W9-AEN-200

TJ
163
.2
T87
2009

ENERGY MANAGEMENT HANDBOOK

SEVENTH EDITION

Co-Editors

Steve Doty
Colorado Springs Utilities
Colorado Springs, Colorado

Wayne C. Turner
School of Industrial Engineering and Management
Oklahoma State University
Stillwater, Oklahoma

Contributors

Simon Baker
California Public Utilities
San Francisco, CA

Gary Berngard
Honeywell Building Solutions
Denver CO

Barney L. Capehart
Industrial Engineering
University of Florida
Gainesville, FL

Clint Christenson
Industrial Engineering
Oklahoma State University
Stillwater, OK

David E. Claridge
Mechanical Engineering Department
Texas A&M University
College Station, TX

Bob Cox
Jacobs Carter & Burgess
Cary, NC

Charles Culp
Energy Systems Laboratory
Texas A&M University
College Station, TX

Michael Dipple
Georgia Southern University
Statesboro, GA

Steve Doty
Colorado Springs Utilities
Colorado Springs, CO

Keith Elder
Coffman Engineers, Inc.
Seattle, WA

John L. Fetters
Effective Lighting Solutions, Inc.
Columbus, OH

Scott Frazier
Oklahoma State University
Stillwater, OK

Carol Freedenthal, CEO
Jofree Corporation,
Houston, TX

Dale A. Gustavson
Consultant
Orange, CA

Jeff Haberl
Energy Systems Laboratory
Texas A&M University
College Station, TX

Jack Halliwell
Halliwell Engineering
Aventura, FL

Michael R. Harrison
Johns-Mansfield Corporation
Denver, CO

Russell L. Heiserman
School of Technology
Oklahoma State University
Stillwater, OK

William J. Kennedy, Jr.
Industrial Engineering
Clemson University
Clemson, SC

John M. Kovacik, Retired
GE Industrial & Power System Sales
Schenectady, NY

Mingsheng Liu
Architectural Engineering
University of Nebraska
Lincoln, NB

Konstantin Lobodovsky
Motor Manager
Penn Valley, CA

William Mashburn
Virginia Polytechnic Institute and
    State University
Blacksburg, VA

Paul Mehta
Director of Industrial Assessment Center
Bradley University
Peoria, IL

Javier Mont
Johnson Controls
Chesterfield, MO

George Owens
Energy and Engineering Solutions
Columbia, MD

Les Pace
Lektron Lighting
Tulsa, OK

Jerald D. Parker, Retired
Mechanical & Aerospace Engineering
Oklahoma State University
Stillwater, OK

S.A. Parker
Pacific Northwest National Laboratory
Richland, WA

David Pratt
Industrial Engineering and Management
Oklahoma State University
Stillwater, OK

Philip S. Schmidt
Department of Mechanical Engineering
University of Texas
Austin, TX

R.B. Scollon
Allied Chemical Corporation
Morristown, NJ

R.D. Smith
Allied Chemical Corporation
Morristown, NJ

Mark B. Spiller
Gainesville Regional Utilities
Gainesville, FL

Nick Stecky
NJS Associates
Denville, NJ

Albert Thumann
Association of Energy Engineers
Atlanta, GA

W.D. Turner
Mechanical Engineering Department
Texas A&M University
College Station, TX

Alfred R. Williams
Ventana Corporation
Bethel, CT

Larry C. Witte
Department of Mechanical Engineering
University of Houston
Houston, TX

Jorge B. Wong Kcomt
Covidien
Wallingford, CT

Eric Woodroof
Profitable Green Solutions.com

Energy Management Handbook

Seventh Edition

BY

Steve Doty

Colorado Springs Utilities

Colorado Springs, Colorado

AND

Wayne C. Turner

School of Industrial Engineering and Management

Oklahoma State University

LIBRARY
NSCC - STRAIT AREA CAMPUS
226 REEVES ST
PORT HAWKESBURY NS
B9A 2A2 CANADA

THE FAIRMONT PRESS, INC.

CRC Press
Taylor & Francis Group

Library of Congress Cataloging-in-Publication Data

Doty, Steve.
Energy management handbook / by Steve Doty and Wayne C. Turner. -- 7th ed.
    p. cm.
Includes bibliographical references and index.
ISBN-10: 0-88173-609-0 (alk. paper)
ISBN-10: 0-88173-610-4 (electronic)
ISBN-13: 978-1-4200-8870-0 (alk. paper)
1. Power resources--Handbooks, manuals, etc. 2. Energy conservation--Handbooks,
manuals, etc. I. Turner, Wayne C., 1942- II. Turner, Wayne C., 1942- Energy management
handbook. III. Title

TJ163.2.T87 2009
658.2'6--dc22

                                    2009015129

Energy management handbook / by Steve Doty and Wayne C. Turner
©2009 by The Fairmont Press, Inc. All rights reserved. No part of this publication may
be reproduced or transmitted in any form or by any means, electronic or mechanical, including photocopy, recording, or any information storage and retrieval system, without
permission in writing from the publisher.

Published by The Fairmont Press, Inc.
700 Indian Trail
Lilburn, GA 30047
tel: 770-925-9388; fax: 770-381-9865
http://www.fairmontpress.com

Distributed by Taylor & Francis Ltd.
6000 Broken Sound Parkway NW, Suite 300
Boca Raton, FL 33487, USA
E-mail: orders@crcpress.com

Distributed by Taylor & Francis Ltd.
23-25 Blades Court
Deodar Road
London SW15 2NU, UK
E-mail: uk.tandf@thomsonpublishingservices.co.uk

Printed in the United States of America
10 9 8 7 6 5 4 3 2 1

10: 0-88173-609-0 (The Fairmont Press, Inc.)
13: 978-1-4200-8870-0 (Taylor & Francis Ltd.)

While every effort is made to provide dependable information, the publisher, authors, and editors
cannot be held responsible for any errors or omissions.

Contents

Foreword to the Seventh Edition

The publishing of the seventh edition of the *Energy Management Handbook* coincides with a new awareness for applying energy efficient technologies. The Energy Independence and Security Act of 2007, the growing concern for global warming, and the spiraling oil prices have again placed energy management at the forefront.

The energy management journey began with the founding of the Association of Energy Engineers (AEE) more than 30 years ago. The Association of Energy Engineers is a non profit professional society which provides continuing education programs to the energy efficiency community. During this period, the Association of Energy Engineers has grown to 8500 members in 77 countries.

No other publication has been as influential in defining the energy management profession. Like the Association of Energy Engineers, the *Energy Management Handbook* was originally launched in the late 1970s. In 1981 the Association of Energy Engineers started its most successful program, the Certified Energy Manager (CEM). To date, over 6500 professionals have been recognized as CEMs. The *Energy Management Handbook* has served as the official reference book for the CEM program.

Today, the energy efficiency industry is poised for new growth, and the *Energy Management Handbook* is more important than ever before. Facility managers today must deal with rising energy costs. For example, June 2008 oil prices leapt above $140 a barrel, setting a record high. The *Energy Management Handbook* offers real solutions to reduce energy costs by utilizing the latest energy efficient technologies and alternative energy strategies, as well as incorporating principles of green buildings and sustainable design.

The seventh edition of the *Energy Management Handbook* will continue to be the indispensable reference required to assist energy managers meet the challenges ahead.

Albert Thumann, P.E., CEM
Executive Director
Association of Energy Engineers
Atlanta, Georgia
www.aeecenter.org

Preface to the Seventh Edition

Good publications always have a successor plan. I (Wayne Turner) am getting older, and there are so many trout to catch that I have to give them more time. The *Energy Management Handbook* is now in its 7th edition, extending the run of publication to more than 25 years. Few books last that long, and I am proud of its success; however, it is time for me to begin stepping aside.

Mr. Steve Doty, of Colorado Springs Utilities, knows energy management perhaps better than I, and he is my chosen successor. He is one of the most thorough and professional people with whom I have ever worked. I will stay on and work with Steve for several years, but eventually this will be his book. He has made changes in this edition that have impressed me; I sincerely think you (the most important people in the world for this book) will also be pleased and impressed. Aren't fresh opinions always helpful?

Meet Steve and the new edition; just don't forget about me YET!

Tight lines.

Wayne Turner
wayne.turner@okstate.edu

Wayne is the pioneer of this book and the reputation it has enjoyed since the first edition. Yet I know he will join me in saying the credit belongs to the authors. The concept of multiple expert authors is truly the strength of the book.

In this edition we have several new chapter authors, each bringing new expertise to the book—and we say *thanks* to the former authors who brought us this far. Future editions will continue this natural evolution, keeping the book vital and current. (And maybe someday I'll get to go fishing too!)

Technical books like this are a continuous work in progress. With the authors' help (and their continuing patience with me), chapters have been carefully reviewed, updated, and improved. I suppose it was inevitable that some of my experiences in engineering would rub off while editing, thus my contribution. We think it is the right mix of fundamentals and application, and hope you agree. Your comments are always welcome.

Steve Doty
dotyranch@msn.com

Chapter 1

Introduction

STEVE DOTY
Colorado Springs Utilities
Colorado Springs, CO

DR. BARNEY L. CAPEHART, PROFESSOR
University of Florida
Gainesville, FL

STEVEN A. PARKER
Pacific Northwest National Laboratory
Richland, WA

DR. WAYNE C. TURNER, REGENTS PROFESSOR
Oklahoma State University
Stillwater, OK

1.1   BACKGROUND

Mr. Al Thumann, executive director of the Association of Energy Engineers, said it well in the 3rd edition foreword of this book: "The energy 'roller coaster' never ceases with new turns and spirals which make for a challenging ride." Those professionals who boarded the ride in the late 70s and stayed on board have experienced several ups and downs. First, being an energy manager was like being a mother, John Wayne, and a slice of apple pie all in one. Everyone supported the concept, and success was around every bend. Then the mid-80s plunge in energy prices caused some to wonder, "Do we really need to continue energy management?" Sometime in the late 80s, the decision was made. Energy management is good business, but it needs to be run by professionals. The Certified Energy Manager (CEM) program of the Association of Energy Engineers became popular, starting a very steep growth curve. AEE continues to grow in membership and stature.

Throughout the years, federal regulation has played an important role in the energy industry. Chapter 20 is devoted to this subject, and a few of the significant regulatory actions are shown in Table 1-1. While energy policy legislation dates back to the industrial revolution, things as we know them probably started with the National Energy Act (NEA) of 1978, which was a legislative response by the U.S. Congress to the 1973 energy crisis. It includes the following statutes:

- Public Utility Regulatory Policies Act (PURPA) (Pub. L. 95-617)
- Energy Tax Act (Pub.L. 95-618)
- National Energy Conservation Policy Act (NECPA) (Pub.L. 95-619)
- Power Plant and Industrial Fuel Use Act (Pub.L. 95-620)
- Natural Gas Policy Act (Pub.L. 95-621).

In 1978, amidst natural gas supply shortages, Congress enacted the Natural Gas Policy Act (NGPA), as part of the National Energy Act (NEA). It had become apparent that price controls put in place to protect consumers from monopoly pricing had begun to hurt consumers by creating natural gas shortages. The NGPA had three central goals:
- Create a single U.S. market for natural gas
- Match natural gas supply and demand
- Allow market pressure to establish the wellhead price of natural gas.[15]

The NGPA granted the Federal Energy Regulatory Commission (FERC) authority over intrastate, as well as interstate, natural gas production. FERC, as the successor to the Federal Power Commission (FPC), was granted jurisdictional authority over virtually all natural gas production, both interstate and intrastate.[16] An informative history of natural gas regulation has been prepared by the Natural Gas Supply Association.[15]

All the legislation we think of since then (EPAct 1992, EPAct 2005, and EISA 2007) is really addenda/modifications to this original legislation, and each had a focus related to the needs of the period. Energy-related mandates for federal facilities are found in many of the energy policy acts. In terms of the number of pages, the federal facility mandates are a small portion of the total, but the concept of leading by example has been consistent; the Federal Energy Management Program (FEMP), charted in 1973, is still alive and well. Another common theme is the endorsement of the energy savings performance contracting delivery method, where a third party provides implementation services in exchange for payment via energy savings (See chapter 25). In addition to these regulations, many presidents have signed executive orders that furthered energy management as good government business practices; however, in reality, only the current ones are relevant. Other federal government activities provide valuable support to the private sector for energy conservation. A couple of examples show how these efforts are enablers to the energy management industry:

- The Department of Energy (DOE) posts bulletins and white papers that are resources to energy professionals researching new subjects. These include everything from industrial topic fact sheets to emissions reporting and other trends, and they are great resources.

- The Energy Information Administration (EIA), part of DOE, provides a wealth of statistical data, including typical energy use intensity (kBtu/SF-yr) figures for business segments—an easy litmus test for determining where energy saving potential exists (your use compared to national average).

- Energy Star® is run by EPA and DOE with private partners. Their appliance branding program has helped make energy usage part of the consumer buying decision process, a market transformation. Also, their Portfolio Manager building rating system provides a bonze plaque with the Energy Star logo to building owners who demonstrate a commitment to

raising the bar of energy conservation.

Several of the federal regulation actions were put in place to allow de-regulation, a fundamental change in how energy is bought and sold. Several states moved toward electrical deregulation, with some successes. But there were side effects. The prospect of electric deregulation and sharing grid infrastructure caused utilities to change their business view of their portion of the grid. Investment in expanding or upgrading this infrastructure became risky business for individual utilities, and so many chose a wait-and-see approach. One (now famous) energy trading company manipulated pricing in the new deregulated electric business environment, and this event gave many states and consumers pause. To regain the confidence of the consumers, a greater degree of oversight of business practices and the sharing of vital U.S. grid infrastructure may be necessary. Other concerns that exist with the U.S.

Table 1-1. Some Key Federal Regulations Related to the Energy Industry

Regulation	Time Period	Action Taken
Natural Gas Policy Act NGPA	1978	— Granted the Federal Energy Regulatory Commission (FERC) authority over intrastate, as well as interstate, natural gas production. — Set wellhead price ceilings by category. — Established rules for allocating the costs of certain high-cost gas to industrial customers served by interstate pipeline companies. — Provided authority to allocate gas to high priority users in times of gas supply emergency. — Put limits on curtailments of sales to high priority agricultural and industrial feedstock sectors. (A key date in the NGPA was January 1, 1985, when price ceilings on most new gas were removed.)
National Energy Conservation Policy Act NECPA	1978	— Provided for the regulation of interstate commerce, to reduce the growth in demand for energy in the United States, and to conserve nonrenewable energy resources produced in this nation and elsewhere, without inhibiting beneficial economic growth. — Prompted residential energy conservation. — Promoted energy conservation programs for schools, hospitals, buildings owned by units of local governments, and public care institutions. — Improved energy efficiency of certain products and processes. — Provided federal energy initiatives and additional energy-related measures.
Energy Policy Act EPAct-1992	1992	— Allowed states to choose de-regulation of electricity purchasing and wheeling through the grid. — Created aggressive efficiency goals for federal facilities. — Created higher motor and appliance efficiency standards. — Allowed federal facilities to utilize the energy services performance contracting (ESPC) project delivery method.
Energy Policy Act EPAct-2005	2005	— Provided tax incentives for efficiency and renewable measures, as well we for investments in electric transmission grid systems. — Created clean coal technology funding. — Required new federal facilities to achieve 30% better efficiency than required by ASHRAE 90.1 (better than energy code levels). — Established renewable energy use amounts.
Energy Independence and Security Act EISA-2007	2007	— Provided loans for battery development. — Increased emphasis on bio fuels. — Created measures intended to phase out the use of incandescent lighting. — Extended life cycle cost periods to 40 years for federal facility projects. — Provided grants to determine viable options for carbon capture and sequestering. — Increased efficiency goals for federal buildings, with milestones to transition energy use away from fossil fuels.

electrical grid infrastructure system include susceptibility to failure and terrorism. Even with the bumps as electricity deregulation was first tried, some states are now deregulated, and wider-scale electric deregulation remains an exciting concept.

As private sector businesses and the federal government expand their needs for energy management programs, opportunities are created for ESCOs (energy service companies), shared savings providers, performance contractors, and other similar organizations. These groups are providing the auditing, energy/economic analyses, capital, and monitoring to help other organizations reduce their energy consumption, thus reducing their expenditures for energy services. By guaranteeing and sharing the savings from improved energy efficiency and improved productivity, both groups benefit and prosper.

Market transformation is a useful by-product of effective energy management. Since the emergence of energy management into the main stream, major associations have taken energy into primary consideration. New associations have emerged that are dedicated to specific aspects of energy (such as wind, solar, biomass, etc.) and are linked to energy conservation goals and laws. This prompts adoption of state-of-the-art practices and standards which, in turn, drive market transformation in the construction industry.

Professional Associations That Have Reacted to Energy

EDITOR'S NOTE: The following lists are intended to show the numerous organizations that are involved in the field of energy. Try as we may, the list will never be complete, and so our apologies go out to any organizations not mentioned. Some of these are industry trade groups, while others are professional societies that offer energy professionals membership and participation opportunities.

Most primary engineering associations now have an energy sub-organization or have integrated energy into their missions. Examples are:
- The Association of Energy Engineers (AEE)
- The American Institute of Architects (AIA)
- The American Society of Mechanical Engineers (ASME)
- The Association of Heating Refrigeration and Air Conditioning Engineers (ASHRAE)
- The Illumination Engineers Society of North America (IESNA)
- The Institute of Electrical and Electronic Engineers (IEEE)

Products and trade groups that include a focus on energy include:

- National Electrical Manufacturers Association (NEMA)
- Air-Conditioning, Heating, and Refrigeration Institute (AHRI, formed when ARI and GAMA merged)
- American Gas Association (AGA)
- Edison Electric Institute (EEI)
- Electric Power Research Institute (EPRI)
- National Insulation Association (NIA)
- North American Insulation Manufacturers Association (NAIMA)

New associations have emerged related to energy policy, energy supply, energy conservation, renewable energy, energy financing, including:
- American Wind Energy Association (AWEA)
- International District Energy Association (IDEA)
- American Solar Energy Society (ASES)
- Association of Energy Service Professionals (AESP)
- Geothermal Heat Pump Consortium (GHPC)
- International Ground Source Heat Pump Association (IGSHPA)
- National Association of State Energy Officials (NASEO)
- World Alliance for Decentralized Energy (WADE)
- US Combined Heat and Power Association (US-CHPA)
- Alliance to Save Energy (ASE)
- American Council for an Energy Efficient Economy (ACEEE)
- Council of American Building Officials (CABO)
- Building Owners and Managers Association (BOMA)
- Biomass Energy Research Association (BERA)
- Green Building Council (USGBC)

Some business segments include inherent properties that resist market change. Consider commercial leased office buildings, for example. These represent a huge business sector, but one with barriers to energy improvement opportunities. In leased office space arrangements, tenants have little interest in making capital improvements to building systems since it is not their building, while landlords have little incentive to make efficiency improvements, as long as they are able to pass the utility costs along. All building costs, including energy, are ultimately paid for by the tenant, but the "built-in utilities" concept sounds like "free utilities" and encourages complacency from the tenant. Furthermore, these buildings are often bought and sold ("flipped"), so large capital improvement measures with payback periods longer than the owner's business horizon will not be considered. It will take creativity by energy professionals to overcome such barriers to change. Focusing on the business case of energy savings is required. For example, when selling a building, its operating expenses subtract from the value of the property, so energy savings can boost the value of a sale. Also, when there is a surplus of rental space in a local market, prospective tenants may choose facilities that boast energy efficiency features. Conscientious tenants may demand sub-metered spaces, so their frugal habits can be rewarded and not washed away by the use of other tenants or that in common areas.

Throughout it all, energy managers have proven time and time again that energy management is cost effective. Furthermore, energy management is vital to our national security, environmental welfare, and economic productivity. This will be discussed in the next section.

1.2   THE VALUE OF ENERGY MANAGEMENT

Business, industry and government organizations are under tremendous economic and environmental pressures. Being economically competitive in the global marketplace and meeting increasing environmental standards to reduce air and water pollution have been the major driving factors in most of the recent operational cost and capital cost investment decisions for all organizations. Energy management has been an important tool to help organizations meet these critical objectives for their short-term survival and long-term success.

The problems that organizations face from both their individual and national perspectives include:

- Meeting more stringent environmental quality standards, primarily related to reducing global warming and reducing acid rain.

    Energy management helps improve environmental quality. Reduced energy consumption directly reduces upstream power plant emissions. For example, the primary culprit in global warming is carbon dioxide, CO_2. The chemistry is not hard, but the math is even easier: With the nation's current mix of power sources, most of our electricity comes from fossil fuel and is generated by equipment that is roughly 35% efficient. Each pound of coal, natural gas, or fuel oil burned creates a predictable amount of carbon dioxide, with coal producing the highest amount of the three. Thus, energy conservation can be equated directly to reducing carbon dioxide emissions. The same is true for other pollutants related to fuel combustion, and energy management efforts are an effective method of reducing NO_x, CO_2, mercury, and particulates.

    Other environmental benefits of the energy

management industry are easy to find: Less energy consumption means less petroleum field development and subsequent on-site pollution.

Less energy consumption means less thermal pollution and emissions at power plants and less cooling water discharge. Less energy consumption extends the capacity of energy distribution infrastructure and prolongs the life span of fossil fuel resources. The list could go on, but the bottom line is that energy management helps improve environmental quality. With increased emphasis on environmental impact, the energy manager serves a vital role in quantifying both the dollar benefit and the emissions reduction for projects, as well as relating them to cost of implementation. The choices of how to proceed will remain with the customer; however advice from a qualified energy manager will provide the necessary input for key business decisions; thus the Certified Energy Manager becomes a valuable business ally.

For facilities focusing on sustainability, the energy manager can serve by linking improvements to current practice and facilities with associated carbon or other emission reductions, providing a path and options for the customer to achieve their goal. Carbon inventory, ranking of options, and other practical measures are brought to the customer by the energy manager as ideas are matched with solutions.

- Becoming—or continuing to be—economically competitive in the global marketplace, which requires reducing the cost of production or services, reducing industrial energy intensiveness, and meeting customer service needs for quality and delivery times.

    Significant energy and dollar savings are available through energy management. Most facilities (manufacturing plants, schools, hospitals, office buildings, etc.) can save according to the profile shown in Figure 1-1. Even more savings have been accomplished by some programs. Part of gaining support for energy projects is the accountability for achieving project success. *"You can't manage what you can't measure"* is a mantra that forms an essential ingredient in any energy management program.

Thus, large savings can be accomplished often, with high returns on investments and rapid paybacks. Energy management can make the difference between profit and loss and can establish real competitive enhancements for most companies.

Energy management helps companies improve their productivity and increase their product or service quality. This is done through implementing new energy efficiency

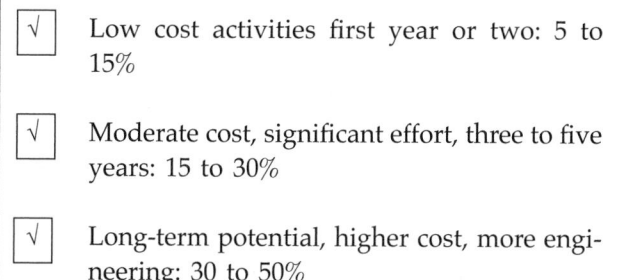

√ Low cost activities first year or two: 5 to 15%

√ Moderate cost, significant effort, three to five years: 15 to 30%

√ Long-term potential, higher cost, more engineering: 30 to 50%

Figure 1-1. Typical Savings through Energy Management

technologies; new materials and new manufacturing processes; and the use of new technologies in equipment and materials for business and industry. Energy cost savings have an amplified effect on the company bottom line profits, and this is more pronounced the larger the fraction of total business expense that comes from energy use and the lower the profit margin. See Figure 1-2. Note that a facility with energy cost at 8% of total operating cost and operating on a 5% net profit margin will experience a profit increase from 5% to 6.7% with a 20% reduction in energy use—*a 34% profit increase.*

Seeing the Big Picture

Well-trained energy professionals bring added value to customers when they can see the bigger picture. Consider a commercial building ripe for a lighting replacement. The quick answer may be to replace lighting one-for-one, and this will in fact produce predictable savings. However, it may pay to pause and do a little homework. If, for example, the project building was designed in an era where twice the lighting per-SF was popular, re-designing the lighting to

current illumination standards and lighting power budgets may produce savings well beyond the efficiency differential of the lighting hardware, amplifying the savings. Likewise, retrofitting water use fixtures, like shower heads, may produce more savings alone than a high efficiency water heater—of course both collectively are best. Evaluating a series of manufacturing processes for opportunities to inherently use less energy would be the logical first step before focusing on equipment. The golden rule for energy conservation measures is to *begin by using less.*

Energy conservation is effective at offsetting the need for increased generation capacity, including renewable energy use. It is almost always true that it is more cost effective to reduce the load through conservation measures than to increase the size of the generator, PV panel, etc. The term "negawatts" was coined and introduced by Amory Lovins, in a 1989 speech, and is effective at describing the symbiotic effects of energy conservation.

A useful principle (and paradigm shift for many customers) is to control energy functions as a direct controllable cost rather than an overhead cost. Tracking energy cost as a component of a manufactured item allows the energy cost to be compared directly to other ingredients; once in this light, management approaches will focus differently upon energy. In buildings, the indoor climate conditions can be equated to energy cost, and the management approach changes. Once energy is believed to be a controllable expense, awareness will be raised, and there will be new incentives to find improvements. Thus, the energy manager has a role in customer education as well as the sciences.

While we may be tempted, ours is not to save the world as energy managers. While employed by a customer,

Original Profit Margin	20% Energy Savings Table shows revised profit value Energy Cost % of Total Operating Cost									
	1%	2%	3%	4%	5%	6%	7%	8%	9%	10%
1%	1.2%	1.4%	1.6%	1.8%	2.0%	2.2%	2.4%	2.6%	2.9%	3.1%
2%	2.2%	2.4%	2.6%	2.8%	3.0%	32%	3.4%	3.7%	3.9%	4.1%
3%	3.2%	3.4%	3.6%	3.8%	4.0%	4.3%	4.5%	4.7%	4.9%	5.1%
5%	5.2%	5.4%	5.6%	5.8%	6.1%	6.3%	6.5%	6.7%	6.9%	7.1%
10%	10.2%	10.4%	10.7%	10.9%	11.1%	11.3%	11.6%	11.8%	12.0%	12.2%
20%	20.2%	20.5%	20.7%	21.0%	21.2%	21.5%	21.7%	22.0%	22.2%	22.4%
30%	30.3%	30.5%	30.8%	31.0%	31.3%	31.6%	31.8%	32.1%	32.4%	32.7%

Figure 1-2. Energy Savings Effect on Profit (14)

their focus becomes our focus—and usually that focus is profit or productivity. Energy consumption is merely one more tool that allows a business to function and thrive. Usually the expectation is to achieve energy savings transparently to existing processes or comfort, and one criterion for a customer selecting an energy professional for a task will be sufficient skill and experience to avoid creating new problems in the process. Thus the successful energy manager will necessarily become a jack of many trades and will be wise to work in teams and connect with other professionals. While we may find it interesting and rewarding that energy savings measures simultaneously reduce emissions and prolong non-renewable energy supplies, the reality for most customers is that energy conservation makes business sense when it brings results to the business bottom line. This careful balancing act of priorities is an ongoing challenge to the energy manager, and finding the business case for energy projects is a special skill all its own.

Often, the energy savings is not the main driving factor when companies decide to purchase new equipment, use new processes, or use new high-tech materials. However, the combination of increased productivity, increased quality, reduced environmental emissions, and reduced energy costs provides a powerful incentive for companies and organizations to implement these new technologies.

Total quality management (TQM) is another emphasis that many businesses and other organizations have developed. TQM is an integrated approach to operating a facility, and energy cost control should be included in the overall TQM program. TQM is based on the principle that front-line employees should have the authority to make changes and other decisions at the lowest operating levels of a facility. If employees have energy management training, they can make informed decisions and recommendations about energy operating costs.

- Maintaining energy supplies that are:
  — Available without significant interruption, and
  — Available at costs that do not fluctuate too rapidly.

Energy management helps reduce the U.S. dependence upon imported oil. During the 1979 oil price crisis, the U.S. was importing almost 50% of our total oil consumption. By 1995, the U.S. was again importing 50% of our consumption. In 2007 about 58% of the petroleum consumed in the U.S. was imported from foreign countries[16]. Sharp increases in crude oil demand from developing countries has pushed crude oil prices to all time highs. Thus, the U.S. is once again vulnerable to an oil embargo or other disruption of supply. The trade balance would be much more favorable if we imported less oil.

- Helping solve other national concerns which include:
  — Need to create new jobs
  — Need to improve the trade balance by reducing costs of imported energy
  — Need to minimize the effects of a potential limited energy supply interruption

None of these concerns can be satisfactorily met without having an energy efficient economy. Energy management plays a key role in helping move toward this.

1.3 THE ENERGY MANAGEMENT PROFESSION

Energy management skills are important to people in many organizations, and certainly to people who perform duties such as energy auditing, facility or building management, energy and economic analysis, and maintenance. The number of companies employing professionally trained energy managers is large and growing. A partial list of job titles is given in Figure 1-3. Even though this is only a partial list, the breadth shows the robustness of the profession.

For some of these people, energy management will be their primary duty, and they will need to acquire in-depth skills in energy analysis, as well as knowledge about existing and new energy using equipment and technologies. For others, such as maintenance managers, energy management skills are simply one more area to cover in an already full plate of duties and expectations. The authors are writing this *Energy Management Handbook* for both of these groups of readers and users.

In the 1980s, few university faculty members would have stated their primary interest was energy management, yet today there are numerous faculty who prominently list energy management as their principal specialty. In 2006, there were 26 universities throughout the country listed by DOE as industrial assessment centers (IAC).[17] Other universities offer coursework and/or do research in energy management but do not have one of the above centers. Finally, several professional journals and magazines now publish exclusively for energy managers.

Utility company demand-side management (DSM) programs have had their ups and downs. DSM efforts peaked in the late 80s and early 90s, then retrenched significantly as utility deregulation and the movement to retail wheeling caused utilities to reduce staff and cut costs—including DSM programs—as much as possible. This short-term cost cutting was seen by many utilities as their only way to become a competitive low-cost supplier of electric power and thereby hold onto their

- Plant Energy Manager
- Utility Energy Auditor
- State Agency Energy Analyst
- Consulting Energy Manager
- DSM Auditor/Manager

- Building/Facility Energy Manager
- Utility Energy Analyst
- Federal Energy Analyst
- Consulting Energy Engineer

Figure 1-3. Tyical Energy Management Job Titles

large customers. Not all utility programs are in a state of reduction. Utilities facing growth and high costs of additional generating capacity may create incentives that curb peak demand and prolong the expense of the next plant. For those, once the next plant is finally built, incentives to continue curbing the use may disappear in a familiar cycle. With national awareness focused on global warming, utility emissions are a certain first target; if emissions become taxed or traded, a new business case will appear for energy conservation programs since the two are closely related.

When there is a reduction in electric utility incentive and rebate programs—and associated customer support—the gap in energy service assistance is met by equipment supply companies and energy service consulting firms that are willing and able to provide the necessary technical and financial assistance. Energy management skills are extremely important to those companies that are in the business of identifying energy savings and providing a guarantee of the savings results.

Thus, the future for energy management is extremely promising. It is cost effective, it improves environmental quality, it helps reduce the trade deficit, and it helps reduce dependence on foreign fuel supplies. Energy management will continue to grow in size and importance.

1.4 SOME SUGGESTED PRINCIPLES OF ENERGY MANAGEMENT

EDITOR'S NOTE: The material in this section is repeated from the first editions of this handbook published in 1982. Mr. Roger Sant, who was then director of the Energy Productivity Center of the Carnegie-Mellon Institute of Research in Arlington, VA, wrote this section for the first edition. It was unchanged for the second edition. Some of the numbers quoted may now be a little old, but the principles are still sound. Amazing, but what was right then for energy management is still right today! The game has changed, the playing field has moved; but the principles stay the same.

If energy productivity is an important opportunity for the nation as a whole, it is a necessity for the individual

company. It represents a real chance for creative management to reduce the component of product cost that has risen the most since 1973.

Those who have taken advantage of these opportunities have done so because of the clear intent and commitment of the top executive. Once that commitment is understood, managers at all levels of the organization can and do respond seriously to the opportunities at hand. Without that leadership, the best designed energy management programs produce few results. In addition, we would like to suggest *four basic principles* which, if adopted, may expand the effectiveness of existing energy management programs or provide the starting point of new efforts.

The *first principle* is to *control the costs of the energy function or service provided, but not the Btu of energy*. As most operating people have noticed, energy is just a means of providing some service or benefit. With the possible exception of feedstocks for petrochemical production, energy is not consumed directly. It is always converted into some useful function. The existing data are not as complete as one would like, but they do indicate some surprises. In 1978, for instance, the aggregate industrial expenditure for energy was $55 billion. Of that, 35% was spent for machine drive from electric motors, 29% for feedstocks, 27% for process heat, 7% for electrolytic functions, and 2% for space conditioning and light. As shown in Table 1-2, this is in blunt contrast to measuring these functions in Btu. Machine drive, for example, instead of 35% of the dollars, required only 12% of the Btu.

In most organizations it will pay to be even more specific about the function provided. For instance, evaporation, distillation, drying, and reheating are all typical of the uses to which process heat is put. In some cases it has also been useful to break down the heat in terms of temperature so the opportunities for matching the heat source to the work requirement can be utilized.

In addition to energy costs, it is useful to measure the depreciation, maintenance, labor, and other operating costs involved in providing the conversion equipment necessary to deliver required services. These costs add as much as 50% to the fuel cost.

It is the total cost of these functions that must be managed and controlled, not the Btu of energy. The large

Table 1-2. Industrial Energy Functions by Expenditure and Btu, 1978
Source: Technical Appendix, *The Least-Cost Energy Strategy,* Carnegie-Mellon University Press, Pittsburgh, Pa., 1979, Tables 1.2.1 and 11.3.2.

Function	Dollar Expenditure (billions)	Percent of Expenditure	Percent of Total Btu
Machine drive	19	35	12
Feedstocks	16	29	35
Process steam	7	13	23
Direct heat	4	7	13
Indirect heat	4	7	13
Electroysi	4	7	3
Space conditioning and lighting	1	1	1
Total	55	100	100

difference in cost of the various Btu of energy can make the commonly used Btu measure extremely misleading. In November 1979, the cost of 1 Btu of electricity was nine times that of 1 Btu of steam coal.

EDITOR'S NOTE: One of the most desirable and least reliable skills for an energy manager is to predict the future cost of energy. Table 1-3 shows the cost of energy in 1979. To the extent that energy costs escalate in price beyond the rate of general inflation, investment paybacks will be shortened, but of course the reverse is also true. Figure 1-4 shows the pattern of energy prices over time. Even the popular conception that energy prices always go up is shown to be false when normalized to constant dollars. This volatility in energy pricing may account for some business decisions that appear overly conservative in establishing rate of return or payback period hurdles.

Table 1-3 Cost of Industrial Energy per Million Btu, 1979

Fuel	Cost
Steam coal	$1.11
Natural gas	$2.75
Residual oil	$2.95
Distillate oil	$4.51
Electricity	$10.31

Availabilities also differ, and the cost of maintaining fuel flexibility can affect the cost of the product. And as shown before, the average annual price increase of natural gas has been almost three times that of electricity. Therefore, an energy management system that controls Btu per unit of product may completely miss the effect of the changing economics and availabilities of energy alternatives and the major differences in usability of each fuel. Controlling

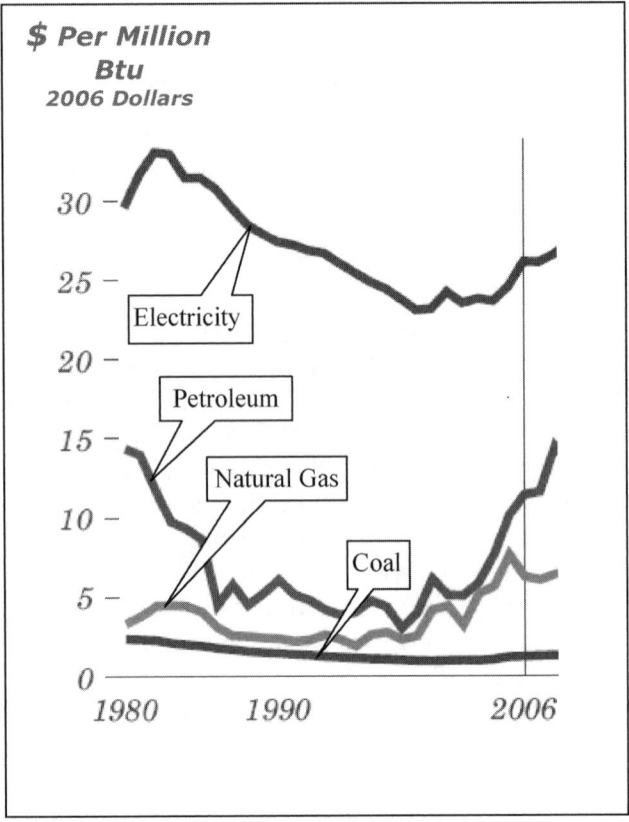

Figure 1-4. Energy prices, 1980-2006, Dollars per Million Btu (2006 dollars)
Source: Energy Information Administration / Annual Energy Outlook 2008

the total cost of energy functions is much more closely attuned to one of the principal interests of the executives of an organization—controlling costs.

EDITOR'S NOTE: Dollars are the bottom line; however, some energy management tasks are better done with Btus. For example, measuring building performance for energy efficiency improvements or for a building efficiency certification is best done with Btus since doing so negates the effect of energy price volatility. Similarly, comparing the heating use of a commercial facility against an industry segment benchmark using cost alone can yield wild results if, for example, one building uses natural gas to heat, while another uses electric resistance; this is another case where using Btus yields more meaningful results.

A *second principle* of energy management is to *control energy functions as a product cost, not as a part of manufacturing or general overhead.* It is surprising how many companies still lump all energy costs into one general or manufacturing overhead account without identifying those products with the highest energy function cost. In most cases, energy functions must become part of the standard cost system so that each function can be assessed as to its specific impact on the product cost.

The minimum theoretical energy expenditure to produce a given product can usually be determined en route to establishing a standard energy cost for that product. The seconds of 25-hp motor drive, the minutes necessary in a 2200°F furnace to heat a steel part for fabrication, or the minutes of 5-V electricity needed to make an electrolytic separation, for example, can be determined as theoretical minimums and compared with the actual figures. As in all production cost functions, the minimum standard is often difficult to meet, but it can serve as an indicator of the size of the opportunity.

In comparing actual values with minimum values, four possible approaches can be taken to reduce the variance, usually in this order:

1. An hourly or daily control system can be installed to keep the function cost at the desired level.
2. Fuel requirements can be switched to a cheaper and more available form.
3. A change can be made in the process methodology to reduce the need for the function.
4. New equipment can be installed to reduce the cost of the function.

The starting point for reducing costs should be in achieving the minimum cost possible with the present equipment and processes. Installing management control systems can indicate what the lowest possible energy use is in a well-controlled situation. It is only at that point that a change in process or equipment configuration should be considered. An equipment change prior to actually minimizing the expenditure under the present system may lead to oversizing new equipment or replacing equipment for unnecessary functions.

The *third principle* is to *control and meter only the main energy functions*—the roughly 20% that make up 80% of the costs. As Peter Drucker pointed out some time ago, a few functions usually account for a majority of the costs. It is important to focus controls on those that represent the meaningful costs and aggregate the remaining items in a general category. Many manufacturing plants in the United States have only one meter, that leading from the gas main or electric main into the plant from the outside source. Regardless of the reasonableness of the standard cost established, the inability to measure actual consumption against that standard will render such a system useless. Sub metering the main functions can provide the information not only to measure but to control costs in a short time interval. The cost of metering and sub metering is usually incidental to the potential for realizing significant cost improvements in the main energy functions of a production system.

The *fourth principle* is to put *the major effort of an energy management program into installing controls and achieving results*. It is common to find general knowledge about how large amounts of energy could be saved in a plant. The missing ingredient is the discipline necessary to achieve these potential savings. Each step in saving energy needs to be monitored frequently enough by the manager or first-line supervisor to see noticeable changes. Logging of important fuel usage or behavioral observations are almost always necessary before any particular savings results can be realized. Therefore, it is critical that an energy director or committee have the authority from the chief executive to install controls, not just advise line management. Those energy managers who have achieved the largest cost reductions actually install systems and controls; they do not just provide good advice.

As suggested earlier, the overall potential for increasing energy productivity and reducing the cost of energy services is substantial. The 20% or so improvement in industrial energy productivity since 1972 is just the beginning. To quote the energy director of a large chemical company: "Long-term results will be much greater."

Although no one knows exactly how much we can improve productivity in practice, the American Physical Society indicated in their 1974 energy conservation study that it is theoretically possible to achieve an eightfold improvement of the 1972 energy/production ratio.[9] Most certainly, we are a long way from an economic saturation of the opportunities (see, e.g., Ref. 10). The common argument that not much can be done after a 15 or 20% improvement has been realized ought to be dismissed as baseless. Energy productivity provides an expanding opportunity, not a last resort. The chapters in this book provide the information that is necessary to make the most of that opportunity in each organization.

CONCLUSION

The energy management industry is integral to the workings of a nations economy and the environment the world shares. For newcomers and seasoned veterans alike, it holds exciting opportunities for the professionals that pursue it. Through professional advancement and texts like this one, new challenges and developments will be met effectively.

References

1. *Statistical Abstract of the United States,* U.S. Government Printing Office, Washington, D.C., 1999.
2. *Energy User News,* Jan. 14, 1980.
3. JOHN G. WINGER et al., *Outlook for Energy in the United States to 1985,* The Chase Manhattan Bank, New York, 1972, p 52.

4. DONELLA H. MEADOWS et al., *The Limits to Growth,* Universe Books, New York, 1972, pp. 153-154.

5. JIMMY E. CARTER, July 15, 1979, "Address to the Nation," *Washington Post,* July 16, 1979, p. A14.

6. *Monthly Energy Review,* Jan. 1980, U.S. Department of Energy, Washington, D.C., p. 16.

7. *Monthly Energy Review,* Jan. 1980, U.S. Department of Energy, Washington D.C., p. 8; *Statistical Abstract of the United States, U.S.* Government Printing Office, Washington, D.C., 1979, Table 1409; *Energy User News,* Jan. 20, 1980, p. 14.

8. American Association for the Advancement of Science, "U.S. Energy Demand: Some Low Energy Futures," *Science,* Apr. 14, 1978, p. 143.

9. American Physical Society Summer Study on Technical Aspects of Efficient Energy Utilization, 1974. Available as W.H. CARNAHAN et al., *Efficient Use of Energy, a Physics Perspective,* from NTIS PB- 242-773, or in *Efficient Energy Use, Vol.* 25 of the American Institute of Physics Conference Proceedings.

10. R.W. SANT, *The Least-Cost Energy Strategy,* Carnegie-Mellon University Press, Pittsburgh, Pa., 1979

11. U.S. Congress Office of Technology Assessment (OTA). Energy Efficiency in the Federal Government: Government by Good Example? OTA-E-492, U.S. Government Printing Office, Washington D.C., May 1991.

12. U.S. Air Force. *DOD Energy Manager's Handbook* Volume 1: Installation Energy Management. Washington D.C., April 1993.

13. Department of Energy Greening Federal Facilities, Second Ed, May 2001, p. vii

14. Commercial Energy Auditing Reference Handbook, Doty, S., Fairmont Press, 2008

15. Natural Gas Supply Association http://www.naturalgas.org/regulation/history.asp

16. U.S. Energy Information Administration (EIA) www.eia.doe.gov

17. U.S. Department of Energy (DOE) Industrial Technology Program, www. eere1.energy.gov

Chapter 2

Effective Energy Management

WILLIAM H. MASHBURN, P.E., CEM

Professor Emeritus
Mechanical Engineering Department
Virginia Polytechnic Institute & State University
Blacksburg, Virginia

2.1 INTRODUCTION

Some years ago, a newspaper headline stated, "Lower energy use leaves experts pleased but puzzled." The article went on to state, "Although the data are preliminary, experts are baffled that the country appears to have broken the decades-old link between economic growth and energy consumption."

For those involved in energy management, this comes as no surprise. We have seen companies becoming more efficient in their use of energy, and that's showing in the data. Those that have extracted all possible savings from downsizing are now looking for other ways to become more competitive. Better management of energy is a viable way, so there is an upward trend in the number of companies that are establishing an energy management program. Management is now beginning to realize they are leaving a lot of money on the table when they do not instigate a good energy management plan.

With the new technologies and alternative energy sources now available, this country could possibly reduce its energy consumption by 50%—if there were no barriers to the implementation. But of course there are barriers, mostly economic. Therefore, we might conclude that **managing energy is not a just technical challenge, but one of how to best implement those technical changes within economic limits, and with a minimum of disruption.**

Unlike other management fads that have come and gone, such as value analysis and quality circles, the need to manage energy will be permanent within our society.

There are several reasons for this:

- There is a direct economic return. Many opportunities found in an energy survey have less than a two-year payback. Some are immediate, such as load shifting or going to a new electric rate schedule.

- Most manufacturing companies are looking for a competitive edge. A reduction in energy costs to manufacture the product can be immediate and permanent. In addition, products that use energy, such as motor driven machinery, are being evaluated to make them more energy efficient, and therefore more marketable. Many foreign countries, where energy is more critical, now want to know the maximum power required to operate a piece of equipment.

- Energy technology is changing so rapidly that state-of-the-art techniques have a half life of ten years at the most. Someone in the organization must be in a position to constantly evaluate and update this technology.

- Energy security is a part of energy management. Without a contingency plan for temporary shortages or outages, and a strategic plan for long range plans, organizations run a risk of major problems without immediate solutions.

- Future price shocks will occur. When world energy markets swing wildly with only a five percent decrease in supply, as they did in 1979, it is reasonable to expect that such occurrences will happen again.

Those people then who choose—or in many cases are drafted—to manage energy will do well to recognize this continuing need and to exert the extra effort to become skilled in this emerging and dynamic profession.

The purpose of this chapter is to provide the fundamentals of an energy management program that can be, and have been, adapted to organizations large and small. Developing a working organizational structure may be the most important thing an energy manager can do.

2.2 ENERGY MANAGEMENT PROGRAM

All the components of a comprehensive energy management program are depicted in Figure 2-1. These components are the organizational structure, a policy, and plans for audits, education, reporting, and strategy. It is hoped that by understanding the fundamentals of managing energy, the energy manager can then adapt a good working program to the existing organizational structure. Each component is discussed in detail below.

2.3   ORGANIZATIONAL STRUCTURE

The organizational chart for energy management shown in Figure 2-1 is generic. It must be adapted to fit into an existing structure for each organization. For example, the presidential block may be the general manager, and VP blocks may be division managers, but the fundamental principles are the same. The main feature of the chart is the location of the energy manager. This position should be high enough in the organizational structure to have access to key players in management, and to have a knowledge of current events within the company. For example, the timing for presenting energy projects can be critical. Funding availability and other management priorities should be known and understood. The organizational level of the energy manager is also indicative of the support management is willing to give to the position.

2.3.1 Energy Manager

One very important part of an energy management program is to have top management support. More important, however, is the selection of the energy manager, who can, among other things, secure this support. The person selected for this position should be one with a vision of what managing energy can do for the com-

pany. Every successful program has had this one thing in common—one person who is a shaker and mover that makes things happen. The program is then built around this person.

There is a great tendency for the energy manager to become an energy engineer and attempt to conduct the whole effort alone. Much has been accomplished in the past with such individuals working alone, but for the long haul, managing the program by involving everyone at the facility is much more productive and permanent. **Developing a working organizational structure may be the most important thing an energy manager can do**.

The role and qualifications of the energy manager have changed substantially in the past few years, affected by required certification of federal energy managers, deregulation of the electric utility industry (bringing both opportunity and uncertainty), and performance contracting requiring more business skills than engineering. In her book titled *Performance Contracting: Expanded Horizons*, Shirley Hansen gives the following requirements for an energy management:

- Create and maintain an energy management plan
- Establish energy records
- Identify outside assistance
- Assess future energy needs
- Identify financing sources
- Make energy recommendations
- Implement recommendations

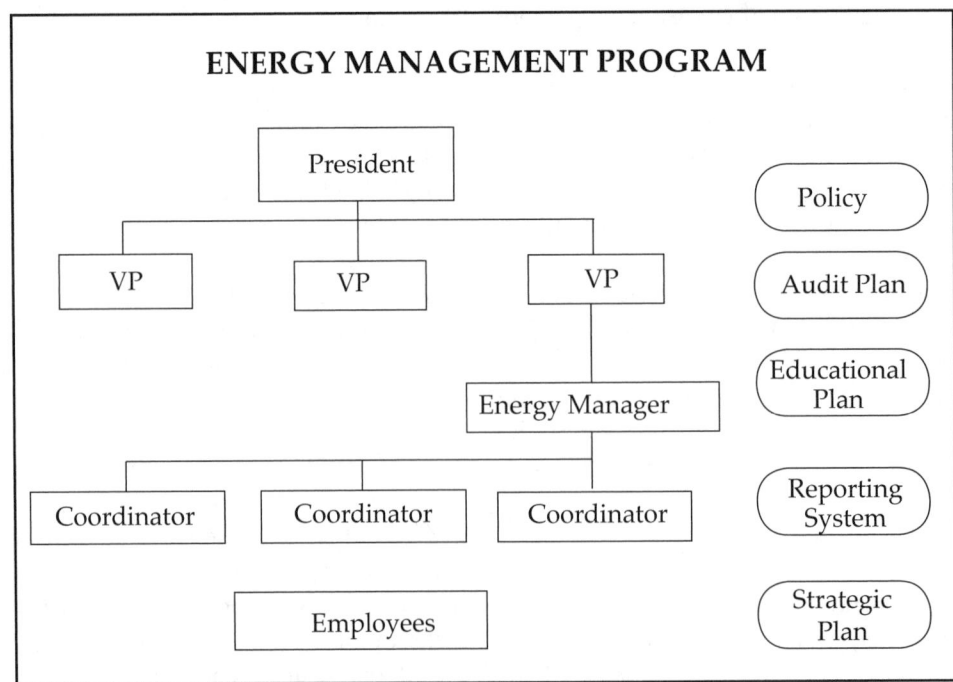

Figure 2-1

- Provide liaison for the energy committee
- Plan communication strategies
- Evaluate program effectiveness

Energy management programs can and have originated within one division of a large corporation. The division, by example and savings, motivates people at corporate level to pick up on the program and make energy management corporate wide. Many programs also originate at corporate level with people who have facilities responsibility and have implemented a good corporate facilities program. They then see the importance and potential of an energy management program and take a leadership role in implementing one. In every case observed by the author, **good programs have been instigated by one individual who has recognized the potential, is willing to put forth the effort (in addition to regular duties), will take the risk of pushing new concepts, and is motivated by a seemingly higher calling to save energy**.

If initiated at corporate level, there are some advantages and some precautions. Some advantages are:

- More resources are available to implement the program, such as budget, staff, and facilities.

- If top management support is secured at corporate level, getting management support at division level is easier.

- Total personnel expertise throughout the corporation is better known and can be identified and made known to division energy managers.

- Expensive test equipment can be purchased and maintained at corporate level for use by divisions as needed.

- A unified reporting system can be put in place.

- Creative financing may be the most needed and the most important assistance to be provided from corporate level.

- Impacts of energy and environmental legislation can best be determined at corporate level.

- Electrical utility rates and structures, as well as effects of unbundling of electric utilities, can be evaluated at corporate level.

Some precautions are:

- Many people at division level may have already done a good job of saving energy and may be cautious about corporate level staff coming in and taking credit for their work.

- All divisions don't progress at the same speed. Work with those who are most interested first, then through the reporting system to top management give them credit. Others will then request assistance.

2.3.2 Energy Team

The coordinators shown in Figure 2-1 represent the energy management team within one given organizational structure, such as one company within a corporation. This group is the core of the program. The main criteria for membership should be an indication of interest. There should be a representative from the administrative group, such as accounting or purchasing, someone from facilities and/or maintenance, and a representative from each major department.

This energy team of coordinators should be appointed for a specific time period, such as one year. Rotation can then bring new people with new ideas, provide a mechanism for tactfully removing non-performers, and involve greater numbers of people in the program in a meaningful way.

Coordinators should be selected to supplement skills lacking in the energy manager since, as pointed out above, it is unrealistic to think one energy manager can have all the qualifications outlined. So total skills needed for the team, including the energy manager, may be defined as follows:

- Have enough technical knowledge within the group to either understand the technology used by the organization or be trainable in that technology.

- Have a knowledge of potentially new technology that may be applicable to the program.

- Have planning skills that will help establish the organizational structure, plan energy surveys, determine educational needs, and develop a strategic energy management plan.

- Understand the economic evaluation system used by the organization, particularly payback and life cycle cost analysis.

- Have good communication and motivational skills since energy management involves everyone within the organization.

The strengths of each team member should be evaluated in light of the above desired skills, and their assignments should be made accordingly.

2.3.3 Employees

Employees are shown as a part of the organizational structure, and are perhaps the greatest untapped resource in an energy management program. A structured method of soliciting their ideas for more efficient use of energy will likely prove to be the most productive effort of the energy management program. A good energy manager will devote 20% of total time working with employees. Too many times employee involvement is limited to posters that say "Save Energy."

Employees in manufacturing plants generally know more about the equipment than anyone else in the facility, because they operate it. They know how to make it run more efficiently, but because there is no mechanism in place for them to have input, their ideas go unsolicited.

An understanding of the psychology of motivation is necessary before an employee involvement program can be successfully conducted. Motivation may be defined as the amount of physical and mental energy that a worker is willing to invest in his or her job. Three key factors of motivation are listed below:

- Motivation is already within people. The task of the supervisor is not to provide motivation, but to know how to release it.

- The amount of energy and enthusiasm people are willing to invest in their work varies with the individual. Not all are over-achievers, but not all are lazy either.

- The amount of personal satisfaction to be derived determines the amount of energy an employee will invest in the job.

Achieving personal satisfaction has been the subject of much research by industrial psychologists, and some revealing facts have emerged. For example, they have learned that most actions taken by people are done to satisfy a physical need, such as the need for food, or an emotional need, such as the need for acceptance, recognition, or achievement.

Research has also shown that many efforts to motivate employees deal almost exclusively with trying to satisfy physical needs, such as raises, bonuses, or fringe benefits. These methods are effective only for the short term, so we must look beyond these to other needs that

may be sources of releasing motivation,

A study done by Heresy and Blanchard [1] in 1977 asked workers to rank job related factors listed below. The results were as follows:

1. Full appreciation for work done
2. Feeling "in" on things
3. Understanding of personal problems
4. Job security
5. Good wages
6. Interesting work
7. Promoting and growth in the company
8. Management loyalty to workers
9. Good working conditions
10. Tactful discipline of workers

This priority list would no doubt change with time and with individual companies, but the rankings of what supervisors thought employees wanted were almost diametrically opposed. They ranked good wages as first.

It becomes obvious from this that **job enrichment is a key to motivation**. Knowing this, the energy manager can plan a program involving employees that can provide job enrichment by some simple and inexpensive recognitions.

Some things to consider in employee motivation are as follows:

- There appears to be a positive relationship between fear arousal and persuasion, if the fear appeals deal with topics primarily of significance to the individual; e.g., personal well being.

- The success of persuasive communication is directly related to the credibility of the source of communication, and it may be reduced if recommended changes deviate too far from existing beliefs and practices.

- When directing attention to conservation, display the reminder at the point of action at the appropriate time for action, and specify who is responsible for taking the action and when it should occur. Generic posters located in the work area are not effective.

- Studies have shown that pro-conservation attitudes and actions will be enhanced through associations with others with similar attitudes, such as being part of an energy committee.

- Positive effects are achieved with financial incentives, if the reward is in proportion to the savings and represents respectable increments of spendable income.

- Consumers place considerable importance on the potential discomfort in reducing their consumption of energy. Changing thermostat settings from the comfort zone should be the last desperate act for an energy manager.

- Social recognition and approval is important and can occur through such things as the award of medals, designation of employee of the month, and selection to membership in elite sub-groups. Note that the dollar cost of such recognitions is minimal.

- The potentially most powerful source of social incentives for conservation behavior—but the least used—is the commitment to others that occurs in the course of group decisions.

Before entering seriously into a program involving employees, be prepared to give a heavy commitment of time and resources. In particular, have the resources to respond quickly to their suggestions.

2.4   ENERGY POLICY

A well-written energy policy that has been authorized by management is as good as the proverbial license to steal. It provides the energy manager with the authority to be involved in business planning, new facility location and planning, selection of production equipment, purchase of measuring equipment, energy reporting, and training—things that are sometimes difficult to do.

If you already have an energy policy, chances are that it is too long and cumbersome. To be effective, the policy should be short—two pages at most. Many people confuse the policy with a procedures manual. It should be bare bones but contain the following items as a minimum:

- Objectives—This can contain the standard motherhood and flag statements about energy, but most important is that the organization will incorporate energy efficiency into facilities and new equipment, with emphasis on life cycle cost analysis rather than lowest initial cost.

- Accountability—This should establish the organizational structure and the authority for the energy

manager, coordinators, and any committees or task groups.

- Reporting—Without authority from top management, it is often difficult for the energy manager to require others within the organization to comply with reporting requirements necessary to properly manage energy. The policy is the place to establish this. It also provides a legitimate reason for requesting funds for instrumentation to measure energy usage.

- Training—If training requirements are established in the policy, it is again easier to include this in budgets. It should include training at all levels within the organization.

Many companies, rather than adopt a comprehensive policy encompassing all the features described above, choose to go with a simpler policy statement.

Appendices A and B give two sample energy policies. Appendix A is generic and covers the items discussed above. Appendix B is a policy statement of a multinational corporation.

2.5   PLANNING

Planning is one of the most important parts of the energy management program, and for most technical people it is the least desirable. It has two major functions in the program. First, a good plan can be a shield from disruptions. Second, by scheduling events throughout the year, continuous emphasis can be applied to the energy management program, and this will play a major role in keeping the program active.

Almost everyone from top management to the custodial level will be happy to give an opinion on what can be done to save energy. Most suggestions are worthless. It is not always wise from a job security standpoint to say this to top management. However, if you inform people—especially top management—that you will evaluate their suggestion, and then assign a priority to it in your plan, not only will you not be disrupted, but you may be considered effective because you do have a plan.

Many programs were started when the fear of energy shortages was greater but have since declined into oblivion. By planning to have events periodically through the year, a continued emphasis will be placed on energy management. Such events can be training programs, audits, planning sessions, demonstrations, research projects, lectures, etc.

The secret to a workable plan is to have people who are required to implement the plan involved in the planning process. People feel a commitment to making things work if they have been a part of the design. This is fundamental to any management planning, but more often that not is overlooked. However, in order to prevent the most outspoken members of a committee from dominating with their ideas and rejecting ideas from less outspoken members, a technique for managing committees must be used. A favorite of the author is the nominal group technique developed at the University of Wisconsin in the late 1980's by Andre Delbecq and Andrea Van de Ven [2]. This technique consists of the following basic steps:

1. Problem definition—The problem is clearly defined to members of the group.

2. Grouping—Divide large groups into smaller groups of seven to ten, then have the group elect a recording secretary.

3. Silent generation of ideas—Each person silently and independently writes as many answers to the problem as can be generated within a specified time.

4. Round-robin listing—the secretary lists each idea individually on an easel until all have been recorded.

5. Discussion—Ideas are discussed for clarification, elaboration, evaluation and combining.

6. Ranking—Each person ranks the five most important items. The total number of points received for each idea will determine the first choice of the group.

2.6   AUDIT PLANNING

The details of conducting audits are discussed in a comprehensive manner in Chapter 4, but planning should be conducted prior to the actual audits. The planning should include types of audits to be performed, team makeup, and dates.

By making the audits specific rather than general in nature, much more energy can be saved. Examples of some types of audits that might be considered are:

- Tuning-Operation-Maintenance (TOM)
- Compressed air

- Motors
- Lighting
- Steam system
- Water
- Controls
- HVAC
- Employee suggestions

By defining individual audits in this manner, it is easy to identify the proper team for the audit. Don't neglect to bring in outside people such as electric utility and natural gas representatives to be team members. Scheduling the audits, then, can contribute to the events that will keep the program active.

With the maturing of performance contracting, energy managers have two choices for the energy audit process. They may go through the contracting process to select and define the work of a performance contractor, or they can set up their own team and conduct audits. In some cases, such as a corporate energy manager, performance contracting may be selected for one facility and energy auditing for another. Each has advantages and disadvantages.

Advantages of performance contracting are:

- No investment is required of the company, other than that involved in the contracting process, which can be very time consuming.

- A minimum of in-house people are involved, namely the energy manager and financial people.

Disadvantages are:

- Technical resources are generally limited to the contracting organization.

- Many firms underestimate the work required.

- The contractor may not have the full spectrum of skills needed.

- The contractor may not have an interest in low/cost no/cost projects.

- High markups are likely.

Advantages of setting up an audit team are:

- The team can be selected to match the equipment to be audited, and it can be made up of in-house personnel, outside specialists, or best, a combination of both.

- They can identify all potential energy conservation projects, both low-cost/no-cost and large capital investments.

- The audit can be an excellent training tool by involving others in the process, and by adding a training component as a part of the audit.

Disadvantages of an audit team approach:

- Financing identified projects becomes a separate issue for the energy manager.

- It takes a well-organized energy management structure to take full advantage of the work of the audit team.

2.7 EDUCATIONAL PLANNING

A major part of the energy manager's job is to provide some energy education to persons within the organization. In spite of the fact that we have been concerned with it since the 70s, **there is still a sea of ignorance concerning energy**.

Raising the energy education level throughout the organization can have big dividends. The program will operate much more effectively if management understands the complexities of energy—and particularly the potential for economic benefit; the coordinators will be more effective is they are able to prioritize energy conservation measures, and are aware of the latest technology; the quality and quantity of employee suggestions will improve significantly with training.

Educational training should be considered for three distinct groups—management, the energy team, and employees.

2.7.1 Management Training

It is difficult to gain much of management's time, so subtle ways must be developed to get them up to speed. Getting time on a regular meeting to provide updates on the program is one way. When the momentum of the program gets going, it may be advantageous to have a half- or one-day presentation for management.

A good, concise report periodically can be a tool to educate management. Short articles that are pertinent to your educational goals, taken from magazines and newspapers, can be attached to reports and sent selectively. Having management be a part of a training program for either the energy team or employees, or

both, can be an educational experience since we learn best when we have to make a presentation.

Ultimately, the energy manager should aspire to be a part of business planning for the organization. A strategic plan for energy should be a part of every business plan. This puts the energy manager into a position for more contact with management people and thus the opportunity to inform and teach.

2.7.2 Energy Team Training

Since the energy team is the core group of the energy management program, proper and thorough training for them should have the highest priority. Training is available from many sources and in many forms.

- Self study—This necessitates having a good library of energy related materials from which coordinators can select.

- In-house training—This may be done by a qualified member of the team, usually the energy manager, or someone from outside.

- Short courses—These are offered by associations such as the Association of Energy Engineers [3], by individual consultants, by corporations, and by colleges and universities.

- Comprehensive courses—Such courses of one to four weeks duration are offered by universities, including Virginia Tech and N.C. State University.

For large decentralized organizations, an annual two- or three-day seminar can be the base for the educational program for energy managers. Such a program should be planned carefully. The following suggestions should be incorporated into such a program:

- Select quality speakers from both inside and outside the organization.

- This is an opportunity to get top management support. Invite a top level executive from the organization to give opening remarks. It may be wise to offer to write the remarks, or at least to provide some material for inclusion.

- Involve the participants in workshop activities so they have an opportunity to have input into the program. Also, provide some practical tips on energy savings that they might go back and implement immediately. One or two good ideas can sometimes pay for their time in the seminar.

- Make the seminar first class, with professional speakers. Consider a banquet with an entertaining—not technical—after dinner speaker and a manual that includes a schedule of events, biosketches of speakers, list of attendees, information on each topic presented, and other things that will help pull the whole seminar together. Vendors will contribute door prizes.

- You may wish to develop a logo for the program and include it on small favors such as cups, carrying cases, etc.

2.7.3 Employee Training

A systematic approach for involving employees should start with some basic training in energy. This will produce a much higher quality of ideas from them. Employees place a high value on training, so a side benefit is that morale goes up. Simply teaching the difference between electrical demand and kilowatt hours of energy, and that compressed air is very expensive, is a start. Short training sessions on energy can be injected into other ongoing training for employees, such as safety. A more comprehensive training program should include:

- Energy conservation in the home
- Fundamentals of electric energy
- Fundamentals of energy systems
- How energy surveys are conducted and what to look for

2.8   STRATEGIC PLANNING

Developing objectives, strategies, programs, and action items constitutes strategic planning for the energy management program. It is the last but perhaps most important step in the process of developing the program, and unfortunately it is where many stop. The very term "strategic planning" has an ominous sound to those who are more technically inclined. However, by using a simplified approach and involving the energy management team in the process, a plan can be developed using a flow chart that will define the program for the next five years.

If the team is involved in developing each of the components of objectives, strategies, programs, and action items, using the nominal group technique, the result will be a simplified flow chart that can be used for many purposes. First, it is a protective plan that discourages

intrusion into the program, once it is established and approved. It provides the basis for resources such as funding and personnel for implementation. It projects strategic planning into overall planning by the organization, and hence it legitimizes the program at top management level. By involving the implementers in the planning process, there is a strong commitment to make it work.

Appendix C contains flow charts depicting a strategic plan developed in a workshop conducted by the author for a large defense organization. It is a model plan in that it deals not only with the technical aspects of energy management, but also with funding, communications, education, and behavior modification.

2.9   REPORTING

There is no generic form to that can be used for reporting. There are too many variables, such as organization size, product, project requirements, and procedures already in existence. The ultimate reporting system is one used by a chemical company making a textile product. The Btu/lb of product is calculated on a computer system that gives an instantaneous reading. This is not only a reporting system, but one that detects maintenance problems. Very few companies are set up to do this, but many do have some type of energy index for monthly reporting.

When energy prices fluctuate wildly, the best energy index is usually based on Btus, but when energy prices are stable, the best index is dollars. However, there are still many factors that will influence any index, such as weather, production, expansion or contraction of facilities, new technologies, etc.

The bottom line is that any reporting system has to be customized to suit individual circumstances. And, while reporting is not always the most glamorous part of managing energy, it can make a contribution to the program by providing the bottom line on its effectiveness. It is also a straight pipeline into management and can be a tool for promoting the program.

The report is probably of most value to the one who prepares it. It is a forcing function that requires all information to be pulled together in a coherent manner. This requires much thought and analysis that might not otherwise take place.

By making reporting a requirement of the energy policy, getting the necessary support can be easier. In many cases, the data may already be collected on a periodic basis and put into a computer. It may simply require combining production data and energy data to develop an energy index.

Keep the reporting requirements as simple as possible. The monthly report could be something as simple as adding to an ongoing graph that compares present usage to some baseline year. Any narrative should be short, with data kept in a file that can be provided for any supporting in-depth information.

With all the above considered, the best way to report is to do it against an audit that has been performed at the facility. One large corporation has its facilities report in this manner and then has an award for those that complete all energy conservation measures listed on the audit.

2.10 OWNERSHIP

The key to a successful energy management program is one word—ownership. This extends to everyone within the organization. Employees that operate a machine "own" that machine. Any attempt to modify their "baby" without their participation will not succeed. They have the knowledge to make or break the attempt. Members of the energy team are not going to be interested in seeing one person—the energy manger—get all the fame and glory for their efforts. Management people that invest in energy projects want to share in the recognition for their risk taking. A corporate energy team that goes into a division for an energy audit must help put a person from the division in the energy management position, then make sure the audit belongs to the division. Below are more tips for success that have been compiled from observing successful energy management programs.

- Have a plan. A plan dealing with organization, surveys, training, and strategic planning—with scheduled events —has two advantages. It prevents disruptions by non-productive ideas, and it keeps the program active.

- Give away—or at least share—ideas for saving energy. The surest way to kill a project is to be possessive. If others have a vested interest they will help make it work.

- Be aggressive. The energy team, after some training, will be the most energy knowledgeable group within the company. Too many management decisions are made with a meager knowledge of the effects on energy.

- Use proven technology. Many programs get bogged down trying to make a new technology work and lose sight of the easy projects with good payback. Don't buy serial number one. In spite of price breaks and promise of vendor support, it can be all too consuming to make the system work.

- Go with the winners. Not every department within a company will be enthused about the energy program. Make those who are look good to top management through the reporting system, and all will follow.

- A final major tip—Ask machine operators what should be done to reduce energy. Then make sure they get proper recognition for ideas.

2.11 SUMMARY

Let's now summarize by assuming you have just been appointed energy manager of a fairly large company. What are the steps you might consider in setting up an energy management program? Here is a suggested procedure.

2.11.1 Situation Analysis

Determine what has been done before. Was there a previous attempt to establish an energy management program? What were the results of this effort? Next, plot the energy usage for all fuels for the past two—or more years. Then project the usage and cost for the next five years, at the present rate. This will not only help you sell your program but will identify areas of concentration for reducing energy.

2.11.2 Policy

Develop some kind of acceptable policy that gives authority to the program. This will help later on with such things as reporting requirements and the need for measurement instrumentation.

2.11.3 Organization

Set up the energy committee and/or coordinators.

2.11.4 Training

With the committee involvement, develop a training plan for the first year.

2.11.5 Audits

Again with the committee involvement, develop an auditing plan for the first year.

2.11.6 Reporting

Develop a simple reporting system.

2.11.7 Schedule

From the above information, develop a schedule of events for the next year, timing them so as to give periodic actions from the program, which will help keep the program active and visible.

2.11.8 Implement the program

2.12 CONCLUSION

Energy management has now matured to the point that it offers outstanding opportunities for those willing to invest time and effort to learn the fundamentals. It requires technical and management skills that broadens educational needs for both technical and management people desiring to enter this field. Because of the economic return of energy management, it is attractive to top management, so exposure of the energy manager at this level brings added opportunity for recognition and advancement. Managing energy will be a continuous need, so persons with this skill will have personal job security, even as we are caught up in the downsizing fad now permeating our society.

References

1. Hersey, Paul and Kenneth H. Blanchard, *Management of Organizational Behavior: Utilizing Human Resources*, Harper and Row, 1970
2. Delbecq, Andre L., Andrew H. Van de Ven, and David H. Gustafson, *Group Techniques for Program Planning*, Green Briar Press, 1986.
3. Mashburn, William H., *Managing Energy Resources in Times of Dynamic Change*, Fairmont Press, 1992
4. Turner, Wayne, *Energy Management Handbook*, 2nd edition, Chapter 2, Fairmont Press, 1993.

Appendix A

ENERGY POLICY

Acme Manufacturing Company

Policy and Procedures Manual

Subject: Energy Management Program

I.   Policy

Energy management shall be practiced in all areas of the company's operation.

II.   Energy Management Program Objectives

It is the company's objective to use energy efficiently and to provide energy security for the organization, both immediate and long range, by:

*   Utilizing energy efficiently throughout the company's operations.

*   Incorporating energy efficiency into existing equipment and facilities, as well as in the selection and purchase of new equipment.

*   Complying with government regulations—federal, state, and local.

*   Putting in place an energy management program to accomplish the above objectives.

III.   Implementation

A. Organization

The company's energy management program shall be administered through the facilities department.

1.   Energy Manager

The energy manager shall report directly to the vice president of facilities and shall have overall responsibility for carrying out the energy management program.

2.   Energy Committee

The energy manager may appoint an energy committee, to be comprised of representatives from various departments. Members will serve for a specified period of time. The purpose of the energy committee is to advise the energy manager on the operation of the energy management program and to provide assistance on specific tasks when needed.

3.   Energy Coordinators

Energy coordinators shall be appointed to represent a specific department or division. The energy manager shall establish minimum qualification standards for coordinators and shall have joint approval authority for each coordinator appointed.

Coordinators shall be responsible for maintaining an ongoing awareness of energy consumption and expenditures in their assigned areas. They shall recommend and implement energy conservation projects and energy management practices.

Coordinators shall provide necessary information for reporting from their specific areas.

They may be assigned on a full-time or part-time

basis, as required to implement programs in their areas.

B.   Reporting

The energy coordinator shall keep the energy office advised of all efforts to increase energy efficiency in their areas. A summary of energy cost savings shall be submitted each quarter to the energy office.

The energy manager shall be responsible for consolidating these reports for top management.

C.   Training

The energy manager shall provide energy training at all levels of the company.

IV.   Policy Updating

The energy manager and the energy advisory committee shall review this policy annually and make recommendations for updating or changes.

Appendix B

POLICY STATEMENT

Acme International Corporation is committed to the efficient, cost effective, and environmentally responsible use of energy throughout its worldwide operations. Acme will promote energy efficiency by implementing cost-effective programs that will maintain or improve the quality of the work environment, optimize service reliability, increase productivity, and enhance the safety of our workplace.

Appendix C

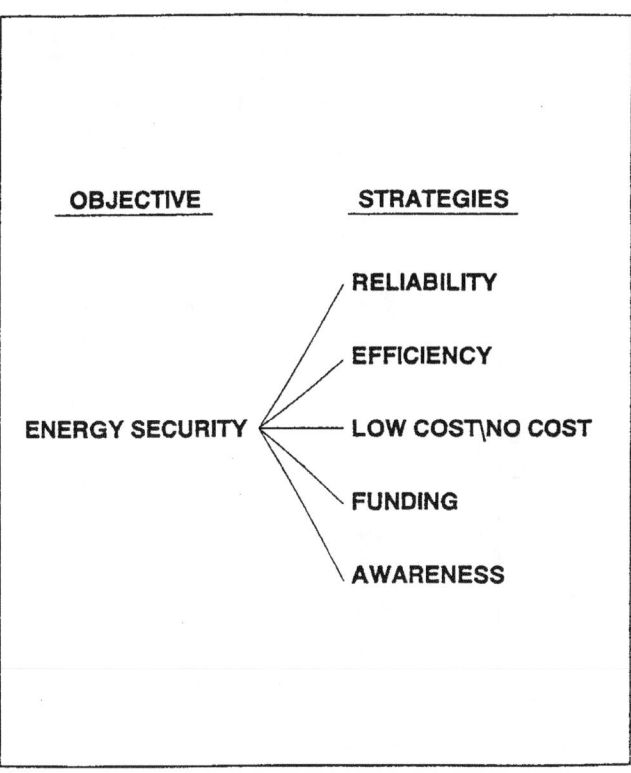

Figure 2-2

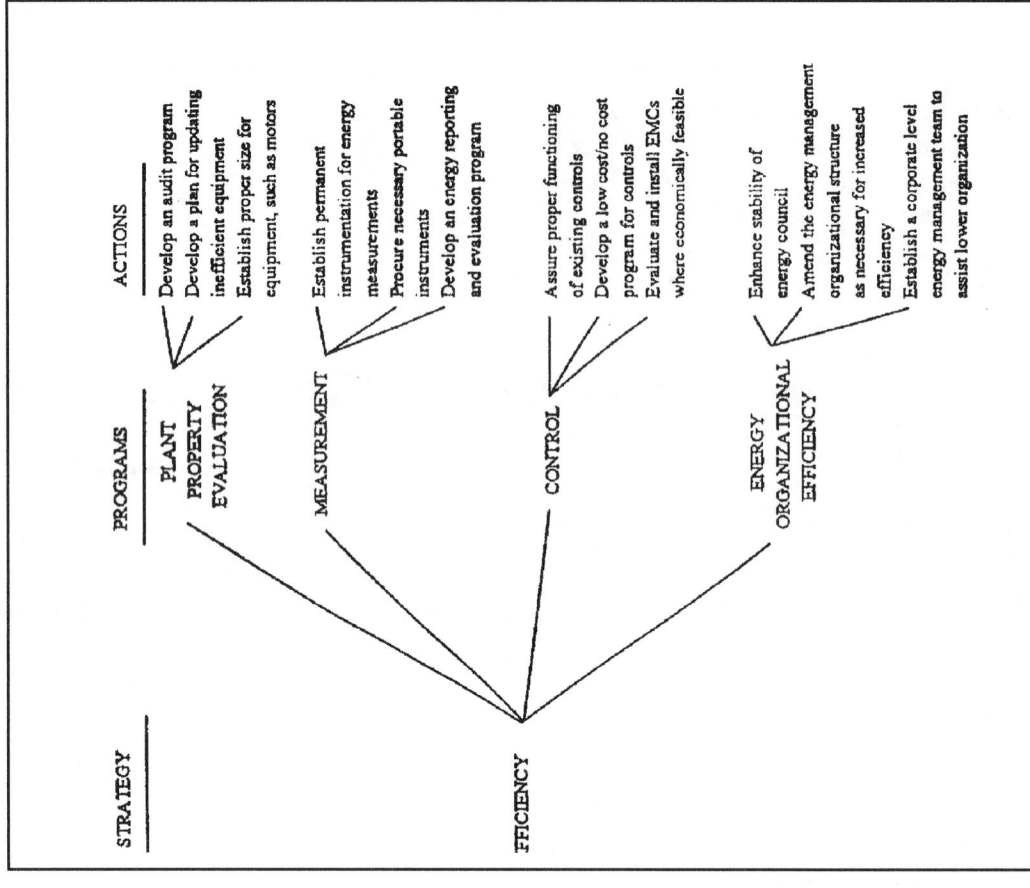

Figure 2-4

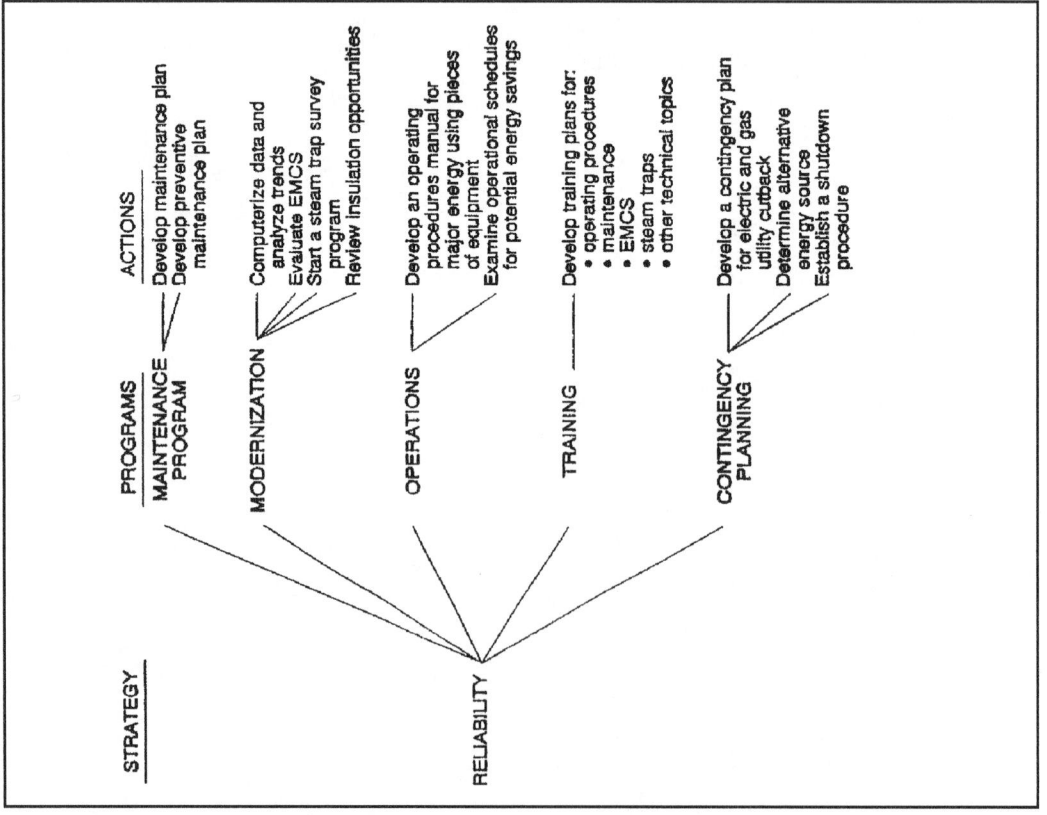

Figure 2-3

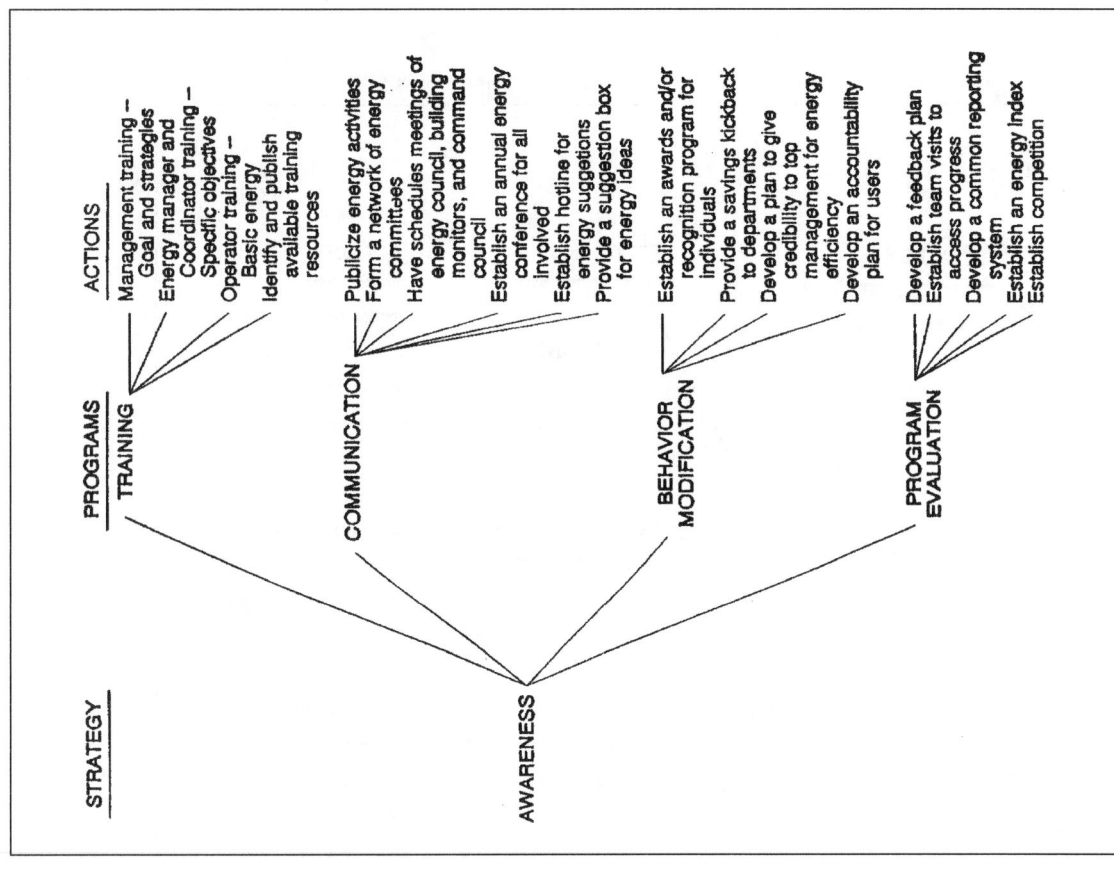

STRATEGY

PROGRAMS

ACTIONS

TRAINING
- Management training — Goal and strategies
- Energy manager and Coordinator training — Specific objectives
- Operator training — Basic energy
- Identify and publish available training resources

COMMUNICATION
- Publicize energy activities
- Form a network of energy committees
- Have schedules meetings of energy council, building monitors, and command council
- Establish an annual energy conference for all involved
- Establish hotline for energy suggestions
- Provide a suggestion box for energy ideas

BEHAVIOR MODIFICATION
- Establish an awards and/or recognition program for individuals
- Provide a savings kickback to departments
- Develop a plan to give credibility to top management for energy efficiency
- Develop an accountability plan for users

PROGRAM EVALUATION
- Develop a feedback plan
- Establish team visits to access progress
- Develop a common reporting system
- Establish an energy index
- Establish competition

AWARENESS

Figure 2-6

STRATEGY

PROGRAMS

ACTIONS

ALTERNATIVES — Develop fuel switching capability for oil and gas

NEGOTIATIONS
- Electric utility
- Gas utility
- Purchase coal by BTU
- Purchase in bulk
- Combine purchases with other facilities

TIME-OF-USE
- Establish an electrical demand control system
- Maximize use of off peak electrical energy

ELIMINATION
- Eliminate excess lighting
- Reduce steam pressure where psi provided not required
- Audit for equipment using energy, but no longer needed, such as transformers, motors
- Audit hot water temperatures for possible reduction

LOW COST/NO COST

Figure 2-5

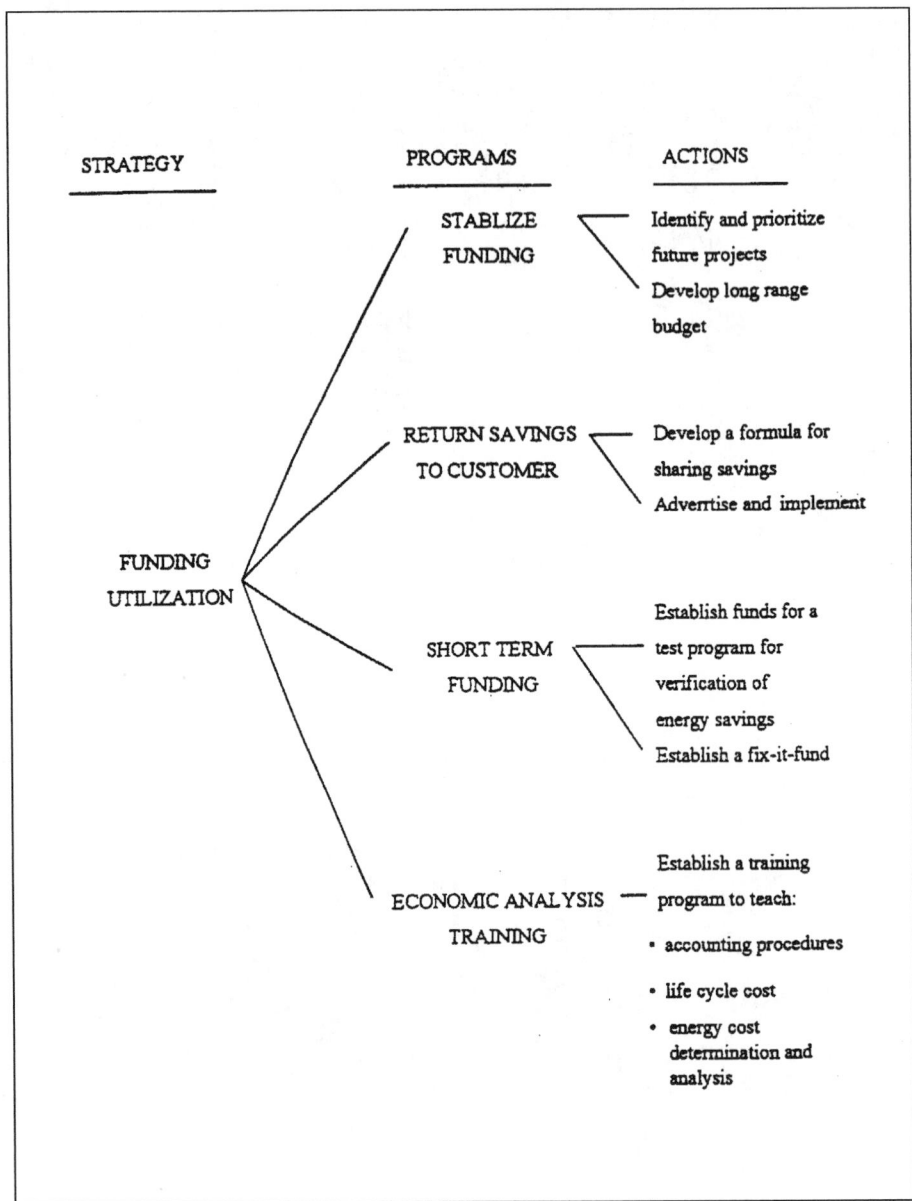

Figure 2-7

CHAPTER 3
ENERGY AUDITING

BARNEY L. CAPEHART AND MARK B. SPILLER
University of Florida
Spiller Consulting
Gainesville, FL

SCOTT FRAZIER
Oklahoma State University

3.1 INTRODUCTION

Saving money on energy bills is attractive to businesses, industries, and individuals alike. Customers whose energy bills use up a large part of their income, and especially those customers whose energy bills represent a substantial fraction of their company's operating costs, have a strong motivation to initiate and continue an on-going energy cost-control program. No-cost or very low-cost operational changes can often save a customer or an industry 10-20% on utility bills; capital cost programs with payback times of two years or less can often save an additional 20-30%. In many cases these energy cost control programs will also result in both reuced energy consumption and reduced emissions of environmental pollutants.

The energy audit is one of the first tasks to be performed in the accomplishment of an effective energy cost control program. An energy audit consists of a detailed examination of how a facility uses energy, what the facility pays for that energy, and finally, a recommended program for changes in operating practices or energy-consuming equipment that will cost-effectively save dollars on energy bills. The energy audit is sometimes called an energy survey or an energy analysis, so that it is not hampered with the negative connotation of an audit in the sense of an IRS audit. The energy audit is a positive experience with significant benefits to the business or individual, and the term "audit" should be avoided if it clearly produces a negative image in the mind of a particular business or individual.

3.2 ENERGY AUDITING SERVICES

Energy audits are performed by several different groups. Electric and gas utilities throughout the country offer free residential energy audits. A utility's residential energy auditors analyze the monthly bills, inspect the construction of the dwelling unit, and inspect all of the energy-consuming appliances in a house or an apartment. Ceiling and wall insulation is measured, ducts are inspected, appliances such as heaters, air conditioners, water heaters, refrigerators, and freezers are examined, and the lighting system is checked.

Some utilities also perform audits for their industrial and commercial customers. They have professional engineers on their staff to perform the detailed audits needed by companies with complex process equipment and operations. When utilities offer free or low-cost energy audits for commercial customers, they usually only provide walk-through audits rather than detailed audits. Even so, they generally consider lighting, HVAC systems, water heating, insulation and some motors.

Large commercial or industrial customers may hire an engineering consulting firm to perform a complete energy audit. Other companies may elect to hire an energy manager or set up an energy management team whose job is to conduct periodic audits and to keep up with the available energy efficiency technology.

The U.S. Department of Energy (U.S. DOE) funds a program where universities around the country operate industrial assessment centers which perform free energy audits for small and medium sized manufacturing companies. There are currently 26 IACs funded by the Industrial Division of the U.S. DOE.

The state energy program is another energy audit service funded by the U.S. Department of Energy. It is usually administered through state energy offices. This program pays for audits of schools, hospitals, and other institutions, and it has some funding assistance for energy conservation improvements.

3.3 BASIC COMPONENTS OF AN ENERGY AUDIT

An initial summary of the basic steps involved in conducting a successful energy audit is provided here, and these steps are explained more fully in the sections that follow. This audit description primarily addresses the steps in an industrial or large-scale commercial audit; not all of the procedures described in this section are required for every type of audit.

The audit process starts by collecting information about a facility's operation and past record of utility bills. These data are then analyzed to get a picture of how the facility uses—and possibly wastes—energy, as well as to help the auditor learn what areas to examine to reduce energy costs. Specific changes—called energy conservation opportunities (ECOs)—are identified and evaluated to determine their benefits and their cost-effectiveness. These ECOs are assessed in terms of their costs and benefits, and an economic comparison is made to rank the various ECOs. Finally, an action plan is created where certain ECOs are selected for implementation, and the actual process of saving energy and saving money begins.

3.3.1 The Auditor's Toolbox

To obtain the best information for a successful energy cost control program, the auditor must make some measurements during the audit visit. The amount of equipment needed depends on the type of energy-consuming equipment used at the facility and on the range of potential ECOs that might be considered. For example, if waste heat recovery is being considered, then the auditor must take substantial temperature measurement data from potential heat sources. Tools commonly needed for energy audits are listed below:

Tape Measures

The most basic measuring device needed is the tape measure. A 25-foot tape measure (1" wide) and a 100-foot tape measure are used to check the dimensions of walls, ceilings, windows, and distances between pieces of equipment for purposes such as determining the length of a pipe for transferring waste heat from one piece of equipment to the other.

Lightmeter

One simple and useful instrument is the lightmeter, which is used to measure illumination levels in facilities. A lightmeter that reads in footcandles allows direct analysis of lighting systems and comparison with recommended light levels specified by the Illuminating Engineering Society. A digital lightmeter that is portable and can be hand carried is the most useful. Many areas in buildings and plants are still significantly over-lighted, and measuring this excess illumination then allows the auditor to recommend a reduction in lighting levels through lamp removal programs, or by replacing inefficient lamps with high efficiency lamps that may not supply the same amount of illumination as the old inefficient lamps.

Thermometers

Several thermometers are generally needed to measure temperatures in offices and other work areas, and to measure the temperature of operating equipment.

Knowing process temperatures allows the auditor to determine process equipment efficiencies and to also identify waste heat sources for potential heat recovery programs. Inexpensive electronic thermometers with interchangeable probes are now available to measure temperatures in both these areas. Some common types include an immersion probe, a surface temperature probe, and a radiation shielded probe for measuring true air temperature. Other types of infra-red thermometers and thermographic equipment are also available. An infrared "gun" is valuable for measuring temperatures of surfaces or steam lines that are not readily reached without a ladder.

Humidity sensors are useful to measure indoor humidity levels. Excessive humidification or dehumidification is expensive and is easily spotted in this way. It is also useful to verify the performance of some economizer control operations that make control choices based on dew point, wet bulb, or enthalpy of air, and for cooling tower optimization from wet bulb temperature.

Infrared Cameras

Infrared cameras have come down in price substantially by 2008, but they are still rather expensive pieces of equipment. An investment of at least $10,000 to $15,000 is needed to have a good quality infrared camera. However, these are very versatile pieces of equipment and can be used to find overheated electrical wires, connections, neutrals, circuit breakers, transformers, motors and other pieces of electrical equipment. They can also be used to find wet insulation, missing insulation, roof leaks, and cold spots. Thus, infrared cameras are excellent tools for both safety related diagnostics and energy savings diagnostics. A good rule of thumb is that if one safety hazard is found during an infrared scan of a facility, then that has paid for the cost of the scan for the entire facility. Many insurers require infrared scans of buildings for facilities once a year.

Voltmeter

An inexpensive digital voltmeter is useful for determining operating voltages on electrical equipment, especially when the nameplate has worn off of a piece of equipment or is otherwise unreadable or missing. The most versatile instrument is a digital combined volt-ohm-ammeter with a clamp-on feature for measuring currents in conductors that are easily accessible. This type of multi-meter is convenient and relatively inexpensive. Any newly purchased voltmeter or multimeter should be a true RMS meter, for greatest accuracy where harmonics might be involved.

Clamp-on Ammeter

These are very useful instruments for measuring current in a wire without having to make any live electrical connections. The clamp is opened up and put around one insulated conductor, and the meter reads the current in that conductor. New clamp-on ammeters can be purchased rather inexpensively that read true RMS values. This is important because of the level of harmonics in many of our facilities. An idea of the level of harmonics in a load can be estimated from using an old non-RMS ammeter and then using a true RMS ammeter to measure the current. If there is more than a five to ten percent difference between the two readings, there is a significant harmonic content to that load.

Wattmeter/Power Factor Meter

A portable hand-held wattmeter and power factor meter is very handy for determining the power consumption and power factor of individual motors and other inductive devices. This meter typically has a clamp-on feature that allows an easy connection to the current-carrying conductor, as well as probes for voltage connections. Any newly purchased wattmeter or power factor meter should be a true RMS meter for greatest accuracy where harmonics might be involved

Combustion Analyzer

Combustion analyzers are portable devices capable of estimating the combustion efficiency of furnaces, boilers, or other fossil fuel burning machines. Electronic digital combustion perform the measurements and readout in percent combustion efficiency. Today these instruments are hand-held devices that are very accurate, and they are also quite inexpensive at $800-$1,000 for most heaters and boilers.

Airflow Measurement Devices

Measuring air flow from heating, air conditioning, or ventilating ducts, or from other sources of air flow, is one of the energy auditor's tasks. Airflow measurement devices can be used to identify problems with air flows, such as whether the combustion air flow into a gas heater is correct. Typical airflow measuring devices include a velometer, an anemometer, or an airflow hood. See section 3.4.3 for more detail on airflow measurement devices.

Blower Door Attachment

Building or structure tightness can be measured with a blower door attachment. This device is frequently used in residences and in small office buildings to determine the air leakage rate or the number of air chang-es per hour in the facility. This often helps determine whether the facility has substantial structural or duct leaks that need to be found and sealed. See section 3.4.2 for additional information on blower doors.

Smoke Generator

A simple smoke generator can also be used in residences, offices, and other buildings to find air infiltration and leakage around doors, windows, ducts, and other structural features. Care must be taken in using this device since the chemical "smoke" produced may be hazardous, and breathing protection masks may be needed. See section 3.4.1 for additional information on the smoke generation process and the use of smoke generators.

Safety Equipment

The use of safety equipment is a vital precaution for any energy auditor. A good pair of safety glasses is an absolute necessity for almost any manufacturing facility audit visit. Hearing protectors may also be required on audit visits to noisy plants or areas with high horsepower motors driving fans and pumps. Electrical insulated gloves should be used if electrical measurements will be taken, and thermally insulated gloves should be used for working around boilers and heaters. Breathing masks may also be needed when hazardous fumes are present from processes or materials used. Steel-toe and steel-shank safety shoes may be needed on audits of plants where heavy materials, hot or sharp materials, or hazardous materials are being used. (See section 3.3.3 for an additional discussion of safety procedures.)

Miniature Data Loggers

Miniature ("mini") data loggers have appeared in low cost models in the last five years. These are often devices that can be held in the palm of the hand and are electronic instruments that record measurements of temperature, relative humidity, light intensity, light on/off, and motor on/off. If they have an external sensor input jack, these little boxes are actually general purpose data loggers. With external sensors they can record measurements of current, voltage, apparent power (kVA), pressure, and CO_2.

These data loggers have a microcomputer control chip and a memory chip, so they can be initialized and then record data for periods of time from days to weeks. They can record data on a 24-hour-a-day basis, without any attention or intervention on the part of the energy auditor. Most of these data loggers interface with a digital computer PC and can transfer data into a spreadsheet of the user's choice, or they can use the software provided by the suppliers of the loggers.

Collecting audit data with these small data loggers gives a more complete and accurate picture of an energy system's overall performance, because some conditions may change over long periods of time or when no one is present.

Vibration Analysis Gear

Relatively new in the energy manager's tool box is vibration analysis equipment. The correlation between machine condition (bearings, pulley alignment, etc.) and energy consumption is related, and this equipment monitors such machine health. This equipment comes in various levels of sophistication and price. At the lower end of the spectrum are vibration pens (or probes) that simply give real-time amplitude readings of vibrating equipment in in/sec or mm/sec. This type of equipment can cost under $1,000. The engineer compares the measured vibration amplitude to a list of vibration levels (ISO2372) and is able to determine if the vibration is excessive for that particular piece of equipment.

The more typical type of vibration equipment will measure and log the vibration into a database (on-board and downloadable). In addition to simply measuring vibration amplitude, the machine vibration can be displayed in time or frequency domains. The graphs of vibration in the frequency domain will normally exhibit spikes at certain frequencies. These spikes can be interpreted by a trained individual to determine the relative health of the machine monitored.

The more sophisticated machines are capable of trend analysis so that facility equipment can be monitored on a schedule and changes in vibration (amplitudes and frequencies) can be noted. Such trending can be used to schedule maintenance based on observations of change. This type of equipment starts at about $3,000 and goes up, depending on features desired.

3.3.2 Preparing for the Audit Visit

Some preliminary work must be done before the auditor makes the actual energy audit visit to a facility. Data should be collected on the facility's use of energy through examination of utility bills, and some preliminary information should be compiled on the physical description and operation of the facility. This data should then be analyzed so that the auditor can do the most complete job of identifying energy conservation opportunities during the actual site visit to the facility.

Energy Use Data

The energy auditor should start by collecting data on energy use, power demand, and cost for at least the previous 12 months. Twenty-four months of data might be necessary to adequately understand some types of billing methods. Bills for gas, oil, coal, electricity, etc. should be compiled and examined to determine both the amount of energy used and the cost of that energy. This data should then be put into tabular and graphic form to see what kind of patterns or problems appear from the tables or graphs. Any anomaly in the pattern of energy use raises the possibility for some significant energy or cost savings by identifying and controlling that anomalous behavior. Sometimes an anomaly on the graph or in the table reflects an error in billing, but generally the deviation shows that some activity is going on that has not been noticed or is not completely understood by the customer.

Rate Structures

To fully understand the cost of energy, the auditor must determine the rate structure under which that energy use is billed. Energy rate structures may go from the extremely simple ones—for example, $2.00 per gallon of Number 2 fuel oil—to very complex ones, such as electricity consumption which may have a customer charge, on- and off-peak charge, energy charge, demand charge, power factor charge, and other miscellaneous charges that vary from month to month. Few customers or businesses really understand the various rate structures that control the cost of the energy they consume. The auditor can help here because the customer must know the basis for the costs in order to control them successfully.

• Energy charges: For electrical use, this is in terms of kWh and is often different for on- and off-peak use. For fuel, this is in terms of gallons of oil, therms of gas, etc. and usually does not differentiate by time of use, although there may be seasonal adjustments (e.g. higher in winter).

• Electrical Demand Charges: The demand charge is based on a reading of the maximum power in kW that a customer demands in one month. Power is the rate at which energy is used, and it varies quite rapidly for many facilities. Electric utilities average the power reading over intervals from fifteen minutes to one hour, so that very short fluctuations do not adversely affect customers. Thus, a customer might be billed for demand for a month based on a maximum value of a fifteen minute integrated average of their power use. Demand charges are often different for on- and off-peak times.

- Ratchet Clauses: Some utilities have a ratchet clause in their rate structure which stipulates that the minimum power demand charge will be the highest demand recorded in the last billing period or some percentage (i.e., typically 70-75%) of the highest power demand recorded in the last year. The ratchet clause can increase utility charges for facilities during periods of low activity or where power demand is tied to extreme weather.

- Discounts/Penalties: Utilities generally provide discounts on their energy and power rates for customers who accept power at high voltage and provide transformers on site. They also commonly assess penalties when a customer has a power factor less than 0.9-0.95. Inductive loads (e.g., lightly loaded electric motors, old fluorescent lighting ballasts, etc.) reduce the power factor. Improvement can be made by adding capacitance to correct for lagging power factor, and variable capacitor banks are most useful for improving the power factor at the service drop. Capacitance added near the loads can effectively increase the electrical system capacity. Turning off idling or lightly loaded motors can also help.

- Water and wastewater charges: The energy auditor also looks at water and wastewater use and costs as part of the audit visit. These costs are often related to the energy costs at a facility. Wastewater charges are usually based on some proportion of the metered water use since the solids are difficult to meter. This can needlessly result in substantial increases in the utility bill for processes which do not contribute to the wastewater stream (e.g., makeup water for cooling towers and other evaporative devices, irrigation, etc.). For many utilities a water sub meter can be installed on branch lines that supply the loads not returning water to the sewer system. This can reduce the sewer charges for these branch water flows by up to 75%.

NOTE: Understanding the relationship between the cost of energy compared to water/waste water is important for effective audit recommendations. For example, in areas where electricity cost is low and water cost is high, suggested HVAC measures that convert air-cooled equipment to evaporatively cooled may find most of the energy savings negated by water and waste water charges, even with the sewer consumptive use credit.

Figure 3-1. Sample Summary of Energy Usage and Costs

Month	kWh Used (kWh)	kWh Cost ($)	Demand (kW)	Demand Cost ($)	Total Cost ($)
Mar	44960	1581	213	1495	3076
Apr	47920	1859	213	1495	3354
May	56000	2318	231	1621	3939
Jun	56320	2423	222	1558	3981
Jul	45120	1908	222	1558	3466
Aug	54240	2410	231	1621	4032
Sept	50720	2260	222	1558	3819
Oct	52080	2312	231	1621	3933
Nov	44480	1954	213	1495	3449
Dec	38640	1715	213	1495	3210
Jan	36000	1591	204	1432	3023
Feb	42880	1908	204	1432	3340
Totals	569,360	24,243	2,619	18,385	42,628
Monthly Averages	47,447	2,020	218	1,532	3,552

This example is simplified for the sake of illustration. Most rate structures that include demand charges also include time of use charges for on/off peak, and power factor charges.

Energy bills should be broken down into the components that can be controlled by the facility. These cost components can be listed individually in tables and then plotted. For example, electricity bills should be broken down into power demand costs per kW per month, and energy costs per kWh. The following example illustrates the parts of a rate structure for an industry in Florida.

Example: A company that fabricates metal products gets electricity from its electric utility at the following general service demand rate structure.

Rate structure:

Customer cost	=	$21.00 per month
Energy cost	=	$0.051 per kWh
Demand cost	=	$6.50 per kW per month
Taxes	=	Total of 8%
Fuel adjustment	=	A variable amount per kWh each month

The energy use and costs for that company for a year are summarized below:

The auditor must be sure to account for all the taxes, the fuel adjustment costs, the fixed charges, and any other costs so that the true cost of the controllable energy cost components can be determined. In the electric rate structure described above, the quoted costs for a kW of demand and a kWh of energy are not complete until all these additional costs are added. Although the rate structure says that there is a basic charge of $6.50 per kW per month, the actual cost including all taxes is $7.02 per kW per month. The average cost per kWh is most easily obtained by taking the data for the twelve month period and calculating the cost over this period of time. Using the numbers from the table, one can see that this company has an average energy cost of $0.075 per kWh.

These data are used initially to analyze potential ECOs and will ultimately influence which ECOs are recommended. For example, an ECO that reduces peak demand during a month would save $7.02 per kW per month. Therefore, the auditor should consider ECOs that would involve using certain equipment during the night shift, when the peak load is significantly less than the first shift peak load. ECOs that save both energy and demand on the first shift would save costs at a rate of $0.075 per kWh. Finally, ECOs that save electrical energy during the off-peak shift should be examined too, but they may not be as advantageous; they would only save at the rate of $0.043 per kWh because they are already used off-peak and there would not be any additional demand cost savings.

Physical and Operational Data for the Facility

The auditor must gather information on factors likely to affect energy use in the facility. Geographic location, weather data, facility layout and construction, operating hours, and equipment can all influence energy use.

*   **Geographic Location/Weather Data:** The geographic location of the facility should be noted, together with the weather data for that location. Contact the local weather station, the local utility, or the state energy office to obtain the average degree days for heating and cooling for that location for the past twelve months. This degree-day data will be very useful in analyzing the need for energy for heating or cooling the facility. Bin weather data would also be useful if a thermal envelope simulation of the facility were going to be performed as part of the audit.

*   **Facility Layout:** Next, the facility layout or plan should be obtained and reviewed to determine the facility size, floor plan, and construction features such as wall and roof material and insulation levels, as well as door and window sizes and construction. A set of building plans could supply this information in sufficient detail. It is important to make sure the plans reflect the "as-built" features of the facility since many original building plans do not get updated after building alterations.

*   **Operating Hours:** Operating hours for the facility should also be obtained. Is there only a single shift? Are there two shifts? Three? Knowing the operating hours in advance allows some determination as to whether some loads could be shifted to off-peak times. Adding a second shift can often be cost effective from an energy cost view since the demand charge can then be spread over a greater amount of kWh.

    NOTE: Shifting production to off-peak hours may create labor costs that negate the demand charge savings.

*   **Equipment List:** Finally, the auditor should get an equipment list for the facility and review it before conducting the audit. All large pieces of energy-consuming equipment such as heaters, air conditioners, water heaters, and specific process-related equipment should be identified. This list, together with data on operational uses of the equipment, allows a good understanding of the major energy-consuming tasks or equipment at the facility. As a general rule,

Table 3-1. Typical Metrics to Extract From Customer Utility Bills (1)

Metric	How to Calculate	How Used
Overall $/kWh	Total electric cost / Total kWh used	Measures defined in terms of kWh savings are converted to $ savings with this factor. When high compared to other facilities in the region, this prompts the question "why?"
Overall $/therm for gas	Total gas cost / Total therms used	Measures defined in terms of therm savings are converted to $ savings with this factor.
Energy Use Index (EUI) in kBtu/SF-yr	Convert electric energy use to kBtu with kWh * 3.413 Convert gas energy use to kBtu with Therms/100 Total Gas and Elec kBtu and divide by building SF	EUI values are benchmarked for common building uses. Similar strategy for manufacturing, where the EUI is in terms of Btu/lb of milk, Btu per ton of concrete, Btu/gallon of beer, etc. Where benchmarks are available, this simple comparison of the customer to their peers establishes whether existing energy use, in general, is high, low, or average and suggests reasonable targets for improvement
Overall $/1000 gallons for water and waste	Total water+waste cost / total water gallons used (thousands)	Measures defined in terms of water savings are converted to $ savings with this factor. These costs also subtract from measures that use water to save energy such as cooling towers and evaporative cooling.
Differential between on-off peak kWh charges		The higher the differential, the greater the incentive to shift loads to off peak. This is a key parameter for economic viability of Thermal Energy Storage (TES) systems.
Differential between on-off peak demand charges		The higher the differential, the greater the incentive to shift loads to off peak. This is a key parameter for economic viability of Thermal Energy Storage (TES) systems.
Load factor	Avg. demand / max demand Or (Total kWh per month / days per month / 24hours) / max demand	Customers with low load factors will almost always have high overall $/kWh, since demand charges are a greater portion of the total costs. Low load factors are a prompt to suggest ways to level the load – spacing out equipment use to avoid setting peaks.
Overall $/therm for electric heating, compared to gas	$/therm electric heating = $/kWh * (100,000/3413) $/therm for gas, from above, divide by firing efficiency (e.g. 0.80).	Provides relative benefit of fuel switching options. For example, if gas heating cost per therm is 40% less than electric heating, then an electric boiler conversion to gas fired will save 40% in dollars by fuel switching.
Fraction of electric bill that is demand	Demand charges / total electric charges	Establishes relative importance of demand charges. For example, if demand charges are two thirds of the bill, demand will get more focus than if it is 25% of the bill.
Magnitude of power factor charges	Usually denoted on the bill.	Establishes a budget for power factor correction measures. For example, with a 3 year payback hurdle, power factor charges of $10,000 per year mean that up to $30,000 in corrective measures would constitute a viable alternative for the customer.

the largest energy and cost activities should be examined first to see what savings could be achieved. The greatest effort should be devoted to the ECOs that show the greatest savings, and the least effort to those with the smallest savings potential.

The equipment found at an audit location will depend greatly on the type of facility involved. Residential audits for single-family dwellings generally involve smaller-sized lighting, heating, air conditioning, and refrigeration systems. Commercial operations such as grocery stores, office buildings, and shopping centers usually have equipment similar to residences, but much larger in size and in energy use. However, large residential structures such as apartment buildings have heating, air conditioning, and lighting that is very similar to many commercial facilities. Business operations is the area where commercial audits begin to involve equipment substantially different from that found in residences.

Industrial auditors encounter the most complex equipment. Commercial-scale lighting, heating, air conditioning, and refrigeration, as well as office business equipment, is generally used at most industrial facilities. The major difference is in the highly specialized equipment used for the industrial production processes. This can include equipment for chemical mixing and blending, metal plating and treatment, welding, plastic injection molding, paper making and printing, metal refining, electronic assembly, and making glass, for example.

3.3.3 Energy Audit Safety Considerations

Safety is a critical part of any energy audit. The audit person or team should be thoroughly briefed on safety equipment and procedures and should never place themselves in a position where they could injure themselves or other people at the facility. Adequate safety equipment should be worn at all appropriate times. Auditors should be extremely careful making any measurements on electrical systems or high temperature devices such as boilers, heaters, cookers, etc. Electrical gloves or heat-resistant gloves should be worn as appropriate.

The auditor should be careful when examining any operating piece of equipment, especially those with open drive shafts, belts or gears, or any form of rotating machinery. The equipment operator or supervisor should be notified that the auditor is going to look at that piece of equipment and might need to get information from some part of the device. If necessary, the auditor may need to come back when the machine or device is idle in order to safely get the data. The auditor should never approach a piece of equipment and inspect it without the operator or supervisor being notified first.

Safety Checklist
1. General:
   a. Decline any task that does not appear safe. Safety is more important than savings.
   b. Do not enter confined spaces or areas where a respiratory breathing hazard exists, without being properly trained and equipped to do so.
   c. Use two hands on ladders; use shoulder straps to carry tools and note pads when climbing.
   d. Conduct the field work with a helper or with the customer rather than alone.
   e. Do not operate switches, disconnects, valves, or open equipment panels; let the customer do this for you.

2. Electrical:
   a. Avoid working on live circuits, if possible.
   b. Securely lock off circuits and switches before working on a piece of equipment.
   c. Always keep one hand in your pocket while making measurements on live circuits to help prevent cardiac arrest.

3. Hearing:
   a. Use foam insert plugs while working around loud machinery to reduce sound levels up to 30 decibels.

4. Clothing:
   a. Avoid loose clothing, especially neck ties.
   b. Remove rings, bracelets, watches, etc. if working near exposed electrical connections.
   c. Wear steel toed shoes in mechanical and machinery areas.

3.3.4 Conducting the Audit Visit

Once the information on energy bills, facility equipment, and facility operation has been obtained, the audit equipment can be gathered up, and the actual visit to the facility can be made.

Introductory Meeting

The audit person (or team) should meet with the facility manager and the maintenance supervisor and briefly discuss the purpose of the audit and indicate the kind of information that is to be obtained during the facility visit. If possible, a facility employee who is in a position to authorize expenditures or make operating policy decisions should also be at this initial meeting.

Audit Interviews

Getting the correct information on facility equipment and operation is important if the audit is going to be most successful in identifying ways to save money on energy bills. The company philosophy towards investments, the impetus behind requesting the audit, and the expectations from the audit can be determined by interviewing the general manager, chief operating officer, or other executives. The facility manager or plant manager is one person that should have access to much of the operational data on the facility, as well as a file of data on facility equipment. The finance officer can provide any necessary financial records (e.g., utility bills for electric, gas, oil, other fuels, water and wastewater, expenditures for maintenance and repair, etc.).

The auditor must also interview the floor supervisors and equipment operators to understand the building and process problems. Line or area supervisors usually have the best information on times their equipment is used. The maintenance supervisor is often the primary person to talk with about types of lighting and lamps, sizes of motors, sizes of air conditioners and space heaters, and electrical loads of specialized process equipment. Finally, the maintenance staff must be interviewed to find equipment and performance problems.

The auditor should write down these people's names, job functions and telephone numbers since it is frequently necessary to get additional information after the initial audit visit.

Walk-through Tour

A walk-through tour of the facility or plant tour should be conducted by the facility/plant manager, and it should be arranged so the auditor or audit team can see the major operational and equipment features of the facility. The main purpose of the walk-through tour is to obtain general information. More specific information should be obtained from the maintenance and operational people after the tour.

Getting Detailed Data

Following the facility or plant tour, the auditor or audit team should acquire the detailed data on facility equipment and operation that will lead to identifying the significant Energy Conservation Opportunities (ECOs) that may be appropriate for this facility. This includes data on lighting, HVAC equipment, motors, water heating, and specialized equipment such as refrigerators, ovens, mixers, boilers, heaters, etc. This data is most easily recorded on individualized data sheets that have been prepared in advance.

Energy Audits: What to Look for

- **Lighting:** Making a detailed inventory of all lighting is important. Data should be recorded on numbers of each type of light fixtures and lamps, wattages of lamps, and hours of operation of groups of lights. A lighting inventory data sheet should be used to record this data. Using a lightmeter, the auditor should also record light intensity readings for each area. Taking notes on types of tasks performed in each area will help the auditor select alternative lighting technologies that might be more energy efficient. Other items to note are the areas that may be infrequently used and may be candidates for occupancy sensor controls of lighting, or areas where daylighting may be feasible.

- **HVAC Equipment:** All heating, air conditioning and ventilating equipment should be inventoried. Prepared data sheets can be used to record type, size, model numbers, age, electrical specifications or fuel use specifications, and estimated hours of operation. The equipment should be inspected to determine the condition of the evaporator and condenser coils, the air filters, and the insulation on the refrigerant lines. Air velocity measurement may also be made and recorded to assess operating efficiencies or to discover conditioned air leaks. This data will allow later analysis to examine alternative equipment and operations that would reduce energy costs for heating, ventilating, and air conditioning.

- **Electric Motors:** An inventory of all electric motors over 1 horsepower should also be taken. Prepared data sheets can be used to record motor size, use, age, model number, estimated hours of operation, other electrical characteristics, and possibly the operating power factor. Measurement of voltages, currents, and power factors may be appropriate for some motors. Notes should be taken on the use of motors, recording particularly those that are infrequently used and might be candidates for peak load control or shifting use to off-peak times. All motors over 1 hp and with usage of 2000 hours per year or greater are likely candidates for replacement by high efficiency motors, at least when they fail and must be replaced.

- **Water Heaters:** All water heaters should be examined and data recorded on their type, size, age, model number, electrical characteristics, or fuel use. What the hot water is used for, how much is used, and what time it is used should all be noted. Temperature of the hot water should be measured. Pipe insulation and control of circulation pumps are other opportunities.

- **Waste Heat Sources:** Most facilities have several sources of waste heat, providing possible opportunities for waste heat recovery to be used as the substantial or total source of heat for needed hot water, make-up air, combustion air, feed water, and other beneficial uses. Waste heat sources are air conditioners, air compressors, heaters and boilers, process cooling systems, ovens, furnaces, cookers, and many others. Temperature measurements for these waste heat sources are necessary to analyze them for replacing the operation of the existing water heaters.

- **Peak Equipment Loads:** The auditor should particularly look for any piece of electrically powered equipment that is used infrequently or whose use could be controlled and shifted to off-peak times. Examples of infrequently used equipment include trash compactors, fire sprinkler system pumps (testing), certain types of welders, drying ovens, or any type of back-up machine. Some production machines might be able to be scheduled for off-peak. Water heating could be done off-peak if a storage system is available, and off-peak thermal storage can be accomplished for on-peak heating or cooling of buildings. Electrical measurements of voltages, currents, and wattages may be helpful. Any information which leads to a piece of equipment being used off-peak is valuable and could result in substantial savings on electric bills. The auditor should be especially alert for those infrequent on-peak uses that might help explain anomalies on the energy demand bills.

- **Other Energy-Consuming Equipment:** Finally, an inventory of all other equipment that consumes a substantial amount of energy should be taken. Commercial facilities may have extensive computer and copying equipment, refrigeration and cooling equipment, cooking devices, printing equipment, water heaters, etc. Industrial facilities will have many highly specialized process and production operations and machines. Data on types, sizes, capacities, fuel use, electrical characteristics, age, and operating hours should be recorded for all of this equipment.

Preliminary Identification of ECOs

As the audit is being conducted, the auditor should take notes on potential ECOs that are evident. Identifying ECOs requires a good knowledge of the available energy efficiency technologies that can accomplish the same job with less energy and less cost. For example, overlighting indicates a potential lamp removal or lamp change ECO, and inefficient lamps indicate a potential lamp technology change. Motors with high use times are potential ECOs for high efficiency replacements. Notes on waste heat sources should indicate what other heating sources they might replace, as well as how far away they are from the end use point. Identifying any potential ECOs during the walk-through will make it easier later on to analyze the data and to determine the final ECO recommendations.

3.3.5 Post-Audit Analysis

Following the audit visit to the facility, the data collected should be examined, organized and reviewed for completeness. Any missing data should be obtained from the facility personnel or from a re-visit to the facility. The preliminary ECOs identified during the audit visit should now be reviewed, and the actual analysis of the equipment or operational change should be conducted. This involves determining the costs and the benefits of the potential ECO and making a judgment on the cost-effectiveness of that potential ECO.

Cost-effectiveness involves a judgment decision that is viewed differently by different people and different companies. Often, simple payback period (SPP) is used to measure cost-effectiveness, and most facilities want a SPP of two years or less. The SPP for an ECO is found by taking the initial cost and dividing it by the annual savings. This results in finding a period of time in which the savings will repay the initial investment, without using the time value of money. One other common measure of cost-effectiveness is the discounted benefit-cost ratio. In this method, the annual savings are discounted when they occur in future years and are added together to find the present value of the annual savings over a specified period of time. The benefit-cost ratio is then calculated by dividing the present value of the savings by the initial cost. A ratio greater than one means the investment will more than repay itself, even when the discounted future savings are taken into account.

Several ECO examples are given here in order to illustrate the relationship between the audit information obtained and the technology and operational changes recommended to save on energy bills.

Lighting ECO

First, an ECO technology is selected, such as replacing an existing 400-watt mercury vapor lamp with a 325-watt multi-vapor (metal halide) lamp when it burns out. The cost of the replacement lamp must be determined. Product catalogs can be used to get typical prices for the new lamp—about $10 more than the 400 watt mercury vapor lamp. The new lamp is a direct screw-in replacement, and no change is needed in the fixture or ballast. Labor cost is assumed to be the same to install either lamp. The benefits, or cost savings, must be calculated next. The power savings is 400-325 = 75 watts. If the lamp operates for 4000 hours per year and electric energy costs $0.075/kWh, then the savings is (.075 kW)(4000 hr/year)($0.075/kWh) = $22.50/year. This gives a SPP = $10/$22.50/yr =.4 years, or about 5 months. This would be considered an extremely cost-effective ECO. (For illustration purposes, ballast wattage has been ignored and average cost has been used to find the savings.)

Motor ECO

A ventilating fan at a fiberglass boat manufacturing company has a standard efficiency 5 hp motor that runs at full load two shifts a day (4160 hours per year). When this motor wears out, the company will have an ECO of using a high efficiency motor. A premium efficiency 5 hp motor costs around $80 more to purchase than the standard efficiency motor. The standard motor is 83% efficient and the premium efficiency model is 88.5% efficient. The cost savings for this fully loaded motor is found by calculating (5 hp)(4160 hr/yr)(.746 kW/hp)[(1/.83) –(1/.885)]($.075/kWh) = (1162 kWh)*($0.075) = $87.15/year. The SPP = $80/$87.15/yr = .9 years (about 11 months). This is also a very attractive ECO when evaluated by this economic measure.

The discounted benefit-cost ratio can be found once a motor life is determined and a discount rate is selected. Companies generally have a corporate standard for the discount rate used in determining measures used to make investment decisions. For a 10 year assumed life, and a 10% discount rate, the present worth factor is found as 6.144 (see Chapter 4, Appendix 4A). The benefit-cost ratio is found as B/C = ($87.15)(6.144)/$80 = 6.7. This example shows an extremely attractive benefit-cost ratio!

Peak Load Control ECO

A metals fabrication plant has a large shot-blast cleaner that is used to remove the rust from heavy steel blocks before being machined and welded. The cleaner shoots out a stream of small metal balls (like shotgun pellets) to clean the metal blocks. A 150 hp motor provides the primary motive force for this cleaner. If turned on during the first shift, this machine requires a total electrical load of about 180 kW, which adds directly to the peak load billed by the electric utility. At $7.02/kW/month, this costs (180 kW)*($7.02/kW/month) = $1263.60/month. Discussions with line operating people resulted in the information that the need for the metal blocks was known well in advance, and that cleaning could easily be done on the evening shift before the blocks were needed. Based on this information, the recommended ECO is to restrict the shot-blast cleaner use to the evening shift, saving the company $15,163.20 per year. Since there is no cost to implement this ECO, the SPP = $0; that is, the payback is immediate.

3.3.6 The Energy Audit Report

The next step in the energy audit process is to prepare a report which details the final results and recommendations. The length and detail of this report will vary depending on the type of facility audited. A residential audit may result in a computer printout from the utility. An industrial audit is more likely to have a detailed explanation of the ECOs and benefit-cost analyses. The following discussion covers the more detailed audit reports.

The report should begin with an executive summary that provides the owners/managers of the audited facility with a brief synopsis of the total savings available and the highlights of each ECO. The report should then describe the facility that has been audited, and provide information on the operation of the facility that relates to its energy costs. The energy bills should be presented, with tables and plots showing the costs and consumption. Following the energy cost analysis, the recommended ECOs should be presented, along with the calculations for the costs and benefits, and the cost-effectiveness criterion. Measures are usually ranked in terms of simple payback period (SPP) and are sometimes bundled into groups of measures.

Regardless of the audience for the audit report, it should be written in a clear, concise and easy-to-understand format and style. The executive summary should be tailored to non-technical personnel, and technical jargon should be minimized. A client who understands the report is more likely to implement the recommended ECOs. An outline for a complete energy audit report is shown below.

Energy Audit Report Format

Executive Summary
    A brief summary of the recommendations showing costs and savings, with a table of ECOs ranked by simple payback.
Table of Contents
Introduction
    Purpose of the energy audit
    Need for a continuing energy cost control program
Facility Description
    Product or service, and materials flow
    Size, construction, facility layout, and hours of operation
    Equipment list, with specifications
Energy Bill Analysis
    Utility rate structures
    Tables and graphs of energy consumptions and costs
    Discussion of energy costs and energy bills
Energy Conservation Opportunities
    Listing of potential ECOs
    Cost and savings analysis
    Economic evaluation
Action Plan
    Recommended ECOs and an implementation schedule
    Designation of an energy monitor and ongoing program
Conclusion
    Additional comments not otherwise covered

3.3.7 The Energy Action Plan

The last step in the energy audit process is to recommend an action plan for the facility. Some companies will have an energy audit conducted by their electric utility or by an independent consulting firm and will then make changes to reduce their energy bills. They may not spend any further effort in the energy cost control area until several years in the future, when another energy audit is conducted. In contrast to this is the company that establishes a permanent energy cost control program and assigns one person (or a team of people) to continually monitor and improve the energy efficiency and energy productivity of the company. Similar to a total quality management program whereby a company seeks to continually improve the quality of its products, services and operation, an energy cost control program seeks continual improvement in the amount of product produced for a given expenditure for energy.

The energy action plan lists the ECOs which should be implemented first, and suggests an overall implementation schedule. Often, one or more of the recommended ECOs provides an immediate or very short payback period, so savings from the ECO(s) can be used to generate capital to pay for implementing the other ECOs. In addition, the action plan also suggests that a company designate one person as the energy monitor for the facility. This person can look at the monthly energy bills and see whether any unusual costs are occurring and can verify that the energy savings from ECOs is really being seen. Finally, this person can continue to look for other ways the company can save on energy costs, as well as be seen as evidence that the company is interested in a future program of energy cost control.

3.4 SPECIALIZED AUDIT TOOLS

3.4.1 Smoke Sources

Smoke is useful in determining airflow characteristics in buildings, air distribution systems, exhaust hoods and systems, cooling towers, and air intakes. There are several ways to produce smoke. Ideally, the smoke should be neutrally buoyant with the air mass around it so that no motion will be detected unless a force is applied. Cigarette and incense stick smoke, although inexpensive, do not meet this requirement.

Smoke generators using titanium tetrachloride ($TiCl_4$) provide an inexpensive and convenient way to produce and apply smoke. The smoke is a combination of hydrochloric acid (HCl) fumes and titanium oxides produced by the reaction of $TiCl_4$ and atmospheric water vapor. This smoke is both corrosive and toxic, so the use of a respirator mask utilizing activated carbon is strongly recommended. Commercial units typically use either glass or plastic cases. Glass has excellent longevity but is subject to breakage since smoke generators are often used in difficult-to-reach areas. Most types of plastic containers will quickly degrade from the action of hydrochloric acid.

Small Teflon® squeeze bottles (i.e., 30 ml), with attached caps designed for laboratory reagent use, resist degradation and are easy to use. The bottle should be stuffed with 2-3 real cotton balls, then filled with about 0.15 fluid ounces of liquid $TiCl_4$. Synthetic cotton balls typically disintegrate if used with titanium tetrachloride. This bottle should yield over a year of service with regular use. The neck will clog with debris but can be cleaned with a paper clip.

Some smoke generators are designed for short-time use. These bottles are inexpensive and useful for a day of smoke generation, but they will quickly degrade. Smoke bombs are incendiary devices designed to emit a large volume of smoke over a short period of time. The smoke is available in various colors to provide good visibility. These are useful in determining airflow capabilities of exhaust air systems and large-scale ventilation systems. A crude smoke bomb can be constructed by placing a stick of elemental phosphorus in a metal pan and igniting it. A large volume of white smoke will be released. This is an inexpensive way of testing laboratory exhaust hoods since many labs have phosphorus in stock.

More accurate results can be obtained by measuring the chemical composition of the airstream after injecting a known quantity of tracer gas (such as sulphur hexafluoride) into an area. The efficiency of an exhaust system can be determined by measuring the rate of tracer gas removal. Building infiltration/exfiltration rates can also be estimated with tracer gas.

3.4.2 Blower Door

The blower door is a device containing a fan, controller, several pressure gauges, and a frame that fits in the doorway of a building. It is used to study the pressurization and leakage rates of a building and its air distribution system under varying pressure conditions. The units currently available are designed for use in residences, although they can be used in small commercial buildings as well. The large quantities of ventilation air limit blower door use in large commercial and industrial buildings.

An air leakage/pressure curve can be developed for the building by measuring the fan flow rate necessary to achieve a pressure differential between the building interior and the ambient atmospheric pressure over a range of values. The natural air infiltration rate of the building under the prevailing pressure conditions can be estimated from the leakage/pressure curve and local air pressure data. Measurements made before and after sealing identified leaks can indicate the effectiveness of the work.

The blower door can help to locate the source of air leaks in the building by depressurizing to 30 Pascals and searching potential leakage areas with a smoke source. The air distribution system typically leaks on both the supply and return air sides. If the duct system is located outside the conditioned space (e.g., attic, under floor, etc.), supply leaks will depressurize the building and increase the air infiltration rate; return air leaks will pressurize the building, causing air to exfiltrate. A combination of supply and return air leaks is difficult to detect

without sealing off the duct system at the registers and measuring the leakage rate of the building compared to that of the unsealed duct system. The difference between the two conditions is a measure of the leakage attributable to the air distribution system.

3.4.3 Airflow Measurement Devices

Two types of anemometers are available for measuring airflow: vane and hot-wire. The volume of air moving through an orifice can be determined by estimating the free area of the opening (e.g., supply air register, exhaust hood face, etc.) and multiplying by the air speed. This result is approximate, due to the difficulty in determining the average air speed and the free vent area. Regular calibrations are necessary to assure the accuracy of the instrument. The anemometer can also be used to optimize the face velocity of exhaust hoods by adjusting the door opening until the anemometer indicates the desired airspeed.

Airflow hoods also measure airflow. They contain an airspeed integrating manifold, which averages the velocity across the opening and reads out the airflow volume. The hoods are typically made of nylon fabric supported by an aluminum frame. The instrument is lightweight and easy to hold up against an air vent. The lip of the hood must fit snugly around the opening to assure that all the air volume is measured. Both supply and exhaust airflow can be measured. The result must be adjusted if test conditions fall outside the design range.

3.5 INDUSTRIAL AUDITS

3.5.1 Introduction

Industrial audits are some of the most complex and most interesting audits because of the tremendous variety of equipment and processes found in these facilities. Much of the industrial equipment can be found during commercial audits too. Large chillers, boilers, ventilating fans, water heaters, coolers and freezers, and extensive lighting systems are often the same in most industrial operations as those found in large office buildings or shopping centers. Small cogeneration systems are sometimes found in both large commercial and industrial facilities.

The highly specialized equipment that is used in industrial processes is what differentiates these facilities from large commercial operations. The challenge for the auditor and energy management specialist is to learn how this complex (and often unique) industrial equipment operates, and then to come up with improvements to the processes and the equipment that can save energy and money. The sheer scope of the problem is so great that industrial firms often hire specialized consulting engineers to examine their processes and recommend operational and equipment changes that result in greater energy productivity.

3.5.2 Process Analysis

Many industrial manufacturing processes include sequential steps, each with unique operations, equipment, and materials. One approach to identify opportunities for improvement is to prepare a block diagram of the sequential steps and indicate where energy and other utilities are input. This step can sometimes reveal process steps that can be linked for energy benefit.

A further refinement of this method is to quantify the magnitude of energy use at

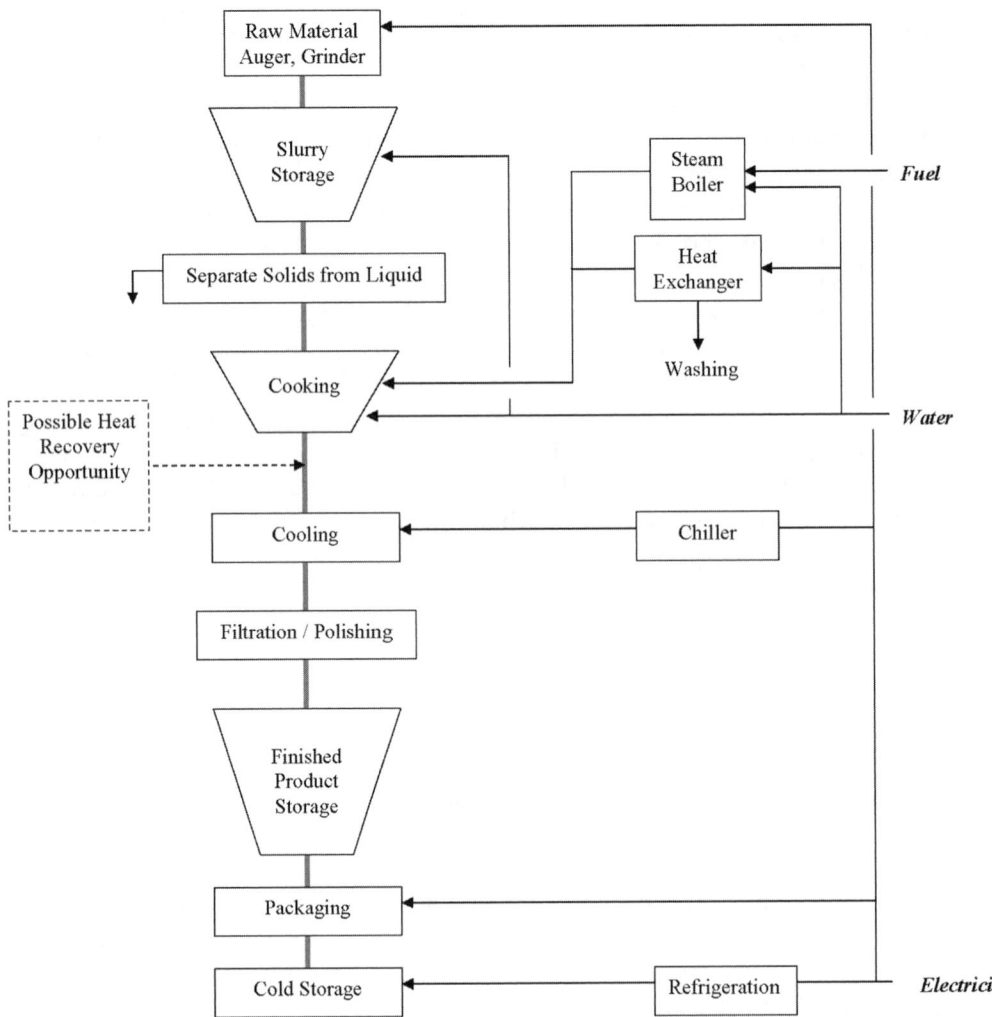

Figure 3-2. Manufacturing Process Flow Diagram with Utility Inputs Added (1)

each process step. The value of this is that it identifies the most significant energy use points and can sometimes be compared to benchmark data for similar businesses.

3.5.3 Industrial Audit Services

A few electric and gas utilities are large enough, and well-enough staffed, that they can offer industrial audits to their customers. These utilities have a trained staff of engineers and process specialists with extensive experience who can recommend operational changes or new equipment to reduce the energy costs in a particular production environment. Many gas and electric utilities, even if they do not offer audits, do offer financial incentives for facilities to install high efficiency lighting, motors, chillers, and other equipment. These incentives can make many ECOs very attractive.

Small and medium-sized industries that fall into

the manufacturing sector (SIC 2000 to 3999) and are in the service area of one of the industrial assessment centers funded by the U.S. Department of Energy can receive free energy audits throughout this program. There are presently 26 IACs operating primarily in the eastern and mid-western areas of the U.S. These IACs are administered by the U.S. Department of Energy. A search vehicle using iIndustrial assessment centers will yield updated locations. Also, information can be found at www.inc.rutgers.edu

3.5.4 Industrial Energy Rate Structures

Except for the smallest industries, facilities will be billed for energy services through a large commercial or industrial rate category. It is important to get this rate structure information for all sources of energy—electricity, gas, oil, coal, steam, etc. Gas, oil and coal are usually billed on a straight cost per unit basis (e.g. $0.90

per gallon of #2 fuel oil). Electricity and steam most often have complex rate structures, with components for a fixed customer charge, a demand charge, and an energy charge. Gas, steam, and electric energy are often available with a time of day rate, or an interruptible rate, that provides much cheaper energy service with the understanding that the customer may have his supply interrupted (stopped) for periods of several hours at a time. Advance notice of the interruption is almost always given, and the number of times a customer can be interrupted in a given period of time is limited.

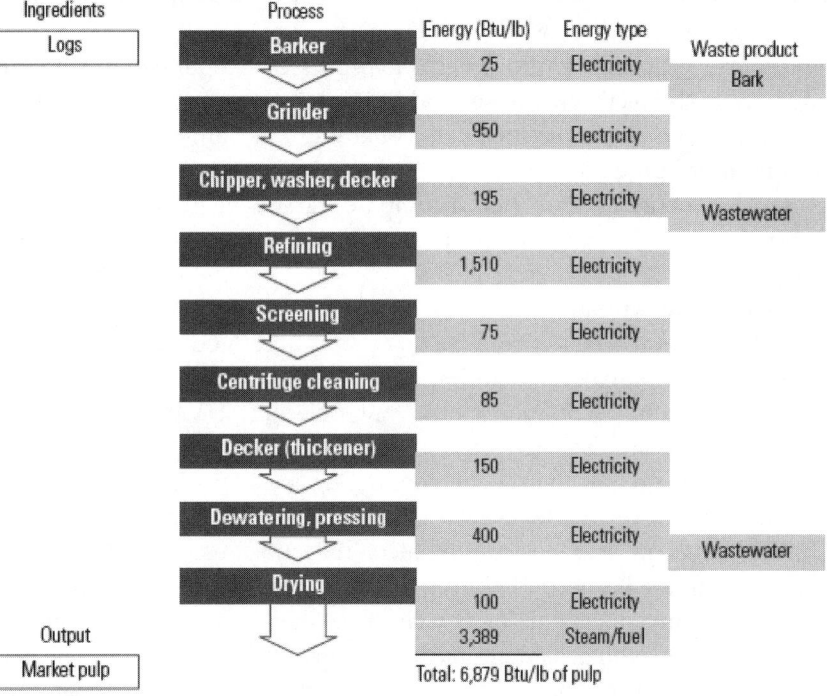

Figure 3-3. Manufacturing Process Flow Chart with Energy Use Identified (1)

3.5.5 Process and Technology Data Sources

For the industrial audit, it is critical to get in advance as much information as possible on the specialized process equipment so that study and research can be performed to understand the particular processes being used and what improvements in operation or technology are available. Data sources are extremely valuable here; auditors should maintain a library of information on processes and technology and should know where to find additional information from research organizations, government facilities, equipment suppliers, and other organizations.

EPRI/GRI

The Electric Power Research Institute (EPRI) and the Gas Research Institute (GRI) are both excellent sources of information on the latest technologies of using electric energy or gas. EPRI has a large number of on-going projects to show the cost-effectiveness of electro-technologies using new processes. GRI also has a large number of projects underway to help promote the use of new cost-effective gas technologies. Both of these organizations provide extensive documentation of their processes and technologies; they also have computer data bases to aid customer inquiries.

U.S. DOE Industrial Division

The U.S. Department of Energy has an Industrial Division that provides a rich source of information on new technologies and new processes. This division funds research into new processes and technologies, and it also funds many demonstration projects to help insure that promising improvements get implemented in appropriate industries. The Industrial Division of US-DOE also maintains a wide network of contacts with government-related research laboratories, such as Oak Ridge National Laboratory, Brookhaven National Laboratory, Lawrence Berkeley National Laboratory, Sandia National Laboratory, and Battelle National Laboratory. These laboratories have many of their own research, development, and demonstration programs for improved industrial and commercial technologies.

State Energy Offices

State energy offices are also good sources of information, as well as good contacts to see what kind of incentive programs might be available in the state. Many states offer programs of free boiler tune-ups, free air conditioning system checks, seminars on energy efficiency for various facilities, and other services. Most state energy offices have well-stocked energy libraries, and they are also tied into other state energy research organizations, national laboratories, and the USDOE.

Equipment Suppliers

Equipment suppliers provide additional sources for data on energy efficiency improvements to processes. Marketing new, cost-effective processes and technologies provides sales for the companies, as well as helping industries to be more productive and more economically competitive. The energy auditor should compare the information from all of the sources described above.

3.5.6 Conducting the Audit
Safety Considerations

Safety is the primary consideration in any industrial audit. The possibility of injury from hot objects, hazardous materials, slippery surfaces, drive belts, and electric shocks is far greater than when conducting residential and commercial audits. Safety glasses, safety shoes, durable clothing, and possibly a safety hat and breathing mask might be needed during some audits. Gloves should be worn while making any electrical measurements as well as while making any measurements around boilers, heaters, furnaces, steam lines, or other very hot pieces of equipment. In all cases, adequate attention to personal safety is a significant feature of any industrial audit.

Lighting

Lighting is not as great a percent of total industrial use as it is in the commercial sector on the average, but lighting is still a big energy use and cost area for many industrial facilities. A complete inventory of all lighting should be taken during the audit visit. Hours of operation of lights are also necessary since lights are commonly left on when they are not needed. Timers, energy management systems, and occupancy sensors are all valuable approaches to insuring that lights that are not needed are not turned on. It is also important to look at the facility's outside lighting for parking and storage areas.

During the lighting inventory, types of tasks being performed should also be noted since light replacement with more efficient lamps often involves changing the color of the resultant light. For example, high pressure sodium lamps are much more efficient than mercury vapor lamps, or even metal halide lamps, but they produce a yellowish light that makes fine color distinction difficult. However, many assembly tasks can still be performed adequately under high pressure sodium lighting. These typically include metal fabrication, wood product fabrication, plastic extrusion, and many others.

Electric Motors

A common characteristic of many industries is their extensive use of electric motors. A complete inventory of all motors over 1 hp should be taken, as well as data being recorded on how long each motor operates during a day. For motors with substantial usage times, replacement with high-efficiency models is almost always cost effective. In addition, consideration should be given to replacement of standard drive belts with synchronous belts that transmit the motor energy more efficiently. For motors which are used infrequently, it may be possible to shift the use to off-peak times and to achieve a kW demand reduction which would reduce energy cost.

HVAC Systems

An inventory of all space heaters and air conditioners should be taken. Btu per hour ratings and efficiencies of all units should be recorded, as well as usage patterns. Although many industries do not heat or air condition the production floor area, they almost always have office areas, cafeterias, and other areas that are normally heated and air conditioned. For these conditioned areas, the construction of the facility should be noted—how much insulation, what the walls and ceilings are made of, how high the ceilings are. Adding additional insulation or high bay anti-stratification fans might be cost effective ECOs for heating needs. Evaporative cooling or spot cooling may provide cooling savings.

Production floors that are not air conditioned often have large numbers of ventilating fans that operate anywhere from one shift per day to 24 hours a day. Plants with high heat loads and plants in the mild climate areas often leave these ventilating fans running all year long. These are good candidates for high efficiency motor replacements. Timers or an energy management system might be used to turn off these ventilating fans when the plant is shut down.

Boilers

All boilers should be checked for efficient operation using a stack gas combustion analyzer. Boiler specifications on Btu per hour ratings, pressures, and temperatures should be recorded. The boiler should be varied between low-fire, normal-fire, and high-fire, with combustion gas and temperature readings taken at each level. Boiler tune-up is one of the most common and most energy-saving operations available to many facilities. The auditor should check to see whether any waste heat from the boiler is being recovered for use in a heat recuperator or for some other use, such as wa-

Library, Nova Scotia Community College

ter heating. If not, this should be noted as a potential ECO. Over-sized boilers and bare hot surfaces create large standby losses that can be reduced with right-size equipment and insulation.

Specialized Equipment

Most of the remaining equipment encountered during the industrial audit will be highly specialized process production equipment and machines. This equipment should all be examined and operational data taken, as well as noting hours and periods of use. All heat sources should be considered carefully as to whether they could be replaced with sources using waste heat, or whether a particular heat source could serve as a provider of waste heat to another application. Operations where both heating and cooling occur periodically, such as a plastic extrusion machine, are good candidates for reclaiming waste heat, or sharing heat from a machine needing cooling with another machine needing heat.

Air Compressors

Air compressors should be examined for size, operating pressures, and type (reciprocating or screw), as well as whether they use outside cool air for intake. Often large air compressors are operated at night when much smaller units are sufficient. Also, screw-type air compressors use a large fraction of their rated power when they are idling, so control valves should be installed to prevent this loss. Efficiency is improved with intake air that is cool, so outside air should be used in most cases, except in extremely cold temperature areas.

The auditor should determine whether there are significant air leaks in air hoses, fittings, and machines. Air leaks are a major source of energy loss in many facilities, and they should be corrected by maintenance action. Finally, air compressors are a good source of waste heat. Nearly 90% of the energy used by an air compressor shows up as waste heat, so this is a large source of low temperature waste heat for heating input air to a heater or boiler, or for heating hot water for process use. Efficiency improvements can also come from reducing the compressed air pressure and converting high volume/low pressure points of use to air blowers, rather than regulating compressed air.

3.6 COMMERCIAL AUDITS

3.6.1 Introduction

Commercial audits span the range from very simple audits for small offices to very complex audits for multi-story office buildings or large shopping centers. Complex commercial audits are performed in substantially the same manner as industrial audits. The following discussion highlights those areas where commercial audits are likely to differ from industrial audits.

Commercial audits generally involve consideration of the structural envelope, as well as lighting, people, equipment, ventilation, and control systems at the facility. Office buildings, shopping centers, and malls all have complex building envelopes that should be examined and evaluated. Building materials, insulation levels, door and window construction, skylights, and many other envelope features must be considered in order to identify candidate ECOs. This step also establishes the relative contribution of envelope loads to overall energy use, and this proportion will vary depending upon the internal loads.

Commercial facilities also have large capacity equipment, such as chillers, space heaters, water heaters, refrigerators, heaters, cookers, and office equipment like computers and copy machines. Small cogeneration systems may be found in commercial facilities and institutions such as schools and hospitals. Much of the equipment in commercial facilities is the same type and size as that found in manufacturing or industrial facilities. Potential ECOs would look at more efficient equipment, use of waste heat, or operational changes to use less expensive energy.

3.6.2 Commercial Audit Services

Electric and gas utilities, as well as many engineering consulting firms, perform audits for commercial facilities. Some utilities offer free walk-through audits for commercial customers and also offer financial incentives for customers who change to more energy efficient equipment. Schools, hospitals and some other government institutions can qualify for free audits under the ICP program described in Section 3.5.3 of this chapter, for those states that still fund the ICP program. Whoever conducts the commercial audit must initiate the ICP process by collecting information on the energy rate structures, the equipment in use at the facility, and the operational procedures used there.

3.6.3 Commercial Energy Rate Structures

Small commercial customers are usually billed for energy on a per energy unit basis, while large commercial customers are billed under complex rate structures containing components related to energy, rate of energy use (power), time of day or season of year, power factor, and numerous other elements. One of the first steps in a commercial audit is to obtain the rate structures for all sources of energy and to analyze at least one to two year's worth of energy bills. This information should be put into a table and also plotted.

3.6.4 Conducting the Audit

A significant difference in industrial and commercial audits arises in the area of lighting. Lighting in commercial facilities is one of the largest energy costs, sometimes accounting for half or more of the entire electric bill. Lighting levels and lighting quality are extremely important to many commercial operations. Retail sales operations in particular want light levels that are far in excess of standard office values. Quality of light in terms of color is also a big concern in retail sales, so finding acceptable ECOs for reducing lighting costs is much more difficult for retail facilities than for office buildings. The challenge is to find new lighting technologies that allow high light levels and warm color while reducing the wattage required. New T8 and T10 fluorescent lamps and metal halide lamp replacements for mercury vapor lamps offer these features, and they usually represent cost-effective ECOs for retail sales and other commercial facilities.

Energy use intensity (EUI) benchmark values are available in terms of Btu/SF and provide a good starting point for audits of commercial facilities. Common business sectors include:

Education
Food sales
Food service
Health care
Lodging
Retail
Office
Public assembly
Churches
Warehouses

The Energy Information Administration (EIA) is a source of such data, specifically their commercial building energy consumption survey (CBECS). Other sources of benchmark data are industry trade associations for particular sectors.

3.7 RESIDENTIAL AUDITS

Audits for large, multi-story apartment buildings can be very similar to commercial audits. (See section 3.6.) Audits of single-family residences, however, are generally fairly simple. For single-family structures, the energy audit focuses on the thermal envelope and appliances such as the heater, air conditioner, water heater, and "plug loads."

The residential auditor should start by obtaining past energy bills and analyzing them to determine any patterns or anomalies, as well as to available benchmark data for energy use intensity per SF (EUI). During the audit visit, the structure is examined to determine the levels of insulation, the conditions of (and seals for) windows and doors, and the integrity of the ducts. The space heater and/or air conditioner is inspected, along with the water heater. Equipment model numbers, age, size, and efficiencies are recorded, as well as equipment condition and evidence of regular maintenance. The post-audit analysis then evaluates potential ECOs, such as adding insulation, adding double-pane windows, window shading or insulated doors, and changing to higher efficiency heaters, air conditioners, and water heaters. The auditor calculates costs, benefits, and simple payback periods and then presents them to the owner or occupant. A simple audit report, often in the form of a computer printout, is given to the owner or occupant.

3.8 INDOOR AIR QUALITY

Implementation of new energy-related standards and practices has sometimes contributed to a degradation of indoor air quality. One example is that homes built tighter with less leakage and no outside ventilation allow off-gassing of construction materials and volatile organic compounds (VOCs) to build up to higher levels than would be expected with more 'loose' construction. For commercial buildings, an example is the reduction of ventilation requirements in the 80s to save energy; this was a classic case of solving one problem and creating a new one. A variety of indoor air quality (IAQ) issues stemmed from this, and the ventilation standards were subsequently repealed. Ventilation standards are a continuing research topic since there are clearly competing interests between reducing ventilation to save energy and increasing ventilation to benefit the indoor occupants.

The energy audit team is generally not tasked with being IAQ experts or to include detailed testing, as would be done by specialists and industrial hygienists. However, an awareness of the phenomena and the potential for some conservation measures to create or exacerbate IAQ issues is important. Any measures that alter exhaust system or ventilation systems have this potential, either by changing the amount of ventilation or the overall air balance of the building. Even changing filters can have an inadvertent negative effect. An example would be a change from flat filters to bag filters to reduce air pressure drop and fan horsepower; bag filters have a tendency to release a puff of dust upon initial start-up and thus occasionally increase airborne particulates. Microbial contamination is also a potential problem indoors, so measures that include water-cooling (cooling tower) or evaporative cooling have this potential. Whenever evaporative cooling measures are suggested, emphasis on routine maintenance and sanitizing such equipment is good practice, to both help educate the customer about the technology and reduce such complications. One other example occurs in dry climates where humidifiers are used. Humidifiers are high energy users, so an ECM may be to lower the rH setting for humidifiers or remove them altogether; however, while this will save energy, there may be occupants that are sensitive to changes in humidity and may experience health complications as a result of the ECM.

If, as part of an energy audit, any IAQ problems are suspected, communicating the concern to the customer is prudent, along with suggestions to consider testing or further review by qualified individuals.

Chapter 17 is dedicated to this important subject.

Ventilation Rates

Recommended ventilation quantities for commercial and institutional buildings are published by the American Society of Heating, Refrigerating, and Air-Conditioning Engineers (ASHRAE) in Standard 62.1-2007, "Ventilation for Acceptable Air Quality for Non-Residential Buildings." These ventilation rates are for effective systems. Many existing systems fail in entraining the air mass efficiently. The density of the contaminants relative to air must be considered in locating the exhaust air intakes and ventilation supply air registers.

Liability

Liability related to indoor air problems appears to be a growing but uncertain issue, because few cases have made it through the court system. However, in retrospect, the asbestos and ureaformaldehyde pollu-

tion problems discovered in the past suggest proceeding with caution and a proactive approach.

3.9 CONCLUSION

Energy audits are an important first step in the overall process of reducing energy costs for any building, company, or industry. A thorough audit identifies and analyzes the changes in equipment and operations that will result in cost-effective energy cost reduction. The energy auditor plays a key role in the successful conduct of an audit, as well as in the implementation of the audit recommendations.

References

1. *Commercial Energy Auditing Reference Handbook*, Doty, S., Fairmont Press, 2008.

Bibliography

Instructions For Energy Auditors, Volumes I and II, U.S. Department of Energy, DOE/CS-0041/12&13, September, 1978. Available through National Technical Information Service, Springfield, VA.

Encyclopedia of Energy Engineering and Technology, Barney L. Capehart, Editor, Taylor and Francis/CRC Publishers, Boca Raton, FL, 2007.

Energy Conservation Guide for Industry and Commerce, National Bureau of Standards Handbook 115 and Supplement, 1978. Available through U.S. Government Printing Office, Washington, DC.

Guide to Energy Management, Sixth Edition, Capehart, B.L., Turner, W.C., and Kennedy, W.J., The Fairmont Press, Atlanta, GA, 2008.

Illuminating Engineering Society, *IES Lighting Handbook, Ninth Edition*, New York, NY, 2000.

Total Energy Management, A Handbook prepared by the National Electrical Contractors Association and the National Electrical Manufacturers Association, Washington, DC.

Handbook of Energy Audits, Thumann, Albert and William J. Younger, Seventh Edition, The Fairmont Press, Atlanta, GA.

Industrial Energy Management and Utilization, Witte, Larry C., Schmidt, Philip S., and Brown, David R., Hemisphere Publishing Corporation, Washington, DC, 1988.

Threshold Limit Values for Chemical Substances and Physical Agents and Biological Exposure Indices, 1990-91 American Conference of Governmental Industrial Hygienists.

Ventilation for Acceptable Indoor Air Quality, ASHRAE 62.1-2007, American Society of Heating, Refrigerating and Air-Conditioning Engineers, Inc., 2007.

Facility Design and Planning Engineering Weather Data, Departments of the Air Force, the Army, and the Navy, 1978.

Handbook of Energy Engineering, Fifth Edition, Thumann, A., and Mehta, D.P., The Fairmont Press, Atlanta, GA, 2004.

CHAPTER 4

ECONOMIC ANALYSIS

DR. DAVID PRATT
Industrial Engineering and Management
Oklahoma State University, Stillwater, OK

4.1 OBJECTIVE

The objective of this chapter is to present a coherent, consistent approach to economic analysis of capital investments (energy related or other). Adherence to the concepts and methods presented will lead to sound investment decisions with respect to time value of money principles. The chapter opens with material designed to motivate the importance of life cycle cost concepts in the economic analysis of projects. The next three sections provide foundational material necessary to fully develop time value of money concepts and techniques. These sections present general characteristics of capital investments, sources of funds for capital investment, and a brief summary of tax considerations which are important for economic analysis. The next two sections introduce time value of money calculations and several approaches for calculating project measures of worth based on time value of money concepts. Following these is a section presenting material to address several special problems that may be encountered in economic analysis. This material includes, among other things, discussions of escalation and inflation, non-annual compounding of interest, and life cycle costing analysis. The chapter closes with a brief summary and a list of references that can provide additional depth in many of the areas covered in the chapter.

4.2 INTRODUCTION

Capital investment decisions arise in many circumstances. The circumstances range from evaluating business opportunities to personal retirement planning. Regardless of circumstances, the basic criterion for evaluating any investment decision is that the revenues (savings) generated by the investment must be greater than the costs incurred. The number of years over which the revenues accumulate and the comparative importance of future dollars (revenues or costs) relative to present dollars are important factors in making sound investment decisions. This consideration of costs over the entire life cycle of the investments gives rise to the name *life cycle cost* analysis that is commonly used to refer to the economic analysis approach presented in this chapter. An example of the importance of life cycle costs is shown in Figure 4-1, which depicts the estimated costs of owning and operating an oil-fired furnace to heat a 2,000 square foot house in the northeast United States. Of particular note is that the initial costs represent only 23% of the total costs incurred over the life of the furnace. The life cycle cost approach provides a significantly better evaluation of long term implications of an investment than methods which focus on first cost or near-term results.

Life cycle cost analysis methods can be applied to virtually any public or private business sector investment decision as well as to personal financial planning decisions. Energy related decisions provide excellent examples for the application of this approach. Such decisions include: evaluation of alternative building designs which have different initial costs, operating and maintenance costs, and perhaps different lives; evaluation of investments to improve the thermal performance of an existing building (wall or roof insulation, window glazing); and evaluation of alternative heating, ventilating, or air conditioning systems. For federal buildings, Congress and the President have mandated, through legislation and executive order, energy conservation goals that must be met using cost-effective measures. This mandate included use of the life cycle cost approach as the means of evaluating cost effectiveness.

4.3 GENERAL CHARACTERISTICS OF CAPITAL INVESTMENTS

4.3.1 Capital Investment Characteristics

When companies spend money, the outlay of cash can be broadly categorized into one of two classifica-

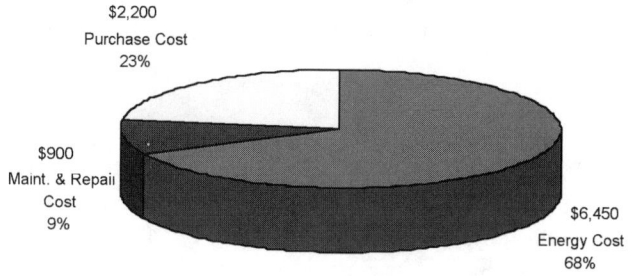

Figure 4-1. 15-year Life Cycle Costs of a Heating System

tions—expenses or capital investments. Expenses are generally those cash expenditures that are routine, on-going, and necessary for the ordinary operation of the business. Capital investments, on the other hand, are generally more strategic and have long term effects. Decisions made regarding capital investments are usually made at higher levels within the organizational hierarchy and carry with them additional tax consequences as compared to expenses.

Three characteristics of capital investments are of concern when performing life cycle cost analysis. First, capital investments usually require a relatively large initial cost. "Relatively large" may mean several hundred dollars to a small company or many millions of dollars to a large company. The initial cost may occur as a single expenditure, such as purchasing a new heating system, or it may occur over a period of several years, such as designing and constructing a new building. It is not uncommon that the funds available for capital investments projects are limited. In other words, the sum of the initial costs of all the viable and attractive projects exceeds the total available funds. This creates a situation known as capital rationing, which imposes special requirements on the investment analysis.

The second important characteristic of a capital investment is that the benefits (revenues or savings) resulting from the initial cost occur in the future, normally over a period of years. The period between the initial cost and the last future cash flow is the life cycle or life of the investment. It is the fact that cash flows occur over the investment's life that requires the introduction of time value of money concepts to properly evaluate investments. If multiple investments are being evaluated and the lives of the investments are not equal, special consideration must be given to the issue of selecting an appropriate planning horizon for the analysis.

The last important characteristic of capital investments is that they are relatively irreversible. Frequently, after the initial investment has been made, terminating or significantly altering the nature of a capital investment has substantial (usually negative) cost consequences. This is one of the reasons that capital investment decisions are usually evaluated at higher levels of the organizational hierarchy than are operating expense decisions.

4.3.2 Capital Investment Cost Categories

In almost every case, the costs which occur over the life of a capital investment can be classified into one of the following categories:

- Initial cost
- Annual expenses and revenues
- Periodic replacement and maintenance
- Salvage value

As a simplifying assumption, the cash flows that occur during a year are generally summed and regarded as a single end-of-year cash flow. While this approach does introduce some inaccuracy in the evaluation, it is generally not regarded as significant relative to the level of estimation associated with projecting future cash flows.

Initial costs include all costs associated with preparing the investment for service. This includes purchase cost as well as installation and preparation costs. Initial costs are usually nonrecurring during the life of an investment. Annual expenses and revenues are the recurring costs and benefits generated throughout the life of the investment. Periodic replacement and maintenance costs are similar to annual expenses and revenues, except that they do not (or are not expected to) occur annually. The salvage value (residual value) of an investment is the revenue (or expense) attributed to disposal of the investment at the end of its useful life.

4.3.3 Cash Flow Diagrams

A convenient way to display the revenues (savings) and costs associated with an investment is a *cash flow diagram*. By using a cash flow diagram, the timing of the cash flows are more apparent and the chances of properly applying time value of money concepts are increased. With practice, different cash flow patterns can be recognized and they, in turn, may suggest the most direct approach for analysis.

It is usually advantageous to determine the time frame over which the cash flows occur first. This establishes the horizontal scale of the cash flow diagram. This scale is divided into time periods which are frequently, but not always, years. Receipts and disbursements are then located on the time scale in accordance with the problem specifications. Individual outlays or receipts are indicated by drawing vertical lines appropriately placed along the time scale. The relative magnitudes can be suggested by the heights, but exact scaling generally does not enhance the meaningfulness of the diagram. Upward directed lines indicate cash inflow (revenues or savings) while downward directed lines indicate cash outflow (costs).

Figure 4-2 illustrates a cash flow diagram. The cash flows depicted represent an economic evaluation of whether to choose a baseboard heating and window air conditioning system or a heat pump for a ranger's house in a national park [Fuller and Petersen, 1994]. The differential costs associated with the decision are:

- The heat pump costs (cash outflow) $1500 more than the baseboard system
- The heat pump saves (cash inflow) $380 annually in electricity costs
- The heat pump has a $50 higher annual maintenance cost (cash outflow)
- The heat pump has a $150 higher salvage value (cash inflow) at the end of 15 years
- The heat pump requires $200 more in replacement maintenance (cash outflow) at the end of year 8.

Although cash flow diagrams are simply graphical representations of income and outlay, they should exhibit as much information as possible. During the analysis phase, it is useful to show the minimum attractive rate of return (an interest rate used to account for the time value of money within the problem) on the cash flow diagram, although this has been omitted in Figure 4-2. The requirements for a good cash flow diagram are completeness, accuracy, and legibility. The measure of a successful diagram is that someone else can understand the problem fully from it.

4.4   SOURCES OF FUNDS

Capital investing requires a source of funds. For large companies, multiple sources may be employed. The process of obtaining funds for capital investment is called *financing*. There are two broad sources of financial funding—debt financing and equity financing. Debt financing involves borrowing and utilizing money which is to be repaid at a later point in time. Interest is paid to the lending party for the privilege of using the money. Debt financing does not create an ownership position for the lender within the borrowing organization. The borrower is simply obligated to repay the borrowed funds, plus accrued interest, according to a repayment schedule.

Car loans and mortgage loans are two examples of this type of financing. The two primary sources of debt capital are loans and bonds. The cost of capital associated with debt financing is relatively easy to calculate, since interest rates and repayment schedules are usually clearly documented in the legal instruments controlling the financing arrangements. An added benefit to debt financing under current U.S. tax law (as of December 2008) is that the interest payments made by corporations on debt capital are tax deductible. This effectively lowers the cost of debt financing since for debt financing, with deductible interest payments, the after-tax cost of capital is given by:

$$\text{Cost of Capital}_{\text{AFTERTAX}} = \text{Cost of Capital}_{\text{BEFORETAX}} * (1 - \text{Tax Rate})$$

where the tax rate is determine by applicable tax law.

The second broad source of funding is equity financing. Under equity financing the lender acquires an ownership (or equity) position within the borrower's organization. As a result of this ownership position, the lender has the right to participate in the financial success of the organization as a whole. The two primary sources of equity financing are stocks and retained earnings. The cost of capital associated with shares of stock is much debated within the financial community. A detailed presentation of the issues and approaches is beyond the scope of this chapter. Additional reference material can be found in Park and Sharp-Bette [1990]. One issue over which there is general agreement is that the cost of capital for stocks is higher than the cost of capital for debt financing. This is at least partially attributable to the fact that interest payments are tax deductible, while stock dividend payments are not.

If any subject is more widely debated in the financial community than the cost of capital for stocks, it is the cost of capital for retained earnings. Retained earnings are the accumulation of annual earnings surpluses that

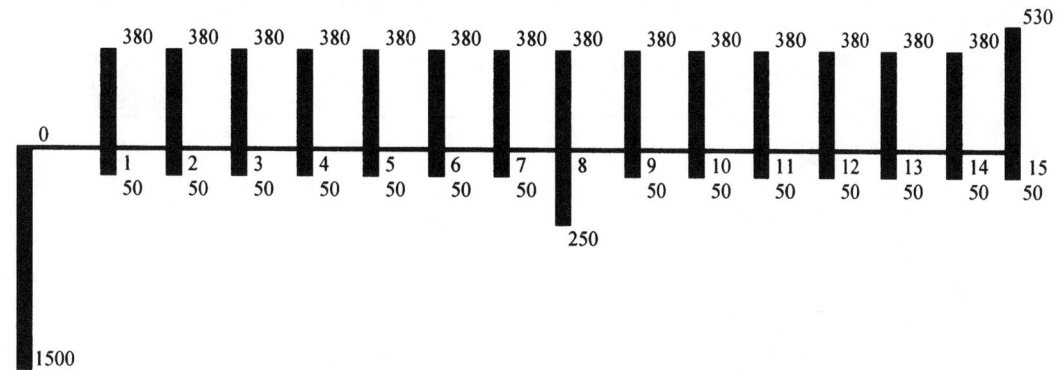

Figure 4-2. Heat Pump and Baseboard System Differential Life Cycle Costs

a company retains within the company's coffers rather than pays out to the stockholders as dividends. Although these earnings are held by the company, they truly belong to the stockholders. In essence the company is establishing the position that by retaining the earnings and investing them in capital projects, the stockholders will achieve at least as high a return through future financial successes as they would have earned if the earnings had be paid out as dividends. Hence, one common approach to valuing the cost of capital for retained earnings is to apply the same cost of capital as for stock. This, therefore, leads to the same generally agreed result. The cost of capital for financing through retained earnings generally exceeds the cost of capital for debt financing.

In many cases the financing for a set of capital investments is obtained by packaging a combination of the above sources to achieve a desired level of available funds. When this approach is taken, the overall cost of capital is generally taken to be the weighted average cost of capital across all sources. The cost of each individual source's funds is weighted by the source's fraction of the total dollar amount available. By summing across all sources, a weighted average cost of capital is calculated.

Example 1

Determine the weighted average cost of capital for financing which is composed of:

25% loans with a before tax cost of capital of 12%/yr and

75% retained earnings with a cost of capital of 10%/yr.

The company's effective tax rate is 34%.

Cost of Capital$_{LOANS}$ = 12% * (1 – 0.34) = 7.92%

Cost of Capital$_{RETAINEDEARNINGS}$ = 10%

Weighted Average Cost of Capital = (0.25)*7.92% + (0.75)*10.00% = 9.48%

4.5 TAX CONSIDERATIONS

4.5.1 After Tax Cash Flows

Taxes are a fact of life in both personal and business decision making. Taxes occur in many forms and are primarily designed to generate revenues for governmental entities ranging from local authorities to the federal government. A few of the most common forms of taxes are income taxes, ad valorem taxes, sales taxes, and excise taxes. Cash flows used for economic analysis should always be adjusted for the combined impact of all relevant taxes. To do otherwise, ignores the significant impact that taxes have on economic decision making. Tax laws and regulations are complex and intricate. A detailed treatment of tax considerations as they apply to economic analysis is beyond the scope of this chapter and generally requires the assistance of a professional with specialized training in the subject. A high level summary of concepts and techniques that concentrate on federal income taxes are presented in the material that follows. The focus is on federal income taxes since they impact most decisions and have relatively wide and general application.

The amount of federal taxes due are determined based on a tax rate multiplied by a taxable income. The rates (as of December 2008) are determined based on tables of rates published under the Omnibus Reconciliation Act of 1993 as shown in Table 4-1. Depending on income range, the marginal tax rates vary from 15% of taxable income to 39% of taxable income. *Taxable income* is calculated by subtracting *allowable deductions* from *gross income*. Gross income is generated when a company sells its product or service. Allowable deductions include salaries and wages, materials, interest payments, and depreciation as well as other costs of doing business as detailed in the tax regulations.

The calculation of taxes owed and after-tax cash

Table 4-1. Federal Tax Rates based on the Omnibus Reconciliation Act of 1993

Taxable Income (TI)	Taxes Due	Marginal Tax Rate
$0 < TI ≤ $50,000	0.15*TI	0.15
$50,000 < TI ≤ $75,000	$7,500+0.25(TI-$50,000)	0.25
$75,000 < TI ≤ $100,000	$13,750+0.34(TI-$75,000)	0.34
$100,000 < TI ≤ $335,000	$22,250+0.39(TI-$100,000)	0.39
$335,000 < TI ≤ $10,000,000	$113,900+0.34(TI-$335,000)	0.34
$10,000,000 < TI ≤ $15,000,000	$3,400,000+0.35(TI-$10,000,000)	0.35
$15,000,000 < TI ≤ $18,333,333	$5,150,000+0.38(TI-$15,000,000)	0.38
$18,333,333 < TI	$6,416,667+0.35(TI-$18,333,333)	0.35

flows (ATCF) require knowledge of:

- Before tax cash flows (BTCF), the net project cash flows before the consideration of taxes due, loan payments, and bond payments
- Total loan payments attributable to the project, including a breakdown of principal and interest components of the payments
- Total bond payments attributable to the project, including a breakdown of the redemption and interest components of the payments
- Depreciation allowances attributable to the project.

Given the availability of the above information, the procedure to determine the ATCF on a year by year basis proceeds, using the following calculation for each year:

- Taxable Income = BTCF – Loan Interest – Bond Interest – Deprecation
- Taxes = Taxable Income * Tax Rate
- ATCF = BTCF – Total Loan Payments – Total Bond Payments – Taxes

An important observation is that depreciation reduces taxable income (hence, taxes) but does not directly enter into the calculation of ATCF since it is not a true cash flow. It is not a true cash flow because no cash changes hands; depreciation is an accounting concept designed to stimulate business by reducing taxes over the life of an asset. The next section provides additional information about depreciation.

4.5.2 Depreciation

Most assets used in the course of a business decrease in value over time. U.S. federal income tax law permits reasonable deductions from taxable income to allow for this. These deductions are called depreciation allowances. To be depreciable, an asset must meet three primary conditions: (1) it must be held by the business for the purpose of producing income; (2) it must wear out or be consumed in the course of its use; and (3) it must have a life longer than a year.

Many methods of depreciation have been allowed under U.S. tax law over the years. Among these methods are straight line, sum-of-the-years digits, declining balance, and the accelerated cost recovery system. Descriptions of these methods can be found in many references, including economic analysis text books [White et al. 1998]. The method currently used for depreciation of assets placed in service after 1986 is the modified accelerated cost recovery system (MACRS). Determination of the allowable MACRS depreciation deduction for an asset is a function of (1) the asset's property class; (2) the asset's basis; and (3) the year within the asset's recovery period for which the deduction is calculated.

Eight property classes are defined for assets which are depreciable under MACRS. The property classes and several examples of property that fall into each class are shown in Table 4-2. Professional tax guidance is recommended to determine the MACRS property class for a specific asset.

The basis of an asset is the cost of placing the asset in service. In most cases, the basis includes the purchase cost of the asset plus the costs necessary to place the asset in service (e.g., installation charges).

Given an asset's property class and its depreciable basis, the depreciation allowance for each year of an asset's life can be determined from tabled values of MACRS

Table 4-2. MACRS Property Classes

Property Class	Example Assets
3-Year Property	special handling devices for food special tools motor vehicle manufacturing
5-Year Property	computers and office machines general purpose trucks
7-Year Property	office furniture most manufacturing machine tools
10-Year Property	tugs & water transport equipment petroleum refining assets
15-Year Property	fencing and landscaping cement manufacturing assets
20-Year Property	farm buildings utility transmission lines and poles
27.5-Year Residential Rental Property	rental houses and apartments
31.5-Year Nonresidential Real Property	business buildings

percentages. The MACRS percentages specify the percentage of an asset's basis that are allowable as deductions during each year of an asset's recovery period. The MACRS percentages by recovery year (age of the asset) and property class are shown in Table 4-3.

Example 2

Determine depreciation allowances during each recovery year for a MACRS 5-year property with a basis of $10,000.

Year 1 deduction: $10,000 * 20.00% = $2,000
Year 2 deduction: $10,000 * 32.00% = $3,200
Year 3 deduction: $10,000 * 19.20% = $1,920
Year 4 deduction: $10,000 * 11.52% = $1,152
Year 5 deduction: $10,000 * 11.52% = $1,152
Year 6 deduction: $10,000 * 5.76% = $576

The sum of the deductions calculated in Example 2 is $10,000, which means that the asset is "fully depreciated" after six years. Though not shown here, tables similar to Table 4-3 are available for the 27.5-Year and 31.5-Year property classes. There usage is similar to that outlined above, except that depreciation is calculated monthly rather than annually.

4.6   TIME VALUE OF MONEY CONCEPTS

4.6.1 Introduction

Most people have an intuitive sense of the time value of money. Given a choice between $100 today and $100 one year from today, almost everyone would prefer the $100 today. Why is this the case? Two primary factors lead to this time preference associated with money—interest and inflation. Interest is the ability to earn a return on money that is loaned rather than consumed. By taking the $100 today and placing it in an interest bearing bank account (i.e., loaning it to the bank), one year from today an amount greater than $100 would be available for withdrawal. Thus, it is the preferred choice. The amount in excess of $100 that would be available depends upon the interest rate being paid by the bank. The next section develops the mathematics of the relationship between interest rates and the timing of cash flows.

The second factor which leads to the time preference associated with money is inflation. Inflation is a complex subject but in general can be described as a decrease in the purchasing power of money. The impact of inflation is that the "basket of goods" a consumer can buy today with $100 contains more than the "basket" the consumer could buy one year from today. This decrease in purchas-

Table 4-3. MACRS Percentages by Recovery Year and Property Class

Recovery Year	3-Year Property	5-Year Property	7-Year Property	10-Year Property	15-Year Property	20-Year Property
1	33.33%	20.00%	14.29%	10.00%	5.00%	3.750%
2	44.45%	32.00%	24.49%	18.00%	9.50%	7.219%
3	14.81%	19.20%	17.49%	14,40%	8.55%	6.677%
4	7.41%	11.52%	12.49%	11.52%	7.70%	6.177%
5		11.52%	8.93%	9.22%	6.93%	5.713%
6		5.76%	8.92%	7.37%	6.23%	5.285%
7			8.93%	6.55%	5.90%	4.888%
8			4.46%	6.55%	5.90%	4.522%
9				6.56%	5.91%	4.462%
10				6.55%	5.90%	4.461%
11				3.28%	5.91%	4.462%
12					5.90%	4.461%
13					5.91%	4.462%
14					5.90%	4.461%
15					5.91%	4.462%
16					2.5%	4.461%
17						4.462%
18						4.461%
19						4.462%
20						4.461%
21						2.231%

ing power is the result of inflation. The subject of inflation is addressed in Section 4.8.4.

4.6.2 The Mathematics of Interest

The mathematics of interest must account for the amount and timing of cash flows. The basic formula for studying and understanding interest calculations is:

$$F_n = P + I_n$$

where: F_n = a future amount of money at the *end* of the nth year,

P = a present amount of money at the *beginning* of the year which is n years prior to F,

I_n = the amount of accumulated interest *over* n years, and

n = the number of years between P and F

The goal of studying the mathematics of interest is to develop a formula for F_n that is expressed only in terms of the present amount P, the annual interest rate i, and the number of years n. The two major approaches for determining the value of I_n are simple interest and compound interest. Under simple interest, interest is earned (charged) only on the original amount loaned (borrowed). Under compound interest, interest is earned (charged) on the original amount loaned (borrowed) plus any interest accumulated from previous periods.

4.6.3 Simple Interest

For simple interest, interest is earned (charged) only on the original principal amount at the rate of i% per year (expressed as i%/yr). Table 4-4 illustrates the annual cal-culation of simple interest. In Table 4-4 and the formulas which follow, the interest rate i is to be expressed as a decimal amount (e.g., 8% interest is expressed as 0.08).

At the beginning of year 1 (end of year 0), P dollars (e.g., $100) are deposited in an account earning i%/yr (e.g., 8%/yr or 0.08) simple interest. Under simple com-pounding, during year 1 the P dollars ($100) earn P*i dol-lars ($100*0.08 = $8) of interest. At the end of the year 1 the balance in the account is obtained by adding P dol-lars (the original principal, $100) plus P*i (the interest earned during year 1, $8) to obtain P+P*i ($100+$8=$108). Through algebraic manipulation, the end of year 1 bal-ance can be expressed mathematically as P*(1+i) dollars ($100*1.08=$108).

The beginning of year 2 is the same point in time as the end of year 1 so the balance in the account is P*(1+i) dollars ($108). During year 2 the account again earns P*i dollars ($8) of interest since under simple compounding, interest is paid only on the original principal amount P ($100). Thus at the end of year 2, the balance in the ac-count is obtained by adding P dollars (the original princi-pal) plus P*i (the interest from year 1) plus P*i (the inter-est from year 2) to obtain P+P*i+P*i ($100+$8+$8=$116). After some algebraic manipulation, this can be writ-ten conveniently mathematically as P*(1+2*i) dollars ($100*1.16=$116).

Table 4-4 extends the above logic to year 3 and then generalizes the approach for year n. If we return our at-tention to our original goal of developing a formula for F_n that is expressed only in terms of the present amount P, the annual interest rate i, and the number of years n, the above development and Table 4-4 results can be sum-marized as follows:

Table 4-4. The Mathematics of Simple Interest

Year (t)	Amount At Beginning Of Year	Interest Earned During Year	Amount At End Of Year (F_t)
0	—	—	P
1	P	Pi	P + Pi = P (1 + i)
2	P (1 + i)	Pi	P (1+ i) + Pi = P (1 + 2i)
3	P (1 + 2i)	Pi	P (1+ 2i) + Pi = P (1 + 3i)
n	P (1 + (n-1)i)	Pi	P (1+ (n-1)i) + Pi = P (1 + ni)

For Simple Interest

$$F_n = P(1+n \cdot i)$$

Example 3

Determine the balance which will accumulate at the end of year 4 in an account which pays 10%/yr simple interest if a deposit of $500 is made today.

$$F_n = P * (1 + n \cdot i)$$
$$F_4 = 500 * (1 + 4 \cdot 0.10)$$
$$F_4 = 500 * (1 + 0.40)$$
$$F_4 = 500 * (1.40)$$
$$F_4 = \$700$$

4.6.4 Compound Interest

For compound interest, interest is earned (charged) on the original principal amount *plus any accumulated interest from previous years* at the rate of i% per year (i%/yr). Table 4-5 illustrates the annual calculation of compound interest. In the Table 4-5 and the formulas which follow, i is expressed as a decimal amount (i.e., 8% interest is expressed as 0.08).

At the beginning of year 1 (end of year 0), P dollars (e.g., $100) are deposited in an account earning i%/yr (e.g., 8%/yr or 0.08) compound interest. Under compound interest, during year 1 the P dollars ($100) earn P*i dollars ($100*0.08 = $8) of interest. Notice that this the same as the amount earned under simple compounding. This result is expected since the interest earned in *previous* years is zero for year 1. At the end of the year 1 the balance in the account is obtained by adding P dollars (the original principal, $100) plus P*i (the interest earned during year 1, $8) to obtain P+P*i ($100+$8=$108). Through algebraic manipulation, the end of year 1 balance can be expressed mathematically as P*(1+i) dollars ($100*1.08=$108).

During year 2 and subsequent years, we begin to see the power (if you are a lender) or penalty (if you are a borrower) of compound interest over simple interest. The beginning of year 2 is the same point in time as the end of year 1 so the balance in the account is P*(1+i) dollars ($108). During year 2 the account earns i% interest on the original principal, P dollars ($100), *and* it earns i% interest on the accumulated interest from year 1, P*i dollars ($8). Thus the interest earned in year 2 is [P+P*i]*i dollars ([$100+$8]*0.08=$8.64). The balance at the end of year 2 is obtained by adding P dollars (the original principal) plus P*i (the interest from year 1) plus [P+P*i]*i (the interest from year 2) to obtain P+P*i+[P+P*i]*i dollars ($100+$8+$8.64=$116.64). After some algebraic manipulation, this can be written conveniently mathematically as P*(1+i)^n dollars ($100*1.08^2 =$116.64).

Table 4-5 extends the above logic to year 3 and then generalizes the approach for year n. If we return our attention to our original goal of developing a formula for F_n that is expressed only in terms of the present amount P, the annual interest rate i, and the number of years n, the above development and Table 4-5 results can be summarized as follows:

For Compound Interest

$$F_n = P(1+i)^n$$

Table 4-5. The Mathematics of Compound Interest

Year (t)	Amount At Beginning Of Year	Interest Earned During Year	Amount At End Of Year (F_t)
0	—	—	P
1	P	Pi	$P + Pi$ $= P(1 + i)$
2	$P(1 + i)$	$P(1 + i)i$	$P(1+i) + P(1+i)i$ $= P(1+i)(1+i)$ $= P(1+i)^2$
3	$P(1+i)^2$	$P(1+i)^2 i$	$P(1+i)^2 + P(1+i)^2 i$ $= P(1+i)^2 (1+i)$ $= P(1+i)^3$
n	$P(1+i)^{n-1}$	$P(1+i)^{n-1} i$	$P(1+i)^{n-1} + P(1+i)^{n-1} i$ $= P(1+i)^{n-1}(1+i)$
i			$= P(1+i)^n$

Example 4

Repeat Example 3 using compound interest rather than simple interest.

$$F_n = P * (1 + i)^n$$
$$F_4 = 500 * (1 + 0.10)^4$$
$$F_4 = 500 * (1.10)^4$$
$$F_4 = 500 * (1.4641)$$
$$F_4 = \$732.05$$

Notice that the balance available for withdrawal is higher under compound interest ($732.05 > $700.00). This is due to earning interest on principal plus interest rather than earning interest on just original principal. Since compound interest is by far more common in practice than simple interest, the remainder of this chapter is based on compound interest unless explicitly stated otherwise.

4.6.5 Single Sum Cash Flows

Time value of money problems involving compound interest are common. Because of this frequent need, tables of compound interest time value of money factors can be found in most books and reference manuals that deal with economic analysis. The factor $(1+i)^n$ is known as the *single sum, future worth factor* or the *single payment, compound amount factor*. This factor is denoted $(F|P,i,n)$ where F denotes a future amount, P denotes a present amount, i is an interest rate (expressed as a percentage amount), and n denotes a number of years. The factor $(F|P,i,n)$ is read, "To find F given P at i% for n years." Tables of values of $(F|P,i,n)$ for selected values of i and n are provided in the chapter appendix. The tables of values in the chapter appendix are organized such that the annual interest rate (i) determines the appropriate page, the time value of money factor $(F|P)$ determines the appropriate column, and the number of years (n) determines the appropriate row.

Example 5

Repeat Example 4 using the single sum, future worth factor.

$$F_n = P * (1 + i)^n$$
$$F_n = P * (F|P,i,n)$$
$$F_4 = 500 * (F|P,10\%,4)$$
$$F_4 = 500 * (1.4641)$$
$$F_4 = 732.05$$

The above formulas for compound interest allow us to solve for an unknown F given P, i, and n. What if we want to determine P with known values of F, i, and n? We can derive this relationship from the compound interest formula above:

$$F_n = P (1+i)^n$$

Dividing both sides by $(1+i)^n$ yields

$$P = \frac{F_n}{(1+I)^n}$$

which can be rewritten as

$$P = F_n (1+i)^{-n}$$

The factor $(1+i)^{-n}$ is known as the single sum, present worth factor or the *single payment, present worth factor*. This factor is denoted $(P|F,i,n)$ and is read, "To find P given F at i% for n years." Tables of $(P|F,i,n)$ are provided in the chapter appendix.

Example 6

To accumulate $1000 five years from today in an account earning 8%/yr compound interest, how much must be deposited today?

$$P = F_n * (1 + i)^{-n}$$
$$P = F_5 * (P|F,i,n)$$
$$P = 1000 * (P|F,8\%,5)$$
$$P = 1000 * (0.6806)$$
$$P = 680.60$$

To verify your solution, try multiplying 680.60 * $(F|P,8\%,5)$. What would you expect for a result? (Answer: $1000.) If you're still not convinced, try building a table like Table 4-5 to calculate the year end balances each year for five years.

4.6.6 Series Cash Flows

Having considered the transformation of a single sum to a future worth when given a present amount and vice versa, let us generalize to a series of cash flows. The future worth of a series of cash flows is simply the sum of the future worths of each individual cash flow. Similarly, the present worth of a series of cash flows is the sum of the present worths of the individual cash flows.

Example 7

Determine the future worth (accumulated total) at the end of seven years in an account that earns 5%/yr if a $600 deposit is made today and a $1000 deposit is made at the end of year two?

For the $600 deposit, n=7 (years between today and end of year 7).

For the $1000 deposit, n=5 (years between end of year 2

and end of year 7).

$F_7 = 600 * (F|P,5\%,7) + 1000 * (F|P,5\%,5)$
$F_7 = 600 * (1.4071) + 1000 * (1.2763)$
$F_7 = 844.26 + 1276.30 = \$2120.56$

Example 8

Determine the amount that would have to be deposited today (present worth) in an account paying 6%/yr interest if you want to withdraw $500 four years from today and $600 eight years from today (leaving zero in the account after the $600 withdrawal).

For the $500 deposit n=4, for the $600 deposit n=8

$P = 500 * (P|F,6\%,4) + 600 * (P|F,6\%,8)$
$P = 500 * (0.7921) + 600 * (0.6274)$
$P = 396.05 + 376.44 = \$772.49$

4.6.7 Uniform Series Cash Flows

A uniform series of cash flows exists when the cash flows in a series occur every year and are all equal in value. Figure 4-3 shows the cash flow diagram of a uniform series of withdrawals. The uniform series has length 4 and amount 2000. If we want to determine the amount of money that would have to be deposited today to support this series of withdrawals starting one year from today, we could use the approach illustrated in Example 8 above to determine a present worth component for each individual cash flow. This approach would require us to sum the following series of factors (assuming the interest rate is 9%/yr):

$P = 2000*(P|F,9\%,1) + 2000*(P|F,9\%,2) + 2000*(P|F,9\%,3) + 2000*(P|F,9\%,4)$

After some algebraic manipulation, this expression can be restated as:

$P = 2000*[(P|F,9\%,1) + (P|F,9\%,2) + (P|F,9\%,3) + (P|F,9\%,4)]$
$P = 2000*[(0.9174) + (0.8417) + (0.7722) + (0.7084)]$
$P = 2000*[3.2397] = \$6479.40$

Fortunately, uniform series occur frequently enough in practice to justify tabulating values to eliminate the need to repeatedly sum a series of $(P|F,i,n)$ factors. To accommodate uniform series factors, we need to add a new symbol to our time value of money terminology in addition to the single sum symbols P and F. The symbol "A" is used to designate a uniform series of cash flows. When dealing with uniform series cash flows, the symbol A represents the amount of each annual cash flow, and the n represents the number of cash flows in the series. The factor $(P|A,i,n)$ is known as the uniform series, present worth factor and is read, "To find P given A at i% for n years." Tables of $(P|A,i,n)$ are provided in the chapter appendix. An algebraic expression can also be derived for the $(P|A,i,n)$ factor that expresses P in terms of A, i, and n. The derivation of this formula is omitted here, but the resulting expression is shown in the summary table (Table 4-6) at the end of this section.

An important observation when using a $(P|A,i,n)$ factor is that the "P" resulting from the calculation occurs one period prior to the first "A" cash flow. In our example the first withdrawal (the first "A") occurred one year after the deposit (the "P"). Restating the example problem above using a $(P|A,i,n)$ factor, it becomes:

$P = A * (P|A,i,n)$
$P = 2000 * (P|A,9\%,4)$
$P = 2000 * (3.2397) = \$6479.40$

This result is identical (as expected) to the result using the $(P|F,i,n)$ factors. In both cases the interpretation of the result is as follows: If we deposit $6479.40 in an account paying 9%/yr interest, we could make withdrawals of $2000 per year for four years (starting one year after the initial deposit) to deplete the account at the end of 4 years.

The reciprocal relationship between P and A is symbolized by the factor $(A|P,i,n)$ and is called the *uniform series, capital recovery factor*. Tables of $(A|P,i,n)$ are provided in the chapter appendix, and the algebraic expression for $(A|P,i,n)$ is shown in Table 4-6 at the end of this section. This factor enables us to determine the amount of the equal annual withdrawals "A" (starting one year after the deposit) that can be made from an initial deposit of "P."

Example 9

Determine the equal annual withdrawals that can be made for 8 years from an initial deposit of $9000 in an account that pays 12%/yr. The first withdrawal is to be made one year after the initial deposit.

$A = P * (A|P,12\%,8)$
$A = 9000 * (0.2013)$
$A = \$1811.70$

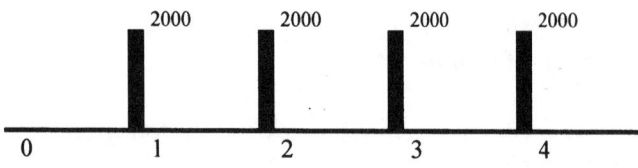

Figure 4-3. Uniform Series Cash Flow

Factors are also available for the relationships between a future worth (accumulated amount) and a uniform series. The factor $(F\,|\,A,i,n)$ is known as the uniform series future worth factor and is read, "To find F given A at i% for n years.." The reciprocal factor, $(A\,|\,F,i,n)$, is known as the *uniform series sinking fund* factor and is read, "To find A given F at i% for n years." An important observation when using an $(F\,|\,A,i,n)$ factor or an $(A\,|\,F,i,n)$ factor is that the "F" resulting from the calculation occurs at the same point in time as to the last "A" cash flow. The algebraic expressions for $(A\,|\,F,i,n)$ and $(F\,|\,A,i,n)$ are shown in Table 6 at the end of this section.

Example 10

If you deposit $2000 per year into an individual retirement account starting on your 24th birthday, how much will have accumulated in the account after your deposit on your 65th birthday? The account pays 6%/yr.

n = 42 (birthdays between 24th and 65th, inclusive)
F = A * (F | A,6%,42)
P = 2000 * (175.9505) = $351,901

Example 11

If you want to be a millionaire on your 65th birthday, what equal annual deposits must be made in an account starting on your 24th birthday? The account pays 10%/yr.

n = 42 (birthdays between 24th and 65th, inclusive)
A = F * (A | F,10%,42)
P = 1000000 * (0.001860) = $1860

4.6.8 Gradient Series

A gradient series of cash flows occurs when the value of a given cash flow is greater than the value of the previous period's cash flow by a constant amount. The symbol used to represent the constant increment is G. The factor $(P\,|\,G,i,n)$ is known as the gradient series, present worth factor. Tables of $(P\,|\,G,i,n)$ are provided in the chapter appendix. An algebraic expression can also be derived for the $(P\,|\,G,i,n)$ factor that expresses P in terms of G, i, and n. The derivation of this formula is omitted here, but the resulting expression is shown in the summary table (Table 4-6) at the end of this section.

It is not uncommon to encounter a cash flow series that is the sum of a uniform series and a gradient series. Figure 4-4 illustrates such a series. The uniform component of this series has a value of 1000 and the gradient series has a value of 500. By convention the first element of a gradient series has a zero value. Therefore, in Figure 4-4, both the uniform series and the gradient series have

length four (n=4). Like the uniform series factor, the "P" calculated by a $(P\,|\,G,i,n)$ factor is located one period before the first element of the series (which is the zero element for a gradient series).

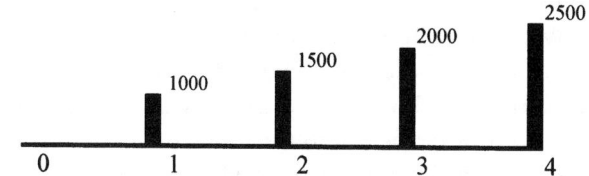

Figure 4-4. Uniform and Gradient Series

Example 12

Assume you wish to make the series of withdrawals illustrated in Figure 4-4 from an account which pays 15%/yr. How much money would you have to deposit today such that the account is depleted at the time of the last withdrawal?

This problem is best solved by recognizing that the cash flows are a combination of a uniform series of value 1000 and length 4 (starting at time=1) plus a gradient series of size 500 and length 4 (starting at time=1).

P = A * (P | A,15%,4) + G * (P | G,15%,4)
P = 1000 * (2.8550) + 500 * (3.7864)
P = 2855.00 + 1893.20 = $4748.20

Occasionally it is useful to convert a gradient series to an equivalent uniform series of the same length. Equivalence in this context means that the present value (P) calculated from the gradient series is numerically equal to the present value (P) calculated from the uniform series. One way to accomplish this task with the time value of money factors we have already considered is to convert the gradient series to a present value using a $(P\,|\,G,i,n)$ factor and then convert this present value to a uniform series using an $(A\,|\,P,i,n)$ factor. In other words:

A = [G * (P | G,i,n)] * (A | P,i,n)

An alternative approach is to use a factor known as the *gradient-to-uniform series conversion factor*, symbolized by $(A\,|\,G,i,n)$. Tables of $(A\,|\,G,i,n)$ are provided in the chapter appendix. An algebraic expression can also be derived for the $(A\,|\,G,i,n)$ factor that expresses A in terms of G, i, and n. The derivation of this formula is omitted here, but the resulting expression is shown in the summary table (Table 4-6) at the end of this section.

4.6.9 Summary of Time Value of Money Factors

Table 4-6 summarizes the time value of money factors introduced in this section. Time value of money factors are useful in economic analysis because they provide a mechanism to accomplish two primary functions: (1) they allow us to replace a cash flow at one point in time with an equivalent cash flow (in a time value of money sense) at a different point in time, and (2) they allow us to convert one cash flow pattern to another (e.g., convert a single sum of money to an equivalent cash flow series or convert a cash flow series to an equivalent single sum). The usefulness of these two functions when performing economic analysis of alternatives will become apparent in Sections 4.7 and 4.8 which follow.

4.6.10 The Concepts of Equivalence and Indifference

Up to this point the term "equivalence" has been used several times but never fully defined. It is appropriate at this point to formally define equivalence, as well as a related term, indifference.

In economic analysis, "equivalence" means "the state of being equal in value." The concept is primarily applied to the comparison of two or more cash flow profiles. Specifically, two (or more) cash flow profiles are equivalent if their time value of money worths at a common point in time are equal.

Question: Are the following two cash flows equivalent at 15%/yr?

Cash Flow 1: Receive $1,322.50 two years from today.

Cash Flow 2: Receive $1,000.00 today.

Analysis Approach 1: Compare worths at t=0 (present worth).

$PW(1) = 1,322.50*(P|F,15,2) = 1322.50*0.756147 = 1,000.$

$PW(2) = 1,000$

Answer: Cash Flow 1 and Cash Flow 2 are equivalent.

Analysis Approach 2: Compare worths at t=2 (future worth).

$FW(1) = 1,322.50$

To Find	Given	Factor	Symbol	Name
P	F	$(1+i)^{-n}$ or $\dfrac{1}{(1+i)^n}$	$(P\|F,i,n)$	Single Payment, Present Worth Factor
F	P	$(1+i)^n$	$(F\|P,i,n)$	Single Payment, Compound Amount Factor
P	A	$\dfrac{(1+i)^n - 1}{i(1+i)^n}$	$(P\|A,i,n)$	Uniform Series, Present Worth Factor
A	P	$\dfrac{i(1+i)^n}{(1+i)^n - 1}$	$(A\|P,i,n)$	Uniform Series, Capital Recovery Factor
F	A	$\dfrac{(1+i)^n - 1}{i}$	$(F\|A,i,n)$	Uniform Series, Compound Amount Factor
A	F	$\dfrac{i}{(1+i)^n - 1}$	$(A\|F,i,n)$	Uniform Series, Sinking Fund Factor
P	G	$\dfrac{1 - (1+ni)(1+i)^{-n}}{i^2}$ or $\dfrac{(1+i)^n - ni - 1}{i^2(1+i)^n}$	$(P\|G,i,n)$	Gradient Series, Present Worth Factor
A	G	$\dfrac{(1+i)^n - (1+ni)}{i[(1+i)^n - 1]}$	$(A\|G,i,n)$	Gradient Series, Uniform Series Factor

Table 4-6. Summary of Discrete Compounding Time Value of Money Factors

FW(2) = 1,000*(F | P,15,2) = 1,000*1.3225 = 1,322.50
Answer: Cash Flow 1 and Cash Flow 2 are equivalent.

Generally the comparison (hence the determination of equivalence) for the two cash flow series in this example would be made as present worths (t=0) or future worths (t=2), but the equivalence definition holds regardless of the point in time chosen. For example:

Analysis Approach 3: Compare worths at t=1
  $W_1(1)$ = 1,322.50*(P | F,15,1) = 1,322.50*0.869565 = 1,150.00
  $W_1(2)$ = 1,000*(F | P,15,1) = 1,000*1.15 = 1,150.00
Answer: Cash Flow 1 and Cash Flow 2 are equivalent.

Thus, the selection of the point in time, t, at which to make the comparison is completely arbitrary. Clearly however, some choices are more intuitively appealing than others (t= 0 and t=2 in the above example).

In economic analysis, "indifference" means "to have no preference." The concept is primarily applied in the comparison of two or more cash flow profiles. Specifically, a potential investor is indifferent between two (or more) cash flow profiles if they are equivalent.

Question: Given the following two cash flows at 15%/yr, which do you prefer?
  Cash Flow 1: Receive $1,322.50 two years from today.
  Cash Flow 2: Receive $1,000.00 today.

Answer: Based on the equivalence calculations above, given these two choices, an investor is indifferent.

The concept of equivalence can be used to break a large, complex problem into a series of smaller more manageable ones. This is done by taking advantage of the fact that, in calculating the economic worth of a cash flow profile, any part of the profile can be replaced by an equivalent representation without altering the worth of the profile at an arbitrary point in time.

Question: You are given a choice between (1) receiving P dollars today or (2) receiving the cash flow series illustrated in Figure 4-5. What must the value of P be for you to be indifferent between the two choices if i=12%/yr?

Analysis Approach: To be indifferent between the choices, P must have a value such that the two alternatives are equivalent at 12%/yr. If we select t=0 as the common point in time upon which to base the analysis (present worth approach), then the analysis proceeds as follows.

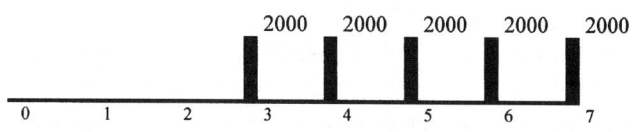

Figure 4-5. A Cash Flow Series

PW(Alt 1) = PW(Alt 2)

Since P is already at t=0 (today), no time value of money factors are involved.

For PW(Alt 2):

Step 1 – Replace the uniform series (t=3 to 7) with an equivalent single sum, V_2, at t=2 (one period before the first element of the series).
  V_2 = 2,000 * (P | A,12%,5) = 2,000 * 3.6048 = 7,209.60

Step 2 – Replace the single sum V_2, with an equivalent value V_0 at t=0.
  PW(Alt 2) = V_0 = V_2 * (P | F,12,2) = 7,209.60 * 0.7972 = 5,747.49

Answer: To be indifferent between the two alternatives, they must be equivalent at t=0. To be equivalent, P must have a value of $5,747.49

4.7 PROJECT MEASURES OF WORTH

4.7.1 Introduction

In this section measures of worth for investment projects are introduced. The measures are used to evaluate the attractiveness of a single investment opportunity. The measures to be presented are (1) present worth, (2) annual worth, (3) internal rate of return, (4) savings investment ratio, and (5) payback period. All but one of these measures of worth require an interest rate to calculate the worth of an investment. This interest rate is commonly referred to as the minimum attractive rate of return (MARR). There are many ways to determine a value of MARR for investment analysis, and no one way is proper for all applications. One principle is, however, generally accepted. MARR should always exceed the cost of capital as described in Section 4.4, Sources of Funds, presented earlier in this chapter.

In all of the following measures of worth, the following conventions are used for defining cash flows: At any given point in time (t = 0, 1, 2, ..., n), there may exist both revenue (positive) cash flows, Rt, and cost (negative) cash flows, Ct. The net cash flow at t, At, is defined as Rt – Ct.

4.7.2 Present Worth (PW)

Consider again the cash flow series illustrated in Figure 4-5. If you were given the opportunity to "buy" that cash flow series for $5,747.49, would you be interested in purchasing it? If you expected to earn a 12%/yr return on your money (MARR=12%), based on the analysis in the previous section, your conclusion should be that you are indifferent between (1) retaining your $5,747.49 and (2) giving up your $5,747.49 in favor of the cash flow series. Figure 4-6 illustrates the net cash flows of this second investment opportunity.

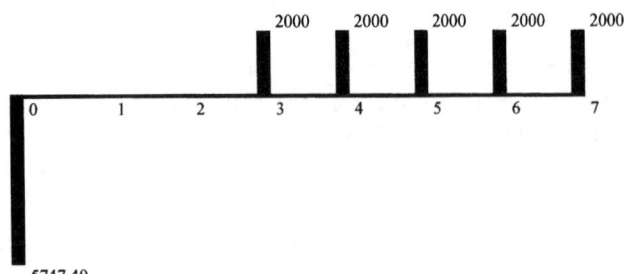

Figure 4-6. An Investment Opportunity

What value would you expect if we calculated the present worth (equivalent value of all cash flows at t=0) of Figure 4-6? We must be careful with the signs (directions) of the cash flows in this analysis since some represent cash outflows (downward) and some represent cash inflows (upward).

PW = –5747.49 + 2000*(P|A,12%,5)*(P|F,12%,2)
PW = –5747.49 + 2000*(3.6048)*(0.7972)
PW = –5747.49 + 5747.49 = $0.00

The value of zero for present worth indicates indifference regarding the investment opportunity. We would just as soon do nothing (i.e., retain our $5747.49) as invest in the opportunity.

What if the same returns (future cash inflows) where offered for a $5000 investment (t=0 outflow); would this be more or less attractive? Hopefully, after a little reflection, it is apparent that this would be a more attractive investment, because you are getting the same returns but paying less than the indifference amount for them. What happens if we calculate the present worth of this new opportunity?

PW = –5000 + 2000*(P|A,12%,5)*(P|F,12%,2)
PW = –5000 + 2000*(3.6048)*(0.7972)
PW = –5000.00 + 5747.49 = $747.49

The positive value of present worth indicates an attractive investment. If we repeat the process with an initial cost greater than $5747.49, it should come as no sur-

prise that the present worth will be negative indicating an unattractive investment.

The concept of present worth as a measure of investment worth can be generalized as follows:

Measure of Worth: Present Worth
Description: All cash flows are converted to a single sum equivalent at time zero using i=MARR.

Calculation Approach: $PW = \sum^{n} A_t (P|F,i,t)$

Decision Rule: If PW ≥ 0, then the investment is attractive.

Example 13

Installing thermal windows on a small office building is estimated to cost $10,000. The windows are expected to last six years and have no salvage value at that time. The energy savings from the windows are expected to be $2525 each year for the first three years and $3840 for each of the remaining three years. If MARR is 15%/yr and the present worth measure of worth is to be used, is this an attractive investment?

The cash flow diagram for the thermal windows is shown in Figure 4-7.

PW =   –10000 + 2525*(P|F,15%,1) + 2525*(P|F,15%,2) + 2525*(P|F,15%,3) + 3840*(P|F,15%,4) + 3840*(P|F,15%,5) + 3840*(P|F,15%,6)
PW =   –10000 + 2525*(0.8696) + 2525*(0.7561) + 2525*(0.6575) + 3840*(0.5718) + 3840*(0.4972) + 3840*(0.4323)
PW =   –10000 + 2195.74 + 1909.15 + 1660.19 + 2195.71 + 1909.25 + 1660.03
PW = $1530.07

Decision: PW≥0 ($1530.07≥0.0); therefore the window investment is attractive.

An alternative (and simpler) approach to calculating PW is obtained by recognizing that the savings cash flows are two uniform series, one of value $2525 and length 3 starting at t= 1, and one of value $3840 and length 3 starting at t= 4.

PW =   –10000 + 2525*(P|A,15%,3) + 3840*(P|A,15%,3)*(P|F,15%,3)
PW =   –10000 + 2525*(2.2832) + 3840*(2.2832)*(0.6575) = $1529.70

Decision: PW≥0 ($1529.70>0.0); therefore, the window investment is attractive.

The slight difference in the PW values is caused by the accumulation of round off errors as the various factors are rounded to four places to the right of the decimal point.

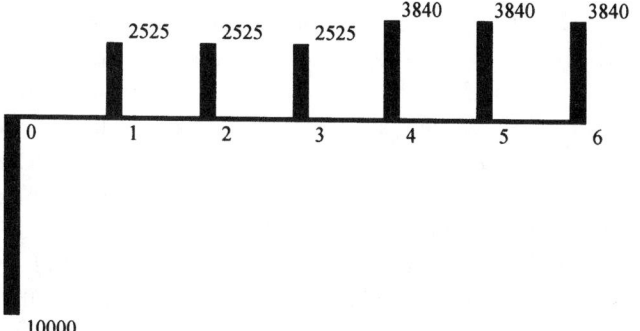

Figure 4-7. Thermal Windows Investment

4.7.3   Annual Worth (AW)

An alternative to present worth is annual worth. The annual worth measure converts all cash flows to an equivalent uniform annual series of cash flows over the investment life, using i=MARR. The annual worth measure is generally calculated by first calculating the present worth measure and then multiplying this by the appropriate $(A|P,i,n)$ factor. A thorough review of the tables in the chapter appendix or the equations in Table 4-6 leads to the conclusion that for all values of i (i>0) and n (n>0), the value of $(A|P,i,n)$ is greater than zero. Hence,

if PW>0, then AW>0;
if PW<0, then AW<0; and
if PW=0, then AW=0.

This is because the only difference between PW and AW is multiplication by a positive, non-zero value, namely $(A|P,i,n)$. The decision rules for investment attractiveness for PW and AW are identical. Positive values indicate an attractive investment, negative values indicate an unattractive investment, and zero indicates indifference. Frequently the only reason for choosing between AW and PW as a measure of worth in an analysis is the preference of the decision maker.

The concept of annual worth as a measure of investment worth can be generalized as follows:

Measure of Worth: Annual Worth

Description: All cash flows are converted to an equivalent uniform annual series of cash flows over the planning horizon, using i=MARR.

Calculation Approach: $AW = PW (A|P,i,n)$

Decision Rule: If AW ≥ 0, then the investment is attractive.

Example 14

Reconsider the thermal window data of Example 13. If the annual worth measure of worth is to be used, is this an attractive investment?

$AW = PW (A|P,15\%,6)$
$AW = 1529.70 (0.2642) = \$404.15/yr$
Decision: AW ≥ 0 ($404.15>0.0); therefore the window investment is attractive.

4.7.4   Internal Rate of Return (IRR)

One of the problems associated with using the present worth or the annual worth measures is that they depend upon knowing a value for MARR. As mentioned in the introduction to this section, the "proper" value for MARR is a much debated topic and tends to vary from company to company and decision maker to decision maker. If the value of MARR changes, the value of PW or AW *must be recalculated* to determine whether the attractiveness/unattractiveness of an investment has changed.

The internal rate of return (IRR) approach is designed to calculate a rate of return that is "internal" to the project. That is,

if IRR > MARR, the project is attractive,
if IRR < MARR, the project is unattractive,
if IRR = MARR, there is indifference.

Thus, if MARR changes, no new calculations are required. We simply compare the calculated IRR for the project to the new value of MARR, and we have our decision.

The value of IRR is typically determined through a trial and error process. An expression for the present worth of an investment is written without specifying a value for i in the time value of money factors. Then various values of i are substituted until a value is found that sets the present worth (PW) equal to zero. The value of i found in this way is the IRR.

As appealing as the flexibility of this approach is, there are two major drawbacks. First, the iterations required to solve using the trial and error approach can be time consuming. This factor is mitigated by the fact that most spreadsheets and financial calculators are pre-programmed to solve for an IRR value, given a cash flow series. The second, and more serious, drawback to the IRR approach is that some cash flow series have more than one value of IRR (i.e., more than one value of i sets the PW expression to zero). A detailed discussion of this multiple solution issue is beyond the scope of this chapter, but one can be found in White, Agee, and Case [1989], as well as most other economic analysis references. However, it can be shown that if a cash flow series consists of an initial investment (negative cash flow at t=0), followed by a series of future returns (positive or zero cash flows for all t>0), then a unique IRR exists. If these conditions are not satisfied a unique IRR is not guaranteed and caution should be exercised in making decisions based on IRR.

The concept of internal rate of return as a measure of investment worth can be generalized as follows:

Measure of Worth: Internal Rate of Return
Description: An interest rate, IRR, is determined that yields a present of zero. IRR implicitly assumes the reinvestment of recovered funds at IRR.
Calculation Approach:

$$\text{find IRR such that PW} = \sum_{}^{n} A_t (P|F,IRR,t) = 0.$$

Important Note: Depending upon the cash flow series, multiple IRRs may exist! If the cash flow series consists of an initial investment (net negative cash flow) followed by a series of future returns (net non-negative cash flows), then a unique IRR exists.

Decision Rule: If IRR is unique and IRR ≥ MARR, then the investment is attractive.

Example 15

Reconsider the thermal window data of Example 13. If the internal rate of return measure of worth is to be used, is this an attractive investment?

First we note that the cash flow series has a single negative investment followed by all positive returns; therefore, it has a unique value for IRR. For such a cash flow series it can also be shown that as i increases PW decreases.

From example 11, we know that for i=15%:
PW = -10000+2525*(P|A,15%,3)+3840*
    (P|A,15%,3)*(P|F,15%,3)
PW = -10000+2525*(2.2832)+3840*
    (2.2832)*(0.6575) = $1529.70
Since PW>0, we must increase i to decrease PW
    toward zero for i=18%:
PW = -10000+2525*(P|A,18%,3)+3840*
    (P|A,18%,3)*(P|F,18%,3)
PW = -10000+2525*(2.1743)+3840*(2.1743)*
    (0.6086) = $571.50
Since PW>0, we must increase i to decrease PW
    toward zero for i=20%:
PW = -10000+2525*(P|A,20%,3)+3840*
    (P|A,20%,3)*(P|F,20%,3)
PW = -10000+2525*(2.1065)+3840*
    (2.1065)*(0.5787) = --$0.01

Although we could interpolate a value of i for which PW=0 (rather than -0.01), for practical purposes PW=0 at i=20%; therefore, IRR=20%.

Decision: IRR≥MARR (20%>15%); therefore, the window investment is attractive.

If a project has a single initial investment at t=0 and savings that are represented as a single uniform series over the life of the project, then an alternate method can be used to calculate the IRR. The alternate method involves calculating the ratio P/A, where P represents the initial investment and A represents the uniform annual savings. The interest tables are then searched for a value of i% such that (P|A,i%,life) = P/A. Consider the following example. A project has a single initial investment of $200,000 and generates annual savings for 10 years of $40,000 per year. Based on this data, the ratio of P/A is 200,000/40,000, or 5.0. We now search the interest tables for a value i% such that (P|A,i%,10) = 5.0. The value of (P|A,15%,10)=5.0188; therefore, the IRR for this project is approximately 15%. It should be noted that this approach is related to a measure of worth which has not yet been introduced. The simple payback period (SPP) will be introduced in Section 4.7.6, but for projects with the cash flow pattern described above, the ratio of P/A is equal to the value of the SPP.

4.7.5 Saving Investment Ratio (SIR)

Many companies are accustomed to working with benefit cost ratios. An investment measure of worth which is consistent with the present worth measure and has the form of a benefit cost ratio is the savings investment ratio (SIR). The SIR decision rule can be derived from the present worth decision rule as follows:

Starting with the PW decision rule PW ≥ 0, replacing PW with its calculation expression,

$$\sum_{}^{n} A_t (P|F,i,t) \geq 0$$

which, using the relationship At = Rt – Ct, can be restated

$$\sum_{}^{n} (R_t - C_t) (P|F,i,t) \geq 0$$

which can be algebraically separated into

$$\sum_{}^{n} R_t (P|F,i,t) - \sum_{}^{n} C_t (P|F,i,t) \geq 0.$$

Adding the second term to both sides of the inequality,

$$\sum_{}^{n} R_t (P|F,i,t) \geq \sum_{}^{n} C_t (P|F,i,t)$$

Dividing both sides of the inequality by the right side term,

$$\frac{\sum_{}^{n} R_t \, (P \mid F,i,t)}{\sum_{}^{n} C_t \, (P \mid F,i,t)} \geq 1$$

which is the decision rule for SIR.

The SIR represents the ratio of the present worth of the revenues to the present worth of the costs. If this ratio exceeds one, the investment is attractive.

The concept of savings investment ratio as a measure of investment worth can be generalized as follows:

Measure of Worth: Savings Investment Ratio

Description: The ratio of the present worth of positive cash flows to the present worth of (the absolute value of) negative cash flows is formed using i=MARR.

Calculation Approach: SIR = $\dfrac{\displaystyle\sum_{t=0}^{n} R_t \, (P \mid F,i,t)}{\displaystyle\sum_{}^{n} C_t \, (P \mid F,i,t)}$

Decision Rule: If SIR $\geq$ 1, then the investment is attractive.

Example 16

Reconsider the thermal window data of Example 13. If the savings investment ratio measure of worth is to be used, is this an attractive investment?

From example 13, we know that for i=15%:

$$SIR = \frac{\sum_{}^{n} R_t \, (P \mid F,i,t)}{\sum_{}^{n} C_t \, (P \mid F,i,t)}$$

$$SIR = \frac{2525^*(P \mid A,15\%,3)+3840^*(P \mid A,15\%,3)^*(P \mid F,15\%,3)}{10,000}$$

$$SIR = \frac{11,529.70}{10,000.00} = 1.15297$$

Decision: SIR$\geq$1.0 (1.15297>1.0); therefore, the window investment is attractive.

An important observation regarding the four measures of worth presented to this point (PW, AW, IRR, and SIR) is that they are all consistent and equivalent. In other words, an investment that is attractive under one mea-sure of worth will be attractive under each of the other measures of worth. A review of the decisions determined in Examples 13 through 16 will confirm the observation. Because of their consistency, it is not necessary to calculate more than one measure of investment worth to determine the attractiveness of a project. The rationale for presenting multiple measures which are essentially identical for de-cision making is that various individuals and companies may have a preference for one approach over another.

4.7.6 Simple Payback Period (SPP)

The simple payback period of an investment is gen-erally taken to mean the number of years required to re-cover the initial investment through net project returns. The payback period is a popular measure of invest-ment worth, and it appears in many forms in economic analysis literature and company procedure manuals. Unfortunately, all too frequently, payback period is used inappropriately and leads to decisions which focus ex-clusively on short-term results while ignoring time value of money concepts. After presenting a common form of payback period, these shortcomings will be discussed.

Measure of Worth: Simple Payback Period

Description: The number of years required to recover the initial investment by accumulating net project returns is determined.

Calculation Approach:

SPP = the smallest m such that $\displaystyle\sum_{t=1}^{m} A_t \geq C_0$

Decision Rule: If SPP is less than or equal to a prede-termined limit (often called a hurdle rate), then the investment is attractive.

Important Note: This form of payback period ignores the time value of money and ignores returns beyond the predetermined limit.

The fact that this approach ignores time value of money concepts is apparent by the fact that no time value of money factors are included in the determination of m. This implicitly assumes that the applicable interest rate for converting future amounts to present amounts is zero. This implies that people are indifferent between \$100 to-day and \$100 one year from today, which is an implica-tion that is highly inconsistent with observable behavior.

The short-term focus of the payback period measure of worth can be illustrated using the cash flow diagrams of Figure 4-8. Applying the SPP approach above yields a payback period for investment (a) of SPP=2 (1200>1000 @ t=2) and a payback period for investment (b) of SPP=4

(1,000,300>1000) @ t=4). If the decision hurdle rate is 3 years (a very common rate), then investment (a) is attractive but investment (b) is not. Hopefully, it is obvious that judging (b) unattractive is not good decision making since a $1,000,000 return four years after a $1,000 investment is attractive under almost any value of MARR. In point of fact, the IRR for (b) is 465%, so for any value of MARR less than 465% investment (b) is attractive.

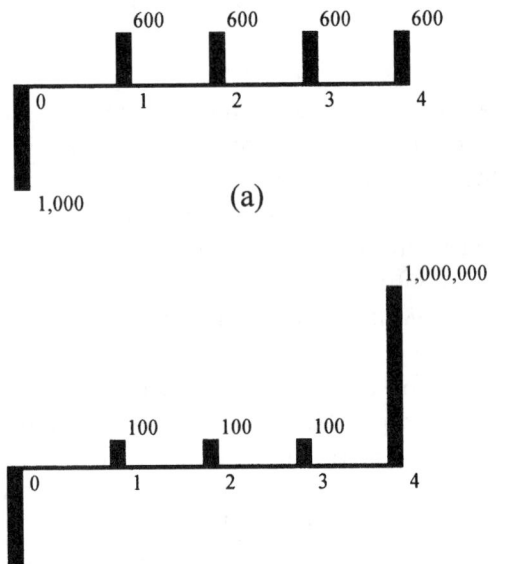

Figure 4-8. Two Investments Evaluated Using Payback Period

4.8   SPECIAL TOPICS

4.8.1 Escalation

The economic analysis methods shown all consider costs and savings. Usually the savings accumulate over time. A standard assumption when estimating energy savings is that utility rate increases are matched with inflation. When utility rate increases outpace the rate of inflation, escalation occurs. In this case, the value of the savings increases over time, and assuming no escalation understates actual savings. Of course, predicting the future is risky business, but trends are estimated by the Department of Energy for the foreseeable future. When escalation of energy savings is anticipated, making adjustments in the calculated savings and cost/benefit ratios helps the measure's chances of success improve since the financial benefit is more accurately represented. Judgment is advised and care should be taken to never overstate savings. De-rating the predicted escalation may be in order. Using 50% of the predicted escalation rate is a conservative approach that should prevent overstating the savings. For engineering economic calculations, the energy savings estimates become gradient series or other forms that are no longer uniform.

4.8.2 Interpolating Interest Tables

All of the examples previously presented in this chapter conveniently used interest rates whose time value of money factors were tabulated in the chapter appendix. How does one proceed if non-tabulated time value of money factors are needed? There are two viable approaches—calculation of the exact values and interpolation. The best and theoretically correct approach is to calculate the exact values of needed factors based on the formulas in Table 4-6.

Example 17

Determine the exact value for (F | P,13%,7).
From Table 4-6,
$(F \mid P,i,n) = (1+i)^n = (1+.13)^7 = 2.3526$.

Interpolation is often used instead of calculation of exact values because, with practice, interpolated values can be calculated quickly. Interpolated values are not "exact" but for most practical problems are "close enough," particularly if the range of interpolation is kept as narrow as possible. Interpolation of some factors, for instance (P | A,i,n), also tends to be less error prone than the exact calculation due to simpler mathematical operations.

Interpolation involves determining an unknown time value of money factor using two known values that bracket the value of interest. An assumption is made that the values of the time value of money factor vary linearly between the known values. Ratios are then used to estimate the unknown value. The example below illustrates the process.

Example 18

Determine an interpolated value for (F | P,13%,7).
The narrowest range of interest rates that bracket 13%, and for which time value of money factor tables are provided in the chapter appendix, is 12% to 15%.
The values necessary for this interpolation are:

i values	(F | P,i%,7)
12%	2.2107
13%	(F | P,13%,7)
15%	2.6600

The interpolation proceeds by setting up ratios and solving for the unknown value, (F | P,13%,7), as shown in the box at the top of the following page.

The interpolated value for (F | P,13%,7), 2.3605, differs from the exact value, 2.3526, by 0.0079. This would

$$\frac{\text{change between rows 2 \& 1 of left column}}{\text{change between rows 3 \& 1 of left column}} = \frac{\text{change between rose 2 \& 1 of right column}}{\text{change between rows 3 \& 1 of right column}}$$

$$\frac{0.13 - 0.12}{0.15 - 0.12} = \frac{(F \mid P, 13\%, 7) - 2.2107}{2.6600 - 2.2107}$$

$$\frac{0.01}{0.03} = \frac{(F \mid P, 13\%, 7) - 2.2107}{0.4493}$$

$$0.1498 = (F \mid P, 13\%, 7) - 2.2107$$
$$(F \mid P, 13\%, 7) = 2.3605$$

imply a $7.90 difference in present worth for every thousand dollars of return at t=7. The relative importance of this interpolation error can be judged only in the context of a specific problem.

4.8.3 Non-annual Interest Compounding

Many practical economic analysis problems involve interest that is not compounded annually. It is common practice to express a non-annually compounded interest rate as follows:

12% per year compounded monthly or 12%/yr/mo.

When expressed in this form, 12%/yr/mo is known as the *nominal* annual interest rate The techniques covered in this chapter up to this point can not be used directly to solve an economic analysis problem of this type, because the interest period (per year) and compounding period (monthly) are not the same. Two approaches can be used to solve problems of this type. One approach involves determining a *period* interest rate; the other involves determining an *effective* interest rate.

To solve this type of problem using a period interest rate approach, we must define the period interest rate:

$$\text{Period Interest Rate} = \frac{\text{Nominal Annual Interest Rate}}{\text{Number of Interest Periods per Year}}$$

In our example,

$$\text{Period Interest Rate} = \frac{12\%/yr/mo}{12 \text{ mo}/yr} = 1\%/mo/mo$$

Since the interest period and the compounding period are now the same, the time value of money factors in the chapter appendix can be applied directly. Note however, that the number of interest periods (n) must be adjusted to match the new frequency.

Example 19

$2,000 is invested in an account which pays 12% per year compounded monthly. What is the balance in the account after 3 years?

$$\text{Nominal Annual Interest Rate} = 12\%/yr/mo$$

$$\text{Period Interest Rate} = \frac{12\%/yr/mo}{12 \text{ mo}/yr} = 1\%/mo/mo$$

Number of Interest Periods = 3 years $\times$ 12 mo/yr = 36 interest periods (months)

$F = P (F \mid P, i, n) = \$2,000 (F \mid P, 1, 36) = \$2,000 (1.4308) = \$2,861.60$

Example 20

What are the monthly payments on a 5 year car loan of $12,500 at 6% per year compounded monthly.

Nominal Annual Interest Rate = 6%/yr/mo

$$\text{Period Interest Rate} = \frac{6\%/yr/mo}{12 \text{ mo}/yr} = 0.5\%/mo/mo$$

Number of Interest Periods = 5 years $\times$

12 mo/yr = 60 interest periods

$A = P (A \mid P, i, n) = \$12,500 (A \mid P, 0.5, 60) =$

$\$12,500 (0.0193) = \$241.25$

To solve this type of problem using an effective interest rate approach, we must define the effective interest rate. The effective annual interest rate is the annualized interest rate that would yield results equivalent to the period interest rate as previously calculated. Note, however, that the effective annual interest rate approach should not be used if the cash flows are more frequent than annual (e.g., monthly). In general, the interest rate for time value of money factors should match the frequency of the cash flows. (If the cash flows are monthly, use the period interest rate approach with monthly periods).

As an example of the calculation of an effective interest rate, assume that the nominal interest rate is 12%/

yr/qtr; therefore, the period interest rate is 3%/qtr/qtr. One dollar invested for 1 year at 3%/qtr/qtr would have a future worth of:

$$F = P (F \mid P,i,n) = \$1 (F \mid P,3,4) = \$1 (1.03)^4$$
$$= \$1 (1.1255) = \$1.1255$$

To get this same value in 1 year with an annual rate, the annual rate would have to be of 12.55%/yr/yr. This value is called the effective annual interest rate. The effective annual interest rate is given by $(1.03)^4 - 1 = 0.1255$ or 12.55%.

The general equation for the effective annual interest rate is:

Effective Annual Interest Rate $= (1 + (r/m))^m - 1$
where: r = nominal annual interest rate
          m = number of interest periods per year

Example 21

What is the effective annual interest rate if the nominal rate is 12%/yr compounded monthly?
    nominal annual interest rate = 12%/yr/mo
    period interest rate = 1%/mo/mo
    effective annual interest rate $= (1+0.12/12)^{12} -1$
        $= 0.1268$ or 12.68%

4.8.4   Economic Analysis Under Inflation

Inflation is characterized by a decrease in the purchasing power of money caused by an increase in general price levels of goods and services without an accompanying increase in the value of the goods and services. Inflationary pressure is created when more dollars are put into an economy without an accompanying increase in goods and services. In other words, printing more money without an increase in economic output generates inflation. A complete treatment of inflation is beyond the scope of this chapter. A good summary can be found in Sullivan and Bontadelli [1980].

When consideration of inflation is introduced into economic analysis, future cash flows can be stated in terms of either constant-worth dollars or then-current dollars. Then-current cash flows are expressed in terms of the *face amount* of dollars (actual number of dollars) that will change hands when the cash flow occurs. Alternatively, *constant-worth* cash flows are expressed in terms of the purchasing power of dollars relative to a fixed point in time known as the base period.

Example 22

For the next 4 years, a family anticipates buying $1000 worth of groceries each year. If inflation is expected to be 3%/yr, what are the then-current cash flows required to purchase the groceries?

To buy the groceries, the family will need to take the following face amount of dollars to the store. We will somewhat artificially assume that the family only shops once per year, buys the same set of items each year, and that the first trip to the store will be one year from today.

Year 1: dollars required $1000.00*(1.03)=$1030.00
Year 2: dollars required $1030.00*(1.03)=$1060.90
Year 3: dollars required $1060.90*(1.03)=$1092.73
Year 4: dollars required $1092.73*(1.03)=$1125.51

What are the constant-worth cash flows, if today's dollars are used as the base year?
The constant worth dollars are inflation-free dollars; therefore, the $1000 of groceries costs $1000 each year.
    Year 1: $1000.00
    Year 2: $1000.00
    Year 3: $1000.00
    Year 4: $1000.00

The key to proper economic analysis under inflation is to base the value of MARR on the types of cash flows. If the cash flows contain inflation, then the value of MARR should also be adjusted for inflation. Alternatively, if the cash flows do not contain inflation, then the value of MARR should be inflation-free. When MARR does not contain an adjustment for inflation, it is referred to as a *real* value for MARR. If it contains an inflation adjustment, it is referred to as a *combined* value for MARR. The relationship between inflation rate, the real value of MARR, and the combined value of MARR is given by:

$$1 + MARR_{COMBINED}= (1 + \text{inflation rate}) * (1 + MARR_{REAL})$$

Example 23

If the inflation rate is 3%/yr and the real value of MARR is 15%/yr, what is the combined value of MARR?

$$1 + MARR_{COMBINED}= (1 + \text{inflation rate}) * (1 + MARR_{REAL})$$
$$1 + MARR_{COMBINED}= (1 + 0.03) * (1 + 0.15)$$
$$1 + MARR_{COMBINED}= (1.03) * (1.15)$$
$$1 + MARR_{COMBINED}= 1.1845$$
$$MARR_{COMBINED}= 1.1845 - 1 = 0.1845 = 18.45\%$$

If the cash flows of a project are stated in terms of then-current dollars, the appropriate value of MARR is the combined value of MARR. Analysis done in this way is referred to as *then current analysis*. If the cash flows of a project are stated in terms of constant-worth dollars, the appropriate value of MARR is the real value of MARR. Analysis done in this way is referred to as then *constant worth analysis*.

Example 24

Using the cash flows of Examples 22 and interest rates of Example 23, determine the present worth of the grocery purchases using a constant worth analysis.

Constant worth analysis requires constant worth cash flows and the real value of MARR.

PW = 1000 * (P | A,15%,4) = 1000 * (2.8550) = $2855.00

Example 25

Using the cash flows of Examples 22 and interest rates of Example 23, determine the present worth of the grocery purchases using a then current analysis.

Then current analysis requires then current cash flows and the combined value of MARR.

PW = 1030.00 * (P | F,18.45%,1) + 1060.90*
   (P | F,18.45%,2)+ 1092.73 * (P | F,18.45%,3)
   +1125.51 * (P | F,18.45%,4)

PW = 1030.00 * (0.8442) + 1060.90 * (0.7127) +
   1092.73 * (0.6017) +1125.51 * (0.5080)

PW = 869.53 + 756.10 + 657.50 + 571.76 = 2854.89

The notable result of Examples 24 and 25 is that the present worths determined by the constant-worth approach ($2855.00) and the then-current approach ($2854.89) are equal. (The $0.11 difference is due to rounding). This result is often unexpected but it is mathematically sound. The important conclusion is that if care is taken to appropriately match the cash flows and value of MARR, the level of general price inflation is not a determining factor in the acceptability of projects. To make this important result hold, inflation must either (1) be included in both the cash flows and MARR (the then-current approach) or (2) be included in neither the cash flows nor MARR (the constant-worth approach).

4.8.5 Life Cycle Cost Analysis (LCCA)

A life cycle costs analysis calculates the cost of a system or product over its entire life span, sometimes referred to as "cradle to grave." The purpose of life cycle cost analysis is to consider all facets of the project for their merit and, specifically, to compare multiple project options using the same method to find the overall best choice. The advantage of using LCCA is that all associated costs become visible, and some designs will be able to compete better economically using this method.

While this method provides a definitive answer for "best choice" based on complete and repeating life cycles of equipment, it may not be used by businesses with a finite planning horizon. In other words, if alternative B wins in the LCCA test, it still may not be chosen if the business has a 5-year planning horizon and alternative B has high initial cost. LCCA results may be overridden if the LCCA winner costs more initially and capital is limited.

For energy related projects, an LCCA study should include consideration of costs for at least the following categories: design, initial construction, energy and utilities, operations and maintenance, repairs, replacement, and salvage. Other relevant costs may also be included. All of the identified costs are normalized to a point in time, often the present worth, in which case the lowest present worth of costs indicates the preferred alternative in the LCCA.

To use this method, a point in time must be identified where all systems are at the end of life at the same time. Life spans of systems and equipment are a matter of opinion and influenced by usage patterns and maintenance and so, in reality, are somewhat subjective. Where LCCA is used for government projects, this subjectivity is often removed by mandated tables that stipulate system life, maintenance cost assumptions per year, etc. The remaining variables are first cost and utility cost, which are easier to estimate and verify.

4.9 SUMMARY AND ADDITIONAL EXAMPLE APPLICATIONS

In this chapter a coherent, consistent approach to economic analysis of capital investments (energy related or other) has been presented. To conclude, this section provides several additional examples to illustrate the use of time value of money concepts for energy related problems. Additional example applications, as well as a more in depth presentation of conceptual details, can be found in the references listed at the end of the chapter. These references are by no means exclusive; many other excellent presentations of the subject matter are also available. Adherence to the concepts and methods presented here and in the references will lead to sound investment decisions with respect to time value of money principles.

Example 26

In Section 4.3.3 an example involving the evaluation of a baseboard heating and window air conditioner versus a heat pump was introduced to illustrate cash flow diagramming (Figure 4-2). A summary of the differential costs is repeated here for convenience.

- The heat pump costs $1500 more than the baseboard system
- The heat pump saves $380 annually in electricity costs
- The heat pump has a $50 higher annual maintenance costs
- The heat pump has a $150 higher salvage value at the end of 15 years
- The heat pump requires $200 more in replacement maintenance at the end of year 8.

If MARR is 18%, is the additional investment in the heat pump attractive?

Using present worth as the measure of worth:

PW = −1500 + 380*(P | A,18%,15) − 50*(P | A,18%,15) + 150*(P | F,18%,15) − 200*(P | F,18%,8)

PW = −1500 + 380*(5.0916) − 50*(5.0916) + 150*(0.0835) − 200*(0.2660)

PW = −1500.00 + 1934.81 − 254.58 + 12.53 − 53.20 = $139.56

Decision: PW≥0 ($139.56>0.0); therefore, the additional investment for the heat pump is attractive.

Example 27

A homeowner needs to decide whether to install R-11 or R-19 insulation in the attic of her home. The R-19 insulation costs $150 more to install and will save approximately 400 kWh per year. If the planning horizon is 20 years and electricity costs $0.08/kWh, is the additional investment attractive at MARR of 10%?

At $0.08/kWh, the annual savings are: 400 kWh * $0.08/kWh = $32.00.

Using present worth as the measure of worth:
PW = -150 + 32*(P | A,10%,20)
PW = -150 + 32*(8.5136) = -150 + 272.44 = $122.44

Decision: PW≥0 ($122.44>0.0); therefore, the R-19 insulation is attractive.

Example 28

The homeowner from Example 27 can install R-30 insulation in the attic of her home for $200 more than the R-19 insulation. The R-30 will save approximately 250 kWh per year over the R-19 insulation. Is the additional investment attractive?

Assuming the same MARR, electricity cost, and planning horizon, the additional annual savings are: 250 kWh * $0.08/kWh = $20.00

Using present worth as the measure of worth:

PW = -200 + 20*(P | A,10%,20)
PW = -200 + 20*(8.5136) = -200 + 170.27 = -$29.73

Decision: PW<0 (-$29.73<0.0); therefore, the R-30 insulation is not attractive.

Example 29

An economizer costs $20,000 and will last 10 years. It will generate savings of $3,500 per year, with maintenance costs of $500 per year. If MARR is 10% is the economizer an attractive investment?

Using present worth as the measure of worth:

PW = −20000 + 3500*(P | A,10%,10) −500*(P | A,10%,10)
PW = −20000 + 3500*(6.1446) − 500*(6.1446)
PW = −20000.00 + 21506.10 − 3072.30 = -$1566.20

Decision: PW<0 (-$1566.20<0.0); therefore, the economizer is not attractive.

Example 30

If the economizer from Example 29 has a salvage value of $5,000 at the end of 10 years is the investment attractive?

Using present worth as the measure of worth:

PW = −20000 + 3500*(P | A,10%,10) 500*(P | A,10%,10) + 5000*(P | F,10%,10)
PW = −20000 + 3500*(6.1446) − 500*(6.1446) + 5000*(0.3855)
PW = −20000.00 + 21506.10 − 3072.30 + 1927.50 = $361.30

Decision: PW≥0 ($361.30≥0.0); therefore, the economizer is now attractive.

References

Brown, R.J. and R.R. Yanuck, 1980, *Life Cycle Costing: A Practical Guide for Energy Managers*, The Fairmont Press, Inc., Atlanta, GA.

Fuller, S.K. and S.R. Petersen, 1994, NISTIR 5165: Life-Cycle Costing Workshop for Energy Conservation in Buildings: Student Manual, U.S. Department of Commerce, Office of Applied Economics, Gaithersburg, MD.

Fuller, S.K. and S.R. Petersen, 1995, NIST Handbook 135: Life-Cycle Costing Manual for the Federal Energy Management Program, National Technical Information Service, Springfield, VA.

Park, C.S. and G.P. Sharp-Bette, 1990, *Advanced Engineering Economics*, John Wiley & Sons, New York, NY.

Sullivan, W.G. and J.A. Bontadelli, 1980, "The Industrial Engineer and Inflation," *Industrial Engineering*, Vol. 12, No. 3, 24-33.

Fabrycky, W.J., G.J. Thuesen, and D. Verma, 1998, *Economic Decision Analysis*, 3rd Edition, Prentice Hall, Upper Saddle River, NJ.

White, J.A., K.E. Case, D.B. Pratt, and M.H. Agee, 1998, *Principles of Engineering Economic Analysis*, 4th Edition, John Wiley & Sons, New York, NY.

Appendix

Time Value of Money Factors—Discrete Compounding
   i = 1%

	Single Sums		Uniform Series				Gradient Series	
	To Find F Given P (F|P,i%,n)	To Find P Given F (P|F,i%,n)	To Find F Given A (F|A,i%,n)	To Find A Given F (A|F,i%,n)	To Find P Given A (P|A,i%,n)	To Find A Given P (A|P,i%,n)	To Find P Given G (P|G,i%,n)	To Find A Given G (A|G,i%,n)
n								
1	1.0100	0.9901	1.0000	1.0000	0.9901	1.0100	0.0000	0.0000
2	1.0201	0.9803	2.0100	0.4975	1.9704	0.5075	0.9803	0.4975
3	1.0303	0.9706	3.0301	0.3300	2.9410	0.3400	2.9215	0.9934
4	1.0406	0.9610	4.0604	0.2463	3.9020	0.2563	5.8044	1.4876
5	1.0510	0.9515	5.1010	0.1960	4.8534	0.2060	9.6103	1.9801
6	1.0615	0.9420	6.1520	0.1625	5.7955	0.1725	14.3205	2.4710
7	1.0721	0.9327	7.2135	0.1386	6.7282	0.1486	19.9168	2.9602
8	1.0829	0.9235	8.2857	0.1207	7.6517	0.1307	26.3812	3.4478
9	1.0937	0.9143	9.3685	0.1067	8.5660	0.1167	33.6959	3.9337
10	1.1046	0.9053	10.4622	0.0956	9.4713	0.1056	41.8435	4.4179
11	1.1157	0.8963	11.5668	0.0865	10.3676	0.0965	50.8067	4.9005
12	1.1268	0.8874	12.6825	0.0788	11.2551	0.0888	60.5687	5.3815
13	1.1381	0.8787	13.8093	0.0724	12.1337	0.0824	71.1126	5.8607
14	1.1495	0.8700	14.9474	0.0669	13.0037	0.0769	82.4221	6.3384
15	1.1610	0.8613	16.0969	0.0621	13.8651	0.0721	94.4810	6.8143
16	1.1726	0.8528	17.2579	0.0579	14.7179	0.0679	107.2734	7.2886
17	1.1843	0.8444	18.4304	0.0543	15.5623	0.0643	120.7834	7.7613
18	1.1961	0.8360	19.6147	0.0510	16.3983	0.0610	134.9957	8.2323
19	1.2081	0.8277	20.8109	0.0481	17.2260	0.0581	149.8950	8.7017
20	1.2202	0.8195	22.0190	0.0454	18.0456	0.0554	165.4664	9.1694
21	1.2324	0.8114	23.2392	0.0430	18.8570	0.0530	181.6950	9.6354
22	1.2447	0.8034	24.4716	0.0409	19.6604	0.0509	198.5663	10.0998
23	1.2572	0.7954	25.7163	0.0389	20.4558	0.0489	216.0660	10.5626
24	1.2697	0.7876	26.9735	0.0371	21.2434	0.0471	234.1800	11.0237
25	1.2824	0.7798	28.2432	0.0354	22.0232	0.0454	252.8945	11.4831
26	1.2953	0.7720	29.5256	0.0339	22.7952	0.0439	272.1957	11.9409
27	1.3082	0.7644	30.8209	0.0324	23.5596	0.0424	292.0702	12.3971
28	1.3213	0.7568	32.1291	0.0311	24.3164	0.0411	312.5047	12.8516
29	1.3345	0.7493	33.4504	0.0299	25.0658	0.0399	333.4863	13.3044
30	1.3478	0.7419	34.7849	0.0287	25.8077	0.0387	355.0021	13.7557
36	1.4308	0.6989	43.0769	0.0232	30.1075	0.0332	494.6207	16.4285
42	1.5188	0.6584	51.8790	0.0193	34.1581	0.0293	650.4514	19.0424
48	1.6122	0.6203	61.2226	0.0163	37.9740	0.0263	820.1460	21.5976
54	1.7114	0.5843	71.1410	0.0141	41.5687	0.0241	1.002E+03	24.0945
60	1.8167	0.5504	81.6697	0.0122	44.9550	0.0222	1.193E+03	26.5333
66	1.9285	0.5185	92.8460	0.0108	48.1452	0.0208	1.392E+03	28.9146
72	2.0471	0.4885	104.7099	9.550E-03	51.1504	0.0196	1.598E+03	31.2386
120	3.3004	0.3030	230.0387	4.347E-03	69.7005	0.0143	3.334E+03	47.8349
180	5.9958	0.1668	499.5802	2.002E-03	83.3217	0.0120	5.330E+03	63.9697
360	35.9496	0.0278	3.495E+03	2.861E-04	97.2183	0.0103	8.720E+03	89.6995

Time Value of Money Factors—Discrete Compounding
  i = 2%

n	Single Sums		Uniform Series				Gradient Series	
	To Find F Given P (F|P,i%,n)	To Find P Given F (P|F,i%,n)	To Find F Given A (F|A,i%,n)	To Find A Given F (A|F,i%,n)	To Find P Given A (P|A,i%,n)	To Find A Given P (A|P,i%,n)	To Find P Given G (P|G,i%,n)	To Find A Given G (A|G,i%,n)
1	1.0200	0.9804	1.0000	1.0000	0.9804	1.0200	0.0000	0.0000
2	1.0404	0.9612	2.0200	0.4950	1.9416	0.5150	0.9612	0.4950
3	1.0612	0.9423	3.0604	0.3268	2.8839	0.3468	2.8458	0.9868
4	1.0824	0.9238	4.1216	0.2426	3.8077	0.2626	5.6173	1.4752
5	1.1041	0.9057	5.2040	0.1922	4.7135	0.2122	9.2403	1.9604
6	1.1262	0.8880	6.3081	0.1585	5.6014	0.1785	13.6801	2.4423
7	1.1487	0.8706	7.4343	0.1345	6.4720	0.1545	18.9035	2.9208
8	1.1717	0.8535	8.5830	0.1165	7.3255	0.1365	24.8779	3.3961
9	1.1951	0.8368	9.7546	0.1025	8.1622	0.1225	31.5720	3.8681
10	1.2190	0.8203	10.9497	0.0913	8.9826	0.1113	38.9551	4.3367
11	1.2434	0.8043	12.1687	0.0822	9.7868	0.1022	46.9977	4.8021
12	1.2682	0.7885	13.4121	0.0746	10.5753	0.0946	55.6712	5.2642
13	1.2936	0.7730	14.6803	0.0681	11.3484	0.0881	64.9475	5.7231
14	1.3195	0.7579	15.9739	0.0626	12.1062	0.0826	74.7999	6.1786
15	1.3459	0.7430	17.2934	0.0578	12.8493	0.0778	85.2021	6.6309
16	1.3728	0.7284	18.6393	0.0537	13.5777	0.0737	96.1288	7.0799
17	1.4002	0.7142	20.0121	0.0500	14.2919	0.0700	107.5554	7.5256
18	1.4282	0.7002	21.4123	0.0467	14.9920	0.0667	119.4581	7.9681
19	1.4568	0.6864	22.8406	0.0438	15.6785	0.0638	131.8139	8.4073
20	1.4859	0.6730	24.2974	0.0412	16.3514	0.0612	144.6003	8.8433
21	1.5157	0.6598	25.7833	0.0388	17.0112	0.0588	157.7959	9.2760
22	1.5460	0.6468	27.2990	0.0366	17.6580	0.0566	171.3795	9.7055
23	1.5769	0.6342	28.8450	0.0347	18.2922	0.0547	185.3309	10.1317
24	1.6084	0.6217	30.4219	0.0329	18.9139	0.0529	199.6305	10.5547
25	1.6406	0.6095	32.0303	0.0312	19.5235	0.0512	214.2592	10.9745
26	1.6734	0.5976	33.6709	0.0297	20.1210	0.0497	229.1987	11.3910
27	1.7069	0.5859	35.3443	0.0283	20.7069	0.0483	244.4311	11.8043
28	1.7410	0.5744	37.0512	0.0270	21.2813	0.0470	259.9392	12.2145
29	1.7758	0.5631	38.7922	0.0258	21.8444	0.0458	275.7064	12.6214
30	1.8114	0.5521	40.5681	0.0246	22.3965	0.0446	291.7164	13.0251
36	2.0399	0.4902	51.9944	0.0192	25.4888	0.0392	392.0405	15.3809
42	2.2972	0.4353	64.8622	0.0154	28.2348	0.0354	497.6010	17.6237
48	2.5871	0.3865	79.3535	0.0126	30.6731	0.0326	605.9657	19.7556
54	2.9135	0.3432	95.6731	0.0105	32.8383	0.0305	715.1815	21.7789
60	3.2810	0.3048	114.0515	8.768E-03	34.7609	0.0288	823.6975	23.6961
66	3.6950	0.2706	134.7487	7.421E-03	36.4681	0.0274	930.3000	25.5100
72	4.1611	0.2403	158.0570	6.327E-03	37.9841	0.0263	1.034E+03	27.2234
120	10.7652	0.0929	488.2582	2.048E-03	45.3554	0.0220	1.710E+03	37.7114
180	35.3208	0.0283	1.716E+03	5.827E-04	48.5844	0.0206	2.174E+03	44.7554
360	1.248E+03	8.016E-04	6.233E+04	1.604E-05	49.9599	0.0200	2.484E+03	49.7112

Time Value of Money Factors—Discrete Compounding
i = 3%

	Single Sums		Uniform Series				Gradient Series	
	To Find F Given P (F|P,i%,n)	To Find P Given F (P|F,i%,n)	To Find F Given A (F|A,i%,n)	To Find A Given F (A|F,i%,n)	To Find P Given A (P|A,i%,n)	To Find A Given P (A|P,i%,n)	To Find P Given G (P|G,i%,n)	To Find A Given G (A|G,i%,n)
n								
1	1.0300	0.9709	1.0000	1.0000	0.9709	1.0300	0.0000	0.0000
2	1.0609	0.9426	2.0300	0.4926	1.9135	0.5226	0.9426	0.4926
3	1.0927	0.9151	3.0909	0.3235	2.8286	0.3535	2.7729	0.9803
4	1.1255	0.8885	4.1836	0.2390	3.7171	0.2690	5.4383	1.4631
5	1.1593	0.8626	5.3091	0.1884	4.5797	0.2184	8.8888	1.9409
6	1.1941	0.8375	6.4684	0.1546	5.4172	0.1846	13.0762	2.4138
7	1.2299	0.8131	7.6625	0.1305	6.2303	0.1605	17.9547	2.8819
8	1.2668	0.7894	8.8923	0.1125	7.0197	0.1425	23.4806	3.3450
9	1.3048	0.7664	10.1591	0.0984	7.7861	0.1284	29.6119	3.8032
10	1.3439	0.7441	11.4639	0.0872	8.5302	0.1172	36.3088	4.2565
11	1.3842	0.7224	12.8078	0.0781	9.2526	0.1081	43.5330	4.7049
12	1.4258	0.7014	14.1920	0.0705	9.9540	0.1005	51.2482	5.1485
13	1.4685	0.6810	15.6178	0.0640	10.6350	0.0940	59.4196	5.5872
14	1.5126	0.6611	17.0863	0.0585	11.2961	0.0885	68.0141	6.0210
15	1.5580	0.6419	18.5989	0.0538	11.9379	0.0838	77.0002	6.4500
16	1.6047	0.6232	20.1569	0.0496	12.5611	0.0796	86.3477	6.8742
17	1.6528	0.6050	21.7616	0.0460	13.1661	0.0760	96.0280	7.2936
18	1.7024	0.5874	23.4144	0.0427	13.7535	0.0727	106.0137	7.7081
19	1.7535	0.5703	25.1169	0.0398	14.3238	0.0698	116.2788	8.1179
20	1.8061	0.5537	26.8704	0.0372	14.8775	0.0672	126.7987	8.5229
21	1.8603	0.5375	28.6765	0.0349	15.4150	0.0649	137.5496	8.9231
22	1.9161	0.5219	30.5368	0.0327	15.9369	0.0627	148.5094	9.3186
23	1.9736	0.5067	32.4529	0.0308	16.4436	0.0608	159.6566	9.7093
24	2.0328	0.4919	34.4265	0.0290	16.9355	0.0590	170.9711	10.0954
25	2.0938	0.4776	36.4593	0.0274	17.4131	0.0574	182.4336	10.4768
26	2.1566	0.4637	38.5530	0.0259	17.8768	0.0559	194.0260	10.8535
27	2.2213	0.4502	40.7096	0.0246	18.3270	0.0546	205.7309	11.2255
28	2.2879	0.4371	42.9309	0.0233	18.7641	0.0533	217.5320	11.5930
29	2.3566	0.4243	45.2189	0.0221	19.1885	0.0521	229.4137	11.9558
30	2.4273	0.4120	47.5754	0.0210	19.6004	0.0510	241.3613	12.3141
36	2.8983	0.3450	63.2759	0.0158	21.8323	0.0458	313.7028	14.3688
42	3.4607	0.2890	82.0232	0.0122	23.7014	0.0422	385.5024	16.2650
48	4.1323	0.2420	104.4084	9.578E-03	25.2667	0.0396	455.0255	18.0089
54	4.9341	0.2027	131.1375	7.626E-03	26.5777	0.0376	521.1157	19.6073
60	5.8916	0.1697	163.0534	6.133E-03	27.6756	0.0361	583.0526	21.0674
66	7.0349	0.1421	201.1627	4.971E-03	28.5950	0.0350	640.4407	22.3969
72	8.4000	0.1190	246.6672	4.054E-03	29.3651	0.0341	693.1226	23.6036
120	34.7110	0.0288	1.124E+03	8.899E-04	32.3730	0.0309	963.8635	29.7737
180	204.5034	4.890E-03	6.783E+03	1.474E-04	33.1703	0.0301	1.076E+03	32.4488
360	4.182E+04	2.391E-05	1.394E+06	7.173E-07	33.3325	0.0300	1.111E+03	33.3247

Time Value of Money Factors—Discrete Compounding
    i = 4%

n	Single Sums		Uniform Series				Gradient Series	
	To Find F Given P (F|P,i%,n)	To Find P Given F (P|F,i%,n)	To Find F Given A (F|A,i%,n)	To Find A Given F (A|F,i%,n)	To Find P Given A (P|A,i%,n)	To Find A Given P (A|P,i%,n)	To Find P Given G (P|G,i%,n)	To Find A Given G (A|G,i%,n)
1	1.0400	0.9615	1.0000	1.0000	0.9615	1.0400	0.0000	0.0000
2	1.0816	0.9246	2.0400	0.4902	1.8861	0.5302	0.9246	0.4902
3	1.1249	0.8890	3.1216	0.3203	2.7751	0.3603	2.7025	0.9739
4	1.1699	0.8548	4.2465	0.2355	3.6299	0.2755	5.2670	1.4510
5	1.2167	0.8219	5.4163	0.1846	4.4518	0.2246	8.5547	1.9216
6	1.2653	0.7903	6.6330	0.1508	5.2421	0.1908	12.5062	2.3857
7	1.3159	0.7599	7.8983	0.1266	6.0021	0.1666	17.0657	2.8433
8	1.3686	0.7307	9.2142	0.1085	6.7327	0.1485	22.1806	3.2944
9	1.4233	0.7026	10.5828	0.0945	7.4353	0.1345	27.8013	3.7391
10	1.4802	0.6756	12.0061	0.0833	8.1109	0.1233	33.8814	4.1773
11	1.5395	0.6496	13.4864	0.0741	8.7605	0.1141	40.3772	4.6090
12	1.6010	0.6246	15.0258	0.0666	9.3851	0.1066	47.2477	5.0343
13	1.6651	0.6006	16.6268	0.0601	9.9856	0.1001	54.4546	5.4533
14	1.7317	0.5775	18.2919	0.0547	10.5631	0.0947	61.9618	5.8659
15	1.8009	0.5553	20.0236	0.0499	11.1184	0.0899	69.7355	6.2721
16	1.8730	0.5339	21.8245	0.0458	11.6523	0.0858	77.7441	6.6720
17	1.9479	0.5134	23.6975	0.0422	12.1657	0.0822	85.9581	7.0656
18	2.0258	0.4936	25.6454	0.0390	12.6593	0.0790	94.3498	7.4530
19	2.1068	0.4746	27.6712	0.0361	13.1339	0.0761	102.8933	7.8342
20	2.1911	0.4564	29.7781	0.0336	13.5903	0.0736	111.5647	8.2091
21	2.2788	0.4388	31.9692	0.0313	14.0292	0.0713	120.3414	8.5779
22	2.3699	0.4220	34.2480	0.0292	14.4511	0.0692	129.2024	8.9407
23	2.4647	0.4057	36.6179	0.0273	14.8568	0.0673	138.1284	9.2973
24	2.5633	0.3901	39.0826	0.0256	15.2470	0.0656	147.1012	9.6479
25	2.6658	0.3751	41.6459	0.0240	15.6221	0.0640	156.1040	9.9925
26	2.7725	0.3607	44.3117	0.0226	15.9828	0.0626	165.1212	10.3312
27	2.8834	0.3468	47.0842	0.0212	16.3296	0.0612	174.1385	10.6640
28	2.9987	0.3335	49.9676	0.0200	16.6631	0.0600	183.1424	10.9909
29	3.1187	0.3207	52.9663	0.0189	16.9837	0.0589	192.1206	11.3120
30	3.2434	0.3083	56.0849	0.0178	17.2920	0.0578	201.0618	11.6274
36	4.1039	0.2437	77.5983	0.0129	18.9083	0.0529	253.4052	13.4018
42	5.1928	0.1926	104.8196	9.540E-03	20.1856	0.0495	302.4370	14.9828
48	6.5705	0.1522	139.2632	7.181E-03	21.1951	0.0472	347.2446	16.3832
54	8.3138	0.1203	182.8454	5.469E-03	21.9930	0.0455	387.4436	17.6167
60	10.5196	0.0951	237.9907	4.202E-03	22.6235	0.0442	422.9966	18.6972
66	13.3107	0.0751	307.7671	3.249E-03	23.1218	0.0432	454.0847	19.6388
72	16.8423	0.0594	396.0566	2.525E-03	23.5156	0.0425	481.0170	20.4552
120	110.6626	9.036E-03	2.742E+03	3.648E-04	24.7741	0.0404	592.2428	23.9057
180	1.164E+03	8.590E-04	2.908E+04	3.439E-05	24.9785	0.0400	620.5976	24.8452
360	1.355E+06	7.379E-07	3.388E+07	2.952E-08	25.0000	0.0400	624.9929	24.9997

Time Value of Money Factors—Discrete Compounding
   i = 5%

	Single Sums		Uniform Series				Gradient Series	
	To Find F Given P (F|P,i%,n)	To Find P Given F (P|F,i%,n)	To Find F Given A (F|A,i%,n)	To Find A Given F (A|F,i%,n)	To Find P Given A (P|A,i%,n)	To Find A Given P (A|P,i%,n)	To Find P Given G (P|G,i%,n)	To Find A Given G (A|G,i%,n)
n								
1	1.0500	0.9524	1.0000	1.0000	0.9524	1.0500	0.0000	0.0000
2	1.1025	0.9070	2.0500	0.4878	1.8594	0.5378	0.9070	0.4878
3	1.1576	0.8638	3.1525	0.3172	2.7232	0.3672	2.6347	0.9675
4	1.2155	0.8227	4.3101	0.2320	3.5460	0.2820	5.1028	1.4391
5	1.2763	0.7835	5.5256	0.1810	4.3295	0.2310	8.2369	1.9025
6	1.3401	0.7462	6.8019	0.1470	5.0757	0.1970	11.9680	2.3579
7	1.4071	0.7107	8.1420	0.1228	5.7864	0.1728	16.2321	2.8052
8	1.4775	0.6768	9.5491	0.1047	6.4632	0.1547	20.9700	3.2445
9	1.5513	0.6446	11.0266	0.0907	7.1078	0.1407	26.1268	3.6758
10	1.6289	0.6139	12.5779	0.0795	7.7217	0.1295	31.6520	4.0991
11	1.7103	0.5847	14.2068	0.0704	8.3064	0.1204	37.4988	4.5144
12	1.7959	0.5568	15.9171	0.0628	8.8633	0.1128	43.6241	4.9219
13	1.8856	0.5303	17.7130	0.0565	9.3936	0.1065	49.9879	5.3215
14	1.9799	0.5051	19.5986	0.0510	9.8986	0.1010	56.5538	5.7133
15	2.0789	0.4810	21.5786	0.0463	10.3797	0.0963	63.2880	6.0973
16	2.1829	0.4581	23.6575	0.0423	10.8378	0.0923	70.1597	6.4736
17	2.2920	0.4363	25.8404	0.0387	11.2741	0.0887	77.1405	6.8423
18	2.4066	0.4155	28.1324	0.0355	11.6896	0.0855	84.2043	7.2034
19	2.5270	0.3957	30.5390	0.0327	12.0853	0.0827	91.3275	7.5569
20	2.6533	0.3769	33.0660	0.0302	12.4622	0.0802	98.4884	7.9030
21	2.7860	0.3589	35.7193	0.0280	12.8212	0.0780	105.6673	8.2416
22	2.9253	0.3418	38.5052	0.0260	13.1630	0.0760	112.8461	8.5730
23	3.0715	0.3256	41.4305	0.0241	13.4886	0.0741	120.0087	8.8971
24	3.2251	0.3101	44.5020	0.0225	13.7986	0.0725	127.1402	9.2140
25	3.3864	0.2953	47.7271	0.0210	14.0939	0.0710	134.2275	9.5238
26	3.5557	0.2812	51.1135	0.0196	14.3752	0.0696	141.2585	9.8266
27	3.7335	0.2678	54.6691	0.0183	14.6430	0.0683	148.2226	10.1224
28	3.9201	0.2551	58.4026	0.0171	14.8981	0.0671	155.1101	10.4114
29	4.1161	0.2429	62.3227	0.0160	15.1411	0.0660	161.9126	10.6936
30	4.3219	0.2314	66.4388	0.0151	15.3725	0.0651	168.6226	10.9691
36	5.7918	0.1727	95.8363	0.0104	16.5469	0.0604	206.6237	12.4872
42	7.7616	0.1288	135.2318	7.395E-03	17.4232	0.0574	240.2389	13.7884
48	10.4013	0.0961	188.0254	5.318E-03	18.0772	0.0553	269.2467	14.8943
54	13.9387	0.0717	258.7739	3.864E-03	18.5651	0.0539	293.8208	15.8265
60	18.6792	0.0535	353.5837	2.828E-03	18.9293	0.0528	314.3432	16.6062
66	25.0319	0.0399	480.6379	2.081E-03	19.2010	0.0521	331.2877	17.2536
72	33.5451	0.0298	650.9027	1.536E-03	19.4038	0.0515	345.1485	17.7877
120	348.9120	2.866E-03	6.958E+03	1.437E-04	19.9427	0.0501	391.9751	19.6551
180	6.517E+03	1.534E-04	1.303E+05	7.673E-06	19.9969	0.0500	399.3863	19.9724
360	4.248E+07	2.354E-08	8.495E+08	1.177E-09	20.0000	0.0500	399.9998	20.0000

Time Value of Money Factors—Discrete Compounding
    i = 6%

	Single Sums		Uniform Series				Gradient Series	
	To Find F Given P (F|P,i%,n)	To Find P Given F (P|F,i%,n)	To Find F Given A (F|A,i%,n)	To Find A Given F (A|F,i%,n)	To Find P Given A (P|A,i%,n)	To Find A Given P (A|P,i%,n)	To Find P Given G (P|G,i%,n)	To Find A Given G (A|G,i%,n)
n								
1	1.0600	0.9434	1.0000	1.0000	0.9434	1.0600	0.0000	0.0000
2	1.1236	0.8900	2.0600	0.4854	1.8334	0.5454	0.8900	0.4854
3	1.1910	0.8396	3.1836	0.3141	2.6730	0.3741	2.5692	0.9612
4	1.2625	0.7921	4.3746	0.2286	3.4651	0.2886	4.9455	1.4272
5	1.3382	0.7473	5.6371	0.1774	4.2124	0.2374	7.9345	1.8836
6	1.4185	0.7050	6.9753	0.1434	4.9173	0.2034	11.4594	2.3304
7	1.5036	0.6651	8.3938	0.1191	5.5824	0.1791	15.4497	2.7676
8	1.5938	0.6274	9.8975	0.1010	6.2098	0.1610	19.8416	3.1952
9	1.6895	0.5919	11.4913	0.0870	6.8017	0.1470	24.5768	3.6133
10	1.7908	0.5584	13.1808	0.0759	7.3601	0.1359	29.6023	4.0220
11	1.8983	0.5268	14.9716	0.0668	7.8869	0.1268	34.8702	4.4213
12	2.0122	0.4970	16.8699	0.0593	8.3838	0.1193	40.3369	4.8113
13	2.1329	0.4688	18.8821	0.0530	8.8527	0.1130	45.9629	5.1920
14	2.2609	0.4423	21.0151	0.0476	9.2950	0.1076	51.7128	5.5635
15	2.3966	0.4173	23.2760	0.0430	9.7122	0.1030	57.5546	5.9260
16	2.5404	0.3936	25.6725	0.0390	10.1059	0.0990	63.4592	6.2794
17	2.6928	0.3714	28.2129	0.0354	10.4773	0.0954	69.4011	6.6240
18	2.8543	0.3503	30.9057	0.0324	10.8276	0.0924	75.3569	6.9597
19	3.0256	0.3305	33.7600	0.0296	11.1581	0.0896	81.3062	7.2867
20	3.2071	0.3118	36.7856	0.0272	11.4699	0.0872	87.2304	7.6051
21	3.3996	0.2942	39.9927	0.0250	11.7641	0.0850	93.1136	7.9151
22	3.6035	0.2775	43.3923	0.0230	12.0416	0.0830	98.9412	8.2166
23	3.8197	0.2618	46.9958	0.0213	12.3034	0.0813	104.7007	8.5099
24	4.0489	0.2470	50.8156	0.0197	12.5504	0.0797	110.3812	8.7951
25	4.2919	0.2330	54.8645	0.0182	12.7834	0.0782	115.9732	9.0722
26	4.5494	0.2198	59.1564	0.0169	13.0032	0.0769	121.4684	9.3414
27	4.8223	0.2074	63.7058	0.0157	13.2105	0.0757	126.8600	9.6029
28	5.1117	0.1956	68.5281	0.0146	13.4062	0.0746	132.1420	9.8568
29	5.4184	0.1846	73.6398	0.0136	13.5907	0.0736	137.3096	10.1032
30	5.7435	0.1741	79.0582	0.0126	13.7648	0.0726	142.3588	10.3422
36	8.1473	0.1227	119.1209	8.395E-03	14.6210	0.0684	170.0387	11.6298
42	11.5570	0.0865	175.9505	5.683E-03	15.2245	0.0657	193.1732	12.6883
48	16.3939	0.0610	256.5645	3.898E-03	15.6500	0.0639	212.0351	13.5485
54	23.2550	0.0430	370.9170	2.696E-03	15.9500	0.0627	227.1316	14.2402
60	32.9877	0.0303	533.1282	1.876E-03	16.1614	0.0619	239.0428	14.7909
66	46.7937	0.0214	763.2278	1.310E-03	16.3105	0.0613	248.3341	15.2254
72	66.3777	0.0151	1.090E+03	9.177E-04	16.4156	0.0609	255.5146	15.5654
120	1.088E+03	9.190E-04	1.812E+04	5.519E-05	16.6514	0.0601	275.6846	16.5563
180	3.590E+04	2.786E-05	5.983E+05	1.672E-06	16.6662	0.0600	277.6865	16.6617
360	1.289E+09	7.760E-10	2.148E+10	4.656E-11	16.6667	0.0600	277.7778	16.6667

Time Value of Money Factors—Discrete Compounding
i = 7%

n	Single Sums		Uniform Series				Gradient Series	
	To Find F Given P (F|P,i%,n)	To Find P Given F (P|F,i%,n)	To Find F Given A (F|A,i%,n)	To Find A Given F (A|F,i%,n)	To Find P Given A (P|A,i%,n)	To Find A Given P (A|P,i%,n)	To Find P Given G (P|G,i%,n)	To Find A Given G (A|G,i%,n)
1	1.0700	0.9346	1.0000	1.0000	0.9346	1.0700	0.0000	0.0000
2	1.1449	0.8734	2.0700	0.4831	1.8080	0.5531	0.8734	0.4831
3	1.2250	0.8163	3.2149	0.3111	2.6243	0.3811	2.5060	0.9549
4	1.3108	0.7629	4.4399	0.2252	3.3872	0.2952	4.7947	1.4155
5	1.4026	0.7130	5.7507	0.1739	4.1002	0.2439	7.6467	1.8650
6	1.5007	0.6663	7.1533	0.1398	4.7665	0.2098	10.9784	2.3032
7	1.6058	0.6227	8.6540	0.1156	5.3893	0.1856	14.7149	2.7304
8	1.7182	0.5820	10.2598	0.0975	5.9713	0.1675	18.7889	3.1465
9	1.8385	0.5439	11.9780	0.0835	6.5152	0.1535	23.1404	3.5517
10	1.9672	0.5083	13.8164	0.0724	7.0236	0.1424	27.7156	3.9461
11	2.1049	0.4751	15.7836	0.0634	7.4987	0.1334	32.4665	4.3296
12	2.2522	0.4440	17.8885	0.0559	7.9427	0.1259	37.3506	4.7025
13	2.4098	0.4150	20.1406	0.0497	8.3577	0.1197	42.3302	5.0648
14	2.5785	0.3878	22.5505	0.0443	8.7455	0.1143	47.3718	5.4167
15	2.7590	0.3624	25.1290	0.0398	9.1079	0.1098	52.4461	5.7583
16	2.9522	0.3387	27.8881	0.0359	9.4466	0.1059	57.5271	6.0897
17	3.1588	0.3166	30.8402	0.0324	9.7632	0.1024	62.5923	6.4110
18	3.3799	0.2959	33.9990	0.0294	10.0591	0.0994	67.6219	6.7225
19	3.6165	0.2765	37.3790	0.0268	10.3356	0.0968	72.5991	7.0242
20	3.8697	0.2584	40.9955	0.0244	10.5940	0.0944	77.5091	7.3163
21	4.1406	0.2415	44.8652	0.0223	10.8355	0.0923	82.3393	7.5990
22	4.4304	0.2257	49.0057	0.0204	11.0612	0.0904	87.0793	7.8725
23	4.7405	0.2109	53.4361	0.0187	11.2722	0.0887	91.7201	8.1369
24	5.0724	0.1971	58.1767	0.0172	11.4693	0.0872	96.2545	8.3923
25	5.4274	0.1842	63.2490	0.0158	11.6536	0.0858	100.6765	8.6391
26	5.8074	0.1722	68.6765	0.0146	11.8258	0.0846	104.9814	8.8773
27	6.2139	0.1609	74.4838	0.0134	11.9867	0.0834	109.1656	9.1072
28	6.6488	0.1504	80.6977	0.0124	12.1371	0.0824	113.2264	9.3289
29	7.1143	0.1406	87.3465	0.0114	12.2777	0.0814	117.1622	9.5427
30	7.6123	0.1314	94.4608	0.0106	12.4090	0.0806	120.9718	9.7487
36	11.4239	0.0875	148.9135	6.715E-03	13.0352	0.0767	141.1990	10.8321
42	17.1443	0.0583	230.6322	4.336E-03	13.4524	0.0743	157.1807	11.6842
48	25.7289	0.0389	353.2701	2.831E-03	13.7305	0.0728	169.4981	12.3447
54	38.6122	0.0259	537.3164	1.861E-03	13.9157	0.0719	178.8173	12.8500
60	57.9464	0.0173	813.5204	1.229E-03	14.0392	0.0712	185.7677	13.2321
66	86.9620	0.0115	1.228E+03	8.143E-04	14.1214	0.0708	190.8927	13.5179
72	130.5065	7.662E-03	1.850E+03	5.405E-04	14.1763	0.0705	194.6365	13.7298
120	3.358E+03	2.978E-04	4.795E+04	2.085E-05	14.2815	0.0700	203.5103	14.2500
180	1.946E+05	5.139E-06	2.780E+06	3.598E-07	14.2856	0.0700	204.0674	14.2848
360	3.786E+10	2.641E-11	5.408E+11	1.849E-12	14.2857	0.0700	204.0816	14.2857

Time Value of Money Factors—Discrete Compounding
    i = 8%

	Single Sums		Uniform Series				Gradient Series	
	To Find F Given P (F|P,i%,n)	To Find P Given F (P|F,i%,n)	To Find F Given A (F|A,i%,n)	To Find A Given F (A|F,i%,n)	To Find P Given A (P|A,i%,n)	To Find A Given P (A|P,i%,n)	To Find P Given G (P|G,i%,n)	To Find A Given G (A|G,i%,n)
n								
1	1.0800	0.9259	1.0000	1.0000	0.9259	1.0800	0.0000	0.0000
2	1.1664	0.8573	2.0800	0.4808	1.7833	0.5608	0.8573	0.4808
3	1.2597	0.7938	3.2464	0.3080	2.5771	0.3880	2.4450	0.9487
4	1.3605	0.7350	4.5061	0.2219	3.3121	0.3019	4.6501	1.4040
5	1.4693	0.6806	5.8666	0.1705	3.9927	0.2505	7.3724	1.8465
6	1.5869	0.6302	7.3359	0.1363	4.6229	0.2163	10.5233	2.2763
7	1.7138	0.5835	8.9228	0.1121	5.2064	0.1921	14.0242	2.6937
8	1.8509	0.5403	10.6366	0.0940	5.7466	0.1740	17.8061	3.0985
9	1.9990	0.5002	12.4876	0.0801	6.2469	0.1601	21.8081	3.4910
10	2.1589	0.4632	14.4866	0.0690	6.7101	0.1490	25.9768	3.8713
11	2.3316	0.4289	16.6455	0.0601	7.1390	0.1401	30.2657	4.2395
12	2.5182	0.3971	18.9771	0.0527	7.5361	0.1327	34.6339	4.5957
13	2.7196	0.3677	21.4953	0.0465	7.9038	0.1265	39.0463	4.9402
14	2.9372	0.3405	24.2149	0.0413	8.2442	0.1213	43.4723	5.2731
15	3.1722	0.3152	27.1521	0.0368	8.5595	0.1168	47.8857	5.5945
16	3.4259	0.2919	30.3243	0.0330	8.8514	0.1130	52.2640	5.9046
17	3.7000	0.2703	33.7502	0.0296	9.1216	0.1096	56.5883	6.2037
18	3.9960	0.2502	37.4502	0.0267	9.3719	0.1067	60.8426	6.4920
19	4.3157	0.2317	41.4463	0.0241	9.6036	0.1041	65.0134	6.7697
20	4.6610	0.2145	45.7620	0.0219	9.8181	0.1019	69.0898	7.0369
21	5.0338	0.1987	50.4229	0.0198	10.0168	0.0998	73.0629	7.2940
22	5.4365	0.1839	55.4568	0.0180	10.2007	0.0980	76.9257	7.5412
23	5.8715	0.1703	60.8933	0.0164	10.3711	0.0964	80.6726	7.7786
24	6.3412	0.1577	66.7648	0.0150	10.5288	0.0950	84.2997	8.0066
25	6.8485	0.1460	73.1059	0.0137	10.6748	0.0937	87.8041	8.2254
26	7.3964	0.1352	79.9544	0.0125	10.8100	0.0925	91.1842	8.4352
27	7.9881	0.1252	87.3508	0.0114	10.9352	0.0914	94.4390	8.6363
28	8.6271	0.1159	95.3388	0.0105	11.0511	0.0905	97.5687	8.8289
29	9.3173	0.1073	103.9659	9.619E-03	11.1584	0.0896	100.5738	9.0133
30	10.0627	0.0994	113.2832	8.827E-03	11.2578	0.0888	103.4558	9.1897
36	15.9682	0.0626	187.1021	5.345E-03	11.7172	0.0853	118.2839	10.0949
42	25.3395	0.0395	304.2435	3.287E-03	12.0067	0.0833	129.3651	10.7744
48	40.2106	0.0249	490.1322	2.040E-03	12.1891	0.0820	137.4428	11.2758
54	63.8091	0.0157	785.1141	1.274E-03	12.3041	0.0813	143.2229	11.6403
60	101.2571	9.876E-03	1.253E+03	7.979E-04	12.3766	0.0808	147.3000	11.9015
66	160.6822	6.223E-03	1.996E+03	5.010E-04	12.4222	0.0805	150.1432	12.0867
72	254.9825	3.922E-03	3.175E+03	3.150E-04	12.4510	0.0803	152.1076	12.2165
120	1.025E+04	9.753E-05	1.281E+05	7.803E-06	12.4988	0.0800	156.0885	12.4883
180	1.038E+06	9.632E-07	1.298E+07	7.706E-08	12.5000	0.0800	156.2477	12.4998
360	1.078E+12	9.278E-13	1.347E+13	7.422E-14	12.5000	0.0800	156.2500	12.5000

Time Value of Money Factors—Discrete Compounding
i = 9%

n	Single Sums		Uniform Series				Gradient Series	
	To Find F Given P (F|P,i%,n)	To Find P Given F (P|F,i%,n)	To Find F Given A (F|A,i%,n)	To Find A Given F (A|F,i%,n)	To Find P Given A (P|A,i%,n)	To Find A Given P (A|P,i%,n)	To Find P Given G (P|G,i%,n)	To Find A Given G (A|G,i%,n)
1	1.0900	0.9174	1.0000	1.0000	0.9174	1.0900	0.0000	0.0000
2	1.1881	0.8417	2.0900	0.4785	1.7591	0.5685	0.8417	0.4785
3	1.2950	0.7722	3.2781	0.3051	2.5313	0.3951	2.3860	0.9426
4	1.4116	0.7084	4.5731	0.2187	3.2397	0.3087	4.5113	1.3925
5	1.5386	0.6499	5.9847	0.1671	3.8897	0.2571	7.1110	1.8282
6	1.6771	0.5963	7.5233	0.1329	4.4859	0.2229	10.0924	2.2498
7	1.8280	0.5470	9.2004	0.1087	5.0330	0.1987	13.3746	2.6574
8	1.9926	0.5019	11.0285	0.0907	5.5348	0.1807	16.8877	3.0512
9	2.1719	0.4604	13.0210	0.0768	5.9952	0.1668	20.5711	3.4312
10	2.3674	0.4224	15.1929	0.0658	6.4177	0.1558	24.3728	3.7978
11	2.5804	0.3875	17.5603	0.0569	6.8052	0.1469	28.2481	4.1510
12	2.8127	0.3555	20.1407	0.0497	7.1607	0.1397	32.1590	4.4910
13	3.0658	0.3262	22.9534	0.0436	7.4869	0.1336	36.0731	4.8182
14	3.3417	0.2992	26.0192	0.0384	7.7862	0.1284	39.9633	5.1326
15	3.6425	0.2745	29.3609	0.0341	8.0607	0.1241	43.8069	5.4346
16	3.9703	0.2519	33.0034	0.0303	8.3126	0.1203	47.5849	5.7245
17	4.3276	0.2311	36.9737	0.0270	8.5436	0.1170	51.2821	6.0024
18	4.7171	0.2120	41.3013	0.0242	8.7556	0.1142	54.8860	6.2687
19	5.1417	0.1945	46.0185	0.0217	8.9501	0.1117	58.3868	6.5236
20	5.6044	0.1784	51.1601	0.0195	9.1285	0.1095	61.7770	6.7674
21	6.1088	0.1637	56.7645	0.0176	9.2922	0.1076	65.0509	7.0006
22	6.6586	0.1502	62.8733	0.0159	9.4424	0.1059	68.2048	7.2232
23	7.2579	0.1378	69.5319	0.0144	9.5802	0.1044	71.2359	7.4357
24	7.9111	0.1264	76.7898	0.0130	9.7066	0.1030	74.1433	7.6384
25	8.6231	0.1160	84.7009	0.0118	9.8226	0.1018	76.9265	7.8316
26	9.3992	0.1064	93.3240	0.0107	9.9290	0.1007	79.5863	8.0156
27	10.2451	0.0976	102.7231	9.735E-03	10.0266	0.0997	82.1241	8.1906
28	11.1671	0.0895	112.9682	8.852E-03	10.1161	0.0989	84.5419	8.3571
29	12.1722	0.0822	124.1354	8.056E-03	10.1983	0.0981	86.8422	8.5154
30	13.2677	0.0754	136.3075	7.336E-03	10.2737	0.0973	89.0280	8.6657
36	22.2512	0.0449	236.1247	4.235E-03	10.6118	0.0942	99.9319	9.4171
42	37.3175	0.0268	403.5281	2.478E-03	10.8134	0.0925	107.6432	9.9546
48	62.5852	0.0160	684.2804	1.461E-03	10.9336	0.0915	112.9625	10.3317
54	104.9617	9.527E-03	1.155E+03	8.657E-04	11.0053	0.0909	116.5642	10.5917
60	176.0313	5.681E-03	1.945E+03	5.142E-04	11.0480	0.0905	118.9683	10.7683
66	295.2221	3.387E-03	3.269E+03	3.059E-04	11.0735	0.0903	120.5546	10.8868
72	495.1170	2.020E-03	5.490E+03	1.821E-04	11.0887	0.0902	121.5917	10.9654
120	3.099E+04	3.227E-05	3.443E+05	2.905E-06	11.1108	0.0900	123.4098	11.1072
180	5.455E+06	1.833E-07	6.061E+07	1.650E-08	11.1111	0.0900	123.4564	11.1111
360	2.975E+13	3.361E-14	3.306E+14	3.025E-15	11.1111	0.0900	123.4568	11.1111

Time Value of Money Factors—Discrete Compounding
     i = 10%

	Single Sums		Uniform Series				Gradient Series	
	To Find F Given P (F|P,i%,n)	To Find P Given F (P|F,i%,n)	To Find F Given A (F|A,i%,n)	To Find A Given F (A|F,i%,n)	To Find P Given A (P|A,i%,n)	To Find A Given P (A|P,i%,n)	To Find P Given G (P|G,i%,n)	To Find A Given G (A|G,i%,n)
n								
1	1.1000	0.9091	1.0000	1.0000	0.9091	1.1000	0.0000	0.0000
2	1.2100	0.8264	2.1000	0.4762	1.7355	0.5762	0.8264	0.4762
3	1.3310	0.7513	3.3100	0.3021	2.4869	0.4021	2.3291	0.9366
4	1.4641	0.6830	4.6410	0.2155	3.1699	0.3155	4.3781	1.3812
5	1.6105	0.6209	6.1051	0.1638	3.7908	0.2638	6.8618	1.8101
6	1.7716	0.5645	7.7156	0.1296	4.3553	0.2296	9.6842	2.2236
7	1.9487	0.5132	9.4872	0.1054	4.8684	0.2054	12.7631	2.6216
8	2.1436	0.4665	11.4359	0.0874	5.3349	0.1874	16.0287	3.0045
9	2.3579	0.4241	13.5795	0.0736	5.7590	0.1736	19.4215	3.3724
10	2.5937	0.3855	15.9374	0.0627	6.1446	0.1627	22.8913	3.7255
11	2.8531	0.3505	18.5312	0.0540	6.4951	0.1540	26.3963	4.0641
12	3.1384	0.3186	21.3843	0.0468	6.8137	0.1468	29.9012	4.3884
13	3.4523	0.2897	24.5227	0.0408	7.1034	0.1408	33.3772	4.6988
14	3.7975	0.2633	27.9750	0.0357	7.3667	0.1357	36.8005	4.9955
15	4.1772	0.2394	31.7725	0.0315	7.6061	0.1315	40.1520	5.2789
16	4.5950	0.2176	35.9497	0.0278	7.8237	0.1278	43.4164	5.5493
17	5.0545	0.1978	40.5447	0.0247	8.0216	0.1247	46.5819	5.8071
18	5.5599	0.1799	45.5992	0.0219	8.2014	0.1219	49.6395	6.0526
19	6.1159	0.1635	51.1591	0.0195	8.3649	0.1195	52.5827	6.2861
20	6.7275	0.1486	57.2750	0.0175	8.5136	0.1175	55.4069	6.5081
21	7.4002	0.1351	64.0025	0.0156	8.6487	0.1156	58.1095	6.7189
22	8.1403	0.1228	71.4027	0.0140	8.7715	0.1140	60.6893	6.9189
23	8.9543	0.1117	79.5430	0.0126	8.8832	0.1126	63.1462	7.1085
24	9.8497	0.1015	88.4973	0.0113	8.9847	0.1113	65.4813	7.2881
25	10.8347	0.0923	98.3471	0.0102	9.0770	0.1102	67.6964	7.4580
26	11.9182	0.0839	109.1818	9.159E-03	9.1609	0.1092	69.7940	7.6186
27	13.1100	0.0763	121.0999	8.258E-03	9.2372	0.1083	71.7773	7.7704
28	14.4210	0.0693	134.2099	7.451E-03	9.3066	0.1075	73.6495	7.9137
29	15.8631	0.0630	148.6309	6.728E-03	9.3696	0.1067	75.4146	8.0489
30	17.4494	0.0573	164.4940	6.079E-03	9.4269	0.1061	77.0766	8.1762
36	30.9127	0.0323	299.1268	3.343E-03	9.6765	0.1033	85.1194	8.7965
42	54.7637	0.0183	537.6370	1.860E-03	9.8174	0.1019	90.5047	9.2188
48	97.0172	0.0103	960.1723	1.041E-03	9.8969	0.1010	94.0217	9.5001
54	171.8719	5.818E-03	1.709E+03	5.852E-04	9.9418	0.1006	96.2763	9.6840
60	304.4816	3.284E-03	3.035E+03	3.295E-04	9.9672	0.1003	97.7010	9.8023
66	539.4078	1.854E-03	5.384E+03	1.857E-04	9.9815	0.1002	98.5910	9.8774
72	955.5938	1.046E-03	9.546E+03	1.048E-04	9.9895	0.1001	99.1419	9.9246
120	9.271E+04	1.079E-05	9.271E+05	1.079E-06	9.9999	0.1000	99.9860	9.9987
180	2.823E+07	3.543E-08	2.823E+08	3.543E-09	10.0000	0.1000	99.9999	10.0000
360	7.968E+14	1.255E-15	7.968E+15	1.255E-16	10.0000	0.1000	100.0000	10.0000

Time Value of Money Factors—Discrete Compounding
i = 12%

	Single Sums		Uniform Series				Gradient Series	
	To Find F Given P (F|P,i%,n)	To Find P Given F (P|F,i%,n)	To Find F Given A (F|A,i%,n)	To Find A Given F (A|F,i%,n)	To Find P Given A (P|A,i%,n)	To Find A Given P (A|P,i%,n)	To Find P Given G (P|G,i%,n)	To Find A Given G (A|G,i%,n)
n								
1	1.1200	0.8929	1.0000	1.0000	0.8929	1.1200	0.0000	0.0000
2	1.2544	0.7972	2.1200	0.4717	1.6901	0.5917	0.7972	0.4717
3	1.4049	0.7118	3.3744	0.2963	2.4018	0.4163	2.2208	0.9246
4	1.5735	0.6355	4.7793	0.2092	3.0373	0.3292	4.1273	1.3589
5	1.7623	0.5674	6.3528	0.1574	3.6048	0.2774	6.3970	1.7746
6	1.9738	0.5066	8.1152	0.1232	4.1114	0.2432	8.9302	2.1720
7	2.2107	0.4523	10.0890	0.0991	4.5638	0.2191	11.6443	2.5515
8	2.4760	0.4039	12.2997	0.0813	4.9676	0.2013	14.4714	2.9131
9	2.7731	0.3606	14.7757	0.0677	5.3282	0.1877	17.3563	3.2574
10	3.1058	0.3220	17.5487	0.0570	5.6502	0.1770	20.2541	3.5847
11	3.4785	0.2875	20.6546	0.0484	5.9377	0.1684	23.1288	3.8953
12	3.8960	0.2567	24.1331	0.0414	6.1944	0.1614	25.9523	4.1897
13	4.3635	0.2292	28.0291	0.0357	6.4235	0.1557	28.7024	4.4683
14	4.8871	0.2046	32.3926	0.0309	6.6282	0.1509	31.3624	4.7317
15	5.4736	0.1827	37.2797	0.0268	6.8109	0.1468	33.9202	4.9803
16	6.1304	0.1631	42.7533	0.0234	6.9740	0.1434	36.3670	5.2147
17	6.8660	0.1456	48.8837	0.0205	7.1196	0.1405	38.6973	5.4353
18	7.6900	0.1300	55.7497	0.0179	7.2497	0.1379	40.9080	5.6427
19	8.6128	0.1161	63.4397	0.0158	7.3658	0.1358	42.9979	5.8375
20	9.6463	0.1037	72.0524	0.0139	7.4694	0.1339	44.9676	6.0202
21	10.8038	0.0926	81.6987	0.0122	7.5620	0.1322	46.8188	6.1913
22	12.1003	0.0826	92.5026	0.0108	7.6446	0.1308	48.5543	6.3514
23	13.5523	0.0738	104.6029	9.560E-03	7.7184	0.1296	50.1776	6.5010
24	15.1786	0.0659	118.1552	8.463E-03	7.7843	0.1285	51.6929	6.6406
25	17.0001	0.0588	133.3339	7.500E-03	7.8431	0.1275	53.1046	6.7708
26	19.0401	0.0525	150.3339	6.652E-03	7.8957	0.1267	54.4177	6.8921
27	21.3249	0.0469	169.3740	5.904E-03	7.9426	0.1259	55.6369	7.0049
28	23.8839	0.0419	190.6989	5.244E-03	7.9844	0.1252	56.7674	7.1098
29	26.7499	0.0374	214.5828	4.660E-03	8.0218	0.1247	57.8141	7.2071
30	29.9599	0.0334	241.3327	4.144E-03	8.0552	0.1241	58.7821	7.2974
36	59.1356	0.0169	484.4631	2.064E-03	8.1924	0.1221	63.1970	7.7141
42	116.7231	8.567E-03	964.3595	1.037E-03	8.2619	0.1210	65.8509	7.9704
48	230.3908	4.340E-03	1.912E+03	5.231E-04	8.2972	0.1205	67.4068	8.1241
54	454.7505	2.199E-03	3.781E+03	2.645E-04	8.3150	0.1203	68.3022	8.2143
60	897.5969	1.114E-03	7.472E+03	1.338E-04	8.3240	0.1201	68.8100	8.2664
66	1.772E+03	5.644E-04	1.476E+04	6.777E-05	8.3286	0.1201	69.0948	8.2961
72	3.497E+03	2.860E-04	2.913E+04	3.432E-05	8.3310	0.1200	69.2530	8.3127
120	8.057E+05	1.241E-06	6.714E+06	1.489E-07	8.3333	0.1200	69.4431	8.3332
180	7.232E+08	1.383E-09	6.026E+09	1.659E-10	8.3333	0.1200	69.4444	8.3333
360	5.230E+17	1.912E-18	4.358E+18	2.295E-19	8.3333	0.1200	69.4444	8.3333

Time Value of Money Factors—Discrete Compounding
    i = 15%

	Single Sums		Uniform Series				Gradient Series	
	To Find F Given P (F|P,i%,n)	To Find P Given F (P|F,i%,n)	To Find F Given A (F|A,i%,n)	To Find A Given F (A|F,i%,n)	To Find P Given A (P|A,i%,n)	To Find A Given P (A|P,i%,n)	To Find P Given G (P|G,i%,n)	To Find A Given G (A|G,i%,n)
n								
1	1.1500	0.8696	1.0000	1.0000	0.8696	1.1500	0.0000	0.0000
2	1.3225	0.7561	2.1500	0.4651	1.6257	0.6151	0.7561	0.4651
3	1.5209	0.6575	3.4725	0.2880	2.2832	0.4380	2.0712	0.9071
4	1.7490	0.5718	4.9934	0.2003	2.8550	0.3503	3.7864	1.3263
5	2.0114	0.4972	6.7424	0.1483	3.3522	0.2983	5.7751	1.7228
6	2.3131	0.4323	8.7537	0.1142	3.7845	0.2642	7.9368	2.0972
7	2.6600	0.3759	11.0668	0.0904	4.1604	0.2404	10.1924	2.4498
8	3.0590	0.3269	13.7268	0.0729	4.4873	0.2229	12.4807	2.7813
9	3.5179	0.2843	16.7858	0.0596	4.7716	0.2096	14.7548	3.0922
10	4.0456	0.2472	20.3037	0.0493	5.0188	0.1993	16.9795	3.3832
11	4.6524	0.2149	24.3493	0.0411	5.2337	0.1911	19.1289	3.6549
12	5.3503	0.1869	29.0017	0.0345	5.4206	0.1845	21.1849	3.9082
13	6.1528	0.1625	34.3519	0.0291	5.5831	0.1791	23.1352	4.1438
14	7.0757	0.1413	40.5047	0.0247	5.7245	0.1747	24.9725	4.3624
15	8.1371	0.1229	47.5804	0.0210	5.8474	0.1710	26.6930	4.5650
16	9.3576	0.1069	55.7175	0.0179	5.9542	0.1679	28.2960	4.7522
17	10.7613	0.0929	65.0751	0.0154	6.0472	0.1654	29.7828	4.9251
18	12.3755	0.0808	75.8364	0.0132	6.1280	0.1632	31.1565	5.0843
19	14.2318	0.0703	88.2118	0.0113	6.1982	0.1613	32.4213	5.2307
20	16.3665	0.0611	102.4436	9.761E-03	6.2593	0.1598	33.5822	5.3651
21	18.8215	0.0531	118.8101	8.417E-03	6.3125	0.1584	34.6448	5.4883
22	21.6447	0.0462	137.6316	7.266E-03	6.3587	0.1573	35.6150	5.6010
23	24.8915	0.0402	159.2764	6.278E-03	6.3988	0.1563	36.4988	5.7040
24	28.6252	0.0349	184.1678	5.430E-03	6.4338	0.1554	37.3023	5.7979
25	32.9190	0.0304	212.7930	4.699E-03	6.4641	0.1547	38.0314	5.8834
26	37.8568	0.0264	245.7120	4.070E-03	6.4906	0.1541	38.6918	5.9612
27	43.5353	0.0230	283.5688	3.526E-03	6.5135	0.1535	39.2890	6.0319
28	50.0656	0.0200	327.1041	3.057E-03	6.5335	0.1531	39.8283	6.0960
29	57.5755	0.0174	377.1697	2.651E-03	6.5509	0.1527	40.3146	6.1541
30	66.2118	0.0151	434.7451	2.300E-03	6.5660	0.1523	40.7526	6.2066
36	153.1519	6.529E-03	1.014E+03	9.859E-04	6.6231	0.1510	42.5872	6.4301
42	354.2495	2.823E-03	2.355E+03	4.246E-04	6.6478	0.1504	43.5286	6.5478
48	819.4007	1.220E-03	5.456E+03	1.833E-04	6.6585	0.1502	43.9997	6.6080
54	1.895E+03	5.276E-04	1.263E+04	7.918E-05	6.6631	0.1501	44.2311	6.6382
60	4.384E+03	2.281E-04	2.922E+04	3.422E-05	6.6651	0.1500	44.3431	6.6530
66	1.014E+04	9.861E-05	6.760E+04	1.479E-05	6.6660	0.1500	44.3967	6.6602
72	2.346E+04	4.263E-05	1.564E+05	6.395E-06	6.6664	0.1500	44.4221	6.6636
120	1.922E+07	5.203E-08	1.281E+08	7.805E-09	6.6667	0.1500	44.4444	6.6667
180	8.426E+10	1.187E-11	5.617E+11	1.780E-12	6.6667	0.1500	44.4444	6.6667
360	7.099E+21	1.409E-22	4.733E+22	2.113E-23	6.6667	0.1500	44.4444	6.6667

Time Value of Money Factors—Discrete Compounding
i = 18%

n	Single Sums		Uniform Series				Gradient Series	
	To Find F Given P (F|P,i%,n)	To Find P Given F (P|F,i%,n)	To Find F Given A (F|A,i%,n)	To Find A Given F (A|F,i%,n)	To Find P Given A (P|A,i%,n)	To Find A Given P (A|P,i%,n)	To Find P Given G (P|G,i%,n)	To Find A Given G (A|G,i%,n)
1	1.1800	0.8475	1.0000	1.0000	0.8475	1.1800	0.0000	0.0000
2	1.3924	0.7182	2.1800	0.4587	1.5656	0.6387	0.7182	0.4587
3	1.6430	0.6086	3.5724	0.2799	2.1743	0.4599	1.9354	0.8902
4	1.9388	0.5158	5.2154	0.1917	2.6901	0.3717	3.4828	1.2947
5	2.2878	0.4371	7.1542	0.1398	3.1272	0.3198	5.2312	1.6728
6	2.6996	0.3704	9.4420	0.1059	3.4976	0.2859	7.0834	2.0252
7	3.1855	0.3139	12.1415	0.0824	3.8115	0.2624	8.9670	2.3526
8	3.7589	0.2660	15.3270	0.0652	4.0776	0.2452	10.8292	2.6558
9	4.4355	0.2255	19.0859	0.0524	4.3030	0.2324	12.6329	2.9358
10	5.2338	0.1911	23.5213	0.0425	4.4941	0.2225	14.3525	3.1936
11	6.1759	0.1619	28.7551	0.0348	4.6560	0.2148	15.9716	3.4303
12	7.2876	0.1372	34.9311	0.0286	4.7932	0.2086	17.4811	3.6470
13	8.5994	0.1163	42.2187	0.0237	4.9095	0.2037	18.8765	3.8449
14	10.1472	0.0985	50.8180	0.0197	5.0081	0.1997	20.1576	4.0250
15	11.9737	0.0835	60.9653	0.0164	5.0916	0.1964	21.3269	4.1887
16	14.1290	0.0708	72.9390	0.0137	5.1624	0.1937	22.3885	4.3369
17	16.6722	0.0600	87.0680	0.0115	5.2223	0.1915	23.3482	4.4708
18	19.6733	0.0508	103.7403	9.639E-03	5.2732	0.1896	24.2123	4.5916
19	23.2144	0.0431	123.4135	8.103E-03	5.3162	0.1881	24.9877	4.7003
20	27.3930	0.0365	146.6280	6.820E-03	5.3527	0.1868	25.6813	4.7978
21	32.3238	0.0309	174.0210	5.746E-03	5.3837	0.1857	26.3000	4.8851
22	38.1421	0.0262	206.3448	4.846E-03	5.4099	0.1848	26.8506	4.9632
23	45.0076	0.0222	244.4868	4.090E-03	5.4321	0.1841	27.3394	5.0329
24	53.1090	0.0188	289.4945	3.454E-03	5.4509	0.1835	27.7725	5.0950
25	62.6686	0.0160	342.6035	2.919E-03	5.4669	0.1829	28.1555	5.1502
26	73.9490	0.0135	405.2721	2.467E-03	5.4804	0.1825	28.4935	5.1991
27	87.2598	0.0115	479.2211	2.087E-03	5.4919	0.1821	28.7915	5.2425
28	102.9666	9.712E-03	566.4809	1.765E-03	5.5016	0.1818	29.0537	5.2810
29	121.5005	8.230E-03	669.4475	1.494E-03	5.5098	0.1815	29.2842	5.3149
30	143.3706	6.975E-03	790.9480	1.264E-03	5.5168	0.1813	29.4864	5.3448
36	387.0368	2.584E-03	2.145E+03	4.663E-04	5.5412	0.1805	30.2677	5.4623
42	1.045E+03	9.571E-04	5.799E+03	1.724E-04	5.5502	0.1802	30.6113	5.5153
48	2.821E+03	3.545E-04	1.566E+04	6.384E-05	5.5536	0.1801	30.7587	5.5385
54	7.614E+03	1.313E-04	4.230E+04	2.364E-05	5.5548	0.1800	30.8207	5.5485
60	2.056E+04	4.865E-05	1.142E+05	8.757E-06	5.5553	0.1800	30.8465	5.5526
66	5.549E+04	1.802E-05	3.083E+05	3.244E-06	5.5555	0.1800	30.8570	5.5544
72	1.498E+05	6.676E-06	8.322E+05	1.202E-06	5.5555	0.1800	30.8613	5.5551
120	4.225E+08	2.367E-09	2.347E+09	4.260E-10	5.5556	0.1800	30.8642	5.5556
180	8.685E+12	1.151E-13	4.825E+13	2.073E-14	5.5556	0.1800	30.8642	5.5556
360	7.543E+25	1.326E-26	4.190E+26	2.386E-27	5.5556	0.1800	30.8642	5.5556

Time Value of Money Factors—Discrete Compounding
    i = 20%

n	Single Sums		Uniform Series				Gradient Series	
	To Find F Given P (F|P,i%,n)	To Find P Given F (P|F,i%,n)	To Find F Given A (F|A,i%,n)	To Find A Given F (A|F,i%,n)	To Find P Given A (P|A,i%,n)	To Find A Given P (A|P,i%,n)	To Find P Given G (P|G,i%,n)	To Find A Given G (A|G,i%,n)
1	1.2000	0.8333	1.0000	1.0000	0.8333	1.2000	0.0000	0.0000
2	1.4400	0.6944	2.2000	0.4545	1.5278	0.6545	0.6944	0.4545
3	1.7280	0.5787	3.6400	0.2747	2.1065	0.4747	1.8519	0.8791
4	2.0736	0.4823	5.3680	0.1863	2.5887	0.3863	3.2986	1.2742
5	2.4883	0.4019	7.4416	0.1344	2.9906	0.3344	4.9061	1.6405
6	2.9860	0.3349	9.9299	0.1007	3.3255	0.3007	6.5806	1.9788
7	3.5832	0.2791	12.9159	0.0774	3.6046	0.2774	8.2551	2.2902
8	4.2998	0.2326	16.4991	0.0606	3.8372	0.2606	9.8831	2.5756
9	5.1598	0.1938	20.7989	0.0481	4.0310	0.2481	11.4335	2.8364
10	6.1917	0.1615	25.9587	0.0385	4.1925	0.2385	12.8871	3.0739
11	7.4301	0.1346	32.1504	0.0311	4.3271	0.2311	14.2330	3.2893
12	8.9161	0.1122	39.5805	0.0253	4.4392	0.2253	15.4667	3.4841
13	10.6993	0.0935	48.4966	0.0206	4.5327	0.2206	16.5883	3.6597
14	12.8392	0.0779	59.1959	0.0169	4.6106	0.2169	17.6008	3.8175
15	15.4070	0.0649	72.0351	0.0139	4.6755	0.2139	18.5095	3.9588
16	18.4884	0.0541	87.4421	0.0114	4.7296	0.2114	19.3208	4.0851
17	22.1861	0.0451	105.9306	9.440E-03	4.7746	0.2094	20.0419	4.1976
18	26.6233	0.0376	128.1167	7.805E-03	4.8122	0.2078	20.6805	4.2975
19	31.9480	0.0313	154.7400	6.462E-03	4.8435	0.2065	21.2439	4.3861
20	38.3376	0.0261	186.6880	5.357E-03	4.8696	0.2054	21.7395	4.4643
21	46.0051	0.0217	225.0256	4.444E-03	4.8913	0.2044	22.1742	4.5334
22	55.2061	0.0181	271.0307	3.690E-03	4.9094	0.2037	22.5546	4.5941
23	66.2474	0.0151	326.2369	3.065E-03	4.9245	0.2031	22.8867	4.6475
24	79.4968	0.0126	392.4842	2.548E-03	4.9371	0.2025	23.1760	4.6943
25	95.3962	0.0105	471.9811	2.119E-03	4.9476	0.2021	23.4276	4.7352
26	114.4755	8.735E-03	567.3773	1.762E-03	4.9563	0.2018	23.6460	4.7709
27	137.3706	7.280E-03	681.8528	1.467E-03	4.9636	0.2015	23.8353	4.8020
28	164.8447	6.066E-03	819.2233	1.221E-03	4.9697	0.2012	23.9991	4.8291
29	197.8136	5.055E-03	984.0680	1.016E-03	4.9747	0.2010	24.1406	4.8527
30	237.3763	4.213E-03	1.182E+03	8.461E-04	4.9789	0.2008	24.2628	4.8731
36	708.8019	1.411E-03	3.539E+03	2.826E-04	4.9929	0.2003	24.7108	4.9491
42	2.116E+03	4.725E-04	1.058E+04	9.454E-05	4.9976	0.2001	24.8890	4.9801
48	6.320E+03	1.582E-04	3.159E+04	3.165E-05	4.9992	0.2000	24.9581	4.9924
54	1.887E+04	5.299E-05	9.435E+04	1.060E-05	4.9997	0.2000	24.9844	4.9971
60	5.635E+04	1.775E-05	2.817E+05	3.549E-06	4.9999	0.2000	24.9942	4.9989
66	1.683E+05	5.943E-06	8.413E+05	1.189E-06	5.0000	0.2000	24.9979	4.9996
72	5.024E+05	1.990E-06	2.512E+06	3.981E-07	5.0000	0.2000	24.9992	4.9999
120	3.175E+09	3.150E-10	1.588E+10	6.299E-11	5.0000	0.2000	25.0000	5.0000
180	1.789E+14	5.590E-15	8.945E+14	1.118E-15	5.0000	0.2000	25.0000	5.0000
360	3.201E+28	3.124E-29	1.600E+29	6.249E-30	5.0000	0.2000	25.0000	5.0000

Time Value of Money Factors—Discrete Compounding
i = 25%

n	Single Sums		Uniform Series				Gradient Series	
	To Find F Given P (F|P,i%,n)	To Find P Given F (P|F,i%,n)	To Find F Given A (F|A,i%,n)	To Find A Given F (A|F,i%,n)	To Find P Given A (P|A,i%,n)	To Find A Given P (A|P,i%,n)	To Find P Given G (P|G,i%,n)	To Find A Given G (A|G,i%,n)
1	1.2500	0.8000	1.0000	1.0000	0.8000	1.2500	0.0000	0.0000
2	1.5625	0.6400	2.2500	0.4444	1.4400	0.6944	0.6400	0.4444
3	1.9531	0.5120	3.8125	0.2623	1.9520	0.5123	1.6640	0.8525
4	2.4414	0.4096	5.7656	0.1734	2.3616	0.4234	2.8928	1.2249
5	3.0518	0.3277	8.2070	0.1218	2.6893	0.3718	4.2035	1.5631
6	3.8147	0.2621	11.2588	0.0888	2.9514	0.3388	5.5142	1.8683
7	4.7684	0.2097	15.0735	0.0663	3.1611	0.3163	6.7725	2.1424
8	5.9605	0.1678	19.8419	0.0504	3.3289	0.3004	7.9469	2.3872
9	7.4506	0.1342	25.8023	0.0388	3.4631	0.2888	9.0207	2.6048
10	9.3132	0.1074	33.2529	0.0301	3.5705	0.2801	9.9870	2.7971
11	11.6415	0.0859	42.5661	0.0235	3.6564	0.2735	10.8460	2.9663
12	14.5519	0.0687	54.2077	0.0184	3.7251	0.2684	11.6020	3.1145
13	18.1899	0.0550	68.7596	0.0145	3.7801	0.2645	12.2617	3.2437
14	22.7374	0.0440	86.9495	0.0115	3.8241	0.2615	12.8334	3.3559
15	28.4217	0.0352	109.6868	9.117E-03	3.8593	0.2591	13.3260	3.4530
16	35.5271	0.0281	138.1085	7.241E-03	3.8874	0.2572	13.7482	3.5366
17	44.4089	0.0225	173.6357	5.759E-03	3.9099	0.2558	14.1085	3.6084
18	55.5112	0.0180	218.0446	4.586E-03	3.9279	0.2546	14.4147	3.6698
19	69.3889	0.0144	273.5558	3.656E-03	3.9424	0.2537	14.6741	3.7222
20	86.7362	0.0115	342.9447	2.916E-03	3.9539	0.2529	14.8932	3.7667
21	108.4202	9.223E-03	429.6809	2.327E-03	3.9631	0.2523	15.0777	3.8045
22	135.5253	7.379E-03	538.1011	1.858E-03	3.9705	0.2519	15.2326	3.8365
23	169.4066	5.903E-03	673.6264	1.485E-03	3.9764	0.2515	15.3625	3.8634
24	211.7582	4.722E-03	843.0329	1.186E-03	3.9811	0.2512	15.4711	3.8861
25	264.6978	3.778E-03	1.055E+03	9.481E-04	3.9849	0.2509	15.5618	3.9052
26	330.8722	3.022E-03	1.319E+03	7.579E-04	3.9879	0.2508	15.6373	3.9212
27	413.5903	2.418E-03	1.650E+03	6.059E-04	3.9903	0.2506	15.7002	3.9346
28	516.9879	1.934E-03	2.064E+03	4.845E-04	3.9923	0.2505	15.7524	3.9457
29	646.2349	1.547E-03	2.581E+03	3.875E-04	3.9938	0.2504	15.7957	3.9551
30	807.7936	1.238E-03	3.227E+03	3.099E-04	3.9950	0.2503	15.8316	3.9628
36	3.081E+03	3.245E-04	1.232E+04	8.116E-05	3.9987	0.2501	15.9481	3.9883
42	1.175E+04	8.507E-05	4.702E+04	2.127E-05	3.9997	0.2500	15.9843	3.9964
48	4.484E+04	2.230E-05	1.794E+05	5.575E-06	3.9999	0.2500	15.9954	3.9989
54	1.711E+05	5.846E-06	6.842E+05	1.462E-06	4.0000	0.2500	15.9986	3.9997
60	6.525E+05	1.532E-06	2.610E+06	3.831E-07	4.0000	0.2500	15.9996	3.9999
66	2.489E+06	4.017E-07	9.957E+06	1.004E-07	4.0000	0.2500	15.9999	4.0000
72	9.496E+06	1.053E-07	3.798E+07	2.633E-08	4.0000	0.2500	16.0000	4.0000
120	4.258E+11	2.349E-12	1.703E+12	5.871E-13	4.0000	0.2500	16.0000	4.0000
180	2.778E+17	3.599E-18	1.111E+18	8.998E-19	4.0000	0.2500	16.0000	4.0000
360	7.720E+34	1.295E-35	3.088E+35	3.238E-36	4.0000	0.2500	16.0000	4.0000

Time Value of Money Factors—Discrete Compounding
    i = 30%

n	Single Sums		Uniform Series				Gradient Series	
	To Find F Given P (F|P,i%,n)	To Find P Given F (P|F,i%,n)	To Find F Given A (F|A,i%,n)	To Find A Given F (A|F,i%,n)	To Find P Given A (P|A,i%,n)	To Find A Given P (A|P,i%,n)	To Find P Given G (P|G,i%,n)	To Find A Given G (A|G,i%,n)
1	1.3000	0.7692	1.0000	1.0000	0.7692	1.3000	0.0000	0.0000
2	1.6900	0.5917	2.3000	0.4348	1.3609	0.7348	0.5917	0.4348
3	2.1970	0.4552	3.9900	0.2506	1.8161	0.5506	1.5020	0.8271
4	2.8561	0.3501	6.1870	0.1616	2.1662	0.4616	2.5524	1.1783
5	3.7129	0.2693	9.0431	0.1106	2.4356	0.4106	3.6297	1.4903
6	4.8268	0.2072	12.7560	0.0784	2.6427	0.3784	4.6656	1.7654
7	6.2749	0.1594	17.5828	0.0569	2.8021	0.3569	5.6218	2.0063
8	8.1573	0.1226	23.8577	0.0419	2.9247	0.3419	6.4800	2.2156
9	10.6045	0.0943	32.0150	0.0312	3.0190	0.3312	7.2343	2.3963
10	13.7858	0.0725	42.6195	0.0235	3.0915	0.3235	7.8872	2.5512
11	17.9216	0.0558	56.4053	0.0177	3.1473	0.3177	8.4452	2.6833
12	23.2981	0.0429	74.3270	0.0135	3.1903	0.3135	8.9173	2.7952
13	30.2875	0.0330	97.6250	0.0102	3.2233	0.3102	9.3135	2.8895
14	39.3738	0.0254	127.9125	7.818E-03	3.2487	0.3078	9.6437	2.9685
15	51.1859	0.0195	167.2863	5.978E-03	3.2682	0.3060	9.9172	3.0344
16	66.5417	0.0150	218.4722	4.577E-03	3.2832	0.3046	10.1426	3.0892
17	86.5042	0.0116	285.0139	3.509E-03	3.2948	0.3035	10.3276	3.1345
18	112.4554	8.892E-03	371.5180	2.692E-03	3.3037	0.3027	10.4788	3.1718
19	146.1920	6.840E-03	483.9734	2.066E-03	3.3105	0.3021	10.6019	3.2025
20	190.0496	5.262E-03	630.1655	1.587E-03	3.3158	0.3016	10.7019	3.2275
21	247.0645	4.048E-03	820.2151	1.219E-03	3.3198	0.3012	10.7828	3.2480
22	321.1839	3.113E-03	1.067E+03	9.370E-04	3.3230	0.3009	10.8482	3.2646
23	417.5391	2.395E-03	1.388E+03	7.202E-04	3.3254	0.3007	10.9009	3.2781
24	542.8008	1.842E-03	1.806E+03	5.537E-04	3.3272	0.3006	10.9433	3.2890
25	705.6410	1.417E-03	2.349E+03	4.257E-04	3.3286	0.3004	10.9773	3.2979
26	917.3333	1.090E-03	3.054E+03	3.274E-04	3.3297	0.3003	11.0045	3.3050
27	1.193E+03	8.386E-04	3.972E+03	2.518E-04	3.3305	0.3003	11.0263	3.3107
28	1.550E+03	6.450E-04	5.164E+03	1.936E-04	3.3312	0.3002	11.0437	3.3153
29	2.015E+03	4.962E-04	6.715E+03	1.489E-04	3.3317	0.3001	11.0576	3.3189
30	2.620E+03	3.817E-04	8.730E+03	1.145E-04	3.3321	0.3001	11.0687	3.3219
36	1.265E+04	7.908E-05	4.215E+04	2.372E-05	3.3331	0.3000	11.1007	3.3305
42	6.104E+04	1.638E-05	2.035E+05	4.915E-06	3.3333	0.3000	11.1086	3.3326
48	2.946E+05	3.394E-06	9.821E+05	1.018E-06	3.3333	0.3000	11.1105	3.3332
54	1.422E+06	7.032E-07	4.740E+06	2.110E-07	3.3333	0.3000	11.1110	3.3333
60	6.864E+06	1.457E-07	2.288E+07	4.370E-08	3.3333	0.3000	11.1111	3.3333
66	3.313E+07	3.018E-08	1.104E+08	9.054E-09	3.3333	0.3000	11.1111	3.3333
72	1.599E+08	6.253E-09	5.331E+08	1.876E-09	3.3333	0.3000	11.1111	3.3333
120	4.712E+13	2.122E-14	1.571E+14	6.367E-15	3.3333	0.3000	11.1111	3.3333
180	3.234E+20	3.092E-21	1.078E+21	9.275E-22	3.3333	0.3000	11.1111	3.3333
360	1.046E+41	9.559E-42	3.487E+41	2.868E-42	3.3333	0.3000	11.1111	3.3333

CHAPTER 5

BOILERS AND FIRED SYSTEMS

S.A. PARKER
Chief Engineer, Energy and Environment Directorate
Pacific Northwest National Laboratory
Richland, Washington

B.K WALKER
QuikWater, a division of Webco Industries
Sand Springs, Oklahoma

5.1 INTRODUCTION

Boilers and other fired systems are the most significant energy consumers. Almost two-thirds of the fossil-fuel energy consumed in the United States involves the use of a boiler, furnace, or other fired system. Even most electric energy is produced using fuel-fired boilers. Over 68% of the electricity generated in the United States is produced through the combustion of coal, fuel oil, and natural gas. (The remainder is produced through nuclear, 22%; hydro-electric, 10%; and geothermal and others, <1%.) Unlike many electric systems, boilers and fired systems are not inherently energy efficient.

This chapter and the following chapter on steam and condensate systems examine how energy is consumed and wasted, as well as opportunities for reducing energy consumption and costs in the operation of boiler and steam plants. A list of energy and cost reduction measures is presented, categorized as: load reduction, waste heat recovery, efficiency improvement, fuel cost reduction, and other opportunities. Several of the key opportunities for reducing operating costs are presented, ranging from changes in operating procedures to capital improvement opportunities. The topics reflect recurring opportunities identified from numerous in-plant audits. Several examples are presented to demonstrate the methodology for estimating the potential energy savings associated with various opportunities. Many of these examples utilize easy to understand nomographs and charts in the solution techniques.

In addition to energy saving opportunities, this chapter also describes some issues relevant to day-to-day operations, maintenance, and troubleshooting. Considerations relative to fuel comparison and selection are also discussed. Developing technologies relative to alterna-

tive fuels and types of combustion equipment are also discussed. Some of the technologies discussed hold the potential for significant cost reductions while alleviating environmental problems.

The chapter concludes with a brief discussion of some of the major regulations impacting the operation of boilers and fired systems. It is important to emphasize the need to carefully assess the potential impact of federal, state, and local regulations.

5.2 ANALYSIS OF BOILERS AND FIRED SYSTEMS

5.2.1 Boiler Energy Consumption

Boiler and other fired systems, such as furnaces and ovens, combust fuel with air for the purpose of releasing the chemical heat energy. The purpose of the heat energy may be to raise the temperature of an industrial product as part of a manufacturing process, it may be to generate high-temperature high-pressure steam in order to power a turbine, or it may simply be to heat a space so the occupants will be comfortable. The energy consumption of boilers, furnaces, and other fire systems can be determined simply as a function of load and efficiency as expressed in the equation:

$$\text{Energy consumption} = \int (\text{load}) \times (1/\text{efficiency}) \, dt \qquad (5.1)$$

Similarly, the cost of operating a boiler or fired system can be determined as:

$$\text{Energy cost} = \int (\text{load}) \times (1/\text{efficiency}) \times (\text{fuel cost}) \, dt \qquad (5.2)$$

As such, the opportunities for reducing the energy consumption or energy cost of a boiler or fired system can be put into a few categories. In order to reduce boiler energy consumption, one can reduce the load, increase the operating efficiency, reduce the unit fuel energy cost, or use combinations thereof.

Of course equations 5.1 and 5.2 are not always that simple because the variables are not always constant. The *load* varies as a function of the process being supported. The *efficiency* varies as a function of the load and other functions, such as time or weather. In addition, the *fuel cost* may also vary as a function of time (such as in seasonal, time-of-use, or spot market rates) or as a function of load (such as declining block or spot market rates.) Therefore, solving the equation for the energy consumption or energy cost may not always be simplistic.

5.2.2 Balance Equations

Balance equations are used in an analysis of a process which determines inputs and outputs to a system. There are several types of balance equations which may prove useful in the analysis of a boiler or fired-system. These include a heat balance and mass balance.

Heat Balance

A heat balance is used to determine where all the heat energy enters and leaves a system. Assuming that energy can neither be created nor destroyed, all energy can be accounted for in a system analysis. Energy in equals energy out. Whether through measurement or analysis, all energy entering or leaving a system can be determined. In a simple furnace system, energy enters through the combustion air, fuel, and mixed-air duct. Energy leaves the furnace system through the supply-air duct and the exhaust gases.

In a boiler system, the analysis can become more complex. Energy input comes from the following: condensate return, make-up water, combustion air, fuel, and maybe a few others, depending on the complexity of the system. Energy output departs as the following: steam, blowdown, exhaust gases, shell/surface losses, possibly ash, and other discharges depending on the complexity of the system.

Mass Balance

A mass balance is used to determine where all mass enters and leaves a system. There are several methods in which a mass balance can be performed that can be useful in the analysis of a boiler or other fired system. In the case of a steam boiler, a mass balance can be used in the form of a water balance (steam, condensate return, make-up water, blowdown, and feedwater). A mass balance can also be used for water quality or chemical balance (total dissolved solids, or other impurity). The mass balance can also be used in the form of a combustion analysis (fire-side mass balance consisting of air and fuel in and combustion gasses and excess air out). This type of analysis is the foundation for determining combustion efficiency and determining the optimum air-to-fuel ratio.

For analyzing complex systems, the mass and energy balance equations may be used simultaneously, such as in solving multiple equations with multiple unknowns. This type of analysis is particularly useful in determining blowdown losses, waste heat recovery potential, and other interdependent opportunities.

5.2.3 Efficiency

There are several different measures of efficiency used in boilers and fired systems. While this may lead to some confusion, the different measures are used to convey different information. Therefore, it is important to understand what is being implied by a given efficiency measure.

The basis for testing boilers is the American Society of Mechanical Engineers (ASME) Power Test Code 4 (PTC 4-1998.) This procedure defined and established two primary methods of determining efficiency: the input-output method and the heat-loss method. Both of these methods result in what is commonly referred to as the gross thermal efficiency. The efficiencies determined by these methods are "gross" efficiencies as apposed to "net" efficiencies, which would include the additional energy input of auxiliary equipment like combustion air fans, fuel pumps, stoker drives, etc. For more information on these methods, see the ASME PTC 4-1998 or Taplin 1991.

Another efficiency term commonly used for boilers and other fired systems is combustion efficiency. Combustion efficiency is similar to the heat loss method, but only the heat losses due to the exhaust gases are considered. Combustion efficiency can be measured in the field by analyzing the products of combustion, the exhaust gases.

Typically measuring either carbon dioxide (CO_2) or oxygen (O_2) in the exhaust gas can be used to determine the combustion efficiency as long as there is excess air. Excess air is defined as air in excess of the amount required for stoichiometric conditions. In other words, excess air is the amount of air above that which is theoretically required for complete combustion. In the real world, however, it is not possible to get a perfect mixture of air and fuel to achieve complete combustion without some amount of excess air. As excess air is reduced toward the fuel rich side, incomplete combustion begins to occur, resulting in the formation of carbon monoxide, carbon, smoke, and in extreme cases, raw unburned fuel. Incomplete combustion is inefficient, expensive, and frequently unsafe. Therefore, some amount of excess air is required to ensure complete and safe combustion.

However, excess air is also inefficient, as it results in the excess air being heated from ambient air temperatures to exhaust gas temperatures, resulting in a form of heat loss. Therefore while some excess air is required, it is also desirable to minimize the amount of excess air.

As illustrated in Figure 5-1, the amount of carbon

dioxide, percent by volume, in the exhaust gas reaches a maximum with no excess air stoichiometric conditions. While carbon dioxide can be used as a measure of complete combustion, it can not be used to optimally control the air-to-fuel ratio in a fired system. A drop in the level of carbon dioxide would not be sufficient to inform the control system if it were operating in a condition of excess air or insufficient air. However, measuring oxygen in the exhaust gases is a direct measure of the amount of excess air. Therefore, measuring oxygen in the exhaust gas is a more common and preferred method of controlling the air-to-fuel ratio in a fired system.

5.2.4 Energy Conservation Measures

As noted above, energy cost reduction opportunities can generally be placed into one of the following categories: reducing load, increasing efficiency, and reducing unit energy cost. As with most energy conservation and cost reducing measures, there are also a few additional opportunities which are not so easily categorized. Table 5-1 lists several energy conservation measures that have been found to be very cost effective in various boilers and fired-systems.

5.3   KEY ELEMENTS FOR MAXIMUM EFFICIENCY

There are several opportunities for maximizing efficiency and reducing operating costs in a boiler or other fired-system, as noted in Table 5-1. This section examines in more detail several key opportunities for energy and cost reduction, including excess air, stack temperature, load balancing, boiler blowdown, and condensate return.

5.3.1 Excess Air

In combustion processes, excess air is generally defined as air introduced above the stoichiometric or theoretical requirements to effect complete and efficient combustion of the fuel.

There is an optimum level of excess-air operation for each type of burner or furnace design and fuel type. Only enough air should be supplied to ensure complete combustion of the fuel since more than this amount increases the heat rejected to the stack, resulting in greater fuel consumption for a given process output.

To identify the point of minimum excess-air operation for a particular fired system, curves of combustibles

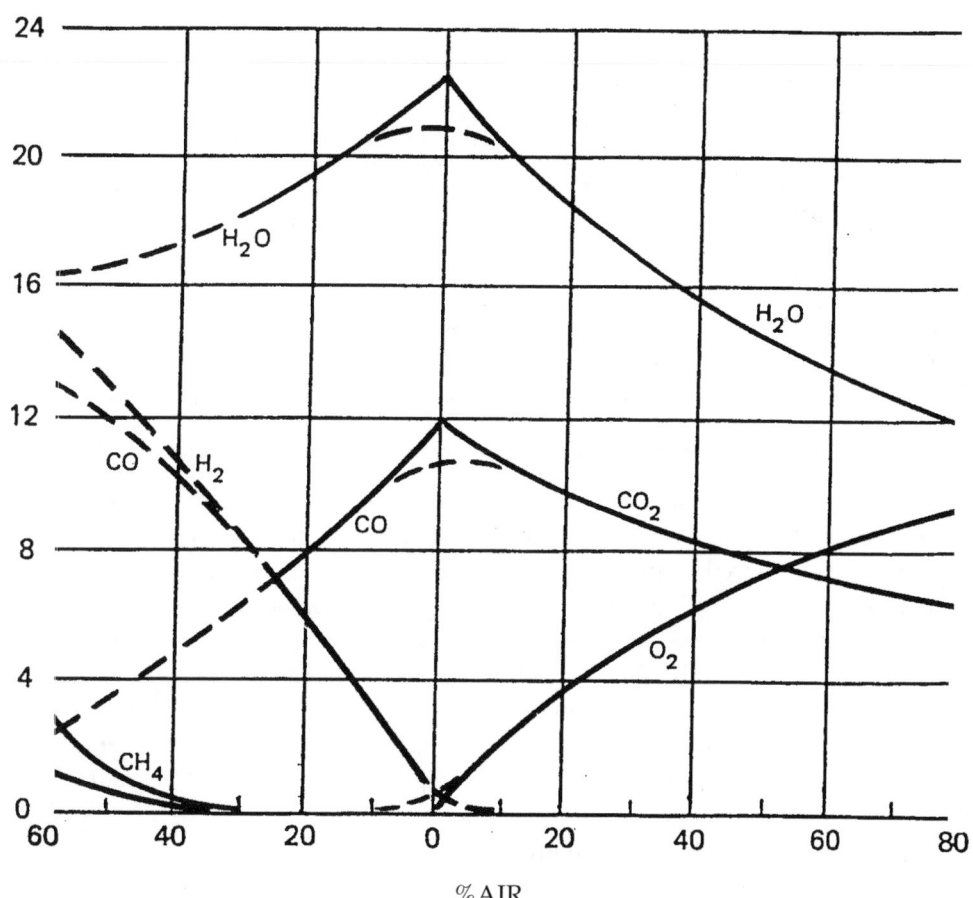

Figure 5-1. Theoretical flue gas analysis versus air percentage for natural gas.

Table 5-1. Energy Conservation measures
for boilers and fired systems[a]

Load Reduction
Insulation
—steam lines and distribution system
—condensate lines and return system
—heat exchangers
—boiler or furnace
Repair steam leaks
Repair failed steam straps
Return condensate to boiler
Reduce boiler blowdown
Improve feedwater treatment
Improve make-up water treatment
Repair condensate leaks
Shut off steam tracers during the summer
Shut off boilers during long periods of no use
Eliminate hot standby
Reduce flash steam loss
Install stack dampers or heat traps in natural draft boilers
Replace continuous pilots with electronic ignition pilots

Waste Heat Recovery (a form of load reduction)
Utilize flash steam
Preheat feedwater with an economizer
Preheat make-up water with an economizer
Preheat combustion air with a recuperator
Recover flue gas heat to supplement other heating system, such as domestic or service hot water, or unit space heater
Recover waste heat from some other system to preheat boiler make-up or feedwater
Install a heat recovery system on incinerator or furnace
Install condensation heat recovery system
—indirect contact heat exchanger
—direct contact heat exchanger

Efficiency Improvement
Reduce excess air
Provide sufficient air for complete combustion
Install combustion efficiency control system
—constant excess air control
—minimum excess air control
—optimum excess air and CO control
Optimize loading of multiple boilers

Shut off unnecessary boilers
Install smaller system for part-load operation
—small boiler for summer loads
—satellite boiler for remote loads
Install low excess air burners
Repair or replace faulty burners
Replace natural draft burners with forced draft burners
Install turbulators in firetube boilers
Install more efficient boiler or furnace system
—high-efficiency, pulse combustion, or condensing boiler or furnace system
Clean heat transfer surfaces to reduce fouling and scale
Improve feedwater treatment to reduce scaling
Improve make-up water treatment to reduce scaling

Fuel Cost Reduction
Switch to alternate utility rate schedule
—interruptible rate schedule
Purchase natural gas from alternate source, self procurement of natural gas
Fuel switching
—switch between alternate fuel sources
—install multiple fuel burning capability
—replace electric boiler with a fuel-fired boiler
Switch to a heat pump
—use heat pump for supplemental heat requirements
—use heat pump for baseline heat requirements

Other Opportunities
Install variable speed drives on feedwater pumps
Install variable speed drives on combustion air fan
Replace boiler with alternative heating system
Replace furnace with alternative heating system
Install more efficient combustion air fan
Install more efficient combustion air fan motor
Install more efficient feedwater pump
Install more efficient feedwater pump motor
Install more efficient condensate pump
Install more efficient condensate pump motor

[a]Reference: F.W. Payne, *Efficient Boiler Operations Sourcebook*, 3rd ed., Fairmont Press, Lilburn, GA, 1991.

as a function of excess O_2 should be constructed similar to that illustrated in Figure 5-2. In the case of a gas-fueled system, the combustible monitored would be carbon monoxide (CO), whereas, in the case of a liquid- or solid-fueled system, the combustible monitored would be the Smoke Spot Number (SSN). The curves should be developed for various firing rates, as the minimal excess-air operating point will also vary as a function of the firing rate (percent load). Figure 5-2 illustrates two potential curves, one for high-fire and the other for low-fire. The optimal excess-air-control set point should be set at some margin (generally 0.5 to 1%) above the minimum O_2 point to allow for response and control variances. It is important to note that some burners may exhibit a gradual or steep

CO-O_2 behavior, and this behavior may even change with various firing rates. It is also important to note that some burners may experience potentially unstable operation with small changes in O_2 (steep CO-O_2 curve behavior). Upper control limits for carbon monoxide vary depending on the referenced source. Points referenced for gas-fired systems are typically 400 ppm, 200 ppm, or 100 ppm. Today, local environmental regulations may dictate acceptable upper limits. Maximum desirable SSN for liquid fuels is typically SSN=1 for No. 2 fuel oil and SSN=4 for No. 6 fuel oil. Again, local environmental regulations may dictate lower acceptable upper limits.

Typical optimum levels of excess air normally attainable for maximum operating efficiency are indicated

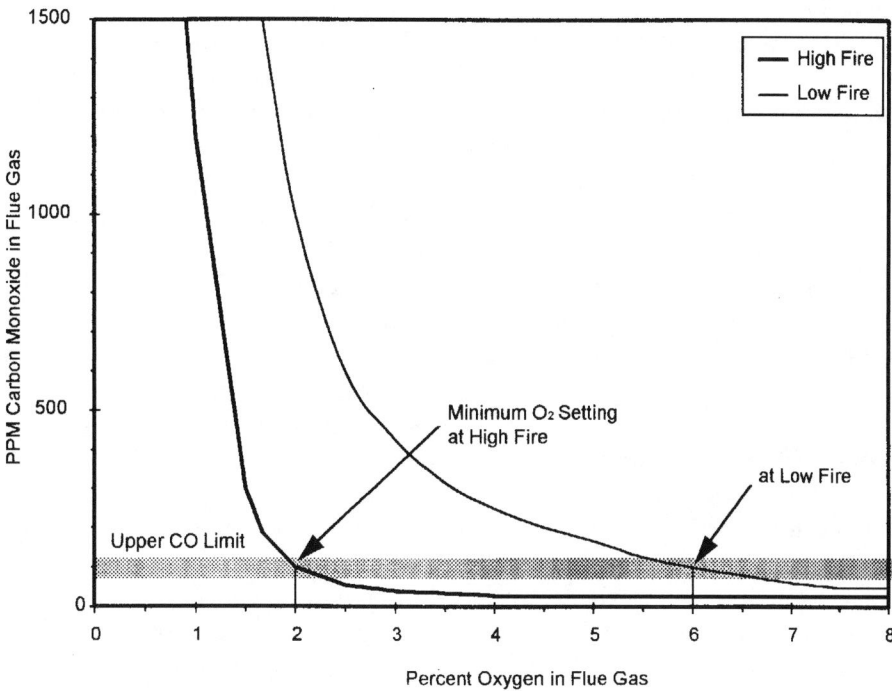

Figure 5-2. Hypothetical CO-O$_2$ characteristic curve for a gas-fired industrial boiler.

in Table 5-2 and classified according to fuel type and firing method.

The amount of excess air (or O$_2$) in the flue gas, unburned combustibles, and the stack temperature rise above the inlet air temperature are significant in defining the efficiency of the combustion process. Excess oxygen (O$_2$) measured in the exhaust stack is the most typical method of controlling the air-to-fuel ratio. However, for more precise control, carbon monoxide (CO) measurements may also be used to control air flow rates in com-

Table 5-2. Typical Optimum Excess Air[a]

Fuel Type	Firing Method	Optimum Excess Air (%)	Equivalent O$_2$ (by Volume)
Natural gas	Natural draft	20-30	4-5
Natural gas	Forced draft	5-10	1-2
Natural gas	Low excess air	.04-0.2	0.1-0.5
Propane	—	5-10	1-2
Coke oven gas	—	5-10	1-2
No. 2 oil	Rotary cup	15-20	3-4
No. 2 oil	Air-atomized	10-15	2-3
No. 2 oil	Steam-atomized	10-15	2-3
No. 6 oil	Steam-atomized	10-15	2-3
Coal	Pulverized	15-20	3-3.5
Coal	Stoker	20-30	3.5-5
Coal	Cyclone	7-15	1.5-3

[a]To maintain safe unit output conditions, excess-air requirements may be greater than the optimum levels indicated. This condition may arise when operating loads are substantially less than the design rating. Where possible, check vendors' predicted performance curves. If unavailable, reduce excess-air operation to minimum levels consistent with satisfactory output.

bination with O_2 monitoring. Careful attention to furnace operation is required to ensure an optimum level of performance.

Figures 5-3, 5-4, and 5-5 can be used to determine the combustion efficiency of a boiler or other fired system burning natural gas, No. 2 fuel oil, or No. 6 fuel oil respectively, so long as the level of unburned combustibles is considered negligible. These figures were derived from H. R. Taplin, Jr., *Combustion Efficiency Tables*, Fairmont Press, Lilburn, GA, 1991. For more information on combustion efficiency, including combustion efficiencies using other fuels, see Taplin 1991.

Where to Look for Conservation Opportunities

Fossil-fuel-fired steam generators, process fired heaters/furnaces, duct heaters, and separately fired superheaters may benefit from an excess-air-control program. Specialized process equipment, such as rotary kilns, fired calciners, etc. can also benefit from an air control program.

How to Test for Relative Efficiency

To determine relative operating efficiency and to establish energy conservation benefits for an excess-air-control program, you must determine: (1) percent oxygen (by volume) in the flue gas (typically dry), (2) stack temperature rise (the difference between the flue gas temperature and the combustion air inlet temperature), and (3) fuel type.

To accomplish optimal control over avoidable losses, continuous measurement of the excess air is a necessity. There are two types of equipment available to measure flue-gas oxygen and corresponding "excess air": (1) portable equipment such as an Orsat flue-gas analyzer, heat prover, electronic gas analyzer, or an equivalent analyzing device; and (2) permanent-type installations such as probe-type continuous oxygen analyzers (available from various manufacturers), which do not require external gas sampling systems.

The major advantage of permanently mounted equipment is that the on-line indication or recording allows remedial action to be taken frequently to ensure

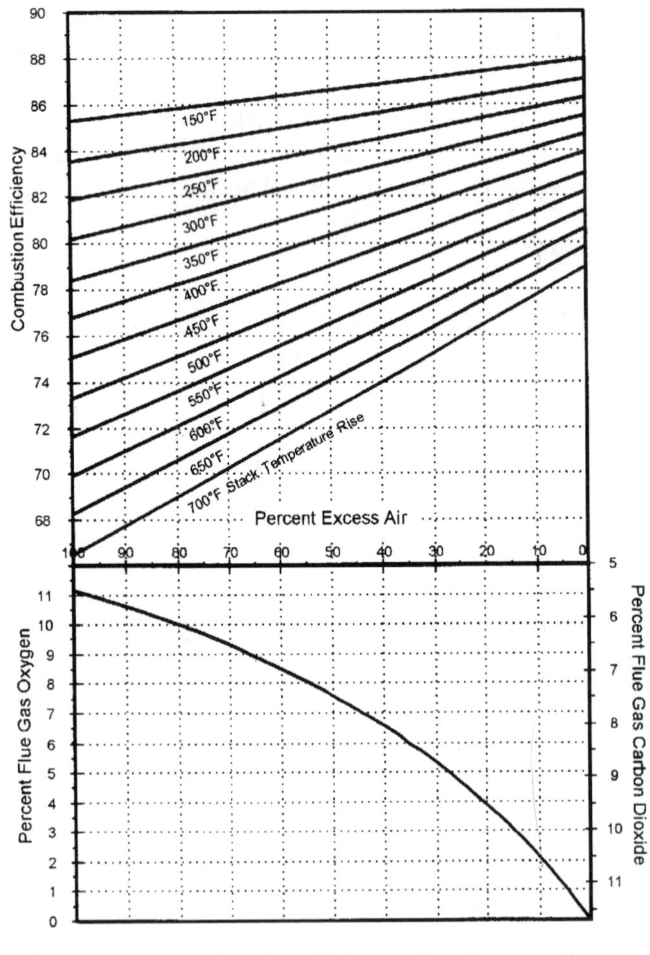

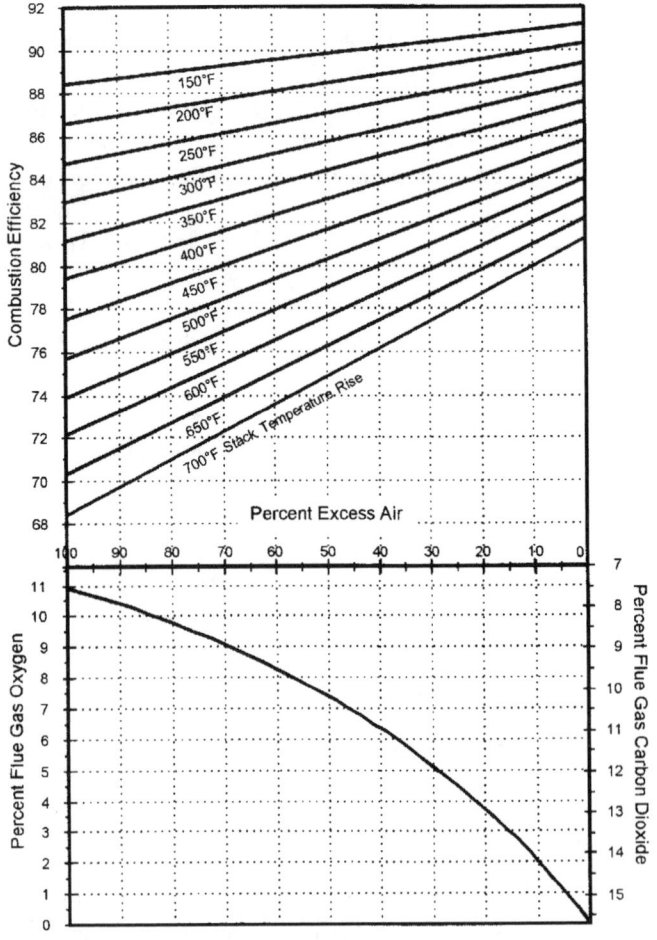

Figure 5-3. Combustion efficiency chart for natural gas.

Figure 5-4. Combustion efficiency chart for number 2 fuel oil.

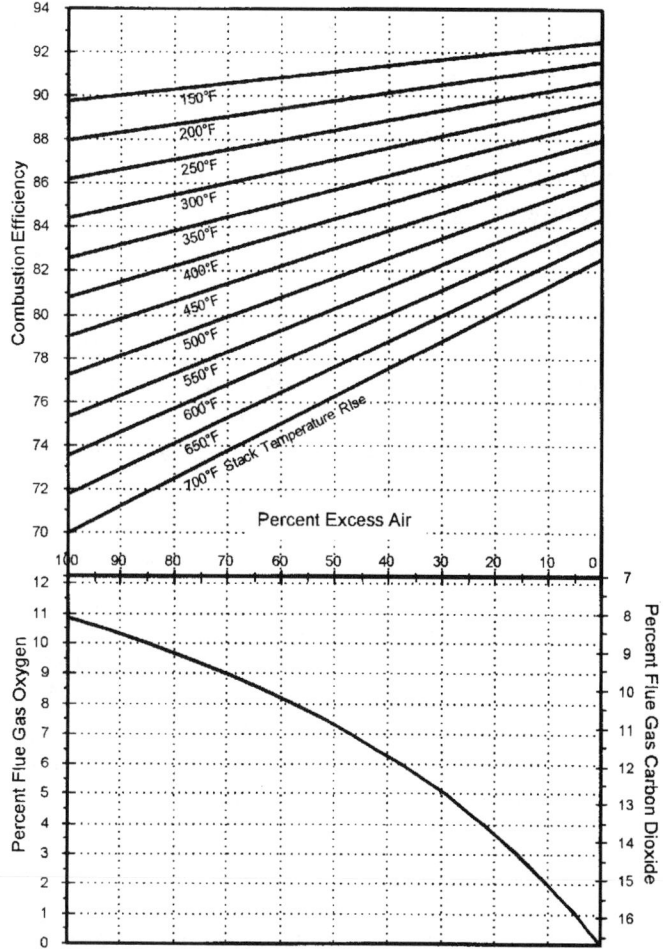

Figure 5-5. Combustion efficiency chart for number 6 fuel oil.

type of equipment provides satisfactory results for the planning and operational results desired.

An analysis to establish performance can be made with the two measurements, percent oxygen and stack temperature rise, in addition to the particular fuel fired. As an illustration, consider the following example.

Example: Determine the potential energy savings associated with reducing the amount of excess air to an optimum level for a natural gas-fired steam boiler.

Operating Data.

Current energy consumption	1,100,000 therms/yr
Boiler rated capacity	600 boiler horsepower
Operating hours	8,500 hr/yr
Current stack gas analysis	9% Oxygen (by volume, dry)
	Minimal CO reading
Combustion air inlet temperature	80°F
Exhaust gas stack temperature	580°F
Proposed operating condition	2% Oxygen (by volume, dry)

Calculation and Analysis.

STEP 1: Determine current boiler combustion efficiency using Figure 5-6 for natural gas. Note that this is the same figure as Figure 5-3.
A) Determine the current stack temperature rise.
    STR = (exhaust stack temperature)
    – (combustion air temperature)
    STR = 580°F - 80°F = 500°F

B) Enter the chart with an oxygen level of 9% and, following a line to the curve, read the percent excess air to be approximately 66%.

C) Continue the line to the curve for a stack temperature rise of 500°F and read the current combustion efficiency to be 76.4%.

STEP 2: Determine the proposed boiler combustion efficiency using the same figure.
D) Repeat steps A through C for the proposed combustion efficiency, assuming the same stack temperature conditions. Read the proposed combustion efficiency to be 81.4%.

Note that in many cases reducing the amount of excess air will tend to reduce the exhaust stack temperature, resulting in an even more efficient operating condition. Unfortunately, it is difficult to predict the extent of this added benefit.

continuous operation at optimum levels. Computerized systems which allow safe control of excess air over the boiler load range have proven economic for many installations. Even carbon monoxide-based monitoring and control systems, which are notably more expensive than simple oxygen-based monitoring and control systems, prove to be cost effective for larger industrial-and utility-sized boiler systems.

Portable equipment only allows performance checking on an intermittent or spot-check basis. Periodic monitoring may be sufficient for smaller boilers or those which do not undergo significant change in operating conditions. However, continuous monitoring and control systems have the ability to respond more rapidly to changing conditions, such as load and inlet air conditions.

The stack temperature rise may be obtained with portable thermocouple probes in conjunction with a potentiometer, or by installing permanent temperature probes within the exhaust stack and combustion air inlet and providing continuous indication or recording. Each

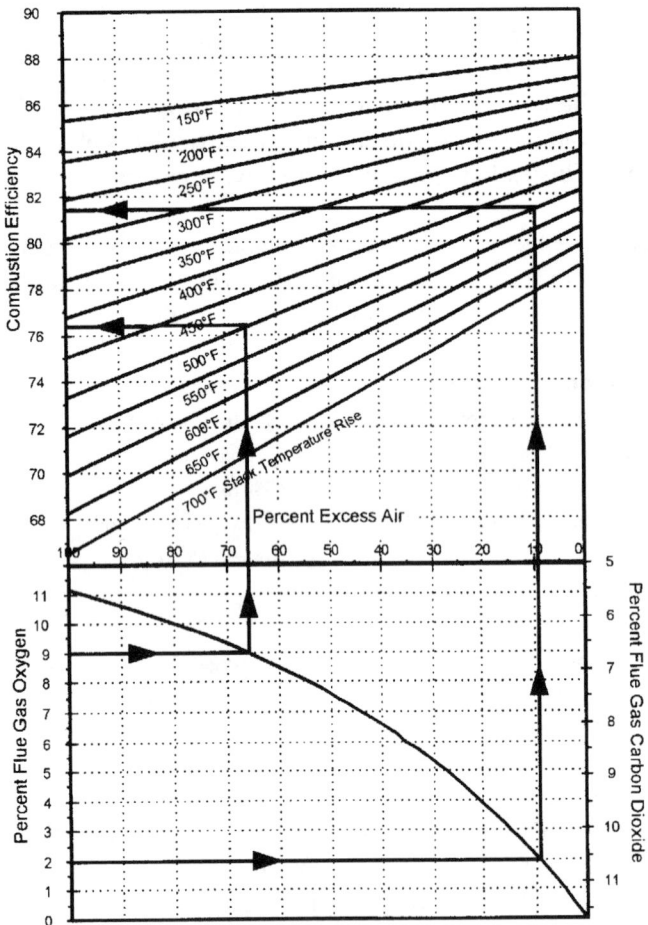

Figure 5-6. Combustion efficiency curve for reducing excess air example.

STEP 3: Determine the fuel savings.

E) Percent fuel savings = [(new efficiency)
    – (old efficiency)]/(new efficiency)
    Percent fuel savings = [(81.4%)
    – (76.4%)]/(81.4%)
    Percent fuel savings = 6.14%

F) Fuel savings =(current fuel consumption)
    × (percent fuel savings)
    Fuel savings = (1,100,000 therms/yr) × (6.14%)
    Fuel savings = 67,540 therms/yr

Conclusions

This example assumes that the results of the combustion analysis and boiler load are constant. Obviously this is an oversimplification of the issue. Because the air-to-fuel ratio (excess air level) is different for different boiler loads, a more thorough analysis should take this into account. One method to accomplish this would be to perform the analysis at various firing rates, such as high-fire and low-fire. For modulating type boilers, which can vary between high- and low-firing rates, a modified bin analysis approach or other bin-type methodology could be employed.

Requirements to Effect Maximum Economy

To obtain the maximum benefits of an excess-air-control program, the following modifications, additions, checks, and procedures should be considered:

Key Elements for Maximum Efficiency

1. Ensure that the furnace boundary walls and flue work are airtight and not a source of air infiltration or exfiltration.

    a. Recognized leakage problem areas include (1) test connection for oxygen analyzer or portable Orsat connection; (2) access doors and ash-pit doors; (3) penetration points passing through furnace setting; (4) air seals on soot-blower elements or sight glasses; (5) seals around boiler drums and header expansion joints; (6) cracks or breaks in brick settings or refractory; (7) operation of the furnace at too negative a pressure; (8) burner penetration points; and (9) deterioration of air preheater radial seals or tube-sheet expansion and cracks on tubular air heater applications.

    b. Tests to locate leakage problem include: (1) a light test whereby a strong spotlight is placed in the furnace and the unit inspected externally; (2) the use of a pyrometer to obtain a temperature profile on the outer casing (generally indicates points where refractory or insulation has deteriorated); (3) a soap-bubble test on suspected penetration points or seal welds; (4) a smoke-bomb test and external examination for traces of smoke; (5) holding a lighted candle along the casing seams (has pinpointed leakage problems on induced- or natural-draft units); (6) operating the forced draft fan on high capacity with the fire out, plus use of liquid chemical smoke producers (has helped identify seal leaks); and (7) use of a thermographic device to locate "hot spots" (may indicate faulty insulation or flue-gas leakage).

2. Ensure optimum burner performance.

    a. Table 5-3 lists common burner difficulties that can be rectified through observation and maintenance.

    b. Ascertain integrity of air volume control: (1) The physical condition of fan vanes, dampers, and operators should be in optimum working

Table 5-3. Malfunctions in Fired Systems

Malfunction	Fuel			Detection	Action
	Coal	Oil	Gas		
Uneven air distribution to burners	x	x	x	Observe flame patterns	Adjust registers (trial and error)
Uneven fuel distribution to burners	x	x	x	Observe fuel pressure gages, or take coal sample and analyze	Consult manufacturer
Improperly positioned guns or impellers	x	x		Observe flame patterns	Adjust guns (trial and error)
Plugged or worn burners	x	x		Visual inspection	Increase frequency of cleaning; install strainers (oil)
Damaged burner throats	x	x	x	Visual inspection	Repair

condition; and (2) positioning air volume controls should be checked for responsiveness and adequacy to maintain optimum air/fuel ratios. Consult operating manual or control manufacturer for test and calibration.

  c. Maintain or purchase high-quality gas analyzing systems; calibrate instrument against a known flue-gas sample.
  d. Purchase or update existing combustion controls to reflect the present state of the art.
  e. Consider adapting "oxygen trim" feature to existing combustion control system.

3. Establish a maintenance program.
  a. Table 5-4 presents a summary of frequent boiler system problems and possible causes.
  b. Perform periodic maintenance as recommended by the manufacturer.
  c. Keep a boiler operator's log and monitor key parameters.
  d. Perform periodic inspections.

Guidelines for Day-to-Day Operation

The following steps must be taken to assure peak boiler efficiency and minimum permissible excess-air operation.

1. Check the calibration of the combustion gas analyzer frequently, and check the zero point daily.
2. If a sampling system is employed, check to assure proper operation of the sampling system.
3. The physical condition of the forced-draft damper should be checked to ensure it is not broken or damaged.
4. Casing leakage must be detected and stopped.
5. Routinely check control drives and instruments.
6. If the combustion gas analyzer is used for monitoring purposes, the excess air must be checked daily. The control may be manually altered to reduce excess air, without shortcutting the safety of operation.
7. The fuel flow and air flow charts should be carefully checked to ensure that the fuel both follows the air on increasing load with proper safety margin and leads the air on decreasing load. This should be compared on a daily shift basis to ensure consistency of safe and efficient operation.
8. Check the burner flame configuration frequently during each shift and note burner register changes in the operator's log.
9. Periodically check flue-gas CO levels to ensure complete combustion. If more than a trace amount of CO is present in the flue gas, investigate burner conditions identified on Table 5-3, or fuel supply quality limits (such as fuel-oil viscosity/temperature or coal fineness and temperature).

Table 5-4. Boiler Performance Troubleshooting

System	Problem	Possible Cause
Heat transfer related	High stack gas temperature	Buildup of gas- or water-side deposits
		Improper water treatment procedure
		Improper soot blower operation
Combustion related	High excess air	Improper control system operation
		Low fuel supply pressure
		Change in fuel heating value
		Change in oil viscosity
		Decrease in inlet air temperature
	Low excess air	Improper control system operation
		Fan limitations
		Increase in inlet air temperature
	High carbon monoxide and combustible emissions	Plugged gas burners
		Unbalanced fuel and air distribution in multiburner furnaces
		Improper air register settings
		Deterioration of burner throat refractory
		Stoker grate condition
		Stoker fuel distribution orientation
		Low fineness on pulverized systems
Miscellaneous	Casing leakage	Damaged casing and insulation
	Air heater leakage	Worn or improper adjusted seals on rotary heaters
		Tube corrosion
	Coal pulverizer power	Pulverizer in poor repair
		Too low classifier setting
	Excessive blowdown	Improper operation
	Steam leaks	Holes in waterwall tube
		Valve packing
	Missing or loose insulation	Overheating
		Weathering
	Excessive soot blower operation	Arbitrary operation schedule that is in excess of requirements

5.3.2 Exhaust Stack Temperature

Another primary factor affecting unit efficiency and ultimately fuel consumption is the temperature of combustion gases rejected to the stack. Increased operating efficiency with a corresponding reduction in fuel input can be achieved by rejecting stack gases at the lowest practical temperature consistent with basic design principles. In general, the application of additional heat recovery equipment can realize this energy conservation objective when the measured flue-gas temperature exceeds approximately 250°F. For a more extensive coverage of waste-heat recovery, see Chapter 8.

Where to Look

Steam boilers, process fired heaters, and other combustion or heat-transfer furnaces can benefit from a heat-recovery program.

The adaptation of heat-recovery equipment to existing units as discussed in this section will be limited to flue gas/liquid and/or flue gas/air preheat exchangers. Specifically, economizers and air preheaters come under this category. Economizers are used to extract heat energy from the flue gas to heat the incoming liquid process feedstream to the furnace. Flue gas/air preheaters lower the flue-gas temperature by transferring heat to the incoming combustion air stream.

Planning-quality guidelines will be presented to determine the final sink temperature, as well as comparative economic benefits to be derived by the installation of heat-recovery equipment. Costs to implement this energy conservation opportunity can then be compared against the potential benefits.

How to Test for Heat-Recovery Potential

In assessing overall efficiency and potential for heat recovery, the parameters of significant importance are temperature and fuel type/sulfur content. To obtain a meaningful operating flue-gas temperature measurement and a basis for heat-recovery selection, the unit under consideration should be operating at, or very close to, design and optimum excess-air values as defined on Table 5-2.

Temperature measurements may be made by mercury or bimetallic element thermometers, optical pyrometers, or an appropriate thermocouple probe. The most adaptable device is the thermocouple probe in which an iron or chromel constantan thermocouple is used. Temperature readout is accomplished by connecting the thermocouple leads to a potentiometer. The output of the potentiometer is a voltage reading which may be correlated with the measured temperature for the particular thermocouple element employed.

To obtain a proper and accurate temperature measurement, the following guidelines should be followed:

1. Locate the probe in an unobstructed flow path and sufficient distance, approximately five diameters downstream or upstream of any major change of direction in the flow path.

2. Ensure that the probe entrance connection is relatively leak free.

3. Take multiple readings by traversing the cross-sectional area of the flue to obtain an average and representative flue-gas temperature.

Modifications or Additions for Maximum Economy

The installation of economizers and/or flue-gas air preheaters on units not presently equipped with heat-recovery devices (and those with minimum heat-recovery equipment) are practical ways of reducing stack temperature while recouping flue-gas sensible heat normally rejected to the stack.

There are no "firm" exit-temperature guidelines that cover all fuel types and process designs. However, certain guiding principles will provide direction to the lowest practical temperature level of heat rejection. The elements that must be considered to make this judgment include (1) fuel type, (2) flue-gas dew-point considerations, (3) heat-transfer criteria, (4) type of heat-recovery surface, and (5) relative economics of heat-recovery equipment.

Tables 5-5 and 5-6 may be used for selecting the lowest practical exit-gas temperature achievable with installation of economizers and/or flue-gas air preheaters.

As an illustration of the potential and methodology for recouping flue-gas sensible heat by the addition of heat-recovery equipment, consider the following example.

Example: Determine the energy savings associated with installing an economizer or flue-gas air preheater on the boiler from the previous example. Assume that the excess-air control system from the previous example has already been implemented.

Available Data

Current energy consumption	1,032,460 therms/yr
Boiler rated capacity	600 boiler horsepower
Operating hours	8,500 hr/yr
Exhaust stack gas analysis	2% Oxygen (by volume, dry)
	Minimal CO reading

Table 5-5. Economizers

	Test for Determination of Exit
Fuel Type	Flue-Gas Temperatures
Gaseous fuel	Heat-transfer criteria:
(minimum percent sulphur)	$T_g = T_1 + 100°F$ (minimum): typically the higher of (a) or (b) below.
Fuel oils and coal	(a) Heat-transfer criteria:
	$T_g = T_1 + 100°F$ (min.)
	(b) Flue-gas dew point
	(from Figure 5-8 for a particular fuel and
	percent sulphur by weight
Where:	T_g = Final stack flue temperature
	T_1 = Process liquid feed temperature

Table 5-6. Flue-Gas/Air Preheaters

	Test for Determination of Exit
Fuel Type	Flue-Gas Temperatures
Gaseous fuel	Historic economic breakpoint:
	T_g (min.) = approximately 250°F
Fuel oils and coal	Average cold-end considerations;
	see Figure 5-9 for determination of T_{ce};
	the exit-gas temperature relationship is $T_g = 2T_{ce} - T_a$
Where:	T_g = Final stack flue temperature
	T_{ce} = Flue gas air preheater recommended average cold end temperature
	T_a = Ambient air temperature

Current operating conditions:

Combustion air inlet temperature	80°F
Exhaust gas stack temperature	580°F
Feedwater temperature	180°F
Operating steam pressure	110 psia
Operating steam temperature	335°F

Proposed operating condition:

Combustion air inlet temperature	80°F
Exhaust gas stack temperature	380°F

Calculation and Analysis

STEP 1: Compare proposed stack temperature

against minimum desired stack temperature.

A) Heat transfer criteria:
$T_g = T1 + 100°F$ (minimum)
$T_g = 180 + 100°F$ (minimum)
$T_g = 280°F$ (minimum)

B) Flue-gas dew point:
$T_g = 120°F$ (from Figure 5-8)

C) Proposed stack temperature
$T_g = 380°F$ is acceptable

STEP 2: Determine current boiler combustion efficiency using Figure 5-7 for natural gas. Note that this is the same figure as Figure 5-3.

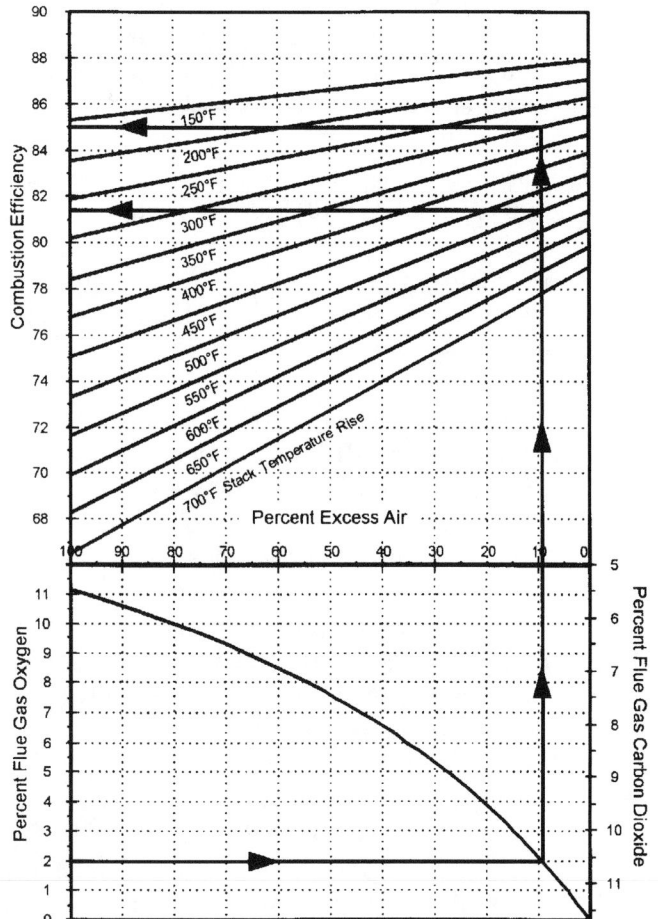

Figure 5-7. Combustion efficiency curve for stack temperature reduction example.

A) Determine the stack temperature rise.
STR = (exhaust stack temperature)
– (combustion air temperature)
STR = 580°F - 80°F = 500°F

B) Enter the chart with an oxygen level of 2% and, following a line to the curve, read the percent excess air to be approximately 9.3%.

C) Continue the line to the curve for a stack temperature rise of 500°F and read the current combustion efficiency to be 81.4%.

STEP 3: Determine the proposed boiler combustion efficiency using the same figure.

D) Repeat steps A through C for the proposed combustion efficiency, assuming the new exhaust stack temperature conditions. Read the proposed combustion efficiency to be 85.0%.

STEP 4: Determine the fuel savings.

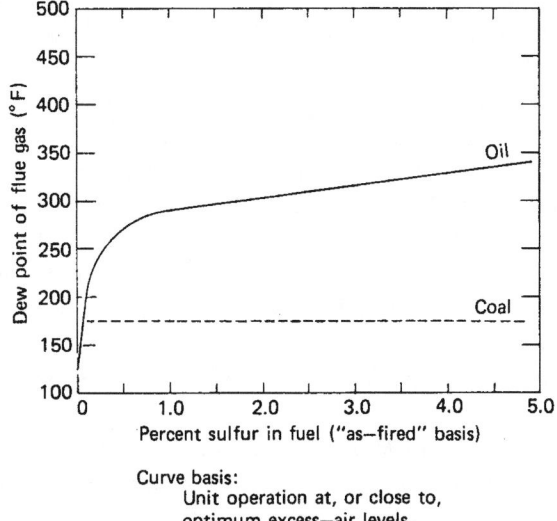

Figure 5-8. Flue-gas dew point. Based on unit operation *at* or *close to* "optimal" excess-air.

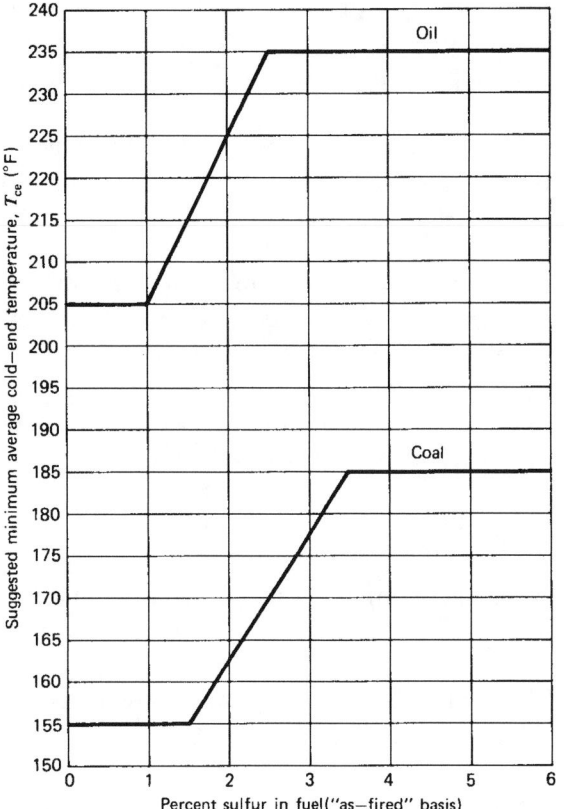

Figure 5-9. Guide for selecting flue-gas air preheaters.

E) Percent fuel savings = [(new efficiency)
– (old efficiency)] / (new efficiency)
Percent fuel savings = [(85.0%) - (81.4%)] / (85.0%)
Percent fuel savings = 4.24%

F) Fuel savings =(current fuel consumption)
× (percent fuel savings)

Fuel savings = (1,032,460 therms/yr) × (4.24%)
Fuel savings = 43,776 therms/yr

Conclusion

As with the earlier example, this analysis methodology assumes that the results of the combustion analysis and boiler load are constant. Obviously this is an oversimplification of the issue. Because the air-to-fuel ratio (excess air level) is different for different boiler loads, a more thorough analysis should take this into account.

Additional considerations in flue-gas heat recovery include:

1. Space availability to accommodate additional heating surface within furnace boundary walls or adjacent area to stack.

2. Adequacy of forced-draft and/or induced-draft fan capacity to overcome increased resistance of heat-recovery equipment. Additional fan power will be required, which will increase electricity consumption, thus reducing net energy savings.

3. Adaptability of soot blowers for maintenance of heat-transfer-surface cleanliness when firing ash- and soot-forming fuels.

4. Design considerations to maintain average cold-end temperatures for flue gas/air preheater applications in cold ambient surroundings.

5. Modifications required of flue and duct work and additional insulation needs.

6. The addition of structural steel supports.

7. Adequate pumping head to overcome increased fluid pressure drop for economizer applications.

8. The need for bypass arrangements around economizers or air preheaters.

9. Corrosive properties of gas, which would require special materials.

10. Direct flame impingement on recovery equipment.

Guidelines for Day-to-Day Operation

1. Maintain operation at goal excess air levels and stack temperature to obtain maximum efficiency and unit thermal performance.

2. Log percent O_2 or equivalent excess air, inlet air temperature, and stack temperatures once per shift or more frequent, noting the unit load and fuel fired.

3. Use oxygen analyzers with recorders for units larger than about 35×10^6 Btu/hr output.

4. Maintain surface cleanliness by soot blowing at least once per shift for ash- and soot-forming fuels.

5. Establish a more frequent cleaning schedule when heat-exchange performance deteriorates due to firing particularly troublesome fuels.

6. External fouling can also cause high excess air operation and higher stack temperatures than normal to achieve desired unit outputs. External fouling can be detected by use of draft loss gauges or water manometers and periodically (once a week) logging the results.

7. For flue gas/air preheaters, oxygen checks should be taken once a month before and after the heating surface to assess condition of circumferential and radial seals. If O_2 between the two readings varies in excess of 1% O_2, air heater leakage is excessive to the detriment of operating efficiency and fan horsepower.

8. Check fan damper operation weekly. Adjust fan damper or operator to correspond to desired excess air levels.

9. Institute daily checks on continuous monitoring equipment measuring flue-gas conditions. Check calibration every other week.

10. Establish an experience guideline on optimum time for cleaning and changing oil guns and tips.

11. Receive the "as-fired" fuel analysis on a monthly basis from the supplier. The fuel base may have changed, dictating a different operating regimen.

12. Analyze boiler blowdown every two months for iron. Internal surface cleanliness is as important as external surface cleanliness to maintaining heat-transfer characteristics and performance.

13. When possible, a sample of coal, both raw and pulverized, should be analyzed to determine if operating changes are warranted and if the design coal fineness is being obtained.

5.3.3 Waste-Heat-Steam Generation

Plants that have fired heaters and/or low-residence-time process furnaces of the type designed during the era of cheap energy may have potentially significant energy-saving opportunities. This section explores an approach to maximize energy efficiency and provide an analysis to determine overall project viability.

The major problem on older units is to determine a practical and economical approach to utilize the sensible heat in the exhaust flue gas. Typically, many vintage units have exhaust-flue-gas temperatures in the range of 1050-1600°F. In this temperature range, a conventional flue-gas air preheater normally is not a practical approach because of materials of construction requirements and significant burner front modifications. Additionally, equipping these units with an air preheater could materially alter the inherent radiant characteristics of the furnace, thus adversely affecting process heat transfer. An alternative approach to utilizing the available flue-gas sensible heat and maximizing overall plant energy efficiency is to consider: (1) waste-heat-steam generation; (2) installing an unfired or supplementary fired recirculating hot-oil loop or ethylene glycol loop to effectively utilize transferred heat to a remote location; and (3) installing a process feed economizer.

Because most industrial process industries have a need for steam, the example is for the application of an unfired waste-heat-steam generator.

The hypothetical plant situation is a reformer furnace installed in the plant in 1963 at a time when it was not considered economical to install a waste-heat-steam generator. As a result, the furnace currently vents hot flue gas (1562°F) to the atmosphere after inspiriting ambient air to reduce the exhaust temperature so that standard materials of construction could be utilized.

The flue-gas temperature of 1562°F is predicated on a measured value by thermocouple and is based on a typical average daily process load on the furnace. This induced-draft furnace fires a No. 2 fuel oil and has been optimized for 20% excess air operation. Flue-gas flow is calculated at 32,800 lb/hr. The plant utilizes approximately 180,000 lb/hr of 300-psig saturated steam from three boilers, each having a nameplate capacity of 75,000 lb/hr. The plant steam load is shared equally by the three operating boilers, each supplying 60,000 lb/hr. Feedwater to the units is supplied at 220°F from a common water-treating facility. The boilers are fired with low-sulfur (0.1% sulphur by weight) No. 2 fuel oil. Boiler efficiency averages 85% at load. Present fuel costs are $0.76/gal or $5.48/$10^6$ Btu basis of No. 2 fuel oil having a heating value of 138,800 Btu/gal. The basic approach to enhancing plant energy efficiency and minimizing cost is to generate maximum quantities of "waste" heat steam by recouping the sensible heat from the furnace exhaust flue gas.

Certain guidelines would provide a "fix" on the amount of steam that could be reasonably generated. The flue-gas temperature drop could practically be reduced to 65-100°F above the boiler feedwater temperature of 220°F. Using an approach temperature of 65°F yields an exit-flue gas temperature of 220 + 65 = 285°F. This assumes that an economizer would be furnished integral with the waste-heat-steam generator.

A heat balance on the flue-gas side (basis of flue-gas temperature drop) would provide the total heat duty available for steam generation. The sensible heat content of the flue gas is derived from Figures 5-10a and 5-10b, based on the flue-gas temperature and percent moisture in the flue gas.

Percentage moisture (by weight) in the flue gas is a function of the type of fuel fired and percentage excess-air operation. Typical values of percentage moisture are indicated in Table 5-7 for various fuels and excess air. For No. 2 fuel oil firing at 20% excess air, percent moisture by weight in flue gas is approximately 6.8%.

Therefore, a flue-gas heat balance becomes:

Flue-Gas Temperature Drop (°F)	Sensible Heat in Flue Gas (Btu/lb W.G.)	
1562	412	(Fig. 5-15)
285	52	(Fig. 5-14)
1277	360	

The total heat available from the flue gas for steam generation becomes:

(32,800 lb.W.G.) × (360 Btu/lb.W.G.) = (11.8 × 10^6 Btu/h)

The amount of steam that may be generated is determined by a thermodynamic heat balance on the steam circuit.

Enthalpy of steam at 300 psig saturated

$$h_3 = 1203 \text{ Btu/lb}$$

Enthalpy of saturated liquid at drum pressure of 300 psig

$$h_f = 400 \text{ Btu/lb}$$

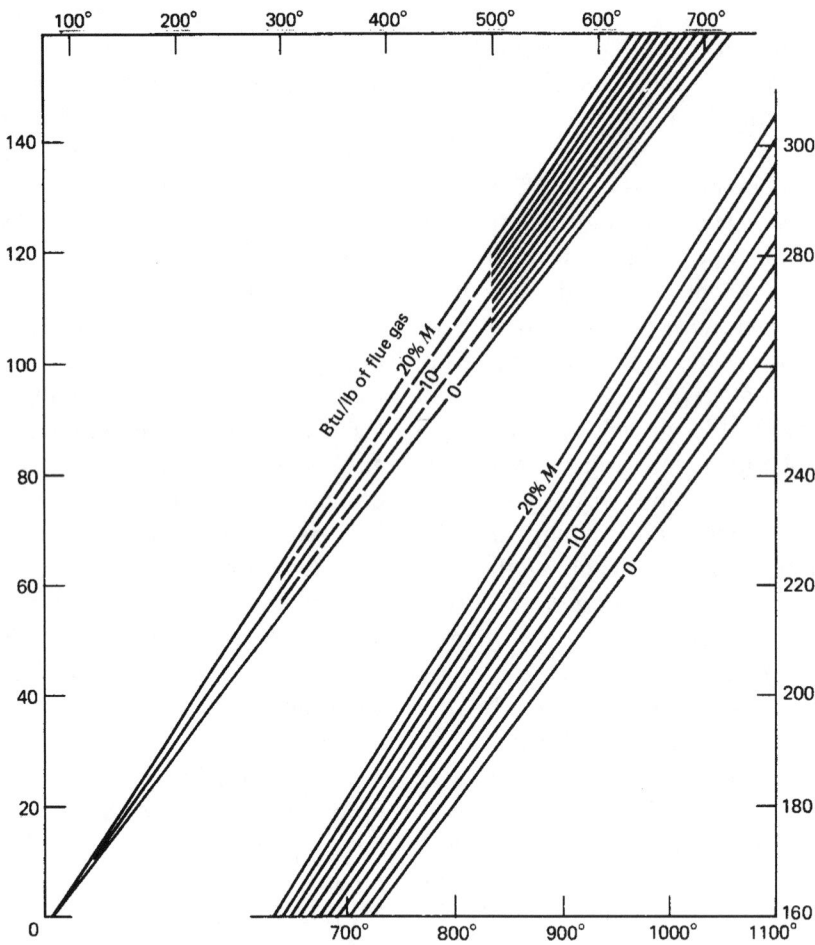

Figure 5-10a. Heat in flue gases vs. percent moisture by weight. (Derived from Keenan and Kayes 1948.)

Enthalpy corresponding to feedwater temperature
of 200°F $h_1 = 188$ Btu/lb

For this example, assume that boiler blowdown is 10% of steam flow. Therefore, feedwater flow through the economizer to the boiler drum will be 1.10 times the steam outflow from the boiler drum. Let the steam outflow be designated as x. Equating heat absorbed by the waste-heat-steam generator to the heat available from reducing the flue-gas temperature (from 1562°F to 285°F) yields the following steam flow:

$$(1.10)(x)(h_f - h_1) + (x)(h_3 - h_f) = 11.8 \times 10^6 \text{ Btu/hr}$$

Therefore,
steam flow, x   = 11,388 lb/hr
feedwater flow   = 1.10(x)= 1.10(11,388)= 12,527 lb/hr
boiler blowdown   = 12,527 – 11,388 = 1,139 lb/hr

Determine the equivalent fuel input in conventional fuel-fired boilers corresponding to the waste heat-steam generator capability. This would be defined as follows:

Fuel input to conventional boilers
        = (output)/(boiler efficiency)

Therefore,
Fuel input   = $(11.8 \times 10^6 \text{ Btu/h})/(0.85)$
                = 13.88×10^6 Btu/h

This suggests that with the installation of the waste-heat-steam generator utilizing the sensible heat of the reformer furnace flue gas, the equivalent of 13.88×10^6 Btu/hr of fossil-fuel input energy could be saved in the firing of the conventional boilers, while still satisfying the overall plant steam demand.

As with other capital projects, the waste-heat-steam generator must compete for capital, and to be viable, it must be profitable. Therefore, the decision to proceed becomes an economic one. For a project to be considered life-cycle cost effective, it must have a net-present value greater than or equal to zero, or an internal rate of return

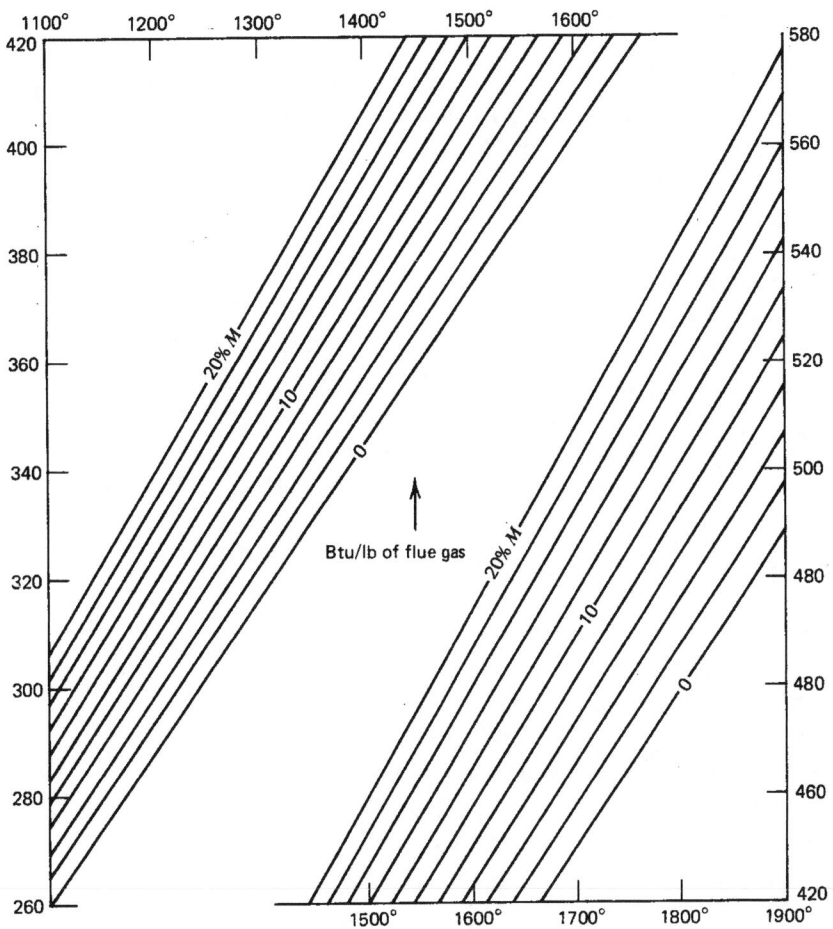

Figure 5-10b. Heat in flue gases vs. percent moisture by weight. (Derived from Keenan and Kayes 1948.)

Table 5-7. Percent Moisture by Weight in Flue Gas

| Fuel Type | Percent Excess Air | | | |
	10	15	20	25
Natural gas	12.1	11.7	11.2	10.8
No. 2 fuel oil	7.3	7.0	6.8	6.6
Coal (varies)	6.7-5.1	6.4-4.9	6.3-4.7	6.1-4.6
Propane	10.1	9.7	9.4	9.1

greater than the company's hurdle rate. For a thorough coverage of economic analysis, see Chapter 4.

5.3.4 Load Balancing
Energy Conservation Opportunities

There is an inherent variation in the energy conversion efficiencies of boilers and their auxiliaries with the operating load imposed on this equipment. It is desirable, therefore, to operate each piece of equipment at the capacity that corresponds to its highest efficiency.

Process plants generally contain multiple boiler units served by common feedwater and condensate return facilities. The constraints imposed by load variations (and the requirement of having excess capacity on line to provide reliability) seldom permit operation of each piece of equipment at optimum conditions. The energy conservation opportunities therefore lie in the establishment of an operating regimen that comes closest to attaining this goal for the overall system in light of operational constraints.

How to Test for Energy Conservation Potential

Information needed to determine energy conservation opportunities through load-balancing techniques requires a plant survey to determine (1) total steam demand and duration at various process throughputs (profile of steam load versus runtime), and (2) equipment efficiency characteristics (profile of efficiency versus load).

Steam Demand

Chart recorders are the best source for this information. Individual boiler steam flowmeters can be totalized for plant output. Demands causing peaks and valleys should be identified and their frequency estimated.

Equipment Efficiency Characteristics

The efficiency of each boiler should be documented at a minimum of four load points between half and maximum load. A fairly accurate method of obtaining unit efficiencies is by measuring stack temperature rise and percent O_2 (or excess air) in the flue gas, or by the input/output method defined in the ASME power test code. Unit efficiencies can be determined with the aid of Figure 5-3, 5-4, or 5-5 for the particular fuel fired. For pump(s) and fan(s) efficiencies, the reader should consult manufacturers' performance curves.

An example of the technique for optimizing boiler loading follows.

Example: A plant has a total installed steam-generating capacity of 500,000 lb/hr and is served by three boilers having a maximum continuous rating of 200,000, 200,000, and 100,000 lb/hr, respectively. Each unit can deliver superheated steam at 620 psig and 700°F with feedwater supplied at 250°F. The fuel fired is natural gas priced into the operation at $3.50/$10^6$ Btu. Total plant steam averages 345,000 lb/hr and is relatively constant.

The boilers are normally operated according to the following loading (top of following page).

Analysis. Determine the savings obtainable with optimum steam plant load-balancing conditions.

STEP 1. Begin with approach (a) or (b).

a) Establish the characteristics of the boiler(s) over the load range suggested through the use of a consultant and translate the results graphically as in Figures 5-11 and 5-12.

b) The plant determines boiler efficiencies for each unit at four load points by measuring unit stack temperature rise and percent O_2 in the flue gas. With these parameters known, efficiencies are obtained from Figures 5-3, 5-4, or 5-5. Tabulate the results and graphically plot unit efficiencies and unit heat inputs as a function of steam load. The results of such an analysis are shown in the tabulation and graphically illustrated in Figures 5-11 and 5-12.

(Unit input) = (unit output)/(efficiency)

STEP 2. Sum up the total unit(s) heat input at the present normal operating steam plant load conditions. From Figure 5-12:

Boiler No.	Steam Load (10^3 lb/hr)	Heat Input (10^6 Btu/hr)
1	140	186
2	140	204
3	65	96
Plant totals	345	486

STEP 3. Optimum steam plant load-balancing conditions are satisfied when the total plant steam demand is met according to Table 5-8.

(Boiler No. 1 input) + (Boiler No. 2 input) + (Boiler No. 3 input) +... = minimum

By trial and error and the use of Figure 5-12, optimum plant heat input is:

Boiler No.	Steam Load (10^3 lb/hr)	Heat Input (10^6 Btu/hr)
1	173	226
2	172	250
3	(Banked standby)	—
Plant totals	345	476

STEP 4. The annual fuel savings realized from optimum load balancing is the difference between the existing boiler input and the optimum boiler input.

Steam plant energy savings
= (existing input) – (optimum input)
= 486 - 476 × 10^6 Btu/hr
= 10 × 10^6 Btu/hr
or annually:
= (10 × 10^6 Btu/hr) × (8500 hr/yr)
× ($3.50/$10^6$ Btu)
= $297,500/yr

Factors that were not considered in the preceding example are the additional energy savings due to more efficient fan operation and the cost of maintaining the third boiler in banked standby.

The cost savings were possible in this example because the plant had been maintaining a high ratio of total capacity in service to actual steam demand. This results in low-load inefficient operation of the boilers. Other operating modes which generally result in inefficient energy usage are:

Boiler No.	Size Boiler (10^3 lb/hr)	Normal Boiler Load (10^3 lb/hr)	Measured Stack Temp. (°F)	O_2 (%)	Unit Eff. (%)
1	200	140	290	5	85.0
2	200	140	540	6	77.4
3	100	65	540	7	76.5
Plant steam demand		345			

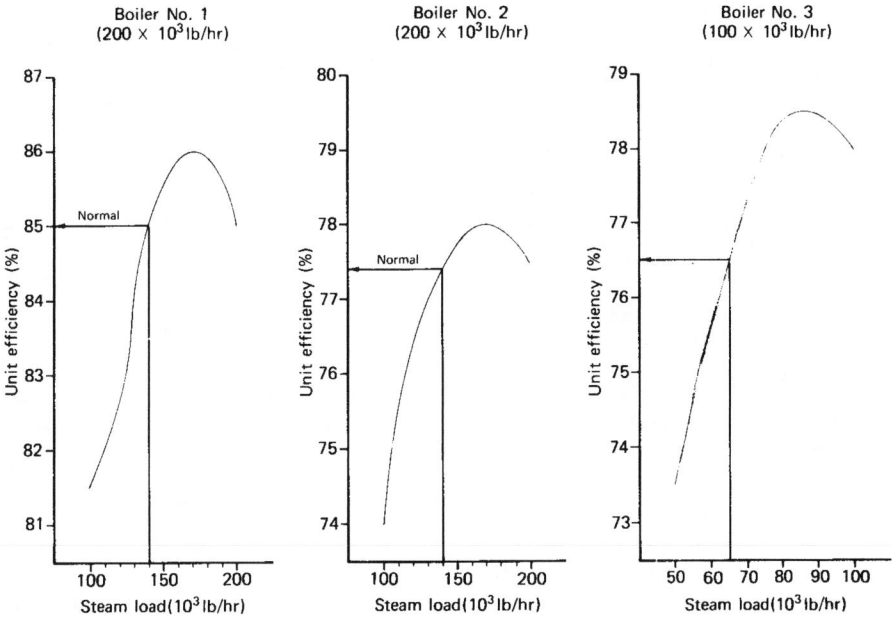

Figure 5-11. Unit efficiency vs. steam load.

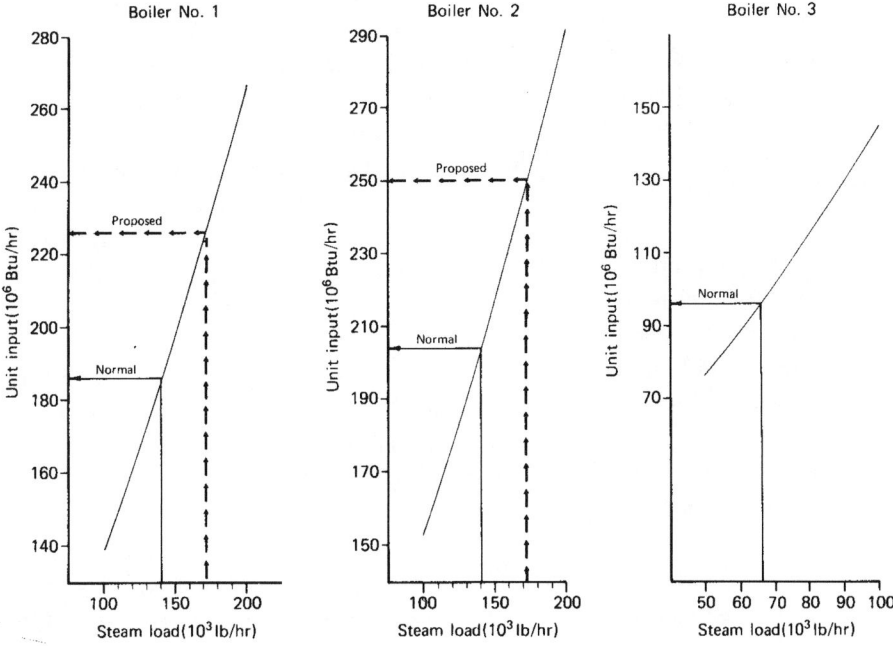

Figure 5-12. Unit input vs. steam load.

Table 5-8. Unit Efficiency and Input Tabulation

Boiler No.	Steam Load (10^3 lb/hr)	Stack Temperature (°F)	Measured Oxygen (%)	Combustion Efficiency (%)	Output (10^6 Btu/hr)	Fuel Input (10^6 Btu/hr)
1	200	305	2	85.0	226.2	266.1
	170	280	2	86.0	192.3	223.6
	130	300	7	84.0	147.0	175.0
	100	280	12	81.5	113.1	138.8
2	200	625	2	77.5	226.2	291.0
	170	570	4	78.0	192.3	246.5
	130	520	7	77.0	147.0	190.9
	100	490	11	74.0	113.1	152.8
3	100	600	2	78.0	113.1	145.0
	85	570	2	78.5	96.1	122.5
	65	540	7	76.5	73.5	96.1
	50	500	11	73.5	56.6	76.9

1. Base-loading boilers at full capacity. This can result in operation of the base-loaded boilers and the swing boilers at less than optimum efficiency unnecessarily.

2. Operation of high-pressure boilers to supply low-pressure steam demands directly via letdown steam.

3. Operation of an excessive number of auxiliary pumps. This results in throttled, inefficient operation.

Requirements for Maximum Economy

Establish a Boiler Loading Schedule. An optimized loading schedule will allow any plant steam demand to be met with the minimum energy input. Some general points to consider when establishing such a schedule are as follows:

1. Boilers generally operate most efficiently at 65 to 85% full-load rating; centrifugal fans at 80 to 90% design rating. Equipment efficiencies fall off at higher or lower load points, with the decrease most pronounced at low-load conditions.

2. It is usually more efficient to operate a lesser number of boilers at higher loads than a larger number at low loads.

3. Boilers should be put into service in order of decreasing efficiency starting with the most efficient unit.

4. Newer units and units with higher capacity are generally more efficient than are older, smaller units.

5. Generally, steam plant load swings should be taken in the smallest and least efficient unit.

Optimize the Use of High-Pressure Boilers. The boilers in a plant that operate at the highest pressure are usually the most efficient. It is, therefore, desirable to supply as much of the plant demand as possible with these units, provided that the high-grade energy in the steam can be effectively used. This is most efficiently done by installation of back-pressure turbines providing useful work output, while providing the exhaust steam for low-pressure consumers.

Degrading high-pressure steam through a pressure reducing and desuperheating station is the least efficient method of supplying low-pressure steam demands. Direct generation at the required pressure is usually more efficient by comparison.

Establish an Auxiliary Loading Schedule. A schedule for cutting plant auxiliaries common to all boilers in and out of service with rising or falling plant load should be established.

Establish Procedures for Maintaining Boilers in Standby Mode. It is generally more economical to run fewer boilers at a higher rating. On the other hand, the integrity of the steam supply must be maintained in the face of forced outage of one of the operating boilers. Both conditions can sometimes be satisfied by maintaining a standby boiler in a "live bank" mode. In this mode the boiler is isolated from the steam system at no load but kept at system operating pressure. The boiler is kept at a pressure by intermittent firing of either the ignitors or a main burner to replace ambient heat losses. Guidelines for live banking of boilers are as follows:

1. Shut all dampers and registers to minimize heat losses from the unit.

2. Establish and follow strict safety procedures for ignitor/burner light-off.

3. For units supplying turbines, take measures to ensure that any condensate which has been formed during banking is not carried through to the turbines. Units with pendant-type superheaters will generally form condensate in these elements.

Operators should familiarize themselves with emergency startup procedures and it should be ascertained that the system pressure decay that will be experienced while bringing the banked boiler(s) up to load can be tolerated.

Guidelines for Day-to-Day Operation

1. Monitor all boiler efficiencies continuously and immediately correct items that detract from performance. Computerized load balancing may prove beneficial.

2. Ensure that load-balancing schedules are followed.

3. Reassess the boiler loading schedule whenever a major change in the system occurs, such as an increase or decrease in steam demand, derating of boilers, addition/decommissioning of boilers, or addition/removal of heat-recovery equipment.

4. Recheck parameters and validity of established operating mode.

5. Measure and record fuel usage and correlate to steam production and flue-gas analysis for determination of the unit heat input relationship.

6. Keep all monitoring instrumentation calibrated and functioning properly.

7. Optimize excess air operation and minimize boiler blowdown.

Computerized Systems Available

There are commercially available direct digital control systems and proprietary sensor devices which accomplish optimal steam/power plant operation, including tie-line purchased power control. These systems control individual boilers to minimum excess air, SO_2, NO_x, CO (and opacity if desired), and they control boiler and cogeneration complexes to reduce and optimize fuel input.

Boiler plant optimization is realized by boiler controls that ensure the plant's steam demands are met in the most cost-effective manner, continuously recognizing boiler efficiencies that differ with time, load, and fuel quality. Similarly, computer control of cogeneration equipment can be cost effective in satisfying plant electrical and process steam demands.

As with power boiler systems, the efficiencies for electrical generation and extraction steam generation can be determined continuously and, as demand changes occur, loading for optimum overall efficiency is determined.

Fully integrated computer systems can also provide electric tie-line control, whereby the utility tie-line load is continuously monitored and controlled within the electrical contract's limits. For example, loads above the peak demand can automatically be avoided by increasing inplant power generation, or in the event the turbines are at full capacity, shedding loads based on previously established priorities.

5.3.5 Boiler Blowdown

In the generation of steam, most water impurities are not evaporated with the steam and thus concentrate in the boiler water. The concentration of the impurities is usually regulated by the adjustment of the continuous blowdown valve, which controls the amount of water (and concentrated impurities) purged from the steam drum.

When the amount of blowdown is not properly established and/or maintained, either of the following may happen:

1. If too little blowdown, sludge deposits and carryover will result.

2. If too much blowdown, excessive hot water is removed, resulting in increased boiler fuel requirements, boiler feedwater requirements, and boiler chemical requirements.

Significant energy savings may be realized by utilizing the guides presented in this section for (1) establishing optimum blowdown levels to maintain acceptable boiler-water quality and to minimize hot-water losses, and (2) the recovery of heat from the hot-water blowdown.

Where to Look For Energy-Saving Opportunities

The continuous blowdown from any steam-generating equipment has the potential for energy savings, whether it is a fired boiler or waste-heat-steam generator. The following items should be carefully considered to maximize savings:

1. Reduce blowdown (BD) by adjustment of the blowdown valve such that the controlling water impurity is held at the maximum allowable level.

2. Maintain blowdown continuously at the minimum acceptable level. This may be achieved by frequent manual adjustments or by the installation of automatic blowdown controls. At current fuel costs, automatic blowdown controls often prove to be economical.

3. Minimize the amount of blowdown required by:
    a. Recovering more clean condensate, which reduces the concentration of impurities coming into the boiler.
    b. Establishing a higher allowable drum solids level than is currently recommended by ABMA standards (see below). This must be done only on recommendation from a reputable water treatment consultant and must be followed up with lab tests for steam purity.
    c. Selecting the raw-water treatment system that has the largest effect on reducing makeup water impurities. This is generally considered applicable only to grass-roots or revamp projects.

4. Recover heat from the hot blowdown water. This is typically accomplished by flashing the water to a low pressure. This produces low-pressure steam (for utilization in an existing steam header) and hot water which may be used to preheat boiler makeup water.

Tests and Evaluations

STEP 1: *Determine Actual Blowdown*. Obtain the following data:

T = ppm of impurities in the makeup water to the deaerator from the treatment plant; obtain average value through lab tests

B = ppm of concentrated impurities in the boiler drum water (blowdown water); obtain average value through lab tests

lb/hr MU = lb/hr of makeup water to the deaerator from the water treatment plant; obtain from flow indicator

lb/hr BFW = lb/hr of boiler feedwater to each

lb/hr STM = lb/hr of steam output from each boiler; obtain from flow indicator

lb/hr CR = lb/hr of condensate return

Note: percentages for BFW, MU, and CR are determined as a percentage of STM.

Calculate the following:

$$\%MU = \text{lb/hr MU} \times 100\% / (\text{total lb/hr BFW})$$
$$= \text{lb/hr MU} \times 100\% / [(\text{boiler no. 1 lb/hr BFW}) + (\text{boiler no. 2 lb/hr BFW}) + ...]$$

$$\%MU = 100\% - \%CR \tag{5.3}$$

$$A = \text{ppm of impurity in BFW} = T \times \%MU \tag{5.4}$$

Now actual blowdown (BD) may be calculated as a function (percentage) of steam output:

$$\%BD = (A \times 100\%) / (B - A) \tag{5.5}$$

Converting to lb/hr BD yields

$$\text{lb/hr BD} = \% BD \times \text{lb/hr STM} \tag{5.6}$$

Note: In using all curves presented in this section, blowdown must be based on steam output from the boiler as calculated above. Boiler blowdown based on boiler feedwater rate (percent BD BFW) to the boiler should not be used. If blowdown is reported as a percent of the boiler feedwater rate, it may be converted to a percent of steam output using

$$\%BD = \%BD_{BFW} \times (1) / (1 - \%BD_{BFW}) \tag{5.7}$$

STEP 2: *Determine Required Blowdown*. The amount of blowdown required for satisfactory boiler operation is normally based on allowable limits for water impurities as established by the American Boiler Manufacturers Association (ABMA).

These limits are presented in Table 5-9. Modifications to these limits are possible as discussed below. The required blowdown may be calculated using the equations presented above by substituting the ABMA limit for B (concentration of impurity in boiler):

$$\% \, BD_{required} = (A)/(B_{required} - A) \times 100\% \qquad (5.8)$$

$$lb/hr \, BD_{required} = \% \, BD_{required} \times lb/hr \, STM \qquad (5.9)$$

STEP 3: *Evaluate the Cost of Excess Blowdown*. The amount of actual boiler blowdown (as calculated in equation 5.4) that is in excess of the amount of required blowdown (as calculated in equation 5.6) is considered as wasting energy since this water has already been heated to the saturation temperature corresponding to the boiler drum pressure. The curves presented in Figure 5-13 provide an easy method of evaluating the cost of excess blowdown as a function of various fuel costs and boiler efficiencies.

As an illustration of the cost of boiler blowdown, consider the following example.

Example: Determine the potential energy savings associated with reducing boiler blowdown from 12% to 10% using Figure 5-13.

Operating Data

Average boiler load	75,000 lb/hr
Steam pressure	150 psig
Make up water temperature	60°F
Operating hours	8,200 hr/yr
Boiler efficiency	80%
Average fuel cost	$2.00/10^6 Btu

Calculation and Analysis

Using the curves in Figure 5-13, enter Chart A at 10% blowdown to the curve for 150 psig boiler drum pressure. Follow the line over to chart B and the curve for a unit efficiency of 80%. Then follow the line down to Chart C and the curve for a fuel cost of $2.00/10^6 Btu. Read the scale for the equivalent fuel value in blowdown. The cost of the blowdown is estimated at $8.00/hr per 100,000 lb/hr of steam generated. Repeat the procedure for the blowdown rate of 12% and find the cost of the blowdown is $10.00/hr per 100,000 lb/hr of steam generated. Potential energy savings then is estimated to be

= ($10.00 - 8.00/hr/100,000 lb/hr)
     × (75,000 lb/hr) × (8,200 hr/yr)
= $12,300/yr.

Energy Conservation Methods

1. **Minimize Blowdown by Manual Adjustment**. This is accomplished by establishing an operating procedure requiring frequent water quality testing and readjustment of blowdown valves so that water impurities in the boiler are held at the allowable limit. Continuous indicating/recording analyzers may be employed, allowing the operator to quickly establish the actual level of water impurity and manually readjust blowdown valves.

2. **Minimize Blowdown by Automatic Adjustment**. The adjustment of blowdown may be automated by the installation of automatic analyzing equipment and the replacement of manual blowdown valves with control valves (see Figure 5-14). The cost of this equipment is frequently justifiable, particularly when there are frequent load changes on the steam-generating equipment, because the automation allows continuous maintenance of the highest allowable level of water contaminants. Literature has

Table 5-9. Recommended Limits for Boiler-Water Concentrations

Drum Pressure (psig)	Total Solids		Alkalinity		Suspended Solids		Silica
	ABMA	Possible	ABMA	Possible	ABMA	Possible	ABMA
0 to 300	3500	6000	700	1000	300	250	125
301 to 450	3000	5000	600	900	250	200	90
451 to 600	2500	4000	500	500	150	100	50
601 to 750	2000	2500	400	400	100	50	35
751 to 900	1500	—	300	300	60	—	20
901 to 1000	1250	—	250	250	40	—	8
1001 to 1500	1000	—	200	200	20	—	2

approximated that the average boiler plant can save about 20% blowdown by changing from manual control to automatic adjustment.

3. **Decrease Blowdown by Recovering More Condensate.** Since clean condensate may be assumed to be essentially free of water impurities, addition of condensate to the makeup water serves to dilute the concentration of impurities. The change in required blowdown may be calculated using equations 5.3 and 5.5.

Example: Determine the effect on boiler blowdown of increasing the rate of condensate return from 50 to 75%:

Operating Data
MU water impurities = 10 ppm = T
Maximum allowable limit in drum
= 100 ppm = B

Calculate:
For 50% return condensate:
MU = 100% − 50% = 50%
A = 10 × 0.5 = 5 ppm in BFW

% BD = A/(B-A) = 5/(100-5) = 5.3%

For 75% return condensate:
MU = 100 − 75% = 25%
A = 10 × 0.25 = 2.5 ppm
% BD = 2.5/(100-2.5) = 2.6%

Conclusion
These values may then be used with the curves on Figure 5-13 to approximate the potential energy savings.

4. **Increase Allowable Drum Solids Level.** In some instances it may be possible to increase the maximum allowable impurity limit without adversely affecting the operation of the steam system. However, it must be emphasized that a water treatment consultant should be contacted for recommendation on changes in the limits as given in Table 5-9. The changes must also be followed by lab tests for steam purity to verify that the system is operating as anticipated.

The energy savings may be evaluated by us-

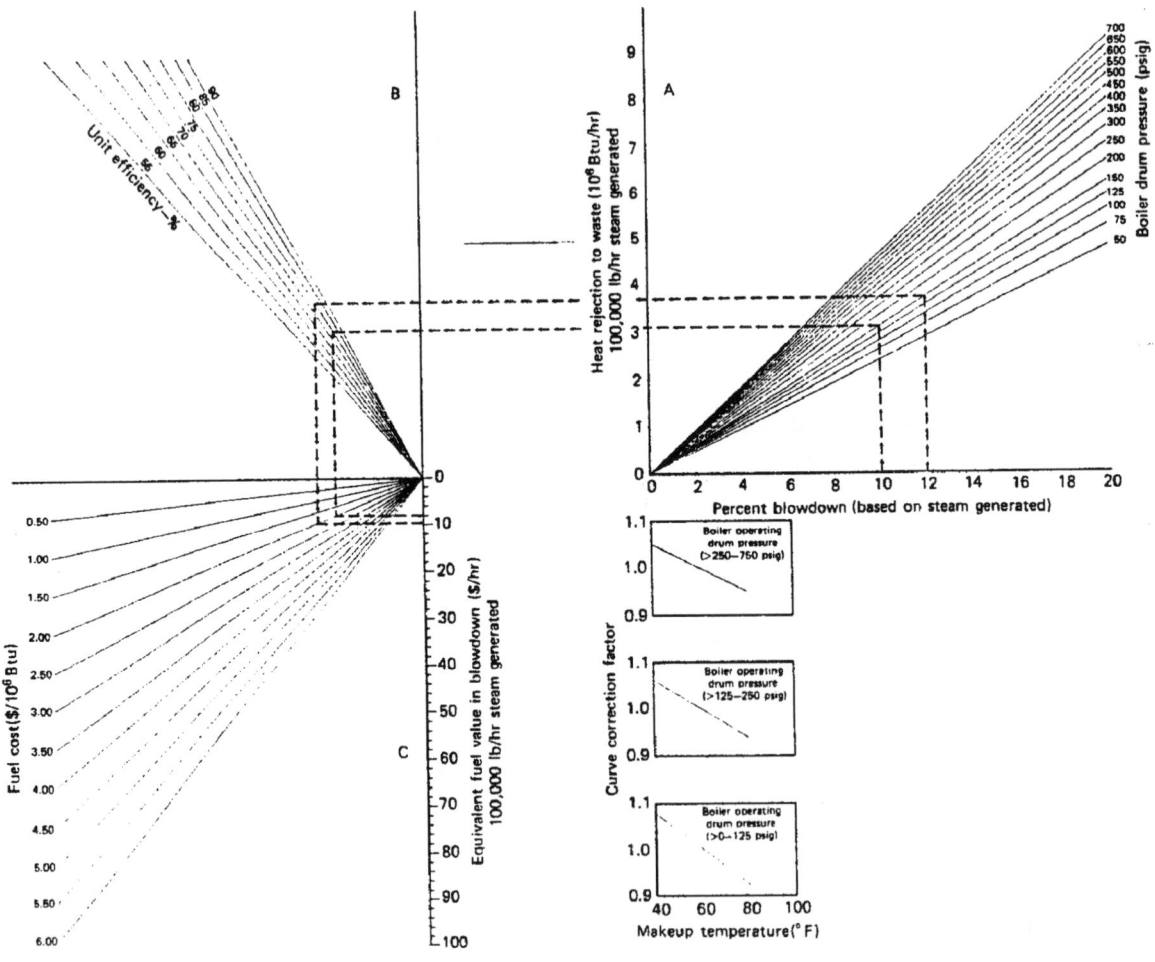

Figure 5-13. Hourly cost of blowdown.

ing the foregoing equations for blowdown and the graphs in Figures 5-13 and 5-15. Consider the following example.

Example: Determine the blowdown rates as a percentage of steam flow required to maintain boiler drum water impurity concentrations at an average of 3000 ppm and 6,000 ppm.

Operating Data
Average makeup water impurity
(measurement)..350 ppm
Condensate return (percent of steam flow)........25%
Assume condensate return free from impurities

Calculation and Analysis
Calculate the impurity concentration in the boiler feedwater (BFW):

$$A = MU \text{ impurity} \times (1.00 - \% \text{ CR})$$
$$A = 350 \text{ ppm} \times (1.00 - 0.25)$$
$$A = 262 \text{ ppm}$$

Mathematical solution.

For drum water impurity level of 3000 ppm:
$$\% \text{ BD} = A / (B - A)$$
$$\% \text{ BD} = 262 / (3000 - 262)$$
$$\% \text{ BD} = 9.6\%$$

Boiler steam drum

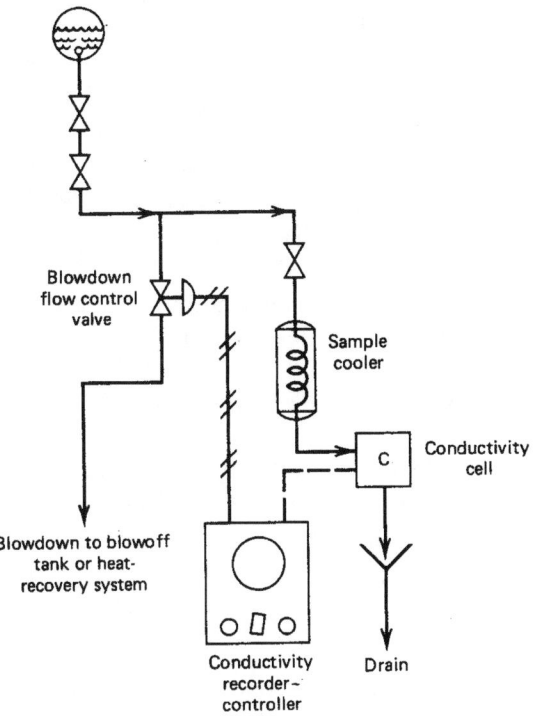

Figure 5-14. Automated continuous blowdown system.

For drum water impurity level of 6000 ppm:
$$\% \text{ BD} = A / (B - A)$$
$$\% \text{ BD} = 262 / (6000 - 262)$$
$$\% \text{ BD} = 4.6\%$$

Graphical Solution. Referring to Figure 5-15 Enter the graph at feedwater impurity level of 262 ppm and follow the line to the curves for 3000 ppm and 6000 ppm boiler drum water impurity level. Then read down to the associated boiler blowdown percentage.

Conclusion
The blowdown percentages may not be used in conjunction with Figure 5-13 to determine the annual cost of blowdown and the potential energy cost savings associated with reducing boiler blowdown.

5. **Select Raw-Water Treatment System for Largest Reduction in Raw-Water Impurities**. Since a large investment would be associated with the installation of new equipment, this energy conservation method is usually applicable to new plants or revamps only. A water treatment consultant should be retained to recommend the type of treatment applicable. An example of how water treatment affects blowdown follows.

Example: Determine the effects on blowdown of using a sodium zeolite softener producing a water quality of 350 ppm solids and of using a demineralization unit producing a water quality of 5 ppm solids. The makeup water rate is 30% and the allowable drum solids level is 3000 ppm.

Solution:
For sodium zeolite:
$$\% \text{ BD} = (350 \times 0.3 \times 100\%) / [3000 - (350 \times 0.3)] = 3.6\%$$

For demineralization unit:
$$\% \text{ BD} = (5 \times 0.3 \times 100\%) / [3000 - (5 \times 0.3)]$$
$$= 0.6\%$$

Therefore, the percentage blowdown would be reduced by 3.0%.

A secondary benefit derived from increasing feedwater quality is the reduced probability of scale formation in the boiler. Internal scale reduces the effectiveness of heat-transfer surfaces and can result in a reduction of as much as 1 to 2% in boiler efficiency in severe cases.

6. **Heat Recovery from Blowdown**. Since a certain amount of continuous blowdown must be main-

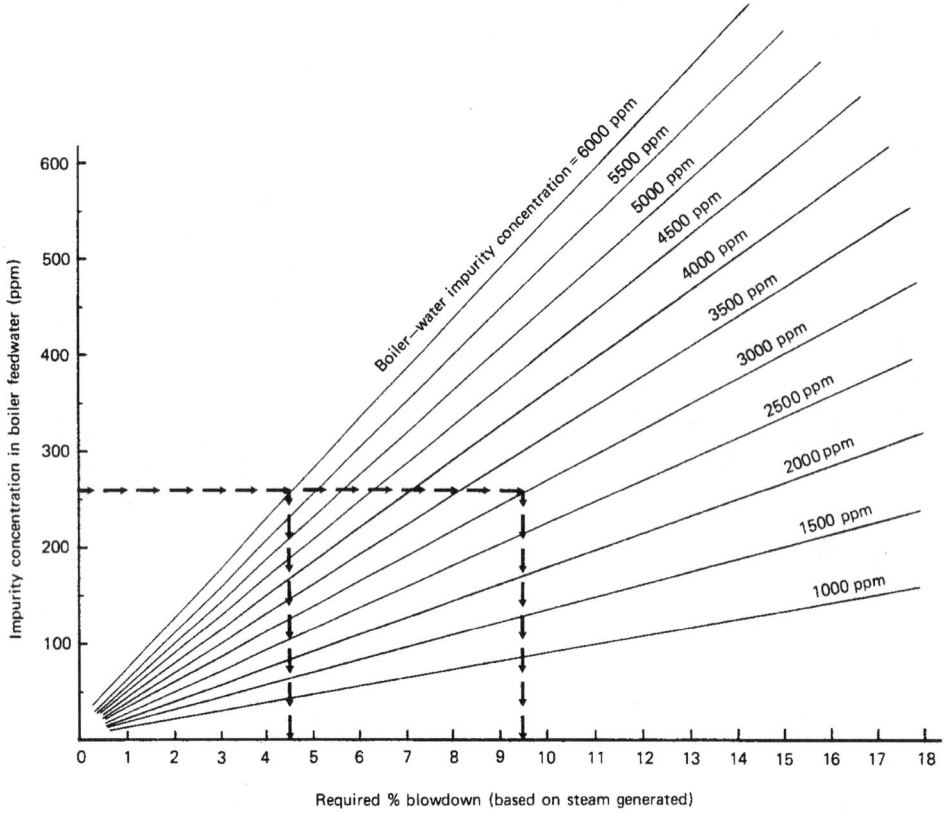

Figure 5-15. Required percent blowdown. Based on equation 5.5.

tained for satisfactory boiler performance, a significant quantity of heat is removed from the boiler. A large amount of the heat in the blowdown is recoverable by using a two-stage heat-recovery system as shown in Figure 5-16 before discharging to the sewer. In this system, blowdown lines from each boiler discharge into a common flash tank. The flashed steam may be tied into an existing header, used directly by process, or used in the deaerator. The remaining hot water may be used to preheat makeup water to the deaerator or preheat other process streams.

The following procedure may be used to calculate the total amount of heat that is recoverable using this system and the associated cost savings.

STEP 1. Determine the annual cost of blowdown using the percent blowdown, steam flow rate (lb/hr), unit efficiency, and fuel cost. This can be accomplished in conjunction with Figure 5-13.

STEP 2. Determine:

Flash % = percent of blowdown that is flashed to steam (using Figure 5-17, curve B, at the flash tank pressure or using equation 5.10a or 5.10b)

COND % = 100% - Flash %

h_{tk} = enthalpy of liquid leaving the flash tank (using Figure 5-17, curve A, at the flash tank pressure)

h_{ex} = enthalpy of liquid leaving the heat exchanger [using Figure 5-17, curve C; for planning purposes, a 30 to 40°F approach temperature (condensate discharge to makeup water temperature) may be used]

STEP 3. Calculate the amount of heat recoverable from the condensate (% QC) using

$$\%QC = [(h_{tk} - h_{ex})/h_{tk}] \times COND \%$$

STEP 4. Since all of the heat in the flashed steam is recoverable, the total percent of heat recoverable (% Q) from the flash tank and heat-exchanger system is found using

$$\% Q = \% QC + Flash \%$$

STEP 5. The annual savings from heat recovery may then be determined by using this percent (% Q) with the annual cost of blowdown found in step 1:

annual savings = (% QC/100) × BD cost

To further illustrate this technique consider the following example.

Example: Determine the percent of heat recoverable (%Q) from a 150 psig boiler blowdown waste stream, if the stream is sent to a 20 psig flash tank and heat exchanger.

Available Data

Boiler drum pressure	150 psig
Flash tank pressure	20 psig
Makeup water temperature	70°F

Assume a 30°F approach temperature between condensate discharge and makeup water temperature

Calculation and Analysis

Referring to Figure 5-17.

Determine Flash % using Chart B:

Entering chart B with a boiler drum pressure of 150 psig and following a live to the curve for a flash tank pressure of 20 psig, read the Steam percentage (Flash %) to be 12.5%.

Determine COND %:

COND % = 100 - Flash %

COND % = 100 - 12.5 %

COND % = 87.5%

Determine h_{tk} using Chart A:

Entering chart A with a flash tank pressure of 20 psig and following a line to the curve for saturated liquid, read the enthalpy of the drum water (h_{tk}) to be 226 Btu/lb.

Determine h_{ex} using Chart C:

Assuming a 30°F approach temperature between condensate discharge and makeup water temperature, the temperature of the blowdown discharge is equal to the makeup water temperature plus the approach temperature, which equals 100°F (70°F + 30°F).

Entering chart C with a blowdown heat exchanger rejection temperature of 100°F and following a line to the curve, read the enthalpy of the blowdown discharge water to be 68 Btu/lb.

Determine the % QC:

% QC = [(h_{tk} - h_{ex})/h_{tk}] × COND %

% QC = [(226 Btu/lb - 68 Btu/lb)/226 Btu/lb]
      × 87.5%

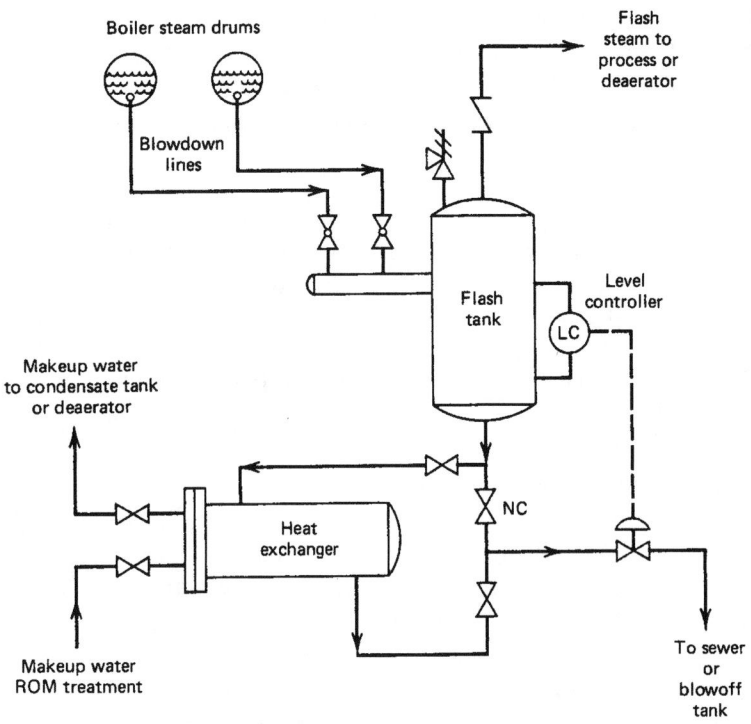

Figure 5-16. Typical two-stage blowdown heat-recovery system.

% QC = 61.2%

Determine the % Q:

% Q = % QC + Flash %

% Q = 61.2% + 12.5%

% Q = 73.7%

Conclusion

Therefore, approximately 73.7% of the heat energy can be recovered using this blowdown heat recovery technique.

More on Flash Steam

To determine the amount of flash steam that is generated by high-pressure, high-temperature condensate being reduced to a lower pressure, you can use the following equation:

$$\text{Flash } \% = (h_{HPl} - h_{LPl}) \times 100\% / (h_{LPv} - h_{LPl}) \qquad (5.10a)$$

or

$$\text{Flash } \% = (h_{HPl} - h_{LPl}) \times 100\% / (h_{LPevp}) \qquad (5.10b)$$

where: Flash % = amount of flash steam as a percent of total mass

h_{HPl} = enthalpy of the high pressure liquid

h_{LPl} = enthalpy of the low pressure liquid

h_{LPv} = enthalpy of the low pressure vapor

h_{LPevap} = evaporation enthalpy of the low pressure liquid = (h_{LPv} - h_{LPl})

Guidelines for Day-to-Day Operation

1. Maintain concentration of impurities in the boiler drum at the highest allowable level. Frequent checks should be made on water quality and blowdown valves adjusted accordingly.

2. Continuous records of impurity concentration in makeup water and boiler drum water will indicate trends in deteriorating water quality so that early corrective actions may be taken.

3. Control instruments should be calibrated on a weekly basis.

5.3.6 Condensate Return

In today's environment of ever-increasing fuel costs, the return and utilization of the heat available in clean steam condensate streams can be a practical and economical energy conservation opportunity. Refer to Chapter 6 for a comprehensive discussion of condensate return. The information below is presented to summarize briefly and emphasize the benefits and major considerations pertinent to optimum steam generator operations. Recognized benefits of return condensate include:

Reduction in steam power plant raw-water makeup and associated treatment costs.

Reduction in boiler blowdown requirements, resulting in direct fuel savings. Refer to section on boiler blowdown.

Reduced steam required for boiler feedwater deaeration.

Raw-water and boiler-water chemical cost reduction.

Opportunities for increased useful work output without additional energy input.

Reduces objectionable environmental discharges from contaminated streams.

Where to Look

Examine and survey all steam-consuming units within a plant to determine the present disposition of any condensate produced or where process modifications can be made to produce "clean" condensate. Address the following:

1. Is the condensate clean and being sewered?

2. Is the stream essentially clean but on occasion becomes contaminated?

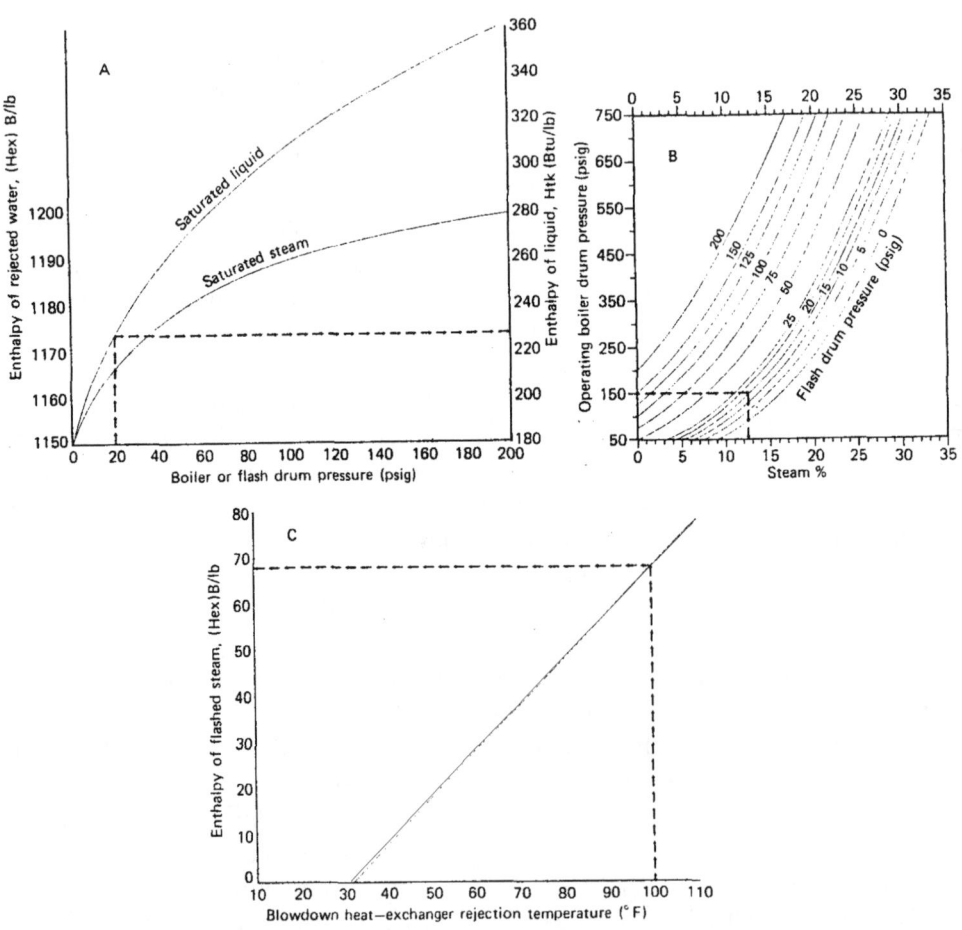

Figure 5-17. Percent of heat recoverable from blowdown.

3. If contaminated, can return to the steam system be justified by polishing the condensate?

4. Can raw makeup or treated water be substituted for condensate presently consumed?

5. Is condensate dumped for operating convenience or lack of chemical purity?

Results from chemical purity tests, establishing battery limit conditions, and analysis of these factors provide the basis of obtaining maximum economy.

Modifications Required for Maximum Economy

Often, the only requirement to gain the benefits of return condensate is to install the necessary piping and/or pumping facilities. Other solutions are more complex and accordingly require a more in-depth analysis. Chemically "clean" or "contaminated" condensate can be effectively utilized by:

Providing single- or multistage flashing for contaminated streams, then recouping the energy of the flashed steam. Recovering additional heat from the flash drum condensate by indirect heat exchange is also a possibility.

Collecting condensate from an atmospheric flash drum with "automatic" provision to dump on indication of stream contamination. This concept, when conditions warrant, allows "normally clean" condensate to be used within the system.

Installing ion-exchange polishing units for condensate streams that may be contaminated but are significant in quantity and heat value.

Providing a centrally located collection tank and pump to return the condensate to the steam system. This avoids a massive and complex network of individual return lines.

Using raw water in lieu of condensate and returning the condensate to the system. An example is the use of condensate to regenerate water treatment units.

Changing barometric condensers or other direct-contact heat exchangers to surface type or indirect exchangers, respectively, and returning the clean condensate to the system.

Collecting condensate from sources normally overlooked, such as space heating, steam tracing, and steam traps.

Providing flexibility to isolate and sewer individual return streams to maintain system integrity. Providing "knockout" or disengaging drum(s) to ensure clean condensate return to the system.

Recovering the heat content from contaminated condensate by indirect heat exchangers. An example is using an exchanger to heat the boiler makeup water.

Returning the contaminated condensate stream to a clarifier or hot lime unit to clean for boiler makeup rather than sewering the stream.

Allowing provision for manual water testing of the condensate stream suspected of becoming contaminated.

Using the contaminated condensate for noncritical applications, such as space heating, tank heating, etc.

Guidelines for Day-to-Day Operation

1. Maintain the system, including leak detection and insulation repair.

2. Periodically test the return water at its source of entry within the steam system for (a) contamination, (b) corrosion, and (c) acceptable purity.

3. Maintain and calibrate monitoring and analyzing equipment.

4. Ensure that the proper operating regimen is followed; that is, the condensate is returned and not sewered.

5.4 CONDENSING BOILERS

As illustrated in Figure 5-1, water is produced in the combustion of a hydrocarbon fuel. The hydrogen in the fuel combines with oxygen in the air to form water. Because the combustion gases are at an elevated temperature, the water is in vapor form. The water vapor, however, contains significant latent energy. For any fuel, the amount of latent heat energy contained by the water vapor is the difference between the higher-heating value of a fuel and the lower-heating value of the fuel. With the combustion of natural gas, which has a high hydrogen-to-carbon ratio, the water vapor holds about 10% of the total heat energy released by the combustion of the fuel.

Conventional boilers are designed to operate with exhaust gas temperatures high enough to keep the water in vapor form until the gases exit the stack. The combustion efficiency charts discussed in Section 5.3 are for noncondensing boilers. Depending on the design and length of the exhaust stack and the outside air temperature, the stack temperature could be quite high, and it may limit combustion efficiency. Table 5-10 illustrates minimum recommended exhaust gas temperatures for various fuels. The reason for this requirement is usually to prevent the formation of acids in the exhaust. When water vapor condenses in an atmosphere of combustion gases (CO_2, NO_x, and SO_2), the moisture forms acids based on those products of combustion, such as carbonic acid, nitric acid, and sulfuric acid. The acids will corrode the exhaust stack, and possibly worse, the boiler itself.

Table 5-10. Recommended Minimum Exhaust Gas Exit Temperature to Avoid Corrosion Problems

Source Fuel	Minimum Temperature (°F)
Fuel Oil, >2.5% sulfur	390
Fuel Oil, <1% sulfur	330
Bituminous coal, >3.5% sulfur	290
Bituminous coal, >1.5% sulfur	230
Pulverized anthracite coal	220
Natural gas	220

Source: Handbook of Energy Engineering, 5th edition. A. Thumann and D.P. Mehta. 2001.

5.4.1 Higher Efficiency

European nations have used condensing boilers for years. Many European energy codes now require the use of condensing boilers for gas-fired, and in some cases oil-fired, central-heating plants. Some natural gas utilities in Canada are incentivizing the use of condensing boilers.

Condensing boilers are designed to extract sufficient heat energy from the combustion gases to lower the temperature below the dew point, causing the moisture to condense within the boiler. A condensing boiler uses very high efficiency heat exchangers designed to recover the majority of the available sensible heat, as well as some of the latent heat generated by the combustion of the fuel. The result is a significant increase in combustion efficiency compared to a non-condensing boiler.

To maximize efficiency, the combustion gases need to be exposed to a heat sink with a low enough temperature to result in condensation. If the distribution system does not provide low enough inlet water temperature to the boiler, the moisture in the combustion gases will only partially condense or may not condense at all. Condensing boilers may not condense all the time or to the same extent. The return water temperature is the primary factor that determines if the boiler operates in condensing mode. Low return water temperatures are essential to obtaining the high efficiencies associated with condensing boilers.

The condensing temperature of the combustion gases is a function of the dew point of the water vapor generated. The amount of hydrogen in the fuel determines how much water vapor is generated during the combustion

process. Natural gas has a higher hydrogen content than fuel oil; therefore, natural gas generates more water per pound of fuel consumed, thus increasing the condensing temperature (allowing heat energy to be recovered at a higher temperature). Excess air lowers the vapor pressure; therefore, it will decrease the dew point (lowering the temperature required to condense the vapor in the combustion gas). This is another reason to optimize the air-to-fuel ratio. Altitude also changes the vapor pressure. Higher altitudes (lower barometric pressure) decrease the dew point.

However, the most influential factor is the return water temperature. Minimizing the return water temperature, by design and operation, will maximize the efficiency of the condensing boiler system, as shown in Figure 5.18. Therefore, condensing boilers are most applicable to applications that allow for low return water temperatures.

5.4.2 Condensing Boiler and System Design

To recover as much latent heat as possible, and because the products of combustion are corrosive, condensing boilers are constructed from high-grade stainless steel or other highly-resistant metals. Because the system operates with lower water temperatures, the heat transfer surface area is greater than an equivalent capacity non-condensing boiler.

Condensing boilers tend to have more advanced controls and higher grade burners. The combustion systems operate under positive pressure (power burner or pulse combustion) with precise-controlled hot surface or spark ignition. In addition to making them more efficient, this also results in lower NO_x emissions. Adequate combustion control is also important because excess air makes it more difficult to achieve high sensible and latent heat recovery.

High efficiency is not only a function of design but also a function of operating conditions. Proper system design and control is important to ensure that the condensing boiler experiences low enough return water temperatures to recover latent heat energy. As noted earlier, condensing boilers may not always condense. The hot water distribution system and operating strategy should be designed to ensure that the condensing boiler receives the lowest return water temperature over the highest number of operating hours to ensure optimizing overall efficiency. For example, hot water temperature reset (see Chapter 12) is a control strategy that reduces the water temperature when the system capacity can still meet the operating conditions (off-design condition optimization).

Most condensing boilers incorporate controls to modulate capacity and avoid cycling losses that result from partial load. When installing large capacity central heating systems, there are advantages in dividing the output over several smaller capacity boilers. Higher overall efficiency, greater reliability, simpler control and generally lower capital cost are among the major advantages that modular systems offer compared with the conventional arrangement of one or two larger boilers.

The capability of the system to maintain tight temperature control, modulate firing rate with partial load, and operate in the condensing mode may eliminate the need for hot water storage, temperature blending valves, and primary-secondary pumping systems.

Several manufacturers make condensing boilers, most under 1 million Btu/h. Only a few manufacturers make units with a capacity in the 1 to 2 million Btu/h range. Condensing boilers are frequently designed for modular configurations. A multiple condensing boiler system has several advantages over a single boiler system. Specifically, they can operate with higher seasonal efficiency. Modular configurations allow less capacity to be online, thus extending the turndown ratio and further avoiding the need for boiler cycling. They also have a higher overall reliability, because if one boiler goes down, the modular configuration still allows heat to be available, although in reduced capacity.

Modern condensing boilers usually have full modulation with high turndown ratios, typically at least 10:1 but possibly as high as 20:1. (Turndown is the ratio of maximum to minimum firing rate.) Firing at reduced input allows more residence time for the combustion gases in the heat transfer area. For this reason, condensing boilers may have higher combustion efficiency at low load than at full load—provided the boiler is modulating and not cycling. As with any boiler, condensing or non-con-

densing, cycling will result in additional heat loss.

Because the boilers condense moisture from the combustion gases, the exhaust stack configuration is different from conventional boilers. The exhaust gases are lower in temperature, thus there is less draft force to draw the gases up the stack. In addition, the boilers also have a drain line to expel the condensed water. As noted earlier, the condensed moisture is acidic, typically with a pH around 4 to 5 for natural gas, but it can be lower. Therefore, drain lines should be PVC, CPVC, ABS or some other corrosion resistant material. Copper drain lines should not be used. There is also typically some concern with sending the condensate to drain. Dilution, or mixing the condensate with other more basic (higher pH) sources, may be acceptable to meet waste water (sewer) requirements. In some cases, however, treatment with an inline neutralizer may be required. Checking with local sources for specific legal requirements as part of the design phase is highly recommended.

5.4.3 Direct-contact condensing boilers and water heaters

Direct-contact water heating is a heat transfer method in which there are no tubes isolating hot combustion gases from the fluid to be heated. The exhaust gases are allowed to come into direct contact with water in a totally non-pressurized environment, so all heating occurs at atmospheric pressure.

Direct-contact water heaters consist of bodies, sometimes made completely from stainless steel, that use a vertical hollow chamber in which water is sprayed at the top of this vertical chamber. The upper portion of this vertical chamber is filled with stainless steel balls (or other fill material), which provides a large heat transfer surface area for the rising exhaust gases to be conducted into the water. This "heat transfer zone" is typically 24 to 36 inches deep, and most heat transfer occurs in this area.

A burner is mounted in the vertical column, below the heat transfer zone, which provides the energy used to heat the water. Depending on the manufacturer, the burner may be substantially removed from the falling water with a dedicated firing chamber, or it may reside directly in the path of the water flow with an overhead metal shield to protect the bulk of the flame from water impingement. The burner is forced draft and is typically fired on natural gas or propane, although some models may be designed for fuel oil.

Because the water falls at atmospheric pressure, the warm water collects at the bottom of the direct-contact heater. This may be a full-sized tank or just a holding area at the base of the heater. The warm water is then circulated, via a mechanical pump, to the needed application.

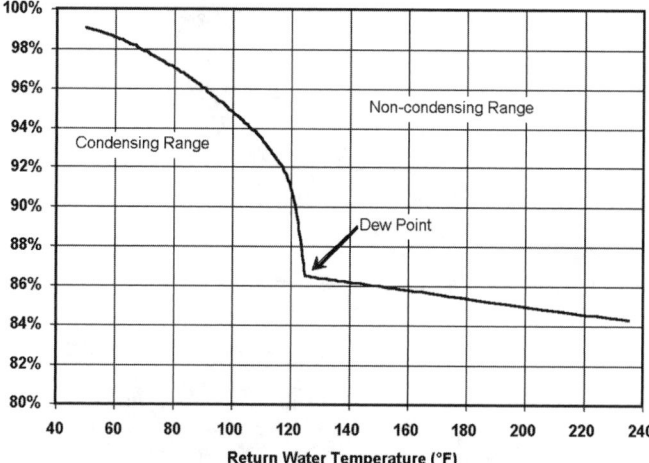

Figure 5-18. Typical Relationship Between Return Water Temperature and Combustion Efficiency (based on natural gas HHV = 1050 Btu/scf)

Like other condensing boilers, direct-contact water heaters are very efficient (hot water energy output ÷ fuel energy input). Direct-contact water heaters sub-cool the combustion gases below the dew point, thereby reclaiming thermal energy that would otherwise be lost through the exhaust stack.

The thermal efficiency of a direct-contact water heater may be determined by the exhaust gas temperature, which should exceed the inlet water temperature by 5 to 10°F. For example, if the inlet water is 55°F, the exhaust gas temperature should be about 60 to 65°F, and the heater efficiency will be about 99%. A direct-contact condensing water heater should operate with an inlet water temperature between 45°F and 80°F, which translates to 99.7% to 97.7%. If the system is recirculating hot water back to the upper portion of a heater, the efficiency of the direct-contact water heater will be lower for the time frame in which the recirculation occurs. For example, if the return water temperature is 160°F, the exhaust gas temperature would be about 165°F, making the direct-contact heater efficiency about 75% for the time that 160°F is being reheated. These dynamics make direct-contact heater system design different from conventional boiler system design. To maximize operating efficiency, the coldest possible return water temperature is desired from the system loop (not the typical 20°F temperature drop across a conventional heat exchanger design).

5.4.4 Water Temperature Limitations with Direct Contact

Direct-contact water heaters are not capable of producing steam and have difficulty in achieving outlet water temperatures higher than 188°F. Direct-contact water heaters can achieve water temperatures up to 193°F. However, the efficiency of the direct-contact condensing boiler drops dramatically because large amounts of water are vaporized carrying away heat energy.

Direct-contact water heaters typically operate with a system pressure of a few inches water column above atmospheric pressure. In theory, direct-contact water heaters should be able to heat water up to 200°F or even higher. The limiting factor occurs when the water is heated as the droplet falls through open air in the heat transfer section of the direct contact. Because the heating occurs while the water is moving through the air, the aerodynamics lower the localized pressure around the water droplets to a point lower than the available atmospheric pressure. This, in turn, lowers the temperature at which the liquid water changes phase to a gaseous form. If this effect is attempting to achieve exit water temperatures higher than 188°F, it becomes inefficient and unstable. The altitude at which a heater is operating will affect the maximum temperature output capability of the direct-contact heater because of the reduction in available atmospheric pressure.

5.4.5 Water Quality and Emissions with Direct Contact

While the affect from the condensing moisture from the combustion gases provides acidity, it is diluted by the water being heated. Therefore, the water quality from a direct-contact water heater is not usually a concern. Different manufacturers use different designs that affect the pH drop. After a single pass through a direct-contact heating unit, water with a neutral pH (7.4) exits the heater with a pH between 6.9 and 6.3. In general, the pH change is minimized in heating units that have a dedicated space to complete the combustion of fuels.

Other than CO_2 injection into the water stream, direct-contact heaters do not negatively affect the consistency of the water stream. There are cases in poor heater design, where combustion has been quenched early and acids created from incomplete combustion, lowering the pH further than is explainable by CO_2 injection.

Exhaust emissions from a direct-contact water heater is typically lower than the emission rates of a boiler firing with the same burner at the same firing rate. This does not make the units low NO_x emitters, but it does allow certain low NO_x technologies to work even more effectively in direct-contact heaters than in comparable boilers.

5.4.6 Applications for Condensing Boilers

Condensing boilers may be used in almost any ap-

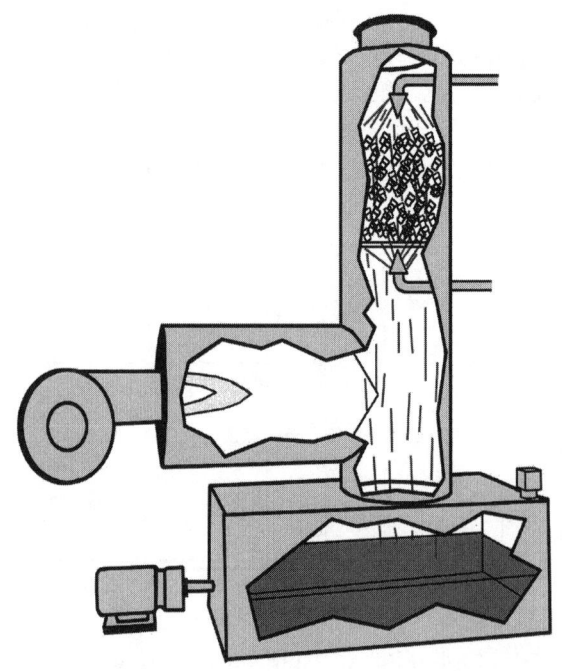

Figure 5-19. Direct-contact condensing hot water boiler

plication that requires water at moderately warm temperatures. Radiant floor heating systems and water-source heat pump systems are two examples of space heating systems that can be designed for low operating water temperatures that can take advantage of the efficiency benefits of condensing boilers. Conventional space heating applications with water temperatures of 150 to 180°F will not result in condensing moisture, even in condensing boilers. However, applying an outdoor air temperature reset control strategy and designing heat transfer coils for larger temperature differentials will increase the combustion efficiency of a condensing boiler system.

A short list of applications is given in Table 5.11, although there are several more possible applications.

5.5   FUEL CONSIDERATIONS

The selection and application of fuels to various combustors are becoming increasingly complex. Most existing units have limited flexibility in their ability to fire alternative fuels, and new units must be carefully planned to assure the lowest first costs without jeopardizing the future capability to switch to a different fuel. This section presents an overview of the important considerations in boiler and fuel selection. Also refer to Section 5.5.

5.5.1 Natural Gas

Natural-gas firing in combustors has traditionally been the most attractive fuel type, because:

1.   Gas costs were low until about 2000, when they started to rise dramatically.

2.   Only limited fuel-handling equipment (typically consisting of pipelines, metering, a liquid knockout drum, and appropriate controls) is required.

3.   Boiler costs are minimized due to smaller boiler sizes that result from highly radiant flame characteristics and higher velocities, resulting in enhanced heat transfer and less heating surface.

4.   Freedom from capital and operating costs associated with pollution control equipment.

Natural gas, being the cleanest readily available conventional form of fuel, also makes gas-fired units the easiest to operate and maintain.

However, as discussed elsewhere, the continued use of natural gas as fuel for most combustors will probably be limited in the future by government regulations, rising

fuel costs, and inadequate supplies. One further disadvantage, which often seems to be overlooked, is the lower boiler efficiency that results from firing gas, particularly when compared to oil or coal.

5.5.2 Fuel Oil
Classifications

Influential in the storage, handling, and combustion efficiency of a liquid fuel are its physical and chemical characteristics.

Fuel oils are graded as No. 1, No. 2, No. 4, No. 5 (light), No. 5 (heavy), and No. 6. Distillates are Nos. 1 and 2, and residual oils are Nos. 4, 5, and 6. Oils are classified according to physical characteristics by the American Society for Testing and Materials (ASTM), according to Standard D-396.

No. 1 oil is used as domestic heating oil and as a light grade of diesel fuel. Kerosene is generally in a lighter class; however, often both are classified the same. No. 2 oil is suitable for industrial use and home heating. The primary advantage of using a distillate oil rather than a residual oil is that it is easier to handle, requiring no heating to transport and no temperature control to lower the viscosity for proper atomization and combustion. However, there are substantial purchase cost penalties between residual and distillate.

It is worth noting that distillates can be divided into two classes: straight-run and cracked. A straight-run distillate is produced from crude oil by heating it and then condensing the vapors. Refining by cracking involves higher temperatures and pressures, or catalysts, to produce the required oil from heavier crudes. The difference between these two methods is that cracked oils contain substantially more aromatic and olefinic hydrocarbons which are more difficult to burn than the paraffinic and naphthenic hydrocarbons from the straight-run process. Sometimes a cracked distillate, called industrial No. 2, is used in fuel-burning installations of medium size (small package boiler or ceramic kilns for example) with suitable equipment.

Because of the viscosity range permitted by ASTM, No. 4 and No. 5 oil can be produced in a variety of ways: blending of No. 2 and No. 6, mixture of refinery by-products, utilization of off-specification products,etc. Because of the potential variations in characteristics, it is important to monitor combustion performance routinely to obtain optimum results. Burner modifications may be required to switch from, say, a No. 4 that is a blend, to a No. 4 that is a distillate.

Light (or cold) No. 5 fuel oil and heavy (or hot) are distinguished primarily by their viscosity ranges: 150 to 300 SUS (Saybolt Universal Seconds) at 100°F and 350 to

750 SUS at 100°F, respectively. The classes normally delineate the need for preheating with heavy No. 5 requiring some heating for proper atomization.

No. 6 fuel oil is also referred to as residual, Bunker C, reduced bottoms, or vacuum bottoms. It is a very heavy oil or residue left after most of the light volatiles have been distilled from crude. Because of its high viscosity, 900 to 9000 SUS at 100°F, it can only be used in systems designed with heated storage and sufficient temperature/viscosity at the burner for atomization.

Heating Value

Fuel oil heating content can be expressed as higher (or gross) heating value and low (or net) heating value. The higher heating value (HHV) includes the water content of the fuel, whereas the lower heating value (LHV) does not. For each gallon of oil burned, approximately 7 to 9 lb of water vapor is produced. This vapor, when condensed to 60°F, releases 1058 Btu. Thus the HHV is about 1000 Btu/lb or 8500 Btu/gal higher than the LHV. While the LHV is representative of the heat produced during combustion, it is seldom used in the United States except for exact combustion calculations.

Viscosity

Viscosity is a measure of the relative flow characteristics of an oil, which is an important factor in the design and operation of oil-handling and -burning equipment, the efficiency of pumps, temperature requirements, and pipe sizing. Distillates typically have low viscosities and can be handled and burned with relative ease. However, No. 5 and No. 6 oils may have a wide range of viscosities, making design and operation more difficult.

Viscosity indicates the time required in seconds for 60 cm³ of oil to flow through a standard-size orifice at a specific temperature. Viscosity in the United States is normally determined with a Saybolt viscosimeter. The Saybolt viscosimeter has two variations (Universal and Furol), with the only difference being the size of orifice

and sample temperature. The Universal has the smallest opening and is used for lighter oils. When stating an oil's viscosity, the type of instrument and temperature must also be stated.

Flash Point

Flash point is the temperature at which oil vapors flash when ignited by an external flame. As heating continues above this point, sufficient vapors are driven off to produce continuous combustion. Since flash point is an indication of volatility, it indicates the maximum temperature for safe handling. Distillate oils normally have flash points from 145 to 200°F, whereas the flash point for heavier oils may be up to 250°F. Thus, under normal ambient conditions, fuel oils are relatively safe to handle (unless contaminated).

Pour Point

Pour point is the lowest temperature at which an oil flows under standard conditions. It is 5°F above the oil's solidification temperature. The wax content of the oil significantly influences the pour point (the more wax, the higher the pour point). Knowledge of an oil's pour point will help determine the need for heated storage, storage temperature, and the need for pour-point depressant. Also, since the oil may cool while being transferred, burner preheat temperatures will be influenced and should be watched.

Sulfur Content

The sulfur content of an oil is dependent upon the source of crude oil. Typically, 70 to 80% of the sulfur in a crude oil is concentrated in the fuel product, unless expensive desulfurization equipment is added to the refining process. Fuel oils normally have sulfur contents of from 0.3 to 3.0%, with distillates at the lower end of the range unless processed from a very high sulfur crude. Often, desulfurized light distillates are blended with high-sulfur residual oil to reduce the residual's sulfur content.

Table 5-11. Condensing Boiler Applications

Single-Pass Applications	Open-Loop Applications	Closed-Loop Applications	Special Applications
Plant clean up	Bottle warming	Hydronic heating	Caustic fluid
Bird scalding	Hospital systems	Indirect heating	Green houses
Car washes	Swimming pools		Hotel showers
Concrete batches	Fruit cleaning		Hotel sinks
Dyeing	Clean-up systems		Demineralized water
Parts cleaning	Car washes		Biogas
Boiler make-up water preheat			
Commercial laundry			
Industrial laundry			

Sulfur content is an important consideration, primarily in meeting environmental regulations.

Ash

During combustion, impurities in oil produce a metallic oxide ash in the furnace. Over 25 different metals can be found in oil ash, the predominant ones being nickel, iron, calcium, sodium, aluminum, and vanadium. These impurities are concentrated from the source crude oil during refining and are difficult to remove since they are chemically bound in the oil. Ash contents vary widely—distillates have about 0 to 0.1% ash and heavier oil 0.2 to 1.5%. Although percentages are small, continuous boiler operations can result in considerable accumulation of ash in the firebox.

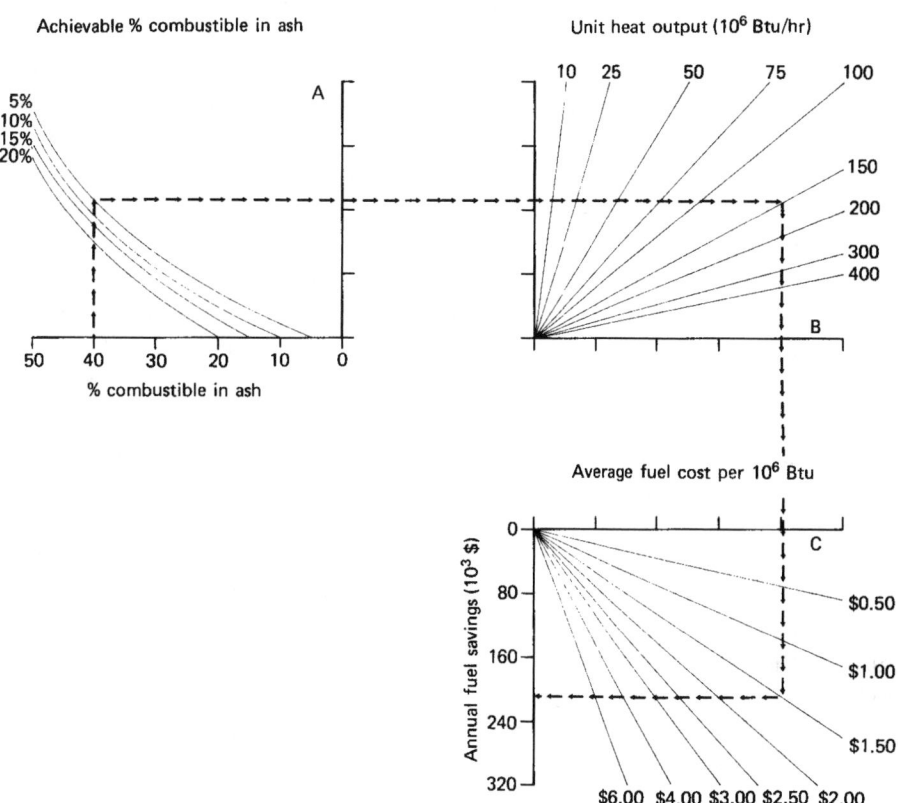

Figure 5-20. Coal annual savings from reducing combustible losses.

Problems associated with ash include reduction in heat-transfer rates through boiler tubes, fouling of superheaters, accelerated corrosion of boiler tubes, and deterioration of refractories. Ashes containing sodium, vanadium, and/or nickel are especially troublesome.

Other Contaminants

Other fuel-oil contaminants include water, sediment, and sludge. Water in fuel oil comes from condensation, leaks in storage equipment, and/or leaking heating coils. Small amounts of water should not cause problems. However, if large concentrations (such as at tank bottoms) are picked up, erratic and inefficient combustion may result. Sediment comes from dirt carried through with the crude during processing and impurities picked up in storage and transportation. Sediment can cause line and strainer plugging, control problems, and burned/nozzle plugging. More frequent filter cleaning may be required.

Sludge is a mixture of organic compounds that have precipitated after different heavy oils are blended. These are normally in the form of waxes or asphaltenese, which can cause plugging problems.

Additives

Fuel-oil additives may be used in boilers to improve combustion efficiency, inhibit high-temperature corro-sion, and minimize cold-end corrosion. In addition, additives may be useful in controlling plugging, corrosion, and the formation of deposits in fuel-handling systems. However, caution should be used in establishing the need and application of any additive program. Before selecting an additive, clearly identify the problem requiring correction and the cause of the problem. In many cases, solutions may be found which would obviate the need and expense of additives. Also, be sure to understand clearly both the benefits and the potential debits of the additive under consideration.

Additives to fuel-handling systems may be warranted if corrosion problems persist due to water which cannot be removed mechanically. Additives are also available that help prevent sludge and/or other deposits from accumulating in equipment, which could result in increased loading from increased pressure drops on pumps and losses in heat-transfer-equipment efficiencies.

Additive vendors claim that excess air can be controlled at lower values when catalysts are used. Although these claims appear to be verifiable, consideration should be given to mechanically controlling O_2 to the lowest possible levels. Accurate O_2 measurement and control should first be implemented and then modifications to burner assemblies considered. Catalysts, consisting of metallic oxides (typically manganese and barium), have demonstrated the capability of reducing carbon carryover in the flue gas and thus would permit lower O_2 levels without smoking. Under steady load conditions, savings can be achieved. However, savings may be negligible under varying loads, which necessitate prevention of fuel-rich mixtures by maintaining higher than optimal air levels.

Other types of combustion additives are available which may be beneficial to specific boiler operating problems. However, these are not discussed here since they are specific in nature and are not necessarily related to improved boiler efficiency. Generally, additives are used when a specific problem exists and when other conventional solutions have been exhausted.

Atomization

Oil-fired burners, kilns, heat-treating furnaces, ovens, process reactors, and process heaters will realize increased efficiency when fuel oil is effectively atomized. The finer the oil is atomized, the more complete combustion and higher overall combustion efficiency. Obtaining the optimum degree of atomization depends on maintaining a precise differential between the pressure of the oil and pressure of the atomizing agent, normally steam or air. The problem usually encountered is that the steam (or air pressure) remains constant while the oil pressure can vary substantially. One solution is the addition of a differential pressure regulator that controls the steam pressure so that the differential pressure to the oil is maintained. Other solutions, including similar arrangements for air-atomized systems, should be reviewed with equipment vendors.

Fuel-Oil Emulsions

In general, fuel-oil emulsifier systems are designed to produce an oil/water emulsion that can be combusted in a furnace or boiler.

The theory of operation is that micro-size droplets of water are injected and evenly dispersed throughout the oil. As combustion takes place, micro explosions of the water droplets take place, producing very fine oil droplets. Thus, more surface area of the fuel is exposed, which then allows for a reduction in excess air level and improved efficiency. Unburned particles are also reduced.

Several types of systems are available. One uses a resonant chamber in which shock waves are started in the fluid, causing the water to cavitate and breakdown into small bubbles. Another system produces an emulsion by injecting water into oil. The primary technical difference among the various emulsifier systems currently marketed is the water particle-size distribu-

tion that is formed. The mean particle size, as well as the particle-size distribution, is very important to fuel combustion performance, and these parameters are the primary reasons for differing optimal water content for various systems (from 3% to over 200%). Manufacturers typically claim improvements in boiler efficiency of from 1 to 6%, with further savings due to reduced particulate emissions that allow burning of higher-sulfur (lower-cost) fuels. However, recent independent testing seems to indicate that while savings are achieved, these are often at the low end of the range, particularly for units that are already operating near optimum conditions. Greater savings are more probable from emulsifiers that are installed on small, inefficient, and poorly instrumented boilers. Tests have confirmed reductions in particulate and NO_x emissions with the ability to keep boiler internals cleaner. Flames exhibit characteristics of more transparency and higher radiance (similar to gas flames).

Before considering installation of an emulsifier system, existing equipment should be tuned up and put in optimum condition for maximum efficiency. A qualified service technician may be helpful in obtaining best results. After performance tests are completed, meaningful tests with an emulsified oil can be run. It is suggested that case histories of a manufacturer's equipment be reviewed and an agreement be based on satisfaction within a specific time frame.

5.5.3 Coal

The following factors have a negative impact on the selection of coal as a fuel:

1. Environmental limitations which necessitate the installation of expensive equipment to control particulates SO_2, and NO_x. These requirements, when combined with the low price of oil and gas in the late 1960s and early 1970s, forced many existing industrial and utility coal-fired units to convert to oil or gas. And new units typically were designed with no coal capability at all.

2. Significantly higher capital investments, not only for pollution abatement but also for coal-receiving equipment, raw-coal storage, coal preparation (crushing, conveying, pulverizing, etc.), prepared-coal storage, and ash handling.

3. Space requirements for equipment and coal storage.

4. Higher maintenance costs associated with the installation of more equipment.

5. Concern over uninterrupted availability of coal resulting from strikes.

6. Increasing transportation costs.

The use of coal is likely to escalate, even in light of the foregoing factors, owing primarily to its relatively secure availability both on a short-term and long-term basis, and its lower cost. The substantial operating cost savings at current fuel prices of coal over oil or gas can economically justify a great portion (if not all) of the significantly higher capital investments required for coal. In addition, most predictions of future fuel costs indicate that oil and gas costs will escalate much faster than coal.

Enhancing Coal-Firing Efficiency

A potentially significant loss from the combustion of coal fuels is the unburned carbon loss. All coal-fired steam generators and coal-fired vessels inherently suffer an efficiency debit attributable to unburned carbon. At a given process output, quantities in excess of acceptable values for the particular unit design and coal rank detract greatly from the unit efficiency, resulting in increased fuel consumption.

It is probable that pulverized coal-fired installations suffer from an inordinately high unburned carbon loss when any of the following conditions are experienced:

A change in the raw-fuel quality from the original design basis.

Deterioration of the fuel burners, burner throats, or burner swirl plates or impellers.

Increased frequency of soot-blowing to maintain heat-transfer and surface cleanliness.

Noted increase in stack gas opacity.

Sluggish operation of combustion controls of antiquated design.

Uneven flame patterns characterized by a particularly bright spot in one sector of the flame and a dark spot in another.

CO formation as determined from a flue-gas analysis.

Frequent smoking observed in the combustion zone.

Increases in refuse quantities in collection devices.

Lack of continued maintenance and/or replacement of critical pulverized internals and classifier assembly.

High incidence of coal "hang-up" in the distribution piping to the burners.

Frequent manipulation of the air/coal primary and secondary air registers.

How to Test for Relative Operating Efficiency

The aforementioned items are general symptoms which are suspect in causing a high unburned carbon loss from the combustion of coal. However, the magnitude of the problem often goes undetected and remedial action never taken, because of the difficulty in establishing and quantifying this loss while relating it to the overall unit operating efficiency.

In light of this, it is recommended that the boiler manufacturer be consulted to establish the magnitude of this operating loss, as well as to review the system equipment and operating methods.

The general test procedure to determine the unburned carbon loss requires manual sampling of the ash (refuse) in the ash pit, boiler hopper(s), air heater hopper, and dust collector hopper, as well as performing a laboratory analysis of the samples. For reference, detailed methods, test procedures, and results are outlined in the ASME Power Test Codes, publications PTC 3.2 (Solid Fuels) and PTC 4 (Fired Steam Generators), respectively. In addition to the sampling of the ash, a laboratory analysis should be performed on a representative raw coal sample and a pulverized coal sample.

Results and analysis of such a testing program will:

1. Quantify the unburned carbon loss.

2. Determine if the unburned carbon loss is high for the type of coal fired, unit design, and operating methods.

3. Reveal a reasonable and attainable value for unburned carbon loss.

4. Provide guidance for any corrective action.

5. Allow the plant, with the aid of Figure 5-18, to assess the annual loss in dollars from operating with a high unburned carbon loss.

6. Suggest an operating mode to reduce the excess air required for combustion.

Figure 5-18 can be used to determine the approximate energy savings resulting from reducing unburned coal fuel loss. The unburned fuel is generally collected with the ash in either the boiler ash hopper(s) or in various collection devices. The quantity of ash collected at various locations is dependent upon and unique to the system design. The boiler manufacturer will generally specify the proportion of ash normally collected in various ash hoppers furnished on the boiler proper. The balance of ash and unburned fuel is either collected in flue-gas cleanup devices or discharged to the atmosphere. A weighted average of total percentage combustibles in the ash must be computed to use Figure 5-18. To further illustrate the potential savings from reducing combustible losses in coal-fired systems using Figure 5-18, consider the following example.

Example: Determine the benefits of reducing the combustible losses for a coal-fired steam generator having a maximum continuous rating of 145,000 lbs/hr at an operating pressure of 210 psig and a temperature of 475°F at the desuperheater outlet; feedwater is supplied at 250°F.

Available Data

Average boiler load	125,000 lb/hr
Superheater pressure	210 psig
Superheater temperature	475°F
Fuel type	Coal
Measured flue gas oxygen	3.5%
Operating combustibles in ash	40%
Obtainable combustibles in ash	5%
Yearly operating time	8500 hr/yr
Design unit heat output	150×10^6 Btu/hr
Average unit heat output	129×10^6 Btu/hr
Average fuel cost	$1.50/$10^6$ Btu

Analysis. Referring to Figure 5-18:
Chart A.

Operating combustibles	40%
Obtainable combustibles	5%

Chart B.

Design Unit output	150×10^6 Btu/hr

Chart C.

Average fuel cost	$1.50/$10^6$ Btu
Annual fuel savings	$210,000

Annual fuel savings corrected to actual operating conditions:

Savings $ = (curve value × [operating heat output]/ (design heat output)] × [actual annual operating hours)/(8,760 hr/yr)]
= ($210,000/yr) × [($129 \times 10^6$ Btu/hr)/(150×10^6 Btu/hr)] × [(8,500 hr/yr)/(8,760 hr/yr)]
= $175,200/yr

Notes: If the unit heat output or average fuel cost exceeds the limit of Figure 5-20, use half the particular value and double the savings obtained from Chart C. The annual fuel savings are based on 8,760 hr/yr. For an operating

factor other than the curve basis, apply a correction factor to curve results. The application of the charts assumes operating at, or close to, optimum excess air values.

**Modifications for
Maximum Economy**

It is very difficult to pinpoint and rectify the major problem detracting from efficiency, as there are many interdependent variables and numerous pieces of equipment required for the combustion of coal. Further testing and/or operating manipulations may be required to "zero in" on a solution. Modification(s) that have been instituted with a fair degree of success to reduce high unburned carbon losses and/or high-excess-air operation are:

Modifying or changing the pulverizer internals to increase the coal fineness and thereby enhance combustion characteristics.

Rerouting or modifying air/coal distribution piping to avoid coal hang-up and slug flow going to the burners.

Installing additional or new combustion controls to smooth out and maintain consistent performance.

Purchasing new coal feeders compatible with and responsive to unit demand fluctuations.

Calibrating air flow and metering devices to ensure correct air/coal mixtures and velocities at the burner throats.

Installing new classifiers to ensure that proper coal fines reach the burners for combustion. Optimally setting air register positions for proper air/fuel mixing and combustion. Replacing worn and abraded burner impeller plates.

Increasing the air/coal mixture temperature exiting the pulverizers to ensure good ignition without coking.

Cleaning the burning throat areas of deposits.

Installing turning vanes or air foils in the secondary air supply duct or air plenum to ensure even distribution and proper air/fuel mixing at each burner.

Purchasing new (or updating existing) combustion controls to reflect the present state of the art.

As can be seen, the solutions can be varied—simple or complex, relatively cheap or quite expensive.

The incentives to correct the problem(s) are offered in Figure 5-18. Compare the expected benefits to the boiler manufacturer's solution and cost to implement and judge the merits.

Guidelines for Day-to-Day Operation

To a great extent, the items listed which detract from operating performance should be those checked on a day-to-day basis. A checklist should be initiated and developed by the plant to correct those conditions that do not require a shutdown, such as:

1. Maintaining integrity of distribution dampers and air registers with defined positions for operating load.

2. Ensuring pulverizer exit temperature by maintaining air calibration devices and air-moving equipment.

3. Checking feeder speeds and coal hopper valves, ensuring an even and steady coal feed.

4. Frequent blending of the raw coal pile to provide some measure of uniformity.

Coal Conversion

Converting existing boilers that originally were designed to fire oil and/or natural gas to coal capability is possible for certain types of units, provided that significant deratings in steam rate capacities are acceptable to existing operations. Since boiler installations are somewhat custom in design, fuel changes should be addressed on an individual basis. In general, however, conversion of field-erected units to coal firing is technically possible, whereas conversion of shop-assembled boilers is not.

Modifications required to convert shop-assembled units would include installation of ash-handling facilities in the furnace and convection pass, soot-blowing equipment, and modifications to tube spacing (each of which is almost physically impossible to do and prohibitive in cost). In addition, even if these modifications could be made, derating of the unit by about two-thirds would be necessary. Alternatives for consideration, rather than converting existing shop-assembled units would be to replace with new coal-fired units or to carefully assess the application of an alternative fuel, such as a coal/oil mixture or ultra-fine coal.

Many field-erected boilers can be converted to coal firing without serious compromises being made to good design. The most serious drawback to converting these boilers is the necessity to downrate by 40 to 50%. Although in some instances downrating may be acceptable, most operating locations cannot tolerate losses in steam supply of this magnitude.

Why is downrating necessary? Generally, there are differences between boilers designed for oil and gas and those designed for coal. Burning of coal requires significantly more combustion volume than that needed for oil or gas, due to flame characteristics and the need to avoid excessive slagging and fouling of heat-transfer surfaces. This means that a boiler originally designed for oil or gas must be downrated to maintain heat-release rates and firebox exit temperatures within acceptable limits for coal. Other factors influencing downrating are acceptable flue-gas velocities to minimize erosion and tube spacing.

A suggested approach for assuring the viability of coal conversion is:

1. Contact the original boiler manufacturer and determine via a complete inspection the actual modifications that would be required for the boiler.

2. Determine if the site can accommodate coal storage and other associated equipment, including unloading facilities, conveyors, dust abatement, ash disposal, and so on.

3. Select the type of coal to burn, considering cost, availability, and pollution restrictions. Before final selection of a coal, it must be analyzed to determine its acceptability and effect on unit rating. For any installation, final selection must be based on the coal's heating value; moisture content; mineral matter content; ash fusion and chemical characteristics; and/or pulverizers' grindability.

4. Assess over economics.

Bibliography

American Society of Mechanical Engineers (ASME), ASME Power Test Code (PTC) 4-1998; Fired Steam Generators, ASME, New York, 1964.

American Society of Heating, Refrigerating, and Air Conditioning Engineers (ASHRAE), 2008 HVAC Systems and Equipment Handbook, Chapter 31 Boilers, ASHRAE, Atlanta, GA., 2008.

D'Antonio, P.C., "Selecting Small Condensing Boilers," *HPAC Magazine*, 2006.

D'Antonio, P.C., "Maximizing Small-Boiler Efficiency," *HPAC Magazine*, 2006.

Ayne, F.W., *Efficient Boiler Operations Source Book*, 3rd ed., Fairmont, Lilburn, GA., 1991.

Consortium for Energy Efficiency (CEE), A Market Assessment for Condensing Boilers in Commercial Heating Applications. Boston, MA., 2001.

Keenan, J.H., and J. Kaye, *The Gas Tables*, Wiley, New York, 1948.

Pilaar, N., "Features, Advantages of Condensing Boilers," *HPAC Magazine*, 2007.

Scollon, R.B., and R.D. Smith, *Energy Conservation Guidebook*, Allied Corporation, Morristown, N.J., 1976

Taplin Jr., H.R., *Combustion Efficiency Tables*, Fairmont, Lilburn, GA., 1991.

CHAPTER 6
STEAM AND CONDENSATE SYSTEMS

PHILIP S. SCHMIDT
Department of Mechanical Engineering
University of Texas at Austin
Austin, Texas

6. 1 INTRODUCTION

Nearly half of the energy used by industry goes into the production of process steam, approximately the same total energy usage as that required to heat all the homes and commercial buildings in America. Why is so much of our energy resources expended for the generation of industrial steam?

Steam is one of the most abundant, least expensive, and most effective heat-transfer media obtainable. Water is found everywhere and requires relatively little modification from its raw state to make it directly usable in process equipment. During boiling and condensation, if the pressure is held constant, and both water and steam are present, the temperature also remains constant. Further, the temperature is uniquely fixed by the pressure, and hence by maintaining constant pressure, which is a relatively easy parameter to control, excellent control of process temperature can also be maintained. The conversion of a liquid to a vapor absorbs large quantities of heat in each pound of water. The resulting steam is easy to transport, and because it is so energetic, relatively small quantities of it can move large amounts of heat. This means that relatively inexpensive pumping and piping can be used, compared to that needed for other heating media.

Finally, the process of heat transfer by condensation, in the jacket of a steam-heated vessel, for example, is extremely efficient. High rates of heat transfer can be obtained with relatively small equipment, saving both space and capital. For these reasons, steam is widely used as the heating medium in thousands of industries.

Prior to the mid-1970s, many steam systems in industry were relatively energy wasteful. This was not necessarily bad design for the times, because energy was so cheap that it was logical to save first cost, even at the expense of a considerable increase in energy requirements. But things have changed, and today it makes good sense to explore every possibility for improving the energy efficiency of steam systems.

This chapter will introduce some of the language commonly used in dealing with steam systems, and it will help energy managers both define basic design constraints and evaluate the applicability of some of the vast array of manufacturer's data and literature that appear in the marketplace. It will discuss the various factors that produce inefficiency in steam system operations and some of the measures that can be taken to improve this situation. Simple calculation methods will be introduced to estimate the quantities of energy that may be lost and may be partially recoverable by the implementation of energy conservation measures. Some important energy conservation areas pertinent to steam systems are covered in other chapters and will not be repeated here. In particular, the reader is referred to Chapters 5 (boilers), 7 (cogeneration), and 15 (industrial insulation).

6.1.1 Components of Steam and Condensate Systems

Figure 6-1 shows a schematic of a typical steam system in an industrial plant. The boiler, or steam generator, produces steam at the highest pressure (and therefore the highest temperature) required by the various processes. This steam is carried from the boiler house through large steam mains to the process equipment area. There, transfer lines distribute the steam to each piece of equipment. If some processes require lower temperatures, the steam may be throttled to lower pressure through a pressure-regulating valve (designated PRV on the diagram) or through a back-pressure turbine. Steam traps located on the equipment allow condensate to drain back into the condensate return line, where it flows back to the condensate receiver. Steam traps also perform other functions, such as venting air from the system on startup, which is discussed in more detail later.

The system shown in Figure 6-1 is, of course, highly idealized. In addition to the components shown, other elements, such as strainers, check valves, and pumping traps may be utilized. In some plants, condensate may be simply released to a drain and not returned. (The potential for energy conservation through recovery of condensate will be discussed later.) Also, in the system shown, the flash steam produced in the process of throt-

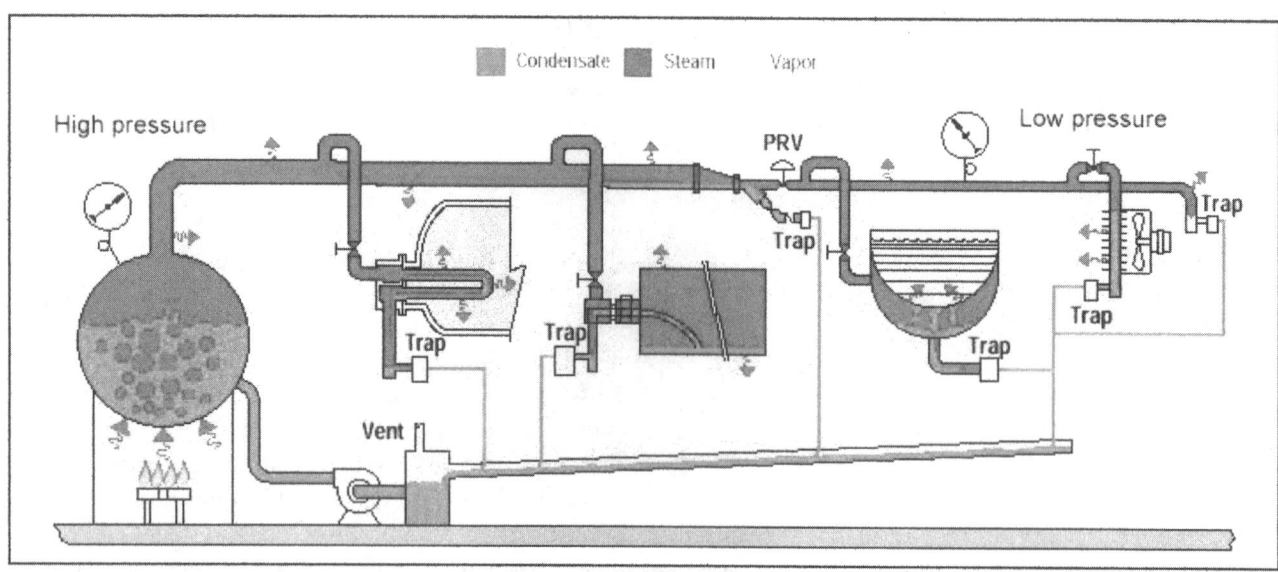

Figure 6-1. Typical steam system components.

tling across the steam traps is vented to the atmosphere. Prevention of this loss represents an excellent opportunity to save Btus and dollars.

6.1.2 Energy Conservation Opportunities in Steam Systems

Many opportunities for energy savings exist in steam system operations, ranging from simple operating procedure modifications to major retrofits requiring significant capital expenditures. Table 6-1 shows a checklist of energy conservation opportunities applicable to most steam systems. It is helpful for energy conservation managers to maintain such a running list applicable to their own situations. Ideas are frequently presented in the technical and trade literature, and plant operators often make valuable contributions since, after all, they are the people closest to the problem.

To "sell" such improvements to plant management and operating personnel, it is necessary to demonstrate the dollars-and-cents value of a project or operating change. The following sections discuss the thermal properties of steam, how to determine the steam requirements of plant equipment and estimate the amount of steam required to make up for system losses, how to assign a dollar value to this steam, and various approaches to alleviating these losses.

6.2 THERMAL PROPERTIES OF STEAM

6.2.1 Definitions and Terminology

Before discussing numerical calculations of steam properties for various applications, it is necessary to establish an understanding of some terms commonly used in the operation of steam systems.

British Thermal Unit (Btu). One Btu is the amount of heat required to raise 1 pound of water 1 degree Fahrenheit in temperature. To get a perspective on this quantity, a cubic foot of natural gas at atmospheric pressure will release about 1000 Btu when burned in a boiler with no losses. This same 1000 Btu will produce a little less than 1 pound of steam at atmospheric conditions, starting from tap water.

Boiling Point. The boiling point is the temperature at which water begins to boil at any given pressure. The boiling point of water at sea-level atmospheric pressure is about 212°F. At high altitude, where the atmospheric pressure is lower, the boiling point is also lower. Conversely, the boiling point of water goes up with increasing pressure. In steam systems, we usually refer to the boiling point by another term, "saturation temperature."

Absolute and Gauge Pressure. In steam system literature, we frequently see two different pressures used. The "absolute pressure," designated psia, is the true force per unit of area (e.g., pounds per square inch) exerted by the steam on the wall of the pipe or vessel containing it. We usually measure pressures, however, with sensing devices that are exposed to the atmosphere outside, and which therefore register an indication, not of the true force inside the vessel, but of the difference between that force and the force exerted by the outside atmosphere. We call this difference the "gauge pressure," designated psig. Since atmospheric pressure at

Table 6-1. Checklist of Energy Conservation Opportunities in Steam and Condensate Systems

General Operations

1. Review operation of long steam lines to remote single-service applications. Consider relocation or conversion of remote equipment, such as steam-heated storage tanks.
2. Review operation of steam systems used only for occasional services, such as winter-only steam tracing lines. Consider use of automatic controls, such as temperature-controlled valves, to assure that the systems are used only when needed.
3. Implement a regular steam leak survey and repair program.
4. Publicize to operators and plant maintenance personnel the annual cost of steam leaks and unnecessary equipment operations.
5. Establish a regular steam-use monitoring program, normalized to production rate, to track progress in reduction of steam consumption. Publicize on a monthly basis the results of this monitoring effort.
6. Consider revision of the plant-wide steam balance in multipressure systems to eliminate venting of low-pressure steam. For example, provide electrical backup for currently steam-driven pumps or compressors to permit shutoff of turbines when excess low-pressure steam exists.
7. Check actual steam usage in various operations against theoretical or design requirement. Where significant disparities exist, determine the cause and correct it.
8. Review pressure-level requirements of steam-driven mechanical equipment to evaluate feasibility of using lower pressure levels.
9. Review temperature requirements of heated storage vessels and reduce to minimum acceptable temperatures.
10. Evaluate production scheduling of batch operations and revise if possible to minimize startups and shutdowns.

Steam Trapping

1. Check sizing of all steam traps to assure they are adequately rated to provide proper condensate drainage. Also review types of traps in various services to assure that the most efficient trap is being used for each application.
2. Implement a regular steam trap survey and maintenance program. Train maintenance personnel in techniques for diagnosing trap failure.

Condensate Recovery

1. Survey condensate sources presently being discharged to waste drains for feasibility of condensate recovery.
2. Where condensate is already being returned, evaluate whether % being recovered can be increased.
3. Consider opportunities for flash steam utilization in low-temperature processes presently using first-generation steam.
4. Consider pressurizing atmospheric condensate return systems to minimize flash losses.
5. Repair leaks in condensate return lines.

Mechanical Drive Turbines

1. Review mechanical drive standby turbines presently left in the idling mode and consider the feasibility of shutting down standby turbines.
2. Implement a steam turbine performance testing program and clean turbines on a regular basis to maximize efficiency.
3. Evaluate the potential for cogeneration in multipressure steam systems presently using large pressure-reducing valves.

Insulation

1. Survey surface temperatures using infrared thermometry (or thermography on insulated equipment and piping) to locate areas of insulation deterioration. Maintain insulation on a regular basis.
2. Evaluate insulation of all uninsulated lines and fittings previously thought to be uneconomic. Recent rises in energy costs have made insulation of valves, flanges, and small lines desirable in many cases where this was previously unattractive.
3. Survey the economics of retrofitting additional insulation on presently insulated lines, and upgrade insulation if economically feasible.

sea level is usually around 14.7 psi, we can obtain the absolute pressure by simply adding 14.7 to the gauge pressure reading. In tables of steam properties, it is more common to see pressures listed in psia, and hence it is necessary to make the appropriate correction to the pressure indicated on a gauge.

Saturated and Superheated Steam. If we put cold water into a boiler and heat it, its temperature will begin to rise until it reaches the boiling point. If we continue to heat the water, rather than continuing to rise in temperature, it begins to boil and produce steam. As long as the pressure remains constant, the temperature will remain at the saturation temperature for the given pressure, and the more heat we add, the more liquid will be converted to steam. We call this boiling liquid a "saturated liquid" and refer to the steam so generated as "saturated vapor." Continuing to add more and more heat will simply generate more saturated vapor (or simply "saturated steam") until the water is completely boiled off. At this point, if we continue to add heat, the steam temperature will begin to rise once more. We call this "superheated steam." This chapter will concentrate on the behavior of saturated steam, because it is the steam condition most commonly encountered in industrial process heating applications. Superheated steam is common in power generation and is often produced in industrial systems when cogeneration of power and process heat is used.

Sensible and Latent Heat. Heat input that is directly registered as a change in temperature of a substance is called "sensible heat," for the simple reason that we can, in fact, "sense" it with our sense of touch or with a thermometer. For example, the heating of the water mentioned above before it reaches the boiling point would be sensible heating. When the heat goes into the conversion of a liquid to a vapor in boiling, or vice versa in the process of condensation, it is termed "latent heat." Thus, when a pound of steam condenses on a heater surface to produce a pound of saturated liquid at the same temperature, we say that it has released its latent heat. If the condensate cools further, it is said to be releasing sensible heat.

Enthalpy. The total energy content of a flowing medium, usually expressed in Btu/lb, is termed its "enthalpy." The enthalpy of steam at any given condition takes into account both latent and sensible heat, and also the "mechanical" energy content reflected in its pressure. Hence, steam at 500 psia and 600°F will have a higher enthalpy than steam at the same temperature but at 300 psia. Also, saturated steam at any temperature

and pressure has a higher enthalpy than condensate at the same conditions, due to the latent heat content of the steam. Enthalpy, as listed in tables of steam properties, does not include the kinetic energy of motion, but this component is insignificant in most energy conservation applications.

Specific Volume. The specific volume of a substance is the amount of space (e.g., cubic feet) occupied by 1 pound of the substance. This term will become important in some of our later discussions, because steam normally occupies a much greater volume for a given mass than water (i.e., it has a much greater specific volume), and this must be taken into account when considering the design of condensate return systems.

Condensate. Condensate is the liquid produced when steam condenses on a heater surface. As shown later, this condensate still contains a significant fraction of its energy, and it can be returned to the boiler to conserve fuel.

Flash Steam. When hot condensate at its saturation temperature (corresponding to the elevated pressure) in a heating vessel rapidly drops in pressure, as, for example, when passing through a steam trap or a valve, it suddenly finds itself at a temperature above the saturation temperature for the new pressure. Steam is thus generated which absorbs sufficient energy to drop the temperature of the condensate to the appropriate saturation level. This is called "flash steam," and the pressure-reduction process is called "flashing." In many condensate return systems, flash steam is simply released to the atmosphere, but it may, in fact, have practical applications in energy conservation.

Boiler Efficiency. The boiler efficiency is the percentage of energy released in the burning of fuel in a boiler that actually goes into the production of steam. The remaining percentage is lost through radiation from the boiler surfaces, blowdown of the boiler water to maintain satisfactory impurity levels, and loss of the hot flue gas up the stack. Although this chapter does go into detail on the subject of boiler efficiency, which is discussed in Chapter 5, it is important to recognize that this parameter relates the energy savings obtainable by conserving steam to the fuel savings obtainable at the boiler, a relation of obvious economic importance. Thus if we save 100 Btu of steam energy and have a boiler with an efficiency of 80%, the actual fuel energy saved would be 100/0.80, or 125 Btu. Because boilers always have an efficiency of less than 100% (more commonly

Table 6-2. Thermodynamic Properties of Saturated Steam

(1) Gauge Pressure	(2) Absolute Pressure (psia)	(3) Steam Temp. (°F)	(4) Enthalpy of Sat. Liquid (Btu/lb)	(5) Latent Heat (Btu/lb)	(6) Enthalpy of Steam (Btu/lb)	(7) Specific Volume (ft³/lb)
In. vacuum						
29.743	0.08854	32.00	0.00	1075.8	1075.8	3306.00
29.515	0.2	53.14	21.21	1063.8	1085.0	1526.00
27.886	1.0	101.74	69.70	1036.3	1106.0	333.60
19.742	5.0	162.24	130.13	1001.0	1131.1	73.52
9.562	10.0	193.21	161.17	982.1	1143.3	38.42
7.536	11.0	197.75	165.73	979.3	1145.0	35.14
5.490	12.0	201.96	169.96	976.6	1146.6	32.40
3.454	13.0	205.88	173.91	974.2	1148.1	30.06
1.418	14.0	209.56	177.61	971.9	1149.5	28.04
Psig						
0.0	14.696	212.00	180.07	970.3	1150.4	26.80
1.3	16.0	216.32	184.42	967.6	1152.0	24.75
2.3	17.0	219.44	187.56	965.5	1153.1	23.39
5.3	20.0	227.96	196.16	960.1	1156.3	20.09
10.3	25.0	240.07	208.42	952.1	1160.6	16.30
15.3	30.0	250.33	218.82	945.3	1164.1	13.75
20.3	35.0	259.28	227.91	939.2	1167.1	11.90
25.3	40.0	267.25	236.03	933.7	1169.7	10.50
30.3	45.0	274.44	243.36	928.6	1172.0	9.40
40.3	55.0	287.07	256.30	919.6	1175.9	7.79
50.3	65.0	297.97	267.50	911.6	1179.1	6.66
60.3	75.0	307.60	277.43	904.5	1181.9	5.82
70.3	85.0	316.25	286.39	897.8	1184.2	5.17
80.3	95.0	324.12	294.56	891.7	1186.2	4.65
90.3	105.0	331.36	302.10	886.0	1188.1	4.23
100.0	114.7	337.90	308.80	880.0	1188.8	3.88
110.3	125.0	344.33	315.68	875.4	1191.1	3.59
120.3	135.0	350.21	321.85	870.6	1192.4	3.33
125.3	140.0	353.02	324.82	868.2	1193.0	3.22
130.3	145.0	355.76	327.70	865.8	1193.5	3.11
140.3	155.0	360.50	333.24	861.3	1194.6	2.92
150.3	165.0	365.99	338.53	857.1	1195.6	2.75
160.3	175.0	370.75	343.57	852.8	1196.5	2.60
180.3	195.0	379.67	353.10	844.9	1198.0	2.34
200.3	215.0	387.89	361.91	837.4	1199.3	2.13
225.3	240.0	397.37	372.12	828.5	1200.6	1.92
250.3	265.0	406.11	381.60	820.1	1201.7	1.74
	300.0	417.33	393.84	809.0	1202.8	1.54
	400.0	444.59	424.00	780.5	1204.5	1.16
	450.0	456.28	437.20	767.4	1204.6	1.03
	500.0	467.01	449.40	755.0	1204.4	0.93
	600.0	486.21	471.60	731.6	1203.2	0.77
	900.0	531.98	526.60	668.8	1195.4	0.50
	1200.0	567.22	571.70	611.7	1183.4	0.36
	1500.0	596.23	611.60	556.3	1167.9	0.28
	1700.0	613.15	636.30	519.6	1155.9	0.24
	2000.0	635.82	671.70	463.4	1135.1	0.19
	2500.0	668.13	730.60	360.5	1091.1	0.13
	2700.0	679.55	756.20	312.1	1068.3	0.11
	3206.2	705.40	902.70	0.0	902.7	0.05

What is the temperature inside the line? Coming down column (1) we find a pressure of 150, and moving over to column (3), we note that the corresponding steam temperature at this pressure is about 366°F.

Column (4) lists the enthalpy of the saturated liquid in Btu/lb of water. We can see at the head of the column that this enthalpy is designated as 0 at a temperature [column (3)] of 32°F.

around 75 to 80%) there is a built-in "amplifier" on any energy savings effected in the steam system.

6.2.2 Properties of Saturated Steam

In calculating the energy savings obtainable through various measures, it is important to understand the quantitative thermal properties of steam and condensate. Table 6-2 shows a typical compilation of the properties of saturated steam.

Columns 1 and 2 list various pressures, either in gauge (psig) or absolute (psia). Note that these two pressures always differ by about 15 psi (14.7 to be more precise). Remember that the former represents the pressure indicated on a normal pressure gauge, while the latter represents the true pressure inside the line. Column 3 shows the saturation temperature corresponding to each of these pressures. Note, for example, that at an absolute pressure of 14.696 (the normal pressure of the atmosphere at sea level), the saturation temperature is 212°F, the figure we are all familiar with. Suppose that we have a pressure of 150 psi indicated on the pressure gauge on a steam line. This is an arbitrary reference point, and therefore the heat indicated at any other temperature tells us the amount of heat added to raise the water from an initial value of 32°F to that temperature. For example, referring back to our 150 psig steam, the water contains about 338.5 Btu/lb; starting from 32°F, 10 lb of water would contain 10 times this number, or about 3385 Btu. We can also subtract one number from another in this column to find the amount of heat necessary to raise the water from one temperature to another. If the water started at 101.74°F, it would contain a heat of 69.7 Btu/lb, and to raise it from this temperature to 366°F would require 338.5—69.7, or about 268.8 Btu for each pound of water. Column (5) shows the latent heat content of a pound of steam for each pressure. For our 150-psig example, we can see that it takes about 857 Btu to convert each pound of saturated water into saturated steam. Note this is a much larger quantity than the heat content of the water alone, confirming the earlier observation that steam is a very effective carrier of heat; each pound can give up, in this case, 857 Btu when condensed on a surface back to saturated liquid. Column (6), the enthalpy of the saturated steam, represents simply the sum of columns (4) and (5) since each pound of steam contains both the latent heat required to vaporize the water and the sensible heat required to raise the water to the boiling point in the first place.

Column (7) shows the specific volume of the saturated steam at each pressure. Note that as the pressure increases, the steam is compressed; that is, it occupies less space per pound. 150-psig steam occupies only

2.75 ft^3/lb; if released to atmospheric pressure (0 psig) it would expand to nearly 10 times this volume. By comparison, saturated liquid at atmospheric pressure has a specific volume of only 0.017 ft^3/lb (not shown in the table), and it changes only a few percent over the entire pressure range of interest here. Thus, 1 lb of saturated liquid condensate at 212°F will expand more than 1600 times in volume when converting to a vapor. This illustrates that piping systems for the return of condensate from steam-heated equipment must be sized primarily to accommodate the large volume of flashed vapor, and that the volume occupied by the condensate itself is relatively small.

The steam tables can be a valuable tool in estimating energy savings, as illustrated in the following example.

Example: A 100-ft run of 6-in. steam piping carries saturated steam at 95 psig. Tables obtained from an insulation manufacturer indicate that the heat loss from this piping run is presently 110,000 Btu/hr. With proper insulation, the manufacturer's tables indicate that this loss could be reduced to 500 Btu/hr. How many pounds per hour of steam savings does this installation represent, and if the boiler is 80% efficient, what would be the resulting fuel savings?

From the insulation manufacturer's data, we can find the reduction in heat loss:

$$\text{heat-loss reduction} = 110,000 - 500 = 109,500 \text{ Btu/hr}$$

From Table 6-2, at 95 psig (halfway between 90 and 100), the total heat of the steam is about 1188.4 Btu/lb. The steam savings is therefore

$$\text{steam savings} = \frac{109,500 \text{ Btu/hr}}{1188.4 \text{ Btu/lb}}$$

Assume that condensate is returned to the boiler at around 212°F; thus, the condensate has a heat content of about 180 Btu/lb. The heat required to generate 95 psig steam from this condensate is 1188.4 − 180.0, or 1008.4 Btu/lb. If the boiler is 80% efficient, then

$$\text{fuel savings} = \frac{1008.4 \text{ Btu/lb} \times 92 \text{ lb/hr}}{0.80}$$

$$= \text{approximately 0.116 million Btu/hr}$$

6.2.3 Properties of Superheated Steam

If additional heat is added to saturated steam with no liquid remaining, it begins to superheat and the tem-

Table 6-3. Thermodynamic Properties of Superheated Steam

Abs. Press (psi)		Temperature (°F)														
		100	200	300	400	500	600	700	800	900	1000	1100	1200	1300	1400	1500
1	v	0.0161	392.5	452.3	511.9	571.5	631.1	690.7								
	h	68.00	1150.2	1195.7	1241.8	1288.6	1336.1	1384.5								
5	v	0.0161	78.14	90.24	102.24	114.21	126.15	138.08	150.01	161.94	173.86	185.78	197.70	209.62	221.53	233.45
	h	68.01	1148.6	1194.8	1241.3	1288.2	1335.9	1384.3	1433.6	1483.7	1534.7	1586.7	1639.6	1693.3	1748.0	1803.5
10	v	0.0161	38.84	44.98	51.03	57.04	63.03	69.00	74.98	80.94	86.91	92.87	98.84	104.80	110.76	116.72
	h	68.02	1146.6	1193.7	1240.6	1287.8	1335.5	1384.0	1433.4	1483.5	1534.6	1586.6	1639.5	1693.3	1747.9	1803.4
15	v	0.0161	0.0166	29.899	33.963	37.985	41.986	45.978	49.964	53.946	57.926	61.905	65.882	69.858	73.833	77.807
	h	68.04	168.09	1192.5	1239.9	1287.3	1335.2	1383.8	1433.2	1483.4	1534.5	1586.5	1639.4	1693.2	1747.8	1803.4
20	v	0.0161	0.0166	22.356	25.428	28.457	31.466	34.465	37.458	40.447	43.435	46.420	49.405	52.388	55.370	58.352
	h	68.05	168.11	1191.4	1239.2	1286.9	1334.9	1383.5	1432.9	1483.2	1534.3	1586.3	1639.3	1693.1	1747.8	1803.3
40	v	0.0161	0.0166	11.036	12.624	14.165	15.685	17.195	18.699	20.199	21.697	23.194	24.689	26.183	27.676	29.168
	h	68.10	168.15	1186.6	1236.4	1285.0	1333.6	1382.5	1432.1	1482.5	1533.7	1585.8	1638.8	1992.7	1747.5	1803.0
60	v	0.0161	0.0166	7.257	8.354	9.400	10.425	11.438	12.466	13.450	14.452	15.452	16.450	17.448	18.445	19.441
	h	68.15	168.20	1181.6	1233.5	1283.2	1332.3	1381.5	1431.3	1481.8	1533.2	1585.3	1638.4	1692.4	1747.1	1802.8
80	v	0.0161	0.0166	0.0175	6.218	7.018	7.794	8.560	9.319	10.075	10.829	11.581	12.331	13.081	13.829	14.577
	h	68.21	168.24	269.74	1230.5	1281.3	1330.9	1380.5	1430.5	1481.1	1532.6	1584.9	1638.0	1692.0	1746.8	1802.5
100	v	0.0161	0.0166	0.0175	4.935	5.558	6.216	6.833	7.443	8.050	8.655	9.258	9.860	10.460	11.060	11.659
	h	68.26	168.29	269.77	1227.4	1279.3	1329.6	1379.5	1429.7	1480.4	1532.0	1584.4	1637.6	1691.6	1746.5	1802.2
120	v	0.0161	0.0166	0.0175	4.0786	4.6341	5.1637	5.6831	6.1928	6.7006	7.2060	7.7096	8.2119	8.7130	9.2134	9.7130
	h	68.31	168.33	269.81	1224.1	1277.4	1328.1	1378.4	1428.8	1479.8	1531.4	1583.9	1637.1	1691.3	1746.2	1802.0
140	v	0.0161	0.0166	0.0175	3.4661	3.9526	4.4119	4.8585	5.2995	5.7364	6.1709	6.6036	7.0349	7.4652	7.8946	8.3233
	h	68.37	168.38	269.85	1220.8	1275.3	1326.8	1377.4	1428.0	1479.1	1530.8	1583.4	1636.7	1690.9	1745.9	1801.7
160	v	0.0161	0.0166	0.0175	3.0060	3.4413	3.8480	4.2420	4.6295	5.0132	5.3945	5.7741	6.1522	6.5293	6.9055	7.2811
	h	68.42	168.42	269.89	1217.4	1273.3	1325.4	1376.4	1427.2	1478.4	1530.3	1582.9	1636.3	1690.5	1745.6	1801.4
180	v	0.0161	0.0166	0.0174	2.6474	3.0433	3.4093	3.7621	4.1084	4.4505	4.7907	5.1289	5.4657	5.8014	6.1363	6.4704
	h	68.47	168.47	269.92	1213.8	1271.2	1324.0	1375.3	1426.3	1477.7	1529.7	1582.4	1635.9	1690.2	1745.3	1801.2
200	v	0.0161	0.0166	0.0174	2.3598	2.7247	3.0583	3.3783	3.6915	4.0008	4.3077	4.6128	4.9165	5.2191	5.5209	5.8219
	h	68.52	168.51	269.96	1210.1	1269.0	1322.6	1374.3	1425.5	1477.0	1529.1	1581.9	1635.4	1689.8	1745.0	1800.9
250	v	0.0161	0.0166	0.0174	0.0186	2.1504	2.4662	2.6872	2.9410	3.1909	3.4382	3.6837	3.9278	4.1709	4.4131	4.6546
	h	68.66	168.63	270.05	375.10	1263.5	1319.0	1371.6	1423.4	1475.3	1527.6	1580.6	1634.4	1688.9	1744.2	1800.2
300	v	0.0161	0.0166	0.0174	0.0186	1.7665	2.0044	2.2263	2.4407	2.6509	2.8585	3.0643	3.2688	3.4721	3.6746	3.8764
	h	68.79	168.74	270.14	375.15	1257.7	1315.2	1368.9	1421.3	1473.6	1526.2	1579.4	1633.3	1688.0	1743.4	1799.6
350	v	0.0161	0.0166	0.0174	0.0186	1.4913	1.7028	1.8970	2.0832	2.2652	2.4445	2.6219	2.7980	2.9730	3.1471	3.3205
	h	68.92	168.85	270.24	375.21	1251.5	1311.4	1366.2	1419.2	1471.8	1524.7	1578.2	1632.3	1687.1	1742.6	1798.9
400	v	0.0161	0.0166	0.0174	00162	1.2841	1.4763	1.6499	1.8151	1.9759	2.1339	2.2901	2.4450	2.5987	2.7515	2.9037
	h	69.05	168.97	270.33	375.27	1245.1	1307.4	1363.4	1417.0	1470.1	1523.3	1576.9	1631.2	1686.2	1741.9	1798.2
500	v	0.0161	0.0166	0.0174	0.0186	0.9919	1.1584	1.3037	1.4397	1.5708	1.6992	1.8256	1.9507	2.0746	2.1977	2.3200
	h	69.32	169.19	270.51	375.38	1231.2	1299.1	1357.7	1412.7	1466.6	1520.3	1574.4	1629.1	1684.4	1740.3	1796.9
600	v	0.0161	0.0166	0.0174	0.0186	0.7944	0.9456	1.0726	1.1892	1.3008	1.4093	1.5160	1.6211	1.7252	1.8284	1.9309
	h	69.58	169.42	270.70	375.49	1215.9	1290.3	1351.8	1408.3	1463.0	1517.4	1571.9	1627.0	1682.6	1738.8	1795.6
700	v	0.0161	0.0166	0.0174	0.0186	0.0204	0.7928	0.9072	1.0102	1.1078	1.2023	1.2948	1.3858	1.4757	1.5647	1.6530
	h	69.84	169.65	270.89	375.61	487.93	1281.0	1345.6	1403.7	1459.4	1514.4	1569.4	1624.8	1680.7	1737.2	1794.3
800	v	0.0161	0.0166	0.0174	0.0186	0.0204	0.6774	0.7828	0.8759	0.9631	1.0470	1.1289	1.2093	1.2885	1.3669	1.4446
	h	70.11	169.88	271.07	375.73	487.88	1271.1	1339.2	1399.1	1455.8	1511.4	1566.9	1622.7	1678.9	1736.0	1792.9
900	v	0.0161	0.0166	0.0174	0.0186	0.0204	0.5869	0.6858	0.7713	0.8504	0.9262	0.9998	1.0720	1.1430	1.2131	1.2825
	h	70.37	170.10	271.26	375.84	487.83	1260.6	1332.7	1394.4	1452.2	1508.5	1564.4	1620.6	1677.1	1734.1	1791.6
1000	v	0.0161	0.0166	0.0174	0.0186	0.0204	0.5137	0.6080	0.6875	0.7603	0.8295	0.8966	0.9622	1.0266	1.0901	1.1529
	h	70.63	170.33	271.44	375.96	487.79	1249.3	1325.9	1389.6	1448.5	1504.4	1561.9	1618.4	167~.3	1732.5	1790.3

perature will rise. Table 6-3 shows the thermodynamic properties of superheated steam. Unlike the saturated steam of Table 6-2, in which each pressure had only a single temperature associated with it (the saturation temperature), superheated steam may exist, for a given pressure, at any temperature above the saturation temperature. Thus the properties must be tabulated as a function of both temperature and pressure, rather than

pressure alone. With this exception, the values in the superheated steam table may be used exactly like those in Table 6-2.

Example: Suppose, in the preceding example, that the steam line is carrying superheated steam at 250 psia (235 psig) and 500°F. (Note that both temperature and pressure must be specified for superheated steam.) For the same reduction in heat loss (109,500 Btu/hr), how many pounds per hour of steam is saved?

From Table 6-3 the enthalpy of steam at 250 psia and 500°F is 1263.5 Btu/lb. Thus,

$$\text{steam savings} = \frac{109,500 \text{ Btu/hr}}{1263.5 \text{ Btu/lb}}$$

$$= 86.6 \text{ lb/hr}$$

Table 6-4. Orders of Magnitude of Convective Conductances

Heating Process	Order of Magnitude of h (Btu/hr ft² • F)
Free convection, air	1
Forced convection, air	5-10
Forced convection, water	250-1000
Condensation, steam	5000-10,000

6.2.4 Heat-Transfer Characteristics of Steam

As mentioned in Section 6.1, steam is one of the most effective heat-transfer media available. The rate of heat transfer from a fluid medium to a solid surface (such as the surface of a heat-exchanger tube or a jacketed heating vessel) can be expressed by Newton's law of cooling:

$$q = h(T_{f} - T_{s})$$

where q is the rate of heat transfer per unit of surface area (e.g., Btu/hr/ft²), h is a proportionality factor called the "convective conductance," T_f is the temperature of the fluid medium, and T_s is the temperature of the surface.

Table 6-4 shows the order of magnitudes of h for several heat-transfer media. Condensation of steam can be several times as effective as the flow of water over a surface for the transfer of heat, and it may be 1000 times more effective than a gaseous heating medium, such as air.

In a heat exchanger, the overall effectiveness must take into account the fluid resistance on both sides of the exchanger and the conduction of heat through the tube wall. These effects are generally lumped into a single "overall conductance," U, defined by the equation

$$q = U(T_{f1} - T_{f2})$$

where q is defined as before, U is the conductance, and T_{f1} and T_{f2} are the temperatures of the two fluids. In addition, there is a tendency for fluids to deposit "fouling layers" of crystalline, particulate, or organic matter on transfer surfaces, which further impede the flow of heat. This impediment is characterized by a "fouling resistance," which, for design purposes, is usually incorporated as an additional factor in determining the overall conductance.

Table 6-5 illustrates typical values of U (not including fouling) and the fouling resistance for exchangers employing steam on the shell side versus exchangers using a light organic liquid (such as a typical heat-transfer oil). The 30 to 50% higher U values for steam translate directly into a proportionate reduction in required heat-exchanger area for the same fluid temperatures. Furthermore, fouling resistances for the steam-heated exchangers are 50 to 100% lower than for a similar service using an organic heating medium since pure steam contains no contaminants to deposit on the exchanger surface. From the design standpoint, this means that the additional heating surface incorporated to allow for fouling need not be as great. From the operating viewpoint, it translates into energy conservation since more heat can be transferred per hour in the exchanger for the same fluid conditions, or the same heating duty can be met with a lower fluid temperature difference if the fouling resistance is lower.

Table 6-5. Comparison of Steam and Light Organics as Heat-Exchange Media

Shell-Side Fluid	Tube-Side Fluid	Typical U (Btu/hr ft² . °F)	Typical Fouling Resistance (hr ft² . °F/Btu)
Steam	Light organic liquid	135-190	0.001
Steam	Heavy organic liquid	45-80	0.002
Light organic liquid	Light organic liquid	100-130	0.002
Light organic liquid	Heavy organic liquid	35-70	0.003

6.3 ESTIMATING STEAM USAGE AND ITS VALUE

To properly assess the worth of energy conservation improvements in steam systems, it is first necessary to determine how much steam is actually required to carry out a desired process, how much energy is being wasted through various system losses, and the dollar value of these losses. Such information will be needed to determine the potential gains achievable with insulation, repair or improvement of steam traps, and condensate recovery systems.

6.3.1 Determining Steam Requirements

Several approaches can be used to determine process steam requirements. In order of increasing reliability, they include the use of steam consumption tables for typical equipment, detailed system energy balances, and direct measurement of steam and/or condensate flows. The choice of which method is to be used depends on how critical the steam-using process is to the plant's overall energy consumption and how the data are to be used. It is often helpful to normalize the total steam and energy use figures to a "per unit of output" basis. This a useful metric in that consumption estimates can easily be scaled as production rates change.

For applications in which a high degree of accuracy is not required, such as developing rough estimates of the distribution of energy within a plant, steam consumption tables have been developed for various kinds of process equipment. Table 6-6 shows steam consumption tables for a number of typical industrial and commercial applications. To illustrate, suppose that we wish to estimate the steam usage in a soft-drink bottling plant for washing 2000 bottles/min. From the table, we see that, typically, a bottle washer uses about 310 lb of 5-psig steam per hour for each 100 bottles/min of capacity. For a washer with a 2000-bottle/min capacity we would, of course, use about 20 times this value, or 6200 lb/hr of steam. Referring to Table 6-2, we see that 5-psig steam has a total enthalpy of about 1156 Btu/lb. The hourly heat usage of this machine, therefore, would be approximately 1156 × 6200, or a little over 7 million Btu/hr. Remember that this is the heat content of the steam itself, not the fuel heat content required at the boiler since the boiler efficiency has not yet been taken into account.

Note that most of the entries in Table 6-6 show the steam consumption "in use," not the peak steam consumption during all phases of operation. These figures are fairly reliable for equipment that operates on a more-or-less steady basis; however, they may be quite low for batch-processing operations or operations where the load on the equipment fluctuates significantly during its operation. For this reason, steam equipment manufacturers recommend that estimated steam consumption values be multiplied by a "factor of safety," typically between 2 and 5, to assure that the equipment will operate properly under peak-load conditions. This can be quite important from the standpoint of energy efficiency. For example, if a steam trap is sized for average load conditions only, during startup or heavy-load operations, condensate will tend to back up into the heating vessel, reducing the effective area for condensation and thus reducing its heating capacity. For steam traps and condensate return lines, the incremental cost of slight oversizing is small, and factors of safety are used to assure that the design is adequately conservative to guarantee rapid removal of the condensate. Gross oversizing of steam traps, however, can also cause excessive steam loss. (This point is discussed in more detail in the section on steam traps, where appropriate factors of safety for specific applications are given.)

A second, and generally more accurate, approach to estimating steam requirements is by direct energy-balance calculations on the process. A comprehensive discussion of energy balances is beyond the scope of this section; analysis of complex equipment should be undertaken by a specialist. It is, however, possible to determine simple energy balances on equipment involving the heating of a single product.

The energy-balance concept simply states that any energy put into a system with steam must be absorbed by the product and/or the equipment itself, dissipated to the environment, carried out with the product, or carried out in the condensate. Recall that the concepts of sensible and latent heat were discussed in Section 6.2, in the context of heat absorption by water in the production of steam. We can extend this concept to consider heat absorption of any material, such as the heating of air in a dryer or the evaporation of water in production of concentrated fruit juice.

The *sensible* heat requirement of any process is defined in terms of the "specific heat" of the material being heated. Table 6-7 gives specific heats for a number of common substances. The specific heat specifies the number of Btu required to raise 1 pound of a substance through a temperature rise of 1°F. Remember that, for water, we stated that 1 Btu was, by definition, the amount of heat required to raise 1 lb by 1°F. The specific heat of water, therefore, is exactly 1 (at least near normal ambient conditions). To see how the specific heat can be used to calculate steam requirements for the sensible heating of products, consider the following example.

Table 6-6. Typical Steam Consumption Rates for Industrial and Commercial Equipment

Type of Installation	Description	Typical Pressure (psig)	Steam Consumption in Use (lb/hr)
Bakeries	Dough-room trough, 8 ft long	10	4
	Oven, white bread, 120-ft^2 surface	10	29
Bottle washing	Soft drinks, per 100 bottles/min	5	310
Dairies	Pasteurizer, per 100 gal heated/20 min	15-75	232 (max)
Dishwashers	Dishwashing machine	15-20	60-70
Hospitals	Sterilizers, instrument, per 100 in.^{3}, approx.	40-50	3
	Sterilizers, water, per 10 gal, approx.	40-50	6
	Disinfecting ovens, double door, 50-100 ft^3, per 10 ft^3, approx.	40-50	21
Laundries	Steam irons, each	100	4
	Starch cooker, per 10 gal capacity	100	7
	Laundry presses, per 10-in. length, approx.	100	7
	Tumblers, 40 in., per 10-in. length, approx.	100	38
Plastic molding	Each 12-15 ft^2 platen surface	125	29
Paper manufacture	Corrugators, per 1,000 ft^2	175	29
	Wood pulp paper, per 100 lb of paper	50	372
Restaurants	Standard steam tables, per ft of length	5-20	36
	Steam-jacketed kettles, 25 gal of stock	5-20	29
	Steam-jacketed kettles, 60 gal of stock	5-20	58
	Warming ovens, per 20 ft^3	5-20	29
Silver mirroring	Average steam tables	5	102
Tire vulcanizing	Truck molds, large	100	87
	Passenger molds	100	29

Example: A paint dryer requires about 3000 cfm of 200°F air, which is heated in a steam-coil unit. How many pounds of 50-psig steam does this unit require per hour?

The density of air at temperatures of several hundred degrees or below is about 0.075 lb/ft^3. The number of pounds of air passing through the dryer is then

3000 ft^3/min × 60 min/hr × 0.075 lb/ft^3= 13,500 lb/hr

Suppose that the air enters the steam-coil unit at 70°F. Its temperature will then be raised by 200 − 70 = 130°F. From Table 6-7 the specific heat of air is 0.24 Btu/lb °F. The energy required to provide this temperature rise is, therefore,

13,500 lb/hr × 0.24 Btu/lb °F × 130°F = 421,200 Btu/hr

Employing the energy-balance principle, whatever energy is absorbed by the product (air) must be provided by an equal quantity of steam energy, less the energy contained in the condensate.

From Table 6-2 the total enthalpy per pound of 50-psig steam is about 1179.1 Btu, and the condensate (saturated liquid) has an enthalpy of 267.5 Btu/lb. Each pound of steam therefore gives up 1179.1 − 267.5, or 911.6 Btu (i.e., its latent heat) to the air. The steam required is, therefore,

$$\frac{\text{heat required by air per hour}}{\text{heat released per pound of steam}} = \frac{421,200}{911.6} = 462 \text{ lb/hr}$$

To illustrate how latent heat comes into play when considering the steam requirements of a process, consider another example, this time involving a steam-heat evaporator.

Example: A milk evaporator uses a steam-jacketed kettle in which milk is batch-processed at atmospheric pressure. The kettle has a 1500-lb per batch capacity. Milk is heated from a temperature of 80°F to 212°F, where 25% of its mass is then driven off as vapor. Determine the

Table 6-7. Specific Heats of Common Materials

Material	Btu/lb·°F	Material	Btu/lb·°F
		Solids	
Aluminum	0.22	Iron, cast	0.49
Asbestos	0.20	Lead	0.03
Cement, dry	0.37	Magnesium	0.25
Clay	0.22	Porcelain	0.26
Concrete, stone	0.19	Rubber	0.48
Concrete, cinder	0.18	Silver	0.06
Copper	0.09	Steel	0.12
Glass, common	0.20	Tin	0.05
Ice, 32°F	0.49	Wood	0.32-0.48
		Liquids	
Acetone	0.51	Milk	0.90
Alcohol, methyl, 60-70°F	0.60	Naphthalene	0.41
Ammonia, 104°F	1.16	Petroleum	0.51
Ethylene glycol	0.53	Soybean oil	0.47
Fuel oil, sp. gr. 86	0.45	Tomato juice	0.95
Glycerine	0.58	Water	1.00
		Gases	
		(Constant-Pressure Specific Heats)	
Acetone	0.35	Carbon dioxide	0.20
Air, dry, 32-392°F	0.24	Methane	0.59
Alcohol	0.45	Nitrogen	0.24
Ammonia	0.54	Oxygen	0.22

amount of 15-psig steam required per batch, not including the heating of the kettle itself.

We must first heat the milk from 80°F to 212°F (sensible heating) and then evaporate off 375 lb (0.25 × 1500) of water.

From Table 6-7, the specific heat of milk is 0.90 Btu/lb °F. The sensible heat requirement is, therefore,

$$0.90 \times 1500 \text{ lb} \times (212 - 80)°F = 178,200 \text{ Btu/batch}$$

In addition, we must provide the latent heat to vaporize 375 lb of water at 212°F. From Table 6-2, 970.3 Btu/lb is required. The total latent heat is, therefore,

$$375 \times 970.3 = 363,863 \text{ Btu/batch}$$

and the total heat input is

$$363,863 + 178,200 = 542,063 \text{ Btu/batch}$$

This heat must be supplied as the latent heat of 15-psig steam, which, from Table 6-2, is about 945 Btu/lb.

The total steam requirement is, then,

$$\frac{542,063 \text{ Btu/batch}}{945 \text{ Btu/lb}} = 574 \text{ lb of 15-psig steam per batch}$$

We could also determine the startup requirement to heat the steel kettle, if we could estimate its weight, using the specific heat of 0.12 Btu/lb °F for steel, as shown in Table 6-7.

6.3.2 Estimating Surface and Leakage Losses

In addition to the steam required to actually carry out a process, heat is lost through the surfaces of pipes, storage tanks, and jacketed heater surfaces, as well as through malfunctioning steam traps and leaks in flanges, valves, and other fittings. Estimation of these losses is important, because fixing them can often be the most cost-effective energy conservation measure available.

Figure 6-2 illustrates the annual heat loss, based on 24-hr/day, 365-day/yr operation, for bare steam lines at

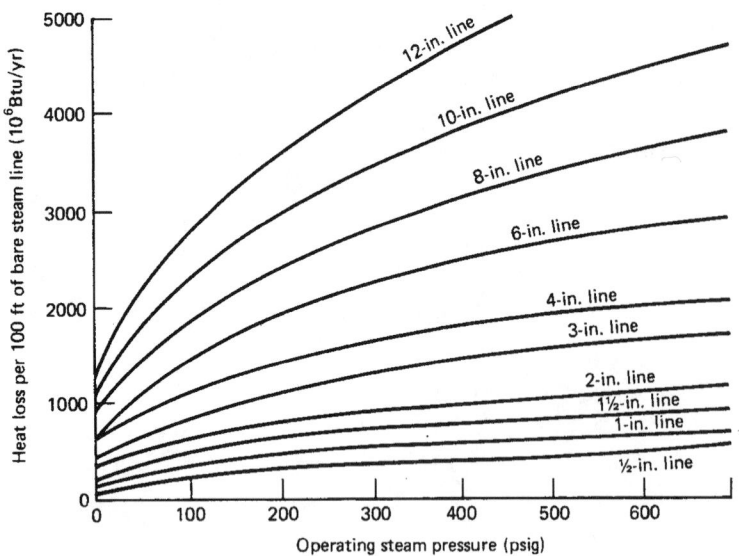

Figure 6-2. Heat loss from bare steam lines.

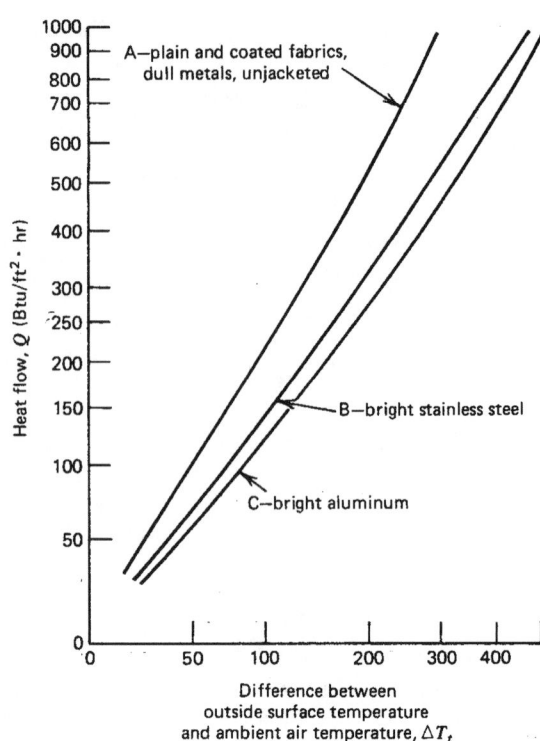

Figure 6-3. Heat losses from surfaces at elevated temperatures.

various pressures. The figure shows, for example, that a 100-ft run of 6-in. line operating at 100 psig will lose about 1400 million Btu/yr. The economic return on an insulation retrofit can easily be determined with price data obtained from an insulation contractor.

Figure 6-3 can be used to estimate heat losses from flat surfaces at elevated temperatures, or from already insulated piping runs for which the outside jacket surface temperature is known. The figure shows the heat flow per hour per square foot of exposed surface area as a function of the difference in temperature between the surface and the surrounding air. It will be noted that the nature of the surface significantly affects the magnitude of the heat loss. This is because thermal radiation, which is strongly dependent on the character of the radiating surface, plays an important role in heat loss at elevated temperature, as does convective heat loss to the air.

Another important source of energy loss in steam systems is leakage from components such as loose flanges, malfunctioning steam traps, and leaking condensate return lines. Figure 6-4 permits estimation of this loss of steam at various pressures leaking through holes. The heat losses are represented in million Btu/yr, based on full-time operation. Using the figure, we can see that a stuck-open steam trap with a 1/8-in. orifice would waste about 600 million Btu/yr of steam energy when leaking from a 100-psig line. This figure can also be used to estimate magnitudes of leakage from other sources of more complicated geometry. It is necessary to first determine an approximate area of leakage (in square inches) and then calculate the equivalent hole diameter represented by that area. The following ex-

ample illustrates this calculation.

Example: A flange on a 200-psig steam line has a leaking gasket. The maintenance crew, looking at the gasket, estimates that it is about 0.020 in. thick and is leaking from about 1/8-in. of the periphery of the flange. Estimate the annual heat loss in the steam if the line is operational 8000 hr/yr.

The area of the leak is a rectangle 0.020 in. wide and 1/8-in. in length:

$$\text{leak area} = 0.020 \times 1/8 = 0.0025 \text{ in.}^2$$

An equivalent circle will have an area of $\pi D_2/4$, so if $\pi D_2/4 = 0.0025$, then $D = 0.056$ in. From Figure 6-4, this leak, if occurring year-round (8760 hr), would waste about 200 million Btu/yr of steam energy. For actual operation

$$\text{heat lost} = \frac{8000}{8760} \times 200{,}000{,}000 = 182.6 \text{ million Btu/hr}$$

Since the boiler efficiency has not been considered, the actual fuel waste would be about 25% greater.

6.3.3 Measuring Steam and Condensate Rates

In maintaining steam systems at peak efficiency, it is often desirable to monitor the rate of steam flow through the system continuously, particularly at points

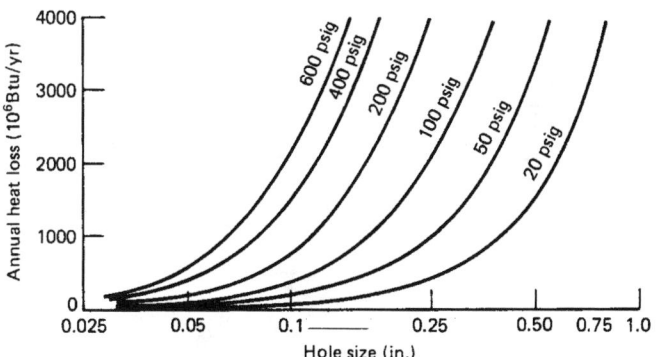

Figure 6-4. Heat loss from steam leaks.

Figure 6-5 Orifice flowmeter.

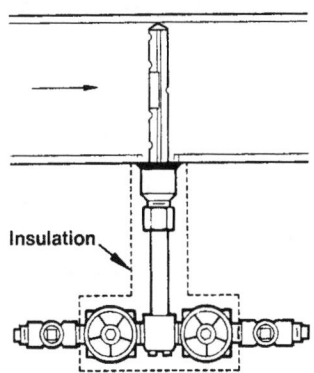

Figure 6-6. Annular averaging element.

of major usage, such as steam mains. Figure 6-5 shows one of the most common types of flowmetering devices, the calibrated orifice. This is a sharp-edged restriction that causes the steam flow to "neck down" and then re-expand after passing through the orifice. As the steam accelerates to pass through the restriction, its pressure drops, and this pressure drop, if measured, can be easily related to the flow rate. The calibrated orifice is one of a class of devices known as obstruction flowmeters, all of which work on the same principle of restricting the flow and producing a measurable pressure drop. Other types of obstruction meters are the ASME standard nozzle and the venturi. Orifices, although simple to manufacture and relatively easy to install (between flanges for example), are also subject to wear that causes them, over a period of time, to give unreliable readings. Nozzles and venturis, although more expensive initially, tend to be more resistant to erosion and wear and also produce less permanent pressure drop, once the steam re-expands to fill the pipe. With all of these devices, care must be exercised in installation since turbulence and flow irregularities produced by valves, elbows, and fittings immediately upstream of the obstruction will produce erroneous readings.

Figure 6-6 shows another type of flowmetering device used for steam, called an annular averaging element. The annular element is somewhat different in principle from the devices discussed above. It averages the pressure produced when steam impacts on the holes facing into the flow direction, and it subtracts from this average impact pressure a static pressure sensed by a tube facing downstream. As with obstruction-type flowmeters, the flow must be related to this pressure difference.

A device that does not utilize pressure drop for steam metering applications is the vortex shedding flowmeter, illustrated in Figure 6-7. A solid bar extends through the flow, and as steam flows around the bar, vortices are shed alternately from one side to

the other in its wake. As the vortices shift from side to side, the frequency of shedding can be detected with a thermal or magnetic detector, and this frequency varies directly with the rate of flow. The vortex shedding meter is quite rugged since the only function of the object extending into the flow stream is to provide an obstruction to generate vortices; hence it can be made of heavy-duty stainless steel. Also, vortex shedding meters tend to be relatively insensitive to variations in the steam properties since they produce a pulsed output rather than an analog signal.

The target flowmeter, not shown, is also suitable for some steam applications. This type of meter uses a "target," such as a small cylinder, mounted on the end of a metal strut that extends into the flow line. The strut is gauged to measure the force on the target, and if the properties of the fluid are accurately known, this force can be related to flow velocity. The target meter is especially useful when only intermittent measurements are needed, as the unit can be "hot-tapped" in a pipe through a ball valve and withdrawn when not in use. The requirement of accurate property data limits the usefulness of this type of meter in situations where steam conditions vary considerably, especially where high moisture is present.

The devices discussed above are useful in perma-

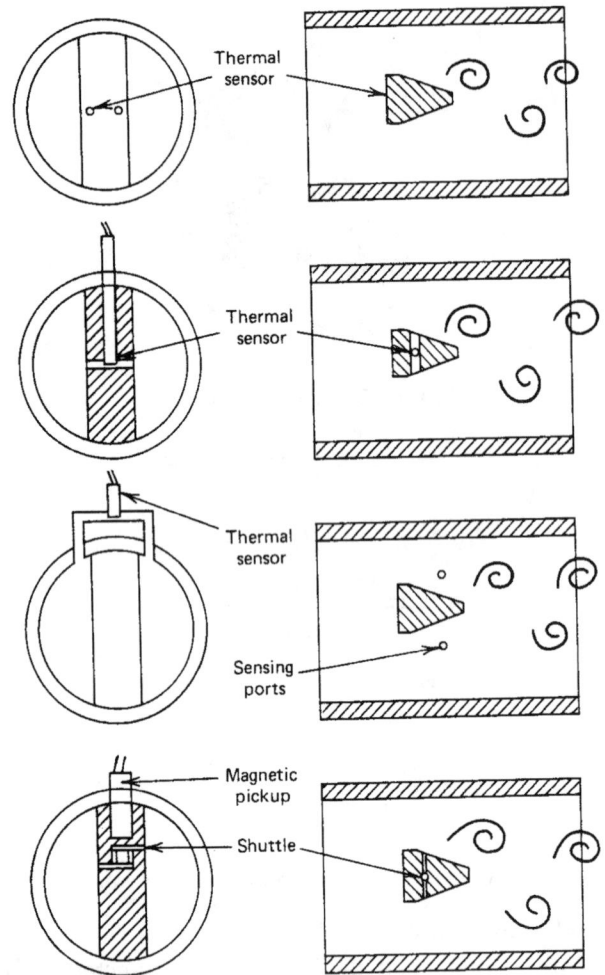

Figure 6-7. Vortex shedding flowmeter with various methods for sensing fluctuations.

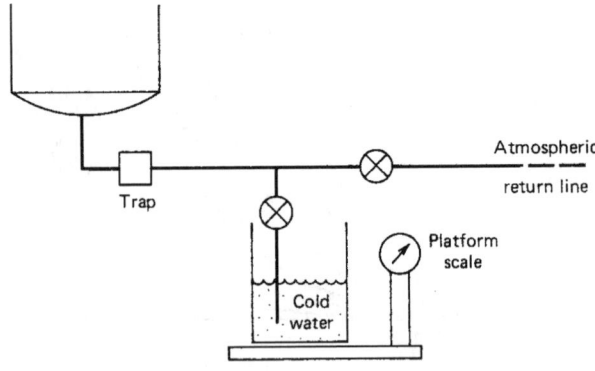

Figure 6-8.
Weigh bucket technique for condensate measurement.

nent installations where it is desired to continuously or periodically measure steam flow; there is no simple way to directly measure steam flow on a spotcheck basis without cutting into the system. There is, however, a relatively simple indirect method, illustrated in Figure 6-8, for determining the rate of steam usage in systems with unpressurized condensate return lines, or in open systems in which condensate is dumped to a drain. If a drain line is installed after the trap, condensate may be caught in a barrel and the weight of a sample measured over a given period of time. Precautions must be taken when using this technique to assure that the flash steam, generated when the condensate drops in pressure as it passes through the trap, does not bubble out of the barrel. This can represent both a safety hazard and an error in the measurements due to the loss of mass in vapor form. The barrel should be partially filled with cold water prior to the test so that flash steam will condense as it bubbles through the water. An energy balance can also be made on the water at

the beginning and end of the test by measuring its temperature, and with proper application of the steam tables, a check can be made to assure that the trap is not blowing through.

Figure 6-9 illustrates another instrument that can be used to monitor condensate flow on a regular basis, the rotameter. A rotameter indicates the flow rate of the liquid by the level of a specially shaped float, which rises in a calibrated glass tube such that its weight exactly balances the drag force of the flowing condensate.

Measurements of this type can be very useful in monitoring system performance since any unusual change in steam or condensate rate not associated with a corresponding change in production rate would tend to indicate an equipment malfunction, producing poor efficiency.

6.3.4 Computing the Dollar Value of Steam

In analyzing energy conservation measures for steam systems, it is important to establish a steam value in dollars per pound; this value will depend on the steam pressure, the boiler efficiency, and the price of fuel.

Steam may be valued from two points of view. The more common approach, termed the "enthalpy method," takes into account only the heating capability of the steam and is most

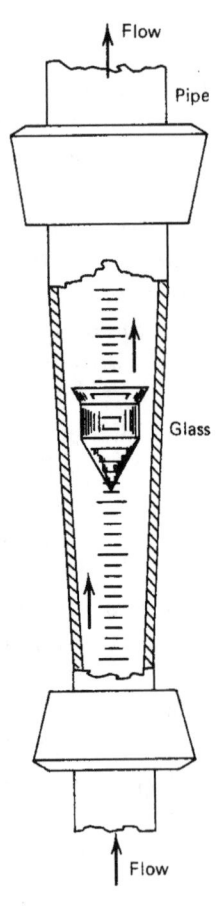

Figure 6-9.
Liquid rotameter.

appropriate when steam is used primarily for process heating. The enthalpy method can be illustrated with an example.

Example: An oil refinery produces 200-psig saturated steam in a large boiler, some of which is used directly in high-temperature processes, and some of which is let down to 30 psig through regulating valves for use at lower temperatures. The feedwater is added to the boiler at about 160°F. The boiler efficiency has been determined to be 82%, and boiler fuel is priced at $2.20 million Btu. Establish the values of 200-psig and 30-psig steam ($/lb).

Using the enthalpy method, we determine the increased heat content for each steam pressure required from the boiler if feedwater enters at 160°F. From Table 6-2:

total enthalpy of steam at 200 psig
    = 1199.3 Btu/lb
total enthalpy of steam at 30 psig (ignoring superheat
        after PRV)
    = 1172 Btu/lb
enthalpy of feedwater at 160°F
            = approx. 130 Btu/lb (about the same
            as saturated liquid at 160°F)
heat added per lb of 200-psig steam
    = 1199.3 – 130 = 1069.3 Btu/lb
heat added per lb of 30-psig steam
    = 1042 Btu/lb
fuel Btu required per pound at 200 psig
    = 1069.3/0.82 = 1304 Btu
fuel Btu required per lb at 30 psig
    = 1042/0.82 = 1271 Btu

With boiler fuel priced at $2.20/million Btu,
    or .22 per 1000 Btu:

value of 200-psig steam
        = 0.22 × 1.304 =.29/lb or $2.90/1000 lb
value of 30-psig steam
        = 0.22 × 1.271 =.28/lb or $2.80/1000 lb

Another approach to the valuation of steam is termed the "availability" or "entropy" method that takes into account not only the heat content of the steam, but also its power-producing potential if it were expanded through a steam turbine. This method is most applicable in plants where cogeneration (the sequential generation of power and use of steam for process heat) is practiced. The availability method involves some fairly complex thermodynamic reasoning and will not be covered here.

The reader should, however, be aware of its existence in analyzing the economics of cogeneration systems, where there is an interchangeability between purchased electric power and in-plant power and steam generation.

6.4 STEAM TRAPS AND THEIR APPLICATION

6.4.1 Functions of Steam Traps

Steam traps are important elements of steam and condensate systems, and they may represent a major energy conservation opportunity, or problem, as the case may be. The basic function of a steam trap is to allow condensate formed in the heating process to be drained from the equipment. This must be done speedily to prevent backup of condensate in the system.

Inefficient removal of condensate produces two adverse effects. First, if condensate is allowed to back up in the steam chamber, it cools below the steam temperature as it gives up sensible heat to the process and thereby reduces the effective potential for heat transfer. Since condensing steam is a much more effective heat-transfer medium than stagnant liquid, the area for condensation is reduced, and the efficiency of the heat-transfer process is deteriorated. This results in longer cycle times for batch processes, or lower throughput rates in continuous heating processes. In either case, inefficient condensate removal almost always increases the amount of energy required by the process.

A second reason for efficient removal of condensate is the avoidance of "water hammer" in steam systems. This phenomenon occurs when slugs of liquid become trapped between steam packets in a line. The steam, which has a much larger specific volume, can accelerate these slugs to high velocity, and when they impact an obstruction, such as a valve or an elbow, they produce an impact force not unlike hitting the element with a hammer (hence the term). Water hammer can be extremely damaging to equipment, and proper design of trapping systems to avoid it is necessary.

The second crucial function of a steam trap is to facilitate the removal of air from the steam space. Air can leak into the steam system when it is shut down, and some gas is always liberated from the water in the boiling process and carried through the steam lines. Air mixed with steam occupies some of the volume that would otherwise be filled by the steam itself. Each of these components, air and steam, contributes its share to the total pressure exerted in the system, as it is a fundamental thermodynamic principle that, in a mixture of gases, each component contributes to the pressure in the same proportion as its share of the volume of the space. For example, consider a steam system at 100 psia, with 10% of the volume air instead of steam. (Note that in this case it is necessary to use *absolute* pressures.) Therefore, from thermodynamics, 10% of the pressure, or 10 psia, is contributed by the air, and only 90%, or 90 psia, by the steam. Referring to Table 6-2, the corresponding steam temperature is between 316 and 324°F, or approximately 320°F. If the air were not present, the steam pressure would be 100 psia, corresponding to a temperature of about 328°F, so the presence of air in the system reduces the temperature for heat transfer. This means that more steam must be generated to do a given heating job. Table 6-8 shows the temperature reduction caused by the presence of air in various quantities at given pressures (shown in psig), indicating that the effective temperature may be seriously degraded.

In actual operation the situation is usually even worse than indicated in Table 6-8. We have considered the temperature reduction on the assumption that the air and steam are uniformly mixed. In fact, on a real heating surface, as air and steam move adjacent to the surface, the steam is condensed out into a liquid, while the air stays behind in the form of vapor. In the region very near the surface, therefore, the air occupies an even larger fraction of the volume than in the steam space as a whole, acting effectively as an insulating blanket on the surface. Suffice it to say that air is an undesirable parasite in steam systems, and its removal is important for proper operation.

Oxygen and carbon dioxide, in particular, have another adverse effect—corrosion in condensate and steam lines. Oxygen in condensate produces pitting or rusting of the surface, which can contaminate the water, making it undesirable as boiler feed; CO_2 in solution with water forms carbonic acid, which is highly corrosive to metallic surfaces. These components must be removed from the system, partially by good steam trapping and partially by proper deaeration of condensate, as is discussed in a subsequent section.

6.4.2 Types of Steam Traps and Their Selection

Various types of steam traps are available on the market, and the selection of the best trap for a given application is an important one. Many manufacturers produce several types of traps for specific applications, and manufacturers' representatives should be consulted in arriving at a choice. This section will give a brief introduction to the subject and comment on its relevance to improved energy utilization in steam systems.

Steam traps may be generally classified into three groups: mechanical, which work on the basis of the density difference between condensate and steam or air; thermostatic, which use the difference in temperature between steam (which stays close to its saturation temperature) and condensate (which cools rapidly); and thermodynamic, which functions on the difference in flow properties between liquids and vapors.

Figures 6.10 and 6.11 show two types of mechanical traps in common use for industrial applications. Figure 6-10 illustrates the principle of the "bucket trap. " In the trap illustrated, an inverted bucket is placed over the inlet line, inside an external chamber. The bucket is attached to a lever arm that opens and closes a valve as the bucket rises and falls in the chamber. As long as condensate flows through the system, the bucket has a negative buoyancy since liquid is present both inside and outside the bucket. The valve is open, and condensate is allowed to drain continuously to the return line. Steam entering the trap fills the bucket, displacing condensate, and the bucket rises, closing off the valve. Noncondensable gases, such as air and CO_2, bubble through a small vent hole and collect at the top of the trap (to be swept out with flash steam the next time the valve opens). Steam may also leak through the vent, but it is condensed on contact with the cool chamber walls and collects as condensate in the chamber. The vent hole is quite small, so the rate of steam loss through this leakage action is not excessive. As condensate again begins to enter the bucket, it loses buoyancy and begins to drop until the valve opens and again discharges condensate and trapped air.

Table 6-8. Temperature Reduction Caused by Air in Steam Systems

Pressure (psig)	Temp. of Steam, No Air Present	Temp. of Steam Mixed with Various Percentages of Air (by Volume)		
		10	20	30
10	240.1	234.3	228.0	220.9
25	267.3	261.0	254.1	246.4
50	298.0	291.0	283.5	275.1
75	320.3	312.9	204.8	295.9
100	338.1	330.3	321.8	312.4

The float-and-thermostatic (F&T) trap, illustrated in Figure 6-11, works on a similar principle. In this case, instead of a bucket, a buoyant float rises and falls in the chamber as condensate enters or is discharged. The float is attached to a valve, similar to the one

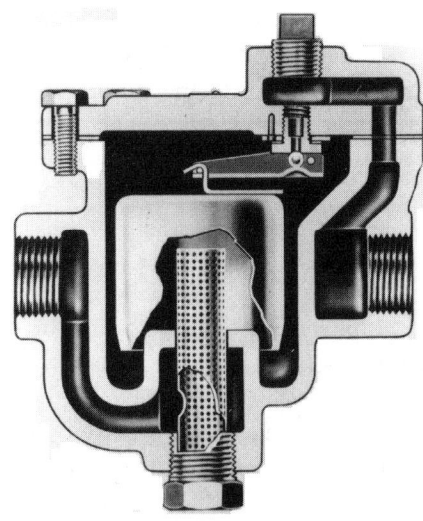

Figure 6-10. Inverted bucket trap.

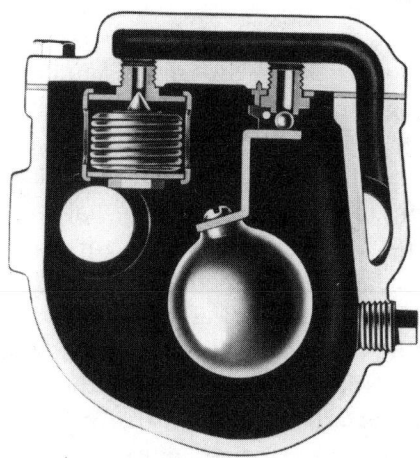

Figure 6-11. Float and thermostatic steam trap.

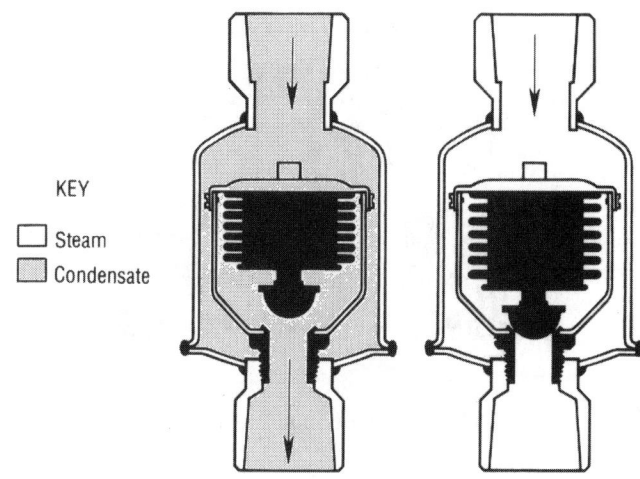

KEY

☐ Steam
☐ Condensate

Figure 6-12. Thermostatic steam trap.

in a bucket trap, which opens and closes as the ball rises and falls. Since there is no natural vent in this trap, and the ball cannot distinguish between air and steam (which have similar densities), special provision must be made to remove air and other gases from the system. This is usually done by incorporating a small thermostatically actuated valve in the top of the trap. At low temperature, the valve bellows contracts, opening the vent and allowing air to be discharged to the return line. When steam enters the chamber, the bellows expands, sealing the vent. Some float traps are also available without this thermostatic air-vent feature; external provision must then be provided to permit proper air removal from the system. The F&T-type trap permits continuous discharge of condensate, unlike the bucket trap, which is intermittent. This can be an advantage in certain applications.

Figure 6-12 illustrates a thermostatic steam trap. In this trap, a temperature-sensitive bellows expands and

contracts in response to the temperature of the fluid in the chamber surrounding the bellows. When condensate surrounds the bellows, it contracts, opening the drain port. As steam enters the chamber, the elevated temperature causes the bellows to expand and seal the drain. Since air also enters the chamber at a temperature lower than that of steam, the thermostatic trap is naturally self-venting; it is also a continuous-drain-type trap. The bellows in the trap can be partially filled with a fluid and sealed, such that an internal pressure is produced that counterbalances the external pressure imposed by the steam. This feature makes the bellows-type thermostatic trap somewhat self-compensating for variations in steam pressure. Another type of thermostatic trap, which uses a bimetallic element, is also available. This type of trap is not well-suited for applications in which significant variations in steam pressure might be expected since it is responsive only to temperature changes in the system.

The thermodynamic, or controlled disk steam trap is shown in Figure 6-13. This type of trap is very simple in construction and can be made quite compact and resistant to damage from water hammer. In a thermodynamic trap, a small disk covers the inlet orifice. Condensate or air, moving at relatively low velocity, lifts the disk off its seat and is passed through to the outlet drain. When steam enters the trap, it passes through at high velocity because of its large volume. As the steam passes through the space between the disk and its seat, it impacts on the walls of the control chamber to produce a rise in pressure. This pressure imbalance between the outside of the disk and the side facing the seat causes it to snap shut, sealing off the chamber and preventing the further passage of steam to the outlet. When condensate again enters the inlet side, the disk lifts off the seat and permits its release.

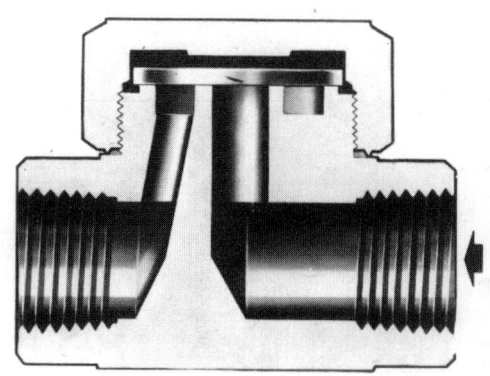

Figure 6-13. Disk or thermodynamic steam trap.

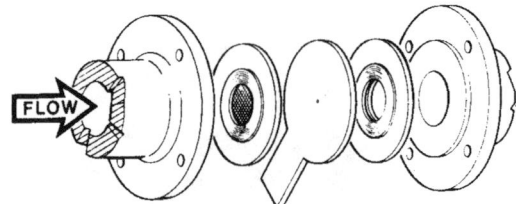

Figure 6-14. Drain orifice.

An alternative to conventional steam traps, the drain orifice is illustrated in Figure 6-14. This device consists simply of an obstruction to the flow of condensate, similar to the orifice flowmeter described in an earlier section but much smaller. This small hole allows the pressure in the steam system to force condensate to drain continuously into the lower-pressure return system. Obviously, if steam enters, rather than condensate, it will also pass through the orifice and be lost. The strategy of using drain orifices is to select an orifice size that permits condensate to drain at such a rate that live steam seldom enters the system. Even if steam does occasionally pass through, the small size of the orifice limits the steam leakage rate to a value much less than would be lost due to a "stuck-open" malfunction of one of the types of traps discussed above. Drain orifices can be successfully applied in systems that have a well-defined and relatively constant condensate load. They are not suited for use where condensate load may vary widely with operating conditions.

As mentioned above, a number of operating requirements must be taken into consideration in selecting the appropriate trap for a given application. Table 6-9 lists these application considerations and presents one manufacturer's ratings on the performance of the various traps discussed above. In selecting a trap for a given application, assistance from manufacturers' representatives should be obtained since a great body of experience in actual service has been accumulated over the years.

6.4.3 Considerations in Steam Trap Sizing

As mentioned earlier in this section, good energy conservation practice demands the efficient removal of condensate from process equipment. It is thus necessary to assure that traps are properly sized for the given condensate load. Grossly oversized traps waste steam by excessive surface heat loss and internal venting, while undersized traps permit accumulation of condensate with resultant loss in equipment heat-transfer effectiveness.

Steam traps are sized based in two specifications, the condensate load (e.g., in lbs/hr or gal/min) and the pressure differential across the trap (in psig). Section 6.3 discussed various methods for estimating condensate loads expected under normal operating conditions.

It is good practice to size the capacity of the trap based on this expected load times a factor of safety to account for peaks at startup and fluctuations in normal operating conditions. It is not unusual for startup condensate loads to be three to four times higher than steady operational loads, and in some applications they may range up to 10 times the steady-state load.

Table 6-10 presents typical factors of safety for condensate capacity recommended by steam trap manufacturers. This indicates typical ranges of factor of safety to consider in various applications. Although there is considerable variation in the recommended values, in both energy and economic terms the cost of oversizing is ordinarily not prohibitive, and conservative safety factors are usually used. The exception to this rule of thumb is in the sizing of disk-type traps, which may not function properly if loaded considerably below design. Drain orifices also must be sized close to normal operating loads. Again, the advice of the manufacturer should be solicited for the specific application in mind.

The other important design specification is the pressure differential over which the trap will operate. Since pressure is the driving force that moves condensate through the trap and on to the receiver, trap capacity will increase, for a given trap size, as the pressure increases. The trap operating-pressure differential is not simply the boiler pressure. On the upstream side of the trap, steam pressure may drop through valves and fittings and through heat-transfer passages in the process equipment. Thus, the appropriate upstream pressure is the pressure at the trap inlet, which to a reasonable approximation, can usually be considered to be the process steam pressure at the equipment. Back pressure on the outlet side of the trap must also be considered. This includes the receiver pressure (if the condensate

Table 6-9. Comparison of Steam Trap Characteristics

Characteristic	Inverted Bucket	Float & Thermostataic	Disk	Bellows Thermostatic
Method of operation	Intermittent	Continuous	Intermittent	Continuous[a]
Energy conservation (time in service)	Excellent	Good	Poor	Fair
Resistance to wear	Excellent	Good	Poor	Fair
Corrosion resistance	Excellent	Good	Excellent	Good
Resistance to Hydraulic shock	Excellent	Poor	Excellent	Poor
Vents air and CO_2 at steam temperature	Yes	No	No	No
Ability to vent air at very low pressure (1/4 psig)	Poor	Excellent	NR[b]	Good
Ability to handle start-up air loads	Fair	Excellent	Poor	Excellent
Operation against back pressure	Excellent	Excellent	Poor	Excellent
Resistance to damage from freezing[c]	Good	Poor	Good	Good
Ability to purge system	Excellent	Fair	Excellent	Good
Performance on very light loads	Excellent	Excellent	Poor	Excellent
Responsiveness to slugs of condensate	Immediate	Immediate	Delayed	Delayed
Ability to handle dirt	Excellent	Poor	Poor	Fair
Comparative physical size	Large[d]	Large	Small	Small
Ability to handle "flash steam"	Fair	Poor	Poor	Poor
Mechanical failure (open-closed)	Open	Closed	Open[e]	Closed[f]

[a]Can be intermittent on low load.
[b]Not recommended for low-pressure operations.
[c]Cast iron traps not recommended.
[d]In welded stainless steel construction—medium..
[e]Can fail closed due to dirt.
[f]Can fail open due to wear.

return system is pressurized), the pressure drop associated with flash steam and condensate flow through the return lines, and the head of water associated with risers if the trap is located at a point below the condensate receiver. Condensate return lines are usually sized for a given capacity to maintain a velocity no greater than 5000 ft/min of the flash steam. Table 6-11 shows the expected pressure drop per 100 ft of return line that can be expected under design conditions. Referring to the table, a 60-psig system, for example, returning condensate to an unpressurized receiver (0 psig) through a 2-in. line would have a return-line pressure drop of just under 1/2 psi/100-ft run, and the condensate capacity of the line would be about 2600 lb/hr. The pressure head produced by a vertical column of water is about 1 psi/2 ft of rise. These components can be summed to estimate the back pressure on the system, and the appropriate pressure for sizing the trap is then the difference between the upstream pressure and the back pressure.

Table 6-10. Typical Factors of Safety for Steam Traps (Condensate Flow Basis)

Application	Factor of Safety
Autoclaves	3-4
Blast coils	3-4
Dry cans	2-3
Dryers	3-4
Dry kilns	3-4
Fan system heating service	3-4
Greenhouse coils	3-4
Hospital equipment	2-3
Hot-water heaters	4-6
Kitchen equipment	2-3
Paper machines	3-4
Pipe coils (in still air)	3-4
Platen presses	2-3
Purifiers	3-4
Separators	3-4
Steam-jacketed kettles	4-5
Steam mains	3-4
Submerged surfaces	5-6
Tracer lines	2-3
Unit heaters	3-4

Table 6-11. Condensate Capacities and Pressure Drops for Return Lines[a]

Supply Press. (psig)	5	15		30			60				100					200					
Return Press. (Psig)	0	0	5	0	5	10	0	5	10	20	0	5	10	20	30	0	5	10	20	30	50
Pipe size (in.), schedule 40 pipe																					
1/2	1,425	590	1,335	360	640	1,065	235	370	535	1,010	180	270	370	615	955	115	165	215	325	450	760
	4.0	4.0	5.3	4.0	5.3	6.5	4.0	5.3	6.5	8.9	4.0	5.3	6.5	8.9	11.3	4.0	5.3	6.5	8.9	11.3	15.9
3/4	2,495	1,035	2,340	635	1,125	1,855	415	650	940	1 770	310	470	645	1,085	1,675	200	285	375	570	795	1,330
	2.35	2.35	3.14	2.35	3.14	3.88	2.35	3.14	3.88	5.32	2.35	3.14	3.88	5.32	6.72	2.35	3.14	3.88	5.32	6.72	9.40
1	4,045	1,880	3,790	1,030	1,820	3,005	670	1,055	1,520	2,865	505	765	1,045	1,755	2,715	325	465	605	925	1. 285	2,155
	1.53	1.53	2.04	1.53	2.04	2.51	1.53	2.04	2.51	3.44	1.63	2.04	2.51	3.44	4.36	1.53	2.04	2.51	3.44	4.36	6.15
1-1/4	7,000	2,905	6,565	1,780	3,150	5,200	1,155	1,830	2,635	4,960	875	1,320	1,810	3,035	4,695	560	800	1,050	1,600	2,225	3.735
	0.95	0.95	1.26	.95	1.26	1.55	.95	1.26	1.55	2.13	95	1.26	1.55	2.13	2.69	0.95	1.26	1.55	2.13	2.69	3.80
1-1/2	9,530	3,955	8,935	2,425	4,290	7.080	1,575	2,490	3,585	6,750	2,190	1,795	2,465	4,135	6,395	760	1,090	1,430	2,175	3,025	5,080
	0.73	0.73	0.97	0.73	0.97	1.20	0.73	0.97	1.20	1.64	0.73	0.97	1.20	1.64	2.07	0.73	0.97	1.20	1.64	2.07	2.93
2	15,710	6,525	14,725	3,995	7,070	11,670	2,595	4,105	5,910	11,125	1,985	2.960	4,060	6,810	10,540	1,255	1,800	2,355	3,585	4,990	8,375
	0.48	0.48	0.64	0.48	0.64	0.79	0.48	0.64	0.79	1.08	0.48	0.64	0.79	1.08	1.37	0.48	0.64	0.79	1.08	1.37	1.93
2-1/2	22,415	9,305	21,005	5,700	10,085	16,650	3,705	5,855	8,430	15,875	2,800	4,225	5,795	9,720	15,035	1,790	2,565	3,380	5,115	7,120	11,950
	0.36	0.36	0.48	0.36	0.48	0.59	0.36	0.48	0.69	0.81	0.36	0.48	0.59	0.81	1.03	0.36	0.48	0.59	0.91	1.03	1.45
3	34,610	14,370	32,435	8,800	15,570	25,710	5,720	9,045	13,020	24 515	4,325	6,525	8,950	15,005	23,220	2,765	3,965	5.185	7,900	10,990	18,450
	0.26	0.26	0.34	0.26	0.34	0.42	0.26	0.34	0.42	0.58	0.26	0.34	0.42	0.58	0.73	0.26	0.34	0.42	0.58	0.73	1.03
3-1/2	46,285	19,220	43,380	11,765	20,825	34,385	7,650	12,095	17,410	32,785	5,785	6,725	11,970	20,070	31,050	3,695	5,300	6,940	10,565	14,700	24,675
	0.21	0.21	0.27	0.21	0.27	0.34	0.21	0.27	0.34	0.46	0.21	0.27	0.34	0.46	0.59	0.21	0.27	0.314	0.46	0.59	0.83
4	59,595	24,745	55,855	15,150	26,815	44,275	9,850	15,575	22,415	42,210	7,450	11,235	15,410	25.840	39,960	4,780	6,825	8,935	13,600	18,925	31,770
	0.17	0.17	0.23	0.17	0.23	0.28	0.17	0.23	0.28	0.38	0.17	0.23	0.28	0.38	0.49	0.17	0.23	0.28	0.36	0.49	0.25
5	93,655	38,890	87,780	23,810	42,140	69,580	15,480	24,475	35,230	66,335	11,705	17.660	24.220	40,610	62,830	7,475	10,725	14,040	21,375	29,745	49,930
	0.12	0.12	0.16	0.12	0.16	0.20	0.12	0.16	0.20	0.05	0.12	0.16	0.20	0.05	0.17	0.12	0.16	0.20	0.05	0.17	0.11
6	135,245	58,160	126,760	34,385	60,855	100,480	22,350	35,345	50,875	95,795	16,905	25,500	34,975	58,645	90,735	10,800	15,490	20,270	30,865	42,950	72,105
	0.10	0.10	0.13	0.10	0.13	0.04	0.10	0.13	0.04	0.05	0.10	0.13	0.04	0.05	0.01	0.10	0.13	0.04	0.05	0.01	0.01
8	234,195	97,245	219,505	59,540	105,380	173,995	38,705	61,205	88,095	165,880	29,270	44,160	60,565	101,650	157,115	18,700	26,820	35,105	53,450	74,175	124,855
	0.02	0.02	0.02	0.02	0.02	0.01	0.02	0.02	0.01	0.01	0.02	0.02	0.01	0.01	0.01	0.02	0.02	0.01	0.01	0.01	0.01

[a]Return-line capacity (lb/hr) with pressure drop (psi) for 100 ft of pipe at a velocity of 5000 ft/min.

Table 6-12 shows a typical pressure-capacity table extracted from a manufacturer's catalog. The use of such a table can be illustrated with the following example.

Example: A steam-jacketed platen press in a plastic lamination operation uses about 500 lb/hr of 30-psig steam in normal operation. A 100-ft run of 1-in. pipe returns condensate to a receiver pressurized to 5 psig; the receiver is located 15 ft above the level of the trap. From the capacity differential pressure specifications in Table 6-12, select a suitable trap for this application.

Using a factor of safety of 3 from Table 6-10, a trap capable of handling $3 \times 500 = 1500$ lb/hr of condensate will be selected.

To determine the system back pressure, add the receiver pressure, the piping pressure drop, and the hydraulic head due to the elevation of the receiver.

Entering Table 6-11 at 30-psig supply pressure and 5-psig return pressure, the pipe pressure drop for a 1-in. pipe is just slightly over 2 psi at a condensate rate somewhat higher than our 1500 lb/hr; a 2-psi pressure drop is a reasonable estimate.

The hydraulic head due to the 15-ft riser is 2 psi/ft $\times 15 = 7.5$ psi. The total back pressure is, therefore,

Table 6-12. Typical Pressure-Capacity Specifications for Steam Traps

	Capacities (lb/hr)			
Model:	A	B	C	D
Differential pressure (psi)				
5	450	830	1600	2900
10	560	950	1900	3500
15	640	1060	2100	3900
20	680	880	1800	3500
25	460	950	1900	3800
30	500	1000	2050	4000
40	550	770	1700	3800
50	580	840	1 900	4100
60	635	900	2000	4400
70	660	950	2200	3800
80	690	800	1650	4000
100	640	860	1800	3600
125	680	950	2000	3900
150	570	810	1500	3500
200	—	860	1600	3200
250	—	760	1300	3500
300	—	510	1400	2700
400	—	590	1120	3100
450	—	—	1200	3200

5 psi (receiver) + 7.5 psi (riser) + 2 psi (pipe) = 14.5 psi

or the differential pressure driving the condensate flow through the trap is 30 – 14.5 = 15.5 psi.

From Table 6-12 we see that a Model C trap will handle 2100 lb/hr at 15-psi differential pressure; this would then be the correct choice.

6.4.4 Maintaining Steam Traps for Efficient Operation

Steam traps can and do malfunction in two ways. They may stick in the closed position, causing condensate to back up into the steam system, or they may stick open, allowing live steam to discharge into the condensate system. The former type of malfunction is usually quickly detectable since flooding of a process heater with condensate will usually so degrade its performance that the failure is soon evidence by a significant change in operating conditions. This type of failure can have disastrous effects on equipment by producing damaging water hammer and causing process streams to back up into other equipment. Because of these potential problems, steam traps are often designed to fail in the open position; for this reason, they are among the biggest energy wasters in an industrial plant. Broad experience in large process plants using thousands of steam traps has shown that typically from 15 to 60% of the traps in a plant may be blowing through, wasting enormous amounts of energy. Table 6-13 shows the cost of wasted 100-psig steam (typical of many process plant conditions) for leak diameters characteristic of steam trap orifices. At higher steam pressures, the leakage would be even greater; the loss rate does not go down in direct proportion at lower steam pressures but declines at a rate proportional to the square root of the pressure. For example, a 1/8-in. leak in a system at 60 psig, instead of the 100 psig shown in the table, would still waste over 75% of the steam rate shown (the square root of 60/100). The cost of wasted steam far outweighs the cost of proper maintenance to repair the malfunctions, and comprehensive steam trap maintenance programs have proven to be among the most attractive energy conservation investments available in large process plants. Most types of steam traps can be repaired, and some have inexpensive replaceable elements for rapid turnaround.

Table 6-13. Annual Cost of Steam Leaks

Leak Diameter (in.)	Steam Wasted per Month (lb)[a]	Cost per Month[b]	Cost per Year[b]
1/16	13,300	$40	$480
1/8	52,200	156	1,890
1/4	209,000	626	7,800
1/2	833,000	2,500	30,000

[a]Based on 100-psig differential pressure across the orifice.
[b]Based on steam value of $3/1000 lb. Cost will scale in direct proportion for other steam values.

A major problem facing the energy conservation manager is diagnosis of open traps. The fact that a trap is blowing through can often be detected by a rise in temperature at the condensate receiver, and it is quite easy to monitor this simple parameter. There are also several direct methods for checking trap operation. Figure 6-15 shows the simplest approach for open condensate systems where traps drain directly to atmospheric pressure. In proper normal operation, a stream of condensate drains from the line, together with a lazy cloud of flash steam produced as the condensate throttles across the trap. When the trap is blow-

Table 6-14. Operating Sounds of Various Types of Steam Traps

Trap	Proper Operation	Malfunctioning
Disk type (impulse of thermodynamic)	Opening and snap-closing of disk several times per minute	Rapid chattering of disk as steam blows through
Mechanical type (bucket)	Cycling sound of the bucket as it opens and closes	Fails open—sound of steam blowing through Fails closed—no sound
Thermostatic type	Sound of periodic discharge if medium to high load; possibly no sound if light load; throttled discharge	Fails closed—no sound

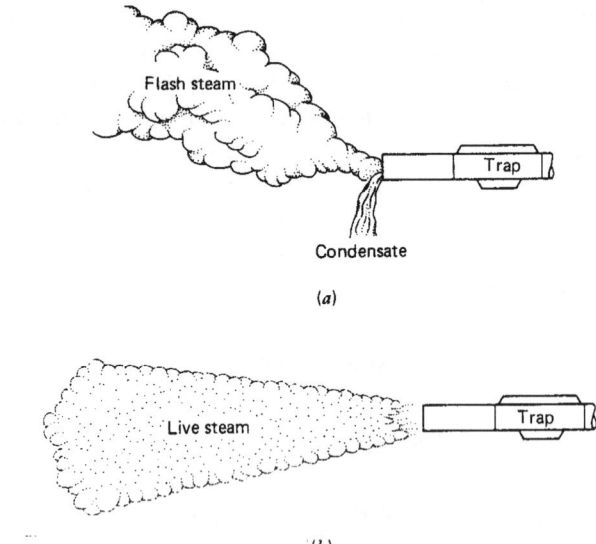

Figure 6-15. Visual observation of steam trap operation in open system: (a) Proper operation, (b) Improper operation.

ing through, a well-defined jet of live steam will issue from the line with either no condensate, or perhaps a condensate mist associated with steam condensation at the periphery of the jet.

Visual observation is less convenient in a closed condensate system, but it can be utilized if a test valve is placed in the return line just downstream of the trap, as shown in Figure 6-16. This system has the added advantage that the test line may be used to actually measure condensate discharge rate as a check on equipment efficiency, as discussed earlier. An alternative in closed condensate return systems is to install a sight glass just downstream of the trap. These are relatively inexpensive and permit quick visual observation of trap operation without interfering with normal production.

Another approach to steam trap testing is to observe the sound of the trap during operation. Table 6-14 describes the sounds made by various types of traps

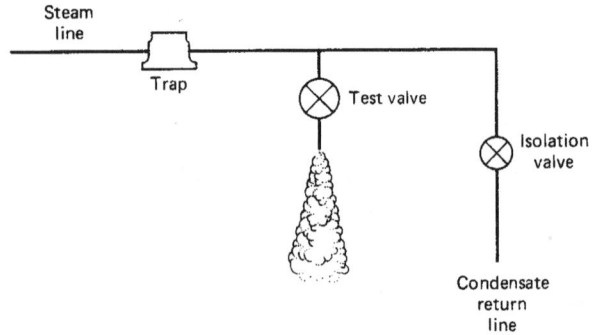

Figure 6-16. Visual observation of steam trap operation in closed systems.

during normal and abnormal operation. This method is most effective with disk-type traps, although it can be used to some extent with the other types as well. An industrial stethoscope can be used to listen to the trap, although under many conditions, the characteristic sound will be masked by noises transmitted from other parts of the system. Ultrasonic detectors may be used effectively in such cases; these devices are, in effect, electronic stethoscopes with acoustic filtering to make them sensitive to sound and vibration only in the very high frequency range. Steam blowing through a trap emits a very high-pitched sound, produced by intense turbulence at the trap orifice, as contrasted with the lower-pitched and lower-intensity sound of liquid flowing through. Ultrasonic methods can, therefore, give a more reliable measure of steam trap performance than conventional "listening" devices.

A third approach to steam trap testing makes use of the drop in saturation temperature and associated pressure drop across the trap. Condensate tends to cool rapidly in contact with uninsulated portions of the return line, accentuating the temperature difference. If the temperature on each side of the trap is measured, a sharp temperature drop should be evident. Table 6-15 shows typical temperatures that can be expected on the condensate side for various condensate pressures. In practice, the temperature drop method can be rather uncertain, because of the range of temperatures the condensate may exhibit, and because in blowing through a stuck-open trap, live steam will, itself, undergo some temperature drop. For example, 85-psig saturated steam blowing through an orifice to 15 psig will drop from 328°F to about 300°F, and it may then cool further by radiation and convection from uninsulated surfaces. From Table 6-15, the expected condensate-side temperature is about 215 to 238°F for this pressure. Thus, although the difference is still substantial, misinterpretation is possible, particularly if accurate measurements of the steam and condensate pressures on each side of the trap are not available.

The most successful programs of steam trap diagnosis utilize a combination of these methods, coupled with a regular maintenance program, to assure that traps are kept in proper operating condition.

This section has discussed the reasons why good steam trap performance can be crucial to successful energy conservation in steam systems. Traps must be properly selected and installed for the given service as well as appropriately sized, to assure efficient removal of condensate and gases. Once in service, expenditures for regular monitoring and maintenance easily pay for themselves in fuel savings.

Table 6-15. Typical Pipe Surface Temperatures for Various Operating Pressures

Operating Pressure (psig)	Typical Line Temperatures (°F)
0	190-210
15	215-238
45	248-278
115	295-330
135	304-340
450	395-437

6.5 CONDENSATE RECOVERY

Condensate from steam systems is wasted, or at least used inefficiently, in many industrial operations. Yet improvements in the condensate system can offer the greatest savings of any of the measures discussed in this chapter. In this section, methods are presented for estimating the potential energy and mass savings achievable through good condensate recovery, and the considerations involved in system design are discussed. It is not possible in a brief survey to provide a comprehensive guide to the detailed design of such systems. Condensate return systems can, in fact, be quite complex, and proper design usually requires a careful engineering analysis. The energy manager can, however, define the type of system best suited to the requirements and determine whether sufficient justification exists for a comprehensive design study.

6.5.1 Estimation of Heat and Mass Losses in Condensate Systems

The saturated liquid condensate produced when steam condenses on a heating surface still retains a significant fraction of the energy contained in the steam itself. Referring to Table 6-2, for example, it is seen that at a pressure of about 80 psig, *each pound of saturated liquid contains about 295 Btu, or nearly 25% of the original energy contained in the steam at the same pressure.* In some plants, this condensate is simply discharged to a waste-water system, which is wasteful not only of energy, but also of water and the expense of boiler feedwater treatment. Even if condensate is returned to an atmospheric pressure receiver, a considerable fraction of it is lost in the form of flash steam. Table 6-16 shows the percent of condensate loss due to flashing from systems at the given steam pressure to a flash tank at a lower pressure. For example, if in the 80-psig system discussed above, condensate is returned to a vented receiver instead of discharging it to a drain, nearly 12% is vented to the atmosphere as flash steam. Thus, about 3% of the origi-

Table 6-16. Percent of Mass Converted to Flash Steam in a Flash Tank

Steam Pressure (psig)	Flash Tank Pressure (psig)										
	0	2	5	10	15	20	30	40	60	80	100
5	1.7	1.0	0								
10	2.9	2.2	1.4	0							
15	4.0	3.2	2.4	1.1	0						
20	4.9	4.2	3.4	2.1	1.1	0					
30	6.5	5.8	5.0	3.8	2.6	1.7	0				
40	7.8	7.1	6.4	5.1	4.0	3.1	1.3	0			
60	10.0	9.3	8.6	7.3	6.3	5.4	3.6	2.2	0		
80	11.7	11.1	10.3	9.0	8.1	7.1	5.5	4.0	1.9	0	
100	13.3	12.6	11.8	10.6	9.7	8.8	7.0	5.7	3.5	1.7	0
125	14.8	14.2	13.4	12.2	11.3	10.3	8.6	7.4	5.2	3.4	1.8
160	16.8	16.2	15.4	14.1	13.2	12.4	10.6	9.5	7.4	5.6	4.0
200	18.6	18.0	17.3	16.1	15.2	14.3	12.8	11.5	9.3	7.5	5.9
250	20.6	20.0	19.3	18.1	17.2	16.3	14.7	13.6	11.2	9.8	8.2
300	22.7	21.8	21.1	19.9	19.0	18.2	16.7	15.4	13.4	11.8	10.1
350	24.0	23.3	22.6	21.6	20.5	19.8	18.3	17.2	15.1	13.5	11.9
400	25.3	24.7	24.0	22.9	22.0	21.1	19.7	18.5	16.5	15.0	13.4

nal steam energy (0.12 × 0.25) goes up the vent pipe. The table shows that this loss could be reduced by half by operating the flash tank at a pressure of 30 psig, providing a low-temperature steam source for potential use in other parts of the process. (Flash steam recovery is discussed later in more detail.)

Even when condensate is fully recovered using one of the methods to be described below, heat losses can still occur from uninsulated or poorly insulated return lines. These losses can be recovered very cost effectively by the proper application of thermal insulation as discussed in Chapter 15.

6.5.2 Methods of Condensate Heat Recovery

Several options are available for recovery of condensate, ranging in cost and complexity from simple and inexpensive to elaborate and costly. The choice of which option is best depends on the amount of condensate to be recovered, other uses for its energy, and the potential cost savings relative to other possible investments.

The simplest system, which can be utilized if condensate is presently being discharged, is the installation of a vented flash tank, which collects condensate from various points of formation and cools it sufficiently to allow it to be delivered back to the boiler feed tank. Figure 6-17 schematically illustrates such a system. It consists of a series of collection lines tying the points of condensate generation to the flash tank, which allows the liquid to separate from the flash steam; the flash steam is vented to the atmosphere through an open pipe. Condensate may be gravity-drained through a strainer and a trap. To avoid further generation of flash steam, a cooling leg may be incorporated to cool the liquid below its saturation temperature.

Flash tanks must be sized to produce proper separation of the flash steam from the liquid. As condensate is flashed, steam will be generated rather violently, and as vapor bubbles burst at the surface, liquid may be entrained and carried out through the vent. This represents a nuisance, and in some cases a safety hazard, if the vent is located in proximity to personnel or equipment. Table 6-17 permits the estimation of flash tank size required for a given application. Strictly speaking, flash tanks must be sized on the basis of volume. However, if a typical length to diameter of about 3:1 is assumed, flash tank dimensions can be represented as the product of diameter times length (having units of square feet/area), even though this particular product has no direct physical significance. Consider, for example, the sizing of a vented flash tank for collection of 80-psig condensate at a rate of about 3000 lb/hr. For a flash tank pressure of

0 psig (atmospheric pressure), the diameter-length product is about 2.5 per 1000 lb. Therefore, a diameter times length of 7.5 ft² would be needed for this application. A tank 1.5 ft in diameter by 5 ft long would be satisfactory. Of course, for flash tanks as with other condensate equipment, conservative design would suggest the use of an appropriate safety factor.

As noted above, venting of flash steam to the atmosphere is a wasteful process, and if significant amounts of condensate are to be recovered, it may be desirable to attempt to utilize this flash steam. Figure 6-18 shows a modification of the simple flash tank system to accomplish this. Rather than venting to the atmosphere, the flash tank is pressurized and flash steam is piped to a low-pressure steam main, where it can be utilized for process purposes. From Table 6-17 it will be noted that the flash tank can be smaller in physical size at elevated pressure, although, of course, it must be properly designed for pressure containment. If in the example above the 80-psig condensate were flashed in a 15-psig tank, only about 2.7 ft² of diameter times length would be required. A tank 1 ft in diameter by 3 ft long could be utilized. Atmospheric vents are usually provided for automatic pressure relief and to allow manual venting if desired.

If pressurized flash steam is to be used, the cost

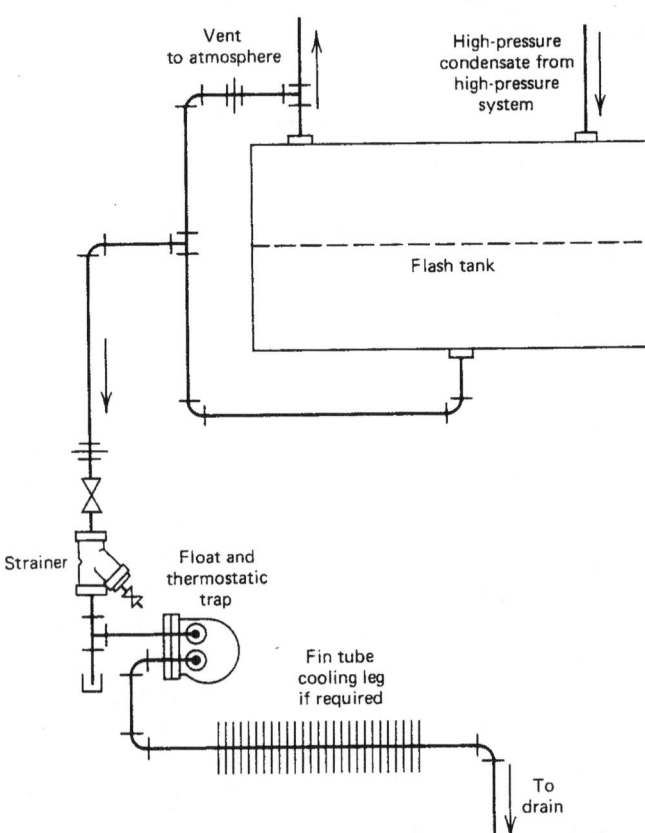

Figure 6-17. Flash tank vented to atmosphere.

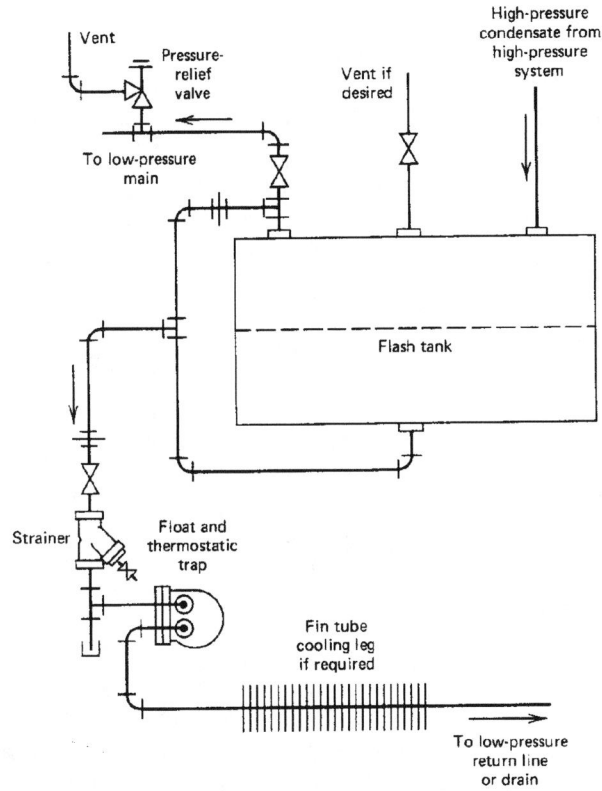

Figure 6-18. Pressurized flash tank discharging to low-pressure steam system.

of piping to set up a low-pressure steam system may be significant, particularly if the flash tank is remote from the potential low-pressure steam applications. Thus it is desirable to plan such a system to minimize these piping costs by generating the flash steam near its point of use. Figure 6-19 illustrates such an application. Here an air heater having four sections formerly utilized 100-psi steam in all sections. Because the temperature difference between the steam and the cold incoming air is larger than the difference at the exit end, the condensate load would be unevenly balanced among the four sections, with the heaviest load in the first section; lower-temperature steam could be utilized here. In the revised arrangement shown, 100-psi steam is used in the last three sections; condensate is drained to a 5-psi flash tank, where low-pressure steam is generated and piped to the first section, substantially reducing the overall steam load to the heater. Note that for backup purposes, a pressure-controlled reducing valve has been incorporated to supplement the low-pressure flash steam at light-load conditions. This example shows how flash steam can be used directly, without an expensive piping system to distribute it. A similar approach could apply to adjacent pieces of equipment in a multiple-batch operation.

Table 6-17. Flash Tank Sizing[a]

Steam Pressure (psig)	Flash Tank Pressure (psig)										
	0	2	5	10	15	20	30	40	60	80	100
400	5.41	4.70	3.89	3.01	2.44	2.03	1.49	1.15	0.77	0.56	0.42
350	5.14	4.45	3.66	2.84	2.28	1.91	1.38	1.07	0.70	0.51	0.37
300	4.86	4.15	3.42	2.62	2.11	1.75	1.26	0.96	0.62	0.44	0.31
250	4.41	3.82	3.12	2.39	1.91	1.56	1.11	0.85	0.52	0.37	0.25
200	3.98	3.40	2.80	2.12	1.68	1.37	0.97	0.72	0.43	0.28	0.18
175	3.75	3.20	2.61	1.95	1.57	1.26	0.87	0.64	0.38	0.23	0.15
160	3.60	3.08	2.50	1.86	1.46	1.19	0.80	0.59	0.34	0.21	0.12
150	3.48	2.98	2.41	1.80	1.40	1.14	0.77	0.56	0.31	0.19	0.10
140	3.36	2.86	2.31	1.72	1.35	1.08	0.72	0.52	0.29	0.16	0.08
130	3.24	2.76	2.23	1.65	1.29	1.02	0.67	0.49	0.26	0.14	0.07
120	3.12	2.65	2.15	1.57	1.22	0.97	0.64	0.44	0.23	0.12	0.04
110	2.99	2.52	2.05	1.50	1.15	0.91	0.58	0.40	0.20	0.09	0.02
100	2.85	2.41	1.92	1.40	1.07	0.85	0.53	0.36	0.16	0.06	
90	2.68	2.26	1.81	1.30	0.99	0.77	0.48	0.31	0.13	0.05	
80	2.52	2.12	1.67	1.18	0.90	0.68	0.42	0.25	0.09		
70	2.34	1.95	1.55	1.08	0.81	0.61	0.35	0.20	0.04		
60	2.14	1.77	1.39	0.96	0.70	0.52	0.27	0.14			
50	1.94	1.59	1.22	0.81	0.58	0.41	0.20	0.08			
40	1.68	1.36	1.02	0.67	0.44	0.30	0.11				
30	1.40	1.10	0.81	0.50	0.29	0.16					
20	1.06	0.81	0.55	0.28	0.12						
12	0.75	0.48	0.28								
10	0.62	0.42	0.23								

[a]Flash tank area (ft²) = diameter X length of horizontal tank for discharge of 1000 lb/hr of condensate.

Although the utilization of flash steam in a low-pressure system appears to offer an almost "free" energy source, its practical application involves a number of problems that must be carefully considered. These are all essentially economic in nature.

As mentioned above, the quantity of condensate and its pressure (thus yielding a given quantity of flash steam) must be sufficiently large to provide a significant amount of available energy at the desired pressure. System costs do not go up in simple proportion to capacity. Rather, there is a large initial cost for piping and installation of the flash tank and other system components; therefore, the overall cost per unit of heat recovered becomes significantly less as the system becomes larger. The nature of the condensate-producing system itself is also important. For example, if condensate is produced at only two or three points from large steam users, the cost of the condensate collection system will be considerably less than that of a system in which there are many small users.

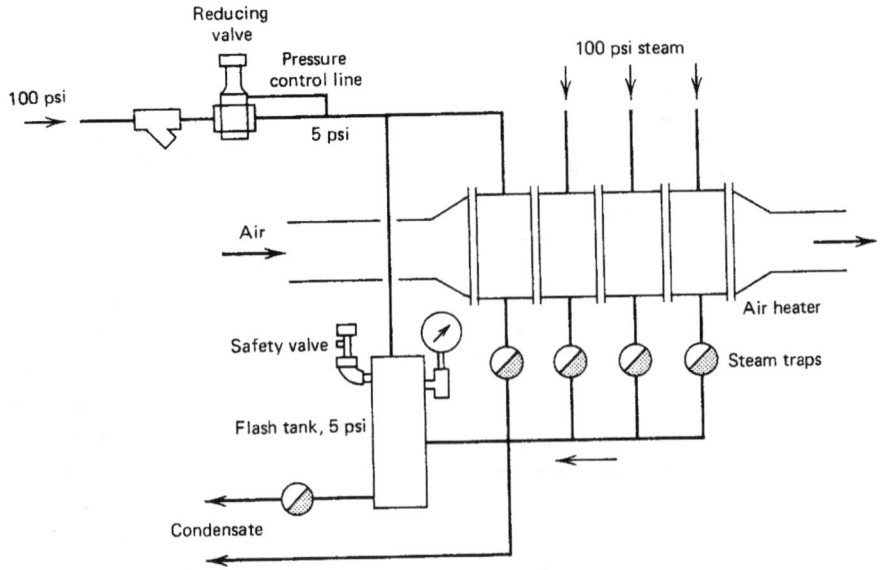

Figure 6-19. Flash steam utilization within a process unit.

Another important consideration is the potential for application of the flash steam. The availability of 5000 lb/hr of 15-psig steam is meaningless unless there is a need for a heat source of this magnitude in the 250°F temperature range. Thus, potential uses must be properly matched to the available supply. Flash steam is most effectively utilized when it can supplement an existing low-pressure steam supply, rather than providing the sole source of heat to equipment. Not only must the total average quantity of flash steam match the needs of the process, but the time variations of source and user must be taken into account since steam cannot be economically stored for use at a later time. Thus, flash steam might not be a suitable heat source for sequential batch processes in which the number of operating units is small, where significant fluctuations in steam demand exist.

When considering the possible conversion to low-pressure steam of an existing piece of equipment presently operating on high-pressure steam, it is important to recognize that steam pressure can have a significant effect on equipment operation. Since a reduction in steam pressure also means a reduction in temperature, a unit may not have adequate heating-surface area to provide the necessary heat capacity to the process at reduced pressure. Existing steam distribution piping may

not be adequate since steam is lower in density at low pressure than at high pressure. Typically, larger piping is required to transport the low-pressure vapor at acceptable velocities. Although one might expect that the heat losses from the pipe surface might be lower with low-pressure steam because of its lower temperature, in fact, this may not be the case if a large pipe (and hence larger surface area) is needed to handle the lower-pressure vapor. This requirement will also make insulation more expensive.

When flash steam is used in a piece of equipment, the resulting low-pressure condensate must still be returned to a receiver for delivery back to the boiler. Flash steam will again be produced if the receiver is vented, although somewhat less so than in the flashing of high-pressure condensate. This flash steam and that produced from the flash tank condensate draining into the receiver will be lost unless some additional provision is made for its recovery, as shown in Figure 6-20. In this system, rather than venting to the atmosphere, the steam rises through a cold-water spray, which condenses it. This spray might be boiler makeup water, for example, and hence the energy of the flash steam is used for makeup preheat. Not only is the heat content of the flash steam saved, but its mass as well, reducing makeup-water requirements and saving the incremental costs of makeup-water treatment. This system has the added advantage that it produces a deaerating effect on the condensate and feedwater. If the cold spray is metered so as to produce a temperature in the tank above about 190°F, dissolved gases in the condensate and feedwater, particularly oxygen and CO_2, will come out of solution, and

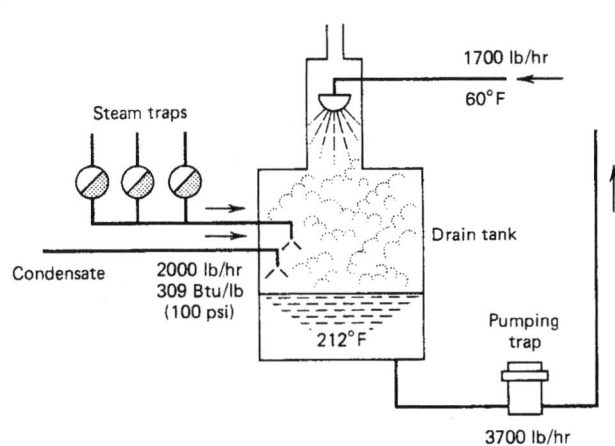

Figure 6-20. Flash steam recovery in spray tank.

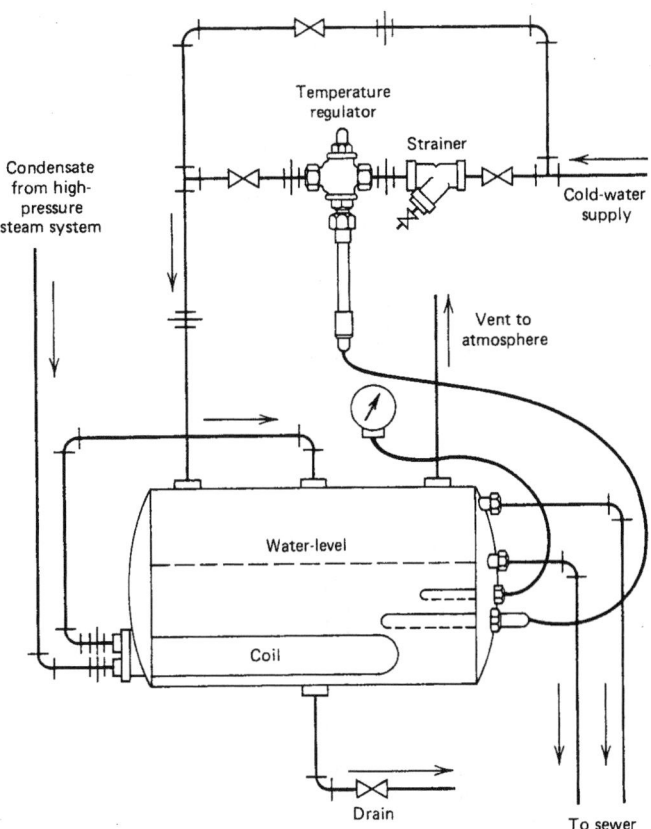

Figure 6-21. Flash tank with condensate precooling.

since they are not condensed by the cold-water spray, they will be released through the atmospheric vent. As with flash steam systems, this system (usually termed a "barometric condenser" or "spray deaerator") requires careful consideration to assure its proper application. The system must be compatible with the boiler feedwater system, and controls must be provided to coordinate boiler makeup demands with the condensate load.

An alternative approach to the barometric condenser is shown in Figure 6-21. In this system, condensate is cooled by passing it through a submerged coil in the flash tank before it is flashed. This reduces the amount of flash steam generated. Cold-water makeup (possibly boiler feedwater) is regulated by a temperature-controlled valve.

The systems described above have one feature in common. In all cases, the final condensate state is atmospheric pressure, which may be required to permit return of the condensate to the existing boiler-feedwater makeup tank. If condensate can be returned at elevated pressure, a number of advantages may be realized.

Figure 6-22 shows schematics of two pressurized condensate return systems. Condensate is returned, in some cases without the need for a steam trap, to a high-pressure receiver, which routes the condensate directly back to the boiler. The boiler makeup unit and/or deaerator feeds the boiler in parallel with the condensate return unit, and appropriate controls must be incorporated to coordinate the operation of the two units. Systems such as the one shown in Figure 6-22a are available for condensate pressures up to about 15 psig. For higher pressures, the unit can be used in conjunction with a flash tank, as shown in Figure 6-22b. This system would be suitable where an application for 15-psig steam is available. Systems of this type are attractive in relatively low-pressure applications, such as steam-driven absorption chillers. When considering them, care

must be exercised to assure that dissolved gases in the boiler makeup are at suitable levels to avoid corrosion since the natural deaeration effect of atmospheric venting is lost.

One of the key engineering considerations that must be accounted for in the design of all the systems described above is the problem of pumping high-temperature condensate. To understand the nature of the problem, it is necessary to introduce the concept of "net positive suction head" (NPSH) for a pump. This term means the amount of static fluid pressure that must be provided at the inlet side of the pump to assure that no vapor will be formed as the liquid passes through the pump mechanism, a phenomenon known as cavitation. As liquid moves into the pump inlet from an initially static condition, it accelerates and its pressure drops rather suddenly. If the liquid is at or near its saturation temperature in the stationary condition, this sudden drop in pressure will produce boiling and the generation of vapor bubbles. Vapor can also be generated by air coming out of solution at reduced pressure. These bubbles travel through the pump impeller, where the fluid pressure rises, causing the bubbles to collapse. The inrush of liquid into the vapor space produces an impact on the impeller surface that can have an effect comparable to sandblasting. Clearly, this is deleterious to the impeller

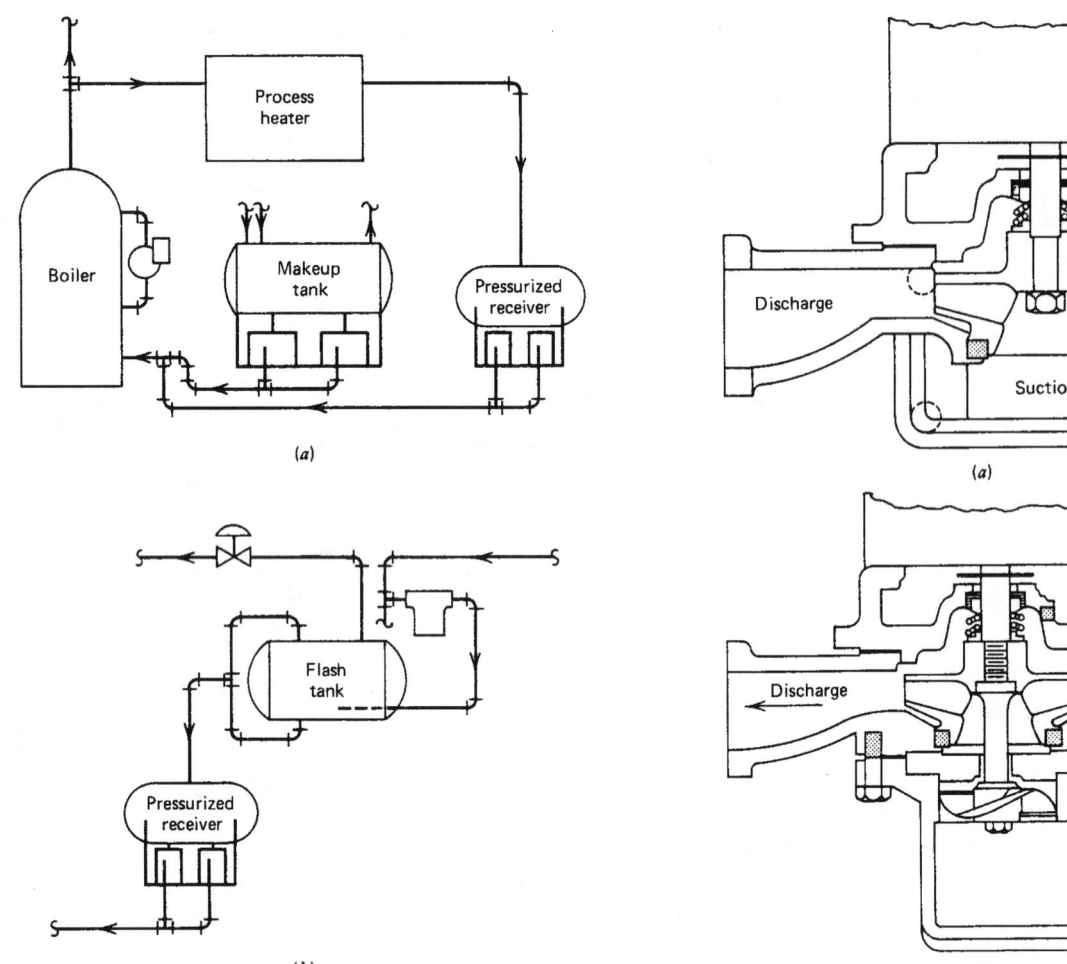

Figure 6-22. Pressurized condensate receiver systems, (a) Low-pressure process requirements, (b) Flash tank for use with high-pressure systems.

Figure 6-23. Conventional and low-NPSH pumps, (a) Conventional centrifugal pump, (b) Low-NPSH centrifugal pump.

and can cause rapid wear. Most equipment operators are familiar with the characteristic "grinding" sound of cavitation in pumps when air is advertently allowed to enter the system, and the same effect can occur due to steam generation in high-temperature condensate pumping.

To avoid cavitation, manufacturers specify a minimum pressure above saturation that must be maintained on the inlet side of the pump, such that, even when the pressure drops through the inlet port, saturation or deaeration conditions will not occur. This minimum-pressure requirement is termed the net positive suction head.

For condensate applications, special low-NPSH pumps have been designed. Figure 6-23 illustrates the difference between a conventional pump and a low-NPSH pump. In the conventional pump (Figure 6-23a), fluid on the suction side is drawn directly into the impeller, where the rapid pressure drop occurs in the entry passage. In the low-NPSH pump (Figure 6-23b), a small "preimpeller" provides an initial pressure boost to the incoming fluid,

with relatively little drop in pressure at the entrance. This extra stage of pumping essentially provides a greater head to the entry passage of the main impeller, so that the system pressure at the suction side of the pump can be much closer to saturation conditions than that required for a conventional pump. Low-NPSH pumps are higher in price than conventional centrifugal pumps, but they can greatly simplify the problem of design for high-temperature condensate return, and they can, in some cases, actually reduce overall system costs.

An alternative device for the pumping of condensate, called a "pumping trap," utilizes the pressure of the steam itself as the driving medium. Figure 6-24 illustrates the mechanism of a pumping trap. Condensate enters the inlet side and rises in the body until it activates a float-operated valve, which admits steam or compressed air into the chamber. A check valve prevents condensate from being pushed back through the inlet port, and another check valve allows the steam or air pressure to drive it out through the exit side. When

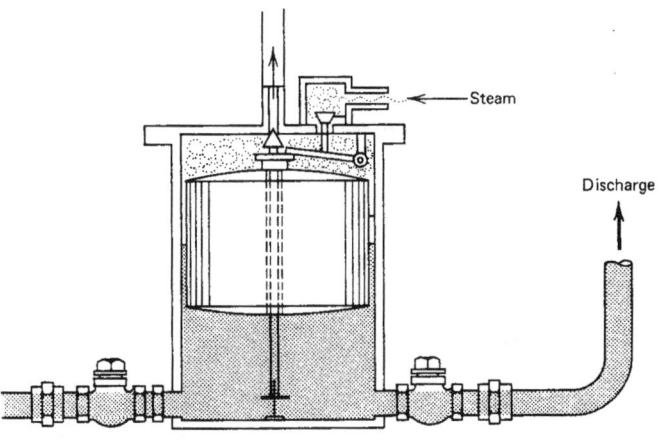

Figure 6-24. Pumping trap.

6.5.3 Overall Planning Considerations in Condensate Recovery Systems

As mentioned earlier, condensate recovery systems require careful engineering to assure that they are compatible with overall plant operations, that they are safe and reliable, and that they can actually achieve their energy-efficiency potential. In this section a few of the overall planning factors that should be considered are enumerated and discussed.

1. **Availability of Adequate Condensate Sources.** An energy audit should be performed to collect detailed information on the quantity of condensate available from all the various steam using sources in the plant, as well as the relevant data associated with these sources. Such

Figure 6-25. Pressurized condensate return systems.

the condensate level drops to a predetermined position, the steam or air valve is closed, allowing the pumping cycle to start again. Pumping traps have certain inherent advantages for condensate return applications over electrically driven pumps. They have no NPSH requirement, and hence can handle condensate at virtually any temperature without regard to pressure conditions. They are essentially self-regulating since the condensate level itself determines when the trap pumps; thus no auxiliary electrical controls are required for the system. This has another advantage in environments where explosion-proofing is required, such as refineries and chemical plants. Electrical lines need not be run to the system since it utilizes steam as a driving force, and the steam line is usually close at hand. Pumping traps operate more efficiently using compressed air, if available, because when steam is introduced to the chamber, some of it condenses before its pressure can drive the condensate out. Thus, for the same pressure, more steam is required to give the same pumping capacity as compressed air. The disadvantages of pumping traps are their mechanical complexity, resulting in a susceptibility to maintenance problems, as well as the fact that they are available only in limited capacities.

The engineering of a complete condensate recovery system from scratch can be a rather involved process, requiring the design of tanks, plumbing, controls, and pumping devices. For large systems, there is little alternative to engineering and fabricating the system to the specific plant requirements. For small- to moderate-capacity applications, however, packaged systems incorporating all of the foregoing components are commercially available. Figures 6.25a and b show examples of two such systems, the former using electrically driven low-NPSH pumps, and the latter utilizing a pumping trap (the lower unit in the figure).

data include, for example, condensate pressure, quantity, and source location relative to other steam-using equipment and the boiler room. Certain other information may also be pertinent. For example, if the condensate is contaminated by contact with other process streams, it may be unsuitable for recovery. It is not valid to assume that steam used in the process automatically results in recoverable condensate. Stripping steam used in refining of petroleum and other separation processes is a good example. In batch processing operations, it is important to classify steam loads as steady or intermittent, and if intermittent, whether or not they are coincident (and hence additive) with other loads.

2. **Survey of Possible Flash Steam Applications.** The process should be surveyed in detail to assess what applications presently using first-generation steam might be adaptable to the use of flash steam or to heat recovered from condensate. Temperatures and typical heat loads are necessary but not sufficient. Heat-transfer characteristics of the equipment itself may be important. A skilled heat-transfer engineer, given the present operating characteristics of a process heater, can make reasonable estimates of that heater's capability to operate at a lower pressure.

3. **Analysis of Condensate and Boiler Feedwater Chemistry.** To assure that conditions in the boiler are kept in a satisfactory state to avoid scaling and corrosion, a change in feedwater treatment may be required when condensate is recovered and recycled. Water samples from the present condensate drain—from the boiler blow-down, from the incoming water source, and from the outlet of the present feedwater treatment system—should be obtained. With this information, a water treatment specialist can analyze the overall water chemical balance and assure that the treatment system is properly configured to maintain good boiler-water conditions.

4. **Piping Systems.** A layout of the present steam and condensate piping system is helpful to assess the need for new piping and the adequacy of existing runs. It is also useful in identifying suitable locations for flash tanks as discussed in Section 6.5.2.

5. **Economic Data.** The bottom line on any energy recovery project, including condensate recovery, is its profitability. In order to analyze the profitability of the system, it is necessary to estimate the quantity of heat recovered (converted into its equivalent fuel usage), the costs and cost savings associated with the water treat-

ment system, and the savings in water cost. In addition to the basic capital and installation costs of the condensate recovery equipment, there may be additional costs associated with modification of existing equipment to make it suitable for flash steam utilization. And there will almost certainly be a cost of lost production during installation and checkout of the new system.

6.6   SUMMARY

This chapter has discussed a number of considerations in effecting energy conservation in industrial and commercial steam systems. Good energy management begins by improving the operation of existing systems and then progresses to evaluation of system modifications to maximize energy efficiency. Methodology and data have been presented to assist the energy manager in estimating the potential for savings by improving steam system operations and implementing system design changes.

Bibliography

Armstrong International Corporation, *Steam Trap Testing Guide for Energy Conservation*, 2007. www.armstronginternational.com/files/products/traps/pdf/310.pdf

Armstrong International Corporation, *Steam Conservation Guidelines Handbook*, 2005. http://steamix.com/common/allproductscatalog/consguidelines.pdf

Bhatt, M. Siddhartha, "Energy Audit Case Studies. I - Steam Systems," *Applied Thermal Engineering*, v 20, n 3, Feb, 2000, p 285-296.

Deacon, W.T., "Successful Steam System Operation Strategies," ASHRAE Transactions, v 99, n pt 1, 1993, p 1259-1264.

French, Scott A. "Lighting a Fire Under Steam Trap Maintenance," *Heating, Piping, Air Conditioning*, v 72, n 3, 2000, p 33-38.

Garcia, E., "Discussion on Steam Trap Maintenance Programs and Their Return," American Society of Mechanical Engineers, Advanced Energy Systems Division Publication AES, v 2-2, 1986, p 113-119.

Hahn, Glenn E., "Effective Condensate Management Can Cut Energy, Maintenance Costs," *Pulp & Paper*, v 62, n 11, Nov, 1988, p 130-133.

Kenney, W.F., *Energy Conservation in the Process Industries*, Academic Press, Orlando, 1984.

Kerr, R.N., "Managing the Steam Trap System," *Strategic Planning for Cogeneration and Energy Management*, 8th World Energy Engineering Congress, Assoc of Energy Engineers, Atlanta, GA,1986, p 263-264.

Lawson, R.L. "How to Minimize Waste in Steam Tracing Systems and Unit Heaters," New Directions in Energy Technology, Proceedings of the 7th World Energy Engineering Congress, Associaton of Energy Engineers, Atlanta, GA, 1985, p 287-289.

Risko, James R., " Handle Steam More Intelligently," *Chemical Engineering*, v 113, n 11, November, 2006, p 38-43

Spirax Sarco Corporation, Steam Engineering Tutorials http://www.spiraxsarco.com/resources/steam-engineering-tutorials.asp

Stultz, S.C. and J.B. Kitto, Eds., *Steam: Its Generation and Use*, 40th Edition, Babcock and Wilcox/American Boiler Manufacturers Association, 1992.

CHAPTER 7

COGENERATION AND DISTRIBUTED GENERATION

JORGE B. WONG
Cogen Masters
Wallingford, CT

JOHN M. KOVACIK
General Electric Company
Retired

7.1 INTRODUCTION

Cogeneration is broadly defined as the coincident or simultaneous generation of combined heat and power (CHP). In a true cogeneration system a "significant" portion of the generated or recovered heat must be used in a thermal process as steam, hot air, hot water, etc. The cogenerated power is typically in the form of mechanical or electrical energy.

The power may be totally used in the industrial plant that serves as the "host" of the cogeneration system, or it may be partially/totally exported to a utility grid. Figure 7-1 illustrates the potential of saving in primary energy when separate generation of heat and electrical energy is replaced by a cogeneration system. This chapter presents an overview of current design, analysis and evaluation procedures.

The combined generation of useful heat and power is not a new concept. The U.S. Department of Energy (1978) reported that in the early 1900s, 58% of the total power produced by on-site industrial plants was cogenerated. However, Polymeros (1981) stated that by 1950, on-site CHP generation accounted for only 15 percent of total U.S. electrical generation and that by 1974, this figure had dropped to about 5 percent. In Europe, the experience has been very different. The US Department of Energy, US-DOE (1978), reported that "historically, industrial cogeneration has been five to six times more common in some parts of Europe than in the U.S." In 1972, "16% of West Germany's total power was cogenerated by industries; in Italy, 18%; in France, 16%; and in the Netherlands, 10%."

Since the promulgation of the Public Utilities Regulatory Policies Act of 1978 (PURPA), however, U.S. cogeneration design, operation and marketing activities have dramatically increased, and that have received much more attention from industry, government and academia. As a result of this incentive, newer technologies such as various combined cycles have achieved industrial maturity. Also, new or improved cogeneration-related processes and equipment have been developed and are actively marketed; e.g., heat recovery steam generator and duct burners, gas-engine driven chillers, direct and indirect fired two-stage absorption chiller-heaters, etc.

The Energy Policy Act of 2005 (EPAct 2005) updated some aspects of the previous law, EPAct 1992. The following summary from the US Department of Energy contains the main updates. For more details see: http://www.oe.energy.gov/purpa.htm.

The Energy Policy Act of 2005 (EPACT 2005) Subtitle E contains four sections, three of which address additional Public Utility Regulatory Policies Act of 1978 (PURPA) Title I Standards: Sections 1251, 1252, 1253 and 1254.

Specifically, EPACT 2005 adds five new federal standards to PURPA Section 111(d):

- (11) NET METERING (see EPACT 2005 Sec. 1251 for details)
- (12) FUEL SOURCES (see EPACT 2005 Sec. 1251 for details)
- (13) FOSSIL FUEL GENERATION EFFICIENCY (see EPACT 2005 Sec. 1251 for details)
- (14) TIME-BASED METERING AND COMMUNICATIONS, also known as "smart metering (see EPACT 2005 Sec. 1252 for details), and
- (15) INTERCONNECTION (see EPACT 2005 Sec. 1254 for details).

Three of these new standards (11, 14, and 15) apply to Title I. In addition, Section 1253, "Cogeneration and Small Power Production and Sale Requirements" establishes the following:

(a) "Termination of Mandatory Purchase and Sale Requirements" (originally mandated by PURPA of 1978), not requiring utilities to purchase cogenerated power from existing qualified facilities (QF) whenever the corresponding state utilities commission ("the commission") finds that said QF has a nondiscriminatory (i.e. 'fair') access to:

1. independent long-term electricity markets or
2. transmission and interconnection services to competitive wholesale markets
3. wholesale markets comparable to 1 and 2 above.

This means that if the commission determines an existing QF has no access or has discriminatory access to any of the three electricity markets listed above, said QF may be able to sell cogenerated power to a local power utility at a non-discriminatory rate schedule.

A) Separate generation of heat and electricity in boiler and conventional condensation power plant.

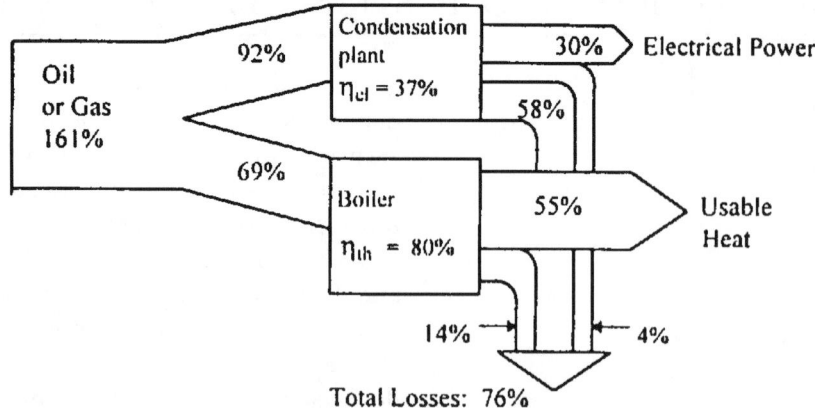

B) Cogeneration of heat and power.

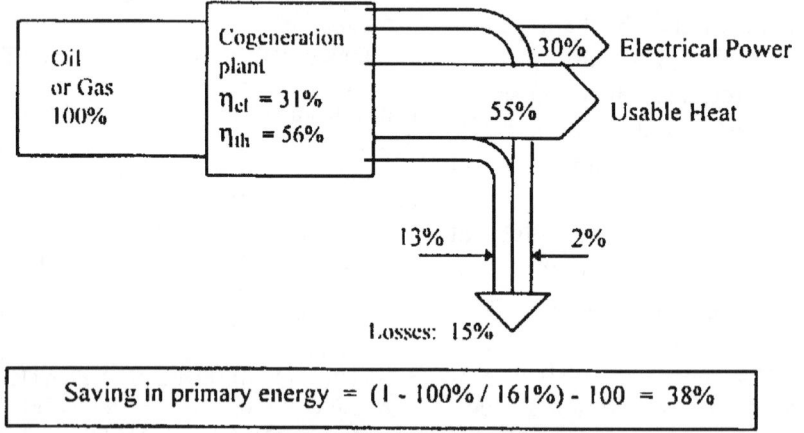

C) Integration of a cogeneration system and an industrial process.

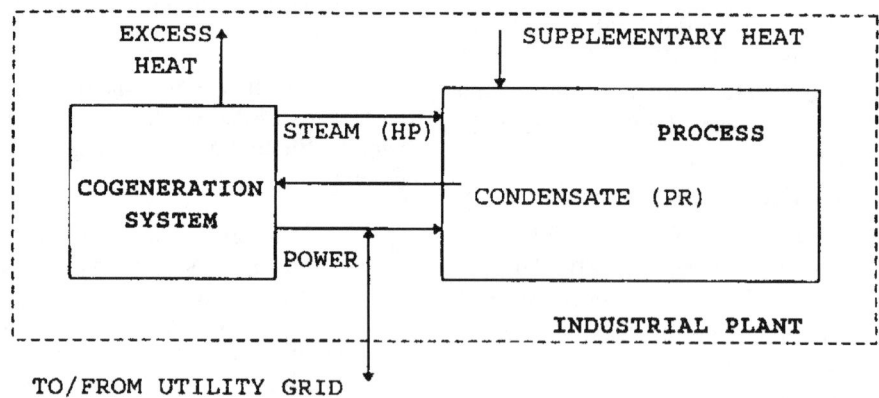

Figure 7-1. Potential of saving in primary energy when separate generation of heat and electrical energy is replaced by an industrial cogeneration system.

(b) "Revised Purchase and Sale Obligation for New Facilities," whereby no electric utility is required to purchase from or sell electric energy to a facility that is not an exisiting QF, "unless the facility meets the criteria for qualifying generation facilities established by the Commision pursuant the rulemaking required by subsection (n)," below.

"(n) Rulemaking for New Qualifying Facilities" establishes that the commission shall issue a rule revising the criteria for new QF's seeking to sell electricity to ensure:

- "The thermal energy output of a new QF is used in a productive and beneficial manner."
- "The electrical, thermal and chemical output of the cogeneration facility is used fundamentally for industrial, commercial, or institutional purposes and is not intended fundamentally for sale to an electric utility, taking into account technological, efficiency, economic, and variable thermal energy requirements, as well as state laws applicable to sales of electric energy from a QF to its host facility"; and
- "Continuing progress in the development of efficient electric energy egenrating technology."

EPACT 2005 also requires state regulatory agencies to make a determination regarding net metering, smart metering, cogeneration, small power production purchase/sale requirements, and interconnection of small, distributed, and renewable energy systems. For example, the Public Utilities Commission of Ohio approved in 2007 a set of recommendations that address net metering, advanced metering infrastructure and demand response, interconnection, stand-by rates, and renewable energy portfolio standards. The commission received comments from Ohio's electric distribution utilities, consumer groups, energy marketers, industrial energy users, manufacturing associations, environmental councils, alternative energy corporations, farming associations, universities, and state and federal agencies. Note that all this information is very relevant when planning and executing cogeneration projects.

The impact of cogeneration in the U.S., both as an energy conservation measure and as a means to contribute to the overall electrical power generation capacity, cannot be overemphasized. SFA Pacific Inc. (1990) estimated that non-utility generators (NUGs) produce 6-7% of the power generated in the US and will account for about half of the 90,000 MW that will be added during the next decade. More recently, Makansi (1991) reported that around 40,000 MW of independently produced and/or cogenerated (IPP/COGEN) power has been put on line since the establishment of PURPA. This constitutes a significant part of new capacity. Makansi also reports that another 60,000

MW of IPP/COGEN power is in construction and development. This trend is likely to increase since CHP is a supplementary way of increasing the existing U.S. power generation capacity. Thus, distributed generation (DG) is arising as a serious CHP alternative. (See Section 7.2.2.4).

PURPA is considered the foremost regulatory instrument in promoting cogeneration, IPPs and/or NUGs. Thus, to promote industrial energy efficiency and resource conservation through cogeneration, PURPA requires electric utilities to purchase cogenerated power at fair rates to both the utility and the generator. PURPA also orders utilities to provide supplementary and back-up power to qualifying facilities (QFs).

The promulgation of the Energy Policy Act of 1992 (EPAct 1992) has brought a "deregulated" and competitive structure to the power industry. EPAct-1992 establishes the framework that allows the possibility of "retail wheeling" (i.e., anyone can purchase power from any generator, utility or otherwise). This is in addition to the existing wholesale wheeling or electricity trade that normally occurs among U.S. utilities. EPAct 1992's main objective is to stimulate competition in the electricity generation sector and reduce electricity costs. The impact of EPAct 1992, deregulation, and retailed wheeling on cogeneration can be tremendous since it creates a new class of generating facilities called exempt wholesale generators (EWGs). Economies of scale and scope—i.e. the natural higher efficiency of larger CHP plants—could trigger the growth of cogeneration-based EWGs. In addition, EPAct 1992 opened the transmission grid to utlities and NUGs by ordering FERC to allow open transmission access to all approved EWGs. This provision has been amended by EPACT 2005; however, specific retail wheeling rules have not been legislated at the time of this writing. So far, only a few retail wheeling cases have been started. Thus, the actual impact of deregulation on cogeneration is considered to be uncertain. Nevertheless, some consider that retail wheeling/deregulation and cogeneration could have a synergistic effect and should be able to support each other.

7.2. COGENERATION SYSTEM DESIGN AND ANALYSIS

The process of designing and evaluating a cogeneration system has so many factors that it has been compared to the Rubik's cube, the ingenious game to arrange a multi-colored cube. The change in one of the cube faces will likely affect some other face. The most important faces are: fuel security, regulations, economics, technology, contract negotiation and financing.

7.2.1 General Considerations and Definitions

Kovacik (1982) indicates that although cogeneration should be evaluated as a part of any energy management plan, the main prerequisite is that a plant shows a significant and concurrent demand for heat and power. Once this scenario is identified, he states that cogeneration systems can be explored under the following circumstances:

1. Development of new facilities

2. Major expansions to existing facilities which increase process heat demands and/or process energy rejection.

3. When old process and/or power plant equipment is being replaced, offering the opportunity to upgrade the energy supply system.

The following terms and definitions are regularly used in the discussion of CHP systems.

Industrial Plant: the facility requiring process heat and electric and/or shaft power. It can be a process plant, a manufacturing facility, a college campus, etc. See Figure 7-1c.

Process Heat (PH): the thermal energy required in the industrial plant. This energy is supplied as steam, hot water, hot air, etc.

Process Returns (PR): the fluid returned from the industrial plant to the cogeneration system. For systems where the process heat is supplied as steam, the process returns are condensate.

Net Heat to Process (NHP): the difference between the thermal energy supplied to the industrial plant and the energy returned to the cogeneration system. Thus, NHP = PH − PR. The NHP may or may not be equal to the actual process heat demand (PH).

Plant Power Demand (PPD): the electrical power or load demanded (kW or MW) by the industrial plant. It includes the power required in for indus-

Table 7-1. Basic cogeneration systems.

Cogeneration Systems	Unit Elec. Capacity (kW)	Heat Rate[2] (Btu/kWh)	Electrical[3] Efficiency (%)	Thermal Efficiency (%)	Total Efficiency (%)	Exhaust Temperature (°F)	125-psig Steam Generation (lbs/hr)
Small reciprocating Gas Engines	1-500	25,000 to 10,000	14-34	52	66-86	600-1200	0-200[1]
Large reciprocating Gas Engines	500-17,000	13,000 to 9,500	26-36	52	78-88	600-1200	100-10,000[1]
Diesel Engines	100-4,000 9,500	14,000 to	24-36	50	74-86	700-1500	100-1500[1]
Industrial Gas Turbines	800-10,000	14,000 to 11,000	24-31	50	74-81	800-1000	3,000-30,000
Utility Size Gas Turbines	10,000-150,000	13,000 to 9000	26-35	50	76-85	700-800	30,000 to 300,000
Steam-Turbine Cycles	5,000-200,000	30,000 to 10,000	10-35	28	38-63	350-1000	10,000 to 200,000

NOTES

[1]Hot water @ 250° is available at 10 times the flow of steam

[2]Heat rate is the fuel heat input (Btu—higher heating value) to the cycle per kWh of electrical output at design (full load) and ISO conditions (60°F ambient temperature and sea level operation)

[3]The electrical generation efficiency in percent of a prime mover can be determined by the formula Efficiency = (3413/Heat Rate)100

Sources: Adapted from Limaye (1985) and Manufacturers Data.

trial processes, air-conditioning, lighting, etc.

Heat/Power Ratio (H/P): The heat-to-power ratio of the industrial plant (demand), or the rated heat-to-power ratio of the cogeneration system or cycle (capacity).

Topping Cycle: Thermal cycle where power is produced prior to the delivery of heat to the industrial plant. One example is the case of heat recovered from a diesel-engine generator to produce steam and hot water. Figure 7-2 shows a diesel engine topping cycle.

Bottoming Cycle: Power production from the recovery of heat that would "normally" be rejected to a heat sink. Examples include the generation of power using the heat from various exothermic chemical processes and the heat rejected from kilns used in various industries. Figure 7-3 illustrates a bottoming cycle.

Combined Cycle: This is a combination of the two cycles described above. Power is produced in a topping cycle—typically a gas-turbine generator. Then heat exhausted from the turbine is used to produce steam, which is subsequently expanded in a steam turbine to generate more electric or shaft power. Steam can also be extracted from the cycle to be used as process heat. Figure 7-4 depicts a combined cycle.

Prime Mover: A unit of the CHP system that generates electric or shaft power. Typically, it is a gas turbine generator, a steam turbine drive, or a diesel-engine generator.

7.2.2 Basic Cogeneration Systems

Most cogeneration systems are based on prime movers such as steam turbines, gas turbines, internal combustion engines, and packaged cogeneration. Table 7-1 shows typical performance data for various cogeneration systems. Figures in this table (and in this chapter) are based on higher heating values, unless stated otherwise.

7.2.2.1 Steam Turbine Systems

Steam turbines are currently used as prime movers in topping, bottoming, and combined cycles. There are many types of steam turbines to accommodate various heat/power ratios and loads. For limited expansion (pressure drop) and smaller loads (<4000 HP), lower cost single stage backpressure turbines are used. When several pressure levels are required (and usually for larger loads), multi-stage condensing and non-condensing turbines with induction and/or extraction of steam at intermediate pressures are generally used. Figure 7-5 shows a variety of condensing and non-condensing turbines.

Four factors must be examined to assure that the maximum amount of power from a CHP steam plant is economically generated, based on the process heat required. These factors are: (1) prime-mover size, (2) initial steam conditions, (3) process pressure levels, and (4) feedwater heating cycle.

1. Prime-mover Type and Size. Process heat and plant electric requirements define the type and size of the steam generator. The type of CHP system and its corresponding prime mover are selected by matching the CHP system heat output to the process heat load.

If process heat demands are such that the plant power requirements can be satisfied by cogenerated power, then the size of the prime mover is selected to meet or exceed the "peak" power demand. However, cogeneration may supply only a portion of the total plant power needs. The balance has to be imported through a utility tie. In isolated plants, the balance is generated by additional conventional units. This discussion assumes that both heat and power demands remain constant all times. Hence, the design problem becomes one of specifying two variables: (1) how much power should be cogenerated on-site and (2) how much power should be imported. Thus, given the technological, economical, and legal constraints for a particular plant, and assuming the CHP system must be constructed at a minimum overall cost, it becomes a constrained optimization problem.

2. Initial Steam Conditions. Many industrial plants do not have adequate process steam demands to generate all the power required. Thus, designers should examine those variables that can be safely adjusted or manipulated to optimize the power that can be economically generated. One set of these variables are the initial steam conditions, i.e. the initial pressure and temperature of the steam generated. In general, an increase in initial pressure and/or temperature will increase the amount of energy available for power generation. But the prime mover construction and cost, as well as the heat demand, impose economical limits for the initial steam conditions. Thus, higher initial steam conditions can be economically justified in industrial plants having relatively large process steam demands.

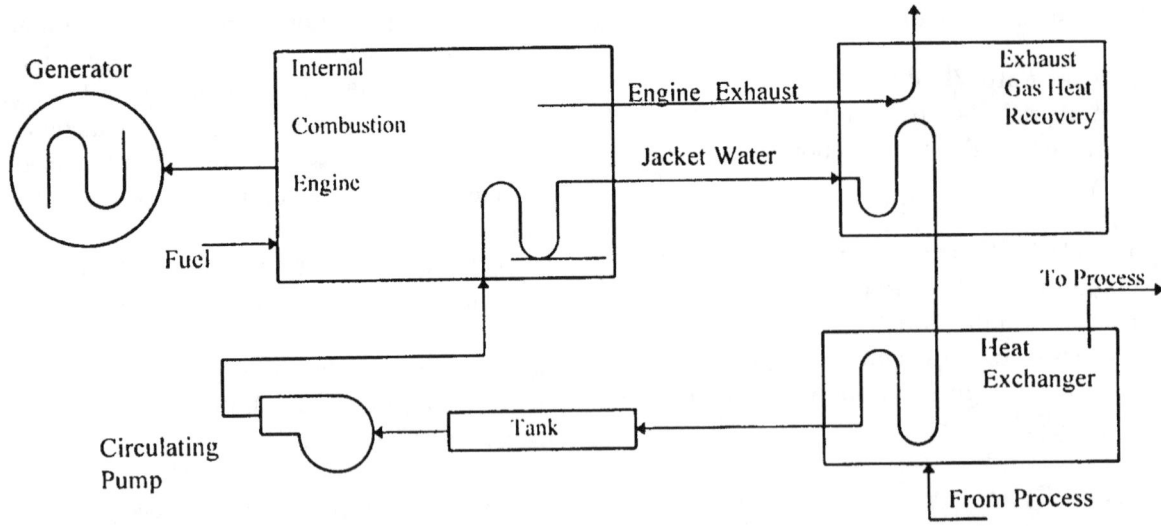

Figure 7-2. Diesel engine topping cycle.

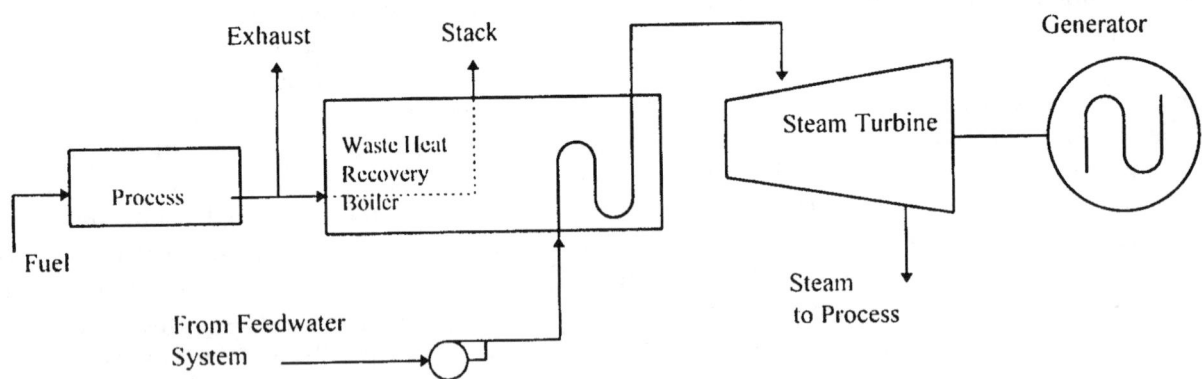

Figure 7-3. Steam turbine bottoming cycle.

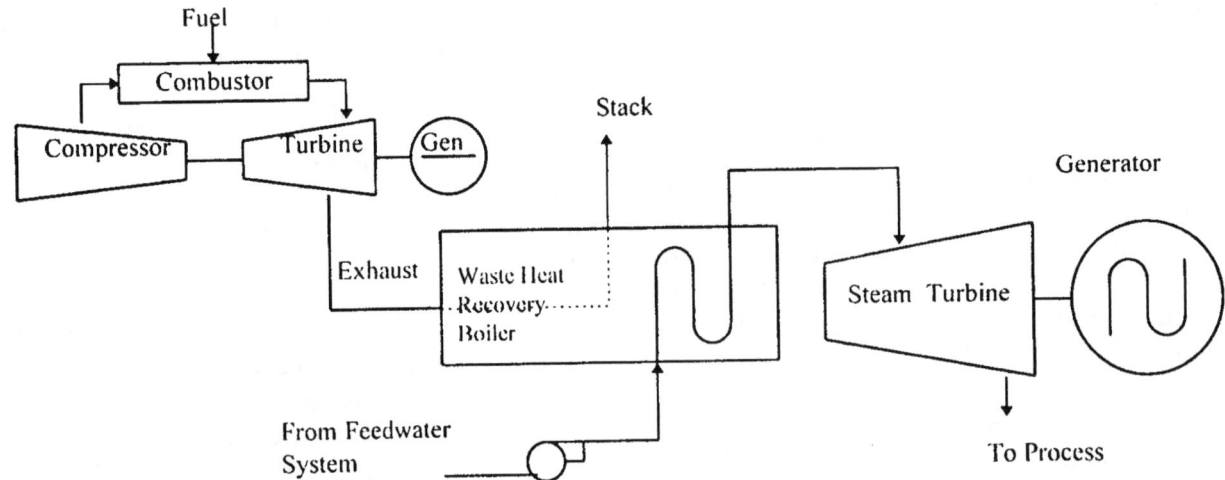

Figure 7-4. Combined cycle: gas-turbine/generator set,
waste heat recovery boiler, and steam-turbine/generator set.

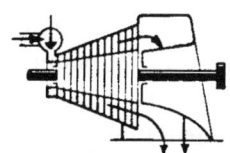

CONDENSING

STRAIGHT FLOW
For continuous and standby power exhausting to a condenser.

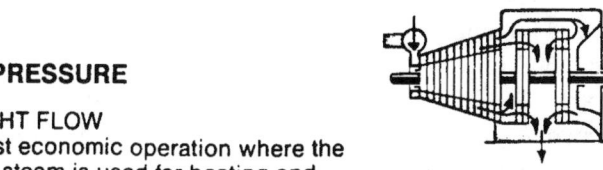

DUAL-FLOW OPPOSED EXHAUST
For large steam flow exhausting to condenser to minimize blade stresses and optimize efficiency.

BACK PRESSURE

STRAIGHT FLOW
The most economic operation where the exhaust steam is used for heating and process purposes.

NON-AUTOMATIC EXTRACTION
All the backpressure benefits plus additional higher pressure bleed for preheating.

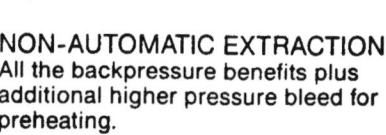

NON-AUTOMATIC EXTRACTION
To provide low flow, up to 15% of throttle flow for heating or process requirements.

CONTROLLED EXTRACTION/ INDUCTION
Where two low pressure steam headers are required.

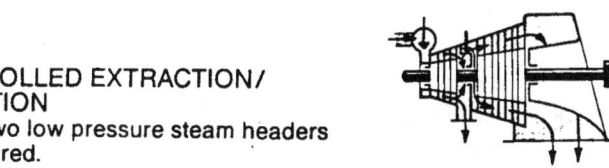

CONTROLLED EXTRACTION
Used when there is a demand for power and low pressure process steam which may be less than the steam flow required to make the power.

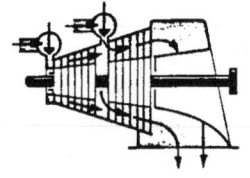

CONTROLLED INDUCTION
Enables the user to produce power from two steam pressure levels with controls which favor the less expensive lower pressure steam flow.

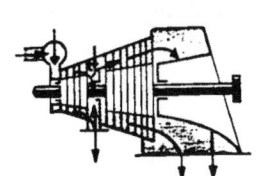

CONTROLLED EXTRACTION/ INDUCTION
Used where there is a variable demand for low pressure process steam which at times results in excess low pressure steam available to do work in the turbine.

Figure 7-5. Various multi-stage steam turbine systems. *Courtesy of Coppus Engineering/Murray Turbomachinery Corporation.*

3. Process Steam Pressure. For a given set of initial steam conditions, lowering the exhaust pressure also increases the energy available for power generation. However, this pressure is limited (totally, with non-condensing turbines, or partially, with extraction turbines) by the maximum pressure required in the industrial process. For instance, in paper mills, various pressure levels are needed to satisfy various process temperature requirements.

4. Feedwater Heating. Feedwater heating through use of steam exhausted and/or extracted from a turbine increases the power that can be generated.

Calculation of Steam Turbine Power

Given the initial steam conditions (psig, °F) and the exhaust saturated pressure (psig), theoretical steam rates (TSR) specify the amount of steam heat input required to generate a kWh in an ideal turbine. The TSR is defined by

$$\text{TSR (lb/kWh)} = \frac{3412 \text{ Btu/kWh}}{h_i - h_o \text{ Btu/lb}} \qquad (7.1)$$

where $h_i - h_o$ is the difference in enthalpy from the initial steam conditions to the exhaust pressure based on an isentropic (ideal) expansion. These values can be obtained from steam tables or a Mollier chart*. However, they are conveniently tabulated by the American Society of Mechanical Engineers. Note that to cogenerate addi-

*See Appendix 1 at the end of the book, "Mollier Diagram for Steam."

tional turbine power, some incremental make-up steam needs to be generated in the boiler. The upper bound (max) for such incremental steam is equivalent to the actual steam rate below. Note that, depending in the turbine and system efficiency, the value of incremental steam required will be exceeded many times by the value of the generated power (3X to 5X). And for smaller turbines (<1 MW) and large steam plants (>50 MW), the actual make-up fuel (or fuel chargeable to power) will be negligible relative to the overall steam generator output. In fact, concurrent steam system efficiency measures and condensate /heat recovery implemented with cogeneration steam turbines can often more than compensate for the additional steam and fuel needed.

The TSR can be converted to the actual steam rate (ASR):

$$\text{ASR (lb/kWh)} = \frac{\text{TSR}}{\eta_g} \qquad (7.2)$$

where η_g is the turbine-generator overall efficiency, stated or specified at "design" or full-load conditions. Some of the factors that define the overall efficiency of a turbine-generator set are: the inlet volume flow, pressure ratio, speed, geometry of turbine staging, throttling losses, friction losses, generator losses, and kinetic losses associated with the turbine exhaust. Most turbine manufacturers provide charts specifying either ASR or η_g values. Once the ASR has been established, the net enthalpy of the steam supplied to process (NEP) can be calculated:

$$
\begin{aligned}
&\text{NEP (Btu/lb)}\\
&= \text{ Hi} - 3500/\text{ASR} - \text{Hc}(x) - \text{Hm}(1-x) \qquad (7.3)
\end{aligned}
$$

where Hi = enthalpy at the turbine inlet conditions (Btu/lb)

3500 = conversion from heat to power (Btu/ kWh), including the effect of 2.6% radiation, mechanical, and generator losses

Hc = enthalpy of condensate return (Btu/lb)

Hm = enthalpy of make-up water (Btu/lb)

x = condensate flow fraction in boiler feedwater

(1–x) = make-up water flow fraction in boiler feed water

Hence, assuming a straight flow turbine (See Figure 7-5), the net heat to process (Btu/hr) defined in Section 7.2.1 can be obtained by multiplying equation 7.3 by the flow rate in lb/hr. The analysis of the overall cycle would require the replication of complete heat and mass balance calculations at part-load efficiencies. To expedite these computations, there are a number of commercially available software packages, which also produce mass/ heat balance tables. See Example 8 for a cogeneration software application.

Selection of Smaller Single-stage Steam Turbines

There exist many applications for smaller units (condensing and noncondensing), especially in mechanical drives or auxiliaries (fans, pumps, etc.). However, a typical application is the replacement (or bypass) of a pressure reducing valve (PRV) by a single-stage back-pressure turbine.

After obtaining the TSR from inlet and outlet steam conditions, Figure 7-6 helps in determining the approximate steam rate (ASR) for smaller (<3000HP) single-stage steam turbines. Figure 7-7 is a sample of size and speed ranges available from a turbine manufacturer. It should be noticed that there is an overlap of capacities. (For example, an ET-30 unit can operate in the ET-25 range and in a portion of the ET-15/ET-20 range.) Thus, the graph can help in selecting a unit subject to variable steam flows and/or loads.

Example 1. A stream of 15,500 lb of saturated steam at 250 psig (406°F) is being expanded through a PRV to obtain process steam at 50 psig. Determine the potential for electricity generation if the steam is expanded using a single-stage back-pressure 3,600 RPM turbine-generator.

Data:

Steam flow (W_s) : 15,500 lb/hr

Inlet Steam : 250 psig sat (264.7 psia), 406°F

Enthalpy, hi : 1201.7 Btu/lb (from Mollier chart)

Outlet Steam : 50 psig subcooled (9.67% moisture).

Enthalpy, ho : 1090.8 Btu/lb (from Mollier chart)

Turbine Speed : 3600 RPM

a) Calculate TSR using equation 7.1

$$
\begin{aligned}
\text{TSR} &= \frac{3412 \text{ Btu/kWh}}{h_i - h_o \text{ Btu/lb}}\\[2mm]
&= \frac{3412 \text{ Btu/kWh}}{1201.7 - 1090.8 \text{ Btu/lb}}\\[2mm]
&= 30.77 \text{ lb/kWh}
\end{aligned}
$$

b) Obtain steam rate from Figure 7-6, ASR = 38.5 lbs/ HP-hr.

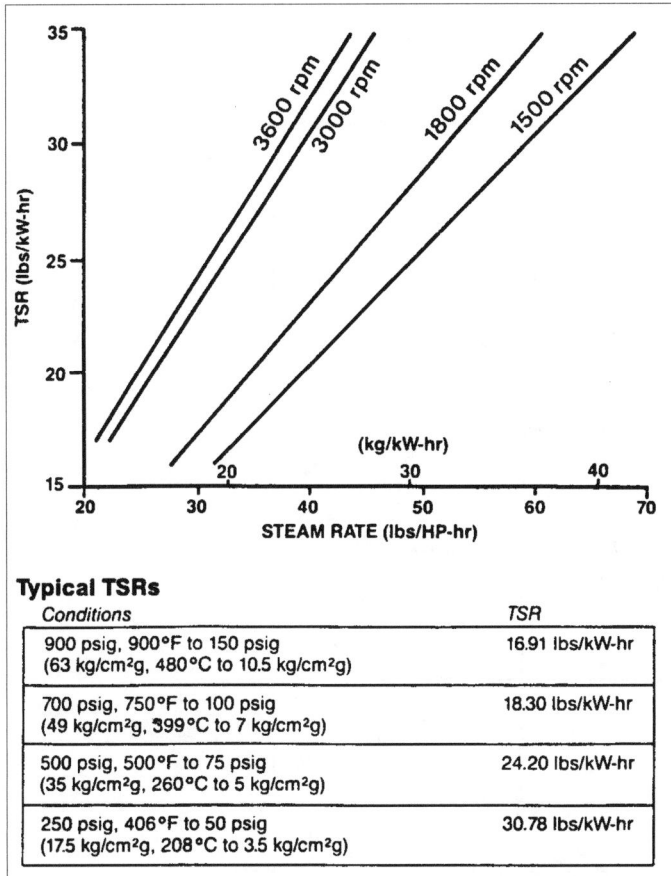

Typical TSRs

Conditions	TSR
900 psig, 900°F to 150 psig (63 kg/cm²g, 480°C to 10.5 kg/cm²g)	16.91 lbs/kW-hr
700 psig, 750°F to 100 psig (49 kg/cm²g, 399°C to 7 kg/cm²g)	18.30 lbs/kW-hr
500 psig, 500°F to 75 psig (35 kg/cm²g, 260°C to 5 kg/cm²g)	24.20 lbs/kW-hr
250 psig, 406°F to 50 psig (17.5 kg/cm²g, 208°C to 3.5 kg/cm²g)	30.78 lbs/kW-hr

Figure 7-6. Estimation of steam rates for smaller (<3,000 HP) single-stage turbines. Courtesy Skinner Engine Co.

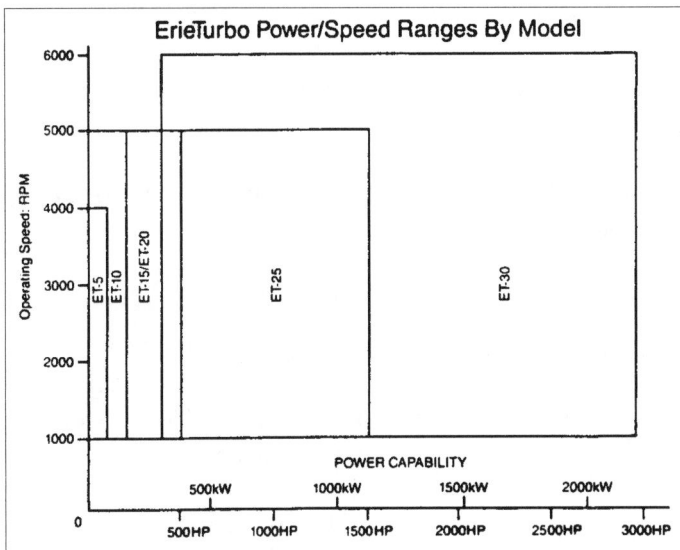

Figure 7-7. Power/speed ranges for single stage turbines. Courtesy Skinner Engine Co.

c) Calculate potential generation capacity (PGC):

$$PGC = \frac{W_s(0.746 \text{ kW/hp})}{ASR}$$

$$= \frac{(15{,}500 \text{ lb/hr}) (0.746 \text{ kW/hp})}{38.5 \text{ lbs/hp-hr}} = 300.3 \text{ kW}$$

Figure 7-7 shows that units ET-15 or ET-20 better match the required PGC. Next, the generator would have to be sized according to a commercially available unit size, e.g. 300 kW.

<u>Selection of Multi-stage Steam Turbines</u>

Multistage steam turbines provide more flexibility to match various pressure levels and variable flow rates in larger cogeneration applications. Figure 7-5 describes a variety of back-pressure and condensing multi-stage steam turbines.

For turbine selection, a Mollier diagram should be used to explore various multi-stage turbine alternatives. In general, a preliminary analysis should include the following:

— Approximation of actual steam rates.

— Defining number of stages.

— Estimation of stage pressure and temperatures.

— Calculation of full-load and part-load steam rates.

— Estimation of induction and/or extraction pressures, temperatures, and flow rates for various power outputs.

For larger units (>3000 kW), Figure 7-8 shows a chart to determine the approximate turbine efficiency when the power range, speed, and steam conditions are known. Figure 7-9 gives steam rate correction factors for off-design loads and speeds. However, manufacturers of multistage turbines advise that stage selection and other thermodynamic parameters must be evaluated taking into account other important factors such as speed limitations, mechanical stresses, leakage and throttling losses, windage, bearing friction, and reheat. Thus, after a preliminary evaluation, it is important to compare notes with the engineers of a turbine manufacturer.

Example 2. Steam flow rate must be estimated to design a heat recovery steam generator (a bottoming cycle). The steam is to be used for power generation and is to be expanded in a 5,000 RPM multistage condensing turbine to produce a maximum of 18,500 kW. Steam inlet conditions are 600 psig/750°F, and exhaust pressure is 4" HGA (absolute). Additional data are given below.

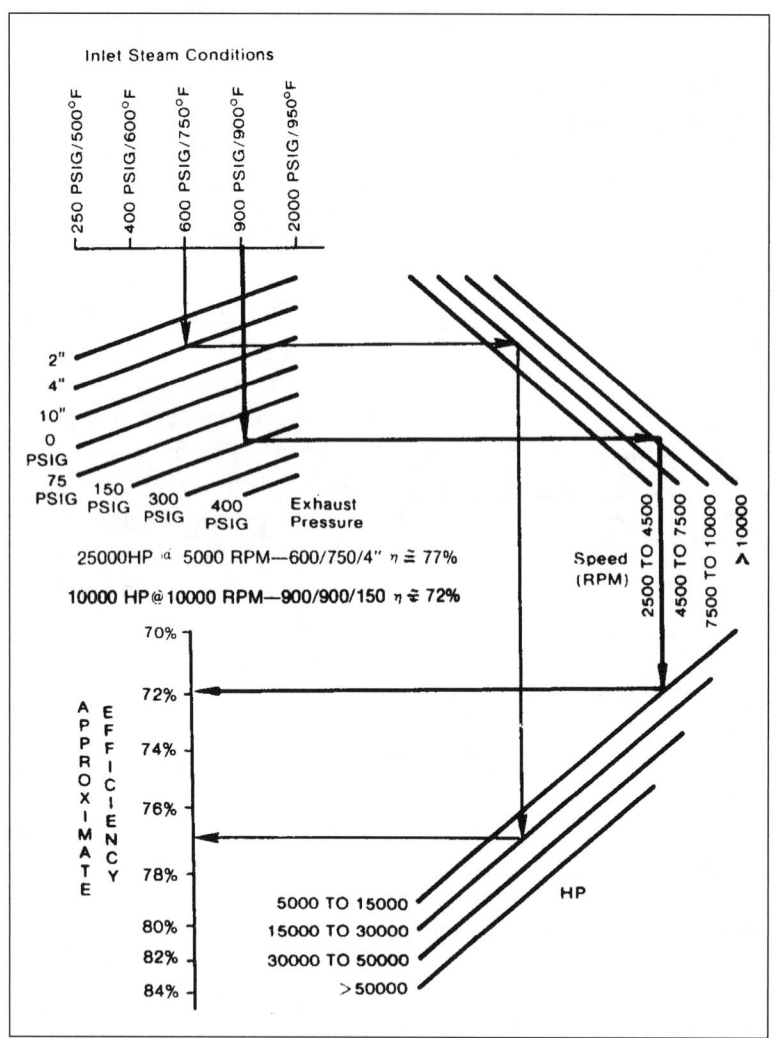

Figure 7-8. Approximate steam turbine efficiency chart for multistage steam turbines (>3,000 kW). *Courtesy Elliot Co.*

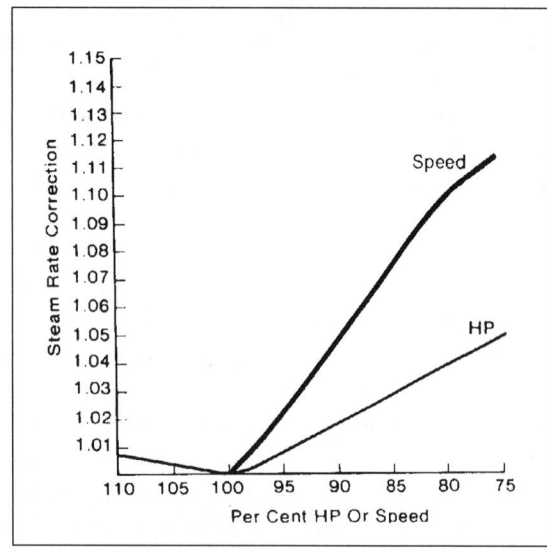

Figure 7-9. Steam turbine part-load/speed correction curves. *Courtesy Elliot Co.*

<u>Data (for a constant entropy steam expansion @ S = 1.61 Btu/lb/°R)</u>

PGC	:	8,500 kW
Inlet Steam:	:	600 psig (615 psia), 750°F
Enthalpy, h_i	:	1378.9 Btu/lb (from Mollier chart or steam tables)
Outlet Steam	:	4″ HG absolute. (2 psia)
Enthalpy, h_o	:	935.0 Btu/lb (from Mollier chart or ASME steam tables)
Turbine Speed	:	5000 RPM

a) Calculate TSR using equation 7.1 (TSR can also be obtained from ASME Tables or the Mollier chart):

$$\text{TSE (lb/kWh} = \frac{3412 \text{ Btu/kWh}}{h_i - h_o \text{ Btu/lb}}$$

$$= \frac{3412 \text{ Btu/kWh}}{1378.9 - 935.0 \text{ Btu/lb}}$$

$$= 7.68 \text{ lb/kWh}$$

b) Using the data above, Figure 7-8 gives $\eta_{tg} = 77\%$.

c) Combining equations 7.2 and 7.4 and solving for Ws, the total steam flow required is

$$W_s = \text{PCG} \times \text{TSR}/\eta_{tg}$$

$$= \frac{18,500 \text{ kW} \times 7.68 \text{ lb/kWh}}{0.77}$$

$$= 184,760 \text{ lb/hr}$$

d) Hence, the waste heat recovery steam generator should be able to generate about 185,000 lb/hr. A unit with a 200,000 lb/hr nominal capacity will likely be specified.

Example 3. Find the ASR and steam flow required when the turbine of Example 2 is operated at 14,800 kW and 4,500 RPM.

% Power variation = 14,800 kW/185,500 kW = 80%

% Speed variation = 4500 RPM/5000 RPM = 90%

From Figure 7-9 the power correction factor is 1.04 and the speed correction factor is 1.05. Then the total correction factor is:

1.04 × 1.05 = 1.09.

Therefore, the part-load ASR is

$$= \frac{7.68 \text{ lb/kWh}}{0.77} \times 1.09 = 10.9 \text{ lb/kWh}$$

The steam flow required (@ 600 psig, 750°F) is

= 10.9 lb/kWh × 14,500 kW

= 158,050 lb/hr.

7.2.2.2 Gas Turbine Systems

Gas turbines are extensively applied in industrial plants. Two types of gas turbines are utilized: one is the lighter aircraft derivative turbine and the other is the heavier industrial gas turbine. Both industrial and aircraft engines have demonstrated excellent reliability/availability in the base load service. Due to the nature of the unit designs, aircraft derivative units usually have higher maintenance costs ($/kWh) than industrial units.

Since gas turbines can burn a variety of liquid and gas fuels and run long times unattended, they are considered to be versatile and reliable. For a fixed capacity, they have the smallest relative foot-print (sq-ft per kW).

Gas Turbine Based CHP Systems

Exhaust gases from gas turbines (from 600 to 1200°F) offer a large heat recovery potential. The exhaust has been used directly, as in drying processes. Topping cycles have also been developed by using the exhaust gases to generate process steam in heat recovery steam generators (HRSGs). Where larger power loads exist, high pressure steam is generated to be subsequently expanded in a steam turbine-generator; this constitutes the so called combined cycle. (See Figure 7-4.)

If the demand for steam and/or power is even higher, the exhaust gases are used (1) as preheated combustion air of a combustion process or (2) are additionally fired by a "duct burner" to increase their heat content and temperature.

Recent developments include combined cycle systems with steam injection or STIG—from the HRSG to the gas turbine (Cheng Cycle)—to augment and modulate the electrical output of the system. The Cheng Cycle allows the cogeneration system to handle a wider range of varying heat and power loads.

All these options present a greater degree of CHP generation flexibility, allowing a gas turbine system to match a wider variety of heat-to-power demand ratios and variable loads.

Gas Turbine Ratings and Performance

There is a wide range of gas turbine sizes and drives. Available turbines have ratings that vary in discrete sizes from 50 kW to 160,000 kW. Kovacik (1982) and Hay (1988) list the following gas turbine data required for design and off-design conditions:

1. **Unit Fuel Consumption/Output Characteristics.** These depend on the unit design and manufacturer. The actual specific fuel consumption or efficiency and output also depend on (a) ambient temperature, (b) pressure ratio, and (c) part-load operation. Figure 7-11 shows performance data of a gas turbine. Vendors usually provide this kind of information.

2. Exhaust Flow Temperature. This data item allows the development of the exhaust heat recovery system. The most common recovery system are HRSGs, which are classified as unfired, supplementary fired and fired units. The amount of steam that can be generated in an unfired or supplementary fired HRSG can be estimated by the following relationship:

$$W_s = \frac{W_g C_p (T_i - T_3)\, e\, L\, f}{h_{sh} - h_{sat}} \qquad (7.5)$$

where

W_s = steam flow rate
W_g = exhaust flow rate to HRSG
C_p = specific heat of products of combustion
T_1 = gas temperature-after burner, if applicable
T_3 = saturation temperature in steam drum
L = a factor to account radiation and other losses, 0.985
h_{sh} = enthalpy of steam leaving superheater
h_{sat} = saturated liquid enthalpy in the steam drum
e = HRSG effectiveness = $(T_1 - T_2)/(T_1 - T_3)$, defined by Figure 7-10.
f = fuel factor, 1.0 for fuel oil, 1.015 for gas.

3. Parametric Studies for Off-design Conditions. Varying the amount of primary or supplementary firing will change the gas flow rate or temperature and the HRSG steam output. Thus, according to the varying temperatures, several iterations of equation [7.5] are required to evaluate off-design or part load conditions. When this evaluation is carried over a range of loads, firing rates, and temperatures, it is called a parametric study. Models can be constructed of off-design conditions using gas turbine performance data provided by manufacturers (See Figure 7-11).

4. Exhaust Pressure Effects on Output and Exhaust Temperature. Heat recovery systems increase the exhaust backpressure, reducing the turbine output in relation to simple operation (without HRSG). Turbine manufacturers provide test data about inlet and back-pressure effects, as well as elevation effects, on turbine output and efficiency (Figure 7-11). In addition, ambient temperature and altitude can impact output capacity of turbine-based plants. As rules of thumb, consider a 3% capacity loss per 10°F of ambient temperature increase above ISO condition and 3% loss per 1000 ft of elevation rise above sea level.

Example 4. In a combined cycle (Figure 7-4), the steam for the turbine of Example 2 must be generated by several HRSGs. (See Figure 7-10.) To follow variable CHP loads and to optimize overall system reliability, each HRSG will be connected to a dedicated 10-MWe gas turbine-generator (GTG) set. The GTG sets burn fuel-oil and each unit exhausts 140,000 kg/hr of gas at 900°F. Estimate the total gas flow to the HRSGs, if the system must produce a maximum of 160,000 lb/hr of 615 psia/750°F steam. How many HRSG-gas turbine sets are needed?

DATA (see notation of Equation 7.5 and Figure 7-10):

Gas W_g = 140,000 kg/hr per gas-turbine unit
Turbines C_p = 0.26 Btu/lb/°F, average specific heat of gases between T_1 and T_4
          T_1 = 900°F, hot gas temperature
          f = fuel factor, 1 for fuel oil.

The inlet steam conditions required in the steam turbine of Example 3 are: (615 psia/750°F) and h_i = 1378.9 Btu/lb (from Mollier chart). Thus,

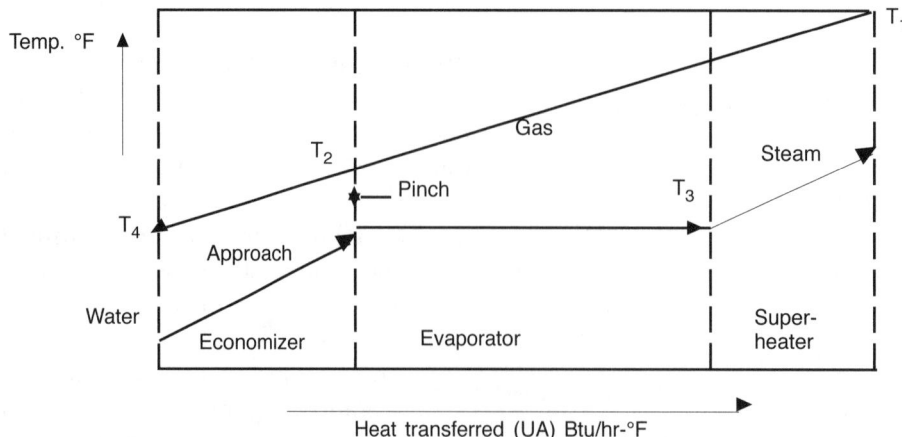

Figure 7-10. Heat recovery steam generator diagram.

FUEL	ISO RATING	POWER kWe	HEAT RATE kJ/kWh (Btu/kWh)		EXHAUST FLOW kg/hr (lb/hr)		EXHAUST TEMP °C (°F)	
Natural Gas	Continuous	4370	12 870	(12 200)	75 050	(165 120)	497	(927)
	Peak	4800	12 865	(12 195)	74 970	(165 280)	534	(994)
Distillate	Continuous	4280	13 135	(12 450)	75 050	(165 120)	497	(927)
	Peak	4700	13 130	(12 445)	74 970	(165 280)	534	(994)

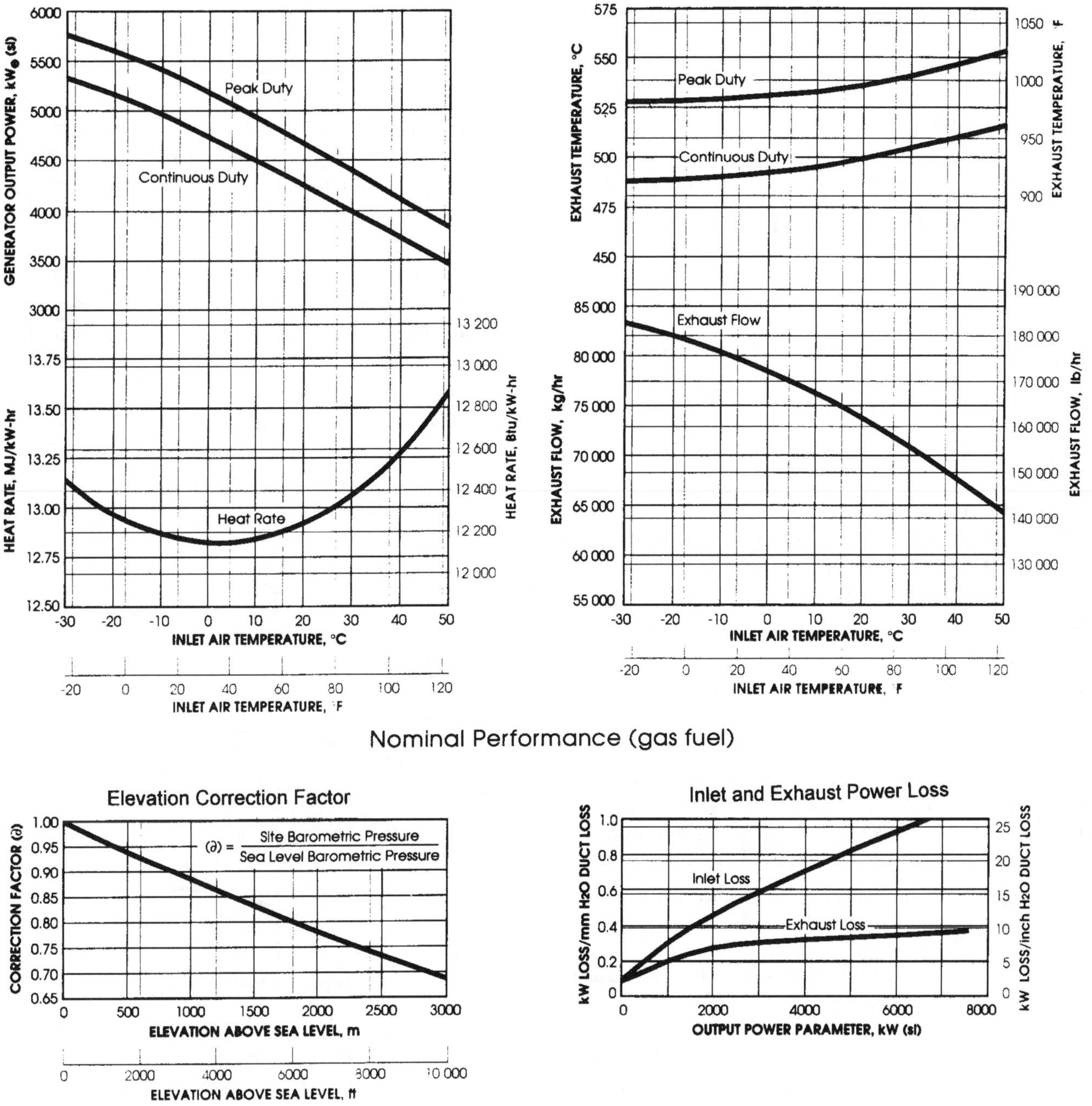

Nominal Performance (gas fuel)

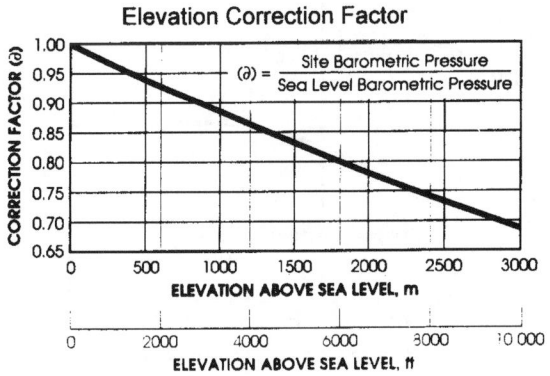

Elevation Correction Factor

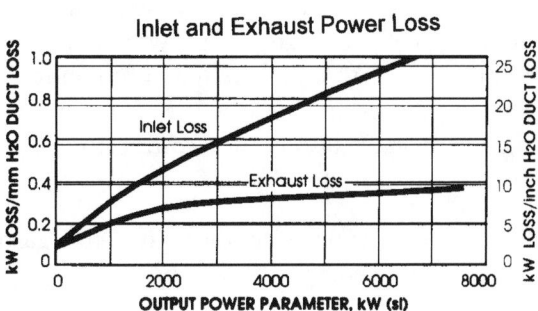

Inlet and Exhaust Power Loss

Figure 7-11. Performance ratings and curves of a gas turbine-generator set. Nominal ratings are given at ISO conditions: 59°F, sea level, 60% relative humidity and no external pressure losses. *Courtesy of Solar Turbines, Inc.*

<u>HRSGs</u> Ws = 160,000 lb/hr from all HRSGs
       T3 = 488.8°F (temp of 615 psia sat steam)
       L   = 0.98
       e   = HRSG effectiveness, 0.9

Therefore, $h_{Sh} = h_i = 1378.9$ Btu/lb and,
              $h_{sat} = 474.7$ Btu/lb (Sat. Water @
                        615 psia)

From equation 7.5, the total combustion gas flow required is

$$W_s = \frac{W_g(h_{sh} - h_{sat})}{Cp\,(T_1 - T_3)\,e\,L\,f}$$

$$= \frac{160,000\ \text{lb/hr}\ (1381.1 - 474.7\ \text{Btu/lb})}{0.26\ \text{Btu/lb/°F}\ (900\text{–}488.8\text{°F})\ 0.9 \times 0.98 \times 1}$$

$$= 1,537,959.3\ \text{lb/hr or } 427.2\ \text{lb/sec.}$$

Finally, the number of required gas turbine/HRSG sets is

= [(1,537,959 lb/hr) (1kg/2.21b)/(140,000 kg/hr/set))

= 4.99, i.e. 5 sets.

7.2.2.3 Reciprocating Engine Systems

Reciprocating engines include a variety of internally fired, piston driven engines. Their sizes range from 10 bhp to 50,000 bhp. According to Kovacik (1982), the largest unit supplied by a U.S. manufacturer is rated at 13,500 bhp. In larger plants, several units are used to accommodate part load and to provide redundancy and better availability.

In these engines, combustion heat rejected through the jacket water, lube oil and exhaust gases, can be recovered through heat exchangers to generate hot water and/or steam. Figure 7-13 shows an internal combustion engine cogeneration system.

Exhaust gases have also been used directly. Reciprocating engines are classified by:

— the thermodynamic cycle, diesel or Otto (gasoline, propane, methane or lean-burn gas)

— the rotation speed: high-speed (1200-1800 rpm), medium-speed (500-900 rpm) or low speed (450 rpm or less)

— the aspiration type: naturally aspirated or turbocharged

— the operating cycle: two-cycle or four-cycle

— the fuel burned: fuel-oil fired, or natural-gas fired.

Reciprocating engines are widely used to move vehicles, generators, and a variety of shaft loads. Larger engines are associated with lower speeds, increased torque, and heavier duties. The total heat utilization of CHP systems based on gas-fired or fuel-oil fired engines approach 60-75%. Figure 7-12 shows the CHP balance vs. load of a diesel engine.

Example 5. Estimate the amount of 180 F water that can be produced by recovering heat, first from the jacket water and then from the exhaust of a 1200 kW diesel generator. On average, the engine runs at a 75% load and the inlet water temperature is 70°F. The effectiveness of the jacket water heat exchanger is 90%, and exhaust heat exchanger is 80%.

From Figure 7-12, at 75% load, the exhaust heat and the jacket water heat are 22% and 33%, respectively. Thus, the flow rate of water heated from 70 to 180°F is:

(1200 kW × 75%) [(90% × 22%) + (80% × 33%)]
(3412 Btu/kW) (1 lb.°F/Btu)(1 lb/8.33 gal)
(1 hr/60 min)/(180 – 70°F)
= 25.8 gal per minute.

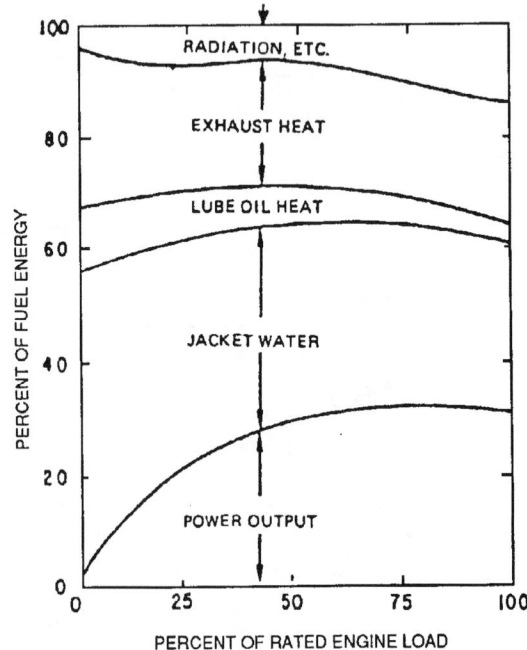

Figure 7-12. Diesel engine heat and power balance.

7.2.2.4 Distributed Generation

Distributed Generation (DG) is emerging as the generation of heat and power (CHP) through relatively small distributed units (from 25 MW to a few kilowatts). Typically, these are smaller, self contained, power generation systems located close to, adjacent to, or within the boundaries of a CHP user or consumer facility. Typical fuels include natural gas, liquefied petroleum gas (LPG), kerosene, and diesel fuel. DG plants may or may not be interconnected to a utility grid. Earlier DG systems have also been known as packaged cogeneration units.

This section focuses on current developments of the following DG technologies:

- Combustion or Gas Turbines
  (See also Section 7.2.2.2)
- Reciprocating Engines
  (See also Section 7.2.2.3)
- Fuel Cells
  (See also Chapter 16)
- Photovoltaics or solar cells
  (See also Chapter 16)
- Microturbines
  (See also Section 7.2.2.2)

Previous sections in this chapter discuss some of the underlying technologies. Chapter 16, on "Alternative Energy," provides a fundamental and renewable-energy approach to Photovoltaics and Fuel Cells. Next, Table 7-2 shows a comparison of current DG technologies.

DG Economics

In addition to the economic factors listed in Table 7-2, i.e. turn key (installed), heat recovery (installed) and operation and maintenance (O&M) costs, there are other evaluation considerations which are often site specific. These include local fuel availability and cost, size and weight limitations, emission and noise regulations, and other factors.

DG can be used for various purposes and applications with varying economic merit:

1. As a prime mover to supply base-load electrical demand with or without heat recovery.
2. As a peak shaving generator.
3. As an uninterruptible power supply or emergency back-up power unit.
4. As a combination emergency power and peak-shaving generator.

In general, DG as a prime mover for continuous operation does not show feasible economics unless there is significant heat recovery. Thus, for 5000 hrs/yr or more, with heat recovery (60-70% CHP system efficiency), our experience with several projects in the US and the Caribbean shows that for the 200 to 800 kW- range of gas- or diesel fired engine generators,

Table 7-2. Comparison of DG Technologies

Comparison Factor	Diesel Engine	Gas Engine	Simple Cycle Gas Turbine	Microturbine	Fuel Cells	Photovoltaics
Product Availability	Commercial	Commercial	Commercial	1999-2005	1996-2010	Commercial
Size Range (kW/unit)	20 - 10,000+	50 - 5,000+	1,000 -30,000	20 - 200	50 - 1000+	1+
Typical DG Range (kW/unit)	200 - 2,000	300-3,000	1,000 - 10,000	20-100	50 - 200	1 - 5
Efficiency (HHV)	36 - 43%	28 - 42%	21 - 40%	25 - 30%	35 - 54%	n.a.
Genset Package Cost ($/kW)	125 – 300	250 - 600	300 - 600	350 - 750	1500 -3000	n.a.
Turnkey Cost - With no heat recovery ($/kW)	350 – 500	600 - 1000	650 - 900	600 - 1100	1900 - 3500	5000 - 10000
Heat Recovery Added Cost ($/kW)	100- 200	75- 150	100 - 200	75 - 350	Incl.	n.a.
O&M Cost ($/kWh)	0.005 - 0.010	0.007 - 0.015	0.003 - 0.008	0.005 - 0.010	0.005 - 0.010	0.001 - 0.004

Source: Gas Research Institute (2000): www.gri.org

Table 7-3. DG Economics and Sensitivity Analysis

Simple Payback in Years for 5,000 hrs/yr operation and $500/kW installed cost							
	DG With Heat Recovery		Assumes Recovered Heat = Power Output			68%	CHP efficiency
Gas Cost	Electrical Cost ($/kWh)						
($/MMBTU)	0.06	0.07	0.08	0.09	0.10	0.11	0.12
5.00	6.13	3.80	2.75	2.16	1.78	1.51	1.31
6.00	9.44	4.86	3.27	2.46	1.98	1.65	1.42
7.00	20.60	6.73	4.02	2.87	2.23	1.82	1.54
8.00	N/F	10.96	5.23	3.43	2.56	2.04	1.69
9.00	N/F	29.54	7.47	4.28	3.00	2.30	1.87
10.00	N/F	N/F	13.07	5.67	3.62	2.66	2.10
11.00	N/F	N/F	52.22	8.39	4.56	3.13	2.39
12.00	N/F	N/F	N/F	16.18	6.18	3.82	2.76
N/F: Not Feasible	DG system generates CHP and includes a heat recovery module ($100/kW installed).						

Simple Payback in Years for 5,000 hrs/yr operation and $400/kW installed cost							
			DG Without Heat Recovery			34%	efficiency
5.00	N/F	16.00	5.33	3.20	2.29	1.78	1.45
6.00	N/F	N/F	16.00	5.33	3.20	2.29	1.78
7.00	N/F	N/F	N/F	16.00	5.33	3.20	2.29
8.00	N/F	N/F	N/F	N/F	16.00	5.33	3.20
9.00	N/F	N/F	N/F	N/F	N/F	16.00	5.33
10.00	N/F	N/F	N/F	N/F	N/F	N/F	16.00
11.00	N/F	N/F	N/F	N/F	N/F	N/F	N/F
12.00	N/F	N/F	N/F	N/F	N/F	N/F	N/F
N/F: Not Feasible							

there is an acceptable payback (3 years or less) only if the displaced electricity costs more than $0.10 per kWh and the natural gas fuel is $6.00 per million Btu or less. The evaluation considers a $0.015/ kWh CHP plant operation and maintenance fee, in addition to fuel consumption. (See upper part of Table 7-3. The shadowed cells on the upper-right part of the tables list paybacks of three years or less.)

Without heat Recovery, DG makes little economic sense for continuous operation, unless an inexpensive and plentiful source of fuel is available (e.g. bio-diesel from waste cooking oil or bio-gas from obtained from agricultural waste, sewer treatment plants and landfills). DG is also feasible in the cases where surplus or pipeline natural or liquefied petroleum gas exists. To have a payback of three years or less, the bottom part of Table 7-3 shows the DG fuel must cost $6/million Btu or less and the displaced power must cost $0.11/kWh or more. This sensitivity analysis assumes $0.010/kWh for operation and maintenance, in addition to fuel cost.

Peak Shaving Optimization Model and Case Study

While DG applications for continuous power generation (with no heat recovery) are rarely economically feasible, DG for peak shaving, in conjunction with back-up or emergency power generation, is generally an attractive business proposition. See Appendix A at the end of this chapter for a detailed case study which shows there is generally a good business case for peak shaving using diesel or gas fired generators. Such a case study illustrates an optimization model to estimate the optimal peak shaving generator size (in kW or MW).

Combustion or Gas Turbines*

Combustion turbine (CT) sizes for distributed generation vary from 1 to 30 MW. CT's are used to power aircraft, marine vessels, gas compressors, and utility and industrial generators. In 1998, over 500 CT's were shipped from the US to worldwide facilities totaling 3,500 MW of power capacity. Most of these were sold overseas. The North American market represents only an 11% share of the total. The primary application of CT's is as prime mover for continuous power, particularly in a combined cycle arrangement, or as a peaking unit to generate during peak demand periods.

Low maintenance, high reliability, and high quality exhaust heat make CT's an excellent choice for industrial and commercial CHP applications larger than 3 MW. CT's can burn natural gas, liquid fuels such as diesel oil, or both gas and liquid fuels (dual-fuel operation). Thus, they contribute to the fuel security of the DG plant. CT emissions can be controlled by using dry low NO_x com-

---

*The updates on DG technologies have been obtained from "Fundamentals of Distributed Generation" and "Distributed Generation: A Primer," from the Gas Research Institute web site: *http://www.gri.org/pub/solutions/dg/index.html*.

bustors, water or steam injection, or exhaust treatments such as selective catalytic reduction (SCR). Due to their inherent reliability and remote diagnostic capability, CTs tend to have one of the lowest maintenance costs among DG technologies.

Reciprocating Engines

Reciprocating internal combustion (IC) engines are a widespread and well-known technology. North American production tops 35 million units per year for automobiles, trucks, construction and mining equipment, lawn care, marine propulsion, and of course, all types of power generation from small portable gen-sets to engines the size of a house, powering generators of several megawatts. Spark ignition engines for power generation use natural gas as the preferred fuel, though they can be set up to run on propane or gasoline. Diesel cycle, compression ignition engines can operate on diesel fuel or heavy oil, or they can be set up in a dual-fuel configuration that burns primarily natural gas with a small amount of diesel pilot fuel (and can be switched to 100% diesel).

Current generation IC engines offer low first cost, easy start-up, proven reliability when properly maintained, good load-following characteristics, and heat recovery potential. IC engine systems with heat recovery have become a popular form of DG in Europe. Emissions of IC engines have been reduced significantly in the last several years by exhaust catalysts and through better design and control of the combustion process. IC engines are well suited for standby, peaking, and intermediate applications for combined heat and power (CHP) in commercial and light industrial applications of less than 10 MW.

Microturbines

Microturbines or turbogenerators are small combustion turbines with outputs of 30-200 kW. Individual units can be packaged together to serve larger loads. Turbogenerator technology has evolved from automotive and truck turbochargers to auxiliary power units for airplanes and small jet engines used for pilot military aircraft.

Recent development of these microturbines has been focused on this technology as the prime mover for hybrid electric vehicles and as a stationary power source for the DG market. In most configurations, the turbine shaft, spinning at up to 100,000 rpm, drives a high speed generator. This high frequency output is first rectified and then converted to 60 Hz (or 50 Hz). The systems are capable of producing power at around 25 to 30% efficiency by employing a recuperator that transfers heat energy from the exhaust stream back into the incoming air stream. Like larger turbines, these units are capable of operating on a variety of fuels. The systems are air-cooled, and some even use air bearings, thereby eliminating both water and oil systems. Low-emission combustion systems are being demonstrated that provide emissions performance comparable to larger CTS. Turbogenerators are appropriately sized for cogeneration or power-only applications in commercial buildings or light industrial markets.

Fuel Cells

Fuel cells (Figure 7-13) produce power electrochemically like a battery rather than like a conventional generating system that converts fuel to heat to shaft-power and finally to electricity. Unlike a storage battery, however, which produces power from stored chemicals, fuel cells produce power when hydrogen fuel is delivered to the negative pole (cathode) of the cell and oxygen in air is delivered to the positive pole (anode). The hydrogen fuel can come from a variety of sources, but the most economic one is steam reforming of natural gas, a chemical process that strips the hydrogen from both the fuel and the steam. Several different liquid and solid media can be used to create the fuel cell's electrochemical reaction a phosphoric acid fuel cell (PAFC), a molten carbonate fuel cell (MCFC), a solid oxide fuel cell (SOFC), and a proton exchange membrane (PEM). Each of these media comprises a distinct fuel cell technology with its own performance characteristics and development schedule. PAFCs are in early commercial market development now, with 200 kW units delivered to over 120 customers.

The SOFC and MCFC technologies are now in field test or demonstration. PEM units are in early development and testing. Direct electrochemical reactions are generally more efficient than using fuel to drive a heat

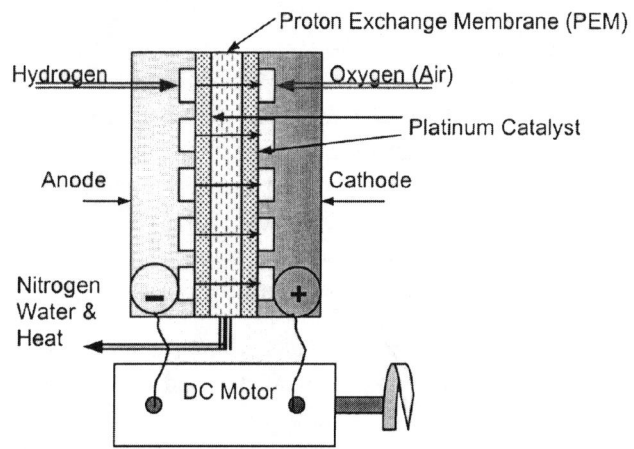

Figure 7-13. Proton Exchange Membrane (PEM) fuel cells use platinum catalysts to promote the flow of anions (positive ions) through the membrane, thus creating a direct current (DC). An inverter can be used to convert DC to AC.

engine to produce electricity. Fuel cell efficiencies range from 35-40% for the PAFC up to 60% with MCFC and SOFC systems under development. PEM unit efficiencies are as high as 50%. Fuel cells are inherently quiet and extremely clean running. Like a battery, fuel cells produce direct current (DC) that must be run through an inverter to get 60 Hz alternating current (AC). These power electronics components can be integrated with other components as part of a power quality control strategy for sensitive customers. Because of current high costs, fuel cells are best suited to environmentally sensitive areas and customers with power quality concerns. Some fuel cell technology is modular and capable of application in small commercial and even residential markets; other technology utilizes high temperatures in larger sized systems that would be well suited to industrial cogeneration applications.

Photovoltaics

Photovoltaic power cells use solar energy to produce power. Photovoltaic power is modular and can be sited wherever the sun shines. These systems have been commercially demonstrated in extremely sensitive environmental areas and remote (grid-isolated) applications. Battery banks are needed to store the energy harnessed during daytime. High costs make these systems a niche technology that is able to compete more on the basis of environmental benefits than on economics. Isolated facilities with need for limited but critical power are typical applications.

7.2.3 The Cogeneration Design Process

The following evaluation steps are suggested to carry out cogeneration system design.

1.   Develop the profile of the various process steam (heat) demands at the appropriate steam pressures for the applications being studied. Also, collect data with regard to condensate returned from the process and its temperature. Data must include daily fluctuations due to normal variations in process needs, as well as seasonal weather effects; include the influence of not-working periods such as weekends, vacation periods, and holidays.

2.   Develop a profile for electric power in the same manner as the process heat demand profile. These profiles typically include hour-by-hour heat and power demands for "typical" days (or weeks) for each season or month of the year.

3.   Determine fuel availability and present-day cost as well as projected future costs. The study should also factor process by-product fuels into the development of the energy supply system.

4.   Determine purchased power availability and its present and expected future cost.

5.   Collect and evaluate plant discharge stream data in the same degree of detail as the process heat demand data.

6.   Determine number and rating of major (demand and generation) equipment items. This evaluation usually establish whether spare capacity and/or supplementary firing should be installed.

7.   Evaluate plant, process and CHP system economic lives.

Once this initial data bank has been established, the various alternatives that can satisfy plant heat and power demands can be identified. Subsequently, detailed technical analyses are conducted. Thus, energy balances are made, investment cost estimated, and the economic merit of each alternative evaluated. Some approaches for evaluation are discussed next.

7.2.4 Economic Feasibility Evaluation Methods

Cogeneration feasibility evaluation is an iterative process; further evaluations generally require more data. There are a number of evaluation methods using various approaches and different levels of technical detail. Most of them consider seasonal loads and equipment performance characteristics. Some of the most representative methods are discussed as follows.

**7.2.4.1 General Approaches For
        Design and Evaluation**

Hay (1988) presents a structured approach for system design and evaluation. It is a sequence of evaluation iterations, each greater than the previous and each producing information as to whether the costs of the next step is warranted. His suggested design process is based on the following steps.

Step 1: Site Walkthrough and Technical Screening
Step 2: Preliminary Economic Screening
Step 3: Detailed Engineering Design

Similarly, Butler (1984) considers three steps to perform the study, engineering, and construction of cogeneration projects. These are discussed as follow.

Step 1. Preliminary studies and conceptual engineering. This is achieved by performing a technical feasibility and economic cost-benefit study to rank and recommend alternatives. The determination of technical feasibility includes a realistic assessment with respect to environmental impact, regulatory compliance, and interface with a utility. Then an economic analysis-based on the simple payback period-serves as a basis for more refined evaluations.

Step 2. Engineering and Construction Planning. Once an alternative has been selected and approved by the owner, preliminary engineering is started to develop the general design criteria. These criteria include specific site information such as process heat and power requirements, fuel availability and pricing, system type definition, modes of operation, system interface, a review of alternatives under more detailed load and equipment data, confirmation of a selected alternative, and finally, the size of plant equipment and systems to match the application.

Step 3. Design Documentation. This includes the preparation of project flow charts; piping and instrument diagrams; general arrangement drawings; equipment layouts; process interface layouts; building, structural and foundation drawings; electrical diagrams; and specification of an energy management system, if required.

Several methodologies and manuals have been developed to carry out Step 1, i.e. screening analysis and preliminary feasibility studies. Some of them are briefly discussed in the next sections. Steps 2 and 3 usually require ad-hoc approaches according to the characteristics of each particular site. Therefore, a general methodology is not applicable for such activities.

7.2.4.2 Preliminary Feasibility Study Approaches

AGA Manual—GKCO Consultants (1982) developed a cogeneration feasibility (technical and economical) evaluation manual for the American Gas Association, AGA. It contains a "Cogeneration Conceptual Design Guide" that provides guidelines for the development of plant designs. It specifies the following steps to conduct the site feasibility study:

a) Select the type of prime mover or cycle (piston engine, gas turbine or steam turbine);

b) Determine the total installed capacity;

c) Determine the size and number of prime movers;

d) Determine the required standby capacity.

According to its authors, "the approach taken (in the manual) is to develop the minimal amount of information required for the feasibility analysis, deferring more rigorous and comprehensive analyses to the actual concept study." The approach includes the discussion of the following "Design Options," or design criteria, to determine (1) the size and (2) the operation mode of the CHP system.

Isolated Operation, Electric Load Following—The facility is independent of the electric utility grid, and it is required to produce all power required on-site and to provide all required reserves for scheduled and unscheduled maintenance.

Baseloaded, Electrically Sized—The facility is sized for baseloaded operation based on the minimum historic billing demand. Supplemental power is purchased from the utility grid. This facility concept generally results in a shorter payback period than that from the isolated site.

Baseloaded, Thermally Sized—The facility is sized to provide most of the site's required thermal energy using recovered heat. The engines are operated to follow the thermal demand with supplemental boiler fired as required. The authors point out that: "this option frequently results in the production of more power than is required on-site and this power is sold to the electric utility."

In addition, the AGA manual includes a description of sources of information or processes by which background data can be developed for the specific gas distribution service area. Such information can be used to adapt the feasibility screening procedures to a specific utility.

7.2.4.3 Cogeneration System Selection and Sizing.

The selection of a set of "candidate" cogeneration systems endeavors to tentatively specify the most appropriate prime mover technology, which will be further evaluated in the course of the study. Often, two or more alternative systems that meet the technical requirements are pre-selected for further evaluation. For instance, a plant's CHP requirements can be met by either a reciprocating engine system or a combustion turbine system. Thus, the two system technologies are pre-selected for a more detailed economic analysis.

To evaluate specific technologies, there exist a vast number of technology-specific manuals and references. A representative sample is listed as follows. Mackay (1983) has developed a manual titled "Gas Turbine Cogeneration: Design, Evaluation and Installation." Kovacik (1984) reviews application considerations for both steam turbine and gas turbine cogeneration systems.

Limaye (1987) has compiled several case studies on industrial cogeneration applications. Hay (1988) discusses technical and economic considerations for cogeneration application of gas engines, gas turbines, steam engines and packaged systems. Keklhofer (1991) has written a treatise on technical and economic analysis of combined-cycle gas and steam turbine power plants. Ganapathy (1991) has produced a manual on waste heat boilers.

Usually, system selection is assumed to be separate from sizing the cogeneration equipment (kWe). However, since performance, reliability, and cost are very dependent on equipment size and number, technology selection and system size are very intertwined evaluation factors. In addition to the system design criteria given by the AGA manual, several approaches for cogeneration system selection and/or sizing are discussed as follows.

Heat-to-power Ratio

Canton et al (1987), of The Combustion and Fuels Research Group at Texas A&M University, has devel-

oped a methodology to select a cogeneration system for a given industrial application using the heat to power ratio (HPR). The methodology includes a series of graphs used (1) to define the load HPR and (2) to compare and match the load HPR to the HPRs of existing equipment. Consideration is then given to either heat or power load matching and modulation.

Sizing Procedures

Hay (1987) considers the use of the load duration curve to model variable thermal and electrical loads in system sizing, along with four different scenarios described in Figure 7-14. Each one of these scenarios defines an operating alternative associated with a system size.

Oven (1991) discusses the use of the load duration curve to model variable thermal and electrical loads in system sizing in conjunction with required thermal and electrical load factors. Given the thermal load duration and electrical load duration curves for a particular facility, different sizing alternatives can be defined for various load factors.

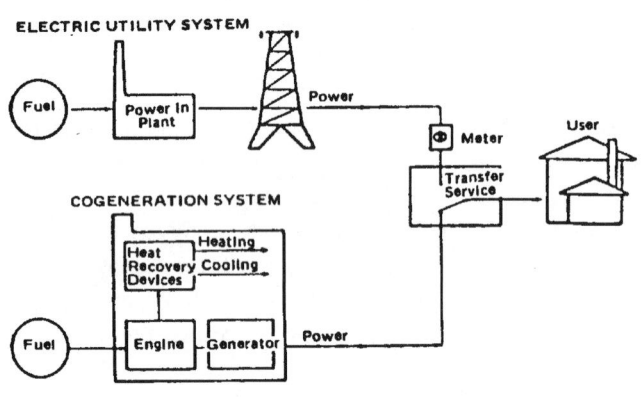

Isolated Cogenerator: User receives all power from cogeneration system *or* electric utility system.

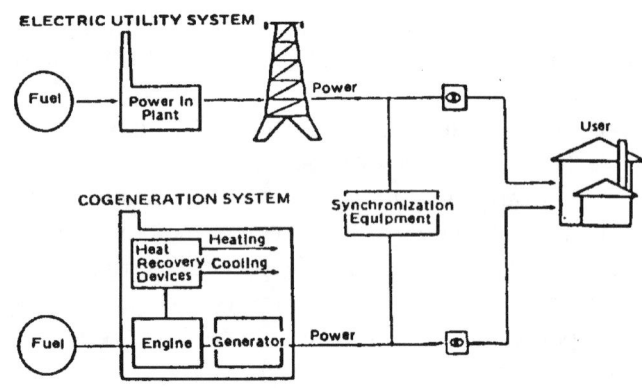

Electrically Base-Loaded Cogeneration: User receives power from cogeneration system and purchases power from the electric utility system.

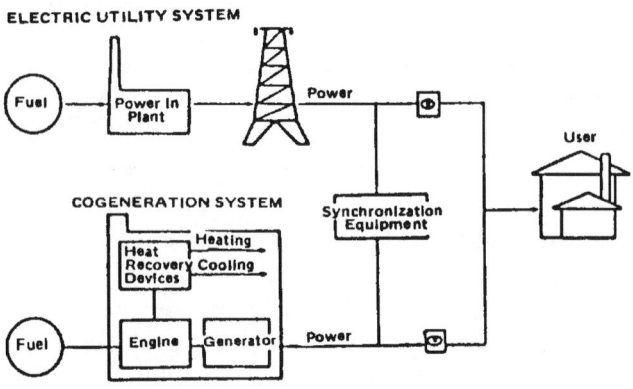

Thermally Base-Loaded Cogeneration: User receives power from cogeneration system *and* cogenerator sells power to the electric utility.

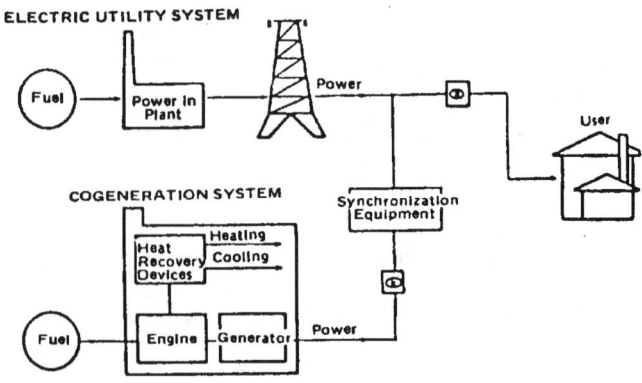

Maximum System: User receives power from the utility. Cogenerator sells output to the utility grid.

Figure 7-14. Each operation mode defines a sizing alternative. *Source: Hay (1987).*

Eastey et al. (1984) discusses a model (CO-GENOPT) for sizing cogeneration systems. The basic inputs to the model are a set of thermal and electric profiles, the cost of fuels and electricity, equipment cost, and performance for a particular technology. The model calculates the operating costs and the number of units for different system sizes. Then it estimates the net present value for each one of them. Based on the maximum net present value, the "optimum" system is selected. The model includes cost and load escalation.

Wong, Ganesh and Turner (1991) have developed two statistical computer models to optimize cogeneration system size subject to varying capacities/loads and to meet an availability requirement. One model is for internal combustion engines, the other for unfired gas turbine cogeneration systems. Once the user defines a required availability, the models determine the system size or capacity that both meets the required availability and maximizes the expected annual worth of its life cycle cost.

7.3   COMPUTER PROGRAMS

There are several computer programs (mainly PC based) available for detailed evaluation of cogeneration systems. In opposition to the rather simple methods discussed above, CHP programs are intended for system configuration or detailed design and analysis. For these reasons, they require a vast amount of input data. The more involved examples in this chapter have been developed using various programs. See Section 7.5.

7.4   U.S. COGENERATION LEGISLATION: PURPA

In 1978 the U.S. Congress amended the Federal Power Act by promulgation of the Public Utilities Regulatory Act (PURPA). The Act recognized the energy saving potential of industrial cogeneration and small power plants, the need for real and significant incentives for development of these facilities, and the private sector requirement to remain unregulated.

PURPA of 1978 eliminated several obstacles to cogeneration so cogenerators can count on "fair" treatment by the local electric utility with regard to interconnection, back-up power supplies, and the sale of excess power. PURPA contains the major federal initiatives regarding cogeneration and small power production. These initiatives are stated as rules and regulations pertaining to PURPA Sections 210 and 201, which were issued in final form in February and March of 1980, respectively. These rules and regulations are discussed in the following sections.

Initially, several utilities—especially those with excess capacity—were reticent to buy cogenerated power and have in the past contested PURPA. Power (1980) magazine reported several cases in which opposition persisted in some utilities to private cogeneration. But after the Supreme Court ruling in favor of PURPA, more and more utilities are finding that PURPA can work to their advantage. Polsky and Landry (1987) report that some utilities are changing attitudes and are even investing in cogeneration projects.

7.4.1 PURPA 201*

Section 201 of PURPA requires the Federal Energy Regulatory Commission (FERC) to define the criteria and procedures by which small power producers (SPPs) and cogeneration facilities can obtain qualifying status to receive the rate benefits and exemptions set forth in Section 210 of PURPA. Some PURPA 201 definitions are stated below.

Small Power Production Facility

A "Small Power Production Facility" is a facility that uses biomass, waste, or renewable resources, including wind, solar and water, to produce electric power and is not greater than 80 megawatts.

Facilities less than 30 MW are exempt from the Public Utility Holding Co. Act and certain state law and regulation. Plants of 30 to 80 MW that use biomass may be exempted from the above but may not be exempted from certain sections of the Federal Power Act.

Cogeneration Facility

A "Cogeneration Facility" is a facility which produces electric energy and forms of useful thermal energy (such as heat or steam) used for industrial, commercial, heating or cooling purposes, through the sequential use of energy. A qualifying facility (QF) must meet certain minimum efficiency standards as described later. Cogeneration facilities are generally classified as "topping cycle" or "bottoming cycle" facilities.

7.4.2   Qualification of a "Cogeneration Facility" or a "Small Power Production Facility" under PURPA

Cogeneration Facilities

To distinguish new cogeneration facilities which will achieve meaningful energy conservation from those

---

*Most of the following sections have been adapted from CFR18 (1990) and Harkins (1980), unless quoted otherwise.

which would be "token" facilities producing trivial amounts of either useful heat or power, the FERC rules establish operating and efficiency standards for both topping-cycle and bottoming-cycle NEW cogeneration facilities. No efficiency standards are required for *existing* cogeneration facilities, regardless of energy source or type of facility. The following fuel utilization effectiveness (FUE) values—based on the lower heating value (LHV) of the fuel—are required from QFs.

- For a new topping-cycle facility:

  — No less than 5% of the total annual energy output of the facility must be useful thermal energy.

- For any new topping-cycle facility that uses any natural gas or oil:

  — All the useful electric power and half the useful thermal energy must equal at least 42.5% of the total annual natural gas and oil energy input; and

  — If the useful thermal output of a facility is less than 15% of the total energy output of the facility, the useful power output plus one-half the useful thermal energy output must be no less than 45% of the total energy input of natural gas and oil for the calendar.

For a new bottoming-cycle facility:

- If supplementary firing (heating of water or steam before entering the electricity generation cycle from the thermal energy cycle) is done with oil or gas, the useful power output of the bottoming cycle must, during any calendar year, be no less than 45% of the energy input of natural gas and oil for supplementary firing.

Small Power Production Facilities

To qualify as a small power production facility under PURPA, the facility must have production capacity of under 80 MW and must get more than 50% of its total energy input from biomass, waste, or renewable resources. Also, use of oil, coal, or natural gas by the facility may not exceed 25% of total annual energy input to the facility.

Ownership Rules Applying to
Cogeneration and Small Power Producers

A qualifying facility may not have more than 50% of the equal interest in the facility held by an electric utility.

7.4.3 PURPA 210

Section 210 of PURPA directs the Federal Energy Regulatory Commission (FERC) to establish the rules and regulations requiring electric utilities to purchase electric power from and sell electric power to qualifying cogeneration and small power production facilities and to provide for the exemption qualifying facilities (QF) from certain federal and state regulations.

Thus, FERC issued in 1980 a series of rules to relax obstacles to cogeneration. Such rules implement sections of the 1978 PURPA and include detailed instructions to state utility commissions that all utilities must purchase electricity from cogenerators and small power producers at the utilities' "avoided" cost. In a nutshell, this means that rates paid by utilities for such electricity must reflect the cost savings they realize by being able to avoid capacity additions and fuel usage of their own.

Tuttle (1980) states that prior to PURPA 210, cogeneration facilities wishing to sell their power were faced with three major obstacles:

- Utilities had no obligation to purchase power and contended that cogeneration facilities were too small and unreliable. As a result, even those cogenerators able to sell power had difficulty getting an equitable price.

- Utility rates for backup power were high and often discriminatory.

- Cogenerators often were subject to the same strict state and federal regulations as the utility.

PURPA was designed to remove these obstacles, by requiring utilities to develop an equitable program of integrating cogenerated power into their loads.

Avoided Costs

The costs avoided by a utility when a cogeneration plant displaces generation capacity and/or fuel usage are the basis for setting the rates paid by utilities for cogenerated power sold back to the utility grid. In some circumstances, the actual rates may be higher or lower than the avoided costs, depending on the need of the utility for additional power and on the outcomes of the negotiations between the parties involved in the cogeneration development process.

All utilities are now required by PURPA to provide data regarding present and future electricity costs on a cent-per-kWh basis during daily, seasonal, peak and off-peak periods for the next five years. This information must also include estimates on planned utility capacity additions and retirements, as well as cost of new capacity and energy costs.

Tuttle (1980) points out that utilities may agree to pay greater price for power if a cogeneration facility can:

- Furnish information on demonstrated reliability and term of commitment.

- Allow the utility to regulate the power production for better control of its load and demand changes.

- Schedule maintenance outages for low-demand periods.

- Provide energy during utility-system daily and seasonal peaks and emergencies.

- Reduce in-house on-site load usage during emergencies.

- Avoid line losses the utility otherwise would have incurred.

In conclusion, a utility is willing to pay better "buyback" rates for cogenerated power if it is short in capacity, if it can exercise a level of control on the CHP plant and load, and if the cogenerator can provide and/or demonstrate a "high" system availability.

PURPA further states that the utility is not obligated to purchase electricity from a QF during periods that would result in net increases in its operating costs. Thus, low demand periods must be identified by the utility and the cogenerator must be notified in advance. During emergencies (utility outages), the QF is not required to provide more power than its contract requires, but a utility has the right to discontinue power purchases if they contribute to the outage.

7.4.4. Emissions and Permits

The main cogeneration regulations about emissions are given by federal laws such as the Clean Air Act (CAA) and the Clean Water Act (CWA), which are enforced according to specific regulations given by the US Environmental Protection Agency. In addition, there is a wide variety of regulation and permitting requirement that vary by state and municipality. However, most environmentally related regulations and permits, whether federal or state, are administered by state agencies. In addition, local approvals may also be required

in some cases.

For example for major power and cogeneration plants (greater than 350 MW) the Washington state Energy Facility Site Evaluation Council (EFSEC) has established the following permitting process, divided in two major verifications (A and B below). Next, representative summaries of applicable federal and state regulation for larger plants (>350 MW) are given.

(A) 1.1.1. The Prevention of Significant Deterioration Process

The prevention of significant deterioration (PSD) procedure is established in Title 40, Code of Federal Regulations (CFR), Part 52.21. Federal rules require PSD review of all new or modified air pollution sources that meet certain criteria. The objective of the PSD program is to prevent serious adverse environmental impact from emissions into the atmosphere by a proposed new source. The program limits degradation of air quality to that which is not considered "significant." It also sets up a mechanism for evaluating the effect that the proposed emissions might have on environmentally related areas for such parameters as visibility, soils, and vegetation. PSD rules also require the utilization of the most effective air pollution control equipment and procedures, after considering environmental, economic, and energy factors.

(B) 1.1.2. The Notice of Construction Process

The procedure for issuing a notice of construction (NOC) permit is established in Chapter 70.94 Revised Code of Washington. Chapter 173-400 WAC and Chapter 173-460 WAC, require all new or modified stationary sources of air pollution to file a NOC application and receive an order of approval, prior to establishing a new or modified stationary source.

WAC 173-400-110 (new source review) outlines the procedures for permitting criteria pollutants. These procedures are further refined in WAC 173-400-113 (requirements for new sources on attainment or unclassifiable areas).

WAC 173-460-040 (new source review) supplements the requirements contained in Chapter 173-400 WAC by adding additional requirements for sources of toxic air pollutants.

1.1.3. Federal Regulations Summary

A given permit may not contain all the requirements included in the following summary, however after the Title V and Acid Rain permits are issued each of the following regulations will be addressed.

Prevention of Significant Deterioration 40 CFR 52.21

New Source Performance Standards (NSPS) 40 CFR 60, subpart GG

NSPS 40 CFR 60, Subpart Db

NSPS Quality Assurance Procedures 40 CFR 60 Appendix F

NSPS Performance Specifications 40 CFR 60, Appendix B

National Emission Standards for 40 CFR 63 Subpart YYYY

Hazardous Air Pollutants (NESHAP)

Acid Rain Permitting 40 CFR 72

Emissions Monitoring and Permitting 40 CFR 75

NOx Requirements 40 CFR 76

Sulfur Content of Natural Gas to be monitored 40 CFR 60.334(b), 40 CFR 72.2, and

40 CFR 75, Appendix D

1.1.4. State Regulation Summary

A given permit may not contain all the requirements included in the following summary, however after the Title V and Acid Rain permits are issued, each of the following regulations will be addressed.

General and Operating Permit Regulations for Chapter 463-39 WAC

Air Polluting Sources

General Regulations for Air Pollution Sources Chapter 173-400 WAC

Operating Permit Regulations Chapter 173-401 WAC

Acid Rain Regulations Chapter 173-406 WAC

Controls For New Sources of Toxic Air Pollutants Chapter 173-460 WAC

Thus, depending on the size and configuration of the proposed cogeneration facility, several permits may apply. Although not exhaustive, the following list is representative of the permits required for a small-to-medium industrial cogeneration plant (1-50 MW).

- Pipeline permit—when a fluid fuel (liquid or gas) needs to be piped from a remote supply location to the cogeneration plant
- Underground storage tank permit—when a

liquid fuel (either primary or back up such as diesel or fuel oil) needs to be stored in an underground tank.

- Waste water and storm water discharge permits—for example, when blowdown water from boilers and cooling towers needs to be discharged
- Solid waste fuel transportation/storage permit—when a solid fuel such as shredded tires or wood chips needs to be transported or stored in a location.
- Air permits—There is a variety of air pollution control permits, depending on the size of the plant, type of fuel, and whether the CHP plant is to be located in an *attainment* or *non-attainment* area. Of particular importance are the 1990 amendments to the Clean Air Act that involved many significant changes to the federal air quality programs, which in turn caused major updates to all state air permitting programs. "Two of the larger changes included the way hazardous air pollutants are addressed and the addition of the Title V (or Part 70) operating permit program. Title V refers to the section of the Clean Air Act, and Part 70 refers to that part of Title 40 of the Code of Federal Regulations, where you will find the requirements for this program" (Minnesota Pollution Control Agency).

The following guidelines are provided as an example of those provided by the Minnesota Pollution Control Agency with respect to air permits.

Who Needs a Permit?

In general, facilities who have the potential to emit (also known as PTE) any regulated pollutant, in greater than specific threshold amounts, must obtain a total facility permit.

What Does PTE Mean?

PTE is defined as the maximum capacity of an emission unit or source to emit a pollutant under its physical and operational design while operating at the maximum number of hours (usually 8760 hours per year). In just about all cases, PTE calculations are based on the assumption that the facility operates at its maximum design capacity, 24 hours a day, 365 days a year (equals 8,760 hours per year). Many times, a facility will take a limit on a unit (such as on the number of hours of operation or amount of material processed) to reduce its potential to emit so that it can avoid certain requirements that would otherwise have been applicable. These limits are then put into the facility's permit.

What is a Regulated Pollutant?

In general there are two types of regulated pollutants:

1. Criteria pollutants—These pollutants all have human health-based or welfare-based standards that set the maximum concentrations allowed in the ambient air (i.e. the air that the general public is exposed to). They include:
   - nitrogen oxides (NO_x)
   - sulfur dioxides (SO_2)
   - ozone/volatile organic compounds (VOC)
   - particulate matter (PM)
   - particulate matter less than 10 microns in diameter (PM10)
   - carbon monoxide (CO)
   - lead (Pb)

2. Hazardous air pollutants (HAPs)—These are defined by a list of chemicals that are known or suspected of causing cancer or other serious health effects, such as developmental effects or birth defects. There were originally 189 HAPs, but various rulemaking activities have removed and/or redefined some of the HAPs. More information is available from:
   - Environmental Protection Agency Air Rules
   - Air Toxics in Minnesota

What are the Thresholds?

There are both federal and state total facility thresholds. First, a facility calculates the PTE for each regulated pollutant from each of its emission units. Then the total potential emissions for the whole facility, for each regulated pollutant, are aggregated and compared to each of the federal thresholds. If the total facility PTE for any of the regulated pollutants is over any federal threshold, a Title V permit will be needed unless limits can be accepted on its PTE to bring it below the federal thresholds. If the facility's total potential emissions are less than the federal thresholds (or it accepts limits so that this is true) but are still greater than the state thresholds, the facility will need a state permit. If the facility's total emissions are less than the state thresholds (without any limits on its PTE), the facility does not need a permit but should keep records of its calculations.

An important point to keep in mind is that even if a facility doesn't need a permit based on its PTE, it may still need one for other reasons. For example, two federal regulatory programs require some facilities to

US (Federal) and Minnesota (State) Air Pollutant Threshold Limits

Pollutant	Total Facility PTE Thresholds (tons per year)	
	Federal	State
NO_x	100	100
SO_2	100	50
VOC	100	100
PM	100	100
PM10	100	25
CO	100	100
Pb	NA	0.5
1 HAP	10	10
> 1 HAP	25	25

Source: http://www.pca.state.mn.us/air/permits/aboutairpermits.html

apply for permits regardless of how much air pollution they could potentially cause.

Besides total facility operating permits, another general class of permits that the MPCA issues are construction permits. Construction permits are issued for the construction of a new facility where PTE is over the federal or state thresholds, or for the modification of an existing facility. In general, a facility that doesn't already require a total facility permit will not need a construction permit for a modification at its facility, unless the additional PTE of the construction activity increases the PTE of the total facility above a permitting threshold.

Other requirements include power interconnection requirements (both technical and sales agreement) by the local utility that will buy excess power or the distribution/transmission line which accepts electricity for wholesale wheeling.

7.5 EVALUATING COGENERATION OPPORTUNITIES: CASE EXAMPLES

The feasibility evaluation of cogeneration opportunities, for both new construction and facility retrofit, require the comparison and ranking of various options, using a figure of economic merit. The options are usually combinations of different CHP technologies, operating modes, and equipment sizes.

A first step in the evaluation is the determination of the costs of a base-case (or do-nothing) scenario. For new facilities, buying thermal and electrical energy from utility companies is traditionally considered the base case. For retrofits, the present way to buy and/or generate energy is the base case. For many, the base-case scenario is the "actual plant situation" after "basic" energy conservation and management measures have been implemented. That is, cogeneration should be evaluated upon an "efficient" base case plant.

Next, suitable cogeneration alternatives are generated using the methods discussed in Sections 7.2 and 7.3. Then, the comparison and ranking of the base case versus the alternative cases is performed using an economic analysis.

Henceforth, this section addresses a basic approach for the economic analysis of cogeneration. Specifically, it discusses the development of the cash flows for each option including the base case. It also discusses some figures of merit such as the gross pay out period (simple payback) and the discounted or internal rate of return. Finally, it describes two case examples of evaluations in industrial plants. The examples are included for illustrative purposes and do not necessarily reflect the latest available performance levels or capital costs.

7.5.1 General Considerations

A detailed treatise on engineering economy is presented in Chapter 4. Even so, since economic evaluations play the key role in determining whether cogeneration can be justified, a brief discussion of economic considerations and several evaluation techniques follows.

The economic evaluations are based on examining the incremental increase in the investment cost for the alternative being considered relative to the alternative to which it is being compared and determining whether the savings in annual operating cost justify the increased investment. The parameter used to evaluate the economic merit may be a relatively simple parameter such as the "gross payout period." Or one might use more sophisticated techniques that include the time value of money (such as the "discounted rate of return"), on the discretionary investment for the cogeneration systems being evaluated.

Investment cost and operating cost are the expenditure categories involved in an economic evaluation. Operating costs result from the operations of equipment, such as (1) purchased fuel, (2) purchased power, (3) purchased water, (4) operating labor, (5) chemicals, and (6) maintenance. Investment-associated costs are of primary importance when factoring the impact of federal and state income taxes into the economic evaluation. These costs (or credits) include (1) investment tax credits, (2) depreciation, (3) local property taxes, and (4) insurance. The economic evaluation establishes whether the operating and investment cost factors result in sufficient after-tax income to provide the company stockholders an adequate rate of return after the debt obligations with

regard to the investment have been satisfied.

When one has many alternatives to evaluate, the less sophisticated techniques, such as "gross payout," can provide an easy method for quickly ranking alternatives and eliminating alternatives that may be particularly unattractive. However, these techniques are applicable only if annual operating costs do not change significantly with time and if additional investments do not have to be made during the study period.

The techniques that include the time value of money permit evaluations where annual savings can change significantly each year. Also, these evaluation procedures permit additional investments at any time during the study period. Thus these techniques truly reflect the profitability of a cogeneration investment or investments.

7.5.2 Cogeneration Evaluation Case Examples

The following examples illustrate evaluation procedures used for cogeneration studies. Both examples are based on 1980 investment costs for facilities located in the U.S. Gulf Coast area.

For simplicity, the economic merit of each alternative examined is expressed as the "gross payout period" (GPO). The GPO is equal to the incremental investment for cogeneration divided by the resulting first-year annual operating cost savings. The GPO can be converted to a "discounted rate of return" (DRR) using Figure 7-15. However, this curve is valid only for evaluations involving a single investment with fixed annual operating cost savings with time. In most instances, the annual savings due to cogeneration will increase as fuel costs increase to both utilities and industries in the years ahead. These

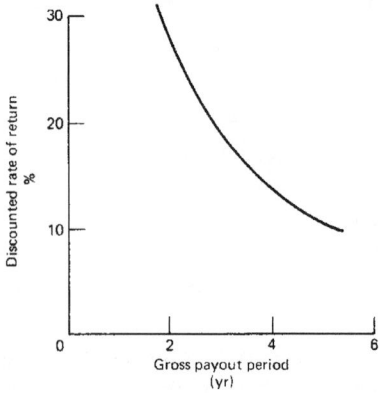

Figure 7-15. Discounted rate of return versus gross payout period. Basis: (1) depreciation period, 28 years; (2) sum-of-the-years'-digits depreciation; (3) economic life, 28 years; (4) constant annual savings with time; (5) local property taxes and insurance, 4% of investment cost; (6) state and federal income taxes, 53%; (7) investment tax credit, 10% of investment cost.

increased future savings enhance the economics of cogeneration. For example, if we assume that a project has a GPO of three years based on the first-year operating cost savings, Figure 7-15 shows a DRR of 18.7%. However, if the savings due to cogeneration increase 10% annually for the first three operating years of the project and are constant thereafter, the DRR increases to 21.6%; if the savings increase 10% annually for the first six years, the DRR would be 24.5%; and if the 10% increase was experienced for the first 10 years, the DRR would be 26.6%.

Example 6: The energy requirements for a large industrial plant are given in Table 7-3. The alternatives considered include:

Base case. Three half-size coal-fired process boilers are installed to supply steam to the plant's 250-psig steam header. All 80-psig steam and steam to the 20-psig deaerating heater is pressure-reduced from the 250-psig steam header. The powerhouse auxiliary power requirements are 3.2 MW. Thus the utility tie must provide 33.2 MW to satisfy the average plant electric power needs.

Case 1. This alternative is based on installation of a noncondensing steam turbine generator. The unit initial steam conditions are 1450 psig, 950°F with automatic extraction at 250 psig, and 80 psig exhaust pressure. The boiler plant has three half-size units providing the same reliability of steam supply as the base case. The feedwater heating system has closed feedwater heaters at 250 psig and 80 psig, with a 20 psig deaerating heater. The 20-psig steam is supplied by noncondensing mechanical drive turbines used as powerhouse auxiliary drives. These units are supplied throttle steam from the 250-psig steam header. For this alternative, the utility tie normally provides 4.95 MW. The simplified schematic and energy balance is given in Figure 7-16.

The results of this cogeneration example are tabulated in Table 7-4. Included are the annual energy requirements, the 1980 investment costs for each case, and the annual operating cost summary. The investment cost data presented are for fully operational plants, including offices, stockrooms, machine shop facilities, and locker rooms, as well as for fire protection and plant security. The cost of land is not included.

The incremental investment cost for Case 1 given in Table 7-4 is $17.2 million. Thus, the incremental cost is $609/kW for the 28.25-MW cogeneration system. This illustrates the favorable per unit cost for cogeneration systems compared to coal-fired facilities designed to provide kilowatts only, which cost in excess of $1000/kW.

The impact of fuel and purchased power costs other than Table 7-3 values on the GPO for this example is shown in Figure 7-17. Equivalent DRR values based on first-year annual operating cost savings can be estimated using Figure 7-15.

Sensitivity analyses often evaluate the impact of uncertainties in the installed cost estimates on the profitability of a project. If the incremental investment cost for cogeneration is 10% greater than the Table 7-4 estimate, the GPO would increase from 3.2 to 3.5 years. Thus the DRR would decrease from 17.5% to about 16%, as shown in Figure 7-15.

Example 7: The energy requirements for a chemical plant are presented in Table 7-5. The alternatives considered include:

Base case. Three half-size oil-fired packaged process boilers are installed to supply process steam at 150 psig. Each unit is fuel-oil-fired and includes a particulate removal system. The plant has a 60-day fuel-oil-storage capacity. A utility tie provides 30.33 MW average to supply process and boiler plant auxiliary power requirements.

Case 1. (Refer to Figure 7-18). This alternative examines the merit of adding a noncondensing steam turbine

Table 7-3. Plant Energy Supply System Considerations: Example 6

Process steam demands
   Net heat to process at 250 psig. 410°F—317 million Btu/hr avg.
   Net heat to process at 80 psig, 330°F—208 million Btu/hr avg. (peak requirements are 10% greater than
      average values)
Process condensate returns: 50% of steam delivered at 280°F
Makeup water at 80°F
Plant fuel is 3.5% sulfur coal
Coal and limestone for SO_2 scrubbing are available at a total cost of $2/million Btu fired
Process area power requirement is 30 MW avg.
Purchased power cost is 3.5 cents/kWh

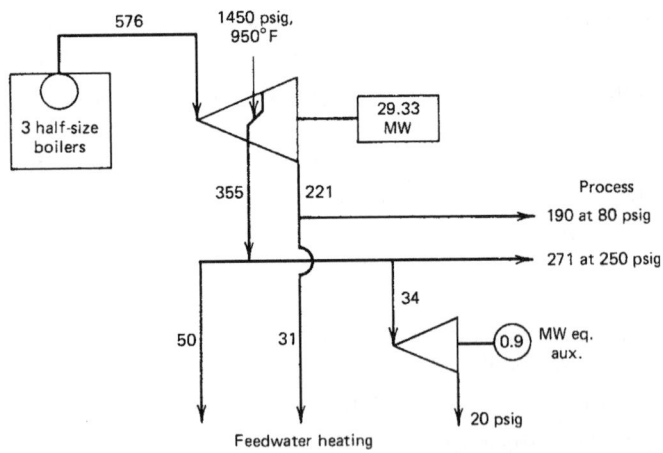

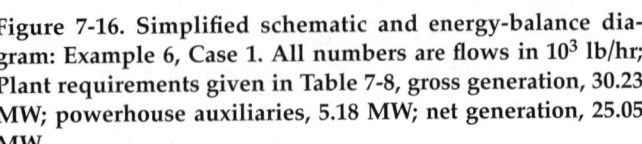

Figure 7-16. Simplified schematic and energy-balance diagram: Example 6, Case 1. All numbers are flows in 10³ lb/hr; Plant requirements given in Table 7-8, gross generation, 30.23 MW; powerhouse auxiliaries, 5.18 MW; net generation, 25.05 MW.

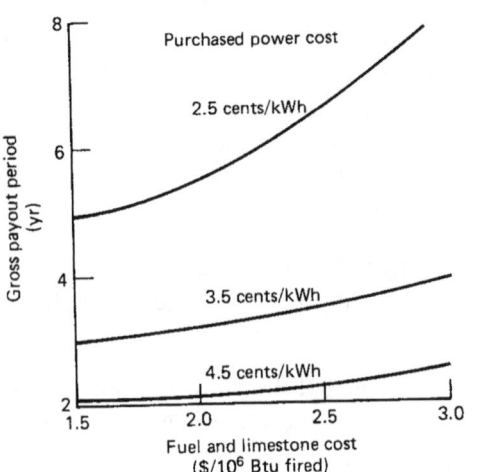

Figure 7-17. Effect of different fuel and power costs on cogeneration profitability: Example 1. *Basis:* Conditions given in Tables 7.3 and 7.4.

Table 7-4. Energy and Economic Summary: Example 6

Alternative	Base Case	Case 1
Energy summary		
Boiler fuel (10⁶ Btu/hr HHV)	599	714
Purchased power (MW)	33.20	4.95
Estimated total installed cost (10⁶ $)	57.6	74.8
Annual operating costs (10⁶ $)		
Fuel and limestone at $2/10⁶ Btu	10.1	12.0
Purchased power at 3.5 cents/kWh	9.8	1.5
Operating labor	0.8	1.1
Maintenance	1.4	1.9
Makeup water	0.3	0.5
Total	22.4	17.0
Annual savings (10⁶ $)	Base	5.4
Gross payout period (yrs)	Base	3.2

Basis: (1) boiler efficiency is 87%; (2) operation equivalent to 8400 hr/yr at Table 7-3 conditions; (3) maintenance is 2.5% of the estimated total installed cost; (4) makeup water cost for case 1 is 80 cents/1000 gal *greater than* Base Case water costs; (5) stack gas scrubbing based on limestone system.

generator with 850 psig, 825°F initial steam conditions, and 150-psig exhaust pressure. Steam is supplied by three half-size packaged boilers. The feedwater heating system is comprised of a 150-psig closed heater and a 20-psig deaerating heater. The steam for the deaerating heater is the exhaust of a mechanical drive turbine (MDT). The MDT is supplied 150-psig steam and drives some of the plant boiler feed pumps. The net generation of this cogeneration system is 6.32 MW when operating at the average 150-psig process heat demand. A utility tie provides the balance of the power required.

Case 2. (Refer to Figure 7-19). This alternative is a combined cycle using the 25,000-kW gas turbine generator whose performance is given in Table 7-7. An unfired HRSG system provides steam at both 850 psig, 825°F and 150 psig sat. Plant steam requirements in excess of that available from the two-pressure level unfired HRSG system are generated in an oil-fired packaged boiler. The steam supplied to the noncondensing turbine is expanded to the 150-psig steam header. The net generation from the overall system is 26.54 MW. A utility tie provides power requirements in excess of that supplied

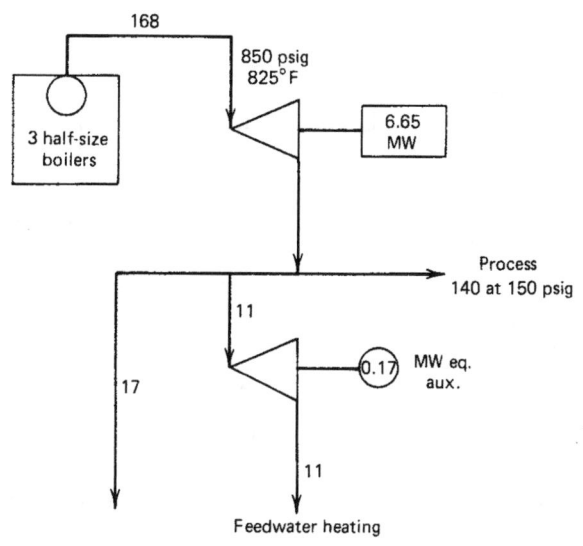

Figure 7-18. Simplified schematic and energy-balance diagram: Example 7, Case 1. All numbers are flows in 1000 lb/hr; gross generation, 6.82 MW; powerhouse auxiliaries, 0.50 MW; net generation; 6.32 MW.

Table 7-5. Plant Energy Supply System Considerations: Example 7

Process steam demands

Net heat to process at 150 psig sat—158.5 million Btu/hr avg. (peak steam requirements are 10% greater than average values)

Process condensate returns: 45% of the steam delivered at 300°F

Makeup water at 80°F

Plant fuel is fuel oil

Fuel cost is $5/million Btu

Process areas require 30 MW

Purchased power cost is 5 cents/kWh

Table 7-6. Energy and Economic Summary: Example 7

Alternative	Base Case	Case 1	Case 2
Energy summary			
Fuel (10^6 Btu/hr HHV)			
Boiler	183	209	34
Gas turbine	—	297	
Total	183	209	331
Purchased power (MW)	30.33	23.77	3.48
Estimated total installed cost (10^6 $)	8.3	12.6	18.9
Annual operating cost (10^6 $)			
Fuel at $5/M Btu HHV	7.7	8.8	13.9
Purchased power at 5 cents/kWh	12.7	10.0	1.5
Operating labor	0.6	0.9	0.9
Maintenance	0.2	0.3	0.5
Makeup water	0.1	0.2	0.2
Total	21.3	20.2	17.0
Annual savings (10^6 $)	Base	1.1	4.3
Gross payout period (yr)	Base	3.9	2.5

Basis: (1) gas turbine performance per Table 7-7; (2) boiler efficiency, 87%; (3) operation equivalent to 8400 hr/yr at Table 7-5 conditions; (4) maintenance, 2.5% of the estimated total installed costs; (5) incremental makeup water cost for cases 1 and 2 relative to the Base Case. $1/1000 gal.

by the cogeneration system. The plant-installed cost estimates for Case 2 include two half-size package boilers. Thus, full steam output can be realized with any steam generator out of service for maintenance.

The energy summary, annual operating costs, and economic results are presented in Table 7-6. The results show that the combined cycle provides a GPO of 2.5 years based on the study fuel and purchased power costs. The incremental cost for Case 2 relative to the

Base Case is $395/kW, compared to $655/kW for Case 1 relative to the Base Case. This favorable incremental investment cost, combined with a FCP of 5510 Btu/kWh, contribute to the low GPO.

The influence of fuel and power costs other than those given in Table 7-5 on the GPO for cases 1 and 2 is shown in Figure 7-20. These GPO values can be translated to DRRs using Figure 7-15.

Figure 7-19. Simplified schematic and energy-balance diagram: Example 7, Case 2. All numbers are flows in 1000 lb/hr; gross generation, 26.77 MW, powerhouse auxiliaries, 0.23 MW: net generation, 26.54 MW.

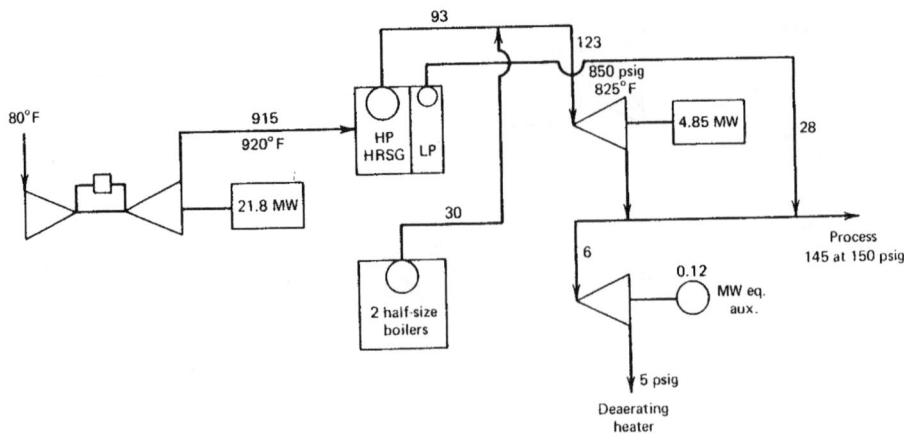

Table 7-7. Steam Generation and Fuel Chargeable to Power: 25,000-kW ISO Gas Turbine and HRSG (Distillate Oil Fuel)[a]

Type HRSG	Unfired		Supplementary Fired		Fully Fired	
Gas Turbine						
Fuel (10^6 Btu/hr HHV)	297				➝	
Output (MW)	21.8		21.6		21.4	
Airflow (10^3 lb/hr)	915				➝	
Exhaust temperature (°F)	920		922		925	
HRSG fuel (10^6 Btu/hr HHV)	NA		190		769	
	Steam (10^3 lb/hr)	FCP (Btu/kWh HHV)	Steam (10^3 lb/hr)	FCP (Btu/kWh HHV)	Steam (10^3 lb/hr)	FCP (Btu/kWh HHV)
Steam conditions						
250 psig sat.	133	6560	317	5620	851	4010
400 psig, 650°F	110	7020	279	5630	751	
600 psig, 750°F	101	7340	268	5660	722	
850 psig, 825°F	93	7650	261	5700	703	
1250 psig, 900°F	—	—	254	5750	687	
1450 psig, 950°F	—	—	250	5750	675	⬇

[a]*Basis*: (1) gas turbine performance given for 80°F ambient temperature, sea-level site; (2) HRSG performance based on 3% blowdown, 1-1/2% radiation and unaccounted losses, 228°F feedwater; (3) no HRSG bypass stack loss; (4) gas turbine exhaust pressure loss is 10 in. H_2O with unfired, 14 in. H_2O with supplementary fired, and 20 in. H_2O with fully fired HRSG; (5) fully fired HRSG based on 10% excess air following the firing system and 300°F stack. (6) fuel chargeable to gas turbine power assumes total fuel credited with equivalent 88% boiler fuel required to generate steam; (7) steam conditions are at utilization equipment; a 5% AP and 5°F AT have been assumed from the outlet of the HRSG.

Example 8. A gas-turbine and HRSG cogeneration system is being considered for a brewery to supply base-load electrical power and part of the steam needed for process. An overview of the proposed system is shown in Figure 7-21. This example shows the use of computer tools in cogeneration design and evaluation.

Base Case. Currently, the plant purchases about 3,500,000 kWh per month at $0.06 per kWh. The brewery uses an average of 24,000 lb/hr of 30 psig saturated steam. Three 300-BHP gas fired boilers produce steam at 35 psig, to allow for pressure losses. The minimum steam demand is 10,000 lb/hr. The plant operates continuously during ten months (7,000 hr/year). The base or minimum electrical load during production is 3,200 kW. The rest of the time (winter) the brewery is down for maintenance. The gas costs $3.50/MMBtu.

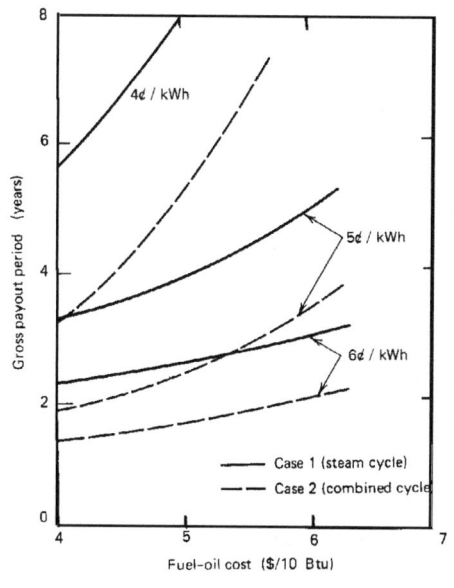

Figure 7-20. Effect of different fuel and power cost on cogeneration profitability: Example 2. *Basis*: **Conditions given in Tables 7.4 and 7.5.**

HRSG and Stack): 12″ H₂O

Location Elevation above sea level: 850 ft

Thus, on a preliminary basis, we assume the turbine will constantly run at full capacity, minus the effect of elevation, the inlet air pressure drop, and exhaust losses. Since the plant will be located at 850 ft above sea level, from Figure 7-11, the elevation correction factor is 0.90. Hence, the corrected continuous power rating (before deducting pressure losses) when firing natural gas and using 70°F inlet air is:

= (Generator Output @ 70°F)
  (Elevation correction @ 850 ft)

= 4,200 kWe × 0.9

= 3,780 kWe

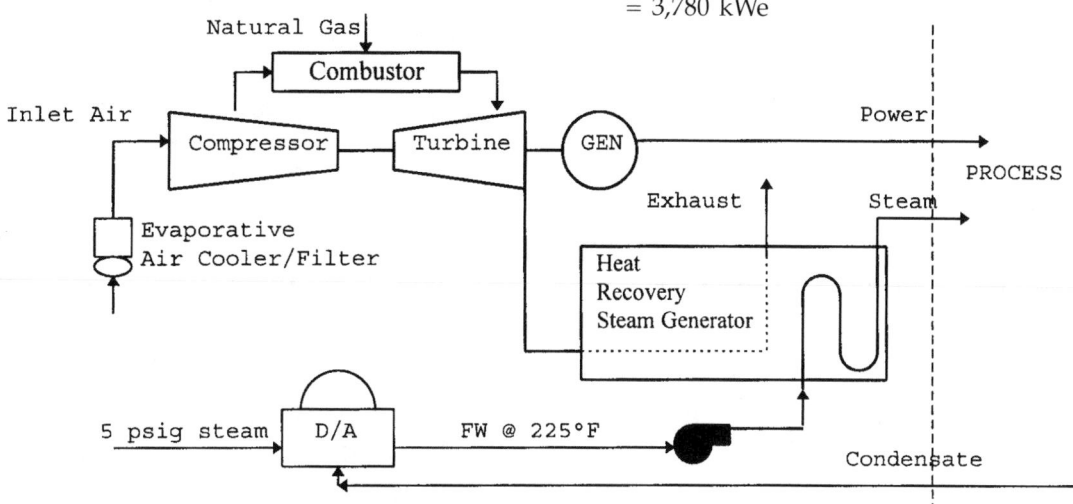

Figure 7-21. Gas turbine/HRSG cogeneration application.

Case 1. Consider the gas turbine whose ratings are given on Figure 7-11. We will evaluate this turbine in conjunction with an unfired water-tube HRSG to supply part of the brewery's heat and power loads. First, we obtain the ratings and performance data for the selected turbine, which has been sized to meet the electrical base load (3.5 MW). An air washer/evaporative cooler will be installed at the turbine inlet to improve (reduce) the overall heat rate by precooling the inlet air to an average 70°F (80°F or less), during the summer production season. Additional operating data are given below.

Operating Data
Inlet air pressure losses (filter and air
    pre-cooler): 5″ H₂O

Exhaust Losses (ducting, by-pass valve,

Next, by using the Inlet and Exhaust Power Loss graphs in Figure 7-11, we get the exhaust and inlet losses (@ 3780 kW output): 17 and 7 kW/inch H₂O, respectively. So, the total power losses due to inlet and exhaust losses are:

= (17 in)(5 kW/in) + (12 in)(8 kW/in)
= 181 kW

Consequently, the net turbine output after elevation and pressure losses is
= 3,780 - 181
= 3,599 kWe

Next, from Figure 7-11 we get the following performance data for 70°F inlet air:

Heat rate   : 12,250 Btu/kWh (LHV)

Exhaust Temperature : 935°F
Exhaust Flow   : 160,000 lb/hr

These figures have been used as input data for HGPRO, a prototype HRSG software program developed by V. Ganesh, W.C. Turner and J.B. Wong in 1992 at Oklahoma State University. The program results are shown in Figure 7-22.

The total installed cost of the complete cogeneration plant, including gas turbine, inlet air precooling, HRSG, auxiliary equipment, and computer based controls is $4,500,000. Fuel for cogeneration is available on a long term contract basis (>5 years) at $2.50/MMBtu. The brewery has a 12% cost of capital. Using a 10-year after tax cash flow analysis with current depreciation and tax rates, *should the brewery invest in this cogeneration option?* For this evaluation, assume: (1) A 1% inflation for power and non-cogen natural gas; (2) an operation and maintenance (O&M) cost of $0.003/kWh for the first year after the project is installed, then escalating at 3% per year; (3) the plant salvage value is neglected.

Economic Analysis

Next, we present system operation assumptions required to conduct a preliminary economic analysis.

1) The cogeneration system will operate during all the production season (7,000 hrs/year).

2) The cogeneration system will supply an average of 3.5 MW of electrical power and 24,000 lb of 35 psig steam per hour. The HRSG will be provided with an inlet gas damper control system to modulate and by-pass hot gas flow. This is to allow for variable steam production or steam load-following operation.

3) The balance of power will be obtained from the existing utility at the current cost ($0.06/kWh).

4) The existing boilers will remain as back-up units. Any steam deficit (considered to be negligible) will be produced by the existing boiler plant.

5) The cogeneration fuel (natural gas) will be metered with a dedicated station and will be available at $2.50/MMBtu during the first five years and at $2.75/MMBtu during the next five-year period. Non cogeneration fuel will be available at the current price of $3.50/MMBtu.

The discounted cash flow analysis was carried out using an electronic spreadsheet (Table 7-8). The results of the spreadsheet show a positive net present value. Therefore, when using the data and assumptions given in this case, the cogeneration project appears to be cost effective. The brewery should consider this project for funding and implementation.

Note: These numbers ignore breakdowns and possible ratchet clause effects.

7.6   CLOSURE

Cogeneration has been used for almost a century to supply both process heat and power in many large industrial plants in the United States. This technology would have been applied to a greater extent if we did not experience a period of plentiful low-cost fuel and reliable low-cost electric power in the 25 years following the end of World War II. Thus, economic rather than technical considerations have limited the application of this energy-saving technology.

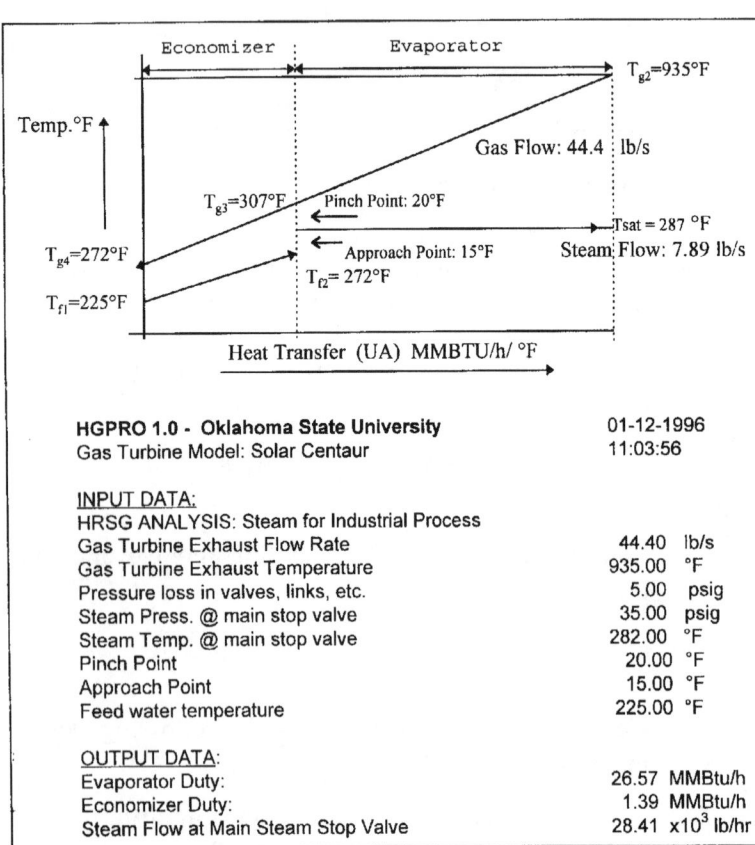

HGPRO 1.0 - **Oklahoma State University**   01-12-1996
Gas Turbine Model: Solar Centaur   11:03:56

INPUT DATA:
HRSG ANALYSIS: Steam for Industrial Process
Gas Turbine Exhaust Flow Rate   44.40   lb/s
Gas Turbine Exhaust Temperature   935.00   °F
Pressure loss in valves, links, etc.   5.00   psig
Steam Press. @ main stop valve   35.00   psig
Steam Temp. @ main stop valve   282.00   °F
Pinch Point   20.00   °F
Approach Point   15.00   °F
Feed water temperature   225.00   °F

OUTPUT DATA:
Evaporator Duty: 26.57   MMBtu/h
Economizer Duty: 1.39   MMBtu/h
Steam Flow at Main Steam Stop Valve   28.41   x10³ lb/hr

Figure 7-22. Results from HGPRO 1.0, a prototype HRSG software.

The continued increase in the cost of energy is the primary factor contributing to the renewed interest in cogeneration and its potential benefits. This chapter was discussed the various prime movers that merit consideration when evaluating this technology. Additionally, approximate performance levels and techniques for developing effective cogeneration systems were presented.

The cost of all forms of energy is rising sharply. Cogeneration should remain an important factor in effectively using our energy supplies and economically providing goods and services in those base-load applications requiring large quantities of process heat and power.

References

1. Butler, C.H., (1984), *Cogeneration: Engineering, Design Financing, and Regulatory Compliance,* McGraw-Hill, Inc., New York, N.Y.
2. Caton, J.A., et al., (1987), *Cogeneration Systems,* Texas A&M University, College Station, TX.
3. CFR-18 (1990): Code of Federal Regulations, Part 292 Regulations Under Sections 201 and 210 of the Public Utility Regulatory Policies Act of 1978 With Regard to Small Power Production and Cogeneration, (4-1-90 Edition).
4. Estey P.N., et al., (1984). "A Model for Sizing Cogeneration Systems," Proceedings of the 19th Intersociety Energy Conversion Engineering Conference" Vol. 2 of 4, August, 1984, San Francisco, CA.
5. Ganapathy, V. (1991). *Waste Heat Boiler Deskbook,* The Fairmont Press, Inc., Lilburn, CA.
6. Harkins H.L., (1981), "PURPA New Horizons for Electric Utilities and Industry," IEEE Transactions, Vol. PAS-100, pp 27842789.
7. Hay, N., (1988), *Guide to Natural Gas Cogeneration,* The Fairmont Press, Lilburn, GA.
8. Kehlhofer, R., (1991). *Combined-Cycle Gas & Steam Turbine Power Plants,* The Fairmont Press, Inc. Lilburn, Ga.
9. Kostrzewa, L.J. & Davidson, K.G., (1988). "Packaged Cogeneration," *ASHRAE Journal,* February 1988.
10. Kovacik, J.M., (1982), "Cogeneration," in *Energy Management Handbook,* ed. by W.C. Turner, Wiley, New York, N.Y.
11. Kovacik, J.M., (1985), "Industrial Cogeneration: System Application Consideration," *Planning Cogeneration Systems,* The Fairmont Press, Lilburn, Ga.
12. Lee, R.T.Y., (1988), "Cogeneration System Selection Using the Navy's CELCAP Code," *Energy Engineering, Vol. 85, No. 5,* 1988.
13. Limaye, D.R. and Balakrishnan, S., (1989), "Technical and Economic Assessment of Packaged Cogeneration Systems Using Cogenmaster," *The Cogeneration Journal, Vol. 5, No. 1,* Winter 1989-90.
14. Limaye, D.R., (1985), *Planning Cogeneration Systems,* The Fairmont Press, Atlanta, CA.
15. Limaye, D.R., (1987), *Industrial Cogeneration Applications,* The Fairmont Press, Atlanta, CA.
16. Mackay, R. (1983). "Gas Turbine Cogeneration: Design, Evaluation and Installation." The Garret Corporation, Los Angeles, CA, The Association Of Energy Engineers, Los Angeles CA, February, 1983.
17. Makansi, J., (1991). "Independent Power/Cogeneration, Success Breeds New Obligation-Delivering on Performance," *Power,* October 1991.
18. Mulloney, et. al., (1988). "Packaged Cogeneration Installation Cost Experience," Proceedings of The 11th World Energy Engineering Congress, October 18-21, 1988.
19. Orlando, J.A., (1991). *Cogeneration Planners Handbook,* The Fairmont Press, Atlanta, GA.
20. Oven, M., (1991), "Factors Affecting the Financial Viability Applications of Cogeneration," XII Seminario Nacional Sobre El Uso Racional de La Energia," Mexico City, November, 1991.
21. Polimeros, G., (1981), *Energy Cogeneration Handbook,* Industrial Press Inc., New, York.
22. Power (1980), "FERC Relaxes Obstacles to Cogeneration," *Power,* September 1980, pp 9-10.
23. SFA Pacific Inc. (1990). "Independent Power/Cogeneration, Trends and Technology Update," *Power,* October 1990.
24. Somasundaram, S., et al., (1988), *A Simplified Self-Help Approach To Sizing of Small-Scale Cogeneration Systems,* Texas A&M University, College Station, TX
25. Spiewak, S.A. and Weiss L., (1994) *Cogeneration & Small Power Production Manual,* 4th Edition, The Fairmont Press, Inc. Lilburn, CA.
26. Turner, W.C. (1982). *Energy Management Handbook,* John Wiley & Sons, New York, N.Y.
27. Tuttle, D.J., (1980), PURPA 210: New Life for Cogenerators," *Power,* July, 1980.
28. Williams, D. and Good, L., (1994) *Guide to the Energy Policy Act of 1992,* The Fairmont Press, Inc. Lilburn, GA.
29. Wong, J.B., Ganesh, V. and Turner, W.C. (1991), "Sizing Cogeneration Systems Under Variable Loads," 14th World Energy Engineering Congress, Atlanta, GA.
30. Wong, J.B. and Turner W.C. (1993), "Linear Optimization of Combined Heat and Power Systems," Industrial Energy Technology Conference, Houston, March, 1993.

Appreciation

Many thanks to Mr. Lew Gelfand for using and testing over the years the contents of this chapter in the evaluation and development of actual cogeneration opportunities, and to Mr. Scott Blaylock for the information provided on fuel cells and microturbines. Messrs. Gelfand and Blaylock are with DukeEnergy/DukeSolutions.

Table 7.8 After tax discounted cash flow economic analysis.

GAS TURBINE / HRSG COGENERATION TEN YEAR ECONOMIC ANALYSIS

INPUT DATA

Power & gas cost escalation rate	1%	(year 1-10)
Labor, operat. & maint. escalation rat	3%	(year 1-10)
Cogen power generation	3,500	kW
Cogen steam production	24,000	lh/hr
Existing boiler efficiency	80%	lh/hr
Present Electricity Cost	$0.06	/kWh
Present Nat. Gas Cost	$3.50	/MMBtu
Turbine plant heat rate, LHV @ 70°F	12,250 MMBtu/kW	93% LHV/HHV
Cogen Nat. Gas Cost	$2.50 /MMBtu	$2.75 /MMBtu
Cogen System O&M Cost	$0.003 /kWh	yrs: 6-10
Cogen System Installed Cost	$4,500,000	
Operation time:	7,000	hr/yr

	NPV @ 12.00%	1996	1997	1998	1999	2000	2001	2002	2003	2004	2005
Savings											
Power cost savings	8610742	1470000	1484700	1499547	1514542	1529688	1544985	1560435	1576039	1591799	1607717
Nat. Gas to steam savings	4305371	735000	742350	749774	757271	764844	772492	780217	788019	795900	803859
Total Savings	12916113	2205000	2227050	2249321	2271814	2294532	2317477	2340652	2364058	2387699	2411576
Costs											
Operating & Maint Costs	-463291	-73500	-75705	-77976	-80315	-82725	-85207	-87763	-90396	-93108	-95901
Cogen Nat. Gas Cost	-4723554	-806788	-806788	-806788	-806788	-806788	-887466	-887466	-887466	-887466	-887466
Depreciation: 5-year MACRS	-3520539	-504000	-1310400	-1290240	-773640	-579600	-435960	-146160	0	0	0
Total Costs	-8707384	-1384288	-2192893	-2175004	-1660743	-1469113	-1408633	-1121389	-977862	-980574	-983367
Net Savings	4208729	820712	34157	74317	611071	825419	908844	1219263	1386196	1407125	1428209
Cashflows											
Post-Tax Income/(Loss)	2569428	501045	20853	45370	373059	503918	554849	744360	846273	859050	871921
Investment	-4258929	-2520000	-2520000	0	0	0	0	0	0	0	0
Depreciation Add Back	3520539	504000	1310400	1290240	773640	579600	435960	146160	0	0	0
Total Cashflows	1831039	-1514955	-1188747	1335610	1146699	1083518	990809	890520	846273	859050	871921
Cumulative Cashflow		-1514955	-2703702	-1368092	-221393	862125	1852934	2743454	3589727	4448777	5320698

NPV : $1,831,039

Appendix A

Statistical Modeling of Electric Demand and
Peak–shaving Generator Economic Optimization
Jorge B. Wong, Ph.D., PE, CEM

ABSTRACT

This chapter shows the development a basic electrical demand statistical model to obtain the optimal kW size and the most cost-effective operating time for an electrical peak shaving generator set. This model considers the most general (and simplified) case of a facility with an even monthly demand charge and a uniformly distributed random demand, which corresponds to a linear load–duration curve. A numerical example and computer spreadsheet output illustrate the model.

INTRODUCTION

Throughout the world, electrical utilities include a hefty charge in a facility's bill for the peak electrical demand incurred during the billing period, usually a month. Such a charge is part of the utility's cost recovery or amortization of newly installed capacity and for operating less efficient power plant capacity during higher load periods.

Demand charge is a good portion of a facility's electrical bill. Typically a demand charge can be as much as 50% of the bill, or more. Thus, to reduce the demand cost, many industrial and commercial facilities try to "manage their loads." One example is by moving some of the electricity–intense operations to "off–peak" hours"—hours when a facility's electrical load is much smaller and the rates ($/kW) are lower. But, when moving electrical loads to "off–peak" hours is not practical or significant, a facility will likely consider a set of engine–driven or fuel cell generators to run in parallel with the utility grid to supply part or all the electrical load demand during "on–peak" hours. We call these peak-shaving generators or PSGs.

While the electric load measurement is instantaneous, the billing demand is typically a 15–to–30– minute average of the instantaneous electrical power demand (kW). To obtain the monthly demand charge, utilities multiply the billing demand by a demand rate. Some utilities charge a flat rate ($/kW–peak per month) for all months of the year. Other utilities have seasonal charges (i.e. different rates for different seasons of the year). Still others use ratchet clauses to account for the highest "on–peak" season demand of the year.

Thus, the model presented in this chapter focuses on the development of a method to obtain the optimal PSG size ($g*$kW) and PSG operation time (hours per year) for a given facility. This model is for the case of a facility with a constant billing demand rate ($/kW/ month) throughout the year. The analysis is based on a linear load-duration curve and uses a simplified life-cycle-cost approach. An example illustrates the underlying approach and optimization method. In addition, the chapter shows an Excel spreadsheet to implement the optimization model. We call this model PSG–1.

ELECTRICAL DEMAND STATISTICAL MODEL

This section develops the statistical–and–math model for the economical sizing of an electrical peak–shaving generator set (PSG) for a given facility. The fundamental question is: *What is the most economical generator set size—$g*$ in kW—for a given site demand profile?* Figure 1 shows a sample record for a facility's electrical demand, which is uniformly distributed between 2000 and 5300 kW. Next, Figure 2 shows the corresponding statistical distributions.

The statistical model of electrical demand is expressed graphically in Figure 2, in terms of two functions:

- The load–duration curve $D(t)$, is the demand as a function of cumulative time t (i.e. the accumulated annual duration in hrs/year of a given $D(t)$ load in kW), and

- The load frequency distribution $f(D)$ (rectangular shaded area in Figures 1 and 2) is the "uniform" probability density function.

MODEL ASSUMPTIONS

This statistical model is based upon the following assumptions:

1. The electrical demand is represented by a linear load–duration curve, as shown in Figure 2. Thus, for a typical year, the facility has a demand D that varies between an upper value D_u (annual maximum) and a lower value D_1 (annual minimum). This implies the electrical load is uniformly distributed between the maximum and minimum demands. The facility operates T hours per year.

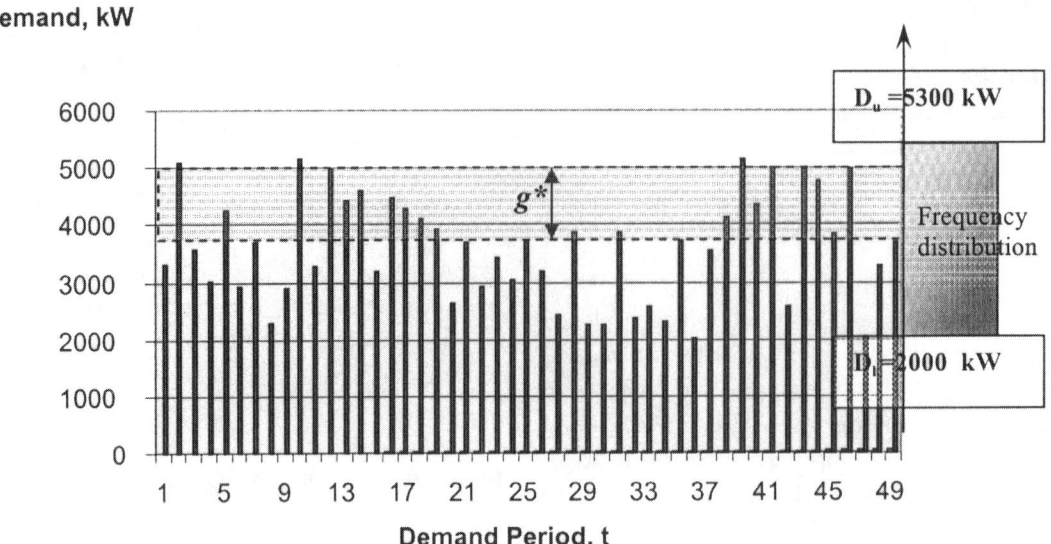

Figure 1. Sample record for a uniformly distributed random demand.

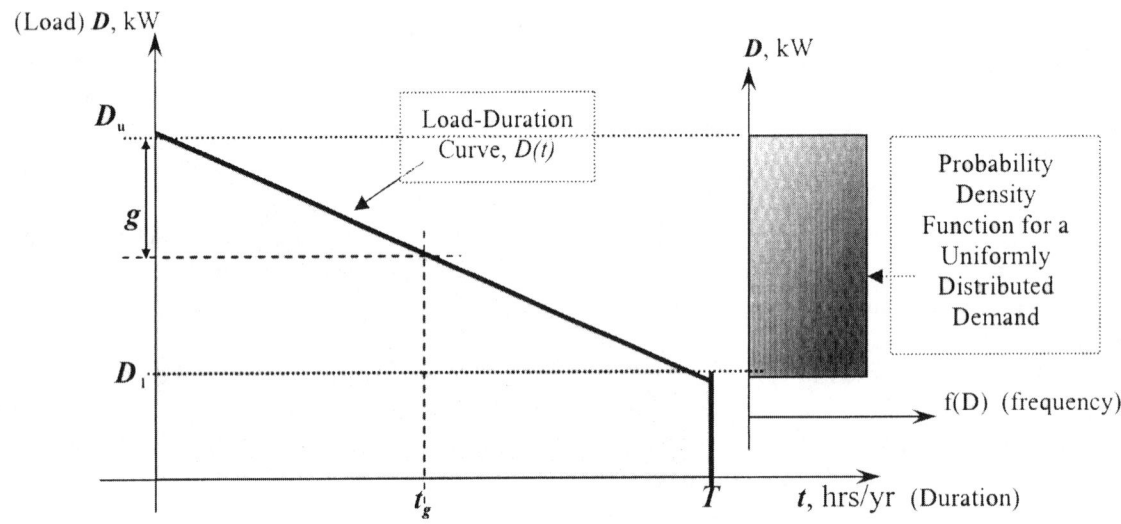

Figure 2. Load—Duration Curve for uniformly distributed demand.

2. There is an even energy or consumption rate Ce ($/kWh) throughout the year.

3. There is an even demand rate Cd ($/kW/month) for every month of the year.

4. There is a same demand peak D_u for every month. (Demand ratchet clauses are not applicable in this case.)

5. The equipment's annual ownership or amortization unit installed cost ($/kW/year) is constant for all sizes of PSGs. The unit ownership or rental cost ($/kW/year) is considered independent of unit size. Ownership, rental, or lease annualized costs are denoted by A_c.

6. A PSG set is installed to reduce the peak demand by a maximum of g kW, operating t_g hours per year.

BASE CASE ELECTRICITY ANNUAL COST—WITHOUT PEAK SHAVING

Consider a facility with the load–duration characteristic shown in Figures 1 and 2. For a unit consumption cost Ce, the *annual energy or consumption cost* (without PSG) for the facility is

$$AEC = T \cdot D_1 \cdot Ce + 1/2\, T\, (D_u - D_1)\, Ce$$

Which is equivalent to

$$AEC = T/2 \cdot Ce \ (D_u + D_1) \qquad [1]$$

Next, considering a peak demand D_u occurs every month, the *annual demand cost* is defined by

$$ADC = 12 \ D_u \cdot Cd \qquad [2]$$

Thus, the total *annual cost* for the facility is

$$TAC = AEC + ADC \qquad [3a]$$

Substituting [1] and [2] in equation [3], we have the base case total annual cost:

$$TAC_1 = T/2 \ (D_u + D_1) \ Ce + 12 \ D_u \cdot Cd \qquad [3b]$$

ELECTRICITY ANNUAL COST WITH PEAK SHAVING

If a peak-shaving generator of size g is installed in the facility to run in parallel with the utility grid during peak–load hours, so the maximum load seen by the utility is $(D_u - g)$, then the *electric bill cost* is

$$EBC = T/2 \ (D_u - g + D_1) \ Ce + 12 \ (D_u - g) \cdot Cd$$

In addition, the facility incurs an ownership (amortization) unit cost Ac ($/kW/yr) and operation and maintenance unit cost $O\&M$ ($/kWh). Hence, the *total annual cost* with demand peak shaving is

$$TAC_2 = [T \cdot D_1 + (T + t_g)/2 \ (D_u - g - D_1)]Ce +$$
$$12 \ (D_u - g) \ Cd + (Ac + 1/2 \ O\&M \cdot t_g)g \qquad [4]$$

ANNUAL WORTH OF THE PEAK-SHAVING GENERATOR

The *annual worth or net savings AW* ($/yr) of the PSG set are obtained by subtracting equation [4] from equation [3]. That is $AW = TAC_1 - TAC_2$. So,

$$AW = 1/2 \ t_g \cdot g \cdot Ce + 12 \cdot g \cdot Cd -$$
$$(Ac + 1/2 \cdot O\&M \cdot t_g)g \qquad [5]$$

From Figure 2 we obtain g: $t_g = (D_u - D_1) : T$

So, the expected PSG operating time is

$$t_g = g \cdot T/(D_u - D_1) \qquad [6]$$

Substituting the value of t_g in equation [5], we have:

$$AW = g^2 \cdot T/[2(D_u - D_1)] \ Ce + 12 \cdot g \cdot Cd -$$
$$\{Ac + O\&M \cdot g \cdot T/[2(D_u - D_1)]\} \ g \qquad [7]$$

OPTIMUM CONDITIONS

We next determine the necessary and sufficient conditions for an optimal PSG size g^* and the corresponding maximum AW to exist.

Necessary Condition

By taking the derivative of AW, Equation [7], with respect to g and equating it to zero we obtain the necessary condition for the maximum annual worth or net saving per year. That is:

$$AW' = g \cdot T \cdot Ce/(D_u - D_1) + 12 \ Cd - Ac -$$
$$g \cdot T \cdot O\&M/(D_u - D_1) = 0 \qquad [8]$$

Sufficient Condition

If the second derivative of AW with respect to g is negative, i.e. $AW'' < 0$, then $AW \ (g)$ is a strictly convex function of g with a global maximum point. So, by taking the second derivative of AW with respect to g and evaluating AW'' as an inequality (< 0) we have:

$$AW'' = T \cdot Ce/(D_u - D_1) - T \cdot O\&M/(D_u - D_1) < 0$$

Multiplying this equation by $(D_u - D_1)/T$ we have the sufficient condition for a maximum AW is

$$Ce - O\&M < 0$$
or
$$Ce < O\&M$$

Therefore, for a global maximum AW to exist, the energy rate Ce must be less than the per unit $O\&M$ cost (including fuel) to operate the peak-shaving generator ($/kWh). Since this is the case for most utility rates Ce and commercial PSGs $O\&M$, we can say there is maximum AW and an optimal g^* for the typical electrical demand case.

OPTIMUM PEAK-SHAVING GENERATOR SIZE

From equation [6] we can solve for g and find the *optimal PSG size, g^** (in kW):

$$g^* = (12Cd - Ac) \ (D_u - D_1)/[T(O\&M - Ce)] \qquad [9]$$

FOR FURTHER RESEARCH

Further research is underway to develop enhanced models which consider:

- Demand profile flexibility. Other load–duration shapes with different underlying frequency distributions (e.g. triangular, normal and auto–correlated loads).

- Economies of Scale. The fact that larger units have better fuel–to–electricity efficiencies (lower heat rates) and lower per unit installed cost ($/kW).

EXAMPLE. A manufacturing plant operates 7500 hours per year and has a fairly constant electrical (billing) peak demand every month (See Figure 1). The actual load, however, varies widely between a minimum of 2000 kW and a maximum of 5300 kW (See Figure 2). The demand charge is $10/kW/month and the energy charge is $0.05/kWh. The installed cost of a diesel generator set, the auxiliary electrical switch gear, and peak–shaving controls is about $300 per kW. Alternatively, the plant can lease a PSG for $50/kW/yr. The operation and maintenance cost (including diesel fuel) is $0.10/kWh.

Assuming the plant leases the PSG, estimate (1) the optimal PSG size, (2) the annual savings, and (3) the PSG annual operation time.

1) The optimal generator size is calculated using equation [9]:

$$g^* = \frac{(12*\$10 - \$50/kWh)(5300 - 2000 \text{ kW})}{7500 \text{ h/yr} (\$0.10/kWh - \$0.05/kWh)}$$
$$= \mathbf{616 \ kW}$$

2) Using a commercially available PSG of size $g^* = 600$ kW, the potential annual savings are estimated using equation [7]:

$$
\begin{aligned}
AW &= g^2 \cdot T \cdot Ce/[2(D_u - D_1)] + 12 \cdot g \cdot Cd \\
&\quad - \{Ac + O\&M \cdot g \cdot T/[2(D_u - D_1)]\} \, g \\
&= 600^2 \times 7500 \times 0.05/(2(5300 - 2000)) + 12 \times 600 \times 10 \\
&\quad - (\$50 + 0.10 \times 600 \times 7500/(2 \, (5300 - 2000))) \, 600 \\
&= \$20{,}455 + \$72{,}000 - \$70{,}909 \\
&= \$21{,}546/\text{year}
\end{aligned}
$$

3) The expected annual operating time for the PSG is estimated using Equation [6]:

g	AWg
(kW)	($/yr)
0	0
66	4373
132	8250
198	11633
264	14520
330	16913
396	18810
462	20213
528	21120
594	21533
660	21450
726	20873
792	19800
858	18233
924	16170
990	13613
1056	10560
1122	7013
1188	2970
1254	-1568

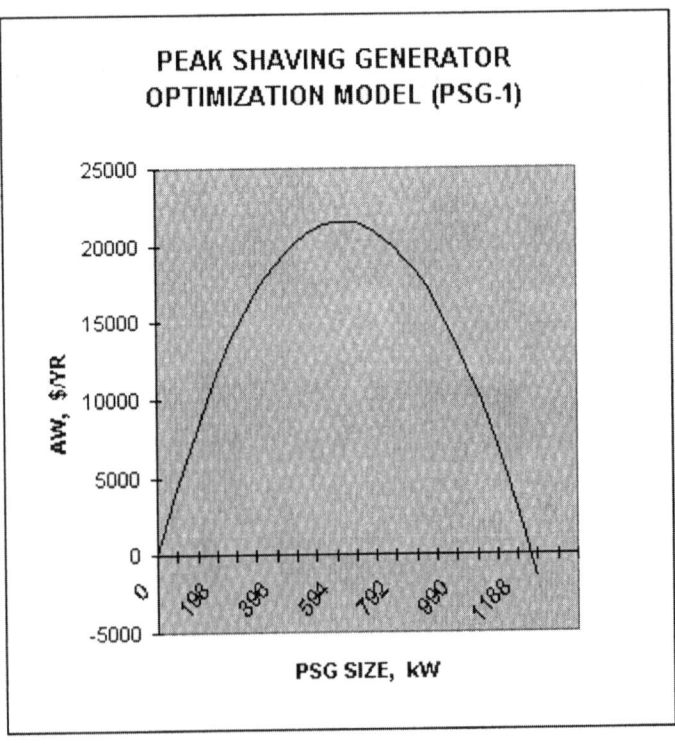

Figure 3. PSG-1 Spreadsheet and Chart

$$t_g = g \cdot T/(D_u - D_1)$$
$$= 600 \times 7500/(5300-2000)$$
$$= 1{,}364 \text{ hours/year}$$

The Excel spreadsheet and chart used to solve this case example is shown in Figure 3.

CONCLUDING REMARKS

The reader should note that the underlying statistical and optimization model is quite "responsive and robust." That is, the underlying methodology can be used in, or adapted to, a variety of demand profiles and rates, while the results remain relatively valid. A forthcoming paper by this author will show how to adapt the linear load– duration models of Figures 1 and 2 to more complex demand profiles. Thus, for example, one typical case is when the electrical load is represented by a Gauss or normal distribution. Also, we will show how to apply equation [9] to more involved industrial cases with multiple billing seasons and demand rates.

Appendix References

Beightler, C.S., Phillips, D.T., and Wilde, D.J., *Foundations of Optimization*, Prentice–Hall, Englewood Cliffs, 1979.

Turner, W.C., *Energy Management Handbook*, 4th Edition, the Fairmont Press, Lilburn, GA, 2001.

Hahn & Shapiro, *Statistical Models in Engineering*, John Wiley 1967, Wiley Classics Library, reprinted in 1994.

Witte, L.C., Schmidt, P.S., and Brown, D.R, *Industrial Energy Management and Utilization*, Hemisphere Publishing Co. and Springer–Verlag, Berlin, 1988.

Appendix Nomenclature

Ac Equipment ownership, lease or rental cost ($/kW/year)

ADC Annual Demand Cost ($/year)

AEC Annual Energy Cost ($/year)

AW Annual Worth ($/year)

Cd Electric demand unit cost ($/kW/month)

Ce Electric energy unit cost ($/kWh)

D Electric demand or load (kW)

D_1 Lower bound of a facility's electric demand or minimum load (kW)

D_u Upper bound of a facility's electric demand or maximum load (kW)

EBC Electric bill cost for a facility with PSG, ($/year)

$f(D)$ Frequency of occurrence of a demand, (unit less)

$O\&M$ Operation and Maintenance cost, including fuel cost ($/kWh)

g Peak shaving generator size or rated output capacity (kW)

g^* Optimal peak shaving generator size or output capacity (kW)

t Time, duration of a given load, (hours/year)

t_g Expected time of operation for a PSG, hours/year

T Facility operation time using power(hours/year)

TAC Total annual electric cost

TAC_1 Total annual cost, base case w/o PSG ($/year)

TAC_2 Total annual cost, with PSG ($/year)

Jorge B. Wong, Ph.D., PE, CEM is an energy management advisor and instructor. Jorge helps facility managers and engineers. Contact Jorge: jorgebwong@att.net

CHAPTER 8
WASTE-HEAT RECOVERY

D. PAUL MEHTA, PH.D.
Professor & Chair, Mechanical Engineering Department
Director of Industrial Assessment Center
Bradley University
Peoria, Illinois

8.1 INTRODUCTION

With the exception of the nuclear bonds of the atom, the molten bowels of the earth, and the tides of the ocean, the sun is the only source of energy for human beings. The fossil fuels of coal, oil, and gas are the locked-in photosynthetic collection of the sun's energy from eons past. The wood we burn and the food we eat are products of the photosynthetic collection of the sun's present energy. The rivers which give hydroelectric power could not exist without the evaporation of the oceans by the sun, and the winds themselves are driven by differential heating of the earth's surface. All these forces are of awesome cyclic complexity, yet strewn with orders we still but dimly perceive. (1)

Alternative Energy Sources, simply defined for our purposes, means energy alternatives to the burning of fossil fuels. For heating and cooling, fossil fuels are burned on-site as primary energy and off-site to generate electricity for secondary energy. At present, for building use, fossil fuels must account as primary or secondary energy for greater than 99% of building demand. But these sources—coal, oil, and gas—which we have so extravagantly drawn upon for the last century or so, are not replenishable in our time and are shockingly limited. We must finally turn directly to the sun for renewable and non- polluting energy. The diagram network in Figure 8-1 nicely encompasses the potential for the sun's capture. (See also Chapter 16, "Use of Alternative Energy.")

A primary concern for the architect and the engineer is the fit of alternative-source technology to new and existing buildings. Alternative sources utilized off-site with an end product of electricity or district heat, a most probable future, require no change. Alternative sources utilization at the building site itself, however, do require a new vision of the construction art, and architects and engineers must henceforth be open to this new vision. The theme of this book—responsible energy management in buildings

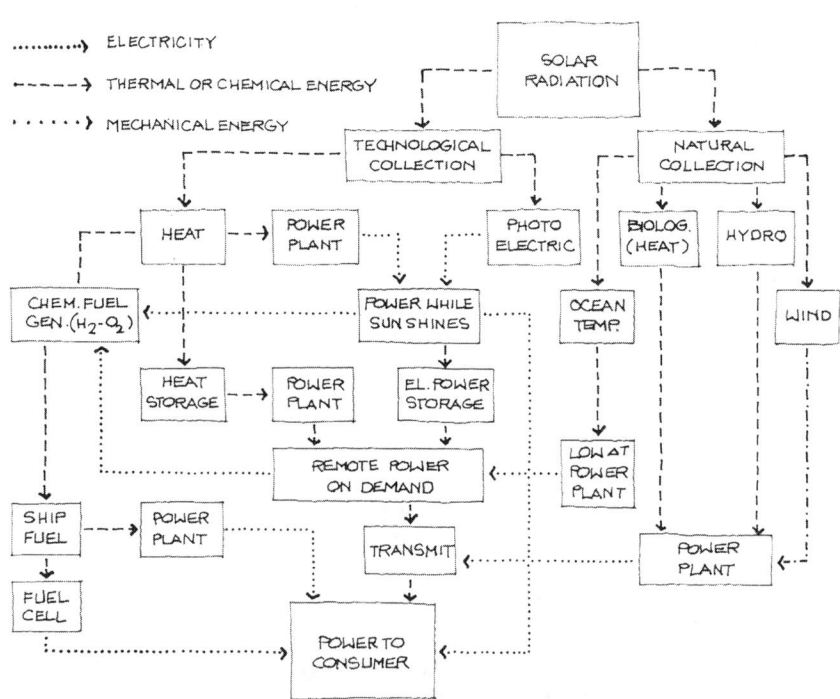

Figure 8-1. The many ways of obtaining electrical power from the sun.

193

and energy efficiency in industrial processes—is a large part of this new vision.

Current promising explorations for direct utilization of alternative energies in buildings include:

- Solar thermal collection
- Solar electrical conversion
- Hydroelectric
- Wind-generating plants
- Methane gas generation
- Solid waste combustion
- Fuel cells

To this listing are added as alternative energy sources:

- Waste heat recovery
- Thermal storage
- Total energy systems (thermal and moisture recovery)

The later items are, not strictly speaking, considered to be alternative sources but are set out here, we think legitimately, as a significant classification of approaches devised to reduce energy demand. The treatment of waste heat recovery is discussed in this chapter.

8.2　WASTE HEAT RECOVERY

Heat recovery is the utilization of heat energy that is otherwise wasted. Properly captured, waste heat can substitute for a portion of the new energy that would normally be required for heating, cooling, and domestic hot-water systems. Heat recovery conserves fuels, reduces operating costs, and reduces peak loads. Applications of heat recovery devices are shown in Table 8-1.

The performance of a heat recovery system will involve some of or all the following factors:

- The temperature difference between the heat source and the heat sink

- The latent heat difference between the heat source and the sink

- The mass flow multiplied by the specific heat of each source and sink

- The efficiency of the heat transfer device

- The extra energy required to operate the heat recovery device

Heat Reclaim Sources	Temper Ventilation Air	Preheat Domestic Hot Water	Space Heating	Terminal Reheat	Temper Makeup Air	Preheat Combustion Air	Heat Heavy Oil	Internal to External Zone Heat Transfer
Exhaust air	1a, 1b, 1c, 2, 3, 4, 5				1a, 1b, 1c, 2, 3, 4, 5	Direct		5, direct
Flue gas	1b, 3, 4		3, 4		1b, 3, 4	1b, 2, 3, 4		
Hot condensate	3, 6	3, 6	3, 5, 6	3, 6	3, 6	3, 6	6	
Refrigerant hot gas	6	6		6	6			
Hot condenser water	3, 6	3, 6	3, 5, 6	6	3, 5, 6	3, 5, 6	6	5
Hot-water drains		6						
Solid waste	7	7	7	7	7	7	7	
Engine exhaust and cooling systems	6, 8	6	6, 8	6	6, 8	6, 8	6	
Lights	5, 9		5, 9	5, 9	5, 9			5, 9
Air-cooled condensers			Direct					

*Heat Reclaim Devices:

1. Thermal wheel	2. Runaround coil	5. Heat pump	8. Waste heat boiler
a. Latent	3. Heat pipe	6. Shell/tube heat exchanger	9. Heat-of-light
b. Sensible	4. Air-to-air heat exchanger	7. Incinerator	10. Thermal storage
c. Combination			

Table 8-1. Applications of heat reclamation devices.

- The fan or pump energy absorbed as heat by the heat transfer device (which either enhances or detracts from performance).

- That the waste heat event and recovered heat event occur at the same time

- Hours of operation per year

8.3 CLASSIFYING WASTE-HEAT QUALITY (2)

For convenience, the total range of waste-heat temperatures, 80 to 3000°F, is divided into three categories: high, medium, and low. These categories are designed to match a similar scale which classifies commercial waste-heat-recovery devices. The two systems of classes allow matches to be made between industrial process waste heat and commercially available recovery equipment. Sub ranges are defined in terms of temperature range as:

High range $1100 \leq T \leq 3000°F$

Medium range $400 \leq T < 1100°F$

Low range $80 \leq T < 400°F$

Waste heat in the high-temperature range is not only the highest quality but is the most useful, and it costs less per unit to transfer than lower-quality heat. However, the equipment needs in the highest part of the range required special engineering and special materials, thus requiring a higher level of investment. All of the applications listed in Table 8-2 result from direct-fired processes. The waste heat in the high range is available to do work through the utilization of steam turbines or gas turbines and thus is a good source of energy for cogeneration plants.*

Table 8-3 gives the temperatures of waste gases primarily from direct-fired process equipment in the medium-temperature range. This is still in the temperature range in which work may be economically extracted using gas turbines in the range 15 to 30 psig, or steam turbines at almost any desired pressure. It is an economic range for direct substitution of process heat since requirements for equipment are reduced from those in the high-temperature range.

*The waste heat generates high-pressure steam in a waste-heat boiler, which is used in a steam turbine generator to generate electricity. The turbine exhaust steam at a lower pressure provides process heat. Alternatively, the high-temperature gases may directly drive a gas turbine generator with the exhaust generating low-pressure steam in a waste-heat boiler for process heating.

Table 8-2.
Waste-heat sources in the high-temperature range

Type of Device	Temperature (°F)
Nickel refining furnace	2500-3000
Aluminum refining furnace	1200-1400
Zinc refining furnace	1400-2000
Copper refining furnace	1400-1500
Steel heating furnaces	1700-1900
Copper reverberatory furnace	1650-2000
Open hearth furnace	1200-1300
Cement kiln (dry process)	1150-1350
Glass melting furnace	1800-2800
Hydrogen plants	1200-1800
Solid waste incinerators	1200-1800
Fume incinerators	1200-2600

Table 8-3.
Waste-heat sources in the medium-temperature range

Type of Device	Temperature (°F)
Steam boiler exhausts	450-900
Gas turbine exhausts	700-1000
Reciprocating engine exhausts	600-1100
Reciprocating engine exhausts (turbocharged)	450-700
Heat treating furnaces	800-1200
Drying and baking ovens	450-1100
Catalytic crackers	800-1200
Annealing furnace cooling systems	800-1200
Selective catalytic reduction systems for NO_x control	525-750

The use of waste heat in the low-temperature range is more problematic. It is ordinarily not practical to extract work directly from the waste-heat source in this temperature range. Practical applications are generally for preheating liquids or gases. At the higher temperatures in this range, air preheaters or economizers can be utilized to preheat combustion air or boiler make-up water, respectively. At the lower end of the range, heat pumps may be required to raise the source temperature to one that is above the load temperature. An example of an application which need not involve heat pump assistance would be the use of 95°F cooling water from an air compressor to preheat domestic hot water from its ground temperature of 50°F to some intermediate temperature less than 95°F. Electric, gas-fired, or steam heaters could then be utilized to heat the water to the temperature desired. Another application could be the use of 90°F cooling water from a battery of spot welders to preheat the ventilating air for winter space heating. Since machinery cooling can not be interrupted or diminished, the waste-heat recovery system, in this latter case, must be designed to be bypassed or supplemented when seasonal load requirements disappear. Table 8-4 lists some waste-heat sources in the low-temperature range.

Generally, the following restrictions must be met before a low temperature source becomes a feasible candidate for waste heat recovery.

Heat Reclaim Sources	Restrictions
Exhaust Air	Flow rate $\geq$ 4,000 CFM
Flue Gas	Reclamation device must be capable of withstanding corrosive force
Hot Condensate	None
Refrigerant	Steady & concurrent
Hot Gas	Demand for refrigeration & waste heat. Refrigeration system is open more than 750 hrs/yr.
Hot Condenser Water	None
Hot-Water Drains	Discharge temp above 120°F and flow rate more than 10,000 gals/wk
Solid Waste	Availability more than 1000 lbs/day
Engine Exhaust &	
Cooling Systems	Exhaust temp greater than 250°F. Size: 50 HP or more
Lights	None
Air-Cooled Condensers	15 HP or more

8.4   STORAGE OF WASTE HEAT

Waste heat can be utilized to adapt otherwise mismatched loads to waste-heat sources by the use of thermal storage. This is possible because of the inherent ability of all materials to absorb energy while undergoing a temperature increase. The absorbed energy is termed stored energy. The quantity that can be stored is dependent upon the temperature rise that can be achieved in the storage material, as well as the intrinsic thermal qualities of the materials, and it can be estimated from the equation

$$Q = \int_{T_1}^{T_2} mCdT = \int_{T_1}^{T_2} \rho VCdT$$

$$= \rho VC \, (T - T_0) \text{ for constant specific heat} \qquad (8.1)$$

Where

m = mass of storage materials, lb_m
ρ = density of storage material, lb/ft^3
V = volume of storage material, ft^3
C = specific heat of storage material, $Btu/lb_m \, °R$
T = temperature in absolute degrees, °R

The specific heat for solids is a function of temperature which can usually be expressed in the form

$$C_o = C_o \, [1 + \alpha \, (T - T_o)] \qquad (8.2)$$

where

C_o = specific heat at temperature T_o
T_o = reference temperature
α = temperature coefficient of specific heat.

It is seen from equation 8.1 that storage materials should have the properties of high density and high specific heat in order to gain maximum heat storage for a given heat. The amount that can be absorbed or given up by the storage material depends upon its thermal conductivity, k, which is defined by the equation

$$\frac{\delta Q}{\delta t} = - kA \frac{dT}{dx} \Big|_{x=0} = \dot{Q} \qquad (8.3)$$

where

t   = time, hr
k   = thermal conductivity, $Btu\text{-}ft/hr \, ft^2 \, °F$
A   = surface area

$\dfrac{dT}{dx} \Big|_{x=0}$ = temperature gradient at the surface.

Table 8-4. Waste-heat sources in the low-temperature range.

Source	Temperature (*F)
Process steam condensate	130-190
Cooling water from:	
Furnace doors	90-130
Bearings	90-190
Welding machines	90-190
Injection molding machines	90-190
Annealing furnaces	150-450
Forming dies	80-190
Air compressors	80-120
Pumps	80-190
Internal combustion engines	150-250
Air conditioning and refrigeration condensers	90-110
Liquid still condensers	90-190
Drying, baking, and curing ovens	200-450
Hot-processed liquids	90-450
Hot-processed solids	200-450
Exhaust	
Commercial building exhaust to make-up	Room Temp.
Paint booths, lab exhaust, etc. to make-up	Room Temp.

Thus, additional desirable properties are high thermal conductivity and large surface area per unit mass (specific area). This latter property is inversely proportional to density but can also be manipulated by designing the shape of the solid particles. Other important properties for storage materials are low cost, high melting temperature, and a resistance to spalling and cracking under conditions of thermal cycling. To summarize, the most desirable properties of thermal storage materials are: (1) high density, (2) high specific heat, (3) high specific area, (4) high thermal conductivity, (5) high melting temperature, (6) low coefficient of thermal expansion, and (7) low cost.

The response of a storage system to a waste-heat stream is given approximately by the following expression due to Rummel. [3]

$$\frac{Q}{A} = \frac{\Delta T_{l,m}/(\theta + \theta'')}{1/h''\theta + 1/h'\theta' + 1/2.5C_s\rho_s R_B/k(\theta' + \theta'')} \qquad (8.4)$$

Where

$T_{l,m}$ = logarithmic mean temperature difference, based upon the uniform inlet temperature of each stream and the average outlet temperatures

C_s = specific heat of storage material, Btu/lb °F

ρ_s = density of storage material, lb/ft^3

k = conductivity of storage material, Btu/hr ft °F

R_B = volume per unit surface area for storage material, ft

h = coefficient of convective heat transfer of gas streams, Btu/hr ft^2 °F

θ = time cycle for gas stream flows, hr.

The primed and double-primed values refer, respectively, to the hot and cold entering streams. In cases where the fourth term in the denominator is large compared to the other three terms, this equation should not be used. This will occur when the cycle times are short and the thermal resistance to heat transfer is large. In those cases there exists insufficient time for the particles to get heated and cooled.

8.5   QUANTIFYING WASTE HEAT

The technical description of waste heat must necessarily include quantification of the following characteristics: (1) quantity, (2) quality, and (3) temporal availability.

The quantity of waste heat available is ordinarily expressed in terms of the enthalpy flow of the waste stream, or

$$\dot{H} = \dot{m}h \qquad (8.5)$$

where

$\dot{H}$ = total enthalpy flow rate of waste stream, Btu/hr

$\dot{m}$ = mass flow rate of waste stsream, lb/hr

$\dot{h}$ = specific enthalpy of waste stream, Btu/lb

The mass flow rate, m, can be calculated from the expression

$$\dot{m} = \rho Q \qquad (8.6)$$

where ρ = density of material, lb/ft^3
         Q = volumetric flow rate, ft^3/hr

The potential for economic waste-heat recovery, however, does not depend as much on the quantity available as it does on whether its quality fits the requirements of the potential heating load which must be supplied and whether the waste heat is available at the times when it is required.

The quality of waste heat can be roughly characterized in terms of the temperatures of the waste stream. The higher the temperature, the more available the waste heat for substitution for purchased energy. The primary sources of energy used in industrial plants are the combustion of fossil fuels and nuclear reaction, both occurring at temperatures approaching 3000°F. Waste heat, of any quantity is ordinarily of little use at temperatures approaching ambient, although the use of a heat pump can improve the quality of waste heat economically over a limited range of temperatures near and even below ambient. As an example, a waste-heat stream at 70°F cannot be used directly to heat a fluid stream whose temperature is 100°F. However, a heat pump might conceivably be used to raise the temperature of the waste heat stream to a temperature above 100°F so that a portion of the waste-heat could then be transferred to the fluid stream at 100°F. Whether this is economically feasible depends upon the final temperature required of the fluid to be heated and the cost of owning and operating the heat pump.

8.6   MATCHING WASTE HEAT SOURCE AND SINK

It cannot be emphasized too strongly that the interruption of a waste heat load, either accidentally or intentionally, may impose severe operating conditions on the source system and might conceivably cause catastrophic failures of that system.

In open system cooling the problem is easier to deal with. Consider the waste heat recovery from the cooling water from an air compressor. In this case the cooling water is city tap water which flows serially through the water jackets and the intercooler and is used as makeup water for several heated treatment baths. Should it become necessary to shut off the flow of makeup water to the baths, it would be necessary to valve the cooling water flow to a drain so that the compressor cooling continues with no interruption. Otherwise, the compressor would become overheated and suffer damage.

Effective utilization of waste heat requires several considerations in addition to a clear understanding of fluid flow patterns in the heat exchangers. Figure 8-2 is the schematic of a refrigeration plant condenser supplying waste heat for space heating during the winter. Since the heating load varies hourly and daily, and disappears in the summer months, it is necessary to provide an auxiliary heat sink which will accommodate the entire condenser discharge when the waste-heat load disappears. In the installation shown, the auxiliary heat sink is a wet cooling tower which is placed in series with the waste-heat exchanger. The series arrangement is preferable to the alternative parallel arrangement for several reasons. One is that fewer additional controls are needed. Using the parallel arrangement would require that the flows through the two paths be carefully controlled to maintain required condenser temperature, while at the same time optimizing the waste-heat recovery.

In the above examples the failure to utilize all of the available waste heat had serious consequences on the system supplying the waste heat. A somewhat different waste-heat using problem occurs when the effect of excessive waste heat availability has an adverse affect on the heat sink. An example would be the use of the cooling air stream from an air-cooled screw-type compressor for space heating in the winter months. During the summer months all of the compressor cooling air would have to be transferred to the outdoors in order to prevent overheating of the work space. This is easily done with sheet metal ducts and diverting dampers.

8.6.1 Open Waste-Heat Exchangers

An open heat exchanger is one where two fluid streams are mixed to form a third exit stream in which energy level (and temperature) is intermediate between the two entering streams. This arrangement has the advantage of extreme simplicity and low fabrication costs with no complex internal parts. The disadvantages are that (1) all flow streams must be at the same pressure, and (2) the contamination of the exit fluids by either of the entrance flows is possible. Several effective applications of open waste-heat exchangers are listed below:

1. The exhaust steam from a turbine-driven feedwater pump in a boiler plant is used to preheat the feedwater in a deaerating feedwater heater.

2. The makeup air for an occupied space is tempered by mixing it with the hot exhaust products from the stack of a gas-fired furnace in a plenum before discharge into the space. This recovery method may be prohibited by codes because of the danger of toxic carbon monoxide; a monitor should be used to test the plenum gases.

3. The continuous blowdown stream from a boiler plant is used to heat the hot wash and rinse water in a commercial laundry. A steam-heated storage heater serves as the open heater.

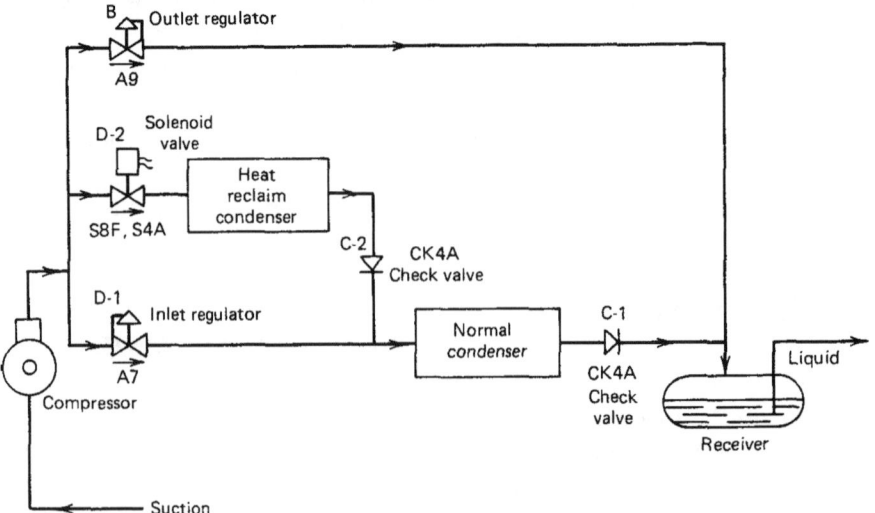

Figure 8-2. System with cold weather condenser pressure control.

8.6.2 Serial Use of Process Air and Water

In some applications, waste streams of process air and water can be directly used for heating without prior mixing with other streams. Some practical applications include:

1. condenser cooling water from batch coolers used directly as wash water in a food-processing plant;
2. steam condensate from wash water heaters added directly to wash water in the bottling section of a brewery;
3. air from the cooling section of a tunnel kiln used as the heating medium in the drying rooms of a refractory;
4. condensate from steam-heated chemical baths returned directly to the baths; and
5. the exhaust gases from a waste-heat boiler used as the heating medium in a lumber kiln.

In all cases, the possibility of contamination from a mixed or a twice-used heat-transport medium must be considered.

8.6.3 Closed Heat Exchangers

As opposed to the open heat exchanger, the closed heat exchanger separates the stream containing the heating fluid from the stream containing the heated fluid, but allows the flow of heat across the separating boundaries. The reasons for separating the streams may be:

1. A pressure difference may exist between the two streams of fluid. The rigid boundaries of the heat exchanger are designed to withstand the pressure differences.
2. One stream could contaminate the other if allowed to mix. The impermeable, separating boundaries of the heat exchanger prevents mixing. Some health or sanitation regulations may require a double wall heat exchange with an intermediate weep hold as further protection from contamination. Theses should only be used when necessary, due to cost and loss of heat transfer efficiency.
3. To permit the use of intermediate fluid better suited than either of the principal exchange media for transporting waste heat through long distances. While the intermediate fluid is often steam, glycol and water mixtures and other substances can be used to take advantage of their special properties.

Closed heat exchangers fall into the general classification of industrial heat exchangers, however they have many pseudonyms related to their specific form or to their specific application. They can be called recuperators, regenerators, waste-heat boilers, condensers, tube-and-shell heat exchangers, plate-type heat exchangers, feedwater heaters, economizers, etc. Whatever name is given, all perform one basic function—the transfer of heat across rigid and impermeable boundaries.

8.6.4 Runaround Systems

Whenever it is necessary to ensure isolation of heating and heated systems, or when it becomes advantageous to use an intermediate transfer medium because of the long distances between the two systems, a runaround heat recovery system is used. Figure 8-3 shows the schematic of a runaround system that recovers heat from the exhaust stream from the heating and ventilating system of a building. The circulating medium is a

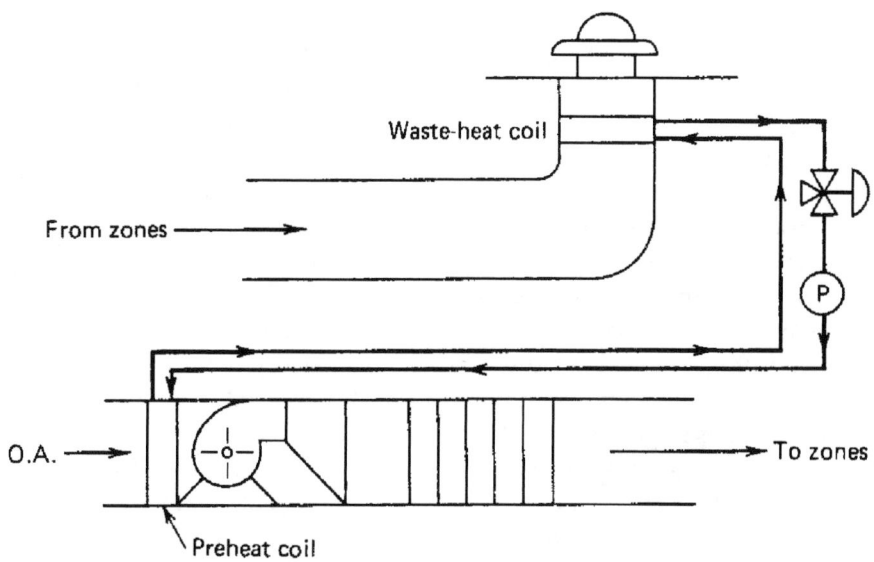

Figure 8-3. Runaround heat-recovery system.

water-glycol mixture selected for its low freezing point. In winter the exhaust air gives up some energy to the glycol in a heat exchanger located in the exhaust air duct. The glycol is circulated by way of a small pump to a second heat exchanger located in the inlet air duct. The outside air is preheated with recovered waste-heat that substitutes for heat that would otherwise be added in the main heating coils of the building's air handler. During the cooling season the heat exchanger in the exhaust duct heats the exhaust air, and the one in the inlet duct precools the outdoor air prior to its passing through

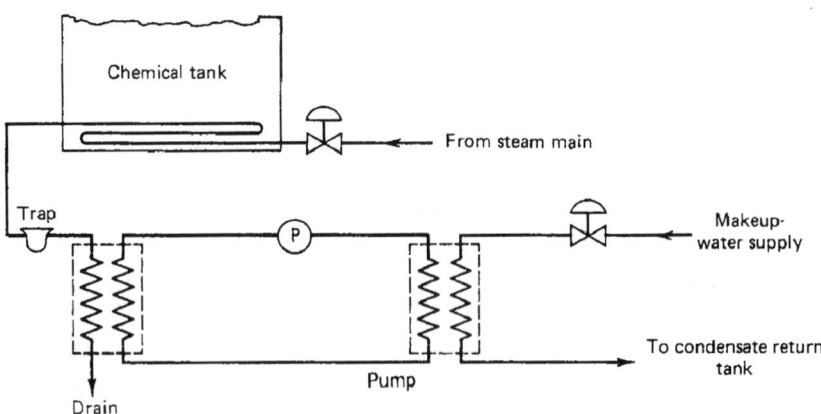

Figure 8-4. Runaround heat-recovery-system process steam source.

the cooling coils of the air handler. The principal reason for using a runaround system in this application is the long separation distance between the inlet air and the exhaust air ducts. Had these been close together, one air-to-air heat exchanger (with appropriate ducting) could have been more economical.

Figure 8-4 is the schematic diagram of a runaround system used to recover the heat of condensation from a chemical bath steam heater. In this case the bath is a highly corrosive liquid. A leak in the heater coils would cause the condensate to become contaminated and thus do damage to the boiler. The intermediate transport fluid isolates the boiler from a potential source of contamination and corrosion. It should be noted that the presence of corrosive chemicals in the bath, which dictated the choice of the runaround system, are also in contact with one side of the condensate heat exchanger. The materials of construction for that heat exchanger should be carefully selected to withstand the corrosion from that chemical.

8.7   WASTE-HEAT EXCHANGERS

8.7.1 Transient Storage Devices

The earliest waste-heat-recovery devices were "regenerators." These consisted of extensive brick work called "checkerwork," located in the exhaust flues and inlet air flues of high-temperature furnaces in the steel industry. Regenerators are still used to a limited extent in open hearth furnaces and other high-temperature furnaces burning low-grade fuels. It is impossible to achieve steel melt temperature with the fuels used unless regenerators are used to boost the inlet air temperature. In the process vast amounts of waste heat are recovered which would otherwise be supplied by expensive high-Btu fuels. Pairs of regenerators are used alternately to store

waste heat from the furnace exhaust gases and then give back that heat to the inlet combustion air. The transfer of exhaust-gas and combustion-air streams from one regenerator to the other is accomplished by using a four-way flapper valve. The design and estimates of the performance of recuperators follows the principles presented in Section 8.4. One disadvantage of this mode of operation is that heat-exchanger effectiveness is maximum only at the beginning of each heating and cooling cycle and falls to almost zero at the end of the cycle. A second disadvantage is that the tremendous mass of the checkerwork and the volume required for its installation raises capital costs above that for the continuous-type air preheaters.

An alternative to the checkerwork regenerator is the heat wheel. This device consists of a permeable flat disk which is placed with its axis parallel to a pair of split ducts and is slowly rotated on an axis parallel to the ducts. The wheel is slowly rotated as it intercepts the gas streams flowing concurrently through the split ducts. Figure 8-5 illustrates those operational features.

As the exhaust-gas stream in the exhaust duct passes through one-half of the disk, it gives up some of its heat that is temporarily stored in the disc material. As the disc is turned, the cold incoming air passes through the heated surfaces of the disk and absorbs the energy. The materials used for the disks include metal alloys, ceramics, and fiber, depending upon the temperature of the exhaust gases. Heat-exchanger efficiency for the heat wheel has been measured as high as 90%, based upon the exhaust stream energy. Some heat wheels are designed to recover sensible heat only, while others with special materials can additionally capture moisture from the exhaust stream. The heat wheel shown in Figure 8-5 recovers both sensible and latent energy from the exhaust.

For commercial buildings in humid climates, desiccant heat wheels (enthalpy wheels) can serve in reverse to de-humidify incoming outside air and thereby reduce the

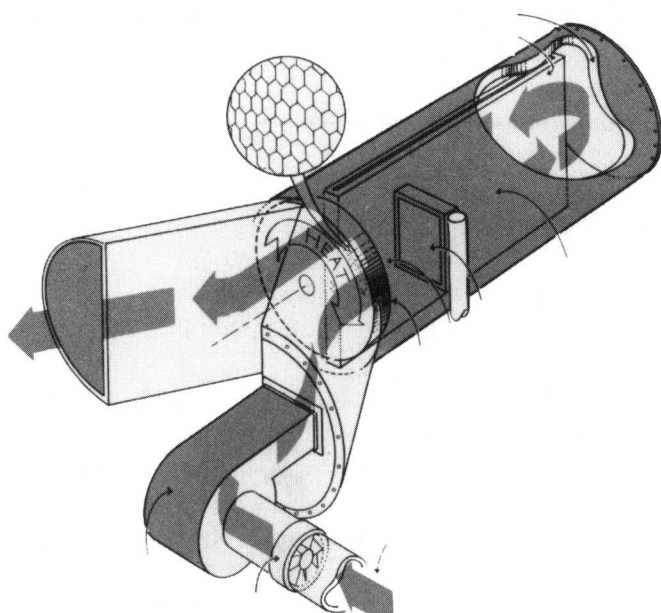

Figure 8-5. Heat wheel.

building dehumidification load considerably.

8.7.2 Steady-State Heat Exchangers

Section 8.8 treats heat exchangers in some detail. However, several important criteria for selection are listed below.

1. Flow Arrangements. These are characterized as:

    Parallel flow   Crossflow
    Counterflow   Mixed flow

    The flow arrangement helps to determine the overall effectiveness, the cost, and the highest achievable temperature in the heated stream. The latter effect most often dictates the choice of flow arrangement in crossflow configuration. Figure 8-6 indicates the temperature profiles for the heating and heated streams, respectively. If the waste heat stream is to be cooled below the cold stream exit, a counterflow heat exchanger must be used.

2. Character of the Exchange Fluids. It is necessary to specify the heated and cooled fluids as to:

    *   Chemical composition
    *   Physical phase (i.e., gaseous, liquid, solid, or multiphase)
    *   Change of phase, if any, such as evaporating or condensing

    These specifications may affect the optimum flow arrangement and/or the materials of construction.

8.7.3 Heat-Exchanger Effectiveness

The effectiveness of a heat exchanger is defined as a ratio of the actual heat transferred to the maximum possible heat transfer considering the temperatures of two streams entering the heat exchanger. For a given flow arrangement, the effectiveness of a heat exchanger is di-

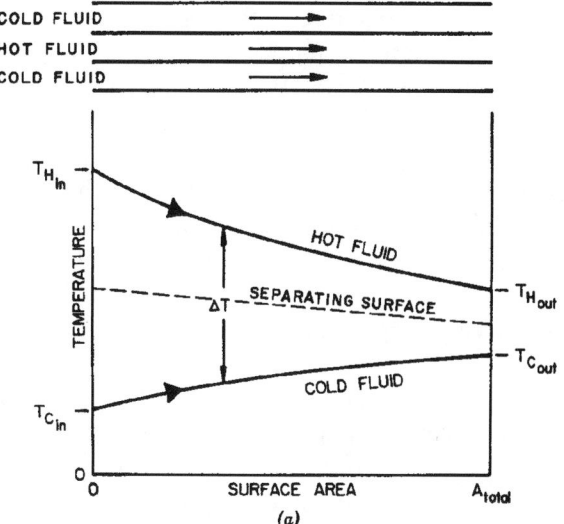

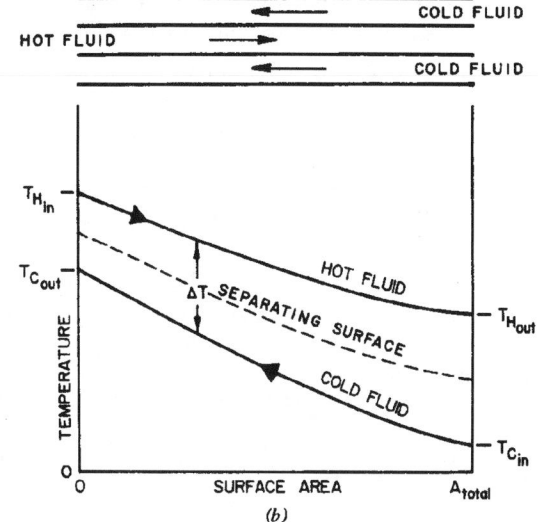

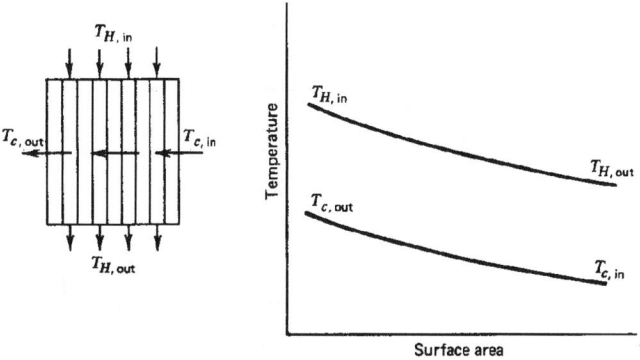

Figure 8-6. Cross-flow heat exchangers.

rectly proportional to the surface area that separates the heated and cooled fluids. The effectiveness of typical heat exchangers is given in Figure 8-7a-c in terms of the parameter AU/C_{min} where A is the effective heat-transfer area, U the effective overall heat conductance, and C_{min} the mass flow rate times the specific heat of the fluid with minimum mc. The conductance is the heat rate per unit area per unit temperature difference. Note that as AU/C_{min} increases, a linear relation exists with the effectiveness until the value of AU/C_{min} approaches 1.0. At this point the curve begins to knee over and the increase in effectiveness with AU is drastically reduced. Thus, one sees a relatively early onset of the law of diminishing returns for heat-exchanger design, and it is implied that one pays heavily for exchangers with high effectiveness.

8.7.4 Filtering or Fouling

One of the important heat-exchanger parameters related to surface conditions is termed the fouling factor. The fouling of the surfaces can occur because of film deposits, such as oil films; because of surface scaling due to the precipitation of solid compounds from solution; because of corrosion of the surfaces; or because of the deposit of solids or liquids from two-phase flow streams. The fouling factor increases with increased fouling and causes a drop in heat exchanger effectiveness. If heavy fouling is anticipated, it may call for the filtering of con-

taminated streams, special construction materials, or a mechanical design that permits easy access to surfaces for frequent cleaning. It is common to specify a 'fouling factor' when specifying a heat exchanger that requires the manufacturer to build the equipment with additional heat transfer surface area due to anticipating a measure of fouling in service.

8.7.5 Materials and Construction

These topics have been reviewed in previous sections. In summary:

1. High or low temperatures may require the use of special materials.
2. The chemical and physical properties of exchange fluids may require the use of special materials.
3. Contaminated fluids may require special materials and/or special construction.
4. The additions of tube fins on the outside, grooved surfaces or swaged fins on the inside, and treated or coated surfaces inside or outside may be required to achieve compactness or unusually high effectiveness.

8.7.6 Corrosion Control

The standard material of construction for heat exchangers is mild steel. Heat exchangers made of steel are the cheapest to buy because the material is the least

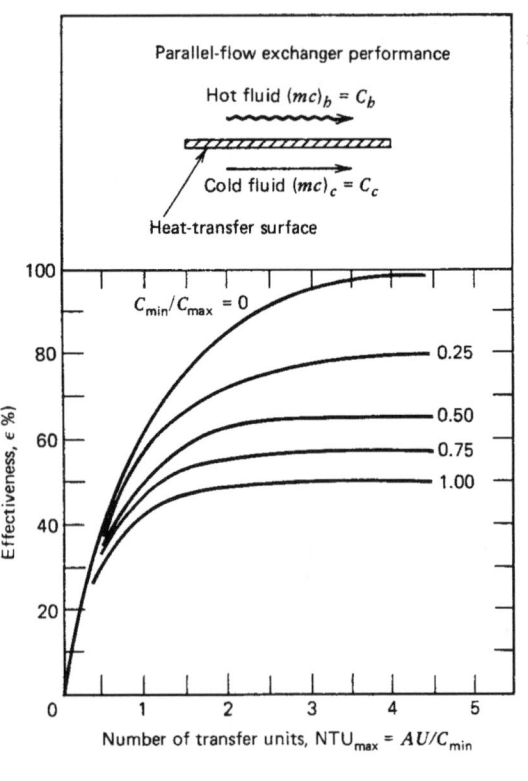

Figure 8-7a.

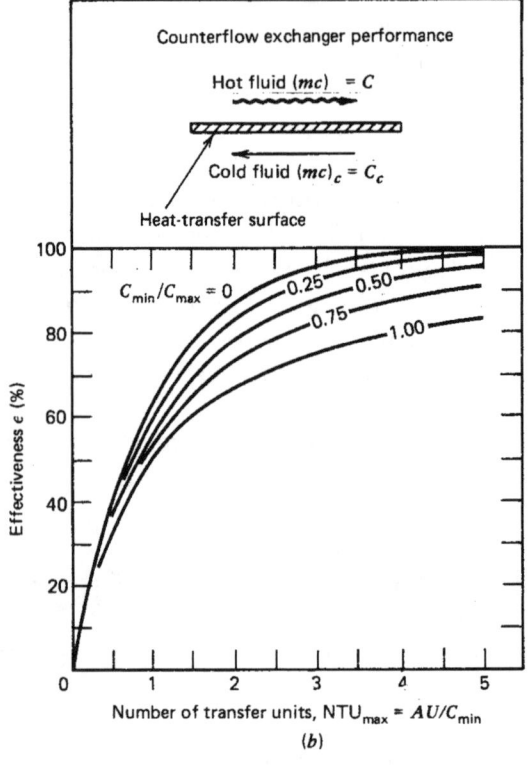

Figure 8-7b.

expensive of all construction materials and is so easy to fabricate. However, when the heat transfer media are corrosive liquids and/or gases, more exotic materials may have to be used. Corrosion tables give the information necessary to estimate the life of the heat exchanger, and life-cycle-costing studies allow valid comparisons of the costs of the steel heat exchanger versus one constructed of exotic materials. The problem is whether it will be cheaper to replace the steel heat exchanger at more frequent intervals or to buy a unit made of more expensive materials that requires less frequent replacement. Mechanical designs which permit easy tube replacement lower the cost of rebuilding and favor the use of mild steel heat exchangers.

Corrosion-resisting coatings, such as TFE plastics, are used to withstand extremely aggressive liquids and gases. However, the high cost of coating and the danger of damaging the coatings during assembly (and during subsequent operation) limit their use. One disadvantage of using coatings is that they almost invariably decrease the overall conductance of the tube walls and thus necessitate an increase in size of the heat exchanger. The decision to use coatings depends upon the availability of alternate materials to withstand the corrosion as well as the comparative life-cycle costs, assuming that alternative materials can be found.

Among the most corrosive and widely used materials flowing in heat exchangers are chlorides, such as hydrochloric acid and saltwater. Steel and most steel alloys have extremely short lives in such service. One class of steel alloys that has shown remarkable resistance to chlorides and other corrosive chemicals is called duplex steels, consisting of half-and-half ferrite and austenitic microstructures. Because of their high tensile strength, thinner tube walls can be used, and this offsets some of the higher cost of the materials.

8.7.7 Maintainability

Provisions for gaining access to the internals may be worth the additional cost so that surfaces may be easily cleaned, or tubes replaced when corroded. A shell and tube heat exchanger with flanged and bolted end caps that are easily removed for maintenance is shown in Figure 8-8. Economizers are available with removable panels and multiple one-piece, finned, serpentine tube elements that are connected to the heaters with standard compression fittings. The tubes can be removed and replaced on site, in a matter of minutes, using only a crescent wrench.

8.8   COMMERCIAL OPTIONS IN WASTE-HEAT-RECOVERY EQUIPMENT

8.8.1 Introduction

It is necessary to completely specify all of the operating parameters as well as the heat exchange capacity for the proper design of heat exchangers, or for the selection of an off-the-shelf item. These specifications will determine the construction parameters and thus the cost of the heat exchanger. The final design will be a compromise among pressure drop (which fixes pump or fan capital and operating costs), maintainability (which strongly af-

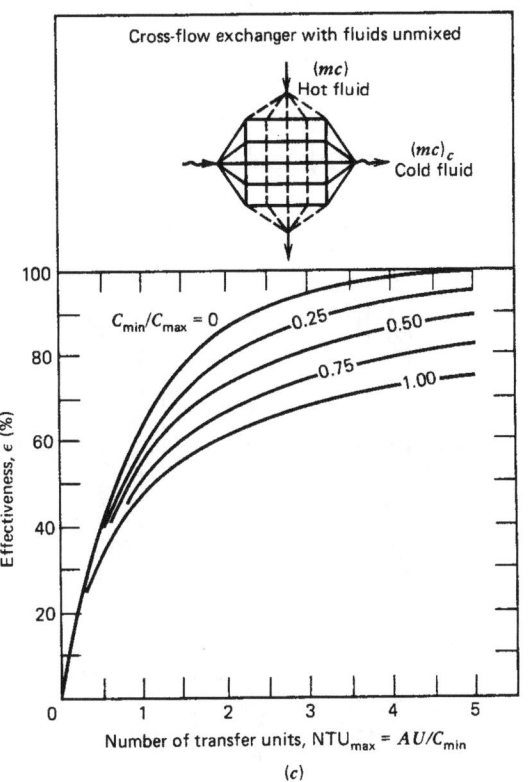

Figure 8-7c. Typical heat-exchanger effectiveness.

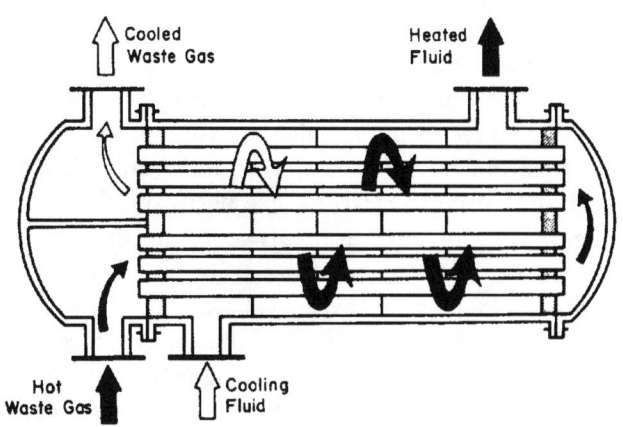

Figure 8-8. Shell and tube heat exchangers

fects maintenance costs), heat exchanger effectiveness, and life-cycle cost. Additional features, such as the on-site use of exotic materials or special designs for enhanced maintainability, may add to the initial cost. That design will balance the costs of operation and maintenance with the fixed costs in order to minimize the life-cycle costs. Advice on selection and design of heat exchangers is available from manufacturers and from T.E.M.A. *Industrial Heat Exchangers*[4] is an excellent guide to heat exchanger selection and includes a director of heat exchanger manufacturers.

The essential parameters that should be known and specified in order to make an optimum choice of waste-heat recovery devices are:

— Temperature of waste-heat fluid
— Chemical composition of waste-heat fluid
— Minimum allowable temperature of waste-heat fluid
— Amount and type of contaminants in the waste-heat fluid
— Allowable pressure drop for the waste-heat fluid
— Temperature of heated fluid
— Flow rate of heated fluid
— Chemical composition of heated fluid
— Maximum allowable temperature of heated fluid
— Allowable pressure drop in the heated fluid
— Control temperature, if control required.

In the remainder of this section, some common types of commercially available waste-heat recovery devices are discussed in detail.

8.8.2 Gas-to-gas Heat Exchangers: Recuperators

Recuperators are used in recovering waste heat to be used for heating gases in the medium- to high-temperature range. Some typical applications are soaking ovens, annealing ovens, melting furnaces, reheat furnaces, afterburners, incinerators, and radiant-heat burners. The simplest configuration for a heat exchanger is the metallic radiation recuperator, which consists of two concentric lengths of metal tubing, as shown in Figure 8-9. This is most often used to extract waste heat from the exhaust gases of a high-temperature furnace for heating the combustion air for the same furnace. The assembly is often designed to replace the exhaust stack.

The inner tube carries the hot exhaust gases, while the external annulus carries the combustion air from the atmosphere to the air inlets of the furnace burners. The hot gases are cooled by the incoming combustion air, which then carries additional energy into the combustion cham-

ber. This is energy that does not have to be supplied by the fuel; consequently, less fuel is burned for a given furnace loading. The saving in fuel also means a decrease in combustion air, and therefore stack losses are decreased, not only by lowering the stack exit temperatures but also by discharging smaller quantities of exhaust gas. This particular recuperator gets its name from the fact that a substantial portion of the heat transfer (from the hot exhaust gases to the surface of the inner tube) takes place by radiative heat transfer. The cold air in the annulus, however, is almost transparent to infrared radiation, so only convection heat transfer takes place to the incoming combustion air. As shown in the diagram, the two gas flows are usually parallel, although the configuration would be simpler and the heat transfer more efficient if counterflow were used. The reason for the use of parallel flow is that the cold air often serves the function of cooling the hottest part of the exhaust duct and consequently extends its service life.

The inner tube is often fabricated from high-temperature materials such as high-nickel stainless steels. The large temperature differential at the inlet causes differential expansion since the outer shell is usually of a different and less expensive material. The mechanical design must take this effect into account. More elaborate designs of radiation recuperators incorporate two sections, the bottom operating in parallel flow, and the upper section using the more efficient counterflow arrangement. Because of the

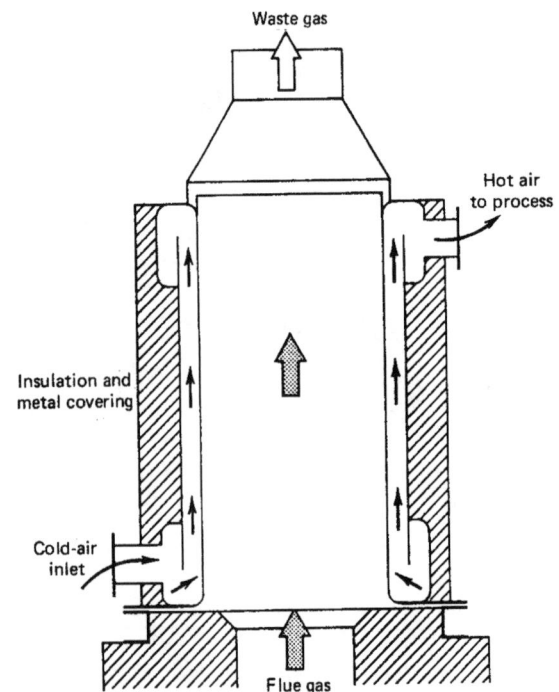

Figure 8-9. Metallic radiation recuperator

large axial expansions experienced and the difficult stress conditions that can occur at the bottom of the recuperator, the unit is often supported at the top by a freestanding support frame joined at the bottom to the furnace by an expansion joint.

A second common form for recuperators is called a tube-type or convective recuperator. As seen in the schematic diagram of a combined radiation and convective type recuperator in Figure 8-10, the hot gases are carried through a number of small-diameter parallel tubes; the combustion air enters a shell surrounding the tubes and is heated as it passes over the outside of the tubes (one or more times) in directions normal to the tubes. If the tubes are baffled as shown so as to allow the air to pass over them twice, the heat exchanger is termed a two-pass convective recuperator; if two baffles are used, it is known as a three-pass recuperator, etc. Although baffling increases the cost of manufacture, as well as the pressure drop in the air path, it also increases the effectiveness of heat exchange. Tube-type recuperators are generally more compact and have a higher effectiveness than do radiation recuperators, because of the larger effective heat-transfer area made possible through the use of multiple tubes and multiple passes of the air. For maximum effectiveness of heat transfer, combinations of the two types of recuperators are used, with the convection type always following the high-temperature radiation recuperator.

The principal limitation on the heat recovery possible with metal recuperators is the reduced life of the liner at inlet temperatures exceeding 2000°F. This limitation forces the use of parallel flow to protect the bottom of the liner. The temperature problem is compounded when furnace combustion air flow is reduced as the furnace loading is reduced. Thus the cooling of the inner shell is reduced and the resulting temperature rise causes rapid surface deterioration. To counteract this effect, it is necessary to provide an ambient air bypass to reduce the temperature of the exhaust gases. The destruction of a radiation recuperator by overheating is a costly accident. Costs for rebuilding one are about 90% of the cost of a new unit.

To overcome the temperature limitations of metal recuperators, ceramic-tube recuperators have been developed with materials that permit operation to temperatures of 2800°F and on the preheated air side to 2200°F, although practical designs yield air temperatures of 1800°F. Early ceramic recuperators were built of tile and joined with furnace cement. Thermal cycling caused cracking of the joints and early deterioration of the units. Leakage rates as high as 60% were common after short service periods. Later developments featured silicon carbide tubes joined by flexible seals in the air headers. This kind of de-

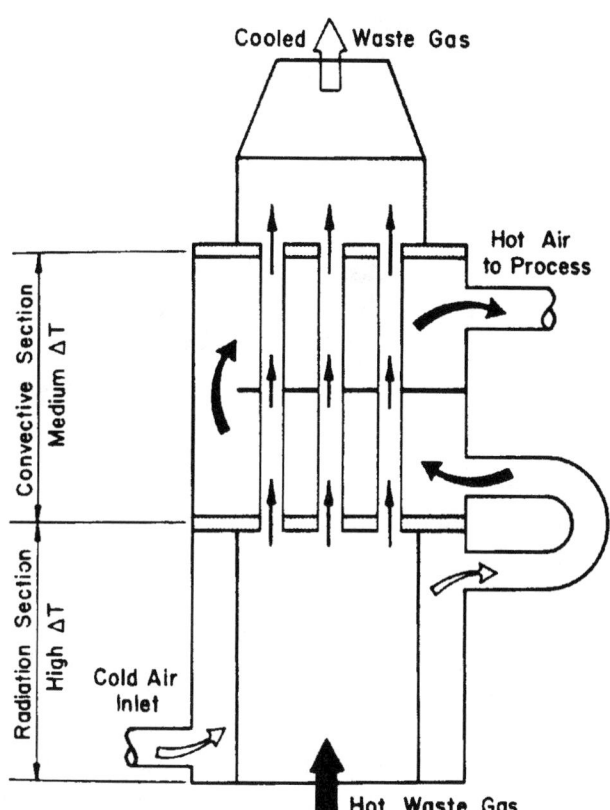

Figure 8-10. Combined radiation and convective recuperator

sign, illustrated in Figure 8-11, maintained the seals at a relatively low temperature and the life of seals was much improved, as evidenced by leakage rates of only a few percent after two years of service.

An alternative design for the convective recuperator is one in which the cold combustion air is heated in a bank of parallel tubes extending into the flue-gas stream normal to the axis of flow. This arrangement is shown in Figure 8-12. The advantages of this configuration are compactness and the ease of replacing individual units. This can be done during full-load operation and minimizes the cost, inconvenience, and possible furnace damage due to a forced shutdown from recuperator failure.

Recuperators are relatively inexpensive, and they do reduce fuel consumption. However, their use may require extensive capital improvements. Higher combustion air temperatures may require:

- burner replacement
- larger-diameter air lines with flexible expansion fittings
- cold-air piping for cooling high-temperature burners
- modified combustion controls
- stack dampers
- cold air bleeds

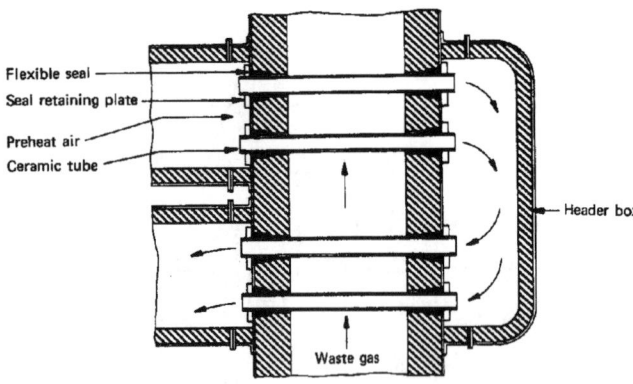

Figure 8-11. Silicon-carbine-tube ceramic recuperator

- recuperator protection systems
- larger combustion air fans to overcome the additional pressure drops in the system and to compensate for higher volumes needed for equivalent mass flow, with the reduction in air density.

8.8.3 Thermal Heat Wheels

A rotary regenerator, also called an air preheater or heat wheel, is used for low-to moderately high-temperature waste-heat recovery. Typical applications are for space heating, curing, drying ovens, and heat-treat furnaces. Originally developed as an air preheater for utility steam boilers, it was later adapted, in small sizes, as a regenerator for automotive turbine applications. It has been used for temperatures ranging from 68°F to 2500 °F.

Figure 8-13 illustrates the operation of a heat wheel in an air conditioning application. It consists of a porous disk, fabricated of material having a substantial specific heat. The disk is driven to rotate between two side-by-side ducts. One is a cold-gas duct and the other is a hot-gas duct. Although the diagram shows a counterflow configuration, parallel flow can also be used. The axis of the disk is located parallel to and on the plane of the partition between the ducts. As the disk slowly rotates, sensible heat (and in some cases, moisture-containing latent heat) is transferred to the disk by the hot exhaust gas. As the disk moves into the area of the cold duct, the heat is transferred from the disk to the cold air. The overall efficiency of heat transfer (including latent heat) can be as high as 90%.

Heat wheels have been built as large as 70 ft in diameter with air capacities to 40,000 cfm. Multiple units can be used in parallel. This modular approach may be used to overcome a mismatch between capacity requirements and the limited number of sizes available in commercial units.

The limitations on the high-temperature range for the heat wheel are primarily due to mechanical difficul-

ties introduced by uneven thermal expansion of the rotating wheel. Uneven expansion can cause excessive deformations of the wheel that result in the loss of adequate gas seals between the ducts and the wheel. The deformation can also result in damage due to the wheel rubbing against its retaining enclosure.

Heat wheels are available in at least four types: (1) a metal frame packed with a core of knitted mesh stainless steel, brass, or aluminum wire, (2) a so-called laminar wheel fabricated from corrugated materials that form many small diameter parallel-flow passages, (3) a laminar wheel constructed from a high-temperature ceramic honeycomb, and (4) a laminar wheel constructed of a fibrous material coated with a hygroscopic so that latent heat can be recovered.

Most gases contain some water vapor since it is a natural component of air and is also a product of hydrocarbon combustion. Water vapor, as a component of a gas mixture, carries with it its latent heat of evaporation. This latent heat may be a substantial part of the energy contained within the exit-gas steams from air-conditioned

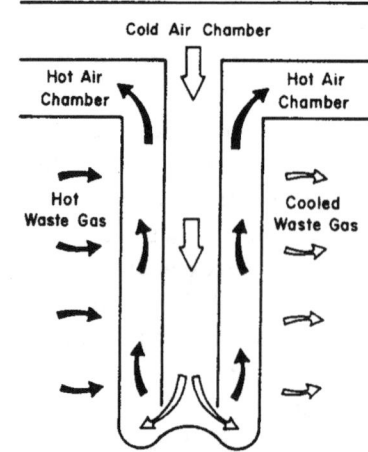

Figure 8-12. Parallel-tube recuperator

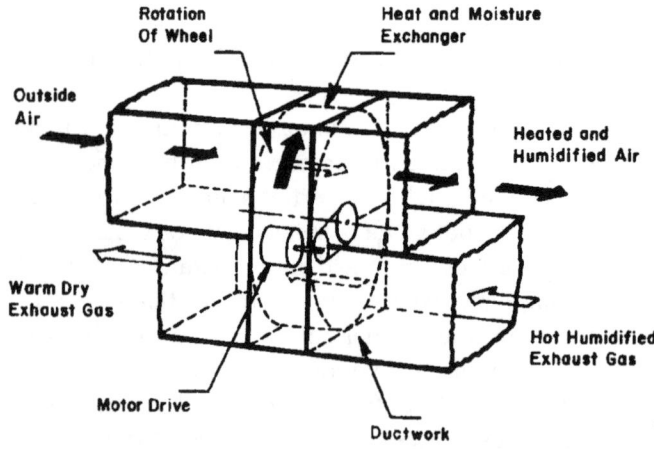

Figure 8-13. Rotary regenerator (heat wheel)

spaces, or from industrial processes. To recover some of the latent heat in the gas stream using a heat wheel, the sheet must be coated with a hygroscopic material such as lithium chloride (LiCl) that readily absorbs water vapor to form a hydrate, which in the case of lithium chloride is the hydrate $LiCl \cdot H_2O$. This hydrate consists of one mole of lithium chloride chemically combined with one mole of water vapor. Thus the weight ratio of water to lithium-chloride is 3:7. In a hygroscopic heat wheel, the hot gas stream gives up some part of its water vapor to the lithium-chloride coating; the gases to be heated are dry and absorb some of the water held in the hydrate. The latent heat in that water vapor adds directly to the total quantity of recovered heat. The efficiency of recovery of the water vapor in the exit stream may be as high as 50%. Other desiccant materials may be used.

Because the pores or passages of heat wheels carry small amounts of gas from the exhaust duct to the intake duct, cross-contamination of the intake gas can occur. If the contamination is undesirable, the carryover of exhaust gas can be partially eliminated by the addition of a purge section located between the intake and exhaust ducts, as shown in Figure 8-14. The purge section allows the passage in the wheel to be cleared of the exhaust gases by introducing clean air, which discharges the contaminant to the atmosphere. Note that additional gas seals are required to separate the purge ducts from the intake and exhaust ducts, and they consequently add to the cost of the heat wheel.

Common practice is to use six air changes of clean air for purging. This results in a reduction of cross-contamination to a value as little as 0.04% for the gas, 0.2% for particulates in laminar wheels, and less than 1.0% total contaminants in packed wheels. If no cross contamination is permitted at all, such as with some laboratory exhaust, the heat recovery device may need to be of a different type with impermeable membranes separating the air flows.

If the heated gas temperatures are to be held constant, regardless of heating loads and exhaust gas temperatures, the heat wheel must be driven at variable speed. This requires a variable-speed drive and a speed-controller with an air temperature sensor as the control element. When operating with outside air to periods of sub-zero temperatures and high humidity, heat wheels may frost up, requiring the protection of an air-preheat system. When handling gases containing water-soluble, greasy, or large concentrations of particulates, air filters may be required in the exhaust system upstream from the heat wheel. These features, however, add to the complexity and the cost of owning and operating the system.

Contaminant buildup on ceramic heat wheels can often be removed by raising the temperature of the exhaust steam to exceed the ignition temperature of the contaminant. However, heat wheels are inherently self-cleaning, because materials entering the wheel from the hot-gas steam tend to be swept out by the reverse flow of the cold-gas steam.

8.8.3 Passive Air Preheaters

Passive gas-to-gas regenerators are available for applications where cross-contamination cannot be tolerated. One such type of regenerator, the plate-type, is shown in Figure 8-15. A second type, the heat pipe array is shown in Figure 8-15. Passive air preheaters are used in the low- and medium-temperature applications. Those include drying, curing, and baking ovens; air preheaters in steam boilers; air dryers; water heat recovery from exhaust steam; secondary recovery from refractory kilns and reverbatory furnaces; and waste heat recovery from conditioned air such as swimming pools and commercial building general exhaust.

The plate-type regenerator is constructed of alternate channels that separate adjacent flows of heated and heating gases by a thin wall of conducting metal. Although

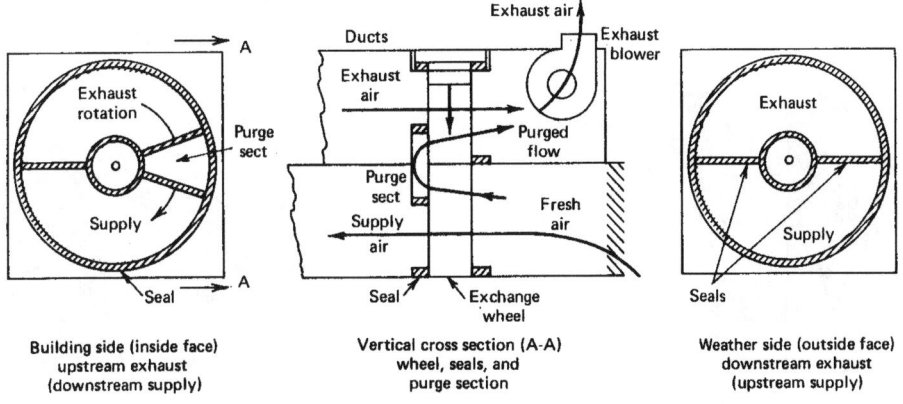

Figure 8-14. Heat wheel with purge section

their use eliminates cross-contamination, they are bulkier, heavier, and more expensive than a heat wheel of similar heat-recovery and flow capacities. Furthermore, it is difficult to achieve temperature control of the heated gas, and fouling may be a more serious problem. Because of the depth of the heat exchanger, cleaning may be impractical and reduce effectiveness significantly. Maintenance provisions are important for all heat exchangers, especially this type. If there will be any appreciable deposits within either of the air streams, filtration may be necessary for reasonable service life.

The heat pipe is a heat-transfer element that is assembled into arrays used to compact and efficient passive gas-to-gas heat exchangers. Figure 8-16 shows how the bundle of finned heat pipes extend through the wall separating the inlet and exhaust ducts in a pattern that resembles the conventional finned tube heat exchangers. Each of the separate pipes, however is a separate sealed element. Each consists of an annular wick on the inside of the full length of the tube, in which an appropriate heat-transfer fluid is absorbed. Figure 8-17 shows how the heat transferred from the hot exhaust gases evaporates the fluid in the wick. This causes the vapor to expand into the center core of the heat pipe. The latent heat of evaporation is carried with the vapor to the cold end of the tube. There it is removed by transferal to the cold gas as the vapor is recondensed. The condensate is then carried back in the wick to the hot end of the tube. This takes place by capillary action, and by gravitational forces if the axis of the tube is tilted from the horizontal. At the hot end of the tube the fluid is then recycled.

The heat pipe is compact and efficient for two reasons. The finned-tube bundle is inherently a good configuration for convective heat transfer between the gases and the outside of the tubes in both ducts. The evaporation-condensing cycle within the heat tubes is a highly efficient method of transferring heat internally. This design is also free of cross-contamination. However, the temperature range over which waste heat can be recovered is severely limited by the thermal and physical properties of the fluid used within the heat pipes. Table 8-5 lists some of the transfer fluids and the temperature ranges in which they are applicable.

8.8.4. Gas or Liquid-to-Liquid Regenerators: The Boiler Economizer

The economizer is ordinarily constructed as a bundle of finned tubes, installed in the boiler's breeching. Boiler feedwater flows through the tubes to be heated by the hot exhaust gases. Such an arrangement is shown in Figure 8-18. The tubes are usually connected in a series arrangement, but can also be arranged in series-parallel to control the liquid-side pressure drop. The air-side pressure drop is controlled by the spacing of the tubes and the number of rows of tubes. Economizers are available both prepackaged in modular sizes and designed and fabricated to custom specifications from standard components. Materials for the tubes and fins can be selected to withstand corrosive liquids and/or exhaust gases.

Temperature control of the boiler feedwater is necessary to prevent boiling in the economizer during low-steam demand, or in case of a feedwater pump failure. This is usually obtained by controlling the amount of exhaust gases flowing through the economizer with a damper, which diverts a portion of the gas flow through a bypass duct.

The extent of heat recovery in the economizer may be limited by the lowest allowable exhaust gas temperature in the exhaust stack. The exhaust gases contain water vapor both from the combustion air and from the combustion of the hydrogen that is contained in the fuel. If the exhaust gasses are cooled below the dew point of the water vapor, condensation will occur and cause damage to the structural materials. If the fuel also contains sulfur, the sulfur-dioxide will be absorbed by the condensed water to form sulfuric acid. This is very corrosive and will

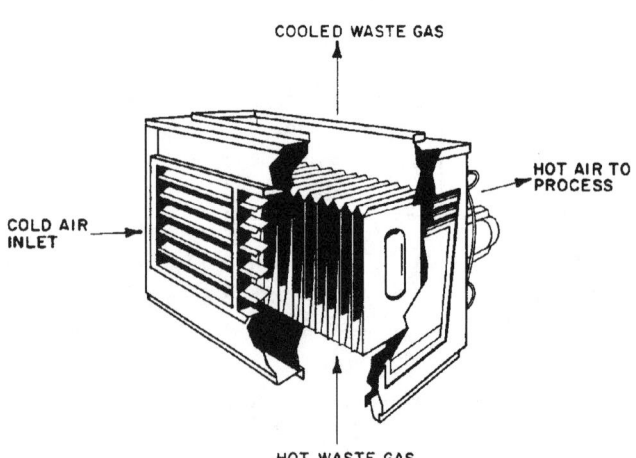

Figure 8-15. Passive gas to gas regenerator

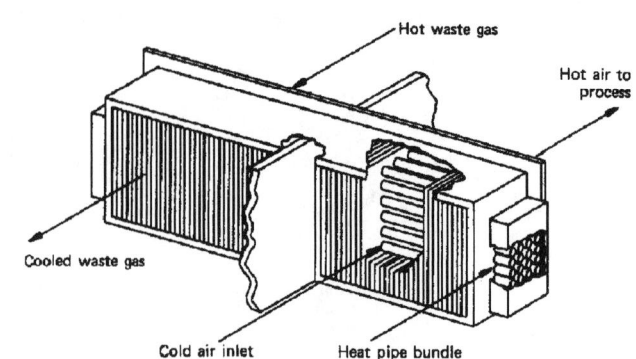

Figure 8-16. Heat pipe

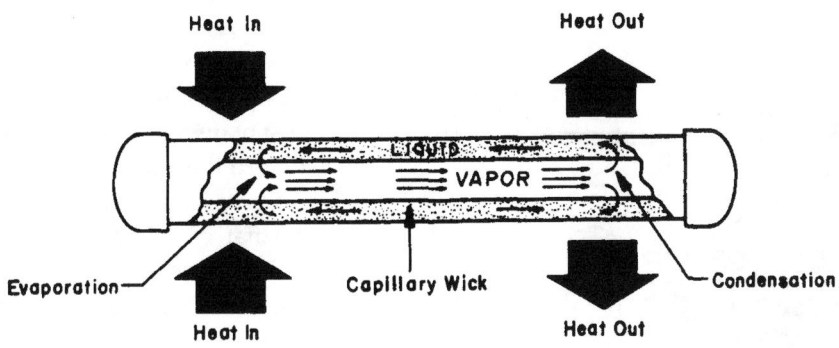

Figure 8-17. Heat pipe operation

Table 8-5. Temperature Ranges for Heat-transfer
Fluids Used in Heat Pipes

Fluid	Temperature Range (°F)	Compatible Metals
Nitrogen	-300 to -110	Stainless steel
Ammonia	-95 to +140	Nickel, aluminum, stainless steel
Methanol	-50 to +200	Nickel, copper, stainless steel
Water	40 to 425	Nickel, copper
Mercury	375 to 1000	Stainless steel
Sodium	950 to 1600	Nickel, stainless steel
Lithium	1600 to 2700	Alloy of niobium and zirconium
Silver	2700 to 3600	Alloy of tantalum and tungsten

attack the breeching downstream of the economizer and the stack lines. The dew point of the exhaust gases from a natural-gas-fired boiler varies from approximately 138°F for a stoichiometric fuel/air mixture, to 113° F for 100% excess air. Because heat-transmission losses through the stack cause axial temperature gradients from 0.2 to 2°F/ft, and because the stack liner may exist at a temperature 50 to 75°F lower than the gas bulk temperature, it is considered prudent to limit minimum stack temperatures to 300°F, or no lower than 250°F when burning natural gas without the use of special materials. When using the fuels containing sulfur, even greater caution is necessary. This means that the effectiveness of an economizer is limited unless the exhaust gases from the boiler are relatively hot. Figure 8-19 is a graph of the percent fuel saved plotted against the percent excess air for a number of stack gas temperatures using natural gas as a boiler fuel. The plots are based on a 300°F hot-gas temperature leaving the economizer.

8.8.5 Shell-and-Tube or Concentric-Tube Heat Exchangers

Shell-and-tube and concentric-tube heat exchangers are used to recover heat in the low and medium range

from process liquids, coolants, and condensates of all kinds for heating liquids.

When the medium containing waste heat is either a liquid or a vapor that heats a liquid at a different pressure, a totally enclosed heat exchanger must be used. The two fluid streams must be separated so as to contain their respective pressures. In the shell-and-tube heat exchanger, the shell is a cylinder that contains the tube bundle. Internal baffles may be used to direct the fluid in the shell over the tubes in multiple passes. Because the shell is inherently weaker than the tubes, the higher-pressure fluid is usually circulated in the tubes, while the lower-pressure fluid circulates in the shell. However, when the heating fluid is a condensing vapor, it is almost invariably contained within the shell. If the reverse were attempted, the condensation of the vapor within the small-diameter parallel tubes would cause flow instabilities. Shell and tube heat exchangers are produced in a wide range of standard sizes with many combinations of materials for the tubes and the shells. The overall conductance of these heat exchangers range to a maximum of several hundred Btu/hr ft²°F.

A concentric-tube exchanger is used when the fluid pressures are so high that a shell design is uneconomi-

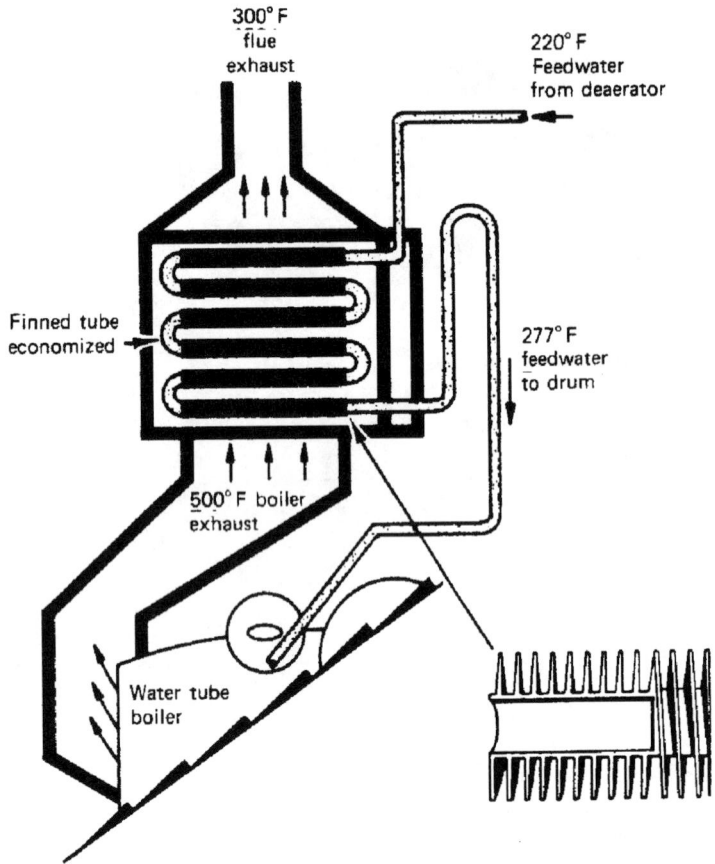

Figure 8-18. Boiler economizer

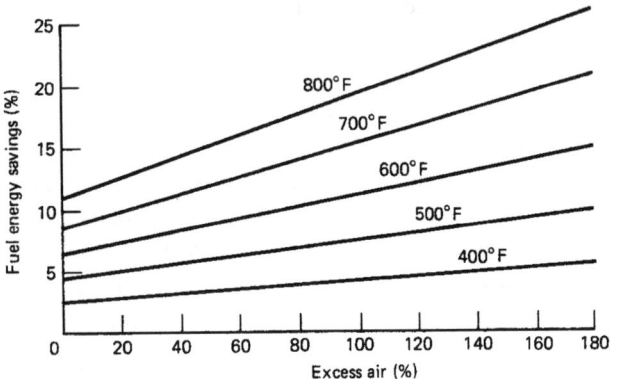

Figure 8-19. Fuel savings from a gas-fired boiler using economizer

cal, or when ease of disassembly is paramount. The hotter fluid is almost invariably contained in the inner tube to minimize surface heat losses. The concentric-tube exchanger may consist of a single straight length, a spiral coil, or a bundle of concentric tubes with hairpin bends.

Shell-and-tube and concentric-tube heat exchangers are used to recover heat in the low and medium range from process liquids, coolants, and condensates of all kinds for heating liquids.

8.8.7   Heat Pumps

Heat pumps offer only limited opportunities for waste-heat recovery, simply because the cost of owning and operating the heat pump may exceed the value of the waste heat recovered.

A heat pump is a device that operates cyclically so that energy absorbed at low temperature is transformed through the application of external work to energy at a higher-temperature that can be absorbed by an existing load. The commercial mechanical refrigeration plant can be utilized as a heat pump with small modifications, as indicated in Figure 8-20. The coefficient of performance (COP) of the heat pump cycle is the simple ratio of heat delivered to work required:

$$COP_{HP} = \frac{Q_h}{Q_{net}} = \frac{Q_h}{W_{net}} \qquad (8.7)$$

Since the work requirement must be met by a prime mover that is either an electric motor or a liquid-fueled engine, the COP must be considerably greater than 3.0 in order to be an economically attractive energy source. That is true because the efficiency of the prime movers used to drive the heat pump, or to generate the electrical energy for the motor drive, have efficiencies less than 33%. The maximum theoretical COP for an ideal heat pump is given by:

$$COP_H = \frac{1}{1 - T_L/T_H} \qquad (8.7)$$

where T_L = temperature of energy source
        T_H = temperature of energy load

The ideal cycle, however, uses an ideal turbine as a vapor expander instead of the usual throttle valve in the expansion line of the mechanical refrigeration plant.

Figure 8-21 is a graph of the theoretical COP versus load temperature for a number of source temperatures. Several factors prevent the actual heat pump from approaching the ideal:

1. The compressor efficiency is not 100%, but is instead in the range 65 to 85%.

2. A turbine expander is too expensive to use in any but the largest units. Thus the irreversible throttling process is used instead of an ideal expansion through a turbine. All of the potential turbine work is lost to the cycle.

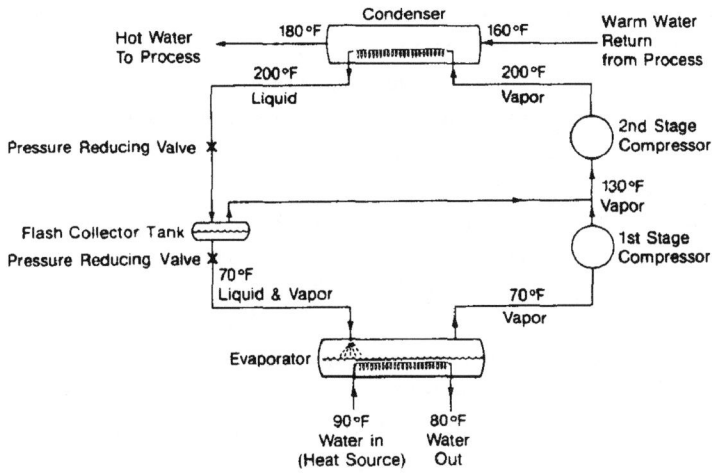

Figure 8-20. Heat Pump

3. Losses occur from fluid friction in lines, compressors, and valving.

4. Higher condenser temperatures and lower evaporator temperatures than the theoretical are required to achieve practical heat flow rates from the source and into the load.

An actual two-stage industrial heat pump installation showed an annual average COP at 3.3 for an average source temperature of 78°F and a load temperature of 190°F. The theoretical COP is 5.8. Except for very carefully designed industrial units, one can expect to achieve actual COP valves ranging from 50% to 65% of the theoretical.

An additional constraint on the use of heat pumps is that high-temperature waste heat above 230°F cannot be supplied directly to the heat pump because of the limits imposed by present compressor and refrigerant technol-

ogy. The development of new refrigerants might raise the limit of heat pump use to 400°F.

8.8.8 Waste-heat Boilers

Waste-heat boilers are water tube boilers in which hot exhaust gases are used to generate steam. The exhaust gases may be from a gas turbine, an incinerator, a diesel engine, or any other source of medium- to high-temperature waste-heat. Figure 8-22 shows a conventional, two-pass waste-heat boiler. When the heat source is in the medium-temperature range, the boiler tends to become bulky. The use of finned tubes extends the heat transfer areas and allows a more compact size. If the quantity of waste heat is insufficient (or intermittent) for generating a needed quantity of steam, it is possible to add auxiliary burners to the boiler, or an afterburner to the ducting upstream of the boiler. The conventional waste-heat boiler cannot generate super-heated steam, so an external superheater is required if superheat is needed.

A more recently designed waste-heat boiler utilizes a finned-tube bundle for the evaporator, an external drum, and forced recirculation of the feedwater. The design, which is modular, makes for a compact unit with high boiler efficiency. Additional tube bundles can be added for superheating the steam and for preheating the feedwater. The degree of superheat that can be achieved is limited by the waste-heat temperature.

Waste-heat boilers are commercially available in capacities from less than 1000 up to 1 million cfm of exhaust gas intake.

8.9 EMERGING TECHNOLOGIES FOR WASTE HEAT RECOVERY (EXPERIMENTAL STAGE)

8.9.1 Transport Membrane Condenser (TMC)

An innovative transport membrane condenser (TMC) concept has been demonstrated by the Gas Technology Institute (GTI) for efficient energy and water recovery from low temperature and low pressure boiler flue gases using the new generation of ceramic membranes as shown in Figure 8-23. The TMC can selectively extract the low pressure water vapor from flue gas into the membrane micropores in the liquid form via the pore condensation mechanism.[7] The recovered ultra-pure hot water is then transported away from the membrane in liquid form for recovery and reuse. This innovative process concept has been demonstrated by GTI to be far superior to conventional condensing economizers. Specifically, due to the low temperature gradient across conventional heat exchanger

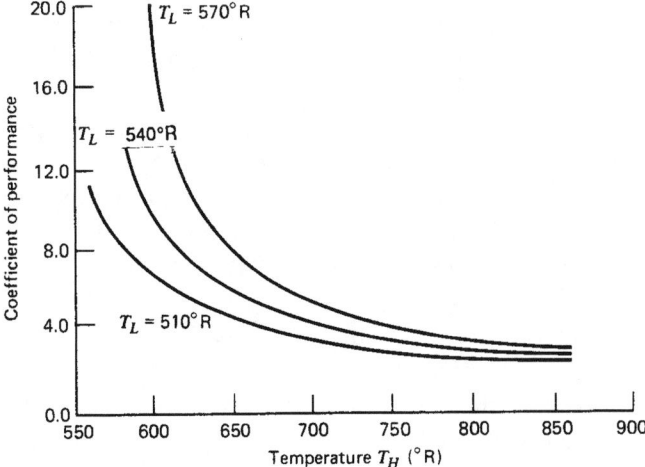

Figure 8-21. Theoretical COP vs. load temperature

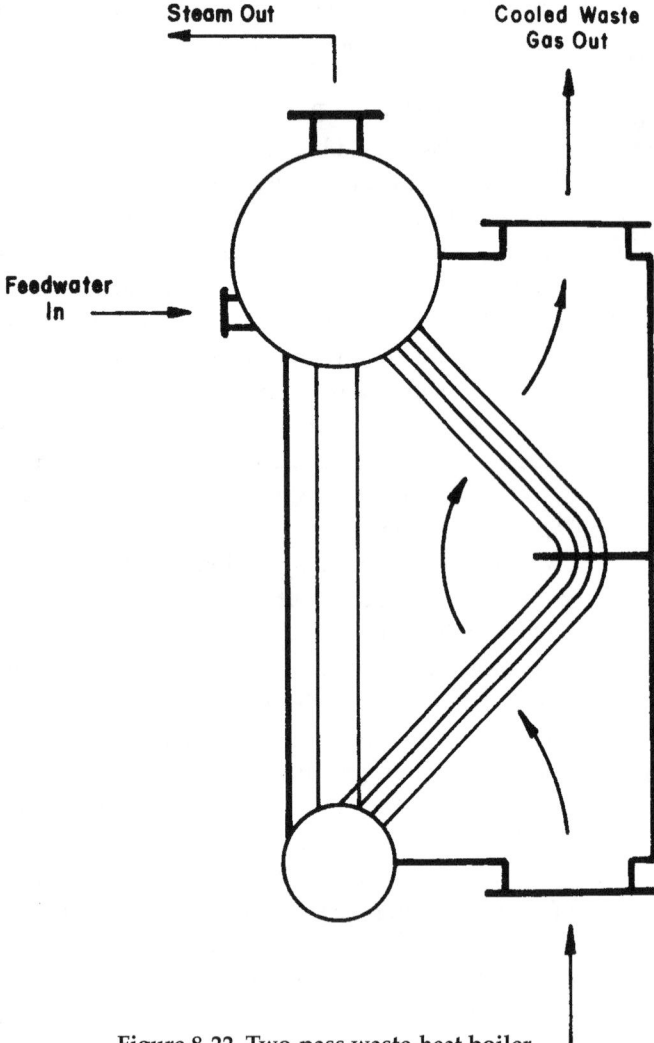

Figure 8-22. Two-pass waste-heat boiler

surfaces, heat transfer is inefficient, leading to a large condensing economizer surface area requirement. These factors make it cost prohibitive, so condensing economizers are not commonly used in the North American market [8] and the energy and water contained in the boiler flue is usually not recovered. [9]

If energy recovery from flue gas can be efficiently achieved, the energy savings potential is enormous. U.S. industry is the largest single energy consuming sector in the nation, accounting for 35% (about 35quads, 1 quad = 10^{15} Btu = 1.055×10^{15} kJ) annually, of which about 15 quads is used for on-site fuel combustion by boilers/ steam systems and process heating. Flue gas heat recovery and water reuse via the proposed TMC can increase the boiler efficiency by about 5% for state-of-the-art boilers (energy efficiency boost from ~89% to ~94%, HHV basis, based upon the feasibility test results) and will be significantly higher for lower efficiency older installed boiler capacity. Annual savings well in excess of one quad can be expected. In addition, this proposed TMC

concept is not limited to boiler applications; it can also economically recover a wide range of the "low grade" waste heat available in industrial processing streams, such as high humidity drying process exhaust in the paper and chemical industry. The total waste energy available in this industry is ~1.8 quad/year. [9] Finally, the fuel consumption reduction realized by the proposed process concept leads directly to a significant reduction in CO_2, CO, and NO_x emissions.

8.9.2 Conventional Technologies Vs TMC

Flue gas is typically available at low pressure, low temperature, and high volume; recovery of water vapor (and it latent heat) in the flue via conventional gas separation technologies, such as membranes or pressure swing adsorption (PSA), is not economical at all. The commercially available condensing economizer offers an avenue to recover this water vapor via condensation (as opposed to gas separation); however, the recovered water is contaminated with the particulate matter (PM) present in the flue. More importantly, it relies on a second fluid as a coolant, to transfer the latent heat away from the condensate. Due to the inefficient conductive heat transfer, surface area requires leads to large units that are very expensive; therefore, the condensing economizer simply has not achieved any appreciable acceptance or market penetration in the US.

Although water vapor separation via pore condensation with a porous membrane as an interphase has been discussed in the literature,[10] the technology was explored for a dehumidification. In summary, the tremendous energy and water resources available in the flue cannot be recovered economically, practically, and "greenly" with conventional gas separation or condensing devices. [10] The TMC is capable of recovering "low grade" waste energy efficiently and effectively in an environment where the water vapor pressure and temperature gradient is too low to be practical for conventional technology. Figure 8-24 shows a TMC based condenser, which can accomplish exactly that recovery.

GTI has recently demonstrated a new unit operation concept for latent heat recovery from the flue gas of packaged boilers, the TMC as shown in Figure 8-24. Figure 8-25 shows standard M&P ceramic membrane elements configured as prototype TMC elements. Water-vapor-laden flue gas near its dew point is passed through the tubes of the TMC and condensed in the pores of the membrane surface layer. The water is then convected away from the surface under a small negative pressure applied to the water on the shell side of the elements. Condensation results from two sources, namely, (i) the cooling effect of the cold water on the shell side and (ii)

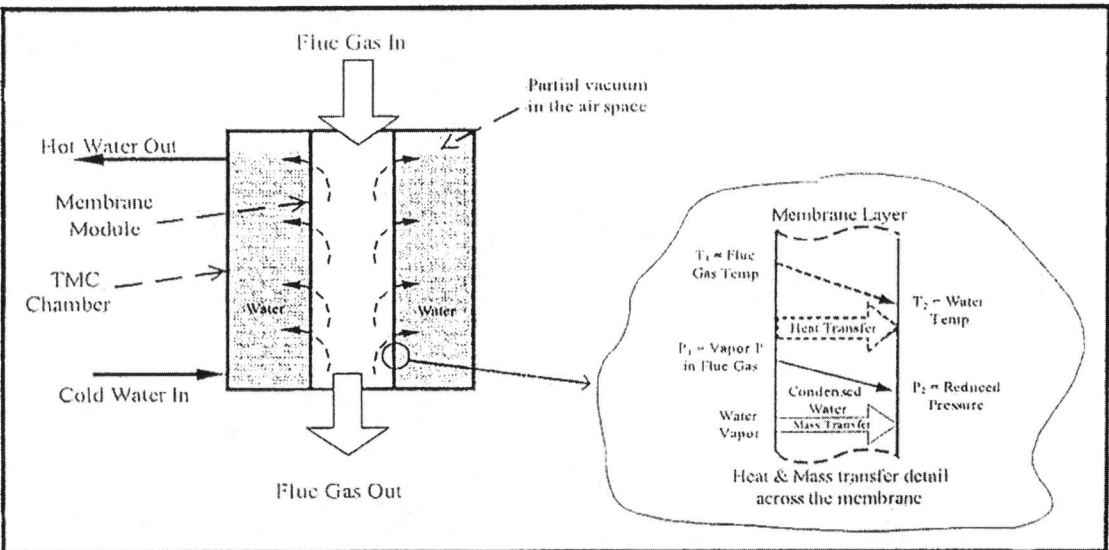

Figure 8-23. Schematic of the transport membrane condenser mechanism[6] (TMC)

capillary condensation in the micropores at the surface of the membrane. Capillary condensation is important here because the water vapor can be condensed in the pores of the membrane, even if its vapor pressure is below the saturation pressure at the same temperature (based upon Kelvin equation). Hence, by using the TMC membrane, it is possible to extract water from flue gas that is well above its dew point via enhanced condensation. In contrast, using the membrane to selectively separate water vapor in the flue via a conventional pressure-driven

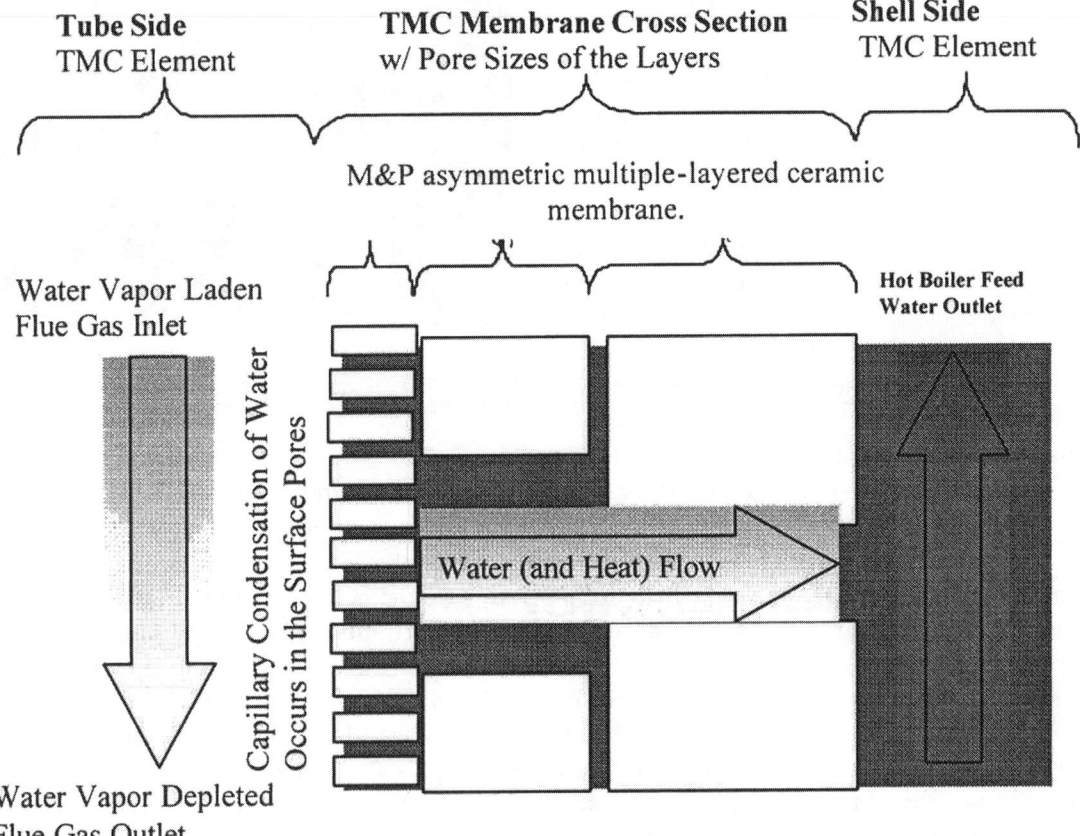

Figure 8-24. The transport membrane condenser concept. Schematic shows the cross section of a membrane tube in a TMC bundle.[7]

process is practically impossible, due to the low vapor pressure difference across the membrane. There are two primary advantages of the TMC over the conventional condensing economizer, specifically, (i) heat transfer between the hot and cold streams occurs not only via conventional heat conduction but also more importantly via convective heat transfer by the condensed water through the membrane, and (ii) the water vapor in the emission stream is condensed directly into the unit. These advantages greatly enhance the heat transfer performance and yield a significant reduction in the size of the unit relative to the condensing economizer. [7]

In addition, the flue gas exiting the TMC unit will always be below its dew point, so it is unnecessary to worry about stack corrosion due to condensation and the unfavorable perception of the plume. Practically speaking, a plastic pipe is all you need as a stack since the flue gas exit from the TMC is lower than 50°C. The TMC concept has been demonstrated successfully at the GTI lab with a prototype unit.

8.9.3 Experimental Results and Discussion

A prototype TMC unit has been fabricated using M&P ceramic membrane products. This prototype unit was installed (see Figure 8-26) and tested at GTI's facility on an 880kW industrial boiler. A typical test result is illustrated in Figure 8-27. [7]

About 45% of the water vapor in the flue gas was recovered. This results in a thermal efficiency improvement from ~89% (current state of the art boiler efficiency with economizer, HHV basis) to ~94%. Based upon the % water recovered, it is believed that the efficiency improvement could be significantly higher, or the required membrane surface area significantly reduced, with improved membrane properties. In summary, the preliminary testing has demonstrated the technical feasibility of the TMC to recover energy and water from a low temperature flue gas stream, but the TMC performance using an existing M&P commercial water filtration ceramic membrane was well below the potential performance of the proposed technology. [7]

8.9.4 Economic Analysis For Prototype

Since water is recovered via condensation and transport through an ultrafiltration (UF) membrane, the recovered water is ultra pure and can be reused for most industrial processes. As a result of the vapor latent heat recovered from the flue gas, a 5% boiler efficiency improvement can be achieved, besides savings in water consumption and treatment in a state-of-the-art boiler. Higher savings can be realized for less efficient installed (legacy) boilers. The new generation of ceramic mem-

Figure 8-25. Photo of several prototype TMC elements[7]

Figure 8-26. Installed TMC elements for water vapor removal from the flue gas of an 880 kW boiler[7]

brane, although not originally developed for this specific application, was evaluated in a prototype, which demonstrated its technical and process viability. Presently, M&P are working actively with GTI and a commercialization partner to optimize the unit performance via a theoretical and empirical approach for membrane/modules design uniquely suited for this selected application. According to the economic analysis, a payback period of less than one year is achievable. [7]

8.9.5 CASE STUDY—Energy Savings
Assessment (ESA) for Prototype TMC

An ESA was conducted for the Granite City Steel

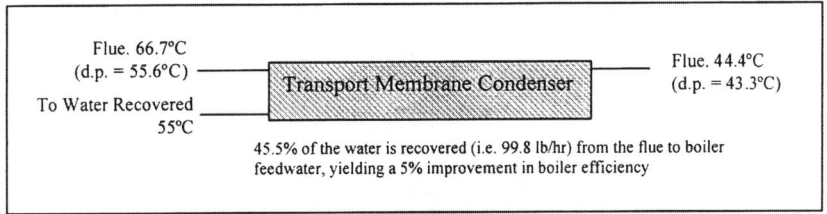

Figure 8-27. Feed and Exit stream temperatures of flue, and water through the proposed transport membrane condenser. A pilot scale testing unit contains a total of 9.75 m^2 of membrane surface area was used[7]

Works (US STEEL) at their Granite City Plant in Illinois in June, 2007. [11] Granite Steel Works has 12 steam boilers, 2 blast oven furnaces (BOF) to make iron using iron pellets, four hot strip furnaces to make carbon steel, two blast furnaces stoves, two coke oven batteries to convert coal to coke and to produce coke oven gas, two BOF Ladle dryers, and several triple-G Galvanizing furnaces. Blast furnaces make iron slabs, which are moved to hot strip furnaces for re-heating in order to make steel coils. Some coils are galvanized before shipping to the customers.

It is estimated that the Granite City Steel Works Plant uses 20,686,000 kWh of electricity at a cost of $1,199,788 per year, and about 18762799 MBtus of natural gas at a cost of $139,162,087 per year. The plant is more than 80 years old and is energy inefficient in several processes and equipments. The most energy-intensive equipments are boilers 1-10 and "B" blast oven furnaces. The steam generated by the boilers is transported above ground to the galvanizing furnaces. Several energy savings strategies were developed using the results of the process heating assessment and survey rool (PHAST) analysis of the plant.

8.9.5.1 Objective of ESA:
The main objective of the ESA was to understand the energy consumption patterns of the direct-fired heating equipments at the Granite City Steel Works.

8.9.5.2 Focus of Assessment:
The focus was to apply PHAST to all the boilers, blast furnaces, stoves, hot strip furnaces, and galvanizing furnaces. The results from the PHAST analysis were used to develop several energy saving recommendations for the plant.

8.9.5.3 Estimated Savings from Installing TMCs
Boilers 1-10 are more than 80 years old and operate 6,552 hours per year. The combustion intake air is taken by natural draft. There are no flue-gas sensors and no air-fuel controls. There is a lot of air-infiltration into

the combustion chamber of each boiler. There is no heat recovery from the flue gas. Intake combustion air is not pre-heated. From the PHAST analysis, it was found that these boilers consume about 3,439,363 MMBtu at an annual cost of $23,634,678. It is recommended that TMCs be installed on these boilers. This application alone is estimated to save about 171,988 MMBtus, saving $1.18 million per year. This measure, when combined with air-fuel ratio control, preheating combustion air, and using most efficient burners can result in an annual savings of approximately $8 million/year.

8.10 SUMMARY

Table 8-6 presents the collation of a number of significant attributes of the most common types of industrial heat exchangers in matrix form. This matrix allows rapid comparisons to be made in selecting competing types of heat exchanges. The characteristics given in the table for each type of heat exchanger are allowable temperature range, ability to transfer moisture, ability to withstand large temperature differentials, availability as packaged units, suitability for retrofitting, compactness, and the allowable combinations of heat transfer fluids. Needless to say that waste heat recovery is a technology that makes sense, and it has a high and bright potential in addressing our energy problem.

References
1. Fred S. Dubin and Chalmers G. Long Jr., *Energy Conservation Standards*, McGraw-Hill Book Company. New York, NY 10020
2. Wayne C. Turner and Steve Doty, eds. *Energy Management Handbook*, 6th Edition. Fairmont Press, Lilburn, GA 30247
3. Annual Energy Review, 1991, U.S. Energy Information Administration, Washington, D.C.
4. G. Walker, *Industrial Heat Exchangers*, 2nd Ed., Hemisphere, New York, NY 1990.
5. Richard J. Ciora, Jr. and Paul K. T. Lie, "Ceramic Membranes for Environmental Related Applications." *Fluid/Particle Separation Journal*, Vol. 15, No 1.
6. Energy Efficiency and Renewable Energy U. S. Department of Energy. "Advanced Membrane Separation Technologies

Table 8-6. Operation and application characteristics of industrial heat exchangers

Commercial Heat-Transfer Equipment	Low temperature: subzero –250°F	Intermediate temp: 250-1200°F	High temperature: 1200-2000°F	Recovers moisture	Large temperature differentials permitted	Packaged units available	Can be retrofit	No cross-contamination	Compact size	Gas-to-gas heat exchange	Gas-to-liquid heat exchanger	Liquid-to-liquid heat exchanger	Corrosive gases permitted with special construction
Radiation recuperator			×		×	a	×	×		×			×
Convection recuperator		×	×		×	×	×	×		×			×
Metallic heat wheel	×	×		b		×	×	c	×	×			×
Hygroscopic heat wheel	×			×		×	×	c	×	×			
Ceramic heat wheel		×	×		×	×	×		×	×			×
Passive regenerator	×	×			×	×	×	×		×			×
Finned-tube heat exchanger	×	×			×	×	×	×	×		×		d
Tube shell-and-tube exchanger	×	×			×	×	×	×	×		×	×	
Waste-heat boilers	×	×	×			×	×	×			×		d
Heat pipes	×	×	×		e	×	×	×	×	×			×

[a]Off-the-shelf items available in small capacities only.
[b]Controversial subject. Some authorities claim moisture recovery. Do not advise depending on it.
[c]With a purge section added, cross-contamination can be limited to less than 1% by mass.
[d]Can be constructed of corrosion-resistant materials, but consider possible extensive damage to equipment caused by leaks or tube ruptures.
[e]Allowable temperatures and temperature differential limited by the phase-equilibrium properties of the internal fluid.

for Energy Recovery." Industrial Technologies Program Fact Sheets.

7. Paul K.T. Liu. "Gas Separations Using Ceramic Membranes" Final Project Report by Media and Process Technologies Inc. Submitted to the U.S. Department of Energy. (DOE) January 2006.

8. "A Market Assessment for Condensing Boilers in Commercial Heating Applications," Consortium for Energy Efficiency, 2001.

9. "Energy Use, Loss and Opportunities Analysis: US Manufacturing & Mining," Report prepared by E3M, Inc., December 2004, for US Department of Energy.

10. Scovazzo, P., J. Burgos, A. Hoehn and P. Todd, "Hydrophilic Membrane-Based Humidity Control.", *J. Memb. Sci., 149, 69* (1998).

11. Mehta, D. Paul. Save Energy Now (SEN) ESA-056-2 Report. U.S. Department of Energy. June 2007.

CHAPTER 9

BUILDING ENVELOPE

KEITH E. ELDER, P.E.

9.1 INTRODUCTION

Building "Envelope" generally refers to those building components that enclose conditioned spaces and through which thermal energy is transferred to or from the outdoor environment. The thermal energy transfer rate is generally referred to as "heat loss" when we are trying to maintain an indoor temperature that is greater than the outdoor temperature. The thermal energy transfer rate is referred to as "heat gain" when we are trying to maintain an indoor temperature that is lower than the outdoor temperature. While many principles to be discussed will apply to both phenomena, the emphasis of this chapter will be upon heat loss.

Ultimately the success of any facility-wide energy management program requires an accurate assessment of the performance of the building envelope. This is true even when no envelope-related improvements are anticipated. Without a good understanding of how the envelope performs, a complete understanding of the interactive relationships of lighting and mechanical systems cannot be obtained.

In addition to a good understanding of basic principles, seasoned engineers and analysts have become aware of additional issues that have a significant impact upon their ability to accurately assess the performance of the building envelope:

1. The actual conditions under which products and components are installed, compared to how they are depicted on architectural drawings.

2. The impact on performance of highly conductive elements within the building envelope; and

3. The extent to which the energy consumption of a building is influenced by the outdoor weather conditions, a characteristic referred to as *thermal mass*.

It is the goal of this chapter to help the reader develop a good qualitative and analytical understanding of the thermal performance of major building envelope components. This understanding will be invaluable in better understanding the overall performance of the facility as well as developing *appropriate* energy management projects to improve performance.

9.1.1 Characteristics of Building Energy Consumption

Figure 9-1 below shows superimposed plots of average monthly temperature and fuel consumption for a natural gas-heated facility in the northwest region of the United States.

Experienced energy analysts recognize the distinc-

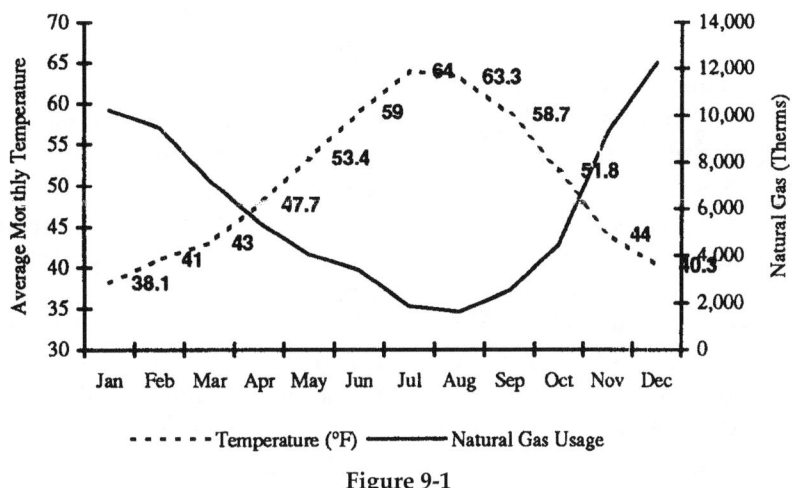

Figure 9-1

tive shape of the monthly fuel consumption profile and can often learn quite a bit about the facility just from inspection of this data. For example, the monthly energy consumption is inversely proportional to the average monthly temperature. The lower the average monthly temperature, the more natural gas appears to be consumed.

Figure 9-1 also indicates that there is a period during the summer months when it appears no heating should be required, yet the facility continues to consume some energy. For natural gas, this is most likely that which is consumed for the heating of domestic hot water, but it could also be due to other sources, such as a gas range in a kitchen. This lower threshold of monthly energy consumption is often referred to as the "base," and is characterized by the fact that its magnitude is independent of outdoor weather trends. The monthly fuel consumption which exceeds the "base" is often referred to as the "variable" consumption, and it is characterized by the fact that its magnitude is dependent upon the severity of outdoor environmental conditions. The distinction between the base and the variable fuel consumption is depicted below in Figure 9-2.

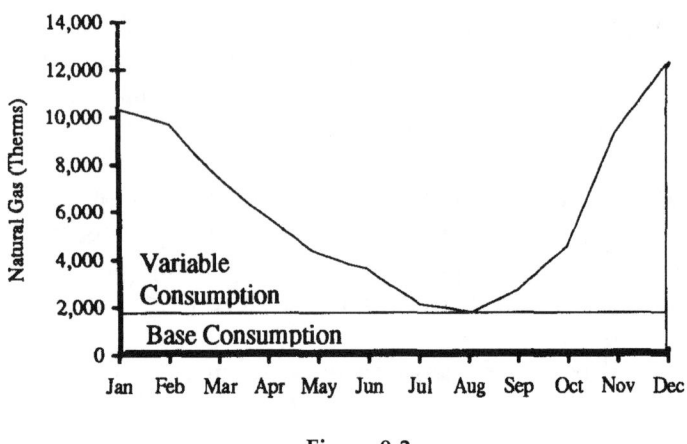

Figure 9-2

Often the base consumption can be distinguished from the variable by inspection. The lowest monthly consumption often is a good indicator of the base consumption. However there are times when a more accurate assessment is necessary. Section 9.9 describes one technique for improving the analyst's discrimination between base and variable consumption.

The distinction between base and variable consumption is an important one in that it is the only variable component of annual fuel consumption that can be saved by building envelope improvements. The more accurate the assessment of the fuel consumption, the more accurate energy savings projections will be.

9.1.2 Quantifying Building Envelope Performance

The rate of heat transfer through the building envelope will be found to be related to the following important variables:

1.   Indoor and outdoor temperature;

2.   Conductivity of the individual envelope components; and

3.   The square footage of each of the envelope components.

For a particular building component exposed to a set of indoor and outdoor temperature conditions, these variables are often expressed in equation form by the following:

$$q = UA(T_i - T_o) \qquad (9.1)$$

Where:

q =   the component heat loss, Btu/hr

U =   the overall heat transfer coefficient, Btu/ (hr-ft^2-°F)

A =   the area of the component, ft^2

T_i =   the indoor temperature, °F

T_o =   the outdoor temperature, °F

9.1.3 Temperatures for Instantaneous Calculations

The indoor and outdoor temperatures for equation (9.1) are those conditions for which the heat loss needs to be known. Traditionally interest has been in the design heat loss for a building or component and is determined by using so-called "design-day" temperature assumptions.

Outdoor temperature selection should be made from the ASHRAE Handbook of Fundamentals, for the geographic location of interest. The values published in this chapter are those that are statistically known not to be exceeded more than a prescribed number of hours (such as 2-1/2 %) of the respective heating or cooling season.

For heating conditions in the winter, indoor temperatures maintained between 68 and 72°F result in comfort to the greatest number of people. Indoor temperatures maintained between 74 and 76°F result in the greatest comfort to the most people during the summer (cooling) period.

We will see in later sections that other forms of temperature data collected on an annual basis are useful for evaluating envelope performance on an annual basis.

The next section will take up the topic of how the U-factors for various envelope components are determined.

9.2   PRINCIPLES OF ENVELOPE ANALYSIS

The successful evaluation of building envelope performance first requires that the analyst be well-versed in the use of a host of analytical tools that adequately address the unique way heat is transferred through each component. While the heat loss principles are similar, the calculation will vary somewhat from component to component.

9.2.1 Heat Loss Through Opaque
Envelope Components

We have seen from Equation (9.1) that the heat loss through a component, such as a wall, is proportional to the area of the component, the indoor-outdoor temperature difference, and U, the proportionality constant which describes the temperature-dependent heat transfer through that component. U, currently described as the "U-factor" by ASHRAE[2], is the reciprocal of the total thermal resistance of the component of interest. If the thermal resistance of the component is known, U can be calculated by dividing the total thermal resistance into "1" as shown:

$$U = 1/R_t \tag{9.2}$$

The thermal resistance, R_t, is the sum of the individual resistances of the various layers of material that comprise the envelope component. R_t is calculated by adding them up as follows:

$$R_t = R_1 + R_2 + R_n + ... \tag{9.3}$$

R_1, R_2 and R_n represent the thermal resistance of each of the elements in the path of the "heat flow." The thermal resistance of common construction materials can be obtained from the ASHRAE Handbook of Fundamentals. Other physical phenomenon, such as convection and radiation, are typically included as well. For instance, free and forced convection are treated as another form of resistance to heat transfer, and the "resistance" values are tabulated in the ASHRAE Fundamentals Manual for various surface orientations and wind velocities. For example, the outdoor resistance due to forced convection (winter) is usually taken as 0.17 hr/(Btu–ft²–°F), and the indoor resistance due to free convection of a vertical surface is usually taken to be 0.68 hr/(Btu–ft²–°F).

To calculate the overall U-factor, one typically draws a cross-sectional sketch of the building component of interest, assigns resistance values to the various material layers, sums the resistances, and uses the reciprocal of that sum to represent U. The total calculation of a wall U-factor is demonstrated in the example below.

Example

Calculate the heat loss for 10,000 square feet of wall with 4-inch face brick, R-11 insulation, and 5/8-inch sheet rock when the outdoor temperature is 20°F and the indoor temperature is 70°F.

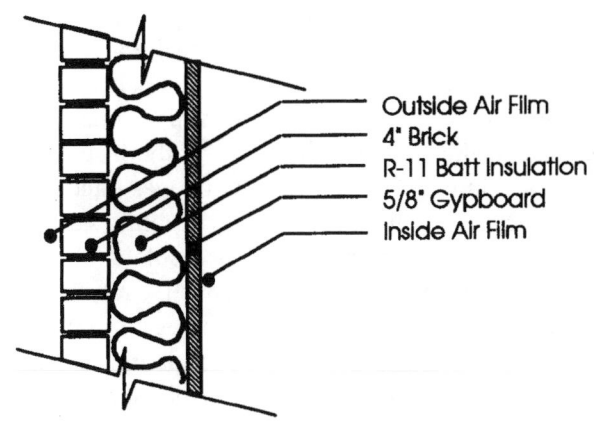

- Outside Air Film
- 4" Brick
- R-11 Batt Insulation
- 5/8" Gypboard
- Inside Air Film

Figure 9-3

In the ASHRAE Handbook we find that a conservative resistance for brick is 0.10 per inch. Four inches of brick would therefore have a resistance of 0.40. Sheet rock (called gypsum board by ASHRAE) has a resistance of 0.90 per inch, which would be 0.56 for 5/8-inch sheet rock. Batt insulation with a rating of R-11 will have a resistance of 11.0, if expanded to its full rated depth.

The first step is to add all the resistances:

	R_i
Outdoor Air Film (15 m.p.h.)	0.17
4-inch Face Brick	0.40
R-11 Batt Insulation	11.00
5/8-inch Gypsum Board	0.56
Indoor Air Film (still air)	0.68

$$R_t = 12.81 \quad \frac{hr\text{–}ft^2\ F}{Btu}$$

The U-factor is the reciprocal of the total resistance:

$$U_o = \frac{1}{R_t} = \frac{1}{12.81} = 0.078 \quad \frac{Btu}{hr\text{–}ft^2\ °F}$$

The heat loss is calculated by multiplying the U-factor by the area and the indoor-outdoor temperature difference:

$$q = 0.078 \times 10,000 \times (70 - 20)$$

$$= 39,000 \text{ Btu/hr}$$

The accuracy of the previous calculation is dependent on at least two important assumptions. The calculation assumes:

1.   The insulation is not compressed; and

2.   The layer(s) of insulation has not been compromised by penetrations of more highly conductive building materials.

9.2.2 Compression of Insulation

The example above assumed that the insulation is installed according to the manufacturer's instructions. Insulation is always assigned its R-value rating according to a specific standard thickness. If the insulation is compressed into a smaller space than it was rated under, the performance will be less than that published by the manufacturer. For example, R-19 batt insulation installed in a 3-1/2-inch wall might have an effective rating as low as R-13. Table 9-1[7] is a summary of the performance that can be expected from various levels of fiberglass batt insulation types installed in different envelope cavities.

9.2.3 Insulation Penetrations

One of the assumptions necessary to justify the use of the one-dimensional heat transfer technique used in equation 9-1 is that the component must be thermally homogeneous. Heat is transferred from the warm side of the component to the colder side and through each individual layer in a series path, much like current flow through simple electrical circuit with the resistances in series. No lateral or sideways heat transfer is assumed to take place within the layers. For this to be true, the materials in each layer must be continuous and not penetrated by more highly conductive elements.

Unfortunately, there are very few walls in the real world where heat transfer can truly be said to be one-dimensional. Most common construction has wood or metal studs penetrating the insulation, and the presence of these other materials must be taken into consideration.

Traditionally studs are accounted for by performing separate U-factor calculations through both wall sections, the stud and the cavity. These two separate U-factors are then combined in parallel by "weighting" them by their respective wall areas. The following example (Figure 9-4) shows how this would typically be done for a wall with studs, plates and headers constituting 23% of the total gross wall area.

	R-Cavity	R-Frame
Outdoor Air Film (15 m.p.h.)	0.17	0.17
4-Inch Face Brick	0.40	0.40
R-11 Batt Insulation	11.00	—
3-1/2-Inch Wood Framing	—	3.59
5/8-Inch Gypsum Board	0.56	0.56
Indoor Air Film (still air)	0.6	0.68

$$R_t = 12.81 \quad 5.40 \quad \frac{\text{hr–ft}^2\text{–°F}}{\text{Btu}}$$

$$U_i = 0.078 \quad 0.185 \quad \frac{\text{hr–ft}^2\text{–°F}}{\text{Btu}}$$

Combining the two U-factors by weighted fractions:

$$U_o = 0.77 \times 0.078 + 023 \times 0.185 = 0.103 \quad \frac{\text{hr–ft}^2\text{–°F}}{\text{Btu}}$$

In the absence of advanced framing construction techniques, wood studs installed 16 inches on center com-

Table 9-1. R-Value of fiberglass batts compressed within various depth cavities.

Insulation R-Value at Standard Thickness									
R-Value		38	30	22	21	19	15	13	11
Standard Thickness		12"	9-½"	6-¾"	5-½"	6-¼"	3-½"	3-5/8"	3-½"
Nominal Lumber Sizes, Inches	Actual Depth of Cavity, Inches	Insulation R-Values when Installed in a Confined Cavity							
2 x 12	11-1/4	37	--	--	--	--	--	--	--
2 x 10	9-1/4	32	30	--	--	--	--	--	--
2 x 8	7-1/4	27	26	--	--	--	--	--	--
2 x 6	5-1/2	--	21	20	21	18	--	--	--
2 x 4	3-1/2	--	--	14	--	13	15	13	11

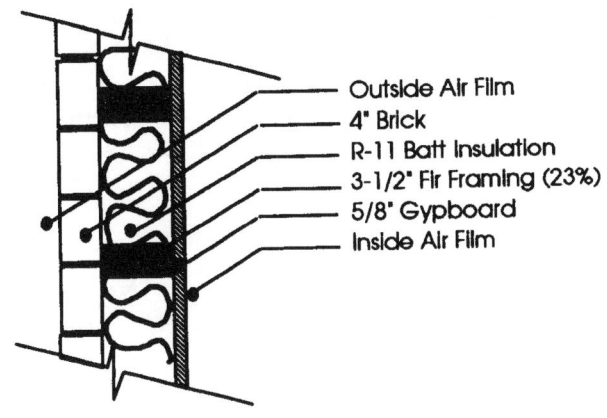

Figure 9-4

Outside Air Film
4" Brick
R-11 Batt Insulation
3-1/2" Flr Framing (23%)
5/8" Gypboard
Inside Air Film

prise 20-25% of a typical wall (including window framing, sill, etc.). Wood studs installed 24 inches on center comprise approximately 15-20% of the gross wall.

The use of this method is appropriate for situations where the materials in the wall section are sufficiently similar that little or no lateral, or "sideways" heat transfer takes place. Of course a certain amount of lateral heat transfer does take place in every wall, resulting in some error in the above calculation procedure. The amount of error depends on how thermally dissimilar the various elements of the wall are. The application of this procedure to walls in which penetrating members' conductivities deviate from the insulation conductivities by less than an order of magnitude (factor of 10) should provide sufficiently accurate results for most analysis of construction materials.

Because the unit R-value for wood is approximately 1.0 per inch and fiberglass batt insulation is R 3.1 per inch, this approach is justified for wood-framed building components.

9.3   METAL ELEMENTS
IN ENVELOPE COMPONENTS

Most commercial building construction is not wood-framed. Economics and the need for fire-rated assemblies have increased the popularity of metal-framing systems over the years. The conductivity of metal framing is significantly more than an order of magnitude greater than the insulation it penetrates. In some instances it is several thousand times greater. However, until recent years, the impact of this type of construction on envelope thermal performance has been ignored by much of the design industry. Yet infrared photography in the field and hot-box tests in the laboratory have demonstrated the severe performance penalty paid for this type of construction.

The introduction of the metal stud framing system into a wall has the potential to nearly double its heat loss! Just how is this possible, given that a typical metal stud is only about 1/20 of an inch thick? The magnitude of this effect is counter-intuitive to many practicing designers, because they have been trained to think of heat transfer through building elements as a one-dimensional phenomenon (as in the previous examples). But when a highly conductive element such as a metal stud is present in an insulated cavity, two and three-dimensional considerations become extremely important, as illustrated below.

Figure 9-5 shows the temperature distribution centered about a metal stud in a section of insulated wall. The lines, called "isotherms," represent regions with the same temperature. Each line denotes a region that is one degree Fahrenheit different from the adjacent line. These lines are of interest because they help us to visualize the characteristics of the heat flow through the wall section. The direction of heat flow is perpendicular to the lines of

Warm Side

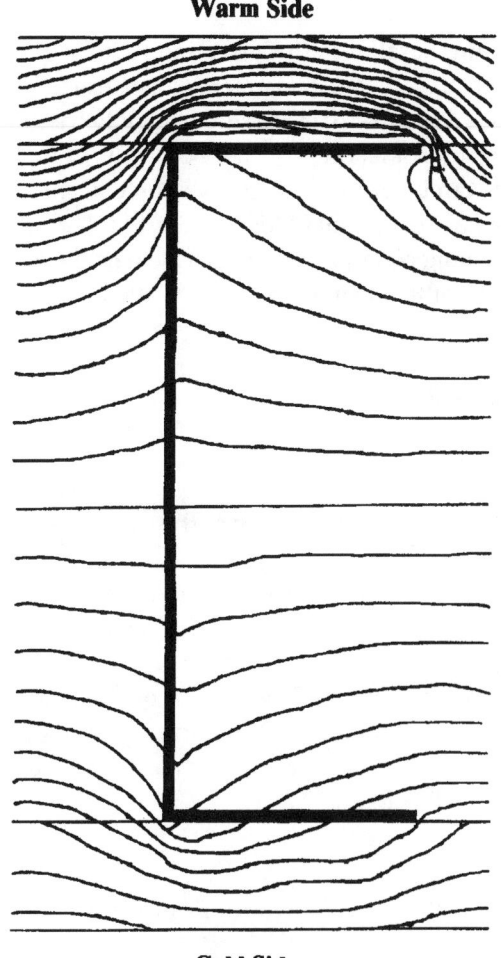

Cold Side

Figure 9-5. Metal stud wall section temperature distribution.

constant temperature, and the heat flow is more intense through regions where the isotherms are closer together. It is possible to visualize both the direction and intensity of heat flow through the wall section by observing the isotherms alone.

If the heat flow was indeed one-dimensional and occurring through a thermally homogeneous material, the lines would be horizontal and would show an even and linear temperature-drop progression from the warm side of the wall to the cold side. Notice that the isotherms are far from horizontal in certain regions around the metal stud. Figure 9-5 shows that the heat flow is not parallel, nor does it move directly through the wall section. Also note that the area with the greatest amount of heat flow is not necessarily restricted to the metal part of the assembly. The metal stud has had a negative influence on the insulation in the adjacent region as well. This is the reason for the significant increase in heat loss reported above for metal studs.

Clearly a different approach is required to determine more accurate U-factors for walls with highly conductive elements.

9.3.1 Using Parallel-Path Correction Factors

Fortunately, most commercial construction consists of a limited number of metal stud assembly combinations, so the results of laboratory "hot box" tests of typical wall sections can be utilized to estimate the U-factors of walls with metal studs. ASHRAE Standard 90.1 recommends the following equation be used to calculate the equivalent resistance of the insulation layer installed between metal studs:

$$R_t = R_i + R_e \qquad (9.4)$$

Where:

R_t = the total resistance of the envelope assembly

R_i = the resistance of all series elements except the layer of the insulation & studs

R_e = the equivalent resistance of the layer containing the insulation and studs

$\quad = R_{insulation} \times F_c$

Where:

F_c = the correction factor from Table 9-2

The use of the above multipliers in wall U-factor calculation is demonstrated in Figure 9-6.

Table 9-2. Parallel path correction factors.

Size of Members	Framing	Insulation R-Value	Correction Factor, F_c
2 × 4	16 in. O.C.	R-11	0.50
2 × 4	24 in. O.C.	R-11	0.60
2 × 6	16 in. O.C.	R-19	0.40
2 × 6	24 in. O.C.	R-19	0.45

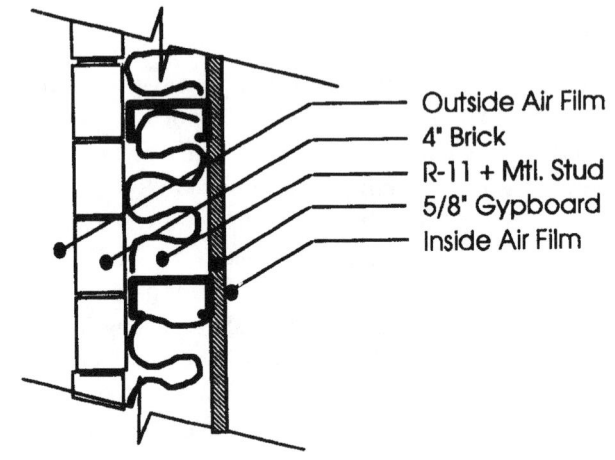

- Outside Air Film
- 4" Brick
- R-11 + Mtl. Stud
- 5/8" Gypboard
- Inside Air Film

Figure 9-6

	R_i
Outdoor Air Film (15 m.p.h.)	0.17
4-inch Face Brick	0.40
R-11/Metal Stud 16" O.C.	
$\quad = 11.0 \times 0.50 =$	5.50
5/8-inch Gypsum Board	0.56
Indoor Air Film (sill air)	0.68

$$R_t = 7.31 \quad \frac{hr - ft^2 - {}^\circ F}{Btu}$$

$$U_o = \frac{1}{R_t} = \frac{1}{7.31} = 0.137 \frac{Btu}{hr - ft^2 - {}^\circ F}$$

Notice that only one path is calculated through the assembly, rather than two. The insulation layer is simply corrected by the ASHRAE multiplier and the calculation is complete. Also notice that it is only the metal stud/insulation layer that is corrected, not the entire assembly. This does not mean that only this layer is affected by the metal studs, but rather that this is the approach ASHRAE Standard 90.1 intends for the factors to give results consistent with tested performance.

The above example shows that the presence of the

metal stud has increased the wall heat loss by 75%! The impact is even more severe for R-19 walls. The importance of accounting for the impact of metal studs in envelope U-factor calculations cannot be overstated.

9.3.3 The ASHRAE "Zone Method"

The ASHRAE Zone Method should be used for U-factor calculation when highly conductive elements are present in the wall that do not fit the geometry or spacing criteria in the above parallel path correction factor table. The Zone Method, described in the ASHRAE Handbook of Fundamentals, is an empirically derived procedure which has been shown to give reasonably accurate answers for simple wall geometries. It is a structured way to calculate the heat transmission through a wall using both series and parallel paths.

The Zone Method takes its name from the fact that the conductive element within the wall influences the heat transmission of a particular region or "zone." The zone of influence is typically denoted as "Zone A," which experiences a significant amount of lateral conduction. The remaining wall section that remains unaffected is called "Zone B." The width of Zone A, which is denoted "W," can be calculated using the following equation:

$$W = m + 2d \qquad (9.5)$$

Where "m" is the width of the metal element at its widest point and "d" is the shortest distance from the metal surface to the outside surface of the entire component. If the metal surface <u>is</u> the outside surface of the component, then "d" is given a minimum value of 0.5 (in English units) to account for the air film.

For example a 3-1/2-inch R-11 insulation is installed between 1.25-inch wide metal studs 16 inches on center. The assembly is sheathed on both sides with 1/2-inch gypsum board. The variable "m" is the widest part of the stud, which is 1.25 inches. The variable "d" is the distance from the metal element to the outer surface of the wall, which in this case is the thickness of the gypsum wall board, 0.5 inches. The width of Zone A is:

$$WA = 1.25 + (2 \times 0.5) = \underline{2.25 \text{ inches}}$$

The width of Zone B is:

$$WB = 16 - 2.25 = \underline{13.75 \text{ inches}}$$

Figure 9-7 shows the boundaries of the zones described above.

Actual step-by-step procedures for performing the calculations for Zone A and Zone B can be found in the

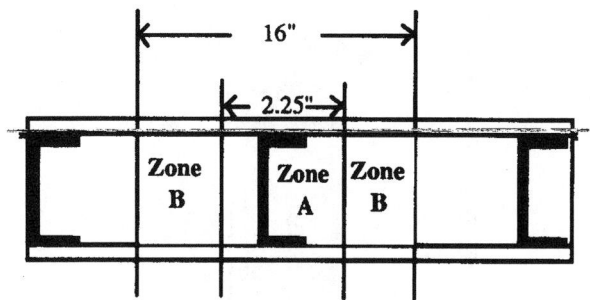

Figure 9-7. Zones A and B.

ASHRAE Handbook of Fundamentals, as well as in other publications.[5]

9.3.4 Improving the Performance of Envelope Components with Metal Elements

The impact of metal elements can be mitigated by:

1. Installing the insulation outside of the layer containing metal studs.

2. Installing interior and exterior finished materials on horizontal "hat" sections, rather than directly on the studs themselves.

3. Using expanded channel (thermally improved) metal studs.

4. Using non-conductive thermal breaks at least 0.40 to 0.50 inches thick between the metal element and the inside and outside sheathing. The value of the thermal break conductivity should be at least a factor of 10 less than the metal element in the envelope component.

9.3.5 Metal Elements in Metal Building Walls

Many metal building walls are constructed from a corrugated sheet steel exterior skin, a layer of insulation, and sometimes a sheet steel V-rib inner liner. The inner liner can be fastened directly to the steel frame of the building. The exterior cladding is attached to the inner liner and the structural steel of the building through cold formed sheet steel elements called Z-girts. Fiberglass batt insulation is sandwiched between the inner liner and the exterior cladding. Figures 9.8 and 9.9 show details of a typical sheet steel wall.

The steel framing and metal siding materials provide additional opportunities for thermal short circuits to occur. The highly conductive path created by the metal in the girts, purlins, and frames connected directly to the metal siding can result in even greater thermal short-cir-

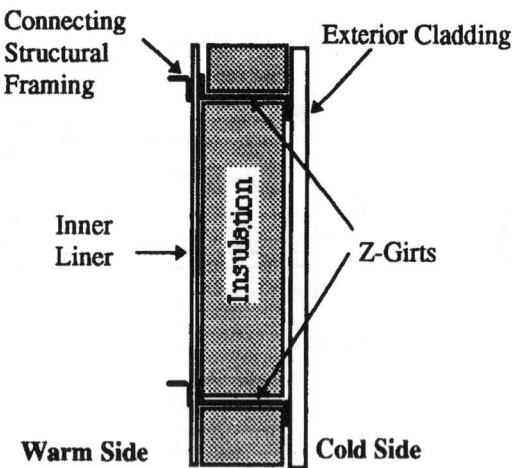

Figure 9-8 Vertical section view.

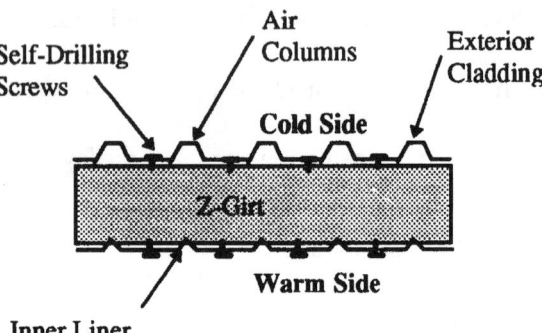

Figure 9-9. Horizontal section view.

cuiting than that discussed in Section 9.3.4 for metal studs in insulated walls. Because of this unobstructed high conductivity path, the temperature of the metal element is nearly the same through the entire assembly, rather than varying as heat flows through the assembly.

In laboratory tests, the introduction of 16 gauge Z-Girts spaced 8 feet apart reduced the R-value of a wall with no girts from 23.4 to 16.5 hr–ft^2–°F/Btu.[4] Substitution of Z-girts made of 12-gauge steel reduced the R-value even further, to 15.2 hr–ft^2–°F/Btu. The extent of this

heat loss will be dictated by the spacing of the Z-girts, the gauge of metal used, and the contact between the metal girts and the metal wall panels.

9.3.6 Metal Wall Performance

While the performance of metal building walls will vary significantly, depending on construction features, Tables 9-3—9-5 will provide a starting point in predicting the performance of this type of construction.[6]

9.3.7   Strategies for Reducing Heat Loss in Metal Building Walls

9.3.7.1 Maintain Maximum Spacing Between Z-girts

Tests on a 10-inch wall[8] with a theoretical R-value of 36 hr-ft^2-°F/Btu demonstrated a measured R-value of 10.2 hr-ft^2-°F/Btu with the girts spaced 2 feet apart. Increasing the spacing to 8 feet between the girts had a corresponding effect of increasing the R-value to 16.5 hr-ft^2-°F/Btu. While typical girt spacing in the wall of a metal building is 6 feet on center, this study does emphasize the desirability of maximizing the girt spacing wherever possible.

9.3.7.2 Use "Thermal" Girts

The thermal performance of metal building walls can be improved by the substitution of thermally improved Z-girts. The use of expanded "thermal" Z-girts has the potential to increase the effective R-value by 15-20% for a wall with girts spaced 8 feet apart.[4] A significant amount of the metal has been removed from these thermal girts, which increases the length of the heat flow path and reduces the girt's heat flow area. Figure 9-10 indicates the basic structure of such a girt.

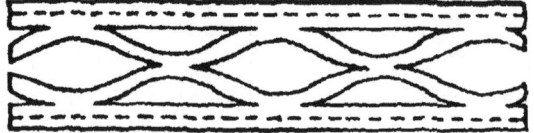

Figure 9-10. Expanded "thermal" girt.

Table 9-3. Wall insulation installed between 16 gauge Z-girts.

Unbridged R-Value	Nominal Thickness (Inches)	R-Value 2' O.C.	R-Value 4' O.C.	R-Value 5' O.C.	R-Value 6' O.C.	R-Value 8' O.C.
6	2	4.9	5.7	5.8	5.9	6.0
10	3	6.4	7.9	8.3	8.5	8.7
13	4	7.4	9.4	9.9	10.2	10.6
19	5-1/2	9.0	12.0	12.9	13.4	14.2
23	7	10.2	13.8	14.9	15.6	16.5
36	10	14.1	19.8	21.5	22.8	24.4

Table 9-4. Wall insulation installed between 12 gauge Z-girts.

Unbridged R-Value	Nominal Thickness (Inches)	R-Value 2' O.C.	R-Value 4' O.C.	R-Value 5' O.C.	R-Value 6' O.C.	R-Value 8' O.C.
6	2	4.8	5.6	5.8	5.9	6.0
10	3	6. 1	7.7	8. 1	8.3	8.6
13	4	6.9	9.0	9.5	9.9	10.3
19	5-1/2	8.1	11.1	12.0	12.6	13.4
23	7	8.7	12.3	13.4	14. 1	15.2

Table 9-5 combines test results with calculations to estimate the benefits that may be realized utilizing "thermal" girts.

9.3.7.3 Increase Z-girt-to-Metal Skin Contact Resistance

While the Z-girt itself does not offer much resistance to heat flow, testing has found that up to half the resistance of the girt-to-metal skin assembly is attributable to "poor" contact between the girt and the metal skin. Quantitative estimates of the benefit of using less conductive materials to "break" the interface between the two metals are not readily available, but laboratory tests have shown some improvement in thermal performance just by using half the number of self-tapping screws to fasten the metal skin to the girts.

9.3.7.4 Install Thermal Break Between Metal Elements

The impact of this strategy will vary with the flexibility offered by geometry of the individual wall component, the on-center spacing of the wall girts, thermal break material, and contact resistance between the girt and metal wall before consideration of the thermal break. Figure 9-11 illustrates the impact of thermal breaks of varying types and thicknesses installed between a corrugated metal wall and Z-girts mounted 6 feet on center.

To be effective, reduction by a factor of 10 or more in thermal conductivity between the thermal break and the metal may be required for a significant improvement in performance. The nominal thickness of the break also plays an important role. Tests of metal panels indicate that, regardless of the insulator, inserts less than 0.40 inches will not normally be sufficient to prevent substantial loss of insulation value.

9.3.8 Calculating U-factors for Metal Building Walls

ASHRAE Standard 90.1 recommends the Johannesson-Vinberg Method[7] for determining the U-factor for sheet metal construction internally insulated with a metal structure bonded on one or both sides with a metal skin. This method is useful for the prediction of U-factors for metal girt/purlin constructions, which can be defined by the geometry described in Figure 9-12.

The dimensions of the elements shown, as well as their conductivities, are evaluated in a series of formulations, which are in turn combined in two parallel paths to determine an overall component resistance. The advantage of this method is that metal gauge, as well as thermal break thickness and conductivity can be taken into account in the calculation. Caution is advised when using this method to evaluate components with metal-to-metal contact. The method assumes complete thermal contact between all components, which is not always the case in metal building construction. To account for this effect, values for contact resistance can be introduced into the calculation in place of or in addition to the thermal break layer.

Table 9-5. Wall insulation installed between 16 gauge "thermal" Z-girts.

Unbridged R-Value	Nominal Thickness (Inches)	R-Value 2' O.C.	R-Value 4' O.C.	R-Value 5' O.C.	R-Value 6' O.C.	R-Value 8' O.C.
6	2	5.6	6.2	6.3	6.4	6.4
10	3	8.3	9.4	9.6	9.8	10.0
13	4	10.2	1 1.7	12. 1	12.3	12.6
19	5-1/2	14.0	16.4	16.9	17.3	17.8
23	7	16.4	19.5	20.2	20.7	21.3

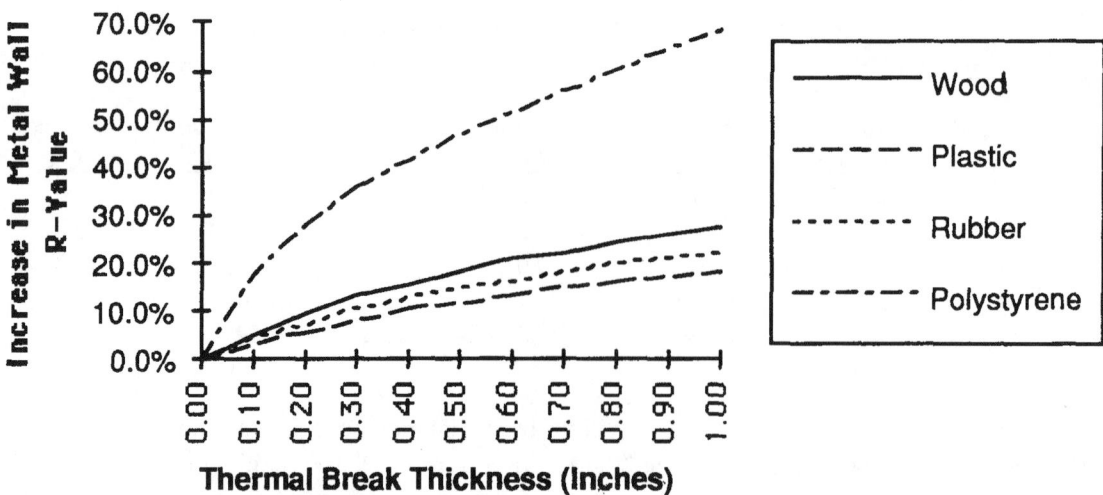

Figure 9-11[5]. Impact of thermal break on metal wall R-value corrugated metal with Z-girts 6'0" on-center.

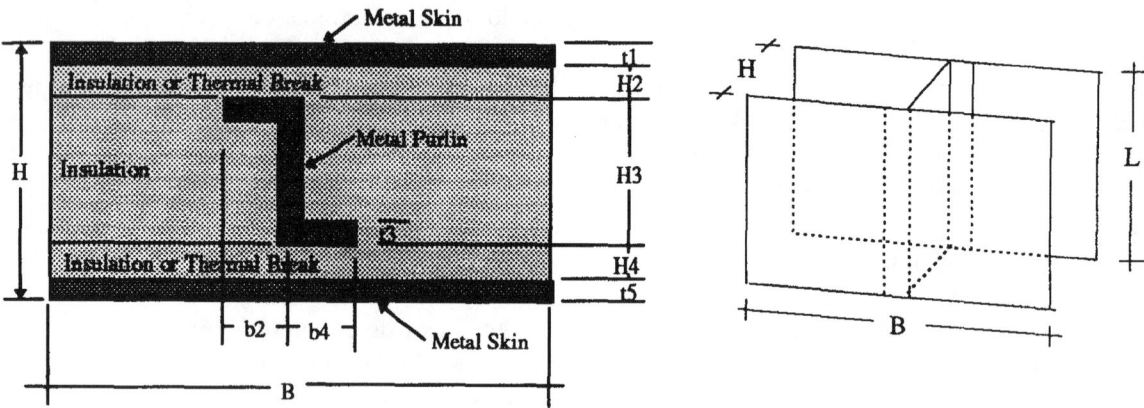

Figure 9-12. Metal skinned structure w/internal metal element.

9.4 ROOFS

In many cases the thermal performance of roof structures is similar to that of walls, and calculations can be performed in a similar way to that described in the previous examples. As was the case with walls, metal penetrations through the insulation will exact a penalty. With roofs, these penetrations will usually take one of several forms. The first form is the installation of batt insulation between z-purlins, similar to that discussed in Section 9.3. The performance of these assemblies will be much like that of similarly constructed walls.

9.4.1 Insulation Between Structural Trusses

Another common thermal short circuit occurs as a result of installing the insulation between structural trusses with metal components (Figure 9-13).

Table 9-6 is a summary of the derated R-values that might be expected as a result of this type of installation compared to the unbridged R-values published by insulation manufacturers.

9.4.2 Insulation Installed "Over-the-Purlin"

One of the most economical methods of insulating the roof of a metal building, shown in Figure 9-14, is to stretch glass fiber blanket insulation over the purlins or trusses, prior to mounting the metal roof panels on top.

The roof purlin is a steel Z-shaped member with a flange width of approximately 2-1/2 inches. The standard purlin spacing for girts supporting roofs in the metal building industry is 5 feet. Using such a method, the insulation is compressed at the purlin/panel interface, resulting in a thermal short circuit. The insulation thickness averages 2-4 inches but can range up to 6 inches. While thicker blanket insulation can be specified to mitigate this, the thickness is limited by structural considerations. With the reduced insulation thickness at the purlin, there is a diminishing return on the investment in additional insulation. Table 9-7 is a summary of the installed R-values that might be expected using this installation technique.

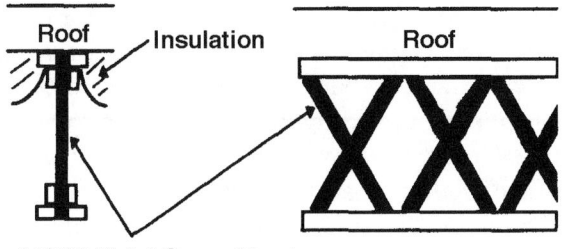

21/32" Metal Cross Members
4, 6, or 8 feet on center

Figure 9-13

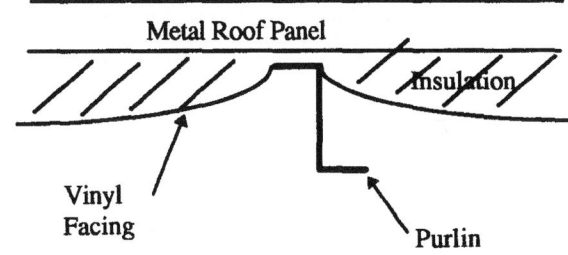

Figure 9-14

**Table 9-6. Roof insulation installed
between metal trusses.[5]**

Unbridged R-Value	R-Value 4 Ft O.C.[10]	R-Value 6 Ft O.C.[11]	R-Value 8 Ft O.C.[11]
5	4.8	4.9	4.9
10	9.2	9.6	9.7
15	13.2	14.0	14.3
20	17.0	18.4	18.9
25	20.3	22.4	23.2
30	23.7	26.5	27.6

9.4.2.1 Calculating Roof Performance for "Over-the-Purlin" Installations

Traditional methods of calculating assembly R-values will not result in accurate estimates of thermal performance for "over-the-purlin" installations. The insulation is not installed uniformly, nor at the thickness required to perform at the rated R-value, and it is penetrated with multiple metal fasteners. The Thermal Insulation Manufacturers Association (TIMA) has developed an empirical formula, based on testing of insulation, meeting the TIMA 202 specification. Insulation not designed to be laminated (such as filler insulation) will show performance up to 15% less than indicated by the formula below. Assuming the insulation meets the TIMA specification, the U-factor of the completed assembly can be estimated as:

$$U = 0.012 + \frac{0.255}{(0.31 \times R_f + t)} \times \left(1 - \frac{N}{L}\right) + N \times \frac{0.198 + 0.065 \times n}{L} \qquad (9.6)$$

Where:

- L = Length of Building Section, feet
- N = Number of Purlins or Girts in the L Dimension
- n = Fastener Population per Linear Foot of Purlin
- R_f = Sum of Inside and Outside Air Film R-Values, $\dfrac{hr - ft^2 {}^\circ F}{Btu}$
- t = Pre-installed insulation thickness (for TIMA 202 type insulation), inches

Example Calculation Using the TIMA Formula

Assume a metal building roof structure with the following characteristics:

Length of Building Section	= 100 feet
Number of Purlins	= 21 purlins
Fastener Population per foot of Purlin	= 1 per foot
Sum of Air Film R-Values = 0.61 + 0.17 = 0.78	$\dfrac{hr\text{-}ft^2 {}^\circ F}{Btu}$
Insulation thickness	= 5 inches

Table 9-7. Roof insulation installed compressed over-the-purlin[5].

Unbridged R-Value	Nominal Thickness (Inches)	R-Value 2' O.C.	R-Value 3' O.C.	R-Value 4' O.C.	R-Value 5' O.C.	R-Value 6' O.C.
10	3	5.4	6.5	7.2	7.7	8.1
13	4	5.7	7.1	8.0	8.7	9.3
16	5-1/2	5.9	7.4	8.6	9.5	10.2
19	6	6.0	7.7	9.0	10.0	10.9

Example Calculation Using the TIMA Formula

Assume a metal building roof structure with the following characteristics:

Length of Building Section	= 100 feet
Number of Purlins	= 21 purlins
Fastener Population per Foot of Purlin	= 1 per foot

$$\text{Sum of Air Film R-Values} = 0.61 + 0.17 = 0.78 \; \frac{\text{hr-ft}^2\text{°F}}{\text{Btu}}$$

Insulation thickness   = 5 inches

$$U = 0.012 + \frac{0.255}{(0.31 \times 0.78 + 6)} \times \left(1 - \frac{21}{105}\right) + 21$$

$$\times \frac{0.198 + 0.065 \times 1}{105} = 0.0973 \; \frac{\text{Btu}}{\text{hr} - \text{ft}^2\text{°F}}$$

9.4.3 Strategies for Reducing Heat Loss in Metal Building Roofs

Many of the strategies suggested for metal walls are applicable to metal roofs. In addition to strategies that minimize thermal bridging due to metal elements, other features of metal roof construction can have a significant impact on the thermal performance.

Tests have shown significant variation in the thermal performance of insulation depending upon the facing used, even though the permeability of the facing itself has no significant thermal effect. The cause of this difference is in the flexibility of the facing itself. The improvement in permeability rating produces a higher U-factor rating, due to the draping characteristics. During installation, the insulation is pulled tightly from eave to eave of the building. The flexibility of vinyl facing allows it to stretch and drape more fully at the purlins than reinforced facing. Figures 9.15 and 9.16 show a generalized illustration of the difference in drape of the vapor retarders and the effect on effective insulation thickness.

The thermal bridging and insulation compression issues discussed above make the following recommendations worth considering.

9.4.3.1 Use the "Roll-Runner" Method of Installing "Over-the-Purlin" Insulation

The "roll-runner" installation technique helps to mitigate the effect described above, achieving less insulation compression by improving the draping characteristics of the insulation. This is done using bands or straps to support the suspended batt insulation, as illustrated in Figure 9-17.

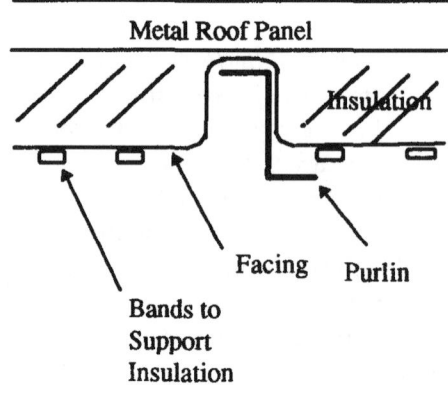

Figure 9-17. Roll Runner Method

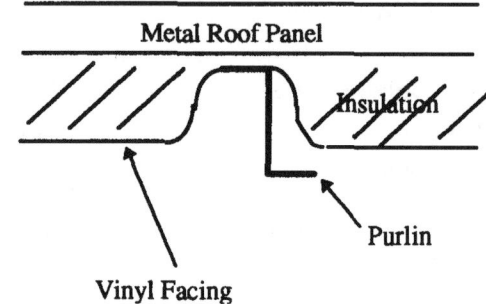

Figure 9-15. Vinyl facing.

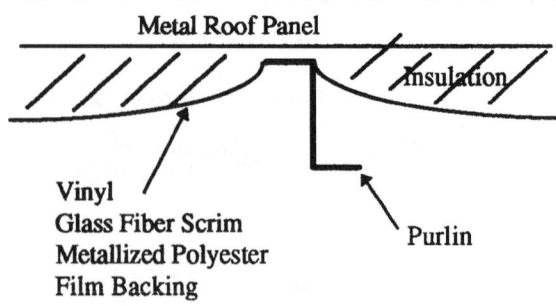

Figure 9-16. Vinyl facing with glass fiber scrim.

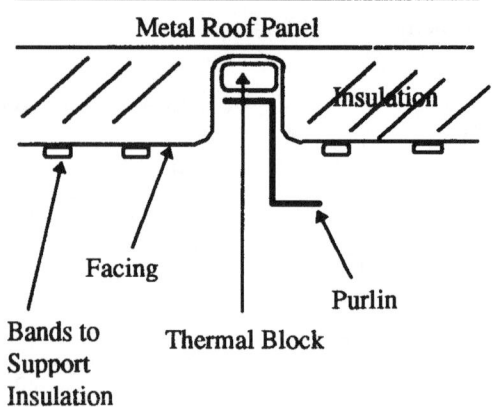

Figure 9-18. Insulated purlin with "roll-runner" installation.

9.4.3.2 Use Thermal Spacers Between Purlin and Standing Seam Roof Deck

Figure 9-18 depicts a thermal spacer installed between the top of the purlin and the metal roof deck. The impact of this strategy will vary with the geometry of the individual component, as well as the location of the spacer.

Table 9-8 summarizes the comparative performance of the two installation techniques discussed above. In all cases, purlins were installed 5 feet on center. While improvements of 8 to 19 percent in effective R-value are achieved using the roll-runner method, performance of 60 to 80 percent can be realized when the purlins are also insulated from direct contact with the metal structure.

9.4.3.3 Install Additional Uncompressed Insulation Between Purlins or Bar Joists

This insulation system illustrated in Figure 9-19 is sometimes referred to as "full depth/sealed cavity," referring to the additional layer of insulation that is installed between purlins or bar joists, which is not compressed as is the over-the-purlin insulation above it. In this configuration, the main function of the over-the-purlin insulation above is to act as a thermal break.

9.4.3.4 Add Rigid Insulation Outside of Purlins

The greatest benefit will be derived where insulation can be added that is neither compressed nor penetrated by conductive elements, as shown in Figure 9-20.

Table 9-9 comparatively summarizes the performance of the above installation for varying levels of insulation installed over purlins with varying on-center spacing.

9.5   FLOORS

Heat loss for floors above grade and exposed to outdoor air can be calculated much the same way as illustrated previously for walls, except that the percentages assumed for floor joists will vary somewhat from that assumed for typical wall constructions.

9.5.1 Floors Over Crawl Spaces

The situation is a little different for a floor directly over a crawl space. The problem is that knowledge of the temperature of the crawl space is necessary to perform the calculation. But the temperature of the crawl space is dependent on the number of exposed crawl space surfaces and their U-factors, as well as the impact of crawl space venting, if any. For design-day heat loss calculations, it is usually most expedient to assume a crawl space temperature equal to the outdoor design temperature. This will very nearly be the case for poorly insulated or vented crawl spaces.

When actual, rather than worst-case heat loss is needed, it is necessary to perform a heat balance on the crawl space. The process is described in the ASHRAE Handbook of Fundamentals. Below is a brief summary of the approach.

Table 9-8. "Roll-runner" method

Mfgr's Rated R-Value	Nominal Thickness (Inches)	Over the Purlin Method	Roll Runner Method	Roll-Runner with Insulated Purlin
10	3	7.1	7.7	12.5
13	4	8.3	9.1	14.3
19	6	11.1	12.5	20.0

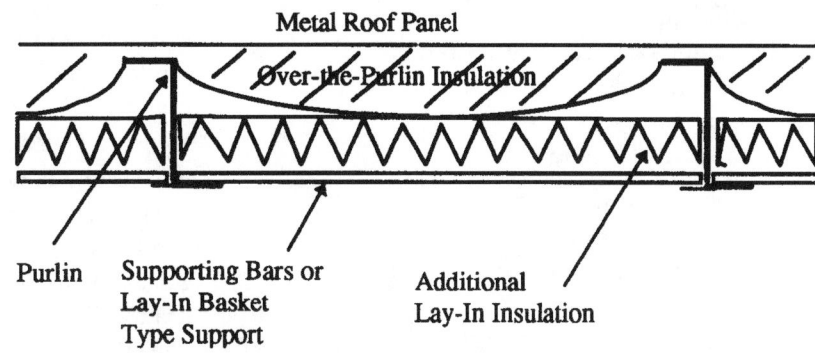

Figure 9-19. Insulation suspension system.

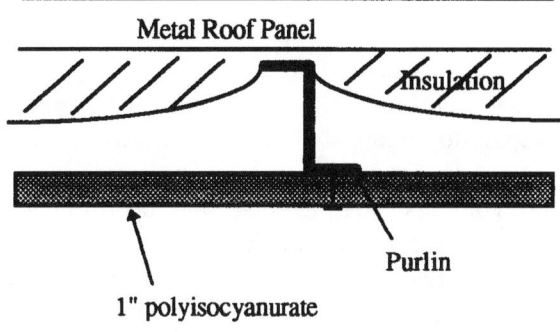

Figure 9-20. Insulation suspension system.

Table 9-9. Insulation "over-the-purlin" w/rigid insulation outside of purlin[14].

Uninstalled Unbridged R-Value	Thickness (Inches)	R-Value 2' O.C.	R-Value 3' O.C.	R-Value 4' O.C.	R-Value 5' O.C.	R-Value 6' O.C.
13	2	12.2	12.9	13.3	12.5	13.8
17	3	12.6	13.7	14.4	14.3	15.3
20	4	12.9	14.3	15.2	15.5	16.5
23	5-1/2	13.1	14.6	15.8	16.7	17.4
26	6	13.2	14.9	16.2	18.3	18.1

Heat loss, q_{floor} from a floor to a crawl space:

$$q_{floor} = q_{perimeter} + q_{ground} + q_{air\,exchange}$$

$$U_f A_f (t_i - t_c) = U_p A_p (t_c - t_o) + U_g A_g (t_c - t_g)$$
$$+ 0.67 \rho c_p V_c (t_c - t_o) \qquad (9.7)$$

Where:

t_i = indoor temperature, °F
t_o = outdoor temperature, °F
t_g = ground temperature, °F
t_c = crawl space temperature, °F
A_f = floor area, ft^2
A_p = exposed perimeter area, ft^2
A_g = ground area ($A_g = A_f$), ft^2
U_f = floor heat transfer coefficient,
    Btu/(hr–ft^2–°F)
U_g = ground coefficient, Btu/hr–ft^2°F
U_p = perimeter heat transfer coefficient,
    Btu/(hr–ft^2-°F)
V_c = volume of the crawl space, ft^3
ρc_p = volumetric air heat capacity
    (0.018 Btu/(ft^3–°F)
0.67 = assumed air exchange rate (volume/hour)

The above equation must be solved for t_c, the crawl space temperature. Then the heat loss from the space above to the crawl space, using the floor U-factor, can be calculated.

9.5.2 Floors On Grade

A common construction technique for commercial buildings is to situate the building on a concrete slab right on grade. The actual physics of the situation can be quite complex, but methods have been developed to simplify the problem. In the case of slab-on-grade construction, it has been found that the heat loss is proportional to the perimeter length of the slab, rather than the floor area. Rather than using a U-factor, which is normally associated with a wall or roof *area*, we use an "F-factor," which is associated with the number of linear *feet* of slab perimeter. The heat loss is given by the equation:

$$q_{slab} = F \times Perimeter \times (T_{inside} - T_{outside}) \qquad (9.8)$$

F-factors are published by ASHRAE and are also available in many state energy codes. As an example, the F-factor for an uninsulated slab is 0.73 Btu/(hr-ft-°F). A slab with 24 inches of R-10 insulation installed inside the foundation wall would be 0.54 Btu/(hr-ft-°F).

9.5.3 Floors Below Grade

Very little performance information exists on the performance of basement floors. What does exist is more relevant to residential construction than commercial. Fortunately, basement floor loss is usually an extremely small component of the overall envelope performance. For every foot the floor is located below grade, the magnitude of the heat loss is diminished dramatically. Floor heat loss is also affected somewhat by the shortest dimension of the basement. Heat loss is not directly proportional to the outside ambient temperature, as other above grade envelope components. Rather, basement floor loss has been correlated to the temperature of the soil four inches below grade. This temperature is found to vary sinusoidally over the heating season, rather than on a daily cycle, as the air temperature does.

A method for determining basement floor and wall heat loss is described, with accompanying calculations, in the ASHRAE Handbook of Fundamentals.

9.6   FENESTRATION

The terms "fenestration," "window," and "glazing" are often used interchangeably. To describe the important aspects of performance in this area requires that terms be defined carefully. "Fenestration" refers to the design and position of windows, doors, and other structural openings in a building. When we speak of *windows*, we are actually describing a system of several components. *Glazing* is the transparent component of glass or plastic windows, doors, clerestories, or skylights. The *sash* is a frame in which the glass panes of a window are set. The *frame* is the complete structural enclosure of the glazing and sash system. *Window* is the term we give to an entire assembly comprised of the sash, glazing, and frame.

Because a window is a thermally nonhomogeneous system of components with varying conductive properties, the thermal performance cannot be accurately approximated by the one-dimensional techniques used to evaluate common opaque building envelope components. The thermal performance of a window system will vary significantly, depending on the following characteristics:

- The number of panes
- The dimension of the space between panes
- The type of gas between the panes
- The emissivity of the glass
- The frame in which the glass is installed
- The type of spacers that separate the panes of glass

9.6.1 Multiple Glass Panes

Because of the low resistance provided by the glazing itself, the major contribution to thermal resistance in single pane glazing is from the indoor and outdoor air films. Assuming 0.17 outdoor and 0.68 indoor air film resistances, a single paned glazing unit might be expected to have an overall resistance of better than 0.85 (hr-ft^2-°F)/Btu, or a U-factor of 1.18 Btu/(hr-ft^2-°F). The addition of a second pane of glass creates an additional space in the assembly, increasing the glazing R-value to 1.85, which results in a U-factor of approximately 0.54 Btu/(hr-ft^2-°F). Similarly, the addition of a third pane of glass might increase the overall R-value to 2.85 (hr-ft^2°F)/Btu, yielding a U-factor of 0.35 Btu/(hr-ft^2-°F). This one-dimensional estimate does not hold true for the entire window unit, but only in the region described by ASHRAE as the "center-of-glass" (see Figure 9-21). Highly conductive framing or spacers will create thermal bridging in much the same fashion as metal studs in the insulated wall evaluated in Section 9.31.

9.6.2 Gas Space Between Panes

Most multiple-paned windows are filled with dry air. The thermal performance can be improved by the substitution of gases with lower thermal conductivities. Other gases and gas mixtures used besides dry air are Argon, Krypton, Carbon Dioxide, and Sulfur Hexafluoride. The use of Argon instead of dry air can improve the "center-of-glass" U-factor by 6-9%, depending on the

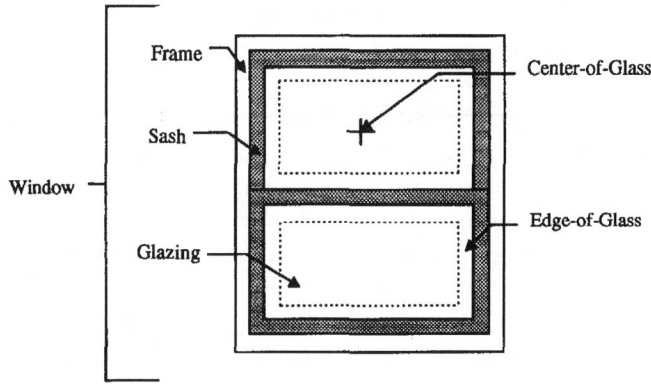

Figure 9-21. Window components.

distance between panes; CO_2 filled units achieve similar performance to Argon gas. For spaces up to 0.5 inches, the mixture of Argon and SF_6 gas can produce the same performance as Argon; Krypton can provide superior performance to that of Argon.

9.6.3 Emissivity

Emissivity describes the ability of a surface to give off thermal radiation. The lower the emissivity of a warm surface, the less heat loss it will experience due to radiation. Glass performance can be substantially improved by the application of special low emissivity coatings. The resulting product has come to be known as "Low-E" glass.

Two techniques for applying the Low-E film are sputter and pyrolytic coating. The lowest emissivities are achieved with a sputtering process by magnetically depositing silver to the glass inside a vacuum chamber. Sputter coated surfaces must be protected within an insulated glass unit and are often called "soft coat." Pyrolytic coating is a newer method which applies tin oxide to the glass while it is still somewhat molten. The pyrolytic process results in higher emissivities than sputter coating, but surfaces are more durable and can be used for single glazed windows. While normal glass has an emissivity of approximately 0.84, pyrolytic coatings can achieve emissivities of approximately 0.40, and sputter coating can achieve emissivities of 0.10 and lower. The emittance of various Low-E glasses will vary considerably between manufacturers.

9.6.4 Window Frames

The type of frame used for the window unit will also have a significant impact on the performance. In general, wood or vinyl frames are thermally superior to metal. Metal frame performance can be improved significantly by the incorporation of a "thermal-break." This usually consists of the thermal isolation of the cold side of the frame from the warm side by means of some low-conducting material. Estimation of the performance of a window, due to the framing elements, is complicated by the variety of configurations and combinations of materials used for sash and frames. Many manufacturers combine materials for structural or aesthetic purposes. Figure 9-22 below illustrates the impact that various framing schemes can have on glass performance schemes.

The center-of-glass curve illustrates the performance of the glazing system without any framing. Notice that single pane glass is almost unaffected by the framing scheme utilized. This is due to the similar order-of-magnitude thermal conductivity between glass and metal. As additional panes are added, emissivities are lowered, and low conductivity gases are introduced, the impact of the framing becomes more pronounced, as shown by the increasing performance "spread" toward the right hand side of the chart. Notice that a plain double pane window with a wood or vinyl frame actually has similar performance to that of Low-E glass (hard coating) with Argon gas fill and a metal frame. Also note how flat the curve is for metal-framed windows with no thermal break. It should be clear that first priority should be given to framing systems before consideration is given to Low-E coatings or low conductivity gases.

9.6.5 Spacers

Double and triple pane window units usually have continuous members around the glass perimeter to separate the glazing lites and provide an edge seal.

These spacers are often made of metal, which increases the heat transfer between the panes and degrades the performance of the glazing near the perimeter. ASHRAE reports this conductive region to be limited to a 2.5-inch band around the perimeter of the glazing unit and appropriately describes it as the "edge of glass." Obviously, low conductivity spacers, such as plastic, fiberglass, or even glass, are to be preferred over high conductivity spacers such as metal. The impact of highly conductive spacers is only felt to the extent that other high performance strategies are incorporated into the window. For example, ASHRAE reports no significant performance difference between spacers types when incorporated into window systems with thermally unimproved frames. When thermally improved or thermally broken frames are incorporated, the spacer material becomes a factor in overall window performance.

The best performing windows will incorporate thermally broken frames, non-metallic spacers between the panes, low-emissivity coatings, or low conductivity gases, such as argon or krypton, between the panes.

9.6.6 Advanced Window Technologies

9.6.6.1 Suspended Interstitial Films

High performance glazing is available that utilizes one or more Low-E coated mylar or polyester films between the inner and outer glazing of the window unit. High performance is achieved by the addition of a second air space (another Low-E surface) without the weight of an additional pane of glass.

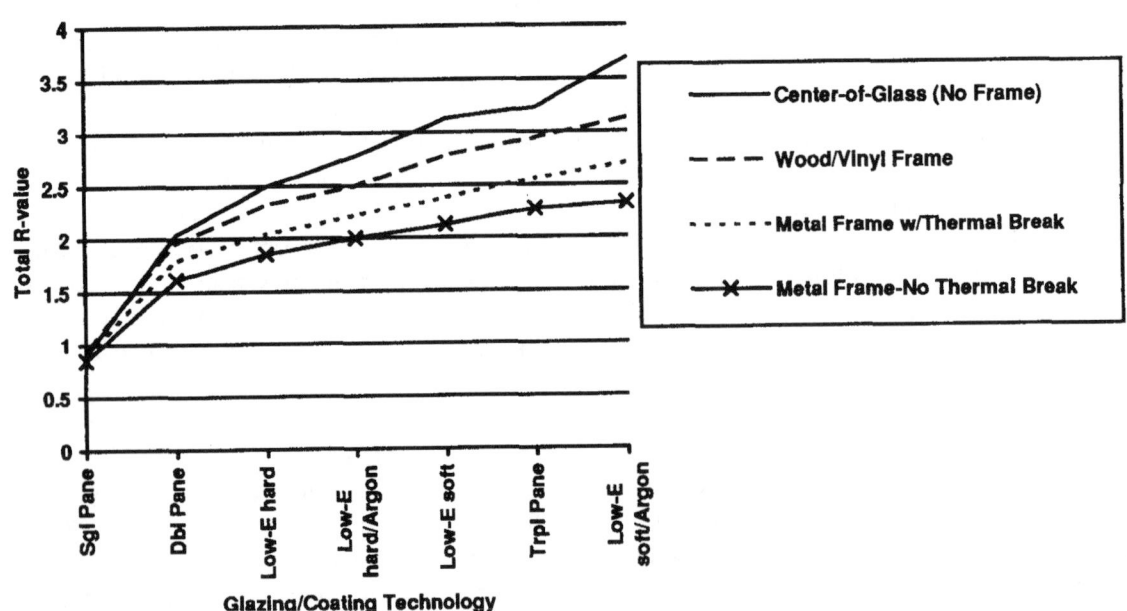

Figure 9-22. Window system performance comparison.

9.6.6.2 Advanced Spacers

Significant improvements in spacer design have been made in the past 10-15 years. The "swiggle strip" spacer sandwiches a thin piece of corrugated metal between two thicker layers of butyl rubber combined with a desiccant. The conductivity is reported to be one-quarter to one-half of the conductivity of a normal aluminum spacer. Other spacers, with conductivities only 6 to 11 percent as great as aluminum, have been designed using silicon foam spacers with foil backing, or by separating two conventional aluminum spacers with a separate strip of polyurethane foam.

9.6.6.3 Future Window Enhancement Technologies

Electrochromic glass systems, referred to as "smart windows," sandwich indium tin oxide, amorphous tungsten trioxide, magnesium fluoride, and gold between layers of heat-resistant Pyrex. Imposing a small voltage to the window changes its light and heat transmission characteristics, allowing radiation to be reflected in the summer and admitted in the winter.

Windows with R-values of R-16 are theoretically possible if the air or other insulating gas between window panes is replaced with a vacuum. Historically, a permanent vacuum has not been achievable, due to the difficulty of forming an airtight seal around the edges of the unit. Work is presently under way on special techniques to create a leakproof seal.

9.6.7 Rating the Performance of Window Products

Window energy performance information has not been made available in any consistent form until recently. Some manufacturers publish R-values that rival many insulation materials. Usually the center-of-glass U-factor is published for glazing, independent of the impact of the framing system that will eventually be used. While the center-of-glass rating alone may be impressive, it does not describe the performance of the entire window. Other manufacturers provide performance data for their framing systems, but they do not reflect how the entire product performs. Even when manufacturers provide ratings for the whole product different methods are used to determine these ratings, both analytical and in laboratory hot box tests. This has been a source of confusion in the industry, requiring some comprehensive standard for the reporting of window thermal performance.

The National Fenestration Rating Council (NFRC), sanctioned by the federal government under the Energy Policy Act of 1992, was established to develop a national performance rating system for fenestration products. The NFRC has established a program in which the factors that affect window performance are included in published performance ratings. NFRC maintains a directory of certified products. Currently many local energy codes require that NFRC ratings be used to demonstrate compliance with their window performance standards.

9.6.8 Doors

In general, door U-factors can be determined in a similar manner as with the walls, roofs and exposed floor demonstrated above. Softwoods have R-values around 1.0 to 1.3 per inch, while hardwoods have R-values that range from 0.80 to 0.95 per inch. A 1-3/8-inch panel door has a U-factor of approximately 0.57 Btu/(Hr-ft^2-°F), while a 1-3/4 solid core flush door has a U-factor of approximately 0.40.

As has been shown to be the case with some walls and windows, metal doors are a different story. The same issues affecting windows, such as framing and thermal break, apply to metal doors as well. The U-factor of a metal door can vary from 0.20 to 0.60 Btu/(Hr-ft^2-°F), depending on the extent to which the metal-to-metal contact can be "broken." In the absence of tested door U-factor data, Table 6, page 24.13 in Chapter 24 of the 1997 ASHRAE Handbook of Fundamentals, can be used to estimate the U-factor of typical doors used in residential and commercial construction.

9.7 INFILTRATION

Infiltration is the uncontrolled inward air leakage through cracks and interstices in a building element and around windows and doors of a building, caused by the effects of wind pressure and the differences in the outdoor/indoor air density or pressure differentials. The heat loss due to infiltration is described by the following equation:

$$q_{infiltration} = 0.019 \times Q \times (T_{inside} - T_{outside}) \qquad (9.9)$$

Where Q is the infiltration air flow in cubic feet per hour.

The determination of Q is an extremely imprecise undertaking, in that the actual infiltration for similar buildings can vary significantly, even though observable parameters appear to be the same.

9.7.1 Estimating Infiltration for Residential Buildings

ASHRAE suggests that the infiltration rate for a residence can be estimated as:

$$Q = L \left[(A(T_i - T_o) + (Bv^2) \right]^{1/2} \qquad (9.10)$$

Where:

Q = The infiltration rate, ft^3/hr
A = Stack Coefficient, $CFM^2/[in^4\text{-}°F]$
T_i = Average indoor Temperature, °F
T_o = Average outdoor Temperature, °F
B = Wind Coefficient, $CFM^2/[in^4\text{-}(mph)^2]$
v = Average wind speed, mph
L = Crack leakage area, in^2

The method is difficult to apply because:

1. L, the total crack area in the building is difficult to determine accurately;

2. The determination of the stack and wind coefficient is subjective;

3. The average wind speed is extremely variable from one micro-climate to another; and

4. Real buildings, built to the same standards, do not experience similar infiltration rates.

ASHRAE reports on the analysis of several hundred public housing units where infiltration varied from 0.5 air changes per hour to 3.5 air changes per hour. If the real buildings experience this much variation, we cannot expect a high degree of accuracy from calculations, unless a significant amount of data are available. However, the studies reported provided some useful guidelines.

In general, older residential buildings without weather-stripping experienced a median infiltration rate of 0.9 air changes per hour. Newer buildings, presumably built to more modern, tighter construction standards, demonstrated median infiltration rates of 0.5 air changes per hour. The structures were unoccupied during tests. It has been estimated that occupants add an estimated 0.10 to 0.15 air changes per hour to the above results.

9.7.2 Estimating Infiltration for Complex Commercial Buildings

Infiltration in large commercial buildings is considerably more complex than small commercial buildings or residential buildings. It is affected by both wind speed and "stack effect." Local wind speed is influenced by distance from the reporting meteorological station, elevation, and the shape of the surrounding terrain. The pressure resulting from the wind is influenced by the local wind velocity, the angle of the wind, the aspect ratio of the building, and which face of the building is im-

pinged and the particular location on the building. This in turn is affected by temperature, distance from the building "neutral plane," the geometry of the building exterior envelope elements and interior partitions, and all of their relationships to each other. If infiltration due to both wind velocity (Q_w) and stack effect (Q_s) can be determined, they are combined as follows to determine the overall infiltration rate.

$$Q_{ws} = \sqrt{Q_w^2 + Q_s^2} \qquad (9.11)$$

Where:

Q_{ws} = The combined infiltration rate due to wind and stack effect, ft^3/hr
Q_w = The infiltration rate due to wind, ft^3/hr
Q_s = The infiltration rate due to stack effect, ft^3/hr

While recent research has increased our understanding of the basic physical mechanisms of infiltration, it is all but impossible to accurately calculate anything but a "worst-case" design value for a particular building. Techniques such as the air-change method and general anecdotal findings from the literature are often the most practical approaches to the evaluation of infiltration for a particular building.

9.7.2.1 Anecdotal Infiltration Findings

General studies have shown that office buildings have air exchange rates ranging from 0.10 to 0.6 air changes per hour, with no outdoor intake. To the extent outdoor air is introduced to the building and it is pressurized relative to the local outdoor pressure, the above rates will be reduced.

Infiltration through modern curtain wall construction is a different situation than that through windows "punched" into walls. Studies of office buildings in the United States and Canada have suggested the following approximate leakage rates per unit wall area for conditions of 0.30 inches of pressure (water gauge):

Tight Construction	$0.10 \, CFM/ft^2$ of curtainwall area
Average Construction	$0.30 \, CFM/ft^2$ of curtainwall area
Leaky Construction	$0.60 \, CFM/ft^2$ of curtainwall area

9.7.2.2 The Air Change Method

While it is difficult to quantitatively predict actual infiltration, it is possible to conservatively predict infiltration that will give us a sense of "worst-case." This requires experience and judgment on the part of the engineer or analyst. ASHRAE Fundamentals Manual published guidelines for estimating infiltration on the basis of the number of "air changes." That is, based on the volume of the space, how many complete changes of air are likely to occur within the space of an hour?

Table 9-10 is a summary of the recommended air changes originally published by ASHRAE in 1972.

These guidelines have been found to be sound over the years and are still widely used by many practicing professionals. The values represent a good starting point for estimating infiltration. They can be modified as required for local conditions, such as wind velocity or excessive building stack effect.

The above values are based on doors and operable windows that are not weather-stripped. ASHRAE originally recommended that the above factors be reduced by 1/3 for weather-stripped windows and doors. This would be a good guideline applicable to modern buildings, which normally have well-sealed windows and doors. Fully conditioned commercial spaces are slightly pressurized and often do not have operable sash. This will also tend to reduce infiltration.

9.7.3 An Infiltration Estimate Example

Assume a 20' × 30' room with a 10' ceiling and one exposed perimeter wall with windows. What would the worst-case infiltration rate be?

The total air volume is $20 \times 30 \times 10 = 6,000$ ft^3. The table above recommends using 1.0 air changes per hour. However, assuming modern construction techniques, we follow ASHRAE's guideline of using 2/3 of the table values:

$$2/3 \times 6,000 \text{ ft}^3/\text{hr} = 4,000 \text{ ft}^3/\text{hr}$$

The heat loss due to infiltration for this space, if the indoor temperature was 70°F and the outdoor temperature was 25°F, would be:

$$q_{\text{infiltration}} \quad \begin{aligned} &= 0.019 \times 4,000 \times (70-25) \\ &= 3,420 \text{ Btu}/\text{hr} \end{aligned}$$

9.8 SUMMARIZING ENVELOPE PERFORMANCE WITH THE BUILDING LOAD COEFFICIENT

In Section 9.1.2 it was shown that building heat loss is proportional to the indoor-outdoor temperature difference. Our review of different building components, including infiltration has demonstrated that each individual component can be assigned a proportionality constant that describes that particular component's behavior with respect to the temperature difference imposed across it. For a wall or window, the proportionality constant is the U-factor times the component area, or "UA." For a slab-on-grade floor, the proportionality constant is the F-factor times the slab perimeter, or "FP." For infiltration, the proportionality constant is 0.019 times the air flow rate in cubic feet per hour. While not attributed to the building envelope, the effect of ventilation must be accounted for in the building's overall temperature-dependent behavior. The proportionality constant for ventilation is the same for infiltration, except that it is commonly expressed as 1.10 times the ventilation air flow in cubic feet per <u>minute</u>.

The instantaneous temperature-dependent performance of the total building envelope is simply the sum of all the individual component terms. This is sometimes referred to as the *building load coefficient* (BLC). The BLC can be expressed as:

$$BLC = \Sigma UA + \Sigma FP + 0.018\, Q_{INF} + 1.1\, Q_{VENT} \quad (9.12)$$

Where:

ΣUA = the sum of all individual component "UA" products

ΣFP = the sum of all individual component "FP" products

Q_{INF} = the building infiltration volume flow rate, in cubic feet per <u>hour</u>

Table 9-10. Recommended air changes due to infiltration.

Type of Room	Number of Air Changes Taking Place per Hour
Rooms with no windows or exterior doors	1/2
Rooms with windows or exterior doors on one side	1
Rooms with windows or exterior doors on two sides	1-1/2
Rooms with windows or exterior doors on three sides	2
Entrance Halls	2

Q_{VENT} = the building ventilation volume flow rate, in cubic feet per <u>minute</u>

From the above and Equation (9.1), it follows that the total instantaneous building heat loss can be expressed as:

$$q = BLC(T_{indoor} - T_{outdoor}) \qquad (9.13)$$

9.9   THERMAL "WEIGHT"

Thermal weight is a qualitative description of the extent to which the building energy consumption occurs in "lock-step" with local weather conditions. Thermal "weight is characterized by the mass and specific heat (heat capacity) of the various components that make up the structure, as well as the unique combination of internal loads and solar exposures which have the potential to offset the temperature-dependent heat loss or heat gain of the facility.

Thermally "light" buildings are those whose heating and cooling requirements *are* proportional to the weather. Thermally "heavy" buildings are those buildings whose heating and cooling requirements are *not* proportional to the weather. The "heavier" the building, the less temperature-dependent the building's energy consumption appears to be, and the less accuracy that can be expected from simple temperature-dependent energy consumption calculation schemes.

The concept of thermal weight is an important one when it comes to determining the energy-saving potential of envelope improvements. The "heavier" the building, the less savings per square foot of improved envelope can be expected for the same "UA" improvement. The "lighter" the building, the greater the savings that can be expected. A means for characterizing the thermal "weight" of buildings, as well as analyzing the energy savings potential of "light" and "heavy" buildings, will be taken up in Section 9.10.

9.10   ENVELOPE ANALYSIS FOR EXISTING BUILDINGS

9.10.1 Degree Days

In theory, if one wanted to predict the heat lost by a building over an extended period of time, Equation 9.12 could be solved for each individual hour, taking into account the relevant changes of the variables. This is possible because the change in the value BLC with respect to temperature is not significant and the indoor temperature (T_i) is normally controlled to a constant value (such as 70°F in winter). That being the case, the total energy transfer could

be predicted by knowing the summation of the individual deviations of outdoor temperature (T_o) from the indoor condition (T_i) over an extended period of time.

The summation described has come to be known as "degree-days," and annual tabulations of degree-days for various climates are published by NOAA, ASHRAE, and various other public and military organizations. Historically an indoor reference point of 65°F has been used to account for the fact that even the most poorly constructed building is capable of maintaining comfort conditions without heating when the temperature is at least 65°F.

Because of the impracticality of obtaining hour-by-hour temperature data for a wide variety of locations, *daily* temperature averages are often used to represent 24-hour blocks of time. The daily averages are calculated by taking the average of daily maximum and minimum temperature recordings, which in turn are converted to degree days. Quantitatively this is calculated as:

$$Degree - Days = \frac{(T_{reference} - T_{average})n}{24} \qquad (9.14)$$

Where "n" represents the number of hours in the period for which the degree days are being reported, and $T_{reference}$ is a reference temperature at which no heating is assumed to occur. Typically, a reference temperature of 65°F is used. Units of degree hours can be obtained by multiplying the results of any degree-day tabulation by 24.

Figure 9-23 shows the monthly natural gas consumption of a metropolitan newspaper building overlaid on a plot of local heating degree days.

Because of the clear relationship between heating degree days and heating fuel consumption for many buildings, such as the above, the following formula has been used since the 1930's to predict the future heating fuel consumption.

$$E = \frac{q \times DD \times 24}{\Delta t \times H \times Eff} \qquad (9.15)$$

Where:

E   = Fuel consumption, in appropriate units, such as Therms natural gas

q   = The design-day heat loss, Btu/hr

DD = The annual heating degree-days (usually referenced to 65°F)

24 = Converts degree-days to degree-hours

Δt = The design day indoor-outdoor temperature difference, °F

H   = Conversion factor for the type of fuel used

Eff = The annualized efficiency of the heating combustion process

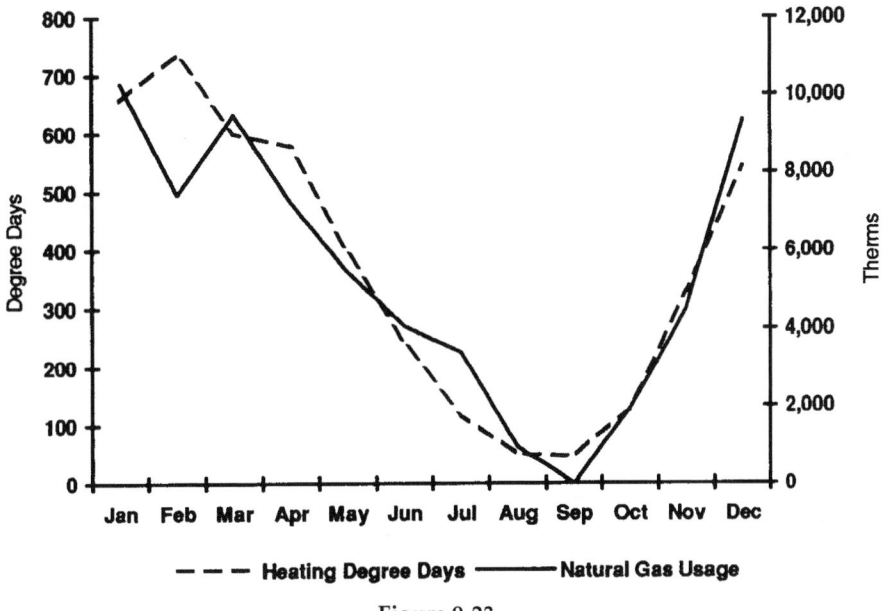

Figure 9-23.

Note that if he BLC is substituted for q/Δt, Equation 9.15 becomes:

$$E = \frac{BLC \times DD \times 24}{H \times Eff} \tag{9.16}$$

Ignoring the conversion terms, we can see that this equation has the potential to describe the relationship between monthly weather trends and the building heating fuel consumption shown in Figure 9-23.

As building construction techniques have improved over the years, and more and more heat-producing equipment has found its way into commercial and even residential buildings, a variety of correction factors have been introduced to accommodate these influences. The degree-day method, in its simpler form, is not presently considered a precise method of estimating future building energy consumption. However, it is useful for indicating the severity of the heating season for a region and can prove to be a very powerful tool in the analysis of existing buildings (i.e. ones whose energy consumption is already known).

It is possible to determine an effective building load doefficient for a building by analyzing the monthly fuel consumption and corresponding monthly degree days using the technique of linear regression.

9.10.2 Analyzing Utility Billings

If we were to plot the monthly natural gas consumption and corresponding degree-days for a building, the result might look something like Figure 9-24. It can be seen that the monthly energy consumption appears to follow the weather in an indirect fashion.

If we could draw a line that represented the closest "fit" to the data, it might look like the Figure 9-25.

As a general rule, the simpler the building and its heating (or cooling) system, the better the correlation. Buildings that are not mechanically cooled will show an even better correlation. To put it in terms of our earlier discussion regarding thermal "weight," the "lighter" the building, the better the correlation between building energy consumption and degree days. To the extent this line fits the data, it gives us two important pieces of information discussed in Section 9.1.1:

- The monthly base consumption

- The relationship between the weather and energy consumption beyond the monthly base, which we have been calling the building load coefficient.

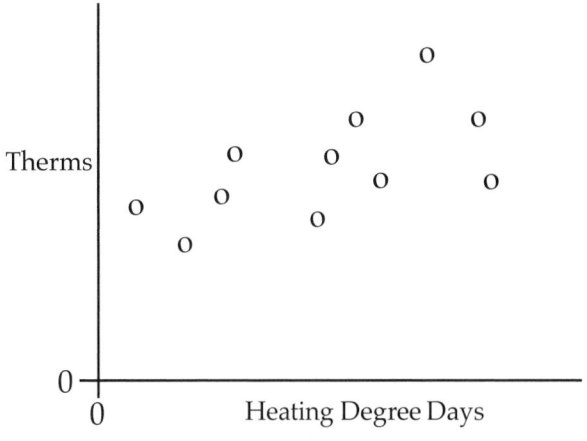

Figure 9-24.

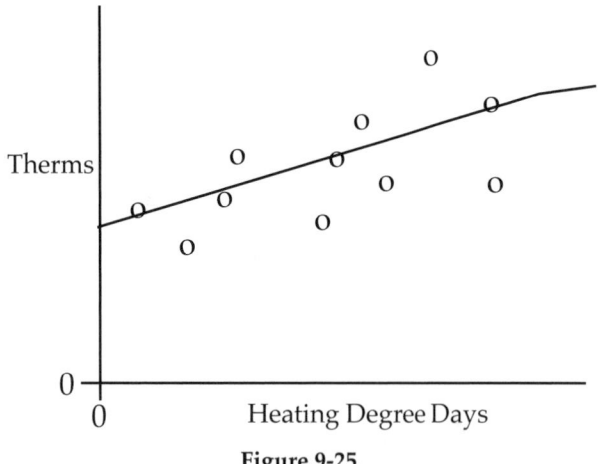

Figure 9-25.

9.10.3 Using Linear Regression for Envelope Analysis

The basic idea behind linear regression is that any physical relationship where the value of a result is linearly dependent on the value of an independent quantity can be described by the relationship:

$$y = mx + b \qquad (9\text{-}17)$$

where "y" is the dependent variable, x is the independent variable, m is the slope of the line that describes the relationship, and b is the y-intercept, which is the value of y when x = 0.

In the case of monthly building energy consumption, the monthly degree days can be taken as the independent variable, and the monthly fuel consumption as the dependent variable. The slope of the line that relates the monthly consumption to monthly degree days is the building load coefficient (BLC). The monthly base fuel consumption is the intercept on the fuel axis (or the monthly fuel consumption when there are no heating degree days). The fuel consumption for any month can be determined by multiplying the monthly degree days by the building load coefficient and adding it to the monthly base fuel consumption.

$$\text{Fuel Consumption} = \text{BLC} \times \text{DD} + \text{base energy} \qquad (9.18)$$

The value of the above is that once the equation is determined, projected fuel savings can be made by recalculating the monthly fuel consumption using a "conservation-modified" BLC.

The determination of the building load coefficient is useful in the analysis of building envelope for a number of reasons:

• Often the information necessary for the analysis of building envelope components is not available.

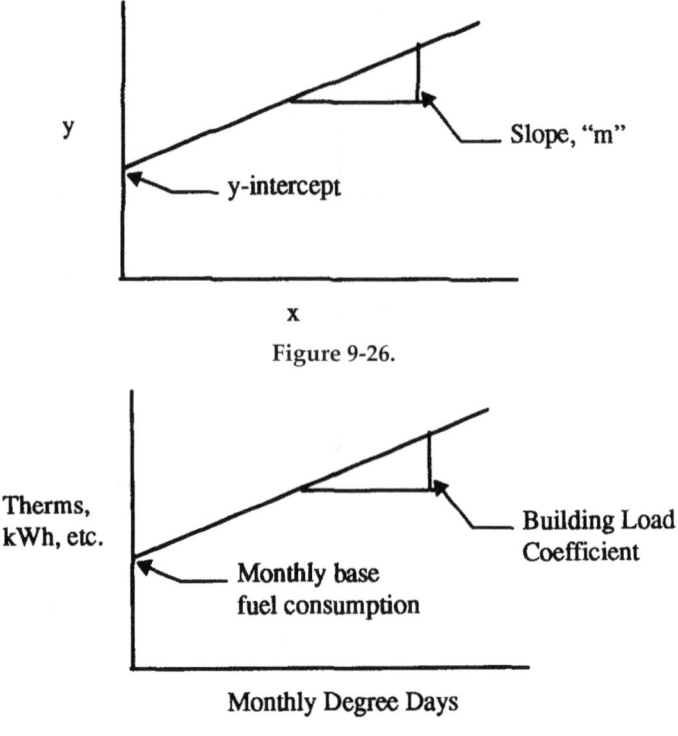

Figure 9-26.

Figure 9-27.

• If envelope information is readily available and the building load coefficient has been determined by analysis of all the individual components, the regression derived BLC gives us a "real-world" check on the calculated BLC.

• Linear regression also gives feedback in terms of how "good" the relationship is between fuel consumption and local climate, giving a good indication as to the thermal "weight" of the building (discussed in Section 9.9).

The following is a brief overview of the regression procedure as it can be applied to corresponding pairs of monthly degree-day and fuel consumption data. The building load coefficient, which describes the actual performance of the building, can be determined with the following relationship:

$$BLC = \frac{n\sum D_i E_i - \sum D_i \sum E_i}{n\sum D_i^2 - (\sum D_i)^2} \qquad (9.19)$$

Where:

n   =   the number of degree-day/fuel consumption pairs

D_i =   the degree days accumulated for an individual month

E_i =   the energy or fuel consumed for an individual month. The monthly base fuel consumption can be calculated as:

$$Base = \frac{\sum E_i}{n} - BLC \frac{\sum D_i}{n} \qquad (9.20)$$

The units of the BLC will be in terms of the fuel units per degree day. The BLC will be most valuable for additional analysis if it is converted to units of Btu/(hr-°F) or Watts/°C.

The following example demonstrates how the above might be used to evaluate the potential for envelope improvement in a building.

Example

A proposal has been made to replace 5,000 ft^2 of windows in an electrically heated building. The building envelope has been analyzed on a component-by-component basis and found to have an overall analytical BLC = 15,400 Btu/(Hr-°F). Of this total, 6,000 Btu/(Hr-°F) is attributable to the single pane windows, which are assumed to have a U-factor of 1.2 Btu/(hr-ft^2-°F). A linear regression is performed on the electric utility data and available monthly degree-day data (65°F reference). The BLC is found to be 83.96 kWh/degree-day, which can be converted to more convenient units by the following:

$$BLC = 83.96 \ \frac{kWh}{°F - Days} \times 3413 \ \frac{Btu}{kWh} \times \frac{1 \ Day}{24 \ hours}$$

$$= 11,940 \ Btu/(hr–°F)$$

What is the reason for the discrepancy between the BLC calculated component-by-component and the BLC derived from linear regression of the building's performance?

The explanation comes back to the concept of thermal weight. Remember, the more thermally "heavy" the building, the more independent of outdoor conditions it is. The difference between the value above and the calculated BLC is due in part to internal heat gains in the building that offset some of the heating that would normally be required.

This has significant consequences for envelope retrofit projects. If the potential savings of a window conservation retrofit is calculated on the basis of the direct "UA" improvement, the savings will be overstated in a building such as the above. A more conservative (and realistic) estimate of savings can be made by de-rating the theoretical UA improvement by the ratio of the regression UA to the calculated UA.

9.10.4 Evaluating the Usefulness of the Regression Results

The correlation coefficient, R, is a useful term that describes how well the derived linear equation accounts for the variation in the monthly fuel consumption of the building. The correlation coefficient is calculated as follows:

$$R = BLC = \sqrt{\frac{\sum D_i^2 - \frac{(\sum D_i)^2}{n}}{\sum E_i^2 - \frac{(\sum E_i)^2}{n}}} \qquad (9.21)$$

The square of the correlation coefficient, R,^{2} provides an estimate of the number of independent values (fuel consumption) whose variation is explained by the regression relationship. For example, an R^2 value of 0.75 tells us that 75% of the monthly fuel consumption data points evaluated can be accounted for by the linear equation given by the regression analysis. As a general rule of thumb, an R^2 value of 0.80 or above describes a thermally "light" building, and a building whose fuel consumption indicates an R^2 value significantly less than 0.80 can be considered a thermally "heavy" building.

9.10.5 Improving the Accuracy of the Building Load Coefficient Estimate

As discussed elsewhere, in Section 9.9, one of the reasons a building might be classified as thermally "heavy" would be the presence of significant internal heat gains, such as lighting, equipment, and people. To the extent these gains occur in the perimeter of the building, they will tend to offset the heat loss predicted, using the tools described previously in this chapter. The greater the internal heat, relative to the overall building temperature dependence (BLC), the lower the outside temperature has to be before heating is required.

The so-called "building balance temperature" is the theoretical outdoor temperature where the total building heat loss is equal to the internal gain. We discussed earlier that the basis of most published degree-day data is a reference (or balance) temperature of 65°F. If data of this sort are used for analysis, the lower the actual building balance temperature than 65°F, the more error will be introduced into the analysis. This accounts for the low R^2 values encountered in thermally heavy buildings.

While most degree data are published with 65°F as the reference point, it is possible to find compiled sources with 55°F and even 45°F reference temperatures. The use of this monthly data can improve the accuracy of the analysis significantly if the appropriate data are utilized.

Example

A regression is performed for a library building with electric resistance heating. The results of the analysis indicate an R^2 of 0.457. In other words, only 46% of the variation in monthly electricity usage is explained by the regression. The figure below is a plot of the monthly electric consumption versus degree days calculated for a 65°F reference.

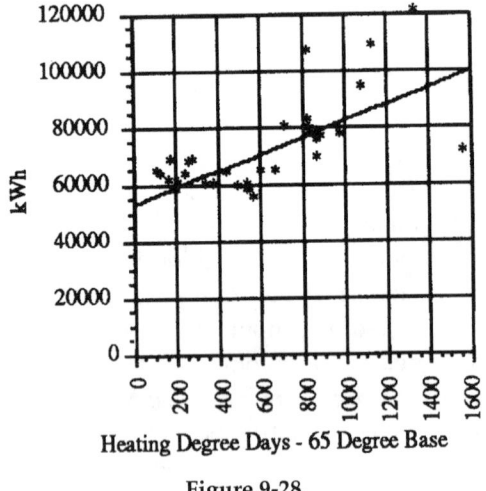

Figure 9-28.

The following plot shows the result of the regression run using monthly degree day referenced to 55°F. Notice in the figure that there are less data points showing. This is explained by the observation that all the days with average temperatures greater than 55°F are not included in the data set. The resulting R^2 has increased to 0.656.

The next figure shows the result of the regression run using monthly degree day referenced to 45°F. The resulting R^2 has increased to 0.884. Notice how few data points are left. Normally in statistical analysis great emphasis is placed on the importance of having an adequate data sample for the results to be meaningful. While we would like to have as many points as possible in our analysis, the concern is not so great with this type of analysis. Statistical analysis usually concerns itself with *whether* there is a relationship to be found. The type of analysis we are advocating here assumes that the relationship *does* exist, and we are merely attempting to discover the most accurate form of the relationship. Another way of saying this is that we are using a trial-and-error technique to discover the building balance point.

The final figure shows the result of the regression run with monthly degree days referenced to 40°F. The resulting R^2 has *decreased* to 0.535.

We have learned from the above that the balance point of the building analyzed is somewhere between 40° and 50°F, with 45°F probably being pretty close. Additional iterations can be made if more precision is desired, and if the referenced degree day data are available.

If the published degree day data desired are not available, it is a relatively simple matter to construct it

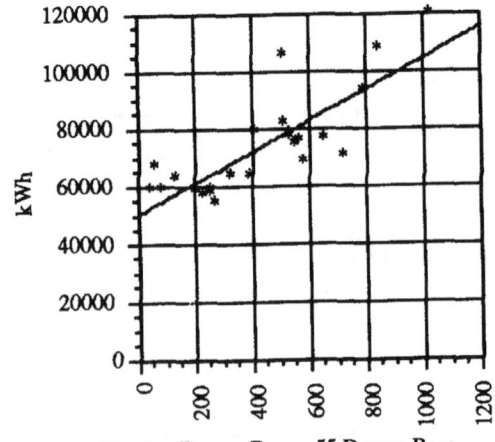

Figure 9-29.

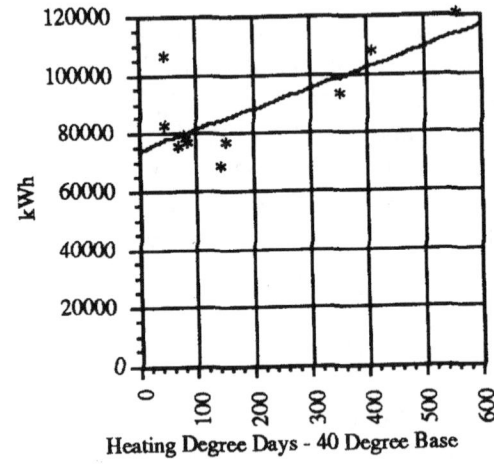

Figure 9-30.

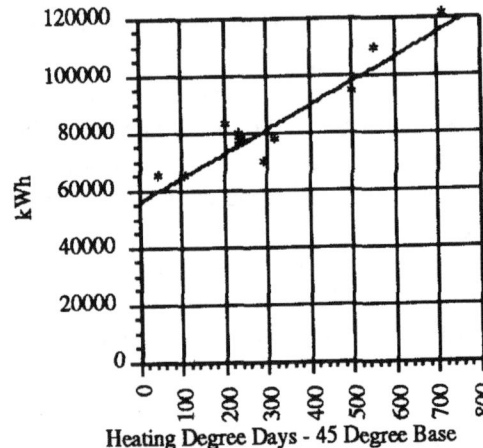

Figure 9-31.

from the average daily temperatures published for the local climate. If this is done with a spreadsheet, a table can be constructed in such a way that the degree days, BLC, monthly base, and R^2 values are all linked to an assumed balance temperature. This balance temperature is modified iteratively until R^2 is maximized. Of course any energy saving calculations should consistently use both the derived BLC and degree days that accompany the correlation with the highest R^2.

9.11 ENVELOPE ANALYSIS FOR NEW BUILDINGS

Envelope analysis of new buildings offers a different kind of challenge to the analyst. While new construction offers much greater opportunity for economical improvements to the envelope, the analysis is much more open-ended than that for existing buildings. In other words, we do not know the monthly energy consumption (the answer), and are left totally at the mercy of tools designed to assist us in *predicting* the future energy consumption of an as-yet unconstructed building, the necessary details of which constitute thousands of unknowns.

We say the above to emphasize the extreme difficulty of the task and the inadvisability of harboring any illusions that this can be done reliably, short of using hour-by-hour computer simulation techniques. Even with the powerful programs currently available, accuracy no better than 10 to 20 percent should be expected.

Nevertheless, we are often called upon to quantify the benefits of using one envelope strategy in lieu of another. We are able to quantify the *difference* in annual energy consumption between two options much more reliably than the absolute consumption. A number of techniques have been developed to assist in this process. One of the more popular and useful tools is the *temperature bin method*.

9.11.1 The Temperature Bin Method

The temperature bin method requires that instantaneous energy calculations be performed at many different outdoor temperature conditions, with the results multiplied by the number of hours expected at each temperature condition. The "bins" referred to represent the number of hours associated with groups of temperatures and are compiled in 5°F increments. The hour tabulations are available in annual, monthly, and sometimes 8-hour shift totals. All hourly occurrences in a bin are assumed to take place at the bin "center temperature." For example, it is assumed that 42°F quantitatively represents the 40-44°F bin.

The basic methodology of this method requires calculating the unique heat loss at each bin by multiplying the BLC by the difference between the indoor temperature assumed and the center temperature for each respective bin. This result is in turn multiplied by the number of hours in the bin. The products of all the

bins are summed to arrive at the total predicted annual heating energy for the building.

The advantage of this "multiple-measure" method over a "single-measure" method, such as the degree day method, is the ability to accommodate other temperature-dependent phenomena in the analysis. For example, the power requirement and capacity of an air-to-air heat pump are extremely temperature dependent. This dependency is easily accommodated by the bin method.

However, just as was the case with the degree-day method, energy savings predicted by the bin method may vary significantly from the actual, depending on the building balance temperature (building thermal weight). While the balance temperature cannot be predicted for a new building with the techniques previously discussed, it can be estimated with the following equation.

$$T_{balance} = T_{indoor} - \frac{q_{internal}}{BLC} \qquad (9.22)$$

Where:

$T_{balance}$ = The predicted balance temperature, °F

T_{indoor} = The assumed indoor conditioned space temperature, °F

$q_{internal}$ = The assumed internal heat gain in building temperature control zones adjacent to the envelope, Btu/hr

BLC = The building load coefficient, BLC, Btu/(hr-°F)

More accurate energy predictions will result by omitting calculations for bins whose center temperature exceeds the assumed balance temperature. For example, no calculation should be performed for the 50-54°F bin if the predicted balance temperature is 50°F. The center temperature of 47°F for the 45-49°F bin indicates that a 3°F temperature difference is appropriate for that bin (50-47°F).

A complete description of the bin method can be found in the ASHRAE Handbook of Fundamentals.

9.12 ENVELOPE STANDARDS FOR NEW & EXISTING CONSTRUCTION

The ASHRAE/IESNA Standard 90.1, *Energy Standard for Buildings Except Low-Rise Residential Buildings,* is the de-facto standard for commercial building efficiency and is the basis of most local energy codes.

9.12.1 ASHRAE 90.1 Compliance Requirements

Compliance with the Standard may be demonstrated utilizing the *prescriptive approach* or by performing calculations utilizing the *building envelope trade-off option.*

9.12.1.1 The Prescriptive Approach

Component criteria necessary for local compliance with the prescriptive approach are tabulated in tables for 8 separate climate zones. One of the features of the standard is the inclusion of many precalculated building components in the Appendix. A variety of common (and some uncommon) envelope assemblies can be referenced and utilized for demonstrating compliance with the prescriptive approach. In most cases, this will mean that no calculations will be required by the designer to demonstrate compliance.

9.12.1.2 The Building Envelope Trade-off Option

The building envelope trade-off option provides more compliance flexibility than might be found in the prescriptive approach. This option requires the designer to demonstrate that the proposed building envelope results in an envelope performance factor that is lower than the budgeted one for the project. Because of the impact of the mechanical and lighting systems on heating and cooling energy consumption, the building envelope trade-off option requires information for the mechanical and lighting systems, as well as for the envelope components.

9.13 SUMMARY

While the above discussion of envelope components has emphasized the information needed to perform rudimentary heat loss calculations, you'll find that the more you understand these basics, the more you begin to understand what makes an efficient building envelope. This same understanding will also guide you in deciding how to prioritize envelope improvement projects in existing buildings.

9.14 ADDITIONAL READING

As you can see from this brief introduction, the best source of comprehensive information on building envelope issues is the *ASHRAE Handbook of Fundamentals.* You are encouraged to continue your study of building envelope by reading the following chapters in the 2001 *ASHRAE Handbook of Fundamentals.*

Chapter	Topic
23,24	Thermal Insulation and Vapor Retarders
25	Thermal and Water Vapor Transmission Data
26	Ventilation and Infiltration
27	Climatic Design Information
28	Residential Cooling and Heating Load Calculations
29	Nonresidential Cooling and Heating Load Calculations
30	Fenestration

9.14.1 References

1. American Society of Heating, Refrigerating, and Air Conditioning Engineers, Inc., *Heat Transmission Coefficients for Walls, Roofs, Ceilings, and Floors,* 1993.
2. ASHRAE Handbook of Fundamentals, American Society of Heating, Refrigerating and Air Conditioning Engineers, Inc., Atlanta, GA, 2005.
3. ASHRAE Standard 901.-2007, American Society of Heating, Refrigerating, and Air Conditioning Engineers, Inc., Atlanta, GA, 2007.
4. Brown, W.C., *Heat-Transmission Tests on Sheet Steel Walls,* ASHRAE Transactions, 1986, Vol. 92, part 2B.
5. Elder, Keith E., *Metal Buildings: A Thermal Performance Compendium,* Bonneville Power Administration, Electric Ideas Clearinghouse, August 1994.
6. Elder, Keith E., *Metal Elements in the Building Envelope. A Practitioner's Guide,* Bonneville Power Administration, Electric Ideas Clearinghouse, October 1993.
7. Johannesson, Gudni, *Thermal Bridges in Sheet Metal Construction,* Division of Building Technology, Lund Institute of Technology, Lund, Sweden, Report TVHB-3007, 1981 .
8. Loss, W., *Metal Buildings Study: Performance of Materials and Field Validation,* Brookhaven National Laboratory, December 1987.
9. Washington State Energy Code, Washington Association of Building Officials April 1, 1994

CHAPTER 10

HVAC SYSTEMS

ROBERT COX, PE CEM
Jacobs, Engineering, Inc.
Raleigh, NC

ERIC NEIL ANGEVINE, P.E.
Associate Professor
School of Architecture
Oklahoma State University
Stillwater, Oklahoma

10.1 INTRODUCTION

The heating ventilation and air conditioning (HVAC) systems in most buildings are normally the largest consumer of energy in the building. Fan energy, pumping energy, mechanical or thermal cooling energy, and thermal heating energy may account for half or more of the annual energy consumption and annual energy cost for the building. In some building types, usually with large mandated ventilation requirements or with stringent temperature and humidity requirements (hospitals, clean rooms, museums, laboratories and data centers), the percentage may be up to 70% of the building's energy consumption.

The total energy consumed by HVAC systems will depend on several factors:

1) Climate (Miami or Minneapolis climates will require more of the HVAC systems than San Diego or Colorado Springs)
2) Type and efficiency of building envelope
   a) Percentage of fenestration area of the total wall area and energy efficiency factors of the glass, U-value, solar heat gain coefficient, etc.
   b) Quality of wall and roof insulation
   c) Orientation of the building envelope
   d) Shading
3) Amount of internal heat gain requiring cooling— from computers, servers, lighting, miscellaneous electrical components, etc. (A data center or heavily loaded office space will impose much greater load than a simple office or a lightly loaded classroom.)
4) Amount of fresh air which must be introduced to the spaces in the building to meet code, good practice, or exhaust requirements (general exhaust, laboratory exhaust, fume hoods, etc.).

5) Amount of minimum air changes required for good indoor air quality and ventilation effectiveness.
6) Requirement for simultaneous heating and cooling due to:
   a) Minimum air change requirements
   b) Dehumidification control
7) Requirements for humidification (cold climates, infection control, humidity sensitive processes)
8) Space temperature and humidity requirements for heating and cooling (e.g. 65°F—40% RH (operating room) versus 76°F—50% RH (office or classroom)
9) Types of HVAC systems selected to serve the building loads
10) Hours of operation of the systems
11) Actual occupied hours of the building spaces
12) Mechanical equipment efficiencies
13) Distribution energy requirements (energy required to move heating or cooling throughout the building)—fan energy or pumping energy
14) System thermal losses (heat loss/gain from or heat gain to insulated or uninsulated equipment or piping or ductwork systems)
15) Equipment condition, including cleanliness of heat transfer elements, duct leaks, etc.

The energy professional will need to consider all of these factors when investigating the energy efficiency of the current building HVAC systems and when evaluating the opportunities for improving the energy efficiency of the building through implementation of energy conservation opportunity (ECO) projects.

10.1.1 Mechanical Equipment Efficiency

Efficiency, by definition, is the ratio of the energy output of a piece of equipment to its energy input, in like units to produce a dimensionless ratio. Since no equipment known can produce energy, efficiency will always be a value less than 1.0 (100%).

The relative efficiency of cooling equipment is usually expressed as a coefficient of performance (COP), which is defined as the ratio of the heat energy extracted to the mechanical energy input in similar units. Since the heat energy extracted by modern air conditioning equipment far exceeds its mechanical energy input, a COP of up to 8 is achievable.

Air-conditioning equipment is also commonly rated by its energy efficiency ratio (EER) or seasonal energy efficiency ratio (SEER). EER is defined as the ratio of heat energy extracted (in Btu/hr) to the mechanical energy input in watts. Although it should have dimensions of Btu/hr/watt, it is expressed as a dimensionless ratio and is therefore related to COP by the equation

$$EER = 3.413 \cdot COP \qquad \text{eq. 10.1}$$

Other useful equipment efficiency conversions:
$$kW/ton = 12/EER \qquad \text{eq. 10.2}$$

$$COP = 3.517/(kW/ton) \qquad \text{eq. 10.3}$$
(The constant 3.517 is 12000/3413)

Although neither COP nor EER is the efficiency of a chiller or air conditioner, both are measures which allow comparison between different pieces of equipment. The term air-conditioning efficiency is commonly understood to indicate the extent to which a given air conditioner performs to its maximum capacity. As discussed below, most equipment does not operate at its peak efficiency all of the time. For this reason, the seasonal energy efficiency ratio (SEER), which takes varying efficiency at partial load into account, is a more accurate measure of air-conditioning efficiency than COP or EER. The SEER rating is applied to residential and light commercial air-cooled refrigeration systems, and EER is used for larger equipment. For chilled water equipment, kW/ton or COP are common, with integrated part load value (IPLV) representing a measure of part load energy efficiency. Both the SEER and IPLV rating system are standardized to permit comparison and specification of part load energy performance between manufacturers.

These energy related terms will be used throughout discussions in this chapter.

10.1.2   HVAC System Goals

The HVAC systems in a building must accomplish several goals:

A.   Provide sensible heating to each of several spaces in the building to offset heat loss from the building envelope and to maintain thermal comfort at some desired space temperature. "Sensible" heating or cooling describes a change in space air temperature, separate from any humidification or dehumidification process.

B.   Provide humidification at the system level, or at the space level if required to maintain space relative humidity setpoints.

C.   Provide sensible and latent cooling to each of several spaces in the building to offset heat gain from the building envelope and internal gains (from lighting, equipment, people) and maintain thermal comfort at some desired space temperature and humidity. In some cases, the desired space conditions may be based on the process requirements or on experiments (labs/classrooms/computer rooms), or on the desired temperature and humidity to facilitate healing (hospitals/clinics/skilled nursing facilities); not just that required for human comfort.

D.   Provide ventilation for each space to maintain good ventilation effectiveness for human comfort (to enhance evaporation) and to meet mandated ventilation needs for process, dilution, infection control or other requirement (labs, classrooms, healthcare spaces, clean rooms, etc.)

E.   Provide pressurization control for the building to the outside elements and in some cases pressurization control of some spaces with respect to each other for safety, process, or infection control reasons (labs, healthcare, clean rooms).

F.   Provide outside air for the building for dilution of odors, to makeup for building exhaust and to provide desired indoor air quality. This outdoor air must be heated, cooled, dehumidified, or humidified, depending upon climate so as to not effect space temperature or humidity control.

G.   Provide filtration of air to maintain good indoor air quality and/or to meet specific process (lab, classroom, computer room) and infection control (healthcare facilities) requirements.

H.   Provide regulation and automated control of system components to maintain desired space temperatures as environmental and operating conditions change. This is further discussed in Chapter 22.

The goal for this chapter is to briefly describe the different types of HVAC systems commonly used in most buildings and the energy consumed by these systems, as well as to illustrate which systems are more or less energy efficient than others. The chapter will also provide direction to the energy professional as to which energy conservation opportunities (ECOs) can be effectively implemented to make each system more energy efficient.

10.2 SURVEYING EXISTING HVAC SYSTEM CONDITIONS

The first stage of any effective energy management program is an audit of the facility in question, which will usually include an HVAC survey. The goal for the energy professional is to understand how to maximize energy conservation and develop energy conservation opportunities (ECOs) without impacting any of the other basic goals the HVAC system is trying to achieve. Regardless of the type of HVAC system being used in the facility, several important criteria need to be documented concerning how the facility is operated:

1. What are the current desired or required temperature and humidity setpoints for each space? (These conditions may be different from original design.)

2. What are actual occupancy hours of each space? Are they different from the hours the HVAC system is operated?

3. What are all the current temperature, humidity, pressure, or other setpoints attempted to be maintained by the HVAC systems? Are these setpoints being achieved by the HVAC systems? Are they capable of being achieved with the capacity of the existing equipment?

4. How maintainable are the systems? Can they be serviced? What are the current preventive and proactive maintenance procedures?

5. What changes have occurred to the facility since its original design and what changes are expected?

6. When might these changes occur?

7. What are all the currently programmed sequences of operation for the HVAC systems?

8. What is the age and condition of the equipment?

As part of the survey, try to document all the HVAC systems and equipment in the facility, including associated air flows, water flows, steam flows, and other capacity and efficiency features of the equipment. Review of design documents, one line diagrams, test and balance reports, and control diagrams are excellent clues as to how the system is intended to operate.

Compare the equipment and system cooling or heating capacities to what is required to meet the current basis of design for the building. Identify equipment that is too small or radically oversized.

The energy professional may obtain capacities from original design drawings. It is usually, however, very important to determine actual performance of the systems through measurement of airflows, water-flows, amperages of fans, pumps, chillers, and other equipment, as well as recordings of temperature and humidity.

If a building automation system is installed in the building, it can often be used to trend performance of many significant points of measurement for the systems. If a building automation system is not available, the energy professional can use a variety of individual data logging devices which can be brought to the building to log temperatures, amperages, or other values over time.

The age and condition of the existing equipment should also be evaluated as part of the audit. Additions or modifications to reduce energy consumption (ECOs) to equipment already in poor condition, out of its service life (compared to current equipment), or with poor efficiency would be a poor life cycle cost decision. The owner should be consulted regarding normal life cycle replacement of equipment to be included in the ECOs.

Systemic Issues

Review the design for appropriate application, including zoning. Simply introducing energy conservation measures into a poorly designed system may produce additional problems.

When design application issues exist, correcting those should be a prerequisite to energy conservation. Some common design related issues include:

• Oversized equipment or motors

• Undersized ducts or pipes and resulting high transport losses

• Duct leakage in unconditioned spaces

• Negative pressurization from poor air balance

• Air economizers with no return/relief fan control

Zoning Problems

HVAC zoning is usually identified by a space temperature control point sensor or thermostat. The area cooled or heated by the HVAC system responding to this control point delineates a "zone." If a zone serves both east and west glass exposures, there are built-in comfort issues. If the thermostat is located in the east exposure, the west areas will be over-cooled in the morning and over-heated in the afternoon; the reverse would be true if the thermostat were located in the west portion of the building. Zones that serve both perimeter and interior areas have similar incompatibilities. In addition to the obvious comfort complaints that accompany such designs, energy use is almost always increased as a result.

Finally, all ECOs must be implemented with maintainability in mind. If a certain ECO is implemented in a manner which cannot be maintained over time, it will be abandoned in the future and will fail to achieve the life cycle savings it was intended to achieve. (Refer to Chapter 14.

10.3 HUMAN THERMAL COMFORT/FACILITY DESIGN CRITERIA

The ultimate objective of any heating, cooling and ventilation system is to maintain each space at its desired temperature, air quality and humidity conditions.

Human thermal comfort is maximized by establishing a heat balance between the occupant and the environment. Since the body can exchange heat energy with its environment by conduction, convection and radiation, it is necessary to look at the factors which affect these heat transfer processes, along with the body's ability to cool itself by the evaporation of perspiration.

In addition to air temperature, humidity and air motion, the surface temperature of surroundings also has a significant influence on the rate at which the human body can dissipate heat. At temperatures below about 80°F (27°C), most of the body's heat loss is by convection and radiation. Convection is affected by air temperature and air velocity. Radiation is primarily a function of the relative surface temperature of the body and its surroundings. Heat transfer by conduction is negligible since we make minimal physical contact with our surroundings, which are not insulated by clothing. At temperatures above 80°F (27°C), the primary heat loss mechanism is evaporation. The rate of evaporation is dependent on the temperature and humidity of the air, as well as the velocity of air that passes over the body carrying away evaporated moisture.

In addition to these environmental factors, the rate of heat loss by all means is affected by the amount of clothing, which acts as thermal insulation. Similarly, the amount of heat which must be dissipated is strongly influenced by activity level. Therefore, the degree of thermal comfort achieved is a function of air temperature, humidity and air velocity, the temperature of surrounding surfaces, the level of activity, and the amount of clothing worn.

In general, when environmental conditions are cool the most important determinant of human thermal comfort is the radiant temperature of the surroundings.

When conditions are warm, air velocity and humidity are more important. It is not by accident that the natural response to being too warm is to increase air motion. Similarly, a reduction in humidity will offset an increase in air temperature, although it is usually necessary to limit relative humidity to no more than 60% in summer and no less than 30% in winter.

There is, of course, a human response to air temperature. The most noticeable comfort response to air temperature is the reaction to drift, the change of temperature over time. A temperature drift of more than 1°F per hour (0.5°C/hr) will result in discomfort under otherwise comfortable conditions. Temperature stratification can also cause discomfort, and temperature variation within the occupied space of a building should not be allowed to vary by more than 5°F (3°C).

The location and type of air distribution devices play a role equal in importance to that of effective controls in achieving thermal comfort. The discomfort caused by stratification can be reduced or eliminated by proper distribution of air within the space.

In general terms, thermal comfort can be achieved at air temperatures between about 68°F and 80°F, and relative humidity's between 30% and 60%, under varying air velocities and radiant surface temperatures. Figure 10-1 shows the generalized "comfort zone" of dry bulb temperatures and humidity levels plotted on the psychrometric chart. (A blank psychrometric chart is provided in the back appendix of this book.) However, human thermal comfort is a complex function of temperature, humidity, air motion, and thermal radiation from local surroundings, activity level and amount of clothing.

10.3.1 Special HVAC System Applications

There are several unique HVAC applications where ventilation rates, pressurization control, desired space temperatures and humidity setpoints, and fresh air requirements may be dictated by code or other concerns (infection control, heat dissipation, lab experiment success, equipment performance, safety). In these cases, the design criteria may greatly exceed that required for human thermal comfort. Some examples of these types of spaces and facilities include:

A.   Laboratory spaces
B.   Healthcare spaces
C.   Critical mission spaces (data centers, computer cen-

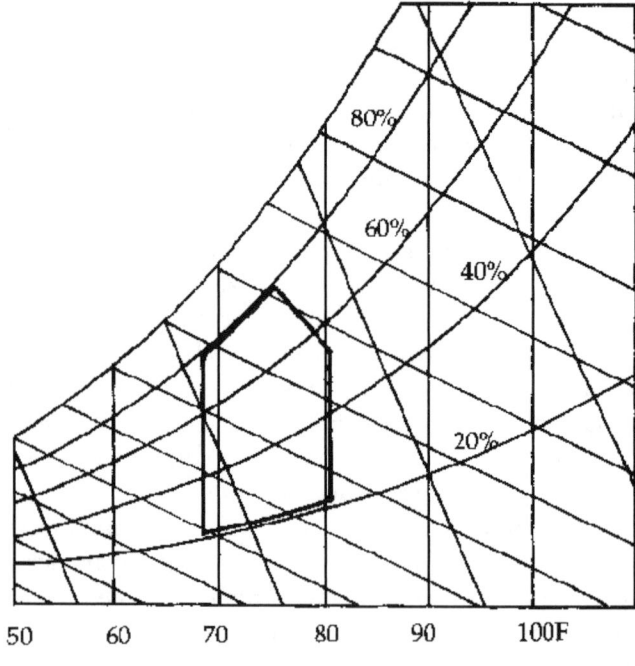

Figure 10-1. Comfort zone.

ters, UPS, emergency power)

D.   Museums, libraries, archival spaces

E.   Clean rooms

The energy professional must be knowledgeable with regard to the unique requirements of these facilities before proposing ECO projects that conflict with the basis of design. However, these types of facilities generally use 2-10 times the amount of energy of a typical office building or school building and are therefore attractive candidates for energy efficiency improvements.

10.4 INTERACTIONS WITH HVAC ECO PROJECTS

Care must be taken during implementation of any ECO Project that it does not cause an impact which will not allow the HVAC system to function as originally designed. Some examples are listed in Table 10-1.

10.5 HVAC SYSTEM TYPES

10.5.1 General Efficiency Discussion

The energy efficiency of systems used to heat and cool buildings varies widely, but is generally a function of the details of the system organization. On the most simplistic level, the amount of energy consumed is a function of the efficiency of the source of the heating or cooling energy (boilers, chillers, refrigeration, electricity, etc.) the amount of energy consumed in distribution of heating and cooling fluids (fan energy or pumping energy) energy required for ventilation (fan energy), and whether the system operation requires simultaneously heating and cooling of the delivery fluid (usually air). System ef-

Table 10-1. ECO Project Interactions
Source: *Commercial Energy Auditing Reference Handbook*, Doty, S., 2008, Fairmont Press

Action	Reaction
Lighting retrofit	Existing heating system was marginally sized and now is inadequate.
Refrigeration retrofit for high efficiency	Waste heat was used for heating, and now auxiliary heating elements are needed, increasing energy use.
Chiller savings from reduced head pressure, via lower condenser water temperature	Low efficiency cooling tower requires large increase in tower fan kW, eroding most of the chiller savings.
	Too much reset can create operational problems for the chiller, Some chillers cannot be reset below 75 degrees. 65-70°F is usually safe, and some can accept colder water with excellent reductions in kW.
Chiller savings from increased suction pressure, from increased chilled water temperature setpoint.	Higher chilled water temperature creates higher apparatus dew point temperature in air handlers and a loss of dehumidification.
Condensing boiler retrofit for 90+ efficiency.	High efficiency not achieved because boiler rating depends on low return water temperature; existing hydronic design is not set up for this, so intended efficiencies are not realized.
Addition of air economizer	Building problems with pressurization, allowing air to infiltrate or exfiltrate.
	For buildings with very low break even points, using very cold air for cooling encourages stratification and nuisance freeze stat tripping.
	Relative humidity swings inside the building for humidity sensitive activities, including electronics, books, and artifacts.
Constant volume HVAC converted to VAV	Low pressure ductwork may not be suitable. Keep pressures low and protect with static pressure switches.
Thermostats turned off in unoccupied areas	Frozen pipes at the perimeter

ficiency is also highly dependent upon the programming of automated control systems.

HVAC system types can be typically classified according to their energy efficiency as highly efficient, moderately efficient, or generally inefficient. This terminology indicates only the comparative energy consumption of typical systems when compared to each other. Using these terms, those system types classified as generally inefficient will result in higher energy bills for the building in which they are installed, while an equivalent building with a system classified as highly efficient will usually have lowest energy bills. However, it is important to recognize that there is a wide range of efficiencies within each category, and that a specific energy-efficient example of a typically inefficient system might have lower energy bills than the least efficient example of a moderately, or even highly efficient type of system.

Figure 10-2 shows the relative efficiency of the more commonly used types of HVAC systems discussed in this chapter. The range of actual energy consumption for each system type is a function of other design variables, including how the system is configured and installed in a particular building and how it is controlled and operated.

To maximize the efficiency of any type of HVAC system, it is important to select efficient equipment, minimize the energy consumed in distribution, and avoid simultaneous heating and cooling of the working fluid (usually air) wherever possible. It is equally important that the control system directly control the variable pa-

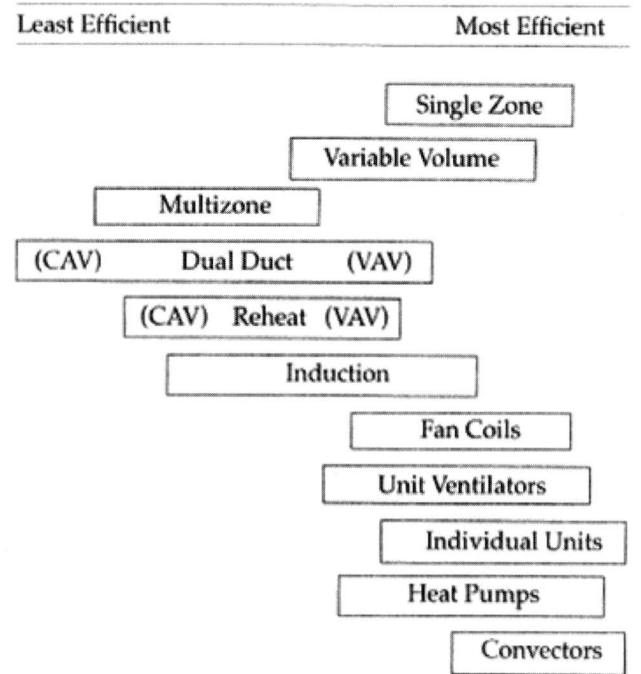

Figure 10-2. Relative Energy Efficiency of HVAC Systems.

rameters of the system and also have occupancy control.

Most HVAC systems serve various zones in the facility that may have different climatic and/or internal thermal loads and for which heat can be supplied or extracted independent of other zones.

The four-volume ASHRAE Handbook, published sequentially in a four-year cycle by the American Society of Heating, Refrigerating and Air-Conditioning Engineers, Inc., provides the most comprehensive and authoritative reference for HVAC systems and equipment as well as applications to particular buildings.

10.5.2 General HVAC System Discussion

The HVAC systems in most buildings can be classified into several categories that will be discussed in this chapter. The HVAC systems serving a facility are generally designed and distributed throughout the building based on a certain number of control zones of operation. The zones of operation are determined by the design engineer based on different envelope exposures and climate conditions, occupancy requirements and schedules, internal and thermal loads, and different zone temperature, humidity, and ventilation requirements.

The zones are provided with HVAC equipment and systems to meet their space needs by adding or removing heat and humidity from the space. The zones are served by the following types of HVAC equipment and systems (unitary equipment):

10.5.3 Decentralized Heating and Cooling Systems
1. Window air-conditioning equipment
2. Through-the-wall room HVAC units (packaged terminal AC units)
3. Air-cooled heat pump systems (self-contained or with remote condenser)
4. Self-contained air handling systems
5. Packaged special procedure units (computer rooms)
6. Light commercial split systems (compressor in the evaporator section or the condensing section)

These systems serve individual zones and have their own local cooling capability in the form of a cooling coil (normally direct expansion refrigerant coil), sometimes a heating coil (normally electric heat, but occasionally hot water heating from a remote source), their own refrigerant compressor or compressors, and local or remote condensers. Split systems have a remote condenser. Some self-contained units may have water cooled compressors, with water provided from a remote heat sink (city water, or water from a well or ground source, or water from a central cooling tower system).

Generally, the heating systems for these units are decentralized, but some may be provided with heating water from a central system. These systems have their own fans, self-contained controls, filters, and self-contained refrigerant systems. See Figure 10-3 for an illustration of a self-contained unit and Figure 10-4 for an illustration of a PTAC unit.

These systems have several equipment and system energy advantages and disadvantages:

Advantages
1. The systems serve single zones; therefore, simultaneous heating and cooling is usually eliminated. These systems normally operate in a heating, cooling, or fan-only mode, or are turned off.
2. Heating or cooling can be provided to individual zones having different operating hours than the remainder of building zones. This is especially advantageous when a majority of building zones can be turned off.
3. Each building zone can have different operating temperatures and humidity conditions, with no effect on other zones.
4. Simple operation.

Disadvantages
1. Cooling and heating equipment has a much poorer coefficient of performance (COP) or energy efficient ratios (EER) than central cooling or heating equipment.
2. Generally air-cooled or water cooled economizers are not provided with this type of equipment. Therefore, the equipment is required to operate for cooling when ambient dry-bulb or wet-bulb temperatures would be adequate to provide space cooling with no mechanical refrigeration.
3. These systems generally cannot adequately dehumidify or filter outdoor air (especially in humid climates), therefore a separate system to provide outdoor air and ventilation is usually required.
4. Normal temperature control of these systems will result in poor space humidity control if outdoor air is introduced to these systems. They are really designed to perform only sensible cooling.
5. These systems, which are through the wall or window systems, provide poor ventilation effectiveness to the space.

Further descriptions of decentralized systems can be found in the current volume of ASHRAE Systems and Equipment (Chapters 2, 48 and 49).

10.5.4 Air Handlings and Distribution Systems
10.5.4.1 General
Air handling systems for heating and cooling buildings are systems which moderate the air temperature of the occupied space by providing a supply of heated or cooled air from a central source via a network of air ducts. These systems, referred to as all-air systems, increase or decrease the space temperature by altering either the volume or temperature of the air supplied to the individual zones connected to the air handling system.

Recalling that the most important determinant of thermal comfort in a warm environment is air velocity, the majority of larger buildings which require cooling

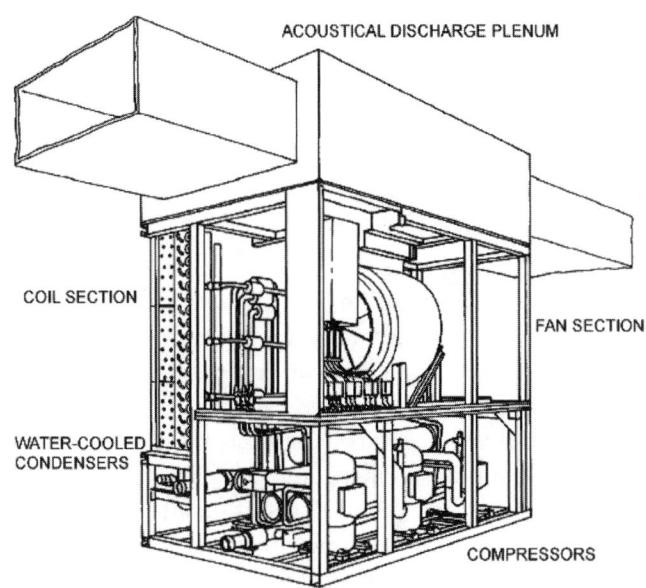

Figure 10-3. Vertical Self-Contained Unit
Source: ASHRAE Systems and Equipment Handbook, 2008, © American Society of Heating, Refrigerating and Air-Conditioning Engineers, Inc., www.ashrae.org.

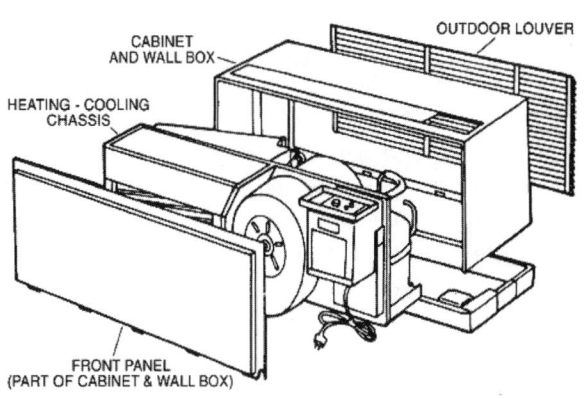

**Figure 10-4. Packaged
Terminal Air Conditioner (PTAC)**
Source: ASHRAE Systems and Equipment Handbook, 2008, © American Society of Heating, Refrigerating and Air-Conditioning Engineers, Inc., www.ashrae.org.

employ all-air systems. Consequently, all-air systems are normally the system of choice when cooling is required. All-air systems can also provide control of outside fresh air, air quality, and humidity control. These systems can also use outside air free-cooling (air-side economizer) when ambient conditions are favorable for cooling spaces. The advantages of all-air systems are offset somewhat by the additional energy consumed in distribution of the air and the inefficiency of using air as the heat transfer medium.

All-air systems tend to be selected when significant ventilation rates are required for reasons other than comfort cooling, when multiple zones are required, and for buildings which have significant internal cooling loads that coincide with heating loads imposed by heat loss through the building envelope.

The components of an all-air HVAC system include an air-handling unit (AHU) with a fan or fans (sometimes both supply air and return or relief air fans), heat transfer coils to preheat, heat, or cool the air passing through them, filters to clean the air, and often elements to humidify the air. Dehumidification, when required, is accomplished by cooling the air below the dew-point temperature.

The air handling system may be provided with the ability to introduce fresh outdoor air into the air handling unit. The outdoor air is used to enhance indoor air quality through dilution of odors, meet minimum code and safety requirements, meet mandated pressurization requirements, if any, and to provide air to offset makeup exhaust from the building. In certain buildings, like laboratory facilities, kitchens, clean rooms, and healthcare facilities, the makeup air requirements may be significant.

The air handling system can be designed to bring in only the minimum amount of outdoor air required for the items stated above, or it may be capable of bringing in the entire amount of air the air handling unit will use for air-side economizer operation. During the air-side economizer operation, at appropriate ambient dry bulb and wet bulb temperatures, the outdoor air can be used to cool the zones served by the air handling system without using any mechanically driven cooling source. If the air-side economizer operation is used, return air, relief air, and outdoor air dampers will be provided, as well as mixed air temperature control, to modulate the dampers to supply the correct temperature of air to the system.

When an air-side economizer system is used, a return fan(s) or a relief fan(s) is provided in the air handling system for building pressurization control.

The air handling system may have been designed to provide a constant amount of air to each zone as dictated by cooling and heating load, space temperature, and ventilation requirements. This type of system is known as a constant volume (CAV) system. A CAV system will then modify the temperature of the air to each space to maintain space temperature setpoint as loads and occupancy vary.

The air handling system may have been designed with the capability to vary the amount of air to each zone as loads and occupancy vary to maintain space temperature and ventilation setpoints. This type of system is known as a variable air volume (VAV) system. The variable air volume system will vary the airflow to a zone, from a maximum airflow required to sensibly cool the zone or required to meet minimum ventilation or exhaust requirements, to a minimum airflow required to meet ventilation or exhaust requirements. If the space is still too cool at the minimum airflow requirements, a reheat coil may be needed to increase air temperature to keep the space at space temperature setpoint, or to heat the space.

VAV systems normally use a variable frequency controller for each fan motor to vary the motor and fan speed. The varying fan speed will increase or decrease airflow to the system. The speed is normally controlled to maintain a static pressure setpoint in the supply or return ductwork. The system is also normally provided with airflow monitoring devices to measure supply and return airflow.

Figure 10-5 illustrates typical features in a single zone air handling unit when an air-side economizer is used, and it identifies a number of common air handler components. Figure 10-6 illustrates a constant volume reheat system, and while Figure 10-7 illustrates a variable volume system. Figure 10-8 illustrates the multizone and dual-duct system concept. Note: Each of these figures show "cooling coil" and "heating coil" symbols. Cooling coils can be chilled water or direct expansion (DX); heating coils can be hydronic heating water, steam, or electric resistance type.

In most air handling and distribution systems, terminal units and associated controls are provided for each zone in either a constant volume or variable volume system, unless it is a single zone system. The terminal units and associated controllers consist of an air flow monitoring station, to measure zone airflow, and a controlled damper which modulates its position (regardless of upstream ductwork static pressure) to maintain an airflow setpoint value as sensed by the air flow monitor.

The required airflow setpoint is varied as required to meet a space temperature setpoint that is sensed by a space temperature sensor. Many of the terminal units and controllers are also provided with a heating coil

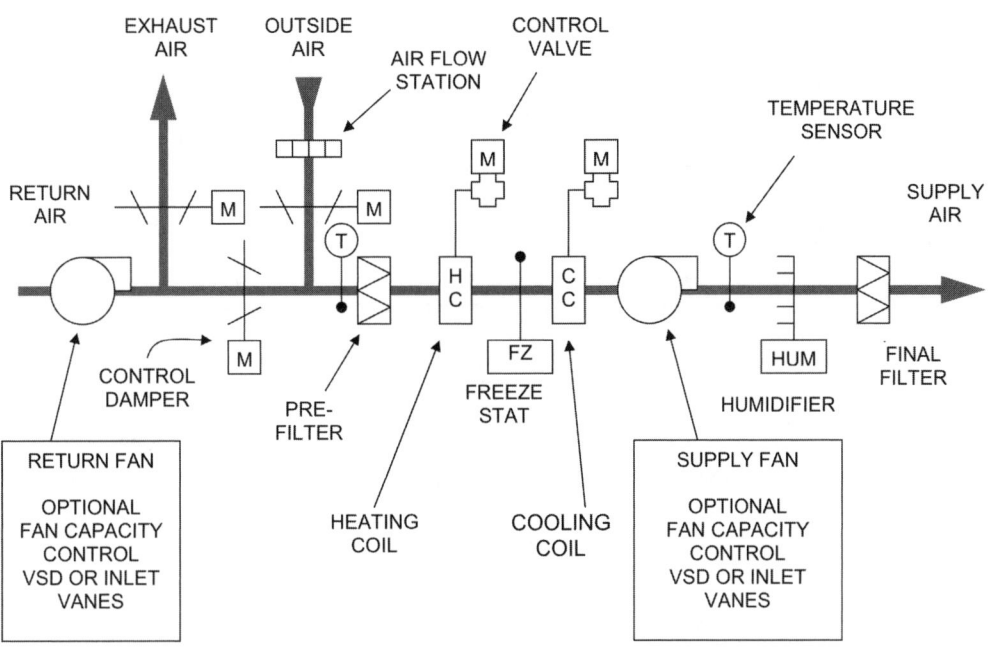

Figure 10-5. Single Zone HVAC System with Air Economizer

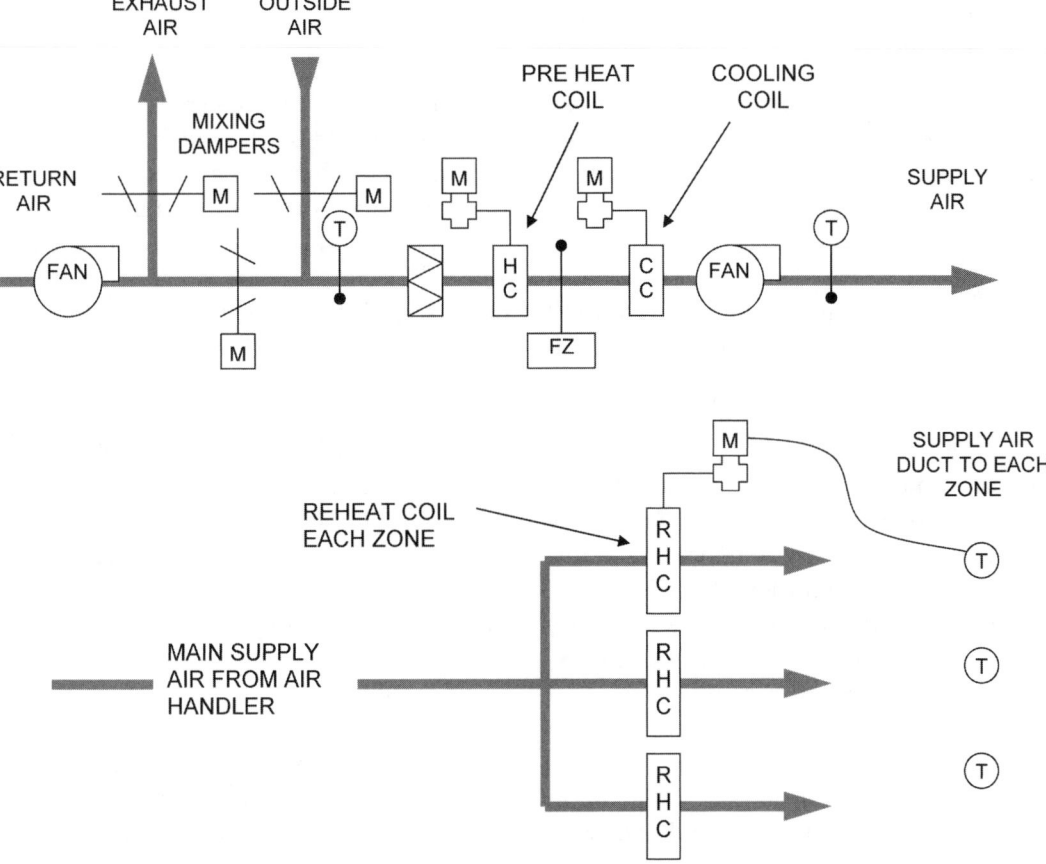

Figure 10-6. Constant Volume Multiple Zone Air Handling System Schematic (Constant Volume Terminal Reheat System)

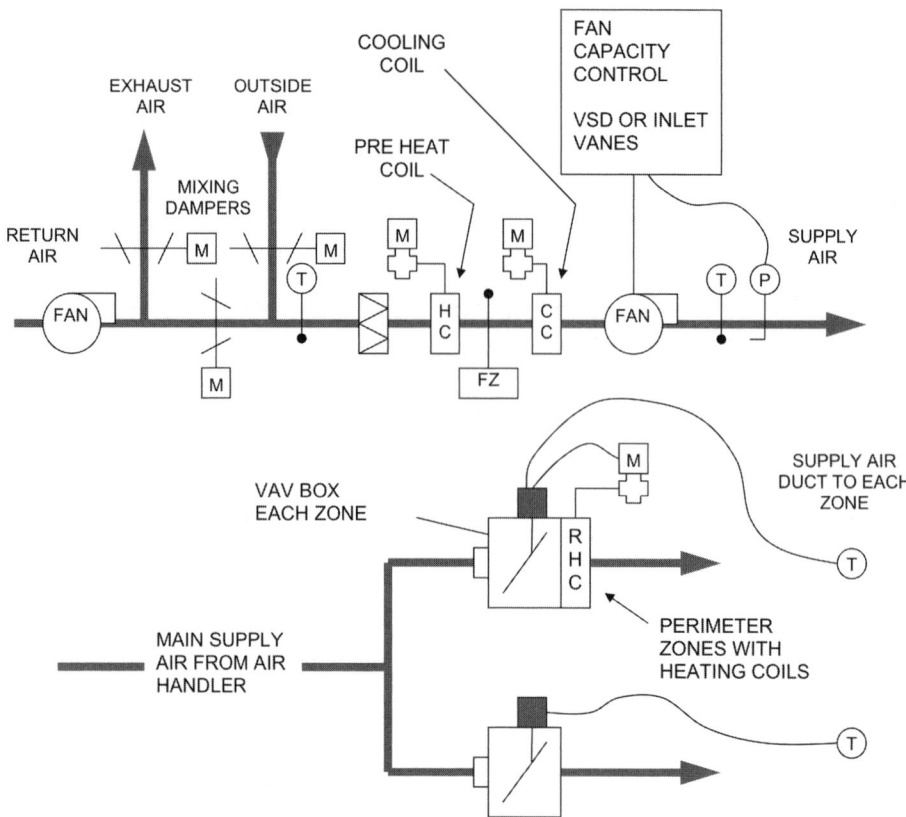

Figure 10-7. Variable Volume Multiple Zone Air Handling System Schematic (VAV Reheat System)

downstream of the damper (reheat coil) to provide additional heat to the air supplied to the zone if the air volume cannot be reduced beyond its minimum airflow requirements and the space is still too cold. The reheat coil is sized to reheat the air or heat the space with air as required. The reheat coil is provided with a heat source of either heating water or electricity, and its output temperature is modulated by varying stages of electrical capacity, or by varying the amount of heating water proceeding through the coil by use of a heating water control valve.

The conditioned air from the AHU is supplied to the terminal units and occupied spaces by a network of supply-air ducts, and air is returned from the conditioned spaces by a parallel network of return-air ducts. The AHU and its duct system also normally include a duct that supplies fresh outside air to the AHU and one which can exhaust some or all of the return air to the outside.

10.5.4.2 Conditioning Outdoor Air

The cost of tempering outdoor air is important to energy professionals. One common ECO for buildings with variable occupancy is to vary the outside air accordingly. (e.g. When there are only half the occupants, only half the ventilation air is needed.) Similar strategies can be used

to reduce outdoor air when not needed as makeup for exhaust (exhaust fans off, fume hoods modulate, etc.). Depending on the climate, heating, cooling, humidifying, or dehumidifying of outdoor air will be required. Table 10-2 provides an example of the cost of heating outdoor air:

10.5.4.3 Single Duct Systems

The majority of all-air HVAC systems (1980s to the present) employ a single network of supply air ducts which provides a continuous supply of either warmed or cooled air to the occupied zones of the building.

Single duct systems can either be single zone systems or multiple zone systems. A single zone system serves a single zone of temperature and humidity control, while multiple zone systems serve multiple zones of temperature and humidity control.

System Components

Air Handling Unit

Single zone and multizone have a central air handling unit which contains most of the following components as illustrated in Figures 10-5 through 10-7:

1.   Supply fan or fans distribute air to the space or spaces and back to the air handling unit.

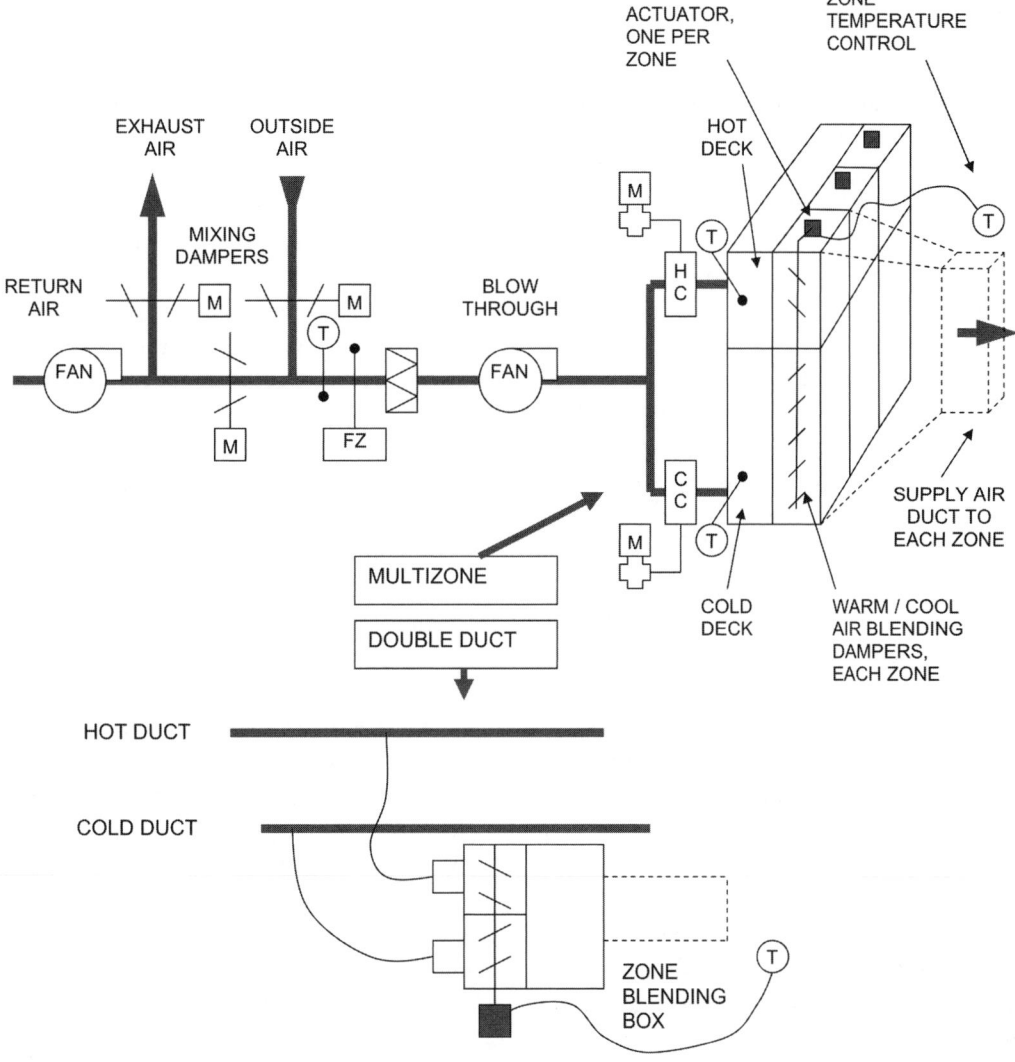

Figure 10-8. Multizone or Dual Duct Air Handling System.

Editor's Note: The diagram shows a multizone air handler. Functionally the standard dual duct is equivalent. Instead of hot and cold "decks," blending dampers at the air handler and low pressure ductwork to each zone, the dual duct system has hot and cold "ducts" under high pressure. Two high pressure ducts are routed through the building and blending boxes at each zone provide temperature control.

Table 10-2. Outdoor Air Heating

Source: *Commercial Energy Auditing Reference Handbook*

		Bldg Space Temp, degF		
		65	**70**	**75**
OA	**30**	39	44	50
Temp	**35**	33	39	44
degF	**40**	28	33	39
	45	22	28	33
	50	17	22	28
	55	11	17	22
	60	6	11	17

Units = therms
Table is based on 100 cfm of outside air at an average temperature being heated for 1000 hours to final temperatures shown.
Formula used: 1.1 * Altitude Factor * cfm * dT.
    Q = Btu/hr
    cfm = cubic feet per minute air flow
    dT = differential temperature, degF
    Fa = altitude correction factor
For gas heating, divide output by efficiency to obtain input.
For table use, determine average outside temperature during the period of interest.

2. Cooling coil or coils (supplied by either direct expansion refrigeration or chilled water) cool and dehumidify the return air from the space to a delivery air temperature. They cool and dehumidify any outdoor air introduced into the air handling unit. Often these coils are designed to provide a discharge temperature setpoint between 50°F and 55°F.

3. Preheat coil or coils (supplied with steam or heating water or electricity) pre-heat outdoor air introduced into the air handling system. These are usually provided only when the amount of minimum outdoor air required to maintain indoor air quality, or to makeup for exhaust from the space, is a large percentage of the total air. They are also normally not provided in warmer climates (minimum winter temperatures greater than 30°F). These coils are usually designed to maintain a discharge temperature between 45°F and 55°F.

4. Reheat coil or coils (supplied with steam, or heating water or electricity) reheat the previously cooled and dehumidified air to maintain space temperature setpoints. In multiple zone systems, these coils are located farther down the duct system, near the zone.

5. Humidifiers (supplied with central steam or locally generated steam) add humidity to the air to maintain space humidity setpoints if required. These are normally provided only in climates with dry cold winter conditions or when spaces have unusual humidity requirements, such as healthcare areas, laboratories, clean rooms, and pharmaceutical areas.

6. Return or relief fan(s) to bring air back from the space and relief air to the outside. These fans are normally only used when return ductwork distribution is very long or when a "free-cooling" air-side economizer system is used.

7. Dampers to control the amounts of outdoor air, return air, and relief air introduced to the air handling unit and relieved from the space.

Terminal Control Units

In multizone single duct systems, each zone of control is normally provided with a terminal control unit. These terminal control units may actually be provided with their own fans to distribute air to the space (fan-powered), or they may use the central air handling unit's fan energy to push the supply air to the space as required.

The terminal units are sized and controlled to provide a designed airflow to meet the space's peak cooling load requirement or the space's peak ventilation rate requirement. At any time when that amount of air is not required, the terminal unit adjusts to meet the space's cooling needs or a minimum airflow amount, whichever is less.

If required, the terminal units are also frequently provided with reheat coils (normally provided with heating hot water or electricity) that reheat the cooler dehumidified air from the air handling unit to a higher delivery temperature when the space is too cool.

Reheat Coils and Control

Both single-zone and multizone air handling systems can provide simultaneous heating and cooling of the air by reheating the air temperature at each zone.

The local space temperature sensor for each zone controls the discharge temperature downstream of the reheat coil, providing excellent control of the zone space temperature. Constant air volume (CAV) or variable air volume (VAV) reheat systems are used to provide precise control of room temperature and/or humidity. However, these systems are energy intensive and costly to operate; thus they are always good energy retrofit candidates.

In a reheat system, energy is consumed to cool the supply air from return air or mixed air temperature; then additional energy is consumed at the terminal control device to reheat the air to a higher temperature. In VAV reheat systems, the reheat coil is not activated until the VAV controls have modulated the airflow to a minimum airflow setpoint, making them more energy efficient than CAV reheat systems.

CAV and VAV reheat systems are frequently used to condition spaces with extremely rigid requirements for humidity control, such as museums, printing plants, textile mills, and industrial process settings.
The energy professional has many opportunities in most facilities to create ECOs which can dramatically reduce reheat energy consumption.

10.5.4.4 Single Zone Systems

The single duct, single-zone system is the simplest of the all-air HVAC systems. It is one of the most energy-efficient systems, as well as one of the least expensive to install. It uses a minimum of distribution fan energy since equipment is typically located within or immediately adjacent to the area that it conditions. The system is directly controlled by a zone temperature sensor, which adjusts to provide cooling or heating as required to maintain space temperature and humidity requirements.

Single zone systems can provide either cooling or heating to a space or both. These systems can be configured to operate in a heating, cooling, fan only, or fan off type of control sequence, or in a reheat control sequence. Single zone systems may be provided with the ability to introduce outdoor air into the space. The outdoor air provided may only meet the minimum outdoor air requirement, or the system may be provided with the capability to provide the entire airflow capacity of the unit from outside air if ambient conditions are desirable.

Typical applications of single-zone systems are large spaces with fairly uniform loads, such as retail stores, public assembly spaces, exhibit halls, auditoriums, lecture halls, warehouse spaces, arenas, etc.

10.5.4.5 Multiple Zone Systems

The multiple zone HVAC system functions like the single zone system, with the exception that the temperature of individual zones is controlled by a zone temperature sensor which regulates the volume and temperature of air discharged into the space. This arrangement allows a high degree of local temperature control at a moderate cost. Installation costs and operating costs are slightly greater than with single-zone systems.

The distribution energy consumed is greater than a single-zone system due to the friction losses in VAV control devices, as well as the fact that the fan in the AHU must be regulated to balance the overall air volume requirements of the system. In most multiple zone systems, the air from the air handling unit is supplied at a constant temperature to the zones, usually about 55°F (13°C). In these cases dedicated heat recovery chillers should be examined

10.5.4.6 Multizone Systems

A multizone air distribution system is a specific type of HVAC system that is a variation of the single-duct CAV reheat system. In a multizone system, each zone is served by a dedicated supply duct which connects it directly to the air handling unit. In the most common type of multizone system, the AHU produces warm air at a temperature of 100°F (38°C) and cool air at about 55°F (13°C), which are blended with dampers to adjust the supply air temperature. The supply air temperature is set to a temperature required to satisfy the zone heating or cooling needs. (See Figure 10-8.)

Multizone systems are among the least energy ef-

ficient systems, sharing the inherent inefficiency of reheat systems since energy is consumed to simultaneously heat and cool air. Normally, since a constant volume of air is supplied to each zone, blended conditioned air must be supplied even when no heating or cooling is required.

Multizone systems also consume a great deal of energy in distribution, due to the large quantity of constant volume air required to meet space loads. Multizone systems were used much more frequently before the 1990s, but they are not frequently used in current design.

Retrofitting multizone systems is always a cost effective ECO.

10.5.4.7 Dual Duct Systems

Dual duct systems are similar to the multizone concept in that both cool supply air and warm supply air are produced by a central AHU. However, instead of blending the air in the fan room, separate warm-air ducts and cold-air ducts run parallel throughout the building, and air is mixed at dual-duct terminal units serving each zone. (See Figure 10-8.)

Dual duct systems require the greatest amount of space for distribution ductwork. In order to offset the spatial limitations imposed by this problem, dual duct systems often employ high velocity/high pressure supply ducts, which reduce the size and cost of the ductwork, as well as the required floor-to-floor height. This option increases the fan energy required for distribution. These systems were used extensively before the 1980s but are generally not used anymore, due to high first installation costs and high annual energy costs. The dual-duct system exhibits the greatest energy consumption of any all-air system.

Retrofitting dual duct systems is always a cost effective ECO.

10.5.4.8 Air and Water Systems and All-water Systems

Air and water systems and all-water systems are

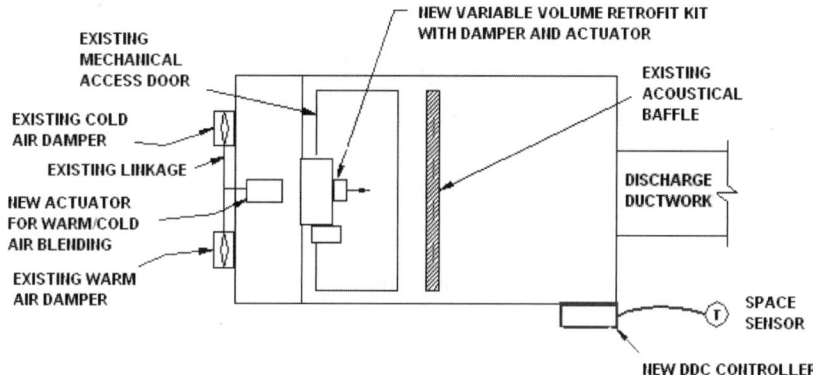

Figure 10-9. Constant Volume Dual Duct Terminal Unit Modified for VAV Operation

Editor's Note: Additional energy savings can be achieved by splitting the linkage between hot/cold duct dampers and supplying a separate actuator for each one.

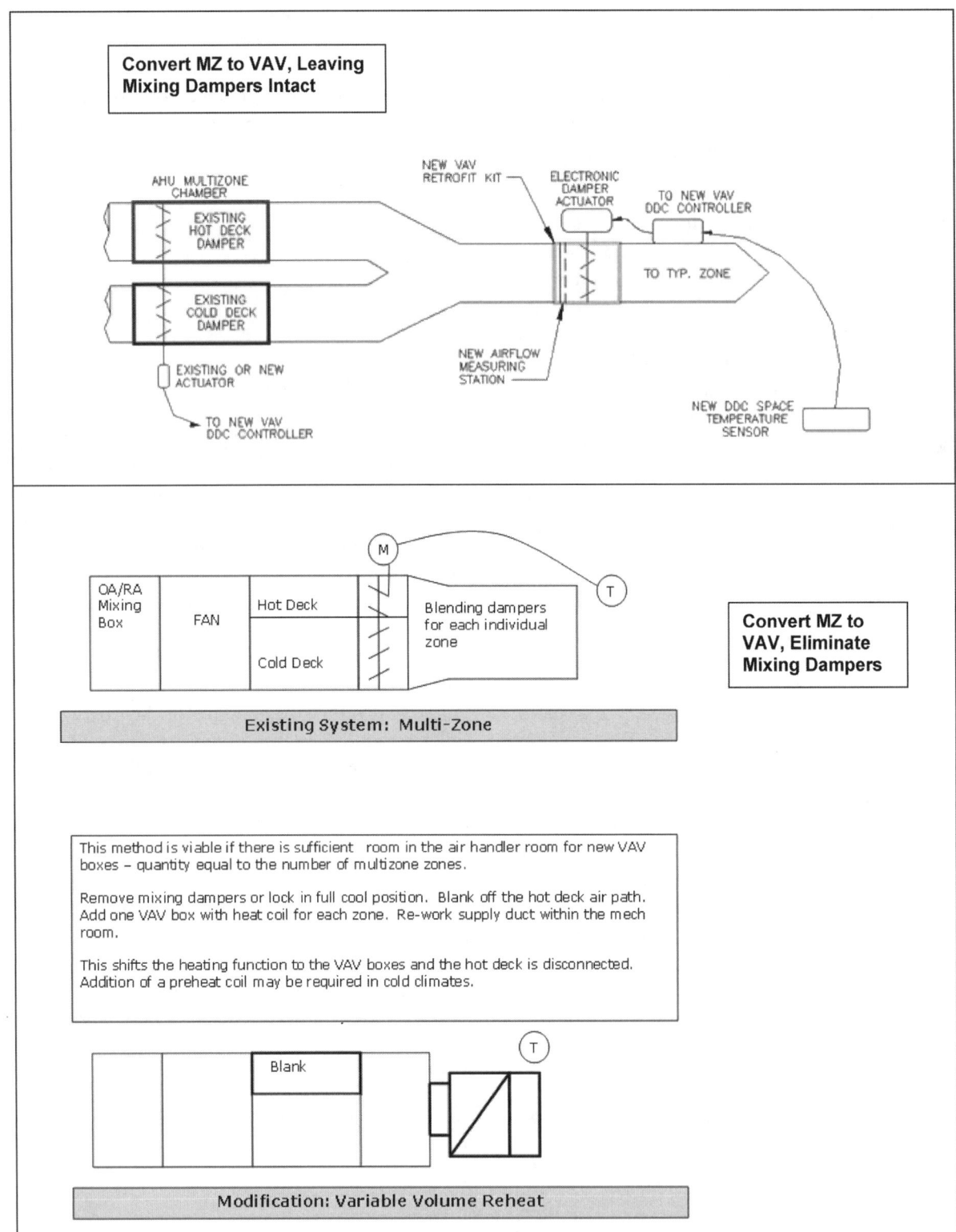

Figure 10-10. Multizone to Variable Air Volume Retrofit Options

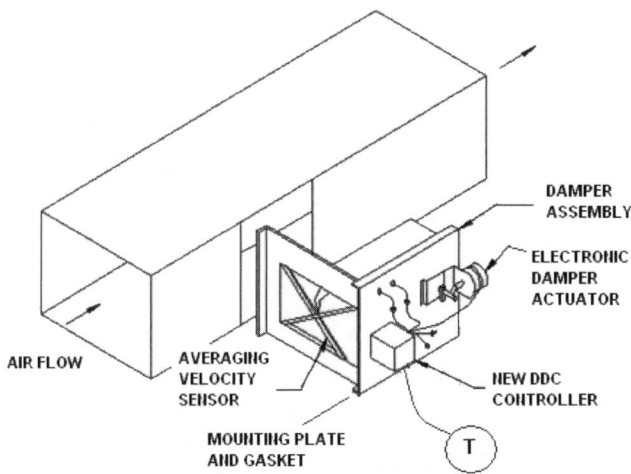

Figure 10-11. Insert-Type VAV Box for Modifying a Constant Volume System for Variable Volume Operation

HVAC systems that serve individual zones for heating and cooling and which are provided with chilled water, heating water, or condenser water from a central chilled water, heating water, or condenser water source. Diagrams of these system types are shown in Figures 10-12, 10-13, and 10-14. These systems include:

A. Water source heat pumps (with condenser water from a cooling tower or boiler loop, ground source or well water, or another heat source or heat sink)
B. Fan coil units (2 pipe and 4 pipe)
C. Unit ventilators
D. Chilled water radiant panels and chilled beams
E. Radiant heating panels
F. Finned tube radiation
G. Unit heaters

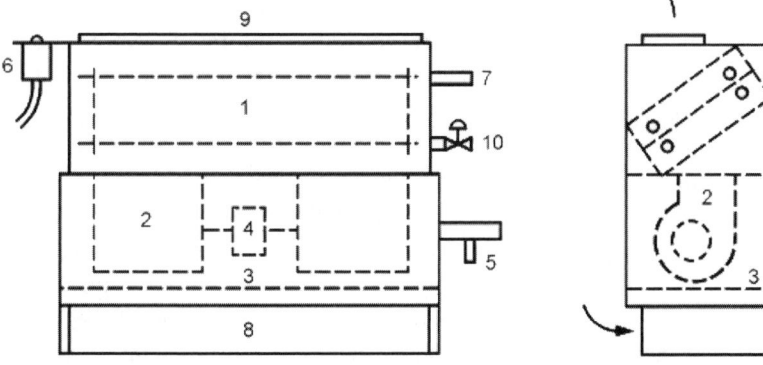

1. FINNED TUBE COILS
2. FAN SCROLLS
3. FILTER
4. FAN MOTOR
5. AUXILIARY CONDENSATE PAN

6. FAN SPEED CONTROL SWITCH
7. COIL CONNECTIONS
8. RETURN AIR OPENING
9. DISCHARGE AIR OPENING
10. WATER CONTROL VALVES

Figure 10-12. Fan Coil Unit

Source: ASHRAE Systems and Equipment Handbook, 2008, © American Society of Heating, Refrigerating and Air-Conditioning Engineers, Inc., www.ashrae.org.

These systems generally are not capable of heating, cooling, dehumidifying, or humidifying outdoor air. A separate source of fresh air is normally provided to the zones when these systems are used.

These systems are inherently more efficient than all-air systems, because the heating and cooling energy is transferred by water.

Air is not the most effective medium for transporting thermal energy. A cubic foot of air weighs only about 0.074 pounds (0.34 kg) at standard conditions (70°F; 1 atm.). With a specific heat of about 0.24 Btu/lb°F (0.14 joule/C), one cubic foot can carry less than 0.02 Btu per degree Fahrenheit temperature difference. By comparison, a cubic foot of water weighs 62.4 pounds and can carry 62.4 Btu/ft^3.

Water can be used for transporting thermal energy in both heating and cooling systems. It can be heated in a boiler to a temperature of 140 to 250°F (60-120°C) or cooled by a chiller to 40 to 50°F (4-10°C), and it can then be pumped and piped throughout the building to terminal devices which take in or extract heat energy.

Steam can also be used to transport heat energy. Steam provides most of its energy by releasing the latent heat of vaporization (about 1000 Btu/lb or 2.3 joules/kg). Thus, one mass unit of steam provides as much heating as fifty units of water that undergo a 20°F (11°C) temperature change.

Air and water and all-water distribution systems provide flexible zoning for comfort heating and cooling and they have a relatively low installed cost when compared to all-air systems. The minimal space required for distribution piping makes them an excellent choice for retrofit installation in existing buildings or buildings with significant spatial constraints. The disadvantage of all-water systems is that a separate OA ventilation system must be supplied, little or no control over air quality or humidity is provided, and normally only sensible or radiant heating and cooling for the space is provided.

Water distribution piping systems are usually described in terms of the number of pipes which are attached to each terminal device:

Two-pipe systems provide a supply pipe and a return pipe to each terminal unit, connected in parallel so that each unit (zone) can draw as much water as needed. Efficiency and thermal control are both high, but the system cannot provide heating in one zone while cooling another.

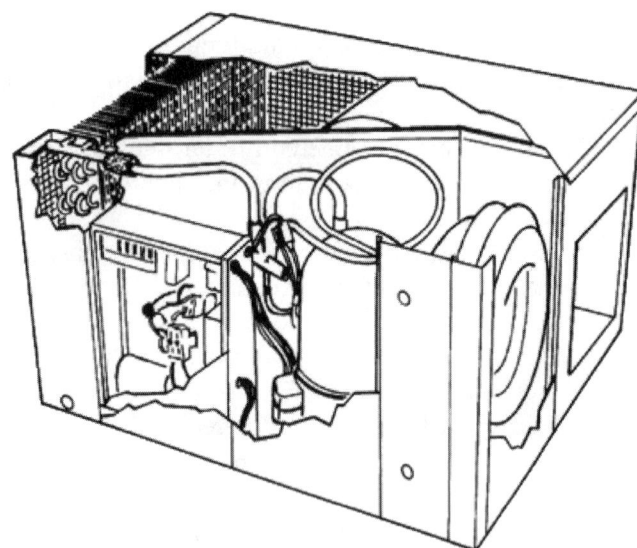

Figure 10-13. Typical Horizontal Water-source Heat Pump
Source: ASHRAE Systems and Equipment Handbook, 2008, © American Society of Heating, Refrigerating and Air-Conditioning Engineers, Inc., www.ashrae.org.

Three-pipe systems employ separate supply pipes for heating and cooling but provide only a single, common return pipe. Mixing the returned hot water, at perhaps 140°F (60°C), with the chilled water return, at 55°F (13°C), is highly inefficient and wastes energy required to reheat or re-cool the water.

Four-pipe systems provide a supply and return pipe for both heating water and chilled water, allowing simultaneous heating and cooling, along with relatively high efficiency and excellent thermal control. Control measures are necessary to prevent simultaneous heating and cooling in these systems which, when it occurs, will lower the efficiency of the system.

10.5.4.9 Radiant Heating

Radiant energy is undoubtedly the oldest method of centrally heating buildings. Recalling that the most important determinant of thermal comfort when environmental conditions are too cool is the radiant temperature of the physical surroundings, radiant heating systems are among the most economical.

The efficiency of radiant heating is a function primarily of the temperature, area and emissivity of the heat source, and the distance between the radiant source and the observer. It is therefore essential that radiant heat sources be located so that they are not obstructed by other objects. Emissivity is an object's ability to absorb and emit thermal radiation and is primarily related to color. Dark objects absorb and emit radiation better than light colored objects.

Because they are not dependent upon maintaining a

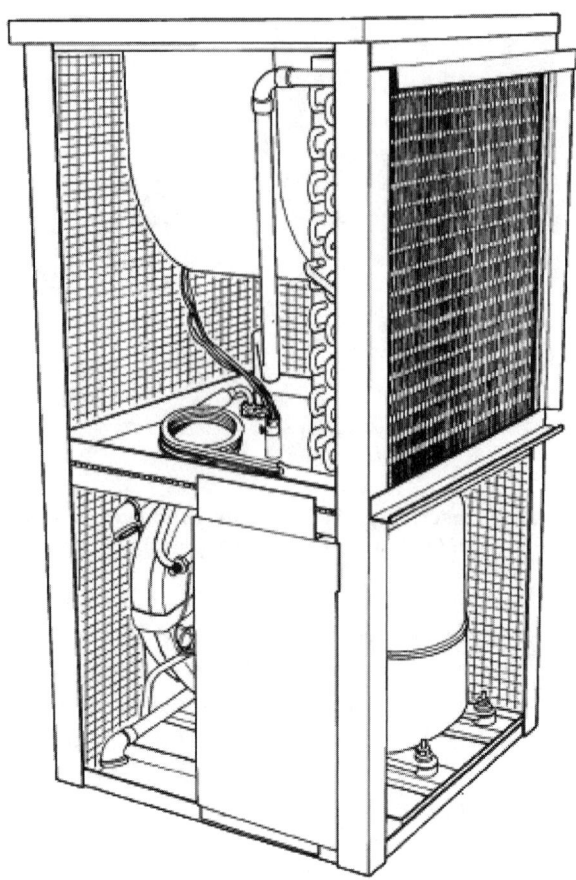

Figure 10-14. Typical Vertical Water-source Heat Pump
Source: ASHRAE Systems and Equipment Handbook, 2008, © American Society of Heating, Refrigerating and Air-Conditioning Engineers, Inc., www.ashrae.org.

static room air temperature, radiant heating systems provide excellent thermal comfort and efficiency in spaces subject to large influxes of outside air, such as factories and warehouses. However they are slow to respond to sudden changes in thermal requirements. Another drawback to radiant systems is that they promote the stratification of room air, concentrating warm air near the ceiling.

Radiant Heating By Natural Convection

The simplest all-water system is a system of hydronic (hot-water) convectors or electric convectors. In this system hot water from a boiler or steam/hot water converter is circulated through a finned tube, usually mounted horizontally behind a simple metal cover which provides an air inlet opening below the tube and an outlet above. Room air is drawn through the convector by natural convection, where it is warmed in passing over the hot finned tube.

A variation on the horizontal finned-tube hydronic convector is the cabinet convector, which occupies less perimeter space. A cabinet convector would have several finned tubes in order to transfer additional heat to the air passing through it. When this is still insufficient, a small electric fan can be added, converting the convector to a unit heater. An electric resistance element can be used in place of the finned tube.

Hydronic convectors are among the least expensive heating systems to operate as well as to install. Their use is limited, however, to heating only, and they do not provide ventilation, air filtration, or humidity control.

Hydronic convectors and unit heaters may be used alone in buildings where cooling and mechanical ventilation is not required, or to provide heating of perimeter spaces in combination with an all-air cooling system. They are the most suitable type of system for providing heat to control condensation on large expanses of glass on exterior wall systems.

10.5.4.10 Fan-coil Systems

A fan-coil terminal is essentially a small air-handling unit serving a single space without a ducted distribution system; the main difference other than size is that fan coils generally do not have outside air and exhaust provisions. One or more independent terminals are typically located in each room connected to a supply of hot and/or chilled water. At each terminal, a fan in the unit draws room air through a filter and blows it across a coil of hot or chilled water and back into the room. Condensate which forms on the cooling coil must be collected in a drip pan and removed by a drain.

Although most fan-coil units are located beneath windows on exterior walls, they may also be mounted horizontally at the ceiling, particularly for installations where cooling is the primary concern.

Technically, a fan-coil unit with an outside air inlet is called a unit ventilator. Unit ventilators provide the capability of using cool outside air during cold weather to provide free cooling when internal loads exceed the heat lost through the building envelope.

Fan-coil units and unit ventilators are directly controlled by local thermostats, often located within the unit, making this system one of the most energy efficient. These systems lack humidity control, and all maintenance must occur within the occupied space.

Fan-coil units are typically used in buildings that have many zones located primarily along exterior walls, such as schools, hotels, apartments and office buildings. They are also an excellent choice for retrofitting air-conditioning into buildings with low floor-to-floor heights. Although a four-pipe fan-coil system can be used for a thermally heavy building with high internal loads, it suffers the drawback that the cooling of interior zones in warm weather must be carried out through active air-conditioning since there is no supply of fresh (cool) outside air to

provide free cooling separately. They are also utilized to control the space temperature in laboratories where constant temperature make-up air is supplied to all spaces.

10.5.4.11 Closed-loop Heat Pump Systems

Closed-loop heat pump systems offer an efficient option for heating and cooling large buildings. Each room or zone contains a water-source heat pump that can provide heating or cooling, along with air filtration and the dehumidification associated with all air systems.

There are two basic types of closed loop heat pump systems. The conventional type of system, discussed here, has central heating and cooling equipment to add or remove heat from the loop and stabilize temperature (usually a hot water boiler and a cooling tower). A variation of this system couples the loop to an earth or water body static heat exchanger that alternates by season to be a heat source and heat sink. One such system is the ground source heat pump that is coupled to a water-to-earth heat exchanger. These systems are especially energy efficient since they do not require the auxiliary heating and cooling equipment energy input to maintain loop temperature. (See Chapter 28.)

The efficiency of the conventional closed loop heat pump is excellent during "swing" weather when some zones call for heat while others call for some cooling. When equally balanced, there is very little cooling or heating energy input into the system and the heat pumps literally move heat throughout the building to where it is needed. During peak cooling season, energy efficiency is good since the heat rejection is via evaporative cooling and requires no supplemental mechanical cooling to sustain the loop. Energy use for these systems is highest during heating season, when the auxiliary boiler must run to sustain the loop *and* the heat pumps run to transfer the heat across the heat exchanger boundary.

The water source for all of the heat pumps in the building circulates in a closed piping loop, connected to a cooling tower for summer cooling and a boiler for winter heating, or to a remote heat source or sink as discussed earlier. Control valves allow the water to bypass either or both of these elements when they are not needed. The primary energy benefit of closed-loop heat pumps is that heat removed from overheated interior spaces is used to provide heat for under-heated perimeter spaces during cold weather.

Since the closed-loop heat pump system is an all-water, piped system, distribution energy is low, and since direct, local control is used in each zone, control energy is also minimized, making this system one of the most efficient.

Heat pump systems have the disadvantage of having many compressors distributed throughout the building to maintain, often located in occupied spaces. Careful economic analysis is necessary to be sure that the energy savings will be great enough to offset the added maintenance costs. Closed-loop heat pumps are most applicable to buildings which exhibit a wide variety of cooling requirements, along with simultaneous heating requirements in perimeter zones and large internal loads.

10.5.4.12 Unit Heaters

Packaged heating-only unit heaters often utilize electricity or natural gas as their primary source of energy, but they can also utilize heating hot water or steam from a central plant.

Unit heaters are an effective source of heating, with minimal fan energy used to distribute the heating energy.

10.5.4.13 Infrared Radiant Heating

High temperature infrared radiant heaters utilize a gas flame to produce a high-temperature (over 500°F, 260°C) source of radiant energy. The heat is distributed through tubes throughout the space. Although they do not respond rapidly to changes in heating requirements, they are very effectively used in areas with significant intrusions of cold outside air (aircraft hangars, garages, warehouses). The radiant heating warms room surfaces and physical objects in the space, and thermal comfort returns within minutes of an influx of cold air. These systems have an added advantage of being able to maintain thermal comfort at lower space temperatures, making them very efficient.

Conversion to infrared radiant heating in the types of spaces listed above is always an attractive ECO.

10.6 CENTRAL COOLING EQUIPMENT, HEAT REJECTION EQUIPMENT, AND DISTRIBUTION

The most common process for producing cooling for buildings is the vapor-compression refrigeration cycle, which essentially moves heat from a controlled environment to a warmer, uncontrolled environment through the evaporation of a refrigerant driven through the refrigeration cycle by a compressor.

Vapor compression refrigeration machines are typically classified according to the method of operation of the compressor. Small air-to-air units most commonly employ a reciprocating or scroll compressor, combined with an air-cooled condenser. This is used in conjunction with a direct-expansion (DX) evaporator coil placed within the air-handling unit.

Cooling systems for large non-residential buildings typically employ chilled water as the medium which transfers heat from occupied spaces to the outdoors through the use of chillers and cooling towers or other heat rejection equipment.

10.6.1 Mechanically Driven Chillers

The most common types of water chillers for large buildings are rotary screw or centrifugal chillers that employ either a screw chiller configuration or a centrifugal compressor to compress the refrigerant. The refrigerant circuit extracts heat from a closed loop of water which is pumped through coils in air-handling or terminal to cool the building. Heat is then rejected from the refrigerant to a condenser. The condenser may reject the heat into the atmosphere directly (air-cooled) or indirectly into a secondary condenser water loop (water-cooled) that is ultimately rejected to the environment by a cooling tower, spray cooler (fluid cooler), evaporative condenser, or dry cooler.

The operating fluid used in older chillers may be either a CFC or HCFC type refrigerant. Some older centrifugal chillers may use CFC-11 or HCFC-12 refrigerants, the manufacture and use of which has been eliminated under the terms of the Montreal Protocol. New refrigerants HCFC-123 and HCFC-134a are being used to replace the CFC refrigerants, but refrigerant modifications to existing equipment will often reduce the overall capacity of this equipment by 15 to 25 percent.

Centrifugal chillers are normally driven by open or hermetic electric motors. Centrifugal chillers are available as low as 80 tons of capacity and rotary screw machines as low as 75 tons of capacity. These machines, when used in a water cooled mode with a fluid cooler or cooling tower, can be as efficient as 0.6 kW/ton of cooling produced, or a COP of approximately 6.0. If chillers have reached the end of their useful life, an ECO should be considered to add newer, more efficient machines.

Larger chillers over 250 tons in capacity can be provided with multiple compressors and variable frequency control of chiller speed. These controls, in coordination with the variable speed control of the condenser water pumps and chilled water pumps, can result in a chiller capable of operating in a 0.4-0.5 kW/ton of cooling production range. The variable speed compressor technology provides superior part load performance. Constant speed centrifugal chillers normally experience progressively worse efficiency, beginning below 50% load.

Small water chillers, up to about 200 tons of capacity, may utilize reciprocating or screw compressors and may be air-cooled using an air cooled condenser instead of using cooling towers. An air-cooled chiller uses a single or multiple compressors to operate a DX liquid cooler. Air-cooled chillers are widely used in smaller commercial and large-scale residential buildings. Air-cooled equipment is much less efficient than water-cooled equipment. An air-cooled machine has a normal operating efficiency of approximately 1.3 kW/ton of cooling produced, or a COP of approximately 3.0 at summer outside design conditions. Efficiencies are somewhat better in cooler weather.

10.6.2 Absorption Chillers

An alternative to vapor-compression refrigeration is absorption refrigeration, which uses heat energy to drive a refrigerant cycle, extracting heat from a controlled environment and rejecting it to the environment. Absorption chillers are rarely used anymore, due to their poor COP and significant use of water for cooling tower evaporation. They also have very high maintenance requirements. Single effect absorbers have a COP of only about 0.6, and double effect absorbers have a COP of approximately 1.0, much worse than any mechanically driven chiller that would have COP values of anywhere from 3 to as high as 7.0.

Absorption chillers are only effectively used when there is a waste heat stream from some other process which would normally be rejected but can be used to power the absorption chiller. Any existing absorption cooling system that operates using a "new energy" heat source is a strong candidate for a replacement ECO. It is generally always an attractive ECO to replace the absorber with a highly efficient electric centrifugal machine, unless it utilizing waste heat.

Anotherl benefit to energy savings when implementing an ECO to replace an absorption chiller with a mechanically driven chiller is the savings in water consumption required to evaporate waste heat (and the associated costs of water and sewer charges). A single effect absorption chiller rejects approximately 30,000 Btuh of heat per ton of cooling produced, and a double effect absorption machine rejects approximately 24,000 Btuh per ton of cooling produced. Correspondingly, a very inefficient water cooled chiller would only reject about 15,400 Btuh per ton of cooling produced, and a highly efficient water-cooled chiller would only reject about 13,500 Btuh per ton of cooling produced. This means as much as 40%-55% of the water required for evaporation in the cooling towers could be saved by replacing the absorption chillers with electric rotary screw or centrifugal chillers.

In locations with high water and sewer charges, experience has shown the annual dollars saved from avoided water consumption can be as great as the annual dollars saved in energy consumption.

10.6.3   Chiller Performance

Most chillers are selected to cool for peak cooling load requirements and then operate at loads less than peak most of the time. Many chiller manufacturers provide data that identifies a chiller's part-load performance as an aid to evaluating energy costs. Ideally a chiller operates at a desired temperature difference (typically 45-55°F; 25-30°C) at a given flow rate to meet a given load. As the load requirement increases or decreases, the chiller will load or unload to meet the need. A reset schedule that allows the chilled water temperature to be adjusted (to meet thermal building loads based on enthalpy) provides an ideal method of reducing energy consumption.

10.6.4   Thermal Storage

Thermal storage can be an effective way of minimizing electrical demand by using stored chilled water or ice to offset peak loads during the peak demand time, and it can often be an effective ECO, depending upon utility rate structure. A detailed knowledge of the building's cooling load profile is essential in determining the effectiveness of thermal storage. (See Chapter 19 for a discussion of thermal storage systems.)

10.6.5   Cooling Towers/Spray Coolers/
Evaporative Condensers

Cooling towers/spray coolers (fluid coolers)/evaporative condensers use atmospheric air to cool the water from a condenser or coil through evaporation.

The use of variable-speed control of cooling tower or spray cooler fan motors is an effective way to optimize the control of the cooling tower in order to reduce fan power consumption. As the required cooling capacity increases or decreases, the fans can be sequenced to maintain the approach temperature difference. For most air-conditioning systems, this approach value usually varies between 8 and 14°F.

When operated in the winter, the quantity of air must be carefully controlled so the water spray is not allowed to freeze. In cold climates it may be necessary to provide a heating element within the tower to prevent freeze-ups. Alternately, some systems utilize an indoor sump, eliminating the freeze concern.

It is generally a very effective ECO to convert constant speed cooling tower fan motors to variable speed operation.

10.6.6   Water-side Economizer

A method of providing "free-cooling" to building systems where no mechanical cooling will be needed is to use the evaporatively cooled cooling tower water to cool supply air or to cool chilled water. This method is referred to as a water-side economizer. The most common and effective way of interconnecting the cooling tower water to the chilled water loop is through the use of a plate-and-frame heat exchanger which transfers heat from the condenser water system to the chilled water system. This method isolates the cooling tower water from the chilled water circuit, maintaining the integrity of the closed chilled water loop. (See Figures 10-16 and 10-18.) Another method is to use a separate circuit and pump that allows cooling tower water to be circulated through a coil located within an air-handling unit.

The water-side economizer can be easily implemented with the plate and frame heat exchanger in series or in parallel with the normal chiller(s). In the case of series operation, the plate and frame heat exchanger can actually pre-cool the return chilled water to some amount, even if not to the total amount required to distribute to the chilled water coils. In this manner, the load on the chillers being used is reduced. The figures referenced indicate the plate and frame heat exchanger in a parallel configuration.

The implementation of a water-side economizer as an ECO is almost always a very cost effective high ROI project if a building already has chilled water and condenser water systems in place with their associated chilled water and condenser water pumping systems.

Key considerations for the energy professional in evaluating use of a water-side economizer are climate (mean coincident wet bulb temperature—great for San Diego and Colorado, poor for Houston and Miami), water costs, electrical costs, and demand charges.

In general, air or water economizers provide benefit when a need for cooling coincides with suitably cool/dry outdoor air conditions, i.e. if it is cool/dry outdoors when cooling is needed indoors, the economizer can be used instead of mechanical cooling. Buildings with high internal heat gains may require cooling year round and are excellent candidates for air or water economizers. However, in some buildings, such as those with very small internal heat gains, the loads must be studied to verify the economizer benefit.

10.6.7 Water Treatment

A good water treatment program is essential to the maintenance of efficient chilled water and condenser water systems. Proper water treatment is also integral to efficient operation of water cooled equipment by keeping heat transfer surfaces clean. If allowed to foul, heat exchanger approach temperatures will increase with a corresponding energy penalty. Evidence of fouling in

the cooling and condenser water systems will be lower suction temperatures and higher condensing temperatures. In all cases, energy use will increase from fouling, and an effective water treatment program is the way to maintain effective heat transfer. Periodic cleaning of water-side heat exchangers and chiller tubes is recommended for sustained efficient operation.

The energy professional can quickly ascertain the efficiency of the chilled water system by measuring the actual approach temperatures and comparing them to values recommended by the chiller manufacturer. This may result in an ECO related to cleaning of the existing systems.

10.6.8 Chilled Water and Condenser Water Pumping Systems

Significant electrical energy is used in chilled water and condenser water distribution systems. Many existing systems are often constant volume pumping systems. Where constant volume pumping is used, the proportion of the total chiller plant load used by the auxiliary equipment increases at part load, thus having a parasitic effect on the overall plant efficiency. Depending upon electric energy rates and the size of existing pumping systems, converting constant volume chilled water and condenser water pumping systems to variable volume operation is almost always a cost effective ECO. (See Figures 10-15 through 10-18.)

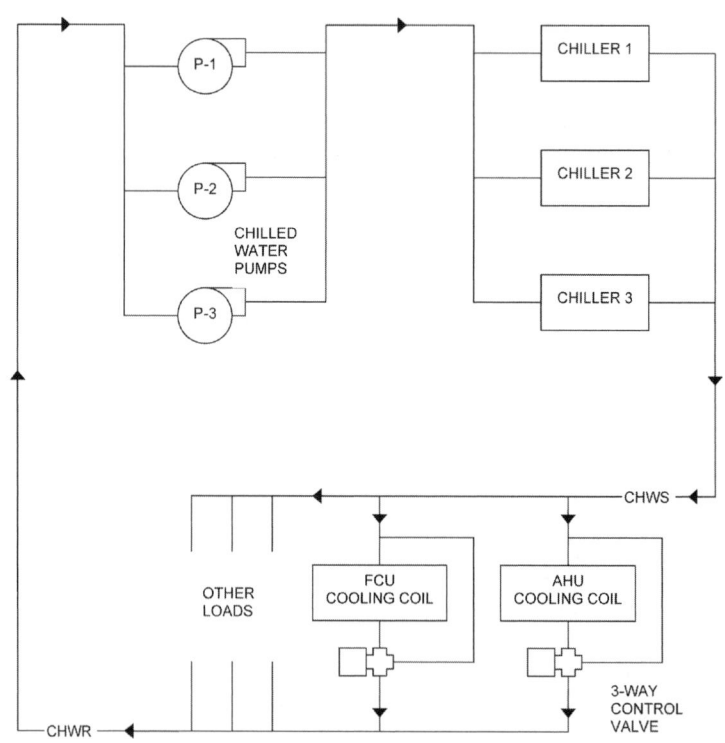

Figure 10-15. Chiller Plant Chilled Water Plant Flow Diagram, Pre-retrofit

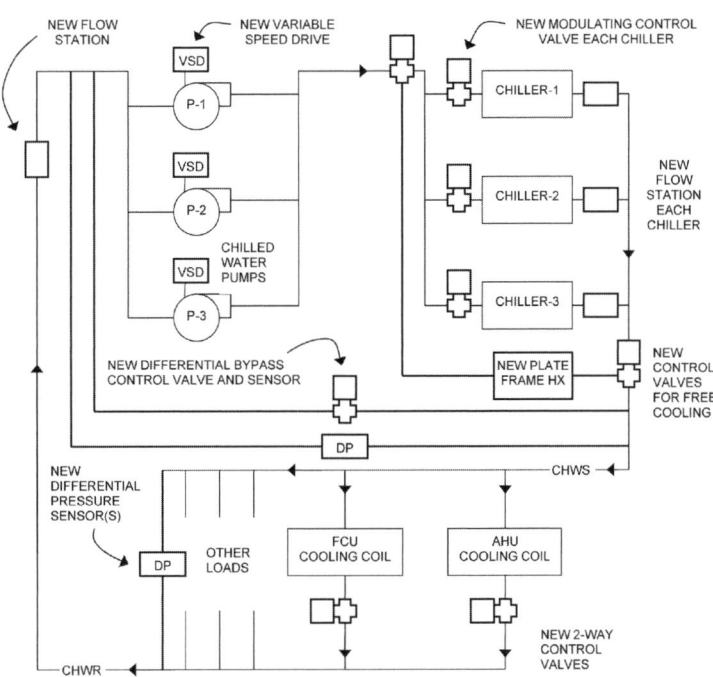

Figure 10-16. Chiller Plant Chilled Water Plant Flow Diagram, Post retrofit

The energy professional can accomplish these modifications by implementing the following modifications:

a. Replace existing control valves with 2-way control valves at AHUs and other chilled water distribution.

b. Add variable frequency controllers to control chilled water pump and condenser water pump motors.

c. If the system is a primary chilled water system, a water-flow measuring station will need to be provided along with a bypass control valve to maintain minimum chilled water flow through the chiller.

d. Similarly, a water flow measuring station will need to be provided on the condenser water system to assist the variable frequency controller in maintaining minimum condenser water flow and head pressure.

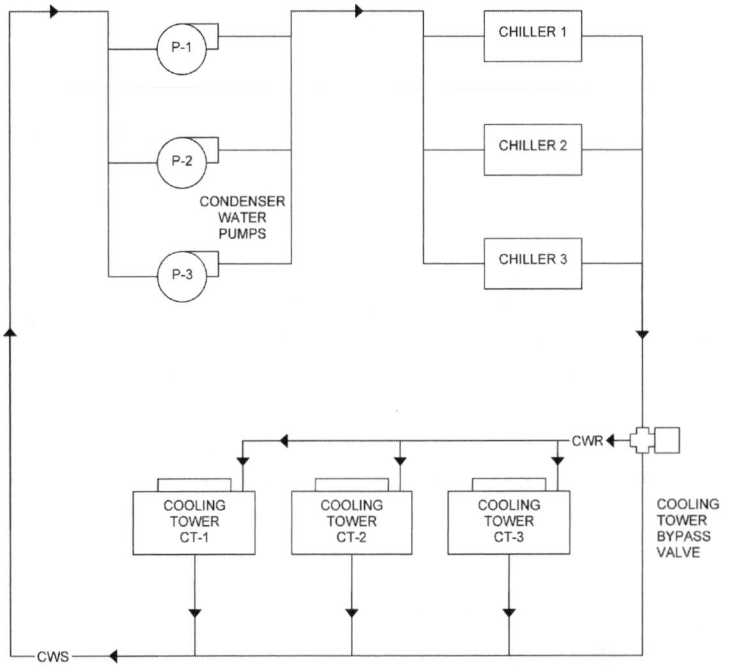

Figure 10-17. Chiller Plant Condenser Water Flow Diagram, Pre-retrofit

10.7 IMPACT OF PART LOAD OPERATION AND OCCUPANCY OF THE BUILDING

All equipment operates at its optimum efficiency when operated at or near its design full-load condition. Both overloading and under-loading of equipment reduces equipment efficiency.

This fact has its greatest impact on HVAC system efficiency when large systems are designed to air condition an entire building or a large segment of a major complex. Since air-conditioning loads vary and the design heating and cooling loads occur only under the most severe

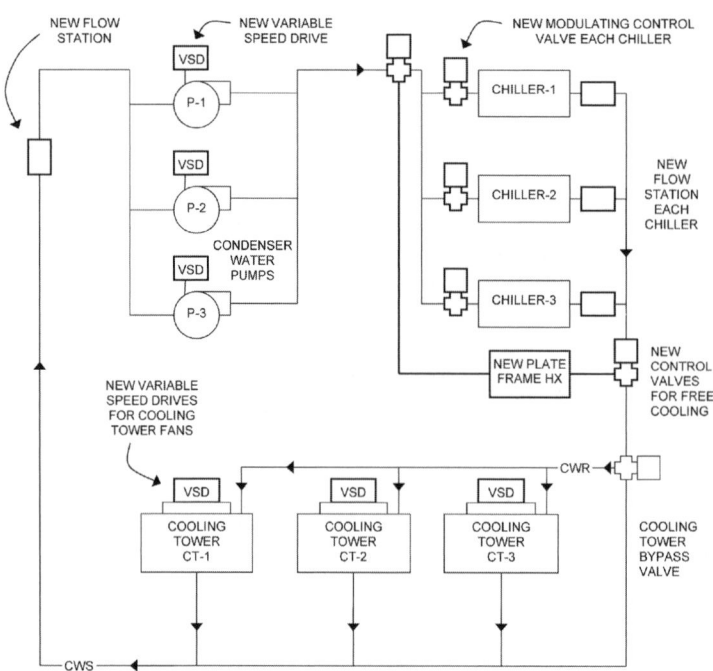

Figure 10-18. Chiller Plant Condenser Water Flow Diagram, Post-retrofit

weather or occupancy conditions, the majority of the time the HVAC system must operate at less than full capacity.

When selected parts of a building are utilized for off-hours operation, this may require that the entire building be conditioned or that the system operate far from its optimum conditions and thus at far less than its optimum efficiency.

The energy professional should evaluate ECOs in those cases which would provide individual heating or cooling equipment for areas of the building with longer or full-time occupancy hours. This will limit use of larger equipment to times when it is operating closer to its full capacity and best efficiency. These type of ECOs have an excellent to moderate ROI, depending on physical constraints for installation and services available.

10.8 HVAC SYSTEM ELECTRICAL DISTRIBUTION ENERGY

Depending upon the types of HVAC systems used, the distribution energy associated with fan systems and pumping systems can often amount to as much as 40-50% of the building's electrical energy usage.

Distribution energy is the electrical energy consumed to operate fans and pumps, with fan energy typically being far greater than pump energy. The performance of similar fans is related by three fan laws that relate fan power,

airflow, pressure and efficiency to fan size, speed and air density. The reader is referred to the ASHRAE Handbook: HVAC Systems and Equipment for additional information on fans and the application of fan laws.

Fan power consumption is a function of the quantity of airflow moved by the fan, the amount of pressure required to be produced by the fan (to overcome the losses through the ductwork distribution system that have to be overcome), and the type of fan and drive system used (which determines the efficiency of the fan mechanical performance). Total power consumption is also effected by the efficiency of the motor driving the fan.

$$\text{FAN Hp} = \frac{\text{CFM} * \text{TSP}}{(6356 * \text{fan eff} * \text{drive eff} * \text{motor eff})}$$

eq. 10.4

Where:

Hp = Horsepower
CFM = Cubic feet per minute
TSP = Total static pressure, in. w.c.

Fan Energy Consumption

(kWh)= FAN Hp *0.746* Hours of Operation   eq. 10.5

Pump power consumption is a function of the quantity of water used, the amount of pressure required to be produced by the pump to overcome system losses, and

Table 10-3. Humidifier Energy User per 100 lbs of Moisture

Source: Commercial Energy Auditing Reference Handbook

Medium	Technology	Energy Consumption	Remarks
Steam	Infrared Lamps	30+ kWh	
	Electric Resistance	30 kWh	
	Gas-Fired Steam	1.25 therms at 80%e	
Adiabatic	Compressed air atomization	3 kWh 12-15 cfm	Note 1
	Ultrasonic	2.5 kWh	Note 1
	Evaporative Pads	---	Note 1

Table Note 1: Adiabatic cooling can either help (if cooling is needed) or will require 1000 Btu of auxiliary heating per lb of evaporated moisture to compensate for the cooling effect.

the type of pump selected (which determines the efficiency of the pump mechanical performance). Total power consumption is also affected similarly by the efficiency of the motor driving the pump.

$$\frac{GPM * HEAD}{(3960 * pump\ eff * drive\ eff * motor\ eff)} \quad \text{eq. 10.6}$$

Where:

Hp = Horsepower

GPM = Gallons per minute

HEAD = Flow resistance, ft. head

Pump Energy Consumption

(kWh)= PUMP Hp *0.746* Hours of Operation

eq. 10.7

Fan and pumping horsepower are proportional to the cube of airflow; reducing airflow or water flow to 75 percent of existing airflow or water-flow will result in a reduction in the fan horsepower by the cube of 75 percent, or about 42 percent.

Therefore, the energy professional can easily see that converting systems to variable flow and reducing airflow or water-flow quantities during any periods when they are not required (unoccupied periods), as discussed throughout this chapter, will have significant impact on energy savings. Similarly, positively affecting any of the other factors in these equations can all be considered as reliable ECOs with moderate payback:

1. Reduce operating hours.

2. Replace motors with higher efficiency motors.

3. Occasionally replace pumps or fans with better selections having higher mechanical efficiencies. (This is especially valuable if the equipment is at the end of its useful life.)

10.9 HUMIDIFICATION SYSTEMS

Some HVAC systems include humidification equipment. In very dry climates these systems may be required to maintain comfort, although humidification is more commonly employed when processes inside are sensitive to humidity changes (e.g. museums, or laboratories, or manufacturing) or where low humidity needs to be avoided (e.g. data centers). Humidification is a focus of energy management because it can be a significant source of energy use.

Each pound of water requires roughly 1000 Btu of energy to evaporate; therefore, an immediate focus when evaluating humidity cost of operation is the "lbs. per hour" humidification rate.

Humidification Formula for Adding Moisture to Air

$$\text{Humidifier Load (lbs/hr)} = 60* Cfm * PCF(air) *delta\text{-}M \quad \text{eq. 10.8}$$

Where:

PCF(air) = lbs/cubic foot air density

delta-M = (lbs moisture/lb dry air)

There are several different types of humidification systems, each with a signature energy use per pound of moisture put into the air stream. Types of systems include:

a. Compressed air atomization

b. Steam injection (This may come from central boiler steam or from a remote electric generator which produces steam.

c. Infrared or pan type humidifiers (usually used in data center equipment)

d. Evaporative pads

For example, compressed air atomization would

include the cost of the compressed air energy input, while simple evaporative pads would not. For winter humidification, each pound or water put into the air as humidification requires about 1000 Btus of heat. However, in some cases there is a simultaneous need for humidification and cooling; in these cases there can be significant energy advantage in the chosen type of humidifier. For data centers with a continuous need for both cooling and humidification, an adiabatic humidifier (pads or atomizer) will do both with very little energy compared to standard-issue infrared or pan heaters. A good way to estimate humidification energy consumption is to meter the make-up water to the humidifier and convert to pounds.

10.10   EXAMPLE HVAC ENERGY CONSERVATION OPPORTUNITIES (ECOs)

10.10.1 Specific Energy Conservation Opportunities for Decentralized HVAC Systems

Several specific energy conservation opportunities can be implemented to improve the energy consumption of these types of systems:

1.   Add local or remote programmable thermostatic and operational control of the systems, with the ability to program times of operation and program unoccupied period setback and setup space temperature setpoints. This ECO will almost always have an excellent ROI.

2.   If the equipment is in need of replacement, or if a water-side free cooling coil can be provided, install new equipment haviang "free-cooling" water-side economizer coils to minimize compressor operation. This ECO will have a much longer ROI, depending upon physical conditions.

3.   If physically feasible with some self-contained equipment, provide ductwork and controls to incorporate a "free-cooling" air-side economizer. This ECO will have a moderate to longer term ROI, depending on physical constraints, and may be unfeasible depending upon physical conditions.

4.   If the equipment is providing introduction of fresh air locally, provide operational control of fresh air with a separate fresh air damper that closes during unoccupied periods, based on an occupancy sensor signal, and can be controlled by a CO_2 sensor during occupied periods. The unoccupied period can be determined either as a programmed scheduled time

or, preferably, as sensed by an occupancy sensor. These controls should be provided as part of the energy conservation opportunity (ECO). This ECO will almost always have an excellent ROI.

5.   Add a water-side economizer "free-cooling" coil in the ductwork associated with self-contained units and a "free-cooling" water distribution system associated with a central "free-cooling" heat sink (cooling tower, ground source water system, lake or other water source). This ECO will have a longer ROI.

6.   Replace any older compressors or condensing units with the highest coefficient of performance (COP) or energy efficiency ratio (EER) equipment as an ECO. This ECO will have a moderate to long ROI, depending upon the condition of existing equipment.

10.10.2 Use of Dedicated Outside Air Systems (DOAS)

A more effective way to treat the minimum outdoor air introduced to the building is by use of a dedicated outdoor air system (DOAS), which consists of a dedicated AHU responsible to temper the incoming air (heat/cool/dehumidify). Often the control setting for such systems is "room neutral." DOAS equipment is normally provided with heat recovery devices to assist in reducing the energy required to cool, dehumidify, heat, and humidify outdoor air.

The DOAS may supply conditioned OA to each of the other AHU systems or directly to the zones in the facility. The ventilation OA may also then be varied to minimum levels through control by CO_2 sensors. Use of a DOAS may be a very cost effective ECO, due to its reduction in energy required to treat minimum OA requirements and the ability to better control the remaining AHUs to maximize energy performance.

When coupled with a VAV system, a DOAS provides further energy advantages. By separating the ventilation function from the heating and cooling function, it is possible to achieve better optimization of each. The main (VAV) system becomes a recirculating system only, and VAV terminal unit minimum settings can be set to zero to allow full closure without a loss of ventilation. In turn, the VAV system terminal units with reheat (perimeter) lose the "reheat penalty" inherent in all single duct VAV air systems. The reheat penalty exists when the terminal unit controls would tend to close the damper based on space temperature but must stay partially open since the supply air contains the ventilation air. This requires the reheat coil to activate to keep the space from getting too cold. This ECO usually has a moderate to high ROI, depending upon the amount of OA to be conditioned and the physical constraints of installing the DOAS. It is a particularly effective ECO in facilities where operating hours are continuous and OA quantities are high, such as healthcare facilities, correctional facilities, dormitories, nursing or skilled nursing facilities, etc.

10.10.3 Specific Energy Conservation Opportunities for Single-zone Systems

1. If not already provided, the unit should be converted to VAV operation and have new BAS sequences incorporated. This ECO will almost always have a very attractive ROI.
    A. New variable frequency controllers should be added to vary the speed of supply and return fan motors.
    B. The fan speed should be varied to meet space temperature requirements as sensed by the space temperature sensor.
    C. New airflow monitoring stations should be added in fan inlets to monitor volume of air. The minimum airflow setting and fan speeds will need to be carefully examined, based on the existing supply diffuser distribution. Excessively low air volume will cause poor space air distribution and drafts.
    D. Cooling coil discharge temperature should be reset upwards after the airflow reaches its minimum airflow setpoint, if the space is indicated to still be too cool.
    E. No heating should be provided until the AHU reaches its minimum airflow setting.

2. If not already provided, install a minimum outdoor air damper separate from the economizer damper control, with separate actuation and control. This ECO will almost always have a very attractive ROI.
    A. This new minimum outdoor air quantity should be measured with a newly installed airflow monitoring station. The new minimum outdoor damper should be scheduled to be open only during occupied periods of the spaces, whether or not the AHU is operating to cooling.
    B. Occupancy can be determined by time of day scheduling or occupancy sensors in the space. During occupied periods, the minimum outdoor air quantity should be varied to meet maximum CO_2 quantities. A CO_2 sensor should be installed in the return ductwork to control the damper. This will produce significant savings from unnecessary heating, cooling, dehumidification and humidification of outdoor air when not required.

3. If not already provided, and if physically feasible, an air-side "free-cooling" economizer or water-side "free-cooling" economizer should be installed to serve the AHU and provide free cooling when ambient conditions are acceptable. This ECO will normally have a moderate to longer term ROI, depending on physical conditions.

10.10.4 Specific Energy Conservation Opportunities for Multizone Systems

The same ECOs related to control of minimum OA and economizer control (air-side or water-side), with and the required zone airflow quantities, should be accomplished for multizone systems as discussed for earlier systems. In addition, the following opportunities should be investigated:

1. Convert the system to VAV multizone operation. (See Figures 10-8, 10-10, and 10-11.) This will always be an attractive ROI ECO. This conversion can be accomplished by adding a zone terminal unit in the ductwork downstream of each set of zone warm deck and cooling deck dampers. This terminal unit should contain an airflow monitoring station and a damper, damper operator, and controls. The new terminal unit should be programmed with new maximum and minimum zone airflow values. The terminal unit damper should modulate the airflow quantity to maintain space temperature until the zone is at minimum airflow. The multizone cooling damper should remain open until the zone is at minimum airflow, and the warm air damper should remain closed. Once heating is required, the zone dampers should modulate from the cooling air damper open to the warm air damper open to maintain zone temperature. This type of ECO will dramatically reduce cooling energy, heating energy, and fan energy requirements.

2. Warm air duct heating coil discharge temperature and cold air duct discharge temperature should be reset based on the zones needing the most heating or cooling. This will result in a cold air duct discharge temperature setup to 60°F or more and a heating duct temperature of 75°F or more, depending on what the zones need.

3. Add new variable frequency controllers to control supply and return fan motor speeds; add new static pressure controllers to control the variable frequency controllers.

10.10.5 Specific Energy Conservation Opportunities for Dual-duct Systems

The same ECOs related to control of minimum OA and economizer control (air-side or water-side), with the required zone airflow quantities, should be accomplished for dual duct systems as discussed for earlier systems. In addition, the following opportunities should be investigated:

1. Convert the system to VAV dual duct operation. (See Figures 10-8 and 10-9.) This ECO will always have an attractive ROI. This conversion can be accomplished by adding a VAV damper retrofit kit to each dual duct terminal unit in the ductwork downstream of each set of terminal unit warm deck and cooling deck dampers. This retrofit kit will replace the existing constant volume regulator. This terminal unit should contain an airflow monitoring station and a damper, damper operator, and controls. The new terminal unit should be programmed with new maximum and minimum zone airflow values. The terminal unit damper should modulate the airflow quantity to maintain space temperature until the zone is at minimum airflow. The terminal unit cooling damper should remain open until the zone is at minimum airflow, and the warm air damper should remain closed. Once heating is required, the zone dampers should modulate from the cooling air damper open to the warm air damper open to maintain zone temperature. This type of ECO will dramatically reduce cooling energy, heating energy, and fan energy requirements.

2. Warm air duct heating coil discharge temperature and cold air duct discharge temperature should be reset based on the zones needing the most heating or cooling. This will result in a cold air duct discharge temperature setup to 60°F or more and a heating duct temperature of 75°F or more, depending on what the zones need.

3. Add new variable frequency controllers to control supply and return fan motor speeds; add new static pressure controllers to control the variable frequency controllers.

10.10.6 Air-to-air Heat Recovery Systems

Air-to-air heat recovery systems can be used during the heating season to extract waste heat and humidity from exhaust air that is used to preheat cold fresh air from outside. In warm weather, they can be used to extract heat and humidity from warm fresh air and dispel it into the exhaust air stream from the building.

Their use is particularly appropriate to buildings with high outside air requirements, like healthcare facilities, laboratories, correctional facilities, dormitories, etc. However, they can be used for any building to condition OA. They can be provided as a separate unit to feed OA to other AHUs, like a DOAS or can be incorporated into existing AHU configurations. (See Figure 10-19 for different types of air-to-air heat recovery devices.)

These devices can often be added to existing systems as an ECO with a moderate to longer term ROI, depending upon the physical constraints of the location of AHUs and exhaust ductwork.

The energy professional should be careful to evaluate this ECO after other ECOs are implemented that reduce AHU operating hours and AHU/OA requirements with CO_2 demand control or other control methods. After these previous ECOs are implemented, the amount of OA required and the hours it is required may have been reduced dramatically, leaving much lower heating and cooling requirements for OA, therefore making this ECO have a poor ROI.

Air-side Economizers

One advantage of all-air HVAC systems is that they can utilize outside air to condition interior spaces when it is at an appropriate temperature and humidity. The use of outside air to actively cool spaces is referred to as an economizer, or economizer cycle.

When the outside air is at desirable enthalpy conditions, only distribution energy is required to provide cooling with outside air. The cooling coil and associated mechanical cooling do not have to be used. When the outside air temperature has a more attractive enthalpy than the return air, it requires less cooling energy to utilize 100 percent outdoor air for supply air than to condition recycled indoor air. An economizer cycle is simply a control sequence that adjusts outside air and exhaust dampers to utilize 100 percent outside air when conditions make it advantageous to do so.

Economizer cycles will provide the greatest benefit in climates having more than 2000 heating degree days per year since warmer climates will have few days cold enough to permit the use of outside air for cooling. Economizer cycles will not be very effective in climates with high coincident wet bulb conditions, even when the outdoor air temperatures are lower (e.g., Houston, Miami).

The simplest type of economizer utilizes a dry-bulb temperature control that activates the economizer at a predetermined outside dry-bulb temperature, usually around the normal supply air temperature, or about 55°F (13°C). Above 55°F (13°C), minimum outdoor air is supplied for ventilation. Below 55°F (13°C), the quantity of

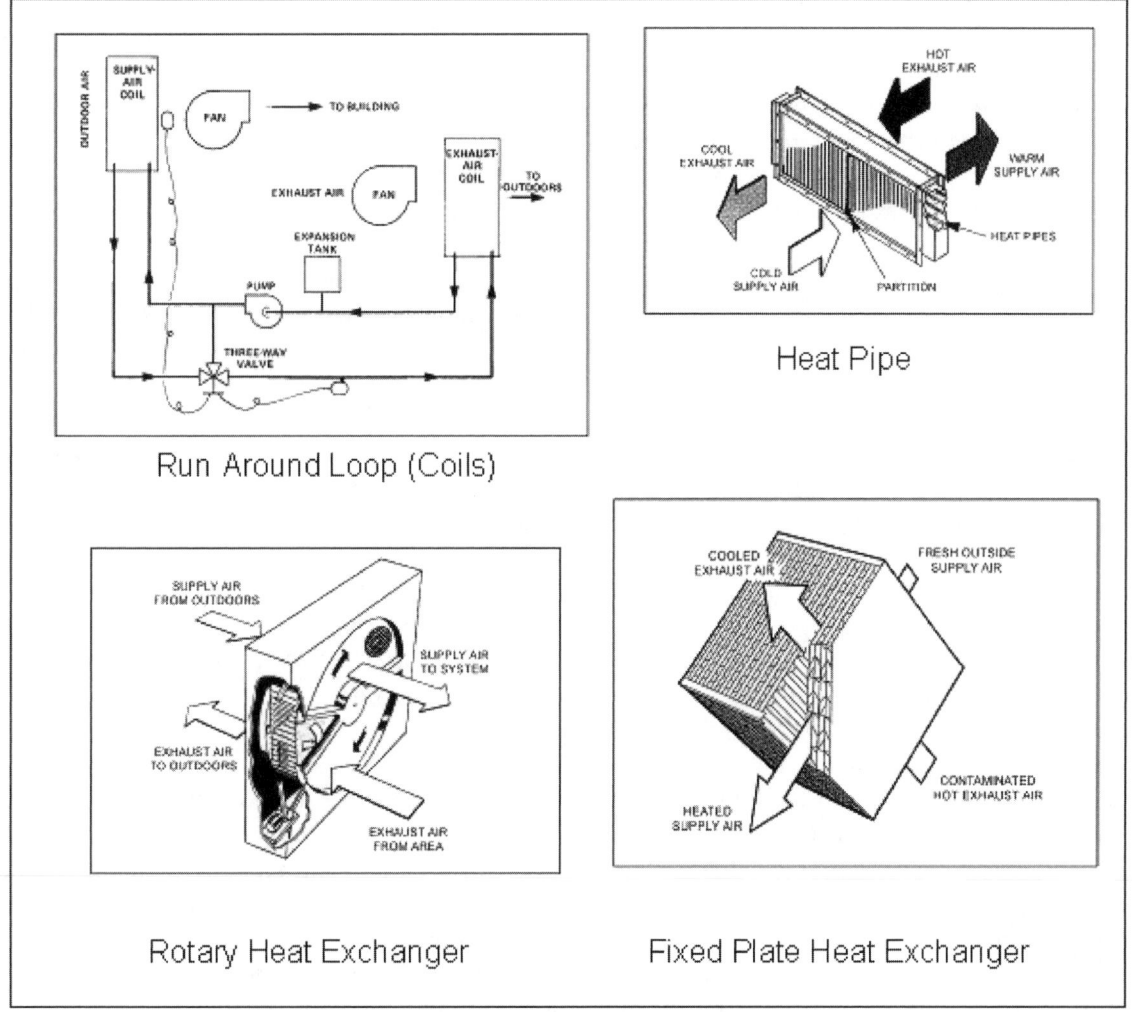

Figure 10-19. Energy Recovery Technologies

Source: ASHRAE Systems and Equipment Handbook, 2008, © American Society of Heating, Refrigerating and Air-Conditioning Engineers, Inc., www.ashrae.org.

outdoor air is gradually reduced from 100 percent and blended with return air to make 55°F (13°C) supply air.

Because it is possible to utilize outdoor air at temperatures above 55°F (13°C) to save cooling energy when wet bulb conditions are low, a modified dry-bulb temperature control can be used. This is identical to the simpler dry-bulb temperature control, except that when the outside temperature is between 55°F (13°C) and a pre-selected higher temperature based on the typical humidity, 100 percent outdoor air is used (but cooled to 55°F for supply air).

The third and most efficient type of control for an economizer cycle is enthalpy control. This control sequence determines and compares the amount of energy required to cool 100 percent outdoor air with that required to cool the normal blend of return air. It then selects the source which requires the least energy for cooling.

Air-handling units which lack adequate provision for 100 percent outside air can utilize an alternative "water-side" economizer discussed earlier in this chapter.

Building Automation System ECOs

1. *Operating Schedule Modifications*
    One of the greatest causes of excess energy consumption in most buildings is unnecessary cooling or heating of spaces or air quantities during unoccupied periods. Turning off equipment or reducing airflow and heating and cooling requirements for central air-conditioning systems in zones which are unoccupied is always a very attractive ROI ECO.

2. *Unoccupied Period Set-back and Set-up Temperatures*
    The principle of establishing unoccupied period space temperature setpoints is used to reduce the

amount of conditioning provided by allowing the interior temperature to drift naturally to a marginal temperature and then recondition it to normal conditions before the space becomes occupied.

Many electric utility companies offer time-of-day rates whereby energy used during off-peak periods is charged at a rate much lower than for electricity consumed during peak periods, adding to the savings achieved by merely reducing consumption.

3. *Warm-up and Cool-down Cycles*
    When a building is operated with unoccupied period control sequence, the system must be restarted in advance so that optimum conditions are reached before the beginning of the next occupied period.

    One of the best approaches to reconditioning a building after unoccupied period set-back is with the use of an optimum-time start program. This program compares the outside and inside temperatures, along with the desired setpoint during the operating cycle. It determines how long it will take to recondition the building to the setpoint, based on previous data, and turns the system on at the appropriate time to reach the setpoint temperature just prior to the start of the occupied period.

    A second method of saving energy during warm-up or cool down is with a warm-up/cool-down cycle. During the warm-up/cool-down cycle the system recirculates building return air until a temperature within one or two degrees of the setpoint is reached, saving the energy which would be required to heat or cool outside ventilation air. Since outside ventilation air is provided to meet human fresh-air needs, it is not normally introduced during warm-up and cool down cycles.

4. *Additional Control Strategies*
    More extensive discussion of building automation systems may be found in Chapters 12 and 22.

10.11 REDUCING SYSTEM LOADS

The loads the HVAC systems and equipment serve should always be addressed in conjunction with any retrofit projects for HVAC systems and equipment. Improvements to lighting systems, building envelope improvements, and other operational changes will reduce the loads and the energy consumption of the HVAC sys-

tems. These improvements will also reduce the size of any future HVAC equipment to be provided, reducing capital costs of ECO projects.

10.12 ESTIMATING HVAC SYSTEMS ENERGY CONSUMPTION

Calculating energy use, baseline and proposed, for justification of HVAC ECO projects can be challenging. A key evaluation factor for determining how to calculate energy consumption and savings is to examine the current HVAC systems and the considered ECO projects.

For systems that are single zone or decentralized systems and do not have any significant simultaneous heating and cooling present, spreadsheet calculations and bin method calculations can be sufficient. If the current systems in the building are large systems with multiple zones and simultaneous heating and cooling requirements, and if VAV conversions are contemplated, a computerized energy simulation program is probably necessary due to the complexity of the interaction of the HVAC load profiles and the HVAC systems. A computerized simulation is also generally required for buildings that have large central plants with multiple chillers and boilers, or remote central plants. Significant detail on this subject can be found in the *ASHRAE Fundamentals Handbook*, Chapter 32.

HVAC energy use profiles are dynamic, coming from a mix of internal loads (people, lights, equipment), ventilation, and envelope. Estimating energy use of HVAC systems requires a good understanding of the load profile and corresponding efficiencies of attendant systems, including seasonal modes, economizers, occupancy patterns, day/night temperature setbacks, etc. In some cases, spreadsheets and bin weather data can provide a reasonable estimation of operating cost and savings. For other cases energy modeling is warranted.

Equipment efficiencies vary, depending upon percent load on the equipment. These changes in part load efficiency must be accounted for in the energy estimating calculations.

10.12.1 HVAC Cooling and Heating Load Profiles
With few exceptions, the HVAC heating and cooling load will be dynamic, and equipment loading will increase and decrease with the changing load. This pattern of change and the hours of operation at different loads is known as the load profile.

For existing systems, actual measurements are an excellent way to determine load profile. This may be through portable data loggers (e.g. measuring chiller load in kW) or long-term trends from the energy management system. Where available, these "real" data are always more accurate than any estimated value.

Computer programs that allow hourly analysis accept inputs on many contributing factors, such as envelope, occupancy and activity schedules, lighting, and internal loads; they overlay this data with weather patterns, solar loads, utility rate structures, and thermal flywheel affect of building materials. Hourly aggregate load profiles are provided. For estimating purposes energy modeling is ideal because of its accuracy and also, once created, is a good tool for evaluating options.

A less accurate but much quicker method uses bin weather data and the general assumptions that:

- The cooling will be its greatest on the hottest day.
- The heating load will be its greatest on the coldest day.
- Between minimum and maximum heating/cooling load values, an approximately linear relationship exists.

This assumption is easily handled in a spreadsheet using bin weather data; it is demonstrated in Tables 10-4 and 10-5. Limitations of this method are:

- Only buildings whose HVAC loads are a strong function of outside temperature should be estimated in this way. These are known as "thermally light" buildings. (See Chapter 9, Building Envelope.)
- Buildings with HVAC loads that are weather independent cannot be predicted from changes in outside air temperature.

For accurate energy consumption estimates, an efficiency curve for HVAC equipment at various loads must also be identified. For example, if the hours of operation are established at different loads, accurate equipment efficiencies at each load will be necessary to evaluate energy use accurately. Part load efficiency is available from the equipment manufacturer but is itself dynamic. At the reduced load, it is usually assumed that outdoor ambient temperatures are lower, or for water cooled equipment, the condenser water temperature is lower. As long as the assumptions made are reasonable for the area and consistent for the baseline and comparison system (for savings of an ECO), the calculation will be valid.

Example:

(Note: This example is for cooling equipment, although the same method applies equally to heating or other HVAC equipment that operates at variable capacity.)

Had this calculation been done with nameplate full load and part load efficiency, instead of a true load pro-file, results would be much different.

Full load (100 hrs @ 100 tons @ 1.19 kW/ton)
$$= 119,000 \text{ kWh}$$
Part load (4000 hrs @ 60.5 tons @ 1.04 kW/ton)
$$= 251,680 \text{ kWh}$$
Total energy = 370,680 kWh

10.13   SUMMARY

The HVAC systems encountered in commercial, industrial, and institutional buildings are numerous in type, personality, and characteristic energy use. The energy use is driven by weather and multiple additional factors, and it is dynamic. The dynamic nature of most HVAC system makes it difficult to quantify energy use and savings potential, so energy professionals engaged in this work must possess good estimating techniques or utilize computer modeling. Understanding the system fundamentals gives insight to characteristic energy use footprints and also suggests proven energy improvement opportunities. Any ECOs implemented must maintain the design intent of the system, which must be understood at the onset. Consideration of system condition, age, and any existing operational issues will allow HVAC ECOs to bring maximum benefit to the customer.

10.14   ITEMS FOR FURTHER INVESTIGATION

This chapter was necessarily brief, and in its brevity cannot be a complete treatment of the subject. There are many excellent texts on the prolific industry of HVAC design and application, most notably the guidelines and handbooks produced by ASHRAE. Related material within this handbook includes:

- Boilers (Chapter 5)
- Heat recovery (Chapter 8)
- Building envelope (Chapter 9)
- Energy systems maintenance (Chapter 14)
- Indoor air quality (Chapter 17)
- Thermal energy storage (Chapter 19)
- Automatic controls (Chapter 22)
- Ground source heat pumps (Chapter 28)

The following topics were not addressed, and are listed for the interested reader to pursue through additional study:

- HVAC load calculation methods
- HVAC energy modeling methods
- Desiccant dehumidification
- Data center HVAC

Table 10-4. Sample Spread sheet of Linear HVAC Load Profile Assumption. *Source:* Commercial Energy Auditing Reference Handbook

	Mid-pts	DB (F)	Total Bin Hours for each value of percent load	Load	Pct Load. This is the load curve
24 X 7 **5 Degree Bins**					
Observed Maximum	92.5	90 to 95	37	**250**	100%
Cooling Load	87.5	85 to 90	100	214	86%
250 tons	82.5	80 to 85	285	179	71%
	77.5	75 to 80	369	143	57%
	72.5	70 to 75	461	107	43%
	67.5	65 to 70	539	71	29%
	62.5	60 to 65	865	36	14%
CROSS OVER ZONE	57.5	55 to 60	813	**0**	0%
ASSUME NO SAVINGS	52.5	50 to 55	744	**0**	0%
	47.5	45 to 50	729	167	8%
	42.5	40 to 45	657	333	17%
	37.5	35 to 40	869	500	25%
	32.5	30 to 35	693	667	33%
	27.5	25 to 30	561	833	42%
Observed Maximum	22.5	20 to 25	399	1000	50%
Heating Load	17.5	15 to 20	302	1167	58%
2000 Mbh	12.5	10 to 15	134	1333	67%
	7.5	5 to 10	95	1500	75%
	2.5	0 to 5	81	1667	83%
	-2.5	-5 to 0	24	1833	92%
	-7.5	-10 to -5	3	**2000**	100%

Table 10-5. Sample Pictorial Representation of Linear HVAC Load Profile Assumption
Dashed line shows assumed linear heating and cooling load variation. *Source:* Commercial Energy Auditing Reference Handbook

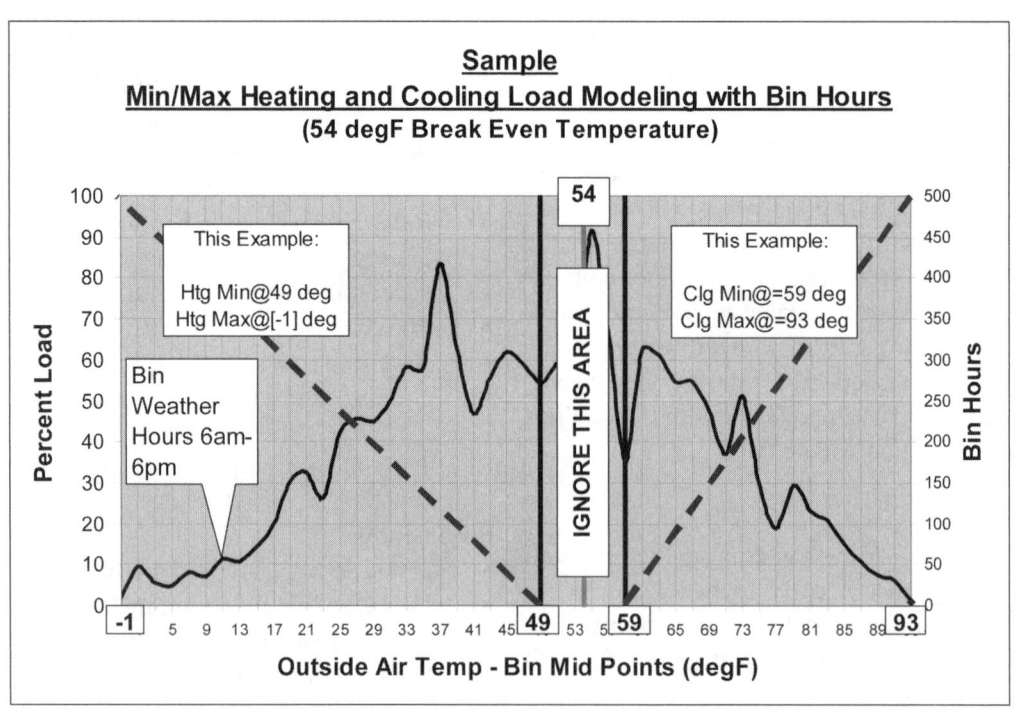

Nominal full load efficiency (EER):   10.1 (1.19 kW/ton)

Nominal part load efficiency (SEER)   11.5 (1.04 kW/ton)

Design load: 100 tons

Pct Load	Tons	Hours	Ton-Hours	EER	Efficiency (kW/ton)	Energy (kWh)
100	100	100	10,000	10.1	1.19	11,881
90	90	200	18,000	10.9	1.10	19,817
80	80	400	32,000	11.5	1.04	33,391
70	70	800	56,000	11.0	1.09	61,091
60	60	1200	72,000	11.0	1.09	78,545
50	50	800	40,000	10.8	1.11	44,444
40	40	600	24,000	10.5	1.14	27,429
						276,598

- Health care facilities HVAC
- Laboratory facilities HVAC
- Refrigeration cycle analysis
- Psychrometric analysis
- Geothermal, solar, or other renewable energy Sources and impacts on HVAC systems
- HVAC system commissioning
- Variable refrigerant flow systems

Bibliography

ASHRAE 2006 Refrigeration Handbook

ASHRAE 2007 HVAC Applications Handbook

ASHRAE 2008 HVAC Systems and Equipment Handbook

ASHRAE 2009 Fundamentals Handbook

CHAPTER 11

MOTORS, DRIVES, AND
ELECTRIC ENERGY MANAGEMENT

K.K. LOBODOVSKY
BSEE & BSME
Certified Energy Auditor
State of California

STEVE DOTY, PE, CEM
Colorado Springs Utilities
Colorado Springs, CO

11.1 INTRODUCTION

Efficient use of electric energy enables commercial, industrial and institutional facilities to minimize operating costs, and increase profits to stay competitive.

The majority of electrical energy in the United States is used to run electric motor driven systems. Generally, systems consist of several components: the electrical power supply, the electric motor, the motor control, the mechanical transmission system, and the driven load.

There are several ways to improve the systems' efficiency. The cost effective way is to check each component of the system for an opportunity to reduce electrical losses. A qualified individual should oversee the electrical system since poor power distribution within a facility is a common cause of energy losses.

Technology Update, Ch. 18[1], lists 20 items to help facility management staff identify opportunities to improve drive system efficiency:

1. Maintain voltage levels.
2. Minimize phase imbalance.
3. Maintain power factor.
4. Maintain good power quality.
5. Select efficient transformers.
6. Identify and fix distribution system losses.
7. Minimize distribution system resistance.
8. Use adjustable speed drives (ASDs) or 2-speed motors where appropriate.
9. Consider load shedding.
10. Choose replacement before a motor fails.
11. Choose energy-efficient motors.
12. Match motor operating speeds.
13. Size motors for efficiency.
14. Choose 200 volt motors for 208 volt electrical systems.
15. Minimize motor rewind losses.
16. Optimize transmission efficiency.
17. Perform periodic checks.
18. Control temperatures.
19. Lubricate correctly.
20. Maintain motor records.

Some of these steps require the one-time involvement of an electrical engineer or technician. Some steps can be implemented when motors fail or major capital changes are made in the facility. Others involve development of a motor monitoring and maintenance program.

11.2 POWER SUPPLY

Much of this information consists of standards defined by the National Electrical Manufacturers Association (NEMA).

The power supply is one of the major factors affecting selection, installation, operation, and maintenance of an electrical motor driven system. Usual service conditions, defined in NEMA Standards Publication MG1, *Motors and Generators*,[2] include:

- Motors designed for rated voltage, frequency, and number of phases.

- The supply voltage must be known to select the proper motor.

- Motor nameplate voltage will normally be less then nominal power system voltage.

Nominal Power System Voltage (Volts)	Motor Utilization (Nameplate) Voltage Volts
208	200
240	230
480	460
600	575
2400	2300
4160	4000
6900	6600
13800	13200

- Operation within tolerance of ±10 percent of the rated voltage.

- Operation from a sine wave of voltage source (not to exceed 10 percent deviation factor).

- Operation within a tolerance of ±5 percent of rated frequency.

- Operation within a voltage unbalance of 1 percent or less.

Operation at other than usual service conditions may result in the consumption of additional energy.

11.3 EFFECTS OF UNBALANCED VOLTAGES ON THE PERFORMANCE OF POLYPHASE SQUIRREL-CAGE INDUCTION MOTORS (MG 1-20.56)

When the line voltages applied to a polyphase induction motor are not equal, unbalanced currents in the stator windings result. A small percentage of voltage unbalance results in a much larger percentage current unbalance. Consequently, the temperature rise of the motor operating at a particular load and percentage voltage unbalance will be greater than for the motor operating under the same conditions with balanced voltages.

Voltages should be evenly balanced as closely as they can be read on a voltmeter. If the voltages are unbalanced, the rated horsepower of polyphase squirrel-cage induction motors should be multiplied by the factor shown in Figure 11-1 to reduce the possibility of damage to the motor. Operation of the motor with more than a 5-percent voltage unbalance is not recommended.

When the derating curve of Figure 11-1 is applied for operation on unbalanced voltages, the selection and setting of the overload device should take into account the combination of the derating factor applied to the motor and the increase in current resulting from the unbalanced voltages. This is a complex problem involving the variation in motor current as a function of load and voltage unbalance, in addition to the characteristics of the overload device relative to $I_{MAXIMUM}$ or $I_{AVERAGE}$. In the absence of specific information it is recommended that overload devices be selected and/or adjusted at the minimum value that does not result in tripping for the derating factor and voltage unbalance that applies. When the unbalanced

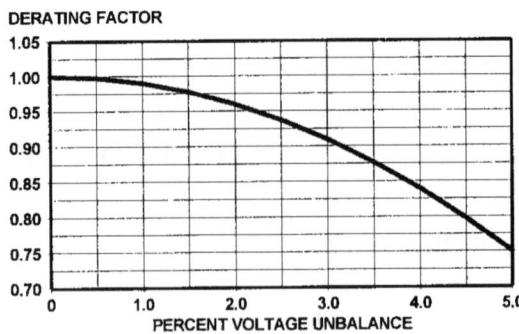

Figure 11-1. Polyphase squirrel-cage induction motors derating factor due to unbalanced voltage.

voltages are unanticipated, it is recommended that the overload devices be selected so as to be responsive to $I_{MAXIMUM}$ in preference to overload devices responsive to $I_{AVERAGE}$.

Notes:
1. Due to motor heating, motors with unbalanced voltage are required to be reduced from nameplate maximum horsepower. This chart reflects nameplate motor Hp de-rate, not efficiency loss.
2. This chart and the concept of voltage imbalance does not apply to single-phase motors.

11.4 EFFECT ON PERFORMANCE— GENERAL (MG 1 20.56.1)

The effect of unbalanced voltages on polyphase induction motors is equivalent to the introduction of a "negative-sequence voltage" having a rotation opposite to that occurring with balanced voltages. This negative-sequence voltage produces an air gap flux rotating against the rotation of the rotor, tending to produce high currents. A small negative-sequence voltage may produce current in the windings considerably in excess of those present under balanced voltage conditions.

11.4.1 Unbalanced Defined (MG 1 20.56.2)

The voltage unbalance in percent may be defined as follows:

$$\begin{array}{l} \text{Percent} \\ \text{voltage} \\ \text{unbalance} \end{array} = 100 \times \frac{\text{Maximum voltage deviation from average voltage}}{\text{average voltage}}$$

Example: With voltages of 220, 215 and 210, the average is 215; the maximum deviation from the average is 5.

Percent Voltage Unbalance
= 100 * 5/215 = 2.3 Percent

11.4.2 Torque (MG 1 20.56.3)

The locked-rotor torque and breakdown torque are decreased when the voltage is unbalanced. If the voltage unbalance is extremely severe, the torque might not be adequate for the application.

11.4.3 Full-Load Speed (MG 1 20.56.4)

The full-load speed is reduced slightly when the motor operates at unbalanced voltages.

11.4.4 Currents (MG 1 20.56.5)

The locked-rotor current will be unbalanced, but the locked rotor kVA will increase only slightly.

11.4.5. Voltage Imbalance Effect on Motor Efficiency

Unbalanced currents and reverse rotation from voltage imbalance produce energy waste and reduced motor efficiency. The reduction in motor efficiency is proportional to the imbalance and is more pronounced at reduced motor load.

11.4.6 Correcting Voltage Imbalance

The cause of voltage imbalance is almost always

Table 11-1. Motor Efficiency Loss from Voltage Imbalance at Reduced Load (10)

% Motor Load	Balanced	1% Imbalance	2.5% Imbalance
100	0.0%	0.0%	1.5%
75	0.0%	0.1%	1.4%
50	0.0%	0.6%	2.1%

single-phase loads that are unevenly distributed with the three-phase loads. These loads commonly include lighting and receptacles, along with smaller motor loads, air conditioners, and electric heaters. An effort is made during initial design to balance these loads, but usage patterns are not always known and changes in the building or plant over time can allow voltage imbalance to creep up. It is good plant practice to periodically measure voltage imbalance at equipment panels and at motor terminals, to assure the best balance possible. When imbalance is discovered, it can be improved by shifting load from one electrical phase or circuit panel to another.

11.5 MOTOR

The origin of the electric motor can be traced back to 1831 when Michael Faraday demonstrated the fundamental principles of electromagnetism. The purpose of an electric motor is to convert electrical energy into mechanical energy.

Electric motors are efficient at converting electric energy into mechanical energy. If the efficiency of an electric motor is 80%, it means that 80% of electrical energy delivered to the motor is directly converted to mechanical energy at the motor shaft. The portion lost within the motor is the difference between electrical energy input and mechanical energy output.

11.6 GLOSSARY OF FREQUENTLY OCCURRING MOTOR TERMS[4]

Motor terms are used quite frequently, usually on the assumption that every one knows what they mean or imply. Such is far too often not the case. The following section is a list of motor terms.

Amps

Full Load Amps

The amount of current the motor can be expected to draw under full load (torque) conditions is called full

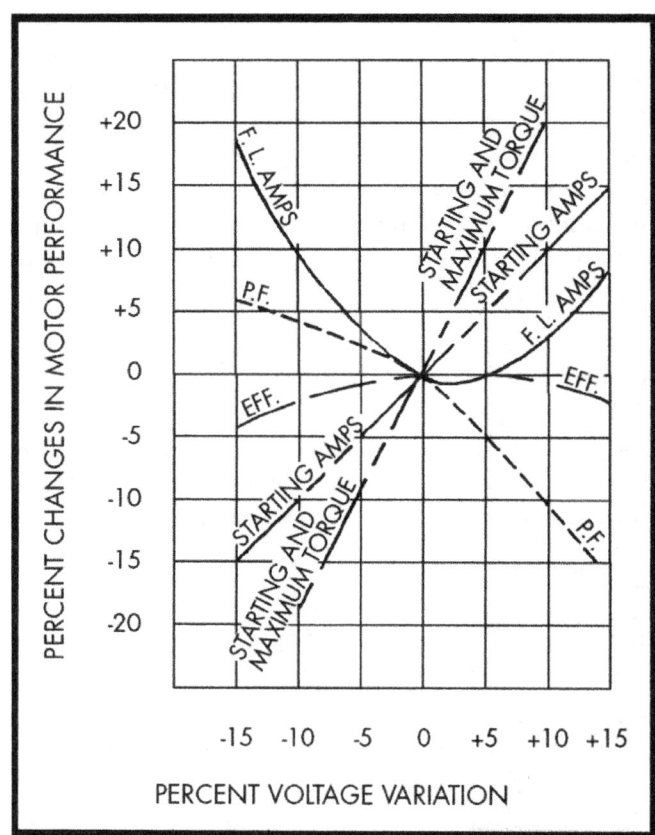

Figure 11-2. Voltage Variation Effect on Motor Performance (9)

load amps. It is also known as nameplate amps.

Locked Rotor Amps

Also known as starting inrush, this is the amount of current the motor can be expected to draw under starting conditions when full voltage is applied.

Service Factor Amps

This is the amount of current the motor will draw when it is subjected to a percentage of overload equal to the service factor on the nameplate of the motor. For example, many motors will have a service factor of 1.15, meaning the motor can handle a 15% overload. The service factor amperage is the amount of current the motor will draw under the service factor load condition.

Note: Extended operation of a motor in its service factor will shorten its life. The motor design allows for intermittent and temporary operation in this overloaded state.

Design

The design letter is an indication of the shape of the torque speed curve. Figure 11-3 shows the typical shape of the most commonly used design letters. They are A, B, C, and D. Design B is the standard industrial duty motor, which has reasonable starting torque with moderate starting current and good overall performance for most industrial applications. Design C is used for hard to start loads and is specifically designed to have high starting torque. Design D is the so-called high slip motor, which tends to have very high starting torque but has high slip RPM at full load torque. In some respects, this motor can be said to have a "spongy" characteristic when loads

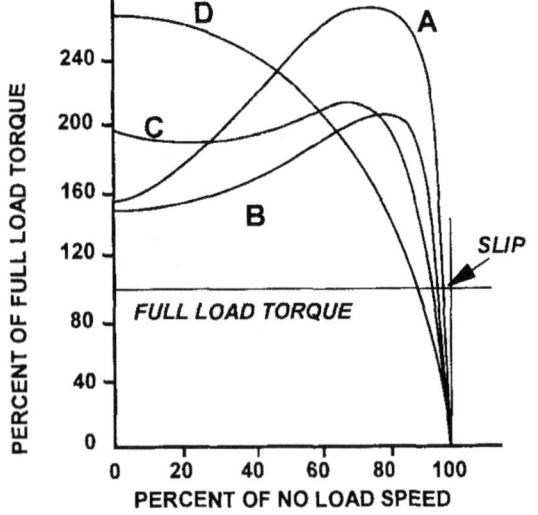

are changing. Design D motors are particularly suited for low speed punch press, hoist and elevator applications. Generally, the efficiency of design D motors at full load is rather poor; thus, they are normally used on those applications where the torque characteristics are of primary importance. Design A motors are not commonly specified, but specialized motors used on injection molding applications have characteristics similar to design B. The most important characteristic of design A is the high pull out torque.

Efficiency

Efficiency is the percentage of the input power that is actually converted to work output from the motor shaft. Efficiency is stamped on the nameplate of most domestically produced electric motors.

$$\text{Efficiency} = \text{EFF} = \frac{746 \times \text{HP Output}}{\text{Watts input}}$$

(See Section 11.9 for discussion of efficiency loss at reduced load.)

Frame Size

Motors, like suits of clothes, shoes, and hats come in various sizes to match the requirements of the applications. In general, the frame size gets larger with increasing horsepower or with decreasing speeds. In order to promote standardization in the motor industry, NEMA (National Electrical Manufacturers Association) prescribes standard frame sizes for certain horsepower, speed, and enclosure combinations. Frame specifically determines the mounting and shaft dimension of standard motors. For example, a motor with a frame size of 56, will always have a shaft height (above the base) of 3- 1/2 inches.

Frequency

This is the frequency for which the motor is designed. The most commonly occurring frequency in the United States is 60 cycles, but internationally other frequencies such as 25, 40, and 50 cycles can be found.

Full Load Speed

An indication of the approximate speed at which the motor will run when putting out full rated output torque or horsepower is called full load speed.

High Inertia Load

These are loads that have a relatively high fly wheel effect. Large fans, blowers, punch presses, centrifuges, industrial washing machines, and other similar loads can be classified as high inertia loads.

Insulation Class

The motor insulation class is a measure of the resistance of the insulating components of a motor to degradation from heat. Four major classifications of insulation are used in motors. They are, in order of increasing thermal capabilities, A, B, F, and H.

Class of Insulation System	Temperature, Degrees C
A	105
B	130
F	155
H	180

Load Types

Constant Horsepower

The term constant horsepower is used in certain types of loads where the torque requirement is reduced as the speed is increased, and vice-versa. The constant horsepower load is usually associated with metal removal applications, such as drill presses, lathes, milling machines, and similar applications.

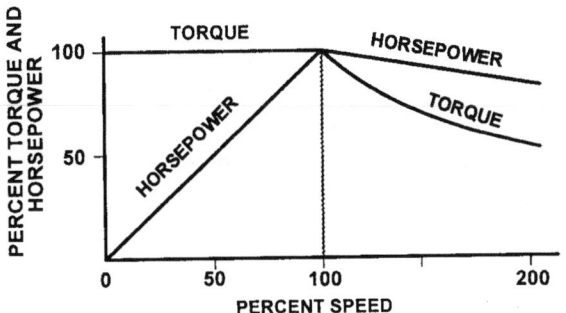

Constant Torque

Constant torque is a term used to define a load characteristic where the amount of torque required to drive the machine is constant, regardless of the speed at which it is driven. For example, the torque requirement of most conveyors is constant.

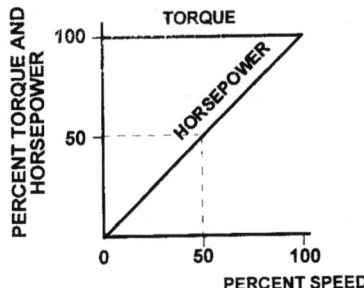

Variable Torque

Variable torque is found in loads having characteristics requiring low torque at low speeds, with increasing values of torque required as the speed is increased. Typi-

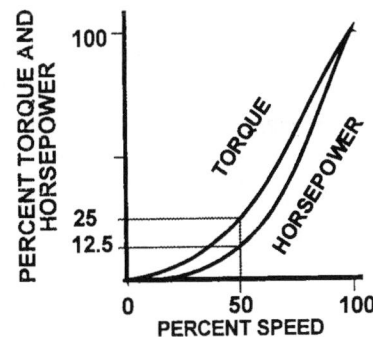

cal examples of variable torque loads are centrifugal fans and centrifugal pumps.

Phase

Phase is the indication of the type of power supply for which the motor is designed. Two major categories exist—single-phase and three-phase. There are some very spotty areas where two-phase power is available, but these are very insignificant.

Poles

This is the number of magnetic poles within the motor when power is applied. Poles are always an even number such as 2, 4, 6. In an AC motor, the number of poles work in conjunction with the frequency to determine the synchronous speed of the motor. At 50 and 60 cycles, common arrangements are:

Synchronous speed		
Poles	60 Cycles	50 Cycles
2	3600	3000
4	1800	1500
6	1200	1000
8	900	750
10	720	600

Power Factor

Percent power factor is a measure of a particular motor's requirements for magnetizing amperage. (For more information see Section 11.7.)

$$\text{Power factor} \atop \text{(3 phase)} = pf = \frac{\text{Watts input}}{\text{Volts} \times \text{amps} \times 1.73}$$

(See Section 11.9 for discussion of power factor reduction at reduced load.)

Service Factor

The service factor is a multiplier that indicates the

amount of overload a motor can be expected to handle. For example, a motor with a 1.0 service factor cannot be expected to handle more than its nameplate horsepower on a continuous basis. Similarly, a motor with a 1.15 service factor can be expected to safely handle intermittent loads amounting to 15% beyond its nameplate horsepower.

Slip

Slip is used in two forms. One is the slip RPM, which is the difference between the synchronous speed and the full load speed. When this slip RPM is expressed as a percentage of the synchronous speed, then it is called percent slip or just "slip." Most standard motors run with a full loadslip of 2% to 5%.

Synchronous Speed

This is the speed at which the magnetic field within the motor is rotating. It is also the approximate speed the motor will run under no load condition. For example, a 4-pole motor running in 60 cycles would have a magnetic field speed of 1800 RPM. The no load speed of that motor shaft would be very close to 1800, probably 1798 or 1799 RPM. The full load speed of the same motor might be 1745 RPM. The difference between the synchronous speed and the full load speed is called the slip RPM of the motor.

Temperature
Ambient Temperature.

Ambient temperature is the maximum safe room temperature surrounding the motor if it is going to be operated continuously at full load. In most cases, the standardized ambient temperature rating is 40°C (104°F). This is a very warm room. Certain types of applications, such as on board ships and in boiler rooms, may require motors with a higher ambient temperature capability, such as 50°C or 60°C.

Temperature Rise.

Temperature rise is the amount of temperature change that can be expected within the winding of the motor from non-operating (cool) condition to its temperature at full load continuous operating condition. Temperature rise is normally expressed in degrees centigrade.

Time Rating

Most motors are rated in continuous duty, which means that they can operate at full load torque continuously without overheating. Motors used on certain types of applications, such as waste disposal, valve actuators, hoists, and other types of intermittent loads, will fre-

quently be rated in short term duty, such as 5 minutes, 15 minutes, 30 minutes, or 1 hour. Just like a human being, a motor can be asked to handle very strenuous work as long as it is not required on a continuous basis.

Torque

Torque is the twisting force exerted by the shaft or a motor. Torque is measured in inch pounds or foot pounds—and on small motors, in terms of inch ounces.

Full Load Torque

Full load torque is the rated continuous torque that the motor can support without overheating within its time rating.

Peak Torque

Many types of loads, such as reciprocating compressors, have cycling torque where the amount of torque required varies depending on the position of the machine. The actual maximum torque requirement at any point is called the peak torque requirement. Peak torque is involved in things such as punch presses and other types of loads where an oscillating torque requirement occurs.

Pull Out Torque

Also known as breakdown torque, this is the maximum amount of torque available from the motor shaft when the motor is operating at full voltage and running at full speed. The load is then increased until the maximum point is reached. Refer to Figure 11-4.

Pull Up Torque

The lowest point on the torque speed curve for a motor accelerating a load up to full speed is called pull up torque. Some motors are designed without a value of pull up torque because the lowest point may occur at the locked rotor point. In this case, pull up torque is the same as locked rotor torque.

Starting Torque

The amount of torque the motor produces when it is energized at full voltage and with the shaft locked in

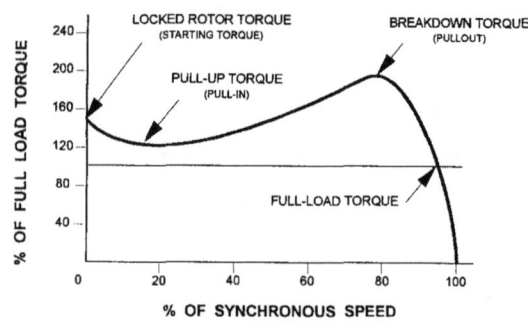

Figure 11-3. Typical speed—torque curve.

place is called starting torque. This value is also frequently expressed as "locked rotor torque." It is the amount of torque available when power is applied to break the load away and start accelerating it up to speed.

Voltage

This is the voltage rating for which the motor is designed. (See Section 11.2.)

11.7 POWER FACTOR

WHAT IS POWER FACTOR (pf)?

It is the mathematical ratio of *ACTIVE POWER* (W) to *APPARENT POWER* (VA)

$$pf = \frac{\text{Active power}}{\text{Apparent power}} = W * \text{Cos } \theta$$

pf angle in degrees = $\cos^{-1} \theta$

ACTIVE POWER = **W** = "real power" = supplied by the power system to actually turn the motor.

REACTIVE POWER = **VAR** = **(W)tan** θ = is used strictly to develop a magnetic field within the motor.

or **(VA)² = (W)² + (VAR)²**

NOTE: Power factor may be "leading" or "lagging," depending on the direction of VAR flow.

CAPACITORS can be used to improve the power factor of a circuit with a large inductive load. Current through capacitor LEADS the applied voltage by 90 electrical degrees (VAC), and has the effect of "opposing" the inductive "LAGGING" current on a "one-for-one" (VAR) basis.

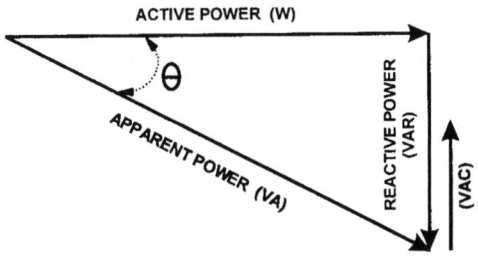

Power Factor Triangle

WHY RAISE POWER FACTOR (pf)?

Low (or "unsatisfactory") power factor is caused by the use of inductive (magnetic) devices and can indicate possible low system electrical operating efficiency.

These devices are:

- non-power factor corrected fluorescent and high intensity discharge lighting fixture ballasts (40%-80% pf)

- arc welders (50%-70% pf)

- solenoids (20%-50% pf)

- induction heating equipment (60%-90% pf)

- lifting magnets (20%-50% pf)

- small "dry-pack" transformers (30%-95% pf)

- and most significantly, induction motors (55%-90% pf)

Induction motors are generally the principal cause of low power factor because there are so many in use, and they are usually *not fully loaded*. The correction of the condition of LOW power factor is a problem of vital economic importance in the generation, distribution and utilization of a-c power.

MAJOR BENEFITS OF POWER FACTOR IMPROVEMENT ARE:

- increased plant capacity,

- reduced power factor "penalty" charges from the electric utility,

- improvement of voltage supply,

- less power losses in feeders, transformers and distribution equipment.

WHERE TO CORRECT POWER FACTOR?

Capacitor correction is relatively inexpensive, both in material and installation costs. Capacitors can be installed at any point in the electrical system and will improve the power factor between the point of application and the power source. However, the power factor between the utilization equipment and the capacitor will remain unchanged. Capacitors are usually added at each piece of offending equipment, ahead of groups of small motors (ahead of motor control centers or distribution panels) or at main services. Refer to the National Electrical Code for installation requirements.

The advantages and disadvantages of each type of capacitor installation are listed below.

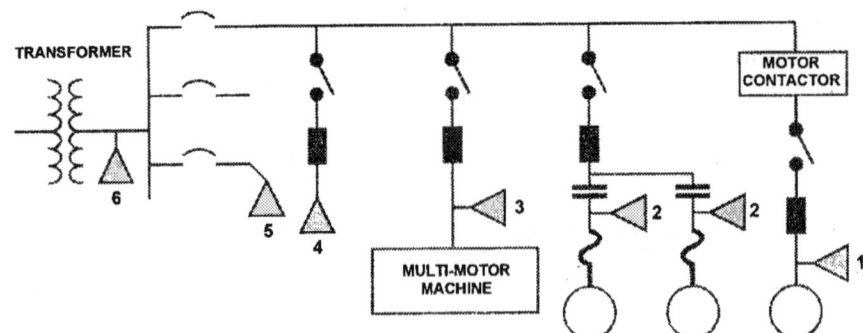

Figure 11-4. Power Factor Correction Locations

Capacitor on each piece of equipment (1,2)

ADVANTAGES

- increases load capabilities of distribution system

- can be switched with equipment; no additional switching is required

- better voltage regulation because capacitor use follows load

- capacitor sizing is simplified

- capacitors are coupled with equipment and move with equipment if rearrangements are instituted

DISADVANTAGES

- small capacitors cost more per KVAC than larger units (economic break point for individual correction is generally at 10 HP)

Capacitor with equipment group (3)

ADVANTAGES

- increased load capabilities of the service

- reduced material costs relative to individual correction

- reduced installation costs relative to individual correction

DISADVANTAGES

- switching means may be required to control amount of capacitance used

Capacitor at main service (4,5, & 6)

ADVANTAGES

- low material installation costs

DISADVANTAGES

- switching usually required to control the amount of capacitance used

- does not improve load capabilities of the distribution system

OTHER CONSIDERATIONS

Where the loads contributing to power factor are relatively constant, and system load capabilities are not a factor, correcting at the main service could provide a cost advantage. When the low power factor is derived from a few selected pieces of equipment, individual equipment correction would be cost effective. Most capacitors used for power factor correction have built-in fusing; if not, fusing must be provided.

The growing use of ASDs (nonlinear loads) has increased the complexity of system power factor and its corrections. The application of pf correction capacitors without a thorough analysis of the system can aggravate rather than correct the problem, particularly if the fifth and seventh harmonics are present.

Harmonics are electric power in wave forms that occur at different frequencies than utility power, but they are in exact multiples of the utility power frequency and overlap the basic utility power sine wave. For example, a 600 cycle frequency wave would align itself with the 60 cycle frequency wave every 10th cycle. A discussion of harmonics is beyond the scope of this text, but they are worth mentioning due to the potential effect on capacitors used for power factor correction in an energy project. Harmonics "look like" a short circuit to a capacitor and, if sufficient harmonics exist, fixed capacitors can suffer premature failures, i.e. within a few years. If the facility in question includes only motors, fixed capacitors are a viable solution; however, in facilities with substantial amounts of electronic equipment that generate harmonics, standard capacitors may not be compatible. Testing is required to determine the existence and degree of harmonics and whether standard capacitors will or will not be appropriate. Common sources of harmonics are variable frequency drives and switching power supplies.

Where harmonics exist in problematic amounts, power factor correction requires special electronic equipment designed to work normally in their midst. Fixed capacitors are used for fixed loads such as constantly loaded motors, or at main electrical services serving large groups of motors or magnetic loads. Where the power factor varies, such as motor with motor loads or different activities in plant shifts, variable capacitor correction equipment or digital "active" power factor/power quality correction equipment may be required.

(See Appendix III in the back of the book for power factor correction capacitor selection method.)

POWER QUALITY REQUIREMENTS[6]

The electronic circuits used in ASDs may be susceptible to power quality related problems if care is not taken during application, specification, and installation. The most common problems include transient overvoltages, voltage sags, and harmonic distortion. These power quality problems are usually manifested in the form of nuisance tripping.

TRANSIENT OVERVOLTAGES—Capacitors are devices used in the utility power system to provide power factor correction and voltage stability during periods of heavy loading. Customers may also use capacitors for power factor correction within their facility. When capacitors are energized, a large transient overvoltage may develop, causing the ASD to trip.

VOLTAGE SAGS—ASDs (VFDs) are very sensitive to temporary reductions in nominal voltage. Typically, voltage sags are caused by faults on the electrical system of either the customer or the utility.

HARMONIC DISTORTION—ASDs (VFDs) introduce harmonics into the power system due to non-linear characteristics of power electronics operation. Harmonics are components of current and voltage that are multiples of the normal 60Hz ac sine wave. ASDs produce harmonics which, if severe, can cause motor, transformer and conductor overheating; capacitor failures; misoperation of relays and controls; and reduce system efficiencies.

Compliance with IEEE-519 "Recommended Practices and Requirements for Harmonic Control in Electrical Power Systems" is strongly recommended.

11.8 SPECIAL HIGH EFFICIENCY MOTOR DESIGNS

Far and away the most common motor style in use in commercial and industrial facilities is the AC induction or "squirrel cage" motor. However some other motor designs are worth mentioning since they have potential application for energy conservation projects.

11.8.1 Electrically Commutated Motors (ECM)

Fractional horsepower motors applied in packaged furnaces, etc. are often low cost single-phase type, such as permanent split capacitor (PSC) or shaded pole (SP) motors. While small, when in sufficient quantity (such as a building full of fan powered VAV boxes) they add up to the equivalent of a large inefficient motor. The ECM technology increases fractional motor efficiency significantly. The ECM uses a brushless DC motor with an electronic inverter controller to drive it. ECM motor technology also has application in refrigerated cases and other coolers/freezers where fans move air across coils. In this case the inefficiency of conventional motors releases heat that adds directly to the cooling load; thus the added cost of the motor can be justified by a reduction in cooling capacity and associated machinery cost, as well as from energy savings. Note that energy waste within a refrigerated environment is a double cost—both the cost of the inefficient operation of the fan *and* the cost of the refrigeration work to remove it.

11.8.2 Emerging Technology: Permanent Magnet Motors (PM)

Permanent magnet motors have been used historically to create compact sized motors where space constraints required it. Another application of the technology is for variable speed motors; the permanent magnet technology resists efficiency loss at reduced speed better than standard induction AC motors, thereby returning greater overall efficiency for motors so operated.

Permanent magnet motor technology is more expensive than conventional motors but can be warranted when equipment has high run time. While still an emerging technology, these motors are seeing ap-

Table 11-2. Characteristic Efficiency Comparison of Fractional Hp Motors

	Shaded Pole (SP)	Permanent Split Capacitor (PSC)	Electrically Commutated Motor (ECM)
Typical Efficiency	25-40%	55-65%	80-88%

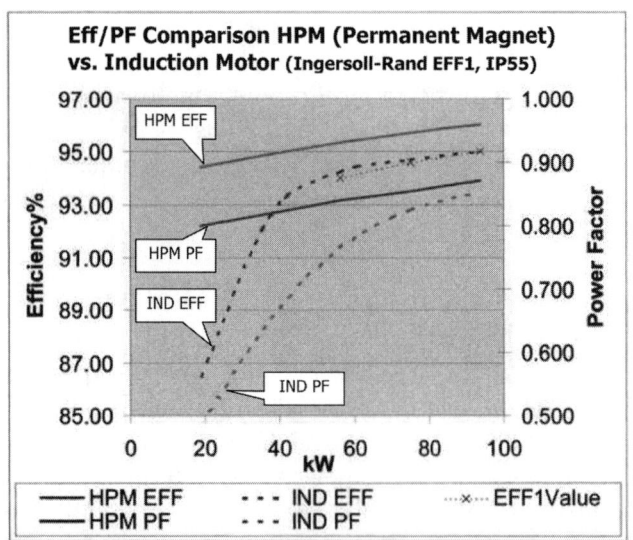

Figure 11-5. Permanent Magnet/Induction Motor Comparison (11)

plication in variable speed screw air compressors and refrigeration compressors where a high percentage of run hours are at part load.

11.9 ELECTRIC MOTOR PERFORMANCE AT PART LOAD

Most electric motors are designed to operate at 50 to 100 percent of their rated load. One reason is the motor's optimum efficiency is generally 75 percent of the rated load, and the other reason is motors are generally sized for the starting requirements.

Several surveys of installed motors reveal that a large portion of motors in use are improperly loaded. Underloaded motors, those loaded below 50 percent of rated load, operate inefficiently and exhibit low power factor. Low power factor increases losses in electrical distribution and utilization equipment, such as wiring,

motors, and transformers, and it reduces the load-handling capability and voltage regulation of the building's electrical system. Typical part-load efficiency and power factor characteristics are shown in Figure 11-6.

Additional motor losses are incurred when motor speed is reduced in conjunction with reduced load, such as with variable speed drive use. See Figure 11-7.

POWER SURVEY

Power surveys are conducted to compile meaningful records of energy usage at the service entrance, feeders and individuals loads. These records can be analyzed to prioritize those areas yielding the greatest energy savings. Power surveys also provide information for load scheduling to reduce peak demand and show operational characteristics of loads that may suggest component or system replacement to reduce energy consumption. Only through the measurement of AC power parameters can true cost benefit analysis be performed.[7]

11.10 DETERMINING ELECTRIC MOTOR OPERATING LOADS

Determining if electric motors are properly loaded enables a manager to make informed decisions about when to replace them and which replacement to choose. There are several ways to determine motor loads. The simplest way is by direct electrical measurement using a power meter. Slip measurement or amperage readings methods can be used to *estimate* the actual load.

11.11 POWER METER

To understand the electrical power usage of a facility, load or device, measurements must be taken over a time span to have a profile of the unit's opera-

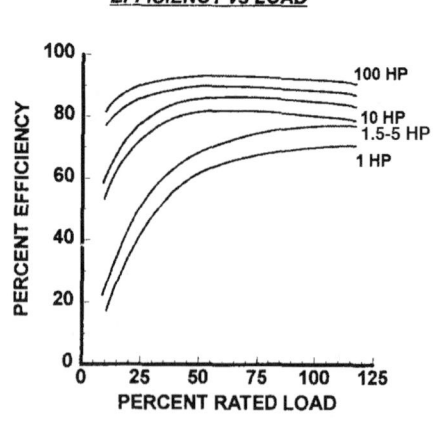
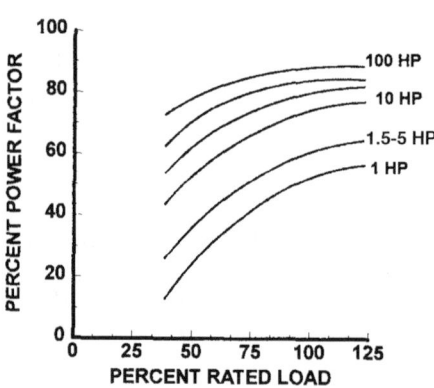

1800 RPM THREE-PHASE DESIGN **B** INDUCTION MOTOR

**Figure 11-6. Typical part-load efficiency and power factor characteristics
(Reduced load at full speed)**

tion. Digital power multimeters measure amps, volts, kWatts, kVars, kVA, power factor, phase angle, and firing angle.

Such measurements should only be performed by trained personnel.

11.11.1 Equipment for Power Measurement or Surveys

When choosing equipment to conduct a power survey, many presentation formats are available, including indicating instruments, chart recorders, and digital devices. For most survey applications, changing loads makes it necessary for data to be compiled over a period of time such as a week or month. Recording devices may be as simple as a data logger to measure amps or watts on a piece of equipment under study. More sophisticated equipment may include portable voltage and current inputs, known as potential transformers (PTs) and current transformers (CTs). The PTs and CTs serve to transform the equipment voltage and current to lower and safer levels, which are then inputs to the machine (transducers). The survey equipment is often attached to the load side terminals of the equipment (or lugs in the electrical distribution point) and left in place to record for the necessary period of time.

Depending upon the depth of measurement, the equipment may be equipped to measure and record:

- volts
- amps
- watts
- VARs
  (volt-amp reactive power)
- kVA

- power factor
- watt-hours
- VAR-hours
- demand
- frequency
- harmonics

Digital power monitoring equipment often includes software that accepts the stored data and formats it into useful tables and graphs.

11.11.2 Loads

When analyzing polyphase motors, it is important to make measurements with equipment suited for the application. Watt measurements or VAR measurements should be taken with a two element device. Power factor

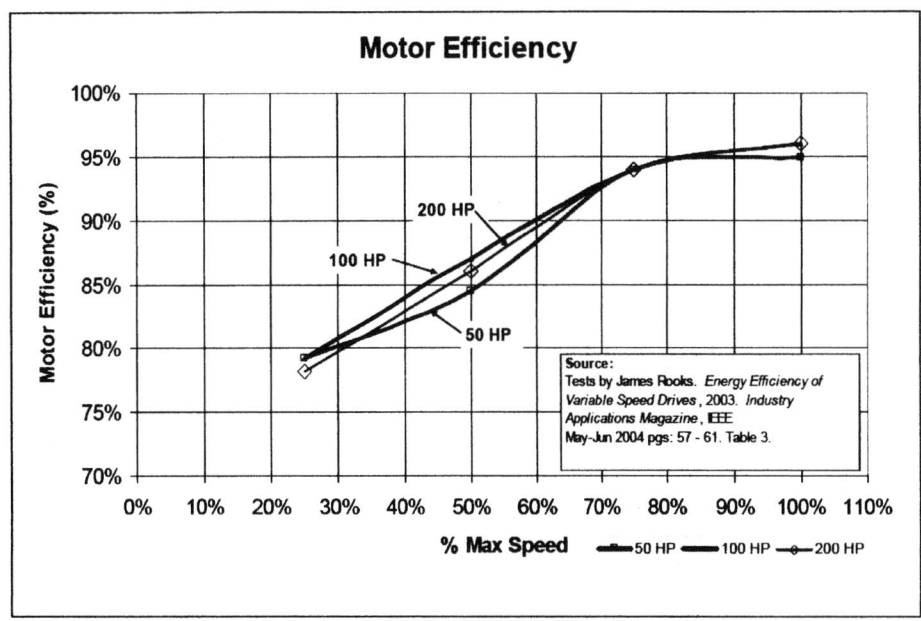

Figure 11-7. Motor efficiency at reduced speed (10)

should be determined from the readings of both measurements. When variable speed drives are encountered, it is always preferable to take measurements on the line side of the controller. When measurements are required on the load side of the controller, the instrument specifications should be reviewed; if there is a question on the application the manufacturer should be contacted.[7]

11.12 APPROXIMATE MOTOR LOAD FROM SLIP MEASUREMENT

Conditions

1. Applied voltage must be within 5% of nameplate rating.

2. This should not be used on rewound motors.

3. Motors should be operating under steady load conditions.

4. This should be performed by trained personnel.

Note: Values used in this analysis are subject to rounding errors. For example, full load speed is often rounded to the nearest 5 RPM.

Procedure

1. Read and record the motors nameplate full load speed (RPM).

2. Determine synchronous/no load speed (RPM 900, 1200, 1800, 3600)

3. Measure and record operating load speed with tachometer. (RPM)

4. Insert the recorded values in the following formula and solve.

$$(\% \text{ Motor load}) = \frac{\text{NLS} - \text{OLS}}{\text{NLS} - \text{FLS}} \times 100$$

Where:

NLS = No load or synchronous speed
OLS = Operating load speed
FLS = Full load speed

Example:

Consider a 100 HP, 1800 RM Motor
FLS = 1775 RPM, OLS = 1786 RPM

$$(\% \text{ Motor load}) = \frac{1800 - 1786}{1800 - 1775} \times 100 = 56$$

Approximate load on motor = **100 HP × 0.56 = 56 HP**

11.13 APPROXIMATE MOTOR LOAD FROM AMPERAGE READINGS[4]

Conditions

1. Applied voltage must be within 5% of nameplate rating.

2. You must be able to disconnect the motor from the load (by removing V-belts or disconnecting a coupling).

3. Motor must be 7-1/2 HP or larger; 3450, 1725, or 1140 RPM (2-, 4-, or 6-pole).

4. The indicated line amperage must be below the full load nameplate rating.

Procedure

1. Measure and record line amperage with load connected and running.

2. Disconnect motor from load. Measure and record the line amperage when the motor is running without load.

3. Read and record the motor's nameplate amperage for the voltage being used.

4. Insert the recorded values in the following formula and solve.

$$(\% \text{ Rated HP}) = \frac{(2 \times \text{LLA}) - \text{NLA}}{(2 \times \text{NPA}) - \text{NLA}} \times 100$$

Where:

LLA = Loaded Line Amps
NLA = No Load Line Amps (Motor disconnected from load)
NPA = Nameplate Amperage (For operating voltage)

Example:

*   A 20 HP motor driving a pump is operating on 460 volts and has a loaded line amperage of 16.5.

*   When the coupling is disconnected and the motor operated at no load, the amperage is 9.3.

*   The motor nameplate amperage for 460 volts is 24.0.

Therefore we have:

Loaded Line Amps	LLA	= 16.5
No Load Amps	NLA	= 9.3
Nameplate Amps	NPA	= 24.0

$$(\% \text{Rated HP}) = \frac{(2 \times 16.5) - 9.3}{(2 \times 24.0) - 9.3} \times 100 = \frac{23.7}{38.7} \times 100 = 61.2\%$$

Approximate load on motor = 20 HP × 0.612 = 12.24 or slightly over 12 HP

Please note: This procedure will generally yield reasonably accurate results only when motor load is in the 40-100% range, with deteriorating results at loads below 40%, as below this point, power factor drops off sharply and line current is a poor indicator of motor loading. The no-load amps include a great deal of reactive power and very little real power. Power factor at no load is very low. (See Figure 11-6.)

11.14 ELECTRIC MOTOR EFFICIENCY

The efficiency of a motor is the ratio of the mechanical power output to the electrical power input. It may be expressed as:

$$\text{Efficiency} = \frac{\text{Output}}{\text{Input}} = \frac{\text{Input} - \text{Losses}}{\text{Input}} = \frac{\text{Output}}{\text{Output} + \text{Losses}}$$

Design changes, better materials, and manufacturing improvements reduce motor losses, making premium or energy-efficient motors more efficient than standard motors. Reduced losses mean that an energy-efficient motor produces a given amount of work with less energy input than a standard motor.[3]

11.15 COMPARING MOTORS

It is essential that motor comparison be done on the same basis, including type, size, load, cost of energy, operating hours, and most importantly the efficiency values, such as nominal vs. nominal or guaranteed vs. guaranteed.

The following equations are used to compare the two motors.

11.15.1 For loads not sensitive to motor speed
Same horsepower—different efficiency.

$$kW_{saved} = hp \times 0.746 \times \left(\frac{100}{E_{STD}} - \frac{100}{E_{EE}} \right)$$

Same horsepower and % load—different efficiency

$$kW_{saved} = hp \times 0.746 \times L \times \left(\frac{100}{E_{STD}} - \frac{100}{E_{EE}} \right)$$

Annual $ savings due to difference in efficiency

$$S = hp \times 0.746 \times L \times C \times N \times \left(\frac{100}{E_{STD}} - \frac{100}{E_{EE}} \right)$$

Example

S	= $ Savings (annual)	100
hp	= Horsepower	100
L	= % Load	100
C	= Energy cost ($/kWh)	0.08
N	= Operating hours (annual)	4000
E_{STD}	= % Efficiency of standard motor	91.7
E_{EE}	= % Efficiency of energy eff. motor	95.0
RPM_{STD}	= Speed of standard motor	1775
RPM_{EE}	= Speed of energy eff. motor	1790

11.15.2 For loads sensitive to motor speed
The above equations should be multiplied by speed ratio correction factor.

SRCF = Speed Ratio Correction Factor

$$\left(\frac{RPM_{EE}}{RPM_{STD}} \right)^3$$

Example:

$$S = 100 \times 0.746 \times 1 \times 0.080 \times 4000$$
$$\times (100/91.7-100/95.0) = \$904$$

$$S = 100 \times 0.746 \times 1 \times 0.080 \times 4000$$
$$\times (100/91.7-100/95.0) \times (1790/1775)^3 = \$262$$

$$\$642 \text{ reduction in expected savings.}$$

A relatively minor 15 RPM increase in a motor's rotational speed results in a 2.6 percent increase in the load placed upon the motor by the rotating equipment.

11.16 SENSITIVITY OF LOAD TO MOTOR RPM

When employing electric motors for centrifugal fans and pumps, it is important to remember that the performance of fans and blowers is governed by certain rules of physics. These rules are known as "affinity laws" or "fan laws." There are several parts to it, and all are all related to each other in a known manner; when one changes, all others change. For centrifugal loads, even a minor change in the motor's speed translates into significant change in energy consumption, being especially troublesome when the additional air flow is not needed or useful. Awareness of the sensitivity of load and energy requirements to motor speed can help effectively identify motors with specific performance requirements. In most cases we can capture the full energy conservation benefits associated with energy efficient motor retrofits.

Terminology of Load to Motor RPM
CFM Fan capacity (cubic feet per minute). Volume of air moved by the fan per unit of time.

P Pressure. Pressure produced by the fan that can exist whether the air is in motion or confined in a closed duct.

HP Horsepower. The power required to drive an air moving device.

RPM Revolutions Per Minute. The speed at which the shaft of air moving equipment is rotating.

Affinity Laws or Fan Laws

$$\text{Law \#1} \quad \frac{CFM_2}{CFM_1} = \frac{RPM_2}{RPM_1}$$

Quantity (CFM) varies as fan speed (RPM)

Table 11-3. Motor Efficiency Comparison (10)

| hp | Open Drip-Proof | | | | | |
| | 1200 RPM (6-pole) | | 1800 RPM (4-pole) | | 3600 RPM (2-pole) | |
	EPAct-1992	NEMA Premium	EPAct-1992	NEMA Premium	EPAct-1992	NEMA Premium
1	80	82.5	82.5	85.5	N/A	77
1.5	84	86.5	84	86.5	82.5	84
2	85.5	87.5	84	86.5	84	85.5
3	86.5	88.5	86.5	89.5	84	85.5
5	87.5	89.5	87.5	89.5	85.5	86.5
7.5	88.5	90.2	88.5	91	87.5	88.5
10	90.2	91.7	89.5	91.7	88.5	89.5
15	90.2	91.7	91	93	89.5	90.2
20	91	92.4	91	93	90.2	91
25	91.7	93	91.7	93.6	91	91.7
30	92.4	93.6	92.4	94.1	91	91.7
40	93	94.1	93	94.1	91.7	92.4
50	93	94.1	93	94.5	92.4	93
60	93.6	94.5	93.6	95	93	93.6
75	93.6	94.5	94.1	95	93	93.6
100	94.1	95	94.1	95.4	93	93.6
125	94.1	95	94.5	95.4	93.6	94.1
150	94.5	95.4	95	95.8	93.6	94.1
200	94.5	95.4	95	95.8	94.5	95

| hp | Totally Enclosed Fan-Cooled | | | | | |
| | 1200 RPM (6-pole) | | 1800 RPM (4-pole) | | 3600 RPM (2-pole) | |
	EPAct-1992	NEMA Premium	EPAct-1992	NEMA Premium	EPAct-1992	NEMA Premium
1	80	82.5	82.5	85.5	75.5	77
1.5	85.5	87.5	84	86.5	82.5	84
2	86.5	88.5	84	86.5	84	85.5
3	87.5	89.5	87.5	89.5	85.5	86.5
5	87.5	89.5	87.5	89.5	87.5	88.5
7.5	89.5	91	89.5	91.7	88.5	89.5
10	89.5	91	89.5	91.7	89.5	90.2
15	90.2	91.7	91	92.4	90.2	91
20	90.2	91.7	91	93	90.2	91
25	91.7	93	92.4	93.6	91	91.7
30	91.7	93	92.4	93.6	91	91.7
40	93	94.1	93	94.1	91.7	92.4
50	93	94.1	93	94.5	92.4	93
60	93.6	94.5	93.6	95	93	93.6
75	93.6	94.5	94.1	95.4	93	93.6
100	94.1	95	94.5	95.4	93.6	94.1
125	94.1	95	94.5	95.4	94.5	95
150	95	95.8	95	95.8	94.5	95
200	95	95.8	95	96.2	95	95.4

$$\text{Law}\#2 \; \frac{P_2}{P_1} = \left(\frac{RPM_2}{RPM_1}\right)^2$$

Pressure (P) varies as the *square* of fan speed (RPM)

$$\text{Law}\#3 \; \frac{HP_2}{HP_1} = \left(\frac{RPM_2}{RPM_1}\right)^3$$

Horsepower (HP) varies as the <u>cube</u> of fan speed (RPM)

Example (constant speed application)
Fan system <u>32,000</u> CFM
Motor <u>20</u> HP <u>1750</u> RPM (existing)
Motor <u>20</u> HP <u>1790</u> RPM (new EE)

kW = 20 × 0.746 = 14.92 kW
New CFM with new motor = 1790/1750 × 32,000 = 32,731, or 2.3% increase
New HP = (1790/1750)3 × 20 × = 21.4 HP or 7% increase.
New kW = 21.4 × 0.746 = 15.96 kW or 7% increase in work performed by motor.

Replacing a standard motor with an energy efficient motor in a centrifugal pump or a fan application can result in increased energy consumption if the energy efficient motor operates at a higher RPM. In some cases, a 10 RPM increase can negate any savings associated with a high efficiency motor retrofit.

11.17 EFFICIENCY LOSSES OF DRIVEN LOADS

Motor efficiency is very important. However, it should be pointed out that the motor is a servant to a shaft load. In many cases the "driven load" includes losses that may be reduced with energy savings equal to or greater than savings from motor efficiency im-

provements. Focusing on the load as well as the motor represents a thorough and effective approach to energy management of motor loads. An example will illustrate the point.

Fan/Pump Motor Work Equation

Hp-input = (Hp air or water)/
(EFF_fan or pump × EFFdrive × EFFmotor)

Where EFF = Efficiency

Example: after determining the air horsepower is 50 hp, find the input power, given fan efficiency is 75%, drive efficiency is 96%, and motor efficiency is 91%.

Ans: Input = 50/ (0.75 * 0.96 * 0.91) = 76 hp
Thus, system inefficiencies in the motor-driven fan assembly added 26 Hp to the theoretical power requirement.

System efficiency is the product of individual efficiencies. In this example, the losses are:

Fan	17 Hp	65% of total
Drive	3 Hp	11% of total
Motor	6 Hp	24% of total

This example is a reminder that the motor is part of a motor-driven system and that energy reduction gains can come from process improvements (less fluid flow, less resistance), more efficient fans or pumps, drive improvements, or driver improvements (motor).

11.18 VARIABLE FREQUENCY DRIVES

Variable frequency drives (VFD) change the speed of standard AC induction motors by changing the voltage and frequency of the electricity supplied to the mo-

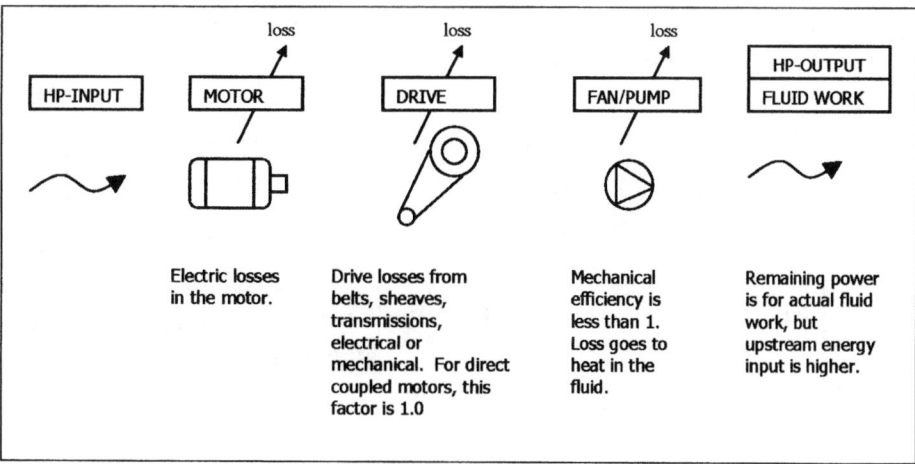

Figure 11-8. Motor-driven system concept (10)

tor. This is accomplished by converting the AC to DC and then electronically re-creating a wave form with the necessary voltage and frequency. This technique allows a motor to be driven at any speed from 0 to full nameplate speed or higher.

The four main components making up AC variable frequency drives (VFDs) are the converter, inverter, the DC circuit linking the two, and a control unit. The converter contains a rectifier and other circuitry that converts the fixed frequency AC to DC. Accessory items include filters for harmonics and other electrical disturbances created by the VFD, safety controls for the VFD and motor (overloads, etc.), and operating controls and user inputs that the VFD will respond to. There are various types of inverters used for VFDs, each with pros and cons and beyond the scope of this chapter. However, each has the same common purpose of motor speed regulation.

Most VFD units sold for commercial use include power factor correction, although this cannot be assumed and should be verified. Without this PF correction, operating a motor at low load (e.g. an idling air compressor) may create significant low power factor conditions, resulting in utility charges.

Through their own internal circuitry, namely the re-creation of a sine wave through high speed switching, VFDs are a source of harmonics. Some designs are more offending than others. VFDs applied in large quantity, large size, or in areas susceptible to harmonic disturbance warrant deeper study of harmonics and options to mitigate them. Sometimes simple harmonic filters upstream and downstream of the drive will provide suitable protection. Other times specific drive manufacturers or product lines of VFDs will be required.

The three common types of adjustable speed loads are variable torque, constant torque, and constant horsepower loads. These are application-specific and the distinction is important.

- A variable torque load requires much lower torque at low speeds than at high speeds, and it is characteristic of centrifugal fans and pumps. With this type of load, horsepower varies approximately as the cube of the speed, and the torque varies approximately as the square of the speed.

- A constant torque load requires the same amount of torque at low speed as at high speed. The torque remains constant throughout the speed range, and the horsepower increases or decreases in direct proportion to the speed. A constant torque load is used in applications such as conveyors, positive displacement pumps, and some extruders, as well as for shock loads, overloads, or high inertia loads.

- A constant horsepower load requires high torque at low speeds, and low torque at high speeds, and therefore constant horsepower at any speed. Constant horsepower loads are encountered in most metal cutting operations and with some extruders .

The savings available from constant torque or constant horsepower loads (non-centrifugal) are mostly from the high efficiency of the VFD as a speed controller (e.g. compared to a mechanical gear reducer), increased power factor, and reduced maintenance costs.

11.19 MOTOR POWER CONSUMPTION WHEN THROTTLING

Most motor applications operate at full speed and do not throttle. Even when connected to variable loads, many systems use start-stop control or two-speed control to match capacity to load. However, many motor-driven systems involve the control of flow or pressure by means of throttling or bypass devices. Throttling and bypass valves are in effect series and parallel power regulators that perform their function by dissipating the difference between source energy supplied and the desired sink energy. In these cases, dramatic savings are often possible by controlling the flow rate or pressure by controlling

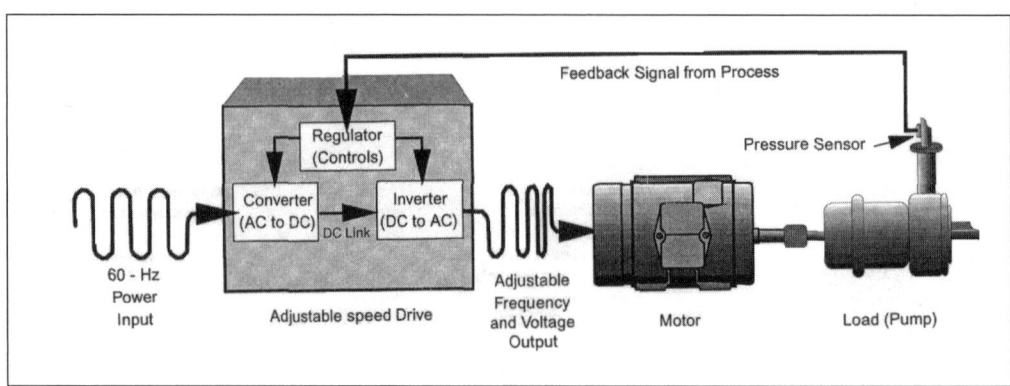

Figure 11-9. Variable Frequency Drive System

motor speed with a variable speed drive. Some examples of variable speed conversion applications include conveyors, compressors, and HVAC fans/pumps.

Some variable speed conversion projects require a system alteration in addition to the motor and drive.

Example 1

Consider a constant flow chilled water system with three-way control valves at each point of use. At full system capacity, most of the valves are open to their zone heat exchanger to extract the cooling energy, but at reduced loads the valves divert more of the chilled water to return. This application is common for heating, cooling, domestic hot water, and fuel oil pumping.

Solution 1

To implement variable pumping on this example, the system would require evaluation and, as a minimum, the three-way control valves would be replaced with two-way control valves. Then, as the array of valves began to close in response to reduced load, the flow from the pump would reduce and create a rise in pump pressure (known as "riding the curve"); the rise in pressure would be sensed by automatic controls and an ensuing signal would slow the pump. In practice, the pump pressure rise is minimal and the system becomes one of constant pressure. With a constant pressure, the energy is reduced with load (termed "load following") in accordance with the affinity laws.

Example 2

Consider a boiler forced draft fan with a 10:1 modulating burner. Burner capacity includes simultaneous throttling of the inlet fuel valve and inlet air dampers. The air control is effective, but the energy to deliver the air does not reduce proportionally to the burner turndown. This application is common for boiler intake and draft fans, pump stations using pressure regulators, and other cases where end use points "push back."

Solution 2

Allowing the same control signal for fuel to control the ASD/VFD will allow it to track the fuel valve change without a linkage. This type of project should always be done with the review and approval of the boiler/burner manufacturers and following any application advice they provide. Note that additional benefits from this particular application can be achieved from "O2 Trim" where measurements of stack gas residual oxygen are used to assure optimum air/fuel mixture ratios at all loads (a shortcoming of all single actuator boilers that have mechanical linkage between air and fuel adjustment points).

Example 3

Consider a constant volume HVAC system. Here, the heating and cooling load swings are compensated by varying heating and cooling equipment capacities, but the air flow rates are established by the greater of the two seasonal demands at maximum design load. The excess air flow at all other times is basically along for the ride and goes around and around in circles.

Solution 3

Identify zones of temperature control and install variable air volume boxes or dampers. The air flow to each zone will then be proportional to need, and the mass flow of circulated air (and the energy to move it around) will follow the seasonal loads rather than be a constant overhead energy expense.

Consider the System, Not Just the Motor

In all cases, conversion to variable speed service requires careful consideration of the system. The constant flow design may be partially (but not totally) out of convenience. Examples of system considerations for variable speed motor conversion projects are:

- Some HVAC air systems have an "air change" design basis, and reduction in air flow can affect indoor air quality and filtration performance.
- Some HVAC air systems have a "pressurization" design basis, and varying a supply air flow stream may require varying the corresponding return air flow stream to maintain the necessary pressure.
- Duct materials are installed based on "pressure class" and will normally be constructed with light gage metal when constant volume service is intended. Thus, a conversion to VAV (and higher duct pressure) raises the question of whether the ductwork can withstand the higher pressures.
- Electric duct heaters have minimum air flow requirements, associated with maximum temperature rise limitations, for fire safety of the heater. Conversion to VAV may encroach on these limits and cause heaters to trip out on high temperature safety controls.
- Air distribution through branch ducts is affected by the pressure in the upstream duct. In cases like a dust collector system or fume hood system, converting to variable flow at the main fan with dampers that close at each station (when not in use) will alter the exhaust flow at a given point, depending on which adjacent stations are closed off. If safety regulations require a certain minimum amount of exhaust, the energy project may inherit the burden of proof that the air/exhaust flow provisions remain intact.
- Hydronic systems (circulating water to convey heat-

ing and cooling energy) include velocity constraints for pipe sizing, and excessive reduction in flow rates can create sediment deposits and air binding.

• Potable water booster pumps inside buildings experience "no flow" conditions, and a run-around system serves to prevent pump overheating. Variable speed control of these pumps must accommodate the no-flow scenario one way or another.

• Reducing water flow rates through heat exchangers will lower the Reynolds number of the fluid (ratio of inertia forces to viscous forces) and reduce heat transfer rate within the heat exchanger. In the case of a water chiller, reducing water flow through variable speed pumping will usually slightly increase the specific power requirements of the chiller in kW/ton, subtracting from the apparent pump savings.

The message is clear but bears repeating: Motors are servants to the systems they are connected to and cannot be evaluated alone. Variable speed conversions are common and provide excellent savings. With words of caution and a sample of nightmares to avoid, a high percentage of these projects can be delivered successfully. Additionally, not all rotating equipment is amenable to reduced speeds. Consider, for example, machines that use splash lubrication or those with critical frequencies. Other complications exist for certain machines considered for variable speed service, including those with hard starting characteristics such as positive displacement compressors. While it may be possible to implement variable speed control in these cases, prudence in application and limited range of operation may be required. Also, motors applied for constant speed service may not be adaptable to electronic AFD/VFD service, and in some cases they will fail prematurely because of it. For these systems, the conversion would include the cost of a new motor that is "VFD rated." The economic penalty of a new motor for retrofit projects can be an incentive to consider an eddy coupling or magnetic coupling solution that reduces the speed of the driven load without reducing the motor speed.

Economic Evaluation of Variable Speed Conversions

Economic evaluation of variable speed system conversions must include the motor drive equipment and controls, but also any required system changes. The economic justification of such projects is energy savings, and determining the energy savings is a critical task. The dollar savings comes directly from motor Hp savings from the conversion, but determining the motor savings requires a load profile and knowledge

of the system. Savings are usually tabulated at 100% load, 90% load, 80% load, etc., with hours associated with each category. Errors in the Hp savings and/or the hours they occur will translate directly to errors in the project payback period, so this is a step that bears close scrutiny. A commonly overlooked fact is that the motor and driven devices do *not* have fixed efficiencies; they have a characteristic droop at low loads. For this reason, savings below 40% load are usually ignored. Additionally, while the affinity laws are accurate for full speed equipment (predicting Hp changes for system changes using existing equipment) the newly introduced variables of changing fan, pump, speed controller (AFD/VFD), and motor efficiency at reduced load will somewhat reduce the actual savings. These effects can be calculated, but the task becomes laborious; another solution is to use a *modified version of the fan law that uses a square instead of cube relationship*. This, along with ignoring any savings below 40% speed, will usually return good energy saving predictions. Experience has shown that motor loads on variable speed service seldom achieve less than 30% of full speed motor load, so placing a "30% minimum" on ASD/VFD calculations is an additional recommended conservative step for savings calculations.

HP2 = HP1 X $(^{N2}/_{N1})^{2.0}$ where N = RPM

Since air flow is proportional to rpm for a centrifugal fan/pump, this may be re-written:

HP2 = HP1 X $(^{Q2}/_{Q1})^{2.0}$ where Q = Flow Rate
Units are cubic feet per minute (CFM), gallons per minute (GPM), Liters/Min., percent flow, etc. (in constant units).

Affinity Law, Modified for VFD Savings, using square instead of cube (10)

Common methods of throttling (varying) the capacity of fans and pumps are shown in Figure 11-10. The principles of capacity modulation for such equipment are shown in Table 11-4.

From the standpoint of maximum energy conservation, the optimal method to reduce capacity of centrifugal equipment (fans, pumps, compressors) is to reduce the speed of the equipment. This can be accomplished in one of two ways:

• By changing the motor speed and its connected driven device, or

• By changing the driven device speed alone by an intermediate speed transformation, leaving the motor speed constant.

Table 11-4. Common Fan/Pump Capacity Modulation Methods

Capacity modulation method	Principle	Remarks
Discharge Damper	Energy dissipation through head loss or added resistance least	Least cost and energy benefit for capacity reduction
Inlet Vane	Combination of inlet pre-rotation and added resistance	Usually limited to 40% turndown
Eddy Current Coupling	Variable speed of the driven device, constant motor speed	0-100% turndown
Magnetic Coupling	Variable speed of the driven device, constant motor speed	0-100% turndown
Adjustable Speed Drive or Variable Frequency Drive	Variable motor speed. Usually variable torque through adjustable volts/ frequency ratio at different speeds	0-100% turndown

Table 11-4. Other Capacity Modulation Devices

Capacity modulation device(s)	Principle	Remarks
Inlet Slide Valves (Screw Compressors)	Slide mechanism uncovers part of the rotor, shunting part of the mass flow back to the suction.	20-100% capacity usually corresponds to a 50-100% power consumption.
Variable Pitch Fan Blades (Axial Fans)	Reduced rake of the impeller unloads the fan.	Unloading curve very similar to Adjustable Speed Drive (or VFD). 0-100% capacity usually corresponds to a 25-100% power consumption.
Modulating Inlet Fan Sleeve (Centrifugal Fans)	A sliding sleeve exposes all or part of the impeller, changing its effective wheel width to load and unload the fan.	0-100% modulation for constant speed fans. More efficient than inlet vanes and no energy penalty in wide-open position. Proprietary.
Variable Pulley Systems	Modulate the speed of the driven device by moving the sides of an adjustable drive pulley in or out, changing the diameter of the pulley.	Once relatively common, this technology is now generally replaced by other methods. These were nicknamed "belt squeezers" or "pulley pinchers." While capable of modulating speeds, the devices were noisy, hard on belts, and had relatively high mechanical losses.

11.20 THEORETICAL POWER CONSUMPTION

Figure 11-10 illustrates the energy saving potential of the application of *adjustable speed drive* to an application that traditionally uses throttling control, such as a *discharge damper, variable inlet vane,* or *eddy current drive.*

From the standpoint of maximum energy conservation, the optimal method to reduce fan CFM is to reduce the fan's speed (RPM). This can be accomplished by changing either the sheaves of the motor, or the sheaves of the fan, or by varying fan motor speed.

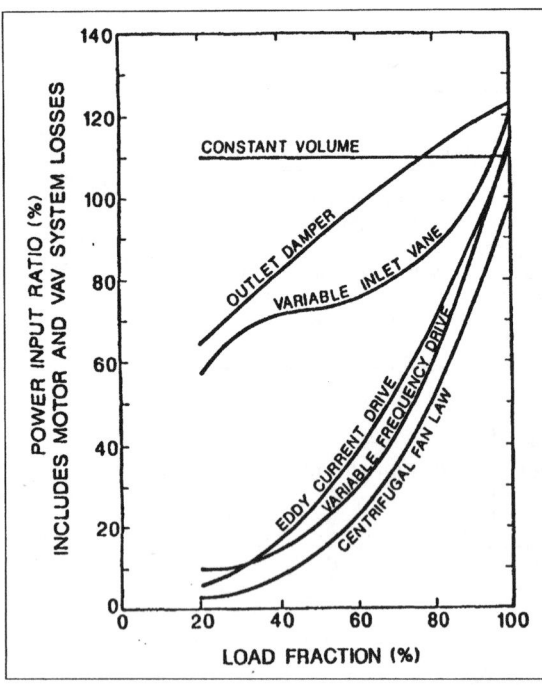

Figure 11-10. Typical Power Consumption of Various Fan Control Systems (Source: Moses et al., 1989)

Figure 11-11. Fan and Pump Capacity Control Methods

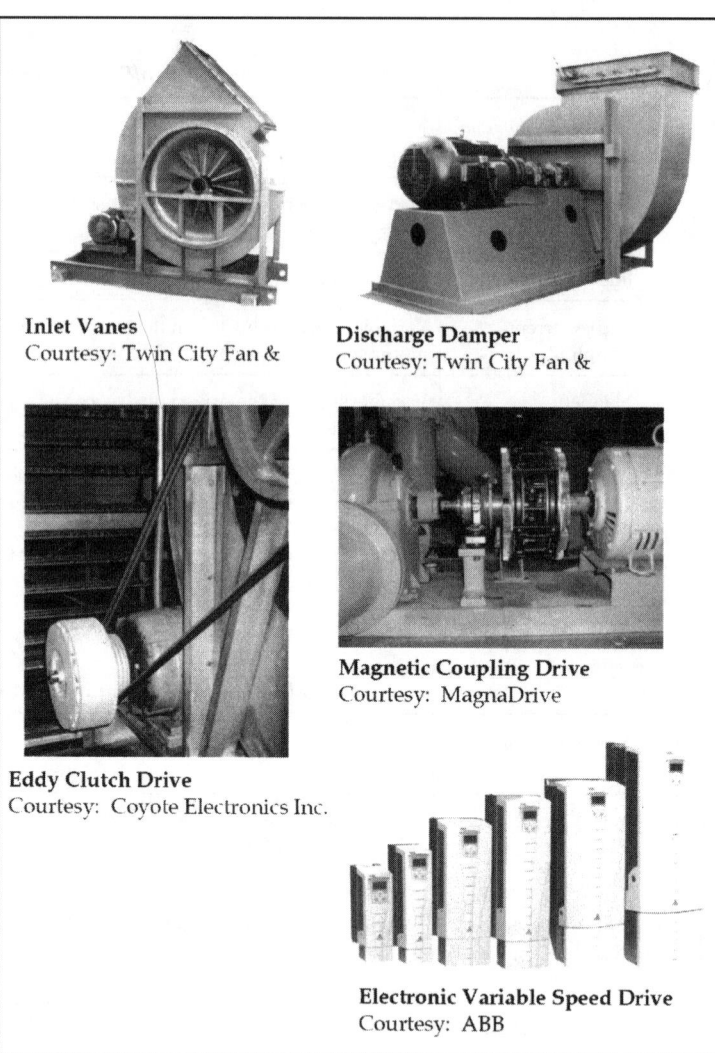

Inlet Vanes
Courtesy: Twin City Fan &

Discharge Damper
Courtesy: Twin City Fan &

Magnetic Coupling Drive
Courtesy: MagnaDrive

Eddy Clutch Drive
Courtesy: Coyote Electronics Inc.

Electronic Variable Speed Drive
Courtesy: ABB

Figure 11-12. Screw Compressor Slide Valve Capacity Modulation (12)

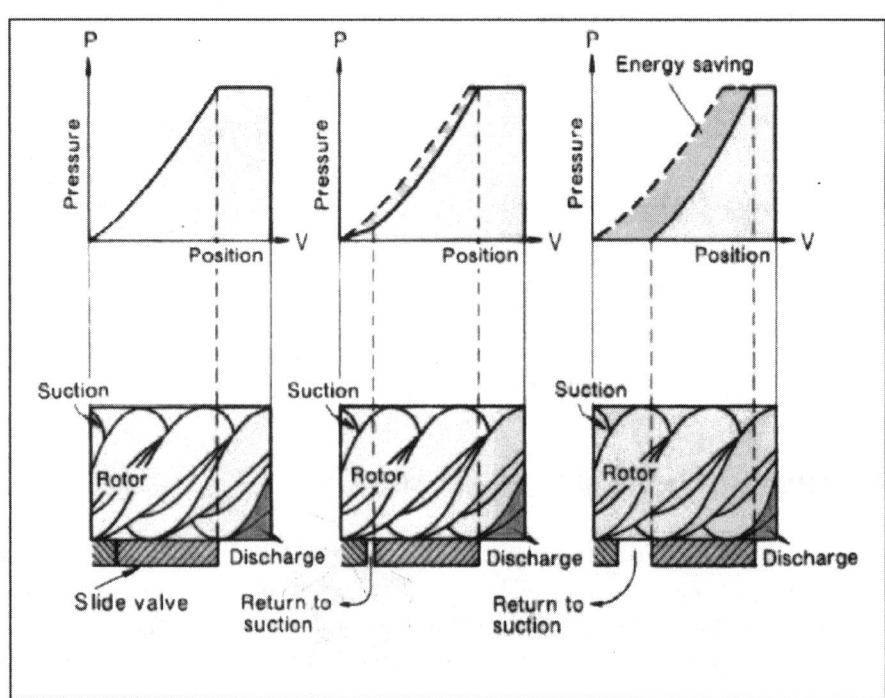

Savings Example

Existing condition: 6000 hours per year, 100 HP constant volume fan
Proposed: Variable Speed Drive (VFD)
Cost of energy: $0.08 per kWh
Motor efficiency = 90% (assumed constant)
Full load VFD efficiency loss assumed to be 5%
Use Square Law instead of Cube Law: HP2 = HP1 X (Q2/Q1) 2.0
Minimum Flow 50%, minimum load 30% regardless of calculation.
Baseline consumption: 6000 hours * 100 HP = 600,000 Hp-Hrs
Load Profile and calculation:

Pct Air Flow	Hours Per Year		Q1	Q2	Hp1	Hp2	HP-Hrs existing	HP-Hrs w/VFD	Hp-Hr savings
100%	300		100%	100%	100	105	30,000	31,500	
90%	500		100%	90%	100	81	50,000	40,500	
80%	1500		100%	80%	100	64	150,000	96,000	
70%	1500		100%	70%	100	49	150,000	73,500	
60%	1000		100%	60%	100	36	100,000	36,000	
50%	1200		100%	50%	100	30 (min)	120,000	36,000	
							600,000	313,500	286,500

Energy savings: 286,500 Hp-Hr.
Conversion to kWh: kWh = (Hp-Hr) * 0.746 * (1/ Motor EFF)
286,500 * 0.746 * (1/ 0.90) = 237,477 kWh
237,477 kWh * 0.08 = $18,998 per year savings (ans.)

Note:
The same calculation method can be done by converting Hp1 and Hp2 to kW, using kW = (Hp) * 0.746 * (1/ Motor EFF).
The added two columns and the final three columns would then look like this:

Pct Air Flow	Hours Per Year		Q1	Q2	Hp1	Hp2	kW1	kW2	kWh existing	kWh w/VFD	kWh savings
100%	300		100%	100%	100	105	82..9	87.0	24,867	26,110	
90%	500		100%	90%	100	81	82..9	67.1	41,444	33,570	
80%	1500		100%	80%	100	64	82..9	53.0	124,333	79,573	
70%	1500		100%	70%	100	49	82..9	40.6	124,333	60,923	
60%	1000		100%	60%	100	36	82..9	29.8	82,889	29,840	
50%	1200		100%	50%	100	30 (min)	82..9	24.9	99,467	29,840	
									497,333	259,857	237,477

Either approach lends itself easily to a standard spreadsheet. Reminder: The math is easy, but the accuracy of the prediction depends upon the accuracy of the load profile.

CASE STUDIES

The following case studies are included as examples of possible VFD applications and the analysis procedures undertaken in the preliminary systems analysis.

Boiler Combustion Fan

This application involves the use of a VFD to vary the speed of a centrifugal combustion air intake fan (50 nameplate horsepower) on a scotch marine-type high pressure steam boiler. The existing system utilizes an actuator to simultaneously vary the amount of gas and air that enters the burner. The air is controlled with the use of inlet dampers (not guide vanes). As the amount of "fire" is reduced, the damper opening is reduced and visa versa. The centrifugal fan and continuous variation in fire rate make this a feasible VFD application. The load profile of the boiler and corresponding motor demand (measured with demand metering equipment) is listed in Table 11-5. This table also includes the cor-

Table 11-5. Example boiler combustion fan load profile and VFD savings

Operating Time at load (hours)	Percent Full Load Power[2]	Existing Load w/ Damper (kW)[1]	Load with VFD Control (kW)[2]	Kilowatt Savings	Kilowatt-Hour Savings
1000	100	37	37	0	0
2000	105	40	30	10	20,000
3000	95	35	20	15	45,000
1000	90	33	10	23	23,000

[1]Measured with a Demand Meter

[2]Approximated using Figure 11-4

                                                                                        88,000 kWh

responding VFD demand requirements (approximated from Figure 11-4), kW savings, and kWh savings.

The annual savings for this example totaled 88,000 kWh, which would equate to an annual savings of $4,400 (based upon a cost of energy of $0.05/kWh). Lobodovsky provides an average estimated installed cost of VFDs in this size range at around $350 per horsepower, or an installed cost of $17,500 (50 hp * $350/hp). This would yield a simple payback of around 4 years. This example does not take into account demand savings which may result if the demand reduction corresponds with the plant peak demand. The control of the fan VFD would be able to utilize the same output signal that the existing actuator does.

Industrial Chiller Plant

A different calculation procedure will be used in the following example. A malting plant in Wisconsin uses seven 550-ton chillers to provide cold water for process cooling. Three of the chillers work all of the time and the other four are operated according to the plant's varying demand for cooling. The chillers are currently controlled by variable inlet vanes (VIV).

The typical load diversity and the power input required for one of the chillers under varying load were obtained from the manufacturer. In addition, the data in Table 11-6 give the power input of a proposed variable frequency drive (also referred to as adjustable speed drive [ASD]) for one chiller that operates about 6,000 hours per year.

A weighted average of the percent load and percent run time fractions gives a load diversity factor (ldf) of 64.2 percent. This means that, on the average, this chiller operates at 64 percent of its full load capacity. A duty-cycle fraction weighted average of the savings attainable by a VFD can be estimated, which is a 26.6 percent savings per year. The energy savings due to avoided cost

Table 11-6. Example centrifugal chiller load and power input (Source: updated from Moses et Al., 1989)

Percent Load (1)	VIV % kW (2)	ASD % kW (3)	Savings % kW (4)	Percent Run Time (5)
0.3	41	17	24	0.08
0.4	53	22	31	0.10
0.5	66	33	33	0.13
0.6	75	40	35	0.18
0.7	82	54	28	0.22
0.8	87	63	24	0.15
0.9	92	78	14	0.09
1.0	100	100	0	0.05

Column 2.   From typical compressor performance with inlet and vane guide control (York Division of Borg-Warner Corp).

Column 3.   From Carrier Corporation's Handbook of Air Conditioning Design: Comparative Performance of Centrifugal Compressor Capacity Control Methods.

Column 5.   Actual performance data.

of electricity usage are computed as follows: Savings = (0.266)(550 ton)(0.7 kW/ton)(0.642)(6,000 hr/yr) = ldf= 394,483 kWh/year. The dollar savings at $0.04 per kWh are (394,483 kWh/year)($0.04/kWh) = $15,779/year. For a 500 horsepower variable frequency drive, the installed cost is estimated at $75,000, based upon Lobodovsky average installed cost of $150 per ton for units of this size. Therefore, the payback period is:

($75,000)/($15,779/year) = 4.75 years.

11.21 MOTORMASTER

Motors are an identifiable efficiency upgrade target since there are so many of them in use. The Energy Efficiency and Renewable Energy (EERE) branch of the U.S. Department of Energy (DOE) provides and maintains a free software tool for public use called MotorMaster.

This tool helps users identify inefficient or oversized facility motors and calculates the savings that can be achieved with more energy efficient motors. The software includes input screens for site-specific information such as motor size, percent load, hours of operation, and cost of power. In the background is a tabulation of commercially available motor sizes and efficiencies. The input screens are intuitively designed and made for widespread public use.

Go to www.eere.energy.gov and search for "Motor-Master."

11.21 SUMMARY AND SUGGESTIONS FOR FURTHER STUDY

This chapter focused mostly on electric motors and drive systems and provides practical methods for understanding and managing electric motors. Most electrical power equipment, like the utility sources that supply them, is a servant to some other process, i.e the energy flows through it to get to its ultimate destination. In the case of a motor, whatever is connected to the output shaft is the primary determinant in the amount of energy the motor relays. Thus, good motor management goes beyond the motor itself to the concept of the motor-driven assembly or system.

Also discussed was another common electrical concern in facilities, power factor, and how to correct for it.

Some additional electrical concepts related to energy management are shown here for additional study:

- Cogeneration (Chapter 7)
- Electric utility rates (Chapter 18)
- Electric deregulation (Chapter 24)
- Lighting (Chapter 13)
- Use of alternative energy (Chapter 16)
- "Ghost" loads from appliances plugged in.
- Transformer efficiency
- Standby losses from transformers
- Power conditioning
- DC motors
- Harmonics

References

1. *Technology Update* produced by the Washington State Energy Office for the Bonneville Power Administration.
2. NEMA Standards publication No. MG 10. *Energy Management Guide for Selection and Use of Polyphase Motors*. National Electrical Manufacturers Association, Washington, D.C. 1989.
3. McCoy, G., A. Litman, and J. Douglass. *Energy Efficient Electric Motor Selection Handbook*. Bonneville Power Admin. 1993.
4. Ed Cowern, Baldor Electric Motors.
5. MotorMaster software, U.S. Dept of Energy, Energy Efficiency and Renewable Energy (EERE), Industrial Technologies Program.
6. Financing for Energy Efficiency Measures, Virginia Power, POST-AND, REV(0)08-17-93.
7. William E. Lanning and Steven E. Strauss, Power Survey Techniques for Electrical Loads, Esterline Angus Instrument Corp.
8. Konstantin Lobodovsky, "Efficient Application of Adjustable Speed Drives," *Motor Manager*, Penn Valley, CA, 1999 (530).
9. Optimizing Your Motor-Driven System, US. Department of Energy, Motor Challenge Fact Sheet, 1996.
10. Commercial Energy Auditing Reference Handbook, Doty,S., 2008, Fairmont Press.
11. *Energy Efficiency in Motor Driven Systems*, Parasalilti, F., P. Bertoldi, Editors, Springer, ISBN 3-540-00666-4. Original data from technical paper *Hybrid Permanent Motors Allow for the Development of High Efficiency Screw Compressors*, R. Cook.
12. Process Gas Applications Where API 619 Screw Compressors Replaced Reciprocating and Centrifugal Compressors, Ohama,T., Tanaka,H., and Koga,T., 2006, 35th Turbomachinery Symposium.

APPENDIX

ELECTRICAL FORMULAS & RULES OF THUMB

Conversion formulas

REQUIRED	DIRECT CURRENT	ALTERNATING CURRENT	
		SINGLE-PHASE	THREE-PHASE
AMPERES WHEN H.P. IS KNOWN	$\dfrac{\text{H.P.x746}}{\text{ExEFF.}}$	$\dfrac{\text{H.P.x746}}{\text{ExEFF.xP.F.}}$	$\dfrac{\text{H.P.x746}}{\text{1.73xExEFF.xP.F.}}$
AMPERES WHEN KILOWATTS ARE KNOWN	$\dfrac{\text{KWx1000}}{\text{E}}$	$\dfrac{\text{KWx1000}}{\text{ExP.F.}}$	$\dfrac{\text{KWx1000}}{\text{1.73xExP.F.}}$
AMPERES WHEN KVA IS KNOWN		$\dfrac{\text{KVAx1000}}{\text{E}}$	$\dfrac{\text{KVAx1000}}{\text{1.73xE}}$
KILOWATTS	$\dfrac{\text{IxE}}{\text{1000}}$	$\dfrac{\text{IxExP.F.}}{\text{1000}}$	$\dfrac{\text{IxExP.F.x1.73}}{\text{1000}}$
KVA		$\dfrac{\text{IxE}}{\text{1000}}$	$\dfrac{\text{IxEx1.73}}{\text{1000}}$
HORSEPOWER OUTPUT	$\dfrac{\text{IxExEFF.}}{\text{746}}$	$\dfrac{\text{IxExEFF.xP.F.}}{\text{746}}$	$\dfrac{\text{IxEx1.73xEFF.xP.F.}}{\text{746}}$

I = AMPERES   H.P. = HORSEPOWER   EFF. = EFFICIENCY (EXPRESSED IN A DECIMAL)
E = VOLTS   P.F. = POWER FACTOR   KW = KILOWATTS   KVA = KILOVOLT-AMPERES

Rules of thumb for motors.

At 3600 RPM, a motor develops 1.5 lb.-ft. per HP.

At 1800 RPM, a motor develops 3 lb.-ft. per HP.

At 1200 RPM, a motor develops 4.5 lb.-ft. per HP.

At 550 & 575 Volts, a 3 phase motor draws 1 amp per HP.

At 440 & 460 Volts, a 3 phase motor draws 1.25 amp per HP.

At 220 & 230 Volts, a 3 phase motor draws 2.5 amp per HP

TECHNICAL FEASIBILITY CHECKLIST
For AFD/VFD

1. How would changing the speed of the driving motor cause a change in the process or its rate?

2. Will product quality be improved or impaired?

3. What effects will the improvement or impairment have?

4. In what way can the machinery operate at other than its current speed?

5. In what way can any speed-changing mechanism (such as step pulleys, gears, or fluid drives) be installed to provide suitable electrical signal(s)?

6. If the existing process is mechanically modulated (dampers, valves, gates), how would new sensor(s) be installed to provide suitable electrical signal(s)?

7. If the existing process is electrically modulated (de motors or wound-rotor induction motors), how would squirrel cage induction motors be adapted to the equipment?

8. Describe how a process modulating control has been (or would be) applied to a drive system of this type.

9. Describe the physical space for installing a new or additional electrical motor controller, (conventional induction motor ASD electronics need at least twice the space of existing starters: synchronous motor ASDs may need more.)

10. If the existing constant speed motor is a totally enclosed fan cooled induction motor, how would additional ventilation be provided if needed, when operating at lower speeds using an ASD?

11. How much derating of the existing motor would be necessary for heating caused by harmonics? Is this derating acceptable?

12. If the machinery is a pump, fan, or compressor, what data sheets or test means are available to estimate the operating characteristics of the unit?

13. Are data, drawings or other means available to estimate the torque requirements at various speeds for machinery other than pumps, fans, or compressors?

14. What drawings or other means available to validate the construction or installation details of the motor and machinery involved?

ECONOMIC FEASIBILITY CHECKLIST
For AFD/VFD

1. How will a change in the speed of the driving motor result in lower energy requirements?

2. What are your electrical energy costs in terms of your utility bill (consumed kWh, demand charges, etc.) or in terms of product costs?

3. If the existing process is mechanically modulated, what portion of the operating time is at other than maximum flow (or load)?

4. How many hours per week does the equipment operate?

5. How will the use of an ASD improve quality (through better speed control, elimination of waste, product reversion, etc.) and/or ultimately result in lower product costs?

6. What costs associated with drive inefficiencies (friction heat, cooling water, etc.) can be reduced by using an energy efficient ASD system?

7. What are the costs for maintaining existing mechanical speed-changing equipment (transmissions, etc.)? Are they obsolete and in need of replacement?

8. What are the costs of maintaining existing electrical speed-changing equipment (de and wound-rotor motor, or reduced-voltage starting)? Are they obsolete and in need of replacement?

9. Are there other problems of equipment reliability that cause production delays and higher product costs? How can they be eliminated by ASDs with self-diagnostic features?

10. What opportunity is there to create additional space by removing large mechanical equipment (transmissions, etc.) with the installation of an ASD controller?

11. Can plant noise be reduced through lessening of the mechanical noise by installing an ASD control, or will the noise of the ASD be excessive?

12. Describe the shutdown arrangements required to provide time for an ASD to be installed.

13. Will the addition of an ASD require a new inverter duty motor? If yes, will the extra cost of the motor, plus the cost of the ASD, still provide acceptable economic return?

CHAPTER 12
ENERGY MANAGEMENT CONTROL SYSTEMS

MICHAEL P. DIPPLE
Associate Director of Physical Plant
Georgia Southern University

DALE A. GUSTAVSON, C.E.M.
D.A. Gustavson Company
Orange, California
(714) 639-6100

12.1 INTRODUCTION

Successful building owners are constantly challenged to provide the highest level of service and comfort while operating within very constrained budget limits. Sustainability awareness is further challenging traditional design and control of heating, ventilating, air conditioning (HVAC), and lighting functions. Facility owners and operators have strong financial incentives to more closely match HVAC and lighting control and zoning, equipment sizing, operating strategies, resources, and labor management. This must be accomplished without sacrificing comfort and safety. Energy management control systems (EMCS) can play a key roll in meeting these challenges.

12.2 ENERGY MANAGEMENT CONTROL SYSTEMS (EMCS)

Energy Management Control Systems can monitor and control an individual building, groups of buildings, a campus, or any combination from either a centralized or decentralized location anywhere on the planet. Current generation EMCS are designed to utilize copper, fiber, intranet, and internet communication paths. The most common method, in a medium to large installation, will use the company's intranet in a networked system to deliver information to one or many operator workstations.

The goal of early generation EMCS (1970s) was to simply turn on and off devices based on need and ease of remote operation rather than a conscious effort to control energy usage. These early systems usually had an interface panel whereby computer on/off signals were converted to actions through a relay panel. While crude, they represented a significant advance over automatic (gener-

ally pneumatic) control systems. Most first generation systems have been replaced or upgraded by now. The advent of solid state electronic control devices and the increasing power of computers led to the second generation of EMCS. Relatively simple electronic control boards were now able to convert electric signals to pneumatic operations of valves and dampers. This was a significant advance over first generation systems. With earlier systems the control was on/off. Now building operators had devices to perform proportional control. Rudimentary control strategies began to show up as software became better defined.

The evolution continued with third generation systems. Further advances in end device technologies allowed electronic actuators to be cost effective and powerful enough to replace pneumatic devices. Computers went from mini-mainframes to personal computers, and controllers went from electronic to digital. Computing power was pushed down to the controller level. Software developments lead the charge to the future with much improved control strategies. Arguably, the most important change of the era was the advent of graphical representations of data. So called human machine interface (HMI) now allowed less technical operators to be very successful managing the day to day operations of a facility. The software is mostly developed at this point with control strategies in place and available for deployment by the operator. EMCS advances are focused on communication technology. Systems today are mostly designed to operate on a common enterprise backbone utilizing internet protocol (IP). This reduces plant wiring and installation costs, and it truly delivers a network-based system capable of reporting information and executing control demands as the owner desires, including wireless technologies.

12.2.1 Direct Digital Control (DDC)

Manufacturers have greatly enhanced EMCS by incorporating direct digital control (DDC). DDC is defined as a digital computer that measures particular variables, processes these data via control algorithms, and controls a terminal device to maintain a given setpoint or the on/off status of an output device. Inputs and outputs relative to a DDC EMCS can be either digital or analog. Most inputs are analog signals converted to digital signals by the DDC controller, and most outputs are likely to be digital (zero to full voltage or amperage or anywhere in between). Newer end devices can be network compatible. Devices that are network compatible offer the advantages of being able to talk on a given network design and eliminate the need to figure out some other method of integration. Other integration methods include field installed translation devices or even hard wired interlocking. Networking allows much better fault tolerance measuring and trouble shooting opportunities.

DDC systems use software to program microprocessors, thereby providing tremendous flexibility for controlling and modifying sophisticated control applications. Changing control sequences by modifying software allows the user to improve performance of control systems throughout a building. In many current DDC controllers programming takes the form of answering questions and setting parameters rather than writing software. The software is written and verified at the manufacturer level, thereby reducing errors and field labor during installation.

DDC EMCS can be programmed for customized control of HVAC and lighting systems and to perform facility wide energy management routines, such as electrical peak demand limiting, ambient condition lighting control, start/stop time optimization, site-wide chilled water reset and hot water reset, time-of-day scheduling, and outdoor air free cooling control. An EMCS using DDC can integrate automatic temperature control functions with energy management functions to ensure that HVAC systems operate in accord with one another for greater energy savings.

The primary objective of an EMCS using DDC is to optimize the control and sequencing of mechanical systems, thereby reducing energy consumption. A DDC EMCS also allows centralized and remote monitoring, supervision, and programming maintenance of the HVAC and lighting functions. Additionally, such systems can lead to improved indoor environmental comfort and air quality.

Be it new construction or a retrofit, when selecting an EMCS the actual needs and requirements of the particular building or campus of buildings must be considered. While facility layout and construction type will influence EMCS configuration, software and technical support are the most critical elements in any EMCS application.

12.2.2 Hardware

In the broadest sense, an EMCS can be from a simple stand-alone unitary microprocessor based controller to a very sophisticated large building DDC EMCS that interfaces with other building systems. Stand-alone unitary microprocessor based controllers will usually have firm-

ware routines (control software logic that the user cannot modify except for setpoints, also known as "canned software") that provide control of a terminal unit such as a heat pump. This chapter will generally cite examples from applications in medium to large facilities. The reader should bear in mind, however, numerous documented successes with stand-alone controllers in small buildings. Fast food and dinner houses, auto dealerships, retail stores, bowling centers, supermarkets, branch banks, small commercial offices, etc., have all benefited from this technology. Often, equipment manufacturers ship smaller HVAC products with on board controls. The EMCS or equipment manufacturer, distributor, installing contractor, end user, or some combination will select the parameters and input the choices into the controller's firmware. The EMCS user or designer should verify the connectivity of controllers installed at the factory to ensure compatibility with the EMCS network. A good strategy is to control the connectivity with the guide specifications.

DDC Controllers

Digital controls come in a variety of styles, to suit the application. For common applications, mass produced and low cost definite purpose controllers are available. These use factory programming, also called "canned" software, which speeds application and start-up but may limit the opportunities for creating special instructions or optimization (unless sufficient programming is available from the supervisory controller to achieve the intended result). Multi-purpose generic controllers are also available that utilize custom programming. These generally come with a fixed number of input and output points, but the point capacity can often be expanded with multiplexers or other point expander cards. Since the programming is custom, these controllers are the natural choice for optimization routines that require dynamic calculations and adaptability. For control processes with a large number of points, point expander boards are usually preferable to multiple linked controllers because having a single processor brain eliminates some unpredictable operations modes that come from partial failures (e.g. if just one of the controllers fails). Conversely, too many expander boards can create a larger single point of failure.

Field User Interface

For effective operations and maintenance, the user interface should not be limited to operator workstation graphics. Like the pressure gages or test ports of a pneumatic system, some form of user interface is helpful at the point of control. This may be a fixed display-adjusting device that can scroll through a few points or a connection jack for a portable operator workstation that can be connected at will. Without some provision for user interface in the field, the DDC system can introduce a frustration to the operations personnel who cannot "see" what is going on at the equipment locations. Lack of field user interface can result in operators who give up trying to understand the control sequences, resorting to the hands-off approach of "it's all being controlled by the computer."

DDC EMCS hardware configuration varies from manufacturer to manufacturer but has a common hierarchical configuration of microprocessor-based digital controllers as well as a front-end personal computer. Generally speaking, sensors are hard wired to terminal controllers (although wireless is available) and terminal controllers are cabled together for communications to system controllers. Most EMCS manufactures have gravitated towards IP packet technologies to capture the advantages of uniformity, speed, and available enterprise IT backbones. IT technologies have largely replaced modems and telecommunications interfaces.

An important part of a DDC EMCS is the system controllers, because they monitor and control most mechanical equipment. A major feature of system level controllers is their ability to handle multiple control loops and functions, such as proportional-integral-derivative (PID) control, energy management routines, and alarms. The terminal equipment controllers are single control loop controllers with specific firmware. The operator interface level (HMI) is an input/output (I/O) device, generally a personal computer (but may also be a hand-held device), that serves as the primary means to monitor the network. Monitoring may include looking for specific data and alarms, customizing the control software for downloading to specific system controllers, maintaining time-of-day scheduling, and generating management reports. The HMI and software bring into play the operator and interaction with the feel of the system. The "human dimension" is a factor that must be clearly understood and weighed. This is true for small design/build projects as well as for very large, sophisticated, engineered projects. The personnel factor must not be ignored. Different building managers and occupants have different needs and occupancy habits. Also, no matter how well a system is designed, if the building operators and occupants are not properly taught how to use and maintain it, the EMCS will never live up to its full potential.

The specific method of communications within an EMCS is significant because of the amount of data being processed simultaneously. While methods of data transfer between DDC controllers and the operator interface level vary from manufacturer to manufacturer, the technology falls largely into two distinct categories:

- Poll/response
- Peer-to-peer or token-ring-passing network

A peer-to-peer network does not have a communication master or center point as does the poll/response system. Every trunk device, be it a terminal level controller attached to a systems level controller or a system level controller networked to other system level controllers, at some point has a time slot allowing it to operate as the master in its peer grouping. The terminal level controller in a peer-to-peer network is usually a poll/response device. A system using peer-to-peer communication can offer distinct advantages over poll/response communication when redundancy of critical global data is accommodated. These advantages include:

- Communication is not dependent on one device and thus is more failure-resistive.
- Direct communication between controllers does not require communication through the operator interface level.
- Global information can be communicated to all controllers quickly and easily.

The speed of system communication in building control is not usually an issue with DDC technology. The response time for reaction to control parameters can be an issue with the application of a centralized poll/response system where control and monitoring point densities are very high or alarm response time is critical. Generally, peer-to-peer communication distributive systems that network system level controllers and terminal equipment level controllers provide the quickest response time.

Although technological gains in desktop computer processing power lend some advantages to poll/response systems that use a personal computer for centralized control, panelized DDC EMCSs are still very common. It used to be the case that using a host PC for building system automation allowed more complicated sequences of operation to be implemented and greater amounts of trend data to be stored. Reduction in physical size as well as cost for computer memory chips has led to much greater capability for stand-alone control units, and as a result there is not much of a performance gap between centralized poll/response and panelized DDC systems.

The selection of EMCS type and sophistication for any given application should balance management and control needs with first cost installation and predicted lifecycle service support and upgrade expenditures. An all encompassing DDC EMCS will provide the best overall control and management capability, but it is also the most expensive. On the other hand, stand-alone controllers are the least expensive for individual control applications; they also limit control strategy and management capability. The best solution is the system that most closely matches the owners' needs, keeping in mind the parameters mentioned above.

Early EMCS manufacturers offered many different controllers, each somewhat specialized. DDC controllers were offered in many versions to match the application. EMCS often had total connected point limits. Once exceeded, the next larger system would need to be purchased. Current trends in controller hardware are to consolidate the number of product controller options and to increase the firmware and software capabilities. Modern systems are much more modular with a "building blocks" architecture instead of system size categories.

One important note: Many sustainability organizations mandate that certain control processes be included in the design. For instance, LEED certified buildings may require outdoor air flow measurement. Make sure the controllers selected and the EMCS software can support what is to be accomplished.

Most existing buildings that are potential candidates for EMCS installations will have already performed some level of retrofit. The cost of energy and market competitiveness will have forced building owners to have at least "harvested the low fruit" by now. Piecemeal renovations over the years will present many challenges. Bit by bit replacement is more common than one might expect, due largely to budgetary pressures. The challenge increases if individual projects were completed by different vendors using different manufactures equipment that may not integrate well or perhaps not communicate at all. While the individual projects results might be worthwhile, without a clear goal and commitment it is very difficult to institute an EMCS capable of delivering to full potential.

12.2.2.1 DDC—Pneumatic Hybrid Conversion

Some existing pneumatic control systems are renovated in an overlay or hybrid fashion, utilizing the existing pneumatic end devices and tubing. The DDC control system output hardware "stops" at an intermediary device called a transducer that converts the DDC output into a pneumatic output, driving the existing devices. Key sensors are added to effect control and provide supervisory control. This method offers economical advantages compared to a complete new DDC system, and it is a step forward in technology and capability; however, its limitations include:

- Pneumatic transducer drift and error.
- Pneumatic end device problems. Any existing problems with actuators and tubing remain in the system.
- Device overlap. For multiple pneumatic actuators

sequenced off of a common signal (common practice with pneumatic controls), the control of heat/cool equipment in sequence is only as accurate as the spring ranges of the devices and not under the direct control of the DDC system.

First cost of DDC EMCS installations were initially a barrier, but this decreased with popularity, and they are now standard of care, with pneumatics all but obsolete in new commercial buildings. For some very large or difficult actuation applications pneumatics may still be used, but those are the exception. Direct electronic actuation with no intermediary transducers offers the greatest potential for DDC control. Compatible technologies in computers, networking, and communications have been adopted, further supporting DDC technology. Additional advantages of DDC are:

- More precise control
- Customization of control schemes for energy management and comfort
- Centralization and integration control and monitoring of HVAC and lighting
- Easier to maintain
- Easier to expand and "grow" with building size and use.

12.2.3 Software

The effectiveness of the software control logic is what provides the building operator with the benefits of an energy management system. While color graphics and other input/output (I/O) capabilities are important, the control logic in the system and terminal controllers is what improves building systems' efficiency and contributes to energy savings.

Networking DDC devices has provided building operators with the ability to customize the traditional energy management strategies such as time-of-day/holiday scheduling, demand shedding, duty cycling, optimum start/stop, and temperature control. Networking also allows the implementation of energy management techniques, such as occupied/unoccupied scheduling of discrete building areas, resulting in reduced airflow volumes and unoccupied period setback strategies that greatly reduce operating cost. PID control provides for more accurate, precise, and efficient control of building HVAC systems (see Chapter 20). Software that provides for adaptive or self-tuning control not only enhances savings but also addresses environmental quality. Adaptive control software monitors the performance of a particular control loop and automatically adjusts PID parameters to improve performance. This feature improves control loop response to more complicated and dynamic processes.

12.2.4 EMCS Graphics

A graphical user interface (GUI) is an overlay onto most EMCS/DDC control systems. While labor-intensive to create, these are useful for gaining acceptance of the control system and can reduce the skill level needed for digital control system navigation. The GUI uses easily recognized icons and symbols, color-coded messages and alarms, and other visual methods to increase the user-friendliness of the system. EMCS graphic displays can also serve as training media by offering a quick visual representation of the system under control.

12.2.5 Proprietary vs. Open DDC Controls

The original DDC controls were proprietary, which means they connected to only other equipment by the same manufacturer and would not connect (communicate) to any other manufacturer's system. The topic of proprietary controls, captive customers, open protocols, hybrid systems, gateways, etc. is an important topic in the control industry evolution. While the concept is not difficult to grasp, details and implications are very numerous and complex, and they are beyond the scope of this text. The basic argument FOR proprietary systems is the one-stop-shopping ease of purchasing and the security of all-encompassing technical resources for system maintenance and system integration. For large or complex systems this can be an important benefit. The basic argument AGAINST proprietary systems is the proprietary lock on the customer, and how the lack of competition almost certainly results in higher replacement costs that escalate during the ownership life cycle.

Both proprietary systems and "open protocol" systems are available, and both are viable, requiring a customer choice where new systems are proposed. Gateways and translator offerings are hardware go-between solutions that offer limited connectivity between proprietary systems. These are common solutions but can create new problems even as they solve others—and many retain the proprietary nature they are intended to solve.

12.2.6 DDC Information Technology (IT) System Maintenance

Systems that share communication infrastructure with other, more sensitive systems require careful attention to information configuration and maintenance. The ideal approach is to have computer information technology (IT) capabilities on staff, devoting their time to control system integration activities as an in-house expense; however, this may be cost prohibitive for all but the largest systems. The more common "drop-in" approach is to select an industry partner to provide this technical service, although this can gravitate back toward proprietary

solutions unless an independent status from any equipment manufacturer is maintained.

12.2.7 Specifications and Procurement

The building owner or representative who is responsible for system design and specifications is best served by concentrating on a performance-based specification and procurement process. Usually, an owner will retain a mechanical consulting engineer to prepare procurement or bid documents that include plans and specifications. The specifying party will be informed and will probably be guided to some extent by the various EMCS manufacturers. It is important the owner be involved and understands the decisions and compromises taking place as the design develops. The outcome should be a system that meets the buildings needs, is cost effective, and is manufacturer neutral, although systems additions often will become manufacturer specific.

HVAC mechanical equipment operation sequences are becoming increasingly complex. This is a result of stricter criteria established by national codes and standards, such as those published by the American Society of Heating, Refrigerating and Air Conditioning Engineers (ASHRAE), LEED, Energy Star, and energy codes. Because of increasing awareness of environmental sustainability, and because building operators require highly accurate control systems, the demands for energy-use monitoring and accounting are increasing. To meet the challenges posed by these stricter codes and standards, building operators demand the EMCS specifying team be thoroughly knowledgeable about all aspects of EMCS design. It is essential that the EMCS specified be technically advanced, provide maximum value and performance, include appropriate technical support, and be easy to use and maintain. The specifying engineer must be knowledgeable about not only the building construction and mechanical and lighting systems to be controlled, but also the businesses and people which occupy the building. Of equal importance is the need to assess the level of training and ongoing support that the EMCS owner/operator will need to realize the anticipated benefit.

12.2.8 Control Strategies

A DDC EMCS can serve the following basic functions:

- Manage the demand or need for energy at any given time
- Manage the length of time that devices consume energy
- Set alarms when devices fail or malfunction
- Facilitate monitoring of HVAC system performance and the functioning of other building systems, in-

cluding trends, logs and reports
- Assist the building operator to administer equipment maintenance

There are financial benefits with each of these functions, yet traditionally, only the first two are quantified in an economic analysis.

The energy management industry has evolved standard, time-proven software routines that can be used as a starting point to develop effective programming. These routines are commonly referred to as control strategies. While these routines can be applied in virtually all EMCS installations, they must be customized for each building. A description of some of these routines follows. See also Chapter 20.

Daily Scheduling

This routine provides for individual, multiple start and stop schedules for each piece of equipment, for each day of the week.

Holiday Programming

This routine provides for multiple holiday schedules, which can be configured up to a year in advance. Typically, each holiday can be programmed for complete shutdown, where all zones of control are maintained at setback levels, or for special days requiring the partial shutdown of the facility. In addition, each holiday generally can be designated for a single date or a range of dates.

Yearly Scheduling

Typically, any number of control points can be assigned to special yearly scheduling routines. The system operator an usually enter schedules for yearly scheduled control points for any date the year. Depending on the particular EMCS software operating system, yearly scheduled dates may be erased once the dates have passed and the schedules have been implemented, or the scheduled dates may repeat in the following year as with schedule holiday dates.

Demand Limiting or Load Shedding

Demand limiting can be based on a single electric meter or multiple meters. Generally, loads are assigned to the appropriate meters in a specified order in which equipment is to be shed. Usually a control point that is allowed to be shed is given a status as a "round-robin" demand point (first off-first on), a "priority" demand point (first off-last on), or a "temperature" demand shed point (load closest to setpoint is shed first). Other parameters per control point include rated kW, minimum shed time, maximum

shed time and minimum time between shed.

Minimum On/Minimum Off Times

Normally a system turns equipment on and off based on temperatures, schedules, duty cycles, demand limits, and other environmental parameters. However, mechanical equipment is often specified to run for a minimum amount of time once started and/or remain off for a minimum amount of time once shut down. Therefore, this routine gives the operator the ability to enter these minimum times for each piece of equipment controlled.

Duty Cycling

This routine provides the ability for a control point to be designated for either temperature-compensated (cycle a piece of equipment on and off to maintain a setpoint within a dead band) or straight time-dependent (cycle a piece of equipment on and off for distinct time intervals) duty cycling. Control parameters for temperature compensated duty cycling include total cycle lengths, long and short off cycles, and high and low temperatures. There are separate sets of parameters for heating and for cooling. Time-dependent duty cycling uses total-cycle and off-cycle lengths. Cycles for each load can be programmed in specific minute increments and based on a selectable offset from the top of the hour. Care should be exercised when implementing duty cycling to ensure minimum ventilation is maintained.

Optimum Start/Stop

For both heating and cooling seasons, DDC systems can provide customized optimum start/stop routines that take into account outside temperature and inside zone temperatures when preparing the building climate for the occupant or shutting the facility down at the end of the day. During unoccupied hours (typically at night) the software tracks the rate of heat loss or gain and then utilizes these data to determine when equipment will be enabled in order to regain desired climatic conditions by the scheduled time of occupation. The same logic is used in reverse for optimum stop.

Night Setback

This routine allows building low temperature limits for nighttime, weekend, and holiday hours, as well as parameters and limits for normal occupied operation to be user selectable. Night setback is usually programmed to evaluate outside air temperature in the algorithm.

Hot Water Reset

This routine varies the temperature of the hot water in a loop such that the water temperature is reduced as the heating requirement for the building decreases. The reset temperature for night set back usually is less than the day reset temperature. Typically, reset is accomplished by controlling a three-way valve in the hot water loop; however, depending on boiler type, reset can also be accomplished by "floating" the aqua stat setpoint. For pneumatic control systems, the primary variable in the reset algorithm is outside air temperature. For newer systems that have DDC controls on air handling units, it is possible to reset the heating hot water temperature based on the "worst case" hot water control valve position. Under this strategy, the heating hot water temperature is reset downwards until the air handling unit with the greatest need for heating has its hot water control valve fully open. For VAV systems that feature DDC control at the zone level, it is also possible to reset the temperature of hot water feeding zonal reheat coils based on the "worst case" zone. Under this strategy, each DDC zone controller reports the position of its hot water reheat control valve, and the temperature is reset downwards until the zone with the greatest reheat requirements has its valve fully open. This strategy can provide a significant reduction in wasteful reheating of conditioned air.

Boiler Optimization

In a facility that has multiple boilers this routine schedules the boilers to maximize plant efficiency by staging the units to give preference to the most efficient boiler, by controlling the burner firing mode when desirable and by minimizing partial loading.

Chilled Water Reset

This routine varies the temperature of the chilled water in a loop such that the water temperature is increased as the cooling requirement for the building decreases. Chilled water reset also can be accomplished by interfacing the EMCS with the chiller controls to reduce the maximum available cooling capacity, such as during demand limiting. In its simplest form the reset algorithm is based on outside air temperature. For newer systems that have DDC controls on air handling units, it is possible to reset the chilled water temperature based on the "worst case" chilled water control valve position. Under this strategy, the chilled water temperature is reset upwards until the air handling unit with the greatest need for cooling has its chilled water control valve fully open. In this way, the chilled water temperature will be reset to as high as possible—which improves chiller capacity and efficiency without compromising comfort. It must be noted that in humid climates this strategy must be carefully implemented within indoor relative humidity limits to prevent indoor air quality complications.

Chiller Optimization

In a facility that has multiple chillers, this routine schedules the chillers to maximize plant efficiency by staging the units to give preference to the most efficient chiller.

Upstream-downstream HVAC Control Coordination:

An EMCS application that can sense and interact with all levels of HVAC equipment can be leveraged for overall optimization to provide enough (but just enough) air flow, water flow, cooling and heating. This is done through setpoint reset, but instead of resetting from outside air (for example), reset is governed from zone level demand. By tracking "most open valve," "most open damper," or "degrees off setpoint" in each space, upstream temperature can be adjusted automatically and can reduce overlapping heating/cooling losses.

For example, a constant 55°F supply air temperature may over-cool some spaces, creating heat load energy waste (sometimes from personal space heaters) detectable from "all fan heating" zone level signals. By resetting air temperature in response to these signals, the ideal temperature can be approached and energy waste from reheat reduced. Other examples of ENCS optimization are VAV fan pressure reset, chilled water rest, boiler reset, and demand controlled ventilation.

Chiller Demand Limiting

This routine limits chiller demand by interfacing the EMCS with the chiller controls to reduce the maximum available cooling capacity in several fixed steps. The primary variable in the demand limiting algorithm is kW demand.

Free Cooling (Air Economizer)

Free cooling is the use of outside air to augment air conditioning or to ventilate a building when the enthalpy (total heat content) of the outside air is less than the enthalpy of the internal air and there is a desire to cool the building environment. In arid climates, an economizer cycle can work well by measuring dry-bulb temperature only. In climates where humidity is of concern, enthalpy-based controls are preferable, as they can provide greater comfort and increased energy savings. An economizer can be categorized as integrated (meaning it can operate in conjunction with mechanical cooling) or non-integrated (meaning that the outside air damper is fixed at its minimum position when mechanical cooling is required).

Free Cooling (Water Economizer)

The water side economizer uses the chilled water

system but creates the cooling effect without a chiller. This utilizes a heat exchange and requires operation of the cooling tower (at low temperature) and pumps but no chiller. While not as economical as an air economizer, these are common in large systems in dry climates. Optimizing the water economizer operation with the air economizer operation to minimize overall cooling cost is a useful task for the EMCS.

Recirculation

This routine provides for rapid warm-up during heating and rapid cool-down during cooling by keeping outside air dampers fully closed and return air dampers fully opened during system start-up.

Hot Deck/Cold Deck Temperature Reset

This routine selects the zones or areas with the greater heating and cooling requirements and establishes the minimum hot and cold deck temperature differential that will meet the requirements in order to maximize system efficiency. (NOTE: Multi-zone system)

Motor Speed Control

This routine will vary the speed of fan and pump motors to reduce air and water velocity as loads decrease. Speed control can be accomplished either by controlling a two speed motor with a digital output or a variable speed drive with an analog output or network connection. In addition to load following, motor speed control can be implemented in unoccupied times if the system is not allowed to be turned off.

Manual Override

This routine provides separate manual override schedules to allow for direct control of equipment for specific periods of time.

Duty Logs

DDC systems can track and display various types of information, such as last time on, last time off, daily equipment runtimes, temperatures (or other analog inputs), kWh, and even remote panel hardware performance. DDC systems can also accumulate and display monthly scheduled and unscheduled runtimes for each piece of equipment controlled or monitored. Temperature and hardware performance information can be monitored and displayed. Change of state can be logged for all loads and can be reviewed on a daily basis for all loads. In addition, a system can produce line graphics or historical trends of input data and hardware systems information. Energy consumption logs can be kept separately for every pulse meter point. For electricity these energy logs

can record data (by the day), such as daily kilowatt hour consumption; kilowatt demand; time the peak occurred; selected demand limit; time that any load was shed; minimum, maximum and average outside temperature; and degree day information.

Alarm Monitoring and Reporting

DDC systems register and display alarms for conditions such as:
- Manual override of machinery at remote locations and reminders to clear the overrides
- Equipment failures
- High temperatures
- Tenant overrides/after hours billing
- Low temperatures, invalid temperatures (sensor is being tampered with or is defective)
- Communication problems

Demand Controlled Ventilation (DCV)

Ventilation costs are significant and can be automatically adjusted to follow occupancy patterns, either by CO_2 sensing or schedules where there are multiple shifts.

12.3 JUSTIFICATION OF EMCS

A well-designed and properly operated EMCS is an investment which offers the building owner/manager a multitude of benefits, which can often be bundled and quantified. In addition to the immediate financial benefits, there are long-range benefits to managing energy and demand as well. Such benefits can be wide-ranging, positively affecting society in general through the construction and operation of sustainable buildings. When buildings are sold, part of their valuation includes expenses such as energy, so reduced energy use can translate into higher valuation. Consider the following EMCS functions, along with the related benefits:

- Manage energy consumption and demand and Optimize operating efficiencies of energy consuming equipment—Lowers operating expenses, produces higher profits and increased competitiveness, keep energy prices affordable in the long term.
- Predictive maintenance, using key measurements for pro-active maintenance—reduces downtime and extends equipment life.
- Improve comfort—Increases occupant productivity and concentration. Helps occupants feel more alert and rested, make fewer mistakes, and perform better. Contributes to a more vital U.S. economy with better educated and less stressed people.

- Improve indoor air quality—Increases productivity, prevents lawsuits, lowers insurance costs, and reduces absenteeism.
- Activate alarms when equipment malfunctions— Provides for less costly disruptions of productivity due to faster response, increased profitability, extended equipment life, and increased competitiveness.
- Assist on- or off-site operator to administer service and maintenance—Results in better records and histories for more efficient and accurate work, less downtime, more productivity and lower cost for service. This leads to higher profits and increased competitiveness.
- Monitor/log building equipment performance and energy use—Results in accumulated data used to improved performance of all EMCS functions. Confirms cost avoidance benefit of project.

Other Benefits
- Reduces energy usage proportionately and reduce green house gas emissions.
- Presents an image of a conscientious "green" building which is good for business.
- With sufficiently reduced energy use, can lead to building certification such as Energy Star or LEED, bringing further building notoriety and customer appeal.

12.3.1 EMCS Application Opportunities

Many commercial and institutional buildings designed and constructed since the 1980s are of similar concept. The exact numbers and placement of equipment will vary from building to building but often include chillers, boilers, cooling towers, and air handling units connected by a network of pipes, pumps and valves, along with an EMCS controlling the equipment. Air systems are constant or variable volume with terminal units such as variable volume boxes (with or without heating ability), fan coils, and other equipment to accomplish final control. (See Chapter 10 for an in depth discussion of HVAC systems and equipment.) The goal of EMCS application to HVAC systems is to maintain occupant comfort or process conditions while minimizing energy use. It is the energy reduction that creates the business case for the control system installation.

The nuances of mechanical design are important in that one must understand the capabilities and limitations of each HVAC system, specifically that the control system applied has as its purpose to control and optimize that piece of equipment, its moving parts, and modes of operation. Thus, a full understanding of the HVAC equipment to be controlled is essential in applying the controls.

Preparing a sequence of operation at this initial stage may be a preliminary task that is well worth the time spent, because the operators of the existing building, or the original design engineers, may be able to spot operational issues by merely reviewing the paragraphs of text. The peer review that verifies proposed optimization with known constraints and design intent can be good insurance against unhappy surprises during implementation.

EMCS and DDC technology is readily applied to new construction, but also provides opportunities for existing facilities. The discussions below illustrate some of the common opportunities in existing commercial buildings, but the concepts apply equally to new designs. Section 12.2.8 "Control Strategies," in this chapter (and Chapter 20) includes specific HVAC control strategies that can be used in HVAC and other control applications. An overlay of known effective control strategies with the building systems and equipment lead to the development of a control system scope of work. Understanding the interaction between HVAC system components and how different control methods impact operations is essential for a successful control system project. Often a change in one portion of the HVAC system creates unanticipated side effects elsewhere. For example, resetting supply air temperatures or supply water temperatures in cooling mode may introduce humidity and comfort issues, because as temperature increases it can hold more water. EMCS can monitor and alarm these conditions in a "watch dog" fashion if desired.

12.3.1.1 Watch-outs

Where HVAC systems are modified, through control or fundamental operations, the potential exists to negatively impact comfort or other design-intent functions in the building. So proceed with caution and diligence. Simple alterations may be effectively managed with a review of design documents, experience, a call to the HVAC design engineer, or a combination. A few are noted here to get the reader's attention. The control system designer's training, skill, and experience with similar buildings and locale are the chief defense against such complications.
- Of particular concern for HVAC control is indoor air quality and the potential for conservation measures and new control routines to create or aggravate air quality issues. This is discussed in greater detail in Chapter 17 Indoor Air Quality.
- Some HVAC conversions and control methods can also impact building pressurization. For example, if the building is allowed to become negatively pressurized the control system actions can create exte-

rior comfort issues and possibly freezing/bursting water pipe damage.

- Some projects may include VFD retrofit of existing cooling towers. In this instance, a careful start-up commissioning step is to "walk" the fan slowly through each speed, looking for critical speeds where vibration occurs and programming the VFD to skip those frequencies. Also, VFDs can create oiling issues with gear boxes that depend upon splash lubrication; simple "minimum speed" programming can protect the equipment.

12.3.2 Example EMCS/DDC HVAC
System Control Retrofit Opportunities
Stand-alone DDC or EMCS

If the existing system is not networked, it is good practice to install cabling for future networking and EMCS control. For full system integration in EMCS, communication compatibility between the various controllers must be established. EMCS offers distinct advantages over stand-alone DDC, especially:

- Enhanced operations from reporting trouble conditions, alarms, reports, trends, troubleshooting, etc.
- Enhanced energy management since the EMCS can communicate to a network of DDC controllers and provide overall system control.

Hybrid Pneumatic Control

An option exists to maintain a portion of the existing pneumatic system, but the shortcomings of such a system should be considered and reviewed with the customer in advance of the decision. (Refer to section 12.2.2.1 "DDC—Pneumatic Hybrid Conversion.") In general it is preferable to provide an all-electronic DDC solution with new end control devices. Most VAV box manufactures and other third parties have kits to retrofit VAVs to accommodate DDC. If practical considerations insist upon the DDC/pneumatic hybrid model, the project scope should include field testing of all pneumatic devices and tubing to be re-used (especially spring ranges and eradication of heat/cool overlap), air filtration to protect expensive pneumatic transducers, high quality transducers with sufficient output volumetric flow rate to compensate for tube leaks, and compressor/drier/filter repair such that a clean, dry, reliable source of compressed air is assured. Depending upon the condition of the existing pneumatic equipment, this can become costly and steer the control design back to DDC.

Heat-cool Overlap

A primary control function is to assure no buck from overlapping or simultaneous heating and cooling action.

This energy waste is common in old and new systems alike and should be a prime directive in all DDC and EMCS projects. The DDC system will determine if the temperature/humidity levels outside allow the mechanical heating and cooling systems to be reset or turned off. Note that overlap of "free cooling" and boiler operation is still heat-cool overlap. This subject is mentioned early in the example since it can be found in all levels of HVAC systems.

Lighting Control

Integrating HVAC and lighting systems in existing buildings presents special challenges. Generally, the electrical circuits will not coincide with the HVAC zoning. For example, VAV boxes might supply three or four offices, while the lighting zone serves a much larger area. The existing conditions force the building operator into compromises, even though the integration will not be perfect, energy savings are possible.

Most VAV box DDC controllers will accept a signal from an occupancy sensor. The occupancy sensor will turn lighting on or off through a local relay at the light fixture, depending on whether the space is occupied or unoccupied. This is simply local hard wiring of interlocks to the lights. By connecting the sensor to its respective VAV box, the space can be set to the occupied or unoccupied setpoint. Through the EMCS, sensors and rooms can be software interlocked to the HVAC system. By turning off lights and resetting setpoints, the operator can reduce energy consumption.

Note: For new buildings, EMCS control of lighting begins at the design stage and includes zoning of lighting circuits to enable central control at circuit breaker panels. An example of this is circuiting the perimeter fixtures separately from inboard fixtures, which then allows logical switching off of perimeter fixtures when sufficient light occurs near windows. These systems may also include definite purpose lighting controllers, designed specifically for lighting control, and can include features like occupancy sensors, overrides, dimming, and daylight harvesting; these controllers are available to directly communicate to the EMCS to share the user workstation or man-machine interface. Many of the lighting control features available via new designs meet practical barriers in retrofits; however, savings are still viable for some basic lighting control measures such as:

- Occupancy sensors on a per-room basis, wired through DDC controller spare points, such as on VAV boxes
- Local relays for large common areas such as open plan offices, warehouses, etc.
- Where relay control is achieved, occupancy sched-

ules can mirror HVAC schedules.

- Overrides to serve cleaning crew work, or for possible after-hours tenant override
- Exterior lighting

Package Rooftop Equipment

Interface to these is usually restricted to relay logic—start-stop, and staging of equipment. Supply air temperature sensors add a measure of operations benefit for remote troubleshooting.

VAV boxes and other terminal units under DDC and EMCS control allow the building operator to apply energy-saving control strategies while being positioned to assure good occupant comfort. Occupied/unoccupied, night setback, and user-adjustment temperature range limits become possible. Supply (leaving air) temperature sensors offer benefit to operations in troubleshooting, especially to easily verify whether the air valve and heating actuators are functional; by remotely overriding the device positions, operators can quickly determine from supply temperature changes whether the devices are responding or have failed. Spare input/output points on standard VAV box and other unitary controllers can be used for interlocks and lighting control, including after-hours overrides. EMCS supervisory control can monitor after-hours overrides and provide custom tenant billing if desired. The building operator will be able to reset box parameters, change control sequences, monitor conditions, and report alarms and equipment failures.

Air handling units (AHUs) come in many configurations and are the mainstay of applied non-canned DDC control strategies. DDC control of each heating/cooling element and return-outside air mixing dampers are typical. Fan control will be via on-off or variable capacity, depending upon the fan and whether it has a VFD or inlet vanes, or is constant volume. Variations of systems may include additional heat/cool coils, mixing dampers (e.g. multi-zone or double duct), humidifiers, etc. Each of these control points are energy consumption points of use and, if carefully controlled, represent opportunities to provide equal performance with reduced energy consumptions.

Outdoor air control is a primary place to look for energy savings. Many systems, over the years, have had the ODA blocked open or shut, or some manual position in between. In some cases, minimum AND maximum air flow dampers and actuators will be required. In others, air flow monitoring stations may be necessary to fully optimize the outside air energy use. Linking outside air intake rates to occupancy, either from time of day (un-oc-cupied periods or warm-up periods) or partial occupancy (CO_2 and demand controlled ventilation) is a very good EMCS candidate for energy reduction in existing buildings currently under conventional control.

Mixing dampers are also the source of "free cooling" in the air-economizer cycle and should be used to full advantage when in suitable climates. EMCS can provide superior control by coordinating actions of ventilation, air-economizer, and supply air reset routines, all in response to dynamic building usage patterns. In climates where air economizers are used successfully, identify and explore options to convert existing equipment that does not have full economizer dampers in place.

Fan/pump capacity control can be enhanced through EMCS. Before applying controls, identify any constant volume systems and explore opportunities for HVAC system alteration that will allow variable volume service. For pumps, the amount of water delivered will usually be from variable speed motor control from variable speed drives (VFD).

For fans, the amount of air delivered will be usually be controlled by one of two methods—inlet vane actuators or VFDs. Inlet vanes throttle flow through a combination of resistance and pre-rotation of the air to unload the fan, while VFDs throttle air flow by slowing the motor, a more efficient process especially at lower loads. Over the years, the inlet vanes may have been disconnected or otherwise become nonfunctional; thus the system operates wide open or at least sub optimally. If existing inlet vanes are to be re-used, verify that their operation is smooth and suitable for another life cycle of system operation; if repairs are needed, the avoided cost of such repairs can subsidize the conversion to VFD control. When VFDs are added to replace inlet vanes, remove inlet vanes.

Many VFD manufacturers offer their drives with interface capability that can simplify the DDC EMCS input/output connection to the VFD; the drive can be a device on the network allowing many more control variables to be exposed and much easier troubleshooting capabilities. Input/output points related to a VFD can be as simple as on-off and % speed command to rpm and power consumption feedback directly from the drive.

Security—In some facilities, occupant protection from contaminated outside air supply is a concern. In these cases, the EMCS can provide an n emergency mode that will close all outdoor air dampers upon demand.

Demand limiting can provide savings to certain customers, when large discretionary electric loads are present. A good example of this is a building heated with electric resistance entering a Monday morning occupancy period after a cold weekend. A combination of optimal start and electric demand meter inputs may be used to implement this control.

Real-time energy consumption can easily be recorded by trending and graphing, providing excellent feedback to building operators, especially when trying out new control routines. Either separate meters or pulse contact additions to utility meters can provide the input data stream to allow this feature through EMCS.

Control valves are common interface points in a DDC retrofit. Other than with the hybrid pneumatic-DDC system, the valves and actuators will normally be replaced to facilitate direct DDC control. (Types of valves and sizing considerations are discussed in Chapter 20.) In general, there are two- and three-way valves, the difference being that a two-way valve fully stops the flow while a three-way valve diverts or mixes a continuous flow. This is an important consideration in any system modification from constant to variable water flow since a system of three-way valves would require replacement with two-way valves to be compatible with variable flow pumping. A common conversion complication worth mentioning here is that converting to two-way valves will increase the "close off" force necessary and require a stronger actuator; thus, simply cutting off and capping the third pipe on a three-way valve will usually render an existing actuator ineffective.

Boiler and chillers on-off control. The most energy efficient setting is "off." Depending upon building usage, this may be best accomplished through simple outdoor air temperature sensing (above this temperature, turn off the boiler, etc.), through zone demand (upon a call for heating by at least two zones that occur for more than 15 minutes, etc.) or a combination of the two.

Primary equipment optimization. Chillers and boilers often operate in multiples and may or may not be identical. Optimizing and sequencing these large points of energy use is an advanced feature of EMCS control that can save considerable money, compared to more simplistic control methods. Large equipment with modulating capacity can be controlled to quite low loads, but efficiency will vary markedly. Thus a very useful application of EMCS is to analyze and determine optimum times to use different machine combinations for overall best efficiency. Implementing such optimization may require additional instruments, such as flow meters, outdoor weather sta-

tions, motor current transmitters, pressure transmitters, and the like. Whether the expense is warranted will depend upon the size and run time of the equipment. For example, a chiller in a mild climate with low annual run hours would warrant less investment in optimization than the same chiller in a hot humid climate with extensive run hours and energy appetite. In large central heating and cooling plants, equipment may be connected via manifolds with motor operated valves or dedicated pumps, requiring on-off control and matrix logic to utilize the equipment properly. To allow for manual control, position feedback may be needed. Of interest in large equipment staging is deciding when to switch off an increment of capacity, as often sequencing for increased capacity is more difficult than switching for less capacity, allowing un-necessary energy use from leaving equipment on too long. Additional instrumentation may be necessary to fully optimize such heating/cooling plants, such as main supply and return temperatures, motor amps, or individual flow stations. The major chiller and boiler equipment manufacturers have interface hardware (gateways) or can directly connect over the network to EMCS.

Cooling tower control may include optimization for cooling mode, water-side economizer operation, and other modes such as basin heating or de-icing, as well as water level control. Optimal temperature control usually requires outdoor wet bulb temperature inputs, which come from a software algorithm combination of outdoor temperature and humidity. Optimized leaving cooling tower water temperature is valuable because condenser water temperature can greatly affect the amount of energy the chiller uses. Hardware interface will follow the intended function. Of particular note for cooling towers is the common use of two speed motors; control of two speed motors when coupled with drive shaft/gear box arrangement can result in equipment-damaging backlash, without proper sequencing between high and low speeds. The use of time delays and interlocks are needed to do this effectively. Other complications in cooling tower control can arise when converting to variable speed control, such as critical frequency vibration and gear box oiling. (Refer to 12.3.1.1 "Watch outs.")

12.3.3 New Construction EMCS

By combining more aggressive zoning with DDC EMCS in new construction, the building designer has the opportunity to better match control of the HVAC system with building usage and optimize equipment settings to loads thus improving operating efficiency. New construction is better able to interface lighting zones to the

HVAC zones. While this type of design integration may prove very costly in an existing building due to existing wiring and circuitry, it is desirable when starting from scratch. Interfacing occupancy sensors to control lights and HVAC through the EMCS allows the control system to reset the VAV box up a few degrees and turn off the light when the space goes unoccupied. However, the bulk of the EMCS opportunity lies in the thousands of existing buildings which are controlled with pneumatic and electrical/mechanical automatic temperature control systems. Surprisingly enough there are also buildings still being controlled by hand!

12.4 SYSTEMS INTEGRATION

The performance of an EMCS is directly proportional to the quality of the designer's systems integration effort. Systems integration can be defined as those activities required to accomplish the precise monitoring and control desired when incorporating an EMCS into a building's HVAC and lighting systems. This definition implies that systems integration includes not only selecting the appropriate EMCS and detailing monitoring points and control interfaces, but also equipment control interfaces and customization of software algorithms to meet the user's needs and commission the system.

12.4.1 Facility Appraisal

An appraisal of the job site is the first step of EMCS systems integration. The appraisal is generally done by interviewing the building users to determine building usage patterns (be careful to differentiate between normally scheduled usage and after hours usage), to identify any comfort complaints (i.e. "hot" or "cold" spots in the building, adequate ventilation, etc.), to assess the attitude that the building maintenance staff has towards an EMCS, and to determine exactly what the building manager expects the system to accomplish. During the facility appraisal, the following people should be interviewed:

- Tenants (by company, department, job function)
- Prospective tenants
- Delivery services
- Security personnel
- Service vendors (including janitorial, HVAC, lighting, electrical, life safety)
- Building operations, engineers and facility managers
- Corporate energy manger, if any
- Building owner/CEO/CFO
- Property manager and staff
- Leasing agents

Perhaps the most overlooked of the above are leasing agents. These professionals must understand the benefits of energy efficiency, or they will not be able to get buy-in from prospective tenants.

12.4.2 Equipment List

The designer should develop a complete and accurate list of the equipment to be controlled. In a retrofit project the list should include all of the equipment and devices to be controlled, the present method of control for these items, the operating condition of the items and their controls, and the area each item services. Do not forget to include exterior and interior lighting on the list. A cost evaluation at this stage will be useful to validate budget; if cost is too high, prioritize those control points with the best savings potential and consider dropping "monitor only" points.

12.4.3 Input/Output Point Definition

The third step is to define the input and output points required to achieve the monitoring and control desired, in the most cost-effective or efficient manner. This requires in-depth appraisal of such items as the equipment list, layout of building spaces, method of building construction, HVAC physical plant makeup and layout, methods of heating and cooling, types of secondary HVAC distribution systems and controls, and the condition of the mechanical equipment and existing controls. The definition of a control point should include a listing of all inputs that will influence the functioning of that point. The installation specifications should detail this information.

12.4.4 Systems Configuration

Once the inputs and outputs have been defined, the designer needs to identify any constraints that might be placed on the EMCS hardware such as specific manufacturer, performance requirements and limitations, etc. Some of the items to be considered when configuring an EMCS are building size, design, location, and use; design of the physical plant; type and layout of the secondary distribution systems; design of the electrical system; type and condition of existing controls in retrofits; and the nature of the personnel operating the building. Cost of installation can vary significantly depending on the type of control system architecture (centralized versus distributive), the method of transmitting information between system controllers and input/output devices (direct hardwire versus multiplexing versus power line carrier), the types of system controllers and input/output devices selected (two way versus one way communicators), and the amount of programming required (customized algorithms versus firmware). Knowing who is going to maintain the software and perform the necessary administrative functions (i.e. inputting schedule changes, monitoring alarms, and producing reports) not only affects how many but also where the operator interface level personal computers are to be located, as well as the type of monitoring and alarming that will be needed. This will also determine training requirements and the extent of support services needed. Internet based systems can utilize resident user PC's and need not have a dedicated workstation location unless one is preferred.

12.4.5 Specifics of Software Logic

The most critical phase of systems integration is the software specification. The designer must take care to be as explicit as possible when detailing the nature of the algorithms to be installed. This can be done by structuring the software specifications in a manner that clearly defines the algorithms and their interrelationships. The designer who simply includes general hardware specifications, rather than detailing the specifics of software sequencing logic, most assuredly adds roadblocks to successful implementation. Detailed software requirements and control sequences are mandatory. This can be done with sequences of operation in conjunction with pre-defined language for common algorithms such as optimal start/stop, time-of-day scheduling, etc.

12.4.6 Control Point Tie-Ins

To ensure that the EMCS is interfaced to a device to be controlled exactly as desired, the integrator should include sketches detailing the specifics of the interface at the control circuit of that device.

12.4.7 Commissioning

The final step of systems integration is commissioning. Commissioning should include both the hardware and software. All the hardware should be physically inspected for correct installation and the wiring tested for continuity. Inputs should be checked for accuracy against a standard. The operation of all outputs should be demonstrated from the operator interface personal computer, or system controller if no personal computer is installed. This inspection should not be taken lightly. The system will not function as designed if any hardware component is incorrectly installed or does not function as desired. The software should be checked before the system is put on the line. This requires a careful review of all algorithms, setpoints and schedules for compliance with the detailed specifications. Where system graphics are used, a good quality control check includes verifying that each hardware point is correctly

mapped all the way through the system to the graphics screen.

12.4.8 Harmonics

Modern electronic ballasts used in fluorescent lighting systems, as well as other digital switching devices, can generate harmonics in the electrical distribution system. There have been reported cases of such harmonics causing malfunctions in control systems that communicate using power line carrier control (PLC) technology. (Such systems use the electrical distribution system as a means for inter-device communication by superimposing their digital "messages" over the 60 Hz power sine wave.) Modern PLC systems that use spread spectrum technology are much more tolerant of harmonics and encounter very few problems.

12.4.9 Communications Technology

Though most digital control systems rely on hard-wired connections between devices, there are alternative technologies that can be used to simplify installation (particularly in retrofit applications), provide a communication link with remote locations, and increase data transmission rates. These include:

- Power Line Carrier (PLC). PLC systems use the existing electrical wiring system as a means of achieving inter-device communication. This is accomplished when a device generates a high frequency communication signal that is superimposed over the standard sixty hertz power sine wave. Other devices connected to the same electrical system can analyze the spectral content of the incoming power and extract the superimposed "message" from the sixty hertz carrier wave. PLC systems can provide reduced installation cost because of the reducing wiring requirements, but they may be less reliable and slower than hard-wired systems.

- Wireless Technology. For control stations that are located a significant distance away from a building, there is the ability to send and receive control messages using wireless technology. Some systems are based on simple radio signals, which are relatively low-cost but slow. Other systems are based on cellular telephone technology, which can provide greater range and faster data transmission speed but may be more expensive to own and operate.

- Fiber Optic Communication. Due to the desire to have ultra fast data transmission capabilities for computer network applications, many buildings and campuses are now installing fiber optic networks. Such networks—which have sufficient bandwidth to provide telephone, computer network, and EMCS communication simultaneously—provide the fastest data transmission rates but are also expensive to install. Many military bases and college campuses have installed fiber optic "backbones" to allow inter-building computer network communication. Such scenarios present an excellent opportunity to replace archaic dial-up modems with more modern systems that can provide more elaborate control sequences and better accuracy. In multi-building or campus settings, fiber optic communications also offers immunity to lighting strikes between buildings that would otherwise create damage via the communication cabling.

12.4.10 Industry Standard Communication Protocols

In an effort to achieve true interoperability for multiple pieces of equipment from different manufacturers, there has been significant effort focused on developing standard communication protocols to which all manufacturers will adhere. The advantage of so-called "open" protocols is that, for example, a chiller control panel can share data with a variable speed drive controlling a chilled water pump in order to reduce energy use, with near "plug and play" simplicity. The marketplace remains dominated by proprietary control systems, making communication between devices from different manufacturers difficult to accomplish in some cases. Open protocols are intended to shift the industry away from proprietary systems that limit control system sophistication and user-friendliness.

Though two standardized protocols are now widely known, BacNET (developed by ASHRAE) and LonWorks (based on the Neuron chip, developed by Echelon Corporation), a number of others are being developed and implemented. However, their implementation is still not yet at the scale it needs to be to provide true interoperability for a wide range of system types. BacNet is an ASHRAE sponsored protocol that relies on gateways to allow communications between various manufacturers' equipment. Gateways are communications interface devices that translate communications streams of one manufacture to another. The downside is if manufactures do not agree, there will be no gateway, and even if there is a gateway who owns the problem if it doesn't work? LonWorks uses an IP based system that utilizes ethernet packets. IP is the standard that allows commercial computers, printers, and the like to talk with each other. The biggest downside to LonWorks is lack of market penetration in the United States; it is much more common in Europe.

For the time being, inter-device communication is typically accomplished using communication "gateways." A gateway serves as a translator between the communication languages of two different pieces of equipment, allowing them to share data and operate synergistically. A gateway can either be a stand-alone piece of hardware, or it may be a built-in feature of some DDC panels. Systems integration consultants are often involved on sophisticated control system projects to ensure that all devices are able to communicate with one another. In the future, it is likely that use of industry standard "open" communication protocols will achieve critical mass, and as a result nearly all manufacturers will adhere to them.

12.4.11 Building Operators and EMCS Specifications

Training should also be considered a part of the commissioning. In fact, training and long-term support are often the most overlooked aspects of an EMCS project. In what form and for whom training should be provided differs for every project. This is because the people involved are different every time.

In the end, the performance of an EMCS depends on the operating personnel. Those who are to work with the system must understand it. The most common cause for the failure of an EMCS is poor training that leads to ignorance about the system's capabilities. Overly complex controls produce the same result, which is to defeat the system and revert to the old ways. Training is essential, as well as good quality documentation and user friendly provisions in hardware and software. An excellent way to sustain user buy-in is to demonstrate savings and associate user participation with the savings. If a system is over-ridden frequently, controls are disconnected, or parameters and schedules are not properly maintained, the EMCS becomes useless. There is a body of evidence that energy saving measures from EMCS tend to slip, eroding initial savings by as much as 20 to 30% in three years. Continuous diligence is necessary to keep these savings, and it all begins with the operators.

The contractor should orient the training specifically to the system installed rather than offer a general training course. A case can be made for expanding this to include business-specific and facility-specific instructions. In new construction, these would be determined by peering as far into the future as possible to determine what might be required. The best approach here is to ask copious questions about how the facility will be used, staffed, serviced, and marketed (in the case of a for lease building). Inviting the operating staff to participate in system start-up can be an excellent training opportunity.

In retrofit applications, the depth and quality of the questions asked during the facility appraisal (refer back to Section 12.4.1) heavily influence the detail spelled out in the specification. For example, to fully develop the possibilities discussed in Section 12.3.1, a list of facility-specific instructions might also include the following:

- Adopt program user-friendly, menu-prompted, after-hours access setup routines.

- Adopt program user-friendly, menu-prompted, monthly after-hours access bill printing routines.

- Print out bill examples using hourly rates to be billed in this building.

- Provide training sessions in after-hours access code programming and bill printing for building engineer.

- Provide training sessions including system overview, control strategies, and temperature and status data interpretation for property manager.

- Provide training sessions in after-hours access code logic, benefits, options, and programming for property manager.

- Provide training session in after-hours bill printing for property manager and property management secretary.

- Conduct a meeting with all tenants' designated representatives to explain how to request HVAC and/or lighting by telephone and to determine the need for additional access code sub-zones.

- Conduct follow-up meeting with all tenants' designated representatives to review first month's billing and to deliver access code maps documenting sub-zones requested in first meeting.

- Conduct a meeting with janitorial service management personnel to explain the use of access codes (lighting only).

- Provide training sessions for outside leasing agents in EMCS benefits (i.e. temperature control, efficiency, improved maintenance, and after-hours fairness). Include why a building with an EMCS is different and better than a building without an EMCS.

- Conduct meetings to explain EMCS operation, provide basic documentation, and establish communication and service-call protocol with all service vendors (i.e. security, HVAC, lighting, and electrical).

- The above are only a few of what might be a long list of services that could be included in an EMCS specification for a particular building owner to ensure that projected benefits are realized.

- Whether one is a consulting engineer, a contractor doing a design/build project, or a facility director writing a guide specification, every effort should be made to anticipate future needs. EMCS projects priced, negotiated, or bid without the benefit of complete specifications are doomed to less than stellar performance at best and total failure at worse.

12.4.12 Equipment and Contractor Qualification and Selection

Many EMCS projects have gone astray because decisions as to system manufacturer and installer (or bid list) were overly influenced by the features of EMCS equipment. There is a big difference between capabilities and effective application of appropriate control and facility-specific strategies.

Too much emphasis on a particular brand of EMCS overlooks the vital role of the contractor in the success or failure of a project. Even when protected by the most detailed of specifications, it pays to look beyond the low bid and conduct a thorough investigation of the installers being considered. Several of the major EMCS manufacturers also have contracting divisions. Talent and experience often vary from branch office to branch office. The people who will install and program the EMCS and the building owner or operator are critical to project success.

Thorough interviews, inspections of previous jobsites, conversations with references, and visits to potential contractors' offices to meet the delivery team are all good ways to increase the chances of quality installation. The delivery team will have a greater impact on the outcome of a project than the EMCS equipment. A thorough investigation of the entire team should be made, and only the qualified should be allowed to compete for the project. The same argument could easily be made for the selection of a consulting engineer.

12.4.13 Close-out Documentation

Before acceptance of the EMCS work, assure the facility has the following.

- Accurate as-built drawings
- Detailed and clear sequences of operation
- Names and contact information for service, warranty, and support
- Recommended maintenance intervals of equipment and calibration
- Records of initial settings made to serve as a baseline for recommissioning, such as reset schedules, setpoints, occupancy schedules, etc.
- Technical manuals for the system

EMCS technology is a powerful tool for saving energy, but it depends heavily on the support and buy-in of the people using it. By seeking out the crafts-people and the people most focused on saving energy, the chances of a project paying off are significantly greater.

GLOSSARY OF TERMS

ALGORITHM—An assortment of rules or steps presented for solving a problem.

ANALOG INPUT—A variable input that is sensed, like temperature, humidity, or pressure that is sent to a computer for processing.

ANALOG OUTPUT—A variable signal such as voltage; current pneumatic air pressure which would in turn operate a modulating motor that will drive valves; dampers; etc.

ANALOG TO DIGITAL A/D CONVERTER—An integrated circuit that takes input data, such as a temperature sensor, and converts them into digital logic that the microprocessor can recognize. All analog inputs of a DDC system are fed into the microprocessor through an A/D converter.

BACNET—ASHRAE supported open protocol allowing equipment form varying manufactures to communicate with each other utilizing common standards and gateways.

BAUD RATE—The speed, in bits per second, at which information travels over a communications channel.

BIT—The smallest representation of digital data there is. It has two states: 1 (on) or 0 (off).

BITS OF RESOLUTION—A representation of how finely information can be represented. Eight bits of resolution means information may be divided into 256 different states, ten bits allows division into 1024 states, and twelve bits allows division into 4096 states.

BYTE—Eight bits of digital information. Eight bits can represent up to 256 different states of information.

CONTROL LOOP—The strategy that is used to make a piece of equipment operate properly. The loop receives the appropriate inputs and sets the desired condition.

DAISY CHAIN—A wiring scheme where units must follow one another in a specific order.

DEAD BAND—An area around a setpoint where there is no change of state.

DEAD TIME—A delay deliberately placed between two related actions in order to avoid an overlap in operation that could cause equipment problems.

DERIVATIVE CONTROL—A system which changes the output of a controller based on how fast a variable is moving from or to a setpoint.

DIGITAL COMMUNICATION BUS—A set of wiring, usually a twisted pair for DDC controllers. Information is sent over this bus using a digital value to represent the value of the information.

DIGITAL INPUT—An input where there are only two possibilities, such as on/off, open/closed.

DIGITAL OUTPUT—An output that has two states, such as on/off, open/closed.

DIGITAL TO ANALOG D/A CONVERTER—An integrated circuit that takes data from the microprocessor and converts them to analog data that are represented by a voltage or current. All analog outputs of a DDC system go through a D/A converter.

DIRECT DIGITAL CONTROL (DDC)—The term used for implementing control processes with microprocessor hardware and software technology in lieu of pneumatic, on-off, or other conventional control system methods. Inputs and outputs are converted to digital signals and processed in computer fashion. DDC repeatability is limited only by the stability of the input/output devices it is connected to, and thus they are known for close control. A distinctive feature of DDC control is software basis and digital communication, both of which create opportunities for optimization.

DISTRIBUTED CONTROL—A system where all intelligence is not in a single controller. If something fails, other controllers will take over to control a unit.

ENERGY MANAGEMENT CONTROL SYSTEM (EMCS)—A collection (system) of DDC controllers and their control devices, connected and operating under supervisory control equipment or software such that they operate collectively. EMCS is distinguished from stand-alone DDC control by its ability to share data globally, coordinate and optimize individual controllers, link control processes, and perform operations tasks such as alarms, trends, and reports. An EMCS may be configured with or without graphics.

EPROM—Erasable Programmable Read Only Memory.

An integrated circuit that is known as firmware, where software instructions have been "burned" into it.

EEPROM—Electrically erasable programmable read only memory. An integrated circuit like the EPROM above, except that the information may be altered electronically with the chip installed in a circuit.

FCU—Fan coil unit. A type of terminal unit.

FEEDBACK—The signal or signals which are sent back from a controlled process telling the current status.

FIBER OPTIC—Individual glass fibers formed into a "cable." Allows computer communications signals to be sent via light pulses versus traditional electrical pulses. Can provide much better immunity from lightning strikes than traditional cable.

FIRMWARE—Software that has been programmed into an EPROM.

FLOATING POINT—Numerical data displayed or manipulated with automatic positioning of the decimal point.

GATEWAY—Network device designed to "translate" computer languages between devices. Usually specifically designed for the application.

HMI- Human machine interface. Also called man-machine interface. A device that portrays the controller activities in table or graphic form suitable for interaction with an operator.

HP—Heat pump. A type of terminal unit.

INPUT—Data which are supplied to a computer or control system for processing.

INTEGRAL CONTROL—A system which changes the output of a controller based on how long a variable has been offset from a setpoint. This type of control reduces the setpoint/variable offset.

INTELLIGENT BUILDING—A building that is controlled by a DDC system.

I/P TRANSDUCER—A device for converting a digital signal into a proportional pneumatic signal for retrofitting existing pneumatic control systems.

LAG—The delay in response of a system to a controlled change.

LAN—Local area network. A communication line through which a computer can transmit and receive information from other computers.

LON—Echelon Corporation supported open protocol network language allowing devices from varying manufacturers to communicate at the IP level. Usually does not require gateway devices.

MICROPROCESSOR—The "brains" of all DDC systems and personal computers. An integrated circuit that has logic and math functions built into it. If fed the proper instructions, it will perform the defined functions.

MODEM—A device that allows the computer to communicate over phone lines. It allows the DDC system to be viewed, operated, and programmed through a telephone system.

ODA—Outdoor air damper. The air modulating element responsible for varying the amount of outside air introduced into an air handler.

OFFSET—The difference between a variable and its setpoint.

OPERATOR'S TERMINAL—The computer at which the operator can control a motor or a relay to start/stop a unit.

OUTPUT—Processed information sent by a controller to an actuator to control a motor or to a relay to start/stop a unit.

PEER-TO-PEER—Type of network configuration allowing devices to communicate with each other without a supervisory controller or computer.

PI CONTROL—A combination of proportional and integral control. Adequate for almost all HVAC control applications. Control loops using PI control look at both how far an input is from setpoint and how long it has been away from setpoint.

PID CONTROL—Proportional plus integral plus derivative control. A system which directs control loops to look at how far away an input is from setpoint, how long it has been at setpoint, and how fast it is approaching or moving away from setpoint.

PROPORTIONAL CONTROL—A system which linearly varies the output as an input variable changes relative to setpoint. The farther the input is from setpoint, the larger the change in the output.

PULSE WIDTH MODULATION—A means of proportionally modulating an actuator using digital outputs; one output opens the motor, another closes it. Used extensively in VAV boxes for driving the damper motor on the box.

RAM—Random Access Memory. A computer chip that information can be written to and read from. Used to store information needed in control loop calculations.

RELAY—Electro-mechanical device that converts electrical command signal into mechanical action. Used primarily for starting/stopping equipment and lighting.

ROUTER—Network device used to sub-divide larger networks into smaller components. Often employed as network "traffic cops."

SETPOINT—A value that has been assigned to a controlled variable. An example would be a cooling setpoint of 76 degrees F.

SOFTWARE—The list of instructions written by an engineer or programmer that makes a controller or computer operate as it does.

TERMINAL UNIT—Part of the mechanical system that serves an individual zone, such as VAV box, hydronic heat pump, fan coil, etc.

TOKEN RING—Network traffic management scheme that allows one controller at a time to send data onto the network. Token ring is passed to next controller in an orderly manner to prevent network packet transportation problems.

TRANSDUCER—A unit that converts one type of signal to another (an electronic signal to a pneumatic signal).

TWISTED PAIR—Two wires in one cable that are twisted the entire length of the cable. DDC systems often use a twisted pair for communication link between controllers.

TWO POSITION—Something that has only two states. A two-position valve has two states, either open or closed.

USER INTERFACE—The operator's terminal is usually referred to as the user interface. It allows the operator to interrogate the system and perform desired functions. Most interfaces today are personal computers or minicomputers.

VAV—Variable air volume. A terminal unit that varies the amount of air delivered to a space, depending upon the demand for cooling.

WORD—Sixteen bits of digital information. Sixteen bit scans represent up to 65,536 states of information.

CHAPTER 13

LIGHTING

ERIC A. WOODROOF, PH.D., CEM, CRM
ProfitableGreenSolutions.com

JOHN FETTERS, CEM, CLEP
Effective Lighting Solutions, Inc.

13.1 INTRODUCTION

In today's cost-competitive, market-driven economy, everyone is seeking technologies or methods to reduce energy expenses and environmental impact. Because nearly all buildings have lights, lighting retrofits are very common and generally offer an attractive return on investment. Electricity used to operate lighting systems represents a significant portion of total electricity consumed in the United States, where they consume approximately 20% of the electricity generated.[1]

An attractive feature of lighting retrofits is they typically provide savings for both kW and kWh charges. Thus, the potential for dollar savings is increased. Many lighting retrofits can also improve the visual environment and worker productivity. Conversely, if a lighting retrofit reduces lighting quality, worker productivity may drop and the energy savings could be overshadowed by reduced profits. This was the case with the lighting retrofits of the 1970s, when employees were left "in the dark" due to massive de-lamping initiatives. However, due to substantial advances in technologies, today's lighting retrofits can reduce energy expenses while *improving* lighting quality and worker productivity.

This chapter will provide the energy manager with a good understanding of lighting fundamentals so that he/she can oversee successful lighting upgrades. The example section, (near the end of this chapter) contains analyses of a few common lighting retrofits. A section on new technologies also should be helpful. The schematics section contains illustrations of many lamps, ballasts, and lighting systems.

13.2 LIGHTING FUNDAMENTALS

This section will introduce the important concepts about lighting and the two objectives of the lighting designer: (1) to provide the right quantity of light, and (2) provide the right quality of light.

13.2.1 Lighting Quantity

Lighting quantity is the amount of light provided to a room. Unlike light quality, light quantity is easy to measure and describe.

13.2.1.1 Units

Lighting quantity is primarily expressed in three types of units: watts, lumens, and foot-candles (fc). Figure 13-1 shows the relationship between each unit. The watt is the unit for measuring electrical power. It defines the rate of energy consumption by an electrical device when it is in operation. The amount of watts consumed represents the electrical input to the lighting system.

The output of a lamp is measured in lumens. For example, one standard four-foot fluorescent lamp would provide 2,900 lumens in a standard office system. The amount of lumens can also be used to describe the output of an entire fixture (comprising several lamps). Thus, the number of lumens describes how much light is being produced by the lighting system.

The number of foot-candles shows how much light is actually reaching the workplane (or task). Foot-candles are the end result of watts being converted to lumens, the lumens escaping the fixture and traveling through the air to reach the workplane. In an office, the workplane is the desk level. You can measure the amount of foot-candles with a light meter when it is placed on the work surface where tasks are performed. Foot-candle measurements are important because they express the "result" and not

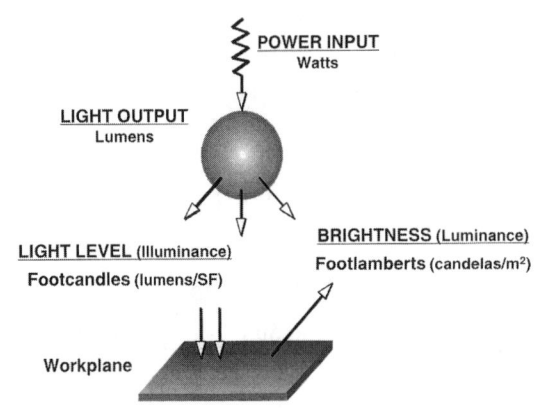

Figure 13-1. Units of measurement.

the "effort" of a lighting system. The Illuminating Engineering Society (IES) recommends light levels for specific tasks using foot-candles, not lumens or watts.

Efficacy

Similar to efficiency, efficacy describes an output/input ratio, the higher the output (while input is kept constant), the greater the efficacy. Efficacy is the amount of lumens per watt from a particular energy source. A common misconception in lighting terminology is that lamps with greater wattage provide more light. However, light sources with high efficacy can provide more light with the same amount of power (watts) when compared to light sources with low efficacy.

Figure 13-2 shows lamp efficacies for various lamp types, based on initial lumen values. These systems will be discussed in greater detail in Section 13.2.3.

13.2.1.2 IES Recommended Light Levels

The Illuminating Engineering Society (IES) is the largest organized group of lighting professionals in the United States. Since 1915, IES has prescribed the appropriate light levels for many kinds of visual tasks. Although IES is highly respected, the appropriate amount of light for a given space can be subjective. For many years, lighting professionals applied the philosophy that "more light is better," and light levels recommended by IES generally increased until the 1970s. However, recently it has been shown that occupant comfort decreases when a space has too much light. Numerous experiments have confirmed that the prescribed light levels were excessive and worker productivity was decreasing due to poor visual comfort. Due to these findings, IES revised their handbook and reduced the recommended light levels for many tasks.

The lighting designer must avoid over-illuminating a space. Unfortunately, this objective can be difficult because over-illuminated spaces have become the "norm" in many buildings. Although not optimal, the tradition of excessive illumination can be continued simply due to habit or an organization's reluctance to change. To correct

Figure 13-2. Lamp Efficacies for Various Lamp Types. (Source: Effective Lighting Solutions, Inc.)

Lamp Family	Lamp Type	Watts (nominal)	Efficacy (LPW)
Incandescent	Incandescent	3 – 1,500	4 – 23
	Halogen	42 – 1,500	14 – 22
CFL	Screw-base	9 – 85	40 – 65
	Twin-tube Pre-heat (2-pin)	5 – 13	50 – 69
	Quad Pre-heat (2-pin)	9 – 26	64 – 69
	Rapid-start (4-pin)	5 – 57	46 – 87
Linear Fluorescent	T2	6 – 13	52 – 66
	T5	14 – 35	96 – 104
	T5HO	24 – 80	83 – 94
	T8 4' Standard	32	53 – 88
	T8 4' Reduced-Wattage	30	94 - 95
	T8 4' High-Performance	32	94 – 98
	T10 4'	40	83
	T12 Reduced-Wattage	34	49 – 85
	T12 4' Standard	40	53 – 83
	T12 8' Slimline Reduced-Wattage	60	64 – 97
	T12 8' Slimline	75	58 – 87
	T12 8' HO Reduced-Wattage	95	81 – 92
	T12 8' HO (110-w)	110	55 – 85
	T12 8' VHO Reduced-Wattage	195	67
	T12 8' VHO Fluorescent	215	65 – 70
Mercury Vapor	Standard	50 – 1,000	32 – 58
	Self-ballasted	160 – 750	14 - 19
Metal Halide	Standard Probe-start	50 – 1,500	69 – 115
	Pulse-start	50 – 1,000	69 – 110
	Ceramic Arc-tube	20 – 400	83 – 95
High-Pressure Sodium	Standard	35 – 1,000	64 – 150
	Improved CRI	70 – 400	63 – 94
Low-Pressure Sodium		18 – 180	100 – 175
Induction	QL	55 – 165	64 – 73
	ICETRON™	70 – 150	70 – 79

this trend, the first step for a lighting retrofit should be to examine the existing system to determine if it is over-illuminated. Appropriate light levels for all types of visual tasks can be found in the most recent IES Handbook. Table 13-1 is a summary of some of the IES recommendations.

It is important to remember that IES light levels correspond to particular visual tasks. In an office, there are many tasks: walking around the office, viewing computer screens, reading and writing on paper. Each task requires a different light level. In the past, lighting designers

Table 13-1. Recommended light levels for visual tasks.

Building/Space Type	Guideline Illuminance Range (footcandles)
Commercial interiors	
Art galleries	30-100
Banks	50-150
Hotels (rooms and lobbies)	10-50
Offices	30-100
-Average reading and writing	50-75
-Hallways	10-20
-Rooms with computers	20-50
Restaurants (dining areas)	20-50
Stores (general)	20-50
Merchandise	100-200
Institutional interiors	
Auditoriums/assembly places	15-30
Hospitals (general areas)	10-15
Labs/treatment areas	50-100
Libraries	30-100
Schools	30-150
Industrial interiors	
Ordinary tasks	50
Stockroom storage	30
Loading and unloading	20
Difficult tasks	100
Highly difficult tasks	200
Very difficult tasks	300-500
Most difficult tasks	500-1000
Exterior	
Building security	1-5
Floodlighting	
(low/high brightness or surroundings)	5-30
Parking	1-5

would identify the task that required the most light and design the lighting system to provide that level of illumination for the entire space. However, as previously stated, these design methods often lead to environments with excessive brightness, glare and poor worker productivity.

If the IES tables are applied for each task ("task lighting"), a superior lighting system is constructed. For example, in an office with computers there should be up to 30 fc for ambient lighting. Small task lights on desks could provide the additional foot-candles needed to achieve a total illuminance of 50 to 75 fc for reading and writing. Task lighting techniques are discussed in greater detail in Section 13.3.

13.2.2 Lighting Quality

Lighting quality can have a dramatic influence on the attitude and performance of occupants. In fact, different "moods" can be created by a lighting system. Consider the behavior of people when they eat in different restaurants. If the restaurant is a fast-food restaurant, the space is usually illuminated by bright white lights, with a significant amount of glare from shiny tables. Occupants rarely spend much time there, partly because the space creates an uncomfortable mood and the atmosphere is "fast" (eat and leave). In contrast, consider an elegant restaurant with a candle-lit tables and a "warm" atmosphere. Occupants tend to relax and take more time to eat. Although occupant behavior is also linked to interior design and other factors, lighting quality represents a significant influence. Occupants perceive and react to a space's light color. It is important that the lighting designer be able to recognize and create the subtle aspects of an environment that define the theme of the space. For example, drug and grocery stores use white lights to create a "cool" and "clean" environment. Imagine if these spaces were illuminated by the same color lights as in an elegant restaurant. How would the perception of the store change?

Occupants can be influenced to work more effectively if they are in an environment that promotes a "work-like" atmosphere. The goal of the lighting designer is to provide the appropriate quality of light for a particular task, to create the right "mood" for the space.

Employee comfort and performance are worth more than energy savings. Although the cost of energy for lighting ($.50-$1.00/year/ft^2) is substantial, it is relatively small compared to the cost of labor ($100-$300/year/ft^2). Improvements in lighting quality can yield high dividends for businesses, because gains in worker productivity are common when lighting quality is improved. Conversely, if a lighting retrofit reduces lighting quality, occupant performance may decrease, quickly off-setting any savings in energy costs. Good energy managers should remember that buildings were not designed to save energy; they exist to create an environment where people can work efficiently. Occupants should be able to see clearly without being distracted by glare, excessive shadows or other uncomfortable features.

Lighting quality can be divided into four main considerations: uniformity, glare, color rendering index, and coordinated color temperature.

13.2.2.1 Uniformity

The uniformity of illuminance describes how evenly light spreads over an area. Creating uniform illumination requires proper fixture spacing. Non-uniform illuminance creates bright and dark spots, which can cause discomfort for some occupants.

Lighting designers have traditionally specified uniform illumination. This option is least risky because it minimizes the problems associated with non-uniform illumination and provides excellent flexibility for changes in the work environment. Unfortunately, uniform lighting applied over large areas can waste large amounts of energy. For example, in a manufacturing building 20% of the floor space may require high levels of illumination (100 fc) for a specific visual task. The remaining 80% of the building may only require 40 foot candles. Uniform illumination over the entire space would require 100 fc at any point in the building. Clearly, this is a tremendous waste of energy and money. Although uniform illumination is not needed throughout the entire facility, uniform illumination should be applied on specific tasks. For example, a person assembling small parts on a table should have uniform illumination across the table top.

13.2.2.2 Glare

Glare is a sensation caused by relatively bright objects in an occupant's field of view. The key word is *relative*, because glare is most probable when bright objects are located in front of dark environments. For example, a car's high beam headlights cause glare to oncoming drivers at night, yet create little discomfort during the day. *Contrast* is the relationship between the brightness of an object and its background. Although most visual tasks generally become easier with increased contrast, too much brightness causes glare and makes the visual task more difficult. Glare in certain work environments is a serious concern because it usually will cause discomfort and reduce worker productivity.

Visual Comfort Probability (VCP)

The visual comfort probability is a rating given to a fixture that indicates the percent of people who are comfortable with the glare. Thus, a fixture with a VCP of 80 means that 80% of occupants are comfortable with the amount of glare from that fixture. A minimum VCP of 70 is recommended for general interior spaces. Fixtures with VCPs exceeding 80 are recommended in computer areas and high-profile executive office environments.

To improve a lighting system that has excessive glare, a lighting designer should be consulted. However there are some basic "rules of thumb" which can assist the energy manager. A high-glare environment is characterized by either excessive illumination and reflection, or the existence of very bright areas, typically around fixtures. To minimize glare, the energy manager can try to obscure the bare lamp from the occupant's field of view, relocate fixtures, or replace the fixtures with ones that have a high VCP.

Reducing glare is commonly achieved by using indirect lighting, deep cell parabolic troffers, or special lenses. Although these measures will reduce glare, fixture efficiency will be decreased, because more light will be "trapped" in the fixture. Alternatively, glare can be minimized by reducing ambient light levels and using task lighting techniques.

Visual Display Terminals (VDTs)

Today's office environment contains a variety of special visual tasks, including the use of computer monitors or visual display terminals (VDTs). Occupants using VDTs are extremely vulnerable to glare and discomfort. When reflections of ceiling lights are visible on the VDT screen, the occupant has difficulty reading the screen. This phenomena is also called "discomfort glare" and is very common in rooms that are uniformly illuminated by fixtures with low a VCP. Therefore, lighting for VDT environments must be carefully designed so that occupants remain comfortable. Because the location VDTs can be frequently changed, lighting upgrades should also be designed to be adjustable. Moveable task lights and fixtures with high VCP are very popular for these types of applications. Because each VDT environment is unique, each upgrade must be evaluated on a case-by-case basis.

13.2.2.3 Color

Color considerations have an incredible influence on lighting quality. Light sources are specified based on two color-related parameters: the color rendering index (CRI) and the coordinated color temperature (CCT).

Color Rendering Index (CRI)

In simple terms, the CRI provides an evaluation of how colors appear under a given light source. The index range is from 0 to 100. The higher the number, the easier to distinguish colors. Generally, sources with a CRI > 75 provide excellent color rendition; sources with a CRI < 55 provide poor color rendition. To provide a "base-case," offices illuminated by most T12 Cool White lamps have a CRI = 62.

It is extremely important that a light source with a high CRI be used with visual tasks that require the occupant to distinguish colors. For example, a room with a color printing press requires illumination with excellent color rendition. In comparison, outdoor security lighting for a building may not need to have a high CRI, but a large quantity of light is desired.

Coordinated Color Temperature (CCT)

The coordinated color temperature (CCT) describes the color of the light source. For example, on a clear day, the sun appears yellow. On an over-cast day, the partially obscured sun appears to be gray. These color differences are indicated by a temperature scale. The CCT (measured in degrees Kelvin) is a close representation of the color that an object (black-body) would radiate at a certain temperature. For example, imagine a wire being heated. First it turns red (CCT = 2000K). As it gets hotter, it turns white (CCT = 5000K) and then blue (CCT = 8000K). Although a wire is different from a light source, the principle is similar.

CCT is not related to CRI, but it can influence the atmosphere of a room. Laboratories, hospitals, and grocery stores generally use "cool" (blue-white) sources, while expensive restaurants may seek a "warm" (yellow-red) source to produce a candle-lit appearance. Traditionally, office environments have been illuminated by Cool White lamps, which have a CCT = 4100K. However, a more recent trend has been to specify 3500K tri-phosphor lamps, which are considered neutral. Table 13-2 illustrates some common specifications for different visual environments.

13.2.3 Lighting System Components

After determining the quantity and quality of illumination required for a particular task, most lighting designers specify the lamp, then the ballast, and finally the fixture to meet the lighting needs. The schematics section (near the end of this chapter) contains illustrations of many of the lamps and systems described in this section.

13.2.3.1 Lamps

The lamp is the first component to consider in the lighting design process. The lamp choice determines the light quantity, CRI, CCT, relamping time interval, and

Table 13-2. Sample design considerations for a commercial building.

Office Areas		Light Levels	CRI	Color Temperature	Glare
Executive	General	100FC	≥80	3000K	VCP≥70
	Task	≥50FC	≥80	3000K	VCP≥70
Private	General	30-50FC	≥70	3000-3500K	VCP≥70
	Task	≥50FC	≥70	3000-3500K	VCP≥70
Open Plan Computers	General	30-50FC	≥70	3000-3500K	VCP≥90
	Task	≥50FC	≥70	3000-3500K	VCP≥90
Hallways	General	10-20FC	≥70	3000-3500K	VCP≥70
	Task	10-20FC	≥70	3000-3500K	VCP≥70
Reception/ Lobby	General	20-50FC	≥80	3000-5000K	VCP≥90
	Task	≥50FC	≥80	3000-3500K	VCP≥90
Conference	General	10-70FC	≥80	3000-4100K	VCP≥90
	Task	10-70FC	≥80	3000-4100K	VCP≥90
Open Plan General	General	30-70FC	≥70	3500-4100K	VCP≥80
	Task	30-70FC	≥70	3500-4100K	VCP≥80
Drafting	General	70-100FC	≥70	4100-5000K	VCP≥90
	Task	100-150FC	≥780	4100-5000K	VCP≥90

Table 13-3. Lamp characteristics

	Incandescent Including Tungsten Halogen 15-1500	Fluorescent 15-219	Compact Fluorescent 4-40	Mercury Vapor (Self-ballasted) 40-1000	Metal Halide 175-1000	High-Pressure Sodium (Improved Color) 70-1000	Low-Pressure Sodium 35-180
Wattages (lamp only)							
Life (hr)	750-12,000	7,500-24,000	10,000-20,000	16,000-15,000	1,500-15,000	24,000 (10,000)	18,000
Efficacy (lumens/W) lamp only	15-25	55-100	50-80	50-60 (20-25)	80-100	75-140 (67-112)	Up to 180
Lumen maintenance	Fair to excellent	Fair to excellent	Fair	Very good (good)	Good	Excellent	Excellent
Color rendition	Excellent	Good to excellent	Good to excellent	Poor to excellent	Very good	Fair	Poor
Light direction control	Very good to excellent	Fair	Fair	Very good	Very good	Very good	Fair
Relight time	Immediate	Immediate	Imm- 3 seconds	3-10 min.	10-20 min.	Less than 1 min.	Immediate
Comparative fixture cost	Low: simple	Moderate	Moderate	Higher than fluorescent	Generally higher than mercury	High	High
Comparative operating cost	High	Lower than incandescent	Lower than incandescent	Lower than incandescent	Lower than mercury	Lowest of HID types	Low

operational costs of the lighting system. This section will cover only the most popular types of lamps. Table 13-3 summarizes the differences between the primary lamps and lighting systems.

Incandescent

The oldest electric lighting technology is the incandescent lamp. Incandescent lamps are also the least efficacious (have the lowest lumens per watt) and have the shortest life. They produce light by passing a current through a tungsten filament, causing it to become hot and glow. As the tungsten emits light, it gradually evaporates, eventually causing the filament to break. When this happens, the lamps is said to be "burned-out."

Although incandescent sources are the least efficacious, they are still sold in great quantities because of economies of scale and market barriers. Consumers still purchase incandescent bulbs because they have low initial costs. However, if life-cycle cost analyses are used, incandescent lamps are usually more expensive than other lighting systems with higher efficacies.

Compact Fluorescent Lamps (CFLs)

Overview of CFLs:

Compact fluorescent lamps (CFLs) are energy efficient, long lasting replacements for some incandescent lamps. CFLs (like all fluorescent lamps) are composed of two parts, the lamp and the ballast. The short tubular lamps can last longer than 8,000 hours. The ballasts (plastic component at the base of tube) usually last longer than 60,000

hours. Some CFLs can be purchased as self-ballasted units that "screw in" to an existing incandescent socket. For simplicity, this chapter refers to a CFL as a lamp and ballast system. CFLs are available in many styles and sizes.

In most applications, CFLs are excellent replacements for incandescent lamps. CFLs provide similar light quantity and quality, while only requiring about 20-30% of the energy of comparable incandescent lamps. In addition, CFLs last 7 to 10 times longer than their incandescent counterparts. In many cases, it is cost-effective to replace an entire incandescent fixture with a fixture specially designed for CFLs.

The "New Technololgies" section contains a more thorough explanation of CFLs.

Fluorescent

Fluorescent lamps are the most common light source for commercial interiors in the U.S. They are repeatedly specified because they are relatively efficient, have long lamp lives, and are available in a wide variety of styles. For many years, the conventional fluorescent lamp used in offices has been the four-foot F40T12 lamp, which is usually used with a magnetic ballast. However, these lamps are being rapidly replaced by T8 or T5 lamps with electronic ballasts.

The labeling system used by manufacturers may appear complex, however it is actually quite simple. For example, with an F34T12 lamp, the "F" stands for fluorescent, the "34" means 34 watts, and the "T12" refers to the tube thickness. Since tube thickness (diameter) is measured in 1/8 inch increments, a T12 is 12/8, or 1.5 inches in diameter. A T8 lamp is 1 inch in diameter. Some lamp labels include additional information, indicating the CRI and CCT. Usually, CRI is indicated with one digit, like "8," meaning CRI = 80. CCT is indicated by the two digits following, "35" meaning 3500K. For example, a F32T8/841 label indicates a lamp with a CRI = 80 and a CCT = 4100K. Alternatively, the lamp manufacturer might label a lamp with a letter code referring to a specific lamp color. For example, "CW" to mean Cool White lamps with a CCT = 4100K.

Some lamps have "ES," "EE," or "EW" printed on the label. These acronyms attached at the end of a lamp label indicate that the lamp is an energy-saving type. These lamps consume less energy than standard lamps; however they also produce less light.

Tri-phosphor lamps have a coating on the inside of the lamp that improves performance. Tri-phosphor lamps usually provide greater color rendition. A bi-phosphor lamp (T12 Cool White) has a CRI= 62. By upgrading to a tri-phosphor lamp with a CRI = 75, occupants will be able to distinguish colors better. Tri-phosphor lamps are commonly specified with systems using electronic ballasts. Lamp flicker and ballast humming are also significantly reduced with electronically ballasted systems. For these reasons, the visual environment and worker productivity is likely to be improved.

There are many options to consider when choosing fluorescent lamps. Carefully check the manufacturers specifications and be sure to match the lamp and ballast to the application. Table 13-4 shows some of the specifications that vary between different lamp types.

High Intensity Discharge (HID)

High-intensity discharge (HID) lamps are similar to fluorescent lamps because they produce light by discharging an electric arc through a tube filled with gases. HID lamps generate much more light, heat, and pressure within the arc tube than fluorescent lamps, hence the title "high intensity" discharge. Like incandescent lamps, HIDs are physically small light sources (point sources), which means that reflectors, refractors and light pipes can be effectively used to direct the light. Although originally developed for outdoor and industrial applications, HIDs are also used in office, retail, and other indoor applications.

With a few exceptions, HIDs require time to warm up and should not be turned ON and OFF for short intervals. They are not ideal for certain applications, because as point sources of light they tend to produce more defined shadows than non-point sources, such as fluorescent tubes, which emit diffuse light.

Most HIDs have relatively high efficacies and long lamp lives (5,000 to 24,000+ hours), reducing maintenance re-lamping costs. In addition to reducing maintenance requirements, HIDs have many unique benefits. There are three popular types of HID sources (listed in order of increasing efficacy): Mercury vapor, metal halide, and high pressure sodium. A fourth source, low pressure sodium, is not technically an HID, but it provides similar quantities of illumination and will be referred to as an HID in this chapter. Table 13-3 shows there are dramatic differences in efficacy, CRI, and CCT between each HID source type.

Mercury Vapor

Mercury vapor systems were the "first generation" HIDs. Today they are relatively inefficient, provide poor CRI, and have the most rapid lumen depreciation rate of all HIDs. Because of these characteristics, other more cost-effective HID sources have replaced mercury vapor lamps in nearly all applications. Mercury vapor lamps provide a white-colored light that turns slightly green over time. A popular lighting upgrade is to replace mer-

Table 13-4.

Fluorescent options at a glance	Low-wattage	T8	Super-T8	T5	T5HO
Watts	32	25 to 30	32	28	54
Lumens	2,700 to 3,200	2,400 to 2,900	3,100+	2,900	4,400 to 5,000
Efficacy (lumens/watts)	92 to 100	89 to 92	88+/90+*	104	93
Rated average life (hours)	20,000	18,000 to 30,000	24,000+	20,000	20,000
CRI	75 to 85	82 to 85	81+	82 to 85	82 to 85

*Programmed rapid start ballasts/instant start ballasts

†This extended life is available from a specific lamp-ballast combination. Normal T10 lamp lives are approximately 24,000 hours. Service life refers to the typical lamp replacement life.

cury vapor systems with metal halide or high pressure sodium systems.

Metal Halide

Metal halide lamps are similar to mercury vapor lamps but contain slightly different metals in the arc tube, providing more lumens per watt with improved color rendition and improved lumen maintenance. With nearly twice the efficacy of mercury vapor lamps, metal halide lamps provide a white light and are commonly used in industrial facilities, sports arenas, and other spaces where good color rendition is required. They are the current best choice for lighting large areas that need good color rendition.

High Pressure Sodium (HPS)

With a higher efficacy than metal halide lamps, HPS systems are an economical choice for most outdoor and some industrial applications where good color rendition is not required. HPS is common in parking lots and produces a light golden color that allows some color rendition. Although HPS lamps do not provide the best color rendition, (or attractiveness) as "white light" sources, they are adequate for indoor applications at some industrial facilities. The key is to apply HPS in an area where there are no other light source types available for comparison. Because occupants usually prefer "white light," HPS installations can result in some occupant complaints. However, when HPS is installed at a great distance from metal halide lamps or fluorescent systems, the occupant will have no reference "white light" and will accept the HPS as "normal." This technique has allowed HPS to be installed in countless indoor gymnasiums and industrial spaces with minimal complaints.

Low Pressure Sodium

Although LPS systems have the highest efficacy of any commercially available HID, this monochromic light source produces the poorest color rendition of all lamp types. With a low CCT, the lamp appears to be "pumpkin orange," and all objects illuminated by its light appear black and white or shades of gray. Applications are limited to security or street lighting. The lamps are physically long (up to 3 feet) and not considered to be point sources. Thus optical control is poor, making LPS less effective for extremely high mounting.

LPS has become popular because of its extremely high efficacy. With up to 60% greater efficacy than HPS, LPS is economically attractive. Several cities, such as San Diego, California, have installed LPS systems on streets. Although there are many successful applications, LPS installations must be carefully considered. Often lighting quality can be improved by supplementing the LPS system with other light sources (having a greater CRI).

13.2.3.2 Ballasts

With the exception of incandescent systems, nearly all lighting systems (fluorescent and HID) require a ballast. A ballast controls the voltage and current that is supplied to lamps. Because ballasts are an integral component of the lighting system, they have a direct impact on light output. The ballast factor is the ratio of a lamp's light output to a reference ballast. General purpose fluorescent ballasts have a ballast factor that is less than one (typically .88 for most electronic ballasts). Special ballasts may have higher ballast factors to increase light output, or lower ballast factors to reduce light output. As can be expected, a ballast with a high ballast factor also consumes more energy than a general purpose ballast.

Fluorescent

Specifying the proper ballast for fluorescent lighting systems has become more complicated than it was 25 years ago, when magnetic ballasts were practically the only option. Electronic ballasts for fluorescent lamps have been available since the early 1980s, and their introduction has resulted in a variety of options.

This section describes the two types of fluorescent ballasts, magnetic and electronic.

Magnetic

Magnetic ballasts are available in three primary types.
- Standard core and coil
- High-efficiency core and coil (energy-efficient ballasts)
- Cathode cut-out or hybrid

Standard core and coil magnetic ballasts are essentially core and coil transformers that are relatively inefficient at operating fluorescent lamps. Although these types of ballasts are no longer sold in the U.S., they still exist in many facilities. The "high-efficiency" magnetic ballast can replace the "standard ballast," improving the system efficiency by approximately 10%.

"Cathode cut-out" or "hybrid" ballasts are high-efficiency core and coil ballasts that incorporate electronic components which cut off power to the lamp cathodes after the lamps are operating, resulting in an additional 2-watt savings per lamp.

Electronic

During the infancy of electronic ballast technology, reliability and harmonic distortion problems hampered their success. However, most electronic ballasts available today have a failure rate of less than one percent, and many distort harmonic current less than their magnetic counterparts. Electronic ballasts are superior to magnetic ballasts, because they are typically 30% more energy efficient and produce less lamp flicker, ballast noise, and waste heat.

In nearly every fluorescent lighting application, electronic ballasts can be used in place of conventional magnetic core and coil ballasts. Electronic ballasts improve fluorescent system efficacy by converting the standard 60 Hz input frequency to a higher frequency, usually 25,000 to 40,000 Hz. Lamps operating on these frequencies produce about the same amount of light, while consuming up to 40% less power than a standard magnetic ballast. Other advantages of electronic ballasts include less audible noise, less weight, virtually no lamp flicker, and dimming capabilities.

T12 and T8 ballasts are the most popular types of electronic ballasts. T12 electronic ballasts are designed for use with conventional (T12) fluorescent lighting systems. T8 ballasts offer some distinct advantages over other types of electronic ballasts. They are generally more efficient, have less lumen depreciation, and are available with more options. T8 ballasts can operate one, two, three, or four lamps. Most T12 ballasts can only operate one, two, or three lamps. Therefore, one T8 ballast can replace two T12 ballasts in a 4 lamp fixture.

Some electronic ballasts are parallel-wired so that when one lamp burns out, the remaining lamps in the fixture will continue to operate. In a typical magnetic, (series-wired system) when one component fails, all lamps in the fixture shut OFF. Before maintenance personnel can relamp, they must first diagnose which lamp failed. Thus the electronically ballasted system will reduce time to diagnose problems, because maintenance personnel can immediately see which lamp failed.

Parallel-wired ballasts also offer the option of reducing lamps per fixture (after the retrofit) if an area is over-illuminated. This option allows the energy manager to experiment with different configurations of lamps in different areas. However, each ballast operates best when controlling the specified number of lamps.

Due to the advantages of electronically ballasted systems, they are produced by many manufacturers, and prices are very competitive. Due to their market penetration, T8 systems (and replacement parts) are more likely to be available, and at lower costs.

HID

As with fluorescent systems, high intensity discharge lamps also require ballasts to operate. Although there are not nearly as many specification options as with fluorescent ballasts, HID ballasts are available in dimmable and bi-level light outputs. Instant restrike systems are also available.

Capacitive Switching HID Fixtures

Capacitive switching or "bi-level" HID fixtures are designed to provide either full or partial light output, based on inputs from occupancy sensors, manual switches, or scheduling systems. Capacitive-switched dimming can be installed as a retrofit to existing fixtures or as a direct fixture replacement. Capacitive switching HID upgrades can be less expensive than installing a panel-level variable voltage control to dim the lights, especially in circuits with relatively few fixtures.

The most common applications of capacitive switching are athletic facilities, occupancy-sensed dimming in parking lots, and warehouse aisles. General purpose transmitters can be used with other control devices such as timers and photosensors to control the bi-level fixtures. Upon detecting motion, the occupancy sensor sends a signal to the bi-level HID ballasts. The system will rapidly bring the light levels from a standby reduced level to about 80 percent of full output, followed by the normal warm-up time between 80 and 100 percent of full light output.

Depending of the lamp type and wattage, the standby lumens are roughly 15-40 percent of full output, and the standby wattage is 30-60 percent of full wattage. When the space is unoccupied and the system is dimmed, you can achieve energy savings of 40-70 percent.

13.2.3.3 Fixtures (aka Luminaires)

A fixture is a unit consisting of the lamps, ballasts, reflectors, lenses or louvers, and housing. The main function is to focus or spread light emanating from the lamp(s). Without fixtures, lighting systems would appear very bright and cause glare.

Fixture Efficiency

Fixtures block or reflect some of the light exiting the lamp. The efficiency of a fixture is the percentage of lamp lumens produced that actually exit the fixture in the intended direction. Efficiency varies greatly among different fixture and lamp configurations. For example, using four T8 lamps in a fixture will be more efficient than using four T12 lamps because the T8 lamps are thinner, allowing more light to "escape" between the lamps and out of the fixture. Understanding fixtures is important because a lighting retrofit may involve changing some components of the fixture to improve the efficiency and deliver more light to the task.

The coefficient of utilization (CU) is the percent of

lumens produced that actually reach the work plane. The CU incorporates the fixture efficiency, mounting height, and reflectances of walls and ceilings. Therefore, improving the fixture efficiency will improve the CU.

Reflectors

Installing reflectors in most fixtures can improve its efficiency because light leaving the lamp is more likely to "reflect" off interior walls and exit the fixture. Because lamps block some of the light reflecting off the fixture interior, reflectors perform better when there are fewer lamps (or smaller lamps) in the fixture. Due to this fact, a common fixture upgrade is to install reflectors and remove some of the lamps in a fixture. Altered light levels and different distributions may not be acceptable; however, these changes need to be considered.

To ensure acceptable performance from reflectors, conduct a trial installation and measure "before" and "after" light levels at various locations in the room. Don't compare an existing system, (which is dirty, old, and contains old lamps) against a new fixture having half the lamps and a clean, dust-free reflector. The light levels may appear to be adequate, or even improved. However, as the new system ages and dirt accumulates on the surfaces, the light levels will drop.

A variety of reflector materials are available: highly reflective white paint, silver film laminate, and anodized aluminum. Silver film laminate usually has the highest reflectance but is considered less durable. Be sure to evaluate the economic benefits of your options to get the most "bang for your buck."

In addition to installing reflectors within fixtures, light levels can be increased by improving the reflectivity of the room's walls, floors, and ceilings. For example, by covering a brown wall with white paint, more light will be reflected back into the workspace, and the coefficient of utilization will be increased.

Lenses and Louvers

Most indoor fixtures use either a lens or louver to prevent occupants from directly seeing the lamps. Light that is emitted in the shielding angle or "glare zone" (angles above 45° from the fixture's vertical axis) can cause glare and visual discomfort, which hinders the occupant's ability to view work surfaces and computer screens. Lenses and louvers are designed to shield the viewer from these uncomfortable, direct beams of light. Lenses and louvers are usually included as part of a fixture when purchased, and they can have a tremendous impact on the VCP of a fixture.

Lenses are sheets of hard plastic (either clear or milky white) that are located on the bottom of a fixture.

Clear, prismatic lenses are very efficient, because they trap less light within the fixture. Milky-white lenses are called "diffusers" and are the least efficient, trapping a lot of the light within the fixture. Although diffusers have been routinely specified for many office environments, they have one of the lowest VCP ratings.

Louvers provide superior glare control and high VCP when compared to most lenses. As Figure 13-3 shows, a louver is a grid of plastic "shields" that blocks some of the horizontal light exiting the fixture. The most common application of louvers is to reduce the fixture glare in sensitive work environments, such as in rooms with computers. Parabolic louvers usually improve the VCP of a fixture; however, efficiency is reduced because more light is blocked by the louver. Generally, the smaller the cell, the greater the VCP and less the efficiency. Deep-cell parabolic louvers offer a better combination of VCP and efficiency; however, deep-cell louvers require deep fixtures, which may not fit into the ceiling plenum space.

Table 13-5 shows the efficiency and VCP for various lenses and louvers. VCP is usually inversely related to fixture efficiency. An exception is with the milky-white diffusers, which have low VCP and low efficiency.

Light Distribution/Mounting Height

Fixtures are designed to direct light where it is needed. Various light distributions are possible to best suit any visual environment. With "direct lighting," 90-100% of the light is directed downward for maximum use. With "indirect lighting," 90-100% of the light is directed to the ceilings and upper walls. A "semi-indirect" system distributes 60-90% down, with the remainder upward. Designing the lighting system should incorporate the different light distributions of different fixtures to maximize comfort and visual quality.

Fixture mounting height and light distribution are presented together since they are interactive. HID systems are preferred for high mounting heights since the lamps are physically small and reflectors can direct light downward with a high degree of control. Fluorescent lamps are physically long and diffuse sources, with less ability to

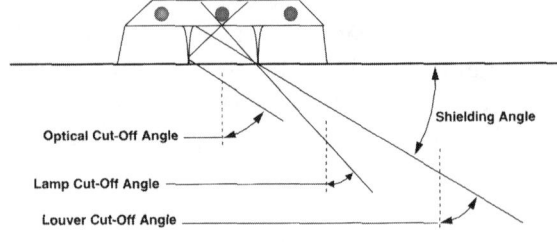

Figure 13-3. Higher shielding angles for improved glare control.

Table 13-5. Luminaire efficiency and VCP.

Shielding Material	Luminaire Efficiency (%)	Visual Comfort Probability (VCP)
Clear Prismatic Lens	60-70	50-70
Low Glare Clear Lens	60-75	75-85
Deep-Cell Parabolic Louver	50-70	75-95
Translucent Diffuser	40-60	40-50
Small-Cell Parabolic Louver	35-45	99

Table 13-6. Minimum mounting heights for HIDs

Lamp Type	feet above ground
400 W Metal Halide	16
1000 W Metal Halide	20
200 W High Pressure Sodium	15
250 W High Pressure Sodium	16
400 W High Pressure Sodium	18
1000 W High Pressure Sodium	26

control light at high mounting heights. Thus fluorescent systems are better for low mounting heights and/or areas that require diffuse light with minimal shadows.

Generally, "high-bay" HID fixtures are designed for mounting heights greater than 20 feet high. "High-bay" fixtures usually have reflectors and focus most of their light downward. "Low-bay" fixtures are designed for mounting heights less than 20 feet and use lenses to direct more light horizontally.

HID sources are potential sources of direct glare since they produce large quantities of light from physically small lamps. The probability of excessive direct glare may be minimized by mounting fixtures at sufficient heights. Table 13-6 shows the minimum mounting height recommended for different types of HID systems.

13.2.3.4 Exit Signs

Recent advances in exit sign systems have created attractive opportunities to reduce energy and maintenance costs. Because emergency exit signs should operate 24 hours per day, energy savings quickly recover retrofit costs. There are generally two options, buying a new exit sign, or retrofitting the existing exit sign with new light sources.

Most retrofit kits available today contain adapters that screw into the existing incandescent sockets. Installation is easy, usually requiring only 15 minutes per sign. However, if a sign is severely discolored or damaged, buying a new sign might be required in order to maintain illuminance as required by fire codes.

Basically, there are five upgrade technologies: compact fluorescent lamps (CFLs), incandescent assemblies, light-emitting diodes (LED), electroluminescent panels, and self-luminous tubes.

Replacing incandescent sources with compact fluorescent lamps was the "first generation" exit sign upgrade. Most CFL kits must be hard-wired and can not simply screw into an existing incandescent socket. Although CFL kits are a great improvement over incandescent exit signs, more technologically advanced upgrades are available that offer reduced maintenance costs, greater efficacy and flexibility for installation in low (sub-zero) temperature environments.

As Table 13-7 shows, LED upgrades are the most cost-effective because they consume very little energy and have an extremely long life, practically eliminating maintenance.

Another low-maintenance upgrade is to install a "rope" of incandescent assemblies. These low-voltage "luminous ropes" are an easy retrofit, because they can screw into existing sockets like LED retrofit kits. However, the incandescent assemblies create bright spots that are visible through the transparent exit sign, and the non-uniform glow is a noticeable change. In addition, the incandescent assemblies don't last nearly as long as LEDs.

Although electroluminescent panels consume less than one watt, light output rapidly depreciates over

Table 13-7. Exit sign upgrades.

Light Source	Watts	Life	Replacement
Incandescent (Long Life)	40	8 months	lamps
Compact Fluorescent	10	1.7 years	lamps
Incandescent Assembly	8	3 + years	light source
Light Emitting Diode (LED)	<4	>25	light source
Electroluminescent	1	8+ years	panel
Self luminous (Tritium)	0	10-20 years	luminous tubes

time. These self-luminous sources are obviously the most energy-efficient, consuming no electricity. However the spent tritium tubes, which illuminate the unit, must be disposed of as a radioactive waste, which will increase over-all costs.

13.2.3.5 Lighting Controls

Lighting controls offer the ability for systems to be turned ON, dimmed, or OFF. There are several control technology upgrades for lighting systems, ranging from simple (installing manual switches in proper locations) to sophisticated (installing occupancy sensors).

Switches

The standard manual, single-pole switch was the first energy conservation device. It is also the simplest device and provides the least options. One negative aspect about manual switches is that people often forget to turn them OFF. If switches are far from room exits or are difficult to find, occupants are more likely to leave lights ON when exiting a room.[1] Occupants do not want to walk through darkness to find exits. However, if switches are located in the right locations, with multiple points of control for a single circuit, occupants find it easier to turn systems OFF. Once occupants get in the habit of turning lights OFF upon exit, more complex systems may not be necessary. The point is that switches can be great energy conservation devices as long as they are convenient to use them.

Another opportunity for upgrading controls exists when lighting systems are designed such that all circuits in an area are controlled from one switch, yet not all circuits need to be activated. For example, a college football stadium's lighting system is designed to provide enough light for TV applications. However, this intense amount of light is not needed for regular practice nights or other non-TV events. Because the lights are all controlled from one switch, every time the facility is used all the lights are turned ON. By dividing the circuits and installing one more switch to allow the football stadium to use only 70% of its lights during practice nights, significant energy savings are possible. Another example includes recircuiting an office fluorescent system such that one of the three lamps can be turned on. When more light is needed, the other two lamps can be switched on.

Generally, if it is not too difficult to re-circuit a poorly designed lighting system; additional switches can be added to optimize the lighting controls. This approach also has applications in "perimeter" lighting near windows, versus interior lighting.

Time Clocks

Time clocks can be used to control lights when their operation is based on a fixed operating schedule. Time clocks are available in electronic or mechanical styles. However, regular check-ups are needed to ensure that the time clock is controlling the system properly. After a power loss, electronic timers without battery backups can get off schedule, cycling ON and OFF at the wrong times. It requires a great deal of maintenance time to reset isolated time clocks if many are installed.

Photocells

For most outdoor lighting applications, photocells (which turn lights ON when it gets dark, and off when sufficient daylight is available) offer a low-maintenance alternative to time clocks. Unlike time clocks, photocells are seasonally self-adjusting and automatically switch ON when light levels are low, such as during rainy days. A photocell is inexpensive and can be installed on each fixture, or it can be installed to control numerous fixtures on one circuit. Photocells can also be effectively used indoors if daylight is available through skylights.

Photocells have worked well in almost any climate; however, they should be aimed north (in the northern hemisphere) to "view" the reflected light of the north sky. This way they are not biased by the directionality of east/west exposure or degraded by intense southern exposure. Photocells should also be cleaned when fixtures are re-lamped. Otherwise, dust will accumulate on the photodiode aperture, causing the controls to always perceive it is a cloudy day, and the lights will stay ON.

The least expensive type of photocell uses a cadmium sulfide cell, but these cells lose sensitivity after being in service for a few years by being degraded from their exposure to sunlight. This decreases savings by keeping exterior lighting on longer than required. To avoid this situation, cadmium sulfide cells can be replaced with electronic types that do not lose sensitivity over time. These electronic photocells use solid-state, silicon phototransistors or photodiodes, which last longer as evidenced by their longer warranties—up to 6 years—and can easily pay back before that time with energy and labor savings.

Photocells combined with Dimmable Ballasts to allow Daylight Harvesting

Daylight harvesting is a control strategy that can be applied where diffuse daylight can be used effectively to light interior spaces. There is a widespread misunderstanding that daylighting can only be done in areas where there is a predominance of sunny, clear days, such as California or Arizona. In fact, many places with over 50% cloudy days can cost-effectively use daylight controls.

Daylight harvesting employs strategically located photo-sensors and electronic dimming ballasts. To effectively apply this strategy requires more knowledge than just plugging a sensor into a dimming ballast. Photo-sensors and dimming ballasts form a control system that controls the light level according to the daylight level. The fluorescent lighting is dimmed to maintain a band of light level when there is sufficient daylight present in the space. The output is changed gradually by a fade control, so occupants are not disturbed by rapid changes in light level.

Lumen Depreciation Compensation
(an additional benefit of a Daylight Harvesting System)

Lighting systems are usually over-designed to compensate for light losses that normally occur during the life time of the system. Alternatively, the "lumen depreciation compensation strategy" allows the design light level to be met without over-designing, thereby providing a more efficient lighting system. The control system works in a way similar to daylight harvesting controls. A photo-sensor detects the actual light level and provides a low-voltage signal to electronic dimming ballasts to adjust the light level. When lamps are new and room surfaces are clean, less power is required to provide the design light level. As lamps depreciate in their light output and surfaces become dirty, the input power and light level is increased gradually to compensate for these sources of light loss. Some building management systems accomplish this control by using a depreciation algorithm to adjust the output of the electronic ballasts instead of relying on photo-sensors.

Occupancy Sensors

Occupancy sensors save energy by turning off lights in spaces that are unoccupied. When the sensor detects motion, it activates a control device that turns ON a lighting system. If no motion is detected within a specified period, the lights are turned OFF until motion

Table 13-8. Estimated % savings from occupancy sensors.

Application	Energy Savings
Offices (Private)	25-50%
Offices (Open Spaces)	20-25%
Rest Rooms	30-75%
Corridors	30-40%
Storage Areas	45-65%
Meeting Rooms	45-65%
Conference Rooms	45-65%
Warehouses	50-75%

is sensed again. With most sensors, sensitivity (the ability to detect motion) and the time delay (difference in time between when sensor detects no motion and lights go OFF) are adjustable. Occupancy sensors are produced in two primary types: Ultrasonic (US) and Passive Infrared (PIR). Dual-Technology (DT) sensors that have both ultrasonic and passive infrared detectors are also available. Table 13-8 shows the estimated percent energy savings from occupancy sensor installation for various locations.

US and PIR sensors are available as wall-switch sensors or remote sensors, such as ceiling mounted or outdoor commercial grade units. With remote sensors, a low-voltage wire connects each sensor to an electrical relay and control module, which operates on common voltages. With wall-switch sensors, the sensor and control module are packaged as one unit. Multiple sensors and/or lighting circuits can be linked to one control module, allowing flexibility for optimum design.

Wall-switch sensors can replace existing manual switches in small areas such as offices, conference rooms, and some classrooms. However, in these applications, a manual override switch should be available so that the lights can be turned OFF for slide presentations and other visual displays. Wall-switch sensors should have an unobstructed coverage pattern (absolutely necessary for PIR sensors) of the room it controls.

Ceiling-mounted units are appropriate in corridors, rest rooms, open office areas with partitions, and any space where objects obstruct the line of sight from a wall-mounted sensor location. Commercial grade outdoor units can also be used in indoor warehouses and large aisles. Sensors designed for outdoor use are typically heavy duty and usually have adjustable sensitivities and coverage patterns for maximum flexibility. Table 13-9 indicates the appropriate sensors for various applications.

Ultrasonic Sensors (US)

Ultrasonic sensors transmit and receive high-frequency sound waves above the range of human hearing. The sound waves bounce around the room and return to the sensor. Any motion within the room distorts the sound waves. The sensor detects this distortion and signals the lights to turn ON. When no motion has been detected over a user-specified time, the sensor sends a signal to turn the lights OFF. Because ultrasonic sensors need enclosed spaces (for good sound wave echo reflection), they can only be used indoors and perform better if room surfaces are hard, where sound wave absorption is minimized. Ultrasonic sensors are most sensitive to motion toward or away from the sensor. Applications include rooms with objects that obstruct the sensor's line of

Table 13-9. Occupancy sensor applications.

Sensor Technology	Private Office	Large Open Office Plan	Partitioned Office Plan	Conference Room	Rest Room	Closets/ Copy Room	Hallways Corridors	Warehouse Aisles Areas
US Wall Switch	•			•	•	•		
US Ceiling Mount	•	•	•	•	•	•		
IR Wall Switch	•			•		•		
IR Ceiling Mount	•	•	•	•		•		
US Narrow View							•	
IR High Mount Narrow View							•	•
Corner Mount Wide-View Technology Type		•		•				

sight coverage of the room, such as restroom stalls, locker rooms, and storage areas.

Passive Infrared Sensors (PIR)

Passive infrared sensors detect differences in infrared energy emanating in the room. When a person moves, the sensor "sees" a heat source move from one zone to the next. PIR sensors require an unobstructed view, and as distance from the sensor increases, larger motions are necessary to trigger the sensor. Applications include open plan offices (without partitions), classrooms, and other areas that allow a clear line of sight from the sensor.

Dual-Technology Sensors (DT)

Dual-technology (DT) sensors combine both US and PIR sensing technologies. DT sensors can improve sensor reliability and minimize false switching; however, these types of sensors are still limited to applications where ultrasonic sensors will work.

Occupancy Sensor Effect on Lamp Life

Occupancy sensors can cause rapid ON/OFF switching that reduces the life of certain fluorescent lamps. Offices without occupancy sensors usually have lights constantly ON for approximately ten hours per day. After occupancy sensors are installed, the lamps may be turned ON and OFF several times per day. Several laboratory tests have shown that some fluorescent lamps lose about 25% of their life if turned OFF and ON every three hours. Although occupancy sensors may cause lamp life to be reduced, the annual burning hours also decreases. Therefore, in most applications, the time period until re-lamp will not increase. However, due to the laboratory results, oc-

cupancy sensors should be carefully evaluated if the lights will be turned ON and OFF rapidly. The longer the lights are left OFF, the longer lamps will last.

The frequency at which occupants enter a room makes a difference in the actual percent time savings possible. Occupancy sensors save the most energy when applied in rooms that are not used for long periods of time. If a room is frequently used and occupants re-enter a room before the lights have had a chance to turn OFF, no energy will be saved. Therefore, a room that is occupied once every three hours will be more appropriate for occupancy sensors than a room occupied once every three minutes, even though the percent vacancy time is the same.

Occupancy Sensors and HIDs

Although occupancy sensors were not primarily developed for HIDs, some special HID ballasts (bi-level) offer the ability to dim and re-light lamps quickly. Another term for bi-level HID technology is capacitive switching HID fixtures, which are discussed in the HID ballast section.

Lighting Controls via a Facility Management System

When lighting systems are connected to a facility management system (FMS), greater control options can be realized. The FMS could control lights (and other equipment, i.e. HVAC) to turn OFF during non-working hours, except when other sensors indicate that a space is occupied. These sensors include standard occupancy sensors or a card access system, which could indicate which employee is in a particular part of the facility. If the facility is "smart," it will know where the employee works and control the lights and other systems in that area. By wiring all systems to the FMS, there is a greater ability to integrate technologies for maximum performance and savings. *For example, an employee can control lights by entering a code into the telephone system or a computer network.*

Specialized controls for individual work environments (offices or cubicles) are also available. These systems use an occupancy sensor to regulate lights, other electronic systems, and even HVAC systems in an energy efficient manner. In some systems, remote controls allow the occupant to regulate individual lighting and HVAC systems. These customized systems have allowed some organizations to realize individual productivity gains via more effective and aesthetic work space environments.

13.3 PROCESS TO IMPROVE LIGHTING EFFICIENCY

The three basic steps to improving the efficiency of lighting systems:
1. Identify necessary light quantity and quality to perform visual task.
2. Increase light source efficiency if occupancy is frequent.
3. Optimize lighting controls if occupancy is infrequent.

Step 1, identifying the proper lighting quantity and quality, is essential to any illuminated space. However, steps 2 & 3 are options that can be explored individually or together. Steps 2 & 3 can both be implemented, but often the two options are economically mutually exclusive. If you can turn OFF a lighting system for the majority of time, the extra expense to upgrade lighting sources is rarely justified. Remember, light source upgrades will only save energy (relative to the existing system) when the lights are ON.

13.3.1 Identify necessary light quantities and qualities to perform tasks.

Identifying the necessary light quantities for a task is the first step of a lighting retrofit. Often this step is overlooked, because most energy managers try to mimic the illumination of an existing system, even if it is over-illuminated and contains many sources of glare. For many years, lighting systems were designed with the belief that no space can be over-illuminated. However, the "more light is better" myth has been dispelled, and light levels recommended by the IES has declined by 15% in hospitals, 17% in schools, 21% in office buildings, and 34% in retail buildings.[2] Even with IES's adjustments, there are

still many excessively illuminated spaces in use today. Energy managers can reap remarkable savings by simply redesigning a lighting system so that the proper illumination levels are produced.

Although the number of workplane footcandles are important, the occupant needs to have a contrast so that he can perform a task. For example, during the daytime your car headlights don't create enough contrast to be noticeable. However, at night, your headlights provide enough contrast for the task. The same amount of light is provided by the headlights during both periods, but daylight "washes out" the contrast of the headlights.

The same principle applies to offices and other illuminated spaces. For a task to appear relatively bright, objects surrounding that task must be relatively dark. For example, if ambient light is excessive (150 fc) the occupant's eyes will adjust to it and perceive it as the "norm." However, when the occupant wants to focus on something he/she may require an additional light to accent the task (at 200 fc). This excessively illuminated space results in unnecessary energy consumption. The occupant would see better if ambient light was reduced to 30-40 fc and the task light was used to accent the task at 50 fc. As discussed earlier, excessive illumination is not only wasteful but can reduce the comfort of the visual environment and decrease worker productivity.

After identifying the proper quantity of light, the proper quality must be chosen. The CRI, CCT, and VCP must be specified to suit the space.

13.3.2 Increase Source Efficacy

Increasing the source efficacy of a lighting system means replacing or modifying the lamps, ballasts, and/or fixtures to become more efficient. In the past, the term "source" has been used to imply only the lamp of a system. However, due to the inter-relationships between components of modern lighting systems, we also consider ballast and fixture retrofits as "source upgrades." Thus increasing the efficacy simply means getting more lumens per watt out of lighting system. For example, to increase the source efficacy of a T12 system with a magnetic ballast, the ballast and lamps could be replaced with T8 lamps and an electronic ballast, which is a more efficacious (efficient) system.

Another retrofit that would increase source efficacy would be to improve the fixture efficiency by installing reflectors and more efficient lenses. This retrofit would increase the lumens per watt, because with reflectors and efficient lenses, more lumens can escape the fixture, while the power supplied remains constant.

Increasing the efficiency of a light source is one of

the most popular types of lighting retrofits, because energy savings can almost be guaranteed if the new system consumes less watts than the old system. With reduced lighting load, electrical demand savings are also usually obtained. In addition, lighting quality can be improved by specifying sources with higher CRI and improved performance. These benefits allow capital improvements for lighting systems that pay for themselves through increased profits.

Task lighting

As a subset of increasing source efficacy, "task lighting" or "task/ambient" lighting techniques involve improving the efficiency of lighting in an entire workplace by replacing and relocating lighting systems. Task lighting means retrofitting lighting systems to provide appropriate illumination for each task. Usually this results in a reduction of ambient light levels, while maintaining or increasing the light levels on a particular task. For example, in an office the light level needed on a desk could be 75 fc. The light needed in aisles is only 20 fc. Traditional uniform lighting design would create a workplace where ambient lighting provides 75 fc throughout the entire workspace. Task lighting would create an environment where each desk is illuminated to 75 fc and the aisles only to 20 fc. Figure 13-4 shows a typical application of task/ambient lighting.

Task lighting upgrades are a model of energy efficiency, because they only illuminate what is necessary. Task lighting designs are best suited for office environments with VDTs and/or where modular furniture can incorporate task lighting under shelves. Alternatively, moveable desk lamps may be used for task illumination. Savings result when the energy saved from reducing ambient light levels exceeds the energy used for task lights.

In most work spaces, a variety of visual tasks are performed, and each employee has lighting preferences. Most workers prefer lighting systems designed with task lighting because it is flexible and allows individual control. For example, older workers may require greater light

Figure 13-4. Task/ambient lighting.

levels than young workers. Identifying task lighting opportunities may require some creativity, but the potential dollar savings can be enormous.

Task lighting techniques are also applicable in industrial facilities. For example, high intensity task lights can be installed on fork trucks (to supplement headlights) for use in rarely occupied warehouses. With this system, the entire warehouse's lighting can be reduced, saving a large amount of energy.

13.3.3 Optimize Lighting Controls

The third step of lighting energy management is to investigate optimizing lighting controls. As shown earlier, improving the efficiency of a lighting system can save a percentage of the energy consumed *while the system is operating*. However, sophisticated controls can turn systems OFF when they are not needed, allowing energy savings to accumulate quickly. The Electric Power Research Institute (EPRI) reports that spaces in an average office building may only be occupied 60-75% of the time, although the lights may be ON for the entire 10 hour day[3]. Lighting controls include switches, time clocks, occupancy sensors and other devices that regulate a lighting system. These systems are discussed in Section 13.2.3, Lighting System Components.

13.4 MAINTENANCE

13.4.1 Isolated Systems

Most lighting manuals prescribe specialized technologies to efficiently provide light for particular tasks. An example is dimmable ballasts. For areas that have sufficient daylight, dimmable ballasts can be used with integrated circuitry to reduce energy consumption during peak periods. Still, though there may be some shedding of lighting load along the perimeter, these energy cost savings may not represent a great percentage of the building's total lighting load. Further, applications of specialized technologies (such as dimmable ballasts) may be dispersed and isolated in several buildings, which can become a complex maintenance challenge, even if lamp types and locations are recorded properly. If maintenance personnel need to make additional site visits to get the right equipment to re-lamp or "fine-tune" special systems, the labor costs may exceed the energy cost savings.

In facilities with low potential for energy cost savings, facility managers may not want to spend a great deal of time monitoring and "fine-tuning" a lighting system if other maintenance concerns need attention. If a specialized lighting system malfunctions, repair may require special components that may be expensive and more difficult to install. If maintenance cannot effectively repair the complex technologies, the systems will fail and occupant complaints will increase. Thus, the isolated, complex technology that appeared to be a unique solution to a particular lighting issue is often replaced with a system that is easy to maintain.

In addition to the often eventual replacement of technologies that are difficult to maintain, well intended repairs to the system may accidentally result in "snapback." "Snap-back" is when a specialized or isolated technology is accidentally replaced with a common technology within the facility. For example, if dimmable ballasts only represent 10% of the building's total ballasts, maintenance personnel might not keep them in stock. When replacement is needed, the maintenance personnel may accidentally install a regular ballast. Thus, the lighting retrofit has "snapped back" to its original condition.

The above arguments are not meant to "shoot down" the application of all new technologies. However, new technologies usually bring new problems. The authors ask that the energy manager carefully consider the maintenance impact when evaluating an isolated technology. Once again, all lighting systems depend on regular maintenance.

13.4.2 Maintaining System Performance

As with most manufactured products, lighting systems lose performance over time. This degradation can be the result of lamp lumen depreciation (LLD), fixture dirt depreciation (LDD), room surface dirt depreciation (RSDD), and many other factors. Several of these factors can be recovered to maintain performance of the lighting system. Figure 13-5 shows the LLD for various types of lighting systems.

Lamp lumen depreciation occurs because as the lamp ages, its performance degrades. LLD can be accelerated if the lamp is operated in harsh environments, or if the system is subjected to conditions for which it was not designed. For example, if a fluorescent system is turned ON and OFF every minute, the lamps and ballasts will not last as long. Light loss due to lamp lumen depreciation can be recovered by re-lamping the fixture.

Fixture dirt depreciation and room surface dirt depreciation block light and can reduce light levels. However, these factors can be minimized by cleaning surfaces and minimizing dust. The magnitude of these factors is dependent on each room, thus recommended cleaning intervals can vary. Generally it is most economical to clean fixtures when re-lamping.

13.4.3 Group Re-lamping

Most companies replace lamps when someone notices a lamp is burned out. In a high rise building, this could become a full-time job, running from floor to floor, office to office, disrupting work to open a fixture and replace a lamp. However, in certain cases, it is less costly to group re-lamp on a pre-determined date. Group relamping can be cost-effective due to economies of scale. Replacing all lamps at one time can be more efficient than relamping "one at a time." In addition, bulk purchasing may also yield savings. The rule of thumb is to group relamp at 50% to 70% of the lamp's rated life. However, depending on site-specific factors and the lumen depreciation of the lighting system, relamping interval may vary.

The facility manager must evaluate their own building and determine the appropriate relamp interval by

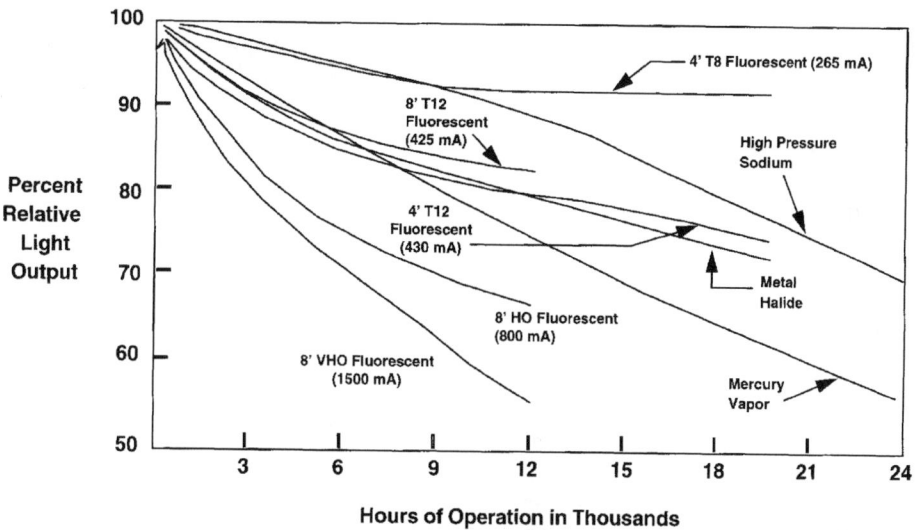

Figure 13-5. Lamp lumen depreciation (LLD).

observing when lamps start to fail. Due to variations in power voltages (spikes, surges and low power), lamps may have different operating characteristics and lives from one facility to another. It is important to maintain records on lamp and ballast replacements and determine the most appropriate relamping interval. This also helps keep track of maintenance costs, labor needs and budgets.

Group relamping is the least costly method to relamp, due to reduced time and labor costs. For example, Table 13-10 shows the benefits of group relamping. As more states adopt legislation requiring special disposal of lighting systems, group relamping in bulk may offer reduced disposal costs due to large volumes of material.

13.4.4 Disposal Costs

Disposal costs and regulations for lighting systems vary from state to state. These expenses should be included in an economic analysis of any retrofit. If proper disposal regulations are not followed, the EPA could impose fines and hold the violating company liable for environmental damage in the future.

13.5 NEW TECHNOLOGIES & PRODUCTS[1]

The energy efficient lighting market is extremely competitive, forcing manufacturers to develop new products to survive. The development is so rapid; that it is challenging to "keep up" with all the latest technologies. This chapter describes the proven technologies; however, it is good idea to evaluate the latest developments before implementing a lighting system.

Smaller fixtures are allowing designers to put lighting fixtures closer to the task, where the light is needed. This breakthrough changes the design because instead of the conventional "blast more light through an efficient fixture, lens, and then the air," we can have a target lit from a close range (or even "back-lit") with much less light. Over the next decade, thin-film "light panels" will be hitting the market with new applications for creating "soft ambient light"—even from walls!

LEDs (light emitting diodes) should provide the next lighting revolution within 5 years. Many lighting manufacturers are actively developing LED white light systems that could replace fluorescent lamps. Today this opportunity already exists in signs, retail displays, medical/dental operating rooms, and other specialty lighting applications. Why? Beyond being low watt, close-range light sources, LEDs can offer multi-colored light without a lot of heat or maintenance. For a retailer, they can light a window display for products with different colors, creating different moods and attracting different buyers. The same principle of more flexibility applies for signs. In operating rooms, smaller lights closer to the patient with less heat being generated means happier patients and doctors! In addition, art museums have applications due

Table 13-10. Group relamping example: 1,000 3-Lamp T8 Lensed troffers

	Spot Relamping (on burn-out)	Group Relamping (@ 70% rated life)
Relamp cycle	20,000 hours	14,000 hours
Avg. relamps/year	525 relamps/yr	750 relamps/yr (group) 52 relamps/yr (spot)
Avg. material cost/year	$1,050/yr	$1,604/yr
Lamp disposal @ 0.50 ea.	$236/yr	$375/yr
Avg. labor cost/year	$3,150/yr	$1,437/yr
TOTAL EXPENSES:	$4,463/yr	$3,416/yr
Assumptions: *Labor:*	*$6.00/lamp*	*$1.50/lamp*
Material:	*$2.00/lamp*	*$2.00/lamp*
Operation:	*3,500 hr/yr*	*3,500 hr/yr*

[1]The majority of this section was provided by John Fetters, Effective Lighting Solutions. ©Effective Lighting Solutions, Inc.

to reduced "light wear and tear" on paintings. In airports, runways, taxiways, and other signaling/marking applications, LEDs are reducing lamp replacement and labor costs.

These are very exciting technologies to watch, as are the remaining ones listed in this section.

13.5.1 Fluorescent Ballasts

Miniaturization of electronic ballasts has been made possible by the use of integrated circuits and surface-mount technologies. The new ballasts are smaller, thinner, and lighter.

Low Profile Housing

The familiar "brick" shape and weight of ballasts will soon be gone. Reduced parts count and surface mount technology have reduced the size of ballasts as well as improved their reliability. These advances have permitted housings of lower profile and smaller cross-section. Today, some ballasts have a dimensional cross-section of 30 × 30 mm. The advantages of smaller ballast packages include lighter weight, less material, and easier handling and installation. In addition, they fit into the new low-profile fixtures, especially indirect and direct-indirect fixtures.

Universal Input Voltage

Many facilities have different lighting system voltages in different parts of their buildings. Maintenance personnel are slowed in their ballast replacement task when they don't know the voltage for a particular area of the building. However, ballasts with the universal voltage feature will automatically use any line voltage applied (between 120-277-v). In addition to saving valuable maintenance time, when the labor cost of identifying the voltage for each ballast to be replaced or the expense of distributor restocking of ballasts ordered with the incorrect voltage is included, any cost difference is very affordable. In addition, fewer replacement ballast models need to be stocked.

Optimizing Ballast Selection

Due to the variety of ballast and lamp combinations available, selecting ballasts by their "ballast factor" can help optimize the illumination. For example, if you require a little more light, use a ballast with a higher ballast factor. If you need less light (and want to save energy), you can specify a lower BF. One thing to consider when using this strategy is maintenance awareness when it is time to replace ballasts. If the maintenance employees don't know about the strategy, they may replace a special ballast with a normal one.

Instant-start ballasts have become the most popular method of starting F32T8/RS rapid-start lamps because of their lower input watts rating compared with rapid-start systems. However, lamp life can be reduced by up to 25% at short burn cycles when lamps are operated instant-start, increasing maintenance costs. In applications where short ON/OFF cycles are common, lamp life increases by using program-start or rapid-start ballasts, instead of instant-start ballasts. Rapid-start operation of rapid-start lamps will ensure normal rated lamp life and program-start ballasts can extend lamp life by up to 50%.

Dimming electronic ballasts for fluorescent lamps.

Electronic ballasts with dimming functions operate fluorescent lamps at high frequency, just like fixed-output electronic ballasts. Most dimmable ballasts now have separate low-voltage control leads that can be grouped together to create control zones which are independent of the power zones. Many dimming ballast designs provide over-voltage protection of the control leads in case line voltage is accidentally applied to the low voltage leads. The control method of choice is 0 to 10vDC, although dimming ballasts are also available that are designed to accept the AC line phase control signals from incandescent wall-box dimmer controls that dim the fluorescent lamps accordingly.

Dimmable ballasts are available for dimming most linear fluorescent lamps (1-, 2- or 3-lamp versions), including T5 HO lamps. Many of these products start the lamps at any dimmer setting and do not have to be ramped up to full-light output before they dim. Most of the models available measure less than 15% total harmonic distortion (THD) throughout the dimming range.

Conference and presentation rooms have traditionally been built with two lighting systems. One, an incandescent system, usually uses recessed cans and is dimmed with wall dimmers. The second is usually a fluorescent non-dimming system for general lighting. The incandescent system requires a lot of maintenance, due to the short life of the incandescent lamps. One solution to this situation is to remove the overhead incandescent system and replace the ballasts in the fluorescent system with line-voltage dimming ballasts that can operate from the existing incandescent wall-box dimmer(s). The main benefit of this improvement is lower maintenance cost for a small investment in ballasts, and the electrical maintenance staff can make the change.

Electronic ballasts for compact fluorescent lamps (CFLs)

Several manufacturers have dimming ballasts for rapid-start (4-pin) compact fluorescent lamps (CFLs).

Most of these offerings are for the higher-wattage CFLs (26 to 57-w). The lowest dimming limit is 5% and the dimming range varies with the manufacturer. There are designs to accept the AC phase-control signals from incandescent wall-box dimmer controls. This makes upgrading an older incandescent downlight system to an energy-efficient CFL system easy, with no new wiring required.

13.5.2 Fluorescent Lamps

Several smaller, yet brighter fluorescent systems (T5, T5HO, and Super T8) can be very useful. Smaller systems have been effective in task lighting environments where less light from a single source is needed. The reduction of unnecessary lighting reduces energy expenses and can reduce installation costs.

T5 lamps

The T5 lamps come in two distinct and different families—standard (high-efficiency) and high output (HO). These recently developed lamps should not be confused with older miniature preheat fluorescent lamps of the same diameter nor with the line of long compact fluorescent lamps of the same diameter.

Standard (high-efficiency) T5 linear lamps

These 5/8" diameter lamps (Figure 13-6) are equipped with miniature bi-pin bases and are powered by electronic ballasts. All the lamps in this family operate on the same current (170 ma) and have the same surface brightness for all wattages. For cove and cornice applications this is a distinct advantage.

Another reason the T5 lamp is suited for these applications is that they are designed to peak in their lumen rating at 35°C (95°F) vs. 25°C (77°F) for T12 and T8 lamps. This characteristic provides higher light output in confined applications where there is little or no air circulation. In indirect fixtures, this thermal characteristic increases efficiency and gives more usable lumens per watt.

Standard T5 lamps are 12-18% more efficient than T8 lamps (96-106 LPW) and 10-15% more efficient than the T5HO. T5s employ rare-earth phosphors with CRI greater than 80 and lamp lumen maintenance rated at 95%. There are 4 sizes of standard T5 lamps as shown in Table 13-11, all rated at 20,000 hours (at 3 hours-per-start).

Note that the 28-watt lamp (not quite 4' long) has an initial lumen rating the same as a 4' T8 lamp. However, the millimeter lengths and miniature bi-pin bases preclude their use in standard length linear fluorescent systems, and the high bulb-wall brightness limits their use to high ceiling applications, because the visible tubes can create too much discomfort glare in low mounting height applications.

**Table 13-11. Standard T5 Lamp Sizes
(Source: Effective Lighting Solutions, Inc.)**

Nominal Watts	Length mm/(in)	Lumens (initial)	Lumens (maintained)
14	549/(21.6)	1,350	1,283
21	849/(33.4)	2,100	1,995
28	1149/(45.2)	2,900	2,755
35	1449/(57.0)	3,650	3,460

High output T5 linear lamps (SEE Table 13-12)

These T5 lamps are physically the same size as standard T5 lamps but provide higher lumen output. T5HO lamps generate from 1.5 to 2 times the light output of the standard T5 and nearly twice the light output (188%) of T8 and T12 systems with the same number of lamps. One-lamp T5 HO fixtures can replace both lamps of 2-lamp T8 fixtures.

They are approximately 10-15% less efficient than standard T5 lamps (83 to 94 lumens per watt) and can be up to 8% less efficient than standard T8 systems. The surface brightness varies among various wattages and since lamps operate on different currents (see Table 13-12), each lamp wattage requires a unique ballast.

T5 HO lamps are available in the three standard fluorescent color temperatures (cool—4100K, warm—3000K and neutral—3500K) and have a color rendering index greater than 80. Lumen maintenance is rated at 95%. These lamps are now rated at 20,000 hours. Similar to standard T5 lamps, T5HO lamps also peak in their lumen rating at 35°C (95°F) vs. 25°C (77°F) for T12 and T8 lamps. This provides higher light output in confined applications where there is little or no air circulation. In indirect fixtures, this thermal characteristic results in increased efficiency with more usable lumens per watt.

T5HO lamps are being used in designs of slim profile indirect fixtures that take advantage of the smaller lamp. Only 1 lamp per 4-ft section is required, replacing

**Table 13-12. T5HO Lamp Sizes
(Source: Effective Lighting Solutions, Inc.)**

Nominal Watts	Length mm/(in)	Lamp Current (ma)	Lumens (initial)	Lumens (maintained)
24	549/(21.6)	300	2,000	1,900
39	849/(33.4)	340	3,500	3,325
54	1149/(45.2)	460	5,000	4,750
80	1449/(57.0)	552	7,500	7,125

designs using 2, 4-ft T8 lamps per 4-ft section. The high bulb wall brightness limits their use in direct applications in low ceiling height conditions due to discomfort glare. Following the trend to fluorescent, T5HO lamps are being used in high ceiling applications, including high-bay industrial fixtures, and are a common replacement for high pressure sodium fixtures due to better light quality and ability to turn on quickly.

T8 lamps

Standard T8 lamps (See Figure 13-6)

T8 lamps (1" dia) were originally imported from Europe in the early 1980s. The lamps now used in the U.S. are different than their European pre-heat cousins, and there are improved models. T8s have been the lamps of choice (along with the high frequency electronic ballasts that drive them) for fluorescent upgrades for several years. T8 lamps are available in 2′, 3′, 4′, and 5′ lengths, at 17, 25, 32, and 40-w respectively. These lamps require ballasts that supply 265 ma. There are also two versions of 8′ retrofit lamps, at 59-w, or 86-w. U-tubes are available in the new 1-5/8" leg spacing and a retrofit U-tube that has 6" leg space used to replace 6" leg-spacing T12 U-tubes.

Recent advances in T8 lamps have been in improvements in color rendering and longer life. Extended performance T8 lamps have a life rating of 24,000 (at 3 hours per start)—20% longer than standard T8 lamps. These extended performance lamps operate on the same electronic ballasts designed to operate standard T8 lamps. Lumen maintenance is rated at 0.94 and it levels off after that. Lumen output is slightly higher at 3,000 lumens and CRI is improved to 8. Standard 3000K, 3500K and 4100K colors are provided.

Reduced-wattage T8 lamps

Sometimes called "energy-saving" T8 lamps, these lamps are available in 28 and 30-w vs. the standard 32-w models. They are designed to replace reduced-wattage,

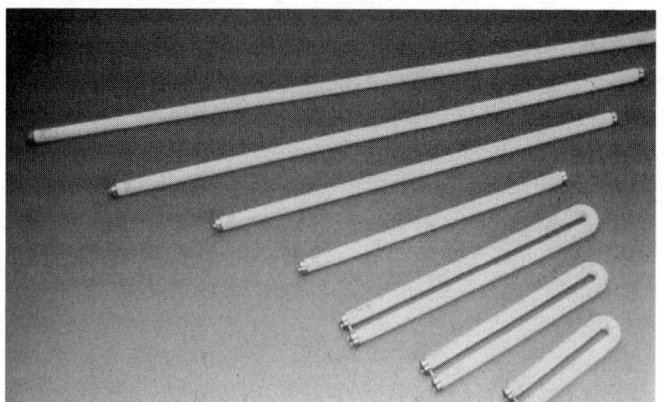

Figure 13-6. Some T8 Lamp Sizes (Source: Sylvania)

34-w T12 lamps when upgrading to electronic ballasts. They are recommended for use only on instant-start electronic ballasts to provide the higher open-circuit voltage required, and they need to be operated above 60°F. In addition, they cost more than standard T8s, but they save about 6% over standard lamps, are TCLP compliant, and have high lumen maintenance (94%).

High Performance "Super" T8 lamps. (See Table 13-4)

These high lumen lamps are a variant of T8 size fluorescent lamps that include special rare earth phosphors and produce higher light output compared to regular T8s. The higher efficacy (lumens per watt) does not translate directly to energy savings but rather translates to more light. However, these can be utilized for energy savings through additional measures such as reduced number of fixtures or tubes, or in combination with reduced ballast factors. For example, if a 4-lamp fixture is a candidate for a retrofit to T8, an option may be to replace the fixture with a 3-lamp Super T8 fixture for equivalent light. (Super T8 cost is higher than standard T8.)

TCLP compliant fluorescent lamps (lamp disposal)

Over 600 million fluorescent lamp tubes are disposed of every year in the U.S. Prior to June 1999, the USEPA required that spent fluorescent lamps that did not pass a toxicity characteristic leaching procedure (TCLP) were to be treated as hazardous waste because they contained more than 0.2 mg/liter (ppm) of mercury.

Standard fluorescent lamps do not pass the TCLP test and were required (prior to the universal waste rule) to be handled as hazardous waste, or recycled by using expensive hazardous waste haulers and massive documentation. Fluorescent lamps are now covered by the universal waste rule (as of this writing). The main result of the inclusion of fluorescent lamps in the universal waste rule is to encourage recycling of spent lamps.

Lamp manufacturers have reduced the mercury content of fluorescent tubes over the past decade to less than half the original content. In response to public concern for mercury in the environment, the major lamp companies started to produce what were originally called "low-mercury" lamps. Philips Lighting, using a proprietary dosing and buffering technology they call ALTO®, produced the first low-mercury fluorescent lamps. 4-foot ALTO fluorescent lamps have less than 10 mg of mercury and therefore will pass the TCLP test. Other lamp companies have followed this trend, and now these lamps are called "TCLP compliant" lamps to indicate the lamps are designed to pass the federal TCLP (toxic characteristic leaching procedure) test.

Low-mercury lamps have distinctive colored end caps, usually green. Use of TCLP compliant lamps provides users with normal lamp performance, light output, and life as an environmentally friendly option to meet their lighting needs. Although they do not need to be recycled, many end-users are avoiding any possible liability for their lamp disposal by recycling their spent TCLP compliant lamps.

13.5.3 Compact Fluorescent Lamps

Improvements to CFL technologies have been occurring every year since they became commercially available. Products available today provide higher efficacies, as well as instant starting, reduced lamp flicker, quiet operation, smaller size, and lighter weight. Dimmable CFLs are now available, and it can be expected that their performance will increase with time. The 2700K color (incandescent appearance) has been replaced by the 3000K for commercial applications. "Pre-heat" models start by blinking before they stay ON. Older lamps blink more than new lamps during starting. Rapid-start models start instantly, with no blinking.

Traditional Problems with CFLs

CFLs suffer from multiple sensitivities that reduce the light output and in some cases lamp life. They are position-sensitive. Gravity determines where the excess mercury "pools," which affects the mercury vapor pressure that determines the lumen output. Lumen ratings published in lamp catalogs are produced according to ANSI testing standards that require the lamp to be in the vertical, "base up" position. In the base-down position some CFLs produce 20% fewer lumens. In the horizontal position, they produce about 15% less light. Lamp lumen depreciation for CFLs is often more accelerated than for incandescent sources. CFLs are also not recommended for wet applications.

Additional sensitivities include temperature sensitivity that reduces the light output when CFLs are operated above or below their optimum temperature rating. The loss due to temperature is approximately 15-20% and is most noticeable in enclosed fixtures, such as recessed downlights, due to self-heating. *However, when the mercury used in a CFL is in the form of an amalgam—an alloy of mercury and other metals—the mercury vapor pressure is reduced without affecting the lamp temperature. This technique makes the lamps less temperature sensitive than conventional CFLs and provides more light at the high and low extremes (above 100°F and below 32°F.) Amalgam lamps are not easily identified, but most "triple-tubes" are amalgam products.*

Screw-base CFLs

Screw-base CFLs have "Edison" bases and are used to replace incandescent lamps. They have an integral ballast built into the base. Early models had magnetic ballasts built into the base; however, most contemporary models have electronic bases, allowing significant size reduction. Some screw-base CFLs are used in commercial applications, but most are used in residential lighting. Higher wattage options are available.

2-pin preheat CFLs (Figure 13-8)

2-pin CFLs require a separate ballast (usually magnetic) located in the fixture. Each lamp has a starter, located in the base, which provides pre-heat starting.

4-pin rapid-start CFLs

4-pin rapid-start lamps are available in 16, 18, 24,

Figure 13-7. Spiral or Spring Shaped CFLs

Table 13-13. Large Screw-base Compact Fluorescent Lamps
(Source: www.maxlite.com)

Shape	Lumens	Watts	LPW	M.O.D	M.O.L.	Replaces
Spiral	3,500	55	64	3.5"	10"	200-w incand
Quad	4,200	65	65	3.5"	11"	250-w incand
Quad	5,500	85	65	3.5"	11.8"	300-w incand

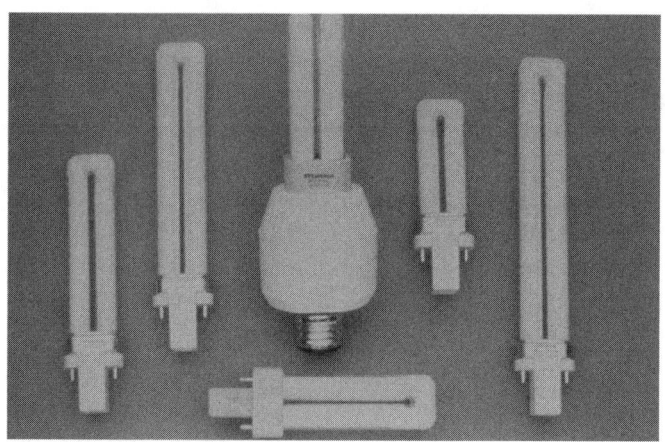

Figure 13-8. Twin-tube Pre-heat CFLs (Source: Sylvania)

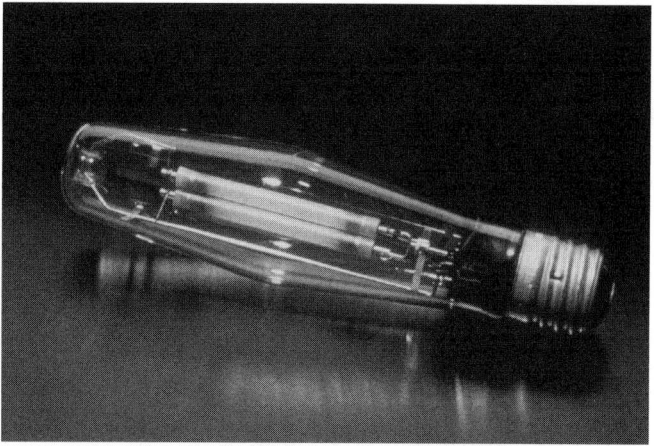

Figure 13-9. Double Arc-Tube HPS (Source: Sylvania)

**Table 13-14. Pre-Heat Compact Fluorescent Lamps
(Source: Effective Lighting Solutions, Inc.)**

Description	Lumens	Watts	Lamp Lumens/Watt
Twin-tube Pre-heat	250	5	50
	400	7	57
	600	9	66
	900	13	69
Quad Pre-heat	575	9	64
	860	13	66
	1200	18	66
	1800	26	69

26, 28, 32, and 42-watt models. All commercial-grade models are rated at 10,000 hours. The majority of these lamps are T5 (5/8" tube diameter), but some are T4 (1/2" tube diameter). Maximum overall length (MOL) ranges from 3.5" to 5.5." The three primary color temperatures are available—3000K (warm), 3500K (neutral), and 4100K (cool). At least one manufacturer also provides the warmer 2700K color. Rapid-start CFLs are designed for operation on electronic ballasts and can be dimmed when operated on a dimming electronic ballast designed for the appropriate lamp wattage.

New generation compact fluorescent lamps (CFL) are a significant improvement over the earlier twin and double twin tube types. Instead of using free mercury, these new CFLs use mercury that has been combined with other metals to form an amalgam. The amalgam makes the lamps less sensitive to the effects of temperature and position.

This is an important advantage over standard CFLs and is the reason that many applications using standard CFLs perform poorly. Amalgam CFLs have stable light output from 23°F to 130°F. Also, amalgam

lamps are not position sensitive and exhibit less color shift than conventional CFLs. They do take slightly longer to warm up, but they are at full brightness in less than 3 minutes. Unfortunately, manufacturers do not always clearly identify their amalgam lamps, but most "triple" tubes are amalgam.

CFLs are available in higher wattage for use in high ceiling downlights. A 32-watt triple-tube amalgam lamp, rated at 2,400 lumens, provides a system replacement for 150-watt incandescent downlights. A 42-watt triple-tube amalgam lamp, rated at 3,200 lumens, is useful as a system replacement for high-wattage incandescent downlights and a 57-watt rapid start, triple-tube amalgam lamp, rated at 4,300 lumens, is equivalent to a 200-w incandescent lamp.

At Lightfair International 2003, Philips Lighting unveiled a new multiple burning position, high lumen-output PL-H lamp. These 4-pin, rapid start lamps are used with high frequency, electronic ballasts. They are composed of 6, T5 limbs joined with bend-and-bridge technology. There are 6 models with wattages ranging from 60 to 120-w. Versatile and powerful, they have a lumen output almost double that of other CFLs, up to 9,000 lumens (120-w model); they provide maximum design freedom in many areas, including high ceiling indoor and outdoor applications. In addition, the white light PL-H range promises stable color rendering, long life, and high lumen maintenance.

Dimmable CFLs

Screw-base dimmable CFLs were introduced in 1996. This lamp is intended to replace incandescent lamps used on wallbox dimming systems. The electronic ballast base in this 1-piece lamp responds to the phase change voltage waveform from most existing dimmers and dims the CFL down to 10% light output.

The dimmable CFL is available in several wattages, the most common being a 23-watt triple-tube amalgam lamp, with a lumen rating of 1500, that will replace 90-watt "A" lamps. The major benefit of this lamp is that dimming is accomplished on existing dimming circuits, with no additional control wiring required.

13.5.4 High Intensity Discharge (HID) Systems:

Metal Halide Systems

Metal halide lamps have become more popular due to technological advancements and consumer preference for "white light." Technologically, the "pulse-start" metal halide systems are a significant improvement in efficiency and performance. Like most electronic ballasts, these operate at high frequency and provide a quicker re-strike time (3-5 minutes) versus standard metal halide systems (6-10 minutes). The pulse-start systems maintain CRI and lumen output better over time.

Pulse-start metal halide

Low-wattage metal halide (< 175-w) and high-pressure sodium lamps have used pulse-start technology for many years, using a high voltage pulse starter to ignite the lamps. What is new is the availability of high-wattage, pulse-start metal halide lamps (175-w to 1000-w) that are quickly replacing standard metal halide lamps. There is a new family of arc tubes called "formed body" that replace the old pinched seal arc tubes and overcome the disadvantages of the old design. The starter electrode, found in standard arc tubes, has been eliminated. The new arc tube design features uniform geometry and higher fill pressures. Improved temperature control is achieved with smaller pinch seals that provide less heat loss, reducing lamp-to-lamp color shift. Formed body arc tube lamps provide a lower ambient temperature limit, -40°F instead of -30°F for standard arc tube lamps. Faster starting and restarting (re-strike) results from the lower mass of the new arc tubes. These changes result in higher lamp efficacy (up to 110 lumens per watt), improved lumen maintenance (up to 80%), consistent lamp-to-lamp color (within 100°K), and 50% faster warm up and re-strike times (three to five minutes vs. eight to 15 minutes).

Ceramic metal halide lamps (CMH)

Ceramic arc tube metal halide lamps use the same ceramic material used in high-pressure sodium arc tubes—polycrystalline alumina (PCA). PCA reduces the sodium loss through the more porous glass arc tube used in standard metal halide lamps. This reduces color shift and spectral variation of standard metal halide lamps

caused as the sodium is depleted. Metal halide lamps with ceramic arc tubes are designated either CDM (ceramic discharge, metal halide) or CMH (ceramic metal halide), and reference may also be made to their constant color in the brand name. They are available from 20 to 400-watt, with color temperature of either 3000K (warm) or 4000K (cool) and an average rated life from 6,000 to 15,000 hours, depending on the wattage. Ceramic metal halide lamps are started by a pulse starter like PS metal halide lamps and operate best on electronic ballasts. The main advantage of the combination of CMH lamps and electronic ballasts is 10-20% higher lumen output (also resulting in a corresponding higher LPW) and the best color stability.

The benefits of CMH lamps include good lamp efficacy (83-95 LPW) in the same range as older, linear fluorescent lamps; high CRI (83-95); limited color shift (from ± 75K to ± 200K CCT); excellent lamp-to-lamp color consistency; and good lumen maintenance (0.70-0.80).

Applications for these improved color metal halide lamps include high ceilings, such as atria, lobbies of hospitality spaces, downlights, and lighting merchandise—anywhere that the higher CRI and color consistency can be justified. Fade-block models with thin-film coatings on the arc-tube shroud are available for merchandise lighting to help reduce the UV fading of materials. There are also HPS replacement lamps that can be used as an interim solution when converting from a high-pressure sodium system to a white light system.

High Pressure Sodium systems

Two new lamp wattages are available to narrow the gap between the 400-w and the 1,000-w standard HPS lamps. Both sizes will probably not survive the market; the 600-watt, 90,000 lumen lamp is not as widely supported by the lighting industry as the 750-w, 105,000-lumen lamp as a good "in between" size. In general, however, all high-pressure sodium lamps are losing ground to white-light sources such as metal halide or fluorescent.

Several improvements have been introduced in "new-generation" HPS lamps. The major improvement is the elimination of end-of-life cycling characteristic of standard high-pressure sodium lamps. However, there are two different design approaches by the three major lamp companies. Two companies have taken the "notification" approach in which the lamp turns a distinctive blue color at end of life. A third company simply shuts off the lamp power at end of life.

New HPS lamps have welded bases that replace the old lead-soldered bases. Several new models have reduced or zero mercury content, qualifying them as TCLP compliant lamps.

These lamps sacrifice efficacy and life to achieve CRI rating up to 65. Lamp efficacy ranges from 63 to 94 LPW, and they have an average rated life of 15,000 hours. They are available in 70, 100, 150, 250, and 400-w models and lumen ratings from 4,400 to 37,500.

Double arc-tube HPS lamps

These HPS lamps are called standby lamps and have two arc tubes welded together in parallel. However only one arc tube operates when the lamp is ignited. Upon loss of power, the second arc tube, hot from being in close proximity to the first arc tube, comes ON at about 50% light output. It then comes up to full light output within the strike time of the lamp (~4 min max). Standby lamps are used for safety and security applications and are popular with prison lighting systems, as well as roadway systems, with a tested life of 40,000 hours, reducing maintenance time and labor cost.

13.5.5 Induction Lighting

Electrodeless Induction Systems

Since the introduction of the first electric light, a search has been on for long-life lighting. The reason for this search is to reduce the cost associated with changing lighting components at or near their end of rated life-maintenance cost.

The lamps used in induction systems have no electrodes to wear out like other lamps, such as fluorescent and HID lamps. The lamps can last much longer without electrodes. Long life is the primary advantage of these systems, and they can provide a good payback where maintenance labor cost is high. When compared with other light sources, electrodeless induction systems will operate 5-8 times longer than fluorescent and metal halide systems and about 4 times longer than HPS systems. In addition, induction lamps come ON relatively quickly and have short re-strike time compared with HID lamps.

Instead of using electrodes to generate electrons as in fluorescent lamps, electrodeless systems produce light by means of induction—the use of an electromagnetic field to induce a plasma gas discharge into a tube or bulb that has a phosphor coating. No electrons are needed since the gas discharge is induced into the bulb or tube by a high-frequency electronic generator that supplies the electromagnetic field. These systems provide white light with a minimum color shift, and CRI and lumen depreciation values are similar to fluorescent lamps.

Each of the two primary electrodeless system lamps has a unique size and shape and require new fixtures that are designed to optically match each unique shape. There is no common electronics package for these products since they operate on much different frequencies. The electronics package must be fairly close to the glass envelopes, and the maximum mounting distance is restricted to the wire length supplied on the electronics package. These are independent systems and are designed so that both the glass envelopes and the electronics are changed out together at end of life.

Genura™ Lamp

GE Lighting developed an electrodeless induction lamp labeled Genura™ and introduced it in the U.S. in 1995. Genura™ is a compact R30 reflector lamp with a standard medium base, that is intended for use as a retrofit lamp (in place of a 100-watt A lamp, a 75-watt R30 lamp, or a 65-watt R30 lamp) in recessed downlights (cans). (This product is a lamp and not a system, so it is covered here before the induction systems.)

QL Induction Lighting System (Figure 13-10)

Philips Lighting developed this induction system and introduced it to the European market in 1991. In 1992 the QL was introduced to the U.S. market. The QL system is comprised of three components: 1) a high-frequency generator, 2) a power coupler, and 3) the glass bulb. The high-frequency generator is in a separate electronics package that provides the 2.65 MHz current to the power coupler (antenna) through a coax cable. The power coupler sits inside the enclosed glass discharge bulb shaped like a large A lamp. The bulb, which contains an inert gas and a small amount of mercury, is attached to the power coupler by a plastic lamp cap that uses a click system. Like fluorescent lamps, the inside walls of the bulb are coated with a phosphor coating. When the high frequency electromagnetic field is applied to the bulb, the gas is ionized and the lamp produces photons (at UV frequency) and visible light in the same manner as a fluorescent tube. The photons collide

Figure 13-10. QL Lighting System (Source: Philips Lighting)

with the phosphor coating and cause the lamp to glow. Full brightness is achieved in 10-15 seconds. The system meets FCC requirements as a low EMI design.

The main advantage of the QL system is its long life, with an average rated life of 100,000 hours. This long life advantage is especially important where maintenance cost is high. The current emphasis in the U.S. is in outdoor lighting systems such as street, roadway, and tunnel lighting. Several fixture manufacturers have incorporated the QL in their designs.

ICETRON™ System

The inductively coupled electrodeless system—ICE-TRON™—was developed by Sylvania. This electrodeless system consists of three parts: 1) a unique rectangular "donut" shaped bulb-filled with an inert gas and a small amount of mercury, 2) two ring-shaped ferrite core couplers, one at each of the short sides of the bulb, and 3) a separate high-frequency (200-300 KHz) generator. A plug-in connector attaches leads from the couplers to the electronic generator. The driver may be mounted up to 66 feet away from the lamp.

When the high-frequency electromagnetic field is applied to the donut-shaped bulb between the ferrite cores at each end, the gas inside the bulb is ionized and produces light by inducing a circulating current in the bulb, which generates photons at UV frequency. These photons collide with the phosphor coating and cause the lamp to glow. ICETRON™ lamps strike and re-strike instantly.

There are three ICETRON™ lamps, 70-w, 100-w, and 150-w, and two drivers. Table 13-15 shows the combinations of lamps and drivers and the resulting system performance.

The mercury in the glass envelope is in the form of an amalgam, providing a universal burn situation. Starting temperatures extend down to –40°F, opening up opportunities for low temperature applications such as freezers and coolers. The ICETRON™ bulbs are available in two color temperatures, 3500K (neutral) and 4100K (cool), and a CRI of 80. Sylvania rates the lumen maintenance at 70% at 60,000 hours (60% of rated life). This is a departure from the standard method of rating lumen maintenance for other light sources (at 40% rated life). The lumen maintenance curve shows a lumen maintenance value of 75 at 40%. At the rated life of 100,000 hours, the lumen maintenance is about 65%.

The ICETRON™ system meets FCC (non-consumer) requirements and has a low EMI design. The principal advantage of this system is long life, 100,000 hours. A comprehensive warranty covers the system for 60 months. Applications where maintenance is difficult and/or costly are prime candidates for these long life systems.

Table 13-15. ICETRON™ System Performance (Source: Sylvania)

Lamp	Driver	System Lumens	System Watts	System LPW
70-w	100-w	6,500	82	79
100-w	100-w	8,000	107	75
100-w	150-w	11,000	157	70
150-w	150-w	12,000	157	76

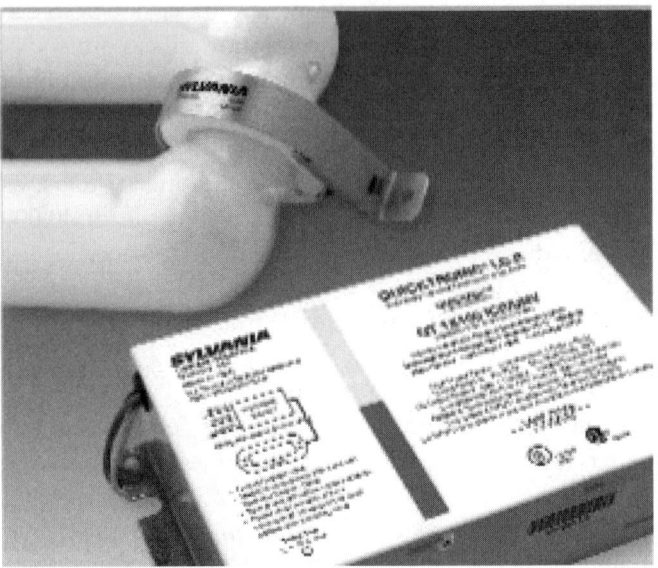

Figure 13-11. ICETRON™ System (Source: Sylvania)

13.5.6 Remote Source Lighting and Fiber Optics

Remote source lighting systems have the lighting source some distance from the point of delivery. Basically, the light source is connected to a light pipe or fiber optics, carrying the light to the point of application. Remote lighting solutions have become more popular because they fill the needs of projects having hazardous or underwater environments, walk-in freezers, architectural restrictions, or special aesthetic objectives. Remote source lighting systems offer reduced maintenance costs, because lamps can be accessed easily and safely. *For example, light pipes can be effective in gymnasiums or swimming pools. The uniform lighting can also result in a lower glare than single bright fixtures.*

Fiber optics can be used to resolve challenges associated with maintaining aesthetics. Light sources can be installed in rooms outside of a viewing area, with the fiber optics routed through walls (or other obscured spaces, like crown molding) to the application. Like miniature flashlights, the fiber optics can be pointed directly at the needed spot. *For example, gallery or church lighting can be achieved without bulky fixtures getting in the way of the occupants' view.*

13.6 SPECIAL CONSIDERATIONS

13.6.1 Rules and Regulations

There are many rules and regulations related to energy projects. These change through the years, and it would easy to be out of date by printing the current regulations here. (A discussion of energy-related regulations can be found in Chapter 20.) When doing a lighting retrofit, you may want to review the latest guidelines and requirements from IESNA, ASHRAE, and especially your local building codes. For example, some cities may pass laws banning certain types of lamps that are inefficient.

There are also opportunities from tax credits and utility incentives. These come and go with the tides of political influences, so take advantage of the good ones while they are available.

Equating energy savings to avoided power plan emissions can easily be calculated and integrated into the report of project savings. (Emission coefficient factors are located in Chapter 16.) Note that the conversion factors vary by state and utility, depending on the local mix of generation technology and fuel used, so the accuracy of these calculations will require using current and local data. An example of how the emissions benefit calculation can be automated for convenience is shown in Figure 13-16.

13.6.2 HVAC Effects

Nearly all energy consumed by lighting systems is converted to light, heat, and noise, which dissipate into the building. Therefore, if the amount of energy consumed by a lighting system is reduced, the amount of heat energy going into the building will also be reduced, and (for air-conditioned buildings) less air-conditioning will be needed. Consequently, the amount of winter-time heating may be increased to compensate for a lighting system that dissipates less heat.

Because most offices use air-conditioning for more months per year than heating, a more efficient lighting system can significantly reduce air-conditioning costs. In addition, air conditioning (usually electric) is much more expensive that heating (usually gas). Therefore, the savings on air-conditioning electricity are usually worth more dollars than the additional gas cost.

13.6.3 The Human Aspect

Regardless of the method selected for achieving energy savings, it is important to consider the human aspect of energy conservation. Buildings and lighting systems should be designed to help occupants work in comfort, safety, and an enjoyable atmosphere. Retrofits that improve the lighting quality (and the performance of workers) should be installed, especially when they save money. The recent advances in electronic ballast technology offer an opportunity for energy conservation that actually improves worker productivity. High frequency electronic ballasts and tri-phosphor lamps offer improved CRI with less audible noise and lamp flicker. These benefits have been shown to improve worker productivity and to reduce headaches, fatigue and absenteeism.

Implementation Tactics

In addition to utilizing the appropriate lighting products, the implementation method of a lighting upgrade can have a serious impact on its success. To ensure favorable reaction and support from employees, they must be involved in the lighting upgrade. Educating employees and allowing them to participate in the decision process of an upgrade will reduce the resistance of change to a new system. Of critical importance is the maintenance department, because it will have an important role in the future upkeep of the system.

Once the decision has been made to upgrade the lighting in a particular area and a trial installation has received approval, a complete retrofit should be completed as soon as possible. Due to economies of scale and minimal employee distraction, an all-at-once retrofit is usually optimal. In some cases, an overnight or over-the-weekend installation might be preferred. This method would avoid possible criticisms from side-by-side comparisons of the old and new systems. For example, a task lighting retrofit may appear darker than a uniformly illuminated space adjacent to it. The average worker who believes "more light is better" might protest the retrofit. However, if the upgrade is done over the weekend, the worker may easily not notice the changes.

13.6.4 Lighting Waste and the Environment

Upgrading any lighting system will require disposal of lamps and ballasts. Some of this waste may be hazardous and/or require special management. Contact the appropriate agency in your state to identify the regulations regarding the proper disposal of lighting equipment in your area.

Mercury

With the exception of incandescent bulbs, nearly all gaseous discharge lamps (fluorescent and HIDs) contain small quantities of mercury that end up in the environment unless recycled. Mercury is also emitted as a by-product of electricity generation from some fossil-fueled power plants. Although compact fluorescent lamps contain the most mercury per lamp, they save a great deal of energy when compared to incandescent sources. Because they reduce energy consumption (and avoid power plant emissions), CFLs introduce to the environment less than half the mercury of incandescents.[5] Mercury sealed in glass lamps is also much less available to ecosystems than mercury dispersed throughout the atmosphere. Nevertheless, mercury is not good for our environment, and the energy manager should check local disposal codes—*you don't want to break the law.* Mercury in lamps can be recycled, and regulations may soon require it.

PCB Ballasts

Ballasts produced prior to 1979 may contain polychlorinated biphenyls (PCBs). Human exposure to these possible carcinogens can cause skin, liver, and reproductive disorders. Older fluorescent and HID ballasts contain high concentrations of PCBs. These chemical compounds were widely used as insulators in electrical equipment such as capacitors, switches, and voltage regulators until 1979. The proper method for disposing used PCB ballasts depends on the regulations in the state where the ballasts are removed or discarded. Generators of PCB containing ballast wastes may be subject to notification and liability provisions under the Comprehensive Environmental Response, Compensation and Liability Act of 1980 (CERCLA)—also known as "Superfund."[6]

Generally, the PCB ballast is considered to be a hazardous waste only when the ballast is leaking PCBs. An indication of possible PCB leakage is an oily tar-like substance emanating from the ballast. If the substance contains PCBs, the ballast and all materials it contacts are considered PCB waste and are subject to state regulations. Leaking PCB ballasts must be incinerated at an EPA approved high-temperature incinerator.

Energy Savings and Reduced Power Plant Emissions

When appliances use less electricity, power plants don't need to produce as much electricity. Because most power plants use fossil fuels, a reduction in electricity generation results in reduced fossil fuel combustion and airborne emissions. Considering the different types of power plants (and the different fuels used) in different geographic regions, the Environmental Protection Agency has calculated the reduced power plant emissions by saving one kWh. Table 13-16 shows the emission factor of CO_2 for each kWh per year saved, using the U.S. average.[7] However, most people can't visualize what a ton of CO_2 looks like, so the table shows equivalent environmental benefits. This spreadsheet can be downloaded from profitablegreensolutions.com.

13.7 DAYLIGHTING

Human beings developed with daylight as their primary light source. For thousands of years humans evolved to the frequency of natural diurnal illumination. Daylight is a flicker-free source, generally with the widest spectral power distribution and highest comfort levels. With the twentieth century's trend towards larger buildings and dense urban environments, the development and wide spread acceptance of fluorescent lighting allowed electric light to become the primary source in offices.

Daylighting interior spaces is making a comeback because it can provide good visual comfort and can save energy if electric light loads can be reduced. New control technologies and improved daylighting methods allow lighting designers to conserve energy and optimize employee productivity.

There are three primary daylighting techniques available for interior spaces: utilizing skylights, building perimeter daylighting, and building core daylighting.

Table 13-16. Emissions Calculator Example

PROFITABLEGREENSOLUTIONS
Complete Emissions Calculator

INSTRUCTIONS: Type in the kWh savings and see the emissions-environmental benefits in green-shaded areas. Insert your own $$ values for the Strategic Benefits in blue text.

Type the amount of electricity your program will save ⟹ 750,000 kWh/year

Emissions Reductions:	Annual Reductions	Reductions over 10 years
Conversion Factor: 1 kWh is worth 1.37 lbs of CO2 (Source: EPA)		
GreenHouse Gas Reduction (in pounds of CO2)	1,022,250 lbs	10,222,500 lbs
or when converted to Metric Tons of CO2 >>>	464 Metric Tons	4,637 Metric Tons

Equivalent Environmental Benefits (mutually-exclusive):	Annual Reductions	Reductions over 10 years
Acid Rain Emission Reduction	5,625 lbs of SOx	56,250 lbs of SOx
Smog Emission Reductions	2,700 lbs of NOx	27,000 lbs of NOx
Barrels of Oil Not Consumed	1,079 Barrels	10,785 Barrels
Cars off the Road	100.2 Cars	1,002 Cars
Gallons of Gas not Consumed	52,812 Gallons	528,119 Gallons
Acres of pine trees reducing carbon	386.3 Acres	3,863 Acres

Skylights are the most primitive and the most common in industrial buildings. Perimeter daylighting is defined as using natural daylight (when sufficient) such that electric lights can be dimmed or shut off near windows at the building perimeter. Traditionally, the amount of dimming depends on the interior distance from fenestration. However, ongoing research and application of "core daylighting techniques" can stretch daylight penetration distance farther into a room. Core daylighting techniques include the use of light shelves, light pipes, active daylighting systems, and fiber optics. These technologies will likely become popular in the near future.

Perimeter and core daylighting technologies are technologies being further developed to function in the modern office environment. The modern office has many visual tasks that require special considerations to avoid excess illumination or glare. It is important to properly control daylighting in offices so that excessive glare does not reduce employee comfort or the ability to work on VDTs. A poorly designed or poorly managed daylit space reduces occupant satisfaction, and it can increase energy use if occupants require additional electric light to balance excessive daylight-induced contrast.

Windows and daylighting typically cause an increased solar heat gain and additional cooling load for HVAC systems. However, development of new glazings and high performance windows has allowed designers to use daylighting without severe heat gain penalties. With dynamic controls, most daylit spaces can now have lower cooling loads than non-daylit spaces with identical fenestration. "The reduction in heat from lights, due to daylighting, can represent a 10% downsizing in perimeter zone cooling and fans.[8]" However, because there are several parameters, daylighting does not always reduce cooling loads any time it displaces electric light. As window size increases, the maximum necessary daylight may be exceeded, creating additional cooling loads.

Whether interior daylighting techniques can be economically utilized depends on several factors. However the ability to significantly reduce electric lighting loads during utility "peak periods" is extremely attractive.

13.8 COMMON RETROFITS

Although there are numerous potential combinations of lamps, ballasts, and lighting systems, a few retrofits are very common.

Offices

In office applications, popular and profitable retrofits involve installing electronic ballasts and energy efficient lamps, and in some cases, reflectors. Table 13-17 shows how a typical system changes with the addition of reflectors and the removal or substitution of lamps and ballasts. Notice that thin lamps allow more light to exit the fixture, thereby increasing fixture efficiency. Reflectors improve efficiency by greater amounts when there are less lamps (or thinner lamps) to block exiting light beams.

The expression (lumens)/(fixture watt) is an indicator of the overall efficiency of the lighting system. It is similar to the efficacy of a lamp.

Indoor/Outdoor Industrial

In nearly all applications with significant annual operating hours, mercury vapor systems can be replaced by metal halide (or high output fluorescent systems). This retrofit will improve CRI and reduce operating and relamping costs. In applications where CRI is not critical, HPS systems (which have a higher efficacy than metal halide systems) can be used.

Almost Anywhere

In nearly all applications where incandescent lamps are ON for more than 5 hours per day, switching to CFLs will be cost-effective.

13.8.1 Sample Retrofits

This section provides the equations to calculate savings from several different types of retrofits. For each type of retrofit, the calculations shown are based on average conditions and costs, which vary from location to location. For example, annual air conditioning hours will vary from building to building and from state to state. The energy costs used in the following examples were based on $10/kW-month and $.05/kWh. In most industrial settings, demand is also billed. In the following examples demand savings would likely occur in all except examples # 4 and # 6. To accurately estimate the cost and savings from these types of retrofits, simply insert local values into the equations.

EXAMPLE 1:
UPGRADE T12 LIGHTING SYSTEM TO T8

A hospital had 415 T12 fluorescent fixtures, which operate 24 hours/day, year round. The lamps and ballasts were replaced with T8 lamps and electronic ballasts, which saved about 30% of the energy and provided higher quality light. Although the T8 lamps cost a little more (resulting in additional lamp replacement costs), the energy savings quickly recovered the expense. In ad-

Table 13-17. Fluorescent lighting upgrade options.

ORIGINAL SYSTEM
Energy Efficient Magnetic Ballast

	40W lamps (T-12)	34W lamps (T-12)
Number of lamps	4	4
Total Watts	176	144
Ballast Factor	0.94	0.87
Available Lumens	12000	9700
Luminaire Efficiency	0.65	0.65
Lumens per Luminaire	7800	6300
Lumens/Luminaire watt	44.3	43.8

POTENTIAL RETROFITS

EE Magnetic Ballast

	42W lamps (T-10)
Number of lamps	2
Total Watts	92
Ballast Factor	0.95
Available Lumens	7000
Luminaire Efficiency	0.77
Lumens per Luminaire	5400
Lumens/Luminaire watt	58.7
ADD SILVER REFLECTOR	
Luminaire Efficiency	0.83
Lumens/Luminaire	5800
Lumens/Luminaire watt	63.0

Electronic Rapid Start Ballast

	42W lamps (T-10)	32W lamps (T-8)
Number of lamps	2	4
Total Watts	63	112
Ballast Factor	0.73	0.88
Available Lumens	5400	10200
Luminaire Efficiency	0.77	0.74
Lumens/Luminaire	4200	7500
Lumens/Luminaire watt	66.7	67.0
ADD SILVER REFLECTOR		
Luminaire Efficiency	0.83	0.78
Lumens/Luminaire	4500	8000
Lumens/Luminaire watt	71.4	71.4

Electronic Instant Start Ballast

	32W lamps (T-8)	32W lamps (T-8)
Number of lamps	3	2
Total Watts	90	60
Ballast Factor	0.88	0.88
Available Lumens	7700	5100
Luminaire Efficiency	0.76	0.78
Lumens per Luminaire	5900	4000
Lumens/Luminaire watt	65.6	66.7
ADD SILVER REFLECTOR		
Luminaire Efficiency	0.81	0.84
Lumens per Luminaire	6200	4300
Lumens/Luminaire watt	68.9	71.7

References:
U.S. EPA Green Lights Program, Lighting Upgrade Manual.
Advance Transformer Specification Guide
Magnetek Specification Guide

NOTES: 40W lamps are rated at 3200 lumens per lamp.
             34W lamps are rated at 2800 lumens per lamp.
             42W lamps are rated at 3700 lumens per lamp.
             32W lamps are rated at 2900 lumens per lamp.
             New luminaires may have greater efficiencies, due to highly reflective paints.

dition, because the T8 system produces less heat, air conditioning requirements during summer months will be reduced. Conversely, heating requirements during winter months will be increased.

Calculations

kW Savings

= (# fixtures) [(Present input watts/fixture) – (Proposed input watts/fixture)]

= (415)[(86 watts/T12 fixture)-(60 watts/T8 fixture)]

= 10.8 kW

kWh Savings

= (kW savings)(Annual Operating Hours)

= (10.8 kW)(8,760 hours/year)

= 94,608 kWh/year

Air Conditioning Savings

= (kW savings)(Air Conditioning Hours/year)(1/Air Conditioner's COP)

= (10.8 kW)(2000 hours)(1/2.6)

= 8,308 kWh/year

Additional Gas Cost

= (kW savings)(Heating Hours/year)(.003413 MCF/kWh)(1/Heating Efficiency)(Gas Cost)

= (10.8 kW)(1,500 hours/year)(.003413 MCF/kWh)(1/0.8)($4.00/MCF)

= $ 276/year

Lamp Replacement Cost

= [(# fixtures)(# lamps/fixture)][((annual operational hours/proposed lamp life)(proposed lamp cost)) – ((annual hours operation/present lamp life)(present lamp cost))]

= [(415 fixtures)(2 lamps/fixture)][((8,760 hours/20,000 hours)($ 3.00/T8 lamp)) – ((8,760 hours/20,000 hours)($ 1.50/T12 lamp))]

= $ 545/year

Total Annual Dollar Savings

= (kW Savings)(kW charge)+[(kWh savings)+(Air Conditioning savings)](kWh cost) -(Additional gas cost) – (lamp replacement cost)

= (10.8 kW)($ 120/kW year)+[(94,608 kWh)+(8,308 kWh)]($ 0.05/kWh) -($ 276/year) – ($ 545/year)

= $ 5,621/year

Implementation Cost

= (# fixtures) (Retrofit cost per fixture)

= (415 fixtures) ($ 45/fixture)

= $ 18,675

Simple Payback

= (Implementation Cost)/(Total Annual Dollar Savings)

= ($ 18,675)/($ 5,621/year)

= 3.3 years

EXAMPLE 2: REPLACE INCANDESCENT LIGHTING WITH COMPACT FLUORESCENT LAMPS

A power plant has 111 incandescent fixtures that operate 24 hours/day, year round. The incandescent lamps were replaced with compact fluorescent lamps, which save over 70% of the energy and last over ten times as long. Because the lamp life is so much longer, there is a maintenance relamping labor savings. Air-conditioning savings or heating costs were not included, because these fixtures are located in a high-bay building that is not heated or air-conditioned.

Calculations

Watts Saved Per Fixture

= (Present input watts/fixture) – (Proposed input watts/fixture)

= (150 watts/fixture) – (30 watts/fixture)

= 120 watts saved/fixture

kW Savings

= (# fixtures)(watts saved/fixture)(1 kW/1000 watts)

= (111 fixtures)(120 watts/fixtures)(1/1000)

= 13.3 kW

kWh Savings

= (Demand savings)(annual operating hours)

= (13.3 kW)(8,760 hours/year)

= 116,683 kWh/year

Lamp Replacement Cost

= [(Number of Fixtures)(cost per CFL Lamp)(operating hours/lamp life)] – [(Number of existing incandescent bulbs)(cost per bulb)(operating hours/lamp life)]

= [(111 Fixtures)($10/CFL lamp)(8,760 hours/10,000 hours)] – [(111 bulbs)($1.93/type "A" lamp)(8,760 hours/750 hours)]

= $ – 1,530/year[§]   [§]Negative cost indicates savings.

Maintenance Relamping Labor Savings

= [(# fixtures)(maintenance relamping cost per fixture)] [((annual hours operation/present lamp life))-((annual hours operation/proposed lamp life))]

= [(111 fixtures)($1.7/fixture)][((8,760/750))-((8,760/10,000))]

= $ 2,039/year

Total Annual Dollar Savings

= (kWh savings)(kWh cost)+ (kW savings)(kW cost) – (lamp replacement cost) + (maintenance relamping labor savings)

= (116,683 kWh)($.05/kWh)+(13.3)($120/kW year) (–1,530/year) + (2,039/year)

= $ 10,999/year

Total Implementation Cost

= [(# fixtures)(cost/CFL ballast and lamp)] + (retrofit labor cost)]

= (111 fixtures)($45/fixture)

= $ 4,995

Simple Payback

= (Total Implementation Cost)/(Total Annual Dollar Savings)

= ($ 4,995)/(10,999/year)

= 0.5 years

EXAMPLE 3: INSTALL OCCUPANCY SENSORS

In this example, an office building has many individual offices that are only used during portions of the day. After mounting wall-switch occupancy sensors, the sensitivity and time delay settings were adjusted to optimize the system. The following analysis is based on an average time savings of 35% per room. Air conditioning costs and demand charges would likely be reduced; however, these savings are not included.

Calculations

kWh Savings

= (# rooms)(# fixtures/room)(input watts/fixture) (1 kW/1000 watts) (Total annual operating hours)(estimated % time saved/100)

= (50 rooms)(4 fixtures/room)(144 watts/fixture)(1/1000) (4,000 hours/year)(.35)

= 40,320 kWh/year

Total Annual Dollar Savings ($/Year)

= (kWh savings/year)(kWh cost)

= (40,320 kWh/year)($.05/kWh)

= $ 2,016/year

Implementation Cost

= (# occupancy sensors needed)[(cost of occupancy sensor)+ (installation time/room)(labor cost)]

= (50)[($ 75)+(1 hour/sensor)($20/hour)]

= $ 4,750

Simple Payback

= (Implementation Cost)/(Total Annual Dollar Savings)

= ($ 4,750)/($ 2,016/year)

= 2.4 years

EXAMPLE 4: RETROFIT EXIT SIGNS WITH L.E.D.s

An office building had 117 exit signs that used incandescent bulbs. The exit signs were retrofitted with LED exit kits, which saved 90% of the energy. Even though the existing incandescent bulbs were "long-life" models, which are expensive, material and maintenance savings were significant. Basically, the hospital should not have to relamp exit signs for 25 years!

Calculations

Input Wattage – Incandescent Signs

= (Watt/fixture) (number of fixtures)

= (40 Watts/fix) (117 fix)

= 4.68 kW

Input Wattage – LED Signs

= (Watt/fixture) (number of fixtures)

= (3.6 Watts/fix) (117 fix)

= .421 kW

kW Savings

= (Incandescent Wattage) – (LED Wattage)

= (4.68 kW) – (.421 kW)

= 4.26 kW

kWh Savings

= (kW Savings)(operating hours)

= (4.26 kW)(8,760 hours)

= 37,318 kWh/yr

Lamp Replacement Cost

= [(Number of LED Exit Fixtures)(cost per LED Fixture)(operating hours/Fixture life)] – [(Number of existing Exit lamps)(cost per Exit lamp)(operating hours/lamp life)]

= [(117 Fixtures)($ 60/lamp kit)(8,760 hours/219,000 hours)] – [(234 Exit lamps)($5.00/lamp)(8,760 hours/8,760 hours)]

= -$ 889/year[§]   [§]Negative cost indicates savings.

Maintenance Relamping Labor Savings

= (# signs)(Number of times each fixture is relamped/yr)(time to relamp one fixture)(Labor Cost)

= (117 signs)(1 relamp/yr)(.25 hours/sign)($20/hour)

= $585/year

Annual Dollar Savings

= [(kWh savings)(electrical consumption cost)] + [(kW savings)(kW cost)] + [Maintenance Cost Savings]-[lamp replacement cost]

= [(37,318 kWh)($.05/kWh)] + [(4.26 kW)($120/kW yr)]+ [$585/yr] – [– $889/yr]

= $ 3,851/year

Implementation Cost

= [# Proposed Fixtures][(Cost/fixture + Installation Cost/ fixture)]

= [117][$60/fixture + $5/fixture]

= $ 7,605

Simple Payback

= (Implementation Cost)/(Annual Dollar Savings)

= ($7,605)/($3,851/yr)

= 2 years.

EXAMPLE 5: REPLACE OUTSIDE MERCURY VAPOR LIGHTING SYSTEM WITH HIGH AND LOW PRESSURE SODIUM LIGHTING SYSTEM

A parking lot is illuminated by mercury vapor lamps, which are relatively inefficient. The existing fixtures were replaced with a combination of high pressure sodium (HPS) and low pressure sodium (LPS) lamps. The LPS provides the lowest-cost illumination, while the HPS provides enough color rendering ability to distinguish the colors of cars. By replacing the fifty 400 watt mercury vapor lamps with ten 250 watt HPS and forty 135 watt LPS fixtures, the company saved approximately $ 2,750/year, with an installed cost of $12,500 and a payback of 4.6 years.

EXAMPLE 6:
REPLACE "U" LAMPS WITH STRAIGHT T8 TUBES

The existing fixtures were 2' by 2' lay-in troffers with two F40T12CW "U" lamps and a standard ballast consuming 96 watts per fixture. The retrofit was to remove the "U" lamps and install three F017T8 lamps with an electronic ballast, which had only 47 watts per fixture.

13.9   SUMMARY

In summary, this chapter will help the energy manager make informed decisions about lighting. The following "recipe" reviews some of the main points that influence the effectiveness of lighting retrofits.

A Recipe for Successful Lighting Retrofits

1. Identify visual task: Distinguish between tasks that involve walking and tasks that involve reading small print.

2. Identify lighting needs for each task: Use IES tables to determine target light levels.

3. Research available products and lighting techniques: Talk to lighting manufacturers about your objectives; let them help you select the products. Perhaps they will offer a demonstration or trial installation. Be aware of the relative costs, especially the costs associated with specialized technologies.

4. Identify lamps to fulfill lighting needs: Pick the lamp that has the proper CRI, CCT, lamp life, and lumen output.

5. Identify ballasts and fixtures to fulfill lighting needs: Select the proper ballast factor, % THD, voltage, fixture light distribution, lenses or baffles, and fixture efficiency.

6. Identify the optimal control technology: Decide whether to use IR, US or DT Occupancy Sensors. Know when to use time clocks or install switches.

7. Consider system variations to optimize:
    Employee performance—Incorporate the importance of lighting quality into the retrofit process.
    Energy savings—Pick the most efficient technologies that are cost-effective.
    Maintenance—Install common systems for simple maintenance, group re-lamping, and maintenance training.
    Ancillary effects—Consider effects on the HVAC system, security, safety, etc.

8. Publicize results: As with any energy management program, your job depends on demonstrating progress. By making energy cost savings known to employees and upper-level management, all people contributing to the program will know there is a benefit to their efforts.

9. Continually look for more opportunities: The lighting industry is constantly developing new products that could improve profitability for your company. Keep in touch with new technologies and methods to avoid "missing the boat" on a good opportunity. Table 13-21 offers a more complete listing of energy saving ideas for lighting retrofits.

13.10 SCHEMATICS

Lamps

Fluorescent

Bulb Shapes (Not Actual Sizes)

The size and shape of a bulb is designated by a letter or letters followed by a number. The letter indicates the shape of the bulb while the number indicates the diameter of the bulb in eighths of an inch. For example, "T-12" indicates a tubular shaped bulb having a diameter of ¹²/₈ or 1½ inches. The following illustrations show some of the more popular bulb shapes and sizes.

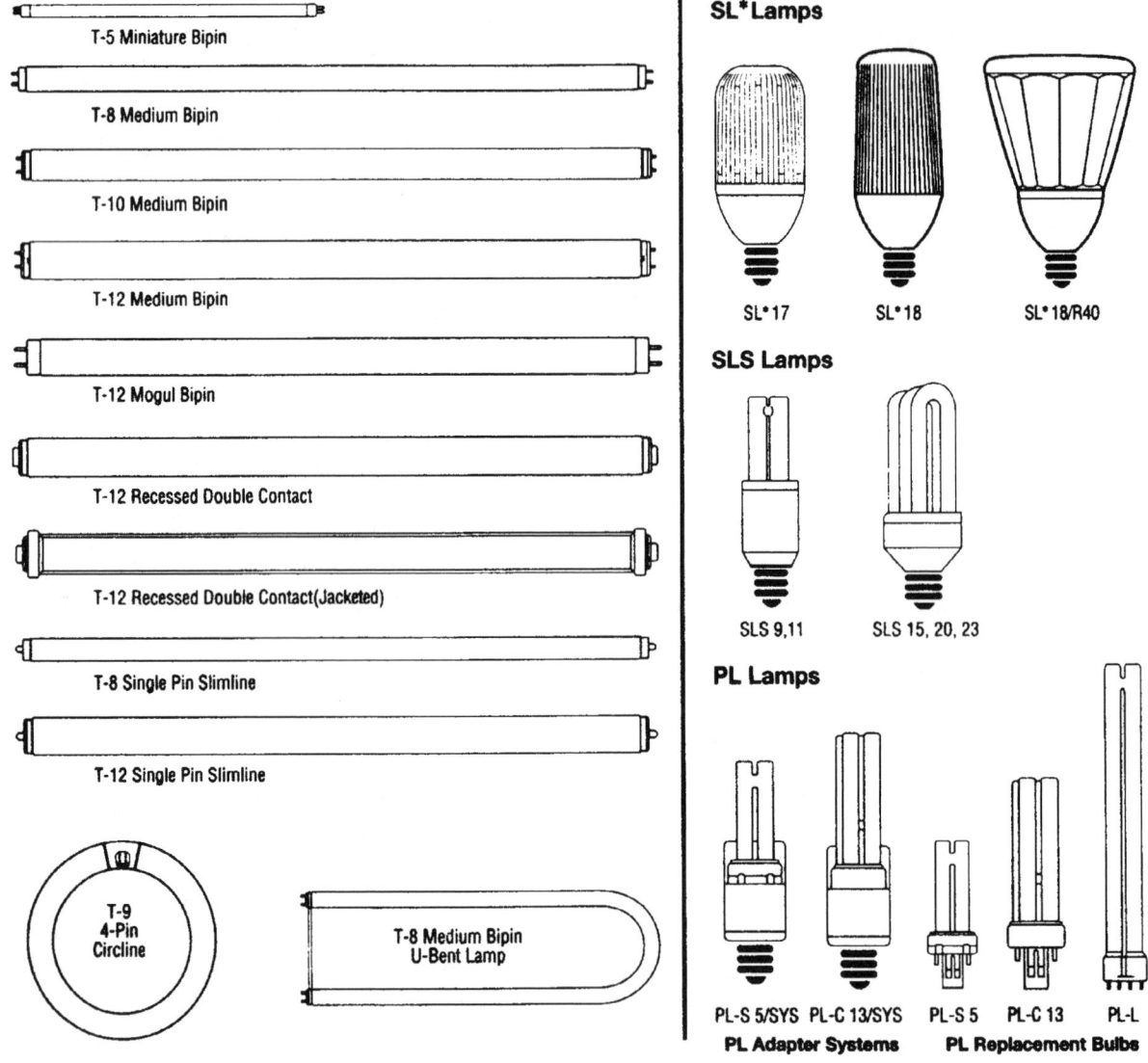

Fluorescent (continued)

Biax Lamps

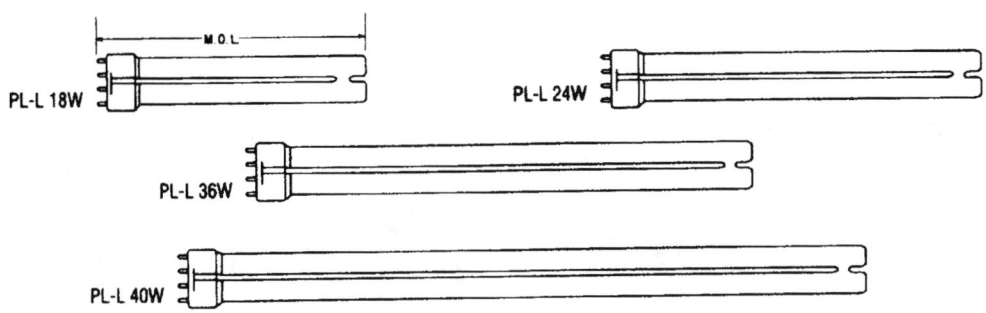

Base Types

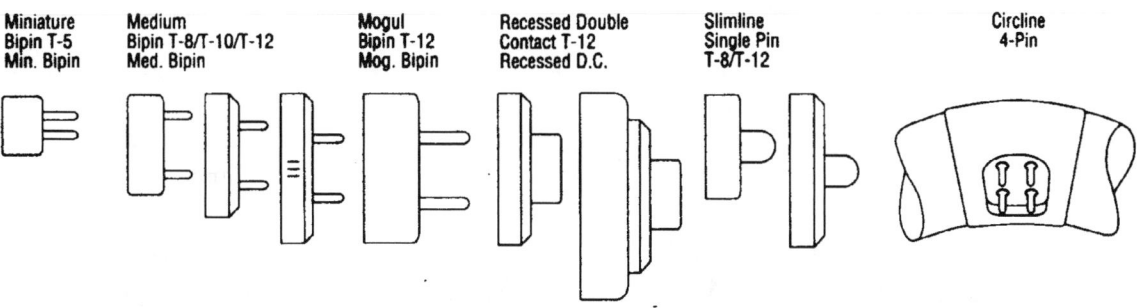

Lamp and Ballast System

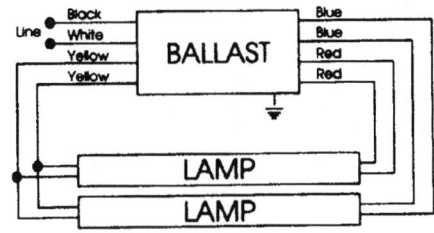

HID

Mercury Vapor

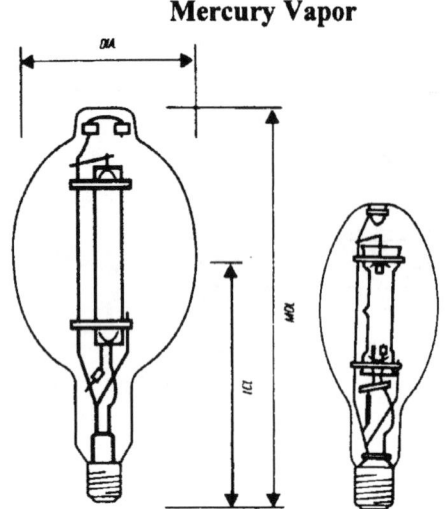

Metal Halide

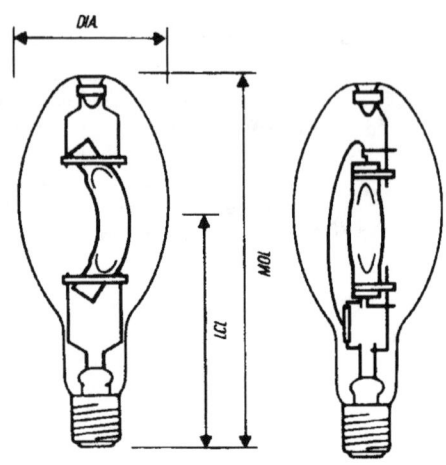

High Pressure Sodium

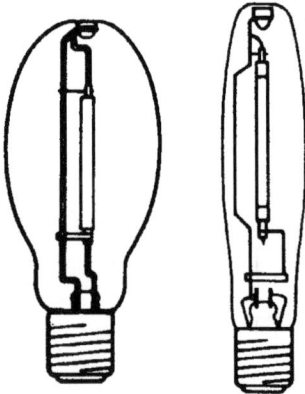

Low Pressure Sodium

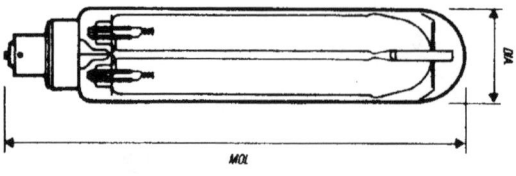

Exit Signs

LED

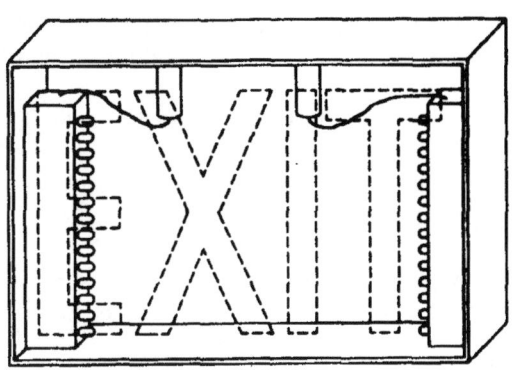

- **1.8-3.6 Input Watts/Fixture. (Replaces standard 20-25 watt lamps.)**
- Convert existing incandescent EXIT signs to use energy efficient LED liehg strips.
- Each kit contains two LED light strips and a reflective backing to provide even light distribution and a new red lens for the fixture.
- Estimated life is 25 years.
- Complies with OSHA and NFPA requirements.
- Available in four base styles to fit existing sockets or as a hard wire kit.
- LED light strips emit a bright red light and are not recommended for use with green signs.
- In addition to DGSC standard warranty, manufacturer's 25-year warranty applies.
- UL approved.

CFL

Quick connecting adapter screws into existing incandescent socket. For use with medium screw base sockets.

Two lamp EXIT sign retrofit system; backup lamp will take over if the primary lamp fails.

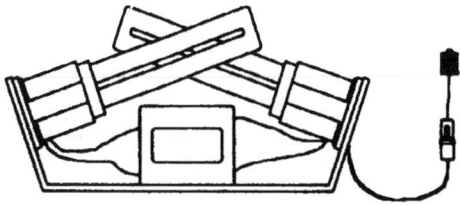

UL approved.
- Lamp: 9 watt, twin tube compact fluorescent
- Lumens: 600
- Lamp Avg Life: 10,000 hours
- Ballast Losses: 2 watts/ballast
- System Input Watts: 11 watts
- Minimum Starting Temperature: 0°F

Unit Dimensions: 8"L X 4-5/8"H X 1"H

Occupancy Sensors
Ceiling Mounted

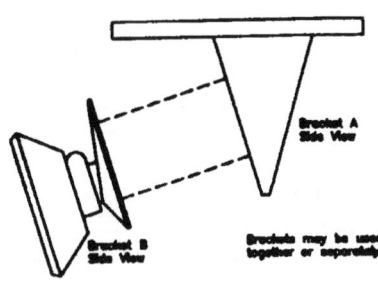

Wall Mounted

Table 13-18. Energy saving checklist.

Lighting Needs

*	Visual tasks: specification	Identify specific visual tasks and locations to determine recommended illuminances for tasks and for surrounding areas.
*	Safety and aesthetics	Review lighting requirements for given applications to satisfy safety and aesthetic criteria.
*	Over-illuminated application	In existing spaces, identify applications where maintained illumination is greater than recommended. Reduce energy by adjusting illuminance to meet recommended levels.
*	Groupings: similar visual tasks	Group visual tasks having the same illuminance requirements and avoid widely separated workstations.
*	Task lighting	Illuminate work surfaces with luminaires properly located in or on furniture; provide lower ambient levels.
*	Luminance ratios	Use wall-washing and lighting of decorative objects to balance brightness.

Space Design and Utilization

*	Space plan	When possible, arrange for occupants working after hours to work in close proximity to one another.
*	Room surfaces	Use light colors for walls, floors, ceilings, and furniture to increase utilization of light; reduce connected lighting power to achieve required illuminances. Avoid glossy finishes on room and work surfaces to limit reflected glare.
*	Space utilization branch circuit wiring	Use modular branch circuit wiring to allow for flexibility in moving, relocating, or adding luminaires to suit changing space configurations.
*	Space utilization:	Light building for occupied periods only, plus when required for security or cleaning purposes occupancy (see chapter 31, Lighting Controls).

Daylighting

*	Daylight compensation	If daylighting can be used to replace some electric lighting near fenestration during substantial periods of the day, lighting in those areas should be circuited so that it may be controlled manually or automatically by switching or dimming.
*	Daylight sensing	Daylight sensors and dimming systems can reduce electric lighting energy.
*	Daylight control	Maximize the effectiveness of existing fenestration-shading controls (interior and exterior) or automatically by switching or dimming.
*	Space utilization	Use daylighting in transition zones, in lounge and recreational areas, and for functions where the variation in color, intensity, and direction may be desirable. Consider applications where daylight can be utilized as ambient lighting, supplemented by local task lights.

Lighting Sources: Lamps and Ballasts

*	Source efficacy	Install lamps with the highest efficacies to provide the desired light source color and distribution requirements.

(Continued)

* Fluorescent lamps — Use T8 fluorescent and high-wattage compact fluorescent systems for improved source efficacy and color quality.

* Ballasts — Use electronic or energy efficient ballasts with fluorescent lamps.

* HID — Use high-efficacy metal halide and high-pressure sodium light sources for exterior floodlighting.

* Incandescent — Where incandescent sources are necessary, use reflector halogen lamps for increased efficacy.

* Compact fluorescent — Use compact fluorescent lamps, where possible, to replace incandescent sources.

* Lamp wattage reduced-wattage lamps — Use reduced-wattage lamps where illuminance is too high.

* Control compatibility — If a control system is used, check compatibility of lamps and ballasts with the control device.

* System change — Substitute metal halide and high-pressure sodium systems for existing mercury vapor lighting systems.

Luminaires

* Maintained efficiency — Select luminaires which do not collect dirt rapidly and which can be easily cleaned.

* Improved maintenance — Improved maintenance procedures may enable a lighting system with reduced wattage to provide adequate illumination throughout systems or component life.

* Luminaire efficiency replacement or relocation — Check luminaire effectiveness for task lighting and for overall efficiency; if ineffective or inefficient, consider replacement or relocation.

* Heat removal — When luminaire temperatures exceed optimal system operating temperatures, consider using special luminaires to improve lamp performance and reduce heat gain to the space.

* Maintained efficiency — Select a lamp replacement schedule for all light sources, to more accurately predict light loss factors and possibly decrease the number of luminaires required.

Lighting controls

* Switching; local control — Install switches for local and convenient control of lighting by occupants. This should be in combination with a building-wide system to turn lights off when the building is unoccupied.

* Selective switching — Install selective switching of luminaires according to groupings of working tasks and different working hours.

* Low-voltage switching systems — Use low-voltage switching systems to obtain maximum switching capability.

* Master control system — Use a programmable low-voltage master switching system for the entire building to turn lights on and off automatically as needed, with overrides at individual areas.

* Multipurpose spaces — Install multi-circuit switching or preset dimming controls to provide flexibility when spaces are used for multiple purposes and require different ranges of illuminance for various activities. Clearly label the control cover plates.

* "Tuning" illuminance — Use switching and dimming systems as a means of adjusting illuminance for variable lighting requirements.

(Continued)

*	Scheduling	Operate lighting according to a predetermined schedule, based on occupancy.
*	Occupant/motion sensors	Use occupant/motion sensors for unpredictable patterns of occupancy.
*	Lumen maintenance	Fluorescent dimming systems may be utilized to maintain illuminance throughout lamp life, thereby saving energy by compensating for lamp-lumen depreciation and other light loss factors.
*	Ballast switching	Use multilevel ballasts and local inboard-outboard lamp switching where a reduction in illuminances is sometimes desired.

Operation and Maintenance

*	Education	Analyze lighting used during working and building cleaning periods, and institute an education program to have personnel turn off incandescent lamps promptly when the space is not in use, fluorescent lamps if the space will not be used for 10 min. or longer, and HID lamps (mercury, metal halide, high-pressure sodium) if the space will not be used for 30 min. or longer.
*	Parking	Restrict parking after hours to specific lots so lighting can be reduced to minimum security requirements in unused parking areas.
*	Custodial service	Schedule routine building cleaning during occupied hours.
*	Reduced illuminance	Reduce illuminance during building cleaning periods.
*	Cleaning schedules	Adjust cleaning schedules to minimize time of operation, by concentrating cleaning activities in fewer spaces at the same time and by turning off lights in unoccupied areas.
*	Program evaluation	Evaluate the present lighting maintenance program, and revise it as necessary to provide the most efficient use of the lighting system.
*	Cleaning and maintenance	Clean luminaires and replace lamps on a regular maintenance schedule to ensure proper illuminance levels are maintained.
*	Regular system checks	Check to see if all components are in good working condition. Transmitting or diffusing media should be examined, and badly discolored or deteriorated media replaced to improve efficiency.
*	Renovation of luminaries	Replace outdated or damaged luminaires with modern ones which have good cleaning capabilities and which use lamps with higher efficacy and good lumen maintenance characteristics.
*	Area maintenance	Trim trees and bushes that may be obstructing outdoor luminaire distribution and creating unwanted shadow.

13.11 GLOSSARY[9]

AMPERE: The standard unit of measurement for electric current that is equal to one coulomb per second. It defines the quantity of electrons moving past a given point in a circuit during a specific period. Amp is an abbreviation.

ANSI: Abbreviation for American National Standards Institute.

ARC TUBE: A tube enclosed by the outer glass envelope of a HID lamp and made of clear quartz or ceramic that contains the arc stream.

ASHRAE: American Society of Heating, Refrigerating and Air-Conditioning Engineers.

AVERAGE RATED LIFE: The number of hours at which half of a large group of product samples have failed.

BAFFLE: A single opaque or translucent element used to control light distribution at certain angles.

BALLAST: A device used to operate fluorescent and HID lamps. The ballast provides the necessary starting voltage, while limiting and regulating the lamp current during operation.

BALLAST CYCLING: Undesirable condition under which the ballast turns lamps ON and OFF (cycles) due to the overheating of the thermal switch inside the ballast. This may be due to incorrect lamps, improper voltage being supplied, high ambient temperature around the fixture, or the early stage of ballast failure.

BALLAST EFFICIENCY FACTOR: The ballast efficiency factor (BEF) is the ballast factor (see below) divided by the input power of the ballast. The higher the BEF—within the same lamp ballast type—the more efficient the ballast.

BALLAST FACTOR: The ballast factor (BF) for a specific lamp-ballast combination represents the percentage of the rated lamp lumens that will be produced by the combination.

CANDELA: Unit of luminous intensity, describing the intensity of a light source in a specific direction.

CANDELA DISTRIBUTION: A curve, often on polar co-ordinates, illustrating the variation of luminous intensity of a lamp or fixture in a plane through the light center.

CANDLEPOWER: A measure of luminous intensity of a light source in a specific direction, measured in candelas (see above).

COEFFICIENT OF UTILIZATION: The ratio of lumens from a fixture received on the work plane to the lumens produced by the lamps alone. (Also called "CU.")

COLOR RENDERING INDEX (CRI): A scale of the effect of a light source on the color appearance of an object compared to its color appearance under a reference light source. It is expressed on a scale of 1 to 100, where 100 indicates no color shift. A low CRI rating suggests that the colors of objects will appear unnatural under that particular light source.

COLOR TEMPERATURE: The color temperature is a specification of the color appearance of a light source, relating the color to a reference source heated to a particular temperature, measured by the thermal unit Kelvin. The measurement can also be described as the "warmth" or "coolness" of a light source. Generally, sources below 3200K are considered "warm," while those above 4000K are considered "cool" sources.

COMPACT FLUORESCENT: A small fluorescent lamp often used as an alternative to incandescent lighting. The lamp life is about 10 times longer than incandescent lamps and is 3-4 times more efficacious. Also called PL, Twin-Tube, CFL, or BIAX lamps.

CONTRAST: The relationship between the luminance of an object and its background.

DIFFUSE: Term describing dispersed light distribution. Refers to the scattering or softening of light.

DIFFUSER: A translucent piece of glass or plastic sheet that shields the light source in a fixture. The light transmitted throughout the diffuser will be directed and scattered.

DIRECT GLARE: Glare produced by a direct view of light sources. Often the result of insufficiently shielded light sources. (SEE GLARE.)

DOWNLIGHT: A type of ceiling fixture, usually fully recessed, where most of the light is directed downward. May feature an open reflector and/or shielding device.

EFFICACY: A metric used to compare light output to energy consumption. Efficacy is measured in lumens per watt. Efficacy is similar to efficiency, but it is expressed in dissimilar units. For example, if a 100-watt source produces 9000 lumens, the efficacy is 90 lumens per watt.

ELECTRONIC BALLAST: A ballast that uses semi-conductor components to increase the frequency of fluorescent lamp operation, typically in the 20-40 kHz range. Smaller inductive components provide the lamp current control. Fluorescent system efficiency is increased due to high frequency lamp operation.

ENERGY-SAVING BALLAST: A type of magnetic ballast designed so the components operate more efficiently, cooler, and longer than a "standard magnetic" ballast. By U.S. law, standard magnetic ballasts can no longer be manufactured.

ENERGY-SAVING LAMP: A lower wattage lamp, generally producing fewer lumens.

FLUORESCENT LAMP: A light source consisting of a tube filled with argon, along with krypton or other inert gas. When electrical current is applied, the resulting arc emits ultraviolet radiation that excites the phosphors inside the lamp wall, causing it to radiate visible light.

FOOTCANDLE (FC): The English unit of measurement of the illuminance (or light level) on a surface. One footcandle is equal to one lumen per square foot.

FOOTLAMBERT: English unit of luminance. One footlambert is equal to $1/p$ candelas per square foot.

GLARE: The effect of brightness or differences in brightness within the visual field sufficiently high to cause annoyance, discomfort, or loss of visual performance.

HARMONIC: For a distorted waveform, a component of the wave with a frequency that is an integer multiple of the fundamental.

HID: Abbreviation for high intensity discharge. Generic term describing mercury vapor, metal halide, high pressure sodium, and (informally) low pressure sodium light sources and fixtures.

HIGH-BAY: Pertains to the type of lighting in an industrial application where the ceiling is 20 feet or higher. Also describes the application itself.

HIGH OUTPUT (HO): A lamp or ballast designed to operate at higher currents (800 mA) and produce more light.

HIGH PRESSURE SODIUM LAMP: A high intensity discharge (HID) lamp in which light is produced by radiation from sodium vapor (and mercury).

HVAC: Heating, ventilating and air conditioning systems.

ILLUMINANCE: A photometric term that quantifies light incident on a surface or plane. Illuminance is commonly called light level. It is expressed as lumens per square foot (footcandles) or lumens per square meter (lux).

INDIRECT GLARE: Glare produced from a reflective surface.

INSTANT START: A fluorescent circuit that ignites the lamp instantly with a very high starting voltage from the ballast. Instant start lamps have single-pin bases.

LAMP LUMEN DEPRECIATION FACTOR (LLD): A factor that represents the reduction of lumen output over time. The factor is commonly used as a multiplier to the initial lumen rating in illuminance calculations, which compensates for the lumen depreciation. The LLD factor is a dimensionless value between 0 and 1.

LAY-IN-TROFFER: A fluorescent fixture, usually a 2' x 4' fixture that sets or "lays" into a specific ceiling grid.

LED: Abbreviation for light emitting diode. An illumination technology used for exit signs. Consumes low wattage and has a rated life of greater than 80 years.

LENS: Transparent or translucent medium that alters the directional characteristics of light passing through it. Usually made of glass or acrylic.

LIGHT LOSS FACTORS (LLF): Factors that allow for a lighting system's operation at less than initial conditions. These factors are used to calculate maintained light levels. LLFs are divided into two categories, recoverable and non-recoverable. Examples are lamp lumen depreciation and fixture surface depreciation.

LOUVER: Grid type of optical assembly used to control light distribution from a fixture. Can range from small-cell plastic to the large-cell anodized aluminum louvers used in parabolic fluorescent fixtures.

LOW-PRESSURE SODIUM: A low-pressure discharge lamp in which light is produced by radiation from sodium vapor. Considered a monochromatic light source (most colors are rendered as gray).

LUMEN: A unit of light flow, or luminous flux. The lumen rating of a lamp is a measure of the total light output of the lamp.

FIXTURE EFFICIENCY: The ratio of total lumen output of a fixture and the lumen output of the lamps, expressed as a percentage. For example, if two fixtures use the same lamps, more light will be emitted from the fixture with the higher efficiency.

FIXTURE: A complete lighting unit consisting of a lamp or lamps, along with the parts designed to distribute the light, hold the lamps, and connect the lamps to a power source. Also called a fixture.

MEAN LIGHT OUTPUT: Light output in lumens at 40% of the rated life.

MERCURY VAPOR LAMP: A type of high intensity discharge (HID) lamp in which most of the light is produced by radiation from mercury vapor. Emits a blue-green cast of light. Available in clear and phosphor-coated lamps.

METAL HALIDE: A type of high intensity discharge (HID) lamp in which most of the light is produced by radiation of metal halide and mercury vapors in the arc tube. Available in clear and phosphor-coated lamps.

OCCUPANCY SENSOR: Control device that turns lights OFF after the space becomes unoccupied. May be ultrasonic, infrared or other type.

PHOTOCELL: A light sensing device used to control fixtures and dimmers in response to detected light levels.

POWER FACTOR: Power factor is a measure of how effectively a device converts input current and voltage into useful electric power. Power factor is the ratio of kW/kVA.

RAPID START (RS): The most popular fluorescent lamp/ballast combination used today. This ballast quickly and efficiently preheats lamp cathodes to start the lamp. Uses a "bi-pin" base.

REACTIVE POWER: Power that creates no useful work; it results when current is not in phase with voltage. Calculated using the equation:

reactive power = V x A x sin ϕ
where ϕ is the phase displacement angle.

RECESSED: The term used to describe the fixture that is flush mounted into a ceiling.

RETROFIT: Refers to upgrading a fixture, room, or building by installing new parts or equipment.

ROOT-MEAN-SQUARE (rms): The effective average value of a periodic quantity, such as an alternating current or voltage wave, calculated by averaging the squared values of the amplitude over one period, then taking the square root of that average.

SPACE CRITERION: A maximum distance that interior fixtures may be spaced that ensures uniform illumination on the work plane. The fixture height above the work plane multiplied by the spacing criterion equals the center-to-center fixture spacing.

SPECULAR: Mirrored or polished surface. The angle of reflection is equal to the angle of incidence. This word describes the finish of the material used in some louvers and reflectors.

T12 LAMP: Industry standard for a fluorescent lamp that is 12 one-eighths (1.5 inches) in diameter.

TANDEM WIRING: A wiring option in which a ballast is shared by two or more fixtures. This reduces labor, materials, and energy costs. Also called "master-slave" wiring.

VCP: Abbreviation for visual comfort provability. A rating system for evaluating direct discomfort glare. This method is a subjective evaluation of visual comfort expressed as the percent of occupants of a space who will be bothered by direct glare. VCP allows for several factors: fixture luminances at different angles of view, fixture size, room size, fixture mounting height, illuminance, and room surface reflectivity. VCP tables are often provided as part of photometric reports for specific fixtures.

TOTAL HARMONIC DISTORTION (THD): For current or voltage, the ratio of a wave's harmonic content to its fundamental component, expressed as a percentage. Also called "harmonic factor," it is a measure of the extent to which a waveform is distorted by harmonic content.

VERY HIGH OUTPUT (VHO): A fluorescent lamp that operates at a "very high" current (1500mA), producing more light output than a "high output" lamp (800 mA) or standard output lamp (430mA).

WATT (W): The unit for measuring electrical power. It defines the rate of energy consumption by an electrical device when it is in operation. The energy cost of operating an electrical device is calculated as its wattage times the hours of use. In single phase circuits, it is related to volts and amps by the formula: Volts x Amps x PF = Watts. (Note: For AC circuits, PF must be included.)

WORK PLANE: The level at which work is done and at which illuminance is specified and measured. For office applications, this is typically a horizontal plane 30 inches above the floor (desk height).

References

1. Mills, E., Piette, M., 1993, "Advanced Energy-Efficient Lighting Systems: Progress and Potential," *Energy—The International Journal*, 18 (2) pp. 75.
2. Woodroof, E., 1995, *A LECO Screening Procedure for Surveyors*, Oklahoma State University Press—Thesis, p. 54.
3. Bartlett, S., 1993, *Energy—The International Journal*, 18 p.99.
4. Flagello, V, 1995, "Evaluating Cost-Efficient P.I.R. Occupancy Sensors: A Guide to Proper Application," *Strategic Planning for Energy and the Environment*, 14 (3) p. 52.
5. Mills, E., Piette, M., 1993, "Advanced Energy-Efficient Lighting Systems: Progress and Potential," *Energy—The International Journal*, 18 (2), p.90.
6. United States EPA Green Lights Program, 1995, *Lighting Upgrade Manual—Lighting Waste Disposal*, p.2.
7. United States EPA 2004-2006.
8. R. Rundquist, 1991, "Daylighting controls: Orphan of HVAC design," *ASHRAE Journal,*, p. 30, November 1991.
9. United States EPA Green Lights Program, 1995, *Lighting Upgrade Manual—Lighting Fundamentals*.

CHAPTER 14
ENERGY SYSTEMS MAINTENANCE

GARY BERNGARD
Honeywell Building Solutions

14.1 INTRODUCTION

"Energy systems maintenance" covered in this chapter generally refers to the processes and procedures associated with efficient operation of major energy consuming equipment and systems in commercial, institutional, and industrial facilities. The intent of this chapter is to provide the basic framework for developing a sustainable approach to optimizing these systems through preventive maintenance in a way that effectively "connects the dots" between mechanical and electrical systems performance and desired business outcomes. In order to achieve this goal it is important to consider maintenance and operations of these systems within the context of the issues, challenges, and goals for the overall maintenance effort—not just energy conservation.

Effective preventive maintenance programs address a broad spectrum of important and desirable outcomes, including:

- Protecting the owners' investment in their building and systems.
- Providing a level of thermal comfort, indoor air quality, and light levels that influence productivity, safety, and health of the building occupants.
- Improve the manufacturing process throughput, quality, and waste minimization.
- Minimize disruptions from equipment failures.
- Minimize the environmental impact from energy consuming processes.
- Accomplish the above with the least amount of energy use and cost.

A solid preventive maintenance plan is complementary and sustains the value of other beneficial strategies, such as building commissioning, retro-commissioning, and continuous commissioning. A good PM program allows for continuous improvement and will also adapt to building uses that change over time. Effective maintenance is assumed in all designs. It is an especially important piece of the business case for energy conservation measures (ECM) projects since those include expectations of return on investment (ROI) performance. An energy-related goal of maintenance is to keep systems and equipment operation as close to "new" performance levels as practical.

It can be said that *energy systems maintenance is essentially a targeted subset of overall effective preventive maintenance* (Doty). A less effective (=non-sustainable) approach would focus purely on conserving energy and all the required maintenance on buildings and systems that influence energy use. While targeting energy influence seems to limit the field of topics to discuss, the systems themselves are very numerous. Within this field are building systems, manufacturing processes galore, cogeneration, alternative energy sources, metering, indoor and outdoor lighting, pools, vats, ovens, conveyors, hydraulics… the list goes on and on. The varied applications of this subject are a challenge as you focus purely on energy systems maintenance.

For many owner/operators, the hope is that the documentation provided when the building was built or renovated can serve as the basis for a preventive maintenance program. Unfortunately, in most cases the documentation left behind—basis of design, as-builts, control sequences, sources of supply, O/M manuals, and training—is less than complete and focuses on tasks that need to be done to protect the manufacturer's warranty, with little attention given to other longer term O&M needs and challenges.

Information is provided in this chapter as a primer to help address the "why," "how," and "where to start" questions relevant to energy efficient O&M practices, including:

1. Why perform energy, facility, and process systems maintenance?
2. How or where should we start developing a plan?
3. Sample preventive maintenance (PM) task lists.
4. Reference material and source data.

It should be noted that this chapter does not address basic PM for building envelopes, for example window caulking, door gaskets, etc. While these building envelope components can have a significant impact on energy efficiency and comfort, our focus is primarily on mechanical and electrical systems. There are significant data resources available that can help with this area, such as the US Department of Energy's "Energy Efficiency and Renewable Energy" website @ http://www1.eere.energy.gov.

14.2 WHY PERFORM ENERGY, FACILITY & PROCESS SYSTEMS PREVENTIVE MAINTENANCE?

One can simply reword the list of desirable outcomes from the introduction to this chapter in the negative to explain why preventive maintenance should be done—and done properly. Lacking a good preventive maintenance program will likely impede a building owner's ability to achieve desired and sustainable outcomes from the building. Without proper maintenance the following are at risk: Investment in building and systems.

- Thermal comfort, indoor air quality, and light levels that influence productivity, safety, and health of the building occupants.
- Manufacturing process throughput, quality, and waste minimization.
- Disruptions from equipment failures.
- Environmental impact from energy consuming processes.
- Controlling energy use and cost.

The "push back" we often times hear (from operation and management personnel) is that, while they would like to have a formalized PM plan, it's not easy to cost-justify when compared to other more pressing operational and capital needs. Perhaps the best response I've heard to this objection comes from a school district superintendent to his board—"Are you suggesting that we ask our taxpayers to fund this bond for capital renewal and that we shouldn't also ask for the proper funding for O&M? If yes, that is tantamount to suggesting that we simply want to restart the deferred maintenance cycle that got us into this dilemma in the first place!"

A solid PM plan, including energy systems maintenance, allows building owners and operators to cost-effectively leverage their building/facility to achieve their business goals, both in terms of financial and other mission critical goals.

14.3 DEVELOPING & IMPLEMENTING A PREVENTIVE MAINTENANCE PLAN

All PM plans should begin with a clear vision for the order of things as it relates to mechanical and electrical systems operations and use. A simple example would include:

- Comfort & safety
- Productivity
- Protect equipment
- Control operating costs (energy, maintenance & repairs, etc.)

Effective PM plans have support from ownership, administration, shareholders, implementers, and stakeholders for the time to develop a plan and for the funds to implement the plan. Building and maintaining alignment with key stakeholders and decision makers ensures success and sustainability of the plan. Refer to Figure 14-1:

A good executive summary should be developed and updated annually to reflect stakeholder and decision maker needs and goals, the resources needed to get the work done, and an overview of the plan to schedule & track maintenance, repairs, labor, supplies, etc.

A few do's and don'ts for a well structured and managed PM plan:

1. *Do* have jointly authored and well-documented standards of service and comfort, including areas such as:
    - Standard hours of operation.
    - Levels of thermal comfort that will be provided during these hours.
    - Levels of process heating, cooling, make-up air, etc. that will be provided.
    - Procedures to obtain these levels of service outside of standard building hours.
2. *Do* identify and address systems that need to be fixed.
3. *Don't* let goals for conservation inadvertently create curtailment of needed indoor environmental outcomes, productivity, occupant health and safety, or code and liability issues.

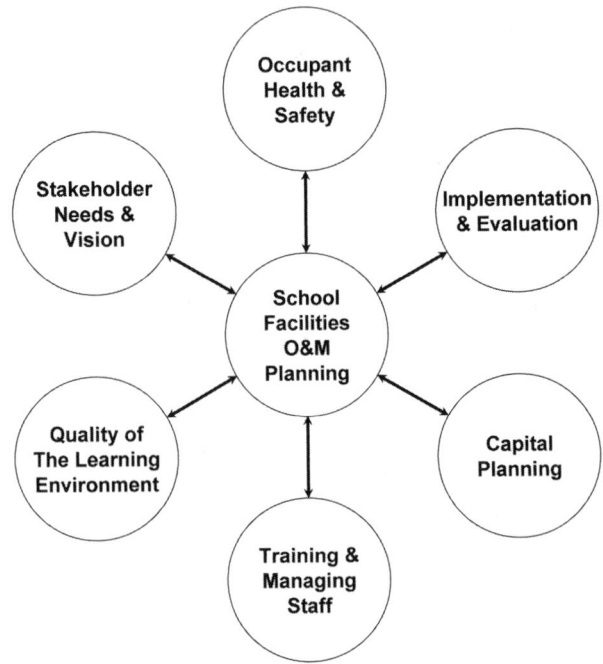

Figure 14-1. Building & maintaining alignment during planning

4. *Do* ensure proper lighting (right amount of light at the right place).

5. *Don't* simply try to make problem systems more efficient; modify and/or replace the systems as needed first, then apply proper O&M practices.

6. *Do* establish a baseline of your current facility use and energy consumption/demand, and trend on-going performance against this baseline.

7. *Do* report operational and financial performance for maintained systems at least annually to ensure continued management/ownership support for your O&M efforts.

8. *Do* have a system for training, orientation, and certification of new O&M employees and stakeholders.

9. *Do* have a system for re-training and re-certification of current O&M staff.

14.3.1. Range of Preventive Maintenance Intensity

The U.S. Department of Education's "Planning Guide for Maintaining School Facilities" suggests that there is a spectrum of maintenance practices for different systems that should be understood when developed, managing, or fine-tuning a PM plan. This is shown graphically in Figure 14-2.

Good business requires a balance of cost and benefit. It is important to pick the right strategy for each major piece of equipment or system to avoid under-maintaining or overspending. A brief discussion of PM strategies follows to help you visualize how and where these might be applied.

1. **Preventive & routine maintenance**
   a. Typically schedule/calendar based.
   b. Can also be based on runtime or other operating parameters.
   c. Manufacturers recommendations are often times greater than needed, because their focus is on trying to protect themselves from warranty period failure and/or trying to sell service agreements.

2. **Diagnostic maintenance**
   a. Diagnose current operation of system and take

necessary preventive or corrective action.
   b. Example: Use oil sample analysis or filter pressure drop to determine if/when to perform maintenance.

3. **Predictive maintenance**
   a. Establish baseline for proper operation and trend operation to predict potential failure.
   b. Then undertake necessary preventive or corrective action.
   c. Example: Vibration baseline and trend analysis can track performance deterioration and be used to forecast needed PM or repairs. Other examples are dirty filters sensed from differential pressure and fouled heat exchangers sensed from approach temperatures.

4. **Run-to-failure (also known as "maintenance by exception")**
   a. Doesn't make sense to spend a lot of time/money maintaining a device that is cheaper to replace if it fails, as long as failure would not create a critical problem.
   b. Example: Spending $100/year on maintenance/supplies for a $25 exhaust fan motor that would only take 15-20 minutes to replace.
   c. Works much better if it's part of a formal plan.
   d. Common application is lighting, where an array of lights are left running until a pre-determined fraction have failed; then all are group relamped.
   e. This approach can create costly repairs when misapplied to large expensive equipment, or for major infrastructure like condenser water treatment or ductwork.

14.3.2 Developing a PM Plan

For owners that do not have a formal PM plan in place but want one and have management support, this nine-step process can be used to develop one:

14.3.3 In-House or Out-Sourced

An example of an O&M labor planning tool is shown

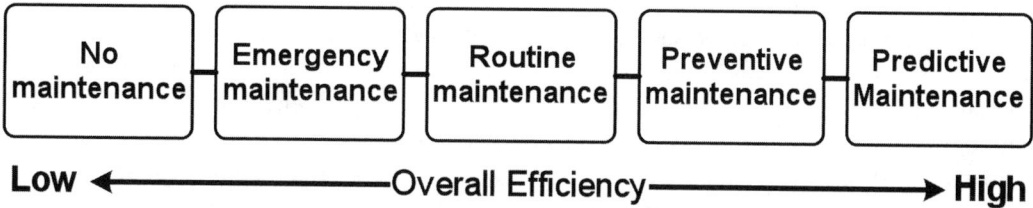

Figure 14-2. Range of Preventive Maintenance Intensity

<u>Nine-Step Process to Develop a Preventive Maintenance (PM) Plan</u>

1. Inventory equipment and systems
   a. Equipment data
   b. Condition assessment
2. Document what needs to be done
   a. Repairs &/or replacements
   b. Maintenance tasks
3. Establish the frequency of the tasks
4. List needed tools & supplies
5. Have a clear and documented explanation of the expertise required to do the tasks
6. Establish manpower plan
   a. Assess current available in-house expertise & available hours. Allow for contingencies, vacations, seasonal work, etc.
   b. Allow for the "unexpected".
   c. Identify current "headache" time & decide if you can reduce through PM. Be realistic
   d. Plan resources by month
7. Decide who's going to do the work
   a. In-house vs. outside
      1) Know capabilities of in-house staff
      2) If you contract with outside service(s), know who's reliable, cost-effective and available in timely fashion
   b. If want to-do in-house, establish "living" training program that includes --
      1) Continuous (annual?) training needs assessment
      2) Formal training plan for all employees
8. Establish a process to schedule & track PM and unscheduled work orders
   a. Create a simple plan and get organized before purchasing any form of computerized maintenance management software ("CMMS")!!!
   b. Then, consider automating. If you don't have a plan before you automate, you'll have automatic chaos with CMMS…
9. Implement, track & adjust parameters (tasking, training, etc.)

---

in Table 14-1. This plan was developed and used by a hospital that was in the process of major renovation and expansion. The management was of the opinion that the timing was right to challenge the established way of doing PM, hence the chart and planning tool. With outside (un-biased) support, the hospital evaluated their readiness level to perform O&M in six areas:

1. Operational logs and inspections
2. Preventive maintenance
3. Problem diagnosis
4. Repairs
5. Energy efficiency/ management
6. Indoor environmental quality

Several key outcomes from this O&M assessment occurred, including the development of individual employee training plans, a re-evaluation of out-sourced services, and a longer-term (budgeted and funded) O&M program with full management support.

Additional itemization of training plan details were provided for other systems:

- Chilled water systems (chillers, cooling towers)
- Miscellaneous cooling equipment (unitary equipment,

  air-cooled condensers)
- Pumps
- Variable Speed Drives
- Water Treatment
- Medical Air Systems
- Pneumatic Tube Carrier Systems

14.3.4 Evaluating Preventive Maintenance Program Effectiveness and Energy Efficiency.

If there is a PM program already in place, consider evaluating the program's effectiveness against owner/operator expectations before suggesting or making any modifications to address energy systems maintenance strategies. An example of an industrial energy plant's assessment follows.

The method shown in this example ranked and prioritized energy management O&M effectiveness. Each category was evaluated with a series of questions. A weighting for each element and the associated sub-category line-items under each element was established. The elements included:

1. Energy management plan
2. Energy procurement
3. Energy consumption—data gathering & analysis

Table 14-1. Sample O&M Sourcing & Training Plan Summary for ABC Hospital (*Continues*)

System	Operational Logs & Inspections			Preventive Maintenance			Problem Diagnosis			Repairs			Energy Efficiency/ Management			Indoor Environmental Quality		
	In-House	Out-Source	Training Needed?	In-House	Out-Source	Training Needed?	In-House	Out-Source	Training Needed?	In-House	Out-Source	Training Needed?	In-House	Out-Source	Training Needed?	In-House	Out-Source	Training Needed?
Heating Hot Water System																		
Boilers																		
Operating & Safety Controls	X		X	X	X	X	X		X	X	X	X	X		X	X		X
Burner Section	X		X	X	X	X	X		X	X	X	X	X		X	X		X
Pressure Vessel	X		X	NA		NA	X	X	X	X	X	X	X		X	NA		NA
Deaerators																		
Operating & Safety Controls	X		X	X		X	X		X	X	X	X	X		X	X		X
Pressure Vessel	X		X	NA		NA	X	X	X	X	X	X	X		X	NA		NA
Steam Condensate Return																		
Operating & Safety Controls	X		X	X		X	X	X	X	X	X	X	X		X	NA		NA
Pressure Vessel	X		X	NA		NA	X	X	X	X	X	X	X		X	NA		NA
Steam Traps	NA		NA	X		X	X		X	X					X	NA		NA
Pump	NA		NA	X		X	X		X	X					X	NA		NA
Pump Motor	NA		NA	X		X	X		X	X					X	NA		NA
Steam driven condensate return assembly	NA		NA	X		X	X	X	X	X		X	X		X	NA		NA
Starter/disconnect	NA		NA	X		X	X	X	X	X	X	X	NA			NA		NA
Steam-to-hot-water Heat Exchangers																		
Operating & Safety Controls	X	X	X	X	X	X	X	X	X	X	X	X	X		X	NA		NA
Pressure Vessel	X		X	NA		NA	X	X	X	X	X	X	X		X	NA		NA
Air Handling Units																		
Fan Sections (squirrel cage, shaft, bearings, etc.)	X		X	X		X	X	X	X	X	X	X	X		X	X		X
Control Dampers	X		X	X		X	X	X	X	X	X	X	NA		NA	X		X
Starters/disconnects	X		X	X		X	X	X	X	X	X	X	NA		NA	NA		NA

4. Energy audits
5. *Energy Consuming Systems Preventive Maintenance*
6. Energy renovation standards
7. Building envelope management
8. Lighting management
9. Air conditioning management
10. Water management
11. Boiler management
12. Electrical management
13. Compressed air management
14. Vehicles management
15. Heating
16. Energy process system optimization

The items listed under Element 5: Energy Consuming Systems Preventive Maintenance are further identified for this chapter to illustrate that each major element includes a subset of plant specific areas of maintenance and energy efficiency:

5a. Areas of PM opportunities:
   1) Steam trap leakage
   2) Compressed air leakage
   3) Thermal insulation damage
   4) Electric Motors PM and Inspection
     a) Electric starter connections

Table 14-1. Sample O&M Sourcing & Training Plan Summary for ABC Hospital (*Concluded*)

System	Operational Logs & Inspections			Preventive Maintenance			Problem Diagnosis			Repairs			Energy Efficiency/ Management			Indoor Environmental Quality		
	In-House	Out-Source	Training Needed?	In-House	Out-Source	Training Needed?	In-House	Out-Source	Training Needed?	In-House	Out-Source	Training Needed?	In-House	Out-Source	Training Needed?	In-House	Out-Source	Training Needed?
Temperature Controls & Energy Management																		
Room Temperature Controls	NA			X		X	X	X	X	X	X	X	X		X	X		X
Air Handling Units Controls	X		X	X	X	X	X	X	X	X	X	X	X		X	X		X
Chiller Plant Controls	X		X	X	X	X	X	X	X	X	X	X	X		X	X		X
Boiler Plant Controls	X		X	X	X	X	X	X	X	X	X	X	X		X	X		X
Heat Exchanger Controls	X		X	X	X	X	X	X	X	X	X	X	X		X	X		X
Miscellaneous Unitary Equipment (e.g. fan coil units, fin-tube radiation, unit heaters, etc.) Controls	NA	NA		X		X	X	X	X	X	X	X	X		X	X		X
Central PC Based Energy Management System																		
Global Energy Management	X		X	X	X	X	X	X	X	X	X	X	X	X	X	X	X	X
Indoor Environmental Quality	X		X	X	X	X	X	X	X	X	X	X	X	X	X	X	X	X
Data Management	X		X	X	X	X	X	X	X	X	X	X	X	X	X	X	X	X

b) Loose wire connections

c) Meter check of starting load and running load against rated loads

d) Belt tension/alignment

e) Bearings (wear, dust, dirt)

f) Internal insulation free of oil

g) Commutator slots and motor housing for dust and good air circulation

h) Fusing and current limiting devices

i) Wear of brushes

5b. Filter inspection and replacement

5c Duct and register inspection/cleanliness

5d. A predictive maintenance system to prevent consumption waste and avoid failures is in place that includes, but is not limited to, the following areas of opportunities

5) Infrared scans

6) Eddy current testing

7) Vibration analysis

8) Oil analysis

9) Ultrasonic testing

10) Megger testing

11) Harmonic testing

12) Multi-amp testing

5e. Cooling tower and water treatment monitoring with special attention to cycles of concentration (important for water conservation)

5f. Coil inspection and cleanliness

A graphical representation of the analysis is presented in Figure 14-3.

As you can see from the above graph, it is apparent that there are a number of areas that fall short of expectations and that a more comprehensive review and refinement should be considered. It is also clear that energy systems maintenance is but one of the facility priorities, although "energy" is embodied in several others. The evaluation was used as the basis for fine-tuning of the plant's O&M program, resulting in approximate savings of 5% in annual O&M cost and 10% in energy savings.

14.4 Examples of Energy Systems Maintenance Procedures

Examples of energy systems maintenance procedures are plentiful and could easily support a full manual in and of themselves. Examples of beneficial energy systems maintenance practices for larger energy consuming boilers, chillers, geo-exchange (ground-source heat pumps), and split-system direct expansion cooling systems are provided below.

Boilers

Boilers generally offer three fuel conservation opportunities that can be captured through effective maintenance procedures:

1. Improvement of combustion efficiency
2. Reduction of losses external to the combustion chamber
3. Better control of the furnace or boiler

Combustion efficiency is a calculated measurement (in percent) of how well the heating equipment is converting a specific fuel into useable heat energy at a specific period of time in the operation of a heating system. Paramount for energy savings on larger boilers (greater than 1 MMBTU output), is the O&M procedure of fine-tuning boiler combustion efficiency.

Complete boiler-efficiency (100%) would extract all the energy available in the fuel and convert it to heat at the point of use. 100% combustion efficiency is not achievable due to incomplete combustion, excess air requirements for safe operation, boiler radiation losses, blow-down losses (steam), and distribution heat losses.

Chapter 5 provides a detailed treatment of boilers and efficiency. O&M procedures can make a positive impact on boiler efficiency in several ways:

1. Adjustment for proper air-fuel mixture (for forced draft boilers with adjustments). This is often outsourced to boiler technical specialists.
2. Maintenance of chemical treatment of water make-up, automatic blow down controls, etc. as applicable.
3. Cleaning of heat exchanger surfaces, both fire side and water side. Part of this measure is also monitoring of stack temperature and heat exchanger approach temperature to predict fouling and schedule cleaning activity. Whenever boilers are opened for jurisdictional inspections, an opportunity exists for inspection and photographing internal parts, usually annually.
4. Maintaining complete insulation systems on hot surfaces to control radiation losses.
5. Sequencing of multiple boilers to reduce part load losses from lightly loaded equipment.

Annual fine-tuning of the combustion process (O_2 + fuel mix, make-up/combustion air control, etc.) can result in energy consumption/cost savings as high as 10-15%.

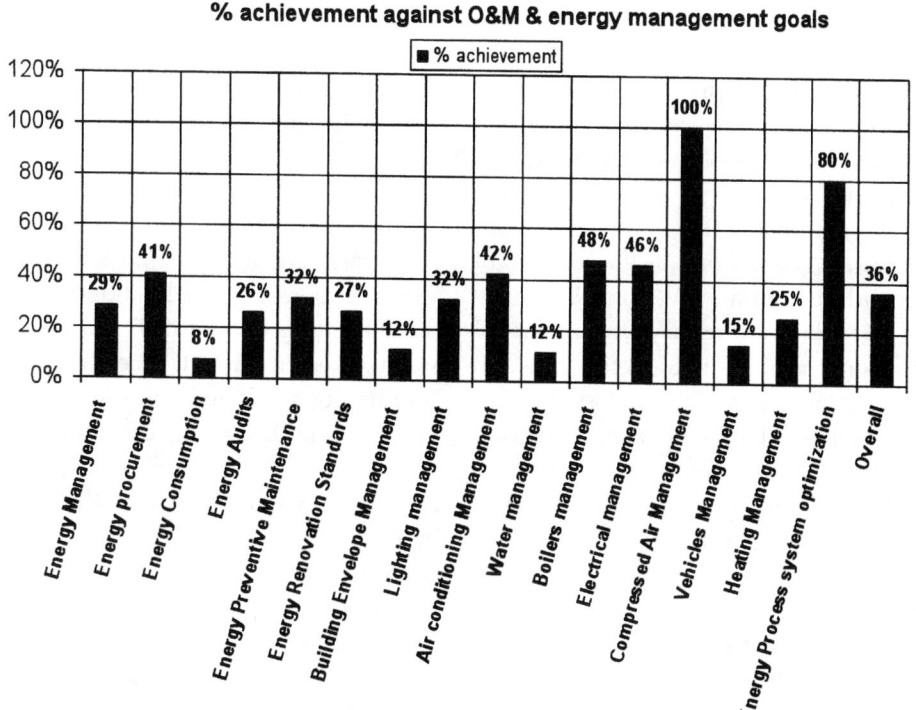

Figure 14-3. Sample Evaluation of an Energy-Focused Preventive Maintenance Program

Chillers

Chiller plant energy savings can be accomplished via O&M procedures in several areas:

1. Reduced compressor head via chilled water and condenser water reset
2. Oil & refrigerant analysis
3. Heat exchanger cleaning
4. Multiple chiller control sequencing to avoid operation at very low loads

The majority of the energy consumed by a chiller is in the process of compressing ("lifting") the incoming hot gas to a higher pressure for subsequent use in the refrigerant cycle. Reductions in lift reduce input power requirements. This compressor head can be reduced by raising chilled water temperature (leaving or entering) or by lowering condenser water temperature (entering or leaving). As a rule of thumb, reducing condenser water in, or chilled water out, will save 1% to 1.5% of the compressor motor energy per 1°F change. Caution should be applied when resetting condenser water temperatures to avoid potential damage to the chiller; each manufacturer has a minimum recommended condenser temperature that should be maintained to assure trouble-free operation. Resetting of chilled water temperatures should be considered in terms of the impact this might have on the ability to meet the cooling loads that the chiller services, including dehumidification performance of the HVAC air systems served by the chilled water plant. In humid climates, over-ambitious reset of chilled water supply temperature can create indoor air quality problems from high humidity. See Chapter 17 for more detail on indoor air quality concepts.

Annual chemical analysis of oil and refrigerant can aid in detecting chiller-contamination problems before they become serious. Testing consists of spectro-chemical analysis to determine contaminants, including moisture, acids, and metals, which hamper performance and efficiency.

In many chillers, oil is entrained in the refrigerant stream for internal lubrication of moving parts. The presence of excess oil will result in efficiency loss, as the oil inhibits the heat transfer capabilities of the refrigerant and the chiller. Generally speaking, there is a 1-2% loss in chiller efficiency for every 1% oil found in the refrigerant. It is not uncommon for poorly maintained chillers to have an oil: refrigerant mix as high as 20:80, resulting in a significant efficiency loss (higher energy use and cost).

Geothermal Heat Pump

One of the greatest contributors to loss of efficiency in heat transfer applications is the build-up of sludge and/or scaling from inadequate water treatment. In a report prepared for the U.S. Department of Energy's Idaho Operations Office titled "Scaling In Geothermal Heat Pump Systems,"

prepared by the Geo-Heat Center at the Oregon Institute of Technology, it was stated that a 0.03" build-up of scale on the heat exchanger surface of the heat pump increased power consumption by 19% versus a clean surface.

Scaling or fouling of a heat transfer surface reduces the cross sectional area for heat to be transferred and causes an increase in the resistance to heat transfer across the heat exchanger. This is because the thermal conductivity of the fouling layer is low. This reduces the overall heat transfer coefficient and efficiency of the heat exchanger. This, in turn, can lead to an increase in pumping and maintenance costs.

In most cases, the formation of scale is a slow process occurring over months or years that could be prevented or minimized through effective water treatment (removal of hardness and proper filtration) and regular maintenance (tube brushing). Figure 14-4 illustrates the impact of scale formation on the performance of a heat pump operating in the cooling mode with 55°F entering water at 2 GPM/ton. Regular O&M attention to heat exchanger surfaces and water treatment will control scale and reduce these losses.

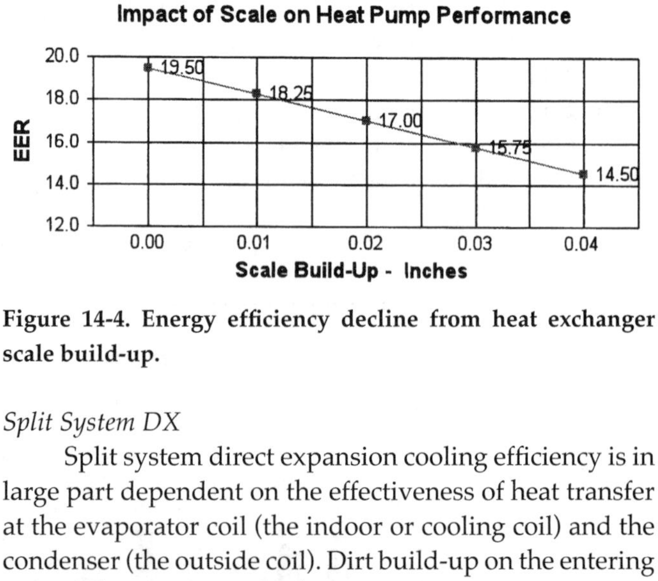

Impact of Scale on Heat Pump Performance

Figure 14-4. Energy efficiency decline from heat exchanger scale build-up.

Split System DX

Split system direct expansion cooling efficiency is in large part dependent on the effectiveness of heat transfer at the evaporator coil (the indoor or cooling coil) and the condenser (the outside coil). Dirt build-up on the entering side of these reduces the heat transfer ability, increasing compressor work and run-times needed to meet cooling needs as well as adding load on the compressor.

Proper filtration of the entering air on the cooling coil, as well as regular changing of this filter media, will reduce build-up on the cooling coil, increasing cooling capacity and decreasing energy use. This coil should also be inspected annually and cleaned if needed. Signs of premature dusting of an indoor coil may be a sign of filter bypassing from an incorrect filter, improper support/sealing, or a loose door and can benefit from O&M intervention, the goal being a filter system that forces all of the HVAC air stream to pass through it.

Dirt and dust build-up on entering condenser coil surfaces is unavoidable. The frequency of inspection and cleaning of these coils during the cooling season should be flexible and allow for more frequent cleanings when there are more airborne elements that might block or coat the coil surface (e.g. cottonwood seeds). Severe outdoor condenser coil blockage and dirt build-up can increase energy consumption by as high as 15-20%. A preventive O&M activity for air-cooled heat-rejection equipment is to maintain the immediate vicinity free from shrubs, grass clippings, leaves, etc. since each of these equipment items acts like a vacuum. Regular inspection of coils is good practice, as well as annual coil cleaning with power-washer equipment.

14-5 Sample PM task lists

The sample preventive maintenance schedules on the following pages outline common energy systems maintenance tasks to be performed. Similar reminder lists can be created for other systems and can also be incorporated into automated work order generating software (CMMS) if desired.

Air Handling Units—
Recommended PM Procedures

Monthly tasks:
- ❏ Visit the equipment personally. Visual / audible inspection. Report unusual noises, vibration, leaks, etc. and repair requirements (if any).
- ❏ Fan and Motor
- ❏ Starter, disconnect & electrical connections
- ❏ Filters
  - ❏ Replace if necessary.

Annual tasks:
- ❏ **Fan and Motor**
  - ❏ Verify operation of system motor, gauges, etc.
  - ❏ Inspect flexible connections and ductwork for damage and leaks.
  - ❏ Inspect the cabinet for cleanliness.
  - ❏ Inspect the cabinet insulation for integrity.
  - ❏ Inspect tension on drive and fan belts. Change and/or adjust as needed.
  - ❏ Lubricate fan shaft and motor bearings.
  - ❏ Lubricate dampers.
  - ❏ Assure minimum damper position and flow.
  - ❏ Clean intake screen on motor.
  - ❏ Inspect fan wheel for free rotation, cracks and alignment.
  - ❏ Inspect for vibrations and unusual noises.
  - ❏ Inspect coils for steam or water leaks.
  - ❏ Report condition of dampers.

- ❏ Test security of guards, doors, and panels.
- ❏ Inspect all structural elements for corrosion and damage.
- ❏ Report condition of coils. Clean if needed.
- ❏ Report condition of air filter(s). Replace if needed.
- ❏ Inspect for excessive loading and bypass around air filter(s).
- ❏ Clean and sanitize drain pan.
- ❏ Verify p-trap is intact and functional.

❏ **Starter, disconnect & electrical connections**
- ❏ Inspect wiring for security and damage.
- ❏ Inspect switch gear, starter, and contactor points.
- ❏ Inspect starter for signs of wear, arcing, overheating, burns, etc.
- ❏ Inspect electrical connections for tightness and absence of moisture.
- ❏ Note if motor surface temperature is unusually hot.
- ❏ For VFDs, verify cooling fan or cooling unit is functional and that any cabinet filters are clean.

❏ **Motors 10 Hp and larger**
- ❏ Measure and record operating voltage. Note any voltage imbalance over 1%.
- ❏ Measure and record operating amperage. Note imbalance or phase amperage above full load amps.

Chiller & Fluid Cooler—
Recommended PM Procedures

Monthly tasks:
- ❏ Visit the equipment personally. Visual / audible inspection. Report unusual noises, vibration, leaks, etc. and repair requirements (if any).
- ❏ Compressors, fans and motors
- ❏ Starter, disconnect & electrical connections
- ❏ Heat transfer surfaces
  - ❏ Clean if necessary.
- ❏ Water Treatment Systems (if applicable)
  - ❏ Verify operation of float / level control and proper level maintained.
  - ❏ Verify level not overflowing.
  - ❏ Verify water treatment controls operable, e.g. proper pH, TDS, as applicable.
  - ❏ Verify bulk chemical are free from leaks and not empty.

Annual tasks:
- ❏ **Compressor(s)**
  - ❏ Verify proper oil level. Record amount of oil

added and date when added.

- ❑ Inspect wiring and connections for signs of wear, overheating, burns, etc.
- ❑ Lubricate motor bearings and shaft bearings.
- ❑ Inspect pulley grooves and belts to alignment, wear, and tension. Adjust to proper tension. Replace belts, if necessary.
- ❑ Oil analysis / vibration analysis if applicable.

❑ **Water-cooled chillers**

- ❑ Inspect condenser tubes and clean annually
- ❑ Inspect chilled water tubes and clean at five years
- ❑ Report any sign of pitting in any water tube.
- ❑ Eddy current testing at five years or when pitting is discovered
- ❑ Condenser Fan Motor(s)
- ❑ Inspect for vibrations and unusual noises in bearings, motors, etc.
- ❑ Inspect fans for vibrations and tightness.
- ❑ Inspect wiring and connections for signs of wear, overheating, burns, etc.
- ❑ Lubricate motor bearings and shaft bearings.
- ❑ Inspect pulley grooves and belts to alignment, wear, and tension.
- ❑ Replace belts if necessary.
- ❑ Heat transfer surfaces—Perform visual inspection & clean if necessary.

❑ **Cooling Tower**

- ❑ Inspect condition of fill (excessive fouling, mechanical damage).
- ❑ Inspect condition of upper and lower sumps (build-up, blockage).
- ❑ Change gearbox oil, if applicable.

❑ **Starter, disconnect & electrical connections**

- ❑ Megger motor at starter and record reading (chiller motor only).
- ❑ Inspect contacts for signs of wear, arcing, overheating, etc.
- ❑ Tighten terminal connections at starter.
- ❑ Inspect wiring for security and damage.
- ❑ For VFDs, verify cooling fan or cooling unit is functional and that any cabinet filters are clean.

❑ **Motors 10 Hp and larger**

- ❑ Measure and record operating voltage. Note any voltage imbalance over 1%.
- ❑ Measure and record operating amperage. Note imbalance or phase amperage above full load amps.

❑ **Operating & Safety Controls**

- ❑ Verify operation of all operating and safety controls, including high/low refrigerant pressure, vibration switches, and oil pressure cutouts.

- ❑ Verify operation of compressor unloaders if applicable.

Unitary and Heat Pump Equipment (Direct Expansion)—Recommended PM Procedures

Annual tasks:

❑ Visit the equipment personally. Visual/audible inspection. Report unusual noises, leaks, etc. and repair requirements (if any).

❑ Fans(s) and Motor(s)

- ❑ Verify operation of system motor, gauges, etc.
- ❑ Inspect flexible connections and ductwork for damage and leaks.
- ❑ Inspect tension on drive and fan belts, and change as needed.
- ❑ Lubricate fan shaft bearings.
- ❑ Lubricate motor bearings.
- ❑ Lubricate dampers.
- ❑ Clean intake screen on motor.
- ❑ Inspect fan wheel for free rotation, cracks, and alignment.
- ❑ Report condition of dampers.
- ❑ Test security of guards, doors, and panels, esp. outdoor units and door seals.
- ❑ Inspect roof curb flashing.*
- ❑ Inspect all structural elements for corrosion and damage.
- ❑ Report condition of coils and clean if needed.

❑ Starter, disconnect and electrical connections

- ❑ Inspect wiring for security and damage
- ❑ Inspect starter for signs of wear, arcing, overheating, burns, etc.
- ❑ Inspect electrical connections for tightness and absence of moisture.
- ❑ Note if motor surface temperature is unusually hot.

❑ Condenser or Outdoor Air Coil Fan Motor(s)

- ❑ Verify all condenser fans are operational.
- ❑ Inspect for vibrations and unusual noises in bearings, motors, etc.
- ❑ Wipe down motor(s) to remove loose dirt and oil buildup.
- ❑ Inspect fans for vibrations and tightness.
- ❑ Inspect wiring and connections for signs of wear, overheating, burns, etc.
- ❑ Lubricate motor bearings.
- ❑ Clean all debris from air inlet louvers.
- ❑ Clean condenser coils with water and coil cleaner.
- ❑ Test for security of guards, doors, and panels.
- ❑ Inspect all structural elements for corrosion and

damage.

❑ Evaporator or Indoor Air Coil
  ❑ Clean condenser coils with water and coil cleaner.
  ❑ Inspect for damage.
❑ Compressor(s)
  ❑ Inspect vibration eliminators for security and damage.
  ❑ Tighten terminal connections at overload elements.*
  ❑ Verify refrigerant charge.
  ❑ Verify superheat adjustment.
  ❑ Inspect moisture indicator for evidence of moisture.*
  ❑ Record compressor suction pressure.
  ❑ Record compressor discharge pressure.
❑ Operating Controls
  ❑ Verify operation of all operating and safety controls including high/low refrigerant pressure.
  ❑ Verify operation of fan speed control or damper control for low ambient operation.*
❑ Heat Pump (Only)
  ❑ Check operation of reversing valve.
  ❑ Verify heating/cooling switchover controls.
  ❑ Verify operation of supplemental heating.
❑ Gas and Oil Burner (Only)
  ❑ Inspect/test for heat exchanger integrity.
  ❑ Verify source of combustion air is unimpeded
  ❑ Inspect and clean all combustion primary air passages.
  ❑ Test all burner linkages for security and/or damage.
  ❑ Test linkage for ease of operation and lubricate as required.
  ❑ Remove, clean and inspect nozzles.*
  ❑ Inspect condition of, and/or clean air filter element.*
  ❑ Inspect ignition assembly and electrode and clean if necessary. *
  ❑ Inspect pilot and clean pilot orifice if necessary.*
  ❑ Inspect high tension wire for deterioration.*
  ❑ Inspect and set spark gap.*
  ❑ Test operation and setting of the gas pressure regulators. *
  ❑ Inspect area around pump seals for seal leakage.*
  Inspect for odors, e.g. gas leaks.
  ❑ Perform combustion analysis and adjust air/fuel mixture, if necessary.
❑ Electric Heat (Only)
  ❑ Inspect elements for hot spots, cracked insula-

tors, other signs of overheating damage
  ❑ Torque heating terminals.
  ❑ Verify staging of heating elements.

*If applicable

Temperature Controls and Energy Management Systems—Recommended PM Procedures

Every Five Years:
❑ Calibrate analog sensors with adjustable ranges, including room, ducts, humidity, and CO_2 and pressure sensors.
❑ Calibrate analog actuators with adjustable zero-span, pneumatic or electronic, including pilot positioners.
❑ Verify accuracy of static (non-adjustable) sensors, and compensate in software or replace.
❑ Verify modulating outputs (actuators) are responsive to commands.
❑ Verify on-off outputs (relays, pressure switches etc.) are responsive to commands.
❑ Perform functional operating sequence test for all equipment, using established sequence of operation or commissioning functional test procedure as the basis of the test.

Annual tasks:
❑ For DDC systems, maintain software for personal computer, EMS network and individual DDC controllers at manufacturer's latest supported revision level.
❑ Perform operating control test for selected equipment
  ❑ Verify heating and cooling valves open and close when told to
  ❑ Verify outdoor air dampers open and close when told to.
  ❑ Verify mixing dampers move freely.
  ❑ Solenoid air valves move when told to.
  ❑ Check pneumatic tubing and devices for leaks.
  ❑ Verify VAV box air dampers and heating valves (if applicable) respond to commands (do this from PC workstation for digital control VAV boxes).
❑ Perform safety operating sequence test for all equipment, and verify proper operation
❑ Boilers
  ❑ Combustion air source unimpeded. If motor actuators used, verify operation and burner interlock that requires damper to be proven open to start.

❏   High water, low water, high/low gas pressure.
❏   Relief valve operation.
❏   Emergency shutdown kill switch.
❏   Chillers
    ❏   Hi/low refrigerant shutdown.
    ❏   Low oil pressure shutdown.
    ❏   Phase loss shutdown, if applicable.
    ❏   Refrigerant monitor.
    ❏   Emergency shutdown kill switch.
    ❏   Self-contained breathing apparatus condition.
❏   Cooling Towers
    ❏   Vibration/low oil shutdown if applicable.
❏   Air Handlers
    ❏   Fire alarm shutdown.
❏   Electric Heaters
    ❏   Inspect high limit safeties for discoloration from overheating; repair if found.
    ❏   Low air flow shutdown control.

Hot Water Boilers—Recommended PM Procedures

Annual tasks:

❏   Visit the equipment personally. Visual/audible inspection. Report unusual noises, leaks, etc. and repair requirements (if any)
❏   Drain boiler as required to perform tests and inspections.
❏   Perform slow drain test of low water cutoff.
❏   Verify operation of makeup water system.
❏   Inspect condition of flues and report.
❏   Inspect refractory and firebrick for defects and report.
❏   Visually inspect boiler exterior for possible leaks and report.
❏   Test boiler room floor drains for proper functioning.
❏   Inspect fireside of boiler and report. Clean fireside heat exchanger surface if any sign of buildup at all
❏   Inspect waterside of boiler for scale buildup and/or oil and report. Clean water-side heat exchanger surface if any sign of buildup at all.
❏   Reassemble and fill boiler. Fire burner to boil off oxygen.
❏   **Boiler Trim**
    ❏   Disassemble, clean, and inspect low water fuel cutoff.
    ❏   Clean or replace sight glass.
    ❏   Disassemble, clean, and inspect water feeder.
❏   **Controls**
    ❏   Clean or replace expansion tank sight glass.*
    ❏   Inspect all electrical connections for tightness.

❏   Verify boiler room supply vents are free from obstructions.
❏   Verify accuracy of temperature gauges.
❏   Inspect air lines for obvious problems.
❏   Inspect wire insulation for signs of overheating, burns, etc.
❏   Verify accuracy of pressure gauges.

*Where applicable

14-6 SUMMARY

Operations and maintenance protects the building investment in many ways. The linkage between maintenance and successful building operation is clear and includes energy conservation, as well as reliability, comfort, and indoor air quality. Data is readily available to show energy efficiency decline from operating most equipment. The default performance degradation, if avoided through targeted maintenance practices, become savings directly attributed to the maintenance effort.

This chapter has focused on building systems. While the details change, many of the fundamental concepts presented would be applicable to other endeavors that include equipment investment, production expectations, wear and tear, and people responsible for maintaining consistent operations. Advanced O&M subjects for additional study include:

•   Process control tuning and optimization (Refer to chapter 22.)
•   Kitchen equipment and systems
•   Power generation and pumping equipment
•   Transportation equipment / fleet operations
•   Manufacturing and specialized equipment

Bibliography

"Energy Conservation with Comfort Manual & Handbook"; by Honeywell Building Solutions.
"Industrial—Energy Plant Self-Assessment"- an interactive Excel based self-assessment tool; by Honeywell Building Solutions
Developing a PM program for K-12 schools—U.S. Department of Education Institution of Educational Sciences http://nces.ed.gov/pubs2003/maintenance/chapter5.asp
Building Energy Statistics—U.S. Energy Information Administration website http://www.eia.doe.gov/emeu/cbecs/
"Scaling In Geothermal Heat Pump Systems"; U.S. Department of Energy's Idaho Operations Office; prepared by the Geo-Heat Center at the Oregon Institute of Technology
"Operation and Maintenance Best Practices for Energy-Efficient Buildings"; produced by Portland Energy Conservation, Inc. (PECI) with funding from the Global Change Division of the U.S. Environmental Protection Agency in cooperation with the U.S. Department of Energy; http://www.wapa.gov/es/pubs/techbrf/oandm.htm
US Department of Energy's "Energy Efficiency and Renewable Energy" website @ http://www1.eere.energy.gov

CHAPTER 15
INSULATION SYSTEMS

JAVIER A. MONT, PH.D., CEM
Johnson Controls, Inc.
Industrial Global Accounts
Chesterfield, Missouri

MICHAEL R. HARRISON
Manager, Engineering and Technical Services
Johns-Manville Sales Corp.
Denver, Colorado

Thermal insulation is a mature technology that has changed significantly in the last few years. What has not changed is the fact that it still plays a key role in the overall energy management picture. In fact, the use of insulation is mandatory for the efficient operation of any hot or cold system. It is interesting to consider that by using insulation, the entire energy requirements of a system are reduced. Most insulation systems reduce the unwanted heat transfer, either loss or gain, by at least 90% as compared to bare surfaces.

Since the insulation system is so vital to energy-efficient operations, the proper selection and application of that system is very important. This chapter describes the various insulation materials commonly used in industrial applications and explores the criteria used in selecting the proper products. In addition, methods for determining the proper insulation thickness are developed, taking into account the economic trade-offs between insulation costs and energy savings.

15.1 FUNDAMENTALS OF THERMAL INSULATION DESIGN THEORY

The basic function of thermal insulation is to retard the flow of unwanted heat energy either to or from a specific location. To accomplish this, insulation products are specifically designed to minimize the three modes of heat transfer. The efficiency of an insulation is measured by an overall property called thermal conductivity.

15.1.1 Thermal Conductivity

The thermal conductivity, or k value, is a measure of the amount of heat that passes through 1 square foot of 1-inch-thick material in 1 hour when there is a temperature difference of 1°F across the insulation thickness. Therefore, the units are Btu-in./hr ft^2 °F. This property relates only to homogeneous materials and has nothing to do with the surfaces of the material. Obviously, the lower the k value, the more efficient the insulation. Since products are often compared by this property, the measurement of thermal conductivity is very critical. The American Society of Testing Materials (ASTM) has developed sophisticated test methods that are the standards in the industry. These methods allow for consistent evaluation and comparison of materials and are frequently used at manufacturing locations in quality control procedures.

Conduction

Heat transfer in this mode results from atomic or molecular motion. Heated molecules are excited and this energy is physically transferred to cooler molecules by vibration. It occurs in both fluids (gas and liquid) and solids, with gas conduction and solid conduction being the primary factors in insulation technology.

Solid conduction can be controlled in two ways: by utilizing a solid material that is less conductive and by utilizing less of the material. For example, glass conducts heat less readily than steel, and a fibrous structure has much less through-conduction than does a solid mass.

Gas conduction does not lend itself to simple modification. Reduction can be achieved by either reducing the gas pressure by evacuation or by replacing the air with a heavy-density gas. In both cases, the insulation must be adequately sealed to prevent reentry of air into the modified system. However, since gas conduction is a major component of the total thermal conductivity, applications requiring very low heat transfer often employ such gas-modified products.

Convection

Heat transfer by convection is a result of hot fluid rising in a system and being replaced by a colder, heavier fluid. This fluid heats, rises, and carries more heat away from the heat source. Convective heat transfer is minimized by the creation of small cells within which the temperature gradients are small. Most thermal insulations are porous structures with enough density to block radiation and provide structural integrity. As such, convection is virtually eliminated within the insulation except for applications where forced convection is being driven through the insulation structure.

Radiation

Electromagnetic radiation is responsible for much of the heat transferred through an insulation and increases in its significance as temperatures increase. The radiant energy will flow even in a vacuum and is governed by the emittance and temperature of the surfaces involved. Radiation can be controlled by utilizing surfaces with low emittance and by inserting absorbers or reflectors within the body of the insulation. The core density of the material is a major factor, with radiation being reduced by increased density. The interplay between the various heat-transfer mechanisms is very important in insulation design. High density reduces radiation but increases solid conduction and material costs. Gas conduction is very significant, but to alter it requires permanent sealing at additional cost. In addition, the temperature in which the insulation is operating changes the relative importance of each mechanism. Figure 15-1 shows the contribution of air conduction, fiber conduction, and radiation in a glass fiber insulation at various densities and mean temperatures.

15.1.2 Heat Transfer

There are many texts dedicated to the physics of heat transfer, some of which are listed in the references.

In its simplest form, however, the basic law of energy flow can be stated as follows:

A steady flow of energy through any medium of transmission is directly proportional to the force causing the flow and inversely proportional to the resistance to that force.

In dealing with heat energy, the forcing function is the temperature difference, and the resistance comes from whatever material is located between the two temperatures.

$$\text{heat flow} = \frac{\text{temperature difference}}{\text{resistance to heat flow}}$$

This is the fundamental equation upon which all heat-transfer calculations are based.

Temperature Difference

By definition, heat transfer will continue to occur until all portions of the system are in thermal equilibrium (i.e., no temperature difference exists). In other words, no amount of insulation is able to provide enough resistance to totally stop the flow of heat as long as a temperature difference exists. For most insulation applications, the two temperatures involved are the operating temperature of the piping or equipment and the surrounding ambient air temperature.

Thermal Resistance

Heat flow is reduced by increasing the thermal resistance of the system. The two types of resistances commonly encountered are mass and surface resistances. Most insulations are homogeneous and as such have a thermal conductivity or k value. Here the insulation resistance, $R_I = tk/k$, where tk represents the thickness of the insulation. In cases of nonhomogeneous products such as multifoil metallic insulations, the thermal properties of the products at their actual finished thicknesses are expressed as conductances rather than conductivities based on a 1-in. thickness. In this case the resistance $R_I = 1/C$, where C represents the measured conductance.

The other component of insulation resistance is the surface resistance, $R_s = 1/f$, where f represents the surface film coefficient. These values

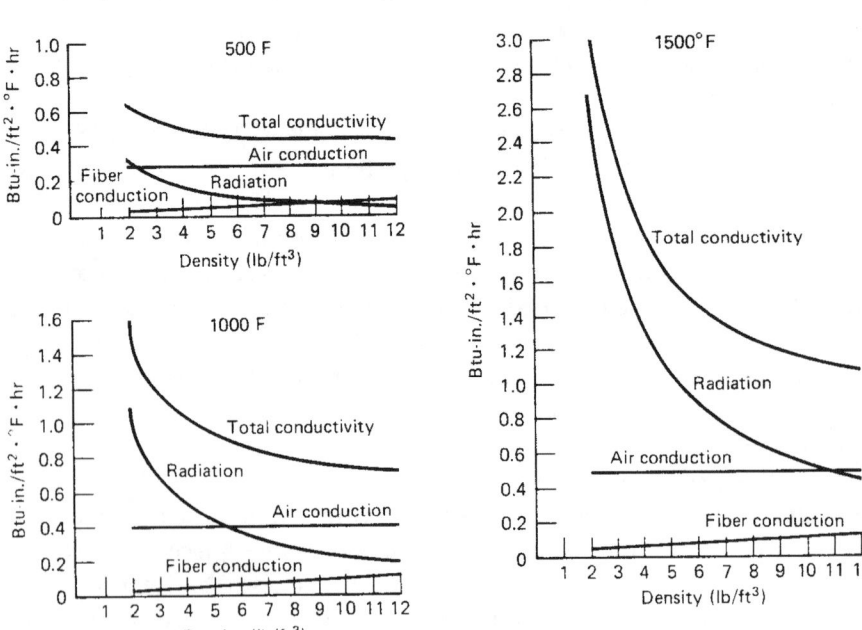

Figure 15-1. Contribution of Each Mode of Heat Transfer. (From Ref. 15.)

are dependent on the emittance of the surface and the temperature difference between the surface and the surrounding environment.

Thermal resistances are additive and as such are the most convenient terms to deal with. Following are several expressions for the heat-transfer equation, showing the relationships between the commonly used R, C, and U values. For a single insulation with an outer film:

$$Q = \frac{\Delta t}{R_I + R_s}$$

$$= \frac{\Delta t}{tk/k + 1/f}$$

$$= \frac{\Delta t}{1/C_I + 1/f}$$

Where the insulation is not homogeneous

$$\text{where } U = \frac{1}{R_{total}} = \frac{1}{R_I + R_s}$$

$$= U \, \Delta t$$

The U-value (U) is termed the overall coefficient of heat transmission of the insulation system.

15.2 INSULATION MATERIALS

Marketplace needs, in conjunction with active research and development programs by manufacturers, are responsible for a continuing change in insulation materials available to industry. Some products have been used for decades, whereas others are relatively new and still being evaluated. The following sections describe the primary insulation materials available today, but first the important physical properties will be discussed.

15.2.1 Important Properties

Each insulation application has a unique set of requirements as it relates to the important insulation properties. However, certain properties emerge as being the most useful for comparing different products and evaluating their fitness for a particular application. Table 15-1 lists the insulation types and product properties that are discussed in detail below. One area that will not be discussed is industrial noise control. Thermal insulations are often used as acoustical insulations for their absorption or attenuation properties. Many texts are available for reference in this area.

Table 15-1. Industrial Insulation Types and Insulation Properties.

Insulation Type and Forms(a)	Temp. Range (°F)	Density (lb/cu.ft)	Thermal Conductivity [Btu-in/hr-ft2-°F at Tmean (°F)]												Compressive Strength (psi) at % Deformation	Fire Hazard Classification Flame-Spread-Smoke Developed	Cell Structure (Permeability and Moisture Absorption)
			-300	-100	0	75	100	200	300	500	700	800	900	1400			
Calcium silicate blocks, shapes and P/C	to 1200	11-15				0.38	0.41	0.44	0.52	0.62			0.72		100-250 at 5%	Non-combustible	Open cell
Glass fiber blankets	to 1000	1				0.24	0.25	0.34	0.46	0.78							
		2				0.22	0.22	0.30	0.36	0.57					0.02-3.5 at 10%	Non-combustible to 25/50	Open cell
Glass fiber boards	to 850	3				0.22	0.23	0.27	0.32	0.49							
Glass fiber P/C	-20 to 850	3				0.22	0.23	0.31	0.39	0.62							
Mineral fiber blocks, boards and P/C	to 1800	15-24					0.32	0.37	0.42	0.52	0.62		0.74		1-18 at 10%	Non-combustible to 25/50	Open cell
Cellular glass blocks, boards and P/C	to 900	8	0.18	0.24	0.29	0.33	0.34	0.41	0.49	0.70					100 at 5%	Non-combustible	Closed cell
Expanded perlite blocks, shapes and P/C	to 1500	13-15					0.4	0.45	0.5	0.6	0.71		0.83		90 at 5%	Non-combustible	Open cell
Urethane foam blocks and P/C	(-100 to -450) to 225	to 1.5				0.16-0.18									16-75 at 10%	25-75 to 140-400	95% closed cell
Polyisocyanurate foam blocks and P/C	to 250	2				0.14	0.15								17-25 at 10%	25-55 to 100	85-90% closed cell
Phenolic foam P/C	-40 to 250	2-3			0.22	0.26									13-22 at 10%	25/50	Open cell
Elastomeric closed cell sheets and P/C	to 400	8.5-9.5			0.29	0.32									40 at 10%	25-75 to 115-490	Closed cell
Ceramic fiber blankets	to 2600	6-8								0.47-0.50		0.70-0.60		1.20-1.80	0.5-1 at 10%	Non-combustible	Open cell

(a) P/C means pipe covering.

Sources: Refs. 17, 18 and manufacturers' literature.

Temperature-Use Range

Since all products have a point at which they become thermally unstable, the upper temperature limit of an insulation is usually quite important. In some cases the physical degradation is gradual and measured by properties such as high-temperature shrinkage or cracking. In such cases, a level is set for the particular property and the product is rated to a temperature at which that performance level is not exceeded. Occasionally, the performance levels are established by industry standards, but frequently, the manufacturers establish their own acceptance levels based on their own research and application knowledge.

In other cases, thermal instability is very rapid rather than gradual. For example, a product containing an organic binder may have a certain temperature at which an exothermic reaction takes place due to a too-rapid binder burnout. Since this type of reaction can be catastrophic, the temperature limit for such a product may be set well below the level at which the problem would occur.

Low-end temperature limits are usually not specified unless the product becomes too brittle or stiff and, as such, unusable at low temperatures. The most serious problem with low-temperature applications is usually vapor transmission, and this is most often related to the vapor-barrier jacket or coating rather than to the insulation. In general, products are eliminated from low-temperature service by a combination of thermal efficiency and cost.

Thermal Conductivity

This property is very important in evaluating insulations, since it is the basic measure of thermal efficiency, as discussed in Section 15.1.1. However, a few points must be emphasized. Since the k value changes with temperature, it is important that the insulation *mean* temperature be used rather than the operating temperature. The mean temperature is the average temperature within the insulation and is calculated by summing the hot and cold surface temperatures and dividing by 2: $(t_h + t_s)/2$. Thermal conductivity data are always published per mean temperature, but many users incorrectly make comparisons at operating temperatures.

A second concern relates to products which have k values that change with time. In particular, foam products often utilize an agent that fills the cells with a gas heavier than air. Shortly after manufacture, some of this gas migrates out, causing an increase in thermal conductivity. This new value is referred to as an "aged k" and is more realistic for design purposes.

Compressive Strength

This property is important for applications where the insulation will see a physical load. It may be a full-time load, such as in buried lines or insulation support saddles, or it may be incidental loading from foot traffic. In either case, this property gives an indication of how much deformation will occur under load. When comparing products it is important to identify the percent compression at which the compressive strength is reported. Five and 10% are the most common; products should be compared at the same level.

Fire Hazard Classification

Insulation materials are involved with fire in two ways—fire hazard and fire protection. Fire protection refers to the ability of a product to withstand fire exposure long enough to protect the column, pipe, or vessel it is covering. This topic is discussed in Section 15.3.1.

Fire hazard relates to the product's contribution to a fire by either flame spread or smoke development. The ASTM E-84 tunnel test is the standard method for rating fire hazard and compares the FS/SD (flame spread/smoke developed) to that of red oak, which has a 100/100 rating. Typically, a 25/50 FHC is specified where fire safety is an important concern. Certain concealed applications allow higher ratings, while the most stringent requirements require a noncombustible classification.

Cell Structure

The internal cell structure of an insulation is a primary factor in determining the amount of moisture the product will absorb, as well as the ease in which vapor will pass through the material. Closed-cell structures tend to resist both actions, but the thickness of the cell walls, as well as the base material, will also influence the long-term performance of a closed-cell product. In mild design conditions such as chilled-water lines in a reasonable ambient, closed-cell products can be used without an additional vapor barrier. However, in severe conditions or colder operating temperatures, an additional vapor barrier is suggested for proper performance.

Available Forms

An insulation material may be just right for a specific application, but if it is not manufactured in a form compatible with the application, it cannot be used. Insulation is available in different types (Ref. 19).

Loose-fill insulation and insulating cements. Loose-fill insulation consists of fibers, powders, granules, or nodules that are poured or blown into walls or other irregular spaces. Insulating cements are mixtures of a loose material with water or other binder that are blown on a surface and dried in place.

Flexible, semirigid and rigid insulation. Flexible and semirigid insulation, which are available in sheets or rolls, are used to insulate pipes and ducts. Rigid insulation is available in rectangular blocks, boards, or sheets and is also used to insulate pipes and other surfaces. The most common forms of insulation are flexible blankets, rigid boards and blocks, pipe insulation half-sections, and full-round pipe sections.

Formed-in-place insulation. This type of insulation can be poured, frothed, or sprayed in place to form rigid or semirigid foam insulation. It is available as liquid components, expandable pellets, or fibrous materials mixed with binders.

Removable-reusable insulation covers. Used to insulate components that require routine maintenance (like valves, flanges, expansion joints, etc.), these covers use belts, Velcro, or stainless steel hooks to reduce the installation time.

Other Properties

For certain applications and thermal calculations, other properties are important. The pH of a material is occasionally important if a potential for chemical reaction exists. Density is important for calculating loads on support structures and occasionally has significance with respect to the ease of installation of the product. The specific heat is used, together with density, in calculating the amount of heat stored in the insulation system, primarily of concern in transient heat-up or cool-down cycles.

15.2.2 Material Description
Calcium Silicate

These products are formed from a mixture of lime and silica and various reinforcing fibers. In general, they contain no organic binders, so they maintain their physical integrity at high temperatures. The calcium silicate products are known for exceptional strength and durability in both intermediate- and high-temperature applications where physical abuse is a problem. In addition, their thermal performance is superior to other products at the higher operating temperatures.

Glass Fiber

Fiberglass insulation is supplied in more forms, sizes, and temperature limits than are other industrial insulations. All of the products are silica-based and range in density from 0.6 to 12 lb/ft^3. The binder systems employed include low-temperature organic binders, high-temperature organic/antipunk binders, and needled mats with no binders at all. The resulting products include flexible blankets, semirigid boards, and preformed one-piece pipe covering, for a very wide range of applications from cryogenic to high temperature. In general, the fiberglass products are not considered load bearing.

Most of the organic binders used begin to oxidize (burn out) in the range 400 to 500°F. The loss of binder somewhat reduces the strength of the product in that area, but the fiber matrix composed of long glass fibers still gives the product good integrity. As a result, many fiberglass products are rated for service above the binder temperature, and successful experience indicates they are completely suitable for numerous applications.

Mineral Fiber/Rock Wool

These products are distinguished from glass fiber in that the fibers are formed from molten rock or slag rather than silica. Most of the products employ organic binders similar to fiberglass, but the very high temperature, high-density blocks use inorganic clay-type binders. The mineral wool fibers are more refractory (heat resistant) than glass fibers, so the products can be used to higher temperatures. However, the mineral wool fiber lengths are much shorter than glass, and the products do contain a high percentage of unfiberized material. As a result, after binder burnout, the products do not retain their physical integrity very well, and long-term vibration or physical abuse will take its toll.

Cellular Glass

This product is composed of millions of completely sealed glass cells, resulting in a rigid insulation that is totally inorganic. Since the product is closed cell, it will not absorb liquids or vapors and thus adds security to cryogenic or buried applications, where moisture is always a problem. Cellular glass is load bearing but also somewhat brittle, making installation more difficult and causing problems in vibrating or flexing applications. At high temperatures, thermal-shock cracking can be a problem, so a cemented multilayer construction is used. The thermal conductivity of cellular glass is higher than for most other products, but it has unique features that make it the best product for certain applications.

Expanded Perlite

These products are made from a naturally occurring mineral, perlite, that has been expanded at a high temperature to form a structure of tiny air cells surrounded by vitrified product. Organic and inorganic binders, together with reinforcing fibers, are used to hold the structure together. As produced, the perlite materials have low moisture absorption, but after heating and oxidizing the organic material, the absorption increases dramatically. The products are rigid and load bearing but have lower compressive strengths and higher thermal conductivities than the calcium silicate products. They are also much more brittle.

Plastic Foams

There are three foam types finding some use in industrial applications, primarily for cold service. They are all produced by foaming various plastic resins.

Polyurethane/Isocyanurate Foams. These two types are rigid and offer the lowest thermal conductivity, since they are expanded with blowing agents. However, sealing is still required to resist the migration of air and water vapor back into the foam cells, particularly under severe conditions with large differentials in vapor pressure. The history of urethanes is plagued with problems of dimensional stability and fire safety. The isocyanurates were developed to improve both conditions, but they still have not achieved the 25/50 FHC (fire hazard classification) for a full range of thickness. As a result, many industrial users will not allow their use except in protected or isolated areas or when covered with another fire-resistant insulation. The advantage of these foam products is their low thermal conductivity, which allows less insulation thickness to be used, of particular importance in very cold service.

Phenolic Foam. These products have achieved the required level of fire safety but do not offer k values much different from fiberglass. They are rigid enough to eliminate the need for special pipe saddle supports on small lines. However, the present temperature limits are so restrictive that the products are primarily limited to plumbing and refrigeration applications.

Polyimide Foams

Polyimide foams are used as thermal and acoustical insulation. This material is fire resistant (FS/SD of 10/10) and lightweight, so it requires fewer mechanical fastening devices. Thermal insulation is available in open-cell structure. Temperature stability limits its application to chilled water lines and systems up to 100°F.

Elastomeric Cellular Plastic

These products combine foamed resins with elastomers to produce a flexible, closed-cell material. Plumbing and refrigeration piping and vessels are the most common applications, and additional vapor-barrier protection is not required for most cold service conditions. Smoke generation has been the biggest problem with the elastomeric products and has restricted their use in 25/50 FHC areas. To reduce installation costs, elastomeric pipe insulation is available in 6-ft long, pre-split tubular sections with a factory-applied adhesive along the longitudinal joint.

Refractories

Insulating refractories consist primarily of two types, fiber and brick.

Ceramic Fiber. These alumina-silica products are available in two basic forms, needled and organically bonded. The needled blankets contain no binders and retain their strength and flexibility to very high temperatures. The organically bonded felts utilize various resins that provide good cold strength and allow the felts to be press cured up to 18 lb/cu.ft. density. However, after the binder burns out, the strength of the felt is substantially reduced. The bulk ceramic fibers are also used in vacuum forming operations where specialty parts are molded to specific shapes.

Insulating Firebrick. These products are manufactured from high-purity refractory clays, with alumina also being added to the higher temperature grades. A finely graded organic filler that burns out during manufacture provides the end product with a well-designed pore structure, adding to the product's insulating efficiency. Insulating firebricks are lighter and therefore store less heat than the dense refractories and are superior in terms of thermal efficiency.

Protective Coatings and Jackets

Any insulation system must employ the proper covering to protect the insulation and ensure long-term performance. Weather barriers, vapor barriers, rigid and soft jackets, and a multitude of coatings exist for all types of applications. It is best to consult literature and representatives of the various coating manufacturers to establish the proper material for a specific application. Jackets with reflective surfaces (like aluminum and stainless steel jackets) have low emissivity (ε). For this reason, reflective jackets have lower heat loss than plain or fabric jackets (high emissivity). In hot applications, this will result in higher surface temperatures and increase the risk of burning personnel. In cold applications, surface temperatures will be lower, which could cause moisture condensation. Regarding jacketing material, existing environment and abuse conditions and desired esthetics usually dictate the proper material. Section 15.3.3 will discuss jacketing systems typically used in industrial work.

15.3 INSULATION SELECTION

The design of a proper insulation system is a two-fold process. First, the most appropriate material must be selected from the many products available. Second, the proper thickness of material to use must be determined. There is a link between these two decisions in that one product with superior thermal performance may require less thickness than another material, and the thickness reduction may reduce the cost. In many cases, however, the thermal values are so close that the same thickness is specified for all the candidate materials. This section deals with the process of material selection. Section 15.4 addresses thickness determination.

15.3.1 Application Requirements

Section 15.2.1 discussed the insulation properties that are of most significance. However, each application will have specific requirements used to weigh the importance of the various properties. There are three items that must always be considered to determine which insulations are suitable for service. They are operating temperature, location or ambient environment, and required form.

Operating Temperature

This parameter refers to the hot or cold service condition that the insulation will be exposed to. In the event of operating design temperatures that may be exceeded during overrun conditions, the potential temperature extremes should be used to assure the insulation's performance.

Cryogenic (–455 to –150°F). Cryogenic service conditions are very critical and require a well-designed insulation system. This is due to the fact that if the system allows water vapor to enter, it will not only condense to a liquid but will subsequently expand and destroy the insulation. Proper vapor barrier design is critical in this temperature range. Closed-cell products are often used since they provide additional vapor resistance in the event that the exterior barrier is damaged or inadequately sealed.

For the lowest temperatures where the maximum thermal resistance is required, vacuum insulations are often employed. These insulations are specially designed to reduce all the modes of heat transfer. Multiple foil sheets (reduced radiation) are separated by a thin mat filler of fiberglass (reduced solid conduction) and are then evacuated (reduced convection and gas conduction). These "super insulations" are very efficient as long as the vacuum is maintained, but if a vacuum failure occurs, the added gas conduction drastically reduces the efficiency.

Finely divided powders are also used for bulk, cavity-fill insulation around cryogenic equipment. With these materials, only a moderate vacuum is required, and in the event that the vacuum fails, the powder still acts as an insulation. It is, however, very important to keep moisture away from the powders, as they are highly absorbent and the ingress of moisture will destroy the system.

Some plastic foams are suitable for cryogenic service, whereas others become too brittle to use. They must all have additional vapor sealing, since high vapor pressures can cause moisture penetration of the cell walls. Closed-cell foamed glass (cellular glass) is quite suitable for this service in all areas except those requiring great thermal efficiency. Since it is not evacuated and has solid structure, the thermal conductivity is relatively high.

Because of the critical nature of much cryogenic work, it is very common to have the insulation system specifically designed for the job. The increased use of liquefied gases (natural and propane), together with cryogenic fluids in manufacturing processes, will require continued use and improvement of these systems.

Low Temperature (–150 to 212°F). This temperature range includes the plumbing, HVAC, and refrigeration systems used in all industries from residential to aerospace. There are many products available in this range, and the cost of the installed thermal efficiency is a large factor. Products typically used are glass fiber, plastic foams, phenolic foam, elastomeric materials, and cellular glass. In below-ambient conditions, a vapor barrier is still required, even though as the service approaches ambient temperature, the necessary vapor resistance becomes less. Above-ambient conditions require little special attention, with the exception of plastic foams that approach their temperature limits around 200°F.

Because of the widespread requirement for the plumbing and HVAC services within residential and commercial buildings, the insulations are subject to a variety of fire codes. Many codes require a flame spread rating less than 25 for exposed material and smoke ratings from 50 to 400, depending on location. A composite rating of 25/50 FHC is suitable for virtually all applications, with a few applications requiring non-combustibility.

Intermediate Temperature (212 to 1000°F). The great majority of steam and hot process applications fall within this operating range. Refineries, power plants, chemical plants, and manufacturing operations all require insulation for piping and equipment at these temperatures. The products generally used are calcium silicate, glass fiber, mineral wool, and expanded perlite. Most of the fiberglass products reach their temperature limit somewhere in this range, with common breakpoints at 450, 650, 850, and 1000°F for various products.

There are two significant elements to insulation selection at these temperatures. First, the thermal conductivity values change dramatically over the range of mean temperatures, especially for light-density products under 18lb/cu.ft. This means, for example, that fiberglass pipe covering will be more efficient than calcium silicate for the lower temperatures, with the calcium silicate having an advantage at the higher temperatures. A thermal conductivity comparison is of value in making sure that the insulation mean temperature is used rather than the operating temperature.

The second item relates to products that use organic binders in their manufacture. All the organics will burn out somewhere within this temperature range, usually between 400 and 500°F. Many products are designed to be used above that temperature, whereas others are not. This is mentioned here only to call attention to the fact that some structural strength is usually lost with organic binder.

High Temperature (1000 to 1600°F). Superheated steam, boiler exhaust ducting, and some process operations deal with temperatures at this level. Calcium silicate, mineral wool, and expanded perlite products are commonly used together with the lower-limit ceramic fibers. Except for a few clay-bonded mineral wool materials, these products reach their temperature limits in this range. Thermal instability, as shown by excessive shrinkage and cracking, is usually the limiting factor.

Refractory (1600 to 3600°F). Furnaces and kilns in steel mills and heat treating and forging shops, as well as in brick and tile ceramic operations, operate in this range. Many types of ceramic fiber are used, with alumina-silica fibers being the most common. Insulating firebrick, castables, and bulk-fill materials are all necessary for meeting the wide variety of conditions that exist in refractory applications. Again, thermal instability is the controlling factor in determining the upper temperature limits of the many products employed.

Location

The second item to consider in insulation selection is the location of the system. Location includes many factors that are critical to choosing the most cost-effective product for the life of the application. Material selection based on initial price only without regard to location can be not only inefficient, but dangerous under certain conditions.

Surrounding Environment. For an insulation to remain effective, it must maintain its thickness and thermal conductivity over time. Therefore, the system must either be able to withstand the rigors of the environment, or be protected from them. An outdoor system needs to keep water from entering the insulation, and in most areas, the jacketing must hold up under radiant solar load. Indoor applications are generally less demanding with regard to weather resistance, but there are washdown areas that see a great deal of moisture. Also, chemical fumes, atmospheres, or spillage may seriously affect certain jacketing materials and should be evaluated prior to specifications. Direct burial applications are normally severe, owing to soil loading, corrosiveness, and moisture. In such applications, it is imperative that the barrier material be sealed from groundwater and resistant to corrosion. Also, the insulation must have a compressive strength sufficient to support the combined weight of the pipe, fluid, soil backfill, and potential wheel loads from ground traffic.

Another concern is insulation application on austenetic stainless steel, a material subject to chloride stress-corrosion cracking. There are two specifications most frequently used to qualify insulations for use on these stainless steels: MIL-1-24244 and Nuclear Regulatory Commission NRC Reg. Guide 1.36. The specifications first require, a stress-corrosion qualification test on actual steel samples; then, on each manufacturing lot to be certified, a chemical analysis must be performed to determine the amount of chlorides, fluorides, sodium, and silicates present in the product. The specific amounts of sodium and silicates required to neutralize the chlorides and fluorides are stated in the specifications.

There are many applications where vibration conditions are severe, such as in gas turbine exhaust stacks. In general, rigid insulations such as calcium silicate withstand this service better than do fibrous materials, especially at elevated temperatures. If the temperature is high enough to oxidize the organic binder, the fibrous products lose much of their compressive strength and resiliency. On horizontal piping, the result can be an oval-shaped pipe insulation that is reduced in thickness on the top of the pipe and sags below the underside of the pipe, thus reducing the thermal efficiency of the system. On vertical piping and equipment with pinned-on insulation, the problem of sag is reduced, but the vibration can still tend to degrade the integrity of the insulation.

Location in a fire-prone area can affect the insulation selection in two ways. First, the insulation system cannot be allowed to carry the fire to another area; this is fire hazard. Second, the insulation can be selected and designed to help protect the piping or equipment from the fire. There are many products available specifically for fireproofing such areas as structural steel columns, but in general they are not very efficient thermal insulations. When an application requires both insulation during operation and protection during a fire, calcium silicate is probably the best selection. This is due to the water of hydration in the product, which must be driven off before the system will rise above the steam temperature. Other high-density, high-temperature products are used as well. With all the products it is important that the jacketing system be designed with stainless steel bands and/ or jacketing since the insulation must be maintained on the piping in order to protect it. Figure 15-2 shows fire test results for three materials per the ASTM E-119 fire curve and indicates the relative level of fire protection provided by each material.

A final concern deals with the transport of volatile fluids through piping systems. When leaks occur around flanges or valves, these fluids can seep into the insulation. Depending on the internal insulation structure, the surface area may be increased significantly, thus reducing the fluid's flash point. If this critical temperature drops below the operating temperature of the system, autoignition can occur, thus creating a fire hazard. In areas where leaks are a problem, either a leakage drain must be provided to remove the fluid, or a closed-cell material such as cellular glass should be used since it will not absorb the fluid.

The previous discussion is intended to draw attention to specific application requirements, not necessarily to determine the correct insulation to be used. Each situation should be evaluated for its own requirements, and in areas of special concern (auto-ignition, fire, etc.) the manufacturer's representative should be called upon to answer questions specific to the product.

Resistance to Physical Abuse. Although this issue is related to location, it is so important that it needs its own discussion. In commercial construction and many light industrial facilities, the pipe and equipment insulation is either hidden or isolated from any significant abuse. In such cases, little attention need be given to this issue. However, in most heavy industrial applications, physical abuse and the problems caused by it are matters of great concern.

Perhaps by definition, physical abuse differs from physical loading in that loading is planned and designed for, whereas abuse is not. For example, with cold piping, pipe support saddles are often located external to the insulation and vapor barrier. This puts the combined weight of the pipe and fluid onto the lower portion of the insulation. This is a designed situation, and a rigid material is inserted between the pipe and the saddle

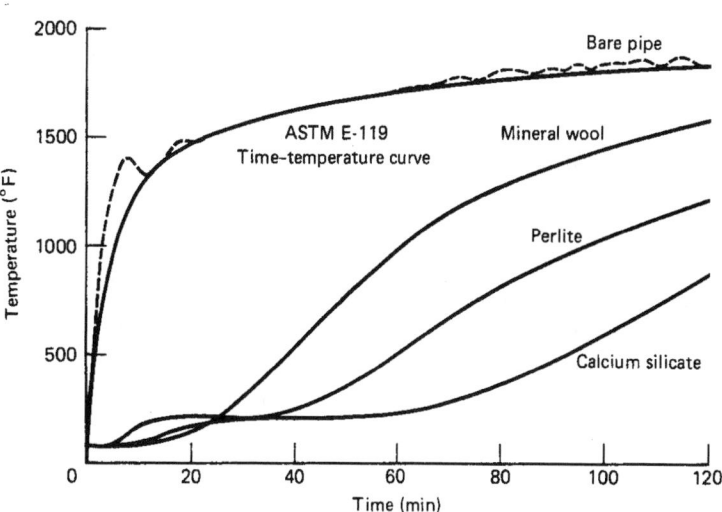

Figure 15-2. Fire Resistance Test Data for Pipe Insulation. (Used by permission from Ref. 6.)

to carry the load. However, if a worker decides to use the insulated pipe as a scaffold support, a walkway, or a hoist support, the insulation may not be designed to support such a load and damage will occur. A quick walk-through of any industrial facility will show much evidence of "unusual" or "unanticipated" abuse. In point of fact, many users have seen so much of this that they now design for the abuse, having determined that it is "usual" for their facility.

The effects of abuse are threefold. First, dented and creased aluminum jacketing is unsightly and lowers the overall appearance of the plant. Nonmetal jackets may become punctured and torn. The second point is that wherever the jacketing is deformed, the material under it is compressed and as a result is a much poorer insulation since the thickness has been reduced. Finally, on outside lines some deformation will undoubtedly occur at the jacketing overlap. This allows for water to enter the system, further degrading the insulation and reducing the thermal efficiency.

In an effort to deal with the physical abuse problem, some specifications call for all horizontal piping to be insulated with rigid material while allowing a fibrous option on the vertical lines. Others modify this specification by requiring rigid insulation to a height of 6 to 10 ft on vertical lines to protect against lateral abuse. Still, in facilities that have a history of a rough environment, it is most common to specify the rigid material for all piping and equipment except that which is totally enclosed or isolated.

As previously mentioned, the primary insulation material choice is between rigid and nonrigid materials. Calcium silicate, cellular glass, and expanded perlite products fit the rigid category, whereas most mineral wool and fiberglass products are nonrigid. Over the years, the calcium silicate products have become the standard for rigorous services, combining good thermal efficiency with exceptional compressive strength and abuse resistance. The maintenance activities associated with rigid insulations are significantly less than the maintenance and replacement needs of softer insulations in abuse areas. The costs associated with this are discussed in a later section. However, it is also recognized that often maintenance activities are lacking, which results in a deteriorated insulation system operating at reduced efficiency for a long period of time.

Form Required

The third general category to consider is the insulation form required for the application. Obviously, pipe insulation and flat sheets are manufactured for specific purposes, and the lack of a specific form eliminates that product from consideration. However, there are subtle differences between form that can make a significant difference in installation costs and system efficiency.

On flat panels, the two significant factors are panel size and the single-layer thickness available. A fibrous 4 × 8 ft sheet is applied much more rapidly than four 2 × 4 ft sheets, and it is possible that the number of pins required might be reduced. In regard to thickness, if one material can be supplied 4 in. thick as a single layer, as opposed to two 2-in.-thick layers, the first option will result in significant labor savings. The same holds true for 18-in.- vs. 12-in.-wide rigid block installation.

Fibrous pipe insulations have three typical forms: one-piece hinged snap-on, two piece halfsections, and flexible blanket wraparound. For most pipe sizes, the one-piece material is the fastest to install and may not require banding if the jacket is attached to the insulation and secured to itself. Two-piece products must be wired in place and then jacketed in a separate operation. Wraparound blankets are becoming more popular, especially for large-diameter pipe and small vessels. They come in standard roll lengths and are cut to length on the job site.

Rigid insulations also have different forms, which vary with the manufacturer. The two-piece half-section pipe insulation is standard. However, these sections can be supplied prejacketed with aluminum, which in effect gives a one-piece hinged section that does not need a separate application of insulation and jacket. Also, thicknesses up to 6 in. are available, eliminating the need for double-layer applications where they are not required for expansion reasons. The greatest diversity comes in the large-diameter pipe sizes. Quads (quarter sections) are available that are both quicker to install and thermally more efficient than scored block bent around the pipe. Similarly, curved radius blocks are available for sizes above quads and provide a better fit than flat beveled block or scored block.

The important point is that the available forms of insulation may well affect the decision as to which material to select and from which manufacturer to purchase that material. It is unwise to assume that all manufacturers offer the same sizes and forms, or that the cost to install the product is not affected by its form.

15.3.2 Cost Factors

Section 15.3.1 dealt with the process of selecting the materials best suited for a specific application. In some cases the requirements are so stringent that only one specific material is acceptable. For most situations, however, more than one insulation material is suitable, even though they may be rank-ordered by anticipated performance. In these cases, several cost factors should be considered to determine which specific material (and/or manufacturer) should be selected to provide the best system for the lowest cost.

Initial Cost

In new construction, the owner is usually interested primarily in the installed cost of the insulation system. As long as the various material options provide similar thermal performance, they can be compared on an equal basis. The contractor, on the other hand, is much more concerned about the insulation form and its effect on installation time. It may be of substantial benefit to the contractor to utilize a more costly material that can be installed more efficiently for reduced labor costs. In a highly competitive market, these savings are often passed through to the owner for the contractor to secure the job. The point is that the lowest-cost material does not necessarily become the lowest-cost installed material, so all acceptable alternatives should be evaluated.

Maintenance Cost

To keep their performance and appearance at acceptable levels, all insulation systems must be maintained. This means, for example, that outdoor weather protection must be replaced when damaged to prevent deterioration of the insulation. If left unattended, the entire system may need to be replaced. In a high-abuse area, a nonrigid insulation may need to be replaced quite frequently in order to maintain performance. Aesthetics often play an important part in maintenance activities, depending on the type of operation and its location. In such cases there is benefit in utilizing an insulation that maintains its form and, if possible, aids the jacketing or coating in resisting abuse.

The trade-off comes between initial cost and maintenance cost in that a less costly system may well require greater maintenance. Unfortunately, the authorities for initial construction and ongoing maintenance are often split, so the owner may not be aware of the future consequences of the initial system selection. It is imperative that both aspects be viewed together.

Lost Heat Cost

If the various suitable insulations are properly evaluated, a more thermally efficient product should require less thickness to meet the design parameters. However, if a common thickness is specified for all products, there can be a substantial difference in heat loss or gain between the systems. In such a case, the more efficient product should receive financial credit for transferring less heat, and this should be considered in the overall cost calculations

Referring to the previous discussion on maintenance costs, there was an underlying assumption that maintenance would be performed to the extent that the original thermal efficiency would be maintained. In reality, maintenance is usually not performed until the situation is significantly deteriorated—and sometimes not even then. The result of this is reduced thermal efficiency for much of the life of a maintenance-intensive system. It is very difficult to assign a figure to the amount of additional heat transfer due to deterioration. In a wet climate, for example, a torn jacket will allow moisture into the system and drastically affect the performance. Conversely, in a dry area, the insulation might maintain its performance for quite some time. Still, when dealing with maintenance costs, it is a valid concern that systems in need of maintenance generally are transferring more heat at greater cost than are systems requiring less maintenance.

Design Life

The anticipated project life is the foundation upon which all costs are compared. Since there are trade-offs between initial cost and ongoing maintenance and heat-loss costs, the design life is important in determining the total level of the ongoing costs. To illustrate, consider the difference between designing a 40-year power plant and a two-year experimental process. Assuming that the insulation in the experimental process will be scrapped at the close of the project, it makes no sense to use a more costly insulation that has lower maintenance requirements since those future benefits will never be re-

alized. Similarly, utilizing a less costly but maintenance-intensive system when the design life is 40 years makes little sense since the additional front-end costs could be regained in only a few years of reduced maintenance costs.

15.3.3 Typical Applications

This section is designed to give a brief overview of commonly used materials and application techniques. For a detailed study of application, techniques, and recommendations, see Ref. 7, as well as the guide specifications supplied by most insulation manufacturers.

The Heat Plant

Boilers are typically insulated with fiberglass or mineral wool boards, with some usage of calcium silicate block when extra durability is desired. Powerhouse boilers are normally insulated on-site, with the fibrous insulation being impaled on pins welded to the boiler. Box-rib aluminum is then fastened to the stiffeners or buckstays as a covering for the insulation. In most commercial and light industrial complexes, package boilers are normally used. These are insulated at the factory, usually with fiberglass or mineral wool.

Breechings and other high-temperature duct work are insulated with calcium silicate (especially where traffic patterns exist), mineral wool, and high-temperature fiberglass. On very large breechings, prefabricated panels are used, as discussed in the following paragraph. H-bar systems supporting the fibrous materials are common, with the aluminum lagging fastened to the outside of the H-bar members. Also, many installations utilize roadmesh over the duct stiffeners, creating an air space, and then wire the insulation to the mesh substrate. Indoors, a finish coat of cement may be used rather than metal lagging.

Precipitators are typically insulated with prefabricated panels filled with mineral wool or fiberglass blankets. For large, flat areas, such panels provide very efficient installations, as the panels are simply secured to the existing structure with self-tapping screws. H-bar and Z-bar systems are also used to contain the fibrous boards.

Steam piping insulation varies with temperature and location, as discussed earlier. Calcium silicate wired in place and then jacketed with corrugated or plain aluminum is very widely used. The jacketing is either screwed at the overlap or banded in place. Fiberglass is used extensively in low-pressure steam work in areas of limited abuse. Mineral wool and expanded perlite can also be used for higher-temperature steam, but calcium silicate is the standard.

Process Work

Hot process piping and vessels are typically insulated with calcium silicate, mineral wool, or high-temperature fiberglass. Horizontal applications are generally subject to more abuse than vertical ones and as such have a higher usage of calcium silicate. Many vessels have the insulation banded in place and then the metal lagging is banded in place, separately. Most vessel heads have a cement finish and may or may not be subsequently covered with metal. Recent product developments have provided a fiberglass wraparound product for large-diameter piping and vessels. This flexible material conforms to the curvature and need only be pinned at the bottom of a horizontal vessel. Banding is then used to secure both the insulation and the jacketing. In areas of chemical contamination, stainless steel jacketing is frequently used.

Cold process vessels and piping also use a variety of insulations, depending on the minimum temperature and the thermal efficiency required. Cellular glass is widely used in areas where the closed-cell structure is an added safeguard. (The material is still applied with a vapor-barrier jacket or coating.) It is also used wherever there is a combined need for closed-cell structure and high compressive strength. However, the polyurethane materials are much more efficient thermally, and in cryogenic work maximum thermal resistance is often required. Multiple vapor barriers are used with the urethanes to prevent the migration of moisture throughout the entire system. In all cold work, the workmanship, particularly on the outer vapor barrier, is extremely critical. There are many other specially engineered systems for cryogenic work, as discussed in Section 15.3.1.

Fluid storage tanks located outdoors are typically insulated with fiberglass insulation. Prefabricated panels are either installed on studs or banded in place. Also, the jacketing can be banded on separately over the insulation. A row of cellular glass is placed along the base of the tank to prevent moisture from wicking up into the fiberglass. Sprayed urethane is also used on tanks that will not exceed 200°F, but a trade-off exists between cost efficiency and the long-term durability of the system.

Tank roofs are a problem because of the need for a rigid walking surface as well as a lagging system that will shed water. Many tops are left bare for this reason, whereas others utilize a spray coating of cork-filled mastic that provides only minimal insulation. Rigid fiberglass systems can be made to work with a well-designed covering system that drains properly. The most secure system is to use a built-up roofing system similar to those used on flat-top buildings. The installation is generally more costly, but acceptable long-term performance is much more probable.

HVAC System

Duct work constructed of sheet metal is usually wrapped with light-density fiberglass with a preapplied foil and kraft facing. The blanket is overlapped and then stapled, with tape or mastic being applied if the duct flow is cold and a vapor barrier is required. Support pins are required to prevent sag on the bottom of horizontal ducts. Fiberglass duct liner is used inside sheet-metal ducts to provide better sound attenuation along the duct; this provides a thermal benefit as well. For exposed duct work, a heavier-density fiberglass board may be used as a wrap to provide a more acceptable appearance. In all cases, the joints in the sheet metal ducts should be sealed with tape or caulking to minimize air leakage and allow the transport of air to the desired location, rather than losing much of it along the run.

Rigid fiberglass duct board and round duct are also used in many low-pressure applications. These products form the duct itself, as well as providing the thermal, acoustical, and vapor-barrier requirements most often needed. The closure system used to join the duct sections also acts to seal the system for minimum air leakage.

Chillers and chilled water expansion tanks are usually insulated with closed-cell elastomeric sheet to prevent condensation on the equipment. The joints are sealed and a finish may or may not be applied to the outside, depending on location.

Piping for both hot and cold service is normally insulated with fiberglass pipe insulation. On cold work, the vapor-barrier jacket is sealed at the overlap with either an adhesive or a factory-applied self-seal lap. If staples are used, they should be dabbed with mastic to secure the vapor resistance. Aluminum jacketing is often used on outside work, with a vapor barrier applied beneath if it is cold service. Domestic hot- and cold-water plumbing and rain leaders are also commonly insulated with fiberglass. Insulation around piping supports takes many forms, depending on the nature of the hanger or support. On cold work, the use of a clevis hanger on the outside of the insulation requires a high-density insert to support the weight of the piping. This system eliminates the problem of adequately sealing around penetrations of the vapor barrier.

15.4 INSULATION THICKNESS DETERMINATION

This section presents formulas and graphical procedures for calculating heat loss, surface temperature, temperature drop, and proper insulation thickness. Over the last few years, computer programs that perform these calculations have become more readily available to customers. (See section 15.5.4.) But still, it is important to understand the basics for their use. Although the overall objective is to determine the right amount of insulation that should be used, some of the equations use thickness as an input variable rather than solving for it. However, all the calculations are simply manipulations or further refinements of the equation in Section 15.1.2:

$$Q = \frac{\Delta t}{R_I + R_s}$$

Following is a list of symbols, definitions, and units to be used in the heat-transfer calculations.

t_a = ambient temperature, °F

t_s = surface temperature of insulation next to ambient, °F

t_h = hot surface temperature, normally operating temperature (cold surface temperature in cold applications), °F

k = thermal conductivity of insulation, always determined at mean temperature, Btu-in./hr ft^2 °F

t_m = $(t_h + t_s)/2$ = mean temperature of insulation, °F

t_h = $(t_{in} + t_{out})/2$ = average hot temperature when fluid enters at one temperature and leaves at another, °F

tk = thickness of insulation, in.

r_1 = actual outer radius of steel pipe or tubing, in.

r_2 = $(r_1 + tk)$ = radius to outside of insulation on piping, in.

Eq tk = $r_2 \, Ln \, (r_2/r_1)$ = equivalent thickness of insulation on a pipe, in.

f = surface air film coefficient, Btu/hr ft^2 °F

R_s = $1/f$ = surface resistance, hr ft^2 °F/Btu

R_I = tk/k = R-value, thermal resistance of insulation, hr ft^2 °F/Btu

Q_F = heat flux through a flat surface, Btu/hr ft^2

Q_p = $Q_f\left(\frac{2\pi_2}{12}\right)$ = heat flux through a pipe, Btu/hr lin. ft

A = area of insulation surface, ft^2

L = length of piping, lin. ft

Q_{T-} = $Q_F \times A$ or $Q_p \times L$ = total heat loss, Btu/hr

H = time, hr

C_p = specific heat of material. Btu/lb. °F

ρ = density, lb/ft^3

M = mass *flow* rate of a material, lb/hr

Δ = difference by subtraction, unitless

RH = relative humidity, %

DP = dew point temperature, °F

U = Btu/hr ft^2 °F

U = U-value, overall heat transfer coefficient, 1/R, Btu/(hr ft^2 °F)

15.4.1 Thermal Design Objective

The first step in determining how much insulation to use is to define what the objective is. There are many reasons for using insulation, and the amount to be used will definitely vary based on the objective chosen. The four broad categories, which include most applications, are (1) personnel protection, (2) condensation control, (3) process control, and (4) economics. Each of these will be discussed in detail, with sample problems leading through the calculation sequence.

15.4.2 Fundamental Concepts
Thermal Equilibrium

A very important law in heat transfer is that under steady-state conditions, the heat *flow* through any portion of the insulation system is the same as the heat *flow* through any other part of the system. Specifically, the heat *flow* through the insulation equals the heat *flow* from the surface to the ambient, so the temperature difference for each section is proportional to the resistance for each section:

$$\text{heat flow} = \frac{\text{temperature difference}}{\text{resistance to heat flow}}$$

$$Q = \frac{t_h - t_a}{R_I + R_s} = \frac{t_h - t_s}{R_I} = \frac{t_s - t_a}{R_s}$$

Because all of the heat flows Q are equal, this relationship is used to check surface temperature or other interface temperatures. For an analysis concerned with the inner surface film coefficient, the same reasoning applies.

$$Q = \frac{t_h - t_a}{R_{s1} + R_1 + R_{s2}} = \frac{t_h - t_{s1}}{R_{s1}} = \frac{t_{s1} - t_{s2}}{R_1} = \frac{t_{s2} - t_a}{R_{s2}}$$

Or for a system with two insulation materials involved, the interface temperature t_{if} between the materials is involved.

$$Q = \frac{t_h - t_a}{R_{I1} + R_{I2} + R_s} = \frac{t_h - t_{if}}{R_{I2}} = \frac{t_{if} - t_s}{R_{I2}} = \frac{t_s - t_a}{R_s} = \frac{t_{if} - t_a}{R_{I2} + R_s}$$

It should be apparent that the heat flow Q is also equal for any combination of Δt and R values, as shown by the last equivalency above, which utilizes two parts of the system instead of just one.

Finally, it is of critical importance to calculate the R_I values using the insulation mean temperature, not the operating temperature. The mean temperature is the sum of the temperatures on either side of the insulation divided by 2. Again, for the last set of equivalencies:

$$t_m \text{ for } R_{I1} = \frac{t_h + t_{if}}{2}$$

$$t_m \text{ for } R_{I2} = \frac{t_{if} - t_s}{2}$$

Pipe vs. Flat Calculations—Equivalent Thickness

Because the radial heat flows in a path from a smaller-diameter pipe, through the insulation, and then off a larger-diameter surface, a phenomenon termed "equivalent thickness" (Eq tk) occurs. Because of the geometry and the dispersion of the heat to a greater area, the pipe really "sees" more insulation than is actually there. When the adjustment is made to enter a greater insulation thickness into the calculation, the standard flat geometry formulas can be used by substituting Eq tk for tk into the equations.

The formula for equivalent thickness is

$$\text{Eq tk} = r_2 \ln \frac{r_2}{r_1}$$

where r_1 and r_2 are the inner and outer radii of the insulation system. For example, an 8-in. IPS with 3-in. insulation would lead to an equivalent thickness as follows (8-in. IPS has 8.625 in. actual outside diameter):

$$r_1 = \frac{8.625}{2} = 4.31 \text{ (Table 15.2)}$$

$$r_2 = r_1 + \text{tk} = 4.31 + 3 = 7.31$$

$$\text{Eq tk} = 7.31 \ln \frac{7.31}{4.31} = 7.31 \ln 1.70$$

$$= 3.86 \quad \textit{actual outside diameter}$$

This Eq tk can hen be used in the flat geometry equation by substituting Eq tk for tk.

$$Q = \frac{t_h - t_k}{\text{Eq tk}/k + R_s}$$

The example above uses an even insulation thickness of 3 in. Some products are manufactured to such even thicknesses, and Table 15-2 lists the Eq tk for such products. However, many products are manufactured to "simplified" thicknesses that allow a proper fit when nesting double-layer materials. ASTM-C-5859 lists these standard dimensions, and Table 15-3 shows Eq tk values for the simplified thicknesses. Figure 15-3 also shows the conversion for any thickness desired and will be used later in the reverse fashion.

Surface Resistance

There is always diversity of opinion when it comes to selecting the proper values for the surface resistance R_s. The surface resistance is affected by surface emittance, surface air velocity, and the surrounding environment. Heat-transfer texts have developed procedures for calculating Rs values, but they are all based on speculated values of emittance and air velocity. In actuality, the emittance of a surface often changes with time, temperature, and surface contamination, such as dust. As a result, it is unnecessary to labor over calculating specific R_s values, when the conditions are estimates at best.

Table 15-4 lists a series of R_s values based on three different surface conditions and the temperature difference between the surface and ambient air. Also included are single-point R_s values for three different surface air velocities. See the note at the bottom of Table 15-4 relating to the effect of R_s on heat-transfer calculations.

15.4.3 Personnel Protection

Workers need to be protected from high-temperature piping and equipment in order to prevent skin burns. Before energy conservation analyses became commonplace, many insulation systems were designed simply to maintain a "safe-touch" temperature on the outer jacket. Now, with energy costs so high, personnel protection calculations are generally limited to temporary installations or waste-heat systems, where the energy being transferred will not be further utilized.

Normally, safe-touch temperatures are specified in the range 130 to 150°F, with 140°F being used most often. It is important to remember that the surface temperature is directly related to the surface resistance R_s, which in turn depends on the emittance of the surface. As a result, an aluminum jacket will be hotter than a dull mastic coating over the same amount of insulation. This is demonstrated below.

Calculation

The objective is to calculate the amount of insulation required to attain a specific surface temperature. As noted earlier,

$$\frac{t_h - t_s}{R_1} = \frac{t_s - t_a}{R_s}$$

Therefore,

$$R_1 = R_s\left(\frac{t_h - t_s}{t_s - t_a}\right) = \frac{tk}{k}(flat) = \frac{\text{Eq tk}}{k}(\text{pipe})$$

Therefore,

$$\text{tk or Eq tk} = kR_s\left(\frac{t_h - t_s}{t_s - t_a}\right)$$

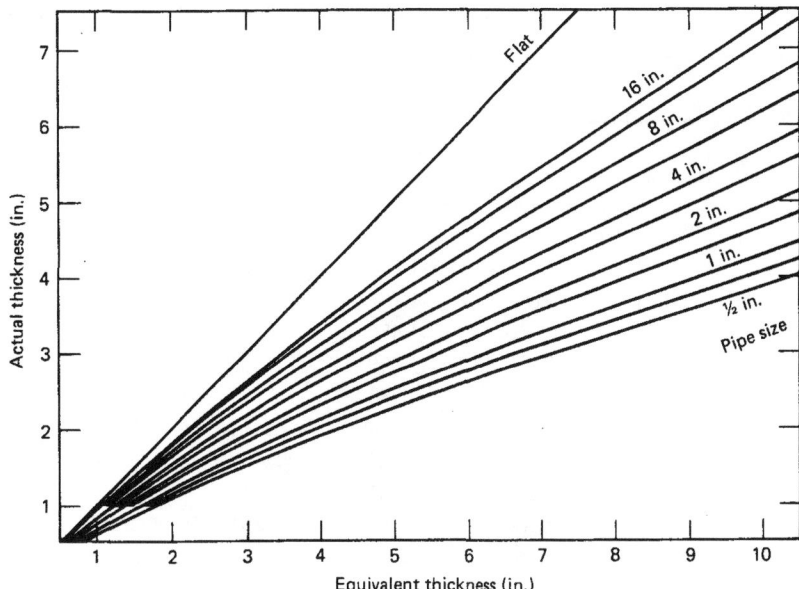

Figure 15-3. Equivalent thickness chart. (From Ref. 16.)

Table 15-2. Equivalent Thickness Values for Even Insulation Thicknesses.

Nominal Pipe Size (in.)	r_1	Actual Thickness (in.)						
		1	1-1/2	2	2-1/2	3	3-1/2	4
1/2	0.420	1.730	2.918	4.238	5.662	7.172	8.755	10.402
3/4	0.525	1.626	2.734	3.966	5.297	6.712	8.199	9.747
1	0.658	1.532	2.563	3.711	4.953	6.275	7.665	9.117
1-1/4	0.830	1.447	2.405	3.472	4.626	5.856	7.153	8.507
1-1/2	0.950	1.403	2.321	3.342	4.449	5.629	6.872	8.171
2	1.188	1.337	2.195	3.148	4.177	5.276	6.436	7.648
2-1/2	1.438	1.287	2.099	2.997	3.968	5.001	6.093	7.234
3	1.750	1.242	2.012	2.858	3.771	4.742	5.768	6.840
3-1/2	2.000	1.217	1.959	2.772	3.649	4.582	5.564	6.592
4	2.250	1.194	1.916	2.704	3.549	4.448	5.396	6.386
4-1/2	2.500	1.178	1.880	2.645	3.464	4.337	5.253	6.211
5	2.781	1.163	1.846	2.590	3.388	4.231	5.118	6.043
6	3.313	1.138	1.799	2.510	3.270	4.071	4.911	5.790
7	3.813	1.120	1.761	2.453	3.184	3.956	4.759	5.604
8	4.313	1.108	1.737	2.407	3.116	3.863	4.644	5.452
9	4.813	1.097	1.714	2.369	3.056	3.783	4.541	5.330
10	5.375	1.088	1.693	2.333	3.007	3.714	4.450	5.214
11	5.875	1.079	1.675	2.305	2.972	3.663	4.383	5.123
12	6.375	1.076	1.662	2.286	2.936	3.619	4.321	5.048
14	7.000	1.069	1.647	2.265	2.900	3.569	4.258	4.969
16	8.000	1.059	1.639	2.231	2.858	3.504	4.178	4.866
18	9.000	1.053	1.622	2.206	2.822	3.449	4.110	4.776
201	0.000	1.048	1.608	2.188	2.789	3.411	4.051	4.711
24	12.000	1.040	1.589	2.163	2.736	3.347	3.971	4.598
30	15.000	1.032	1.572	2.122	2.704	3.281	3.874	4.497

Source: Ref. 1   6.

Table 15-3. Equivalent Thickness Values for Simplified Insulation Thicknesses.

Nominal Pipe Size (in.)	Actual Thickness (in.)							
	$r1$	1	1-1/2	2	2-1/2	3	3-1/2	4
1/2	0.420	1.730	3.053	4.406	6.787	8.253	9.972	12.712
3/4	0.523	1.435	2.660	3.885	5.996	7.447	8.965	10.642
1	0.638	1.715	2.770	4.013	5.358	6.702	8.112	9.581
1-1/4	0.830	1.281	2.727	3.333	4.552	5.777	7.070	8.420
1-1/2	0.950	1.457	2.382	4.025	5.253	6.476	7.759	9.179
2	1.188	1.438	2.367	3.398	4.446	5.561	6.733	8.027
2-1/2	1.438	1.383	2.765	3.657	4.737	5.815	7.015	8.195
3	1.750	1.286	2.114	2.968	3.889	4.868	5.965	7.046
3-1/2	2.000	1.625	2.459	3.258	4.166	5.251	6.266	7.256
4	2.230	1.281	2.010	2.806	3.659	4.059	5.577	6 543
4-1/2	2.300	1.564	2.351	3.152	4.905	4.962	5.907	7 080
5	2.781	1.202	1.893	2.639	3.489	4.339	5.230	6.461
6	3.313	1.138	1.799	2.555	3.317	4.122	5.237	6.015
7	3.813		1.804	2.495	3.230	4.153	4.969	5.821
8	4.313		1.776	2.445	3.391	4.010	4.842	5.768
9	4.813		1.752	2.579	3.232	3.971	4.786	5.583
10	5.375		1.810	2.457	3.108	3.850	4.591	5.361
11	5.875		1.793	2.428	3.140	3.793	4.519	5.271
12	6.375		1.777	2.405	3.103	3.745	4.456	5.241
14	7.000		1.647	2.265	2.900	3.569	4.258	4.969
16	8.000		1.639	2.231	2.858	3.504	4.178	4.866
18	9.000		1.622	2.206	2.822	3.449	4.110	4.776
20	10.000		1.608	2.188	2.789	3.411	4.051	4.711
24	12.000		1.589	2.163	2.736	3.347	3.971	4.598
30	15.000		1.572	2.122	2.704	3.281	3.874	4.497

Source: Ref. 16.

Table 15-4. R_s Values[a] (hr · ft^2 °F/Btu).

$t_s - t_a$ (°F)	*Still Air* Plain, Fabric, Dull Metal: $\varepsilon = 0.95$	Aluminum: $\varepsilon = 0.2$	Stainless Steel: $\varepsilon = 0.4$
10	0.53	0.90	0.81
25	0.52	0.88	0.79
50	0.50	0.86	0.76
75	0.48	0.84	0 75
100	0.46	0.80	0 72
With Wind Velocities Wind Velocity (mph)			
5	0.35	0.41	0.40
10	0.30	0.35	0.34
20	0.24	0.28	0.27

Source: Courtesy of Johns-Manville, Ref. 16.
[a]For heat-loss calculations, the effect of R_s is small compared to R_p, so the accuracy of R_s is not critical. For surface temperature calculations, R_s is the controlling factor and is therefore quite critical. The values presented in Table 15-4 are commonly used values for piping and flat surfaces. More precise values based on surface emittance and wind velocity can be found in the references.

Example. For a 4-in. pipe operating at 700°F in an 85°F ambient temperature with aluminum jacketing over the insulation, determine the thickness of calcium silicate that will keep the surface temperature below 140°F. Since this is a pipe, the equivalent thickness must first be calculated and then converted to actual thickness.

STEP 1. Determine k at $t_m = (700 + 140)/2 = 420°F$. $k = 0.49$ from Table 15-1 or appendix Figure 15-A1 for calcium silicate.

STEP 2. Determine R_s from Table 15-4 for aluminum. $t_s - t_a = 140 - 85 = 55$. So $R_s = 0.85$.

STEP 3. Calculate Eq tk:

$$\text{Eq tk} = (0.49)(0.85) \ \frac{700 - 140}{140 - 85}$$

$$= 4.24 \text{ in.}$$

STEP 4. Determine the actual thickness from Table 15-2. The effect of 4.24 in. on a 4-in. pipe can be accomplished by using 3 in. of insulation.

Note: Thickness recommendations are always increased to the next 1-in. increment. If a surface temperature calculation happens to fall precisely on an even increment (such as 3 in.), it is advisable to be conservative and increase to the next increment (such as 3-1/2 in.). This reduces the criticality of the R_s number used. In the preceding example, it would not be unreasonable to recommend 3-1/2 in. of insulation since it was found to be so close to 3 in.

To illustrate the effect of surface type, consider he same example with a mastic coating.

Example. From Table 15-4, $R_s = 0.50$, so

$$\text{Eq tk} = (0.49)(0.50) \ \frac{700 - 140}{140 - 85}$$

$$= 2.49 \text{ in.}$$

This corresponds to an actual thickness requirement on a 4-in. pipe of 2 in. This compares with 3 in. required for an aluminum-jacketed system. It is of interest to note that even though the aluminum system has a higher surface temperature, the actual heat loss is less because of the higher surface resistance value.

Graphical Method

The calculations illustrated above can also be carried out using graphs which set the heat loss through the insulation equal to the heat loss off the surface, following the discussion in Section 15.4.2.

Figure 15-4 will be used for several different calculations. The following example gives the four-step procedure for achieving the desired surface temperature for personnel protection. The accompanying diagram outlines this procedure.

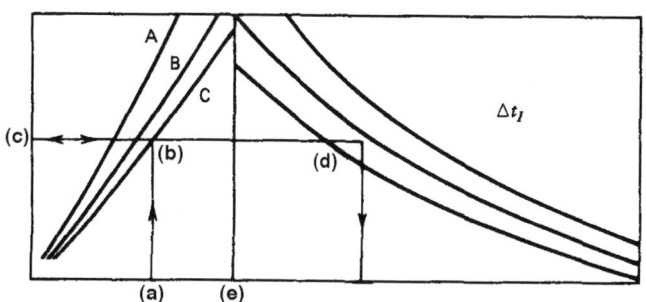

Example. We follow the procedure of the first example, again using aluminum jacketing.

STEP 1. Determine $t_s - t_a$, $140 - 85 = 55°F$.

STEP 2. In the diagram, proceed vertically from (a) of $\Delta t = 55$ to the curve for aluminum jacketing (b).

STEP 2a. Although not required, read the heat loss $Q = 65$ Btu/hr ft² (c).

STEP 3. Proceed to the right to (d), the appropriate curve for $t_h - t_s = 700 - 140 = 560°F$. Interpolate between lines as necessary.

STEP 4. Proceed down to read the required insulation resistance $R_t = 8.6$ at (e). Since $R = $ tk/k or Eq tk/k,

$$\text{tk } or \text{ Eq tk} = R_t k$$

$$t_m = \frac{700 + 140}{2} = 420°F$$

$k = 0.49$ from appendix Figure 15-A1 and

$$\text{tk } or \text{ Eq tk} = (8.6)(0.49) = 4.21 \text{ in.}$$

which compares well with the 4.24 in. from the earlier calculation.

The conversion of Eq tk to actual thickness required for pipe insulation is done in the same manner, using Figure 15-3.

A better understanding of the procedure involved in utilizing this quick graphical method will be obtained after working through the remainder of the calculations in this section.

15.4.4 Condensation Control

On cold systems, either piping or equipment, insulation must be employed to prevent moisture in the warmer surrounding air from condensing on the colder surfaces. The insulation must be of sufficient thickness to keep the insulation surface temperature above the dew point of the surrounding air. Essentially, the calculation procedures are identical to those for personnel protection except that the dew point temperature is substituted for the desired surface temperature. (*Note:* The surface temperature should be kept 1 or 2° above the dew point to prevent condensation at that temperature.)

Dew Point Determination

The condensation (saturation) temperature, or dew point, is dependent on the ambient dry-bulb and wet-bulb temperatures. With these two values and the use of a psychrometric chart, the dew point can be determined. However, for most applications, the relative humidity is more readily attainable, so the dew point is determined using dry-bulb temperature and relative humidity rather than wet-bulb. Table 15-5 is used to find the proper dew-point temperature.

Calculation

This equation is identical to the previous surface-temperature problem except that the surface temperature t_s now takes on the value of the dew point of the ambient air. Also, t_h now represents the cold operating temperature.

$$\text{tk or Eq tk} = kR_s \left(\frac{t_h - t_s}{t_s - t_a} \right)$$

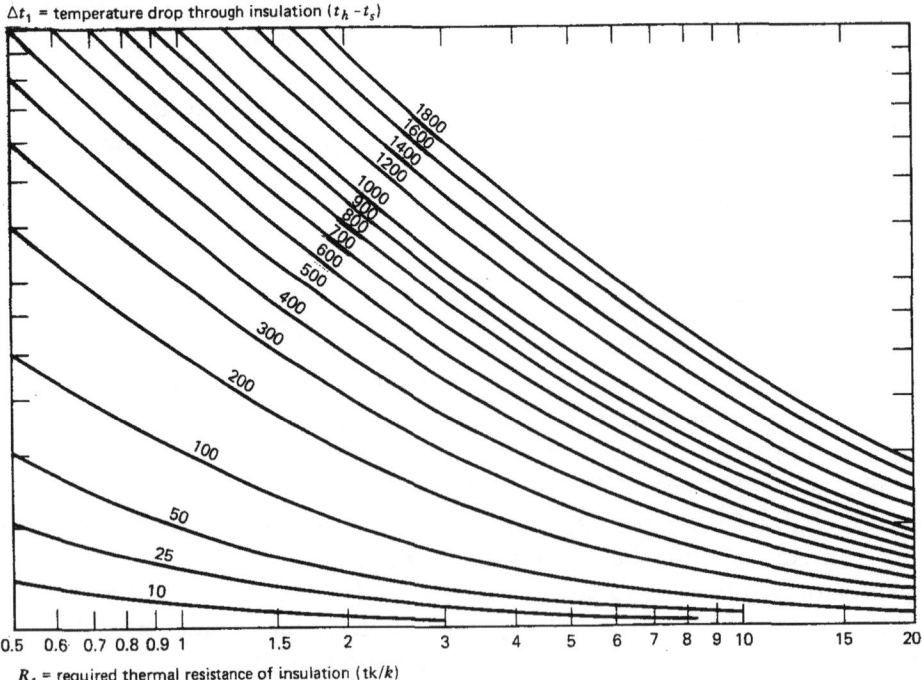

Figure 15-4. Heat Loss and Surface Temperature Graphical Method. (From Ref. 16.)

Example. For a 6-in.-diameter chilled-water line operating at 35°F in an ambient of 90°F and 85% RH, determine the thickness of fiberglass pipe insulation with a composite kraft paper jacket required to prevent condensation.

STEP 1. Determine the dew point (DP) using either a psychrometric chart or Table 15-5. DP at 90°F and 85% RH = 85°F. (In Step 5, the thickness is rounded up, which yields a higher temp.)

STEP 2. Determine k at $t_m = (35 + 85)/2 = 60°F$. k at 60°F = 0.23, from Table 15-1 or appendix Figure 15-A2.

STEP 3. Determine Rs from Table 15-4. Δt here is $(t_{a'} - t_s)$ rater than $(t_s - t_a)$, $t_a - t_s = 90 - 85 = 5°F$, $R_s = 0.54$.

STEP 4. Calculate Eq tk.

$$Eq\ tk = (0.23)(0.54)\frac{35-85}{85-90}$$
$$= 1.24\ in.$$

STEP 5. Determine the actual thickness from Figure 15-2 for 6-in. pipe, 1.24 in. Eq tk. The actual thickness is 1.5 in.

Graphical Method
The graphical procedures are as described in Section 15.4.3. As the applications become colder, it is apparent that the required insulation thicknesses will become larger, with R_I values toward the right side of Figure 15-4. It is suggested that the graphical procedure not be used when the resulting R_I values must be determined from a very flat portion of the $(t_h - t_s)$ curve (anytime the numbers are to the far right of Figure 15-4). It is difficult to read the graph with sufficient accuracy, particularly in light of the simplicity of the mathematical calculation.

Thickness Chart for Fiberglass Pipe Insulation
Table 15-6 gives the thickness requirements for fiberglass pipe insulation with a white, all-purpose jacket in still air. The calculations are based on the lowest

Table 15-5. Dew Point Temperature.

Dry-Bulb Temp. (°F)	Percent Relative Humidity																		
	10	15	20	25	30	35	40	45	50	55	60	65	70	75	80	85	90	95	100
5	−35	−30	−25	−21	−17	−14	−12	−10	−8	−6	−5	−4	−2	−1	1	2	3	4	5
10	−31	−25	−20	−16	−13	−10	−7	−5	−3	−2	0	2	3	4	5	7	8	9	10
15	−28	−21	−16	−12	−8	−5	−3	−1	1	3	5	6	8	9	10	12	13	14	15
20	−24	−16	−8	−4	−2	2	4	6	8	10	11	13	14	15	16	18	19	20	
25	−20	−15	−8	−4	0	3	6	8	10	12	15	16	18	19	20	21	23	24	25
30	−15	−9	−3	2	5	8	11	13	15	17	20	22	23	24	25	27	28	29	30
35	−12	−5	1	5	9	12	15	18	20	22	24	26	27	28	30	32	33	34	35
40	−7	0	5	9	14	16	19	22	24	26	28	29	31	33	35	36	38	39	40
45	−4	3	9	13	17	20	23	25	28	30	32	34	36	38	39	41	43	44	45
50	−1	7	13	17	21	24	27	30	32	34	37	39	41	42	44	45	47	49	50
55	3	11	16	21	25	28	32	34	37	39	41	43	45	47	49	50	52	53	55
60	6	14	20	25	29	32	35	39	42	44	46	48	50	52	54	55	57	59	60
65	10	18	24	28	33	38	40	43	46	49	51	53	55	57	59	60	62	63	65
70	13	21	28	33	37	41	45	48	50	53	55	57	60	62	64	65	67	68	70
75	17	25	32	37	42	46	49	52	55	57	60	62	64	66	69	70	72	74	75
80	20	29	35	41	46	50	54	57	60	62	65	67	69	72	74	75	77	78	80
85	23	32	40	45	50	54	58	61	64	67	69	72	74	76	78	80	82	83	85
90	27	36	44	49	54	58	62	66	69	72	74	77	79	81	83	85	87	89	90
95	30	40	48	54	59	63	67	70	73	76	79	82	84	86	88	90	91	93	95
100	34	44	52	58	63	68	71	75	78	81	84	86	88	91	92	94	96	98	100
105	38	48	56	62	67	72	76	79	82	85	88	90	93	95	97	99	101	103	105
110	41	52	60	66	71	77	80	84	87	90	92	95	98	100	102	104	106	108	110
115	45	56	64	70	75	80	84	88	91	94	97	100	102	105	107	109	111	113	115
120	48	60	68	74	79	85	88	92	96	99	102	105	107	109	112	114	116	118	120
125	52	63	72	78	84	89	93	97	100	104	107	109	111	114	117	119	121	123	125

temperature in each temperature range. Three temperature/humidity conditions are depicted.

15.4.5 Process Control

Included under this heading will be all the calculations other than those for surface temperature and economics. It is often necessary to calculate the heat flow through a given insulation thickness, or conversely, to calculate the thickness required to achieve a certain heat flow rate. The final situation to be addressed deals with temperature drop in both stagnant and flowing systems.

Heat Flow for a Specified Thickness
Calculation Equations. Again, the basic equation for a single insulation material is

$$Q_F = \frac{t_h - t_a}{R_I + R_s}$$

Example. For an 850°F boiler operating indoors in an 80°F ambient temperature insulated with 4 in. of calcium silicate covered with 0.016 in. aluminum jacketing, determine the heat loss per square foot of boiler surface and the surface temperature.

STEP 1. Find k for calcium silicate at t_m. Assume that $t_s = 140°F$. Then $t_m = (850 + 140)/2 = 495°F$, k at 495°F = 0.53, from Table 15-1 or appendix Figure 15-A1.

STEP 2. Determine R_s for aluminum from Table 15-4. $t_s - t_a = 140 - 0 = 60°F$, so $R_s = 0.85$.

STEP 3. Calculate $R_I = 4/0.53 = 7.5$.

STEP 4. Calculate

$$QF = \frac{850 - 80}{7.5 + 0.85} = 92 \text{ Btu/hr} \bullet \text{ft2}$$

STEP 5. Calculate the surface temperature *ts*, as follows:

$$
\begin{aligned}
R_s \times Q_F &= t_s - t_a \\
(R_s \times Q_F) + ta &= ts \\
t_s &= (0.85 \times 92) + 80 \\
&= 158°F
\end{aligned}
$$

STEP 6. Calculate t_m to check assumption and to check the k value used.

$$t_m = \frac{850 - 80}{2}$$

$$= 504°F$$

Since k at 504°F = k at 495°F (assumed) = 0.53, the assumption is okay. A check on R_s can also be made based on the calculated surface temperature.

STEP 7. If the assumption is not okay, recalculate using a new k value based on the new t_m.

The Q_F used above is for flat surfaces. In determining heat flow from a pipe, the same equations are used with Eq tk substituted for tk in the R_I calculation as discussed in Section 15.4.2. Often it is desired to express pipe heat losses in terms of Btu/hr-lin.-ft rather than

Table 15-6. Fiberglass Pipe Insulation: Minimum Thickness to Prevent Condensation[a].

Operating Pipe Temperature (°F)	80°F and 90% RH		80°F and 70% RH		80°F and 50% RH	
	Pipe Size (in.)	Thickness (in.)	Pipe Size (in.)	Thickness (in.)	Pipe Size (in.)	Thickness (in.)
0-34	Up to 1	2	Up to 8	1	Up to 8	1/2
	1-1/4 to 2	2-1/2	9-30	1-1/2	9-30	1
	2-1/2 to 8	3				
	9-30	3-1/2				
35-49	Up to 1-1/2	1-1/2	Up to 4	1/2	Up to 30	1/2
	2-8	2	412-30			
	9-30	3				
50-70	Up to 3	1-1/2	Up to 30	1/2	Up to 30	1/2
	3-1/2 to 20	2				
	21 -30	2-1/2				

Source: Courtesy of Johns-Manville, Ref. 16.
[a]Based on still air and AP Jacket.

Btu/hr ft². This is termed Q_P, with

$$Q_P - Q_F \left(\frac{2\pi r_2}{12} \right)$$

Graphical Method. Figure 15-4 may again be used in lieu of calculations. The main difference from the previous chart usage is that surface temperature is now an unknown, and it must be determined such that thermal equilibrium exists.

Example. Determine the heat loss from the side walls of a vessel operating at 300°F in an 80°F ambient temperature. Two inches of 3-lb/ft³ fiberglass is used with aluminum lagging.

STEP 1. Assume a surface temperature t_s = 120°F.

STEP 2. Calculate

$$t_m = \frac{t_h = t_s}{2} = \frac{300 + 120}{2} = 210°F$$

Determine k from appendix Figure 15-A3 at 210°F. k = 0.27.

STEP 3. Calculate R_I = tk/k = 2/0.27 = 7.41.

STEP 4. Go to position (a) on the chart shown for RI = 7.41 and read vertically to (b), where $t_h - t_s$ = 180°F.

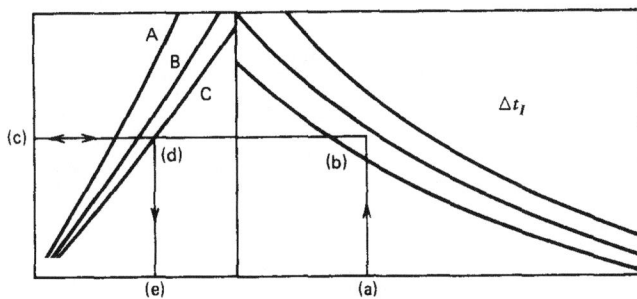

STEP 5. Read to the left to (c) for heat loss Q = 24 Btu/hr ft².

STEP 6. Read down from the proper surface curve from (d) to (e), which represents $ts - ta$, to check the surface temperature assumption. For aluminum, $ts - ta$ (chart) is 21°F, compared with the 120 − 80 = 40°F assumption.

STEP 7. Calculate a new surface temperature, 80 + 21 = 101°F; then calculate a new tm, = (300 + 101)/2

= 200.5°F. Then find a new k = 0.26, which gives a new RI = 2/126 = 7.69.

STEP 8. Return to step 4 with the new RI and proceed. This example shows the insensitivity of heat loss to changes in surface temperature since the new Q = 22 Btu/hr ft².

For pipe insulation, the same procedure is followed except that RI is calculated using the equivalent thickness. Also, conversion to heat loss per linear foot must be done separately after the square-foot loss is determined.

Thickness for a Specified Heat Loss

Again, a surface temperature t_s must first be assumed and then checked for accuracy at the end of the calculation.

From Section 15.4.2,

$$Q = \frac{t_h - t_s}{R_I} = \frac{t_h - t_s}{tk/k}$$

$$tk \text{ (or Eq th)} = k \left(\frac{t_h - t_s}{Q} \right)$$

where k is determined at $t_m = (t_h + t_s)/2$.

Example. How much calcium silicate insulation is required on a 650°F duct in an 80°F ambient temperature if the maximum heat loss is 50 Btu/hr ft²? The insulation will be finished with a mastic coating.

STEP 1. Assume that t_s = 105°F. So t_m = (650 + 105)/2 = 377°F. k from Table 15-1 or appendix Figure 15-A1 at 377°F = 0.46.

STEP 2. Find tk as follows:

$$tk = k \left(\frac{t_h - t_s}{Q} \right) = 0.46 \left(\frac{650 - 105}{50} \right)$$

$$= 5.01 \text{ in.}$$

STEP 3. Check surface temperature assumption by $t_s = (Q \times R_s) + t_a$ using R_s = 0.52. From Table 15-4 for a mastic finish,

$$\begin{aligned} t_s &= 50(0.52) + 80 \\ &= 106°F \end{aligned}$$

(Note that this in turn changes the $t_s - t_a$ from 40 to 25, which changes R_s from 0.49 to 0.51, which is insignificant.)

For a graphical solution to this problem, Figure 15-4 is again used. It is simply a matter of reading across the desired Q level and adjusting the t_s and R_I values to reach equilibrium. Thickness is then determined by $tk = kR_I$.

Temperature Drop in a System

The following discussion is quite simplified and is not intended to replace the service of the process design engineer. The material is presented to illustrate how insulation ties into the process design decision.

Temperature Drop in Stationary Media over Time. The procedure calls for standard heat-flow calculations now tied into the heat content of the fluid. To illustrate, consider the following example.

Example. A water storage tank is calculated to have a surface area of 400 ft^2 and a volume of 790 ft^3. How much will the temperature drop in a 72-hr period with an ambient temperature of 0°F, assuming that the initial water temperature is 50°F? The tank is insulated with 2-in. fiberglass with a mastic coating.

Before proceeding, realize that the maximum heat transfer will occur when the water is at 50°F. As it drops in temperature, the heat-transfer rate is reduced due to a smaller temperature difference. As a first approximation, it is reasonable to use the maximum heat transfer based on 50°F. Then if the temperature drop is significant, an average water temperature can be used in the second iteration.

STEP 1. Assume a surface temperature, calculate the mean temperature, find the k factor from Table 15-1 or appendix Figure 15A.3, and determine R_s from Table 15-4. With $t_s = 10°F$, $_m = 30°F$, $k = 0.22$, and $R_s = 0.53$.

STEP 2. Calculate heat loss with fluid at 50°F.

$$Q_F = \frac{50 - 0}{(2/0.22) + 0.53} = 5.2 \text{ Btu/hr} \cdot \text{ft}^2$$

$$Q_F = Q_F \times A = 5.2 \, (400) = 2080 \text{ Btu/hr}$$

STEP 3. Calculate the amount of heat that must be lost for the entire volume of water to drop 1°F.

Available heat per °F
= volume × density × specific heat
= 790 ft^3 × 62.4 lb/ft^3 × 1 Btu/lb F
= 49,296 Btu/°F

STEP 4. Calculate the temperature drop in 72 hr by determining the total heat flow over the period: Q

= 2080 × 72 = 149,760 Btu. Divide this by the available heat per 1°F drop:

$$\frac{149,760 \text{ Btu}}{49,296 \text{ Btu/°F}} = 3.04°F \text{ drop}$$

This procedure may also be used for fluid lying stationary in a pipeline. In this case it is easiest to do all the calculations for 1 linear foot rather than for the entire length of pipe.

One conservative aspect of this calculation is that the heat capacity of the metal tank or pipe is not included in the calculation. Since the container will have to decrease in temperature with the fluid, there is actually more heat available than was used above.

Temperature Drop in Flowing Media. There are two common situations in this category, the first involving flue gases and the second involving water or other fluids with a thickening or freezing point. This section discusses the flue-gas problem, and the following section, freeze protection.

A problem is encountered with flue gases that have fairly high condensation temperatures. Along the length of a duct run, the temperature will drop, so insulation is added to control the temperature drop. This calculation is actually a heat balance between the mass flow rate of energy input and the heat loss energy outlet.

For a round duct of radius r_1 and length L, gas enters at t_h, and must not drop below t_{min} (the dew point). The flow rate is M lb/hr and the gas has a specific heat of C_p Btu/lb °F. Therefore, the maximum allowable heat loss in Btu/hr is

$$Q_t = MC_p\Delta t = MC_p(t_h - t_{min})$$

Also,

$$Q_T = Q_P \times L = \frac{t_h - t_a}{R_I - R_s} \times \frac{2\pi r_2}{12} \times L$$

where

$$t_h = \left(\frac{t_{in} + t_{out}}{2}\right)$$

(A conservative simplification would be to set $t_h = t_{in}$ since the higher temperature, t_{in}, will cause a greater heat loss.)

To simplify on large ducts, assume that $r_1 = r_2$ (ignore the insulation-thickness addition to the surface area). Therefore,

$$\frac{t_h - t_a}{R_I - R_s} \times \frac{2\pi r_1}{12} \times L = MC_p(t_h - t_{min})$$

and

$$R_I + R_s = \left(\frac{t_h - t_a}{t_h - t_{min}} \right) \times \frac{2\pi r_1}{12} \times \frac{L}{MC_P}$$

Therefore,

$$R_I = \left[\frac{t_h - t_a}{MC_p(t_h - t_{min})} \times \frac{2\pi r_1}{12} \times L \right] - R_s$$

$$= \frac{tk}{k} \therefore tk = R_I \times k$$

Example. A 48-in.-diameter duct 90 ft long in a 60°F ambient temperature has gas entering at 575°F and 15,000 cfm. The gas density standard conditions is 0.178 lb/ft³ and the gas outlet must not be below 555°F. $Cp = 0.18$ Btu/lb °F. Determine the thickness of calcium silicate required to keep the outlet temperature above 565°F, giving a 10°F buffer to account for the interior film coefficient. A more sophisticated approach calculates an interior film resistance R_s instead of using a 10°F or larger buffer. The resulting equation for Qp would be

$$Q_p = \frac{t_h - t_a}{R_s(\text{inferior}) + R_I + R_s} \times \frac{2\pi r_2}{12}$$

This equation, however, will not be used.

STEP 1. Determine t_h the average gas temperature $= (575 + 565)/2 = 570°F$. (A logarithmic mean could be calculated for more accuracy, but it is usually not necessary.)

STEP 2. Determine M lb/hr. The flow rate is 15,000 cfm of hot gas (570°F). At standard conditions 1 atm, (70°F), the flow rate must be determined by the absolute temperature ratio:

$$\frac{t_h + 460}{70 + 460} = \frac{15,000}{\text{std. flow}}$$

$$\text{Std. flow} = 15,000 \left(\frac{70 + 460}{570 + 460} \right)$$

$$= 6262 \text{ cfm std. gas (or scfm)}$$

$$M = 6262 \text{ cfm} \times 0.178 \text{ lb/ft}^3 \times 60 \text{ min/hr}$$

$$= 66,878 \text{ lb/hr}$$

STEP 3. Determine Rs from Table 15-4 assuming t_s $= 80°F$ and a dull surface $R_s = 0.5$.

STEP 4. Calculate R_I.

$$R_I = \frac{570 - 60}{(66,878)\ (0.18)\ (575–565)} \times \frac{2\pi 24}{12} \times 90 - 0.52$$

$$= 4.79 - 0.52$$

$$= 4.27$$

STEP 5. Calculate the thickness. Assume that $t_s = 80°F$.

$$t_m = \frac{570 + 80}{2} = 325°F$$

k at $325°F = \begin{array}{l} 0.45 \text{ for calcium silicate from} \\ \text{Appendix Figure 15-A1.} \end{array}$

$$tk = R_I \times k = 4.27 \times 0.45$$
$$= 1.93 \text{ in.}$$

STEP 6. The thickness required for this application is 2 in. of calcium silicate. Again, a more conservative recommendation would be 2-1/2 in.

Note: The foregoing calculation is quite complex. It is, however, the basis for many process control and freeze-prevention calculations. The two equations for Q, can be manipulated to solve for the following:

Temperature drop, based on a given thickness and flow rate.

Minimum flow rate, based on given thickness and temperature drop.

Minimum length, based on thickness, flow rate, and temperature drop.

Freeze Protection. Four different calculations can be performed with regard to water-line freezing (or the unacceptable thickening of any fluid).

1. Determine the time required for a stagnant, insulated water line to reach 32°F.

2. Determine the amount of heat tracing required to prevent freezing.

3. Determine the flow rate required to prevent freezing of an insulated line.

4. Determine the insulation required to prevent freezing of a line with a given flow rate.

Calculations 1 and 2 relate to Section 15.4.5, where we dealt with stationary media. To apply the same principles to the freeze problems, the following modifications should be made.

a. In calculation 1, the heat transfer should be based on the average water temperature between the starting temperature and freezing:

$$t_h = \frac{t_{start} + 32}{2}$$

b. Rather than solving for temperature drop, given the number of hours, the hours are determined based on

$$\text{hours to freeze} = \frac{\text{available heat}}{\text{heat loss/hr}} \quad \frac{\text{Btu}}{\text{Btu/hr}}$$

where available heat is $WCp\ \Delta t$, with

W= lb of water
Cp= specific heat of water (1 Btu/lb °F)
$\Delta t = t_{start} - 32$

c. In calculation 2, the heat-loss value should be calculated based upon the minimum temperature at which the system should stay, for example, 35°F. The heat tracing should provide enough heat to the system to offset the naturally occurring losses of the pipe. Heat-trace calculations are quite complex and many variables are involved. References 8 and 10 should be consulted for this type of work.

Calculations 3 and 4 relate to Section 4.5.3, dealing with flows. In the case of water, the minimum temperature can beset at 32°F, and the heat-transfer rate is again on an operating average temperature

$$t_h = \frac{t_{start} + 32}{2}$$

The equations given can be manipulated to solve for flow rate or insulation thickness.

As an aid in estimating the amount of insulation for freeze protection, Table 15-7 shows both the hours to freezing and the minimum flow rate to prevent freezing, based on different insulation thicknesses. These figures are based on an initial water temperature of 42°F, an ambient temperature of – 10°F, a surface resistance of 0.54, and a thermal conductivity for fiberglass pipe insulation of k = 0.23.

15.4.6 Operating Conditions

Like all other calculations, heat-transfer equations yield results that are only as accurate as the input variables used. The operating conditions chosen for the heat-transfer calculations are critical to the result, and very misleading conclusions can be drawn if improper conditions are selected.

The term "operating conditions" refers to the environment surrounding the insulation system. Some of the variable conditions are operating temperature, ambient temperature, relative humidity, wind velocity, fluid type, mass flow rate, line length, material volume, and others. Since many of these variables are constantly changing, the selection of a proper value must be made on some logical basis. Following are three suggested methods for determining the appropriate variable values.

1. **Worst Case.** If a severe failure might occur with insufficient insulation, a worst-case approach is probably warranted. For example, freeze protection should obviously be based on the historical temperature extremes rather than on yearly averages. Similarly, exterior condensation control should be based on both ambient temperature and humidity extremes, in addition to the lowest operating temperature. The *ASHRAE Handbook of Fundamentals* as well as U.S. Weather Bureau data give proper design conditions for most locales. In process areas, an appropriate example involves flue-gas condensation. Here the minimum flow rate is the most critical and should be used in the calculation.

As a general rule, worst-case conditions will result in greater insulation thickness than will average conditions. In some cases the difference is very substantial, so it is important to determine initially if a worst-case calculation is required.

2. **Worst Season Average.** When a heating or cooling process is only operating part of the year, it is sensible to consider the average conditions only during that period of time. However, in year-round operations, a seasonal average is also justified in many cases. For example, personnel protection requires a maximum surface temperature that is dependent on the ambient air temperature. Taking the average summer daily maximum temperature is more practical than taking the absolute maximum ambient that could occur. The following example illustrates this.

Example. Consider an 8-in.-diameter, 600°F waste-heat line operating indoors with an average daily high of 80°F (but occasionally it will be 105°F). To maintain the surface below 135°F, 2 in. of calcium silicate is required with the 80°F ambient, whereas 3-1/2 in. is required with the 105°F ambient. The difference is significant and must be weighed against the benefit of the additional insulation in terms of worker safety.

3. **Yearly Average.** Economic calculations for continuously operating equipment should be based

Table 15-7. Hours to Freeze and Flow Rate Required to Prevent Freezing[a].

Nominal Pipe Size (in.)	1 in		2 in.		3 in	
	Hours to Freeze	gpm/100 ft	Hours to Freeze	gpm/100 ft	Hours to Freeze	gpm/100 ft
1/2	0.30	0.087	0.42	0.282	0.50	0.053
3/4	0.47	0.098	0.66	0.070	0.79	0.058
1	0.66	0.113	0.96	0.078	1.16	0.065
1-1/2	0.90	0.144	1.35	0.096	1.67	0.078
2	1.72	0.169	2.64	0.110	3.31	0.088
2-1/2	2.13	0.195	3.33	0.124	4.24	0.098
3	2.81	0.228	4.50	0.142	5.80	0.110
4	3.95	0.279	6.49	0.170	8.49	0.130
5	5.21	0.332	8.69	0.199	11.54	0.150
6	6.48	0.386	10.98	0.228	14.71	0.170
7	7.66	0.437	13.14	0.255	17.75	0.189
8	8.89	0.487	15.37	0.282	20.89	0.207

Source: Ref. 16.
[a]Calculations based on fiberglass pipe insulation with k = 0.23, initial water temperature of 42°F, and ambient air temperature of − 10°F. Flow rate represents the gallons per minute required in a 100-ft pipe and may be prorated for longer or shorter lengths.

on yearly average operating conditions rather than on worst-case design conditions. Since the intent is to maximize the owner's financial return, an average condition will not overstate the savings as the worst case or worst season might. A good approach to process work is to calculate the economic thickness based on yearly averages and then check the sufficiency of that thickness under the worst-case design conditions. That way, both criteria are met.

15.4.7 Bare-surface Heat Loss

It is often desirable to determine if any insulation is required and also to compare bare surface losses with those using insulation. Table 15-8 gives bare-surface losses based on the temperature difference between the surface and ambient air. Actual temperature conditions between those listed can be arrived at by interpolation. To illustrate, consider a bare, 8-in.-diameter pipe operating at 250°F in an 80°F ambient temperature. $\Delta t = 250 - 80 = 170°F$. Q for Δt of 150°F = 812.5 Btu/ hr-lin.-ft; Q for Δt of 200°F = 1203 Btu/hr lin. ft. Interpolating between 150 and 200°F gives

$$Q_{170} = Q_{150} + (2/5)(Q_{200} - Q_{150})$$

$$= 812.5 + 0.4(1203 - 812.5)$$

$$= 968.7 \text{ Btu/hr lin. ft}$$

15.5 INSULATION ECONOMICS

Thermal insulation is a valuable tool in achieving energy conservation. However, to strive for maximum energy conservation without regard for economics is not acceptable. There are many ways to manipulate the cost and savings numbers, and this section explains the various approaches and the pros and cons of each.

15.5.1 Cost Considerations

Simply stated, if the cost of insulation can be recouped by a reduction in total energy costs, the insulation investment is justified. Similarly, if the cost of additional insulation can be recouped by the additional energy-cost reduction, the expenditure is justified. There is a significant difference between the "full thickness" justification and the "incremental" justification. This is discussed in detail in Section 15.5.3. The following discussions will generally use the incremental approach to economic evaluation.

Insulation Costs

The insulation costs should include everything that it takes to apply the material to the pipe or vessel and to properly cover it to finished form. Certainly, it is more costly to install insulation 100 ft in the air than it is from ground level, and metal jackets are more costly than all-purpose indoor jackets. Anticipated maintenance costs

Table 15-8. Heat Loss from Bare Surfaces[a].

Normal Pipe Size (in.)	Temperature Difference (°F)															
	50	100	150	200	250	300	350	400	450	500	550	600	700	800	900	1000
1/2	22	47	79	117	162	215	279	355	442	541	650	772	1,047	1,364	1,723	2,123
3/4	27	59	99	147	203	269	349	444	552	677	812	965	1,309	1,705	2,153	2,654
1	34	75	124	183	254	336	437	555	691	846	1,016	1,207	1,637	2,133	2,694	3,320
1-1/4	42	94	157	232	321	425	552	702	873	1,070	1,285	1,527	2,071	2,697	3,406	4,198
1-1/2	49	107	179	265	367	487	632	804	1,000	1,225	1,471	1,748	2,371	3,088	3,899	4,806
2	61	134	224	332	459	608	790	1,004	1,249	1,530	1,837	2,183	2,961	3,856	4,870	6,002
2-1/2	74	162	271	401	556	736	956	1,215	1,512	1,852	2,224	2,643	3,584	4,669	5,896	7,267
3	89	197	330	489	677	897	1,164	1,480	1,841	2,256	2,708	3,219	4,365	5,685	7,180	8,849
3-1/2	102	225	377	558	773	1,024	1,329	1,690	2,102	2,576	3,092	3,675	4,984	6,491	8,198	10,100
4	115	254	424	628	869	1,152	1,496	1,901	2,365	2,898	3,479	4,135	5,607	7,304	9,224	11,370
4-1/2	128	282	471	698	965	1,280	1,662	2,113	2,628	3,220	3,866	4,595	6,231	8,116	10,250	12,630
5	142	313	524	776	1,074	1,424	1,848	2,350	2,923	3,582	4,300	5,111	6,931	9,027	11,400	14,050
6	169	373	624	924	1,279	1,696	2,201	2,799	3,481	4,266	5,121	6,086	8,254	10,750	13,580	16,730
7	195	430	719	1,064	1,473	1,952	2,534	3,222	4,007	4,910	5,894	7,006	9,501	12,380	15,630	19,260
8	220	486	813	1,203	1,665	2,207	2,865	3,643	4,531	5,552	6,666	7,922	10,740	13,990	17,670	21,780
9	246	542	907	1,343	1,859	2,464	3,198	4,066	5,057	6,197	7,440	8,842	11,990	15,620	19,720	24,310
10	275	606	1,014	1,502	2,078	2,755	3,576	4,547	5,655	6,930	8,320	9,888	13,410	17,470	22,060	27,180
11	300	661	1,106	1,638	2,267	3,005	3,901	4,960	6,169	7,560	9,076	10,790	14,630	19,050	24,060	29,660
12	326	718	1,202	1,779	2,463	3,265	4,238	5,338	6,701	8,212	9,859	11,720	15,890	20,700	26,140	32,210
14	357	783	1,319	1,952	2,703	3,582	4,650	5,912	7,354	9,011	10,820	12,860	17,440	22,710	28,680	35,350
16	408	901	1,508	2,232	3,090	4,096	5,317	6,759	8,407	10,300	12,370	14,700	19,940	25,970	32,790	40,410
18	460	1,015	1,698	2,514	3,480	4,612	5,987	7,612	9,467	11,600	13,930	16,550	22,450	29,240	36,930	45,510
20	510	1,127	1,885	2,790	3,862	5,120	6,646	8,449	10,510	12,880	15,460	18,380	24,920	32,460	40,990	50,520
24	613	1,353	2,263	3,350	4,638	6,148	7,980	10,150	12,620	15,460	18,570	22,060	29,920	38,970	49,220	60,660
30	766	1,690	2,827	4,186	5,795	7,681	9,971	12,680	15,770	19,320	23,200	27,570	37,390	48,700	61,500	75,790
Flat	98	215	360	533	738	978	1,270	1,614	2,008	2,460	2,954	3,510	4,760	6,200	7,830	9,650

Source: Ref. 16.

[a]Losses given in Btu/hr lin. ft of bare pipe at various temperature differences and Btu/hr-ft^2 for flat surfaces. Heat losses were calculated for still air and $\varepsilon = 0.95$ (plain, fabric or dull metals).

should also be included, based on the material and application involved. The variations in labor costs due to both time and base rate should be evaluated for each particular insulation system design and locale. In other words, insulation costs tend to be job specific as well as differentiated by product.

Lost Heat Costs

Reducing the amount of unwanted heat loss is the function of insulation, and the measurement of this is in Btu. The key to economic analyses rests in the dollar value assigned to each Btu that is wasted. At the very least, the energy cost must include the raw-fuel cost, modified by the conversion efficiency of the equipment. For example, if natural gas costs $2.50/million Btu and it is being converted to heat at 70% efficiency, the effective cost of the Btu is 2.50/0.70 = $3.57/million Btu.

The cost of the heat plant is always a point of discussion. Many calculations ignore this capital cost on the basis that a heat plant will be required whether insulation is used or not. On the other hand, the only purpose of the heat plant is to generate usable Btus. So the cost of each Btu should reflect the capital plant cost ammortized over the life of the plant. The recent trend that seems most reasonable is to assign an incremental cost to increases in capital expenditures. This cost is stated as dollars per 1000 Btu per hour. This gives credit to a well-insulated system that requires less Btu/hr capacity.

Other Costs

As the economic calculations become more sophisticated, other costs must be included in the analysis. The major additions are the cost of money and the tax effect of the project. Involving the cost of money recognizes the real fact that many projects are competing for each investment dollar spent.

Therefore, the money used to finance an insulation project must generate a sufficient after-tax return or the money will be invested elsewhere to achieve such a return. This topic, together with an explanation of the use of discount factors, is discussed in detail in Chapter 4.

The effect of taxes can also be included in the analysis as it relates to fuel expense and depreciation. Since both of these items are expensed annually, the after-tax cost is significantly reduced. The final example in Section 15.5.3 illustrates this.

15.5.2. Energy Savings Calculations

The following procedure shows how to estimate the energy cost savings resulting from installing thermal insulation.

Procedure

STEP 1. Calculate present heat losses (Q_{Tpres}). You can use one of the following methods to calculate the heat losses of the present system:

- Heat flow equations. These equations are in Section 15.4.2.

- Graphical method. Consists of Steps 1, 2 and 2a of the graphical method presented in Section 15.4.3.

- Table values. Table 15-8 presents heat losses values for bare surfaces (dull metals).

STEP 2. Determine insulation thickness (tk). Using Section 15.4, you can determine the insulation thickness according to your specific needs. Depending on the pipe diameter and temperature, the first inch of insulation can reduce bare surface heat losses by approximately 85-95% (Ref. 20). Then, for a preliminary economic evaluation, you can use tk = 1-in. If the evaluation is not favorable, you will not be able to justify a thicker insulation. On the other hand, if the evaluation is favorable, you will need to determine the appropriate insulation thickness and reevaluate the investment.

STEP 3. Calculate heat losses with insulation (Q_{Tins}). Use the equations from Section 15.4.5.

STEP 4. Determine heat loss savings ($Q_{Tsavings}$). Subtract the heat losses with insulation from the present heat losses ($Q_{Tsavings} = Q_{Tpres} - Q_{Tins}$).

STEP 5. Estimate fuel cost savings. Estimate the amount of fuel used to generate each Btu wasted and use this value to calculate the energy cost savings. With this savings, you can evaluate the insulation investment using any appropriate financial analysis method (see Section 15.5.3).

Example. For the example presented in section 15.4.3, determine the fuel cost savings resulting from insulating the pipe with 3-1/2 in. of calcium silicate.

Data

- Pipe data: 4-in pipe operating at 700°F in an 85°F ambient temperature.

- Jacket type: Aluminum.

- Pipe length: 100-ft.

- Operating hours: 4,160 hr/yr

- Fuel data: Natural gas, burned to heat the fluid in the pipe at $3/MCF. Efficiency of combustion is approximately 80%.

STEP 1. Determine present heat loss. From Table 15-8 (4-in. pipe, temperature difference = $t_s - t_a$ = 700 − 85 = 615°F), heat loss = 4,356 Btu/hr-lin.ft. Then,

$$Q_{Tpres} = \text{(heat loss/lin.ft)(length)}$$
$$= (4{,}356 \text{ Btu/hr-lin.ft.})(100 \text{ ft})$$
$$= 435{,}600 \text{ Btu/hr}$$

STEP 2. Determine insulation thickness. In this example, the surface temperature has to be below 140°F, which is accomplished with an insulation thickness tk = 3.5-in.

STEP 3. Determine heat losses with insulation. For this example, we need to calculate the heat losses for tk = 3.5-in following the procedure outlined in Section 15.4.5.

1) From the example in Section 15.4.3, t_s = 140 °F, k = 0.49 and R_s = 0.85.

2) From Table 15-2, Eq tk for 3-1/2-in insulation n a 4-in. pipe = 5.396 in. Then,

$$R_I = \text{Eq tk}/k = 5.396/0.49 = 11$$

3) Calculate heat loss Q_F:

$$Q_F = \frac{700 - 85}{11 + 0.85} = 52 \text{ Btu/hr ft}^2$$

4) Calculate surface temperature t_s:
$$t_s = t_a + R_s \times Q_F = 85 + (0.85 \times 52) = 129°F$$

5) Calculate t_m = (700+129)/2 = 415°F. The insulation thermal conductivity at 415°F is 0.49, which is close

enough to the assumed value (see Appendix 15.1). Then, $Q_F = 52$ Btu/hr ft^2.

6) Determine the outside area of insulated pipe. From Table 15-2, pipe radius $rl = 2.25$-in., then outside insulated area (ft^2)

$$= 2\pi \ (rl+\text{tk})(\text{length})/(12 \ \text{in./ft})$$

$$= 2\pi \ (2.25 \ \text{in}+3.5 \ \text{in.})(100 \ \text{ft})/(12 \ \text{in./ft})$$

$$= 301 \ \text{ft}^2$$

7) Calculate heat losses with insulation.

$$Q_{Tins} = (Q_F)(\text{outside area})$$

$$= (52 \ \text{Btu/hr ft}^2)(301 \ \text{ft}^2)$$

$$= 15,652 \ \text{Btu/hr}$$

STEP 4. Determine heat losses savings $Q_{Tsavings}$.

$$Q_{Tsavings} = (Q_{Tpres} - Q_{Tins})(\text{hr/yr})$$

$$= (435,600–15,652 \ \text{Btu/hr})(4,160 \ \text{hr/yr})$$
$$(1 \ \text{MMBtu}/10^6 \ \text{Btu})$$

$$= 1,747 \ \text{MMBtu/yr}$$

STEP 5. Determine fuel cost savings. Assuming 1 MCF = 1 MMBtu.

Fuel savings

$$= \ (Q_{Tsavings})(\text{conversion factor})/$$
$$(\text{combustion efficiency})$$

$$= \ (1,747 \ \text{MMBtu/yr})(1 \ \text{MCF/MMBtu})/(0.8)$$

$$= \ 2,184 \ \text{MCF/yr}$$

Then,

Fuel cost savings $= (\text{fuel savings}) \ (\text{fuel cost})$

$$= (2,184 \ \text{MCF/yr}) \ (\$3/\text{MCF})$$

$$= \$6,552/\text{yr}$$

15.5.3 Financial Analysis Methods— Sample Calculations

Chapter 4 offers a complete discussion of the various types of financial analyses commonly used in industry. A review of that material is suggested here, as the methods discussed below rely on this basic understanding.

To select the proper financial analysis requires an understanding of the degree of sophistication required

by the decision maker. In some cases, a quick estimate of profitability is all that is required. At other times, a very detailed cash flow analysis is in order. The important point is to determine what level of analysis is desired and then seek to communicate at that level. Following is an abbreviated discussion of four primary methods of evaluating an insulation investment: (1) simple payback, (2) discounted payback, (3) minimum annual cost using a level annual equivalent, and (4) present-value cost analysis using discounted cash flows.

Economic Calculations

Basically, a simple payback period is the time required to repay the initial capital investment with the operating savings attributed to that investment. For example, consider the possibility of upgrading a present insulation thickness standard.

	Thickness Current Standard	Upgraded Thickness	Difference
Insulation investment ($)	225,000	275,000	50,000
Annual fuel cost ($)	40,000	30,000	10,000

$$\text{Simple payback} = \frac{\text{investment difference}}{\text{annual fuel saving}} = \frac{50,000}{10,000} = 5.0 \ \text{years}$$

This calculation represents the incremental approach, which determines the amount of time to recover the additional $50,000 of investment.

In the following table, the full thickness analysis is similar except that the upgraded thickness numbers are now compared to an uninsulated system with zero insulation investment.

	Uninsulated System	Upgraded Thickness	Difference
Insulation investment ($)	0	275,000	275,000
Annual fuel cost ($)	340,000	30,000	310,000

$$\text{Simple payback} = \frac{275,000}{310,000} = 0.89 \ \text{year}$$

The magnitude of the difference points out the danger in talking about payback without a proper definition of terms. If in the second example, management had a payback requirement of 3 years, the full insulation investment easily complies, whereas the incremental investment does not. Therefore, it is very important to understand the intent and meaning behind the payback requirement.

Although simple payback is the easiest financial calculation to make, its use is normally limited to rough estimating and the determination of a level of financial risk for a certain investment. The main drawback with this simple analysis is that it does not take into account the time value of money, a very important financial consideration.

Time Value of Money

Again, see Chapter 4. The significance of the cost of money is often ignored or underestimated by those who are not involved in their company's financial mainstream. The following methods of financial analysis are all predicated on the use of discount factors that reflect the cost of money to the firm. Table 15-9 is an abbreviated table of present-value factors for a steady income stream over a number of years. Complete tables are found in Chapter 4.

Discounted Payback

Although similar to simple payback, the utilization of the discount factor makes the savings in future years worth less in present-value terms. For discounted payback, then, the annual savings times the discount factor must now equal the investment to achieve payback in present-value dollars. Using the same example:

	Thickness		
	Current Standard	Upgraded Thickness	Difference
Insulation investment ($)	225,000	275,000	50,000
Annual fuel cost ($)	40,000	30,000	10,000

Now, payback occurs when:

investment = discount factor × annual savings

50,000 = (discount factor) × 10,000

so solving for the discount factor,

$$\text{discount factor} = \frac{\text{investment}}{\text{annual savings}} = \frac{50,000}{10,000} = 5.0$$

For a 15% cost of money, read down the 15% column of Table 15-9 to find a discount factor close to 5. The corresponding number of years is then read to the left, approximately 10 years in this case. For a cost of

money of only 5%, the payback is achieved in about 6 years. Obviously, a 0% cost of money would be the same as the simple payback calculation of 5 years.

Minimum Annual Cost Analysis

As previously discussed, an insulation investment must involve a lump-sum cost for insulation as well as a stream of fuel costs over the many years. One method of putting these two sets of costs into the same terms is to spread out the insulation investment over the life of the project. This is done by dividing the initial investment by the appropriate discount factor in Table 15-9. This produces a "level annual equivalent" of the investment for each year, which can then be added to the annual fuel cost to arrive at a total annual cost.

Utilizing the same example with a 20-year project life and 10% cost of money:

	Thickness	
	Current Standard	Upgraded Thickness
Insulation investment ($)	225,000	275,000

For 20 years at 10%, the discount factor is 8.514 (Table 15-9), so

		Current Standard	Upgraded Thickness
Equivalent annual insulation costs		225,000	275,000
		8.514	8.514
	=	26,427	32,300
Annual fuel cost ($)		40,000	30,000
Total annual cost ($)		66,427	62,300

Therefore, on an annual cost basis, the upgraded thickness is a worthwhile investment because it reduces the

Table 15-9. Present-Value Discount Factors for an Income of $1 Per Year for the Next _n_ Years

	Cost of Money at:				
Years	5%	10%	15%	20%	25%
1	.952	.909	.870	.833	.800
2	1.859	1.736	1.626	1.528	1.440
3	2.723	2.487	2.283	2.106	1.952
4	3.546	3.170	2.855	2.589	2.362
5	4.329	3.791	3.352	2.991	2.689
10	7.722	6.145	5.019	4.192	3.571
15	10.380	7.606	5.847	4.675	3.859
20	12.460	8.514	6.259	4.870	3.954

annual costs by $4127.

Now, to illustrate again the importance of using a proper cost of money, change the 10% to 20% and recompute the annual cost. The 20% discount factor is 4.870.

	Thickness	
	Current Standard	Upgraded Thickness
Equivalent annual insulation cost ($)	225,000	275,000
	4.870	4.870
=	46,201	56,468
Add the annual fuel cost ($)	40,000	30,000
Total annual cost	86,201	86,468

In this case, the higher cost of money causes the upgraded annual cost to be greater than the current cost, so the project is not justified.

Present-Value Cost Analysis

The other method of comparing project costs is to bring all the future costs (i.e., fuel expenditures) back to today's dollars by discounting and then adding this to the initial investment. This provides the total present-value cost of the project over its entire life cycle, and projects can be chosen based on the minimum present-value cost. This discounted cash flow (DCF) technique is used regularly by many companies because it allows the analyst to view a project's total cost rather than just the annual cost and assists in prioritizing among many projects.

	Thickness Current Standard	Upgraded Thickness
Annual fuel cost ($)	40,000	30,000

For 20 years at 10% the discount factor is 8.514 (Table 15-9), so

Present value of fuel cost over 20 years	40,000 × 8.514	30,000 × 8.514
	=340,560	255,420
Insulation investment ($)	225,000	275,000
Total present-value cost of insulation project	$565,460	$530,420

Again, the lower total project cost with the upgraded thickness option justifies that project.

So far, the effect of taxes and depreciation has been ignored so as to concentrate on the fundamentals. However, the tax effects are very significant on the cash flow to the company and should not be ignored. In the case

where the insulation investment is capitalized utilizing a 20-year straight-line depreciation schedule and a 48% tax rate, the following effects are seen (see table at top of next page).

This illustrates the significant impact of both taxes and depreciation. In the preceding analysis, the PV benefit of upgrading was (565,460 – 530,420) = $35,040. In this case, the cash flow benefit is reduced to (356,116 – 351,626) = $4490.

The final area of concern relates to future increases in fuel costs. So far, all the analyses have assumed a constant stream of fuel costs, implying no increase in the base cost of fuel. This assumption allows the use of the PV factor in Table 15-9. To accommodate annual fuel-price increases, either an average fuel cost over the project life is used, or each year's fuel cost is discounted separately to PV terms. Computerized calculations permit this, whereas a manual approach would be extremely laborious.

15.5.4 Economic Thickness (ETI) Calculations

Section 15.5.3 developed the financial analyses often used in evaluating a specific insulation investment. As presented, however, the methods evaluate only two options rather than a series of thickness options. Economic thickness calculations are designed to evaluate each 1/2-in. increment and sum the insulation and operating costs for each increment. Then the option with the lowest total annual cost is selected as the economic thickness. Figure 15-5 graphically illustrates the optimization method. In addition, it shows the effect of additional labor required for double- and triple-layer insulation applications.

Mathematically, the lowest point on the total-cost curve is reached when the incremental insulation cost equals the incremental reduction in energy cost. By definition, the economic thickness is:

that thickness of insulation at which the cost of the next increment is just offset by the energy savings due to that increment over the life of the project.

Historical development

A problem with the McMillan approach was the large number of charts that were needed to deal with all the operating and financial variables. In 1949, Union Carbide Corp., in a cooperation with West Virginia University, established a committee headed by W.C. Turner to establish practical limits for the many variables and develop a manual for performing the calculations. This was done, and in 1961, the manual was published by the National Insulation Manufacturers Association (previously called TIMA and

	Thickness Current Standard	Upgraded Thickness
1. Annual energy cost ($)	40,000	30,000
2. After-tax energy cost ($) ((1) × (1.0 − 0.48))	20,800	15,600
3. Insulation depreciation ($ tax benefit)		
(225,000/20 yr)(0.48)	5,400	
(275,000/20 yr)(0.48)		6,600
4. Net annual cash costs [$; (2) − (3)]	15,400	9,000
5. Present-value factor for 20 years at 10% = 8.514		
6. Present value of annual cash flows [$; (4) × (5)]	131,116	76,626
7. Present value of cash flow for insulation purchase ($)	225,000	275,000
8. Present-value cost of project [$; (6) + (7)]	356,116	351,626

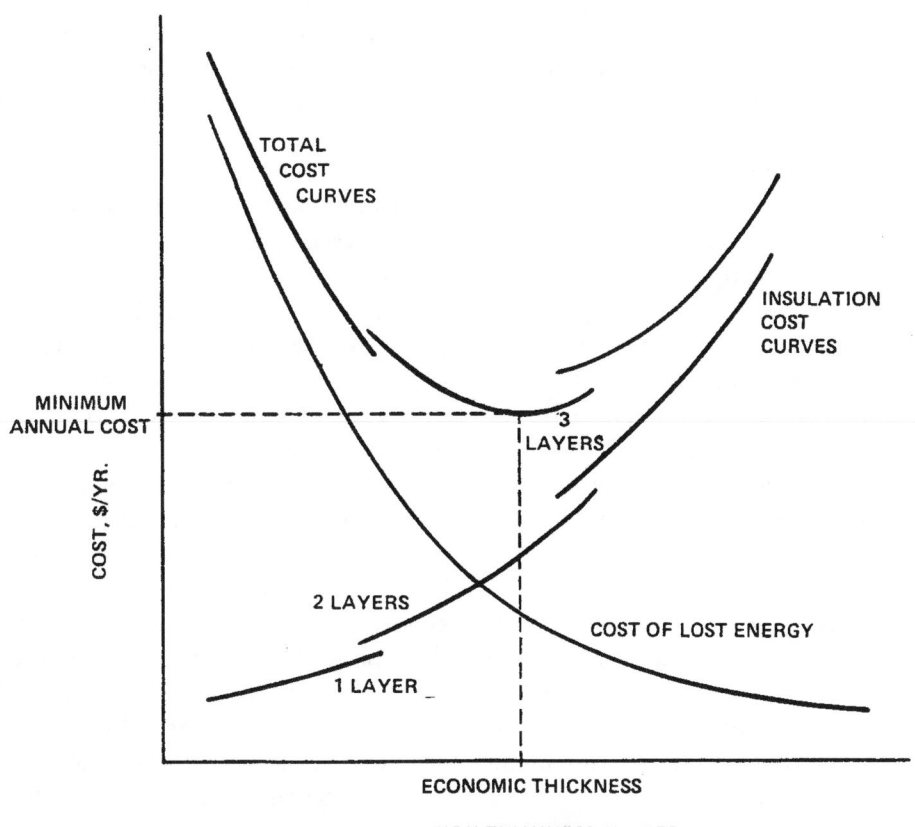

Figure 15-5. Economic Thickness of Insulation (ETI) Concept.

now NAIMA, North American Insulation Manufacturers Association). The manual was entitled *How to Determine Economic Thickness of Insulation* and employed a number of nomographs and charts for manually performing the calculations.

Since that time, the use of computers has greatly changed the method of ETI calculations. In 1973, TIMA released several programs to aid the design engineer in selecting the proper amount of insulation. Then in 1976, the Federal Energy Administration (FEA) published a no-

mograph manual entitled *Economic Thickness of Industrial Insulation* (Conservation Paper #46). In 1980, these manual methods were computerized into the "Economic Thickness of Industrial Insulation for Hot and Cold Surfaces." Through the years, NAIMA developed a version for personal computers; the newest program was renamed 3EPLUS and calculates the ETI thickness of insulation.

Perhaps the most significant change occurring is that most large owners and consulting engineers are developing and using their own economic analysis programs, specifically tailored to their needs. As both heat-transfer and financial calculations become more sophisticated, these programs will continue to be upgraded, and their usefulness in the design phase will increase.

Nomograph Methods

A nomograph method is not presented here, but the interested reader can review the following references:

- FEA manual (Ref. 12). This manual provides a fairly complete but time-consuming nomograph method.

- 1972 ASHRAE Handbook of Fundamentals, Chapter 17 (Ref. 13) provides a simplified, one-page nomograph. This approach is satisfactory for a quick determination, but it lacks the versatility of the more complex approach. The nomograph has been eliminated in the latest edition, with reference made to the computer analyses and the FEA manual.

Computer Programs

Several insulation manufacturers offer to run the analysis for their customers. Also, computer programs such as the 3EPLUS are available for customers who want to run the analysis on their own. The 3EPLUS software is an ETI program developed by the North American Insulation Manufacturers Association and the Steam Challenge Program. The program, available for free download (Ref. 14), calculates heat losses, energy and cost savings, thickness for maximum surface temperature, and optimum thickness of insulation.

All the insulation owning costs are expressed on an equivalent uniform annual cost basis. This program uses the ASTM C680 method for calculating the heat loss and surface temperatures. Each commercially available thickness is analyzed, and the thickness with the lowest annual cost is the economic thickness (ETI).

Figure 15-6 shows the output generated by the NAIMA 3EPLUS program. The first several lines are a readout of the input data. The different variables used in the program allow simulating virtually any job condition. The same program can be used for retrofit analyses and bare-surface calculations. There are two areas of input data that are not fully explained in the output. The first is the installed insulation cost. The user has the option of entering the installed cost for each particular thickness or using an estimating procedure developed by the FEA (now DOE).

The second area that needs explanation is the insulation choice, which relates to the thermal conductivity of the material. The example in Figure 15-6 shows the insulation as glass fiber blanket. The program includes the thermal conductivity equations of several generic types of thermal insulation that were derived from ASTM materials specifications. The user has the option of supplying thermal conductivity data for other materials.

The lower portion of the output supplies seven columns of information. The first and second columns are input data, while the others are calculated output. The program also calculates the reduction in CO_2 emissions by insulating to economic thickness. The meaning of columns two to seven of the output are explained below.

Annual Cost ($/yr). This is the annual operating cost, including both energy cost and the amortized insulation cost. Tax effects are included. This value is the one that determines the economic thickness. As stated under the columns, the lowest annual cost occurs with 2.50 in. of insulation that is the economic thickness.

Payback period (yr). This value represents the discounted payback period of the specific thickness as compared to the reference thickness. In this example, the reference thickness variable was input as zero, so the payback is compared to the uninsulated condition.

Present Value of Heat Saved ($/ft). This gives the energy cost savings in discounted terms as compared to the uninsulated condition. As discussed earlier, the first increment has the most impact on energy savings, but the further incremental savings are still justified, as evidenced by the reduction of annual cost to the 2.50-in. thickness. Heat Loss (Btu/ft). This calculation allows the user to check the expected heat loss with that required for a specific process. It is possible that, under certain conditions, a thickness greater than the economic thickness may be required to achieve a necessary process requirement.

Surface Temperature (°F). This final output allows the user to check the resulting surface temperature to assure the level is within the safe-touch range. The ETI program is very sophisticated. It employs sound methods of both thermal and financial analysis and provides output that is relevant and useful to the design engineer and owner. NAIMA makes this program available to those desiring to have it on their own computer systems. In addition, several of the insulation manufacturers offer to run the analysis for their customers and send them a program output.

Figure 15-6. NAIMA 3E Computer Program Output.

Project Name =	Date = 11-13-1995
Project Number =	Engineer =
System =	Contact =
Location =	Phone =

Fuel Type =	Gas
First Year Price =	3.36 $ per mcf
Heating Value =	1000 Btu per cf
Efficiency =	80.0%
Annual Fuel Inflation Rate =	6.0%
Annual hours of operation =	8320 hours

ECONOMIC DATA

Interest rate or Return on investment =	10.0%
Effective Income Tax Rate =	30%
Physical Plant Depreciation Period =	7 years
New Insulation Depreciation Period =	7 years
Incremental Equipment Investment Rate =	3.47 $/MMBtu/hr
Percent of New Insulation Cost for	
Annual Insulation Maintenance =	2%
Percent of Annual Fuel Bill for	
Physical Plant Maintenance =	1%
Ambient temperature =	75 F
Emittance of outer jacketing =	0.10
Wind speed =	0 mph
Emittance of existing surface =	0.80
Reference thickness for payback calculations =	0.0 inches

Insulation material = GLASS FIBER BLANKET
Horizontal Pipe

Pipe Size =	5 inch
Average Installation Complexity factor =	1.20
Performance Service factor =	1.00

Insulation costs estimated by FEA method

Labor rate =	38.35 $/hr
Productivity factor =	100
Price of 2x2 pipe insulation =	4.97 $/ln ft
Price of 2 inch block =	1.71 $/sqft

Operating Temperature 450 F

Insulation Thick	Cost	Annual Cost	Payback Period	Pres Value Heat Saved	Heat Loss	Surf Temp
Inches	$/ft	$/ft	Years	$/ft	Btu/ft	F
Bare	57.37				1834	450
1.0	10.18	9.25	0.2	1133.90	226	193
1.5	12.10	8.07	0.2	1170.00	174	163
2.0	14.96	7.76	0.3	1190.76	145	146
2.5	17.27	7.60	0.4	1205.83	124	133
3.0	19.68	7.71	0.4	1214.97	111	125
4.0	25.45	8.36	0.6	1228.33	92	113
Double layer						
3.0	22.30	8.27	0.5	1214.97	111	125
4.0	29.22	9.18	0.7	1228.33	92	113
5.0	36.14	10.34	0.9	1235.90	81	106
6.0	43.07	11.63	1.1	1240.32	75	102
Triple layer						
6.0	67.48	16.91	1.7	1240.32	75	102
7.0	77.39	18.86	2.0	1244.51	69	99
8.0	87.00	20.79	2.3	1247.76	64	96

The Economic Thickness is single layer 2.5 inches.

The savings for the economic thickness is 49.77 $/ln ft/yr and the reduction in Carbon Dioxide emissions is 1608 lbs/lnft/yr.

APPENDIX 15.1

Typical Thermal Conductivity Curves Used in Sample Calculations*

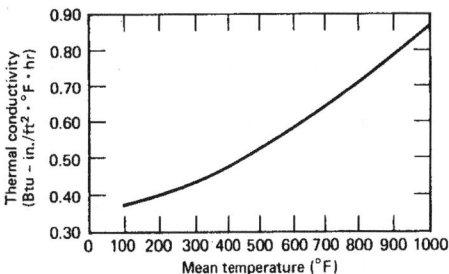

Figure 15-A1. Calcium Silicate.

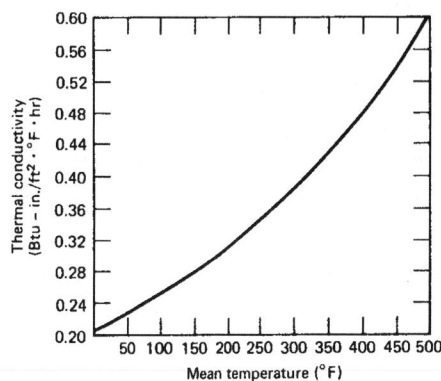

Figure 15-A2. Fiberglass Pipe Insulation.

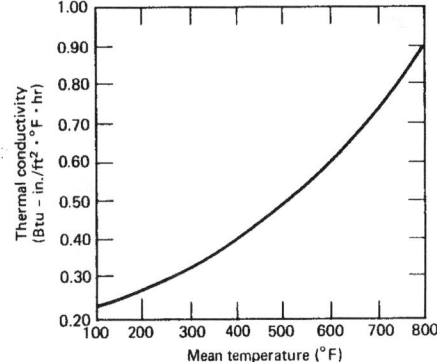

Figure 15-A3. Fiberglass Board, 3 lb/ft³.

*Current manufacturers' data should always be used for calculations.

References

1. American Society for Testing and Materials, Annual Book of ASTM Standards: Part 18—Thermal and Cryogenic Insulating Materials; Building Seals and Sealants; Fire Test; Building Constructions; Environmental Acoustics; Part 17—Refractories, Glass and Other Ceramic Materials; Manufactured Carbon and Graphite Products.
2. W.H. McADAMS, Heat Transmission, McGraw-Hill, New York, 1954.
3. E.M. SPARROW and R.D. CESS, Radiation Heat Transfer, McGraw-Hill, New York, 1978.
4. L.L. BERANEK, Ed., Noise and Vibration Control, McGraw-Hill, New York, 1971.
5. F.A. WHITE, Our Acoustic Environment, Wiley, New York, 1975.
6. M. KANAKIA, W. HERRERA, and F. HUTTO, JR., "Fire Resistance Tests for Thermal Insulation," Journal of Thermal Insulation, Apr. 1978, Technomic, Westport, Conn.
7. Commercial and Industrial Insulation Standards, Midwest Insulation Contractors Association, Inc., Omaha, Neb., 1979.
8. J.F. MALLOY, Thermal Insulation, Reinhold, New York, 1969.
9. American Society for Testing and Materials, Annual Book of ASTM Standards, Part 18, STD C-585.
10. J.F. MALLOY, Thermal Insulation, Reinhold, New York, 1969, pp. 72-77, from Thermon Manufacturing Co. technical data.
11. L.B. McMILLAN, "Heat Transfer through Insulation in the Moderate and High Temperature Fields: A Statement of Existing Data," No. 2034, The American Society of Mechanical Engineers, New York, 1934.
12. Economic Thickness of Industrial Insulation, Conservation Paper No. 46, Federal Energy Administration, Washington, D.C., 1976. Available from Superintendent of Documents, U.S. Government Printing Office, Washington, D.C. 20402 (Stock No. 041-018-00115-8).
13. ASHRAE Handbook of Fundamentals, American Society of Heating, Refrigerating and Air Conditioning Engineers, Inc., New York, 1972, p. 298.
14. NAIMA 3 E's Insulation Thickness Computer Program, North American Insulation Manufacturers Association, 44 Canal Center Plaza, Suite 310, Alexandria, VA 22314.
15. P. Greebler, "Thermal Properties and Applications of High Temperature Aircraft Insulation," American Rocket Society, 1954. Reprinted in Jet Propulsion, Nov.-Dec. 1954.
16. Johns-Manville Sales Corporation, Industrial Products Division, Denver, Colo., Technical Data Sheets.
17. ASHRAE Handbook of Fundamentals, American Society of Heating, Refrigerating and Air Conditioning Engineers, Inc., Atlanta, GA, 1992, p.22.16.
18. W.C. Turner and J.F. Malloy, Thermal Insulation Handbook, Robert E. Krieger Publishing Co. And McGraw Hill, 1981.
19. Ahuja, A., "Thermal Insulation: A Key to Conservation," Consulting-Specifying Engineer, January 1995, p. 100-108.
20. U.S. Department of Energy, "Industrial Insulation for Systems Operating Above Ambient Temperature," Office of Industrial Technologies, Bulletin ORNL/M-4678, Washington, D.C., September 1995.

CHAPTER 16

USE OF ALTERNATIVE ENERGY

SIMON EILIF BAKER, M.E.E.P. *

Lead Regulatory Analyst, California
    Public Utilities Commission
San Francisco, California

JERALD D. PARKER, Ph.D.

Professor Emeritus, Oklahoma State University
Stillwater, Oklahoma

16.1 INTRODUCTION

"Alternative energy" is a widely understood but vaguely defined term for a class of energy resources characterized by what they are not—conventional or fossil-based. These non-traditional resources are distinguished by low environmental impact, in contrast to fossil fuels. Often considered synonymous with "renewable energy," alternative energy is, in fact, a broader class of resources that includes energy efficiency, combined heat and power (or cogeneration), and zero-emissions energy conversion technologies such as fuel cells. Nuclear energy and large hydropower, while not fossil-based, are not normally regarded as alternative technologies due to environmental harm attributed to their use.

Renewable energy, the bulk of alternative energy choices, is energy that comes from natural, non-depletable or rapidly replenished resources—sun, wind, rain, tidal, and geothermal heat. Notwithstanding tidal and geothermal energy, all renewable energy traces back to the sun. Solar energy is the sun's radiation captured in the form of heat or power. Wind energy harnesses the kinetic energy of air movement caused by solar radiation. Biomass energy is the sun's radiation converted through the process of photosynthesis into chemical energy and organic matter, then passed through the food chain. For as long as the sun continues to exist (another 5-6 billion years) renewable energy should be available in sufficient quantities to meet human demands for heat and power.

In 2006, 3.7% of U.S. *primary (total) energy* consumption came from renewable sources (excluding large hydropower), over 80% of which was from biomass. In

the same period, 2.4% of *electrical energy* consumption came from non-hydro renewables, of which 56% was supplied by biomass, 28% from wind, and 0.5% from solar (Ref. 1). Studies of global renewable energy potential have shown that solar, wind, and biomass resources can provide 200–300 x 10^{12} kWh/year of electricity at production costs below 10¢/kWh—enough to serve projected 2050 global electricity demand in most regions (Ref. 2). A key uncertainty in these projections, however, is the rate and extent at which technology costs decline over time. (See Section 16.1.1.) In general, deployment of alternative energy of one sort or another is limited by cost, and to some extent land use, but not by resource availability.

This chapter is organized around four main principles to make it most useful to energy practitioners. (See Table 16-1.) First, the broad range of renewable energy resources is narrowed to three (solar, wind and biomass) for the sake of brevity, but also because at least one of these resources is likely to be available at any location in the U.S. Second, renewable fuels such as ethanol and biodiesel are excluded from this chapter, because they are rarely used in stationary applications. Transportation is a huge component of overall world energy use, but it is beyond the scope of this chapter. Third, examples of technology options are given at two scales, the nature of which varies in terms of how energy managers experience them: (a) *distributed-scale*—small, on-site applications experienced directly but somewhat limited due to economics and resource availability; (b) *utility-scale*—large, bulk power applications with better economics but experienced indirectly through purchases of premium electricity products with renewable energy content. Fourth, a mix of existing and future technologies is presented because rapid changes in the industry require energy practitioners to have a good foundation in established methods of renewable energy utilization, while keeping abreast of technology trends that could change project economics.

16.1.1 Technology Costs

Renewable energy technologies typically have higher investment costs than traditional resources, as shown in Figure 16-1. But costs have dropped by orders of magnitude in recent decades due to sustained

---

*Master of Energy & Environmental Policy

411

Table 16-1. Chapter overview: energy sources, technologies, scale, and commercial status.

Energy Source	Featured Technology	Scale	Status	Notes
Solar	Passive solar	Distributed	Fully commercial	Most cost-effective; featured prominently in green building designs and programs (e.g., LEED)
	Solar thermal (non-electric)	Distributed	Fully commercial	Technologies vary by operating temperature; most common is unglazed collectors in low-temp applications (e.g. pool heating)
	Solar thermal electric – Parabolic trough	Utility	Fully commercial	Competes with wind in some regions for new utility procurement of renewables; well-established track record
	Solar thermal electric – Central receiver	Utility	Commercial	Thermal storage capability yields high capacity factors demonstrated in the 1980s; project agreements with two California utilities to bring new projects on-line.
	Photovoltaic (PV)	Distributed*	Fully commercial	Easy to site and install; low maintenance; expensive, but some experts predict costs to achieve "grid parity" by 2015.
Wind	Horizontal axis wind turbine (HAWT)	Utility	Fully commercial	Dominant technology for bulk power renewables; trend toward bigger turbines; grid integration issues due to variability
	Vertical axis wind turbine (VAWT)	Distributed*	Commercial	Utility-scale VAWT becoming less popular, but commercial designs available for certain niche distributed applications
Biomass	Biomass combined heat & power (CHP)– Steam turbine	Utility (Industrial)	Fully Commercial	75% of biomass energy is CHP, mostly using woody feedstocks from forest products industry; established technologies – back-pressure, condensing-extraction, condensing turbines
N/A	Storage – Several technologies	Utility & distributed	R&D to Commercial	Compliments discussion of big wind and utility integration issues. Many technologies with different attributes.

* These technologies are mostly distributed but can also be scaled up to utility-scale.

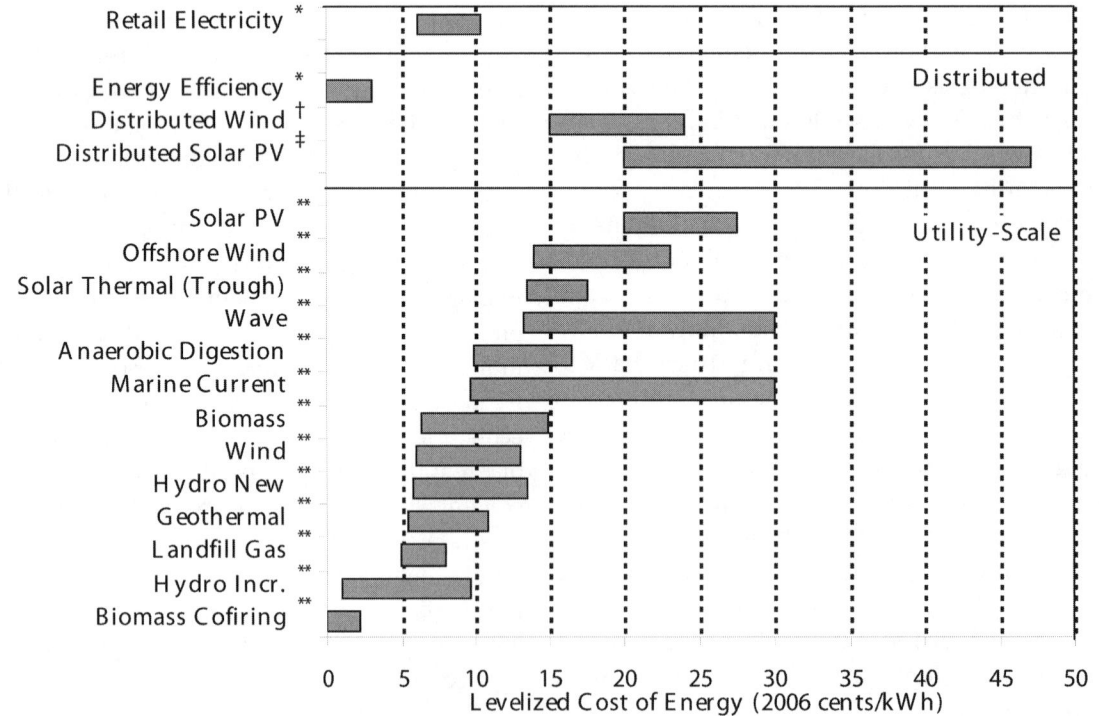

Figure 16-1. Levelized cost of distributed and utility-scale alternative energy technologies compared to national average cost of retail electricity from mostly fossil-fired technologies, U.S. cents. (Compiled from Refs. 6*, 7†, 8‡, and 9**·)

growth in installed capacity of renewables. These so-called *experience curve* effects empirically show that each time cumulative volume doubles, costs fall by a constant and predictable percentage. As a rule of thumb, costs of solar photovoltaics and wind have dropped 17% for each doubling of installed capacity (Refs. 3 and 4). Figure 16-2 shows this log-linear relationship for solar photovoltaics, (See section 16.2.6,) Two inflection points are worth noting in Figure 16-2; the first occurred in the 1980s when federal tax incentives for PV expired, causing manufacturers to sell off inventory at a loss; the second, in 1995, reflects the year Japanese and German stimulus programs drove excess demand for PV caused an uptick in price. These events show the complex relationship between technology costs and government incentive programs. (See section 16.1.3.) Indeed, not shown in Figure 16-2 is the period since 2005, when solar and wind capital costs have, in fact been climbing due to high demand and constraints on supply.

Staggering growth trends indicate that costs will likely continue to decline, assuming a supply-demand equilibrium. Global grid-connected solar PV grew at 60% annually, solar thermal at 16%, and wind at 25%, while conventional resources grew at 3-5% from 2002-2006. In the period 2005-2007, annual global investment in renewable capacity grew over 75% to $71 billion (Ref. 5).

Several factors explain why renewables are typically more expensive than traditional resources, such as coal- and natural gas-fired technologies. First, scale of investment over time has resulted in mature technologies and fossil energy commodities with low production costs. Second, historical, government subsidies of traditional industries have created an uneven playing field for renewables to compete. Third, negative *externalities* of fossil-fuel production and consumption are difficult to monetize and frequently are not reflected in the price of traditional resources. For example, in most cases, the cost of pollution is born by society, but not by the individual or firm responsible (Note 1). Appendix A provides data on emission rates of major pollutants from the electric sector by state. These data can be used to quantify pollution prevention from alternative energy projects.

Despite these disadvantages, existing renewable technologies can be attractive from a lifecycle cost perspective; they are capital intensive but usually have low (or no) fuel and operating costs. Because these technologies are typically long-lived (20-30 years), fuel savings, over the life of the asset recover the capital outlay and pay dividends. Lifecycle cost assessment is most commonly used in government or institutional settings where operating budgets are the main concern. Most private firms' investment bias favors short-term returns with hurdle rates typically less than 5 years. Increasingly, strategic companies are weighing the benefits of long-term price stability and positive "green" image to justify the investment in renewable technologies. These same firms may be concerned with unpredictability of future energy costs, as a result of increasing volatility in energy markets, especially natural gas.

A good guiding principle for any firm is: *efficiency first, then renewables*. As can be seen from Figure 16-1, energy efficiency is the lowest cost alternative to provide energy services. The energy efficiency resource has been dubbed *"negawatts"* on the principle that the cheapest energy resource is one that you don't consume. When packaged with a renewable project, low-cost energy efficiency effectively buys down the total cost of the project, making it more attractive and easier to sell to decision-makers.

16.1.2 Government Policy Incentives

Beginning in the late 1970s, federal, state and local governments have intervened at various times in attempts to level the playing field for renewable energy technologies. The goal of these policies is market transformation to achieve socially efficient outcomes, either to scale-up investment to drive renewable costs down or to produce price signals that reflect the cost of negative externalities. The first significant policy to spur renewable investment was Public Utilities Regulatory Act (PURPA) of 1978, which required utilities to buy wholesale power from *qualifying facilities* such as renewable energy generators at the utilities' avoided

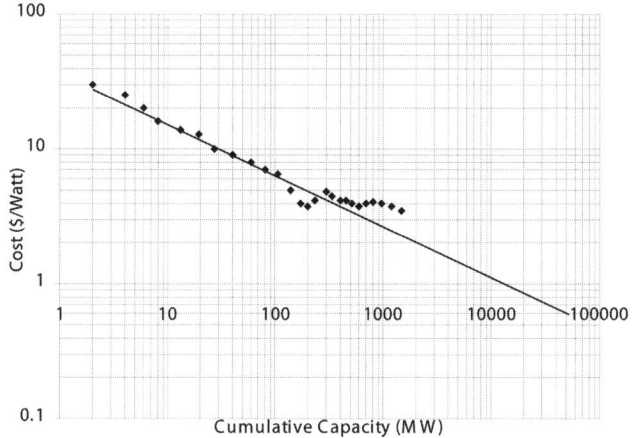

Figure 16-2. Photovoltaic Module Experience Curve: 1975-2000 (Adapted from Ref. 4, Data courtesy of Paul Maycock).

cost of production. Since the late 1990s, an attractive policy environment has been responsible for most (if not all) growth in the renewables industry. A basic knowledge of these policies is fundamental to understanding the economics of technology-specific energy projects. A dizzying array of financial incentives may be available to project owners, depending on technology, project location, ownership structure, utility rate class, and myriad other variables. A recommended source to consult regarding specific projects is the Database on State Incentives for Renewables and Efficiency (DSIRE), www.dsireusa.org, which maintains updated information in summary tables and through an interactive map-based interface.

Five policy mechanisms are highlighted here as being most responsible for growth in utility-scale and distributed renewable energy:
• state renewable portfolio standards (utility-scale)
• federal tax credits (utility-scale and distributed)
• federal accelerated depreciation (utility-scale and distributed)
• state or local net metering tariffs (distributed)
• state and/or local cash rebates (distributed)

In the late 1990's, states began enacting renewable portfolio standards (RPS), which generally place an obligation on electric utilities to supply a specified and increasing fraction of electricity from eligible renewable sources. RPS policies are designed to "pull the market" with policy-driven demand for renewables. As of 2008, 27 states have binding RPS policies in place, which have spurred investment in utility-scale renewables, most prominently wind power.

In 1999, federal tax credits became available for the first time since the early 1980s. For the technologies

profiled in this chapter, the most important of these are the corporate production tax credit (PTC) for wind and biomass and the corporate investment tax credit (ITC) for solar thermal, solar thermal electric, PV, and fuel cells. Not all businesses have sufficient tax liability to maximize the benefits of tax credits. Consequently, projects involving the use of tax credits often involve third-party tax investors. Figure 16-3 underscores the impact of both RPS and PTC policies in particular on the pattern of growth in the wind industry since 1981. Wind capacity investment stimulated, in part by RPS policies, fell sharply in years when the PTC expired (2000, 2002, and 2004), only to pick up again when Congress passed short-term PTC extensions.

Under the federal Modified Accelerated Cost-Recovery System (MACRS), businesses may recover investments in certain property through depreciation deductions. Depreciation schedules generally range from 3 to 7 years for solar thermal, solar thermal electric, PV, wind, biomass, and fuel cell property.

A new tariff structure called *net metering* allows retail customers in some regions to sell excess on-site generation to the electric utility, usually at retail rates (Note 2). Grid-connected, distributed generation benefits greatly from these policies. Prior to the advent of net metering laws, distributed renewable energy systems had to be sized to minimum annual loads or designed with storage in order to avoid "dumping" excess power onto the grid for no compensation to the owner. Net metering enables project owners to size projects to their annual electric load, essentially using the grid as storage during periods of net surplus generation and accumulating credits on their utility bill to be used during periods of net consumption. Net metering rules vary substantially by utility, rate class, technology, and project size, as in

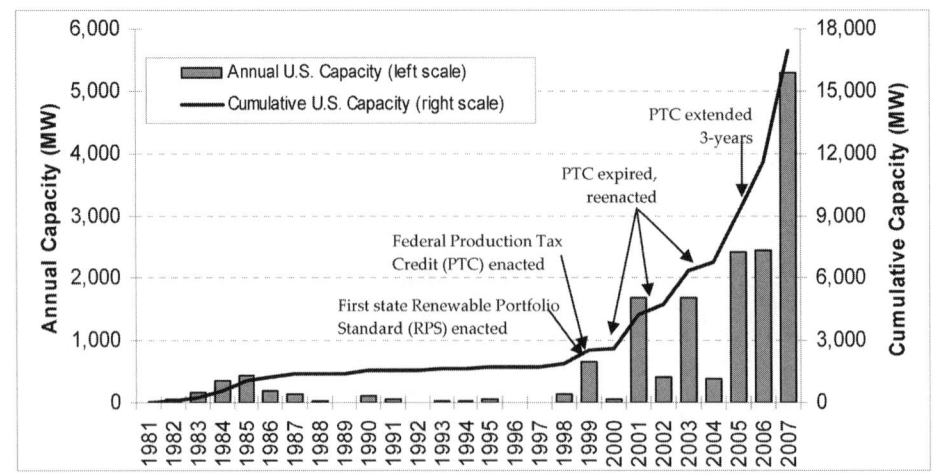

Figure 16-3. Annual and cumulative U.S. wind power capacity, 1981-2007, and the effect of public policy support (Ref. 10).

the case of rules for PV shown in Figure 16-4. In some cases, net metering can be structured under time-of-use (TOU) rates such that net generation during high-value, on-peak (summer, daytime) periods can be credited towards net consumption during low-cost, off-peak (winter, daytime) periods. Energy managers should consult the DSIRE database and contact their local utility to understand the options available to them. Another good resource is the Interstate Renewable Energy Council's (IREC) state-by-state net metering table, which is regularly updated to show tariff variations by state at www.irecusa.org.

Finally, many state and local jurisdictions have implemented cash rebates (or "buydowns") offered by utilities or third-party program administrators. Eligibility rules vary substantially by locale, making it difficult to generalize, but potential financial impacts can be huge and should not be overlooked. Application processes are often prescriptive, lengthy, and at times complex. Again, energy managers are advised to first consult the DSIRE database and then contact their local utility to inquire about specific project eligibility.

The dependence of project economics on incentive programs cannot be overstated. PV project viability, in particular, is highly sensitive to the availability of incentives; in some cases up to 90% of the project benefits stem from incentives. It is not uncommon for the net present value of corporate-owned PV projects to be proportionately distributed as follows: 40% cash rebate,

25% ITC, 20% MACR, and 15% energy sales (or avoided energy purchases).

Given the relatively high cost of renewables in the marketplace, the question arises, "Are subsidy programs cost-effective in terms of net benefits?" The answer is that it depends on the stakeholders' perspective and the scope of benefits included in the benefit-cost analysis. Table 16-2, produced by the National Renewable Energy Laboratory (NREL), illustrates this point and the various benefits that accrue from participant, ratepayer and societal perspectives. Ordinarily, in project evaluations energy managers consider only those values that flow directly to the firm (the participant perspective), as these affect the bottom line. However, the benefits to ratepayers and society can also be demonstrated using Table 16-2. The table introduces an important concept, *avoided cost*, which is the stream of benefits that government agencies and utilities use to evaluate cost-effectiveness of incentive programs. In broad terms, avoided cost is what it would otherwise cost participants, ratepayers, society and the environment if the resource in question (in this case, solar PV) is not procured. Finally, certain benefits, such as market transformation effects, may be real but hard to quantify. A more detailed description of program cost-effectiveness tests is described in the *California Standard Practice Manual: Economic Analysis of Demand-side Programs and Projects*, October 2001, and in Appendix B.

A final important note is that the cumulative effect

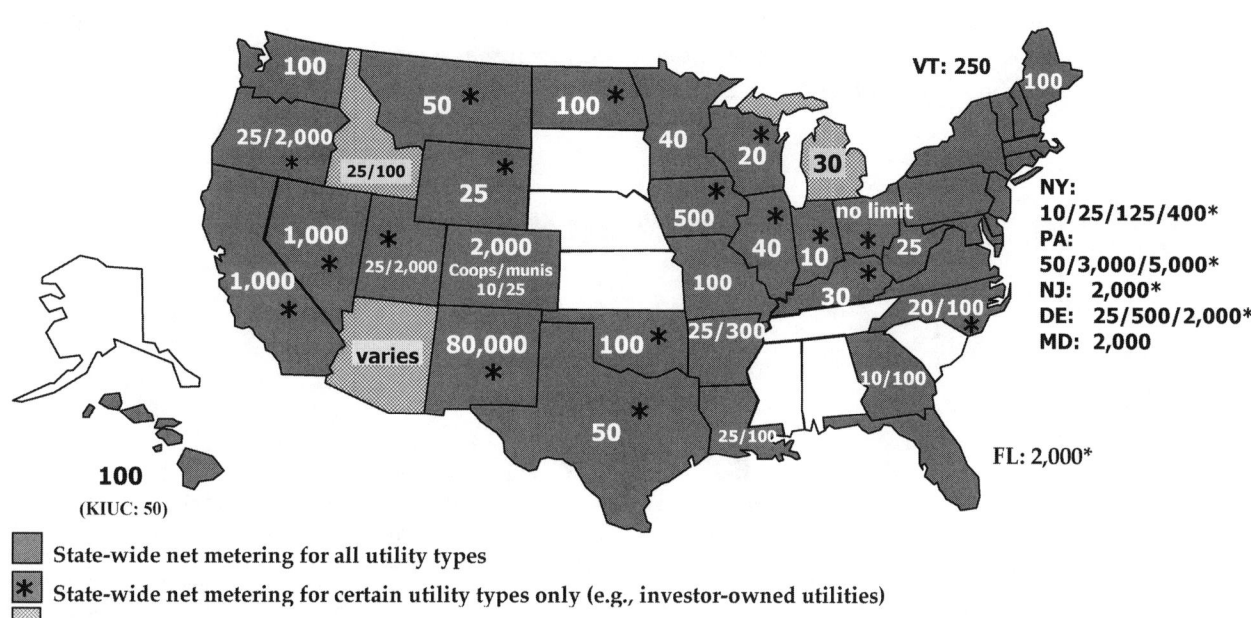

Figure 16-4. PV net metering laws in the United States by utility type and size limit (Ref. 11).

Table 16-2. Photovoltaics (PV) value ranges by stakeholder perspective, 2007, Benefit/(Cost) (Adapted from Ref. 8).

PV Values	Customer/ Participant (¢/kWh)	Utility/ Ratepayers (¢/kWh)	Society/ Environment (¢/kWh)	Net (¢/kWh)	Value Drivers
Benefits (Avoided Costs)					
Generation Energy Cost		3.2 - 9.7		3.2 - 9.7	Fuel price, heat rate
Generation Capacity Cost		1.1 -10.8		1.1 -10.8	Peak capacity value
Transmission & Distribution Costs		0.1 - 10.0		0.1 - 10.0	Location, growth
System Losses		0.5 - 4.3		0.5 - 4.3	Location, time period
Ancillary Services		0 - 1.5		0 - 1.5	Ancillary service prices, voltage support
Hedge Value		0 - 0.9		0 - 0.9	Fuel price forecasts, futures, heat rate
Customer Price Stability	0.5-1.0			0.5-1.0	Calculation method
Criteria Pollutant Emissions			0.02 - 2.0	0.02 - 2.0	Market value of SO_2, NO_x emissions
CO_2 Emissions			0.02 - 4.2	0.02 - 4.2	Abatement costs, market value, discount rate
Implicit Value of PV			0 - 2.0	0 - 2.0	Customer willingness to pay a premium
Costs					
Equipment and Installation	(47) - (19)			(47) - (19)	Technology, system size, location
PV O&M Expenses	(0.15) - (0.05)			(0.15) - (0.05)	Type of system
Benefits Overhead		(0.2) - (0.1)		(0.2) - (0.1)	Infrastructure, marketing & admin. costs
Transfers					
PV Owner Electricity Bill	1.1 - 33.0	(33.0) - (1.1)		-	Customer type, rate structure, load profile
Federal Incentives	1.58 - 7.95		(7.95) - (1.58)	-	Customer type, system size, incentive cap
State Incentives	0 - 17.8		(17.8) - 0	-	State, customer type, size, output, cap
Stakeholder Total	(43.97) - 40.7	(28.3) - 36.0	(25.7) - 6.6	(41.9) - 27.3	

of these policies may have unintended consequences, i.e., higher renewables prices. Since 2005, wind and solar technology prices have actually risen steadily, and in the case of wind, quite dramatically due to high demand (See Figure 16-5). While these trends coincide with increasing power sector costs, generally, due to global demand for cement, steel, copper, and other commodities,

short-term, supply-demand imbalance in the renewable sector is a very likely cause. Most observers expect that, once industry catches up to demand, these trends will reverse and prices will come down, but these predictions are uncertain, given renewables' growth trajectory and prominence as a leading solution to climate change. (See Section 16.1.3.)

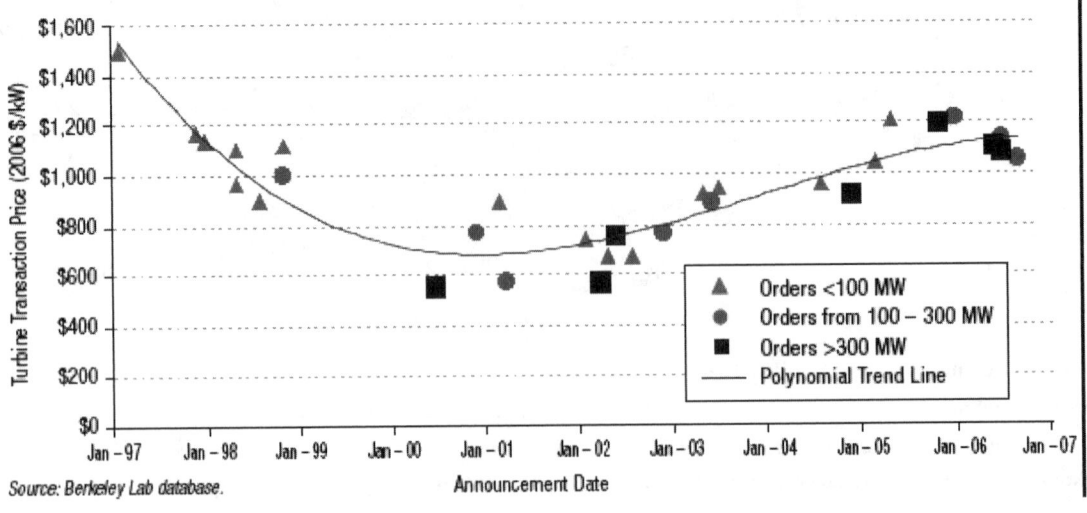

Figure 16-5. Reported U.S. wind turbine transaction prices, 1997-2007 (Ref. 13).

16.1.3 Carbon Footprint

After decades of scientific debate and policy negotiations, the issue of global climate change and what to do about it appears to have reached a tipping point. In 2007, the Intergovernmental Panel on Climate Change (IPCC), a body of more than 1200 scientists and climate experts from 130 countries, issued its Fourth Assessment Report and made its strongest statement to date: "It is likely [better than 2:1 odds] that anthropogenic [human-caused] warming has had a discernible influence on many physical and biological systems (Ref. 14)." Regardless of the veracity of this statement, policy-makers in the U.S. and across the globe have begun to respond with regulations to control carbon dioxide (CO_2) and other greenhouse gases (GHGs) emitted from sources in the electricity, industrial, buildings, and transportation sectors. These policies are already affecting energy prices in European countries. In the U.S., certain states and regions such as California and New England have enacted binding GHG targets. Negotiations on Capital Hill to pass federal limits on GHG emissions have adopted a tone of "not if, but when and how." In the climate change era, energy managers will be increasingly called on to understand and manage carbon risks and opportunities associated with their companies' energy use.

The Kyoto Protocol, an international treaty established in 1997, set binding GHG emissions reduction targets for signatory industrialized nations and instituted an emissions trading (or "cap-and-trade") system to most efficiently reach those targets. Carbon prices in the European Union Emissions Trading Scheme (ETS) have fluctuated between 15 and 30 euros per metric ton (tonne) CO_2 equivalent (CO_2e) for Phase II allowances

corresponding to the Kyoto Protocol's first compliance period, 2008-2012 (Figure 16-6) (Note 3). Hypothetically, if U.S. carbon emissions were covered (which they are not), this would raise the cost of electricity by an average of 1.4 to 2.9 ¢/kWh premium (Note 4). In contrast, carbon prices on the voluntary U.S. Chicago Climate Exchange (CCX) have been trading from 1 to 2 $US (1.6 to 3.2 euros) per tonne CO_2e. The monetization of carbon exposes wasteful carbon-intensive energy users to operating cost risk, which can be managed through alternative energy strategies. Conversely, efficient and low-carbon intensity energy users may position themselves to be sellers of carbon credits, introducing new cash flows towards profits or to pay for renewable energy projects.

Table 16-3 provides data on 16 gases and normalizes their global warming potential (GWP) to the effect of the principal gas, CO_2; this is the concept of CO_2 equivalent (CO_2e). In this comparison table, the effect of CO_2 is "1." Four orders of magnitude separate CO_2 from the most potent gas, sulfur hexafluoride (SF_6). Hydrofluorocarbons (HFCs) and perfluourinated compounds (CFs and SF_6), used in industrial processes, are relatively rare. Methane emissions, on the other hand, are quite common in landfill, wastewater treatment, and agricultural operations. At 21 times GWP of CO_2, methane has been targeted as an early GHG abatement strategy due to relative ease of capture, co-benefits of energy conversion, and high emissions avoidance per investment dollar.

Using the CO_2 emissions factors given in Table 16-4, a *carbon footprint* can be calculated for energy end-use and electricity purchases when the fuel mix of the grid is known. A carbon footprint is the total amount of

Figure 16-6. Trading price of carbon in the European Trading Scheme (ETS): Dec 2004 - Jun 2008 (Source: Point Carbon).

CO_2e attributable to the actions of a firm or individual over a given period (usually one year), or that accumulated over the full lifecycle of a product or service. Using Tables 16-3 and 16-4, it is fairly easy to calculate the annual carbon footprint of a company's operations. Life cycle assessments (LCA) or "cradle-to-grave" analyses are substantially more complex. Energy managers interested in pursuing LCA are encouraged to consult sections 14040:2006 and 14044:2006 of the International Organization on Standards (ISO) 1400 energy management standards.

Initially, carbon footprint methods have been primarily applied by companies utilizing "green" strategies to differentiate themselves in the marketplace. But as the concept has become more popular, bigger companies have taken carbon risk into account in their bond ratings. Wall Street banks have also begun taking notice of carbon exposure when assessing the risks of financing power projects (Note 5).

16.2 SOLAR ENERGY

Solar energy arrives at the outer edge of the earth's atmosphere at a rate of about 428 Btu/hr ft^2 (1,353 W/m^2). This value is referred to as the *solar constant*. Part of this radiation is reflected back to space, part is absorbed by the atmosphere and re-emitted, and part is scattered by atmospheric particles. As a result, only about two-thirds of the sun's energy reaches the surface of the earth. Peak solar radiation on the surface of the earth is approximately 1 kW/m^2. Solar radiation (*insolation or irradiance*) data are often given in kWh/m^2/day, sometimes referred to as *peak sun hours*, a term for the solar insolation that a particular location would receive if the sun were shining at its maximum value for a certain number of hours. Using a peak insolation of 1 kW/m^2, the number of peak sun hours is numerically identical to the average daily solar insolation, as represented in Figure 16-7.

Massive amounts of solar insolation data have been collected over the years by various government and private agencies. Figure 16-8, a solar resource map produced by NREL, gives average insolation data for fixed-plate surfaces tilted at latitude degrees. The solar resource ranges from 6 to 7 peak sun hours in the Southwest to 3 to 4 hours in the Pacific Northwest. For fixed systems, the maximum amount of solar radiation can be captured using a south-facing collector at tilt angle approximately equal to the site's latitude. Fixed solar collectors are usually tilted at some angle from the horizontal so as to provide a maximum amount of total

Table 16-3. Global Warming Potentials (GWPs) of greenhouse gases (Ref. 15).

Gas	GWP*
Carbon dioxide (CO₂)	1
Methane (CH₄)	21
Nitrous oxide (NO₂)	310
HFC-23	11,700
HFC-32	650
HFC-125	2,800
HFC-134a	1,300
HFC-143a	3,800
HFC-152a	140
HFC-227ea	2,900
HFC-236fa	6,300
HFC-4310mee	1,300
CF₄	6,500
C₂F₆	9,200
C₄F₁₀	7,000
C₆F₁₄	7,400
SF₆	23,900

*The GWP of a greenhouse gas (GHG) is the ratio of global warming, or radiative forcing – both direct and indirect– from one unit mass of a GHG to that of one unit mass of carbon dioxide over a period of time.

Table 16-4. Uncontrolled CO2 emissions factors by fuel type: end-use and electricity generation (Ref. 16).

Fuel	CO₂ Emissions Factor (lbs/MMBtu)*
Blast Furnace Gas	116.97
Bituminous Coal	205.45
Distillate Fuel Oil	161.27
Geothermal	0.34
Jet Fuel	159.41
Kerosene	159.41
Landfill Gas	115.12
Lignite Coal	215.53
Municipal Solid Waste	14.63
Natural Gas	116.97
Other Biomass Gas	115.11
Other Gases	141.54
Petroleum Coke	225.13
Propane Gas	139.04
Residual Fuel Oil	173.72
Synthetic Coal	205.45
Subbituminous Coal	212.58
Waste Coal	205.16
Waste Oil	163.61

* CO₂ factors do not vary by boiler type or firing configuration. Emissions factors for natural gas, propane, kerosene, distillate fuel oil and other common end-use fuels are the same in end-use and electricity generation applications.

Area Under Curve = Solar Insolation
A0 = A1
1 kW/m2
1 kW/m2
Solar Radiation
A0
A1
Time of Day (TOD)
Peak Sun Hours
TOD

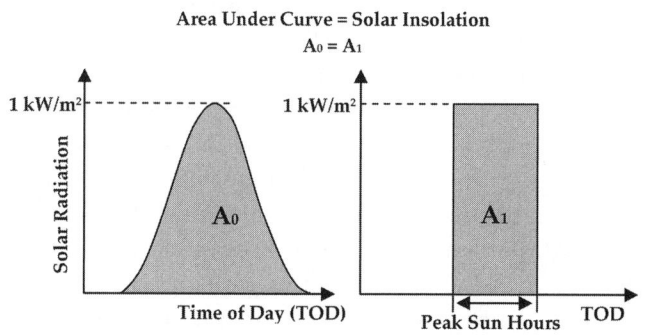

Figure 16-7. Peak sun hours and the relationship to peak solar radiation, 1 kW/m^2 (Ref. 17).

solar energy collected over the year (in the case of solar PV and solar thermal electric), or to provide a maximum amount during a particular season of the year (in the case of solar thermal).

Resource maps provide a first-look at feasibility and indicate better availability in places like the southwest U.S. What is needed in preliminary economic studies is high-resolution, location-specific data on the rate of solar insolations on tilted surfaces. Before the advent of web-based data sets and solar calculators, this involved obtaining solar insolation data and making complex calculations of the sort described in Refs. 18 and 19. Presently, on-line tools such as NREL's PVWATTS allow the user to calculate expected output from solar photovoltaic systems. Behind these calculators is a dataset called a typical meteorological year (TMY), which derives from the National Solar Radiation Data Base (NSRDB). The TMYs are data sets of hourly values of solar radiation and meteorological elements for typical months aggregated into a typical year. Because they represent typical rather than extreme conditions, they are not suited for designing systems to meet the worst-case conditions occurring at a location.

Figure 16-9 shows the procedure for the conversion of horizontal insolation to insolation on a tilted surface. The measured insolation data on a horizontal

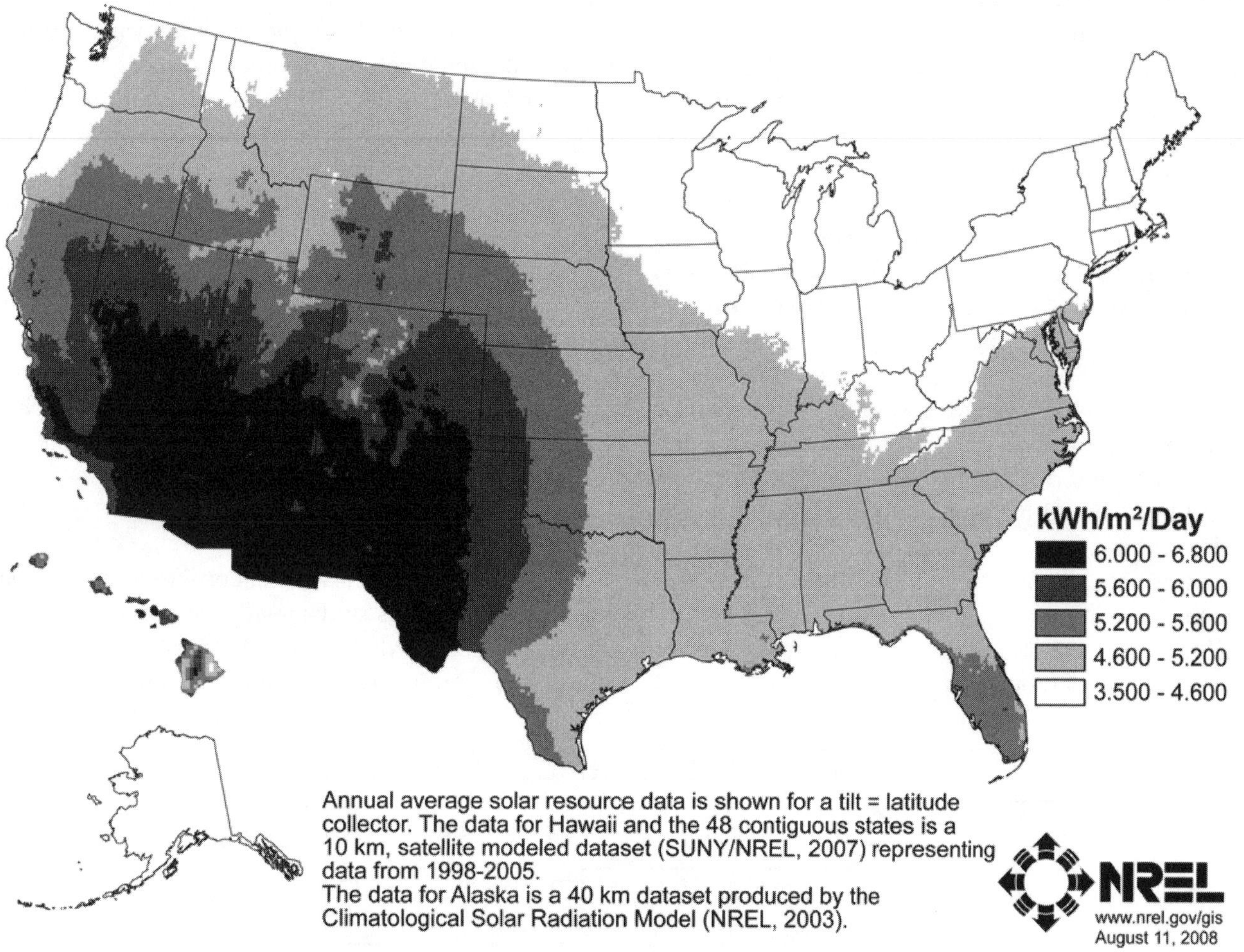

kWh/m²/Day
6.000 - 6.800
5.600 - 6.000
5.200 - 5.600
4.600 - 5.200
3.500 - 4.600
Annual average solar resource data is shown for a tilt = latitude collector. The data for Hawaii and the 48 contiguous states is a 10 km, satellite modeled dataset (SUNY/NREL, 2007) representing data from 1998-2005.
The data for Alaska is a 40 km dataset produced by the Climatological Solar Radiation Model (NREL, 2003).
NREL
www.nrel.gov/gis
August 11, 2008

Figure 16-8. U.S. annual average daily solar radiation for flat-plate collectors tilted at latitude. This map was developed by the National Renewable Energy Laboratory for the U.S. Department of Energy.

surface consist of direct radiation from the sun and diffuse radiation from the sky. The total radiation must be split into these two components (step A) and each component analyzed separately (steps B and C). In addition, the solar energy reflected from the ground and other surroundings must be added into the total (step D). Energy managers can use the global horizontal, direct normal, and diffuse horizontal radiation data given in the TMY dataset and make performance calculations for various system types and configurations, using the procedures previously cited in Refs. 18 and 19.

Four main categories of solar energy use are:
- Passive solar
- Solar thermal
- Solar-thermal electric
- Solar photovoltaic

16.2.1 Passive Solar

Passive solar uses design principles and building materials to take advantage of natural processes of radiation, conduction and convection of the sun's energy to efficiently heat and cool buildings. Passive solar can reduce or even eliminate the need for mechanical cooling and heating and artificial daytime lighting. When included at the design stages of building construction, passive solar is the most economical form of solar energy, because it is nothing more than a common sense planning approach to a project that would have occurred anyway. Knowledge of solar geometry, window technology, and local climate is all that is necessary to design a passively heated (or cooled) structure. In northern latitudes, the basic design principles are as follows:
- The length of the building should be oriented on an east-west axis to maximize southern exposure;
- High-use, interior spaces requiring the most light and heat should be located along the south face of the building;
- Use of an open floor plan to facilitate passive system operation;
- Use of shading structures to prevent summertime heating is desirable.

Figure 16-10 illustrates five basic elements of passive solar design. In cold climates, windows oriented within 30 degrees of true south, exposed to the sun from 9AM to 3PM during the heating season, and having a glazing area up to 7 percent of the building floor area are designed to provide the proper *aperture* through which solar radiation enters the building. The *absorber* is a hard, darkened surface such as brick, concrete, or tile placed in the direct path of sunlight where it can draw heat into a storage element (or thermal mass). *Thermal mass* is any kind of material, such as masonry or water, that retains heat and releases it gradually to the interior space. Whereas the absorber is an exposed surface, thermal mass is the material behind or below that surface. Once the thermal energy is collected and stored, *distribution* is the method by which solar heat circulates to various interior spaces. Finally, a variety of *control* elements are used to prevent over- and/or under-heating; these include roof overhangs to block unwanted summer sun, operable vents and dampers to allow or restrict airflow, low-emissivity blinds, and window awnings.

Passive solar design typically employs one or more of the following heating techniques: direct gain, indirect gain, or isolated gain. Figure 16-10a is an example of direct gain, where direct sunlight is permitted to enter and heat the interior space. Direct gain systems will utilize 60-75% of the solar energy that strikes the aperture. Figure 16-10b shows an indirect gain configuration, also called a Trombe wall, where the thermal mass is located between the sun and the interior space. These systems will typically use 30-45% of the sun's energy striking the glazing. A third configuration, isolated gain, is commonly used in the construction of sunrooms. Isolated gain systems use 15-30% of the sun's energy that strikes the exterior window.

16.2.2 Solar Thermal

Solar thermal uses solar energy for water or space heating applications and usually employs mechanical systems to collect, distribute, and/or store solar-heated fluid or air. After passive solar, it is generally the sec-

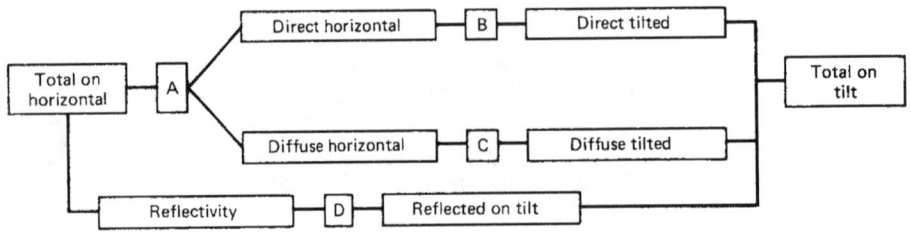

Figure 16-9. Conversion of horizontal insolation to insolation on tilted surface.

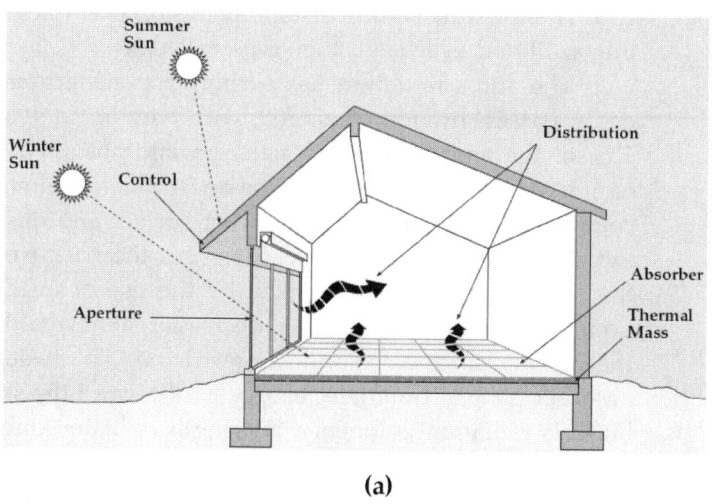
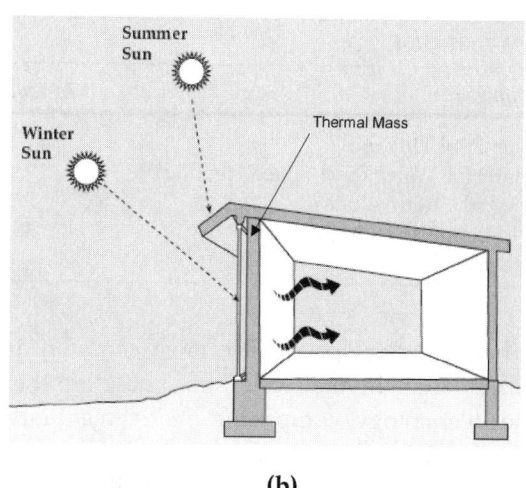

(a)   (b)

Figure 16-10. Five elements of passive solar design in a (a) direct gain configuration, and (b) an example of an indirect gain configuration (Ref. 20).

ond most economical use of solar energy in distributed applications. Indeed, since 1993 over three-quarters of distributed solar systems installed have been for pool heating and other water and space heating, as shown in Table 16-5. Principal factors determining cost-effectiveness are ability to use the solar energy when it is available, combined use for water- and space-heating applications, and local climate. In industrial systems, energy demand will rarely correlate with solar energy availability. In some cases, the energy can be stored until needed, but in most systems, there will be some available solar energy that will not be collected. Because of this factor, particular types of solar energy systems are most likely to be economically viable. Year-round use of large quantities of hot water at laundries, car washes, motels, and restaurants make them good candidates for solar thermal. Systems designed to both water- and space-heating loads can be more economical, because fuel savings accrue from both end-uses to payback capital costs. Cold climates require more expensive design components to prevent damage from nighttime freezing of the heating fluid.

An article on how to identify cost-effective solar-thermal applications is given in the *ASHRAE Journal* (Ref. 22). In almost any solar energy system the largest single expenses are the solar collector panels and support structure. For this reason the system is usually "sized" in terms of collector panel area. Pumps, piping, heat exchangers, and storage tanks are then selected to match. The optimum-size solar system is usually the one that is the most economical on some chosen investment criteria: simple payback, internal rate of return, or positive cash flow.

A typical set of calculations might lead to the results shown in Figure 16-11, the net annual savings per year versus the collector area, with the present cost of fuel as a parameter (Ref. 23). Curve "a" represents a low fuel cost, a negative net savings, and that the system would cost rather than save money. Curve "b" represents a slightly higher fuel cost where a system of about 800 ft² of collectors would break even. Curves "c" and "d," representing even higher fuel costs, show a net savings, with optimum savings occurring at about 1,200 and 2,000 ft², respectively.

A wide variety of devices may be used to collect solar energy. A general classification of types is given in Figure 16-12. These collectors are commonly categorized by operating temperature. *Low-temperature collectors* (~0 to 10°C above ambient temperature) are generally installed to heat swimming pools and, in rare instances,

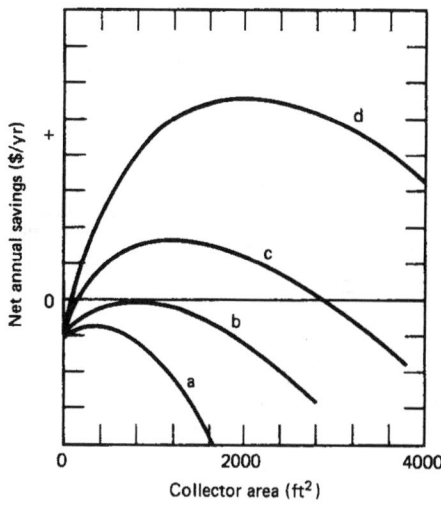

Figure 16-11. Collector area optimization curves for a typical solar heating system (Ref. 23).

Table 16-5. U.S. cumulative solar installations by technology, 1993-2007 (Ref. 21).

Technology	Installations
Solar Pool Heating	317,000
Solar Hot Water (and Space Heating)	193,000
Off-grid Photovoltaics	73,000
Grid-connected Photovoltaics	48,000
Total 631,000	

to heat interior spaces. The most common are the unglazed solar thermal collectors. These are the predominant technology, comprising over three-quarters of the installed solar thermal capacity in the U.S. in 2007 (Ref. 24). *Medium-temperature collectors* (~10°C to 50°C above ambient temperature) make the majority of building applications, and mainly include flat-plate technologies. *High-temperature collectors* (> 50°C above ambient temperature), such as evacuated tubes, are used mainly for industrial process heat, solar air conditioning, and (rarely) water heating in buildings. Flat-plate collectors, because they are most common in building applications, will be discussed first, followed by a discussion of tube-type or mildly concentrating collectors.

The flat-plate collector is a device that is usually faced to the south in the northern hemisphere (north in the southern hemisphere) and usually at some fixed angle of tilt from the horizontal. Its purpose is to use the solar radiation that falls upon it to raise the temperature of some fluid to a level above the ambient conditions. That heated fluid, in turn, may be used to provide hot water or space heat to drive an engine or a refrigerating device, or perhaps to remove moisture from a substance. A typical glazed flat-plate solar collector of the liquid

type is shown in Figure 16-13b, alongside examples of unglazed and evacuated tube-type collectors.

The sun's radiation has a short wavelength and easily passes through glazing(s), with only about 10 to 15% of the energy typically reflected and absorbed in each glazing. The sunlight that passes through is almost completely absorbed by the absorber surface and raises the absorber temperature. Heat loss out the back from the absorber plate is minimized by the use of insulation. Heat loss out the front is decreased somewhat by the glazing since air motion is restricted. The heated absorber plate also radiates energy back toward the sky, but this radiation is longer-wavelength radiation; most of this radiation not reflected back to the absorber by the glazing is absorbed by the glazing. The heated glazing, in turn, converts some of the absorbed energy back to the air space between it and the absorber plate. The trapping of sunlight by the glazing and the consequent heating is known as the *greenhouse effect*.

Energy is removed from the collector by the coolant fluid. A steady condition would be reached when the absorber temperature is such that losses to the coolant and to the surroundings equal the energy gain from the solar input. When no energy is being removed from the collectors by the coolant, the collectors are said to be at *stagnation*. For a well-designed solar collector, that stagnation temperature may be well above 300°F. This must be considered in the design of solar collectors and solar systems since loss of coolant pumping power might be expected to occur sometime during the system lifetime. A typical coolant flow rate for flat-plate collectors is about 0.02 gpm/ft^2 of collector surface (for a 20°F rise).

The fraction of the incident sunlight that is collect-

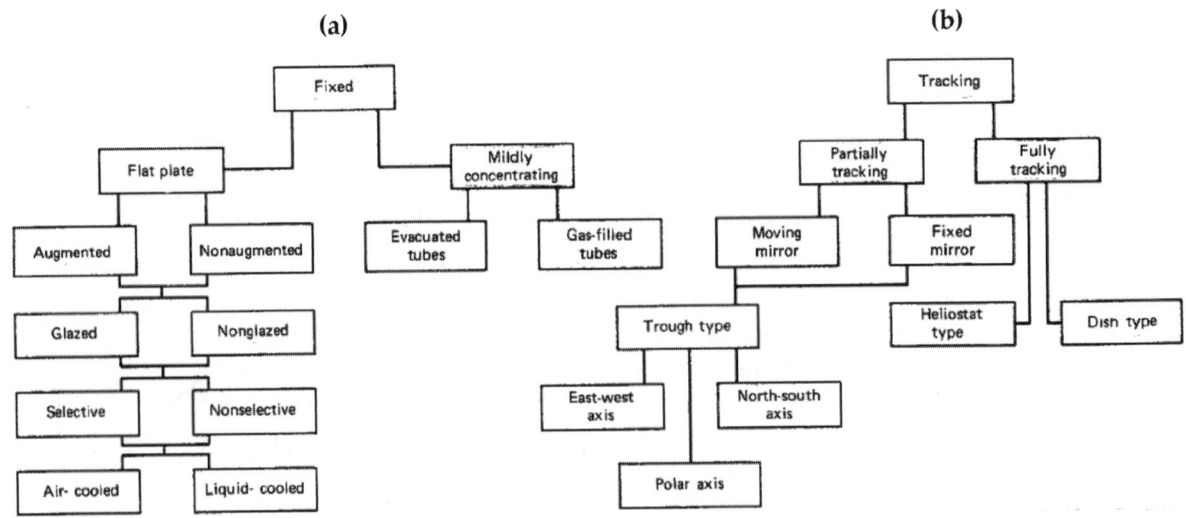

(a)   **(b)**

Figure 16-12. Types of solar collector systems used for (a) solar thermal and (b) solar-thermal electric applications.

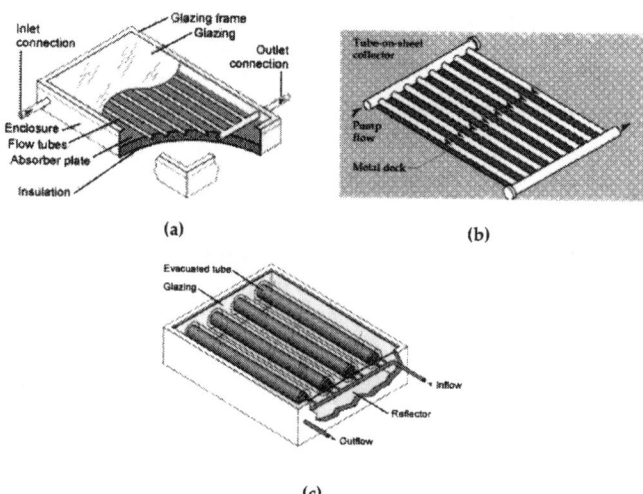

Figure 16-13. Examples solar thermal collectors: (a) glazed flat-plate, (b) unglazed, and (c) evacuated tube.

ed by the solar collector for useful purposes is called the *collector efficiency*. This efficiency depends upon several variables, which might change for a fixed absorber plate design and fixed amount of back and side insulation. These are:

1.   Ratio of insolation

2.   Number and type of glazing

3.   Ambient air temperature

4.   Average (or entering) coolant fluid temperature

A typical single-glazed flat-plate solar collector efficiency curve is given in Figure 16-14. The measured performance can be approximated by a straight line. The left intercept is related to the product $(\tau\alpha)$, where (τ) is the *transmittance* of the glazing and (α) is the *absorbance* of the absorber plate. The slope of the line is related to the magnitude of the heat losses from the collector, a flatter line representing a collector with reduced heat-loss characteristics.

A comparison of collector efficiencies for unglazed, single-glazed, and double-glazed flat-plate collectors is shown in Figure 16-15. Because of the lack of glazing reflections, the unglazed collector has the highest efficiencies at the lower collector temperatures. This factor, combined with its lower cost, makes it useful for swimming pool heating. The single-glazed collector also performs well at lower collector temperatures, but like the unglazed collector, its efficiency drops off at higher collection temperatures because of high front losses. The double-glazed collector, although not a good performer at lower temperatures, is superior at the higher temperatures and might be used for space heating and/or

cooling applications. The efficiency of an evacuated tube collector is also shown in Figure 16-15. It can be seen that it performs very poorly at low temperatures, but because of small heat losses, does very well at higher temperatures. Due to their high operating temperature, evacuated tube-type collectors are good for applications such as air conditioning, power generation, and the furnishing of industrial or process heat above 250°F (121°C).

A very important characteristic of a solar collector surface is its *selectivity*, the ratio of its absorbance (α_σ) for sunlight to its emittance (ϵ) for long-wavelength radiation. A collector surface with a high value of (α_σ/ϵ)

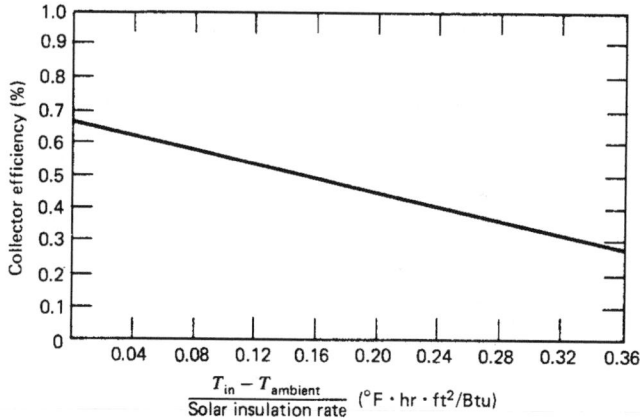

Figure 16-14. Efficiency of a typical liquid-type solar collector panel.

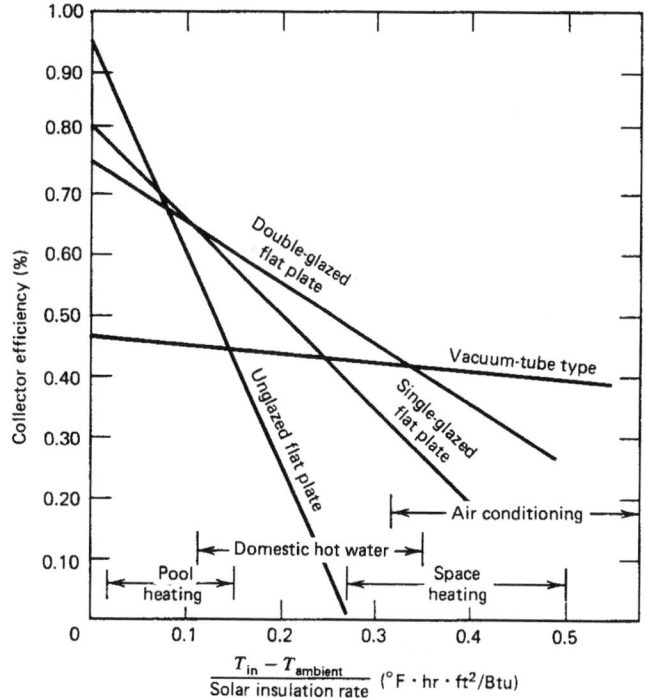

Figure 16-15. Comparison of collector efficiencies for various liquid type collectors, and suitable applications.

is called a selective surface. Since these surfaces are usually formed by a coating process, they are sometimes called *selective coatings*. The most common commercial selective coating is *black chrome*. The characteristics of a typical black chrome surface are shown in Figure 16-16, where $\alpha_\lambda = \varepsilon_\lambda$, the monochromatic absorbance and monochromatic emittance of the surface. Note that at short wavelengths (~ 0.5 μ), typical of sunlight, the absorbance is high. At the longer wavelengths (~2 μ and above), where the absorber plate will emit most of its energy, the emittance is high. Selective surfaces will generally perform better than ordinary blackened surfaces. The performance of a flat black collector and a selective coating collector are compared in Figure 16-17. The single-glazed selective collector performance is very similar to the double-glazed nonselective collector. Economic considerations usually lead one to pick a single-glazed, selective or a double-glazed, nonselective collector over a double-glazed, selective collector, although this decision depends heavily upon quoted or bid prices.

Air-type collectors are particularly useful where hot air is the desired end product. An increasingly popular commercial application for air-type collectors is HVAC pre-heat of intake air to improve furnace efficiency in cold climates; transpired air solar collectors (SolarWall®) represents one such technology. Air collectors have distinct advantages over liquid-type collectors:

1. Freezing is not a concern.
2. Leaks, although undesirable, are not as detrimental as in liquid systems.
3. Corrosion is less likely to occur.

Air systems may require large expenditures of fan power if the distances involved are large, or if the delivery ducts are too small. Heat-transfer rates to air are typically lower than those to liquids, so care must be taken in air collectors and in air heat exchangers to provide sufficient heat-transfer surface. This very often involves the use of extended surfaces or fins on the sides of the surface, where air is to be heated or cooled. Typical air collector designs are shown in Figure 16-18.

Flat-plate collectors usually come in modules about 3 ft wide by 7 ft tall, although there is no standard size.

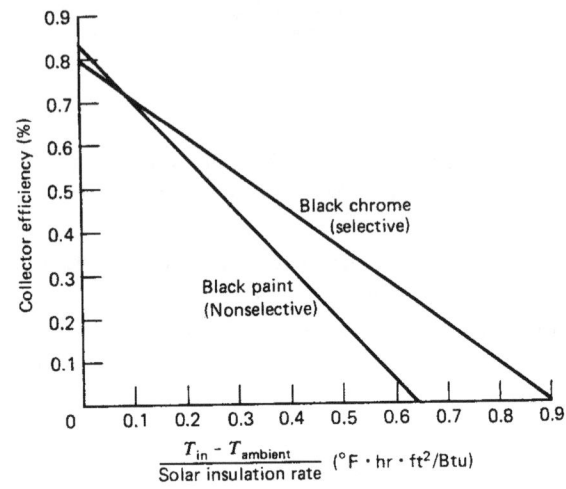

Figure 16-17. Comparison of the efficiencies of selective and nonselective collectors.

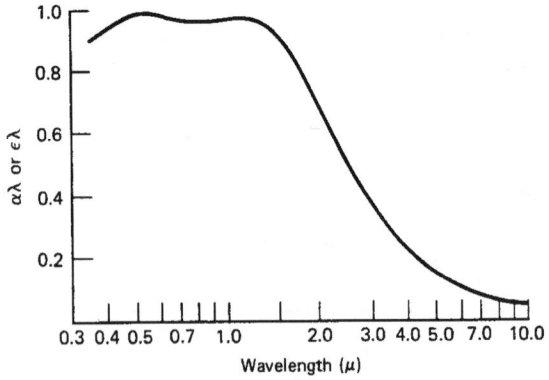

Figure 16-16. Characteristics of a typical selective (black chrome) collector surface.

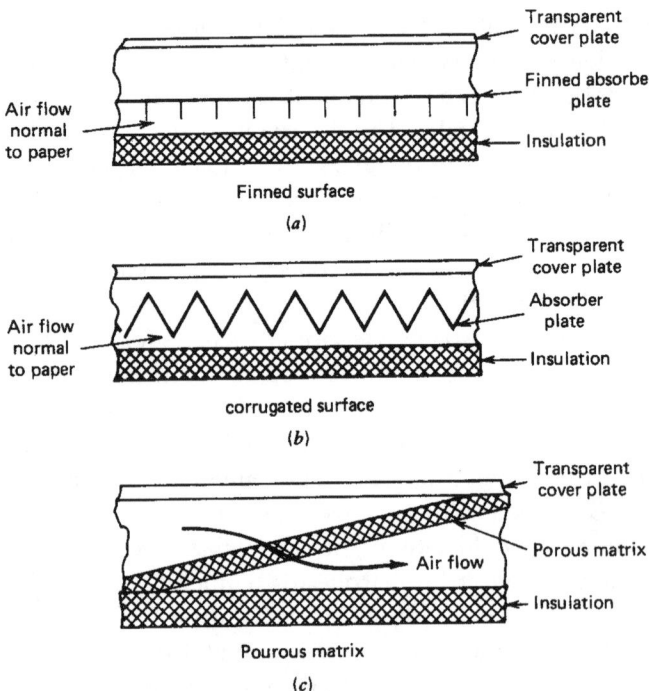

Figure 16-18. Typical air collector designs. (a) Finned surface. (b) Corrugated surface. (c) Porous matrix.

Collectors may have internal manifolds or they may be manifolded externally to form collector arrays. (See Figure 16-19.) Internally manifolded collectors are easily connected together, but only a small number can be hooked together in a single array and still have good flow distribution. Small arrays (5 to 15) are often piped together with similar arrays in various series and parallel arrangements to give the best compromise between nearly uniform flow rates in each collector and as small a pressure drop and total temperature rise as can be attained. Externally manifolded collectors are easily connected in balanced arrays when properly designed. However, these types of arrays require more field connections, have more exposed piping to insulate, and are not as tidy looking.

The overall performance of a collector array, measured in terms of the collector array efficiency, may be quite a bit less than the collector efficiency of the individual collectors. This is due primarily to unequal flow distribution between collectors, larger temperature rises in series connections than in single collectors, and heat losses from the connecting piping. A good array design will minimize these factors together with the pumping requirements for the array.

16.2.2.1 Thermal Storage Systems

Because energy demand is almost never tied to solar energy availability, a storage system is usually a part of the solar heating or cooling system. The type of storage may or may not depend upon the type of collectors used. With air-type collectors, however, a rock-bed type of storage is sometimes used. (See Figure 16-20.) The rocks are usually in the size range of 3/4 to 2 in. in di-

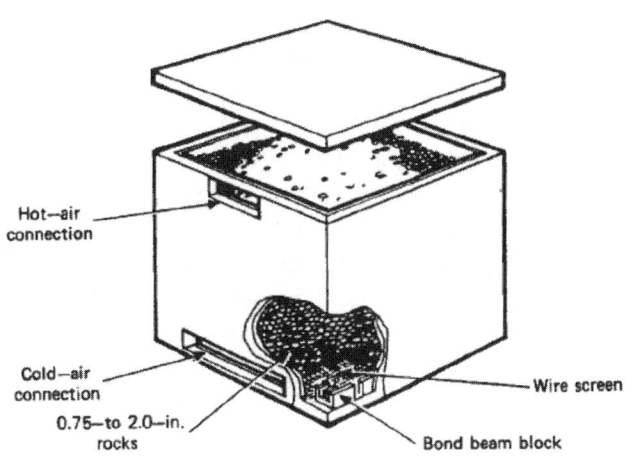

Figure 16-20. Rock-bed-type storage system.

ameter to give the best combination of surface area and pressure drop. Air flow must be down for storing and up for removal if this type system is to perform properly. Horizontal air flow through a storage bed should normally be avoided. An air flow rate of about 2 cfm/ft² of collector is recommended. The amount of storage required in any solar heating system is tied closely to the amount of collector surface area installed, with the optimum amount being determined by a computer calculation. As a rule of thumb, for rough estimates one should use about 75 lb of rock per square foot of air-type collectors. If the storage is too large, the system will not be able to attain sufficiently high temperatures, and in addition, heat losses will be high. If the storage is too small, the system will overheat at times and may not collect and store a large enough fraction of the energy available.

The most common solar thermal storage system is one that uses water, usually in tanks. As a rule the water storage tank should contain about 1.8 gal/ft² of collector surface. Water has the highest thermal storage capability of any common single-phase material per unit mass or per unit volume. It is inexpensive, stable, nontoxic, and easily replaced. Its main disadvantage is its high vapor pressure at high temperatures. This means that high pressures must be used to prevent boiling at high temperatures. Water also freezes; therefore, in most climates, the system must either (1) drain all of the collector fluid back into the storage tank, or (2) use antifreeze in the collectors and separate the collector fluid from the storage fluid by use of a heat exchanger.

Drain-down systems must be used cautiously, because one failure to function properly can cause severe damage to the collectors and piping. It is the more usual practice in large systems to use a common type of heat exchanger, such as a shell-and-tube exchanger, placed

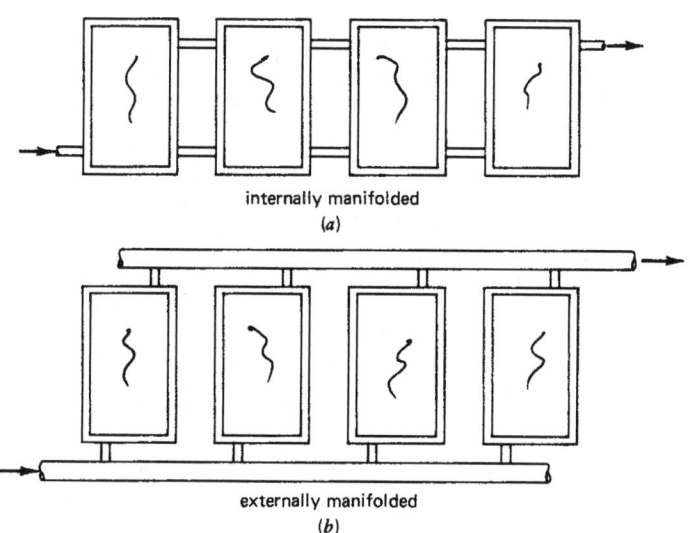

Figure 16-19. Examples of collectors hooked in parallel. (a) Internally manifolded. (b) Externally manifolded.

external to the storage tank, as shown in Figure 16-21. Another method, more common to small solar systems, is to use coils of tubing around or inside the tank, as shown in Figure 16-22.

In any installation using heat exchangers between the collectors and storage, the exchanger must have sufficient surface for heat transfer to prevent impairment of system performance. Too small a surface area in the exchanger causes the collector operating temperature to be higher relative to the storage tank temperature, and the collector array efficiency decreases. As a rough rule of thumb, the exchanger should be sized so as to give an effectiveness of at least 0.60, where the effectiveness is the actual temperature decrease of the collector fluid passing through the exchanger to the maximum possible

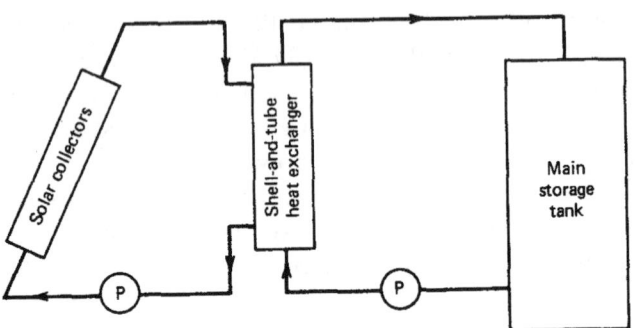

Figure 16-21. External heat exchanger between collectors and main storage

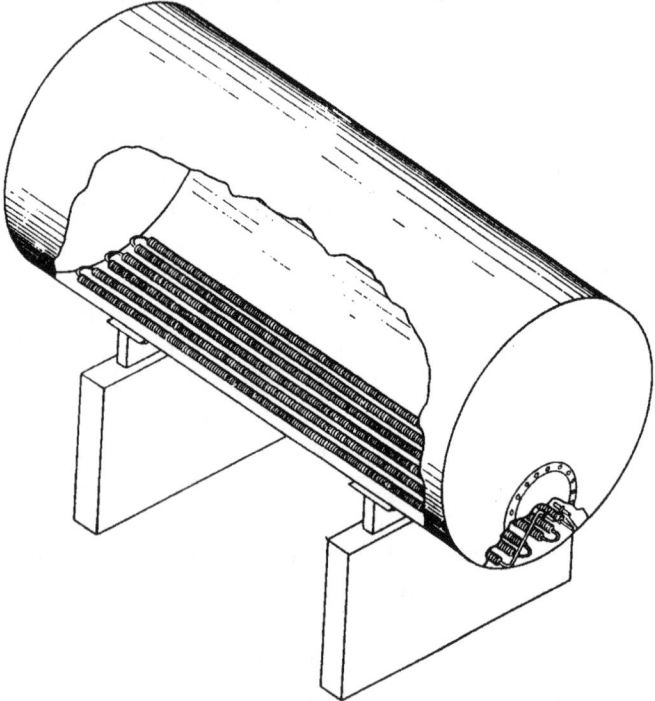

Figure 16-22. Internal heat exchanger between collector and storage medium.

temperature change. The maximum possible would be the difference between the design temperature of the collector fluid entering the exchanger and the temperature entering from the storage tank.

Stratification normally occurs in water storage systems, with the warmest water at the top of the tank. Usually, this is an advantage, and flow inlets to the tank should be designed so as not to destroy this stratification. The colder water at the bottom of the tank is usually pumped to the external heat exchanger and the warmer, returning water is placed at the top or near the center of the tank. Hot water for use is usually removed from the top of the tank.

16.2.2.2 Control Systems

Solar systems should operate automatically with little attention from operating personnel. A good control system will optimize the performance of the system with reliability and a reasonable cost. The heart of any solar thermal collecting system is a device to turn on the collector fluid circulating pump (and other necessary devices) when the sun is providing sufficient insolation so that energy can be used, or collected and stored. With flat-plate collectors it is common to use a differential temperature controller (see Figure 16-23), a device with two temperature sensors. One sensor is normally located on the collector fluid outlet and the other in the storage tank near the outlet to the heat exchanger (or at the level of the internal heat exchanger). When the sun is out, the fluid in the collector is heated. When a prescribed temperature difference (about 20°F) exists between the two sensors, the controller turns on the collector pump and other necessary devices. If the temperature difference

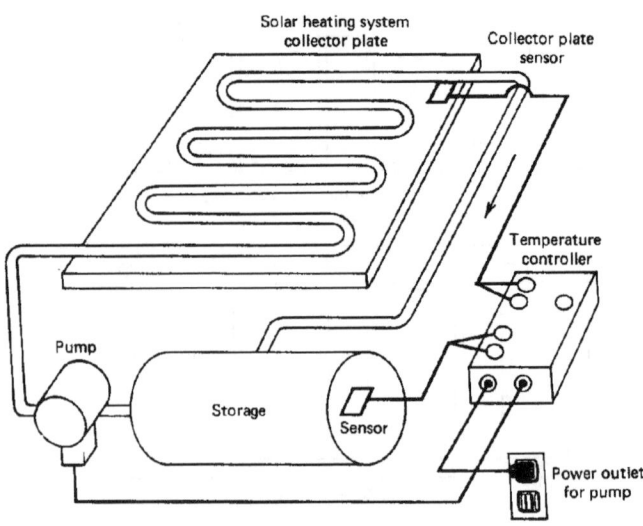

Figure 16-23. Installation of a differential temperature controller in a liquid heating system.

drops below some other prescribed difference (about 3 to 5°F), the controller turns off the necessary devices. Thus clouds or sundown will cause the system to shut down and prevent not only the unnecessary loss of heat to the collectors but also the unnecessary use of electricity. The distinct temperature difference to start and to stop is to prevent excessive cycling.

Differential temperature controllers are available with adjustable temperature difference settings and can also be obtained to modulate the flow of the collector fluid, depending upon the solar energy available. Controllers for high-temperature collectors, such as evacuated tubes and tracking concentrators, sometimes use a light meter to sense the level of sunlight and turn on the pumps. Some concentrating collectors are inverted for protection when light levels go below a predetermined value.

In some systems the storage fluid must be kept above some minimum value (e.g., to prevent freezing). In such cases a *low-temperature controller* is needed to turn on auxiliary heaters if necessary. A *high-temperature controller* may also be needed to bypass the collector fluid or to turn off the system so that the storage fluid is not overheated.

Figure 16-24 shows a control diagram for a solar-heated asphalt storage system (see Figure 16-25) in which the fluid must be kept between two specified temperatures. Solar heat is used whenever it is available (collector pump on). If the storage temperature drops below the specified minimum, the pump *and* an electric heater are turned on to circulate electrically heated fluid

to the tank. If the tank fluid gets too warm, the system shuts off. Almost any required control pattern can be developed for solar systems, using the proper arrangement of a differential temperature controller, high- and low-temperature controllers, relays, and electrically operated valves.

16.2.3 Solar-thermal Electric

An area of growing interest for solar thermal is bulk power applications. Concentrating collectors operated at very high temperatures (>250°C) can produce steam used to generate electricity through a steam-Rankine cycle. Because these technologies utilize a thermal intermediary, they can be hybridized with fossil fuel and, in some cases, can be adapted to utilize thermal storage, making them attractive for utility-scale applications. These technologies make use of the direct normal radiation component of solar radiation. They generally cannot use the diffuse or scattered radiation from the sky (global radiation) and therefore use solar tracking systems to ensure that the sun's direct rays will be concentrated on the receiver. An important point to make is that concentrating collectors do not increase the amount of energy above that which falls on the mirrored surfaces; the energy is merely concentrated to a smaller receiver surface.

In the mid 1980s the first solar thermal power stations were installed in the southwestern U.S. The most famous of these is still operating today. The solar energy generating station (SEGS) facility is the largest solar power installation in the world, consisting of nine plants

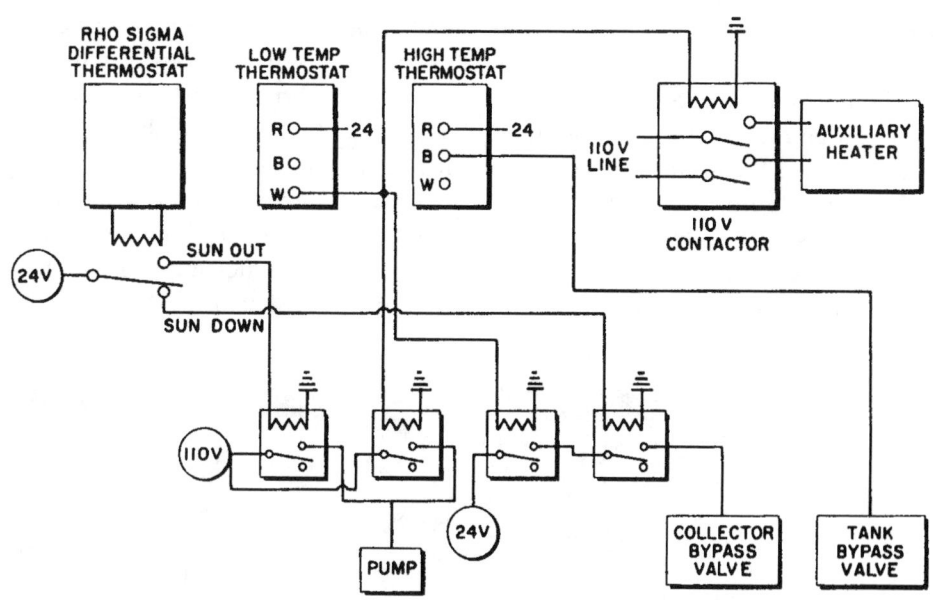

Figure 16-24. Control system for the solar-heated asphalt storage tank of Figure 16-22.

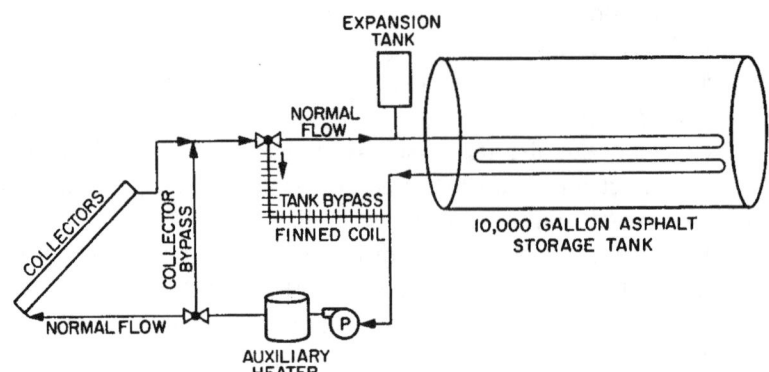

Figure 16-25. Flow schematic of a solar-heated asphalt storage.

totaling 354 MW. For the first time since the 1980s, solar thermal projects of various types are being developed for power production in the U.S. Solar thermal plants are beginning to compete with wind power in utility plans for renewable procurement, in part because their peak output is roughly coincident with the electric system peak (summer, midday), when air conditioning loads are at maximum and power is most expensive to produce.

Solar thermal plants consist of two main sub-components—a collector system and a power block. Three main types of solar thermal power systems are profiled here: parabolic trough, central receiver (or power tower), and dish-engine.

Parabolic trough systems have been the dominant technology to date. They are considered commercially available, with several developers producing the technology. A typical parabolic trough-type solar collector array is shown in Figure 16-26. Here the concentrating surface or mirror is moved, to keep the sun's rays concentrated as much as possible on the receiver pipe or *heat collection element* (HCE), in this case a tube through which a *heat transfer fluid* (HTF) flows. In some systems the tube moves and the mirrored surfaces remain fixed. This type of collector can be mounted on an east-west axis and track the sun by tilting the mirror or receiver in a north-south direction. (See Figure 16-27a.) An alternative is to mount the collectors on a north-south axis and track the sun by rotating in an east-west direction. (See Figure 16-27b.) A third scheme is to use a polar mount,

aligning the trough and receiver parallel to the earth's pole or inclined at some angle to the pole and tracking east to west. (See Figure 16-27c.) Each has its advantages and disadvantages; the selection depends upon the application. All tracking collectors must have some device to locate the sun in the sky, either by sensing or by prediction. Tracking motors, and in some cases flexible or movable line connections, are additional features of tracking systems. A good discussion of concentrating collectors is given in Ref. 25.

Existing US parabolic trough projects include the SEGS and a 64 MW plant installed in 2007 in Nevada, USA. These plants use high-temperature oil as a thermal intermediary that passes through a heat exchanger to produce steam. Natural-gas fired back-up is used to manage thermal flux of the oil HTF. An important concept in solar thermal power is *thermal inertia*: the ability of a given volume of a substance to store (and release) energy without undergoing a phase change. Thermal inertia enables solar thermal systems to produce relatively stable power output, despite variation in the solar resource caused by passing clouds. Parabolic trough systems designed with thermal storage commonly employ molten nitrate salt, which has higher thermal inertia than oil, but these systems require oil/salt heat exchangers and add capital costs. In general, constraints on parabolic trough technology are the high cost of tubes and mirrors, significant heat losses from long runs of hot oil from the collector system to the power block, and

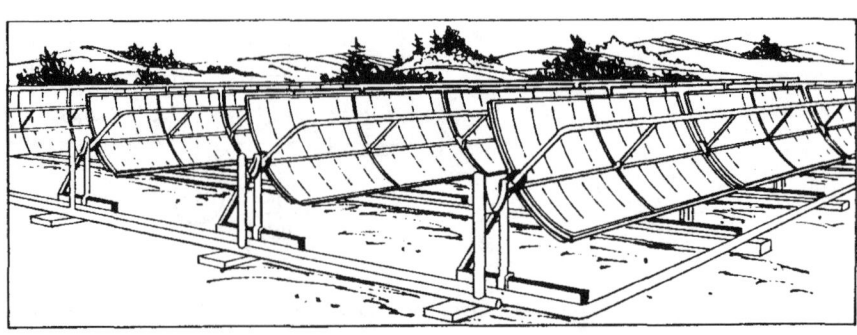

Figure 16-26. Typical parabolic trough-type solar collector array (Suntec, Inc.).

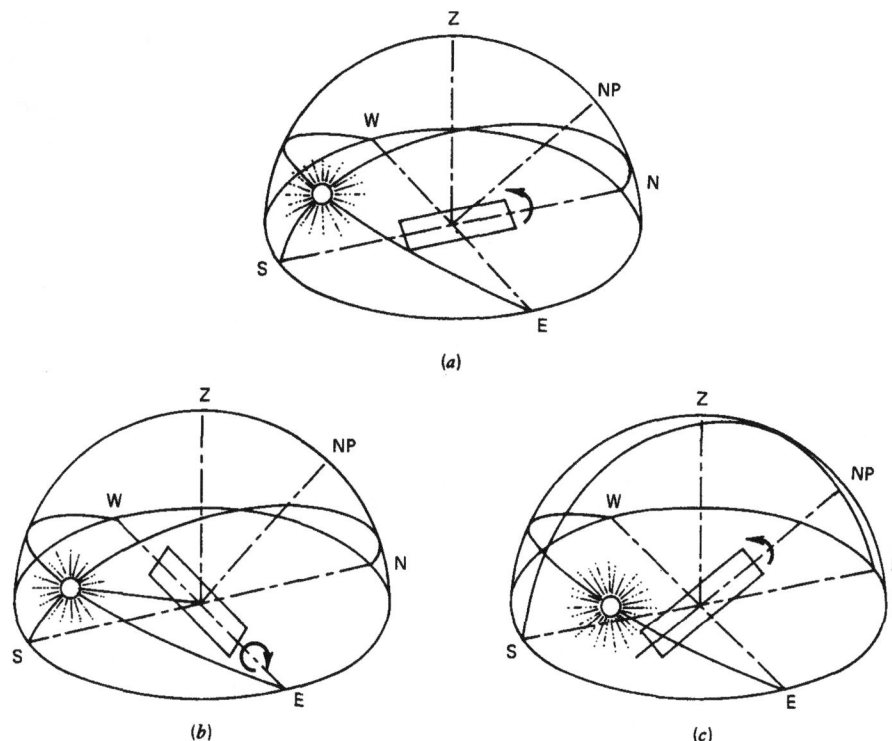

(a)

(b)   *(c)*

Figure 16-27. Trough-type collector arrangements for sun tracking. (a) N-S horizontal E-W tracking. (b) E-W horizontal N-S tracking. (c) Polar axis E-W tracking.

limited worldwide supply of curved mirrors.

Central receiver systems use sun-tracking mirrors, called *heliostats*, to focus sunlight onto a receiver at the top of a tower. A schematic is given in Figure 16-28. Steam is either produced directly in the tower, or indirectly via a molten nitrate salt or other HTF. An advantage to central receiver design is that the molten nitrate salt can be used as both HTF and thermal storage medium because of the shorter run from the collector system to the power block, unlike the long runs in parabolic trough systems. Figure 16-28 denotes molten salt as the HTF at this facility. Despite its advantages, molten salt can be challenging to work with due its high freezing point (Note 7). Research and development are focused on systems that use air as HTF, as well as salt alternatives with lower freezing points.

A 10 MW power tower plant, Solar One, operated in California from 1982 to 1988, and then again (as Solar Two) from 1998 to 1999 after it was retrofitted to improve heat transfer and storage capabilities. These operations demonstrated certain design improvements that are needed for the technology to become commercially viable. In 2007, announcements of new plants under development in California and Spain suggest that the technology is moving into the mainstream market.

Table 16-6 gives a summary of key performance and cost characteristics of solar thermal electric technologies, including dish-Stirling engine, an emerging technology profiled in Section 16.6.1. Due to their high operating temperatures, central receiver and dish-engine technologies are able to achieve higher solar-to-electric efficiencies, compared to parabolic troughs. Central receiver has the highest *capacity factor*—fraction of rated capacity utilized on an annual average basis—because of its thermal storage capability and ability to generate power after dark.

16.2.4 Solar Photovoltaics

Solar cells, or photovoltaics (PV), use the electronic properties of semiconductor material to convert sunlight directly into electricity. They are widely used today in space vehicles and satellites, and in terrestrial applications requiring electricity at remote locations. Since the conversion is direct, solar cells are not limited in efficiency by the Carnot principle. A wide variety of texts are available to give details of the operating principles, technology, and system applications of solar cells.

Most solar cells are very large area p-n junction diodes. (See Figure 16-29a.) A p-n junction has electronic asymmetry. The n-type regions have large electron densities but small hole densities. Electrons flow readily through the material, but holes find it very difficult. P-type material has the opposite characteristic. Excess electron-hole pairs are generated throughout the p-type material when it is illuminated. Electrons flow from the p-type region to the n-type, and a flow of holes occurs in the opposite direction. If the illuminated p-n junction is electrically short circuited, a current will flow in the

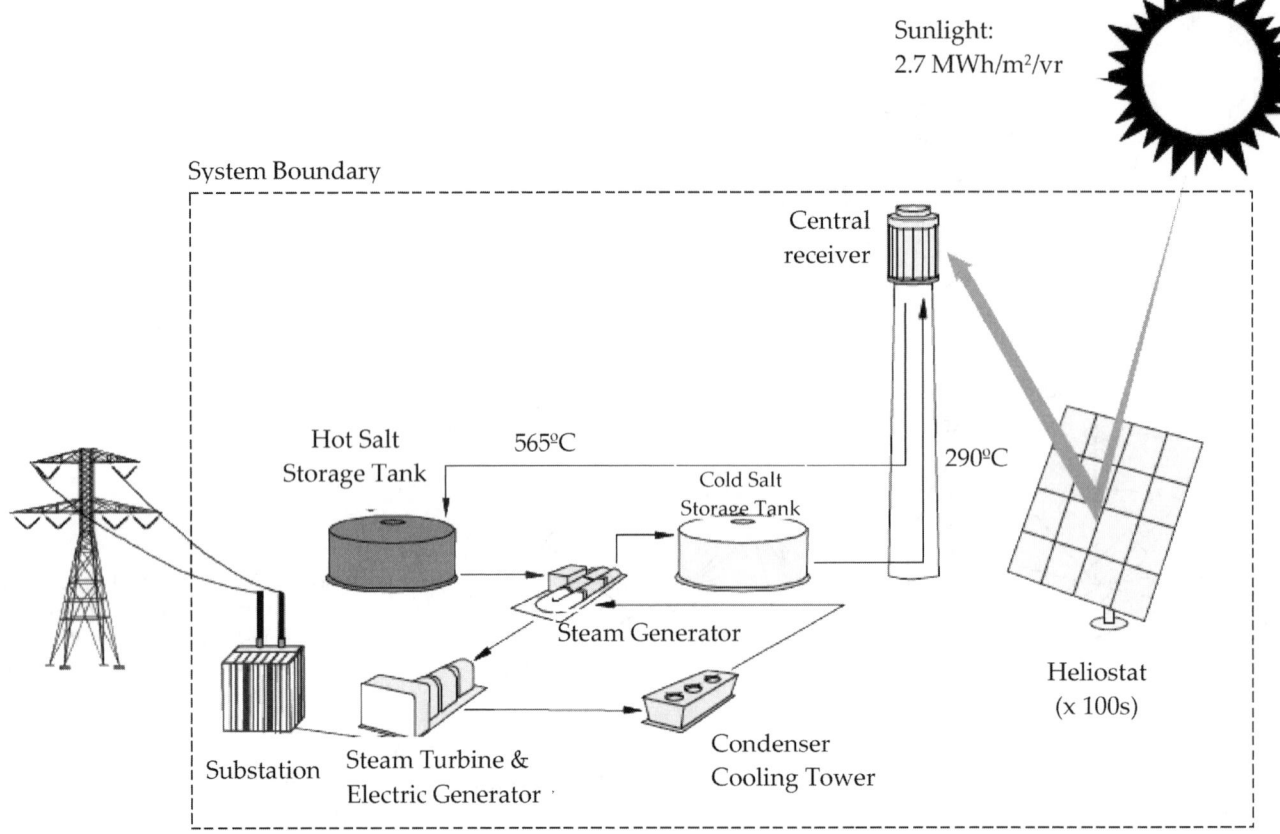

Figure 16-28. Schematic of a central receiver solar power station (Adapted from Ref. 26).

Table 16-6. Characteristics of solar thermal electric power systems (Adapted from Ref. 27 and Ref. 28*).

	Parabolic Trough	Central Receiver	*Dish-Stirling Engine†*
Size	30-320 MW	10-200 MW	*5-25 kW*
Operating temperature (°C/°F)	390/734	565/1,049	*750/1,382*
Annual capacity factor*	28%	78%	*24%*
Peak efficiency	20%	23%	*29%*
Net annual efficiency*	13%	14%	*20%*
Commercial status	Commercial	Scale-up demonstration	*Prototype demonstration*
Technology development risk	Low	Medium	*High*
Storage available	Limited	Yes	*Battery*
Hybrid (fossil) designs	Yes	Yes	*Yes*
Cost			
Capital ($/kW)*	2,805	6,800	*N/A*
O&M ($/kWh)*	0.02	0.04	*N/A*

† See section 16.6.1. Cost information for dish-engine is not given, because, as of 2008, the technology is still in the prototype demonstration phase.

short-circuiting lead. The normal rectifying current-voltage characteristic of the diode is shown in Figure 16-29b. When illuminated (insulated) the current generated by the illumination is superimposed to give a characteristic where power can be extracted.

The characteristic voltage and current parameters of importance to utilizing solar cells are shown in Figure 16-29b. The short-circuit current (I_{sc}) is, ideally, equal to the light generated current (I_L). The open-circuit voltage (V_{oc}) is determined by the properties of the semiconductor. At the particular point on the operating curve where the power is maximum, the rectangle defined by (V_{mp}) and (I_{mp}) will have the greatest area. The fill factor FF is a measure of how "square" the output characteristics are. It is given by:

$$FF = \frac{V_{no}I_m}{V_{oc}I_{sc}}$$

Ideally FF is a function only of the open-circuit voltage and in cells of reasonable efficiency has a value in the range of 0.7 to 0.85.

Most solar cells are made by doping silicon, the second most abundant element in the earth's crust. Sand is reduced to metallurgical-grade silicon, which is then further purified and converted to single-crystal silicon wafers. The wafers are processed into solar cells. Multiple solar cells are interconnected and encapsulated into a weatherproof photovoltaic *module* with ratings from 10 to 350 Watts (DC). Multiple modules are wired into an *array* to achieve the desired system sizing.

In cell manufacturing processes, boron is frequently used as a doping agent to produce p-type wafers, and phosphorus is the most common material used for the n-type impurity. These simple cells have theoretical maximum conversion efficiencies around 21 to 23 percent, because energy absorption is limited by the solar spectrum associated with the doping compound. Figure 16-30 illustrates how different doping materials absorb specific segments of the solar spectrum. Advanced, multi-junction cells have demonstrated efficiencies approaching 40% in the research lab by layering cells with

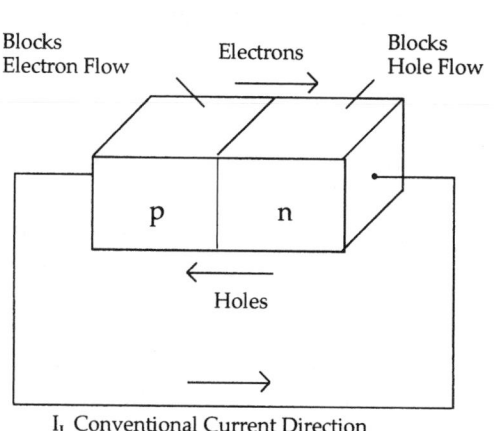

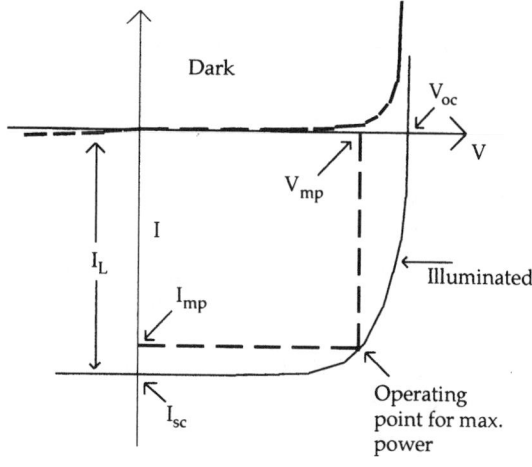

Figure 16-29. Nomenclature of solar cells.

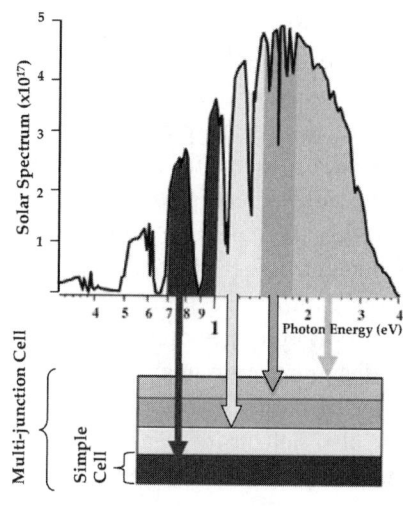

Figure 16-30. Conceptual illustration of solar spectra absorbed by multi-junction and simple cell PV technologies.

multiple doping agents to absorb a broader spectrum of solar energy. Other factors that limit the attainment of theoretical efficiencies include reflection losses, incomplete absorption and partial energy utilization, incomplete collection of electron-hole pairs, a voltage factor, a curve factor, and internal series resistance.

Numerous strains of PV technology have emerged employing various solutions to these limiting factors with the goal of achieving higher efficiencies and lower cost. Figure 16-31 provides a useful overview of various PV technologies and illustrates an increasing trend in research cell efficiencies since 1975. NREL classifies solar cells into four broad categories: multi-junction, crystalline silicon, thin film, and emerging PV. High-efficiency

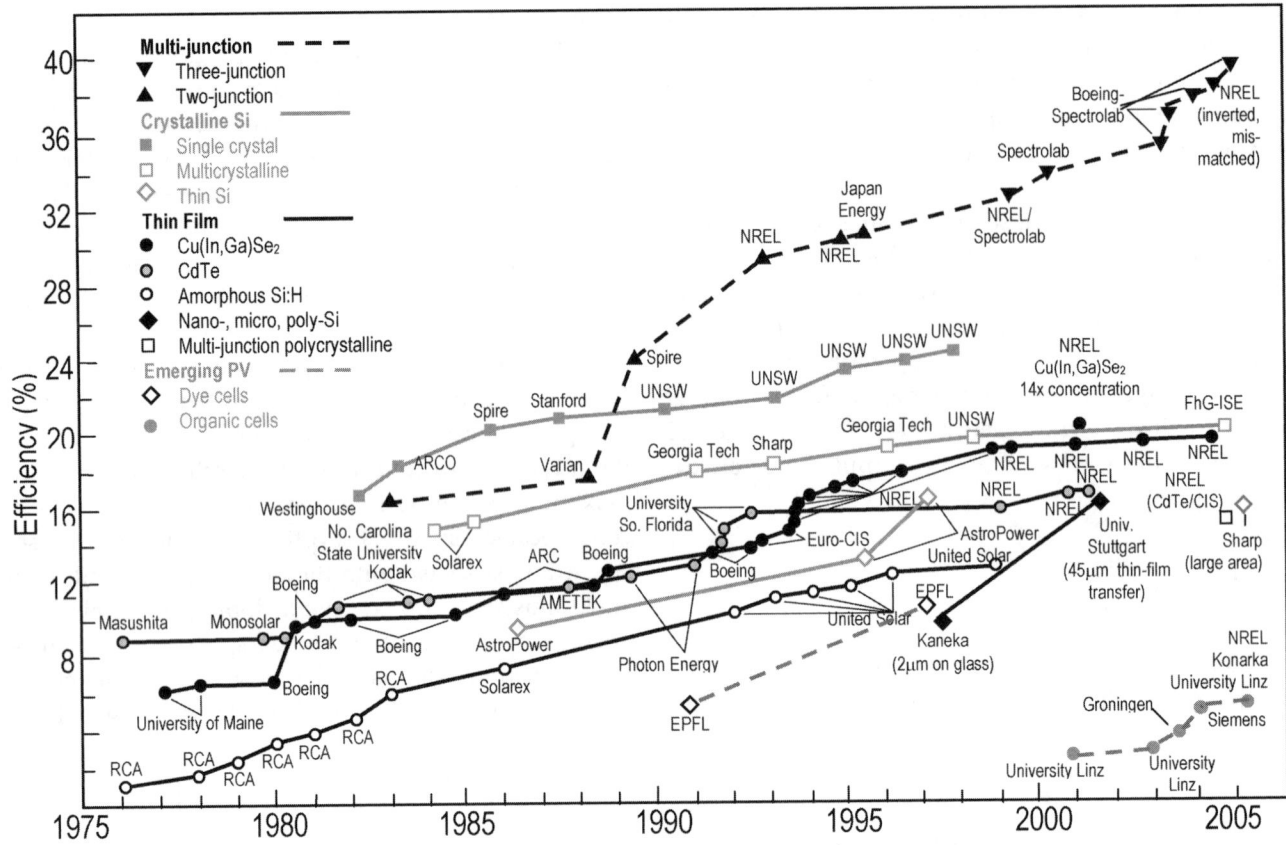

Figure 16-31. Photovoltaic research cell efficiencies, 1975-2005 (Source: NREL).

multi-junction cells are the dominant power generation technology in space applications, but they are not yet commercially available due to their very high cost.

Crystalline silicon cells are the dominant technology in the PV market today, with over 90% global market share in 2007 (Ref. 29). Thin film cells include amorphous silicon (which currently holds 60% of the thin-film market) and higher efficiency technologies such as cadmium telluride (CdTe), copper indium (gallium), and di-selenide (CIS or CIGS). Because they have lower efficiency than crystalline silicon, thin film arrays occupy more square footage for the same rated output. A good rule of thumb for estimating the footprint of crystalline PV is 100 ft^2 per kW. On a dollars per watt basis, crystalline and thin film PV modules are somewhat comparable.

16.2.4.1 Cost, Performance & Sizing

Forecasts of PV module efficiency and production cost are given in Tables 16-7 and 16-8. Energy managers involved with planning PV projects will be primarily concerned with module costs, and secondarily with efficiencies, to the extent that space is a limiting factor. From a practical standpoint, another important factor in selecting a technology is the module's PTC rating,

which is a more "real world" test condition than the manufacturer's STC rating (Note 8). A useful reference for looking up PTC ratings is maintained on the Go Solar California website, www.gosolarcalifornia.gov/equipment/pvmodule. Module costs are typically 50% of the total installed cost of a PV system, with balance-of-system components and installation making up the difference. The module costs shown in Table 16-7 are factory costs; consumer cost is higher. For example, a customer in 2006 might see retail costs in the range of $4 to 5 per watt.

Solar cells are arranged in a variety of series and parallel arrangements to give the voltage-current characteristics desired and to assure reliability in case of individual cell failure. Fixed arrays are placed at some optimal slope and usually faced due south in the northern hemisphere, with some notable exceptions. Latitude tilt is roughly the optimal tilt to maximize annual output from the system. However, lower tilt or horizontal configurations can effectively maximize summer output, if reduced utility demand (kW) charges are the driving consideration. Likewise, designers may wish to increase afternoon solar output by orienting the array slightly east of south, which may also help to decrease demand charges. In effect, these strategies trade off greater an-

Table 16-7. PV Module Efficiency Status in 2006, 2010 and 2015; % DC Efficiencies for Production Modules (Adapted from Ref. 29).

	Cell Technology	2006	2010	2015
Crystalline Silicon	Monocrystalline Silicon	14-19	16-22	22-25
	Multicrystalline Silicon Cast Ingot	13-17	16-18	20+
	a-Si/Monocrystalline slice (HIT)	16-18	18-20	22-24
	Ribbon/Sheet Silicon	14-16	16-18	20+
	Concentrators Silicon Cell/III-V Cell	24/36	26/27	28/38
Thin Film	Amorphous Silicon (a-Si)	6-8	9-10	12
	Silicon Film	10	14	16
	$Cu(In,Ga)Se_2$ (CIS/CIGS)	9-11	10-12	14
	Cadmium Telluride (CdTe)	9-10	11	12

Table 16-8. PV Module Manufacturing Costs and Factory Profitable Price in 2006, 2010, 2015; 2006 US$ per Wall DC (Adapted from Ref. 29).

	Cell Technology	2006	2010	2015
Crystalline Silicon	Monocrystalline Silicon	2.50/3.75	2.00/2.50	1.40/2.20
	Multicrystalline Silicon Cast Ingot	2.40/3.55	1.75/2.20	1.20/2.00
	Ribbon/Sheet Silicon	2.00/3.35	1.60/2.20	1.00/1.70
	Concentrators Silicon Cell	3.00/5.00	1.50/2.90	1.00/1.70
Thin Film	Amorphous Silicon (a-Si)	1.50/2.50	1.25/1.75	0.90/1.40
	$Cu(In,Ga)Se_2$ (CIS/CIGS)	1.50/2.50	1.00/1.75	0.90/1.33
	Cadmium Telluride (CdTe)	1.50/2.50	0.90/1.50	0.65/1.25
	Factory Profitable Price	2.50 - 3.75	1.50 - 2.50	1.25 - 2.20

nual output at lower average value for lower annual output at higher average value to the customer. Another point, illustrated in Table 16-9, is that sub-optimal array orientation and tilt-angle rarely reduce annual expected output by more than 10%. Particularly in roof-mounted situations, orientation may be controlled by the available roof surface, but the penalty is small, even for west or east-facing systems. Single- and double- axis tracking systems can increase PV output by up to 30% and 35%, respectively, but each requires significant increases in capital investment and O&M cost for the tracking devices. Wind loads can also be a serious problem for any solar collector array designed to track.

When sizing a grid-tied PV system for a building application, a good first step is to first estimate the rating required to hypothetically supply 100% of the expected annual load, using Eq. 16.1:

$$\text{Rating} = \frac{\text{Annual Load}}{(8,760)\,(0.18)} \qquad \text{Eq. 16.1}$$

Table 16-9. Typical de-rate factors for PV output as a function of array tilt angle and orientation at latitude 30° in the northern hemisphere (Note: for southern hemisphere, "north" replaces "south").

	De-Rating Factor	Array Tilt Angle From Horizontal					
		0°	18°	30°	45°	60°	90°
Orientation	South	0.89	0.97	1	0.97	0.89	0.58
	SSE or SSW	0.89	0.97	0.99	0.96	0.88	0.59
	Southeast or Southwest	0.89	0.95	0.96	0.93	0.85	0.60
	ESE or WSW	0.89	0.92	0.91	0.87	0.79	0.57
	East or West	0.89	0.88	0.84	0.78	0.70	0.52

where:

Rating = kW overall AC power output from the array/inverter system

Annual

load = kWh

8,760 = the number of hours in a year (or other for partial occupancy)

0.18 = average annual capacity factor for PV (Note 9)

Solar cells produce DC electricity, which in grid-connected applications must be converted to AC electricity at suitable frequency and voltage through a power conditioning device, or *inverter*. (See Figure 16-32.) The inverter causes 5 to 15 % parasitic losses, as do other factors beyond the user's control: array temperature loss (5% to 12%), dust and dirt (2% to 4%), module mismatch (2%), and DC and AC wire losses (2%). Temperatures have a dominant effect on the open circuit voltage of solar cells; for silicon cells the power output decreases by 0.4 to 0.5 % per degree Kelvin increase. Additional de-rate factors are somewhat controllable, including the degree of shading from obstacles, array orientation and tilt, and the PTC rating of module and inverter components. Together, these factors combine to produce an average DC-to-AC de-rate factor of 77%, which needs to be taken into account in DC sizing estimates. Once the full-load DC rating is calculated, the designer can adjust the final design capacity to the project budget and objectives. These procedures for calculating expected performance are automated when using on-line PV performance calculators, such as NREL's PVWATTS.

16.2.4.2 Installation Economic Models

Largely as a result of policy incentives, grid-tied PV is one of the most prominent distributed renewable power technologies in the commercial and industrial marketplace. Availability of some PV incentives varies by location. Table 16-10 gives an indication of which U.S. states are most favorable for PV. But even in locations with favorable policy, PV projects can entail a sizeable capital outlay if they are owner-financed. For example, if a 30 kW commercial installation costs the owner $120,000 after rebates and tax benefits, this poses a sizeable investment for the owner. Few companies have cash reserves lying around to pay large upfront costs unless utility bill savings can be shown to quickly return the investment. When considering a PV project, energy managers must analyze project economics under different project financing options. (See also Chapter 25 for project financing options and Chapter 4 for economic analysis.)

In commercial retrofit situations, the two most common ownership models are (1) end-user owned and (2) third-party owned. The end-user model had been the dominant model in the marketplace, until 2007, when an estimated 50% of commercial and institutional installations were third-party financed (Ref. 30). Several financial analysis methods are possible in the end-user ownership case, including simple payback, total lifecycle payback, rate of return, cash flow with financing, and property appraisal valuation (Note 6). A description of these methods is given below,

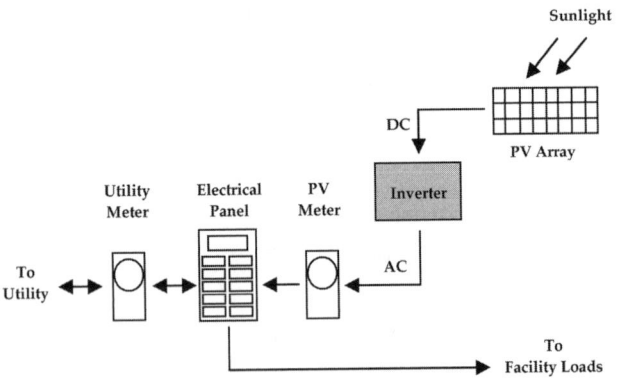

Figure 16-32. Typical grid-connected PV system diagram showing inverter.

Table 16-10. Top-ten states by cumulative grid-connected PV capacity (Ref. 21).

State	MW$_{DC}$	Market Share
1. California	329	69%
2. New Jersey	44	9%
3. Nevada	19	4%
4. Arizona	19	4%
5. New York	15	3%
6. Colorado	15	3%
7. Massachusetts	5	1%
8. Hawaii	4	<1%
9. Texas	3	<1%
10. Oregon	3	<1%
All other states	20	4%
Total	**476**	**100%**

as well as indicative results for illustration purposes, with the caveat that these generalizations and examples do not substitute for properly conducted research and financial analyses of a specific project. For project-level evaluations, a recommended tool is the Clean Power Estimator, a web-based spreadsheet model available for free to the public at www.consumerenergycenter.org/renewables/estimator.

Simple payback, which indicates the number of years it takes to payback the initial investment, is a relatively crude method because it ignores the time value of money. By this method, PV is rarely cost-effective because of high first cost. Total lifecycle payback evaluates cost-effectiveness over the life of the asset. Using the lifecycle method, the PV system shows excellent return since it has a 25-year life with no fuel cost and very low O&M costs. For example: a PV system that has a 5- to 7-year payback and a 25-year life will have an internal rate of return (IRR) rate of nearly 14%.

$$P/A \text{ (payback)} = \frac{(1+i)^n - 1}{(i\,(1+i)^n)} \qquad \text{Eq. 16.2}$$

where in this example:

P/A= payback period = 7 years

N=measure life expectancy = 25 years

i =interest rate; iterating four times to solve for i yields approximately 13.8% IRR

Another way to show PV projects can be economically viable is to consider cash flow, while financing the up front cost. The cash flow method compares the utility bill savings to the cost of financing a PV system. As long as the cost of borrowing is less than the bill savings, cash flow is positive from day one. However, cash flow benefits must be weighed against potentially negative balance sheet impacts affecting debt-equity ratios. Finally, investing in a PV system may improve the value of a commercial property. Because a PV system reduces operating costs, appraisal valuation can sometimes add value to a property on the basis of O&M savings multiplied by a "cap rate" ranging from 4% to 7%. Appraisal valuation is a method that applies only to owner-occupied or owner-managed business properties, which is fairly limited in commercial real estate.

Third-party ownership is becoming increasingly common as PV developers become more sophisticated in their delivery models and as enabling policies spread to new markets. In general, these types of deals are seen on larger installations (>100kW). There are several reasons why third-party ownership may be attractive

when it is available as an alternative:

- To avoid capital investment or debt financing, if cash reserves are limited or balance sheets are tight
- To monetize tax benefits available to for-profit companies, as in the case of institutions or non-profits
- To avoid performance risk and the hassle of installation, operations, maintenance, and repair
- To hedge against future electricity rate escalation or volatility
- To combine solar services with other building services, such as roofing

The main caveats with the third-party model are that end-user ownership is delayed (if it occurs), legal transaction costs are higher, provider access to the site may present problems for high-security facilities, and landlord acceptance is required for leased facilities.

Figure 16-33 gives an overview of the third-party ownership-model and illustrates the parties and transactions involved. It is important to note that in an end-user ownership structure most, if not all, of the same vectors are involved, except that the end-user, rather than a third-party, is at the nexus of the transaction activities. Thus, Figure 16-33 is useful when considering either model.

In the third-party business model, the end-user acts as a "host" and agrees to pay a fixed price for all-inclusive solar electric services under a power purchase agreement (PPA). The contract term is usually 5 to 20 years, with nominal performance guarantees. The third-party is responsible for all aspects of the PV project: purchasing equipment; designing and installing the system; pulling building permits and passing local inspections; obtaining an interconnection agreement and net metering service from the local utility; applying for incentives from the state or local utility; capturing federal (and state) tax benefits, principally the ITC and accelerated depreciation; selling the green attributes or renewable energy credits (RECs) to an aggregator (where such markets exist) or giving them to the utility in exchange for cash rebates; operating the system to maximize performance; and decommissioning the system at the term of the agreement (if ownership does not transfer). Under the terms of the contract, the third-party is usually paid through a monthly payment that covers their cost and a margin.

Third-party arrangements come in several varieties. A more thorough discussion of these and other PV business models is given in Refs. 31 and 32.

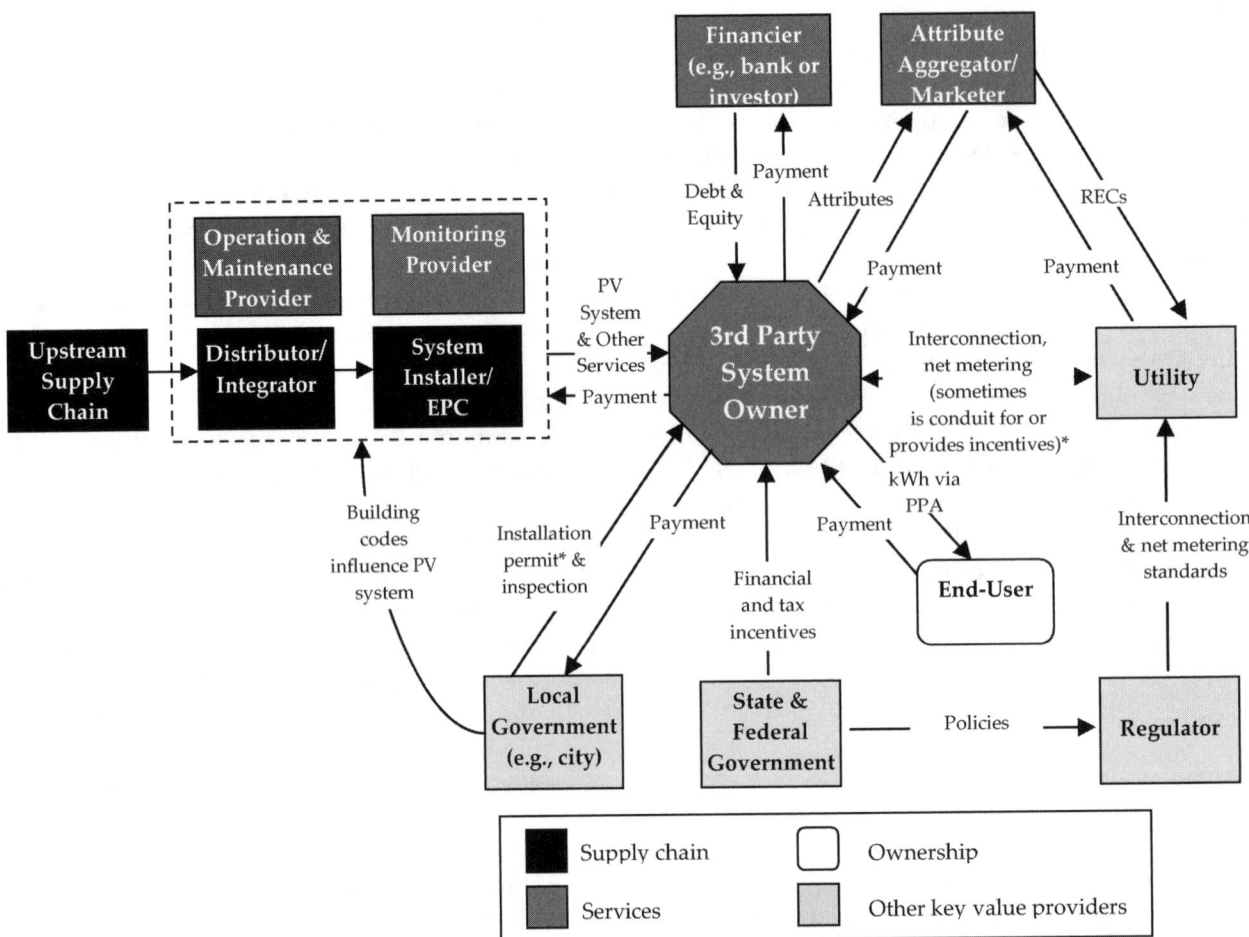

* In order for these transactions to occur special applications and/or agreements need to be filled out, processed and finalized

Figure 16-33. Third-party ownership PV business model for commercial retrofit applications (Adapted from Ref. 31).

16.3 WIND ENERGY

According to the American Wind Energy Association the U.S. added 5,329 MW of new wind power capacity in 2007, a more than 50% increase from the previous year. More wind capacity was installed than coal capacity in that same year. The midwestern U.S. has been called the "Saudi Arabia of wind energy" because of the strength and consistency of the wind over the Great Plains. Cost of wind-powered electricity has fallen by about 80% since the early 1980s, and it is expected to continue to fall as the technology develops. The average cost in 2000 was in the 5 cents/kWh range, although this has risen to around 7 to 10 ¢/kWh since 2005.

Larger turbine designs allow higher density of wind power production. Turbines with 1000MW capacity are available from 80m rotors on 80m towers. Since power output is a function of sweep area (A) and $A = \pi r^2$, each unit of rotor size increase yields exponential gains in power production. Thus, turbine sizes have scaled up to an optimal size limited mainly by logistics

of manufacturing, transporting, and assembling tower components. Off-shore (sea-based) designs are not limited by tower logistics, so they can potentially be larger than land based units. Limiting factors for off-shore wind turbines include seafloor depth for tower foundations and siting opposition due to visual impacts.

A major issue for large-scale wind utilization in the U.S. is the transmission infrastructure needed to bring wind from remote regions where the wind resource is plentiful to population centers where the demand is. These transmission projects will be costly, take time to construct, and may raise controversy over land-use and other environmental impacts. The transmission link raises the cost per installed unit of power, proportionally to the distance it must travel to inject into an existing power grid.

16.3.1 Wind Power Availability

A panel of experts from NSF and NASA estimated that the wind power potentially available across the continental United States, including offshore sites and the Aleutian arc, is equivalent to approximately 10^5 GW of electricity (Ref. 33). At the time, this was about 100 times the electrical generating capacity of the United States. Figure 16-34 shows a wind resource map and legend of wind classes. Economic wind classes are Class 5 and above. As previously mentioned, the middle of the country has a rich wind resource that could potentially supply the entire nation's power needs (if transmission were available to access it). It is interesting to compare and contrast the wind availability shown in Figure 16-34 with the solar energy availability shown in Figure 16-7.

The power that is contained in a moving air stream per unit area normal to the flow is proportional to the cube of the wind velocity. Thus small changes in wind velocity lead to much larger changes in power available. The equation for calculating the *wind power density* is

$$\frac{P}{A} = \frac{1}{2}\rho V^3 \qquad \text{Eq. 16.3}$$

Where

P = power contained in the wind

A = area normal to the wind velocity

ρ = density of air (about 0.07654 lbm/ft^3 or 1.23 kg/m^3)

V = velocity of the air stream

Consistent units should be selected for use in Eq. 16.3. It is convenient to rewrite Eq. 16.3 as

$$\frac{P}{A} = KV^3 \qquad \text{Eq. 16.4}$$

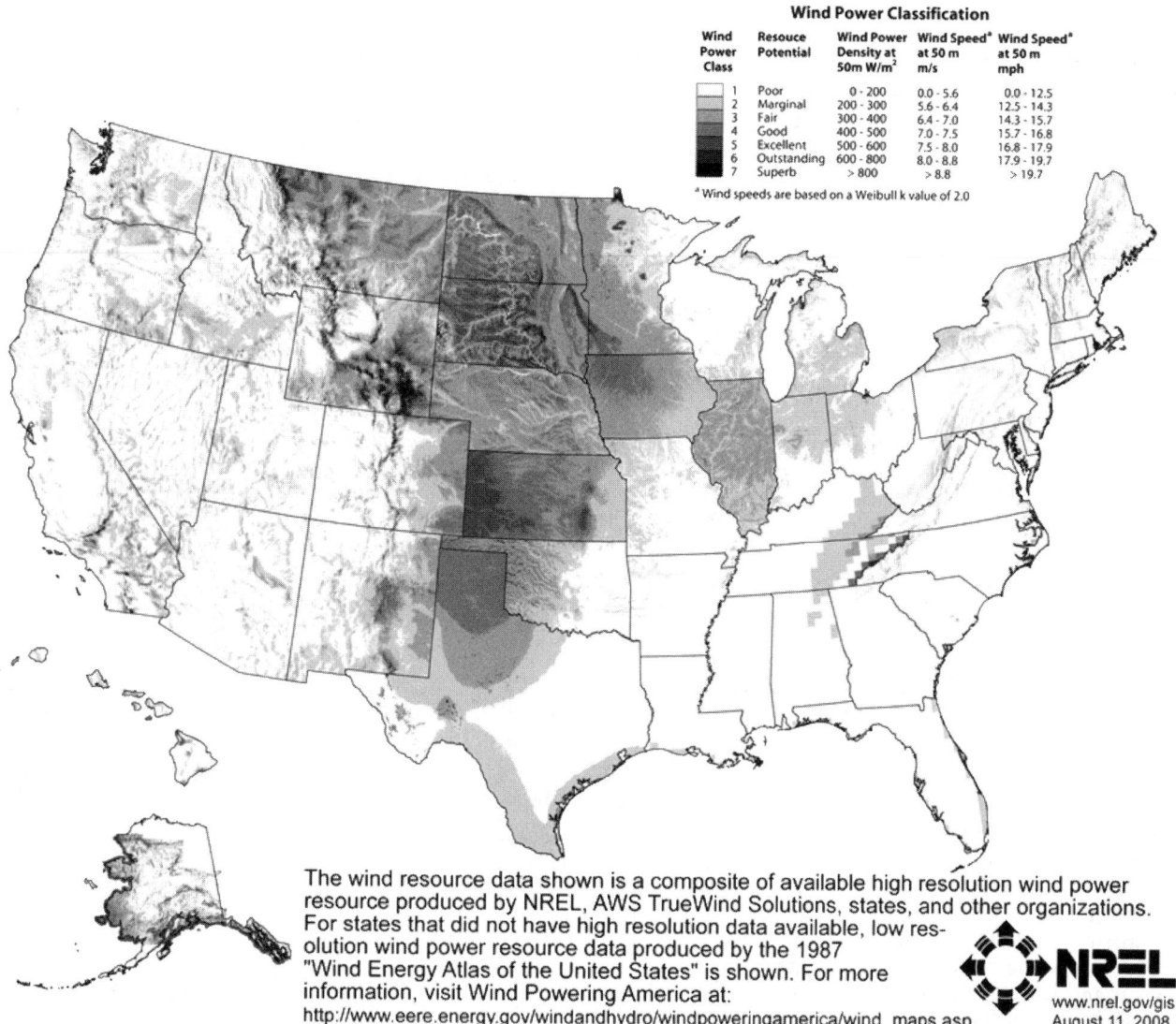

Wind Power Classification

Wind Power Class	Resouce Potential	Wind Power Density at 50m W/m²	Wind Speed[a] at 50 m m/s	Wind Speed[a] at 50 m mph
1	Poor	0 - 200	0.0 - 5.6	0.0 - 12.5
2	Marginal	200 - 300	5.6 - 6.4	12.5 - 14.3
3	Fair	300 - 400	6.4 - 7.0	14.3 - 15.7
4	Good	400 - 500	7.0 - 7.5	15.7 - 16.8
5	Excellent	500 - 600	7.5 - 8.0	16.8 - 17.9
6	Outstanding	600 - 800	8.0 - 8.8	17.9 - 19.7
7	Superb	> 800	> 8.8	> 19.7

[a] Wind speeds are based on a Weibull k value of 2.0

The wind resource data shown is a composite of available high resolution wind power resource produced by NREL, AWS TrueWind Solutions, states, and other organizations. For states that did not have high resolution data available, low resolution wind power resource data produced by the 1987 "Wind Energy Atlas of the United States" is shown. For more information, visit Wind Powering America at: http://www.eere.energy.gov/windandhydro/windpoweringamerica/wind_maps.asp.

NREL
www.nrel.gov/gis
August 11, 2008

Figure 16-34. U.S. wind resource map. This map was developed by the National Renewable Energy Laboratory for the U.S. Department of Energy.

If the power density (P/A) is desired in the units W/ft^2, then the value of K depends upon the units selected for the velocity (V). Values of K for various units of velocity are given in Table 16-11.

The fraction of the power in a wind stream that is converted to mechanical shaft power by a wind device is given by the *power coefficient* (C_p).

It can be shown that only 16/27 or 0.5926 of the power in a wind stream can be extracted by a wind machine since there must be some flow velocity downstream from the device for the air to move out of the way. This upper limit is called the Betz coefficient (or Glauert's limit). No wind device can extract this theoretical maximum. More typically, a device might extract some fraction, such as 70%, of the theoretical limit. Thus a real device might extract approximately (0.5926)(0.70) = 41% of the power available. Such a device would have an *aerodynamic efficiency* of 0.70 and a power coefficient of 0.41. The power conversion capability of such a device could be determined by using Equation 16.3 and Table 16-11. Assume a 20-mile/hr wind. Then

$$\frac{P}{A} = (5.08 \times 10^{-3})(20)^3(0.41) = 16.7 \text{ w/ft}^2$$

Notice that for a 30-mile/hr wind the power conversion capability would be 56.2 W/ft^2, or more than three times as much.

Because the power conversion capability of a wind device varies as the cube of the wind velocity, one cannot predict the annual energy production from a wind device using mean wind velocity. This is a very important concept since such a prediction would tend to underestimate the actual energy available. A good resource for doing wind resource calculations is provided on the Danish Wind Industry Association's website, www.windpower.org.

16.3.2 Wind Devices

Wind conversion devices have been proposed and built in a very wide variety of types. Figure 16-35 gives the generic classification of wind turbines in two types: horizontal axis wind turbine (HAWT) or vertical

Table 16-11. Values of K to Give P/A (W/ft2) in Equation 16.3a.

Units of V	K
ft/sec	1.61×10^{-3}
miles/hr	5.08×10^{-3}
km/hr	1.22×10^{-3}
m/sec	5.69×10^{-2}
knots	7.74×10^{-3}

aTo convert w/ft^2 to W/m^2, multiply by 10.76.

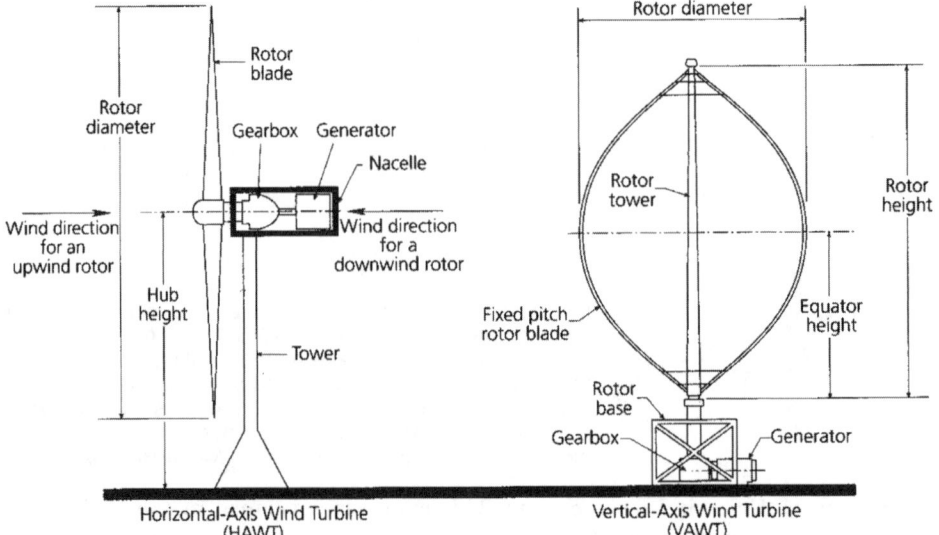

Figure 16-35. Principle wind generator design concepts: (a) Horizontal-axis wind turbine (HAWT) and (b) Vertical-axis wind turbine (VAWT). Adapted from *Renewable Energy* by Thomas B. Johansson, et al., eds. Copyright © 1993 Island Press. Reproduced by permission of Island Press, Washington, DC.

axis wind turbine (VAWT). The most common type for large scale power generation is the HAWT, where the axis of rotation is parallel to the direction of the wind stream. When wind direction is variable, the device must be turned into the wind, usually by a *yaw system*. The rotational speed of these devices can be controlled by feathering of the blades, by flap devices, or by varying the load.

In most HAWTs, the generator is directly coupled to the turbine shaft, sometimes through a gear drive. In the case of the bicycle multi-bladed type, the generator may be belt driven off the rim, or the generator hub may be driven directly off the rim by friction. In the later case there is no rotational speed control except that imposed by the load.

In the case of a VAWT, the direction of the wind is not important, which is an advantage. The system is simpler and there is no ancillary yawing equipment required to turn the unit to align with wind direction. VAWTs are also lighter in weight, require only a short tower base, and can have the generator near the ground. The side wind loads on a VAWT are accommodated by guy wires or cables stretched from the ground to the upper bearing fixture. The problem with VAWTs is that they are difficult to mount on top of towers to access higher wind speeds. Another engineering challenge is design that prevents blade failure due to fatigue caused by forces on each blade which alternate direction with every revolution.

Figure 16-36 illustrates the most common HAWT and VAWT designs and indicates the primary motive

forces that drive them. The Darrieus-type VAWT can have one, two, three, or more blades, but two or three are most common. Like the modern propeller-type HAWT, Darrieus turbines are a lift-type device because the blade cross-section forms an airfoil. Darrieus types have very low starting torque and a high tip-to-wind speed.

The Savonius-type turbine has a very high starting torque but a relatively low tip-to-wind speed. It is primarily a drag-type device, as are the American multi-bladed and Dutch four-arm windmills, although these produce shaft (not electrical) power for water pumping or milling. The Savonius and the Darrieus types are sometimes combined in a single turbine to give good starting torque and yet maintain good performance at high rotational speeds.

Figure 16-37 shows the variation of the power coefficient (Cp) as the ratio of blade tip speed to wind speed varies for different types of wind devices. It can be seen that two-blade types operating at relatively high speed ratios have the highest value of Cp, in the range of 0.45, which is fairly close to the limiting value of the Betz coefficient (0.593). The Darrieus rotor is seen to have a slightly lower maximum value, but like the two-blade type, performs best at high rotational speeds. The American multi-blade type is seen to perform best at lower ratios of tip to wind speed, as is the Savonius.

A three-bladed HAWT has six basic subsystems, as indicated in Figure 16-38 (Ref. 34):

- The rotor, usually consisting of three blades mounted on a hub and including pitch controls and aerodynamic braking system, which protects the unit from over speed in excessively high winds
- The drive train, including gearbox; hydraulic systems; shafts; mechanical braking systems; and a nacelle, which houses the turbine generator
- The yaw system, which orients the rotor perpendicular to the wind stream
- Electrical and electronic systems, including the generator, relays, circuit breakers, droop cables, wiring, controls and electronics, and sensors
- The tower
- Balance-of-station systems including roads, ground-support equipment, and interconnection equipment

16.3.3 Wind Characteristics—Siting

The wind characteristics given in wind resource maps such as Figure 16-33 are simple average values. The wind is almost always quite variable in both speed and direction. It is therefore necessary to characterize

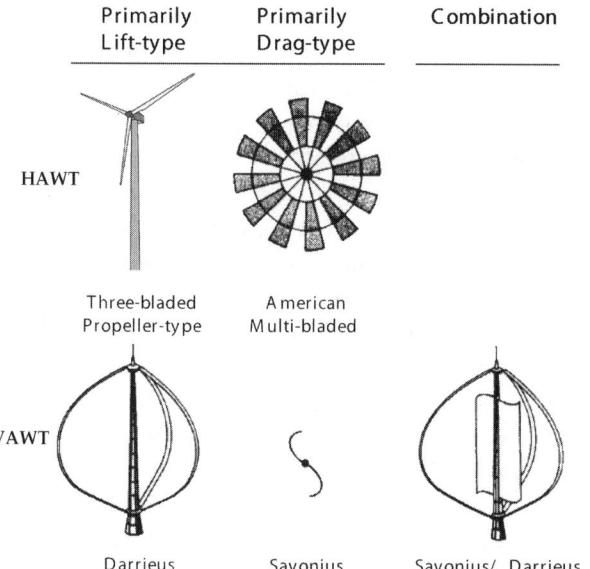

Figure 16-36. Types of wind-conversion devices (Adapted from Ref. 33).

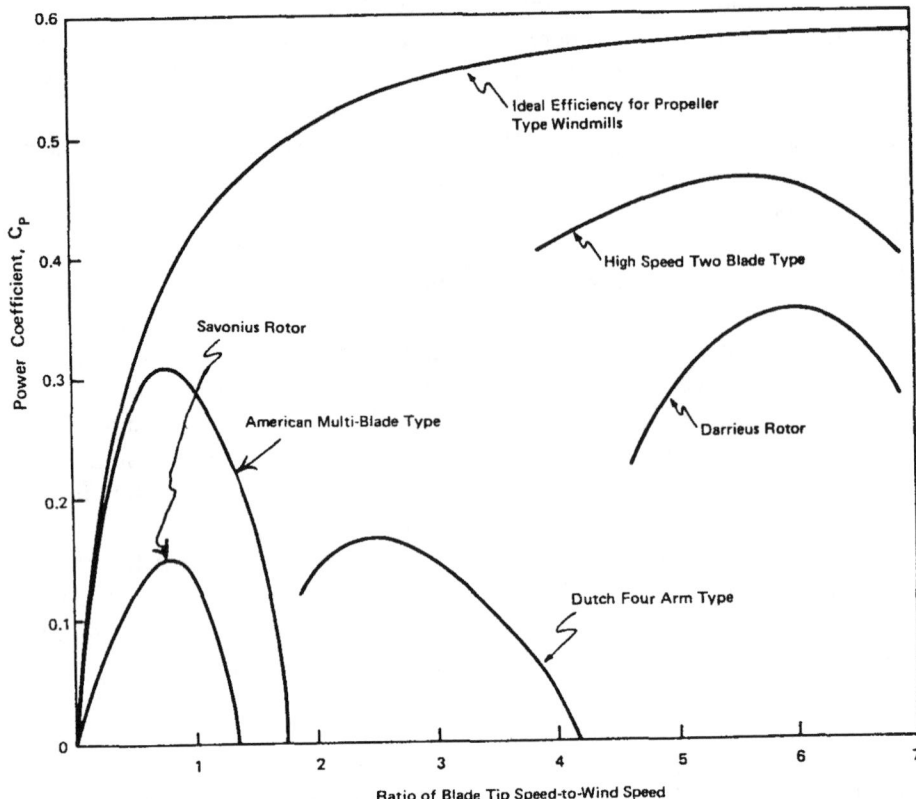

Figure 16-37. Typical pressure coefficients of several wind turbine devices (Ref. 33).

the wind resource at a given site by collecting data over a period of one to two years. A device called an *anemometer* is installed on top of a 60 to 80 m tower to gather wind data. With reliable wind data, an important characteristic of the wind resource becomes the number of hours that the wind exceeds a particular speed. This information can be expressed as speed-duration curves, such as those shown in Figure 16-39 for three sites in the United States. These curves are similar to the load duration curves used by electric utilities.

Because the power density of the wind depends on the cube of the wind speed, the distribution of annual average energy density of winds of various speeds will be quite different for two sites with different average wind speeds. A comparison between sites having average velocities of 13 and 24 miles/hr (5.8 and 10.7 m/sec) is given in Figure 16-40. The area under the curve is indicative of the total energy available per unit area per year for each case.

Sites should be selected where the wind speed is as high and steady as possible. Higher altitude sites tend to have higher average wind speeds; a 5-10% increase in wind speed can be expected for each 100 meters above sea level. But reduced air density at higher elevations offsets the increase in wind power to a small degree; to

maintain the same power density, speed must increase by about 3% per 1,000 meters (Ref. 34). Placement of wind devices at higher altitude also increases *laminar flow*, the flow of air in parallel layers, which reduces turbulence and wear and tear on the rotor. Rough terrain and the presence of trees or buildings should be avoided. The crest of a well-rounded hill is ideal in most cases, whereas a peak with sharp, abrupt sides might be very unsatisfactory because of flow reversals near the ground. Mountain gaps that might produce a funneling effect could be most suitable.

16.3.4 Performance of Turbines and Systems

There are three important wind speeds that might be selected in designing a wind energy conversion system (WECS). They are (1) cut-in wind speed, (2) rated wind speed, and (3) cut-off wind speed. The wind turbine is kept from turning at all by some type of brake as long as the wind speed is below the cut-in value. The wind turbine is shut off-completely at the cut-off wind speed to prevent damage to the turbine. The rated wind speed is the lowest speed at which the system can generate its rated power. If frequency control were not important, a wind turbine would be permitted to rotate at a variable speed as the wind speed changed.

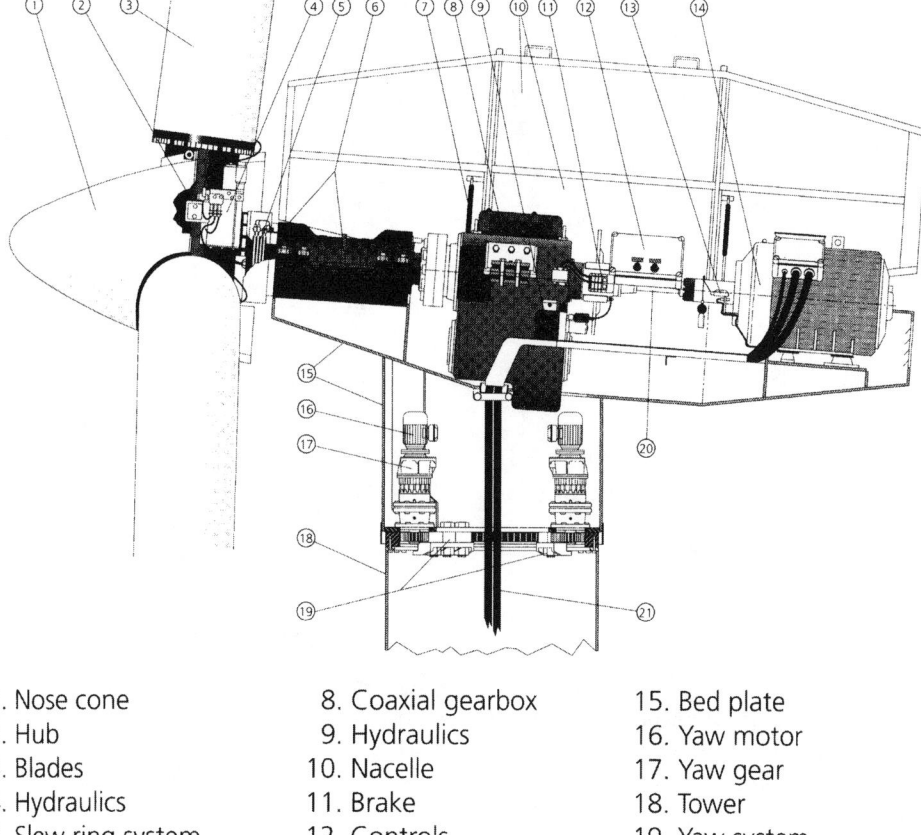

1. Nose cone	8. Coaxial gearbox	15. Bed plate
2. Hub	9. Hydraulics	16. Yaw motor
3. Blades	10. Nacelle	17. Yaw gear
4. Hydraulics	11. Brake	18. Tower
5. Slew ring system	12. Controls	19. Yaw system
6. Main shaft	13. Vibration sensor	20. Transmission shaft
7. Shock absorber	14. Generator	21. Power cables

Figure 16-38. A modern HAWT showing primary components. Adapted from *Renewable Energy* by Thomas B. Johansson, et al., eds. Copyright © 1993 Island Press. Reproduced by permission of Island Press, Washington, DC.

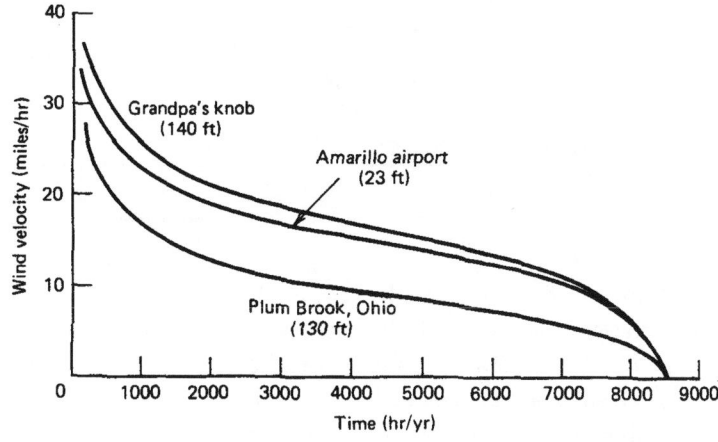

Figure 16-39. Annual average speed-duration curves for three sites (Ref. 33).

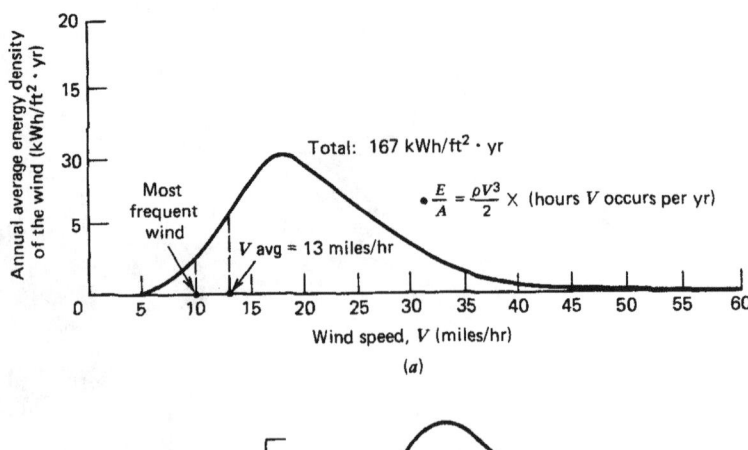

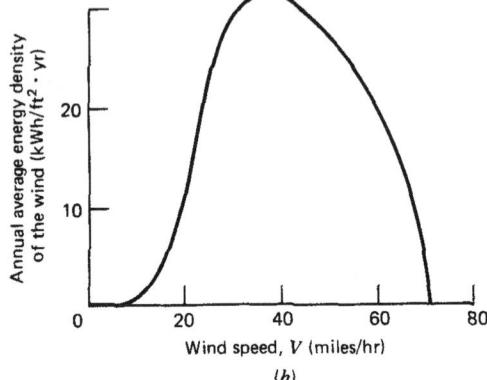

Figure 16-40. Comparison of distribution of annual average wind energy density at two sites: (a) V_{avg} = 13 miles/hr. (b) V_{avg} = 24 miles/hr (Ref. 33).

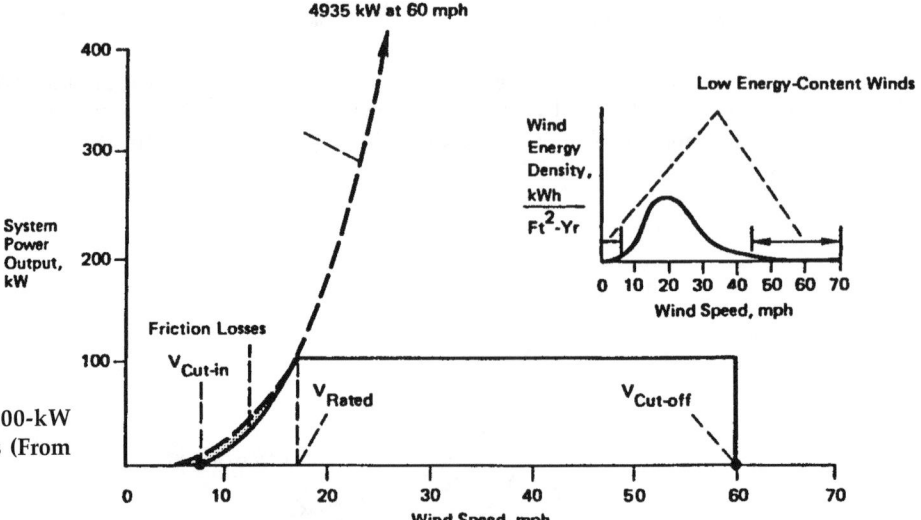

Figure 16-41. Power output of a 100-kW wind turbine at various wind speeds (From Ref. 33).

In practice, however, since frequency control must be maintained, the wind turbine rotational speed might be controlled by varying the load on the generator when the wind speed is between the cut-in and rated speed. When the wind speed is greater than the rated speed but less than cut-out speed, the spin can be controlled by changing the blade pitch on the turbine. This is shown in Figure 16-41 for a 100-kW system. A system such as that shown in Figure 16-41 does not result in large losses

of available wind power if the average energy content of the wind at that site is low for speeds below the cut-in speed and somewhat above the rated speed.

Another useful curve is the actual annual power density output of a wind power system (See Figure 16-42). The curve shows the hours that the device would actually operate and the hours of operation at full rated power. The curve is for a system with a *rated wind speed* of 30 miles/hr (13.4 meters/sec), a *cut-in velocity* of 15

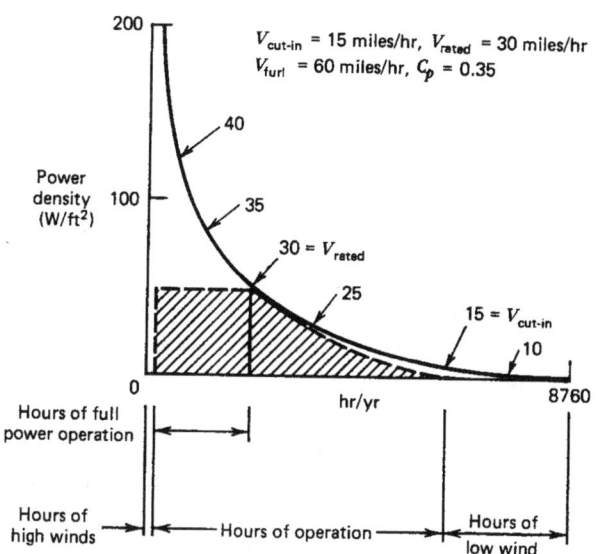

Figure 16-42. Actual annual power density output of a WECS (From Ref. 33).

miles/hr (6.7 meters/sec), and a *cut-off velocity* of 60 miles/hr (26.8 meters/sec), with constant output above 30 miles/hr.

16.3.5 Wind Integration and Storage

Because the typical wind device cannot furnish energy to exactly match the demand, wind energy must be "firmed" and integrated into the utility grid using storage systems and/or backup conventional energy sources. Variations in wind output present challenges to grid operators in two basic dimensions. First, seasonal and diurnal patterns of wind availability are usually characterized by maximum output during "shoulder" months (Spring and Fall) and evening hours—so-called "off-peak" hours when demand for power is lowest. Figure 16-43 illustrates the daily cycles of wind generation as compared to the utility system peak in California during a 2006 heat wave. Although not always the case, the coincident peak availability of wind is usually less than

15% of rated capacity. By comparison, average annual capacity factors for Class 5-6 wind are in the 30-40% range. In order to utilize the wind during peak hours of energy consumption, long-term storage technologies are required to shift delivery of wind power to when it is most needed. Without storage provisions utilized, the excess wind power potential of the turbine may be lost and result in underutilized equipment. The principle technologies for these energy management applications are pumped hydro storage and, to a limited extent, compressed air energy storage (CAES). Each of these incurs mechanical losses but is able to return the majority of the excess energy sent to storage.

Second, wind power varies in hourly and sub-hourly timeframes, requiring other resources on the grid to ramp their output up and down to maintain stable grid voltage and frequency. Short-term storage technologies, mainly battery banks, have been used to provide these integration services. Table 16-12 provides an overview of energy storage technologies, performance characteristics, and typical applications. A recommended reference is Ref. 35. A good source for additional information on storage is at the Electricity Storage Association website, www.electricity storage.org.

16.4 BIOMASS ENERGY

Biomass energy (or *bio-energy*) is broadly defined as any solid, liquid or gaseous fuel source derived from recently dead biological material; as opposed to fossil fuels, which come from long-dead biological material. Biomass constitutes the largest non-hydroelectric renewable source of primary energy and electricity in the U.S. It is the most diverse type of renewable energy, both in terms of fuel sources and end-uses. Bio-energy can be derived from any type of plant, animal or other biological carbon source. Uses include bio-fuels for transportation, direct-use or combined heat and power (CHP) in industry, and direct-use in buildings and rural subsis-

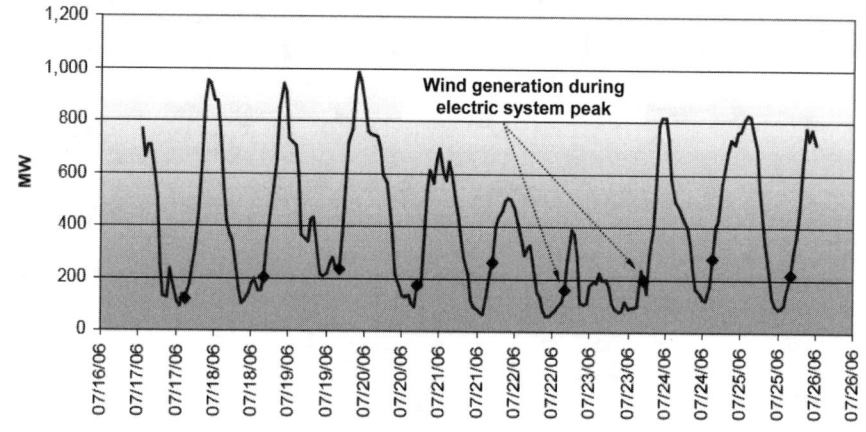

Figure 16-43. Wind generation on the California Independent System Operator (CAISO) power grid and the anti-coincident pattern of wind output during a July 2006 heat wave (Source: CAISO).

Table 16-12. Overview of energy storage technologies (Adapted from Ref. 36).

Storage Duration	Technology	Efficiency (%)	Energy Density (W-h/kg)	Power Density (kW/kg)	Sizes (MWh)	Comments
Days to hours	Pumped hydro	75	0.27/100m	low	5,000-20,000	37 existing in U.S.
	CAES	70	0	low	250-2,200	1 U.S., 1 German
Hours to minutes	Batteries	70-84	30-50	0.2-0.4	17-40	e.g., Lead acid, Li-Ion

tence. Energy crops cultivated for transportation fuels, such as grain or sugar cane feedstocks to produce ethanol, are the most well-known form of biomass. But in the buildings and industrial sectors, biomass is normally consumed as *opportunity fuel*, a waste or byproduct of consumptive or productive activity that is not typically used as fuel but has potential to be economically viable. A broad classification, opportunity fuels include renewable biomass as well as waste or byproducts derived, wholly or in part, from hydrocarbons such as municipal solid waste (MSW) and tire-derived fuels (Ref. 37). This chapter will focus on opportunity fuels for stationary applications.

A biomass energy resource map for the U.S. is given in Figure 16-44. This map provides aggregated tonnage data for the most important feedstocks, but NREL also produces maps for specific feedstocks. To a large extent, climate, hydrology, and soil conditions determine where rural biomass is available; urban centers create their own concentrations from economic activity. Biomass feedstocks can generally be categorized into rural and urban sources. Table 16-13 provides a list of the main feedstocks that energy managers may encounter as opportunity fuels in their field of operations.

Biomass has many potential advantages over fossil fuels for supplying energy. Biomass is carbon neutral,

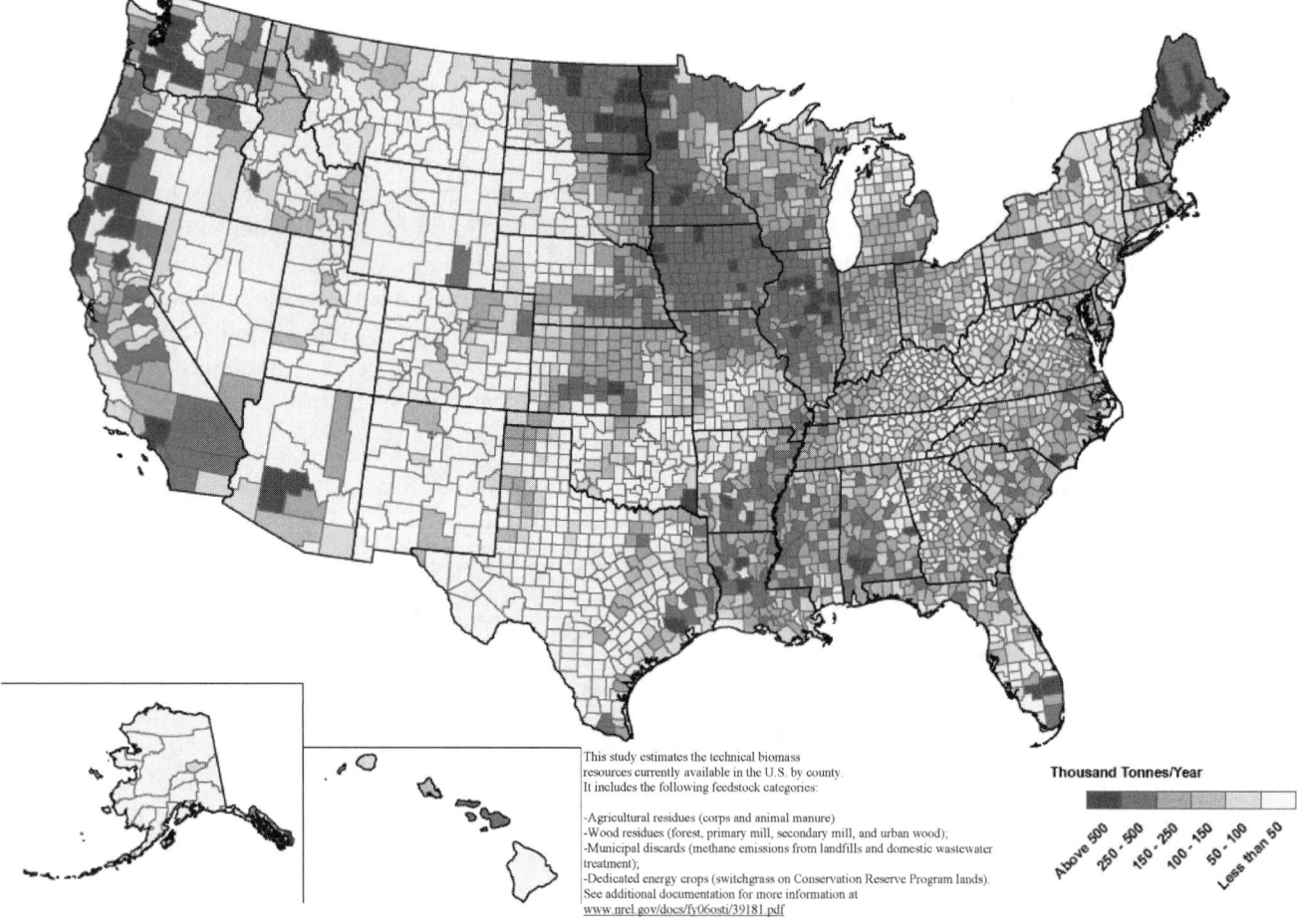

This study estimates the technical biomass resources currently available in the U.S. by county. It includes the following feedstock categories:

-Agricultural residues (corps and animal manure)
-Wood residues (forest, primary mill, secondary mill, and urban wood);
-Municipal discards (methane emissions from landfills and domestic wastewater treatment);
-Dedicated energy crops (switchgrass on Conservation Reserve Program lands).
See additional documentation for more information at www.nrel.gov/docs/fy06osti/39181.pdf

Thousand Tonnes/Year

Above 500   250 - 500   150 - 250   100 - 150   50 - 100   Less than 50

Figure 16-44. U.S. biomass energy resource map. This map was developed by the National Renewable Energy Laboratory for the U.S. Department of Energy.

because CO_2 released during fuel combustion is reabsorbed and "nets out" the carbon cycle in new plant growth. Although variations depend on fuel source and application, biomass use generally produces lower emissions of GHGs and criteria pollutants. Further, it tends to be more flexible (e.g., in multi-fuel or CHP applications) and reliable (as a non-intermittent resources) than many other renewables. Ability to convert waste into savings (or revenues) is perhaps the biggest driver for biomass installations. The dominant stationary applications in the field today rely on opportunity fuels, such as wood waste from the forest product industry, biogas from wastewater treatment facilities in CHP applications, and municipal solid waste (MSW) and landfill gas in power applications. For brevity, this section will focus on woody biomass CHP.

Table 16-13. Common rural and urban biomass feedstocks (Ref. 38).

Rural Resources	Urban Resources
Forest residues, wood wastes and black liquor	Urban wood waste
Crop residues	Wastewater treatment biogas and sludge waste
Energy crops	Municipal solid waste (MSW) and landfill gas (LFG)
Manure biogas	Food processing residue

16.4.1 Biomass Combined Heat and Power (CHP)

Wood-fired systems account for almost 95% of the *biopower* capacity in the U.S. (Ref. 36), although farm, wastewater, and landfill gas systems are increasingly the target of strategies to reduce emissions of high GWP gases such as methane. Of the 75% of non-hydroelectric renewable power that originates from biomass, about two-thirds is generated in combined heat and power operations (Ref. 37). A flow chart showing biomass feedstocks, conversion technologies, intermediate fuels, and prime movers associated with biomass CHP is given in Figure 16-45. Depending on the specific application, biomass generally goes through mechanical and/or thermal preparation and energy conversion to raise high-temperature steam or intermediate fuel, which then drives a prime mover to generate heat and power. Landfill gas and wastewater digester systems rely mostly on reciprocating engines, although, with proper gas clean-up, combustion turbines and even fuel cells have been used. Blending a portion of biomass fuel with conventional fuel, called *co-firing*, can serve to utilize conventional power equipment and reduce fossil fuel consumption rate; an example of this is blending wood chips or saw dust with coal.

It can be seen that processes to convert and utilize biomass energy are nearly as varied and diverse as the types of feedstocks. Despite this diversity, direct combustion to produce boiler steam accounts for 90% of bio-

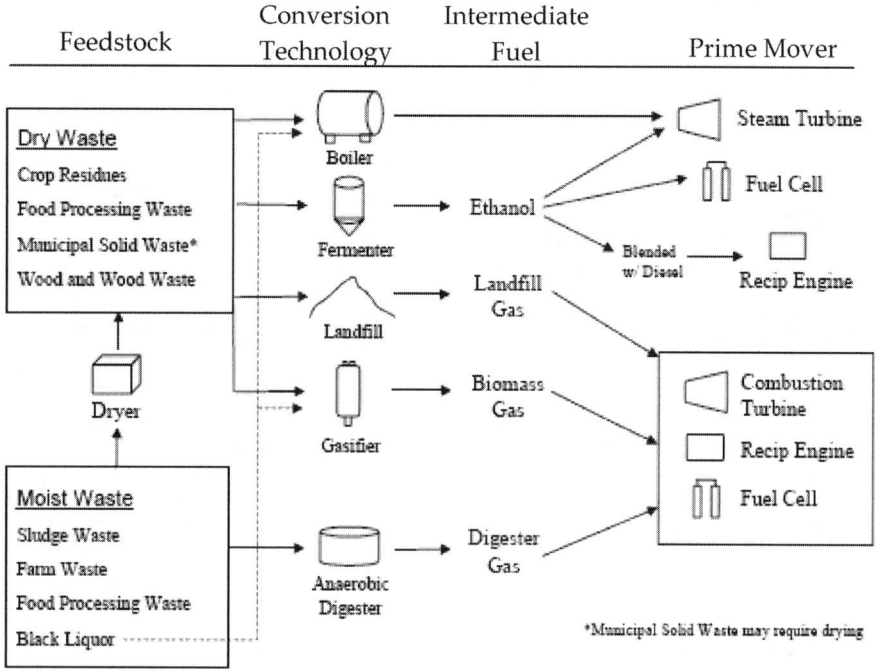

Figure 16-45. Flow chart of biomass energy use in CHP applications (Adapted from Ref. 37).

mass electricity in 2008 (Ref. 36). Gasification technologies developed for coal-fired power plants are generally too costly at distributed scale, despite higher efficiency. Modular gasifiers are at early stages of commercialization. Anaerobic digesters and landfill gas systems are commercial technologies, but as previously mentioned they are the minority. A comprehensive reference on all aspects of biomass CHP technologies, performance, cost, and applications is given in Ref. 38.

In a direct firing biomass CHP configuration, steam generated in the boiler powers a steam turbine generator and waste heat and steam is used to serve process needs at lower pressure and temperature. Process steam can be supplied using an extraction-condensing steam turbine, with partial steam output extracted from the turbine at required pressure for the process. Chapters 7 and 8 of this book are devoted to cogeneration and waste heat recovery, respectively. As shown in Figure 16-46, extraction-condensing steam turbines are designed for variable power and steam production, whereas back-pressure turbines that exhaust all the steam at the pressure required by the process will maximize steam production. Non-CHP (power-only) configurations would divert all steam production to a condensing turbine.

A commonly used conversion technology for a direct-fired biomass CHP system is the fixed stoker boiler. To illustrate the trade-offs between power and steam production, Table 16-14 gives the energy requirements and outputs of representative stoker boiler systems in various power and steam configurations, including back-pressure steam turbines, various extraction turbine

configurations, and power-only condensing turbine configurations. The table also provides moisture and energy content of biomass feedstock, energy conversion efficiencies, and boiler steam conditions at various sizes from 100 to 900 tons per day. A good reference on biomass energy systems is Ref. 39.

16.5 EMERGING TECHNOLOGIES

Editor's Note: The focus of this handbook is energy management, highlighting proven technology and solutions. It also serves as an educational text for professionals and students new to the field. In general, the philosophy of the book is to remain conservative and refrain from prototype products, new developments, and topics still in the research and development phase. If proven in widespread commercial use, they will appear in future editions. In the case of renewable energy, the field is changing very rapidly and decisions on topics to include are not easy. Where select technologies show excellent promise, they are mentioned in this closing section. The purpose is to show some of the recent activity in response to demand for more renewable energy solutions.

16.5.1 Stirling-dish Engine

This arrangement combines an existing proven technology, the Stirling engine, with a solar concentrator system. The Stirling unit is driven by heat, which is provided by the solar concentrators. Like other solar technologies, specific efficiencies are low and large collection of solar energy is required. Simplicity is one

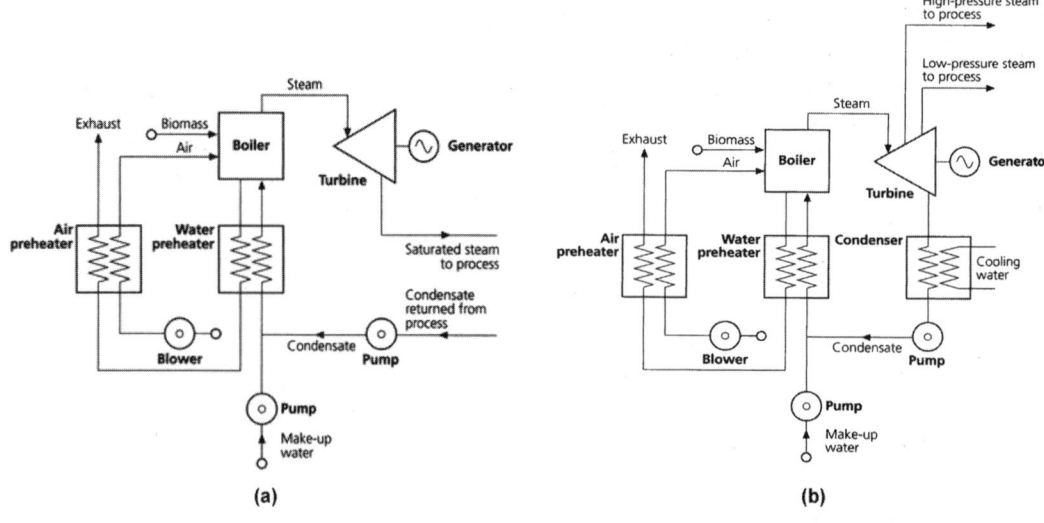

Figure 16-46. Power cycles for direct-fired biomass CHP in two configurations: (a) extraction-condensing turbine for variable power/steam production and (b) back-pressure turbine for maximum steam production. Adapted from *Renewable Energy* by Thomas B. Johansson, et al., eds. Copyright © 1993 Island Press. Reproduced by permission of Island Press, Washington, DC.

Table 16-14. Example biomass stoker boiler power generation system input and output requirements (Ref. 38).

	Size (tons/day)		
	100	600	900
Biomass Fuel Characteristics			
Energy content (dry) (Btu/lb)	8,500	8,500	8,500
Moisture content (%)	50	30	30
Energy content (as received) (Btu/lb)	4,250	5,950	5,950
Biomass Conversion			
Boiler efficiency (zero moisture) (%)	77	77	77
Boiler efficiency (moisture adjusted) (%)	63	71	71
Heat input to boiler (MMBtu/hr)	35.4	297.5	446.3
Heat to the steam (MMBtu/hr)	22.5	212	318
Plant capacity factor	0.9	0.9	0.9
Boiler Steam Conditions			
Boiler output pressure (psig)	275	750	750
Boiler output temperature (° F)	494	750	750
Nominal steam flow (lb/hr)	20,000	165,000	250,000
Steam Turbine Options			
CHP—Back-Pressure Turbine			
Electric output (MW)	0.5	5.6	8.4
Process steam conditions (psig [saturated])	15	150	150
Process steam flow (lb/hr)	19,400	173,000	260,000
CHP efficiency (%)	62.9	70.5	70.5
CHP—Extraction Turbine			
Process steam conditions (psig [saturated])	N/A	150	150
Electric output (MW) (150,000 lb/hr steam)	N/A	6.9	14.7
Electric output (MW) (100,000 lb/hr steam)	N/A	9.8	17.5
Electric output (MW) (50,000 lb/hr steam)	N/A	12.6	20.4
Power Only—Condensing Turbine			
Electric output (MW)	N/A	15.5	23.3
Electric efficiency (%)		17.8	17.8

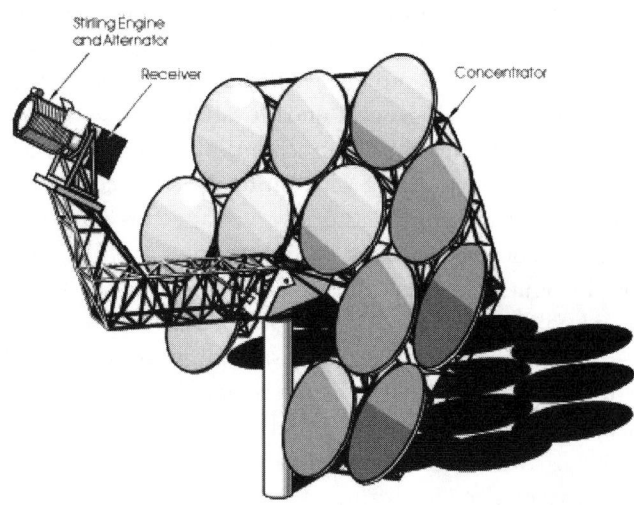

Figure 16-47. Prototype Stirling-dish concentrating solar system (Ref. 40)

advantage of this system.

Dish-engine technology uses a parabolic dish reflector to concentrate direct normal radiation onto a receiver adjoined to a power conversion unit. Figure 16-47 gives a schematic of a dish-type concentrator coupled to a Stirling engine. Because of the small focal region, dish-engine systems track the sun on two axes—azimuth-elevation and polar. Most concentrators approximate the ideal shape with multiple spherically-shaped mirrors supported with a truss structure. Due to a short focal length and the need to pin-point solar rays at the receiver, dish-type systems require robust support structures to prevent efficiency losses from wind vibrations. A receiver transfers solar energy to a high-pressure working gas, usually helium or hydrogen. Stirling engines convert heat to mechanical power in basically the same manner as conventional engines, except the heat is impressed from the outside of the engine. Expansion and contraction of the solar-heated working gas drives a set of pistons and a crankshaft to produce power.

Dish-engine is the oldest of the solar technologies, dating back to the 1800s when solar-powered, Stirling-based systems were first demonstrated. Modern technol-

ogy began developing in the late 1970s. Since the 1990s, prototype demonstrations have attempted to fine-tune the technology, with the goal of developing 5 to 10 kW units for distributed power and 25 kW systems for utility-scale applications. In 2005, two major California utilities signed power purchase agreements totaling over 800 MW; once built, these will be the first commercially operational, dish-engine power plants. Dish-engine is attractive because of its high efficiency, modular design and relatively less stringent siting requirements compared to other solar thermal technologies that require large expanses of flat land area. In general, constraints on wide-scale deployment include the high cost of truss assemblies in a high-cost global commodities market for steel and aluminum, availability and cost of specialized dish-shaped mirrors at scale, and reliability and cost to maintain engine components.

16.5.2 Fuel Cells

As of the date of this edition, fuel cell application remains mostly in niche markets, such as remote power generation, low noise applications, and space flight. Since the thrust of this text is commercial and industrial energy management solutions, fuel cells will be mentioned only briefly so the fundamentals are understood. Much research has been done and continues to be done with fuel cells. (See www.fe.doe.gov.) A fundamental limiting factor for the fuel cell is the fuel it uses; while hydrogen in a fuel cell produces the hallmark "no emissions" exhaust, common fuel cells consume hydrocarbons en route to the hydrogen and so are not renewable. Assuming a renewable source of hydrogen, the fuel cell would be more attractive and would have application in distributed generation (DG) and combined heat and power (CHP). However, fuel cells have important disadvantages when competing for favor with other forms of DG, including high cost and durability.

The fuel cell is an electrochemical device in which the chemical energy of a conventional fuel is converted directly and efficiently into low voltage direct-current electrical energy. It can be thought of as a primary battery in which the fuel and oxidizer are stored external to the battery and are fed to it as needed.

Fundamentals of fuel cell operation are described in texts on direct energy conversion, such as in Refs. 41 and 42. A schematic of a fuel cell is given in Figure 16-48. In the electrochemical reaction, the fuel, which is in gaseous form, diffuses through the anode and is oxidized. This releases electrons to the external circuit. The oxidizer gas diffuses through the cathode, where it is reduced by the electrons coming from the anode through the external circuit. The resulting oxidation products are carried away.

A typical fuel cell system consists of three components: a fuel processor (or reformer), a fuel cell "stack," and a power conditioner (or inverter). Fuel cells can be externally or internally reforming, with the former having an auxiliary fuel processor and the latter being a class of high-temperature fuel cells that reform natural gas within the fuel cell itself. At the heart of a fuel cell is the "stack," which is comprised of many individual fuel cells layered together. Because they produce DC power, fuel cell systems also require an inverter to convert power to AC for grid applications.

Fuel cells may be classified by their electrolytes and the corresponding temperatures at which they operate. Table 16-15 gives an overview of four main types of fuel cells: phosphoric-acid fuel cell (PAFC), molten carbonate fuel cell (MCFC), proton exchange membrane (PEM) and solid-oxide fuel cell.

16.6 TOPICS FOR FURTHER READING

As mentioned at the outset, the subject matter in this chapter is too broad to touch on all aspects of alternative energy in details. The author has attempted to touch on a few of the most applicable and/or promising technologies, but there are many others that could be

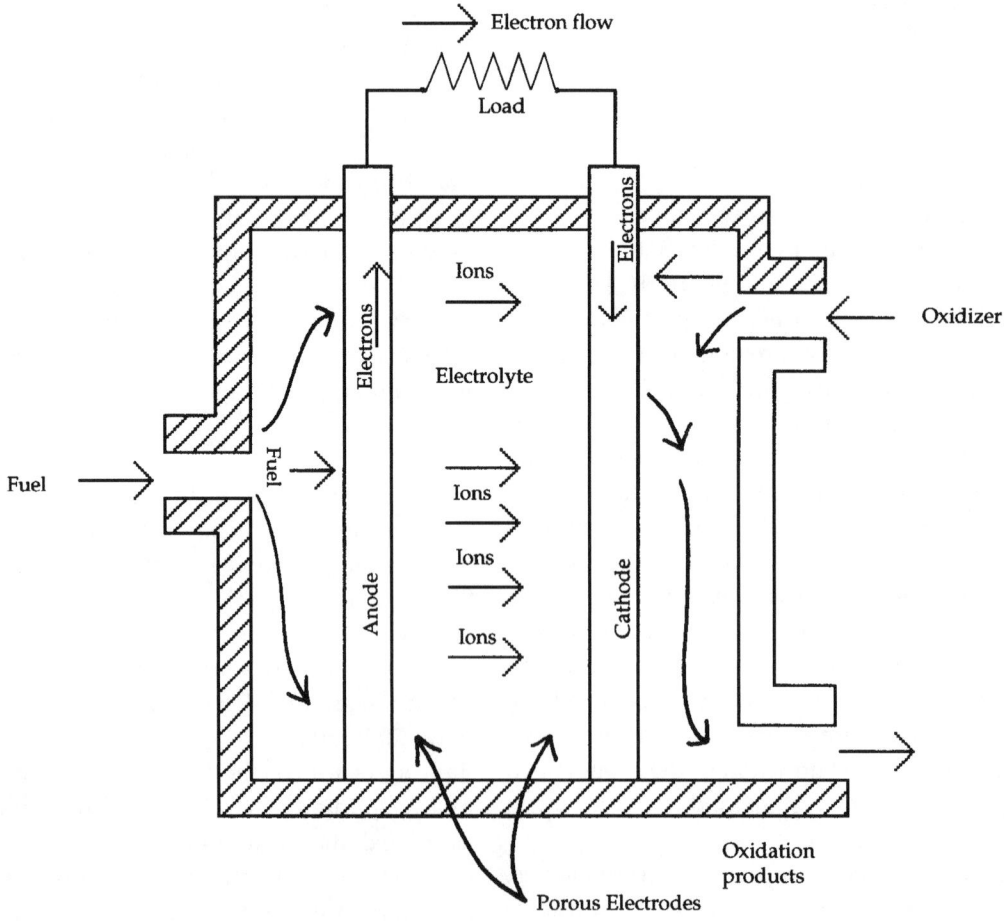

Figure 16-48. Simple schematic of a fuel cell.

Table 16-15. Commercial status, cost, and performance characteristics of stationary fuel cells in a CHP application (Adapted from Ref. 38).

Fuel cell type (Electrolyte)	PAFC	MCFC	PEM[†]	SOFC[†]
Capacity (kW)	200	250	200	100
Commercial status	Commercial	Pre-commercial	Demonstration	Demonstration
Operating temperature	200°C	650°C	80°C	1,000°C
Equipment cost ($/kW)*	5,100	5,100		
O&M costs ($/kW)*	0.03	0.043	0.33	
Heat rate (Btu/kWh)	9,480	7,930	9,750	7,580
Electrical efficiency, HHV (%)	36	43	35	45
Fuel input (MMBtu/hr)	2	2	2	0.8
Fuel reforming capability	External	Internal	External	Internal
CHP Characteristics				
Heat output (MMBtu/kWh)	3,709	1,750	3,592	1,906
CHP efficiency, HHV (%)	75	65	72	70
Power/heat ratio	1	2	0.95	1.79

[†] Cost information for PEM and SOFC are not given, because, as of 2008, the technologies are still in the demonstration phase.

potentially beneficial or relevant in an energy manager's activities. Following is a list of topics for further reading:

Solar thermal
- Passive solar water heating—thermosyphon, integral-collector
- Phase change materials (PCMs) for thermal storage (See Chapter 19)
- Seasonal thermal storage
- Thermal chimney and passive down-draft cooltower
- Solar A/C using desiccants
- Solar thermal (absorption chiller) cooling

Fuel cells and the hydrogen economy

Electricity storage
- Batteries—NiCd, Li-ion, NaS, Flow (Zn-Br)
- Superconducting magnetic energy storage (SMES)
- Flywheels
- Ultracapacitors

Solar thermal electric
- Linear Fresnel reflector
- Solar updraft tower
- Concentrating photovoltaics
- Solar dish engines

Other renewable technologies
- Passive solar cooling
- Small (low-impact) hydro power
- Distributed wind
- Off-shore wind
- Geothermal energy
- Wave energy
- Tidal (marine current) energy
- Anaerobic digester gas
- Waste-to-energy

Notes

1. In the U.S. some negative externalities have been effectively monetized under state and federal air quality controls; for example, the U.S. EPA's National Ambient Air Quality Standards (NAAQS) regulate six "criteria pollutants:" ozone (O_3), particulate matter (PM10 and PM 2.5), carbon monoxide (CO), sulfur dioxide (SO_2), nitrogen oxides (NO_x), and lead (Pb). Under the U.S. EPA's acid rain program, an emissions trading scheme, coal-fired power plants buy and sell pollution permits (called "allowances") for SO_2 and NO_x emissions in some regions. Mercury, another hazardous air pollutant, is regulated under the Clean Air Act. Other negative externalities are not currently monetized in the U.S., most notably green-house gases emissions (GHG) such as carbon-dioxide (CO_2), although CO_2 allowances are traded voluntarily on the Chicago Climate Exchange. In Europe, CO_2 is traded under the European Union Emissions Trading Scheme (EU ETS), a cap-and-trade program pursuing binding GHG emissions reductions targets under the Kyoto Protocol.

2. Net metering tariffs vary by utility. The majority of grid-tied PV systems in the U.S. are interconnected under net metering tariffs that buy back excess power at retail rates, an arrangement that is most favorable to the system owner. Retail buyback involves a cross-subsidy, because the component of retail rates designed to recover fixed costs, such as transmission and distribution infrastructure costs, becomes "stranded" when net metering customers avoid paying these costs and the burden shifts to non-net metering customers. Such inequities are generally justified on the environmental benefits of encouraging distributed solar PV and the as-yet *de minimis* scale of PV market penetration. Some jurisdictions limits the cross-subsidy by restricting net metering to a percentage (say 2%) of electric system capacity, while others avoid it altogether by setting the buyback rate at the cost of wholesale electricity.

3. Phase I, 2005-2007, of the EU ETS was purposely set up to test, verify, and make necessary changes to the cap-and-trade system before launching Phase II, 2008-2012, which corresponds to the binding compliance period under the Kyoto Protocol. The drop in Phase I allowance prices from 2006 to 2007 is the market's response to the fact that Phase I allowance cannot be applied to Phase II compliance, and therefore, beginning in 2008, they have no value.

4. Assuming an exchange rate of 1.58 Euro per $US, and given a 2006 U.S. average electricity sector carbon intensity of 0.6 tonnes CO_2/MWh (Ref. 41), 1.4 to 2.9 ¢/kWh premium is derived from a 15 to 30 Euro/tonne CO_2 range of carbon prices.

5. For example, JP Morgan Chase, Citigroup and Morgan Stanley established the "Carbon Principles," an enhanced due diligence process that evaluates power industry borrowers in terms of their use of energy efficiency, renewables, and low-carbon gen-

erating technologies.

6. However, these traditional evaluation methods do not define a broader evaluation framework that is contained in the *California Standard Practice Manual (SP): Economic Analysis of Demand-side Programs and Projects*, October 2001. This framework is necessary for the evaluation of various conservation and load management programs, which include, for example, energy efficiency (EE), demand response (DR); and distributed generation (DG). In addition, the SPM tests can be used to evaluate fuel substitution programs, like comparing the gas verses electric water heating programs.

7. Molten nitrate salt has desirable properties, such as superior heat transfer and energy storage capabilities, which makes it attractive. However, molten nitrate salt is also difficult to work with. For example, its relatively high freezing point causes problems when salt crystals form inside the equipment and become virtually impossible to clean.

8. PTC stands for PV USA Test Conditions, which were developed at the PV USA test site at Davis, California. PTC are 1,000 W/m^2 solar irradiance, 1.5 Air Mass, and 20 °C ambient temperature at 10 meters above ground level and wind speed of 1 meter/second. PTC is more like "real-world" conditions but does not factor in solar array temperature loss (5% to 12%), dust and dirt (2% to 4%), module mismatch (2%), DC and AC wire losses (2%), or real inverter loss (5% to 15%). STC stands for standard test condition. It is the laboratory testing conditions to measure photovoltaic cells' or modules' nominal output power. Irradiance level is 1,000 W/m^2, 1.5 reference air mass solar spectral irradiance distribution, and cell or module junction temperature of 25°C.

9. PV average annual capacity factor is 0.18, although this varies by location according to degree of cloud cover and other weather variables, as well as and solar-to-electric efficiency of the PV system.

References

1. Energy Information Administration. 2007. *Renewable Energy Annual 2006*.
2. de Vries, B.J.M et al. 2006. Renewable energy sources: Their global potential for the first-half of the 21st century at a global level: An integrated approach, *Energy Policy*: 35(4), pp. 2590-2610.
3. European Wind Energy Association. 2000. *Wind Power Economics: Fact Sheet*.
4. Handleman, C. 2000. An experience curve based model for the projection of PV module costs and its policy implications.
5. REN21. 2008. *Renewables 2007 Global Status Report*, (Paris: REN21 Secretariat and Washington, DC:Worldwatch Institute).
6. Energy Information Administration. 2007. *Annual Energy Outlook 2006*.
7. Forsyth, F. and I. Baring-Gould. 2007. *Distributed Wind Market Applications*, NREL/TP-500-39851, November 2007.
8. Contreras, J.L., L. Frantzis, S. Blazewicz, D. Pinault, and H. Sawyer. 2008. *Photovoltaics Value Analysis*, NREL/SR-581-42303.
9. Black & Veatch. 2008. *Renewable Energy Transmission Initiative Phase 1A Final Report*, prepared for the RETI Coordinating Committee and RETI Stakeholder Steering Committee, May 2008.
10. American Wind Energy Association (AWEA)
11. Database on State Incentives for Renewables and Efficiency (DSIRE). www.dsireusa.org.
12. State of California. 2001. *California Standard Practice Manual: Economic Analysis of Demand-side Programs and Projects*, October 2001.
13. Wiser, R., and M. Bolinger. 2008. *Annual Report on U.S. Wind Power Installation, Cost, and Performance Trends: 2007*, U.S. Department of Energy's Office of Energy Efficiency and Renewable Energy, DOE/GO-102008-259.
14. IPCC. 2007: *Climate Change 2007: Synthesis Report. Contribution of Working Groups I, II and III to the Fourth Assessment Report of the Intergovernmental Panel on Climate Change* [Core Writing Team, Pachauri, R.K and Reisinger, A. (eds.)]. IPCC, Geneva, Switzerland, 104 pp.
15. U.S. Environmental Protection Agency. 2005. *Inventory of U.S. Greenhouse Gas Emissions and Sinks: 1990-2003*, EPA 430-R-05-003 (Final Version: April 2005), Table ES-1.
16. Energy Information Administration. 2005. *Electric Power Annual 2004*, DOE/EIA-0348(2004), November 2005, Table A1.
17. Christiana, H., and S. Bowden, Photovoltaic CDROM.
18. Fuller, S.E. 1976. "The relationship of diffuse, total and extra-terrestrial solar radiation," *Journal of Solar Energy and Technology*, 18(3), pp. 259-263.
19. Solar Radiation Data Manual for Flat-Plate and Concentrating Collectors, http://rredc.nrel.gov/solar/pubs/redbook/HTML/redbook_HTML_index.html.
20. U.S. Department of Energy Office of Energy Efficiency and Renewable Energy. 2001. *Passive Solar Design for the Home*, DOE/GO-102001-1105-FS121, February 2001.
21. Sherwood, L. 2008. *U.S. Solar Market Trends 2007*, prepared for the Interstate Renewable Energy Council, August 2008.
22. Telkes, M. 1974. "Solar Energy Storage," *ASHRAE Journal*, September1974, pp. 38-44.
23. *ERDA Facilities Solar Design Handbook*. 1977. ERDA 77-65, Aug. 1977.
24. Energy Information Administration. 2006. Shipments of Solar Thermal Collectors by Market Sector, End-use and Type.
25. de Laquil III, P., D. Kearney, M. Geyer, and R. Diver. 1994. Solar-thermal electric technology, in *Renewable Energy: Sources for Fuels and Electricity*, Johansson, T.B., H. Kelly, A.K.N. Reddy and R.H. Williams (Eds.), Island Press, Washington, D.C.
26. Sun-Lab, 1997, *Technology Characterization: Solar Power Tower*, Sandia National Laboratory and National Renewable Energy Laboratory.
27. Sun-Lab, 1997, *Technology Characterization: Overview of Solar Thermal Technologies*, Sandia National Laboratory and National Renewable Energy Laboratory.
28. National Renewable Energy Laboratory. 2004. *Solar Energy Technologies Multiyear Technical Plan*, Report No. MP-520-33875; DOE/GO-1020040-1775.
29. Maycock, P. and T. Bradford. 2008. *PV Technology, Performance, and Cost: 2007 Update*, produced by the Prometheus Institute for Sustainable Development and PV Energy Systems.
30. Guice, J. and J. King. 2008. *Solar Power Services: How PPAs are Changing the PV Value Chain*, Greentech Media Research.
31. Frantzis, L, S. Graham, R. Katofsky, and H. Sawyer. 2008. *Photovoltaic Business Models*, prepared for the National Renewable Energy Laboratory, NREL/SR-581-42304.
32. Cory, K., J. Coughlin, T. Jenkin, J. Pater, and B. Swevey. 2008. *Innovations in Wind and Solar PV Financing*, National Renewable Energy Laboratory, NREL/TP-670-42919, February 2008.
33. Eldridge, F.R. 1975. *Wind Machines*, NSF-RA-N-75-051, prepared for NSF by the Mitre Corporation, Oct., U.S. Government Printing Office, Washington, D.C. (Stock No. 038000-00272-4).
34. Cavallo, A.J., S.M. Hock, D.R. Smith. 1993. Wind energy: technology and economics, in *Renewable Energy: Sources for Fuels and Electricity*, Johansson, T.B., H. Kelly, A.K.N. Reddy and R.H. Williams (Eds.), Island Press, Washington, D.C.
35. Electric Power Research Institute. 2003. *EPRI-DOE Handbook of Energy Storage for T&D Applications*.
36. National Renewable Energy Laboratory. 2006. *Power Technologies Energy Data Book, 4th Edition*, NREL/TP-620-39728, August 2006.

37. Resource Dynamics Corporation. 2004. *Combined Heat and Power Market Potential for Opportunity Fuels: A Distributed Energy Program Report*, prepared for the U.S. Department of Energy's (DOE) Office of Energy Efficiency and Renewable Energy (EERE), August 2004.

38. U.S. Environmental Protection Agency. 2007. *Biomass Combined Heat and Power Catalog of Technologies*, prepared by: Energy and Environmental Analysis, Inc., an ICF International Company, and Eastern Research Group, Inc. (ERG) for the U. S. Environmental Protection Agency, Combined Heat and Power Partnership, September 2007.

39. Williams, R.H., and E.D. Larson. 1993. Advanced gasification-based biomass power generation, in *Renewable Energy: Sources for Fuels and Electricity*, Johansson, T.B., H. Kelly, A.K.N. Reddy and R.H. Williams (Eds.), Island Press, Washington, D.C.

40. Sun-Lab, 1997, *Technology Characterization: Solar Dish*, Sandia National Laboratory and National Renewable Energy Laboratory.

41. World Fuel Cell Council, Kroegerstrasse 5, D-60313 Frankfurt/M, Germany.

42. U.S. Fuel Cell Council, 1625 K. Street NW Suite 75, Washington, DC, 2000.

43. Energy Information Administration. 2007. *State Electricity Profiles 2006*, November 2007.

APPENDIX A—EXTERNALITIES

Table 16-16. Emissions coefficients of major air pollutants from electric generation by U.S. state, lbs/MWh (Ref. 43)

State	SOx	NOx	CO2	State	SOx	NOx	CO2	State	SOx	NOx	CO2
Alabama	7.2	1.9	1332	Kentucky	8.7	3.5	2079	Ohio	13.8	3.2	1830
Alaska	1.4	6.0	1514	Louisiana	3	2.2	1312	Oklahoma	3.4	2.6	1631
Arizona	0.9	1.6	1127	Maine	2.2	1.4	739	Oregon	0.5	0.5	293
Arkansas	3.5	1.6	1204	Maryland	12.2	2.8	1373	Pennsylvania	8.5	1.8	1268
California	0.3	0.9	604	Massachusetts	2.4	1.1	1146	Rhode Island	0.3	1.1	928
Colorado	2.5	2.9	1820	Michigan	6.4	2.2	1481	South Carolina	4.9	1.1	907
Connecticut	0.3	0.6	703	Minnesota	3.9	3.5	1556	South Dakota	3.6	4.3	1090
Delaware	9.2	3.5	1806	Mississippi	3.9	2.2	1230	Tennessee	6.3	2.3	1441
District of Columbia	8.8	9.7	2681	Missouri	6.2	2.6	1902	Texas	3.1	1.4	1417
Florida	3.2	2.1	1247	Montana	1.7	3	1490	Utah	1.8	3.7	1947
Georgia	10.9	2.1	1436	Nebraska	4.5	4.3	1552	Vermont [†]	*	0.2	3
Hawaii	4.2	5.5	1723	Nevada	0.6	2.1	1150	Virginia	5.9	1.8	1269
Idaho	0.7	0.3	144	New Hampshire	3.7	0.9	706	Washington	0.2	0.4	211
Illinois	3.5	1.4	1140	New Jersey	2	1	721	West Virginia	10.1	3.3	1999
Indiana	12.8	3.4	2060	New Mexico	1.7	4.3	1955	Wisconsin	7.4	2.8	1726
Iowa	6.4	3.1	1967	New York	1.9	1	790	Wyoming	4.1	4	2196
Kansas	4.9	3.6	1726	North Carolina	7.9	1.8	1288	**US Total**	**5.2**	**2.1**	**1334**
Kentucky	8.7	3.5	2079	North Dakota	8.5	4.9	2232				

* Value is less than half of the smallest unit of measure. For values with no decimals the smallest unit is 1, and values under 0.5 are shown as *

[†] Vermont's power stations are primarily nuclear, hydroelectric, or other renewable.

APPENDIX B
PROGRAM COST-EFFECTIVENESS TESTS

The contents of this appendix are based on the *California Standard Practice Manual: Economic Analysis of Demand-side Programs and Projects* (SPM) (Ref. 12). Table 16-17 provides an overview of various economic tests that utilities and policy-makers use to evaluate the cost-effectiveness of programs to support alternative energy. The formulae in this appendix are abbreviations of those that appear in the SPM and should only be used for evaluation of conservation and load management (not fuel switching) programs.

16.B.1. PARTICIPANT TEST

The participant test is the measure of the quantifiable benefits and costs to the customer due to participation in a program. The benefits of participation include the reduction in the customer's utility bill(s), any incentive paid by the utility or other third parties, and any federal, state, or local tax credit received. The costs to a customer are all out-of-pocket expenses incurred as a result of participating in the program, such as the cost of any equipment or materials purchased as well as any ongoing operation and maintenance costs. The benefit-cost ratio (BCR_P) for the participant test is as follows:

$$BCR_P = B_P/C_P$$

If BCR_P is greater than one, then the program is considered cost effective.

$$B_P = \sum_{t=1}^{N} \frac{BR_t + TC_t + INC_t}{(1+d)^{t-1}}$$

$$C_P = \sum_{t=1}^{n} \frac{PC_t}{(1+d)^{t-1}}$$

Where

B_P = Net present value (NPV) of benefits to participants

C_P = Net present value of costs to participants

BR_t = Bill reductions in year t

TC_t = Tax credits in year t

INC_t = Incentives paid to participant by utility

d = Discount rate

t = year

n = Lifecycle of the DSM measure

PC_t = Participant costs in year t

16.B.2. RATEPAYER IMPACT MEASURE TEST

The ratepayer impact measure (RIM) test measures what happens to customer rates due to changes in utility (or other program administrator) revenues and operating costs caused by the program. The test indicates the direction and magnitude of the expected change in customer rate levels. If DSM causes utility rates to go up, non-participants will see increases in their bills. Participants, on the other hand, encountering the same rate increases, may see still their total utility bills go down since they will consume less energy.

The benefits calculated in the RIM test are the savings from avoided supply costs. These *avoided costs* include the reduction of transmission, distribution, generation, and capacity costs for periods of load reduction. The costs calculated in the RIM test include program costs (incentives paid to the participants, program administrative costs) and decreases in utility revenues.

Table 16-17. Overview of economic test to evaluate cost-effectiveness of alternative energy programs

Cost-Effectiveness Test	Perspective	Benefit-Cost Ratio
Participant Cost	Participant	$\dfrac{\text{Bill Reduction} + \text{Incentive} + \text{Tax Credits}}{\text{Participant Cost}}$
Ratepayer Impact Measure	Ratepayer	$\dfrac{\text{Avoided Supply Cost}}{\text{Program Cost} + \text{Incentive Cost} + \text{Revenue Loss}}$
Program Administrator Test	Program Administrator	$\dfrac{\text{Avoided Supply Cost}}{\text{Program Cost} + \text{Incentive Cost}}$
Total Resource Cost	Society	$\dfrac{\text{Avoided Supply Cost}}{\text{Program Cost} + \text{Net Participant Cost}}$
Societal Cost	Environment	$\dfrac{\text{Avoided Supply Cost} + \text{Avoided Environmental Impact}}{\text{Program Cost} + \text{Net Participant Cost}}$

For a program to be cost-effective using the RIM test, utility rates must not increase as a result of the program, i.e., non-participants will see no increase in their utility bills. The benefit-cost ratio (BCR_{RIM}) for the RIM Test is as follows:

$$BCR_{RIM} = B_{RIM}/C_{RIM}$$

If BCR_{RIM} is greater than one, then the program is considered cost effective.

$$B_{RIM} = \sum_{t=1}^{N} \frac{UAC_t + RG_t}{(1+d)^{t-1}}$$

$$C_{RIM} = \sum_{t=1}^{n} \frac{RL_t + PRC_t + INC_t}{(1+d)^{t-1}}$$

Where

B_{RIM}	=	NPV of benefits to rate levels
C_{RIM}	=	NPV of costs to rate levels
UAC_t	=	Utility avoided supply costs in year t
RG_t	=	Revenue gain from increased sales in year t
RL_t	=	Revenue loss from reduced sales in year t
PRC_t	=	Program costs in year t
INC_t	=	Incentives paid to participant by utility
d	=	Discount rate
t	=	Year
n	=	Lifecycle of the DSM measure

16.B.3. PROGRAM ADMINISTRATOR TEST

The program administrator test measures the net costs of a program as a resource option based on the costs incurred by the program administrator (usually a utility) and excluding any net costs incurred by the participant. The benefits of the program administrator test are the avoided supply costs. The costs are program administration costs associated with running the program (rebates and administrative costs). When benefits exceed costs, the program administrator test is satisfied, indicating a reduction in the total revenue requirements of the utility and resulting in a lower customer bill on average. Even though total utility revenues drop, the addition of program costs may result in higher rates ($/kWh); thus, non-participants' bills may go up even if the average customer bill goes down. The benefit-cost ratio (BCR_{PA}) for the program administrator test is as follows:

$$BCR_{PA} = B_{PA}/C_{PA}$$

If BCR_{PA} is greater than one, then the program is considered cost-effective.

$$B_{PA} = \sum_{t=1}^{N} \frac{UAC_t}{(1+d)^{t-1}}$$

$$C_{PA} = \sum_{t=1}^{n} \frac{PRC_t + INC_t}{(1+d)^{t-1}}$$

Where

B_{PA}	=	NPV of program benefits
C_{PA}	=	NPV of program costs
UAC_t	=	Utility avoided supply costs in year t
PRC_t	=	Program costs in year t
INC_t	=	Incentives paid to participant by utility
d	=	Discount rate
t	=	Year
n	=	Lifecycle of the DSM measure

16.B.4. TOTAL RESOURCE COST TEST

The total resource cost (TRC) test measures the net costs of a program as a resource option based on the total costs of the program, including both the participants' and the utility costs. The TRC test is the most commonly used measure of cost effectiveness since it provides an indication of whether the totality of costs, to utility and ratepayer, is being reduced. The benefits calculated in the TRC test are the avoided supply costs. The costs are the program costs paid by both the utility and the participants. The benefit-cost ratio (BCR_{TRC}) for the TRC Test is as follows:

$$BCR_{TRC} = B_{TRC}/C_{TRC}$$

If BCR_{TRC} is greater than one, then the program is considered cost effective.

$$B_{TRC} = \sum_{t=1}^{N} \frac{UAC_t}{(1+d)^{t-1}}$$

$$C_{TRC} = \sum_{t=1}^{n} \frac{PRC_t + PC_t}{(1+d)^{t-1}}$$

Where

B_{TRC}	=	NPV of benefits to total resources
C_{TRC}	=	NPV of costs to total resources
UAC_t	=	Utility avoided supply costs in year t
PRC_t	=	Program costs in year t
PC_t	=	Participant costs in year t
d	=	Discount rate
t	=	Year
n	=	Lifecycle of the DSM measure

16.B.5. SOCIETAL COST TEST

The societal cost test (SC) is a variant of the TRC test, the difference being that it includes quantified effects of externalities (such as environmental impacts) in the measure of costs and benefits. The benefits calculated in the SC test are the avoided supply costs plus the avoided environmental costs. The costs are the program costs paid by both the utility and the participants. The benefit-cost ratio (BCR_{SC}) for the SC Test is as follows:

$$BCR_{SC} = B_{SC}/C_{SC}$$

If BCR_{SC} is greater than one, then the program is considered cost effective.

$$B_{SC} = \sum_{t=1}^{N} \frac{UAC_t + UEC_t}{(1 + d)^{t-1}}$$

$$C_{SC} = \sum_{t=1}^{n} \frac{PRC_t + PCN_t}{(1 + d)^{t-1}}$$

Where

B_{SC} = NPV of benefits to total resources including environmental effects

C_{SC} = NPV of costs to total resources including environmental effects

UAC_t = Utility avoided supply costs in year t

UEC_t = Avoided environmental costs in year t

PRC_t = Program costs in year t

PCN_t = Net participant costs in year t (net off rebates)

d = Discount rate

t = Year

n = Lifecycle of the DSM measure

CHAPTER 17

INDOOR AIR QUALITY

JACK L. HALLIWELL, P.E.
Halliwell Engineering Associates
Miami, Florida

17.1 INTRODUCTION: WHY IAQ IS IMPORTANT TO CEMs

17.1.1 Statement of Chapter Intent

In commercial buildings, energy management and indoor air quality (IAQ) management often co-exist in a competing interests relationship. Overly aggressive energy management measures can and have resulted in the degradation of IAQ, with subsequent complaints and claims from building occupants. Conversely, excessive IAQ control measures can also bring significant energy penalties to a building's operating costs.

Therefore, balancing these two operational goals can be of significant value to Certified Energy Managers (CEMs), building owners, and their facility managers.

Energy managers are not required to become IAQ experts in order to successfully achieve this balance. What is needed, however, is a clear understanding of IAQ principles: the mechanics of indoor air quality, how problems arise, what causes them, how they can be resolved, and most importantly, how they can be avoided.

The goal of this chapter is to provide the reader with a strong understanding of how IAQ works in buildings and to establish an IAQ foundation in the reader's thinking that will allow most IAQ problems to be avoided or addressed decisively.

The 3 key learning objectives of this chapter are:

1) To understand what IAQ is—its impact on building occupants, managers, owners, and CEMs.
2) To learn IAQ fundamentals, including the sources of IAQ problems, how to investigate and resolve them, and when to call for help.
3) To understand energy management/IAQ equilibrium and how to intuitively avoid problems.

17.1.2 IAQ Defined

Indoor air quality is both quantitative and qualitative, measurable and perceived, objective and subjective. While there are some regulatory standards for IAQ, most of them have industrial applications with airborne pollutant concentrations generally not seen in commercial, institutional, or residential buildings. Therefore, most IAQ complaints in non-industrial buildings result from early detections of increased airborne contaminants by sensitive individuals, long before those concentrations reach regulatory limits. As a result, IAQ can often be difficult to define.

ASHRAE defines "acceptable indoor air quality" in its Standard 62.1-2004 as *air in which there are no known contaminants at harmful concentrations as determined by cognizant authorities <u>and</u> with which a substantial majority (80% or more) of the people exposed do not express dissatisfaction* (underlining added for emphasis). This best effort by ASHRAE's SSPC62 Committee to define IAQ, presents some difficulty in its practical application.

First, this definition provides that IAQ is acceptable as long as the pollutant levels are below regulatory or recommended standards. The problem is that IAQ complaints in commercial and institutional settings usually arise long before those limits are reached. The second problem is that ASHRAE considers the IAQ acceptable if 80% of the occupants find it so. Therefore, in a high-rise office building, or a university structure with 500 occupants, we could have 100 people complaining of IAQ health concerns (by ASHRAE's definition), yet the indoor air could still be deemed acceptable.

It therefore may be easier to define what the *absence* of acceptable IAQ is, i.e. what is an IAQ problem? Here is a definition we have found to be both useful and helpful.

An IAQ problem is a condition that occurs in occupied buildings when the concentrations of pollutants in the indoor air (particles, gases, or bioaerosols) increase to the point where they cause physical discomfort, allergic reactions, or illness to the occupants.

17.1.3 Law of Unintended Consequences

IAQ first became a significant issue for building managers in the late 1970's. During that time, the United States was going through a difficult energy crisis that resulted from an oil embargo created by OPEC. In response, building owners and managers looked for new ways to reduce energy consumption. Some of those measures resulted in significant reductions to outdoor ventilation rates, which reduced operating costs to heat or cool outdoor air.

However, the reduction in ventilation rates also resulted in elevated concentrations of airborne pollutants in the buildings' air, and the energy management/IAQ equilibrium balance became disrupted. The take-away lesson from that experience was that energy management measures must first be reviewed for their potential impacts on a building's IAQ—and ultimately on the comfort and health of the occupants.

17.1.4 Health and Comfort Issues

ASHRAE Standard 62.1-2004, "Ventilation for Acceptable Indoor Air Quality," states: *1.1 The purpose of this standard is to specify minimum ventilation rates and indoor air quality that will be acceptable to human occupants and are intended to minimize the potential for adverse health effects* (underlining added for emphasis).

In other words, this standard seeks to achieve IAQ that will satisfy both the comfort *and* the health of the building occupants. This is an important concept for IAQ investigators, as well as energy managers, because the threshold of occupant displeasure and complaints is often encountered first with discomfort, then with health issues.

By way of example, the US OSHA regulated threshold for carbon dioxide levels is 5,000 parts per million (ppm). However, many building occupants will begin to notice increased CO_2 concentrations at 1,000 to 1,200 ppm (comfort levels). At 1,500 to 2,000 ppm, concerns can turn into complaints of stuffiness, headaches, and lethargy.

The main points are these:

1) Building occupants expect IAQ that will keep them healthy *and* comfortable.

2) The threshold of indoor contaminant concentrations needed to maintain comfort is generally well below regulatory levels.

17.1.5 People are IAQ Monitors

Very few commercial or industrial buildings monitor or measure pollutant concentrations of indoor air. Instead, they generally follow prescriptive or performance guidelines for ventilation rates, filtration, humidity levels, and temperature to achieve a comfortable and healthy building environment. By default, the building occupants serve as biological monitors of indoor air quality parameters.

When airborne concentrations of pollutants increase,

the detection and warning systems that respond are usually the building occupants themselves. Unfortunately, those responses are often negative as an increasing number of the building's population become uncomfortable or experience health symptoms.

17.1.6 The Emotional Aspect of IAQ Problems

IAQ problems are unique. Building occupants who suffer from significant discomfort or health effects from IAQ problems will often perceive them as health threats. If the building management is slow to respond and does not appear to take the problems seriously or is ineffective in solving the problem quickly, occupants will often perceive management as uncaring or inept. The ensuing reactions can be driven by anger and sometimes rage. When this occurs, the scope and costs for correcting what may have been a simple building system problem can increase significantly.

17.1.7 The Unique Nature of IAQ Problems

Once building management has lost the confidence and trust of their occupants, the building-related problem can bifurcate into a building problem *and* a people problem. Attempts to resolve both problems by simply correcting the underlying cause of the elevated pollutant concentrations are usually ineffective. This is because in the occupant's view, acceptable risk has now moved towards zero risk; and acceptable IAQ has now moved toward pristine IAQ.

This is not a theoretical problem, but one that we have seen played out in hundreds of facilities throughout the US. Some of these problems were able to be resolved by addressing the building problem and the occupants' perception problems, while others had gone unresolved for a long period and went to litigation.

The most important IAQ issues for CEMs be aware of in occupied buildings are as follow:

1) Building occupants often consider IAQ problems to be health threats and take them very seriously.

2) As a consequence, the prevention and resolution of IAQ problems are held to a much higher standard by building occupants than most other building-related issues.

3) Management's inability to react quickly or failure to adequately prioritize IAQ problems can result in a significant increase in the scope of work needed to satisfy building occupants.

4) CEMs routinely work with building systems that can

and will impact IAQ. Understanding the mechanics of IAQ and how to avoid or properly respond to IAQ problems is an important part of a CEM's role and responsibility.

17.1.8 IAQ Litigation

When building management is unable to adequately resolve the underlying cause of the IAQ problem and the resulting emotional response of the building occupants, litigation can result. Typically this will occur when a group of occupants join together with a plaintiff's law firm to seek financial restitution for their claimed injuries to health and/or property. Plaintiff's counsel will often seek to obtain class action status for the complaint from the court, which if granted, will put the entire body of building occupants adverse to the owners and management.

This type of litigation is very expensive. Even if the building owners prevail legally, the costs of defending these types of lawsuits, the negative publicity they bring, and the alienation of the building occupants can be difficult to recover from. The main point for CEMs to be aware of is that the high cost of IAQ litigation demands careful consideration of maintaining an energy/IAQ balance.

17.1.9 ASHRAE and IAQ

ASHRAE's primary IAQ standards are 62.1 and 62.2, "Ventilation for Acceptable Indoor Air Quality." Standard 62.1 is applicable to commercial, institutional, and mid/high-rise residential buildings, while Standard 62.2 is designed for low-rise (multi-family) residential buildings.

Both of these standards attempt to take a holistic view of IAQ (causes and solutions), although their primary focus for providing "acceptable indoor air quality" is through the proper quantity and quality of ventilation air. As such, the most recent version of Standard 62 was published in 2004 (Standard 62-2004). This version of the standard provides two primary means of selecting ventilation rates for different types of facilities.

Ventilation Rate Procedure

This procedure provides a prescriptive amount of ventilation air based upon the type of space, the number of people in the space, and the size of the area. It assumes two primary sources and strengths of indoor pollution: people and the building materials/operations. These ventilation rates are provided in prescriptive tables that are easy to use.

IAQ Procedure

This procedure analyzes contaminant sources, strengths, and targeted concentrations of airborne contamination in order to select ventilation rates. This IAQ procedure allows for reductions in ASHRAE's prescribed ventilation rates if the contaminant sources can be adequately reduced or filtered out.

Since cooling or heating outdoor ventilation air can be a significant part of a building's overall energy load, the ventilation rates selected are important for both energy management and IAQ. Most jurisdictions have codified minimum ventilation rates for buildings, and ASHRAE Standard 62 is often used as the benchmark. Caution is required, however, as many jurisdictions still reference earlier versions of Standard 62, which may contain higher ventilation rates than the current (62-2004) version.

In addition, ASHRAE has placed the revision of Standard 62 into a process of "continuous maintenance" that allows for the Standard 62 Committee to update individual sections of the standard on a piecemeal basis. When considering energy management initiatives that could potentially impact a building's ventilation, a CEM should first consult the applicable code and the referenced ventilation standard for that building. Different ventilation rates could apply, based on the date of construction and major renovations that triggered more recent codes and standards.

In addition to Standard 62, ASHRAE has published two other standards that are applicable to IAQ and energy management:

Standard 55-2004: "Thermal Environmental Conditions for Human Occupancy"

The purpose of this standard is to specify the combination of indoor thermal environmental factors (primarily temperature, air speed, and humidity) and personal factors (metabolic rate, activity, and clothing insulation) that will produce thermal environmental conditions acceptable to a majority of occupants in the building.

Standard 52.2-1999: Method of Testing General Ventilation Air Cleaning Devices for Removal Efficiency by Particle Size

This standard is designed to provide users with a simplified method for the selection of HVAC filters of varying efficiencies in the removal of airborne particulates. The rating of filter efficiencies can be complex, with two main variables: (1) the size of the particles to be removed from the airstream and (2) the cleanliness of the filter media (filters perform at their minimum efficiency when they are new). This standard classifies filter media based upon a minimum efficiency rating value (MERV) for different particle sizes that are captured over the range of a filter's life (clean to dirty). The minimum MERV for most commercial applications is MERV 6. That filter will remove particles that range in size from 3 to 10 microns in diameter with a minimum efficiency of 35 to 50 percent. Pleated filters are rated up to MERV 8, box filters range from MERV 10 to 14, and bag filters range from MERV 11 to 16. The highest MERV rating is 20.

Summary

ASHRAE Standard 62 is an excellent guide for selecting minimum ventilation rates for buildings. Those published rates are the result of significant research by ASHRAE's SSPC62 committee members on striking a balance between energy management and acceptable IAQ.

However, CEMs need to be aware that complying with ASHRAE's minimum ventilation rates will not, by itself, guarantee that healthy or comfortable conditions will exist within a building. The mechanics of IAQ are, unfortunately, more complex than that. While the quality and quantity of ventilation is a large part of providing acceptable IAQ, there is more to the IAQ equation than delivering minimum amounts of clean outdoor air, as explained in the following section.

17.2 IAQ FUNDAMENTALS: MECHANICS OF THE PROBLEM, CAUSE(S), INVESTIGATIONS, AND SOLUTIONS

17.2.1 The Root Cause of All IAQ Problems

All IAQ problems have one common denominator, root cause—increased concentrations of airborne contaminants. These airborne pollutants can and will cause reactions in susceptible (sensitive) building occupants first, when some condition(s) exist within the building that promote(s) an increased concentration of those contaminants. Generally, the underlying origin of the problem is either a reduction in the ventilation rate, an increased source strength of airborne pollutants, or both.

An example of a reduction in the ventilation rate would be a VAV damper that did not have a minimum air setting and that shut off ventilation supply when thermal conditions satisfied the thermostatic controls.

An example of increased pollutant source strength would be the conversion of an office space into a conference room, placing 12 people into a space where ventilation was designed for 1 or 2.

In both examples, the bioeffluent discharge of the occupants (mostly CO_2 and water vapor) increased to levels that exceeded some of the occupant's comfort thresholds. This would result (primarily) in comfort complaints. If the contaminant of concern were allergenic or toxic, such as mold or certain solvents, the occupant reactions could be more robust, based upon health concerns.

17.2.2 IAQ Mechanics: People, Pathways, and Pollutants

The three primary factors that determine the acceptability of a building's IAQ are people, pathways, and pollutants ($IAQ = P_1 + P_2 + P_3$). This simple concept is the basic idea behind all IAQ management, problem solving, and problem avoidance. A strong working knowledge of this concept and its applications, as described herein, will allow CEMs to identify most pre-existing IAQ problems, investigate others, and avoid future IAQ issues in the planning and implementation of their energy management programs.

(P₁) People

The most important concept to understand about the first IAQ factor (people) is that *acceptable* IAQ is a subjective measurement. It is only acceptable when the building occupants find it to be acceptable. Engineers can design ventilation rates, filtration levels, humidity, and temperature controls for the building that they know will satisfy most of the people most of the time. In the end however, it is the building occupants who judge the acceptability of the IAQ.

When we start to modify building systems and operational parameters for energy management, any changes that affect IAQ will be judged by that subjective audience. They are the biological IAQ monitors within the building, and there will be a smaller subset of that population who are more sensitive to their environment and will be the first to detect any increases in airborne pollutant concentrations. These are usually people with allergies or compromised immune systems. As the airborne concentrations of contaminants continue to increase, a larger percentage of the building population can be expected to feel the effects and respond accordingly.

ASHRAE's Standard 62-2004 takes full recognition of the subjectivity of acceptable IAQ in Section 2, Scope:

2.5 Acceptable indoor air quality may not be achieved in all buildings meeting the requirements of this standard for one or more of the following reasons:

    a) because of the diversity of sources and contaminants in indoor air;

    b) because of the many other factors that may affect occupants perception and acceptance of indoor air quality, such as air temperature, humidity, noise, lighting, and psychological stress;

    c) because of the range of susceptibility in the population; and

    d) because outdoor air brought into the building may be unacceptable or may not be adequately cleaned.

Because most buildings that CEMs deal with will be

occupied, these subjective IAQ monitors (occupants) represent the most challenging of the three IAQ factors we need to manage.

The key here is to understand that:

a) Of the three primary IAQ factors, P_1 (people) is the one factor we cannot control;

b) Occupants will be the first to tell us when energy management measures start to negatively impact IAQ, and;

c) We need to respond quickly and effectively upon receipt of those complaints.

(P_2) Pathways

This second IAQ factor, pathways, includes the defined HVAC system and as well all other routes that a contaminant could travel from its source to the breathing zone of the building occupants.

From an IAQ perspective, the HVAC system can serve as the cause *or* the cure for problem. Since the HVAC system controls ventilation, filtration, temperature, and humidity in the building, it can serve as the source of many IAQ problems. In fact, in our experience with over 1,200 IAQ investigations, the HVAC system was either the cause or a contributing cause of the problem in over 70% of the buildings we investigated. The point here is that any energy conservation measures related to the HVAC system should be carefully scrutinized for potential IAQ impacts. Below is a brief summary of the main HVAC-related causes of IAQ problems. CEMs that alter any of these HVAC parameters have the potential to create new (or exacerbate existing) IAQ problems within the building.

Ventilation—Interruptions or reductions in the delivery of clean outdoor air can result in increased levels of airborne pollutants, even without corresponding increases in a pollutant source strength. Reference ASHRAE Standard 62.1-2004, "Ventilation for Acceptable Indoor Air Quality," for additional information.

Temperature—Variability in indoor temperature usually does not cause IAQ problems, although it can play a significant role in occupants' perceptions of indoor air quality.

Elevated temperatures outside the occupants' perceived range of acceptability will tend to lower their response threshold for increased levels of indoor contaminants. Conversely, cooler air temperatures are often perceived as cleaner air. Reference ASHRAE Standard 55-2004, "Thermal Environmental Conditions for Human Occupancy," for additional information.

Humidity—Prolonged elevated humidity in buildings can result in mold growth. This can occur through a number of mechanisms, from condensation on surfaces below the surrounding air's dewpoint temperature to elevated surface moisture levels from building materials that take on the same moisture content as the air that surrounds them (equilibrium relative humidity). The availability of a free water source, combined with carbon-based building materials (or dirt) provides the necessary growth mechanisms for mold in a building. Mold growth inside of buildings can release mold spores into the breathing zone, significantly increasing the indoor concentrations of the airborne allergens. Reference ASHRAE Standard 55-2004, "Thermal Environmental Conditions for Human Occupancy," for additional information on acceptable levels of humidity (and avoiding mold growth) inside of buildings.

Filtration—Inadequate filtration can result in increased levels of indoor airborne pollutants by ineffective cleansing of outdoor air or inadequate removal of pollutants from indoor sources. ASHRAE's filtration standard (52.2-1999) recommends a minimum filtration level of MERV 6 in new buildings (pleated filters).

Air Speed—As with temperature, air speed does not directly cause elevations in indoor contaminant levels, but it can affect occupants' perception of comfort and the acceptability of IAQ. Elevations in air speed can improve the perceived acceptability of warmer air temperatures. Conversely, increased air speed of cooler temperatures can create unwanted local over cooling and negative responses from building occupants.

Fresh Air Intake—The location of fresh air intakes can be critical with regard to the ingestion of outdoor air pollutants into the HVAC system and increased levels of indoor contaminants (vehicle exhaust, emergency generator exhaust, trash bin odor, cooling tower effluent, etc.). They can also act as collection points for leaves, bird droppings, snow, and rain, and thereby serve as unintentional pollutant sources themselves.

Building Envelope—In addition to the HVAC system, the building envelope (exterior walls, windows, and roofing system) can serve as a pollutant pathway system. Most building exterior cladding systems allow for the drainage of rainwater and other moisture from the exterior wall cavity. Consequently, the exterior wall cavities can often communicate with other interstitial spaces within the structure to provide an unintended pathway for unfiltered and humid air to enter the building. In addition to these exterior pathways, pressure imbalances can exist (from wind pressures, elevator shafts, return air plenums, and exhaust chases) that serve to transport moisture-laden air and pollutants throughout the building. Service tunnels and crawl spaces can also serve as unintentional air pathways, carrying airborne pollutants where differential pressures exist. These uncontrolled airflows throughout a building can serve as hidden delivery systems of pollutant sources from both outside and inside of the building.

(P₃) Pollutants

This third and final IAQ factor represents the underlying cause of all IAQ complaints. An increased concentration of airborne contaminants in the indoor air is at the heart of all IAQ problems, solutions, and prevention.

To simplify this concept, we will identify three classes of indoor pollutants that we will need to control: particulates, bioaerosols, and gases. The distinction among airborne particles, bioaerosols, and gases is important in determining their cause, origin, and control strategies.

Particulates

Particles dispersed in the air are referred to as aerosols. This class of contaminants covers a large range of particle sizes, from asbestos fibers at .3 microns in diameter to visible dust at over 10 microns in diameter. (Human hair is 100 to 150 microns in diameter) The size of a particle determines where and how deep into the human respiratory system it can be deposited. EPA defines respirable particles that can penetrate into the lungs as having a median diameter of 2.5 microns ($PM_{2.5}$).

Bioaerosols

Bioaerosols are airborne particles with a biological origin. They include viruses, bacteria, fungal spores (mold), pollen, and animal dander.

Most bioaerosol-related problems occur when micro organisms (fungi) grow indoors on wet organic material and disperse spores into the indoor air. The four requirements for mold growth are favorable temperature (same as for humans), mold spores (present in moist environments), a food source (any carbon based building material or dirt), and water. Of these four elements, water (moisture and humidity) is the only one that we can control to prevent mold growth inside of buildings.

Gases and Vapors

The gaseous class covers airborne chemical contaminants. They can exist as gases that are naturally gaseous under ambient conditions, or vapors that are normally in a liquid or solid state, but which can evaporate readily.

Chemical gases of concern in the indoor environment include inorganic gases (ozone, nitrogen dioxide, radon, carbon monoxide, and carbon dioxide) and volatile organic compound gases (VOCs) from solvent-based building materials and cleansers.

Outdoor Air Contaminants

Outdoor air contaminants include both natural and man-made impurities. Natural occurrences such as sea spray, tree pollen, and mold spores can create a baseline of these pollutants in the outdoor air. In addition, manmade outdoor air contaminants are numerous and varied, originating from industrial processes, power plants, transportation, and construction. They can include (in descending order of typical ambient concentrations) carbon dioxide, carbon monoxide, methane, carbon disulphate, and sulfur dioxide. The concern is when these outdoor pollutants become elevated and are carried into the indoor environment.

The US EPA has identified several outdoor contaminants as criteria pollutants. The list includes suspended particulate matter (PM_{10}), lead particulate matter, ozone, nitrogen dioxide, sulfur dioxide, carbon monoxide, and total hydrocarbons. Standards have been set for these contaminants, and levels measured at a large number of locations throughout the US are published by the EPA every year (www.epa.gov/air/criteria.html).

The importance of outdoor air contaminant levels is:

1) Outdoor air contaminants can enter the indoor space and combine with indoor pollutants to create a large total airborne pollutant burden.

2) Outdoor air particulate matter can be filtered effectively if the appropriate filter media is selected. However, gaseous pollutants in the outdoor air are much more problematic and expensive to remove.

3) Outdoor air with high levels of contaminants is not as effective for use in dilution ventilation.

Indoor Air Contaminants

Indoor air pollutant *sources* include people (carbon monoxide, tobacco smoke), building products (volatile organic compounds, formaldehyde), HVAC systems (fiberglass insulation fibers, mold growth, bacteria, including legionella), building systems (gas heaters, appliances, boilers), and cleaning products (solvents).

Pollutants Summary

Understanding the three main classes of indoor air pollutants (particulates, bioaerosols, and gases) and their two main sources (outdoor or indoor) is the key to developing effective control strategies.

17.2.3   Control Strategies for Indoor Pollutants

As previously stated, the underlying cause of all IAQ problems is increased amounts of airborne contaminants from some source(s), either inside or outside of the building. Once that contaminant and its source have been identified, the next step is to select a control strategy to either eliminate (if possible) or reduce that pollutant in the indoor air to acceptable levels. There are three primary strategies available to accomplish this:

Source Removal
Filtration
Dilution Ventilation

Source removal is obviously the best long term approach, as it eliminates the need (and cost) to continually filter or ventilate the airborne pollutants. However, source removal may not always be possible or practical. For example, people themselves are contaminant sources in buildings, as are some building products (finishes and furnishings) and processes (operations and cleaning). Outdoor air with elevated levels of contaminants is also an example of a pollutant that cannot be controlled by source removal. Source removal is, however, the preferred approach when dealing with bioaerosols and many other indoor-generated airborne particulates or gases.

Filtration is the best approach to cleansing airborne particulates from the outdoor air. It will also reduce particulates from indoor sources (including bioaerosols) but not before they enter the breathing zone. Source removal is the preferred particulate removal strategy when practical. ASHRAE Standard 52.2 provides ratings for different filter efficiencies when challenged with varying particle sizes. Particulate filtration is not effective in removing gases from the indoor environment. Gaseous filtration is possible but very expensive. The preferred strategy for gaseous contaminants is source removal, local exhaust, or dilution ventilation.

Dilution ventilation is the control strategy of last resort. It is used when the pollutant sources cannot be removed (such as people or some building products). It is the most expensive control strategy since the outdoor ventilation air needs to be heated or cooled and dehumidified and filtered. ASHRAE Standard 62.1-2004 prescribes minimum ventilation rates for different types of buildings with "typical" indoor pollutant source loadings of people, building products, and cleaning. These ventilation rates take into account the needed balance between energy consumption and IAQ and have been adopted into the building codes of most jurisdictions. Local exhaust ventilation can also be used effectively to remove indoor pollutants at the source location in kitchens, restrooms, and closed parking garages.

In selecting the best indoor pollutant control strategy for your situation, start with source removal (or reduction) as your preferred strategy. If that is not possible or practical, consider filtration for airborne particulates. Ventilation and local exhaust are most applicable to pollutant sources that cannot be removed or adequately reduced.

17.2.4 IAQ Investigations

With a firm understanding of the basic IAQ formula, IAQ = P_1 (People) + P_2 (Pathways) + P_3 (Pollutants), we can use those 3 major factors to help us identify and investigate most IAQ problems we may encounter. Once we have identified the pollutant(s) of concern and their pathway(s), we can select a control strategy to resolve the problem.

There are a number of different approaches to IAQ investigations with two primary strategies that are generally followed:

1) Industrial hygienists, whose professional focus is the health and safety of the building occupants, will often follow a strategy that places its emphasis on people (their health-related reactions to the pollutants) and on identifying the airborne pollutants of concern, often with sampling, in an effort to investigate the problem.

2) Engineers will usually take a different approach, focusing on identifying the airborne pollutant *sources* (as opposed to sampling and analysis) and their pathways to the affected occupants (e.g. HVAC systems).

Both strategies have their place. Engineers utilize this strategy based on its speed of investigation, its lower costs, and the fact that HVAC systems play a major contributing role to IAQ problems.

Others have employed an industrial hygiene approach when the engineering strategy was unable to quickly identify the pollutant source. In these cases, measurements of indoor airborne pollutant sources would come later in the investigation and would be compared to sample results in non-complaint areas of the building and outdoor ambient conditions.

EPA has also developed a hybrid approach to IAQ investigations that recognizes the importance of the interaction among occupants (P_1), transport mechanisms (P_2), and contaminant sources transport mechanisms (P_3). In their IAQ guide, "Building Air Quality," EPA recommends beginning an IAQ investigation by inventorying occupant complaints, contaminant sources, and transport mechanisms. While gathering information, investigators look for patterns that can help them form hypotheses and potential explanations for the complaints.

Regardless of the approach that is taken, here are the main questions that any IAQ investigation must answer:

- What is the IAQ issue of concern? What are the affected occupants' complaints?

- What new or elevated contaminant sources exist?

- Is the HVAC system performing as designed (ventilation, humidity, and temperature control, pressurization)?

- How are the airborne pollutants being transported to the affected occupants' area(s)?

- Are these findings consistent with the occupants' symptoms?

One final note on IAQ investigations. It has been our experience that the more open and transparent the investigation process, the better it will be received by the building occupants. Building managers may incorrectly believe that keeping some or all of this information from the occupants will help protect their liability. Our experience has shown the opposite to be true. In the end, the IAQ problem is not resolved until the building occupants believe it to be resolved. Keeping occupants informed of the investigative findings along the way will greatly help in achieving closure to the emotional aspects of the IAQ problem.

17.2.5 IAQ Solutions

The results of the IAQ investigation will identify the occupants who are affected, their location(s) in the building, the nature of their complaints/symptoms, the offending airborne pollutant(s), their source(s), and their route(s) of travel throughout the building. With this information, we are then ready to select a control strategy. As explained earlier, removal or reduction of the pollutant source is the preferred solution.

A good example of a source removal solution is mold remediation. Since excess moisture is the underlying cause of all mold growth, removal of the mold, wet materials, and the source of the excess moisture is required. (An excellent reference for this work is Standard S520, "Standard and Reference Guide for Professional Mold Remediation." of the Institute of Inspection, Cleaning, and Restoration Certification (IICRC).

When removal of the pollutant source is not possible or practical, then we will look to our other two strategies of filtration or dilution ventilation to resolve the problem.

What's also important to understand in developing IAQ problem solutions is that the underlying cause of the problem has been created somewhere during the design and construction (or renovation) or operations and maintenance of the facility. Design errors and omissions and construction defects have led to many IAQ problems

in properties, sometimes with significant consequences. Construction related IAQ problems are common when renovation work is performed in occupied buildings, with temporary HVAC systems and construction products and activities being contiguous to the occupied section of the building (increased pollutant sources and transport opportunities).

More often however, IAQ problems result from building operations and maintenance or lack of maintenance. Buildings with a significant backlog of deferred maintenance, especially within their HVAC systems, are candidates for IAQ trouble. Finally, energy conservation and management measures can also trigger IAQ-related problems when those programs and processes do not fully consider their impacts on the IAQ equation (IAQ = $P_1 + P_2 + P_3$), *or* the existing IAQ conditions within the building.

17.2.6 When to Call For Help

Asking for help when IAQ complaints arise can be counterintuitive. On one hand, only a small percentage of the building occupants may be claiming to feel symptoms (discomfort, odor detection, or health effects), while the majority of the building population appears to be unaffected. On the other hand, those persons claiming to be affected will become more vocal in their reporting of complaints as time goes on, and they will discuss it with other occupants.

If the initial investigations did not reveal any obvious or significant pollutant sources, what is the prudent course of action?

As noted earlier, IAQ problems are very different from most other building-related issues, primarily because they are often perceived as health threats by the affected persons. For that reason, it is beneficial to ask for help as soon as possible and practical. If initial efforts to resolve the problem have not been successful (generally within a 48 to 72 hour period after initial receipt of the complaint), then bringing in an IAQ-focused HVAC engineer or industrial hygienist would be prudent. The cost of that service compared with the potential downside of not resolving the problem after multiple attempts is small. If the health related complaints are of a serious nature, it may warrant evacuation of the affected area(s) of the building and calling for a professional third party inspection. Any and all persons reporting health-related issues should seek medical attention.

The other benefit of an early response with an outside expert is increased credibility with the affected occupants. If it is conveyed from the start that management takes this problem as seriously as the affected occupants do much of the emotionally charged response can be avoided. In addition, it will help to avoid claims of bias in the findings of the investigation.

Selection of an outside expert and their protocols are important. As explained earlier, both HVAC engineers with specific IAQ training and industrial hygienists are capable of undertaking the inspection, although they typically follow different strategies.

Sampling and measurement of indoor air contaminant levels is *not* recommended as part of an initial IAQ investigation. There are a number of reasons for this, some of which are listed below.

- It is generally more efficient and expeditious to locate the *source* of the indoor air contaminants than to analyze and classify all airborne constituents.

- Sampling for gases, particulates, and bioaerosols all require different equipment and analyses. If you know in advance which class of pollutant is offending, it is easier and quicker to search for that source.

- Most airborne pollutant standards that air sampling results would be compared to relate to industrial settings. Those levels are rarely reached in commercial or institutional buildings.

- Bioaerosol sampling requires an extensive number of samples to be taken inside and out in order for the result to be statistically valid; the results will not have any appreciable effect on the remediation strategy, scope, or requirements.

On a final note, utilizing duct cleaning contractors or other IAQ remediation companies to perform the outside independent IAQ investigation can often result in incomplete or biased outcomes. The services of these firms should be reserved for remediation, once the true cause of the problem has been identified.

17.3 THE ENERGY MANAGEMENT/ IAQ BALANCE: HOW TO AVOID PROBLEMS

Now we are ready to move forward with the main purpose of this chapter, which is to enable CEMs to perform their work without creating new (or exacerbating existing) IAQ problems in their subject buildings.

Let's briefly summarize the main IAQ concepts we have discussed before we put them to work:

1) The Basic IAQ Equation:

$$IAQ = P_1 \text{ (People)} + P_2 \text{ (Pathways)} + P_3 \text{ (Pollutants)}$$

2) The three categories of indoor airborne pollutants:
- Particulates
- Bioaerosols
- Gases

3) The three main IAQ control strategies:
- Source Removal (or reduction)
- Filtration
- Dilution Ventilation (and local exhaust)

4) The underlying causes of IAQ problems:
- Design errors or omissions
- Construction defects
- Inadequate operations and maintenance

Now let's take these concepts and apply them to energy management.

First, before any energy management work is performed in the building, it is recommended that a baseline IAQ survey be performed. This is particularly important for all energy management measures that affect the building envelope, HVAC systems, energy management control systems, lighting systems, control systems, and operations and maintenance of the building. This can be conducted by the CEM and building management together to establish what IAQ conditions (and problems) exist prior to the implementation of any energy management measures. The baseline survey should, at a minimum, review any outstanding (unresolved) IAQ complaints, take baseline temperature and humidity readings, inspect the condition and hygiene of the HVAC system, and identify any potential pollutant sources. This baseline survey should be shared with the building management and agreed upon before moving forward. It will serve as a benchmark for the CEM against any future claims of changed conditions or IAQ degradation. If any energy conservation measures (ECMs) contemplated would change the ventilation rate or overall air exchange rates, baseline HVAC test and balance measurements should also be considered.

Second, individual ECMs should be reviewed to evaluate their potential impact for increasing indoor airborne pollutant loads in the building's air. Sometimes the potential for energy management measures to negatively impact IAQ are obvious, and other times they are more subtle. While it is not possible to cover every condition that could possibly increase indoor pollutant loads, we will provide some examples here to guide your thinking in the key area of energy management and IAQ balance. We will select specific chapters and topics in this book as a guide to illustrate how IAQ problems can be created, unintentionally.

Building Envelope—Chapter 9

Energy management measures designed to improve the performance of the building envelope can have unintended consequences that result in IAQ problems.

ECM

Elimination of infiltration of uncontrolled outdoor air.

Unintended Consequence

The HVAC system has been balanced to the original (infiltration) condition and the total building ventilation rate may be reduced as a result of a tighter building envelope. This may result in less dilution, higher pollutant concentrations, or negatively pressurizing the building and pulling make-up air from drains (sewer gas) or trash chutes.

ECM

Modification to the exterior walls: cladding, insulation, or vapor barrier location.

Unintended Consequence

Most exterior walls allow water to enter and drain. Any interruption to that system could create moisture retention in the wall cavity, promoting mold growth.

Mislocation of vapor barriers can also result in condensation occurring within the exterior wall, supporting mold growth.

HVAC Systems—Chapter 10
ECM

Reduce outdoor ventilation.

Unintended Consequence

In most jurisdictions, ASHRAE Standard 62 has been codified at the minimum ventilation rates allowed, unless operable windows are provided.

Reduced ventilation will increase indoor pollutant concentrations without an increased pollutant source strength.

Reduced outdoor ventilation can also result in depressurizing the building (with control exhaust systems operating), which can result in increased infiltration. In warmer climates, this infiltration of humid air can condense on cooler surfaces, resulting in free water and mold growth.

In addition, the utilization of demand ventilation systems must be considered carefully and with some caution. Systems that utilize CO_2 to measure the need for ventilation can require continual maintenance and calibration in order to operate effectively and avoid IAQ problems. Also, systems that only measure CO_2 do not consider all other airborne pollutants that require ventilation.

ECM

Raise chilled water temperature and/or supply air temperature.

Unintentional Consequence

Raising the chilled water or supply air temperatures in warmer climates, or during the summer months in other climates, can reduce the dehumidification capacity of the HVAC system and increase the moisture load within the building. Continual operation in this mode can result in condensation on cooler surfaces and widespread surface mold growth.

ECM

Reduction in filtration rating to reduce static pressure drop across the filter.

Unintended Consequence

This will increase the dirt burden within the building, as well as the airborne particulate loading. Increased dirt build-up in fiberglass-lined ductwork provides increased opportunity for mold growth inside the HVAC system and downstream of the filter.

Energy Management Control Systems—Chapter 1
Control Systems—Chapter 22

ECM

Reduction in ventilation rates, building pressurization, chilled water flow, or temperature.

Unintended Consequence

Increased levels of indoor airborne pollutants, increased infiltration, increased moisture load, possible mold growth.

Lighting—Chapter 13
ECM

Reductions in lighting levels (watts/sq. ft.) in occupied areas.

Unintended Consequence

In some building populations this can reduce occupant tolerance to previously acceptable levels of indoor pollutant concentrations.

Summary

These examples represent some of the areas where ECMs could result in, or exacerbate existing, IAQ problems. There are many other causes of IAQ problems that can exist in a building, but all come down to the simple principle of an increased level of airborne contaminants, either from a reduction in ventilation or from a new or increased source strength of that pollutant.

Carefully reviewing the contemplated ECMs for their potential to create these unintended effects within the building, along with a solid baseline IAQ survey beforehand, will help CEMs to avoid most IAQ problems.

17.4 SUMMARY AND CONCLUSIONS

Indoor air quality in occupied buildings can be a complex subject. This is primarily because of the diversity of human tolerance to different airborne contaminants, the large number of potential indoor contaminant sources, and the impacts of HVAC system design and operations within the building. As such, finding the balance between an effective energy management plan and acceptable IAQ can sometimes be challenging.

In this chapter we have sought to breakdown a relatively complex subject and frame it with some simple, straightforward operating principles that CEMs can use as a guide in finding their own energy management/IAQ balance.

To put a fine point on this discussion, we have developed the following "Top Ten List" of IAQ pitfalls that should be avoided.

17.5 TOPICS FOR ADDITIONAL STUDY

Air Contaminants
US EPA
"Primary Ambient Air Quality Standards of the United States" (2007)

ASHRAE Fundamentals Handbook (2005)
Chapter 12—*"Air Contaminants"*

Building Envelope
Energy Management Handbook (2006)
Chapter 9—*"Building Envelope"*

ASHRAE Fundamentals Handbook (2005)
Chapter 23—*"Thermal and Moisture Control in Insulated Assemblies"*

Filtration
ASHRAE Standard 52.2-1999
"Method of Testing General Ventilation Air Cleaning Devices

TOP 10 IAQ MISTAKES A CEM SHOULD AVOID

10) Do not underestimate or under-respond to IAQ complaints in your building.

9) Do not delay investigating IAQ complaints or their cause and origin.

8) Do not forget to conduct an IAQ baseline survey *before* you implement your energy management measures.

7) Do not fail to review your energy management program for its potential to elevate indoor airborne contaminants.

6) Do not implement energy management measures that would reduce ventilation, increase humidity, depressurize the building, or reduce filtration without carefully reviewing their potential impacts on IAQ.

5) Do not be intimidated by IAQ. All problems have a common denominator cause—increased airborne contaminants.

4) Understand that many IAQ problems occur, not from their energy management programs, but from operator error.

3) Do not assume that solving the physical IAQ problem will automatically solve the occupants' perceptions of the problem.

2) Do not perform air sampling as an initial IAQ investigative strategy.

1) Do not panic. IAQ problems can be solved quickly and effectively when responded to early.

---

for Removal Efficiency by Particle Size"

Health Effects
Natural Academy of Sciences
"Damp Indoor Spaces and Health" (2004)

IICRC S500 (2006)
Standard and Reference Guide for Professional Water Damage Restoration
Chapter 3—*Health Effects from Exposure to Microbial Contamination in Water Damaged Buildings.*

Mold

Halliwell Engineering Associates, Inc. (June 2004)
"Mold Prevention and Response Programs"

US EPA (2001)
"Mold Remediation in Schools and Commercial Buildings"

IICRC S500 (2006)
"Standard and Reference Guide for Professional Water Damage Restoration"

National Apartment Association (2002)
"Operations and Maintenance Plan for Mold and Moisture in Apartment Properties"

National Academy of Sciences (2004)
"Damp Indoor Spaces and Health"

Humidity, Temperature and IAQ

ASHRAE Standard 55-2004
"Thermal Environmental Conditions for Human Occupancy"

Ventilation

ASHRAE Standard 62.1-2004
"Ventilation for Acceptable Indoor Air Quality"

Bibliography

ASHRAE Standard 62.1-2004, *Ventilation for Acceptable Indoor Air Quality*. American Society of Heating, Refrigerating, and Air-Conditioning Engineers, Inc., 1791 Tullie Circle, NE, Atlanta, GA. 30329

ASHRAE Standard 55-2004, *Thermal Environmental Conditions for Human Occupancy*. American Society of Heating, Refrigerating, and Air-Conditioning Engineers, Inc., 1791 Tullie Circle, NE, Atlanta, GA. 30329

ASHRAE Standard 52.2-1999, *Method of Treating General Ventilation Air Cleaning Devices for removal Efficiency by Particle Size*. American Society of Heating, Refrigerating, and Air-Conditioning Engineers, Inc., 1791 Tullie Circle, NE, Atlanta, GA. 30329

US EPA, Primary Ambient Air Quality Standards for the United States. US EPA, Office of Toxic Substances, Washington, D.C.

Building Air Quality: A Guide for Building Owners and Facility Managers. US EPA, Office of Radiation and Indoor Air and US Department of Health and Human Services, National Institute of Health. 1991.

IICRC S520: Standard and Reference Guide for Professional Mold Remediation. 2003. Institute for Inspection Cleaning and Restoration Certification, 2715 East Mill Plain Blvd, Vancouver, WA. 98661.

Mold Remediation in Schools and Commercial Buildings. US EPA, Office of Air, Radiation, and Indoor Environment Division (66095). March 2001.

CHAPTER 18

ELECTRIC AND GAS UTILITY RATES FOR COMMERCIAL AND INDUSTRIAL CUSTOMERS

R. SCOTT FRAZIER, Ph.D., PE, CEM
Assistant Professor
Oklahoma State University

LYNDA J. WHITE
RICHARD A. WAKEFIELD
JAIRO A. GUTIERREZ
CSA Energy Consultants
Arlington, VA

18.1 INTRODUCTION

Purpose and Limitations

The main focus of this chapter on rates is to provide information on how an average commercial or industrial customer can identify potential rate-related ways of reducing its energy costs. The basic costs incurred by electric and gas utilities are described and discussed. How these costs are reflected in the final rates to commercial and industrial customers is illustrated. Some examples of gas and electric rates and how they are applied are included. In addition, this chapter identifies some innovative rates developed by electric and gas utilities as a response to the increasing pressure for the development of a more competitive industry.

Because of the breadth and complexity of the subject matter, the descriptions, discussions and explanations presented in this chapter can not cover every specific situation. Energy consumption patterns are often unique to a particular commercial and/or industrial activity and therefore case by case evaluations are strongly suggested. The purpose is to present some general cost background and guidelines to better understand how to identify potential energy cost savings measures.

18.1.2 General Information

Historically, electric and gas utility rate structures were developed by the utilities themselves within a much less complex regulatory environment, by simply considering market factors (demand) as well as cost factors (supply). Today the increasing pressures to develop more competitive markets have forced utilities to reconsider their traditional pricing procedures. Other factors affecting today's electric and gas markets include rising fuel prices, environmental concerns, and energy conservation mandates. These factors and pressures have affected gas and electric utility costs and hence their rates to their final customers.

In general, electric and gas rates differ in structure according to the type and class of consumption. Differences in rates may be due to actual differences in the costs incurred by a utility to serve one specific customer vs. another. Utility costs also vary according to the time when the service is used. Customers using service at off-peak hours are less expensive to serve than on-peak users. Since electricity cannot be stored, and since a utility must provide instantaneous and continuous service, the size of a generation plant is determined by the aggregate amount of service taken by all its customers at any particular time.

The main cost elements generally included in rate-making activities are: energy costs, customer costs, and demand costs. Each of these is discussed in the next section.

18.2 UTILITY COSTS

Utilities perform their activities in a manner similar to that of any other privately-owned company. The utility obtains a large portion of its capital in the competitive money market to build its system. It sells a service to the public. It must generate enough revenues to cover its operating expenses and some profit to stay in business and attract capital for future expansions of its system.

In general there are two broad types of costs incurred by a utility in providing its service. First, there are the fixed capital costs associated with the investment in facilities needed to produce (or purchase) and deliver the service. Some of the expenses associated with fixed capital costs include interest on debts, depreciation, insurance, and taxes. Second, there are the expenses associated with the operation and maintenance of those same facilities. These expenses include such things as salaries and benefits, spare parts, and the purchasing, handling, preparing, and transporting of energy resources. The rates paid by utility customers are designed to generate the necessary revenues to recover both types of costs. Both capital and operation and maintenance costs are allocated between the major cost elements incurred by a utility.

18.2.1 Cost Components

The major costs to a utility can be separated into three components. These include customer costs, energy/commodity costs, and demand costs. These cost components are briefly described below.

Customer Costs

Customer costs are those incurred in the connection between customer and utility. They vary with the number of customers, not with the amount of use by the customer. These costs include the operating and capital costs associated with metering (original cost and on-going meter-reading costs), billing, and maintenance of service connections.

Energy/Commodity Costs

Energy and commodity costs consist of costs that vary with changes in consumption of kilowatt-hours (kWh) of electricity or of cubic feet of gas. These are the capital and operating costs that change only with the consumption of energy, such as fuel costs and production supplies. They are not affected by the number of customers or overall system demand.

Demand Costs

Electric utilities must be able to meet the peak demand—the period when the greatest number of customers are simultaneously using service. Gas utilities must be responsive to daily or hourly peak use of gas. In either case, the utility will need to generate or purchase enough power to cover its firm customers' needs at all times. Demand-related costs are dependent upon overall system requirements. Demand costs can be allocated in many different ways, but utilities tend to allocate on-peak load. Included in these costs are the capital and operating costs for production, transmission, equipment (e.g., transformers) and storage (in the case of gas utilities) that vary with demand requirements.

Power Cost Adjustment

This is also sometimes known as energy cost adjustment (ECA). Some utilities use a variable factor in the rate schedule that compensates for the variable fuel costs used to produce energy. The power cost adjustment can be positive or negative and, in the case of electricity, is usually multiplied by the kWh used.

Universal Service Charge

Government bodies such as state public service commissions can require that utilities collect fees to be used in public welfare projects. These projects are funded by a universal service charge fund and can include low income utility bill assistance and weatherization programs, among others. The universal service charge is often collected as a multiplier on the kWh usage.

18.2.2 Allocation of Costs

Once all costs are identified, the utility must decide how to allocate these costs to its various customer classes. How much of each cost component is directly attributable to serving a residential, a lighting, or a manufacturing customer? In answer to this question, each utility performs a cost-of-service study to devise a set of allocation factors that will allow them to equitably divide these costs to the various users. After the costs are allocated, the utility devises a rate structure designed to collect sufficient revenue to cover all its costs, plus a fair rate of return (currently, this is running between 10 and 14% of the owners' equity.)

Impact Fees

Infrastructure costs are high for utilities and must be recovered one way or another. Incremental increases in generator capacity, water storage capacity, waste treatment capacity, and distribution pipes and wires are expensive and are also long term investments. Some rate structures build in the infrastructure costs with the rates, while others rely on impact fees.

Impact fees assess the cost of infrastructure directly instead of recovering it over time with rates. Impact fees, also called "development fees," are usually unpopular, because they represent a large up-front cost and can deter development or raise selling prices of new facilities. One advantage of separating the impact fees is that the money can be identified, escrowed, and ultimately used for the exact purpose it is named for. When buried in the rates, an additional task exists to pare out the "impact" portion or to ignore it and end up down the road with infrastructure costs looming without funding. The magnitude of the impact fee is determined by the connection size, which reflects the amount of infrastructure that must be in place for the new demand. Using pipe size as the analogy, a 6 inch pipe can carry twice as much water as a 4 inch pipe, and so on; thus, the impact fees based on connection size will increase exponentially and can be quite high for large connections. These fees reflect the utility is projected costs of infrastructure capacity increases and the concept that the customers already served by the existing, working infrastructure should not pay for an upgrade for a new neighbor. This argument is countered with the concept that new customers in the area bring jobs and economic vitality that other existing customers share. The regulatory bodies work to balance the two viable points of view.

A related one time charge is a "tap fee." This is commonly applied to water and gas services and is levied on the size of the "tap" to the municipal supply. The basis for these fees are the meter size and the potential for demand that affects the shared infrastructure; the larger the "tap," the larger the fee.

Other one-time fees may include special electric service, such as dedicated transformers, dual feeds for water or electric supply, etc., or extension of service to a remote location. These are generally pass-through fees for the special requirements, but they are sometimes subsidized by the utility if it seems likely that, for example, other customers will soon come along and share a new pipe or wire. Large, up-front expenditures are sometimes amortized over 5-10 years via a special customer contract.

18.3 RATE STRUCTURES

18.3.1 Basic Rate Structure

The rate structure generally follows the major cost component structure. The rates themselves usually consist of a customer charge, an energy charge, and a demand charge. Each type of charge may consist of several individual charges and may be varied by the time or season of use.

Customer Charge

This is generally a flat fee per customer, ranging from zero to $25 for a residential customer to several thousand dollars for a large industrial customer. Some utilities base the customer charge to large industrial customers on the level of maximum annual use.

Energy Charge

This is a charge for the use of energy that is measured in dollars per kilowatt-hour for electricity or in dollars per therm or cubic foot of gas. The energy charge often includes a fuel adjustment factor that allows the utility to change the price allocated for fuel cost recovery on a monthly, quarterly, or annual basis without resorting to a formal rate hearing. This passes the burden of variable fuel costs (either increases or decreases) directly to the consumer. Energy charges are direct charges for the actual use of energy.

Demand Charge

The demand charge is usually not applied to residential or small commercial customers, though it is not always limited to large users. The customer's demand is generally measured with a demand meter that registers the maximum demand or maximum average demand in any 15-, 30-, or 60-minute period in the billing month. For customers who do not have a demand meter, an approximation may be made based on the number of kilowatt hours consumed. Gas demand is determined over an hour or a day and is usually the greatest total use in the stated time period.

Another type of demand charge that may be included is a reactive power factor charge, a charge for kilovolt-amp reactive demand (kVAR). This is a method used to charge for the power lost due to a mismatch between the line and load impedance. Where the power-factor charge is significant, corrective action can be taken; for example, by adding capacitors to electric motors.

Demand may be "ratcheted" back to a period of greater use in order to provide the utility with revenues to maintain the production capabilities to fulfill the greater-use requirement. This is sometimes called a ratchet charge. In other words, if a customer uses a maximum demand of 100 kW or 100 MMBtu one month, then uses 60 kW or 60 MMBtu for the next six months, he/she may have to pay for 100 kW or MMBtu each month until the ratchet period (generally 12 months) is over. This is done to reimburse the utility for the "stranded capacity."

Load Factor

Many individual provisions of the rate structures are addressing load factor, which is defined as the average use / maximum use. For example, if a customer uses power steadily during the day but is closed for 12 hours each day, the load factor would be around 50%. Load factor is usually not billed directly, although it may as well be. Customers with low load factors will pay more than customers with high load factors. For example a customer with a 25% load factor may pay 2-3 times the overall rate ($/kWh) than a customer with a 90% load factor. In the case of residential customers, where load factor is not measured, sample measurements are often taken to establish a good approximation of load factor and the rates set accordingly. This is also done as part of the effort to share the overall utility costs fairly between residential and commercial/ industrial customers.

Load Management Programs

For many utilities there are periods when demand begins to tax supply. Load management (LM) programs are used to alleviate this demand automatically. A typical example uses control of electrical water heaters. If a customer signs up for the LM program, a control will be placed on the water heater power line. This control is an on/off switch that the utility can activate (usually) via carrier wave signal. At any given time, some percentage of water heaters will be in the heating mode and the LM is used to temporarily turn these heaters off. With a large enough number of customers, LM can have an impact during peaking times. Load management is usually "one-way" control, where the utility is not aware of whether the controlled device was actually on at the time of control. In "two-way" load management, the control device reports back to the utility whether the device was on or off at the time of control.

Typical devices that can be controlled by load management programs are HVACs, and some refrigeration and water heating equipment. Load management is also addressed in the 2005 Epact legislation.

18.3.2 Variations

Utilities use a number of methods to tailor their rates to the needs of their customers. Some of the different structures used to accomplish this include seasonal pricing, block pricing, riders, discounts, and innovative rates.

Seasonal Pricing

Costs usually vary by season for most utilities. These variations may be reflected in their rates through different demand and energy charges in the winter and summer. When electric utilities have a seasonal variation in their charges, the summer rates are usually higher than the winter rates, due to high air conditioning use. Gas utilities will generally have winter rates that are higher than summer rates, reflecting increased space-heating use.

Block Pricing

Energy and demand charges may be structured in one of three ways: 1) a declining block structure, 2) an inverted block structure, or 3) a flat rate structure. An inverted block pricing structure increases the rate as the consumption increases. A declining block pricing method decreases the rate as the user's consumption increases. When a rate does not vary with consumption levels it is a "flat" structure. With the declining or inverted block structures, the number of kWh, MMBtu, or therms used is broken into blocks. The unit cost (cents per kWh or cents per MMBtu or therm) is lower or higher for each succeeding block.

A declining block reflects the fact that most utilities can generate additional electricity or provide additional gas for lower and lower costs—up to a point. The capital costs of operation are spread over more usage. The inverted block structure reflects the fact that the incremental cost of production exceeds the average cost of energy. Hence, use of more energy will cause a greater cost to the utility.

Most utilities offer rates with more than one block pricing structure. A utility may offer some combination of inverted, declining and flat block rates, often reflecting seasonal energy cost differentials as well as use differentials. For example, a gas utility may use an inverted block pricing structure in the winter that reflects the higher energy costs in that period but use either a flat or declining block pricing structure in the summer when energy costs are lower.

Riders

A "rider" modifies the structure of a rate based on specific qualifications of the customer. For example, a customer may be on a general service rate and subscribe to a rider that reduces summer energy charges where the utility is granted physical control of the customers air conditioning load.

Discounts

The discount most often available is the voltage discount offered by electric utilities. A voltage discount provides for a reduction in the charge for energy and/or demand if the customer receives service at voltages above the standard voltage. This may require the customer to install, operate and maintain the equipment necessary to reduce the line voltage to the appropriate service voltage. Each customer must evaluate the economics of the discounts against the cost of the required equipment.

Innovative Rates

Increased emphasis on integrated resource planning, demand-side management, and the move to a more competitive energy marketplace has focused utility attention on innovative rates. Those rates designed to change customer load use, help customers maintain or increase market share, or provide the utility with a more efficient operating arena are innovative. Most rates offered today fit into the innovative category.

Net Metering

Customers with alternate electric energy sources, such as photo-voltaics, may be allowed to reverse energy flow into the utility electric grid. The concept is the same for on-site generation but maybe termed "cogeneration" when this is done. Cogeneration has additional restrictions outside the realm of the utility, such as environmental pollution and noise restrictions. Net metering usually allows the meter to turn backwards when the net electric generation exceeds local use; in this case, the excess power flows into the grid for use elsewhere.

For the alternative energy source to see its best economic return, receiving net metered reimbursements from the utility at retail rates is desirable; however, there is disagreement on whether this is truly a fair partnership between customer and utility. On one hand, the net available power is available for sale somewhere else and reduces fuel use needed to generate that unit of energy. On the other hand, the reduction in revenue from the retail point of use leaves the distribution system (wires, transformers) as a stranded investment, underutilized. Typical electric rates are a bundled cost of energy and transmission utility costs (T&D), and if the utility buys back power at a retail rate it may not collect sufficient revenue for the T&D portion. Future growth in renewable electric systems and net metering may ultimately force a revision in rate structures to break out the relatively fixed T&D costs from the more load-dependent generation costs. See also Chapter 16—Renewable Energy.

Wheeling

The key ingredient in any de-regulated energy arrangement is the ability to ship the energy from one location to another by sharing someone else's infrastructure. This method is used for both electric and natural gas purchases. Where electric de-regulation is used, the customer will pay the remote producer for the commodity and pay an additional fee to each link of transmission between supplier and user. Obviously, the farther away and the more territories to traverse, the fees (tolls) will increase. In the electric industry, the term used for passing power through the grid is "wheeling," and each owner of a transmission section used for this purpose can charge for the service. The fees are regulated, but those are generally established at levels appropriate for the maintenance of the infrastructure.

For large-scale wheeling to occur, the cost accounting for infrastructure may ultimately become a separate entity, possibly even resulting in discrete ownership of transmission equipment vs. generation equipment. Until that is done, utilities may hesitate to volunteer the service; the hesitancy may be due to upsetting the monopoly cart, but it may also be due to the unknown of how much the distribution piece is really costing since it has historically been all lumped together. Establishing fair pricing for wheeling through distribution systems will serve to eliminate barriers to more widespread use of the de-regulation concept.

Real-time (Dynamic) Utility Pricing (RTP)

While real-time utility pricing can apply to natural gas and water utilities, this discussion focuses on electrical energy. Real-time electricity pricing (RTP) is a form of utility billing that incorporates a feedback mechanism between the rapidly changing energy costs associated with changing demand (kW) patterns and customer awareness. Utility energy costs can vary hourly, depending on supply and demand characteristics of the system. Effects such as dynamic daily residential and commercial load profiles, changing weather (temperatures), and large local industrial use can influence the demand side of the equation. Supply side effects can include cost and availability of fuels or other market dynamics. Whereas most traditional electrical energy pricing is fairly static except for fuel adjustment charges that can vary per month, RTP pricing and billing can vary in time periods ranging from day-ahead to literally real-time energy market prices.

Traditional electrical energy pricing is fixed for some period (usually months) by an agreement between the utility and a government body such as a state public service commission. On one hand, this makes the cost per unit of energy somewhat predictable; however, from an energy management, demand reduction, or ultimate cost standpoint, there are no effective price signals informing the customer when electricity is more or less expensive.

Real-time electrical pricing is essentially a rapidly changing time-of-use (TOU) rate. As the overall customer demand on the utility electrical system increases during the day, the cost of this energy will also increase. On a hot summer day the utility system might experience a peak electrical demand in the early afternoon as a majority of the customer's air conditioning units simultaneously operate. Peak electrical loads have real additional capacity costs in the form of peaking generators, over-sized generation, transmission and distribution systems, etc.

Demand response (DR) strategies, or demand-side management (DSM) strategies, focus on trying to reduce these variable peak demands. Real-time pricing is the use of economic signals to the customer to encourage usage reduction during peaking periods. Therefore, RTP has two primary goals: reduction of peaking demand loads on the utility system and, an opportunity for the customer to participate and realize economic benefits from these voluntary reductions.

The main challenges for RTP are [11]:

- Will customers shift or reduce demand when they see time-varying price signals?
- How much will the demand costs change (elasticity)?
- Will customers accept dynamic pricing?

• With enabling technology and grid-friendly appliances, to what extent will customers automate their response?

How is RTP Accomplished?

Given that the cost of electricity is changing continuously, how does the consumer become aware of and respond to this? Obviously, a communications interface must exist between the customer and the utility. This is accomplished by using utility meters that can continuously record customer usage over short time intervals. These meters are sometimes called "smart" or "advanced interval" meters. Typically the utility will communicate with the customer via an email notification that the forecast price of energy in the next hour or 24 hours will be a certain amount. The customer then has the option of reducing usage or not. At the end of the billing period, the interval meter will tell the utility billing system what the customer usage was during all periods. On the utility side, new billing systems will be needed. If RTP becomes widespread, this upgrading of the utility infrastructure will be very expensive.

Some forms of RTP use a simple, 24-hour communications rule [11]. Most large-demand price events that are seasonal and weather-related can be predicted with some accuracy within a 24-hour time period. In some cases the sophisticated industrial customer may have automated controls that will reduce usage based on predetermined signals. For example, at a certain price signal the customer may program the HVAC system thermostat to turn up a few degrees in the summer. Industrial and commercial customers could also decide if certain processes could be run at different times. This

can become a complicated economic decision based on the trade-off between production and changing energy costs. For residential RTP, some home appliances, such as electric water heaters, could be programmed to turn off during higher price periods.

In order to smooth out some of the volatility of RTP costs, price hedging can be used. In one study [12], simple RTP hedging by buying blocks of TOU rates eliminated 90% of the price volatility of RTP. Obviously, there is a limit to hedging, as one eventually ends up with a simple TOU rate. Another strategy would be to have a tiered electrical rate, with only the top tier varying with the RTP signal.

Applications

In the United States there have been several RTP pilot programs, and many states offer voluntary RTP [12]. As of 2003, more than half of the states had some form of RTP rates available to customers. Typically, these rates are available to industrial customers above a certain demand load threshold.

The largest applications of RTP to date have been the Georgia Power and Niagara Mohawk programs. The RTP rate schedules for these utilities are shown on their respective websites [13,14].

An example of a voluntary RTP rate schedule is provided by Otter Tail Power Company in Minnesota. Otter Tail Power offers the RTP rate schedule to a limited number of customers in their territory. Shown below are the main points of interest from the RTP rate schedule as of November 30, 2007 (*used by permission and subject to change*) [15].

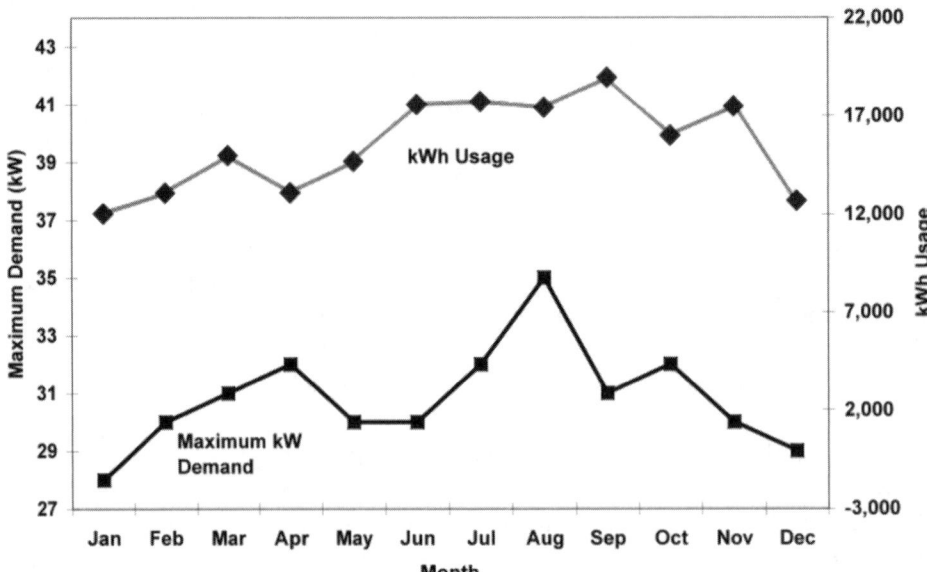

Figure 18-1 Monthly Electric Load Profile of Convenience Store)

Otter Tail Corporation
REAL TIME PRICING RIDER
(Experimental)

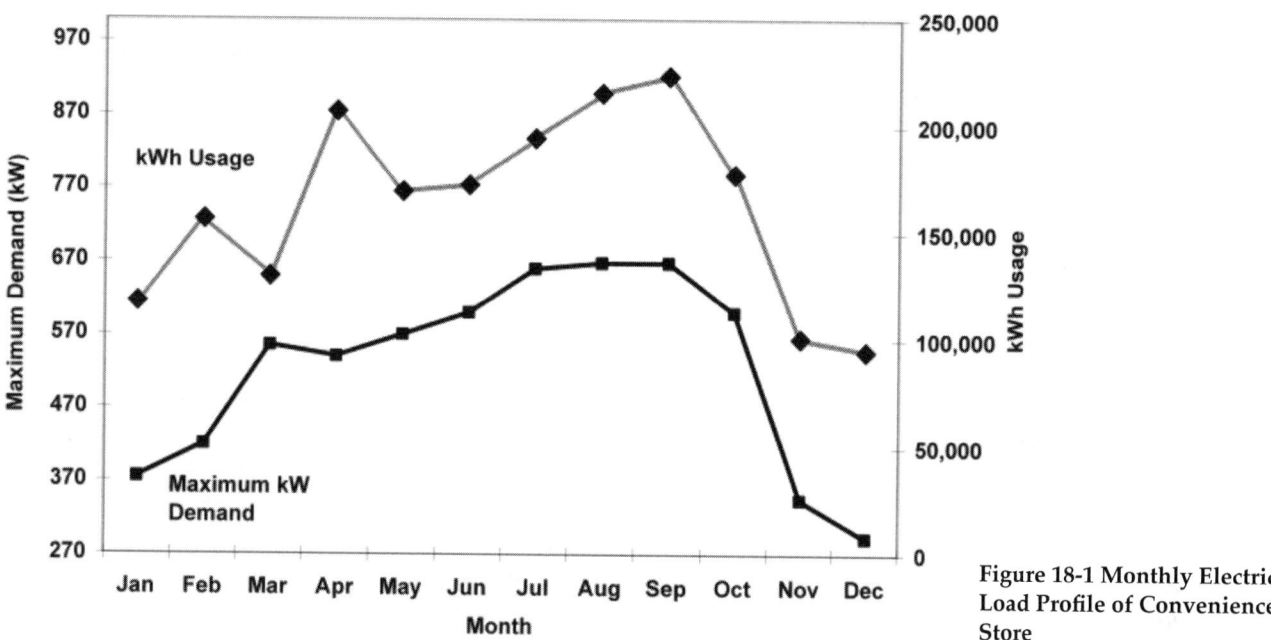

Figure 18-1 Monthly Electric Load Profile of Convenience Store

PRICING METHODOLOGY: Hourly prices are determined for each day, based on projections of the hourly system incremental costs, losses according to voltage level, and hourly outage costs (when applicable).

CUSTOMER ACCESS CHARGE: The customer access charge is specific to each real-time pricing ("RTP") customer and is developed using a 12-month period of hourly (8,760) energy levels (kWh), as well as the corresponding twelve monthly billing demands based on the customer's rate schedule under which it was being billed immediately prior to taking service under the RTP rider. The customer access charge must be agreed to in writing by the customer as a precondition of receiving service under this rider.

The customer access charge will be computed by:

1) Applying the charges in the rate schedule (under which the customer was being billed immediately prior to taking service under the RTP Rider) to the agreed upon 12-month period of hourly energy levels (kWh) and corresponding twelve monthly billing demands;

2) Subtracting from the amount found in step one an amount equal to the sum of each hourly energy level (8760 values) times Otter Tail Power's variable energy costs recorded over the agreed upon 12-month period;

3) Finding the difference between the results of steps one and two, then dividing by 12, will constitute the monthly customer access charge.

BILL DETERMINATION: A real-time pricing bill will be rendered after each monthly billing period. The bill consists of an

administrative charge, a customer access charge, an energy consumption charge, and an excess reactive demand charge/credit. The monthly bill is calculated using the following formula:

RTP Bill Mo = Adm. Charge + Access Chg. + Consumption Chg. + Excess Reactive Demand
Where:
RTP Bill Mo = Customer's monthly bill for service under this Rider
Adm. Chg. = See Administrative Charge section below
Access Charge = See Customer Access Charge section above
Consumption Chg. = 3 {PriceHr x LoadHr }
Excess Reactive Demand = See Excess Reactive Demand section below
3 = Sum over all hours of the monthly billing period
PriceHr = Hourly RTP price as defined under Pricing Methodology
LoadHr = Customer's actual load for each hour of the billing period

ADMINISTRATIVE CHARGE: An administrative charge will be applied to each monthly bill to cover billing, administrative, metering, and communication costs associated with real-time pricing.

EXCESS REACTIVE DEMAND: The reactive demand shall be the maximum kVAR registered over any period of one hour during the month for which the bill is rendered. A separate charge or credit will be made on the bill to reflect incremental changes from the reactive demand used in the standard bill calculation.

PRICE NOTIFICATION: Otter Tail Power shall make available to customers, no later than 4:00 p.m. (Central Time) of the preceding day, hourly RTP prices for the next business day. Except for unusual periods where an outage is at high risk, Otter Tail will make prices for Saturday through Monday available to customers on the previous Friday. More than one-day ahead pricing may also be used for the following holidays: New Year's Day, Memorial Day, Independence Day, Labor Day, Thanksgiving, and Christmas. Because high-outage-risk circumstances prevent Otter Tail from projecting prices more than one day in advance, Otter Tail reserves the right to revise and make available to customers prices for Sunday, Monday, any of the holidays mentioned above, or for the day following a holiday. Any revised prices shall be made available by the usual means no later than 4:00 p.m. of the day prior to the prices taking effect.

Otter Tail is not responsible for a customer's failure to receive or obtain and act upon the hourly RTP prices. If a customer does not receive or obtain the prices made available by Otter Tail, it is the customer's responsibility to notify Otter Tail by 4:30 p.m. (Central Time) of the business day preceding the day that the prices are to take effect. Otter Tail will be responsible for notifying the customer if prices are revised.

18.4 INNOVATIVE RATE TYPES

Utilities have designed a variety of rate types to accomplish different goals. Some influence the customer to use more or less energy or use energy at times that are helpful to the utility. Others are designed to retain or attract customers. Still others are designed to encourage efficient use of energy. The following are some of the innovative rate types that customers should know about.

Time-of-use Rates—For a variety of reasons, it may be more or less expensive to provide the utility service at different times of day. Similar to the discussion on real-time pricing below, utilities can experience different costs throughout the day as supply and demand dynamics occur. A typical example of this is the summer peaking costs. As the aggregate HVAC load in the utility territory increases on a hot day in August, the cost of generating or buying electrical power goes up significantly. Because this effect is somewhat predictable, the utility will try to persuade customers to reduce usage during these times.

When these cost differences are pronounced, the utility may offer time-of-use (TOU) rates. The most common of these are "on-peak" and "off-peak" rates for large commercial users. The same concept can apply to natural gas and water use; one reason this has not been done historically is that it requires special metering which has not been practical to use on other than electric systems. Automated metering advances will make time-of-use rates more prevalent and represents an industry change underway. EPACT-2005 requires time-of-use rates to be made available to all electric customers, including residential customers who request it. Paying attention to "when" electric power is used, as well as other utilities, is a new concept for the residential sector.

To all TOU customers, the differential rates are price signals that encourage customers to defer energy use until costs are lower.

End-use Rates—These rates include air-conditioning, all-electric, compressed natural gas, multi-family, space-heating, thermal energy storage, vehicle fuel and water-heating rates. These rates are all intended to encourage customers to use energy for a specific end-use.

Financial Incentive Rates include rates such as residential assistance, displacement, economic development, and surplus power rates. Assistance rates provide discounts to residential customers who meet specific low-income levels, are senior citizens, or suffer from some physical disability. Displacement rates are offered by electric utilities to customers who are capable of generating their own electricity. The price offered to these customers for utility-provided power is intended to induce the customer to "displace" its own generated electricity with utility-provided electricity. Economic development rates are generally offered by utilities to provide economic incentives for businesses to remain, locate, or expand into areas which are economically distressed. This type of rate is an attempt to attract new customers into the area and to get existing customers to expand until the area is revitalized. Surplus power rates are offered to large commercial and industrial customers. They are offered gas or electric capacity at greatly reduced prices when the utility has an excess available for sale.

Interruptible Rates generally apply to commercial and industrial customers. The utilities often offer several options with respect to the customer's ability to interrupt. Prices vary based on the amount of capacity that is interruptible, the length of the interruption, and the notification time before interruption. Such interruptions are generally, but not always, customer controlled. In addition, the total number of interruptions and the maximum annual hours of interruption may or may not be limited.

Curtailment—Utilities may have economic incentive to reduce peak loads and flatten their load curve. For example, beyond the capacity of base-load equipment, peak load generation may be needed, which, if less efficient or if using higher priced fuel, may raise the incremental cost of produced power substantially. Or, on peak days, some utilities may purchase power off the grid, again at a higher price. In lieu of a pass-through cost adjustment, utilities may offer customers a curtailment or "interruptible" rate. In this agreement, the utility can call an interruption event to the participating customer. The customer then either curtails load as agreed or pays a large surcharge for continued usage; either way the utility has balanced the cost of peak generation with this partnering arrangement. The customer curtailment can be to reduce load (close the facility for the day), scale back production, etc. as they see fit, or to start backup generation.

Curtailable electrical rates are an option for customers with large back-up generators. Typically the utility will send a signal to the customer and the customer must rapidly shed a certain amount of load. In order to minimize this disruption and to satisfy the requirements of the curtailable rate, the customer may switch the load over to the back-up generators.

18.5 CALCULATION OF A MONTHLY BILL

Following is the basic formula used for calculating the monthly bill under a utility rate. The sum of these components will result in the monthly bill.

1) Customer Charge
   • Customer charge = fixed monthly charge
2) Energy Charge
   • Energy charge = dollars × energy use
   • Energy/fuel cost adjustment = dollars × energy use
3) Demand Charge
   • Demand charge = dollars × demand
   • Reactive demand charges (electric only) = dollars × measured kilovolt-ampere reactive demand

4) Tax/Surcharge
   • Tax/surcharge = one or more of items 1-3 above multiplied by tax percentage, dollars × energy use, or dollars × demand

Surcharge for Conservation or DSM—Some utilities, either voluntarily or through regulation, surcharge energy use. This money is collected at a state or regional level and then re-distributed to the utility for disbursement through conservation or demand-side management (DSM) programs. The methods of distribution are stipulated and monitored. The disbursements are made in the form of rebates and other incentives, each aimed at either reducing energy consumption or demand, and all have the common them of public education.

Some examples illustrating the calculation of a monthly bill follow. These examples are actual rate schedules used by the utilities shown. The information regarding the consumption of electricity used in the sample calculations are based on the typical figures shown in Table 18-1.

18.5.1 Commercial General Service with Demand Component

The commercial general service rate often involves the use of demand charges. Table 18-2 provides rate data from Public Service Electric & Gas Company. A sample bill is calculated using the data from Table 18-1 for a convenience store.

Energy Usage – 17,588 kWh; Billing Demand – 30 kW
Season – Summer (June)

Customer Charge:	$300.00
Energy Charge:	$696.48
17,588 kWh x $0.0396/kWh	

Table 18-1. Typical Usage Patterns

Month	Conv. Store kW	Conv. Store kWh	Conv. Store LF	Office Building kW	Office Building kWh	Office Building LF	Shopping Center kW	Shopping Center kWh	Shopping Center LF
Jan	28	12,049	0.60	375	118,000	0.44	1,440	661,000	0.64
Feb	30	13,097	0.61	420	156,500	0.52	1,395	667,000	0.66
Mar	31	15,001	0.67	555	130,000	0.33	1,455	733,000	0.70
Apr	32	13,102	0.57	540	207,000	0.53	1,455	646,000	0.62
May	30	14,698	0.68	570	169,500	0.41	1,545	675,000	0.61
Jun	30	17,588	0.81	600	172,500	0.40	1,740	768,000	0.61
Jul	32	17,739	0.77	660	194,500	0.41	1,935	892,000	0.64
Aug	35	17,437	0.69	668	215,500	0.45	1,905	795,000	0.58
Sep	31	18,963	0.85	668	223,500	0.46	1,740	719,000	0.57
Oct	32	16,003	0.69	600	177,500	0.41	1,515	633,000	0.58
Nov	30	17,490	0.81	345	101,000	0.41	1,425	571,000	0.56
Dec	29	12,684	0.61	293	95,000	0.45	1,455	680,000	0.65

Table 18-2. Sample commercial general service with demand component

Company:	Consolidated Service Electric and Gas Company
Rate Class:	Commercial/Industrial
Rate Type:	Power and Light
Rate Name:	PL-A
Service:	Electric
Effective Date:	12/29/2005
Qualifications:	General purposes where demand is between 10 and 400 kW

	Winter Oct-May	Summer Jun-Sep
Customer Charge:	$300	$300
Energy Cost Adjustment:	-$0.019037	-$0.017192
Tax Rate:	$0.00	$0.00
Surcharge:	$0.00	$0.00
No. of Energy Blocks	2	2
Block 1 Size:	0-1,000,000 kWh	0-1,000,000 kWh
Block 1 Energy Cost:	$0.0396/kWh	$0.0396/kWh
Block 2 Size:	>1,000,000 kWh	>1,000,000 kWh
Block 2 Energy Cost:	$0.0366/kWh	$0.0366/kWh
Demand Charge:	$5.27/kW/Month	$10.97/kW/Month

Energy Cost Adjustment:	-$302.37
17,588 kWh x <$0.017192/kWh>	
Demand Charge:	$329.10
30 kW x $10.97/kW	

Total Monthly Charge:	$1,023.21

Tax Rate:	
25% of first $50 of subtotal	$12.50
12% on remainder of subtotal	$137.66
Total Monthly Charge:	$1,284.79

Energy Usage Measured in kWh per kW Demand—The standard measure of electric energy blocks is in kilowatt hours. An alternative measure used by some electric utilities for commercial and industrial rates is energy per unit of demand (e.g., kWh per kW). This block measurement is illustrated by Costal Electric and Power Company's schedule GS-B rate in Table 18-3. We are calculating the bill for October use in the convenience store of Table 18-1. According to the table, the store used 16,003 kWh in October and had a billing demand of 32 kW. The bill for this use is as follows:

Customer Charge:
$32.00
Energy Charge:

Block 1 – (150 kWh x 32 kW),	
or 4,800 kWh @ $0.06064 =	$291.07
Block 2 – (150 kWh x 32 kW),	
or 4,800 kWh @ $0.03528 =	$169.34
Block 3 – (all remaining kWh),	
or 6,403 kWh @ $0.03167 =	$202.78
Energy Cost Adjustment: (16,003 x $0.01418)	$226.92
Demand Charge: (32 kW x $7.032)	$225.02
Subtotal:	$1,147.13

Time-of-use Rates

Time-of-use rates are calculated very differently from general service rates. The customer's use must be recorded on a time-of-use meter so that billing can be calculated on the use in each time period. In the following sample calculation from Northern Lighting Company (Table 18-4), we assume the customer has responded to the price signal and has relatively little on-peak usage.

Energy Usage - 1,200 kWh; Season - Summer; On-Peak Period - 10:00 a.m.-8:00 p.m., Monday-Friday; On-Peak Usage - 12.5%

Customer Charge:	$40.00
Energy Charge:	$176.53
150 on-peak kWh @ $0.3739 = $56.09	
1050 off-peak kWh @ $0.1147 = $120.44	
Energy Cost Adjustment:	$2.88
Tax: $10.02	
Total Monthly Charge:	**$229.43**

18.5.2 Gas General Service Rates

Gas general service rates are calculated in a similar manner as the electric general service rates. Gas rates are priced in dollars per MMBtu or therm MCF, depending upon the individual utility's unit of measurement. Some gas schedules may include a demand component. The schedule may also have differential seasonal rates.

Table 18-3. Sample energy usage measured in kWh per kW of demand.

Company:	Coastal Electric & Power Company	
Rate Class:	Commercial	
Rate Type:	General Service	
Rate Name:	Schedule GS-B	
Effective Date:	6/01/2006	
Qualifications:	Non-residential use with at least three billing demands => 30 kW in the current and previous 11 billing months, but not more than two billing months of 500 kW or more.	

	OCT-MAY	JUN-SEP
Customer Charge:	$32.00	$32.00
Minimum Charge:	See Note	See Note
Energy Cost Adjustment:	$0.01418	$0.01418
Tax Rate:	See Note	See Note
No. of Energy Blocks:	3	3
Block 1 Size:	150 kWh/kW demand	150 kWh/kW demand
Block 2 Size:	150 kWh/kW demand	150 kWh/kW demand
Block 3 Size:	> 300 kWh/kW demand	> 300 kWh/kW demand
Block 1 Energy Charge:	$0.0640	$0.0640
Block 2 Energy Charge:	$0.03528	$0.03528
Block 3 Energy Charge:	$0.03167	$0.03167
No. of Demand Blocks:	1	1
Block 1 Size:	> 0	> 0
Block 1 Demand Charge:	$7.032	$9.481

MINIMUM CHARGE: Greater of: 1) contract amount or 2) sum of customer charge, energy charge and adjustments, plus $1.604 times the maximum average 30-minute demand measured in the month. TAX: Tax is 25% of the first $50 and 12% of the excess.

BILLING DEMAND: Maximum average 30-minute demand measured in the month, but not less than the maximum demand determined in the current and previous 11 months when measured demand has reached 500 kW or more.

Table 18-4. Electric time-of-use rate.

Company:	Northern Lighting Company	
Rate Class:	Commercial/Industrial	
Rate Type:	Time-of-Use	
Rate Name:	PL-TOU	
Effective Date:	12/29/05	
Qualifications:		

Use for all residential purposes where consumption is 39,000 kWh or less for year ending September 30, or under 12,600 kWh for June through September.

	OCT-MAY	JUN-SEP
Customer Charge:	$40.00	$40.00
Energy Cost Adjustment:	$0.0024	$0.0024
Tax Rate:	5.29389%	5.29389%
No. of Energy Blocks:	1	1
Block 1 Size:	> 0	> 0
Block 1 Energy Charge:		
On-Peak:	$0.1519	$0.3739
Off-Peak:	$0.0978	$0.1147

TAX: applied to total bill
PEAK PERIOD:
On-Peak Hours: 10 a.m.-8 p.m., MON-FRI.
Off-Peak Hours: All remaining hours.

Commercial General Service (Gas) Rate with Delivery (Transportation) and Gas Cost Components

Gas rates for commercial and industrial customers often incorporate both a delivery and actual gas purchased charge. The delivery activity may be contracted to a different business entity than the gas purchase. These two charges may show up on separate bills or may be consolidated into one customer bill. Often some amount of assumed gas line-loss is incorporated into one of these billing components. The example below uses the gas schedule shown in Table 18-5 and is for a small commercial establishment.

Energy Usage:	1,687 Therms
Customer Charge:	$25.73
Purchased Gas Adjustment:	$46.56
Energy Delivery Charge:	$125.38
Environmental Recovery Charge:	$7.59
Natural Gas Cost:	$1,130.29
Taxes:	$84.14
Total Charge:	$1,419.69

Table 18-5.

Company:	Central States Gas Company
Rate Class:	Commercial
Rate Type:	General Service
Effective Date:	12/30/2007
Qualifications:	General commercial use. All charges shown in Therms (100,000 Btu)

Delivery Charges	
Customer Charge:	$25.73
Purchased Gas Adjustment:	$0.0276/Therm
Tax Rate:	6.3%
No. of Energy Blocks:	2
Block 1 Size:	0-150 Therms
Block 1 Energy Charge:	$0.14/Therm
Block 2 Size:	151-1,687 Therms
Block 2 Energy Charge:	$0.068/Therm
Environmental Recovery:	$0.0045/Therm

Purchased Natural Gas	
Natural Gas Cost:	$0.67/Therm

18.6 CONDUCTING A LOAD STUDY

Once a customer understands how utility rates are implemented, he/she can perform a simple load study to make use of this information. A load study will help the energy user to identify load patterns, amount and time of occurrence of maximum load, and the load factor. This information can be used to modify use in ways that can lower electric or gas bills. It can also help the customer determine the most appropriate rate to use.

The basic steps of a utility load study are shown in Table 18-6.

The first step is to collect historical load data. Past bills are one source for this information. One year of data is necessary to identify seasonal patterns; two or more years of data is preferable. Select a study period that is fairly representative of normal consumption conditions.

The next step is to organize the data so that use patterns are evident. One way to analyze the data is to plot the kWh usage, the maximum demand, and the load factor. The load factor is the ratio of the average demand to the maximum demand. The average demand is determined by the usage in kWh divided by the total number of hours (24 × number of days) in the billing period. The number of days in the billing period may vary, depending on how often the meter is read. (See Table 18-7.)

Next, review the data. Seasonal variations will be easily pinpointed. For example, most buildings will show seasonal trends. There may be other peaks due to some aspect of some industrial process, such as a cannery

Table 18-6. Basic steps for conducting a load analysis.

1) Collect historical load data
   - compile data for at least one year
2) Organize data by month for
   - kWh consumption
   - maximum kW demand
   - load factor
3) Review data for
   - seasonal patterns of use
   - peak demands
4) Determine what demand or use can be eliminated or reduced
5) Review load data with utility

Table 18-7. Load factor calculation.

$$\text{Load Factor} = \frac{\text{average demand (kW)}}{\text{maximum demand (kW)}}, \quad \text{where}$$

$$\text{Average Demand} = \frac{\text{kWh usage}}{(24) \times (\text{number of days})}$$

Example: June office building load from Table 18-1

$$\text{Average Demand} = \frac{172,500 \text{ kWh}}{(24) \times (31)} = 231.9 \text{ kW}$$

$$\text{Load Factor} = \frac{231.9 \text{ kW}}{600 \text{ kW}} = 0.39$$

where crops are processed when they are harvested.

In Figures 18-1 and 18-2, kWh and maximum demand (kW) are plotted from the data in table 18-1 for the convenience store and the office building, respectively.

Note that while both buildings seem to use more energy in the late summer, the office building's usage trails off significantly, starting in October. The differences in the electrical energy usage patterns could be due to building construction, orientation, or the timing of activities throughout the year. If the buildings are using gas heating, we would expect to see a corresponding rise in gas usage in colder winter months (not shown).

The load factor calculations for the convenience store vary from 0.57 in April to a high of 0.85 in September. This could be a function of longer operating hours or equipment such as refrigeration or HVAC units running more in the summer.

The office building shows a different load pattern. The load factor varies from 0.41 to a high of 0.52. That is, the load factor is lower and more consistent. This is expected from a building that is probably used only 8-10 hours per day, with little changes in usage patterns depending on the season. The increase in electrical usage in the summer is probably due almost entirely to HVAC units running more in warmer weather.

If the load factor were 1, this would imply uniform levels of use—in effect, a system that was turned on and left running continuously. This may be the case with some manufacturing processes, such as steel mills and refineries.

The fourth step is to determine what demand or use can be eliminated, reduced, or redirected. How can the shopping center reduce its energy costs? By reducing or shifting the peak demand, it can shave demand costs. Although overall consumption is not necessarily reduced, the demand charge is reduced. Where demand ratchets are in place, shaving peak demand may result in savings over a period of several months, not just the month of use. One way to shift peak demand is to install thermal storage units for space cooling purposes; this will shift day time load to night time, giving the cus-

tomer an overall higher load factor. This may qualify the customer for special rates from the utility as well.

Where there may not be much that can be done about the peak demand (in a high load factor situation), more emphasis should be placed on methods to reduce usage. Some examples: turn up the thermostat at night during the summer and down during the winter, install motion detectors to turn off unnecessary lights, and turn off other equipment that is not in use.

Where the customer is charged for electric service on a time-of-use basis, a more sophisticated load study should be performed. The data collected should consist of hourly load data over at least one year. This data can be obtained through the use of recording meters. Once acquired, the data should be organized to show use patterns on a monthly basis, with Monday through Friday (or Saturday, depending on the customer's uses) use plotted separately from weekend use. Review of these data should show where shaving or shifting energy or demand can lower overall electricity bills.

Once the customer has obtained a better understanding of his energy usage patterns, he can discuss with his utility how to best benefit from them. The utility most likely will be interested, because it will also receive some benefits. The customer can consider implementing certain specific measures to better fit in the utility's load pattern and at the same time improve his energy use. The customer's benefit will generally be associated with less energy-related costs. Table 18-8 contains some examples of options that can be taken by commercial and industrial customers and the effect of those options on the utility.

18.7 EFFECTS OF DEREGULATION ON CUSTOMER RATES

18.7.1 Gas and Electric Supply Deregulation

In the period since 1980, many changes have either occurred or begun to occur in the structure of the nation's electric and gas supply industries. These changes have already begun to affect the rate types and structures for U.S. gas and electricity consumers. In the natural gas industry,

Table 18-8. Customer options and their effects on utility.

OPTIONS		
Commercial	*Industrial*	*Utility Effect*
Accept direct control of water heaters	Subscribe to interruptible rates	Reduction of load during peak periods
Store hot water to increase space heating	Add nighttime operations	Builds load during off-peak periods

well-head prices were deregulated as a result of the Natural Gas Policy Act of 1978 and the subsequent Natural Gas Well-Head Decontrol Act of 1989. Subsequently, FERC introduced a number of restructuring rules (Order Nos. 436, 500, and 636) that dramatically changed the regulation of the nation's pipelines and provided access for end-users to transport gas purchased at the well-head. In the electric industry, supply deregulation commenced with passage of the Public Utility Regulatory Policies Act of 1978, which encouraged electric power generation by certain non-utility producers. The Energy Policy Act of 1992 further deregulated production and mandated open transmission access for wholesale transfers of electricity between qualified suppliers and wholesale customers. EPACT 2005 authorized the Federal Regulatory Commission (FERC) to require non-public utilities to provide open-access transmission service, although it did not regulate transmission rates.

These legal and regulatory changes will have a significant and lasting effect on the rate types and rate structures experienced by end-users. In the past, most gas and electric customers paid a single, bundled rate that reflected all costs for capacity and energy, storage, delivery, and administration. Once customers are given the opportunity to purchase their gas and electric resources directly from producers, it then becomes necessary to unbundle the costs associated with production from the costs associated with transportation and delivery to end users. This unbundling process has already resulted in separate rates for many services with costs previously combined in the single unit price for either gas or electricity.

18.7.2 Effect on Gas Rates

Much of the discussion in Section 18.3 of this chapter pertains to bundled rates for gas. However, as a result of unbundling, many utilities are now offering customers four separate services, including balancing, procurement, storage, and transportation of gas. Gas balancing rates provide charges for over- or under-use of customer-owned gas over a specified period of time. When the customer has the utility procure gas for transportation to the customer, gas procurement rates are charged. Gas storage rates are offered to customers for the storage of customer-owned gas. Gas transportation rates are offered to commercial, industrial, and non-utility generator customers for the transportation and delivery of customer-owned gas. In addition to these rates, there is the actual cost of purchasing the gas to be transported. Gas procurement, balancing, storage and transportation rates have increased in usage as the structure of the gas industry has evolved.

Two other types of gas rates are also evolving as a result of industry deregulation. These include negotiated gas rates and variable gas rates. The former refers to rates that are negotiated between individual customers and the utility. Such rates are often subject to market conditions. The latter, variable gas service rates, refer to rates that vary from month to month. A review of all of the gas service rates collected by the Gas Research Institute (GRI) in 1994 indicated that 52% of the gas utilities surveyed offered at least one type of variable pricing. Such rates are often indexed to an outside factor, such as the price of gasoline or the price of an alternative fuel, and they usually vary between established floor and ceiling prices. The most common types of variable rates are those offered for transportation services.

18.7.3 Effect on Electric Rates

In the past, most U.S. electric customers have paid a single bundled rate for electricity. Many of these customers purchased from a utility that produced, transmitted, and delivered the electricity to their premises. In other cases, customers purchased from a distribution utility that had itself purchased the electricity at wholesale from a generating and transmitting utility. In both of these cases, the customer paid for electricity at a single rate that did not distinguish between the various services required to produce and deliver the power. In the future, as a result of the deregulation process already underway, there is a far greater likelihood that initially large customers, and later many smaller customers, will have the ability to select among a number of different suppliers. In most of these cases, however, the transmission and delivery of the purchased electricity will continue to be a regulated monopoly service. Consequently, future electricity consumers are likely to receive separate bills for:

* electric capacity and energy,;
* transmission, and
* distribution.

In some cases, a separate charge may also be made for system control and administrative services, depending on exact industry structure in the given locality. For each such charge, a separate rate structure will apply. At present, it appears likely that there will be significant regional and local differences in the way these rates evolve and are implemented.

GLOSSARY

There are a few terms that the user of this document needs to be familiar with. Below is a listing of common terms and their definitions.

Billing Demand: The billing demand is the demand that is billed to the customer. The electric billing demand is generally the maximum demand or maximum average measured demand in any 15-, 30-, or 60-minute period in the billing month. The gas billing demand is determined over an hour or a day and is usually the greatest total use in the stated time period.

British Thermal Unit (Btu): Quantity of heat needed to raise one pound of water one degree Fahrenheit.

Btu Value: The heat content of natural gas is in Btu per cubic foot. Conversion factors for natural gas are:
- Therm = 100,000 Btu;
- 1 MMBtu = 1,000,000 Btu = 1 Decatherm = 10 therms.

Contract Demand: The demand level specified in a contractual agreement between the customer and the utility. This level of demand is often the minimum demand on which bills will be determined.

Controllable Demand: A portion or all of the customer's demand that is subject to curtailment or interruption directly by the utility.

Cubic Foot: Common unit of measurement of gas volume; the amount of gas required to fill one cubic foot.
- CCF = 100 cubic feet.
- MCF = 1,000 cubic feet

Curtailable Demand: A portion of the customer's demand that may be reduced at the utility's direction. The customer, not the utility, normally implements the reduction.

Customer Charge: The monthly charge to a customer for the provision of the connection to the utility and the metering of energy and/or demand usage.

Demand Charge: The charge levied by a utility for metered demand of the customer. The measurement of demand may be either in kW or kVA.

Demand Response or Demand-side Management: This describes programs that attempt to limit or reduce periods of high customer demand for electricity, gas, or water. These programs can include utility rebates for energy efficient equipment to innovative billing methods such as real time pricing.

Dual-fuel Capability: Some interruptible gas rates require the customer to have the ability to use a fuel other than gas to operate their equipment.

Energy Blocks: Energy block sizes for gas utilities are either in MCFs or in MMBtus. The standard measures of energy block sizes for electric utilities are kWhs. However, several electric utilities also use an energy block size based on the customers' demand level (i.e. kWh per kW). Additionally, some electric utilities combine the standard kWh value with the kWh per kW value.

Energy Cost Adjustment (ECA): A fuel cost factor charged for energy usage. This charge usually varies on a periodic basis, such as monthly or quarterly. It reflects the utilities' need to recover energy related costs in a volatile market. It is often referred to as the fuel cost adjustment, purchased power adjustment, or purchased gas adjustment.

Excess or Non-Coincidental Demand: Some utilities charge for demands in addition to the on- or off-peak demands in time-of-use rates. An excess demand is demand used in off-peak time periods that exceeds usage during on-peak hours. Non-coincidental demand is the maximum demand measured any time in a billing period. This charge is usually in addition to the on- or off-peak demand charges.

Firm Demand: The demand level that the customer can rely on for uninterrupted use.

Interruptible Demand: All of the customer's demand may be completely interrupted at the utility's direction. Either the customer or the utility may implement the interruption.

Load Factor: Average Demand / Maximum Demand. A measure of how level or consistent the energy usage is over time.

Load Profile/Study: The examination of energy usage of a facility over time. Typically this involves examining energy bills for a building over a time period of at least one year. Load profiles highlight seasonal energy usage trends as well as abnormalities such as malfunctioning equipment or billing errors.

MCF: Thousand (1000) cubic feet.

MMCF: Million (1,000,000) cubic feet.

Minimum Charge: The minimum monthly bill that will be charged to a customer. This generally is equal to the customer charge but may include a minimum demand charge as well.

Off-Peak Demand: Greatest demand measured in the off-peak time period.

On-Peak Demand: Greatest demand measured in the on-peak time period.

Power Factor: A term that describes the reactive power component of electrical power usage. Power factor is often a billing component of electrical demand on commercial/industrial energy bills.

Ratchet: A ratchet clause sets a minimum billing demand that applies during peak and/or non-peak months. It is usually applied as a percentage of the peak demand for the preceding season or year.

Rates: This term is synonymous with "schedule" and "tariff" for this chapter. Essentially the term describes the contract between the utility provider and the customer. Rate design is usually controlled by state government agencies.

Reactive Demand: In electric service, some utilities have a special charge for the demand level in kilovolt-amperes reactive (kVAR) that is added to the standard demand charge. This value is a measure of the customer's power factor.

Real-time Pricing: An energy rate schedule that incorporates rapidly changing energy prices and a feedback mechanism to alert customers to high price time periods.

Surcharge: A charge levied by utilities to recover fees or imposts other than taxes.

Therm: A unit of heating value equal to 100,000 Btu.

Transportation Rates: Rates for the transportation of customer-owned gas. These rates do not include purchase or procurement of gas.

Voltage Discounts: Most electric utilities offer discounted rates to customers who will take service at voltages other than the general distribution voltages. The voltages for which discounts are generally offered are secondary, primary, sub-transmission and transmission. The actual voltage of each of these levels vary from utility to utility.

References

1. Acton, J.P., Gelbard, E.H., Hosek, J.R., & Mckay, D.J. (1980, February). *British Industrial Response to the Peak-Load Pricing of Electricity*. The Rand Corporation, R-2508-DOE/DWP.
2. David, A.K., & Li, Y.Z. (1991, November). A Comparison of System Response For Different Types of Real-Time Pricing. IEEE International Conference on Advances In Power System Control, *Operation and Management*. Hong Kong, p. 385-390.
3. Anonymous (1997, August). *Energy User News*. Chilton Co., p. 32.
4. EPRI (1980, October). Industrial Response To Time Of Day Pricing—A Technical and Economic Assessment Of Specific Load Management Strategies. Gordian Associates, EA-1573, Research Project 1212-2.
5. Hanser, P., Wharton, J., & Fox-Penner, P. (March 1, 1997). Real-time Pricing—Restructuring's Big Bang? *Public Utilities Fortnightly*, 135 (5), p. 22-30.
6. Kirsch, L. D., Sullivan, R.L., & Flaim, T.A. (1988, August). Developing Marginal Costs For Real-Time Pricing. IEEE *Transactions on Power Systems*, 3 (3), p. 1133-1138.
7. Mykytyn Consulting group, Inc. (1997). Electric Utilities and Tariffs. PowerRates [Online]. Available: http://www.mcgi.com/pr/samples/utility_list.html [November 4, 1997].
8. O'Sheasy, M. (MIKE.T.OSHEASY@gpc.com). (1998, March 12). RTP. E-mail to Mont, J. (mjavier@okstate.edu).
9. Tabors, R.D., Schweppe, F.C., & Caraminis, M.C. (1989, May). Utility Experience with Real-Time Rates. *Transactions on Power Systems*, 4(2), p. 463-471.
10. Tolley, D.L. (1988, January). Industrial Electricity Tariffs. *Power Engineering Journal*. p. 27-34.
11. Kiesling, Lynne. (2006, April), Differential Retail Pricing of Electricity: The Evolution of Policy, Technology and Markets, Presentation to University of Illinois' Institute of Government and Public Affairs.
12. Goldman, Charles. (2006, April), Customer Experience with Real Time Pricing as the Default Service, Presentation at ISO-NE DR Summit, Sturbridge, Connecticut.
13. Georgia Power (2008), Rate Schedules [Online]. Available: http://www.georgiapower.com/pricing/gpc_rates.asp
14. Niagra Mohawk Power (2008), Rate Schedules [Online]. Available: http://www.nationalgridus.com/niagaramohawk/non_html/rates_psc207.pdf
15. Otter Tail Power Company (2008), RTP Rate Schedule [Online]. Available: http://www.otpco.com/ElectricRates/PDF/MN/c-03m.option2.pdf

CHAPTER 19

THERMAL ENERGY STORAGE

CLINT CHRISTENSON
Noreso, Inc.

19.1 INTRODUCTION

A majority of the technology developed for energy management has dealt with the more efficient *consumption* of electricity, rather than timing the demand for it. Variable frequency drives, energy efficient lights, electronic ballasts and energy efficient motors are a few of these consumption management devices. These techniques often only impact a small portion of the facilities demand (when compared to say the mechanical cooling equipment), which is normally a major portion of the facilities overall annual electric bill. The management of demand charges deals very little with conservation of energy but mainly with the ability of a generator to supply power *when* needed. It is this timing of consumption that is the basis of demand management and the focus of thermal energy storage (TES).

Experts agree that demand management is actually not a form of energy *conservation* but a form of cost *management*. The utility generation and power distribution systems approach capacity planning through many avenues, including construction, acquisition, wholesale wheeling, incentives, and rate structure management. Consumers have the ability to capitalize on these penalties and rewards to reduce their costs and potentially enhance the flexibility of their mechanical systems.

Utilities often charge more for energy and demand during certain periods in the form of on-peak rates and ratchet clauses. The process of managing the generation capacity that a particular utility has "on-line" involves the utilization of those generating units that produce power most efficiently first since these units would have the lowest avoided costs (ultimately the actual cost of energy). When the loads are approaching the connected generation capacity of the utility, additional generating units must be brought on line. Each additional unit has an incrementally higher avoided cost since these "peaking units" units are less efficient and used less often. This has prompted many organizations to implement some form of demand management.

Thermal energy storage (TES) is the concept of gen-

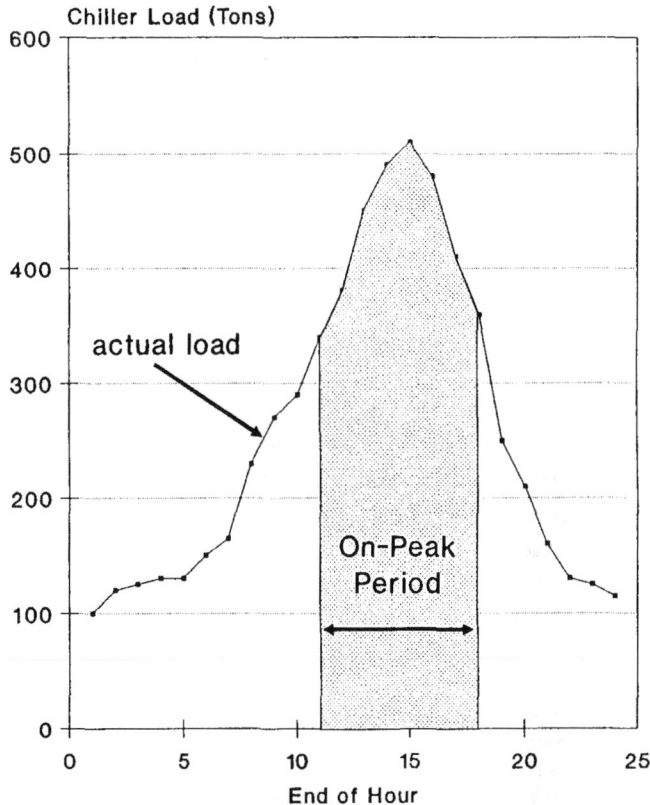

Figure 19-1. Typical office building chiller consumption profile.

erating and storing energy in the form of heat or cold for use during peak periods. Examples in this chapter will focus on cooling, but the principles are similar for heating thermal storage. For the profile in Figure 19-1, a cooling storage system could be implemented to reduce or eliminate the need to run the chillers during the on-peak rate period. *By running the chillers during off-peak hours and storing this capacity for use during the on-peak hours, a reduction in energy costs can be realized.* If this type of system is implemented during new construction or when equipment is being replaced, smaller capacity chillers can be installed since the chiller can spread the production of the total load over the entire day, rather than being sized for peak loads.

Thermal energy storage has been used for centuries, but only recently have large electrical users taken advantage of the technique for cost management. The process involves storing Btus (or lack of Btus) for use

483

when either a heat source or a heat sink is required. The use of eaves, root cellars, ground coupled heat pump systems, and adobe type thermal mass could all be considered forms of thermal storage. Today, the ability to take advantage of a source of inexpensive energy (whether waste heat source or time based rate structure) for use during a later time of more expensive energy has extended the applications of TES. For this particular chapter, the focus of discussion will concentrate on the storage of cooling capacity; the storage of heat will not be considered. The two main driving forces behind the storage of cooling capacity, rate structure and cooling system management, will be discussed in the following paragraphs.

Often the chiller load and efficiency follow the chiller consumption profile, in that the chiller is running at high load, i.e. high efficiency, only a small portion of the day. This is due to the HVAC system having to produce cooling when it is needed as well as to be able to handle instantaneous peak loads. With smaller chiller systems designed to handle the base and peak loads during off-peak hours, the chillers can run at higher average loads and thus higher efficiencies.

Thermal energy storage also has the ability to balance the daily loads on a cooling system. Conventional air conditioning systems must employ a chiller large enough to handle the peak cooling demand as it occurs. This mandates that the cooling system be required to operate in a load following mode, varying the output of the system in response to changes in the cooling requirements. Systems that operate within a one or two shift operation or those that are much more climatically based, can benefit from the smoothing characteristics of TES. A school, for example, that adds a new wing could utilize the existing refrigeration system during the evening to generate cooling capacity to be stored for use during the day. Although additional piping and pumping capacity would need to be added to the addition, new chiller capacity may not have to be added. A new construction project that would have similar single shift cooling demand profile could utilize a smaller chiller in combination with storage to better balance the chiller operation. This could significantly reduce the capital cost of the renovation in addition to any rate based savings as discussed above.

Companies often control the demand of electricity by utilizing some of the techniques listed above and other consumption management actions that also reduce demand. More recently, the ability to shift the *time* when electricity is needed has provided a means of balancing or shifting the demand for electricity to "off-peak" hours. This technique is often called demand

balancing or demand shifting. This demand balancing may best be seen with the use of an example 24-hour chiller consumption plot during the peak day, Figure 19-1 and Table 19-1. This facility exhibits a typical single shift building load profile. Note that the load listed in this table for the end of hour 1 identifies the average load between midnight and 1:00 a.m. and for end of hour 2 the average load between 1 and 2 a.m., etc. This example will employ a utility rate schedule with a summer on-peak demand period from 10 am to 5:59 p.m., an 8-hour period. Moving load from the on-peak rate period to the off-peak period can both balance the demand and reduce residual ratcheted peak charges. Thermal energy storage is one method available to accomplish just that.

Table 19-1. Example chiller consumption profile

Chiller Consumption Profile

End of Hour	Chiller Load (Tons)	Rate
1	100	Reg
2	120	Reg
3	125	Reg
4	130	Reg
5	130	Reg
6	153	Reg
7	165	Reg
8	230	Reg
9	270	Reg
10	290	Reg
11	340	On-Peak
12	380	On-Peak
13	450	On-Peak
14	490	On-Peak
15	510	On-Peak
16	480	On-Peak
17	410	On-Peak
18	360	On-Peak
19	250	Reg
20	210	Reg
21	160	Reg
22	130	Reg
23	125	Reg
24	115	Reg

Daily Total	6123	Ton-Hrs
Daily Avg.	255	Tons
Peak Total	3420	Ton-Hrs
Peak Demand	510	Tons

19.2 STORAGE SYSTEMS

There are two general types of storage systems. Ones that shut the chiller down during on-peak times and run completely off the storage system during that time are known as "full storage systems." Those designed to have the chiller run during the on-peak period supplementing the storage system are known as "partial storage systems." The full storage systems have a higher first cost since the chiller is off during peaking times and the cooling load must be satisfied by a larger chiller running fewer hours, with a larger storage system storing the excess. The full storage systems do realize greater savings than the partial system since the chillers are completely turned off during on-peak periods. Full storage systems are often implemented in retrofit projects since a large chiller system may already be in place.

A partial storage system provides attractive savings with less initial cost and size requirements. New construction projects will often implement a partial storage system so the size of both the chiller and the storage system can be reduced. Figures 19-2 and 19-3 and Tables 19-2 and 19-3 demonstrate the chiller load required to satisfy the cooling needs of the office building presented in Figure 19-1 for the full and partial systems, respectively. Column 2 in these tables represents the building cooling load each hour, and column 3 represents the chiller output for each hour. Discussion of the actual calculations required for sizing these different systems is included in a subsequent section. For simplicity's sake, these numbers do not provide for any system losses, which will also be discussed in a later section.

The full storage system has been designed so that the total daily chiller load is produced during the off-peak hours. This eliminates the need to run the chillers during the on-peak hours, saving the increased rates for demand charges during this period, as well as any future penalties due to ratchet clauses. The partial storage system produces 255 tons per hour during the entire day, storing excess capacity for use when the building demand exceeds the chiller production. This provides the ability to control the chiller load, limit the peak chiller demand to 255 kW,[*] and still take advantage of the off-peak rates for a portion of the on-peak chiller load.

An advantage of partial load systems is that they can provide a means of improving the performance of a system that can handle the cumulative cooling load but not the instantaneous peak demands of the building. In

Table 19-2. Full storage chiller consumption profile.

Chiller Consumption Profile—Full Storage System

1	2	3	4
End of Hour (Tons)	Cooling Load (Tons)	Chiller Load[2]	Rate
1	100	383	Reg
2	120	383	Reg
3	125	383	Reg
4	130	383	Reg
5	130	383	Reg
6	153	383	Reg
7	165	383	Reg
8	230	383	Reg
9	270	383	Reg
10	290	383	Reg
11	340	0	On-Peak
12	380	0	On-Peak
13	450	0	On-Peak
14	490	0	On-Peak
15	510	0	On-Peak
16	480	0	On-Peak
17	410	0	On-Peak
18	360	0	On-Peak
19	250	383	Reg
20	210	383	Reg
21	160	383	Reg
22	130	383	Reg
23	125	383	Reg
24	115	383	Reg

	Without storage	With storage
Daily Total (Ton-Hrs)	6123	6123
Daily Avg (Tons):	255[1]	255
Peak Total (Ton-Hrs)	3420[3]	0[4]
Peak Demand (Tons)	510[3]	0[4]

[1]
$$\frac{6123 \text{ Ton-Hr}}{24 \text{ Hours}} = 255.13 \text{ Avg Tons}$$

[2]
$$\frac{6123 \text{ Ton-Hr}}{16 \text{ Hours}} = 382.69 \text{ Avg Tons}$$

[3]This peak load is supplied by the TES, not the chiller.
[4]This is the chiller load and peak during on-peak periods.

such a system, the chiller could be run nearer optimal load continuously throughout the day, with the excess cooling tonnage being stored for use during the peak periods. An optional method for utilizing partial storage is a system that already utilizes two chillers. The daily cooling load could be satisfied by running both chillers during the off-peak hours, storing any excess cooling

---

[*]assuming COP = 3.5, then kW/ton = $\dfrac{3.517}{COP}$ = 1.0 kW/ton

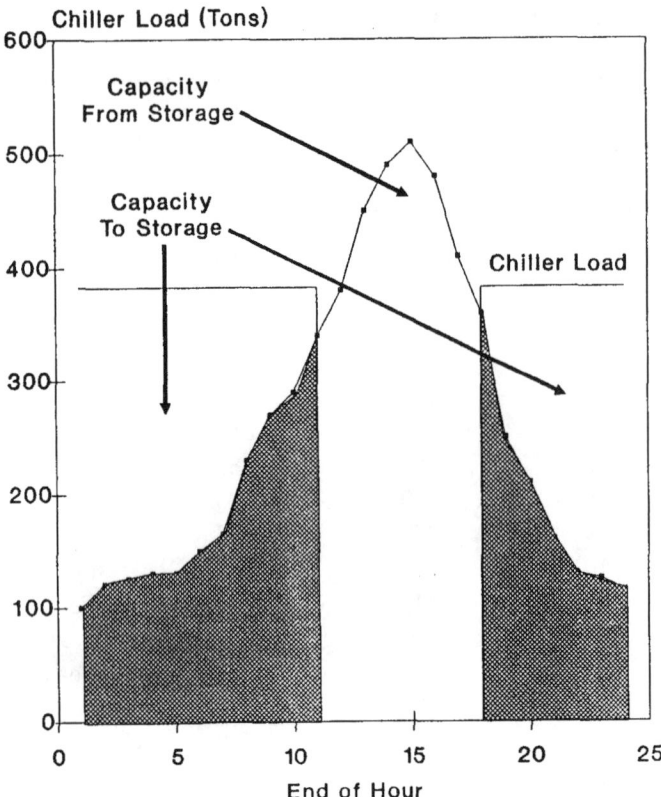

Figure 19-2. Full storage chiller consumption profile.

Table 19-3. Partial storage chiller consumption profile.

Chiller Consumption Profile
Partial Storage System

1	2	3	4
Hour of Day	Cooling Load (Tons)	Chiller Load (Tons)[1]	Rate
1	100	255	Reg
2	120	255	Reg
3	125	255	Reg
4	130	255	Reg
5	130	255	Reg
6	153	255	Reg
7	165	255	Reg
8	230	255	Reg
8	270	255	Reg
10	290	255	Reg
11	340	255	On-Peak
12	380	255	On-Peak
13	450	255	On-Peak
14	490	255	On-Peak
15	510	255	On-Peak
16	480	255	On-Peak
17	410	255	On-Peak
18	360	255	On-Peak
19	250	255	Reg
20	210	255	Reg
21	160	255	Reg
22	130	255	Reg
23	125	255	Reg
24	115	255	Reg

	Without storage	With storage
Daily Total (Ton-Hrs)	6123	6123
Daily Avg (Tons):	255	255
Peak Total (Ton-Hrs):	3420[2]	2041[3]
Peak Demand (Tons):	510[2]	255[3]

[1]
$$\frac{6123 \text{ Ton-Hr}}{24 \text{ Hours}} = 255.13 \text{ Avg Tons}$$

[2]This peak load is supplied by the TES, supplemented by the chiller.
[3]This is the chiller load and peak during on-peak period.

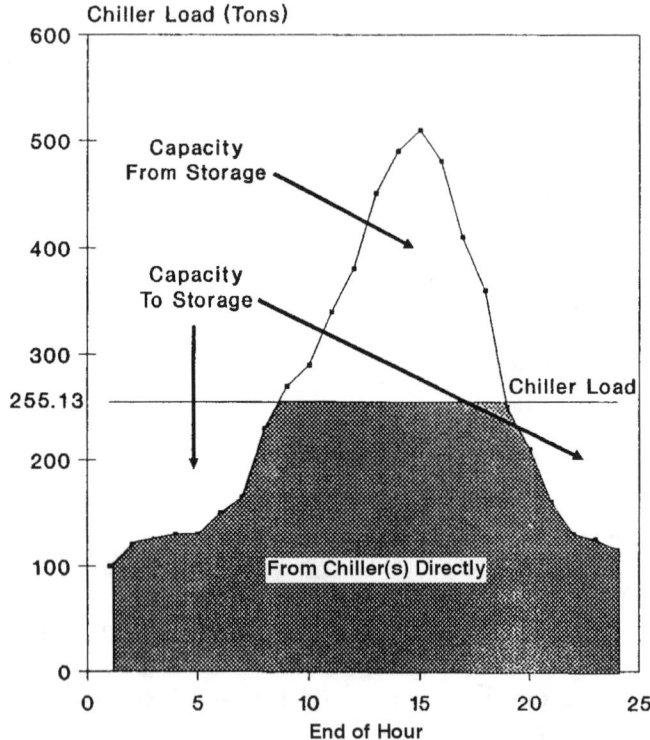

Figure 19-3. Partial storage chiller consumption profile.

capacity, and running only one chiller during the on-peak period to supplement the discharge of the storage system. This also has the important advantage of offering a reserve chiller during peak load times. Figure 19-4 shows the chiller consumption profile for this optional partial storage arrangement, and Table 19-4 lists the consumption values. Early and late in the cooling season, the partial load system could approach the full load system characteristics. As the cooling loads and peaks begin to decline, the storage system will be able to handle more of the on-peak requirement, and eventually the on-peak chiller could also be turned off. A system such as this can be designed to run the chillers at optimum load, increasing efficiency of the system.

Storage systems also provide various operational advantages to mechanical systems. Even a partially charged storage system could provide a certain level of capacity if the primary system failed or if the utility became unavailable during an outage or a curtailment procedure. This "redundancy" could be provided with minimal power, say an emergency generator of small capacity since we could simply start the chilled water pumps rather than a back-up chiller to circulate water through a critical hospital system or data center. With the correct relationships in place, the utility could utilize a large storage system as a virtual generator during high power periods.

19.3 STORAGE MEDIUMS

There are several methods currently in use to store cold in thermal energy storage systems. These are water, ice, and phase change materials. The water systems simply store chilled water for use during on-peak periods. Ice systems produce ice that can be used to cool the actual chilling water, utilizing the high latent heat of fusion. Phase change materials are those materials that exhibit properties (melting points for example) that lend themselves to thermal energy storage. Figure 19-5a represents the configuration of the cooling system with either a water or phase change material thermal storage system, and Figure 19-5b represents a general configuration of a TES utilizing ice as the storage medium The next few sections will discuss these different mediums.

19.3.1 Chilled Water Storage

Chilled water storage is simply a method of storing chilled water generated during off-peak periods in a large tank or series of tanks. These tanks are the most commonly used method of thermal storage. One factor to this popularity is the ease to which these water tanks can be interfaced with the existing HVAC system. The chillers are not required to produce chilled water any colder than presently used in the system, so the system efficiency is not sacrificed. The chiller system draws warmer water from one end of the system, and this is replaced with chilled water in the other. During the off-peak charge cycle, the temperature of the water in the storage will decline until the output temperature of the chiller system is approached or reached. This chilled water is then withdrawn during the on-peak discharge cycle, supplementing or replacing the chiller(s) output.

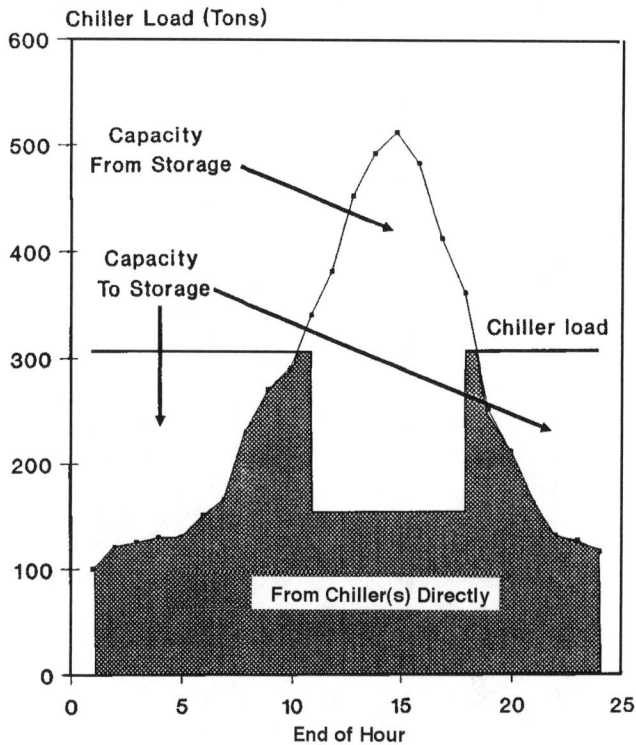

Figure 19-4. Optional partial storage chiller profile.

Facilities that have a system size constraint such as lack of space often install a series of small insulated tanks that are plumbed in series. Other facilities have installed a single, large volume tank either above or below ground. The material and shape of these tanks vary greatly from installation to installation. These large tanks are often designed very similar to municipal water storage tanks. The main performance factors in the design of these tank systems, either large or multiple, is location and insulation. An Electric Power Research Institute's (EPRI) Commercial Cool Storage Field Performance Monitoring Project (RP-2732-05) report states that the storage efficiencies of tanks significantly decreased if tank walls were exposed to sunlight and outdoor ambient conditions and/or had long hold times prior to discharging[7]. To minimize heat gain, tanks should be out of the direct sun whenever possible. The storage efficiency of these tanks is also decreased significantly if the water is stored for extended periods.

One advantage to using a single large tank rather than a series of smaller ones is that the temperature differential between the warm water intake and the chilled water outlet can be maintained. This is achieved utilizing the property of thermal stratification, where the warmer water will migrate to the top of the tank and the colder to the bottom. Proper thermal stratification can only be maintained if the intake and outlet diffusers are located at the top and bottom of the tank, and the flow rates of the water during charge and discharge cycles is kept

Table 19-4. Partial storage chiller consumption profile.

1 Hour of Day	2 Cooling Load (Tons)	3 Chiller Load[1,2] (Tons)	4 Rate
1	100	306	Reg
2	120	306	Reg
3	125	306	Reg
4	130	306	Reg
5	130	306	Reg
6	153	306	Reg
7	165	306	Reg
8	230	306	Reg
9	270	306	Reg
10	290	306	Reg
11	340	153	On-Peak
12	380	153	On-Peak
13	450	153	On-Peak
14	490	153	On-Peak
15	510	153	On-Peak
16	480	153	On-Peak
17	410	153	On-Peak
18	360	153	On-Peak
19	250	306	Reg
20	210	306	Reg
21	160	306	Reg
22	130	306	Reg
23	125	306	Reg
24	115	306	Reg

	Without storage	With storage
Daily Total (Ton-Hrs):	6123	6123
Daily Avg (Tons):	255.13	255.13
On-Peak (Ton-Hrs):	3420[3]	1225[4]
Peak Demand (Tons):	510[3]	153[4]

[1]
$$\frac{(6123 \text{ Ton-Hr}) (2 \text{ Chillers Operating})}{(16 \text{ Hours})(2 \text{ Chillers}) + (8 \text{ Hours})(1 \text{ Chiller})} = 306 \text{ Tons}$$

[2]
$$\frac{(6123 \text{ Ton-Hr}) (1 \text{ Chiller Operating})}{(16 \text{ Hours})(2 \text{ Chillers}) + (8 \text{ Hours})(1 \text{ Chiller})} = 153 \text{ Tons}$$

[3]This peak load is supplied by the TES, supplemented by the chiller.
[4]This is the chiller load and peak during the on-peak period.

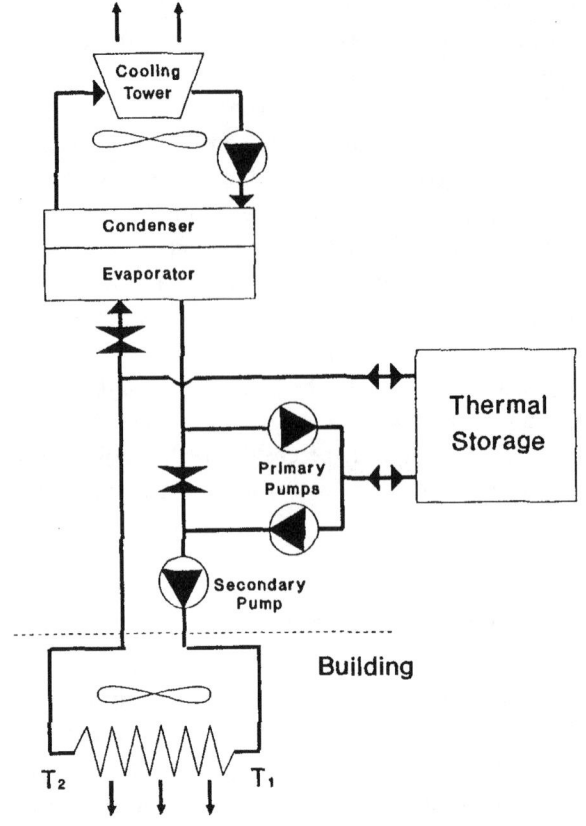

Figure 19-5a. Water & eutectic storage system configuration.

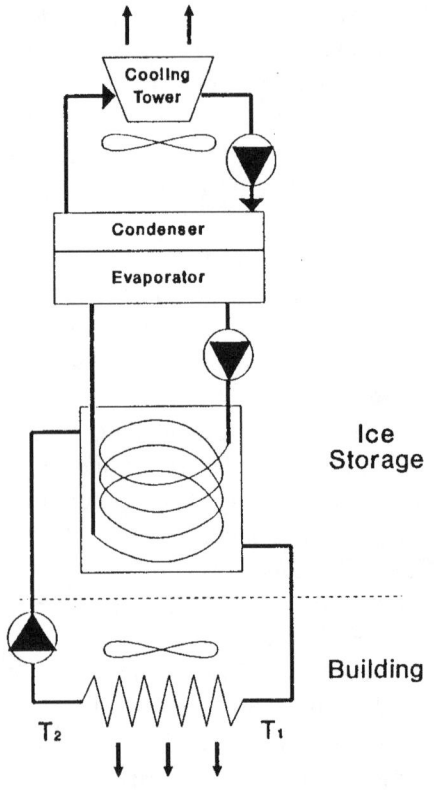

Figure 19-5b. Ice storage system configuration.

low. This will reduce a majority of the mixing of the two temperature waters. Another method used to assure that the two temperature flows remain separated is the use of a movable bladder, creating a physical partition. One top/bottom diffuser tank studied in the EPRI study used a thermocouple array, installed to measure the chilled water temperature at one foot intervals from top to bottom of the tank. This tank had a capacity of 550,000 gallons and was 20 feet deep but had only a 2.5 foot blend zone over which the temperature differential was almost 20 degrees[7].

The advantages of using water as the thermal storage medium are:

1. Retrofitting the storage system with the existing HVAC system is very easy.

2. Water systems utilize normal evaporator temperatures.

3. With proper design, the water tanks have good thermal storage efficiencies.

4. Full thermal stratification maintains chilled water temperature differential, maintaining chiller loading and efficiencies.

5. Water systems have lower auxiliary energy consumption than both ice and phase change materials since the water has unrestricted flow through the storage system.

19.3.2 Ice Storage

Ice storage utilizes water's high latent heat of fusion to store cooling energy. One pound of ice stores 144 Btu's of cooling energy, while chilled water only contains 1 Btu per pound –°F[7,8]. This reduces the required storage volume approximately 75%[7] if ice systems are used rather than water. Ice storage systems form ice with the chiller system during off-peak periods, and this ice is used to generate chilled water during on-peak periods.

There are two main methods in use to utilize ice for on-peak cooling. The first is considered a static system in which serpentine expansion coils are fitted within an insulated tank of cooling water. During the charging cycle, the cooling water forms ice around the direct expansion coil as the cold gases or brine pass through it. (See Figure 19-5b.) The thickness of the ice varies with the ice building time (charge time) and heat transfer area. During the discharge cycle, the cooling water contained in the tank is used to cool the building, and the warmer water returned from the building is circulated through the tank, melting the ice and using its latent heat of fusion for cooling.

The second major category of thermal energy storage systems utilizing ice can be considered a dynamic system. This system has also been labeled a plate ice maker or ice harvester. During the charging cycle the cooling water is pumped over evaporator "plates," where ice is actually produced. These thin sheets of ice are fed into the cooling water tank, dropping the temperature. During on-peak periods, this chilled water is circulated through the building for cooling. This technology is considered dynamic due to the fact that the ice is removed from the evaporator rather than simply remaining on it.

Static ice storage systems are currently available in factory-assembled packaged units which provide ease of installation and can provide a lower initial capital cost. When compared to water storage systems, the size and weight reduction associated with ice systems makes them very attractive to facilities with space constraints. One main disadvantage to ice systems is the fact that the evaporator must be cold enough to produce ice. These evaporator temperatures usually range from 10° to 25°, while most chiller evaporator temperatures range from 42° to 47°[9]. This required decrease in evaporator temperature results in a higher energy demand per ton, causing reduced cooling efficiency. The EPRI Project reported that chillers operating in chilled water or eutectic salt (phase change material) used approximately 20% less energy than chillers operating in ice systems (0.9 vs. 1.1 kW/ton)[7]. The advantages of using ice as the thermal storage medium are:

1. Retrofitting the storage system with the existing HVAC chilled water system is feasible.

2. Ice systems require less space than that required by the water systems.

3. Ice systems have higher storage but lower refrigeration efficiencies than those of water.

4. Ice systems are available in packaged units, due to smaller size requirements.

19.3.3 Phase Change Materials

The benefit of capturing latent heat of fusion while maintaining evaporating temperatures of existing chiller systems can be realized with the use of phase change materials. There are materials that have melting points higher than that of water that have been successfully used in thermal energy storage systems. Several of these materials fall into the general category called "eutectic salts" and are salt hydrates that are mixtures of inorganic salts and water. Some eutectic salts have melting (solidifying) points of 47°[7], providing the opportunity for a direct retrofit using the existing chiller system since this is at or above the existing evaporator temperatures. In a thermal storage system, these salts are placed in plastic containers that are immersed within an insulated chilled water tank. During the charging cycle, the chilled water flows through the gaps between the containers, freezing the salts within them. During the on-peak discharge, the warmer building return water circulates through the tank, melting the salts and utilizing the latent heat of fusion to cool the building. These salt solutions have latent heat of fusion around 40 Btu/lb[9].

This additional latent heat reduces the storage volume by 66% of that required for an equivalent capacity water storage system[9]. Another obvious benefit of using eutectic salts is that the efficiency of the chillers is not sacrificed, as stated earlier since the phase change occurs around normal evaporator suction temperatures. One problem with the eutectic salt systems is that the auxiliary energy consumption is higher since the chilled water must be pumped through the array of eutectic blocks. The auxiliary energy consumption of the ice systems is higher than both the water and eutectic salt systems since the chilled water must be pumped through the ice system coils, nozzles, and heat exchangers. The EPRI study found that the chilled water systems had an average auxiliary energy use of 0.43 kWh/ton-hr compared to the phase change systems' (eutectic and ice) average auxiliary energy use of 0.56 kWh/ton-hr[5]. The advantages of using eutectic salts as the thermal storage medium are that they:

1. can utilize the existing chiller system for generating storage due to evaporator temperature similarity,

2. require less space than that required by the water systems,

3. have higher storage and equivalent refrigeration efficiencies to those of water, and

4. do not suffer the efficiency penalties of ice systems.

19.4 SYSTEM CAPACITY

The performance of thermal storage systems depends upon proper design. If sized too small or too large, the entire system performance will suffer. The following section will explain this sizing procedure for the example office building presented earlier. The facility has a maximum load of 510 tons, a total cooling requirement of 6,123 ton-hours, and an on-peak cooling requirement of 3,420 ton-hours. This information will be analyzed to size a conventional chiller system, a partial storage system, a full storage system, and the optional partial storage system. These results will then be used to determine the actual capacity needed to satisfy the cooling requirements utilizing either a chilled water, a eutectic salt, or an ice thermal storage system. Obviously, some greatly simplifying assumptions are made.

19.4.1 Chiller System Capacity

The conventional system would need to be able to handle the peak load independently, as seen in Figure 19-1. A chiller or series of chillers would be needed to produce the peak cooling load of 510 tons. Unfortunately, packaged chiller units are usually available in increments that mandate excess capacity, but for simplicity one 600-

ton chiller will be used for this comparison. The conventional chiller system will provide cooling as it is needed and will follow the load presented in Figure 19-1 and Table 19-1.

To determine the chiller system requirement of a cooling system utilizing partial load storage, further analysis is needed. Table 19-1 showed that the average cooling load of the office building was 255 tons per hour. The ideal partial load storage system will run at this load. (See Figure 19-3 and Table 19-3.) The chiller system would need to be sized to supply the 255 tons per hour, so one 300-ton chiller will be used for comparison purposes. Table 19-5 shows how the chiller system would operate at 255 tons per hour, providing cooling required for the building directly and charging the storage system with the excess. Although the storage system supplements the cooling system for 2 hours before the peak period, the cooling load is always satisfied.

Comparing the peak demand from the bottoms of columns 2 and 3 of Table 19-5 shows that the partial storage system reduced this peak load almost 50% (510 – 255 = 254.87 tons). Column 4 shows the tonnage that is supplied to the storage system, and column 5 shows the amount of cooling contained in the storage system at the end of each hour of operation. This system was design so there would be zero capacity remaining in the thermal storage tanks after the on-peak period. The values contained at the bottom of Table 19-5 are the total storage required to assure no remaining capacity and the maximum output required from storage. These values will be utilized in the next section to determine the storage capacity required for each of the different storage mediums.

The full storage system also requires some calculations to determine the chiller system size. Since the chillers will not be used during the on-peak period, the entire daily cooling requirement must be generated during the off-peak periods. Table 19-1 listed the total cooling load as 6,123 ton-hours for the peak day. Dividing this load over the 16 off-peak hours yields that the chillers must generate 383 tons of cooling per hour (6,123 ton-hours/16 hours). A 450-ton chiller will be utilized in this situation for comparison purposes. Table 19-6 shows how the chiller system would operate at 383 tons per hour, providing cooling required for the building directly and charging the storage system with the excess.

Comparing the peak demand with and without storage in Table 19-6 shows that the full storage system eliminated all load from the on-peak period. Column 4 shows the tonnage is supplied to the storage system, and column 5 shows the amount of cooling contained in the storage system at the end of each hour of operation.

Table 19-5. Partial storage operation profile.

Thermal Storage Operation Profile
Partial Storage System

1 End of Hour (Tons)	2 Cooling Load (Tons)	3 Chiller Load (Ton-Hrs)	4 Capacity to Storage (Ton-Hrs)	5 Capacity In Storage	6 Storage Cycle
1	100	255.13	155	696	Charge
2	120	255.13	135	831	Charge
3	125	255.13	130	961	Charge
4	130	255.13	125	1086	Charge
5	130	255.13	125	1211	Charge
6	153	255.13	102	1314	Charge
7	165	255.13	90	1404	Charge
8	230	255.13	25	1429	Charge
9	270	255.13	-15	1414	Discharge
10	290	255.13	-35	1379	Discharge
11	340	255.13	-85	1294	Discharge
12	380	255.13	-125	1169	Discharge
13	450	255.13	-195	974	Discharge
14	490	255.13	-235	740	Discharge
15	510	255.13	-255	485	Discharge
16	480	255.13	-225	260	Discharge
17	410	255.13	-155	105	Discharge
18	360	255.13	-105	0	Discharge
19	250	255.13	5	5	Charge
20	210	255.13	45	50	Charge
21	160	255.13	95	145	Charge
22	130	255.13	125	271	Charge
23	125	255.13	130	401	Charge
24	115	255.13	140	541	Charge

	Without storage	With storage		
Daily Total (Ton-Hrs):	6123	6123		
Daily Avg (Tons):	255.13	255.13		
Peak Total (Ton-Hrs):	3420	2041	Storage Total =	1429
Peak Demand (Tons):	510	255.13	Peak Storage Output =	255

Column 4 = Column 3 – Column 2
Column 5(n) = Column 5(n–1) + Column 4(n)

This system was designed so there would be 0 capacity remaining in the thermal storage tanks after the on-peak period, as shown at the bottom of Table 19-6. The values in Table 19-6 will be utilized in the next section to determine the storage capacity required for each of the different storage mediums.

The optional partial storage system is a blend of the two systems presented earlier. Values given in Table 19-7 and Figure 19-4 are one combination of several possibilities that would drop the consumption and peak demand during the on-peak period. Once again this system has been designed to run both chillers during off-peak hours and run only one during on-peak hours. Benefits of this arrangement are that the current chiller system could be used in combination with the storage system and that the storage system does not require as much capacity as the full storage system. Also, a reserve chiller is available during peak-load times.

Comparing the peak demand with and without storage in Table 19-7 shows that the optional partial storage system reduces the peak load from 510 tons to 153 tons, or approximately 70% during the on-peak period.

Table 19-6. Full storage operation profile.

Thermal Storage Operation Profile
Full Storage System

1 Hour of Day	2 Cooling Load (Tons)	3 Chiller Load (Tons)	4 Capacity to Storage (Ton-Hrs)	5 Capacity In Storage (Ton-Hrs)	6 Storage Cycle
1	100	383	283	1589	Charge
2	120	383	263	1852	Charge
3	125	383	258	2109	Charge
4	130	383	253	2362	Charge
5	130	383	253	2615	Charge
6	153	383	230	2844	Charge
7	165	383	218	3062	Charge
8	230	383	153	3215	Charge
9	270	383	113	3327	Charge
10	290	383	93	3420	Charge
11	340	0	-340	3080	Discharge
12	380	0	-380	2700	Discharge
13	450	0	-450	2250	Discharge
14	490	0	-490	1760	Discharge
15	510	0	-510	1250	Discharge
16	480	0	-480	770	Discharge
17	410	0	-410	360	Discharge
18	360	0	-360	0	Discharge
19	250	383	133	133	Charge
20	210	383	173	305	Charge
21	160	383	223	528	Charge
22	130	383	253	781	Charge
23	125	383	258	1038	Charge
24	115	383	268	1306	Charge

	Without storage	With storage		
Daily Total (Ton-Hrs):	6123	6123		
Daily Avg (Tons):	255.13	255.13		
Peak Total (Ton-Hrs):	3420	0	Storage Total =	3420
Peak Demand (Tons):	510	0	Peak Storage Output =	510

Column 4 = Column 3 – Column 2
Column 5(n) = Column 5(n–1) + Column 4(n)

Column 4 shows the tonnage that is supplied to the storage system, and column 5 shows the amount of cooling contained in the storage system at the end of each hour of operation. This system was designed so there would be zero capacity remaining in the thermal storage tanks after the on-peak period. The values contained at the bottom of Table 19-7 are the total storage capacity required and the maximum output required from storage. These values will be utilized in the next section to determine the storage capacity required for each of the different storage mediums. Table 19-8 summarizes the performance parameters for the three configurations discussed above. The next section summarizes the procedure used to determine the size of the storage systems required to handle the office building.

19.4.2 Storage System Capacity

Each of the storage mediums has different size requirements to satisfy the needs of the cooling load. This section will describe the procedure to find the actual volume or size of the storage system for the partial load system for each of the different storage mediums. The

Table 19-7. Optional partial storage operation profile.

Thermal Storage Operation Profile—Optional Partial Storage System					
1	2	3	4	5	6
End of Hour	Cooling Load (Tons)	Chiller Load (Tons)	Capacity to Storage (Ton-Hrs)	Capacity In Storage (Ton-Hrs)	Storage Cycle
1	100	306	206	1053	Charge
2	120	306	186	1239	Charge
3	125	306	181	1420	Charge
4	130	306	176	1597	Charge
5	130	306	176	1773	Charge
6	153	306	153	1926	Charge
7	165	306	141	2067	Charge
8	230	306	76	2143	Charge
9	270	306	36	2179	Charge
10	290	306	16	2195	Charge
11	340	153	-187	2008	Discharge
12	380	153	-227	1782	Discharge
13	450	153	-297	1485	Discharge
14	490	153	-337	1148	Discharge
15	510	153	-357	791	Discharge
16	480	153	-327	464	Discharge
17	410	153	-257	207	Discharge
18	360	153	-207	0	Discharge
19	250	306	56	56	Charge
20	210	306	96	152	Charge
21	160	306	146	298	Charge
22	130	306	176	475	Charge
23	125	306	181	656	Charge
24	115	306	191	847	Charge

	Without storage	With storage			
Daily Total (Ton-Hrs):	6123	6123			
Daily Avg (Tons):	255.13	255.12			
Peak Total (Ton-Hrs):	3420	1225	Storage Total =		2195
Peak Demand (Tons):	510	153	Peak Storage Output =		357

Column 4 = Column 3 - Column 2
Column 5(n) = Column 5(n-1) + Column 4(n)

design of the chiller and thermal storage system must provide enough chilled water to the system to satisfy the peak load, so particular attention should be paid to the pumping and piping. Table 19-9 summarizes the size requirement of each of the three different storage options.

To calculate the capacity of the partial load storage system, the relationship between capacity (C), mass (M), specific heat of material (Cp), and the coil temperature differential $(T_2–T_1)$ shown in Figure 19-5a will be used:

$$C = M Cp (T_2–T_1)$$

where
$$M = lbm$$
$$Cp = Btu/lbm\ °R$$
$$(T_2–T_1) = °R$$

The partial load system required that 1,429 ton-hrs be stored to supplement the output of the chiller during on-peak periods. This value does not allow for any thermal loss that normally occurs. For this discussion, a conservative value of 20% is used, which is an average suggested in the EPRI report[7]. This will increase the storage requirements to 1,715 ton-hrs, and chilled water storage systems

Table 19-8. System performance comparison.

PERFORMANCE PARAMETERS	SYSTEM			
	Conventional No Storage	Partial Storage	Full Storage	Optional Partial
Overall Peak Demand (Tons)	510	255.13	383	306
On-Peak, Peak Demand (Tons)	510	255.13	0	153
On-Peak Chiller Consumption (Ton-Hrs)	3,420	2,041	0	1,225
Required Storage Capacity[1] (Ton-Hrs)	—	1,379	3,420	2,195
MAXIMUM STORAGE OUTPUT[1] (Tons)	—	255	510	357

[1]Values from Table 19-5, 19-6, and 19-7 represent the capacity required to be supplied by the TES.

in this size range cost approximately \$200/ton-hr including piping and installation[5]. Assuming that there are 12,000 Btu's per ton-hr, this yields:

$$C = (1{,}715 \text{ ton-hrs})*(12{,}000 \text{ Btu}/\text{ton-hr})$$
$$= 20.58 \times 10^6 \text{ Btu's.}$$

Assuming $(T_2 - T_1) = 12°$ and $Cp = 1 \text{ Btu}/\text{lb}_m \text{ °R}$, the relation becomes:

$$M = \frac{C}{Cp(T_2 - T_1)} = \frac{20.58 \times 10^6 \text{ Btus}}{1 \text{ Btu}/\text{lb}_m - °R)(12°R)} = 1.72 \times 10^6 \text{ lbm H}_2\text{O}$$

$$\text{Volume of Water} = \text{Mass}/\text{Density} = \frac{1.72 \times 10^6 \text{ lb}_m}{62.5 \text{ lb}_m/\text{Ft}^3}$$

$$\frac{1.72 \times 10^6 \text{ lb}_m}{8.34 \text{ lb}_m/\text{gal}} \quad \begin{aligned} &= 27{,}520 \text{ Ft}^3 \text{ or} \\ &= 206{,}235 \text{ gal.} \end{aligned}$$

Sizing the storage system utilizing ice is completed in a very similar fashion. The EPRI study states that the ice storage tanks had average daily heat gains 3.5 times greater than the chilled water and eutectic systems due to the higher coil temperature differential $(T_2 - T_1)$. To allow for these heat gains a conservative value of 50% will be added to the actual storage capacity, which is an average suggested in the EPRI report.[7] This will increase the stor-

age requirements to 2,144 ton-hrs. Assuming that there are 12,000 Btu's per ton-hr, this yields: (2,144 ton-hrs)*(12,000 Btu's/ton-hr) = 25.73 × 10⁶ Btu's. The ice systems utilize the latent heat of fusion, so the C_1 now becomes

$$C_1 = \text{Latent Heat} = 144 \text{ Btu}/\text{lb}_m.$$

Because the latent heat of fusion, which occurs at 32°F, is so large compared to the sensible heat, the sensible heat (Cp) is not included in the calculation. The mass of water required to be frozen becomes:

$$M = C/C_1 = \frac{25.73 \times 10^6 \text{ Btus}}{(144 \text{ Btu}/\text{lb}_m)} = 1.79 \times 10^5 \text{ lb}_m \text{ H}_2\text{O}$$

$$\text{Volume of Ice} = \frac{\text{Mass}}{\text{Density}} = \frac{1.79 \times 10^5 \text{ lb}_m}{62.5 \text{ lb}_m/\text{Ft}^3}$$
$$= 2{,}864 \text{ Ft}^3$$

This figure is conservative since the sensible heat has been ignored, but it calculates the volume of ice needed to be generated. The actual volume of ice needed will vary, and the total amount of water contained in the tank around the ice coils will vary greatly. The ability to purchase pre-packaged ice storage systems makes their sizing quite easy. For this situation, two 1,080 ton-hr ice storage units will be purchased for approximately \$150/ton-hr, including piping and installation.[4] (Note that this provides 2,160 ton-hrs compared to the needed

Table 19-9. Complete system comparison.

Performance Parameters	SYSTEM			
	Conventional No Storage	Partial Storage	Full Storage	Optional Partial
CHILLER				
SIZE (# and Tons)	1 @ 600	1 @ 300	1 @ 450	2 @ 175
COST($)	180,000	90,000	135,000	105,000
WATER STORAGE				
Capacity (Ton-Hrs)	—	1,715	4,104	2,634
Volume (cubic feet)	—	27,484	65,769	42,212
Volume (gallons)	—	205,635	492,086	315,827
Cost per Ton-Hr ($)	—	200	135	165
Storage cost ($)	—	343,000	554,040	434,610
ICE STORAGE				
Capacity (Ton-Hrs)		2,144	5,130	3,293
# and size (Ton-Hrs)	—	2 @ 1,080	4 @ 1,440	3 @ 1,220
Ice volume (cubic feet)	—	2,859	6,840	4,391
Cost per Ton-Hr (S)	—	150	150	150
Storage cost ($)[1]	—	324,000	864,000	549,000
EUTECTIC STORAGE				
Capacity (Ton-Hrs)		1,715	4,104	2,634
Eutectic vol (cubic feet)	—	8,232	19,699	12,643
Cost per Ton-Hr ($)	—	250	200	230
Storage cost ($)	—	428,750	820,000	605,820

[1](2 units)(1,080 Ton-Hrs/units)($150/Ton-Hr) = $324,000
Note: The values in this table vary slightly from those in the text from additional significant digits.

2,144 ton-hrs.)

Sizing the storage system utilizing the phase change materials or eutectic salts is completed only as the ice storage system. The EPRI study states that the eutectic salt storage tanks had average daily heat gains approximately the same as that of the chilled water systems. To allow for these heat gains a conservative value of 20% is added to the actual storage capacity[5]. This increases storage requirements to 1,715 ton-hrs. Assuming there are 12,000 Btu's per ton-hr, this yields: (1,715 ton-hrs)*(12,000 Btu's/ton-hr) = 20.58×10^6 Btu's.

The eutectic system also utilizes the latent heat of fusion like the ice system; the temperature differential shown in Figure 19-5a is not used in the calculation. The C_1 now becomes:

$$C_1 = \text{Latent Heat} = 40 \text{ Btu/lbm}$$

$$M = C/C_1 = \frac{20.58 \times 10^6 \text{ Btus}}{(40 \text{ Btu/lmb})} = 5.15 \times 10^5 \text{ lbm}$$

$$\begin{array}{l}\text{Volume of} \\ \text{Eutectic Salts (assuming} \\ \text{density = water)}\end{array} = \frac{\text{Mass}}{\text{Density}} = \frac{5.15 \times 10^5 \text{ lbm}}{62.5 \text{ lbm/ft}^3}$$

$$= 8,232 \text{ ft}^3$$

The actual volume of eutectic salts needed would need to be adjusted for density differences in the various combinations of the salts. Eutectic systems have not been studied in great detail, and factory sized units are not yet readily available. The EPRI report[7] studied a system that required 1,600 ton-hrs of storage that utilized approximately 45,000 eutectic "bricks" contained in an 80,600 gallon tank of water. For this situation, a similar eutectic storage unit will be purchased for approximately $250/ton-hr, including piping and installation. The ratio of ton-hrs required for partial storage and the required tank size will be utilized for sizing the full and optional partial storage systems.

Table 19-9 summarizes the sizes and costs of the different storage systems and the actual chiller systems for each of the three storage arrangements. The values presented in this example are for a specific case, and each application should be analyzed thoroughly. The cost per ton hour of a water system drops significantly as the size of the tanks rises, as will the eutectic systems since the engineering and installation costs are spread over more capacity. Also we ignored the sensible heat of the ice and eutectic systems.

19.5 ECONOMIC SUMMARY

Table 19-9 covered the approximate costs of each of the three system configurations utilizing each of the three different storage mediums. Table 19-8 listed the various peak day performance parameters of each of the systems presented. To this point, the peak day chiller consumption has been used to size the system. To analyze the savings potential of the thermal storage systems, much more information is needed to determine daily cooling and chiller loads and the respective storage system performance. To calculate the savings accurately, a daily chiller consumption plot is needed for at least the summer peak period. These values can then be used to determine the chiller load required to satisfy the cooling demands. Only the summer months may be used since most of the cooling takes place and a majority of the utilities "time of use" charges (on-peak rates) are in effect during that time. There are several methods available to estimate or simulate building cooling load. Some of these methods are available in a computer simulation format, or they can also be calculated by hand.

For the office building presented earlier, an alternative method will be used to estimate cooling savings. An estimate of a monthly, average day cooling load will be used to compare the operating costs of the respective cooling configurations. For simplicity, it is assumed that the peak month is July and the average cooling day is 90% of the cooling load of the peak day. The average cooling day for each of the months that make up the summer cooling period are estimated based upon July's average cooling load. These factors are presented in Table 19-10 for June through October[11]. These factors are applied to the hourly chiller load of the average July day to determine the season chiller/TES operation loads. The monthly average day, hourly chiller loads for each of the three systems are presented in Table 19-11. The first column for each month in Table 19-11 lists the hourly cooling demand. The chiller consumption required to satisfy this load utilizing each of the storage systems is also listed. This table does not account for the thermal efficiencies used to size the systems, but for simplicity these values will be used to determine the rate and demand savings that will be achieved after implementing the system. The formulas presented for the peak day thermal storage systems operations have been used for simplicity. These chiller loads do not represent the optimum chiller load since some of partial systems approach full storage systems during the early and late cooling months. The bottom of the table contains the totals for the chiller systems. These totaled average day values will now be used to calculate the savings. The difference between the actual cooling load and the chiller load is the approximate daily savings for each day of that month.

A hypothetical southwest utility rate schedule will be used to apply economic terms to these savings. The electricity consumption rate is $0.04/kWh, and the demand rate during the summer is $3.50/kW per month for the peak demand during the off-peak hours and $5.00/kW per month for the peak demand during the on-peak hours. These summer demand rates are in effect from June through October. This rate schedule only

Table 19-10. Average summer day cooling load factors.

MONTH	kW FACTOR[1]	PEAK TONS[2]	kWh FACTOR[1]	Ton-Hrs/day[3]
JUNE	0.8	360	0.8	4,322
JULY	1	450	1	5,403
AUGUST	0.9	405	0.9	4,863
SEPT	0.7	315	0.7	3,782
OCT	0.5	225	0.5	2,702

[1]kW and kWh factors were estimated to determine utility cost savings.
[2]The average day peak load is estimated to be 90% of the peak day. The kW factor for each month is multiplied by the peak months average tonnage. For JUNE: PEAK TONS = (0.8)*(450) = 360
[3]The average day consumption is estimated to be 90% of the peak day. The kWh factor for each month is multiplied by the peak months average consumption. For JUNE: CONSUMPTION = (0.8)*(5,403) = 4,322

Table 19-11. Monthly average day chiller load profiles.

END OF HOUR	JUNE (in tons)				JULY (in tons)				AUGUST (in tons)				SEPTEMBER (in tons)				OCTOBER (in tons)			
	Actual	partial	full	optional	Actual	partial	full	optional	Actual	partial	full	optional	Actual	partial	full	optional	Actual	partial	full	optional
1	71	180	270	216	88	225	338	270	79	203	304	243	62	158	236	189	44	113	169	135
2	85	180	270	216	106	225	338	270	95	203	304	243	74	158	236	189	53	113	169	135
3	88	180	270	216	110	225	338	270	99	203	304	243	77	158	236	189	55	113	169	135
4	92	180	270	216	115	225	338	270	103	203	304	243	80	158	236	189	57	113	169	135
5	92	180	270	216	115	225	338	270	103	203	304	243	80	158	236	189	57	113	169	135
6	108	180	270	216	135	225	338	270	122	203	304	243	95	158	236	189	68	113	169	135
7	116	180	270	216	146	225	338	270	131	203	304	243	102	158	236	189	73	113	169	135
8	162	180	270	216	203	225	338	270	183	203	304	243	142	158	236	189	101	113	169	135
9	191	180	270	216	238	225	338	270	214	203	304	243	167	158	236	189	119	113	169	135
10	205	180	270	216	256	225	338	270	230	203	304	243	179	158	236	189	128	113	169	135
11	240	180	0	108	300	225	0	135	270	203	0	122	210	158	0	95	150	113	0	68
12	268	180	0	108	335	225	0	135	302	203	0	122	235	158	0	95	168	113	0	68
13	318	180	0	108	397	225	0	135	357	203	0	122	278	158	0	95	199	113	0	68
14	346	180	0	108	432	225	0	135	389	203	0	122	303	158	0	95	216	113	0	68
15	360	180	0	108	450	225	0	135	405	203	0	122	315	158	0	95	225	113	0	68
16	339	180	0	108	424	225	0	135	381	203	0	122	296	158	0	95	212	113	0	68
17	289	180	0	108	362	225	0	135	326	203	0	122	253	158	0	95	181	113	0	68
18	254	180	0	108	318	225	0	135	286	203	0	122	222	158	0	95	159	113	0	68
19	176	180	270	216	221	225	338	270	199	203	304	243	154	158	236	189	110	113	169	135
20	148	180	270	216	185	225	338	270	167	203	304	243	130	158	236	189	93	113	169	135
21	113	180	270	216	141	225	338	270	127	203	304	243	99	158	236	189	71	113	169	135
22	92	180	270	216	115	225	338	270	103	203	304	243	80	158	236	189	57	113	169	135
23	88	180	270	216	110	225	338	270	99	203	304	243	77	158	236	189	55	113	169	135
24	81	180	270	216	101	225	338	270	91	203	304	243	71	158	236	189	51	113	169	135

TOTALS:	JUNE	partial	full	optional	JULY	partial	full	optional	AUG	partial	full	optional	SEPT	partial	full	optional	OCT	partial	full	optional
TOTAL: (ton–hrs)	4,322	4,322	4,322	4,322	5,403	5,403	5,403	5,403	4,862	4,862	4,862	4,862	3,782	3,782	3,782	3,782	2,701	2,701	2,701	2,701
AVG: (tons)	180	180	180	180	225	225	225	225	203	203	203	203	158	158	158	158	113	113	113	113
OFF-PEAK MAX: (tons)	205	180	270	216	256	225	338	270	230	203	304	243	179	158	236	189	128	113	169	135
ON-PEAK MAX: (tons)	360	180	0	108	450	225	0	135	405	203	0	122	315	158	0	95	225	113	0	68
ON-PEAK CONSUMP: (ton–hrs)	2,414	1,441	0	864	3,018	1,801	0	1,081	2,716	1,621	0	972	2,112	1,261	0	756	1,509	900	0	540

For June Partial Storage: (4,322 Ton–Hrs)/(24 Hrs) = 180 Tons
For June Full Storage: (4,322 Ton–Hrs)/(16 Hrs) = 270 Tons

provides savings from balancing the demand, although utilities often have cheaper off-peak consumption rates. It can be seen that the off-peak demand charge assures the demand is leveled and not merely shifted. This rate schedule will be applied to the total values in Table 19-11 and multiplied by the number of days in each month to determine the summer savings. These savings are contained in Table 19-12. The monthly average day loads in Table 19-11 are assumed to be 90% of the actual monthly peak billing demand, and they are adjusted accordingly in Table 19-12. The total monthly savings for each of the chiller/TES systems is determined at the bottom of each monthly column.

These cost savings are not the only monetary justification for implementing TES systems. Utilities often extend rebates and incentives to companies installing thermal energy storage systems to shorten their respective payback period. This helps the utility reduce the need to build new generation plants. The southwest utility serv-

ing the office building studied here offers $200 per design day peak kW shifted to off-peak hours, up to $200,000.

19.6 CONCLUSIONS

Thermal energy storage will play a large role in the future of demand side management programs of both private organizations and utilities. An organization that wishes to employ a system-wide energy management strategy will need to be able to track, predict, and control their load profile in order to minimize utility costs. This management strategy will only become more critical as electricity costs become more variable in a deregulated market. Real-time pricing and multi-facility contracts will further enhance the savings potential of demand management, which thermal energy storage should become a valuable tool.

The success of the thermal storage system and the

HVAC system as a whole depend on many factors:

- The chiller load profile,

- The utility rate schedules and incentive programs,

- The condition of the current chiller system,

- The space available for the various systems,

- The selection of the proper storage medium, and

- The proper design of the system and integration of this system into the current system.

Thermal storage is a very attractive method for an organization to reduce electric costs and improve system management. New installation projects can utilize storage to reduce the initial costs of the chiller system as well as savings in operation. Storage systems will become easier to justify in the future, with increased mass production, and technical advances and more companies switching to storage.

References

1. Cottone, Anthony M., "Featured Performer: Thermal Storage," in *Heating Piping and Air Conditioning*, August 1990, pp. 51-55.
2. Hopkins, Kenneth J., and James W. Schettler, "Thermal Storage Enhances Heat Recovery," in *Heating Piping and Air Conditioning,* March 1990, pp. 45-50.
3. Keeler, Russell M., "Scrap DX for CW with Ice Storage," in *Heating Piping and Air Conditioning,* August 1990, pp. 59-62.
4. Lindemann, Russell, Baltimore Aircoil Company, Personnel Phone Interview, January 7, 1992.
5. Mankivsky, Daniel K., Chicago Bridge and Iron Company, Personnel Phone Interview, January 7, 1992.
6. Pandya, Dilip A., "Retrofit Unitary Cool Storage System," in *Heating Piping and Air Conditioning,* July 1990, pp. 35-37.
7. Science Applications International Corporation, *Operation Performance of Commercial Cool Storage Systems* Vols. 1 & 2, Electric Power Research Institute (EPRI) Palo Alto, September 1989.
8. Tamblyn, Robert T., "Optimizing Storage Savings," in *Heating Piping and Air Conditioning,* August 1990 pp. 43-46.
9. Thumann, Albert, *Optimizing HVAC Systems* The Fairmont Press, Inc., 1988.
10. Thumann, Albert and D. Paul Mehta, *Handbook of Energy Engineering*, The Fairmont Press, Inc., 1991.
11. Wong, Jorge-Kcomt, Dr. Wayne C. Turner, Hemanta Agarwala, and Alpesh Dharia, *A Feasibility Study to Evaluate Different Options for Installation of a New Chiller With/Without Thermal Energy Storage System,* study conducted for the Oklahoma State Office Buildings Energy Cost Reduction Project, Revised 1990.

Table 19-12. Summer monthly system utility costs and TES savings.

	JUNE (30 days)				JULY (31 days)				AUGUST (31 days)				SEPTEMBER (30 days)				OCTOBER (31 days)			
	Actual	Partial	Full	Optional	Actual	Partial	Full	Optional	Actual	Partial	Full	Optional	Actual	Partial	Full	Optional	Actual	Partial	Full	Optional
ON-PEAK, PEAK (kW) [1]	400	200	0	120	500	250	0	150	450	226	0	136	350	176	0	106	250	126	0	76
OFF-PEAK, PEAK (kW) [1]	228	200	300	240	284	250	376	300	256	226	338	270	199	176	262	210	142	126	188	150
CONSUMPTION (kW-Hr) [1]	4,322	4,322	4,322	4,322	5,403	5,403	5,403	5,403	4,862	4,862	4,862	4,862	3,782	3,782	3,782	3,782	2,701	2,701	2,701	2,701
DEMAND COST ($) [2]	2,797	1,700	1,050	1,440	3,496	2,125	1,314	1,800	3,144	1,917	1,182	1,623	2,446	1,492	918	1,263	1,748	1,067	657	903
CONSUMPTION COST ($) [3]	5,186	5,186	5,186	5,186	6,700	6,700	6,700	6,700	6,029	6,029	6,029	6,029	4,538	4,538	4,538	4,538	3,349	3,349	3,349	3,349
TOTAL COST ($) [4]	7,984	6,886	6,236	6,626	10,195	8,825	8,014	8,500	9,173	7,946	7,211	7,652	6,985	6,031	5,456	5,801	5,097	4,416	4,006	4,252
SAVINGS ($) [5]		1,097	1,747	1,357		1,371	2,181	1,696		1,227	1,962	1,522		954	1,528	1,183		681	1,091	845

Table 19-13. Available demand management incentives.

Performance Parameters	System			
	Conventional No Storage	Partial Storage	Full Storage	Optional Partial
Actual On-Peak Demand[1] (kW)	510	255	0	153
On-Peak Demand Shifted[2] (kW)		255	510	357
Utility Subsidy[3] ($)		51,000	102,000	71,400

[1]Yearly design peak demand from Table 19-8.
[2]Demand shifted from design day on-peak period. For partial: 510 kW - 255 kW   = 255 kW.
[3]Based upon $200/kW shifted from design day on-peak period. For partial: 255 kW * $200/kW = $51,000.

CHAPTER 20
CODES, STANDARDS, & LEGISLATION

ALBERT THUMANN, P.E., C.E.M.
Association of Energy Engineers
Atlanta, Georgia

20.1 INTRODUCTION

This chapter presents a historical perspective on key codes, standards, and regulations, that have impacted energy policy and are still playing a major role in shaping energy usage. The context of past standards and legislation must be understood in order to properly implement the proper systems and be able to impact future codes. The Energy Policy Act, for example, has created an environment for retail competition. Electric utilities will drastically change the way they operate in order to provide power and lowest cost. This in turn will drastically reduce utility-sponsored incentive and rebate programs, which have influenced energy conservation adoption. The chapter attempts to cover a majority of the material that currently impacts the energy related industries, with relationship to their respective initial writing.

The main difference between standards, codes, and regulations is an increasing level of enforceability of the various design parameters. A group of interested parties (vendors, trade organizations, engineers, designers, citizens, etc.) may develop a standard in order to assure minimum levels of performance. The standard acts as a suggestion to those parties involved, but it is not enforceable until codified by a governing body (local or state agency), which makes the standard a code. Not meeting this code may prevent continuance of a building permit or result in the ultimate stoppage of work. Once the federal government makes the code part of the federal code, it becomes a regulation. Often this progression involves equipment development and commercialization prior to codification in order to assure that the standards are attainable.

20.2 THE ENERGY INDEPENDENCE AND SECURITY ACT OF 2007 (H.R.6)

The Energy Independence and Security Act of 2007 (H.R.6) was enacted into law December 19, 2007. Key provisions of EISA 2007 are summarized below.

Title I Energy Security through Improved Vehicle Fuel Economy

- Corporate average fuel economy (CAFE). The law sets a target of 35 miles per gallon for the combined fleet of cars and light trucks by 2020.
- The law establishes a loan guarantee program for advanced battery development, a grant program for plug-in hybrid vehicles, incentives for purchasing heavy-duty hybrid vehicles for fleets, and credits for various electric vehicles.

Title II Energy Security Through Increased Production of Biofuels

- The law increases the renewable fuels standard (RFS), which sets annual requirements for the quantity of renewable fuels produced and used in motor vehicles. RFS requires 9 billion gallons of renewable fuels in 2008, increasing to 36 billion gallons in 2022.

Title III Energy Savings Through Improved Standards for Appliances and Lighting

- The law establishes new efficiency standards for motors, external power supplies, residential clothes washers, dishwashers, dehumidifiers, refrigerators, refrigerator freezers, and residential boilers.
- The law contains a set of national standards for light bulbs. The first part of the standard would increase energy efficiency of light bulbs 30% and phase out most common types of incandescent light bulb by 2012-2014.
- Requires the federal government to substitute energy efficient lighting for incandescent bulbs.

Title IV Energy Savings in Buildings and Industry

- The law increases funding for the Department of Energy's weatherization program, providing $3.75 billion over five years.
- The law encourages the development of more energy efficient "green" commercial buildings. The law creates an Office of Commercial High Performance Green Buildings at the Department of Energy.
- A national goal is set to achieve zero-net energy use for new commercial buildings built after 2025. A further goal is to retrofit all pre-construction 2025 buildings to zero-net energy by 2050.

- Requires that total energy use in federal buildings relative to the 2005 level be reduced 30% by 2015.
- Requires federal facilities to conduct a comprehensive energy and water evaluation for each facility at least once every four years.
- Requires new federal buildings and major renovations to reduce fossil fuel energy use 55% relative to 2003 level by 2010 and be eliminated (100 percent reduction) by 2030.
- Requires that each federal agency ensure that major replacements of installed equipment (such as heating and cooling systems) or renovation or expansion of existing space employ the most energy efficient designs, systems, equipment, and controls that are life cycle cost effective. For the purposes of calculating life cycle cost calculations, the time period will increase from 25 years in the prior law to 40 years.
- Directs the Department of Energy to conduct research to develop and demonstrate new process technologies and operating practices to significantly improve the energy efficiency of equipment and processes used by energy-intensive industries.
- Directs the Environmental Protection Agency to establish a recoverable waste energy inventory program. The program must include an ongoing survey of all major industry and large commercial combustion services in the United States.
- Includes new incentives to promote new industrial energy efficiency through the conversion of waste heat into electricity.
- Creates a grant program for healthy high performance schools that aims to encourage states, local governments, and school systems to build green schools.
- Creates a program of grants and loans to support energy efficiency and energy sustainability projects at public institutions.

Title V Energy Savings in Government and Public Institutions

- Promotes energy savings performance contracting in the federal government and provides flexible financing and training of federal contract officers.
- Promotes the purchase of energy efficient products and procurement of alternative fuels with lower carbon emissions for the federal government.
- Reauthorizes state energy grants for renewable energy and energy efficiency technologies through 2012.
- Establishes an energy and environmental block grant program to be used for seed money for innovative local best practices.

Title VI Alternative Research and Development

- Authorizes research and development to expand the use of geothermal energy.
- Improves the cost and effectiveness of thermal energy storage technologies that could improve the operation of concentrating solar power electric generation plants.
- Promotes research and development of technologies that produce electricity from waves, tides, currents, and ocean thermal differences.
- Authorizes a development program on energy storage systems for electric drive vehicles, stationary applications, and electricity transmission and distribution.

Title VII Carbon Capture and Carbon Sequestration

- Provides grants to demonstrate technologies to capture carbon dioxide from industrial sources.
- Authorizes a nationwide assessment of geological formations capable of sequestering carbon dioxide underground.

Title VIII Improved Management of Energy Policy

- Creates a 50% matching grants program for constructing small renewable energy projects that will have an electrical generation capacity less than 15 megawatts.
- Prohibits crude oil and petroleum product wholesalers from using any technique to manipulate the market or provide false information.

Title IX International Energy Programs

- Promotes US exports in clean, efficient technologies to India, China, and other developing countries.
- Authorizes US Agency for International Development (USAID) to increase funding to promote clean energy technologies in developing countries.

Title X Green Jobs

- Creates an energy efficiency and renewable energy worker training program for "green collar" jobs.
- Provides training opportunities for individuals in the energy field who need to update their skills.

Title XI Energy Transportation and Infrastructure

- Establishes an office of climate change and environment to coordinate and implement strategies to reduce transportation-related energy use.

Title XII Small Business Energy Programs

- Loans, grants, and debentures are established to help small businesses develop, invest in, and purchase energy efficient equipment and technologies.

Title XIII Smart Grid

- Promotes a "smart electric grid" to modernize and strengthen the reliability and energy efficiency of the electricity supply. The term "smart grid" refers to a distribution system that allows for flow of information from a customer's meter in two directions: both inside the house to thermostats, appliances, and other devices, and from the house back to the utility.

20.3 THE ENERGY POLICY ACT OF 2005

The first major piece of national energy legislation since the Energy Policy Act of 1992, EPAct 2005 was signed by President George W. Bush on August 8, 2005 and became effective January 1, 2006. The major thrust of EPAct 2005 is energy production. However, there are many important sections of EPAct 2005 that do help promote energy efficiency and energy conservation, as well as provide tax incentives to encourage participation in the private sector. There are also some significant impacts on federal energy management. Highlights are described below:

EPACT 2005 Highlights
Federal Energy Reduction – Existing Buildings
- Baseline changed to year 2003.
- An annual energy reduction goal of 2% is in place from fiscal year 2006 to fiscal year 2015, for a total energy reduction of 20% by year 2015.

Federal Facility Metering
- Electric metering is required in all federal building by the year 2012.

Energy Efficient Products
- Energy efficient specifications are required in procurement bids and evaluations.
- Energy efficient products to be listed in federal catalogs include Energy Star and FEMP products recommended by GSA and Defense Logistics Agency.

Federal—Energy Savings Performance Contracting (ESPC)
- ESPC authority extended through September 30, 2016.
- No caps, limitations or restrictions.
- Impacts all agencies.

Federal—Energy Efficient New Buildings
- New federal buildings will incorporate life cycle costing.
- New federal buildings are required to be designed 30% below ASHRAE standard or the International Energy Code (if life-cycle cost effective). Agencies must identify those that meet or exceed the standard.
- Incorporate sustainable design principles.

Federal Building Renewable Energy: Section 203
- Renewable electricity consumption by the federal government cannot be less than: 3% from fiscal year 2007-2009, 5% from fiscal year 2010-2012, and 7.5% from fiscal year 2013 and beyond.
- The goal for photovoltaic energy is to have 20,000 solar energy systems installed in federal buildings by the year 2012.
- Double credits are earned for renewables produced on the site, on federal lands and used at a federal facility, or produced on Native American lands.
- The goals are based on technical and economic feasibility.

Commercial Buildings
- A tax deduction of up to $1.80 per square foot for energy efficient upgrades to HVAC, lighting, hot water systems, and the building envelope—and 60 cents per square foot for building subsystems that reduce annual power consumption 50 percent compared to the ASHRAE standard.

Residential Buildings
- 30% tax credit for purchase of qualifying residential solar water heating, photovoltaic equipment and fuel cell property. The maximum credit is $2000 (for solar equipment) and $500 for each kilowatt of capacity(fuel cell). The credit applies for property placed in service after 2005 and before 2008.
- Provides a 10% investment tax credit for expenditures with respect to improvements to building envelope.
- Allows tax credits for purchases of high efficiency HVAC systems; advanced main air circulating fans; natural gas, propane or fuel oil furnaces, or hot water boilers; and other energy efficient property. Credit applies to property placed in service after December 31, 2005 and prior to January 1, 2008. The lifetime maximum credit per tax payer is $500.
- Provides $1000 tax credit to eligible contractor for construction of a qualified new energy-efficient home. Tax credit applies to manufactured homes meeting Energy Star standards.

Appliances

- Energy Star Dishwashers (2007)—up to $100 tax credit.

- $75 tax credit for refrigerators that save 15per cent energy; $125 tax credit for refrigerators that save 20 percent or $175 that save 25 percent, based on 2001 standards.

- $100 tax credit for Energy Star Clothes washers (2007).

Fuel Cells, Microturbine Power Plants and Solar Energy

- Provides a 30% tax credit for purchase of qualified fuel cell power plants for businesses.

- Provides a 10 %tax credit for purchase of qualifying stationary micro turbines.

- Provides a 30% tax credit for purchase of qualifying of solar energy property. Tax credits apply to property placed in service after December 31,2005 and before January 1,2008.

Transportation

- Provides tax credits up to $3400 for purchase of hybrid and lean diesel vehicles (capped at 60,000 vehicles per manufacturer for 2006-2010).

Electricity

- Repeals Public Utility Holding Company Act (PUCHA).

- Requires mandatory reliability standards to make the electric power grid more reliable against blackouts.

- Tax incentives to expand investments in electric transmission and generation.

Domestic Production

- Reforms to clarify oil and gas permitting process.

- Authorizes full funding for clean coal research initiative.

- Establishes a new renewable fuel standard that requires the annual use of 7.5 billion gallons of ethanol and biodiesel in the nation's fuel supply by 2012.

20.4 THE ENERGY POLICY ACT OF 1992

The Energy Policy Act of 1992 is substantial and its implementation is impacting electric power deregulation, building codes, and new energy efficient products. Sometimes policy makers do not see the extensive impact of their legislation. This comprehensive legislation is far-reaching and impacts energy conservation, power generation, and alternative fuel vehicles, as well as en-

ergy production. The federal as well as private sectors are impacted by this comprehensive energy act. Highlights of EPACT 1992 are described below.

Energy Efficiency Provisions

Buildings

- Requires states to establish minimum commercial building energy codes and to consider minimum residential codes based on current voluntary codes.

Utilities

- Requires states to consider new regulatory standards that would require utilities to undertake integrated resource planning, allow efficiency programs to be at least as profitable as new supply options, and encourage improvements in supply system efficiency.

Equipment Standards

- Establishes efficiency standards for commercial heating and air-conditioning equipment, electric motors, and lamps.

- Gives the private sector an opportunity to establish voluntary efficiency information/labeling programs for windows, office equipment and luminaries (or the Department of Energy will establish such programs).

Renewable Energy

- Establishes a program for providing federal support on a competitive basis for renewable energy technologies. Expands program to promote export of these renewable energy technologies to emerging markets in developing countries.

Alternative Fuels

- Gives Department of Energy authority to require a private and municipal alternative fuel fleet program, starting in 1998. Provides a federal alternative fuel fleet program with phased-in acquisition schedule; also provides state fleet program for large fleets in large cities.

Electric Vehicles

- Establishes comprehensive program for the research and development, infrastructure promotion, and vehicle demonstration for electric motor vehicles.

Electricity

- Removes obstacles to wholesale power competition in the Public Utilities Holding Company Act by allowing both utilities and non-utilities to form exempt wholesale generators without triggering the

PUHCA restrictions.

Global Climate Change

- Directs the Energy Information Administration to establish a baseline inventory of greenhouse gas emissions and establishes a program for the voluntary reporting of those emissions. Directs the Department of Energy to prepare a report analyzing the strategies for mitigating global climate change and to develop a least-cost energy strategy for reducing the generation of greenhouse gases.

Research and Development

- Directs Dept. of Energy to undertake research and development on a wide range of energy technologies, including energy efficiency technologies, natural gas end-use products, renewable energy resources, heating and cooling products, and electric vehicles.

20.5 CODES AND STANDARDS

Energy codes specify how buildings *must* be constructed or perform, and are written in a mandatory, enforceable language. State and local governments adopt and enforce energy codes for their jurisdictions. Energy standards describe how buildings *should* be constructed to save energy cost effectively. They are published by national organizations such as the American Society of Heating, Refrigerating, and Air Conditioning Engineers (ASHRAE). They are not mandatory but serve as national recommendations, with some variation for regional climate. State and local governments frequently use energy standards as the technical basis for developing their energy codes. Some energy standards are written in a mandatory, enforceable language, making it easy for jurisdictions to incorporate the provisions of the energy standards directly into their laws or regulations. The requirement for the Federal sector to use ASHRAE 90.1 and 90.2 as mandatory standards for all new Federal buildings is specified in the Code of Federal Regulations—10 CFR 435.

Most states use the ASHRAE 90 standard as their basis for the energy component of their building codes. ASHRAE 90.1 is used for commercial buildings and ASHRAE 90.2 is used for residential buildings. Some states have quite comprehensive building codes (for example: California Title 24).

ASHRAE Standard 90.1
ENERGY EFFICIENT DESIGN FOR NEW BUILDINGS

- Sets minimum requirements for the energy-efficient design of new buildings so they will be constructed, operated, and maintained in a manner that minimizes the use of energy without constraining the building function and productivity of the occupants.
- ASHRAE 90.1 addresses building components and systems that affect energy usage.
- Sections 5-10 are the technical sections that specifically address components of the building envelope, HVAC systems and equipment, service water heating, power, lighting, and motors. Each technical section contains general requirements and mandatory provisions. Some sections also include prescriptive and performance requirements.

ASHRAE Standard 90.2
ENERGY EFFICIENT DESIGN FOR NEW LOW-RISE RESIDENTIAL BUILDINGS

When the Department of Energy determines that a revision would improve energy efficiency, each state has two years to review the energy provisions of its residential or commercial building code. For residential buildings, a state has the option of revising its residential code to meet or exceed the residential portion of ASHRAE 90.2. For commercial buildings, a state is required to update its commercial code to meet or exceed the provision of ASHRAE 90.1.

ASHRAE standards 90.1 and 90.2 are developed and revised through voluntary consensus and public hearing processes that are critical to widespread support for their adoption. Both standards are continually maintained by separate, standing, standards projects committees. Committee membership varies from 10 to 60 voting members. Committee membership includes representatives from many groups to ensure balance among all interest categories. After the committee proposes revisions to the standard, it undergoes public review and comment. When a majority of the parties substantially agree, the revised standard is submitted to the ASHRAE board of directors. This entire process can take anywhere from two to ten years to complete. ASHRAE Standards 90.1 and 90.2 are automatically revised and published every three years. Approved interim revisions are posted on the ASHRAE website (www.ashrae.org) and are included in the next published version.

The energy-cost budget method permits trade-offs between building systems (lighting and fenestration, for example) if the annual energy cost estimated for the proposed design does not exceed the annual energy cost of a base design that fulfills the prescriptive requirements. Using the energy-cost budget method approach requires simulation software that can analyze energy consumption in buildings and model the energy features in the proposed design. ASHRAE 90.1 sets minimum requirements for the simulation software; suitable programs include BLAST, eQUEST, and TRACE.

20.6 CLIMATE CHANGE

Kyoto Protocol

The goal of the Kyoto Protocol is to stabilize green house gases in the atmosphere that would prevent human impact on global climate change. The nations that signed the treaty come together to make decisions at meetings called Conferences of the Parties. The 38 parties are grouped into two groups, developed industrialized nations and developing countries. The Kyoto Protocol, an international agreement reached in Kyoto in 1997 by the third Conference of the Parties (COP-3), aims to lower emissions from two groups of three green house gases: 1) carbon dioxide, methane, and nitrous oxide and 2) hydrofluorocarbons (HFCs): sulfur hexafluoride and perfluorocarbons.

20.7 INDOOR AIR QUALITY STANDARDS

Indoor air quality (IAQ) is an emerging issue of concern to building managers, operators, and designers. Recent research has shown that indoor air is often less clean than outdoor air, and federal legislation has been proposed to establish programs to deal with this issue on a national level. This, like the asbestos issue, will have an impact on building design and operations. Americans today spend long hours inside buildings, and building operators, managers, and designers must be aware of potential IAQ problems and how they can be avoided.

IAQ problems, sometimes termed "sick building syndrome," have become an acknowledged health and comfort problem. Buildings are characterized as sick when a significant number of occupants complain of acute symptoms such as headache, eye, nose, and throat irritations, dizziness, nausea, sensitivity odors, and difficulty in concentrating. The complaints may become more clinically defined such that an occupant may develop an actual building-related illness that is believed to be related to IAQ problems.

The most effective means to deal with an IAQ problem is to remove or minimize the pollutant source, when feasible. If not, dilution and filtration may be effective.

The purpose of ASHRAE Standard 62 is to specify minimum ventilation rates and indoor air quality that will be acceptable to human occupants with the intention of minimizing the potential for health effects. ASHRAE defines acceptable indoor air quality as the air in which there are no known contaminants at harmful concentrations, as determined by cognizant authorities and with which a substantial majority (80% or more) of those exposed do not express dissatisfaction.

ASHRAE Standard 55 for thermal environmental conditions for human occupancy covers several environmental parameters, including temperature, radiation, humidity, and air movement. The standard specifies conditions in which 80% of the occupants will find the environment thermally acceptable. This applies to healthy people in normal indoor environments, for winter and summer conditions. Adjustment factors are described for various activity levels and clothing levels.

20.8 MEASUREMENT AND VERIFICATION

The international performance measurement and verification protocol (IPMVP) is used for commercial and industrial facility operators. The IPMVP offers standards for measurement and verification of energy and water efficiency projects. The IPMVP volumes are used to: 1) develop a measurement and verification strategy and plan for quantifying energy and water savings in retrofits and new construction, 2) monitor indoor environmental quality, and 3) quantify emissions reduction (www.evo-world.org).

20.9 REGULATORY AND LEGISLATIVE ISSUES IMPACTING COGENERATION AND INDEPENDENT POWER PRODUCTION

Public Utilities Regulatory Policies Act (PURPA)

This legislation was part of the 1978 National Energy Act and has had perhaps the most significant effect on the development of cogeneration and other forms of alternative energy production in the past decade. Certain provisions of PURPA also apply to the exchange of electric power between utilities and cogenerators. PURPA provides a number of benefits to those cogenerators who can become qualifying facilities (QFs) under this act. Specifically, PURPA:

- Requires utilities to purchase the power made available by cogenerations at reasonable buy-back rates. These rates are typically based on the utilities' cost.
- Guarantees the cogeneration or small power producer interconnection with the electric grid and the availability of backup service from the utility.
- Dictates that supplemental power requirements of

cogeneration must be provided at a reasonable cost.

- Exempts cogenerations and small power producers from federal and state utility regulations and associated reporting requirements of these bodies.

To assure a facility the benefits of PURPA, a cogenerator must become a qualifying facility. To achieve qualifying status, a cogenerator must generate electricity and useful thermal energy from a single fuel source. In addition, a cogeneration facility must be less than 50% owned by an electric utility or an electric utility holding company. Finally, the plant must meet the minimum annual operating efficiency standard established by the Federal Energy Regulatory Commission (FERC) when using oil or natural gas as the principal fuel source. The standard is that the useful electric power output, plus one half of the useful thermal output, of the facility must be no less than 42.5% of the total oil or natural gas energy input. The minimum efficiency standard increases to 45% if the useful thermal energy is less than 15% of the total energy output of the plant.

Natural Gas Policy Act

The Natural Gas Policy Act created a deregulated natural gas market for natural gas, which was the major objective of this regulation. It provides for incremental pricing of higher cost natural gas to fluctuate with the cost of fuel oil. Cogenerators classified as qualifying facilities under PURPA are exempt from the incremental pricing schedule established for industrial customers.

Public Utility Holding Company Act of 1935

The Public Utility Company Holding Act of 1935 authorized the Securities and Exchange Commission (SEC) to regulate certain utility "holding companies" and their subsidiaries in a wide range of corporate transactions.

The utility industry and would-be owners of utilities lobbied Congress heavily to repeal PUHCA, claiming that it was outdated. On August 8, 2005, the Energy Policy Act of 2005 passed both houses of Congress and was signed into law, repealing PUHCA—despite consumer, environmental, union, and credit-rating agency objections. The repeal became effective on February 8, 2006.

20.10 SUMMARY

The dynamic process of revisions to existing codes, plus the introduction of new legislation, will impact the energy industry and bring a dramatic change. Energy conservation and creating new power generation supply options will both be required to meet the energy demands of the twenty-first century.

CHAPTER 21

NATURAL GAS PURCHASING

CAROL FREEDENTHAL
JOFREEnergy Consulting
Houston, Texas

21.1 PREFACE

It's easy to understand why natural gas is an important fuel. Supplies are readily available, and it's clean burning, requires no storage, and is relatively cheap compared to other petroleum products. If fuel requirements allow it and supply sources are in the proximity, natural gas is the ideal fuel. The sources and systems for buying natural gas have increased many ways since federal decontrol concluded in the 1900s.

The purpose of this chapter is to give the fuel buyer, new or experienced, in any operation or industry, the knowledge and information needed to buy natural gas for fuel, or feed stock for chemical operations, at the lowest possible cost and highest security of supply. Knowledge of the industry is a must for getting secure supplies at the lowest possible prices. The buyer may be a large petrochemical plant where natural gas is a major raw material, the commercial user having hundreds of apartments needing gas for heat and hot water, or a plant where the gas is used for process steam and power. It might be a first time experience or an on-going job for the buyer. This chapter gives the background and information to get natural gas supplies and includes, additionally, information on industry history, supply sources, transportation, distribution, storage, contracts, regulatory and financial considerations.

Buying natural gas can be as simple as going to a single marketer for the gas and services needed for delivery. Or—depending on the amount of fuel required, the state where the fuel will be used, and the financial sophistication of the buyer—it can involve dealing with natural gas marketers ranging from producers where gas is essentially bought at the wellhead to a local distribution company, dealing with individual pipelines and others for delivery, and even working with financial people to hedge the risk of buying a volatile commodity. In the decontrolled environment of today's industry, the permutations are many for the knowledgeable buyer to get the best prices and security of supply.

The first edition in 1993 was written just as the natural gas industry was going through its evolution. It went from a government-regulated industry to a strongly market-dominated business. The major industry changes are in place, but a continuing evolution is still occurring since the energy business itself is going through its own metamorphosis. Latest changes are included in this revision.

The changes of going from a federally regulated, price-controlled business to an economically dynamic, open industry were tremendous. Major changes include the ability of anyone to be a natural gas buyer or seller, something impossible under the old system. In addition to trading physical quantities of natural gas, there are now financial markets where gas futures and other financial instruments can be bought and sold. The impact of computerized systems and online trading has had significant effect on the business. In addition to supplies coming from certain areas of the U.S. as recovered resources from the ground, imports by pipeline and as liquefied natural gas (LNG) are playing bigger and bigger roles. Environmental concerns and the fear of global warming are also a factor in the natural gas business. This is truly a new business.

A new business today, but coming from an old business that dates back over 150 years has meant the revision and changing of almost every aspect of the industry. Newly "reformed" energy companies with new marketing organizations to sell and market gas, new systems impacting gas trading and marketing, and even a new industry structure makes it necessary to start from scratch in revising this chapter. This revision, completed in the fall of 2008, comes at a time the total energy business is in transition.

Lead by significant increases in the price of crude oil, the number one source for U.S. energy demand, all energy and commodity products have seen increased prices way above expected incremental growth. Crude oil prices in 2008 increased almost a hundred percent in the last 12 months reaching a high of around $147 per barrel in July 2008 for West Texas Intermediate (WTI). At the Henry Hub market center, natural gas prices went from around $6 per million British thermal units (MMBtu) in September 2007 to over $13/MMBtu. And even though prices fell off for both fuels significantly in a few months later in the year, the exact impact to the consumer and the country's economy as well as the world, is yet unknown. Change will continue in all parts of the energy business to cope with the severe changes caused by increased prices as companies and businesses try different strategies.

A few things are sure. Natural gas is a major fuel for stationary electrical power generation in the United States. Gas trails coal, which supplies over 50% of the fuel for electric generation. Coal use for power generation has reached record levels in recent years, but environmental concerns and the high capital required for new coal burning generating plants might reduce coal's market share. The public's dislike of nuclear power and the high costs to build plants with the desired safety has meant no growth for this industry into the early 2000s. With the increased costs of fossil fuels and the dependency on imports, a new philosophy is developing which recognizes the importance of nuclear power to the country's economy. Increased safety features and the environmental benefits offered by atomic power will open new horizons for atomic use and could impact natural gas use in future years.

Natural gas is the best and, in most locations, the most economic fuel for space heating whether it is residential, commercial, or industrial. For many other heating uses, natural gas is the best fuel. This will not change and only means more and more natural gas will be needed as the country's population increases.

The most important factor for growth of the natural gas business, as with all the energy industries, is the profit motive. It is still the driving force of industries today, even though severely challenged by social and environmental issues. Economics will govern change and remain the basis for decision-making. All the transformations—buying and selling of companies, new marketing companies, new systems for handling the merged assets, etc.—are subject to one metric. Is it profitable?

21.2 INTRODUCTION

Natural gas is predominately the chemical compound "methane," CH4. It has the chemical structure of one carbon atom and four hydrogen atoms and is the simplest of the carbon-based chemicals. In the early days, gas was used for illumination only. No longer used for lighting except for ornamental purposes, natural gas is now used for heating and, to a much lesser extent, for cooling of homes, offices, schools, and factories, as well as for generating electricity and transportation. In addition to fuel uses, natural gas is a major feedstock in the chemical industry in making such products and their derivatives

as ammonia, methanol and many plastics. Natural gas is used in refining and chemical plants as a source for hydrogen needed by these processing operations. Through the reforming process, hydrogen is stripped from the methane, leaving carbon dioxide, which also has usefulness in chemical manufacturing. In addition, carbon dioxide is used for cooling, "dry ice," in carbonated drinks, and in crude oil recovery. A potential large use of natural gas will be as a source of bulk hydrogen for transportation use.

The term "natural gas industry" includes the people, equipment, and systems used to run and operate the business. The natural gas industry starts with exploration in many parts of the U.S. to find areas where natural gas might lie below the surface of the earth in porous beds. From these target locations, wells are located and drilled. Natural gas is produced from gas only wells and from crude oil wells, where it is a by-product of the oil production. Tasks such as gathering, treating, and processing are included in the field operations. Pipelines called "gathering systems" are used to transport the gas from the producing fields to collection points or to interstate or intrastate pipelines to further carry the gas to storage or distribution systems that deliver the gas to the consumer for consumption at the burner tip.

The burner tip might be in a boiler, hot water heater, combustion engine, or as a feedstock for a chemical reactor, to name a few of the many uses of natural gas. In addition to the field production, pipeline transportation, and delivery functions, there is a whole "backroom" of marketing and sales personnel, accounting and bookkeepers, and other financial operatives, as well as general management personnel. In addition to the physical market for natural gas, a complimentary financial market operates affording various instruments for the financial trading of natural gas. All of this makes up the natural gas system.

The gas industry is the oldest utility except for water and sanitation. In the middle of the 19th century, many large cities used a synthetic gas made from passing steam over coal to light downtown areas and provide central heating systems. Big cities like Baltimore, New York, Boston, and many other cities and municipalities used gas for illumination. Many utilities from that period exist today and are still gas and electric suppliers in the areas they serve.

In the early days of the gas business, there was no natural gas, as known today. The synthetic gas they produced, sometimes called "water gas" because of the method of producing it, had many bad attributes. Water gas contained a high amount of hydrogen and carbon monoxide, two bad actors for a gas used in homes, businesses, and factories. The carbon monoxide was poisonous. People die when exposed to it. The hydrogen in the water gas made it very explosive, and many buildings were destroyed when gas from leaks or pipe ruptures was ignited. When what is known today as natural gas came on the scene in the early 1900s, where it was available, it quickly replaced the old manufactured gas.

About the same time, advances were made in electricity, so cities and municipalities changed to electricity for lighting and illumination. Natural gas quickly lost its market for municipal lighting. Even with the loss of these markets, natural gas was an important source of energy for the growing nation. What was originally an unwanted by-product from the oil fields now was a major product of its own. In the early day, the problem was getting rid of the gas at the well site. Flaring was used, but this was a waste of good natural resources. Around the beginning of the century, associated gas from Ohio oil fields was shipped to Cleveland in wooden pipes to replace the then used synthetic gas.

In the early days of the industry, the limitations to greater uses of natural gas were that gas was produced in only certain parts of the country and transportation was available for only very short distances. Market penetration was thwarted by the ability to ship it. Natural gas, from the time it is produced at the well until the time it is consumed at the burner tip, is always in an enclosed container, whether it be a pipe or storage facility. In the beginning of the industry, there were no long distance pipelines. Natural gas was a great replacement for the synthetic counterpart. Methane is essentially safe as far as toxicity and is much safer as far as explosion. Gas' growth was dependent on building long distance pipelines to move the gas from production to consumption.

Not until the 1930s did the industry have the capability of making strong enough, large diameter steel piping needed for the long-distance pipelines. Completion of major interstate pipelines to carry gas from producing regions to consumers was the highlight of a period from the 1930s to the start of World War II in the early 1940s. Pipeline construction came to a halt and was dormant until the war's end. Construction went full force after the war to insure delivering the most economical and easiest fuel to America's homes, commercial facilities, and industrial players. There are still some areas of the U.S. where fully developed natural gas distribution and delivery system are non-existent.

Natural gas is produced by drilling into the earth's crust from onshore or offshore locations, anywhere from a couple of thousand feet to five miles in depth. Once the gas is found and the well completed to bring the gas to the earth's surface, it is treated if necessary to remove acid impurities such as hydrogen sulfide and carbon dioxide. Again, if necessary, it is processed to take out liquid hydrocarbons of longer carbon chains than methane (with its single carbon atom), that might be in the gas stream. These are called "liquid petroleum gases (LPG) or simply "natural gas liquids" (NGL). These are marketed separately from natural gas and serve various industries either as end products or raw materials. Making up the LPG basket are ethane, propane, butanes, and other higher carbon products.

After processing, the gas is transported in pipelines to consuming areas for either direct delivery to a consumer or to local distribution companies (LDC), which handles the delivery to the specific customer, whether it is a residential, commercial, industrial, or electric generator user. In addition to the people and companies directly involved in the production, transportation, storage, and marketing of natural gas, there are countless other businesses and people involved in assisting the gas industry to complete its tasks. There are systems companies, regulatory and legal professionals, financial houses, banks, and a host of other businesses assisting the natural gas industry.

Figure 21-1 shows the many parts of the industry as it is known today.

The just under $200 billion natural gas industry (2007 prices and consumption) only represents the functions and services to get natural gas, the commodity, to market for consumption. Not included in the overall industrial revenues are the moneys generated by the sales and resale of gas before its consumption, the processing and marketing of natural gas liquids coming from the gas, and the financial markets where gas futures and other financial instruments are sold and traded. These are big businesses also. Estimates are that the physical gas is traded three to four times before consumption. In the financial markets, gas volumes 10 to 15 times the amount of gas consumed on an average day are traded daily.

Natural gas goes through many stages after it is produced in the field and goes from the wellhead to the burner tip. It can travel many alternate paths, almost always through pipelines, before coming to its end use as a fuel in various types of equipment or as feedstock for chemical manufacturing. Each one of the stages in its travel gives it an added value. Raw gas coming from the wellhead many times has the quality to go directly into a transporting pipeline for delivery to the consuming area. Sometimes the gas before leaving the field needs treating and/or processing to meet pipeline specifications for acceptance into the pipeline for transportation. Once in the pipeline, the gas can go directly to a consumer, to a "market hub" for transportation, or to the city-gate where a local distribution company will take it to the local con-

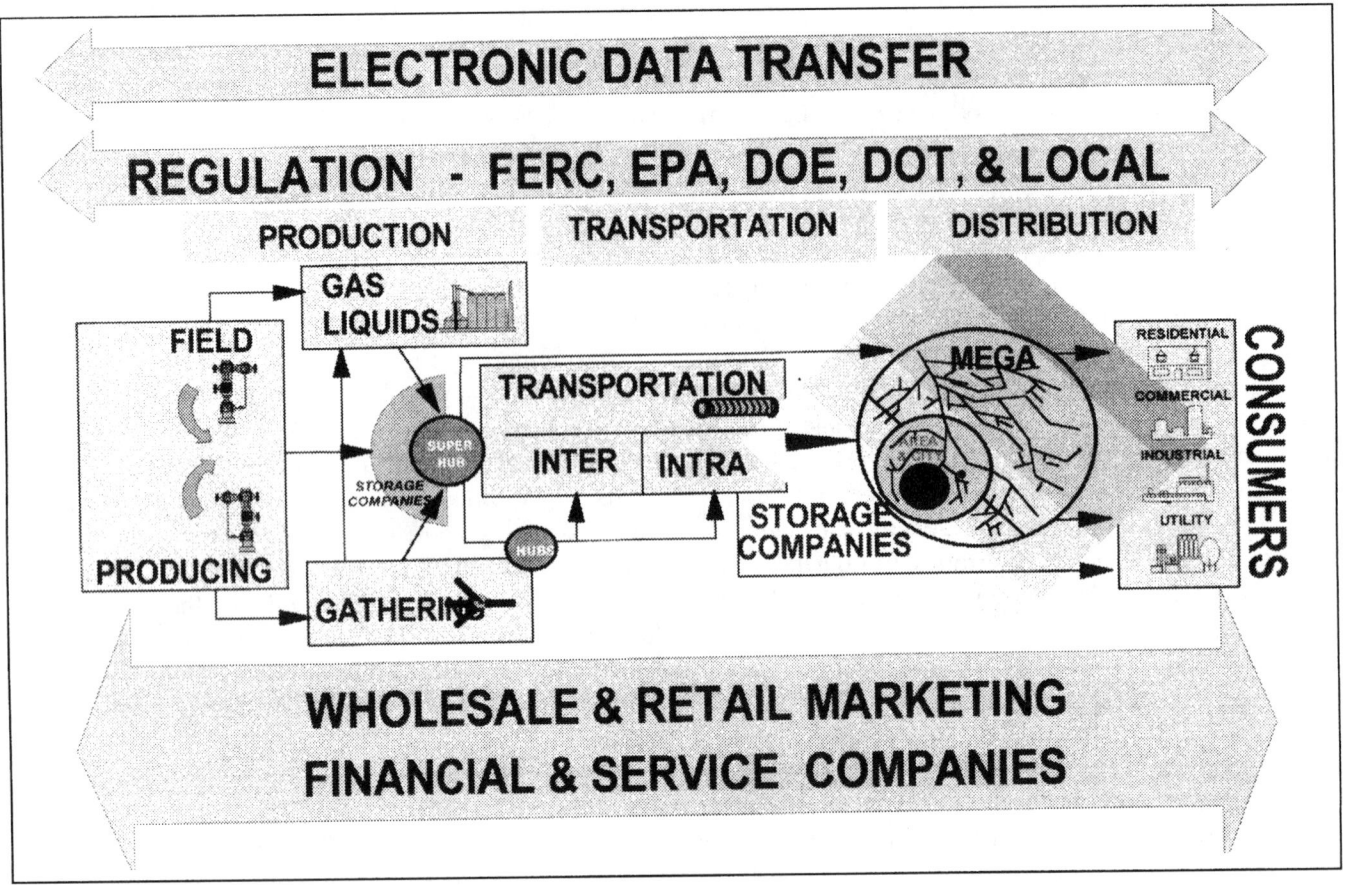

Figure 21-1. Natural Gas Industry Flowsheet.

sumer.

Some areas in the U.S. West where population is sparse, parts of the Northeast where oil prices were too competitive for many years to delivered gas prices, and some other parts of the country where distribution systems cannot be justified based on the market are still deficient in natural gas supplies. Where natural gas is not available, what is called "bottled gas," a mixture of propane and butane or propane only, is used for home heating and other critical uses. In recent years, new natural gas supplies and pipelines were developed to bring natural gas to the northeast U.S. from Canada. Additional distribution systems will bring more gas to more customers throughout the country, from the tip of Florida to the North Central and Northwestern states.

Ever since natural gas became available for fuel, it has been under some form of government economic control. Through government control of pricing passed in the Natural Gas Act in the mid 1930s, the government had the power to make gas prices more or less attractive to competing fuels. Further, with the government controlling wellhead prices and being very slow to make changes in prices as conditions changed, it became difficult and economically undesirable to expand natural

gas production. Government price controls hampered the growth of the U.S. natural gas business. Even as late as the mid-1970s, government price controls stifled expansion and growth just when the country needed new energy supplies. There was no shortage of gas reserves. The only shortages were incentives for producers to develop and supply the gas, because the government dictated low prices. The free market builds its own controls to foster competition and growth; and government controls did nothing but impede market growth.

Congress passed the Natural Gas Policy Act of 1978 (NGPA) to change the government's economic control of the industry and make the gas business a more market oriented, responsive business. A number of years of transition were needed before significant changes began in the industry. Real impact started in 1985 and ended in 1992 with passage of FERC Order 936. The federal decontrol specifically allowed wellhead gas pricing to be market sensitive, interstate pipelines to be transporters only, and open marketing to be available to all parties. Gas at local levels where the state public utility commission or similar local government has control is still heavily regulated except in a few states.

Decontrol at the federal level has slowly filtered

down to local agencies. In the turn of the century, some states began moving to "open transportation" rules. The transition has had many obstacles and is currently somewhat in limbo. An obstacle to the swifter implementation of rules at the state and local levels was the tie of gas and electricity as utilities within state regulatory control. With the electric industry going through its own "decontrol," many wanted to see the much larger electric industry work out the utility problems first. Then gas could follow with less negotiating and discussion. But, the electric timetable has had its own obstacles and is now years behind its planned evolution. Much is still planned and needed in electric and gas decontrol at the local levels. Important in buying natural gas is the degree of decontrol in states where the gas will be used. This is discussed later.

With the price of gas changing each year, the total industry value changes. The industry in nominal annual terms was roughly $191.7 B in 2007 at the retail end. Many electric companies were both gas and electricity utilities even before deregulation and in recent years more electric utilities have bought major natural gas pipelines or gas distributors to expand their utility business. Large electric companies bought into the natural gas industry purchasing transporters, distributors, and marketing companies. Interestingly, in a relatively few years, some of these combinations have come apart because of poor profitability.

Surprisingly few of the expanding companies have sought to buy into the initial phase of the gas business, the oil and gas exploration and production companies (E&P companies). These are the companies looking for natural gas and then producing it. While all of the transporting companies (whether long distance or distribution in nature and, further, whether electric and/or natural gas in business) have shied away from the production companies, other E&P companies have merged or acquired smaller operations to add to the total capability of the company. The significant changes in recent decades saw major E&P companies merge and acquire independent E&P assets.

21.3 NATURAL GAS AS A FUEL

This section will cover in general, the use of natural gas as a fuel. Natural gas demand, supply, transportation, economics, environmental, and regulatory changes will be covered in general. Later in this chapter, when discussing "Buying Natural Gas." more specific discussion will be made on certain areas covered here in general. This section also includes definitions of terms used in discussing the natural gas business.

21.3.1 Definitions & Abbreviations

Certain terms are unique to the natural gas and energy business. A partial listing of important terms and abbreviations are listed below.

Btu—A measure of heat as produced by burning a fuel. Based on the English units of measure, it is the amount of energy (as heat) needed to raise a pound of water one degree at 60 degrees F at standard atmospheric pressure.

Contract—a legally binding agreement between two or more parties, either verbally but more commonly in writing, setting forth the conditions and basis for doing business, such as buying and selling commodities, transporting products, etc.

Cubic foot—A measure of unit volume equal to a container measuring one foot long by one foot high and one foot wide. Since a single cubic foot is relatively small, typically, the units are measured in 1,000 cubic feet (Mcf) or million cubic feet (MMcf). Normally, gas accounted for in cubic measure is referenced to standard temperature and pressure.

Dekatherm—a unit of energy measurement equal to 10 therms, with a therm equal to 100,000 Btu.

DOT—Department of Transportation—government agency responsible for safety of pipeline operations.

EIA—Energy Information Administration—U.S. agency responsible for energy information both domestically and foreign.

EPA—Environmental Protection Agency—government agency responsible for environmental regulation of natural gas production, pipeline operation, and other energy matters.

FERC—Federal Energy Regulatory Commission—government agency responsible for economic regulation of interstate natural gas pipelines and other matters relating to natural gas.

Gas—Any substance that moves freely to fill the space available.

Interstate Pipeline—A pipeline that moves materials from one state to another; crossing state lines in the transportation of goods.

Intrastate Pipeline—A pipeline for carrying goods operating within the boundaries of a given state.

LDC—Local Distribution Company—A pipeline for receiving natural gas at the city gate for transmission to end-users.

Marketer—One who buys or sells a product.

Market Hub—A relatively centrally located terminal where various pipelines can deliver or pick up gas for further transportation and affording various services to buyers and sellers. (See section 21.3.4 for various function afforded by market hubs and further definitions of transportation terms.)

Natural Gas—A fuel in the form of a gas that is produced from reservoirs below the ground's surface, mainly consisting of the chemical compound methane—CH4.

Tariff—A set of fixed charges prepared by a regulatory agency for the movement of goods from one location to another.

Therm—A unit of measure equal to 100,000 Btu.

21.3.2 Demand

Why has natural gas grown in popularity? What makes it a fuel of choice in so many industries as energy sources come under critical review from economic and environmental critics? What shortcomings does it have? The table below shows the change in mix for basic fuels used in the U.S. for the period 2003 through 2007.

Nuclear, which started in 1960, enjoyed a period of rapid growth. The high costs for all the safety engineered into the plants made it an uneconomical system towards the end of the century. There are no nuclear plants scheduled for construction at this time. Smaller supplies of available fossil fuels and increased demand as the world economy expands have brought new interest in atomic

power. A country like France uses it almost totally for power generation, while in the U.S. it accounts for only around 20% of the electricity generated.

Nuclear generation of electricity has changed since the initial surge of plant construction. Existing plants and those still under construction have been made more efficient and safer. A new era of nuclear plants will become a necessity if the country really wants to achieve energy independency. Expanding nuclear-supplied generation would free up gas supplies and reduce the need for coal-fired generators, helping to reduce pollution.

Coal usage in the U.S. has grown in recent years, with it being the major fuel for electric generation. Record coal production occurred in the late 1990s. Coal commands slightly over 50% market share of all fuels currently used for electric generation. It has many negative properties, like the need for railroads for its transportation, high pollution from the burner after-products, and poor handling characteristics, including being dirty, losses on storage, and the difficulties of moving a solid material. Also, disposal of the remaining ash from the burnt coal is a major problem. Still, coal has a few things going for it that will keep it in use for many years to come. The ready availability and abundance are major merits. The stability of low coal prices, at around $1-$2 per million Btu, will always give coal a place in the market. Its domestic availability helps keep U.S. money in the country.

Coal will slowly lose position because of its disadvantages of pollution, higher costs to meet changing standards, and high capital costs for building new generating plants. Petroleum products have lost market share for electric generation in later years because of their costs and the dependence of the U.S. on foreign suppliers for oil products. Oil products used for electric generation include distillate fuel oil, a relatively lightweight oil, that during the refining process can have most of the sulfur removed. Low sulfur fuels are desirable to keep emissions low for environmental reasons.

The other major oil product used in electric generation is residual fuel oil, the bottom of the barrel from the refining process. This is a heavy, hard to transport fuel

Table 21-1. U.S. primary energy consumption by source 2003-2007, quadrillion Btu.

YEAR	FOSSIL FUELS				NUCLEAR	RENEW'L	TOTAL
	COAL	NAT'L GAS	PETROLEUM	TOTAL	NUCLEAR	RENEW'L ENERGY	TOTAL
2003	22.321	22.897	38.809	84.078	7.959	6.150	98.209
2004	22.466	22.931	40.294	85.830	8.222	6.261	100.351
2005	22.797	22.583	40.393	85817	8.160	6.444	100.506
2006	22.447	22.191	39.958	84.657	8.214	6.922	99.856
2007	22.776	23.639	39.773	86.214	8.415	6.835	101.570

Source: EIA September 2008

with many undesirable ingredients that become environmental problems after combustion. Many states have put costly tariffs on using residual fuel oil because of its environmental harm when used.

Natural gas is the nation's second largest source of fuel for electric generation in recent years, as well as the fuel for space heating whether in residential, commercial or industrial applications, and it is a major source of feedstock for chemicals manufacturing. Plentiful supplies at economical prices, a well developed delivery system of pipelines to bring gas from the production areas to the consumer, and its environmental attractiveness has made natural gas the choice of fuel for many applications. Natural gas will remain a popular fuel even though in recent times it too has seen significant price increases. As a fuel for heating and generating electricity, and as a feedstock for chemicals, natural gas is very attractive.

For residential and commercial applications, the security of supply and efficiency in supplying makes it an ideal fuel. Even though natural gas is a fossil fuel, it has the lowest ratio of combustion-produced carbon dioxide to energy released. Carbon dioxide is believed by many to be the biggest culprit in the concern on global warming. Natural gas consumption data are followed in five major areas by the Federal Energy Information Administration (EIA)—residential, commercial, industrial, electric generation, and transportation. In addition, it lists the data for natural gas used in the fields for lease purposes and plant fuel, and as fuel for natural gas pipelines.

Natural gas demand has always, in modern times, determined the amount of gas produced, except for the surplus produced for storage so that sufficient gas will be available during the peak demand winter season. The mid 1970s was a time when the country experienced a severe natural gas supply shortage. In those years, while there were more than sufficient reserves in the ground to meet demand, the control of gas pricing by the federal government stymied the initiative of producers to meet

demand. Potential supply was available, but the lack of profit incentive prevented meeting demand in those years. Demand increased because of changes and shortages in crude oil supplies. The 1970s were the start of the change in crude pricing, with the formation of the Oil Producing Export Countries (OPEC) and the raising of crude oil prices significantly. The U.S. was faced with decreased supplies from foreign producers. Crude prices doubled almost overnight, but because natural gas was price controlled and could not meet the rising prices, supplies in the interstate market suffered.

The major market for natural gas in 2007 was for electric generation, 29.8%. Industrial was close, taking 28.8% of the total natural gas demand. Residential was next with 20.5% of the total 23,056 billion cubic feet consumed. Table 21-2 below gives natural gas consumption by market for 2003-2007.

The residential market is basically for home heating and hot water fuel. The commercial market is pretty much the same, space heating and hot water. Natural gas consumed in industrial plants for space heating is included in this category. The industrial category covers all other industrial uses of natural gas in industry except for power generation. Having started in 2000 power generation in industrial locations is now included in the category of electric generation.

Weather plays a major role in demand for all categories. Residential and commercial consumption are almost entirely affected by weather since these two categories reflect space and water heating. To a much lesser degree, because of the amount of gas used for this purpose, water heating is also impacted by weather. Electric generation is weather sensitive too since, in addition to the load for heating in the winter, the summer electric demand is highly responsive to air conditioning needs in hot weather. Even though the industrial load is not as sensitive to weather as are other categories, it does reflect the additional heating load needed for the process industries

Table 21-2. Natural gas consumed in U.S. 2003-2007, billion cubic feet.

YEAR	LEASE & PLANT FUEL	PIPE-LINE & DISTRI-BUTION USE	DELIVERED TO CUSTOMERS						TOTAL CON-SUMP-TION
			RESID-ENTIAL	COMM-ERCIAL	INDUS-TRIAL	ELEC-TRIC POWER	VEHICLE FUEL	TOTAL	
2003	1,122	591	5,079	3,179	7,150	5,135	18	20,563	22,277
2004	1,098	566	4,869	3,129	7,243	5,464	21	20,725	22,389
2005	1,112	584	4,827	2,999	6,597	5,869	23	20,316	22,011
2006	1,124	584	4,368	2,836	6,496	6,222	25	19,945	21,653
2007	1,168	622	4,724	3,005	6,636	6,874	26	21,265	23,056

Source: EIA/Natural Gas Marketing August 2008

when temperatures fall and raw materials, including process air and/or water, are much colder.

Demand growth in natural gas is dependent on many factors besides increased use as population grows. The biggest growth potential is in electric generation, which continues to grow in the U.S. While some have forecast 30-40% increase in demand of gas for electric generation, its growth is dependent on many things. First, it could be the choice fuel to replace nuclear plants that currently may not be re-certified as they age. However, there are significant indications of the rebirth of nuclear generation. This is an unknown which at the current time is hard to evaluate since political pressure plays as big a role as technical and economic forces.

As coal plants age and need replacement or are replaced for environmental concern, natural gas could be the replacement. It is easier to get to the plant and to handle in the plant. The environmental needs are much smaller, and this and other elements play a role in the capital required for the generating plant and facilities. Natural gas is the fuel of choice among the fuels currently available, including some of the renewable fuels like wind, solar, etc.

Even if the electric systems in effect currently were to change to more "distributive" in nature, such as fuel cells or small, dual cycle gas turbines, natural gas would be the fuel. Planners see fuel cells or turbines being used by residential and commercial units so that each could have its own source of electricity. When additional power is needed, they would draw it from the utility lines. When the fuel cell produces more than needed, the utility would take the excess. Most fuel cell work today involves hydrogen and oxygen as the combined fuels for operation. Natural gas could be the source of hydrogen. Since many homes already have natural gas piped to the house, it would be easy to handle this new fuel to make electric power locally.

21.3.3 Supply

Natural gas is a product coming from the earth. As discussed previously, the major component of natural gas is the chemical compound methane, CH4. Methane is the product formed when organic matter like trees, foliage, and animal matter decays without sufficient oxygen available to completely transform the carbon in these materials to carbon dioxide. The theory is that natural gas deep in the ground is a product of decaying material from the past millions of years of Earth's history. Chemical elements available as the matter decayed gives the methane such contaminants as hydrogen sulfide, carbon dioxide, nitrogen, and many more compounds and elements. Natural gas comes from shallow depths as little as a few thousand feet into the earth and as deep as 20 to 25

thousand feet—almost five miles into the earth's crust.

Natural gas wells are drilled on dry land and on water-covered land such as offshore locations in the Gulf of Mexico. Current drilling in the Gulf of Mexico deep waters is in water depths up to around 3,000 feet. Federal and state laws currently prohibit additional drilling on the East and West coasts. With the increased demand for fuels and so much of the U.S. demand being met by imports, various measures are being proposed to open up all locations to further exploration and production, including some parts of Alaska currently not tapped.

Natural gas quantities are measured using two sets of units; one is a measure of volume and the other is a measure of energy. The volume of the gas at standard conditions is one measure. At standard conditions of temperature and pressure, the number of cubic feet of natural gas is one way to define the amount of gas. Since a cubic foot is a relatively small volume when talking of natural gas, the usual term is a thousand cubic feet (Mcf). As a volume measurement, the next largest unit would be a million cubic feet (MMcf), which is a thousand, thousand cubic feet. A billion cubic feet is expressed as Bcf and a trillion is Tcf.

Since natural gas is not a pure compound but a mixture of many products formed from decaying organic matter, the energy content or heat content of each cubic foot at standard conditions is another method of measuring natural gas quantities. The energy units used in the U.S. are British thermal units (Btu), the amount of heat needed to raise a pound of water one degree Fahrenheit at standard conditions of pressure at 60 degrees Fahrenheit. Again, since a single Btu is a relatively small measure, the amount of energy is expressed as a million Btu or MMBtu. A cubic foot of pure natural gas at standard conditions would have a thousand Btu (MBtu).

For comparison, a barrel of crude oil, which is 42 gallons, is typically 6 MMBtu for a grade like West Texas Intermediate, a relatively clean and mid-weight crude oil. Thus a barrel of crude oil is equivalent to about six thousand cubic feet of pure natural gas. Crude oil energy content varies by source, with the lighter grades and thinner oils having less energy content per barrel, and the heavier and many times, "dirty oils" having more energy per barrel.

Gas coming from wells can range from very low heat contents of 200 to 300 Btu/cf (because of non-combustible contaminants like oxygen, carbon dioxide, nitrogen, water, etc.) to 1500 to 1800 Btu/cf. The additional heat comes from liquid hydrocarbons of higher carbon content entrained in the gas. The higher carbon content molecules are known as "natural gas liquids" (NGLs). Also, other combustible gases like hydrogen sulfide contained in natural gas can raise the heat content of the gas produced. Since a cubic foot of gas can have a varying heat content, most gas sales today are done on an energy or Btu basis.

Data from the EIA show an "average" cubic foot of gas produced in the U.S. as dry natural gas in recent years would have an average of 1,028 Btu/cf. Gas coming from a well having a different heating value is treated and/or processed to remove the contaminants. This is done to lower or raise the Btu quantity per cubic foot to meet pipeline specifications for handling and shipping the gas. Pipeline quality natural gas is 950 to 1150 Btu per cubic foot.

A frequently used term to describe the energy content of natural gas when sold at the local distribution level, such as residential, commercial or small industrial users, is the "therm." A therm is equivalent to 100,000 Btu. Ten therms would make a "dekatherm" (Dt) and would be equivalent to a million Btu (MMBtu). The therm makes it easier when discussing smaller quantities of natural gas.

When exploration and production companies search for gas in the ground, they refer to the quantities located as "reserves." This is a measure of the gas the companies expect to be able to produce economically from the fields where signs of gas were found. Through various exploration methods, from basic geophysical studies of the ground and surrounding areas to the final steps of development wells, more accurate pin-pointing of reserve volumes is achieved. Reserves are the inventory these companies hold and from which gas is produced to fill market needs. In November 2007, the U.S. Department of Energy showed that U.S. natural gas reserves in 2006 had increased 3% that year, rising to over 211 trillion cubic feet, the highest level since 1976. The new reserves replaced 136% of the dry natural gas produced in 2006. In 2008, another group, the American Clean Skies Foundation, said U.S. reserves of natural gas were 2,247 Tcf, which would give 118 years of supply at 2007 demand level.

Without any further replacement by new reserves, this would be a 6- to 7-year life of existing reserves at current consumption rates based on the DOE estimate. U.S. exploration and production companies are continuously looking for new reserves to replace the gas taken from the ground for current consumption. From 1994 to 1997, producers found reserves equal or more in volume to gas produced during that year. Reserve volumes from areas where gas is already being produced represent a very secure number for the amount of gas thought to be in the ground and economically feasible to produce. These are called recoverable reserves based on produced and flowing gas.

The next level of measuring reserves is gas held behind these recoverable, producing reserves. These are a little less secure and a little more speculative but still have a good chance of producing as designated. Using this category, just for the U.S. there are enough gas reserves for 25 to 35, years depending on the amount consumed each year. There are abundant gas reserves in North America to assure a steady supply for the near term and the future.

In addition to the two levels of gas reserves discussed previously, there is an additional category of "possible" or "potential" reserves. These become more speculative but are still an important potential supply for the future. Some of these may become more important sooner than expected. Good examples are the gas supplies coming from coal seam sources and from shale deposits. Considerable gas is produced from both of these sources and has added considerable gas supply to domestic production in recent years. Additional potential supplies (but with long lead times for further development) are gas from hydrates and gas from sources deeper in the Gulf Coast.

In addition to the domestic supply, imports play a big role. Currently, little gas is available from Mexico, and some U.S. gas goes to Mexico. As gas demand and prices increase, Mexico could play an important role as an U.S. supplier. Considerable amounts of gas come from Canada. Pipeline imports have grown significantly from the 845Bcf imported in 1985. Canada supplied in 2007, 3,783 Bcf, or16.4% of U.S. total demand. Canada did much in the late 1990s and early 2000 to expand the pipeline systems bringing gas to the U.S. Most Canadian production is in the provinces of Alberta and British Columbia. New production came from the eastern coast late in the last century and was imported into the U.S. from the maritime provinces. Canadian gas makes up a significant portion of the gas going to the U.S. northeast and the west coast. Major importing locations for gas coming into the U.S. from Canada are on the west coast, at mid-continent near Chicago, and on the east coast.

Canadian imports will diminish in future years as Canada uses more and more of its own supply; natural gas is important to Canada for production of crude oil from tar sands. Imports from Mexico (about 2% in 2007) and Canada are considered pipeline imports.

21.3.3.1 Liquified Natural Gas—LNG

Imports from other countries into the U.S. are transported as liquefied natural gas (LNG). Natural gas in the producing country is cooled to about minus 260 degrees F and compressed until it is liquid. The reduction in volume is roughly 600 times less than the original volume. The liquefied gas with its reduced volume is then economically sized for shipping worldwide and is transported between countries in large vessels that are cryogenically insulated, floating containers. The LNG is received at terminals in the U.S. and worldwide, where it is re-vaporized to gas.

During this step, large quantities of refrigeration are available from the expanding liquid to gas. The cooling "energy" is sold and used in commercial applications to recoup some of the costs in making the gas into LNG. There are currently six LNG terminals in the continental U.S. for receiving and handling LNG. These are in Boston, MA; Lake Charles, LA; Baltimore, MD; off the coast of Georgia at Elba Island; Sabine Pass, TX; and Gulf Gateway, LA. Another terminal is in Puerto Rica.

The Baltimore and Georgia locations were idled years ago when natural gas prices would not justify LNG imports. LNG in world markets is priced in comparison to crude oil prices, while in the U.S. the LNG is priced based on the U.S. market. This leads to a considerable unbalance in pricing between the U.S. and other international LNG buyers. In recent years, European and Asiatic price were in the teens while, with the exception of a short period in mid-2008, U.S. prices were less than $10/MMBtu. This has prevented U.S. LNG supplies from increasing as originally planned. Total LNG imports in 2007 were 770.8 Bcf, or about 17% of total imports and about 3% of total gas demand. This was the record amount of gas imported as LNG to date.

In 2008, European and Asian countries took increased supplies of LNG because of increased demand. European and Asiatic markets get first serve because of pricing as discussed above. Asian and European buyers have paid in the upper teens per million Btu, with a new contract being signed recently between South Korea and Indonesia for roughly $20/MMBtu.

The number of countries supplying LNG to the U.S. increased from few in the early days to six in 2007. Trinidad was the leader, supplying 451 Bcf, followed by Egypt with 115 Bcf and Nigeria with 95 Bcf. While there are seven terminals currently in operation, over 20 prospects are under consideration for future sites to receive LNG.

Natural gas produced from wells where crude oil is the major product is termed "associated gas." Roughly 40% of the gas produced in the U.S. comes from associated wells, while the rest comes from wells drilled specifically for natural gas. The only difference is that the associated wells' gas may contain greater amounts of what has been mentioned previously as "natural gas liquids" (NGL).

These liquids are organic compounds with a higher number of carbon atoms in each of the molecules making up that compound, and they are entrained in the gas as minute liquid droplets. Methane, which is the predominant compound in natural gas, has one carbon and four hydrogen atoms in its molecule. A two-carbon molecule with only hydrogen is called ethane; the three-carbon molecul propane, four-carbon molecule butane, and the fifth pentane. All molecules with more than five carbons are collected with the pentanes and the product is called "pentane plus." It is also known as "natural gasoline," which must be further refined before it can be used as motor fuel. The NGL are removed by physical means either through absorption in an organic solvent or through cryogenically cooling the gas stream so that the liquids can be separated from the methane and each other by distillation. This is done in gas processing plants either near the well sites or at a central field location.

There are markets for the individual NGL products. The ethane is used by the chemical industry for making plastics. Propane is also used in the chemical industry but finds a significant market as fuel. Butanes go to the chemical and fuels market, and the pentanes plus are basically feedstock for motor fuels production from refineries. The overall NGL market is around a $30 billion a year business, depending on the product prices. Prices for NGL vary as the demand varies for each of the specific products, and they have had little relationship to the price of natural gas until recent years when natural gas prices increased to the $10-14/MMBtu range. When gas prices are high and NGL prices are low, profitability on the NGL is very poor. When profitability is poor, the ethane will be re-injected back into the natural gas stream and sold with the gas to boost the heat content of the gas, enhancing its value.

A second difference between associated and gas well gas is strictly of a regulatory nature. Gas from associated wells is produced with no quantity regulations so the maximum amount of crude oil can be produced from the well. Gas from "gas only," wells depending on the state where produced, may be subject to production restrictions based on market, conservation, or other conditions. Major natural gas producing areas in the U.S. are Texas, Wyoming, Oklahoma, New Mexico, and Louisiana. These states, including the offshore areas along the Gulf Coast stretching from Alabama to the southern tip of Texas, account for over 80% of the gas produced in the country in recent years. Table 21-3 gives marketed production of gas for the period 2003-2007.

21.3.4 Transportation

Natural gas in the United States is transported almost exclusively by pipeline. From the moment natural gas leaves the wellhead, whatever route it takes in getting to the burner tip it is through a pipe! Short or long

Table 21-3. U.S. marketed production in selected states and Gulf of Mexico, 2003-2007, billion cubic feet.

YEAR	AS	LA	NM	OK	TX	WY	OTHERS	FED'L G OF MEX	TOTAL
2003	490	1,350	1,604	1,558	5,244	1,539	884	4,406	19,974
2004	472	1,353	1,633	1,656	5,067	1,592	906	3,969	19,517
2005	487	1,296	1,645	1,639	5,276	1,639	910	3,132	18,927
2006	445	1,361	1,609	1,689	5,514	1,816	4,046	2,902	19,382
2007	444	1,327	1,608	1,805	6,093	2,028	4,174	2,771	20,151

Source: EIA/Natural Gas Marketing August 2008

distance, regardless, natural gas is transported in a pipe. The only exceptions are the few times compressed natural gas is transported by truck for short distances. And, in some locations where gas is liquefied (LNG) for storage for use during peak demand times, the LNG is moved by truck also. However, movement of gas through these two means is insignificant in the overall picture of transporting natural gas in the US. Pipeline transportation is basically the only way to move natural gas.

When talking of transporting natural gas through pipelines, there are three main groups of pipelines to be considered:

Gathering System

These are the pipelines in the field for collecting the gas from the individual wells and bringing it to either a central point (sales point or larger collection point called "market center or market hub") for pick up by the long-haul pipeline or to a central treating and/or processing facility.

Long-haul transportation

This is the pipeline picking up the gas at the gathering point or a market center (or if a highly productive well, near a pipeline, from the well itself) and moving the gas to a city-gate for delivery to the distribution company, or to a sales point for a large user where the gas is delivered directly to the consumer. The long-haul pipeline can be either an *interstate pipeline* that crosses from one state into another or an *intrastate pipeline* where the transportation is only within the state where the gas was produced.

The Federal Energy Regulatory Commission (FERC) regulates the economics of interstate pipelines. Operating regulations fall under the Department of Transportation (DOT). The Environmental Protection Agency (EPA) has jurisdiction regardless of the type of pipeline in respect to environmental matters. Even with decontrol, the Federal Energy Regulatory Commission (FERC) is still involved with economic regulation of

the interstate pipelines since these are utilities engaged in interstate commerce. As utilities, the rates for transportation are set through regulatory procedures. The pipeline makes a rate case for presentation to the FERC for authorization to charge the rates shown in the case. The pipeline is allowed to recover all of its costs for transporting the gas and make a return on the invested capital of the pipeline.

Intrastate pipelines are also economically regulated by state agencies with respect to economics. Utilities are granted a license to operate in certain areas and are allowed to make a rate of return on their invested capital. This is different from non-regulated businesses that compete to make profits from the operations.

Local Distribution Companies (LDC)

These are the utilities that take the natural gas from the delivering pipelines at the city-gate for distribution to customers within their franchised area. While many large customers take natural gas directly from a pipeline, even some large users, depending on circumstances and the state involved, also take gas from an LDC. The LDC handles all residential services. Depending on the size of commercial and industrial gas users, and the state where being served, the LDC may be used for direct pipeline sales or service.

For someone buying transportation services from the pipelines, they offer essentially two basic types of rates for transporting natural gas—firm and interruptible. With firm transportation, the transportation buyer is guaranteed a certain volume capacity daily for the gas it wants transported. The buyer is obligated to pay a portion of the transportation charge regardless whether its uses the volume or not on a daily basis. This is called a "demand charge" and is a part of the transportation tariff. The second part of the tariff is the commodity charge, a variable charge the pipeline charges depending on how much gas is transported. Since today's marketing system allows for different types of buying systems, the buyer may pay for the transportation as part of the overall natural gas rate, with the marketer being the one committing with the pipeline for transportation.

Pipelines also offer an "interruptible" tariff where space is on a "first come-first served" basis. Interruptible transportation carries no guarantee to the party buying the transportation that space in the pipeline will be available when needed. The tariff here is usually very close to the commodity rate under firm transportation. Again, the gas buyer may not be the transportation buyer, as the overall gas price could include transportation. A buyer seeking supplies from a marketing company and paying a lump sum price should establish whether the seller has firm or interruptible transportation.

The methodology of the ratemaking procedure used to recover the pipeline's costs and rate of return is such that when a pipeline sells all of its firm transportation, it will make its allowed rate of return. A pipeline can legally exceed its accepted rate of return, based on its handling of the firm and interruptible transportation. Typically, the pipeline has about 80% of its volume contracted in firm transportation. When a firm transporter does not use its full capacity, the pipeline can mitigate the costs to that pipeline by selling its firm transportation to another transporter as interruptible transportation.

The gas buyer at times can use what is called "back hauling" to get a lower rate for gas transportation. An example of this might be gas coming from Canada through a North Central U.S. area such as Chicago. A buyer for this gas might be located in the Southwest, say, in Texas. Rather than ship gas from Chicago to Texas and have to pay the full tariff, a shipper might exchange gas in Texas for the gas to come from Chicago to Texas. In turn, the gas coming from Canada would be sold in the Chicago area as "Texas" gas. Here the shipper would pay the much lower fee for the "paper transportation" of the gas volumes. This would be a back haul arrangement.

The interstate pipeline community is relatively small. Many of the pipelines have merged or been acquired by other utilities since the regulatory changes in the industry took the merchant function from them and made them strictly transporters. There are 20 major interstate pipelines moving gas from the production areas of the country to the consumer. These are owned or controlled by a much smaller number of companies, as there have been many mergers in the industry. The major U.S. interstate pipelines are given in Table 21-4.

Intrastate pipeline companies transport only within the state where the gas is produced. Many of these have miles of pipeline comparable to the interstate systems, but they do not cross state lines. Within the state, these pipelines serve the same mission as the interstate pipelines: bringing the gas from the field, whether the well or gathering point, to the city gate for distribution by the local distributor or directly to large consumers. They also bring gas to sales points for interstate shipment in the interstate pipelines. These pipelines are under state and local regulation for marketing and economics and the federal and state agencies for safety and environmental concerns. Some of the larger ones for the gas-producing states are listed in Table 21-5.

While the pipelines themselves are no longer sellers of natural gas, the buyer should review the pipelines' systems to see if there is a close connection possible so a direct supply might be made from the pipeline to the consumer. In cases where a pipeline is close to a plant or other large user, a marketer or the buyer can make arrangements for the nearby pipeline to bring gas from the producing source or even another transporting pipeline to the facility.

As seen, pipeline transportation might include more than one pipeline to complete the shipment from well to burner tip. Who pays for the transportation at each step is open to negotiation between the gas supplier and the buyer, except when the buyer is buying system gas from an LDC. Usually, the producers are responsible for the gathering and field costs of getting the gas to the transportation pipeline's inlet, which may be on the pipeline or at a terminal point, sometimes designated as a "hub." or "market center." Many times when the transporting pipeline goes through a producing field, the producer will only be responsible for gathering charges to get the gas from the wellhead to the field's central point for discharge into the pipeline's inlet. The gathering and field charges, along with the transportation to the transporting pipeline inlet, are what make the difference between wellhead gas prices and "into pipe" or hub gas prices.

As the industry changed after decontrol and the pipelines totally became transporters, what is known as "marketing hubs" began to grow in importance. These are centers where many pipelines merge or terminate for a given area and where many others take the gas to further destinations—either to a city gate, a consumer, or even to another hub. Currently, there are roughly 30 hubs across the U.S. where pipelines can transfer their supplies to other pipelines. The hubs do more than merely exist as a manifold for gas transfer. They have grown to provide a whole list of services. Much on hubs can be found at the EIA website. According to the EIA, market centers or hubs offer the following services:

Transportation/Wheeling—Transfer gas from one interconnected pipeline to another through a header (hub) by displacement (including exchanges), or by

Table 21-4. Major interstate natural gas pipelines, top 20 by system capacity, 2006 (2,500MMcf/d).

PIPELINE NAME	MARKET REGION SERVED	STATES PIPELINE OPERATES IN	CAPACITY/ TRANSPORTED MMcf/d/Bcf	SYSTEM MILEAGE Miles
Columbia Gas Trans'n Co.	Northeast	DE,PA,MD,KY,NC,NJ,NY,OH, VA,WV,	9,000/1,792	10,318
Transcontinental Gas Pipeline	Northeast, Southeast	AL,GA,LA,MD,MS,NC,NY,SC, TZ,VA,GM	8,161/2,751	10,412
Northern Natural Gas Co.	Central, Midwest	IA,IL,KS,NE,NM,OK,SD,TX,W I,GM	7,200/950	15,743
ANR Pipeline Co.	Midwest	AR,IA,IL,IN,KS,KY,LA,MI,MO, MS,NE,OH,OK,WI,GM	7,129/2,058	10,600
Texas Eastern Transmission Corporation	Northeast	AL,AR,IL,IN,KS,KY,LA,MI,M O,MS,NJ,NY,OH,OK,PA,TX, WV,GM	6,672/1,2411	9,176
Tennessee Pipeline Co.	Northeast, Midwest	AR,KY,LA,MA,NY,OH,PA,TN, TX,WV,GM,AZ,CO,NM,TX,W V,GM	6,329/1,678	14,100
El Paso Nat'l Gas Company	Western, Southwest	AZ,CO,NM,TX	6,152/1,653	10,295
Dominion Trans'n Co.	Northeast	PA,MD,NY,OH,VA,WV	5,934/585	3,392
Natural Gas Pipeline of America	Midwest	AR,IA,IL,KS,LA,MO,NE,OK,T X,GM	4,508/1,709	9,297
Northwest Pipeline Company	Western	CO,ID,OR,UT,WA,WY	4,500/676	3,865
Southern Natural Gas Company	Southeast	AL,GA,LA,MS,SC,TN,TX,GM	3,532/805	7,439
Centerpoint Gas Trans'n Co.	Southwest	AR,KS,LA,OK,TX	3,432/671	6,170
Colorado Interstate Gas Co.	Central	CO,KS,OK,TX,WY	3,170/733	3,979
Texas Gas Trans'n Corp	Midwest	AR,IN,KY,LA,MS,OH,TN	3,098/725	5,609
Great Lakes Trans'n Co.	Midwest	MI,MN,WI	2,958/819	2,115
Gulf South Pln Co.	Southwest	AL,FL,LA,MS,TX,GM	2,946/510	6,532
Panhandle Eastern Pln Co.	Southwest	IL,IN,KS,MI,MO,OH,OK,TX	2,840/579	6,376
Gas Transmission Northwest Corp.	Canada	ID,OR,WA	2,636/805	1,356
Northern Border Pln Co.	Canada	IA,IL,IN,MN,MT,ND,SD	2,626/858	1,400
Southern Star Central Pln Co.	Central	CO,KS,MO,NE,OK,TX,WY	2,526/317	5,725

Source: EIA Oct 2008

Table 21-5. Intrastate pipelines with 1,000 miles of pipeline.

Pipeline Name	Parent Company	Region	State(s) in which it has operations
Central Region			
NorthWestern Energy Co	NorthWestern Corporation	Central	MT WY SD NE
Midwest Region			
Dominion East Ohio Gas Co	Dominion Resources Inc	Midwest	OH
Northern Illinois Gas Co	NICOR Corp	Midwest	IL
Northeast Region			
National Fuel Gas Distribution Co	National Fuel Gas Co	Northeast	NY
Southeast Region			
Southwest Region			
Atmos Pipeline - Texas	Atmos Energy Corp	Southwest	TX
Bridgeline Gas Systems Inc	Bridgeline Holdings LP	Southwest	LA
ET Fuel System	Energy Transfer Partners LP	Southwest	TX
Enbridge Pipelines (East Texas)	Enbridge Energy Partners LP	Southwest	TX
Enogex Inc	OGE Energy Corp	Southwest	OK
Enterprise Texas Pipeline LP	Enterprise Products Partners LP	Southwest	TX
Houston Gas Pipeline System	Energy Transfer Partners LP	Southwest	TX
KM Tejas Pipeline System	KM Texas Intrastate Pipeline Group	Southwest	TX
Kinder Morgan Texas Pipeline Co	KM Texas Intrastate Pipeline Group	Southwest	TX
Louisiana System	Crosstex Energy Services Inc	Southwest	LA
Oklahoma Natural Gas Co	ONEOK Inc	Southwest	OK
Public Service Co of New Mexico	PNM Resources Inc	Southwest	NM
Southern Union Intrastate Pipelines	Southern Union Co	Southwest	TX NM
Westex Pipeline Co	ONEOK Inc	Southwest	TX
Western Region			
PG&E Transmission Co	Pacific Gas & Electric Co	Western	CA
Southern California Gas Co	Sempra Energy Inc	Western	CA

Source: EIA Oct 2008

physical transfer over the transmission of a market center pipeline.

Parking—A short-term transaction where the market center holds the shipper's gas for redelivery at a later date.

Loaning—Another short-term transaction where gas is advanced by the market center to a shipper that is repaid in kind by the shipper a short time later. Also referred to as advancing, drafting, reverse parking, and imbalance resolution.

Storage—Storage is longer than parking, such as seasonal storage. Injection and withdrawal operations may be separately charged.

Peaking—Short-term (usually less than a day and even, hourly) sales of gas to meet unanticipated increases in demand or other shortages experienced by the trader.

Balancing—A short-term interruptible arrangement to cover temporary imbalance situations. The service is often combined with parking and loaning.

Title Transfer—A service where changes in ownership of a specific gas package are recorded by the market center. Title may transfer several times for some gas before leaving the center. The service is an accounting or documentation of title transfers that may be done electronically, by hard copy, or both.

Electronic Trading—Trading systems that either electronically match buyers with sellers or facilitate direct negotiation for legally binding transactions. The market center serves as the location where gas is transferred from buyer to seller. Customers may connect with the hub electronically to enter gas nominations, examine their account position, and access e-mail and bulletin board services.

Administration—Assistance to shippers with the administrative aspects of gas transfers, such as nominations and confirmations.

Compression—Provides compression as a separate service. If compression is bundled with transportation, it is not a separate service.

Risk Management—Services relating to reducing the risk of price changes to gas buyers and sellers; for example, exchange of futures for physicals.

Hub-to-Hub Transfers—Arranging simultaneous receipt of a customer's gas into a connection associated with one center and an instantaneous delivery at a distant connection associated with another center.

Payment for the transportation charges from the pipeline's pick-up to the city gate (LDC transfer point) or user's inlet, even if it includes more than one transporting pipeline, is negotiable between the seller (either a gas producer or marketing company) and the buyer. The marketing company selling the gas might quote a delivered price to the buyer including all transportation charges, especially if the seller is holding transportation rights with the pipeline handling the transportation. If the buyer has transportation rights, he might take the gas FOB (free on board, the point where title transfers and where transportation charges to that point are included in the sales price) at the transportation pipeline's inlet. These are all part of the marketing and negotiating in moving gas from the field to the city gate and/or the consumer.

When buying system gas, such as from an LDC, the transportation and other service charges are part of the sales price. This is the only way the small user can buy and receive gas, but as the size of the user increases the many permutations of buying, transporting, and servicing come into play. Depending on the size of the require-

ments and the state where the gas will be consumed, each buyer should look at the different options to decide which give the best possible price for the delivered gas with the required security of supply.

What are typical prices for transporting natural gas from producing area to consumers in various parts of the country where there is no intrastate gas? The buyer can get detailed information from the pipeline tariffs, which can be gotten from the FERC, the pipeline, and other sources like trade letters and magazines. These are listed under Section 21.3. 6, Information Sources.

Pipeline rates or tariffs are set by the regulatory agencies involved. There is some negotiation possible by the user, based on amount to be transported, firm or interruptible, etc. Still, the gas in different locations will have a value based on market conditions regardless of transportation rates. This is called "basis differential." Some of the trade publications give the different prices for natural gas in different locations based on the basis differential.

For natural gas to be carried in transportation pipelines, it must meet certain conditions of quality and composition. This was previously referred to as "pipeline specifications." These standards include the heating content of the gas per unit volume, i.e. British thermal units per cubic feet (Btu/cf). Gas coming out of the well can range from very low values of 200-300 Btu/cf to over 1,500-1,600 Btu/cf. The lower values come from gas having contaminants like carbon dioxide or nitrogen in the stream, while the higher values come from the gas containing entrained liquid hydrocarbons or hydrogen sulfide.

Contaminants are removed in treating, for the hydrogen sulfide, carbon dioxide and other acid impurities, and in processing for the liquid hydrocarbons such as ethane, propane, etc., as well as other contaminants such as nitrogen, oxygen, etc. Typically, pipeline quality gas will run around 1,000 Btu/cf, with a range of from 950 to 1150 Btu/cf. The exact amount is measured in the stream, as the gas is sold on a Btu basis. Typical specifications for pipeline transmission of natural gas are given in the pipeline's tariff, which is registered with the FERC and can be accessed on line.

Typical pipeline specifications were reviewed for a random group of pipelines based on their tariffs. Specifications were given as follows for the group:

Receipt Point Quality Requirements—All referred to their tariff sheets under category of, "Transportation General Terms and Conditions."

Objectionable Odors, solids, dust, gums—Commercially free for all pipelines observed.

Merchantability—All gas, in the pipeline's judgment, must be free from any toxic or hazardous substance in concentrations, which, in the normal use of the gas, may be hazardous to health, injurious to pipeline facilities, or a limit to merchantability. All were essentially the same language.

Liquids—Free of liquid water and hydrocarbons. In no event will the gas contain water vapor in excess of seven pounds per MMCF. Language and water specification is the same for all pipelilnes in the sample.

Hydrocarbon Dew Point—Hydrocarbon dew point shall not exceed 20 degrees F (range of the five samples were 15 & 20 degrees, with one having a 60 degree dew point) at normal pipeline operating pressures.

Oxygen—No more than 2% by volume, for all except one stating it as no more than 10 parts per million.

Carbon Dioxide—No more than 2% by volume—same for all five pipelines.

Combined non-hydrocarbon gases (diluents)—No more than 3% by volume—same for all five.

Hydrogen sulfide—Not more than 0.25 grain per 100 cubic foot—same for all five.

Total Sulfur—Not more than a range of 0.75 to 20 grains per 100 cubic feet.

Mercapton Sulfur—No more than 0.30 (three at 3.0, one at 0.75 and one at 2.0) grain per 100 cubic feet.

Heating Value—Not less than 967 to 980 Btu per cubic foot.

Temperature—On the low side, not less than 35-50 degrees F and on the high, not more than 120 degrees except for one at 105 degrees F.

Local distribution companies (LDCs) originate services at the city gate where they receive the gas from the transporting pipelines and distribute it to local users, including residential, commercial, and even some small industrial users. There may be more than one city gate in various municipalities. Once the natural gas is moved from the producing area it can travel from a few miles to thousands of miles in getting to its marketing area. The usual terminating point for the gas is at a consumer's

"plant gate," for large industrial and electric generating companies, or at the city gate where the local distribution company (LDC) delivers it to the individual user through a network of pipelines. The LDC has its territory set by a franchise with the local municipality. There are many distribution companies in the country. Some are investor owned utilities, while many are municipality owned and operated. Some are co-ops formed for distributing the gas locally. The American Gas Association is the major trade organization for this group of utilities.

Municipality and state regulatory agencies such as the Public Service Commission usually regulate local distribution companies. This group of natural gas transporters is yet to be deregulated uniformly throughout the country. Some states, Georgia being the most notable, have passed new regulations much like the decontrol of the national interstate pipelines. In these locations where decontrol is in effect, the transporter is strictly a mover of gas and has no merchant function. It may have a subsidiary or affiliated company doing the merchant function, or marketing the gas. The eventual result of deregulation at this level will be for local distribution companies to offer open access to their transportation facilities. Each state will have to make its decision as to whether the LDC is freed from the merchant role or retains it, if only in part, along with offering open transportation for other merchants to move gas to the final consumer.

Natural gas is odorized so that its presence can be detected easily since natural gas as such is an odorless gas. Usually the local distribution company is responsible for this, and it is done before distributing the gas. The odorant is a sulfur-containing hydrocarbon with an obnoxious odor that can be detected by human smell even when it is used in very small, minute quantities in the gas. While it is commonly thought all natural gas must be odorized when it is sold to the user, this is not necessarily true. Gas going to industrial and generating users does not need to be odorized. In some cases, adding the odorizer could be harmful to the process using the natural gas. There are both federal and state regulations governing the odorization. In buying natural gas, the buyer should insure the contract includes provision for adding the odorant and who is responsible for proper addition and monitoring.

21.3.5 Natural Gas Markets & Economics

Until the time natural gas at the national level was well into the decontrol phase, the mid-1980s, the only market for natural gas was as a physical commodity. Natural gas was bought and sold on essentially an "instant basis" even though there were long-term contracts providing for continuing supply by the producer and sales by the current marketer—the pipelines. The buyer essentially bought gas as a commodity for its use as a fuel or feedstock. Once decontrol was underway and the pipelines were strictly transporters, anyone and everyone could be a gas marketer—either a buyer or seller at all levels from the wellhead to the consumer. A new system began to allow for future sales without a specific contract between a producer and consumer. This was the beginning of the financial markets.

Since natural gas is a commodity—it is fungible—and its supply is at times at the mercy of many factors including weather, demand, economics, etc., there is a market for buying gas supplies for the future. Commonly, this is called the "futures market" or financial markets, as opposed to the physical market where the actual commodity goes to the buyer either for resale or consumption. The financial market in volume is many times the size of the physical market. The futures market serves not only the actual user by giving them price protection but can be used by financial groups such as hedge funds, pension funds, etc. to wager on how prices for the commodity will change with time and afford opportunities for making money by playing the market. Currently, financial "speculators" play a large portion of the market.

Many users of natural gas buy or "hedge" on the commodity market to take advantage of prices offered in the future. The New York Commodity Exchange (NYMEX) offers contracts for up to many years and several banks and operators do an over the counter market offering prices even farther out. The consumers or sellers (producers, marketers, users, etc.) using the futures market are usually hedging as a means of price risk protection.

As an example, a fertilizer manufacturer making ammonia and various products where ammonia is the basic raw material is a large user of natural gas. If it takes the ammonia manufacturer an average of 60 to 90 days from the time it buys the raw material natural gas to be ready to sell it as ammonia or another product, it has to worry about the price of both the ammonia and the price of the replacement natural gas changing during the period. If the manufacturer uses $8/MMBtu gas for the ammonia and then, after selling the ammonia, has to buy $12/MMBtu gas to make new ammonia, it could be at a price disadvantage in the ammonia market. To "hedge" against these kinds of price changes, the manufacturer can buy "futures" when it starts making ammonia with the $8 raw material. It can protect its future-buying price for the raw material, which represents 70-80 percent of the manufactured cost of the ammonia, by hedging its future purchases. If the futures price for natural gas is—say for these purposes $9/MMBtu three to four months out—the manufacturer can lock in on the price by buying futures. It can then adjust the price for its product, knowing what the future product will cost.

Since the prices on the futures market move constantly (daily for the near term market and less frequently as time goes out), it makes an ideal medium of wagering what the price will be in the future. "Speculators" who come into the market have no need for the commodity, nor will they most likely ever take actual physical possession of the gas. Their purpose is strictly to wager on where the price will be on a certain date. It can be either up or down from the price on the day they buy "futures." This is the financial commodity market. There are many ways to play the financial markets using different instruments and derivatives.

This is not a small market but one in billions of dollars. In 2008, it was estimated that for every billion cubic feet of gas consumed in an average day, 10 to 15 times that were traded on the NYMEX exchange and other over-the-counter markets. Some of this excess trading went to hedging, but most it went to speculators trading strictly for the sake of profits. The average amount of gas consumed per day in 2007 without regard to seasonality was about 63 billion cubic feet. Using the average wellhead price for the gas (from the EIA) of around $6.39/Mcf, about $400 million was traded on average each day for actual consumption of gas—and much more in speculation.

In the financial trading markets, billions per day of natural gas were traded! That amount is only in the direct buying and selling of gas and does not include derivatives and other financial plays. This is big business and too intricate to be covered thoroughly in this text. Natural gas is only a small part of the commodity trading game. Crude oil and petroleum products are also traded and present extremely large markets and money transfers. When crude oil reached $130/B, which was well below the record set in July 2008 of $147/B, approximately $111B/day went for crude for consumption on a worldwide basis. If crude is being traded roughly 75-100 times the volume consumed, then it is easy to see the magnitude of the commodities market. Big players in these markets are financial houses, hedge funds, pension funds, and anyone else with a lot of money to invest! U.S. hedge funds alone are estimated to have $1.2 trillion dollars, and if they invest on a conservative 10% margin it is easy to see the magnitude of commodity markets.

Other than to have mentioned the financial market and show its significance in the natural gas industry, this chapter is devoted to physical gas buying. The buyers and sellers both need to know about the financial markets and evaluate their own need to participate or not in this type of gas transaction. There are many marketing companies, financial houses, and consultants well versed in the financial markets and how trading in these can lower the over all purchase costs of the commodity. Many books are written on this aspect. Buyers and sellers should become familiar with all sources of information in this area in helping to either maximize the return for the product for the sellers or minimize the purchase costs for the consumers. The comments on buying gas for use do not negate the financial market but leave it to other sources for the users to learn how to work within the financial framework including its benefits and risks. Knowledge of the financial markets is necessary because of the impact the financial market has on the physical market and prices for natural gas.

Natural gas prices were originally set by the federal agency having jurisdiction over natural gas. The original methodology for price setting was much like the rate of return methodology for pipeline transportation tariffs. This was a direct function of the believed costs of finding, developing, and producing natural gas. As discussed previously, the low prices paid at the wellhead prevented the natural gas industry from maintaining the necessary supply and caused the dire gas shortages of the mid-1970s. After natural gas prices were decontrolled and natural gas became a true commodity, prices became a reflection of the normal economic factors impacting commodity pricing. Typical supply/demand factors plus the additional influence of financial players (speculators), money value, and other economic considerations will dictate natural gas prices.

The price for natural gas at the burner tip is dependent on many things—market conditions, supply/demand balances, economic conditions, and many more, including the activity of natural gas financial markets, prices for competing fuels, etc. In the early stages of the industry, because natural gas was considered a burdensome byproduct of the crude oil industry, it was sold for very low prices. When crude oil was around $2/B, or about 30 cents per million British thermal units (MMBtu), natural gas under federal price control sold for a penny or two per thousand cubic feet, or roughly the same per MMBtu. In actual heating value, a thousand cubic feet of natural gas has close to a million Btu. A barrel of fuel oil is 42 gallons of oil and about six million Btu, depending on the grade of oil—the heavier the oil, the more Btu per barrel.

On an economic basis of energy content, natural gas prices for a thousand cubic feet, compared to a 42 gallon barrel of oil, should be close to one-sixth the value of the oil, i.e. a $120/B of oil would be equivalent to $20/MMBtu natural gas. Through the rise of crude oil prices, natural gas prices have seldom matched crude oil. Instead, the value of gas has run about half or about one-twelfth or one-tenth the value of crude oil, and even a lesser ratio compared to finished petroleum products like distillate oil, kerosene, etc. In late summer 2008, when oil prices (West Texas Intermediate, WTI) were around $100/B, natural gas prices were around $7/MMBtu. Gas prices were about one-fifteenth the value of oil in dollars per barrel, far from the equivalent heat ratio of the theoretical one-sixth.

While this comparison used crude oil against wellhead natural gas prices, the real comparison would be between delivered natural gas versus delivered light oil or residual fuel oil, the commercial products from crude oil used in industry for heating. At that time, light oil at the refinery rack was about $160/B (or about $30/MMBtu) and natural gas at the city-gate was about $10/MMBtu in the Eastern markets.

Pricing is not a logical phenomenon. Data and basic considerations can help in predicting prices, but the final price is very dependent on perception—market perception at the time. Too many of the variables are unknown precisely enough for pricing to be a scientific conclusion. Forecasting prices is an art. Perception of the value based on supply/demand parameters sets the price. The market itself will do a lot to raise or lower the price. Further, the large financial market compared to the physical market for natural gas has an immense impact on the prevailing price. Gas prices can "spike" for many reasons, real or perceived. Hurricanes, hot weather spells, changes in the economy, etc. can make prices go up or down quickly and significantly. Short-term changes are always a possibility.

Seasonality at times has little bearing on the current price. Natural gas prices have dropped precipitously in the middle of January and have reached highs for the year in "shoulder months." Eventually, prices come back to reality, but in the time they are moving, large dollar gains or losses can occur. In looking at gas prices, it is necessary to know where the gas is sold, as prices vary according to where the sale is made in the wellhead and consumer path. Unlike crude oil, which is transformed into various commercial products, each with its own value, natural gas is essentially the same for the commodity, once it enters the transportation portion of the journey to the burner tip. Its value does increase as it moves through the system, going from the wellhead to the consumer, because of the added value of the transportation and services bringing the gas to market.

The basic place for pricing natural gas is gas sold at the wellhead or at a marketing hub. Gas priced on a Btu value at the wellhead will accurately reflect the value of the gas further down the chain, even though wellhead gas might need to be gathered, treated, and/or processed and transported to a user area location. Once the gas is pipeline quality, its price reflects where in the delivery system it is at the time, from production to the sales point. Anywhere in the chain, the wellhead price can be estimated by net backing or subtracting the additional costs to get to that point of pricing. In addition to the cost of the commodity (the gas), the cost of transportation and delivery are also susceptible to supply/demand restraints since pipeline capacity is a relatively fixed volume. The price differentials based between locations are affected by the price of the commodity, transportation and delivery, and local market conditions. The term, "basis differential" is the difference of gas prices in a given location versus a common market area like Henry Hub or the wellhead price for a given area.

Gas purchased at the wellhead is done so on a wellhead price. Gas purchased further downstream might be termed "into pipe price." Gas sold at a central marketing center ("market hub") would have the term, "hub price." While currently there are about 30 hubs across the U.S., the best known is the Henry Hub in Louisiana. The New York Mercantile, in making a futures market for natural gas, has its only contract at the Henry Hub because of the hub's central locality and easy accessibility.

Market hubs have the ability to dispatch incoming gas from one pipeline to another in the manifold of pipelines coming into and leaving the hub. Hubs make an ideal site for gas pricing since they carry such large volumes of gas. From hub pricing, gas might then be priced at the "city gate," which is where the gas is transferred to a "local distribution company" for delivery to the consumer. The pricing for the consumer would be based on the "sales point" price, which would be a total price for the gas, including all the transportation and services required to get the gas to the user's receipt point.

The individual price paid by the buyer is dependent on many factors, starting from the wellhead pricing to the price at the meter coming onto the buyer's property. The government and other reporting services report the prices at the major pricing points and at the city gate. The major consuming sectors where prices are reported are the residential, commercial, industrial, and electric generating markets. Since the progression from each of the stages from production to market carries a cost factor, it is important to know where in the delivery chain the price quoted applies.

21.3.6 Information Sources

In recent years, as decontrol has continued, changes in the industry have occurred, and the internet has grown, information on the natural gas industry has literally exploded! Information sources include government agencies, trade publications and organizations, companies in the industry, producers, pipelines, local distribution companies, and financial houses. Everything is covered, including what is natural gas, who are transporters, where consumption occurs, prices, imports, etc. There is no shortage of information; where to find it is the problem!

A major source of information on natural gas is the federal government, with the various agencies dealing with natural gas, energy, transportation, safety, environmental, and energy regulation. Others include industry sources such as producers, pipelines, local distribution companies, trade organizations, and trade publications.

The major government sources of information on natural gas are:

Energy Information Administration (EIA)—www.eia.doe. gov—Includes supply/demand, pricing & other energy sources data. EIA is a division of the **Department of Energy (DOE).**

Federal Energy Regulatory Commission (FERC)—www.ferc. gov—For rules and regulations regarding natural gas marketing, pipeline tariffs and facilities.

Others that have some input applicable to natural gas and are therefore a source of information are:

Department of Transportation (DOT)—www.dot.gov—Regulates the safety of pipelines used in transporting natural gas.

Environmental Protection Agency (EPA)—Reviews and regulates environmental considerations of pipeline operations.

Commodities Futures Trading Commission (CFTC)—Regulatory and enforcement action regarding financial trading of natural gas, including derivatives.

Many states also have information available on natural gas for either production and/or consumption. A complete list of state agencies and the web site for each is given by EIA. Some of the more important ones are listed below.

California Division of Oil & Gas—www.consrv.ca.gov
Colorado Oil & Gas Conservation Commission—www.dnr. state.co.us
Louisiana Department of Natural Resources—www.dnr.state. la.us
New Mexico Department of Energy & Natural Resources— www.emnrd.state.nm.us
Oklahoma Corporation Commission—www.occ. state. ok. us
Texas Railroad Commission—www.rrc.state.tx.us
Wyoming Oil & Gas Commission—http://wogcc.state.wy.us

Canada is a good source for natural gas information; since it is an important supplier of natural gas to the U.S. it makes a relevant source for data. Canadian agencies of interest are the following:

Canadian Energy Pipeline Association—CEPA—www.cepa.com
Canadian Environmental Agency—www.ceaa-acee.gc.ca
Canadian National Energy Board—www.neb.gc.ca
Nova Scotia Department of Energy—www.gov.ns.ca/energy

Further, in Canada, a good non-government source providing an overview of all energy sources in Canada, as well as worldwide, is:

Canadian Center for Energy Information.—www.centreforenergy.com

Industry sources can be broken up into different major areas: those supplying, transporting, and marketing natural gas; financial and banking institutions; trade organizations; and trade publications. A partial listing by these categories follows.

Suppliers, transporters & marketers
    Anadarko Petroleum
    Apache
    BP Energy
    Chesapeake Oil & Gas
    Chevron
    Conoco Energy
    El Paso Corporation
    Exxon Mobil
    Shell Oil Company
    Williams Energy

Financial & Banking Institutions
    Deutschland Bank
    Goldman Sachs
    Louis Dreyfus
    Merrill Lynch
    Raymond James

Trade Organizations
    American Gas Association—AGA
    American Petroleum Institute—API
    Independent Petroleum Producers of America—IPPA
    National Association of Regulatory Utility Commissions—NARUC
    Natural Gas Supply Association—NGSA
    North American Energy Standards Board, Inc.—NAESB

Trade Publications
    Btu WEEKLY—www.Btu.net
    Energy Daily—service@energycentral.com
    Gas Daily—McGraw Hill Publications—support@platts.com
    Inside FERC—www.mcgraw-hill.com
    Natural Gas Intelligence—prices@intelligencepress.com
    Natural Gas Week—www.energyintel.com
    Oil & Gas Journal—Pennwell Publications—www.ogjonline.com
    Pipeline & Gas Journal—www.oildompublishing.com

21.3.7 Environmental

Environmentally, natural gas is the preferred fuel. Even though it is a fossil fuel, the amount of carbon dioxide released is the lowest per unit of energy received from the major fossil fuels. Natural gas is ideal for its handling and transportation qualities, and its environmental advantages make it the most popular fuel. It is the fuel of choice for many applications. It presents no unique environmental concerns to the user, and as long as the supply is pipeline quality, the fuel source is of no concern in regard to environmental purposes, other than the basic combustion exhaust releases. Since any combustion exhaust presents certain environmental concerns, local release laws should be reviewed when going to any gas burning system.

21.3.8 Regulatory

To the average gas buyer, the new natural gas industry presents few regulatory problems or concerns other than those imposed by local or state authorities. This is crucial, as depending on where the buyer or the use of the gas is located, the local and state rules will play a big part in buying natural gas.

The federal regulations on natural gas from prior years have been reduced, even though natural gas is still under federal regulation from the wellhead to city-gate. Natural gas federal rules and regulations can be accessed at the Federal Energy Regulatory Commission (FERC) website if necessary. The small gas user, especially those buying through an LDC, will have no problems. Only for the sake of information, a brief review of natural gas history of regulation follows.

1938—Congress passed the **Natural Gas Act** giving Federal jurisdiction to wholesale sales of natural gas and interstate transportation. The Federal Power Commission (FPC) was the responsible agency.

1954—The Supreme Court rules (**Phillips decision**) that sales prices charged by producers come under FPC jurisdiction.

1977—**Federal Energy Regulatory Commission** (FERC) is created and the FPC abolished.

1978—As part of the **National Energy Act** (NEA), the **Natural Gas Policy Act** (NGPA), with its changes of gas pricing at the wellhead, are passed.

1985/1987—FERC issues **Order 436** and then modifies it with **Order 500, Regulation of Natural Gas Pipelines After Partial Wellhead Deregulation**. This was the Open Access Order that had pipelines offer open access for transportation services. This made it possible for end users to buy gas directly from producers.

1989—Natural Gas Wellhead Decontrol Act called for the gradual elimination of government price control.

1992—Order 636 was issued, calling for interstate pipelines to unbundled all of their non-transportation or non-regulated services from their regulated transportation services. This was the final major decontrol order.

While natural gas pricing is no longer under federal regulations, it is still tied to some of the original federal natural gas laws, and the regulatory agency FERC has jurisdiction over economic and other policies affecting natural gas pricing and marketing. FERC does have oversight control regarding both physical and financial natural gas trading. In today's markets, this presents essentially no interference to commerce, but it does mean that under certain conditions federal regulations could again be imposed on natural gas and certain policies and marketing plays restricted.

In the financial markets, FERC shares responsibility with the Commodities Futures Trading Commission (CFTC) to insure proper trading procedures and that no market manipulation practices are made. In recent times, the regulatory bodies have accused several natural gas traders and brokers with market manipulation and price influencing.

For current conditions, the buyer has to be mainly concerned with local and state rules and regulations. Intrastate transportation, storage, and handling regulations are basically local and state issues. The status of states in following the federal decontrol action is followed by the EIA and can be seen on its website. On the EIA site it is called **natural gas residential choice programs** but is applicable to all gas sales within the state. A summary of the status as of December 2007 from the EIA is given in Table 21.6.

The table is an overview of the status of natural gas restructuring in each state, with the focus on the residential customer. The restructuring applies to all users, and the retail unbundling or restructuring is an indication of the division of services that are required to supply natural gas to consumers that can be separately purchased. In a state with complete unbundling, a buyer can buy gas from any source, and transportation has to be supplied by the various pipelines within the state, including an LDC to deliver the gas to the consumer as a transporter only. Where there is no debundling, the local user would have to buy the gas from the LDC, which would be supplying the gas and transportation and delivery services.

The EIA site is an excellent source of information for the status of unbundling programs by state. It also includes pricing information for various methods of buying gas within the state. EIA contact is possible through the internet. For certain areas such as this, the internet material will usually give a telephone contact at the EIA for further information.

Other resources for information in this area are the following:

National Association of State Utility Consumer Advocates (NASUCA)

EIA Publication—Natural Gas Marketer Prices and Sales To Residential and Commercial Customers: 2002-2005

State Public Utility Commission Web Sites

In addition to the EIA and FERC, the Department of

Table 21-6. Natural gas residential choice programs, 2007

Residential Natural Gas Restructuring Stats – Dec 2007	States
Statewide unbundling - 100% eligibility: Active	DC, NJ, NY, PA
Statewide unbundling – 100% eligibility: Inactive/Limited programs	CA, MA, NH, WV
Statewide unbundling- implementation phase: > 50% eligibility	GA, IL, MO, MI, OH, VA
Pilot programs/partial unbundling	CO, FL, IN, KY, MT, NE, SD, WY
No unbundling	AK, AL, AR, AZ, CT, HI, IA, ID, KS, LA, ME, MN, MO, MS, NC, ND, NH, NV, OK, OR, RI, SC, TN, TX, UT, VT, WA
Pilot program discontinued	DE, WI

Source: EIA, October 2008

Transportation and the Environmental Protection Agency have jurisdiction in the areas of pipeline safety and environment, respectively. Other government agencies do play a role in the natural gas business, but the listed ones are the major ones in connection with natural gas buying. Buyers should insure in their negotiations and contracts with sellers, transporters, and providers that all regulations are covered and that the responsibility for meeting these rules are a part of the supply contract. The contracts for buying and transporting should speak directly as to whom the responsibility for meeting the requirements will fall and which parties will be responsible for the consequences if failure occurs.

Agencies having responsibility for natural gas regulations at the federal and state levels are easily accessible. Federal agencies and websites, as well as some of the more important state agencies, are given in the Information Sources Section, 21.3.6 . All State Public Utility Commissions (PUCs) can easily be located on the Energy Information Administration (EIA) website if more information is necessary.

Further, many law firms and consultants specialize in the regulatory aspects and should be contacted if necessary. Many of the major consulting firms with natural gas expertise are listed on the web by using one or more of the search engines as listed below. Using key words or phrases, such as "Natural Gas Consultants," "Buying Natural Gas," "Natural Gas Marketing," etc. will give many sites for further information.

Table 21-7. Internet major search engines

Google
Netscape
Dig
Yahoo

21.4 BUYING NATURAL GAS

A thorough knowledge of the structure of the natural gas marketing system is essential in buying natural gas, regardless of the size of the operations. The big change in the business from when the industry was price-regulated to the current open market makes knowing the business even more essential. In the old days, gas sales to consumers were essentially through only one route—producers to pipelines—with producers serving as transporter and merchants to local distribution companies, to consumer. Now, buying and selling occurs at all levels, as anyone can be either a buyer or a seller. Transportation pipelines are only transporters, and all other functions they once did have been sold or assigned to other companies.

While many states still have control over how open the distribution systems are, the chain can be as short as producer to consumer, with the buyer or seller handling the transportation. More generally, producers sell to marketers, who then sell to consumers directly or to an LDC that either makes delivery to the local user or sells as system gas to the user. The quantity of gas needed plays a big role in how the gas moves from production to consumption. The large users, usually an industrial consumer, big commercial or industrial users, or generating application, has the gas go from producer or marketer to the consumer. (And sometimes the marketer is the marketing arm of the production company.) Smaller users, like residential and small commercial consumers, almost always make use of the local distribution company for their supply. The approximate cut-off where the buyer goes through the distribution company or sets up its own buying and delivery is about a million Btu/day, or 1,000 cubic feet per day, essentially the same amount. This is the free market for natural gas today.

Buyers are free to pick any marketer or seller to supply gas. Open transportation is available to everyone at the interstate and intrastate level and, depending on the state where the buyer is located, may extend even further down the chain.

The next sections will cover buying natural gas, who the marketers are, pipelines, outlines of contracts for buying, and other necessary areas to help the buyer secure reliable, secure, and economic natural gas supplies.

21.4.1 Actually Buying the Gas

So, how does the gas consumer get down to the basics of buying natural gas? Do they call the local distributor commonly referred to as an LDC, if the consuming facility is in an area served by that distributor? Or does the buyer shop around for a marketer who can supply at the best price and service? Again, information and knowledge are the secrets to success. The buyer must know what is needed to determine what path to follow in buying natural gas, because, depending on two factors, the path goes from one to many!

If the buyer is looking for a source of gas for a new operation, one never before using natural gas, then a little homework must be done before proceeding. There are two primary factors to consider. The first is, "What are the natural gas requirements—the amount needed on a daily basis, including variance due to peak demand, reduced operations, etc.?" The second controlling question is simple—"What state is the user in and what are the current gas marketing rules?"

If the buyer is replacing an expiring contract or having to change vendors, then the historical record is available to help in knowing what is needed to renew the supply sources. The existing information and records can be used to predict with great accuracy what volume of gas will be needed, along with the changes on a daily or other time basis. They will provide the prior costs for the gas supply as a guide to continued buying. With this information, the fuel buyer can look for new sources to meet the needs more efficiently and cheaply.

The very first question to be answered is how much natural gas will be consumed on a daily basis and what will be the range of use on a daily, weekly, monthly and annual basis. The information could even be a factor in the question of an hourly rate, dependent on how large a swing the user anticipates. These are the big questions to answer in making the first step in trying to select a supplier, whether it be a marketer or a local distribution company. The quantity and conditions of the rate and how it will vary are crucial to starting the buying process. Whether the consumer is a large or small user of gas will play a major role in what selections are open to it for purchasing gas.

Typically, the break from a small user to a large one is a rate of about one thousand cubic feet per day, or in energy units, about a million Btu per day. Most local gas distribution companies will talk in "therms" and "dekatherms" rather than Btu or cubic feet. The dekatherm is ten therms. Each therm is 100,000 Btu. Each dekatherm is one million Btu. The line between large and small users is not rigid.

Applications coming close to this approximation may still meet the criteria for going the large user route. If the user is on the small side, depending on the state or location of the use, it still may have an alternative of buying from the local distribution company or using the LDC for transportation only and buying the commodity from a marketing entity.

Making contact with marketing companies, which will be discussed later, and getting information on the local regulatory rules will help in making this decision. Local distribution companies that are in an open access state may have set up their own non-regulated marketing companies to help consumers buy gas at the lowest price with the required service criteria of their own operations. One should not forget the potential of e-commerce and its influence in helping to buy and sell natural gas. A smart buyer will look at all possible sources for meeting the requirements at the lowest price but with reliability and security of service. In buying a commodity like natural gas, price alone should not be the only criteria. Service (security of supply, emergency additional supplies, etc.) equally impacts the buyer's bottom line as does price in

meeting fuel requirements. Having a cheap supply of gas where its availability is so uncertain as to disrupt plant or business operations is really an expensive supply when looked at in the total picture.

The large users—those over the thousand cubic feet level or close to it—should investigate all possible sources for supply and transportation. Their sources may go all the way back to the wellhead or producer and its marketing companies. Depending on how large a supply is needed at a given location, the buyer may include dealing with pipelines and distribution companies for transportation and delivery of the gas. Once the buyer knows in general which direction to go, the big issues then become finding a marketer, transportation, and contracts for the services and commodity.

The next major question is where the gas will be used. This is necessary for two reasons—what pipelines are in the vicinity for delivering the gas and what state regulatory laws apply to gas transportation and marketing.

Pipeline availability: Availability of a transporting pipeline for gas delivery to the user site is important since natural gas is delivered only in pipelines. There are a few exceptions, but they do not apply to the common use of natural gas. Either a pipeline is already in place for the gas delivery, or a line will have to be put in from the closet delivery point to the consuming location. Within a city or a franchised area, the LDC is usually responsible for the pipeline to the using facility. That is a negotiable element with the LDC supplying the service. If the consuming location is not within the given area, or if the state regulations allow for open access, the buyer will have to be responsible for a pipeline from the closet natural gas pipeline in the area to the consuming facility. In those cases where the consumer is too far, or it is too expensive to put in a pipeline for delivery, then the alternative to natural gas would be bottled propane or propane/butane mixtures.

State open access: The state where the consuming location is will determine the availability of open access transportation. Some states have opted to follow the federal government and natural gas interstate pipelines economically within the state and allow them to be transporters only. Some are still using the system where the pipeline is both the transporter and the merchant. Many states are somewhere between the two extremes. The latest status of states with open access is given in Table 21.6 in the Regulatory Section. 21.3.8. Depending on the degree of open access in a given state, the consumer has the choice of buying "system" gas from the local distribution company or buying gas from a marketer and developing its own transportation for delivery to the site. As reviewed previously, the amount of gas to be delivered also plays a role in making this decision.

21.4.2 Natural Gas Marketers

Marketers come in varying forms, sizes, and descriptions. One can look at it much like purchasing gasoline at the local filling station—"full-service" or "self-serve." To add a little more variety or confusion, gas buying and selling has moved to the internet, or e-commerce. When marketing companies started in the mid-1980s to do the merchant function previously performed by the interstate and intrastate pipelines, almost anyone with a telephone and a pencil could be called a natural gas marketer. Through the years, as the number of the marketing companies grew and at the same time took on additional scope and responsibilities, the "fly-by-night," less reliable marketers were pushed out of the business.

Even some of the more reputable, better-financed groups have gone away or merged into other companies. The inability to be profitable in a fast moving, sometimes irrational market place has taken its toll. The scars from the Enron debacle in the early 2000s hurt many gas marketers. It gave birth to a new breed and brought additional responsibilities to the business. The highly volatile gas market, along with the increased volumes of financial trades each day, has made risk analysis and financial planning even more important.

Natural gas marketers can range from independents with no oil and gas, transportation, storage, or any other holding within the industry to subsidiaries of natural gas producers, pipelines, and others in the business. Also, because of the large financial involvement in the gas business, many marketers are subsidiaries of large financial houses and banks. A marketer, regardless of affiliation, buys and sells natural gas for a profit. The marketer performs various services for its customers, including buying natural gas in the appropriate market, handling the needed transportation for the gas delivery, nominating and balancing services, and recording use patterns.

Within each group, there are reputable players who can help get the buyer's natural gas requirements. Some shopping among different marketers is good business thinking. There are also natural gas consultants in the industry who can help the buyer get the best possible supply source.

Marketing natural gas is more than just selling gas to a consumer. The gas business is big business, running into revenues of around $ 200 billion per year, depending on the exact price for the commodity that year. In 2007, it was approximately $192 billion. The $192 billion is only a measure of the actual commodity trading on an idealized basis of direct sells from producers to marketers to consumers. Actually, an average cubic foot of gas most likely gets traded three to four times before coming to the burner tip, where the gas as fuel or feedstock goes out of the market. This is only for the physical side of the trading, the place where the commodity actually is moved to a final destination for consumption.

This pales in comparison to the financial markets, where 10 to 15 times the volume traded each day in the physical market of consumed gas is traded in the financial sector. The money moved in this arena is way beyond the $200 billion discussed previously. At times, the market responds more to financial than to physical drives. While people using gas do some financial buying and selling, much of the financial trading comes from big money entities such as pension funds, financial houses, hedge funds and others.

21.4.3 Hedging

Many gas buyers are speculators who might do more to move the market than the actual users needing the natural gas for fuel or feedstock. Like all commodities, natural gas makes an ideal medium for financial trading. There are those gas consumers who need to play the financial market for the protection or risk-adverse properties the market can offer.

Those who produce natural gas and those using large quantities can buy some protection of the future price by buying financial "futures." This is "hedging." The futures buyer is taking a position for a given month in the future, where the price he pays will be the price for the quantity of gas he purchased now, for future delivery in a given month. The consumer can lock in the price for gas anywhere from a month forward to 10 years or more forward. Whether buying or selling gas, hedging is a good tool to relieve some of the market risk in buying or selling a relatively volatile commodity.

The volatility of natural gas prices (no pun intended) makes it an ideal commodity for speculators and other investors to make a market in for the sheer purpose of making money. The speculator is betting the price will be higher or lower on a given date and is willing to take a position by buying the commodity for trading at that time. Much of the natural gas trading is for speculation. There are arguments on both sides that this can add to the volatility of the gas market.

Most of the physical buyers of natural gas bring a relatively simple mentality to the market place based on supply/demand parameters—the economy, weather, and other pertinent factors. Using an entirely different set of parameters, the investors and speculators might use a different system for their actions. Investors in this group have a "statistics" of their own for analyzing the market. Basically, the investment traders are "market technicians" and play a statistical analysis of the market itself for buying and selling the commodities.

The mentality of the technician is basically, "Who needs to know all the details of the commodity?" The market place itself shows the results, and following the market with its own statistical tools is the way to go." Technicians are also known as "chartists" since they keep a chart of the volume and price on a daily basis for the commodity. Their statistics involve reading the chart of past and current performance to predict future performance for the commodity price. By charting, the technician can predict the expected maximum price for the commodity and other helpful information needed to invest properly. Using the charting method makes any direct analysis of supply/demand, weather, etc. less important in investing wisely. The analysis can be completely dependent on the past and current activities of the commodity, or if a stock, on prices for when to buy or sell the financial holdings.

Many of the market speculators such as pension funds, hedge funds, etc. have very large amounts of money they control. When the signals show it's time to buy or sell, very large sums of money can come into or leave the market. It's easy to see how this can make the price of the commodity very volatile and puts the buyer (and seller) in a very risky position. It's also easy to understand why risk analysis and other protective systems are used by large natural gas players.

21.4.4 Finding the Marketer

Now, where does one go to buy natural gas? As stated earlier, major items in seeking a seller are the customer's gas supply needs, the location where delivery will be made, and the degree of service desired and needed. The buyer might want to prepare a "load profile" for discussions with marketing companies so that the needs for supply can be presented fully. The profile would include the estimated annual gas needs, along with daily and even hourly requirements. It would include varying rates as the business or seasonal changes impact gas requirements. All the elements of demand would be included in the profile.

Once the buyer knows its requirements, then it is in a position to prepare a request for proposal (RFP) to work with various marketing groups in setting up the gas supply. The marketers, in replying to the RFP, can work with the buyer to develop the best possible alternatives to supply. The buyer might want to work with a few marketers to see which make the best offers for supply, including not only pricing but also varying the supply, meeting unexpected increased demands, and other needs.

A large user wanting to hedge prices to insure stronger control on the price paid for the commodity might enlist the aid of gas marketing consultant to learn more about the financial aspects of gas buying. While a marketer can help some in developing financial expertise, certain rules prevent the marketer from taking certain positions. The marketing consultant having no supply ca-

pability is free to completely advise the buyer on various financial plays to hedge gas pricing.

For the buyer needing relatively small amounts, the best approach would be to call the local distribution company. Again, remember that the approximate break point between a small user and large one is a million Btu/day, but this is only an approximate figure. The wise buyer with requirements around this quantity will check with local sources to see what can be developed. Again, the question of financial hedging to protect pricing is an individual matter for the buyer to decide how much it wants to go this route. Depending on the state where the gas will be used, the buyer may have little choice in picking a supplier.

Once the buyer can make a selection, based on supply requirements and location, the selection of a marketer can be difficult, because there are so many alternatives and choices. There are roughly 200 marketing companies handling natural gas at the state level. A complete list of currently registered gas marketers can be found on the EIA web site. These include independent marketers, subsidiaries of producers and transporters, and financial houses doing marketing.

In addition to the regular marketers of natural gas, buyers and sellers of supply, there is also a group who serve the additional function of acting as "market aggregators." Here the marketer will put together various small users to have a bigger package and be able to negotiate from a better position in lining up gas supply. Checking marketers to find an aggregator can be helpful to the small buyer.

Some state utility commissions require natural gas marketers to be certified to operate in the state. Going back to the state commission is one way to find marketing companies in the desired area. In addition to the listing of marketing companies on the EIA website, the National Association of Regulatory Utility Commissions, NARUC, can give state utility commission information. Again, an additional source of marketing companies can be had on the internet itself. Using one of the search engines, such as Google, and the phrase, "natural gas marketers" will give many sources of information listing marketers, as well as marketers themselves.

First in line for the new buyer seeking a seller are the local distribution companies in the area. Those in states where some open access is in operation will, in addition to selling "system gas," have an affiliate or subsidiary selling market-sensitive priced natural gas as well as transportation only. System gas is natural gas which the LDC has purchased for resale to its local customers. Since this customer base includes residential and commercial customers, as well as the industrial sector, the average price will usually be higher, as the financial impact of including these sectors will be in the LDC tariff.

If allowed in the state, the local distribution companies will make available open access transportation so that a large industrial user can bring in its own gas supply and let the local distribution company transport it to the buyer's facility with transportation charges only, again set by the tariff the pipeline has set with the state or local government. As part of its tariff, the LDC will set a minimum amount of gas the buyer uses as a criteria for allowing the buyer to purchase its own gas and use the LDC for transportation. The tariff will set the cost for transportation by the LDC, and in addition to transportation costs the tariff will include a rate of return, amortization of the facilities, and other pipeline costs. Tariffs also include any local taxes or fees made by local governments for transporting natural gas within a franchised area.

If the LDC is in an open access situation and has an affiliate to sell the commodity only, then the buyer has a choice of buying from the LDC marketing affiliate or other marketers. The buyer can "shop" its purchase needs to get the best package of prices, services, and other options. The local distribution company would still be the transporter for the customer. This is all dependent on the degree of open access in the state.

If the buyer has the option of open access and needs a sufficiently large supply of gas, then the marketer of choice would be one of the independent marketers serving the area where the gas will be used or one of the subsidiaries of the producers, transporters, or financial houses doing gas marketing. Table 21-8 lists the major natural gas marketers in the second quarter of 2008 and the volume of gas sold by each on a daily basis. Data are from the publication *Natural Gas Daily*, a subsidiary of McGraw-Hill Companies.

The buyer should sample a large enough group of marketers to insure getting the best price, reliability, and service. Selecting a gas supply source is not an overnight task. The work needed should be in proportion to the value of the gas to the operation. If large supplies of gas are needed, differences of a penny or two and reliability are very important.

Remember, price alone should not be the only consideration in purchasing natural gas. Dependability and service have a definite value. One should always keep this in mind when buying gas supplies. It might seem wonderful to buy the cheapest gas, only to be unable to get it when weather or other problems make the supply scarce!

The marketer can assist the buyer in preparing nominations for the transporter so that it can plan its pipeline capacity needs. Nominations will identify the receipt point, where the gas is delivered to the pipeline, and delivery points, where the gas transfers to the next transporter or consumer. The list of services the marketer can supply the buyer is, to a degree, dependent on the sophistication of the buyer and how much the buyer wants to be involved. They can do the balancing of the account where takes do not match precisely the nominated quantities. Again, these are things to be negotiated with the marketer.

Because of the many parameters to be covered and the need to know the players and the system, many companies seek consultants to help either initially or continually to make better decisions in gas purchases. The difficulty is that unless the buyer is in the buying sector almost continuously, he or she will be at a distinct disadvantage in seeking the optimum natural gas sources. The expense of using a marketing company or a consultant can be a very small price in finding the most effective and efficient source of supply.

The new area of buying gases using e-commerce and the business-to-business internet is growing in importance. Many of the major gas marketers are making markets buying and selling gas through the internet. The internet marketers make it easy for the knowledgeable trader to buy or sell gas without having to use a marketer or broker. How much additional effort the buyer will need to complete the sale and transportation will depend on how this system of marketing grows and prospers. After only a short time of this method of marketing being in existence, large enough volumes have been traded to see the value and potential for e-commerce business in the natural gas industry.

21.4.3 Impact of Gas Pricing on Buying

Natural gas is a commodity. There are many suppliers and the commodity is fungible. Its price is a function of its availability and costs for delivery, which is almost always by pipeline. Simple, but true. When supply is perceived sufficient, current natural gas prices are in the $6-8/MMBtu range. Over-supply will see the price drop significantly, sometimes coming down 20 to 30 percent of this price. Tight supplies can do the same with the cap, based on recent experience, rising close to the mid-teens per MMBtu. Early 2008 prices were in the $6/MMBtu range and went into the mid-teens during the early summer. By early fall, prices were in the $7-8/MMBtu range, as all energy prices declined because of the downturn in the economy, sufficient supply, and other factors. Concern was generated that speculators in the financial markets influenced the high prices during the early summer. While price movement is basically from supply/demand parameters, the problem is two-fold: no one knows the

Table 21-8. Top North American Gas Marketers—Second Quarter 2008, Bcf/d.

Rank	Company	Parent Co.	Volume
1	BP	Energy Co	20.3
2	ConocoPhillips	Energy Co	14.2
	Constellation	Utility Mkt	14.2
4	Shell Energy	Energy Co	13.7
5	Chevron	Energy Co	8.3
6	Louis Dreyfus Energy Service	Financial	7.9
7	RBS Sempra	LDC	7.7
8	Nexen Marketing	Energy Co	7.3
9	Lehman Brothers	Financial	5.8
10	Tenaska	Financial	5.1
11	EnCana	Pipeline	3.8
12	UBS	Financial	3.0
13	Oneok	Pipeline	2.9
14	Devon Energy	Energy Co	2.5
15	Merrill Lynch	Financial	2.4
	Sequent	Utility Mkt	2.4
17	Chesapeake Energy	Energy Co	2.1
18	Anadarko Petroleum	Energy Co	1.9
	ExxonMobil	Energy Co	1.9
20	XTO Energy	Energy Co	1.8
21	Hess	Energy Co	1.7
22	Enserco	Utility Mkt	1.6
23	Canadian Nat'l Resources	Energy Co	1.5
24	CenterPoint Energy	Pipeline	1.3
25	Apache	Energy Co	1.1
26	El Paso Marketing	Pipeline	1.0

supply/demand picture with accuracy, and secondly, fact and perception play unequal roles. In the end, each buyer and seller must make its own decision on where the price will go in the short and long-term futures.

Historically, natural gas prices in the beginning were cents per thousand cubic feet. After crude oil prices became market sensitive in the early 1970s, it was not until natural gas prices were decontrolled that prices for gas in interstate commerce came up to realistic prices ranging from over a dollar to $5-9/MMBtu. Gas prices during the 1970s, before federal decontrol, saw the intrastate market quickly come to market sensitive levels of $3 to $6/MMBtu because supplies on the interstate markets were low. The Natural Gas Policy Act of 1978 ended the difference between interstate and intrastate pricing.

The higher price for natural gas right after decontrol started in 1980 was an effect of the legislation, which set up about a dozen pricing categories. When the gas surpluses of the mid-1980s started, where the legislation had set the "maximum lawful price," it did nothing for a minimum price. The gas merchants of that time, the pipelines, brought the prices down to the $2/MMBtu range quickly. Since 1985, natural gas prices have varied from around a $1/MMBtu to highs above $13/MMBtu at a marketing center or hub. Table 21-9 gives average annual natural gas prices in the field for the major gas producing area and at Henry Hub in Louisiana for 2003 through 2007.

Table 21-9. U.S. Natural gas prices in field and Henry Hub, 2002-2007.

Year	Field $MMBtu	Henry Hub $/MMBtu
2003	5.28	5.50
2004	5.60	5.90
2005	8.01	8.88
2006	6.49	6.76
2007	6.42	6.98

Source: JOFREE ENERGYCONSULTING

There are tools to help in price analysis and forecasting. In addition to the sources for tracking the current gas prices, there are tools for helping in projections of future prices. Services that can supply forecasts based on their interpretation of the future are available from various consulting groups. Many of the financial houses making a market in natural gas and other energy futures have current material on their analysis of gas markets. The federal government has many publications and resources for tracking and estimating gas supply, demand, and pricing. Many of the sources having information on natural gas pricing are free.

A couple of elements are important to keep in mind about price forecasting, regardless of the source of information. Forecasting is an art. There are statistical methods and models to help in making predictions, but many of the assumptions are based on the forecasters' ability and experience. It is still art, not fact or science. Who can predict with accuracy and precision the weather for a week or six months out? Hurricanes occur, blizzards come, and sometimes little is known before hand. There is even a difference if the extremely cold weather comes during the week or only on the weekend. During the week, schools and business facilities need gas for heating; weekends they are closed.

The next point is simple. If the forecaster has an ax to grind, be careful of the conclusions! Since it is an art, the forecaster who has a specific purpose can be prejudiced whether conscious of it or not. Some of the trade sources for natural gas price reporting and forecasting are listed in Table 21-10.

Since gas is a commodity and depends more on the factors of supply/demand for pricing than actual costs, gas prices vary significantly over a short time period. Each month, some of the gas trade publications give what is called the "gas price index." This number is based on the price sellers and buyers are using at the end of the month, and it becomes the index for the next month. The index changes each month, and many contracts use the index from a given publication as the price at which gas will be bought or sold for that month. While the monthly gas price index is well known in the industry, many contracts call for the use of other pricing indexes to set the price for the gas, either on a daily or monthly basis.

Instead of the Henry Hub price index, a contract may call for using the Houston Ship Channel price as the index. Many other locations can be used for pricing purposes, and in some cases the price index can be set against a commodity using natural gas in its production such as ammonia. These are elements to be considered in developing the gas contract with the seller or marketer. Using the gas price index on a monthly basis, or taking each day's published price for a given location such as Henry Hub, is another decision to be made by the buyer. An analysis of current monthly indexed prices versus daily prices makes a good reference to deciding which way to go.

A contract will usually call for a penny or two or some other amount per million Btu above or below the index. The major index used at this time, gas price index, is based on natural gas sold at the Henry Hub in Louisiana, a very important hub used for gas sales and trades and the basis of the NYMEX (New York Mercantile Exchange) futures market. There are many more places where gas is traded, and each of these will have an index of its own or a "basis price," a method for converting from the Henry Hub price to that location's price. It is usually based on the value of the gas at that location versus Henry Hub and the added cost of transportation between the two locations. The basis does not always vary as the value of gas transportation changes. When gas prices are rising, the basis value can increase and vice-versa. Examples of these differences can be found in the trade publications listing natural gas prices.

21.4.4 Natural Gas Purchasing Contracts

As has been said previously, the major change to buying natural gas in the new millennium is the ability to buy from many sources. This can mean buying from the actual producer regardless of where the consumer is located, to buying from local or national marketers or the local distribution company. A consumer might buy from the local distribution company in its area either directly from the utility or from a non-regulated marketing subsidiary of the utility. Another major difference today is that the consumer can buy the commodity and the transportation separately or together depending on the source of the gas, the quantity, the service required, and/or the location of the consumer

These changes in how gas can be purchased have

Table 21-10. Natural gas price reporting and forecasting publications.

Publication	Publisher	Home Office	Frequency
Btu Weekly	Energy Management Institute	New York, NY10128	Weekly
Gas Daily	McGraw-Hill	New York, NY	Daily
Inside FERC	Platts/McGraw-Hill	New York, NY	Weekly
Natural Gas Intelligence	Intelligence Press	Washington, DC	Weekly
Natural Gas Week	Energy Intelligence	Washington, DC 20005	Weekly

brought changes in how contracts are written between the supplier and the consumer. If the buyer were responsible for its own transportation, it would mean having a contract for this transportation as well as contracting for the gas supply. It also opens up new considerations. The buyer wants to make sure it is protected in getting the gas it pays for from the vendor and, if the case is such, the transporter as well. In addition, the buyer must be concerned that he is protected from any liability that might occur because of damage caused by the gas in the sale and delivery to the user.

Contracts are legal documents covering these elements and need to be clear and accurate. After something happens—such as being charged for gas not received or for someone hurt in an accident involving the gas in question—is not the time to start looking at the contract. Who is responsible, or what limits there are for the difference between paying for a volume of gas and receiving a smaller amount, and any other conditions and situations differing from what was expected should be stated in the contract. Recourse and responsibilities should be spelled out in the contract. Even in very short times of delivery or for very small quantities of gas to be purchased and delivered, contracts must accurately and legally cover protection of all parties involved in the transaction. This is where an "ounce of prevention is worth a pound of cure!" With contracts being legal documents, the expense and time to insure that proper legal resources are used in negotiating and drawing up the contracts for buying and delivering natural gas are well worth the effort.

A contract or contracts between the two or more the parties will spell out the details of the transactions needed to purchase and deliver the gas from the source to the consumer. Many of today's gas deals are done over the telephone based upon agreed-to basic terms. Some are being done through computer and cyberspace. Whenever there is an on going relationship of supplying natural gas over a period of time, regardless of the length of the time of delivery, there will be a contract or contracts covering buying and selling conditions, including transportation, delivery, metering, payment, ownership, etc. There might be a contract to supply natural gas for as little as an hour, or as long as a year or two, and up to as long as 10 to 20 years. The long-term contracts of the controlled period, when the transporting company marketed gas and worked on a fixed price schedule, are no longer in use.

Typically today, regardless of the terms, contracts include provisions for price adjustments and for security of supply. Also typically today, contracts longer than six months or a year are considered long-term contracts. Contracts up to three to four years can be for a fixed price

or a market-based price, depending on the whims of the buyer and seller. Most fixed-priced contracts today will be based on the financial market for futures contracts, to protect the buyer and seller from catastrophe due to sudden market peaks where the seller would be obligated to supply gas at a fixed price when prices are rising for the commodity. Further, the contract will protect the seller from the buyer ending the contract prematurely. Likewise, the buyer will want protection, should prices drop significantly, to reopen the pricing provisions.

A contract is an allocation of risks between the buyer and seller. It is the same between the party buying the transportation and the transporter. Every business deal involves risks, and the contract sets the responsibilities of the parties so there should be little argument if something does not go according to plan. The seller is taking the risk of supplying gas and the risk of getting its payment for the gas and services supplied. The buyer is taking the risk of having a reliable, secure source of gas for its business needs. These are the major risks each party is taking, and the contract is a written document to insure both are protected as much as possible. Contracts are written documents to help in allocating these risks.

But even the best contract, written by the best lawyers and negotiators, is really no better than the people offering the commodity and services and the people buying the services and commodity. No contract will help if the party involved is not honorable, trustworthy, and capable of doing what it claims it can do in the contract. Further, signing a contract and then planning on going to court to enforce it is a waste of good time and assets of either party. Contracts are like locks on doors—they are for the benefit of good people to insure no one gets confused or forgets the details of the arrangement. Contracts do little to protect from dishonest or untrustworthy business associates.

Of course, even with good contracts and good intentions of the parties, things go wrong and contract disputes arise. These disputes can involve large loses of management time and company assets. Well-written and negotiated contracts can keep the disputes to minimum occurrences and to minimum losses when the unexpected does occur.

Since one or more contracts may be needed to purchase and deliver natural gas, the buyer should be careful of his actions. Contracts for the purchase of natural gas will usually have the major areas of consideration as listed below. Many of these will apply to the transportation contracts as well, unless the purchase of the gas includes the transportation. Since today sellers and buyers will vary considerably in their position in the respective industries, the contract needs to be tailored specifically between the two or more parties involved in the transactions. A contract for buying natural gas from a local distribution company will be different in many aspects from the contract between a marketing company and the buyer. The local distribution company is a regulated entity, and many elements that will be in a contract are part of the regulatory aspect. Most of the specific items the utility will have to abide by are given in its tariffs, which are filed with local or state regulatory agencies.

A general form natural gas contract for use between a buyer and a marketer is available from the North American Energy Standards Board, Inc. (NAESB), and is called **Base Contract for Sale and Purchase of Natural Gas.** Many marketers use this as a starting point for developing contract trading. It is used for both firm and interruptible sales and covers both written and oral transactions. Section headings in the contract are as follows.

1. Purpose and Procedures
2. Definitions
3. Performance Obligation
4. Transportation, Nominations, and Imbalances
5. Quality and Measurement
6. Taxes
7. Billing, Payment, and Audit
8. Title, Warranty, and Indemnity
9. Notices
10. Financial Responsibility
11. Force Majeure
12. Term
13. Limitations
14. Miscellaneous

Some of the major elements of any typical gas purchase contract are outlined in the discussion below. These are applicable to the NAESB mentioned earlier. But remember, there is no typical contract, as each situation requires its own specific contract to insure security of supply at the best economic costs.

1. Purpose & Scope of the Agreement—What is to be accomplished by the contract. Who will be supplying gas, how will it be transported, and who will receive the gas. Additional comments as to the potential use, whether a sole supplier, etc. might also be included in this section.

2. Definitions—Lists the standard and special terms used in the contract. Especially important in natural gas contracts because of the uniqueness of the commodity, the industry ways of doing business, and the specific parameters of the operations for which the gas is being purchased by this contract.

3. Term of the Agreement—A statement giving how long the agreement will be in force and what conditions will terminate the agreement. Some contracts will include information on methods and options of extending the contract past the initial terms of the agreement.

4. Quantity—Here the details of the total quantity of natural gas to be sold and delivered by the seller and received by the buyer will be stated. Information on the daily contract quantity (DCQ) or even hourly contract quantity will be stated. Any specific deviations from the regular amount, such as swing quantities needed during high production or other causes, are listed here. Penalties the seller is willing to accept for the buyer's failure to take the quantity of gas set in the contract will be listed in this section. Also the converse, penalties the buyer is willing to accept for the seller's nonperformance according to the contract, will be stated in this section. If there is any take-or-pay language, this is the section for it. Take-or-pay is an agreement for the buyer to pay for gas if contracted but not taken. The buyer usually has a period to make up the deficiency. This section will also state whether the gas is being sold on a firm basis, with the buyer and seller obligated as stated to perform, or if the gas is being made available or will be taken on a "best efforts" basis. Very important in this section could be the ways the buyer "nominates" takes for certain periods. The section will include means for balancing the account and give additional penalties for under- or over-quantities of gas taken by the buyer. Other subjects that play a role in the quantity of gas to be supplied, such as well or reserve measurements if buying directly from a producer or other supply considerations if buying from a marketer, can be in this section.

5. Price—Price to be paid for the natural gas to be delivered by this contract, as well as any statements regarding price escalators and/or means to renegotiate the price, will be stated in this section. Omissions of statements to this effect can be construed as a statement of the contract, so care must be exercised that what is said and what is left out is properly covered. Any language needed for agreement on price indexing or other means of adjusting the price to current market conditions will be included. The writing should include provisions

for both price increases as well as decreases, if this is the desired purpose of the statements regarding changes for market or other conditions.

The price section will cover any additional expenses or costs the buyer is willing to undertake in addition to the direct cost of the gas. If the contract is with the producer or an interest owner in a gas well, this section will state who is responsible for any gathering, treating, or processing costs. Again, for a contract with the seller being a producer, provisions will be in this section for who has responsibility for severance and other taxes, royalties, or other charges for which the producer might be liable for payment. Pricing units most commonly used today will be energy units such as British thermal units (Btu). Since a typical volume measurement of natural gas is a thousand cubic feet (Mcf) and typical pipeline quality gas of this volume would have about a million Btu in energy units, the typical unit for gas sales is a million Btu or MMBtu.

The pricing section will also include language in today's contracts protecting both the buyer and the seller from the vagaries of the natural gas markets today. While these in effect reduce the coverage of the contract and change some of the allocations of risks set by the contract, often both parties are willing to have a contract with legal means of changing the pricing conditions of the contract. The long-term, fixed price contracts went out with decontrol. There are still fixed price contracts, but the seller will protect its position by going to the financial market and buying futures to protect his position of supplying long-term, fixed price natural gas. Since the seller is taking steps to insure supplying the gas at a fixed price by buying futures, the seller will protect his actions by putting clauses in the contract to protect this position, should the buyer fail to take the gas as contracted.

6. Transportation—Transportation details as to who has responsibility (the transporter, costs, etc.) to deliver the gas to the buyer must be included in the contract. Crucial items are who is responsible for arranging the transportation, who will pay for the transportation, and whether the transportation will be on a firm or interruptible basis. The transportation must cover the full course of bringing the natural gas from the source to the buyer's location, including bringing it to the accepted delivery point(s) as stated in the next paragraph, Section

7. The buyer must insure they are covered in case transportation is unobtainable or ceases after delivery has started. The contract will include any special conditions on either the buyer's or seller's part to take into account any special situations either party could have that interferes with transporting the gas from the source to the delivery point(s). Further, any regulatory matters pertaining to the gas transportation should be referred to in this section, and in more detail in the regulatory section discussed further in this chapter. This section should cover who has responsibility for overages and balancing of the account, measurement, disagreements on quantities, and payments of transportation and associated charges.

7. Delivery Point(s)—Since the delivery point or points are different in each situation, the contract needs to state delivery or alternate delivery points in very specific language. This clause can become a very crucial one in times where there is a dispute over quantities of gas sold or received. There is also a slightly different interpretation of this clause in light of the new sales methods where there is a separate contract for the sale of the natural gas and one for the transportation.

To the gas buyer, the only real delivery point is when the natural gas crosses the meter and the valve where the gas comes directly into the buyer's system. The buyer wants to be responsible only for the gas received in his system. What gas is presumed sold or dispatched at some other location, such as where the gas might come off of an interstate pipeline into the pipe of the local pipeline delivering the gas to the consumer, is really not the concern of the buyer. This is an argument between the two pipelines or between the pipeline and the seller, depending on the contract for transportation between the significant parties. Delivery point is also crucial in assessing responsibility for problems that might arise from the gas in question. In the case of an explosion or fire resulting from the improper handling of the gas, the ensuing legal action by one or more parties could be influenced by the delivery point as to who had responsibility for the gas at the time and location of the accident.

At times, delivery points may need to be changed to meet requirements of either parties, and the need to change should be included in this section to insure that changing the locations according to the contract do not in any way negate the contract or the terms in the contract.

8. Measurement & Quality—Methods, conditions, timing, and authority for the measurement of the gas volume or energy content and quality are given in this section. Usually, a trade association or other organization's methods and requirements for measurement are called for in this section to insure the proper measurement of the quantity of gas sold or bought and the quality of the gas under the contract. Remedies or alternatives should be included in this section for those cases where the gas fails to meet the quality requirements of the contract, whether on a short-time, unexpected rare or single occurrence or a continuing failure to meet the specifications.

9. Billing—Terms for the billing, who is responsible for payment, manner, and methods of payment, etc., are included in this section.

10. Force Majeure—This the clause in the contract to protect both parties affected by a totally unforeseeable occurrence which is beyond the control of the party seeking protection from the responsibilities of the contract. Many times, this is referred to as an "act of God" and includes severe weather, acts of war or insurrection, strikes, etc.

11. Warranty of Title—The clause guarantees the seller has title to the gas and can sell it. Included are allowances for the buyer to recover damages if there is a failure of title, should another party make claim to ownership of the gas.

12. Regulatory—All necessary permits, licenses, etc. must be obtained according to FERC regulations and any state or local authorities having jurisdiction over the selling and transportation of the natural gas. The party or parties having the responsibility for obtaining these items and the payments required should be fully covered.

13. Assignments—Specifications for the transfer of rights under the contract are covered in this section. This could be an important item in light of the various changes occurring in the gas industry today. The buyer should insure coverage that includes changes that might impact gas transportation as well as the commodity, if the seller was responsible for transportation as well as for the natural gas.

21.5 NEW FRONTIERS FOR THE GAS INDUSTRY

Many challenges face the natural gas industry as energy becomes more and more important and greater reliance is placed on natural gas for the country's energy. Each of these will have an impact on the future buying and trading of the commodity. A summary of these follows.

1. Complete the natural gas decontrol to the final level—local markets in each state.

2. Complete the development of an energy industry that incorporates other energy sources like power, fuel oil, nuclear, wind, biomass, etc. and broaden into one massive energy grid to cover shortages or interruptions in any specific source of energy.

3. Develop new energy sources of both forms already used and new forms to make the U.S. energy independent.

4. Develop the delivery system to insure secure supply of the larger quantities of natural gas as demand forecasts for future years.

5. Develop new and additional natural gas supply sources such a gas hydrates to meet the forecasted demand through the 2030 and forward. Insure sufficient capital is available to develop new sources and systems.

21.6 SUMMARY

Natural gas as a fuel has many advantages over other fuels. Based on quantity and location of use, there are many permutations on how to buy gas. But it is really simple and, once mastered, rewards will be seen. U.S. natural gas supplies are adequate, and marketers and others are available to assist in finding the right sources for natural gas supply.

CHAPTER 22

CONTROL SYSTEMS

STEVE DOTY, PE, CEM

22.1 INTRODUCTION

Control systems are an integral part of many energy related processes. Control systems can be as simple as a residential thermostat, to very complex computer controlled systems for multiple buildings, to industrial process control. Their diligence and repeatability can also serve to maintain the savings of project improvements for years, further justifying their existence by providing economic return to the customer. This chapter will introduce the reader to some concepts of automatic control theory, followed by practical applications useful to the field of the energy professional. Upon completing this chapter the reader will gain a basic understanding of common terms, control technology and control mode categories, basic input and output instrumentation, and the practical need to temper "things possible" with the skill level of the operators who will inherit it. The importance of system controllability and user-friendliness as primary design parameters will be stressed. Basic control strategies will be discussed, as will estimating savings from the use of automatic controls. Finally, there will be an introduction to some complex optimization methods and suggested topics for further study. Examples are used throughout.

The intent of this chapter is to focus on the application of automatic controls as a tool to achieve energy savings. Chapter 12, Energy Management Control Systems, is a partner chapter to this one, and it expands the particulars of automatic controls to their application as a system within buildings. There will be some overlap between chapters, but in general they complement each other. Some background information on controls is obviously a necessary prerequisite; however, this chapter does not attempt to cover all aspects of automatic control theory or application—or to make the reader a controls expert. Less emphasis will be placed on hardware and theory, and more emphasis will be placed on practical applications and tangible benefits. A basic background in energy-related subject matter, common energy units, common energy-consuming machinery and systems, HVAC concepts, and the general field of energy engi-neering is assumed. The field of automatic controls is a busy technology with lots of jargon, hardware and software variations and details, and sheer volume that can create an air of mystery and awe. If this chapter is successful, the reader will be able to separate the fundamental control concepts from the technical details and effectively apply automatic controls to achieve energy savings.

Many of the examples given are for commercial building HVAC and lighting systems since these examples are common and should be familiar to the reader. Similar concepts and considerations apply equally to other fields of endeavor where energy savings are a driving force.

A glossary of terms provides clarification of common terms used in the field of automatic controls.

22.2 WHY AUTOMATIC CONTROL?

- *Regulation:* Many things need attention and adjustment to compensate for changing conditions or varying demands. Examples of this are common in living organisms, such as body temperature, blood pressure, etc. Process control regulation is really just emulating the concepts of such natural processes. The field of automatic control is similar in that we *continually adjust some device to cause a particular measured variable to remain at a desired state.*
  Examples:
  — The need to throttle heating and cooling equipment sized for maximum load that is effectively over-sized at part load conditions.
  — Varying occupancy, and systems attendant to the occupants (lighting, ventilation).
  — Varying product throughput rate through manufacturing facilities.
  — Varying demands, and the need to maintain level or full state for water or fuel reservoirs, feed or coal bins, etc.
- *Coordination:* Organizing or sequencing multiple processes in a logical and efficient manner is an important aspect of automatic control applications.
- *Automation:* Human beings can make very good manual controllers because we can think on our feet and consider many variables together, but most con-

trol tasks are repetitive and suitable for mechanization. Automatic operation allows people to provide oversight of system operations and more effectively utilize their time.

- *Consistency:* Manual control by people can be effective, although we are not all that repeatable and are sometimes forgetful. Using machinery for automatic control adds the improvement of consistent, repeatable operations. The repeatability and consistency feature of automatic control is very important in manufacturing.

- *Conservation:* Supplemental enhancement control routines can be incorporated to reduce energy use while still maintaining good control. It is important to note that control systems do not necessarily reduce energy consumption unless specifically applied and designed for that purpose.

22.3 WHY OPTIMIZATION?

The *80-20 rule* reminds us that we can usually hope to achieve 80 percent of a measure's potential with 20 percent of the difficulty, but the remaining portion requires much more effort. For this text, *optimizing* refers to reducing energy consumption as much as possible, often approaching the barrier of diminishing returns. Optimization can be characterized as taking over where the basic controls left off and working on the remaining opportunities—the ones that aren't as easy to attain. The appropriate use of optimization depends upon the customer's priorities, and these should be tested before the decision to *optimize* is made. Of course, from an environmental standpoint, we would press for that last 20 percent. But if maximum simplicity controls that require only basic skills are a main focus of the customer, optimization may not be a good application. Similarly, projects where reliability is priority #1 may be better served with basic control routines, allowing the extra 20 percent potential to slip away to gain the advantages of simplicity. Economics always comes into play, and some optimization projects (chasing the last 20 percent) may not have the attractive payback periods of their 80 percent counterparts. Most projects represent some balance of these interests, depending upon the needs of the customer. It is important to understand that optimization for maximum benefit will not be for everyone.

A case in point for optimization is the subject of fixed setpoints, which are often a matter of convenience or approximation and usually represent a compromise in optimal energy use. The more factors we can take into account, the closer to optimal will be the result, as stated

by Liptak: "*...multivariable optimization is the approach of common sense. It is the control technique applied by nature, and frequently it is also the simplest and most elegant method of control.*" [1: pp xi]

To summarize, *the desires for maximum simplicity and maximum efficiency are at odds with each other.* **A system that is perceived as being too complex will likely fall into disrepair and be bypassed or unplugged. If the customers are committed to squeezing their energy costs through optimization, they will need to also embrace the technology and be willing to accept additional complication and raise the bar of required operational skill. This concept should be discussed with the customer in advance to be sure the project isn't set up to fail by being unacceptably complex.**

22.4 TECHNOLOGY CLASSIFICATIONS

22.4.1 Introduction

The following is a very important first statement before any discussion about control hardware: "*The type of hardware used in optimization is less important than the understanding of the process and of the control concepts that are to be implemented.*" [1: pp 42]

The main goal should be to become clear about the process fundamentals and what should happen—then the parts and pieces are just details. This discussion of different available hardware types is a familiar but sometimes laborious and dull part of any controls text. Remember that automatic controls are really nothing more than machines that do for us what we would do ourselves if we had nothing better to do; they do work for us like any other tool, and they are only as clever as the people who craft them.

22.4.2 Conventional Electric
(also On-Off or Two-State)

Electricity is used as the power source. Control is discrete (on-off, high-low, cut-in/cut-out, etc.). Contact closures are used to implement control logic. This principle of control has widespread use, varying from simple and familiar control to complex interlocks and Boolean Logic.

Examples:

- A basic home heating thermostat *cuts in* at 67 degrees F and *cuts out* at 69 degrees F, thereby maintaining a room temperature of approximately 68 degrees F.

- A well pump controller *cuts in* when the storage tank pressure is 50 psig and cuts out when the storage tank pressure is 70 psig, thereby maintaining a system pressure of approximately 60 psig.

- An interlock circuit that prevents an exhaust fan from starting until the associated make-up air damper is first proven open.

The differentiator between conventional electric controls and analog electronic controls is the discrete (two-state) nature of the inputs and outputs; analog controls have varying rate inputs and outputs.

22.4.1 Floating Control

A variation of this two-state electric control is "floating control." Technically still "on-off," this unique control method has a system control action similar to analog (modulating) control. Floating control is described in more detail in control modes.

22.4.3 Analog Electronic

Electricity is used as the power source. The key difference between analog and conventional electric control is *modulation*. Analog controls have variable inputs and outputs, not just two states. Minor changes in output positioning of controlled devices make tighter control possible than with two-state (on-off) electric controls.

The hardware for analog electronic controls may include resistors, rheostats, Wheatstone bridges, operational amplifiers, or it may use solid state components to measure the process and modulate the output devices.

Some considerations of analog electronic controls:

- Unless a control dead band is built into the controller, it is likely that the controlled device will be activated by even the minutest deviation from setpoint. The action may be nearly imperceptible, but if this occurs the repeated hovering about the ideal setpoint may cause premature failure of the actuators.
- Analog electronic technology *"connectibility"* options are generally limited to remote adjustments, remote alarm panels, etc. Their "stand alone" nature and modest cost make them a popular choice for basic HVAC manufactured systems that come with factory-installed controls.

22.4.4 Pneumatic

This is the general term for controls that use compressed air as the motive force for control inputs and outputs, instead of electricity. Analog pneumatic pressures are alternately coded and de-coded into control units; for example: 3-15 psig = 0-100 degF. Discrete pneumatic pressures are also coded; for example, 0 psig = off, 20 psig = on. Pneumatic devices can be two-state or modulating, but they are most commonly modulating. Pneumatic controls often have interface devices that communicate pneumatic signals to and from their electric counterparts, such as pressure-to-electric switches (PE switches) and various electric-to-pneumatic solenoids and transducers.

Some considerations for pneumatic controls:

- Air supply quality is critical—no oil or water allowed at the instruments!
- Contamination in the main air system or a loss of air pressure represents a single point of failure for pneumatic controls.

22.4.5 Digital Control (also called Direct Digital Control or DDC)

This control technology uses microprocessors to provide control. A big advantage of DDC controls is the fact that changes to the system are often made with software and do not automatically require physical changes and cost like other technologies. Discrete (on-off) information is readily absorbed as a "1" or "0." For analog processing, interface equipment called digital-to-analog converters (D/A) and analog-to-digital converters (A/D) are used. The higher the resolution of the A/D and D/A conversion, the closer the digital signals resemble true analog signals, allowing smoother control.

One of the great enhancements of digital controls over the last 30 years has been the concept of distributed control. This technology shift occurred in response to customer complaints of excessive dependence on single hardware or software points, and widespread loss of control after a single item failure. It is generally good practice to use multiple smaller controllers at the points of control. These communicate upstream to a supervisory operator workstation, but each single failure point now affects a much smaller area, increasing overall process reliability.

Refer to **Chapter 12—Energy Management Control Systems** for greater detail on DDC control systems and the following:

- DDC Graphics
- Field User Interface
- DDC Controller Hardware
- Proprietary vs. Open DDC Controls
- DDC Information Technology (IT) System Maintenance

22.5 CONTROL MODES

22.5.1 Introduction

Deciding which *control mode* to apply is important, regardless of the technology used. It is important to understand that these modes can be implemented using many of the available technology types. In many cases, simple on-off control is adequate and very appropriate. In other cases, the desired effect can only be achieved

TECHNOLOGY	PROS	CONS
Conventional Electric	• Simple • Proven • Easy to understand and troubleshoot • Inexpensive • Accommodates "and-or" logic with simple series-parallel circuits (relay logic). • Floating control variation approaches analog control quality with reduced cost.	• Definite purpose controls are not flexible without hardware change. • Limited capabilities for optimization.
Analog Electronic	• Lowest cost stand-alone modulating control option. • Often standard equipment on packaged HVAC equipment.	• Long term drift unless properly maintained. • Short-lived components. • Definite purpose controls are not flexible without hardware change. • Often not user-friendly. • Limited capabilities for optimization. • Input/Output hardware items, especially actuators, are more expensive than conventional on-off devices.
Pneumatic	• Simple • Proven • End devices extremely durable • Can be long lived if designed and maintained properly.	• Long term drift unless properly maintained. • Temperature dependent drift. • Definite purpose controls are not flexible without hardware change. • Limited capabilities for optimization. • Susceptible to system-wide failure if compressed air system is contaminated. • Complicated adjustments for pneumatic controllers • First cost includes infrastructure cost (tubing). • Future operator work force may have a reduction in skill level for this technology.
Digital Control	• Excellent flexibility, using software changes with existing hardware. • Best option for optimization due to computing power. • Opportunities for communication linkage to other, compatible equipment for large data gathering / workstation benefit. Convenient method of recording key measurements, baseline information, historical data, trends, run-times, etc. • Convenient remote monitoring.	• Highest first cost. • First cost includes infrastructure cost (cabling). • Often outdated before physically worn out, due to rapid technology advances. • Tendency for proprietary equipment manufacturers to create captive customers and high life cycle costs for repair parts and software upgrades.

Figure 22-1. Pros and Cons of Different Control Technologies

with modulating controls. The following are basic control modes. The accompanying diagrams will illustrate typical system performance.

The term *system capacitance* refers to the rate of response of a system to a stimulus. Systems with a large capacitance tend to resist change, and the effects of control are felt more slowly than with systems of smaller capacitance. Comparing the effect to a flywheel or relative mass is a good way to describe this concept. Another useful example to illustrate system capacitance is an instantaneous electric water heater (small volume of water) compared to a standard residential tank-type water heater. Upon energizing the heater elements, the water temperature in the tank unit changes much more slowly because it has more mass, and we say it has greater system capacitance.

The term *gain* is a control term synonymous with sensitivity and is usually an adjustable amplification value used to tune the controls. If a quicker response is desired for a small input change the gain is increased, in a

stronger output reaction from the controller.

22.5.2 On-off Control

Also called *two position control*, this rudimentary mode is used with equipment that is either *on* or *off*. A nominal setpoint exists but is rarely actually achieved except in passing. A *range* of control values must be tolerated to avoid short-cycling the equipment, and temperature ranges are often fairly wide for this reason. In the case of equipment that cannot be modulated, this is often the only choice. The smoothness of control depends strongly upon the system capacitance; systems with very low capacitance can experience short cycling problems using two position control.

22.5.3 Floating Control

This is a hybrid combination of on-off control and modulating control, also called *incremental* control. As with on-off control, there is a control range (cut-in/cut-out). However, unlike on-off control, floating control systems have the ability to maintain a mid-position of the controlled device instead of being limited to full-on or full-off. Between the cut-in and cut-out thresholds the controlled device merely holds its last position. The process variable is not actually under control within this range, and it is seen to *float* with the load until it crosses a threshold to get another incremental nudge in the correcting direction. This control is tighter than simple on-off control, although not as tight as true modulating control, but it is inexpensive and reliable. Equipment items from small HVAC terminal units to 1000 HP water chiller inlet vanes are controlled in this manner with good success. Note that floating control is limited to processes that

change slowly, and floating control actuators are usually selected to be slow moving.

22.5.4 Proportional-only Control (P)

This is the basic modulating control and what most commercial pneumatic and analog electronic systems utilize. It is essentially an error-sensing device with an adjustable gain or amplification. A control output is issued to regulate a process, and the magnitude of the output is directly *proportional* to the size of the error. This type of control is economical and reliable. A characteristic offset (residual error) is natural with this type of controller, and the size of the offset will increase with load. This offset occurs because an error must increase (further off setpoint) before an output increase can occur.

If the proportional control action is too sensitive (gain set too high) the controller's response will be excessive, and oscillation or *hunting* will occur. When this occurs the controller output (and the equipment connected to it) will oscillate up-and-down, open-and-closed, etc., and the control action will not settle out.

22.5.5 Proportional-Plus-Integral Control (PI)

An *integral* function is added to a proportional-only controller to eliminate the residual error. This control action adjusts the gain to a stronger and stronger value until the error is eliminated. In theory, the integral controller will not rest as long as any error exists; however, it is common to allow a small, acceptable error band around the nominal setpoint to prevent incessant low-level hunting as the controller seeks the perfect "zero" error condition. In practice this integral action has the appearance of slowly but surely building up the output to taper off the error. Since the effect is slow but also relentless, integral control problems can occur in processes that change rapidly. Also, if the controller is left active for long periods while the controlled equipment has been turned off, the integral controller will "wind up" to a 100% output. Upon start-up after a long period of wind-up, the integral function can be strong enough to "stick" at full output for long periods of time with complete loss of control. Therefore, whenever integral control is used, some form of hardware or software interlock must be used as an *anti-wind-up* measure.

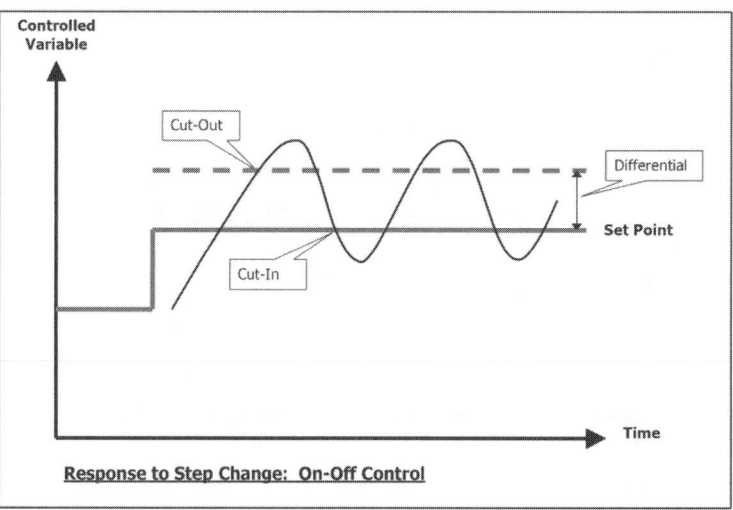

Figure 22-2. On-Off Control Mode Diagram

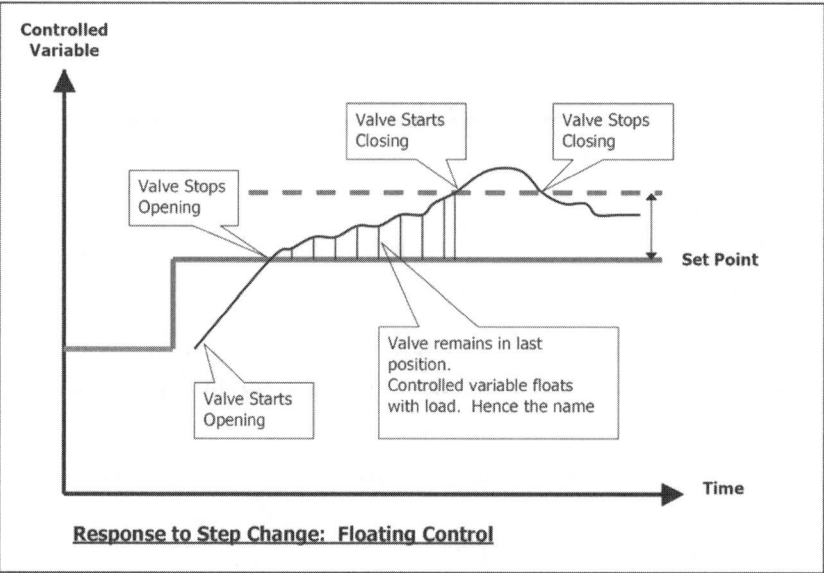

Figure 22-3. Floating Control Mode Diagram

The most common approach to preventing wind-up is to simply turn off the controller when the process is stopped.

22.5.6 Proportional-integral-derivative Control (PID)

PID control is used to accommodate rapid changes in process and minimize overshoot. This is done by reacting to the rate of change of error (derivative) instead of the magnitude of the error (proportional) or the duration of the error (integral). In reviewing the characteristic response curves, PID control looks like the absolute best and, in fact, does provide the tightest control of all the control modes. However, the derivative gain is very touchy to set up properly and can easily cause instability of control, especially at the beginning of a batch process or after a large step change. In HVAC work, the derivative term is seldom used, to avoid the potential for instability and because most HVAC processes are relatively slow acting and tolerant of temporary overshoot of PI control. In many process control applications, PID control is essential since close control is often tied to product quality.

22.5.7 Sequencing

While not actually a separate control mode, this topic deserves special attention since it has great potential for energy savings. A significant blight in many industry processes is the overlapping of adjacent and opposing processes. A common example is HVAC *reheat*, where air that has just been cooled with an energy source is now being heated. This is analogous to driving around with the accelerator fixed and controlling the vehicle speed with the brakes; even if the detriment to the brake system is neglected, the effect upon vehicle fuel economy is obvious. There are many examples of heating/cooling overlap, some deliberate but most inadvertent—but all increase energy consumption. One very effective tool in combating this is effective equipment sequencing. Due to imperfections of real-world control systems, some control systems that appear sequenced will allow overlap either occasionally or over time. Some things that contribute to inadvertent overlap are:

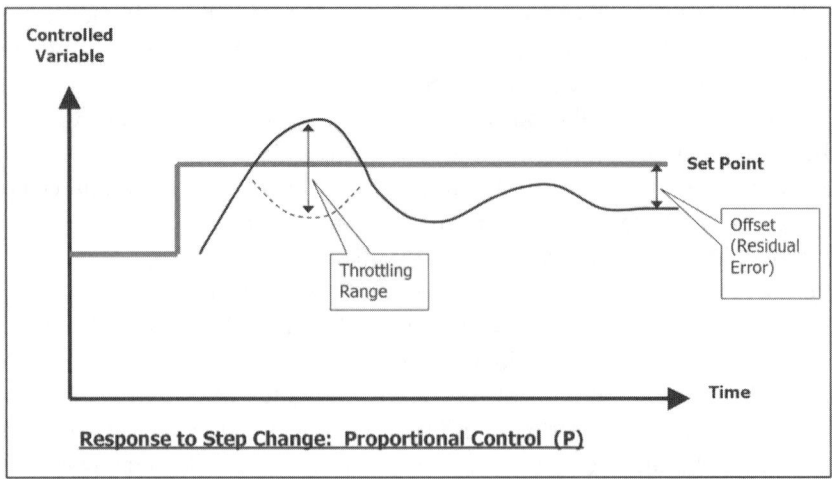

Figure 22-4. Proportional Control Mode Diagram

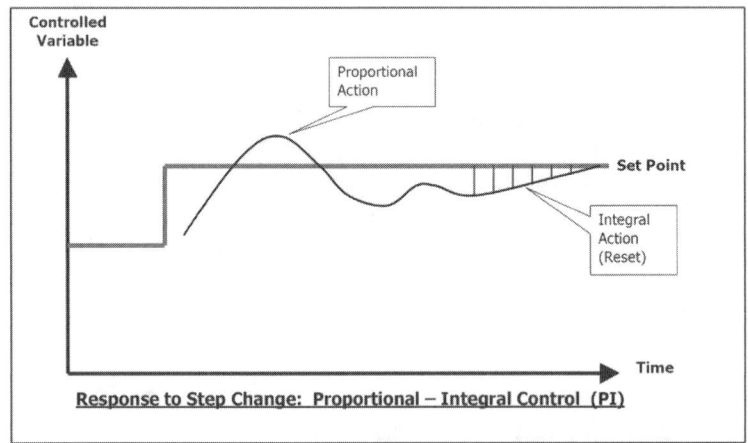

Figure 22-5. Proportional-integral Control Mode Diagram

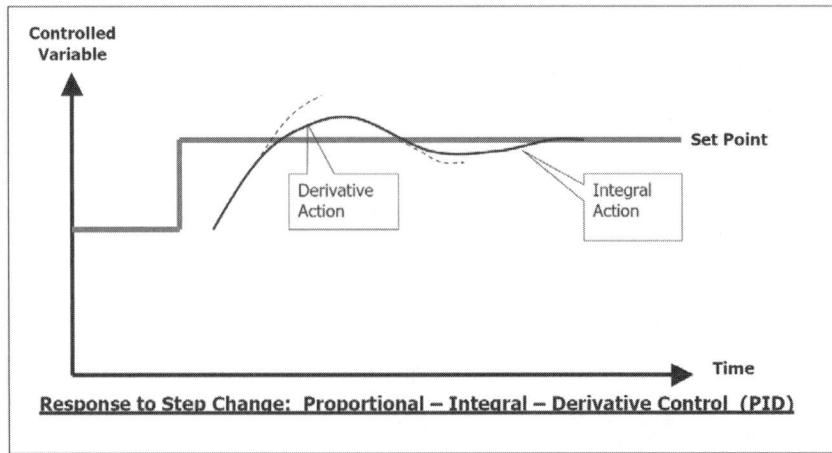

Figure 22-6. Proportional-Integral-Derivative Control Mode Diagram

• Instrument drift
• System influence upon actuator range of motion
• Normal overshoot from on-off controls

- Normal offset from proportional control
- Normal time lag from integral control
- Independent, but sequenced, modulating controllers
- Adjacent processes, such as two comfort zones in an open bay with widely different user setpoints

Where the process does not require absolute tight control, building in a blank spaces or *dead band* between sequenced elements is a simple way of achieving and sustaining energy savings. Within the dead band there is no control; however, the possibility of overlap is eliminated.

22.6 INPUT/OUTPUT DEVICES

22.6.1 Introduction

There are many available input and output devices, serving many basic and specific needs, along with ranges of quality and other features as required for the job. The controls application engineer quickly becomes familiar with many of these in great detail. Unless there is a specific interest, the energy professional does not need to delve into the sea of products; project direction can be very effective by providing only performance-based generic requirements, delegating the product selection to the vendor.

For input and output devices, there are basic distinctions between *transducers, switches, sensors,* and *transmitters* that are useful to understand.

- **Transducers:** These are the core of any instrument and are used to convert the basic physical phenomena of interest into a form more useful to the instrument. Examples:
  — Temperature: bimetal coil, two-phase gas bellows, thermistor, RTD, etc.

  — Pressure: Bourdon tube, diaphragm, strain gage, etc.

The term *transducer* is also commonly applied to output instruments that convert the signal from one form to another, such as converting an analog electronic signal to a corresponding pneumatic signal for use by a pneumatic actuator.

- **Switches:** These are devices that can have two states (on-off, open-close, etc.) used for regulating on-off electrical circuits or to actuate other equipment or devices in a two-position manner. Solenoid valves, relays, etc. are in this class of instruments. Switches are also used as inputs. Examples:
  — Pressure switch proves pump operation.
  — Air flow switch proves fan operation.
  — Current switch proves motor operation.
  — Relay start/stop action controls motors or lights.

- **Sensors:** These are "passive" input devices that can be read directly by the controller with no signal conditioning. They are usually limited to short cable lengths and further limited to transducers with inherent linear outputs. The distance limitation is due to the resistance of the cable itself, which adds to the resistance of the sensor and creates error.
  Example:
  — RTD (resistance temperature detector) that can be wired and read directly to the controller.

- **Transmitters:** These are the typical analog input instruments. A transducer is coupled with some form of pneumatic or electronic signal conditioning. The transmitter output is a linear, standard value easily decoded as an input. Often, the signal conditioning

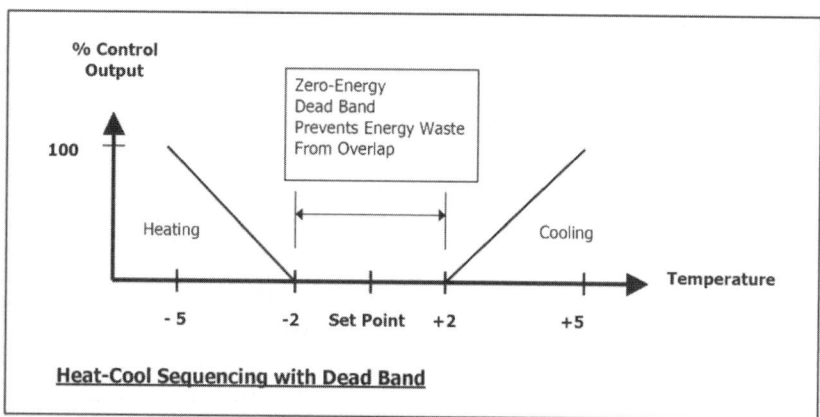

Figure 22-7. Sequencing with Dead Band

allows for stable signal over a large distance to allow remote location of the device, hence the term "transmitter."

Examples:

— 0-100 degF	→ 3-15 psig
— 0-100 degF	→ 4-20 mA (milliamps)
— 0-100 degF	→ 0-10 VDC (Volts DC)
— 4-20mA output	→ 3-15 psig

22.6.2 Conventional Devices and Wiring

Conventional wiring architecture consists of devices in the field, each wired back to the input or output terminals of the controller with a dedicated set of wires (often a pair of wires). Thus, for 100 instruments there would be 100 sets of wires finding their way home to the controller. This type of wiring is very straight forward and is referred to as *home run* wiring. At the control panel the number of wires is the greatest, and the system of wires fans out and disperses as it gets farther from the central point.

22.6.3 Addressable Devices and Wiring

Also called *smart* devices, these devices can be controlled directly by a digital control system by giving the device a unique identification code or "address" to differentiate it from all other similar devices. Each *addressable* device has a means of setting up its own unique address or name to permit uniqueness in communication. *Addressable* devices relay their information or accept their commands digitally and do not rely on transmitters or the like. Addressable devices have the distinct advantage of reduced wiring since a single loop of communication trunk wiring can be shared by multiple devices; on systems with large number of points, the difference can be remarkable. However, addressable devices cost more than conventional (non-addressable) devices since they include on-board communication hardware, as well as A/D and D/A converters for addressable analog devices. Since they cost more their application requires balancing the cost vs. the benefits. The single loop communication can present new failure modes compared to conventional *home run* wiring methods. (i.e., If a single cable is cut how many devices are affected?)

Common Addressable Devices:
• Fire alarm devices—input and output
• Lighting control devices—input and output
• Security devices—input and output

Other Addressable Devices:
• Lighting ballast
• Actuators

• Transmitters and sensors
• Thermostats
• Relays

22.6.4 Linearization

"For simplicity of design, a linear relationship between input and output is highly desirable." [3: pp 28]

Linearity is a high priority in instrumentation for both inputs and outputs. This is because the controller's algorithms or ratios are arithmetic in nature, comparing the input to a standard and producing a derived output. Linearization makes the effect of the controller's decisions predictable and manageable, and makes instruments interchangeable. Understanding how linearization can affect the control system is important to assure project success.

Input/output instruments are considered linear when an incremental change of input value produces an equal increment of output value regardless of the value of the input or location of the device's range. For example, a change of 1 degree F might produce a change of 0.16mA in transmitter output. If truly linear, the device would produce this 0.16mA change in current if reading 0 to 1 degF, or if reading 100 to 101 degF. *To the extent that non-linearity exists in the control loop, errors and unpredictability will also exist.*

Many natural phenomena are linear, and many are not. Linear examples include metal resistance with respect to temperature changes, static pressure with respect to depth of liquid, and volume of a vertical cylinder with respect to level. Non-linear examples include the volume of a conical container or a horizontal cylinder with respect to level, the change in flow with respect to butterfly valve or single blade damper position, and a heat exchanger heat transfer rate with respect to flow rate. Some natural phenomena behave in predictable but non-linear fashion. These include thermocouples and differential pressure flow meters (head loss devices). Thermocouples are linearized by a look-up table or mathematical expression that defines the non-linearity, while head-loss meters are linearized with a square root extractor. Most common input measurements have already been linearized by the instrument manufacturers. Instead of hardware characterization, it is possible to linearize inputs and outputs through software. This "software linearization" is sometimes used with industrial controls but seldom with commercial controls.

A common issue with linearization is control valves and dampers. These are notorious for having non-linear response with respect to travel position. Control valves are generally characterized as either *linear, quick opening* or *equal percentage*. For valves, the flow "character" is achieved by specially contouring the valve plug to influence the flow rates at different valve stem positions. Characterized ball valves are also available to greatly improve the inherent quick-opening flow pattern of these valve trim shapes. Control dampers and butterfly valves are either flat blade or flat disk shapes and do not have selectable characterized flow patterns like control valves do. In the case of control dampers, about all that can be done to reduce non-linear air flow through the damper is to down-size it to create a high pressure drop, which may create other complications or costs, especially in outside air ducts and large ducts without room for transitions. Where linear control of an air stream is important, *air valves* can be used that are available with characterized flow patterns.

Modulation of a heat exchange process is a common automatic control application. The following example applies for most types of heat exchanges, including shell and tube, tube and fin, etc., and all HVAC air coils. The effect is more pronounced in heating applications than cooling, owing to higher temperature differentials. The heat transfer characteristics of a heat exchanger can be likened to a "quick opening valve" since the incremental change in heat transfer for the first fraction of fluid input is much higher than the last fraction. Controlling flow linearly through a heat exchanger will yield a non-linear output with associated control problems, especially tuning issues. In this case, the non-linear flow characteristic of a control valve is deliberately used to improve process control. The standard approach to correct this is to use a control valve with a flow characteristic that is a mirror image of the inherent heat exchanger performance (*equal percentage type*), canceling this inherent phenomenon so the overall control effect is nearly linear. This example not only illustrates how linearity is important to a control system but also points out that *for heat exchanger applications, the control valve selected should almost always be the equal percentage type.*

22.7 VALVES AND DAMPERS

22.7.1 Valve and Damper Selection

Selection criteria usually include a minimum pressure drop for proper *authority* and best linearity over the process.

22.7.1.1 Valve Sizing

The typical sizing procedure for a hydronic control valve is 5 psig wide open pressure drop. The following explains the reason for this. A rule of thumb for reasonable control of a heat transfer coil is for the control valve wide open pressure drop to be at least as high as the coil it controls. So, if an air handler coil is sized for a 5 psig pressure drop at full flow, then the control valve at 5 psig full flow pressure drop would be appropriate. It is common practice to select HVAC water coils at 10 ft. w.c. or so wide open pressure drop, which equates to 10/2.31 = 4.3 psig. Thus the 5-psi valve pressure drop convention is a reflection of the coil sizing convention. An extension of

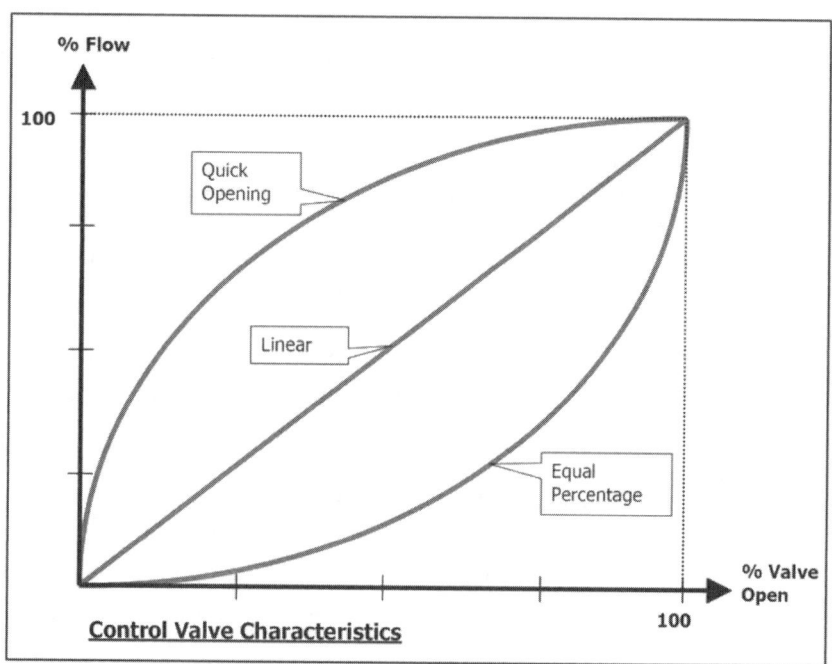

Figure 22-8. Control Valve Characteristics

this logic would be that if the coil were selected at 1 psig wide open pressure drop, then the control valve sizing criteria could also be reduced. This is in fact the case, although seldom done in practice since coils selected at extremely low pressure drop have other issues. However, if the coil were selected at 10 psig wide open pressure drop, a 5 psig control valve would be over-sized and tend to hunt.

Editor's Note: The practice of adding system resistance to achieve good control is counterintuitive and definitely an opportunity for improvement in the industry, because adding circuit resistance to any fluid handling system increases the system energy requirements.

22.7.1.2 Leak-by and Close Off

When specifying control actuators, specifications and close attention are important to achieve reliable close-off performance. This is true for both valves and dampers, especially those with marginal actuator close-off ratings, excessive system pressure, large damper sections, metal-seated valves, undercut butterfly valves, and actuators without a positive seating mechanism to impart a residual tight seating force. Spring-return pneumatic actuators are forgiving in this sense, because they inherently provide residual seating force. Some electronic actuators have mechanisms to provide ample seating force; others rely on simple travel adjustments that define the open and closed positions, which are undesirable since the opportunity for internal leak by, with subsequent heat/cool overlap, is high. Ball valves are inherently tight seating and can be used for modulating service if they have characterized seats.

Actuators that are only marginally strong enough to close off against system flow can rob the system of efficiency over time as system pressures change, valve stems bind, damper axles stiffen, etc. Requiring close-off ratings at least 50% in excess of minimum "new condition" requirements is good practice for long term sustainable operation.

22.7.1.3 Damper Sizing

Just like valves, down-sizing dampers will improve control at the expense of raising system pressure. In practice, dampers are often left duct-sized, even though resulting control is poor. There are several reasons for this:

- Dampers are relatively cheap, compared to the transition costs for a duct fabricator.

- Dampers are often large, and transitions take up space that may not be available.

- At outside air intakes, a down-sized damper can cause rain or snow entrainment.

For air flows controlled by dampers, other than HVAC, proper sizing will yield more linear control and is recommended where practical. For control purposes, the *opposed blade* damper is normally used, since its aperture size and overall resistance varies the most directly with travel. Conversely, "parallel blade" or round dampers are highly non-linear in nature and hard to control unless drastically down-sized or used for two position control only.

22.7.1.4 Other Damper Considerations

Large damper sections are often problematic. For cost reasons, there is often a desire to use fewer, larger actuators and link the dampers together so an adjacent actuator is driven by another damper, not an actuator. In practice, this can easily result in one end of a long section being substantially open even as the actuator-end is closed. This is due to the fact that the damper blades and axles will twist and stretch. Methods to prevent this undesirable condition include multiple actuators or jack-shafting.

22.7.2 Valve and Damper Actuators

Like other instruments, actuators come in a variety of styles and quality levels, and each has its pros and cons. Significant differences exist between manufacturers that make generalizations and rules of thumb difficult. One thing is certain about actuators: they are a moving part, with mechanical components, and will require maintenance; therefore, consideration should be given to life cycle cost and maintainability. Some of the lower cost actuators are not intended to be serviced. For each of the types listed there are *serviceable* and *throwaway* variations, as well as spring-return/non-spring return types. Types of actuators include:

- Pneumatic spring return
- Pneumatic air-to-open/air-to-close
- Electric motor/gear reduction in oil bath
- Electric motor/gear reduction-open air
- Electric hysteresis/stalling motor
- Hydraulic
- Wax motors (thermal expansion)
- System powered actuators, using air or water system pressure as the motive force
- Self-powered actuators, using a capillary bulb and bellows

22.7.2.1 Actuator "Normal" (spring return) Position

For valves and dampers, the phrases *normally open*

(N.O.) and *normally closed* (N.C.) refer to the device position with no power applied, where a spring-returning mechanism exists to drive it to one position or another. Valves required to have a "fail position" are necessary in many applications to provide an increased measure of reliability if control power is lost. In the case of comfort heating and cooling, the choice is made by asking, "*upon a loss of power, control signal or air pressure, would I rather have full heat, full cool, or don't care?*" In other cases, there are other operational issues like freezing, overheating, over-humidifying, etc. that should be considered. Without the spring return feature, the actuator will simply remain at its last position prior to the power interruption. Spring-return actuators are more costly, and should be used prudently so the added cost is justified.

For large pneumatic cylinder actuators, a measure of fail-safe control can be provided without the expense of a spring system. Using an *air-to-open/air-to-close* actuator (no spring) and a small spring return air solenoid valve, the position of the actuator can be relatively assured on power loss, as long as compressed air remains available.

22.8 INSTRUMENT ACCURACY, REPEATABILITY, AND DRIFT

22.8.1 Introduction

Like anything else, there are different grades of instruments with corresponding costs, so the task of the specifier is to separate the needs from the wants, and to balance the performance with the costs. With an awareness of some of the basic considerations and of instrument grades and selection criteria, good decisions are usually evident. Leaving the instrument selection entirely up to the vendor may or may not be the best approach. To the extent that the "standard offering" instrument portfolio has good performance, this can save money; however, a review of the proposed instruments is advised just to be sure. When reviewing product literature for instrumentation, like any other equipment, *it is often as important what is not said as what is said on a component specification sheet.* For example, if long term drift is not mentioned, ask yourself, "Why is that?"

22.8.2 Accuracy

In the jargon of instrument calibration and specifications, there are two important terms: *percent of reading* and *percent of span*. It is not enough to say "±5%" when specifying a calibration tolerance or instrument accuracy rating. The tolerance should be stated as either ±xx degF, ±xx psi, etc., ±xx% of reading, or ±xx% of span. Usually the instruments that are rated in terms of "% of

reading" are the higher quality instruments.

The following example illustrates the important difference between *"% of reading"* and *"% of span"* concepts. Consider a 0-100 psig (span) transmitter that is indicating 10 psig (reading).

Accuracy Spec "+/- 5% of *reading*,"

Acceptable Range 9.5-10.5 psig

Accuracy Spec "+/- 5% of *span*,"

Acceptable Range 5-15 psig

22.8.3 Repeatability

Repeatability is self-defining; it is the ability of an instrument or process to faithfully repeat itself, given identical conditions. For instrumentation, this is synonymous with precision and is the mark of better grade instruments. Regarding instrumentation, a general statement is that *accuracy can be adjusted, but repeatability is a function of the instrument quality and cannot be changed.* Repeatability is determined, in large part, by the stability of the output in the face of environmental changes such as ambient temperature effect, voltage fluctuation, pressure fluctuation, etc. Instruments whose readings are susceptible to changes in ambient temperature can be very problematic if located in areas where the temperature is expected to change; however, this specification (and the cost to mitigate it with higher quality instruments) is of much less concern if located in areas of constant temperature. Commercial grade pneumatic instruments are usually susceptible to drift from ambient temperature changes since thermal expansion changes the volume of tubing, size of orifices, etc. Some lower grade electronic components are also affected by ambient temperature changes.

22.8.4 Drift

Drift is an undesirable but inevitable quality for any instrument. Long-term drift is attributable to many factors, including normal degradation from age. A general rule of thumb is that the long term drift, from all sources combined, should leave the instrument reading within

Measurement	Instrument Tolerance
	Note: Industrial instrument tolerances are much tighter.
Temperature (space or room)	+/- 1.0 degF
Temperature (duct)	+/- 0.5 degF
Pressure (duct static)	+/- 0.5 in. w.c.
Pressure (other)	+/- 5.0 pct of reading
Flow rates for air	+/- 5.0 pct of reading
Flow rates for water	+/- 5.0 pct of reading
Relative Humidity	+/- 5.0 pct of reading
Oxygen or CO2 monitor	+/- 1.0 pct of span

Figure 22-9. Suggested Tolerances For Commercial Instruments.

reason for a period of five (5) years, to reduce the need for O/M activities and constant maintenance of the device, leaving it to serve instead of being served by the facility.

Auto-zero Feature: Some devices, such as stack gas sensors and very-low range differential pressure transmitters, periodically re-establish the zero point of their output span by simultaneously providing a zero input condition and automatically adjusting the output to a zero value.

22.9 BASIC CONTROL BLOCK DIAGRAMS

22.9.1 Introduction

Besides control technology and control mode choices, a basic consideration of control strategy is whether it is open or closed loop.

22.9.2 Closed-loop Control

The controlled system impacts the measured variable, and process measurement provides feedback to the controller.

Examples of Closed Loop Control:
- Room thermostat controls a heating water valve to regulate heat to that room.
- Leaving water temperature sensor for a heat exchanger controls the steam inlet valve to that heat exchanger.

22.9.3 Open-loop Control

An open-loop control system is characterized as one whose output has no impact on the measured variable, so any form of process feedback is impossible.

Examples of Open Loop Control:
- **Automatic reset of hot water temperature from outside air.** A common control strategy, this provides general compensation based on the common sense notion that "the colder it gets outside the more heat we'll need"; however, it is open-loop since the outside air temperature is unaffected by water temperature.
- **Starting the building HVAC system one hour before occupancy.** Another common strategy, this follows the common sense notion that the building will need some time to warm up (or cool down) after being off all night or all weekend. It provides general compensation for the thermal lag in the building mass, but it is open-loop, be-

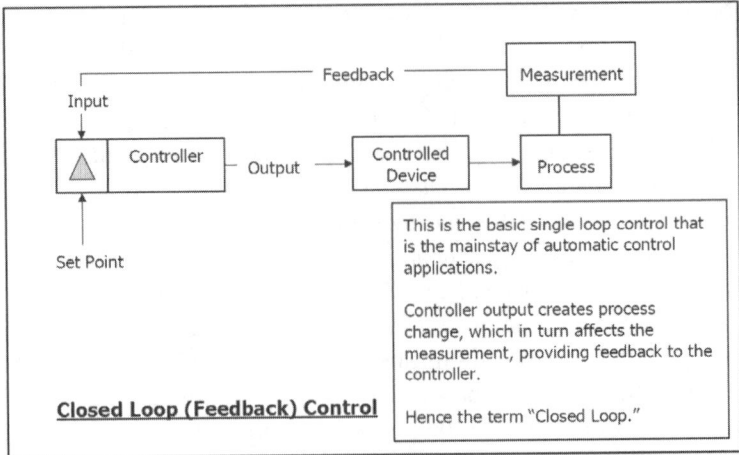

Figure 22-10. Closed-loop Control

cause variations in actual time required in different seasons is not considered. For example it may take (4) hours after a long and cold weekend but only a half-hour after a single night in spring; however, the controller is blind to these facts.

- **Thermostat in room 1 controls the hot water control valve serving room 2.** In this example, closed-loop feedback control is changed into open-loop control due to a design or installation error.

22.10 KEY ELEMENTS OF SUCCESSFULLY APPLIED AUTOMATIC CONTROLS

22.10.1 Examples of Good and Bad Control Applications
Example.
- BAD. Two thermostats in an exterior room, one serving the perimeter heat and the other serving the VAV cooling box overhead air distribution.

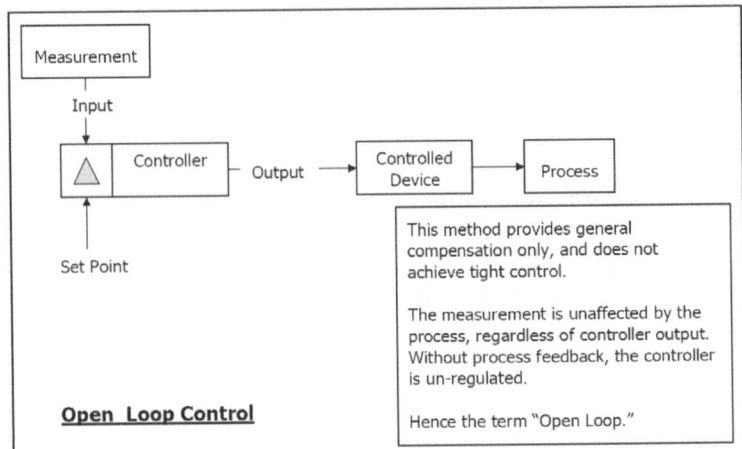

Figure 22-11. Open-loop Control

CONTROL SUCCESS ITEM	REMARKS
Have a good understanding of the process intended to be controlled.	Understanding the system to be put under control is essential to a properly applied control system. See Examples below for some good and bad control applications.
Understand reasonable and unreasonable expectations of the controls.	Maintain proper perspective of controls in the overall list of project priorities: **Project priorities, in ascending order.** [4] Priority 1 Safety Priority 2 Reliability Priority 3 Control and Regulation Priority 4 Energy Efficiency Use controls to automate and optimize, but not to solve inherent, systemic problems. Understand that controls can break, require maintenance, don't last forever. Energy efficiency should only be a focus after the higher priorities have been satisfied, and none of the energy efficiency work should compromise any higher priority.
Verify the system is controllable and lends itself to control.	Without this condition met, the control application will not be successful. See Examples below for good and bad controllability.
Know what should be accomplished, be able to define clearly and in detail what is to be controlled, and how.	The clear and thorough Sequence of Operation in plain English is a very good sanity check for the control application. This exercise is a great check of the proposed control system. If it can be accurately described in words or with diagrams, it can probably be built.
Verify that the basis for savings is sound.	Understand the underlying principles of that which is to be controlled, including how the result constitutes an improvement. Understand the control behavior and energy use as-is, and compare that to the proposed control method. Verify that there will be a savings benefit to justify the work.
Take care not to solve one problem and create a new one at the same time.	Consider each control system for *what could happen if*.... and try to ferret out as many of those as practical. Examples: • If the control system quits, will all the lights be OFF or will they be ON by default? • If power is lost, will anything freeze up? • With all controls in their default states, would it be better to have full heating or full cooling, or don't care? • If the actuator fails in its last commanded state (half way), how will anyone know that has happened? • If the sensing tube falls off and the fan ramps up to full speed, will there be any duct damage? Provide a responsible review to be sure the controls do not create new issues, especially those that could create safety issues or equipment damage. Peer review is always good to double check for pitfalls and opportunities for improvement.

Figure 22-12a. Successful Controls Application Checklist (*Continued*)

• GOOD. One thermostat in a perimeter room sequencing the VAV box and perimeter heat, with a dead band in between to prevent overlap.

22.10.2 Examples of Good and Bad System Controllability

Example.

• BAD. A single step controller on a 15-ton HVAC package unit that has one compressor attempting to maintain a constant 55 degree discharge temperature. Excessive equipment cycling will result, regardless of controls, with the consequence of poor control and likely premature equipment failure.

• GOOD. A two-step controller on a 15-ton HVAC package unit that has two compressors, where 7.5 tons represents a typical part load day for the building. Cut-in and cut-out are set wider than the temperature change of each stage, to prevent short-

CONTROL SUCCESS ITEM	REMARKS
Make sure the controls will provide economic return.	Identify approximate costs and savings attributable to the controls. As a sanity check, almost any control system proposal that costs more than the equipment it controls or has a payback of greater than 5-years is suspect. Prioritize control system expenditures with need. For example, if 2/3 of the savings can be achieved with a very simple and economical system, it may be appropriate to stop there, depending upon the owner's staff or project ambitions for savings. Avoid Bells and Whistles. Be critical of each proposed 'point'. Ask how this will benefit the customer operationally or financially; and which points the customer can do without. Separate the core control features from the 'gee whiz' points that have minimal value. Monitor-only points add project cost and extend payback periods, so use these add-on's with discretion.
Understand the user so that the new control application is a facility-fit.	The control system must survive if it is to remain viable over the long run. Consider the user's point of view: • Identify the level of sophistication the operators will accept. • Make provisions for the system to be user-friendly. • Accommodate customer preferences for materials, system supplier, man-machine interface, etc. • Identify needs for additional staff training. See <u>Examples</u> below for user-friendly design features.
Have sufficient knowledge of control theory and terminology to be conversant with the customer and a controls contractor.	Communication using this trade language is helpful, but remember not to sweat the details. With the basic control expectations made clear, the process fundamentals sound, and the project economically viable, many of the details can be left to the specialty contractor.

Figure 22-12b. Successful Controls Application Checklist (*Concluded*)

cycling. Minimum on-off times further safeguard equipment from short-cycle damage.

22.10.3 Examples of User-friendly Control Design Features

In general, try to provide a control design that will be accepted and will last. Match the system complexity to the user's level of sophistication or their willingness to learn. If it's too complex, they'll probably just unplug it 6-months after start-up. Training will help to raise skill levels and avoid oversimplification. Be patient, especially for anything complex. Any type of measurement that allows the user to see the savings created by the new control system will help spark interest and encourage ownership and buy-in. For the control design to stand the test of time (e.g. to be sustainable), it must be a good fit for the customer. See Figure 22-13.

22.11 EXPECTED LIFE OF CONTROL EQUIPMENT

Be realistic about how long things will last. Nothing lasts forever. For example, a control project based on a 10-year life will need to include cost of repairs to be realistic. Note for the table in Figure 22-14 that life spans shown vary by manufacturer and grade. This information is offered as a prompt for realistic life cycle cost estimating when controls are applied.

22.12 CONTROL APPLICATIONS FOR SAVING ENERGY

Note the common themes throughout the following list of applications:

• Strive to satisfy most of the people most of the time, but not all the people all the time (80-20 rule).

• Try not to run any equipment continuously or off-season.

• Avoid heating, cooling, and lighting areas that are unoccupied.

• Use "just enough" air pressure, water pressure, etc. to satisfy the point of use.

• "Just enough" heating and cooling, as determined by the point of use.

• Use "just enough" ventilation air for the occupants and to make up the exhaust needs.

• Eliminate simultaneous heating and cooling wherever possible; minimize where unavoidable.

22.13 BASIC ENERGY-SAVING CONTROL APPLICATIONS

(See Figures 22-15 and 22-16.)

USER-FRIENDLY ITEM	REMARKS
Be Clear	Without explicitly clear diagrams and operating instructions, the controls can be perceived as a complicated contraption, and quickly fall into disrepair, eroding the intended savings. Clear and thorough documentation is helpful for sustained operation of these systems. Making control intentions clear from the beginning helps ensure customer buy-in and support.
Man-Machine Interface (MMI)	This can take many forms, but has the common theme of convenient communication of key parameters to operations personnel. Consider the following as appropriate: • Graphic control display screens for DDC. • Gages and thermometers by key instruments. • Test wells adjacent to sensors for calibration or portable readings. • Local display-adjust panels (fixed or portable) as a window into the DDC controller, or a connection jack for a portable device.
Training	Essential for operations staff buy-in. The customer needs to understand it and inherit it – and be proud of it. Involvement during the start-up and functional testing phase is a great way to gain familiarity and start training.
Tuning	Each control loop should be rigorously tested by large step changes to verify stability. For DDC controls, time-trends are very useful for verifying that the control loops are tuned to be responsive, without hunting. Be sure the tuning constants are backed up if DDC controls are used. Be realistic for how tight the controls need to be; striving for absolute zero-error control can cause the system to oscillate around the perfect set point, thereby causing excess end device activity and premature failures. Allowing a small variance around the nominal set point can allow the actuators to last a lot longer.
Calibration	Don't accept factory calibration. Check and verify each new instrument and repeat the exercise periodically – e.g. each two years. Instruments drift, especially commercial grade ones. Industrial grade instruments hold tolerance much better but are much more expensive. Some commercial instruments have no calibration adjustment, so the calibration would be more like verification; if not within specified tolerances, adjust or replace it.
Control Diagrams	As much detail as possible. DON'T LOSE THESE! As the years go by, they will become difficult or impossible to replace. If possible, keep the control diagrams and sequences with each control panel, near the equipment it controls. There are almost always changes in the original shop drawings as the installed system molds itself to the project conditions. The as-built version of the shop drawings is more valuable than the original or submittal diagrams, and should be part of the close-out documents provided to the customer. As-built drawings must be accurate to be effective.
Sequence of Operation	Sequences should be painstakingly clear. This cannot be overemphasized. The sequence should serve more than the installing technician; it should provide the basis for functional testing, customer training including new hires, and troubleshooting over the years. A good sequence will explain each of the system's functions in detail, conveying the intent of the sequence and not just stages of operation.
Start-up and Check Out.	Insist on a good QC plan for start-up and check out of the controls. Understand that Quality Control is not governed by any industry standards – some suppliers apply significant effort to this and other so not. For large and complicated systems, requesting a written plan and final report is a good idea. This includes tuning. Examples: • Point-to-point testing (ringing) for all inputs and outputs from the hardware device all the way into the graphic screen. This verifies all points in between and is strongly recommended. • Computer trends and step changes are a very good way to verify control stability and proper tuning. • Points list printout • VAV min/max settings printout In all cases, avoid the QC method of attention by exception which is to say the QC is a reactive process in response to complaints.

Figure 22-13. User-friendly Control System Checklist (*Continued*)

USER-FRIENDLY ITEM	REMARKS
Functional Testing	"Show Me" time. Don't be bashful at this stage. Take it through its paces from top to bottom, and don't be afraid to kick the tires a little bit. Include all pertinent operational modes and related safety controls. The purpose of this step is to see that the system is meeting the project's intentions for functionality, energy saving, and generally if it is doing what it is supposed to. Functional testing also verifies that the customer is getting what they have paid for, and is very appropriate as a final payment contingency item. It's also a great way for the customer to learn, and can be an effective part of customer training. A written functional test plan is useful to follow and get to the point of being done (for payment), and also can be the basis of repeat verification testing over time.
Periodic Exercising	For controls that only act periodically instead of continuously this can make sense. Examples: • The change-over valves for a two pipe system need to move quickly and seat firmly to avoid leak by between the cooling and heating systems. A physical check of these once a year makes sense. • Equipment that has been drained or shut off for a season, as part of the normal start-up work to bring it back on line.
Standardized parts	For additions to facilities, especially large facilities, inventories and familiarization of parts are important issues. Often, there are preferences for or against certain types of parts, features, etc. which should be incorporated to provide a more friendly new system.
Documents for customer's Use	• Sequence of Operation. • Control drawings, as-installed. • Functional testing procedures and results. • Tuning and calibration parameters. • Copy of training materials, or video taping, for re-training and new hires. • Operations and Maintenance manuals, trouble shooting, and emergency procedures. • Equipment catalog data and contact information for repair parts
Other	• User adjustment for comfort control applications. • Devices located in accessible locations for repairs and calibration. • Valve position indicators that are visible from the floor. • Override Provisions for key output points to accommodate emergencies without disconnecting things. • Sensors and Transmitters with long-term stability for minimal need for calibration. • Arrange control wiring/tubing with troubleshooting in mind. • Power on-off switches and air shut-off valves, in key locations to assist the service person. • 3-valve manifolds at flow meters to allow zero-calibration. • Drain and Vent valves at fluid transmitters for routine calibration. • Minimum on-times and off-times to prevent short cycling. • Make normal-use operator parameters adjustable, and critical parameters non adjust-able, to allow user interaction and reduce inadvertent changes for the worse. • Plug-In relays for easy replacement. • Relays with LED lights to indicate at a glance when the relay is on or off.

Figure 22-13. User-friendly Control System Checklist (*Concluded*)

Device	Typical Life
Pneumatic Cylinder Actuator	20 years
Electronic Actuator – oil-filled motor style	10-15 years
Electronic Actuator – mini hydraulic	5 years
Open-air direct coupled electric actuators	5-7 years
Residential style electric "clock motor" valve actuator with stalling motor	5 years
VAV box controller with built-in actuator	5-7 years
Low range differential air pressure sensor, analog	5 years
Specialty sensors (humidity, CO2)	5 years
Economizer dampers	10 years
Globe control valve with adjustable packing	15 years
Analog electronic controls, in general	7-10 years
DDC controls, before technology advances render them obsolete, working or not	7-10 years

Figure 22-14. Typical Control Equipment Life Spans

Lighting Control-Basic	Remarks
Programmed Start-Stop	Timed coincident with occupant schedules. Two features are important for acceptance of these systems: • A warning that the lights are about to go out • Local overrides that allow people to keep the lights on
Photo Cell Control of Outdoor Lights	Outdoor lights off during day-lit hours Time switch to shut off some site lighting after dark, but security lighting stays on.
Occupancy Sensor	Use for individual offices, meeting rooms, classrooms, multi-purpose rooms, warehouses, etc., that are only occasionally occupied. For warehouses, it is good practice to leave a few lights on continuously for safety and security. Not practical for restrooms, since they can leave people in the dark, unless at least one light is one. *Editor's Note: In mechanical / electrical areas, an occupancy sensor may be undesirable if it could leave a worker in darkness in the midst of exposed operating equipment. In such cases, occupancy sensors could be applied to some portion of the lights. Safety always comes before energy savings.*
Timed Tenant Override	When after-hours tenants invoke the override, limit the amount of time this will remain active before the system reverts to unoccupied mode again (e.g. 2 hours), and what actually gets overridden.

Figure 22-15. Basic Lighting Control Applications

22.14 ADVANCED ENERGY-SAVING CONTROL APPLICATIONS

(See Figures 22-17 and 22-18.)

22.15 FACILITIES OPERATIONS CONTROL APPLICATIONS

Automatic controls, especially DDC controls, can be very valuable tools for facilities personnel. Computerized maintenance management, early detection, remote servicing, automatic notification, trends and logs, and other features can be used to improve facility operational quality. See Figure 22-19. *See also Chapter 12–Energy Management Control Systems.*

22.16 CONTROL SYSTEM APPLICATION PITFALLS TO AVOID

Automatic control system choices can affect facility operations; consequently, there are many things to consider when investing in a control system. Figure 2.20 shows some common pitfalls to avoid in applying DDC automatic controls. Refer also to Figure 22-12 and Figure 22-13 for successful and user-friendly features to strive for. Exclusion of any of these items can become a pitfall as well.

22.17 COSTS AND BENEFITS OF AUTOMATIC CONTROL

Refer to Figure 22-21. Automatic controls are unique in that they often provide both tangible and intangible costs and benefits. Tangible benefits, like other energy-related projects, include energy savings and demand savings. Intangible benefits are those that are difficult to quantify or predict. Some of these only apply to digital control systems.

22.18 ESTIMATING SAVINGS FROM APPLIED AUTOMATIC CONTROL SYSTEMS

22.18.1 Introduction

Economic barriers are among the greatest obstacle for the energy professional, and automatic controls are no exception. While it is often easy to visualize that savings will occur from automatic control improvements, estimating them can be a very daunting and intimidating challenge. This presents a dilemma to the energy professional since cost justification is almost always an expectation. Projects that have merit but defy quantification may be overlooked.

Done accurately, the cost saving calculations can be laborious and expensive, posing a barrier to otherwise vi-

HVAC Control - Basics	Remarks
Programmed Start-Stop	Any equipment left to run continuously should be controlled for automatic start-stop based on occupancy schedules. Can be Energy Management System (EMS) control or can be a simple time switch. Outside air dampers are usually closed and exhaust fans off during unoccupied times.
Optimal Start	Use to delay HVAC system start-up as long as possible. During the warm-up mode, the building is normally unoccupied so the outside air damper can remain closed and exhaust fans off until occupied.
Optimal Stop	Use to turn off primary heating and cooling equipment (Chillers / Boilers) shortly before the end of the occupied period, utilizing the thermal lag from the circulating fluid, air, building and furniture mass, and the fact that most people will not detect (or complain about) a temperature change of less than 2 degrees F. Ideally, the comfort conditions will have slipped just 2 degrees F just as the occupied period ends.
Occupied/Unoccupied Mode	Use to set-up and set-back space temperatures, and to reduce or stop outside air intake during unoccupied times. Typical unoccupied temperatures are 55 degF heating and 85 degF cooling.
Timed Tenant Override	When after-hours tenants invoke the override for comfort, limit the amount of time this will remain active before the system reverts to unoccupied mode again (e.g. 2 hours), and what equipment needs to operate.
Dead band Thermostats	Use or adjust thermostats to allow for a zero-energy dead band, a range where no action is taken by the HVAC system. A simple, but effective routine that should be part of every non-critical comfort control application. The dead band separation of heating and cooling controlled devices provides a positive separation to prevent the energy waste associated with heat/cool overlap.
Restrict Tenant Adjustment Limits	Remote lock-out of tenant adjustment for space temperature, or limiting the adjustment to +/- 2 degrees F at most.
VAV Box Control Dead Band	Independent VAV heating and cooling set points with a minimum 5 degree zero-energy dead band in between, in lieu of a single setting with a proportional band. This will increase the zero-energy dead band and reduce simultaneous heating and cooling. Note: For normal VAV systems where the ventilation air comes mixed with the supply air, there is a *minimum stop* setting, for minimum ventilation, that creates some heating-cooling overlap. The dead band application for VAV boxes minimizes, but does not eliminate this overlap.
Air Handler Control	Independent preheat, mixed air, and cooling coil control set points, with a minimum 5 degree zero-energy dead band in between, in lieu of a single setting with a proportional band. This will reduce simultaneous heating and cooling.
Outside Air for Morning Building Cool-Down	Use outside air for pre-occupancy cool-down cycle in the summer, in conjunction with scheduled off time and/or optimal start routines. This routine takes advantage of cool nights and flushes the warm air out of the building that has accumulated and avoids or reduces demand on the primary cooling equipment. Normally limited to dry climates. In humid climates or whenever the outside air humidity is higher than building humidity, this has the potential to add load if subsequent de-humidification (and energy expenditure) occurs as a result.

Figure 22-16. Basic HVAC Control Applications (*Continued*)

able projects. One approach is to produce reasonable estimates using abbreviated estimating methods or "rules of thumb" where possible. As with all estimates, being conservative is a key to success so that project performance is seen to *under-sell and over-deliver*. Even when not a contract requirement to guarantee savings, there is always an expectation that the estimated costs and savings come to pass; further, the credibility of the energy professional is built largely on the accuracy of these estimates. Bearing

in mind that there will always be uncertainties and uncontrolled variables at work, it is usually good practice to *de-rate the calculated savings*. By artificially reducing savings (de-rating) and artificially increasing projected costs (contingency allowance), two things happen.

Effect of de-rating project estimated savings:
* The odds of project economic performance exceeding expectations increases.

HVAC Control - Basics	Remarks
Extended Air Economizer Operation	Extend air-side economizer operation as far as practical, with the use of enthalpy comparison sensors, in lieu of the standard practice of limiting the economizer operation to 55 degrees F or less. With this method, whenever the outside air has less energy in it than the return air, cooling costs will be reduced by using outside air instead of return air. On days when the outside air is significantly drier than inside air, the free cooling range can be extended and outside air temperatures above return air temperatures can be utilized. When the outside air is humid, 55 degrees F is a smart cutoff point.
Boiler Lockout from Outside Air	Generally, prohibit boiler operation above 60 degrees F. Suggested control would be to cut-in at 55 and cut-out at 60 degrees F outside air. This is open loop control, but saves energy compared to "start based on demand" routines, since all it takes is one strong demand point to run the large machine for extended periods, or to start it unnecessarily in warm weather.
Chiller Lockout from Outside Air	Generally, prohibit chiller operation below 50 degrees F, and coordinate with air and water economizer settings. Suggested control would be to 'cut-in' at 55 and 'cut-out' at 50 degrees F outside air. This is open loop control, but saves energy compared to "start based on demand" routines, since all it takes is one strong demand point to run the large machine for extended periods, or to start it unnecessarily in cold weather.
Load Limiting	Reducing demand for primary electrical HVAC equipment can save money on demand costs. This has application for chillers, large variable speed fans, and large variable speed pumps. For <u>example</u>, by load limiting to 90%, this keeps 90% of design capacity available but will reduce electric demand by 20-25%. Since 30-40% of the electric costs for large facilities are often from demand, this will create savings.
Outside Air Reset of Hot Water Converters	Compared to a constant temperature setting, this will reduce standby losses. In the cooling season, these losses are also unnecessary cooling loads.
Mid-Range Vestibule Temperature	For vestibule spaces, temper the space to a mid-range temperature, half way between indoor and outdoor temperatures.
Operable Window Interlock	Where operable windows are used, provide automatic control interlock to disable the HVAC serving that room, to avoid heating and cooling the great outdoors.
Roll-Up Door Interlock	Provide automatic control interlock to disable the HVAC serving interior area, to avoid heating and cooling the great outdoors when the door is up. This will encourage people to keep the door closed.
Staggered Heating and Cooing System Start-Up *(Buildings with demand charges)*	After extended 'off' periods, such as night set back, normal automatic control response will be to drive the heating and cooling equipment in order to reach the occupied set point. To avoid setting the utility *maximum electrical demand* value and subsequent higher demand charges coming out of unoccupied periods, bring large electrically driven heating and cooling equipment loads on in <u>segments</u> for 30 minutes or until the process has "caught up" and is no longer at full load. . NOTE: this is not in reference to inrush currents for starting motors. For <u>example</u>, consider a building heated by electric resistance heat. By default, coming out of unoccupied periods during winter will result in 100% of the heat being energized simultaneously. Controlling in 2 or more zones and allowing ample *pull-up* time for zone 1 to get to temperature and begin cycling normally before starting zone 2 will reduce the maximum demand for that day and, if done diligently through automation, will reduce the overall seasonal demand and demand charges. <u>Examples:</u> • Electric resistance heat (winter) • Packaged HVAC cooling equipment (summer)

Figure 22-16. Basic HVAC Control Applications (*Concluded*)

Lighting Control - Advanced	Remarks
Daylight Harvesting	Control perimeter lighting on/off or modulate, in response to daytime sunlight entering the building. In-board / out-board switching for interior lighting.
Programmable Ballast	Special "addressable" fluorescent ballast are available that can be controlled individually, without conventional relays. Remote control, custom scheduling by room, by area, and *by light fixture* are possible with this system to minimize lighting use.

Figure 22-17. Advanced Lighting Control Applications

HVAC Control-Advanced	Remarks
Occupancy Sensor Control of HVAC *(Where a dedicated VAV box or terminal unit exists)*	When the room is sensed as unoccupied, the space temperatures revert to unoccupied values. For VAV boxes, minimum flow settings are adjusted to zero. Note: perimeter areas strongly influenced by envelop loads may need specially defined "day time unoccupied": settings to avoid comfort complaints and to allow quicker return to occupied temperatures when the room is re-occupied. Examples: • Class room • Hotel room • Conference room
VAV Box Optimal Minimum Setting	The usual method of calculating VAV box minimums is to assume full load conditions and the outside air percent of the total supply air at those conditions. With constant outside air control, the mixture is richer in outside air at supply air flows that are below maximum. Therefore, the VAV minimum stops can be reset automatically based on the actual OA mix as loads change. Example • A 20% VAV minimum stop is required based on 25% OA at design conditions. This same VAV box could have a minimum position of 10% when the supply fan is at 50%, since there is now twice as much OA in that air stream. The effect of this is to reduce over-cooling and reheating when the VAV box is at minimum while the zone thermostat continues to call for less air.
Optimal Supply Air Static Pressure Setting – Variable Air Volume (VAV) Systems	This requires polling of individual VAV boxes for air valve position and reduces system pressure until at least one box is 90% open, thereby providing the optimal system duct pressure (just enough pressure). This reduces fan horsepower.
Optimal Supply Water Pressure Setting – Variable Pumping Systems	This requires polling individual air handlers for control valve position and reduces system pressure until at least one control valve is 90% open, thereby providing the optimal system water pressure (just enough pressure). This reduces pump horsepower.
Optimal Evaporative Cooling Setting	This uses outdoor air wet bulb temperature, which can be measured directly, but is usually calculated from temperature and humidity. The evaporative process can get close to, but never reach or exceed, the wet bulb temperature. By knowing the wet bulb temperature, the control system will know its boundaries and won't try to achieve something it cannot. For water-cooled refrigeration equipment, the low limit for the reset is normally around 55 or 60 degrees F and the chiller mfg needs to be consulted to confirm. Resetting the cooling water temperature down in this way, in lieu of a constant temperature setting, will reduce kW/ton energy use and demand. NOTE: achieving colder condenser water is a trade off between improved chiller kW/ton and increased cooling tower fan kW/ton, so evaluation for diminishing returns is required.

Figure 22-18. Advanced HVAC Control Applications

Note for Advanced Routines: Many optimization routines rely on end-use polling of demand or valve/damper positions such as "most open valve" routines. This can be done cost effectively without actual measurement of position—by polling the individual "percent commanded output." This is referred to as "implied position" and is acceptable in most cases in lieu of actual position. Note also that routines that use polling have the potential to be inefficiently operated if one errant measurement exists. For polling space temperatures, for example, limiting the user adjustment is strongly recommended in conjunction with demand polling of space controls. Additionally, it may make sense to "discard" the high and low values from such polling to prevent errant operation. Some polling techniques wait to react until several "calls" exist; this reduces the chance of an errant signal driving the entire heat/cool plant, but also introduces dissatisfaction if a single and legitimate call exists, since the control system would ignore it.

HVAC Control-Advanced	Remarks
Optimal Evaporative Cooling Setting	This uses outdoor air wet bulb temperature, which can be measured directly, but is usually calculated from temperature and humidity. The evaporative process can get close to, but never reach or exceed, the wet bulb temperature. By knowing the wet bulb temperature, the control system will know its boundaries and won't try to achieve something it cannot. For water-cooled refrigeration equipment, the low limit for the reset is normally around 55 or 60 degrees F and the chiller mfg needs to be consulted to confirm. Resetting the cooling water temperature down in this way, in lieu of a constant temperature setting, will reduce kW/ton energy use and demand. NOTE: achieving colder condenser water is a trade off between improved chiller kW/ton and increased cooling tower fan kW/ton, so evaluation for diminishing returns is required.
Optimal Supply Air Temp Set Point for VAV Air Systems	This requires polling individual VAV box air valve position and reheat valve position. This utilizes a fixed temperature for cooling (e.g. 52 degrees F) with no reset at all until polling of individual boxes indicates that most of the boxes are in heating – only then is the SA temperature allowed to gradually be reset upward to its maximum limit (e.g. 62 degrees F). The reset is accomplished by polling VAV reheat valve positions and the air temperature is gradually reset upwards until at least one VAV box reheat valve is 90% open, thereby providing optimal air temperature (just warm enough). Simultaneously, VAV damper positions are polled to be sure enough cooling is being provided for any zone still calling for cooling, and cooling will prevail if the two are at odds.

A similar, but less accurate, method uses the supply fan VAV output percent command along with outside air to estimate when most of the VAV boxes are at minimum and in the heating mode, signaling the time to begin resetting supply air temperature. An important feature of this reset schedule is that NO reset occurs during summer months, preserving the basis of VAV savings. A simple reset schedule is used for this method as follows:

SA Fan % Capacity	Outside Air Temp	Supply Air Temp Set point
100%	Any	55
40% (*)	40 (**)	55
20%	10	62

(*) The percent of the supply fan capacity at which the reset begins should correspond to the aggregate VAV box minimum settings. For example, if the aggregate (weighed average) VAV box setting is 30%, then a supply fan value of 40% indicates that most, but not all, VAV boxes are at minimum.
(**) The temperature at which the reset begins should correspond to the building thermal **break-even** point, which is the point below which cooling is not required, other than some interior zones.

Another method for supply air reset of VAV systems uses outside air temperature to imply when heating loads will begin to appear. This is easy to implement, but is an open loop control system, i.e. no amount of change in supply temperature can alter outside temperature. A reset schedule for this method would look like the following:

Outside Air Temp	Supply Air Temp Set point
Above 70	55
60 (***)	60
Below 60 (***)	60

(***)The values of the lower outside air reset parameter will vary depending on the break even temperature. |

Figure 22-18. Advanced HVAC Control Applications (*Continued*)

HVAC Control-Advanced	Remarks
CO2 Demand Controlled Ventilation (DCV)	Has application for large open assembly areas, characterized by a single point of CO_2 control in the occupied space of the single zone unit serving that area. Applications include a ballroom, theatre or open plan office area. A separate CO_2 point of control would be required for each dividable meeting area, each assembly area, and each group area, so that areas of great ventilation demand are served appropriately and do not get overlooked by sensor averaging. In essence, this control method uses a CO_2 sensor(s) as a people counter, thereby optimizing the use of outside air, and the energy required to condition it.
Optimal Ventilation by People Count	A combination of number of people and the duration at each location would provide the necessary information to calculate the exact amount of ventilation air needed. This method puts the welfare of the occupant first, but also minimizes the energy use by providing the optimal ventilation quantity (just enough ventilation) at any given time. Note: This routine is conceptual only, but is presented due to the large potential for energy savings. To the author's knowledge, the friendly "people counter" instrument, other than a turnstile, hasn't been developed yet. This may be possible, manually or automatically, based on ticket sales for assembly occupancies.
Optimal Ventilation Effectiveness by Season	Adjust ventilation rates based on improved "**e**" (ventilation effectiveness) in summer (value 1.0) compared to winter (value 0.8), to save outside air conditioning energy costs.
Optimal Sequencing of Multiple Chillers / Boilers	Strategically selecting cut-in and cut-out points to keep the primary equipment operating in its most efficient range. Typically, maintaining this equipment between 50-90% load achieves good efficiency, but verifying the actual best-efficiency points or range for each system, from manufacturer's load profile data, is recommended.
Optimal Reset of Hot Water Converters	This requires polling of individual hot water points of use, for control valve position, such that at least one valve is open 90%. By resetting based on demand, this routine will provide the optimal water temperature (just hot enough).
Multi-Zone System: Reset of Hot Deck and Cold Deck temperatures based on zone demand	This requires polling of individual multi-zone (MZ) mixing dampers to determine the greatest cooling and heating demands, ideally so that at least one MZ damper in cooling and heating is fully open. By resetting hot and cold deck from space demand, optimal heating and cooling (just enough of each) will be provided in the hot and cold decks. Since these systems inherently mix cooled and warmed air, this reduces simultaneous heating and cooling.

Figure 22-18. Advanced HVAC Control Applications (*Continued*)

- Projects with marginal return on investment (ROI) look worse and may be eliminated.

The applied de-rate will depend on the level of uncertainty. A high degree of uncertainty suggests the need for a higher de-rate. A value of 30% is suggested for most applications, although there are cases where no de-rate is needed. One of the big uncertainties for savings associated with controls is whether the savings will last. Controls are different from other "bolted down" changes, because many of the optimizing features are software and can easily be changed.

Examples:
- *Easy to Quantify.* A lighting replacement project with 24-7 operation need not be de-rated at all since there

are few, if any, uncertainties. This is easy to quantify since only one parameter has changed (light fixture energy efficiency).

- *Hard to Quantify.* An Automatic Control project that includes multiple control system improvements implemented at the same time, such as variable pumping, free cooling modes, supply air reset, boiler lock-out, morning warm-up, and new quarter-turn (no leak by) terminal unit control valves. This adds uncertainty for what savings come from individual measures. From a purely simplistic standpoint, a project consisting of several control improvements could be implemented one at a time with six months of post-project measurement and verification (M&V), changing only one thing at a time. This would have the clear advantage of know-

HVAC Control-Advanced	Remarks
Dual Duct Terminal Units	Unless constant volume is critical to the process service, splitting the two dampers for independent control allows optimization. With the dampers split (separate actuators), and ventilation air in one or the other air streams (usually the cold duct), control like a VAV box with a dead band. During a call for cooling, the hot duct damper would remain closed while the cold duct damper modulates to maintain temperature. As cooling demand decreases, the cold duct damper throttles toward closed and finally reaches its minimum position (for ventilation). Here it will float within the dead band. If space temperature falls sufficiently to require heating, the hot duct damper would begin to throttle open with the cold duct damper at minimum. This sequence minimizes heating-cooling overlap and significantly reduces the energy use of the dual duct system that otherwise has simultaneous heating and cooling waste built into it.
District Heating and Cooling Delta-T Control	By actively controlling differential temperatures through buildings points of use to be as high has practical, flow can be reduced and pumping costs minimized. This can be touchy and, if too aggressive, can lead to comfort issues. Many, if not all, of these distribution systems pump more water than they need to. There are a variety of reasons for this, which are beyond the scope of this report, but many of them can be mitigated by imposing a control limit of **minimum delta T** (differential temperature) across the coil, building, or segment of the distribution system. This has the effect of requiring the point of use to extract all the available energy out of the circulated fluid before returning it and thus reduce pumping volumes. Control is implemented by some manner of throttling device, either a control valve or a pump speed, and supervised by differential temperature measurement and/or flow measurement. Applications vary, but the common theme is to wring the heat out of the water and reduce flow when possible, while still maintaining comfort.

Figure 22-18. Advanced HVAC Control Applications (*Concluded*)

ing what savings occurred from what measure, but this would probably not be done in practice due to the protracted length of time for the entire project to be implemented, as well as lost savings from delayed implementation.

22.18.2 Normal Replacement Costs and the New Control System Hurdle

A common practice for existing facilities is to propose equipment replacement using energy efficiency as justification. In general, it is easy to justify the differential cost of upgrades to higher efficiency equipment, but it is often impossible to justify the entire replacement project on energy savings alone. Burdening the project cost with unrelated expenses such as equipment replacement that was due anyway makes the payback look worse and creates an unfair perception of long paybacks. Whenever possible, energy improvement expenses should be fairly separated from normal replacement project costs. Equipment that is near the end of its useful life should be a planned replacement expense, regardless of the desire to reduce energy costs. If replaced early, the remaining value of the equipment may be appropriately "charged" to the energy project, but not the entire project cost since this would need to be done anyway.

Like the normal replacement discussion above, requiring economic justification for installing a new control system in place of an antiquated one is an unfair hurdle. Some equipment requires automatic control in one form or another for simple temperature regulation, and the control system basic infrastructure cost is analogous to the bricks and mortar of the building. For example, if a 20 year old pneumatic system is worn out and no longer working, getting a quotation on a new, complete replacement pneumatic system may be a reasonable baseline of what the "normal replacement" dollar value is worth. The upgrades to controls beyond the basic regulation functions—the energy saving routines, the special optimizing strategies—should pass economic muster, however. Fair is fair.

22.18.3 Barriers to Quantifying Savings from Automatic Controls Projects

- Control parameters are almost always user-adjustable and will usually be fine-tuned during the life of the project, including the post-project measurement period.
- Control algorithms often include multiple variables that do not act independently. Consequently the effects on energy consumption often defy iso-

Facilities Operations Application Examples	Remarks
Just in Time Scheduled Filter Changes (predictive maintenance)	Air mover efficiency is lost if filter replacement is postponed. Filter costs are higher than necessary when filters are changed too early. Identify filter change-out points. For constant volume air systems this is a straight forward application of a filter differential pressure (dP) switch, set to announce the need for service. VAV systems will require special scheduled off-hour events to open the VAV boxes and speed up the fan to a fixed high air flow rate (e.g. 75% fan speed), in order to establish a baseline and determine actual filter loading.

Monitor Heat Exchanger Approach Temperatures (predictive maintenance)	Efficiency is reduced when heat transfer surfaces become fouled. On large equipment, it is practical to monitor the approach temperature and perform maintenance as required to keep equipment running optimally. After 'clean condition' benchmarks are determined, the control system can monitor actual approach and provide alert messages when fouling is indicated.
	Examples:
	  • Boiler stack temperature vs. process temperature.
	  • Chiller condenser water temperature vs. refrigerant temperature.
	  • Hot water heat exchanger leaving "hot side" temp vs. hot water supply temp.
	  • Air-cooling-coil coldest air vs. coldest water differential

VAV Box Controls Health / Sanity Check	VAV box controllers and actuators are inexpensive and have a short life span. It is very common to find systems left to their own accord for a number of years to have 25% of the VAV boxes unresponsive and stuck in some last position, with obvious control and energy ramifications. From the operator workstation, it is relatively easy to set up an automated polling routine to sequentially go from box-to-box and command incremental increases in air flow and heating coil flow, and log whether or not a process change occurred (e.g. is the box alive or not), thereby generating a list of boxes needing attention. Doing this annually is a good idea. NOTE: This is most readily accomplished for VAV boxes with a discharge air temperature sensor, but is possible without it by waiting for confirmation that air flow and space temperature change upon a step change.

Figure 22-19. Facilities Operations Control Applications (*Continued*)

lating and quantifying separately. An uncertainty is whether one measure will interact with another. In many cases, the aggregate savings may be less than the sum of the parts, in which case the savings would be over-stated if calculated independently.

- Control system improvements that improve quality, comfort, ventilation, or other features from some deficient state may actually increase energy use in some ways, eroding the overall savings of the measure itself. This is especially true of ventilation.

22.18.4 General Methodology

The following are general guidelines for estimating savings, and are a good starting point for the specific examples that follow.

- Establish realistic baselines.
- Identify variables and their interactions.
- Identify competing or complementing processes. that will subtract from measure savings.
- Reduce the number variables to simplify analysis.
- Treat uncertainties with contingencies, by either inflating the baseline or de-rating the savings.
- With the baseline established, use experience to

evaluate calculated savings as a percent of total expenses.

- Use all these indicators as sanity checks to avoid over-stating savings.
- After project completion, get after-the-fact measurements where possible to compare actual to estimated savings; collecting this real data will allow improved estimating and reduced contingencies over time.

22.18.5 Quantified Savings Examples

The following are examples of how to approximate savings from automatic controls measures. Many of the methods use rules of thumb and assumptions to simplify the work. More accurate results can be had by using a rigorous computer model, but since time is money the luxury of detailed models is not always available. Without an excessive investment in time, the methods shown below yield results close enough to identify probable savings and to tell if the measure is worthwhile or not. While this is by no means a complete listing, it is hoped to convey the general method of abbreviated energy accounting for multi-faceted processes when improvements are

Facilities Operations Application Examples	Remarks

EMS Monitoring for Energy Use Benchmarking	Using the Energy Management System to monitor the primary energy sources is useful to establish benchmark usage patterns for the building.
	
	Suggested primary energy metering points for a typical building:
	Gas
	    • Whole Building Consumption
	Electric
	    • Whole Building Demand
	    • Whole Building Consumption
	    • Lighting Loads sub-meter
	    • Mechanical Loads sub-meter
	    • Other / Plug Loads sub-meter
	
	Additional sub-metering may be used, as appropriate and trended and logged by the Energy Management System. For example, sub-metering of Data Centers or other high-use areas, or large sub-let areas, may make sense for both energy and business considerations.
	
	Overall Energy Use Index (EUI) values, in kBtu/SF-yr, kBtu/SF-degree day-mo, or similar normalized units, are also very useful as a building energy use watchdog, to provide early detection of things going awry. This provides a reminder mechanism to prompt the operations personnel to keep the building energy use on track.
	
	Once the benchmarks are set, the Energy Management System can easily be set with energy use thresholds to alert the customer if usage creeps up over time. This early detection mechanism serves to prompt investigation and corrective measures. This energy use feedback will help sustain the intended building energy use over time.

EMS for Start-Up Commissioning	The DDC operator workstation is a useful tool for initial start-up and checkout of new controls. With a set of walkie-talkies it can be used to verify each point from the sensor all the way to the graphic screens and reports by disconnecting a wire at the field device. The trending feature can demonstrate proper tuning by recording the control reaction to a set point step change – ultimately returning to stable operation with no hunting.
EMS for Measurement and Verification	The trending and reporting features of Energy Management Systems are well suited for Measurement and Verification activities. The differential **before and after** energy use can be identified using definite purpose energy meters (kWh, kW, flow meters, Btu meters, etc) or calculated values.

Figure 22-19. Facilities Operations Control Applications (*Concluded*)

proposed. Some of the examples show straightforward benefits, and others show benefits with parasitic losses or competing processes. The first example below is condenser water reset, and is discussed in detail. Other abbreviated solutions are provided in Figure 22-24.

22.18.4 Detailed Example: Condenser Water Reset

Note: The cooling tower term *approach* is the difference between leaving condenser water temperature and ambient wet bulb temperature. Cooling towers with closer (smaller) approach ratings either have more heat transfer surface area or larger fan motors.

The Variables

The chiller kW/ton varies according to load, and the cooling tower kW/ton varies according to wet bulb temperature and approach. The cooling tower fan energy varies according to chiller ton hours, tower capacity, and condenser temperature setpoint with respect to ambient wet bulb. Overall savings varies with annual ton-hours, chiller efficiency, the ratio of chiller-to-tower efficiency (kW/ton), chiller low limit in accepting colder condenser water, and coincident wet bulb temperatures. The wet bulb variable means that any rules of thumb developed for annual savings would be area/climate specific.

The more hours the equipment runs, the quicker the payback. Conversely, cooling improvements in moderate climates are often difficult to justify.

Although the chiller energy is reduced directly as the condenser temperature is lowered (1-1.5% per degree), the cooling tower efficiency decreases at lower approach temperatures, and eventually a point is reached where the savings in chiller energy are met by the added cooling tower horsepower. This break-even point varies with the

Pitfall	Remarks
Trying to Use Controls to Fix a Problem	As a general statement, **controls should not be used to compensate for inherent problems.** Better to fix the root issues first.
Not Considering the User!	Get input and buy-in.
Project Scope Not Clear	Avoid run-ins with the vendor and contractor. Be clear on scope and price, and seek verification that <u>all agree the scope is clear</u>, in advance of the work. Avoid money surprises. Make system and performance expectations clear and seek verification that the scope is well understood. Things like wiring responsibility, calibration, quality control, functional testing, training, and documentation are often overlooked. A measure of good business and project management is whether, at the end of the project, all parties are happy, and nobody loses. Depending upon complexity, it may be appropriate for the energy professional to delegate the project implementation to a skilled project manager.
Contractors not Pre-Qualified	Learning this after the work starts is way too late. Build quality into the project by hiring the right people. Things to look for in a controls contractor: • Factory authorized system provider. • Factory trained technicians. • Local support, parts and technical support. • Acceptable quick-response time. • Verifiable experience with similar systems. • Customer-defined expectations for quality control.
Quality Control Expectations not Defined	For a project of any consequence, active quality control is important to achieve success. This topic is a benefit to both the customer and the contractor. NOTE: not all of these apply to every job, and logic must prevail. Quality control properly applied should be a benefit, not a barrier. <u>Examples:</u> • Acceptable instrumentation quality levels • Wire marking, terminal marking, etc. • Detailed and accurate shop drawings and sequences • Submittal review process prior to work • Software demonstration prior to installation • Wiring checkout procedure • Instrument calibration • Check-out and Start-up procedures • System demonstration and testing rigor, and whether a written test plan is required. • As-built documents for hardware and software changes • Terms and details of system guarantee, including various hardware guarantees (and whether labor is included or not), calibration guarantee, etc. • Definition of "substantial completion" (when is it DONE) for final payment

Figure 22-20. Control System Pitfalls (*Continued*)

cooling tower sizing, and towers with high hp/ton (small box, big fan) will hit the wall sooner than more generously sized cooling towers. The cooling tower energy penalty varies, depending upon the approach temperature, and rises sharply for each degree below 7 degrees F approach. Some common values of the energy penalty, in percent increase per degree lowered, are as follows:

12 degF Approach	7.3%
11 degF Approach	7.8%
10 degF Approach	8.3%
9 degF Approach	8.9%
8 degF Approach	9.7%
7 degF Approach	10.6%

8.9% average cooling tower fan energy penalty per degree lowered

The chillers place a physical limit on this control process; some can take colder condenser water than others. Centrifugal chillers can accept from 55 to 70°F entering condenser water, depending upon the manufacturer. Screw chillers are generally limited to 70°F entering condenser water. Many reciprocating chillers are limited to 70°F, but some can operate at lower temperatures. *IMPORTANT: For each application, the limits of the machinery need to be identified, and the control system must stay within those limits to assure no damage or detriment is done to the chillers.*

Refer to Figure 22-22. For a chiller with 0.5kW/ton efficiency, paired with a 0.07 kW/ton cooling tower at 12 degrees approach, a proposed 5 degree reduction (from 12 to 7 degrees F approach) would yield *worse* energy consumption than leaving it at 12 degrees F, due to the high cooling tower fan energy penalty. This same example, with all things equal except a 0.04 kW/ton cooling tower, would save 0.35% per degree lower overall cooling energy. This example shows that *during operation at or near design wet bulb conditions, most or all of the theoretical chiller*

Pitfall	Remarks
System Integrators not Pre-Qualified (DDC)	Unless the digital controls are standalone, integrating the information into the supervisory control system database is an important step for overall system success. • Customer-established standards developed in advance, to define expectations. • Adherence to standards, to promote consistency between successive projects. • IT qualifications and credentials, if any of the integration touches or shares communication infrastructure. • Verifiable experience with similar IT systems. • Verifiable experience with similar control systems.
Too Many Bells and Whistles	Each controlled or monitored item should have a definable need, and justify its existence. Avoid the bells and whistles. Controls are expensive! Apply them prudently. DDC controls cost more than pneumatic or other controls, but can also do more and justify the difference if properly applied.
Too Complex	It will only get unplugged if overly complex. Often simple solutions will work well and stay working longer. Find the balance between complexity and optimization that fits the customer.
Use of Experimental or Beta Site Products	BETA stands for **B**rand new item, and we aren't **E**xactly sure it will work, so please be our **T**est site, **A**nd let us know what does/doesn't work. It doesn't actually stand for that, but that is what it means to the customer. As a rule, don't volunteer the customer to be a guinea pig. Use proven solutions.
Introducing New Single Points of Failure	Evaluate systems for weaknesses. Avoid bottlenecks and central hub issues where possible, and have contingency plans where they are unavoidable. For each element, ask *"If this item quits, what am I left with?"*
Creating a Captive Customer – Proprietary Systems	The customer needs to make this decision, and there is no 'right' answer. There are pros and cons to both proprietary and open systems, and both ways can achieve good control results.
Not expandable	Most facilities will expand their DDC system over time. Know in advance that provisions and requirements for expansion, including hardware and software.
Backward Compatibility Issues	As hardware and software upgrades are introduced, there is a question of how it will work with the existing equipment. Some vendors are very good about this, and some are not.
Gee-Whiz Points	Adding points with little or no value should be avoided. For each hardware point connected to the system, question the value to be sure it is a worthy investment.

Figure 22-20. Control System Pitfalls (*Concluded*)

savings from condenser water reset will usually be negated by the added cooling tower energy use, unless the cooling tower is very efficient (0.04 kW/ton or less).

While the energy savings near design conditions may be marginal, most of the chiller hours will be at wet bulb conditions that are more favorable, and this is where the energy savings are attained. The control strategy capitalizes on this by continuously adjusting the operating setpoint based on ambient wet bulb. When at or near design wet bulb, the controller will stop trying to make colder water, avoiding the penalty. The savings occur on drier days. For those chiller operating hours when the wet bulb is significantly lower than design and the cooling tower can easily produce colder water with little tower energy penalty, the savings will be much more pronounced and closer to the theoretical 1-1.5% per degree. Of course, the chiller load

during these shoulder seasons or overnight periods are usually lower than maximum, so *knowing the chiller load profile and coincident wet bulb is key to quantifying savings.*

Editor's Note: Chillers that can accept very cold condenser water and have cooling needs coincident with reduced wet bulb temperatures such as in the Southwestern US, when it is easy to achieve colder condenser water temperatures, can achieve 10-15% annual overall cooling energy savings using condenser water reset, compared to a fixed temperature setpoint of 70 degrees F.

Figures 22-22 and 22-23 illustrate the pronounced effect the system cooling tower has upon the overall cooling savings of the condenser water reset control routine. Remember, without the cooling tower fan energy expense

Benefits		Costs	
Tangible	Intangible	Tangible	Intangible
Energy Savings	Reliability improvements through increased monitoring.	First Cost.	Higher skill level required for maintenance personnel (DDC).
Demand Savings	Automated equipment operating logs and ease of archiving.	Control System Repair Costs. Often overlooked, these should be budgeted for like any other building system. A rule of thumb is 5% of initial cost per year as an allowance for Operations and Maintenance.	Increased training costs (DDC).
Labor Savings. A touchy subject. Whether job elimination is used as project justification is an individual decision.	Reliability improvements from expedited / remote troubleshooting, early detection and other pro-active benefits from Trending, Reporting, and Alarm features of DDC controls.	Labor costs for the IT (Information Technology) work on the computer-side of the control system, especially when interfacing with other systems or sharing common cabling. Costs may be in-house special personnel or contracted.	Shorter technology obsolescence cycle for DDC controls means earlier system replacement, and increased life cycle costs.
	Predictive / planned maintenance.		Increased time spent in attendance of the controls and operator workstations (DDC).
	Improved comfort and occupant productivity. NOTE: Some attempts have been made to identify associated cost savings from productivity, but these are usually in reference to corrective actions made to areas where very poor conditions existed previously. So, smaller improvements in comfort probably would not yield proportional results.		Higher maintenance costs if proprietary DDC system chosen
	Avoided cost of replacement of existing controls that need replacing anyway.		Electrical consumption of controls and instrumentation.

Figure 22-21. Control System Costs and Benefits

for making colder condenser water, the savings from the chiller would be lowered 1-1.5% per degree.

Figure 22-22 shows how the hp rating of the cooling tower (kW/ton), relative to chiller kW/ton, affects the optimal cooling tower leaving water temperature setpoint. The "approach" value is a parameter of the optimized condenser temperature calculation, and it should be selected based on the cooling tower in use, for best economy. Suggested values of cooling tower approach for best overall cooling efficiency (kW/ton) are shown in Figure 22-23. By using the "ratio" instead of specific combinations, this information can be applied to any combination

of chiller and cooling tower. This is the value inserted in the sequence "...*optimum cooling tower setpoint shall be equal to the calculated wet bulb temperature plus approach...,*" and provides further evidence of the importance of amply sized, low HP cooling towers.

22.18.7 Intangible Control System Savings Examples

There are ancillary savings from control system improvements. These can be discussed but often defy being quantified. Generally, dollar figures are used with these bonus savings; if they are used at all, they should be heavily de-rated (by half), and it should be understood that

Nominal Cooling Tower kW/ton, at 12 degF Approach	Ratio of kW/ton Chiller / Tower , Chiller @ 0.5kW/ton	Overall Cooling Energy Savings % per degF lowered. Compare to rule of thumb 1-1.5% per degF.
0.03	16.7	0.63%
0.04	12.5	0.35%
0.05	10.0	0.08%
0.06	8.3	-0.18%
0.07	7.1	-0.42%
0.08	6.3	-0.66%
0.09	5.6	-0.88%
0.10	5.0	-1.10%

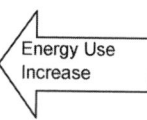

Energy Use Increase

Figure 22-22. Cooling Tower Effect on Condenser Water Reset Savings (Lowering setpoint to reduce approach from 12 to 7 degF). The more hp the cooling tower fan use per ton of cooling, the more the chiller savings are eroded.

Chiller kW/ton	Tower kW/ton	Ratio of Chiller / Tower kW/ton	Typical Lowest Economical Tower Approach For Condenser Water Reset, degF
0.5	0.10	5:1	15
0.5	0.09	6:1	13.5
0.5	0.07	7:1	12
0.5	0.06	8:1	11
0.5	0.05	10:1	10
0.5	0.04	12.5:1	8.5
0.5	0.03	17:1	7

Figure 22-23. Cooling Tower Effect on Optimum Condenser Water Temperature

they are easy to argue.

Cost Benefit from Operations and Maintenance Efficiency Improvements:

DDC reduced labor costs are likely from the reduction in service calls to individual sites when a condition can be analyzed and a setting can be adjusted remotely. This is increased when active computerized maintenance systems are used (CMMS).

Repair costs are reduced by avoiding equipment shutdown or damage from early detection (pre-alarms) for out-of-norm conditions.

Cost Benefit from Improved Comfort and Indoor Air Quality

There are some limited data for this. One study conservatively estimated the following productivity losses [5: pp 4-9]:

Comfort:

Impaired Air and Thermal Quality—1.5%

Indoor Air Quality

3.5% for an unhealthy building. ("Conditions similar to an SBS building,... but with a lower percentage of employees affected.")

6% for SBS/BRI [sick building syndrome/building-related illness]

Total worker productivity costs in terms of $/SF for various

building use [5: pp 4-8, 4-11]:

Values below were derived from the gathered data in the referenced document, in 1995 dollars.

$13 Assembly
$45 Education
$19 Food Service
$110 Healthcare
$19 Lodging
$23 Mercantile and Services
$97 Office

To identify a dollar value of improvements to comfort or IAQ, use the appropriate percentage productivity increase with the total productivity benchmark data, expressed in $/SF (1995 dollars). For example, if a 1% productivity increase is expected for a 25,000 SF office building with a total productivity rate of $97 per SF-yr, the savings (with a 30% de-rate) would be $0.01*97*25,000*0.7, or approximately $17,000 per year. It is suggested that this be used **only** in extreme cases where detriment from poor comfort or poor indoor air quality is a core issue and project driver.

Cost Benefit from Improved Operator/Technician Training

Data are limited. One study concluded [6]:

Improved HVAC preventive maintenance resulting directly from technician training has a significant energy savings po-

Measure	Measure Savings	Other Process or Measure that Subtracts from Savings	Remarks
Programmed HVAC Air System Start-Stop	Rule of thumb: HVAC Air Systems left on continuously without setback instead of being turned off each night have been found to increase annual energy use by **15%.**		Values stated were estimated on one building from envelope losses from not setting back temperatures in winter, plus air horsepower for fans that never stop. Calculated savings were conservatively reduced by 30%.
Chiller - Condenser Water Reset	Rule of thumb: **1-1.5%** chiller energy (kW/ton) reduction per degree <u>lowered</u>.	Additional cooling tower fan horsepower subtracts from measure savings, proportional to the additional cooling tower hours of operation. Increased tower fan energy can consume from 40% to 100% of the chiller theoretical savings at or near design wet bulb conditions.	This is a VERY dynamic calculation. Ignoring cooling tower fan horsepower penalty will over-state savings. Knowing the chiller load profile and coincident wet bulb is essential to quantifying savings. See <u>detailed example</u> in <u>22.18.6</u>
Chiller – Constant Volume Chilled Water Reset	Rule of thumb: **1-1.5%** chiller energy (kW/ton) reduction per degree the chilled water temperature is <u>raised.</u> This can be applied to the annual aggregate load (ton-hours) with little error, and does not need a load profile.		This can also cause loss of humidity control in many climates, and the risk of mold damage may overshadow the potential for energy savings. If done in humid climates, layering this control with indoor and ambient wet bulb temperatures would be suggested, to restrict the reset activity to times of low humidity.
Chiller – Variable Volume Chilled Water Reset	Rule of thumb: **1-1.5%** chiller energy (kW/ton) reduction per degree the chilled water temperature is <u>raised.</u> This can be applied to the annual aggregate load (ton-hours) with little error, and does not need a load profile.	Additional pump horsepower subtracts from measure savings, according to affinity laws: $HP2 = HP1 \left({}^{Q2}/_{Q1}\right)^3$ Energy transport efficiency is <u>high</u>, so pumping costs are usually a small fraction of cooling costs. A load profile is required. Make two identical load profiles in a spreadsheet, and insert volume and static pressure for the existing and proposed systems. **Pump bhp =** gpm * head * sp. Grav $/_{3960 * eff}$	Since pump energy is usually small compared to chiller energy, per ton, there is a net savings, but the added flows negate most (e.g. 2/3) of the chiller savings. See humidity concerns above in "Chiller – Constant Volume Chilled Water Reset"

Figure 22-24. Quantifying Control System Savings (*Continued*)

tential. From a study of nine community colleges in California, savings potential estimates ranged from 6% to 19% of total annual campus energy costs, or $0.09-0.26/SF-yr.

It is suggested for use *only* in extreme cases where detriment from poor maintenance is a core issue and project driver.

Cost Benefit from Calibration

This is very tough to estimate. The question is which direction are the calibrations off and how bad is it? Some control companies have estimated a rule of thumb of *5-10%* energy waste from badly neglected system calibration. Much of this is dependent on whether the primary heating and cooling systems run simultaneously. If they do not, for example a heating valve slightly open in summer, no harm is caused.

The worst case is two adjacent sequenced measure-

ments that drift together and overlap. This can increase cooling load by 20% from false loading, plus the heating energy spent. Note that these same two measurements could also drift apart, thereby creating a nice dead band. Likewise, if space temperatures drift down in winter, indicating they are colder than they really are, the heating system will use more energy to heat the space than designed; this same drift (if it is the same thermostat) will reduce cooling load in summer. If one air handler static pressure sensor drifts down, the fan will work harder to achieve a false value. But the neighboring air handler static sensor may have drifted the other way. Allowing the instruments to get out of calibration will almost surely cause inefficient operation and higher energy costs, mostly due to heating-cooling overlap or other bucking processes.

Measure	Measure Savings	Other Process or Measure that Subtracts from Savings	Remarks
Air Handler – Constant Volume Supply Air Reset in cooling mode	Rule of thumb: **1-1.5%** chiller energy reduction per degree the supply air temperature is raised, for hours when mechanical cooling is used. Reset only occurs in mild weather and so a load profile is required.	Hours when air-economizer is used do not count toward these savings.	See humidity concerns above in "Chiller – Constant Volume Chilled Water Reset"
Air Handler – VAV Supply Air Reset in cooling mode	Rule of thumb is **1-1.5%** chiller energy reduction per degree the supply air temperature is raised, for hours when mechanical cooling is used. Reset only occurs in mild weather and so a load profile is required.	Hours when air-economizer is used do not count toward these savings. Additional fan horsepower subtracts from measure savings, according to affinity laws: $$HP2 = HP1 \left(^{Q2}/_{Q1}\right)^3$$ Energy transport efficiency is <u>low</u>, so air moving costs are usually a large fraction of cooling costs. A load profile is required. Make two identical load profiles in a spreadsheet, and insert volume and static pressure for the existing and proposed systems. **Fan bhp =** $^{cfm \, * \, tsp}/_{6356 \, * \, eff}$	Added fan HP negates most of the cooling savings. For generously sized duct and low air friction losses, net gains have been measured. See humidity concerns above in "Chiller – Constant Volume Chilled Water Reset"
VAV Supply air reset, once most of the VAV boxes are confirmed to be at minimum positions and in heating mode.	Elevating the supply air when most boxes are at minimum positions reduces the reheat coil burden. This is because the VAV reheat system must first heat the primary air to room temperature, before any building heating occurs. This is known as the inherent VAV reheat penalty.	Additional fan horsepower subtracts slightly from measure savings. Energy transport efficiency is low, but at minimum flows the fan laws have reduced baseline horsepower substantially. Example: • If at 40% of max, fan HP will be 20% or less of full load. Heat from the extra fan energy occurs only in heating mode and provides some beneficial heating, but at efficiencies equal to electric resistance heating. The amount subtracted from the measure savings is the added fan Hp, expressed in Btu, less the corresponding HVAC heating input needed for the same amount of heat.	Heating savings dominates slight fan penalty. At minimum air flow conditions (winter), the extra fan hp is a low magnitude penalty. Once the primary air flow is at minimum (with most VAV boxes at minimum), then raising supply air temperature produces savings that outweigh the added fan energy. The aggregate VAV box minimum air flow settings establish the winter supply air flow rate for the fan horsepower penalty calculation.

Figure 22-24. Quantifying Control System Savings (*Continued*)

22.19 CONCLUSION

Automatic controls are useful for basic regulation and quality control of processes and environments. They can also be leveraged for energy savings through optimization. Properly applied, these systems are reliable and cost effective. Returning to the chapter intent, the stated purpose of this chapter was *to focus on the application of automatic controls as a tool to achieve energy savings.* The reader should review the titles of each section, reflect on the key topics taken away, and decide if the stated objective was met. It is hoped that the reader has gained insight into how automatic controls can help achieve energy goals and will endeavor to put these systems to work optimiz-

ing processes and saving energy.

22.20 FURTHER STUDY TOPICS

This chapter was necessarily made brief and does not pretend to be a complete treatment of the subject. The following advanced topics were not addressed and are listed for the interested reader to pursue through additional study:

• Addressable I/O devices, network communications, and equipment interfacing technologies
• Boolean logic
• Cascade control
• Computerized maintenance management systems

Measure	Measure Savings	Other Process or Measure that Subtracts from Savings	Remarks
Boiler lock-out above 60 degrees F, in lieu of continuous operation.	Use weather data to determine hours above 60 degrees F when the boiler will now be off, and estimate savings from boiler standby losses for these times. Thermal losses from distribution piping will add to cooling load if boiler runs in summer. Rule of thumb: Boilers left idling all summer instead of being turned off in warm weather have been found to increase annual gas use by **30%.** Values were taken from an Hourly Analysis Program for one building Climate Zone 2. Calculated savings were conservatively reduced by 30%.		Boiler efficiency drops rapidly below 50% load, due to standby losses. **Efficiency loss is roughly 1% thermal efficiency less for each 6% capacity reduction**, for operation below 50% load. This is an average between 50% and 20% load. The curve is exponentially worse below 20%. Losses are for the boiler only and do not include distribution system loses or pump energy.

Measure	Measure Savings	Other Process or Measure that Subtracts from Savings	Remarks
Chiller lock-out below 55 degrees F, in lieu of continuous operation.	Use weather data to determine hours below 55 degrees F when the chiller will now be off, and cooling loads during these times (if any). Estimate savings from chiller efficiency at reduced load for the hours it should have been off.		Chiller efficiency drops rapidly below 50% load. **Efficiency loss is roughly 0.1kW/ton worse for each 6% capacity reduction**, for operation below 50% load (water-cooled chiller). This is an average between 50% and 20% load. The curve is exponentially worse below 20%. Losses are for the chiller only and do not include distribution system loses or pump energy. For air-cooled chillers, the efficiency may not drop much, due to the reduced ambient temperature beneficial effect. Load for the chiller will be from control valves that don't close, pumping, and heat absorbed into the distribution piping. If air handler winter pre-heat is set to 55 degrees F (sequenced with mixing dampers), simultaneous heating and cooling can exist between the preheat coil and cooling coil if they are each set to the same 55 degrees F. This loss can be estimated at **5%** of the preheat coil load.

Figure 22-24. Quantifying Control System Savings (*Continued*)

- (CMMS)
- Control loop interaction
- Ergonomic considerations: trends, logs, reports, alarms, graphics
- Failure modes, mitigation, and fault tolerance
- Feed forward control
- Fuzzy logic
- IT security
- Ladder logic
- Loop tuning: step changes, dead time, stability, response time
- Measurement and verification with DDC controls

Measure	Measure Savings	Other Process or Measure that Subtracts from Savings	Remarks
Constant Volume Hot water converter reset	Reduced distribution system piping heat losses. Estimate from hours and heat loss tables, using average system temperatures, and take-offs of pipe sizes and lengths. If the heating system runs during cooling season, these losses (now credits) are amplified as new cooling loads, increasing benefit.	Evaluate <u>when</u> (winter or summer?) and <u>where</u> the piping heat losses occur. Wherever the thermal losses are beneficial, they subtract from measure savings. If boiler lockout is used, losses from summer hours must be subtracted from measure savings.	During heating season, heating losses from distribution piping is not a true loss if the piping is within the building envelope. During these periods, consider at least <u>half</u> of the loss as beneficial heat.
Variable Flow Hot water converter reset	Reduced distribution system piping heat losses. Estimate from average systems temperatures and hours, using heat loss tables. If the heating system runs during cooling season, these losses (now credits) are amplified as new cooling loads, increasing benefit.	Variable speed pumping will ramp up as temperature is lowered, increasing pumping costs and tending to offset the savings in standby piping losses. Heat from the extra pump energy occurs only in heating mode and provides some beneficial heating, but at the efficiency of electric resistance heating. This beneficial heat subtracts from measure savings. Evaluate <u>when</u> (winter or summer?) and <u>where</u> the piping heat losses occur. Wherever the thermal losses are beneficial, they subtract from measure savings. If boiler lockout is used, losses from summer hours must be subtracted from measure savings.	For annual hours with the boiler active, the piping losses can be estimated. From these, subtract added pumping costs. Energy transport energy penalty for hydronic systems is low, so this may still produce good savings. During heating season, heating losses from distribution piping is not a true loss if the piping is within the building envelope. During these periods, consider at least <u>half</u> of the loss as beneficial heat.
Control off or reduce lighting in a thermally <u>heavy</u> building	Electric savings is from the hours the lights are now off, compared to existing hours of operation (or reduced wattage if de-lamped) In summer, excess heat from inefficient lighting is amplified since it is also an extra cooling load.		This measure assumes the building is almost always in cooling, so the extra lighting energy is almost never a benefit at all. Cooling load during hours of air economizer are not counted.

Figure 22-24. Quantifying Control System Savings (*Continued*)

- Open protocols, XML, gateways, and translators for DDC networking
- SAMA logic diagrams, multi-element industrial control methods
- SCADA systems
- Self-powered controls
- Wireless controls

22.21 GLOSSARY OF TERMS

Accuracy: A performance measurement of an instrument to produce a value equal to the actual value. Regarding instrumentation, a general statement is that accuracy can be adjusted, but repeatability is a function of the instrument.

Addressable Devices: Also called *smart* devices. A term used to describe equipment that can be given a unique identifying *address* and controlled directly by a digital control system, often in reference to commodity-type equipment that exists in quantity in a facility. By having unique addresses, multiple devices can share a common communication bus and reduce the volume of wiring.

Analog: Variable (input or output), contrasting to discrete.

Bi-Metal: A basic temperature transducer formed by joining two metals with different thermal expansion

Measure	Measure Savings	Other Process or Measure that Subtracts from Savings	Remarks
Control off or reduce lighting in a thermally light building	Electric savings is from the hours the lights are now off, compared to existing hours of operation (or reduced wattage if de-lamped). Excess heat from inefficient lighting also adds cooling load during cooling season.	Excess heat from inefficient lighting is actually part of the building heating but at the efficiency of electric resistance heating. This loss of beneficial waste heat in winter subtracts from measure savings. The winter energy loss is equal to the heat output saved from lighting. The amount subtracted from the measure savings is the avoided lighting waste heat, less the corresponding HVAC heating input needed for the same amount of heat.	The reduced heat loss from better lights means the HVAC heating system must increase output to compensate. Unless the building heating system is electric resistance, the HVAC heating system will be more efficient at producing heat than waste heat from lighting. Note that for buildings with electric resistance heat this would result in no overall savings during heating season. **If heating and cooling seasons are reasonably balanced, it may be acceptable to ignore both effects, as counteracting.**
Use night-time unoccupied mode and morning warm-up in lieu of constant conditioning	During unoccupied periods when the indoor temperature drifts between occupied and ambient temperatures, skin loads reduce proportional to the reduced differential temperature (dT). <u>Rule of thumb:</u> **1%** savings per degree of setback, if kept there for at least 8 hours.	If electric heat is used, a pitfall can exist in the morning warm-up by increasing electric demand, reducing or negating the measure savings, especially if there is a utility ratchet clause. Avoid this by bringing the building up in segments, allowing sufficient time between stages to prevent large demand events.	
Change indoor cooling temperature set points from 74 cooling / 72 heating to 76 cooling and 70 heating.	Similar effect as night set-back, except use **2%** per degree setback for continuous operation instead of 1% for setback.	If local adjustment is available to the tenants, some adjacent areas may have sufficiently different local set points to cause simultaneous heating and cooling, reducing measure savings.	Limiting the local user adjustment to +/- 2 degrees F from nominal set point will help ensure these savings are achieved.

Figure 22-24. Quantifying Control System Savings (*Continued*)

properties. Common shapes are coiled ribbons (that turn upon a temperature change), bars (that bend upon a temperature change), and disks (that warp and "snap" upon a temperature change). Movement is predictable and repeatable and is integral to many temperature control instruments.

Binary: Synonymous with *on-off* or *discrete* (input or output), contrasting to analog.

Boolean Logic: A logic technique used to formulate precise queries using true-false connectors or "operators" between concepts. The primary operators are AND, OR, and NOT. Words or concepts joined with these operators, and parentheses are used to organize the sequence groups of concepts. The true-false nature of Boolean Logic makes it compatible with binary logic used in digital computers since a TRUE or FALSE result can be easily represented by a 0 or 1. Named after George Boole.

Building Automation System (BAS): BAS is the term given

to a computerized automatic control system when the primary focus: "…is on automating as much as possible to save labor." [7:pp 38.10]. BAS is the most common term used to describe computerized control in buildings that provide one or more of these functions:
Energy management system (EMS)
Facility management system (FMS)
Energy management and control system (EMCS), also called energy monitoring and control systems. (See Chapter 12.)

Cascade Control: Also called Master-Sub Master. This is a combination of two controllers in series with a related measured variable. The output of the first controller becomes an input to the second controller, which can then amplify it. Often, the output of the first (master) becomes the setpoint of the second controller (sub-master). Useful in improving modulating control of systems with very slow process response.

Measure	Measure Savings	Other Process or Measure that Subtracts from Savings	Remarks
Air-side economizer	Use weather data and building load characteristics (thermal break even point) to determine when the building needs cooling indoors and free cooling is simultaneously available. Both the number of hours of economizer cooling and the magnitude of displaced cooling load at those hours are required to make this determination.	The actual cooling load will be a mix of envelope loads and internal loads, and will decline at lower temperatures – the slope of the load decline curve will depend on the relative magnitudes of the thermal and internal load constituents. For thermally light buildings, where the thermal break even outdoor temperature is higher than 55 or 60 deg F, savings will be near zero.	For thermally heavy buildings, e.g. those with high internal heat gains that need cooling all the time, the air economizer can be very beneficial since both the hours of benefit and the displaced load (ton-hours) will be high. Rules of thumb for savings are climate dependent, and will vary by region. In dry climates, this can be estimated with fair accuracy using outdoor dry bulb temperature profiles, but in other regions, the coincident OA humidity or wet bulb profile will need to be overlaid as well to establish cooling loads. For thermally heavy buildings cooling benefit is viable whenever OA temp is 55 degF or below, even in humid climates or when raining. Economizer savings occur when cooling loads exist at outside air temperatures low enough that mechanical cooling can be turned off and outside air act as the cooling medium. The mechanical cooling cut off point varies by region but can b certainly be cut off at 55 degrees F. Example: Assume 55 degF is the cut off point for mechanical cooling. Unless cooling loads exist below 55 degF there are no savings from an air economizer. Identifying the ton-hours of load in the striped triangle is the economizer savings.

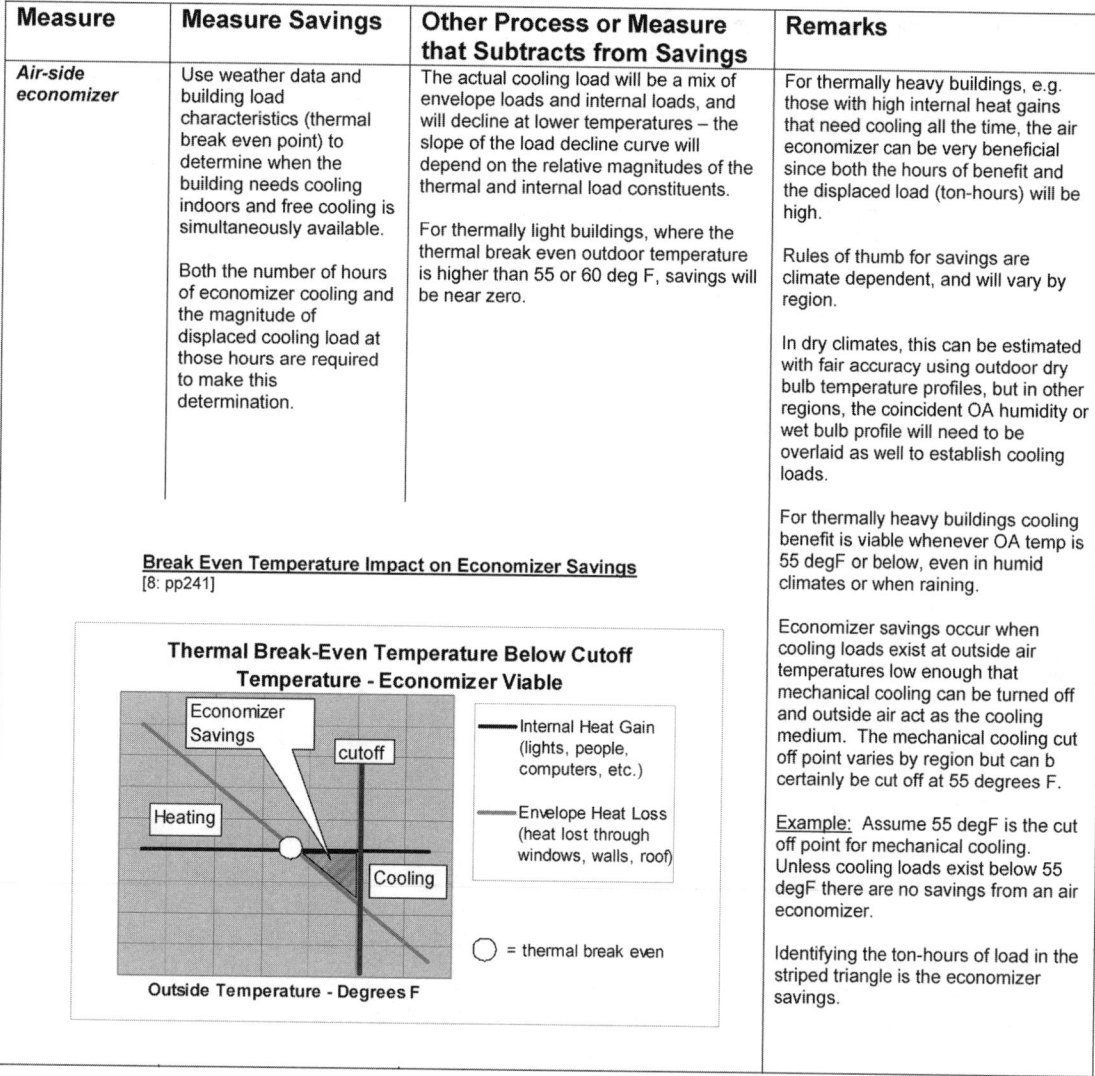

Break Even Temperature Impact on Economizer Savings
[8: pp241]

Figure 22-24. Quantifying Control System Savings (*Continued*)

Cavitation: A problem phenomenon with fluid flow, often with control valves or pumps, where the local pressure drop is sufficiently high to cause temporary boiling of the fluid. When the pressure is sufficiently regained, the vapor bubbles created by boiling collapse and create a shock. If this occurs within the vicinity of fluid handling apparatus, it can damage the equipment.

Closed Loop: A control system which includes related process feedback input(s) and controlled device output(s), collectively forming a regulating process. The defining characteristic is the feedback input point that senses changes in the process, and the effect of the controlled device, to close the loop of communication.

Control Loop: The general term for a collection of control system components used to regulate a process that includes as a minimum a controller, setpoint, control element (controlled device), associated process, and process measurement (measured variable).

Controlled Device: The manipulated device responding to the controller output, which then impacts the process itself. The item being manipulated by the controller.

Controller: A device that compares setpoint to measured values and determines an appropriate output response.

Cut In: A parameter of two-position control. The cut-**in** value is where the controlled variable is sufficiently beyond the setpoint for the controlled equipment to be turned on (cut-in).

Cut Out: A parameter of two-position control. The cut-**out** value is where the controlled variable is sufficiently beyond the setpoint for the controlled equipment to

Measure	Measure Savings	Other Process or Measure that Subtracts from Savings	Remarks
Replace terminal reheat valves with quarter turn 'no leak-by' valves.	Savings are from hot water use when not needed, over-heating spaces from leakage. For heating systems that run in the summer, cooling savings ADD to the hot water savings. Where cooling is available at the same time, the cooling load will increase to compensate for the unintended heat load from leaking control valves.		For low cost light commercial control valves – those with marginal shutoff ratings and characteristic short life spans – leakage is very common. For rooms with only heating the classic symptom is overheating in the spaces during winter. In extreme cases, the occupant will control indoor temperature by opening the windows in winter. For rooms with cooling too, such as a VAV box with reheat, this may go un-noticed for years. In some facilities with large terminal reheat loads (e.g. hospital) this can represent **20%** of summer cooling load, and for most systems with conventional globe-style heating valves that are over 10-years old, assuming a loss equal to **10%** of summer cooling load is reasonable.

Figure 22-24. Quantifying Control System Savings (*Continued*)

be turned off (cut-out).

Dead Band: Also called zero energy dead band. Refers to a deliberate gap in control span between sequenced, usually conflicting processes, to avoid control overlap and subsequent operational or energy consumption issues.

Example:

A single temperature controller that sequences both heating and cooling equipment to a common duct or space may use a dead band to prevent simultaneous heating and cooling.

Dead Time: The time between a change in the process input and when that change is felt in the downstream process measurement. Often a function of the time it takes material to flow from one point to another.

De-Bouncing: A type of discrete input signal conditioning used for digital control systems. Mechanical snap-acting contacts actually open and close many times and can trigger control output problems when monitored by a high speed digital circuit. The de-bouncing signal conditioning has, as its purpose, to filter out the bouncing "noise" so the controller sees a simple open or closed state.

Demand Limiting: A control strategy with the purpose of reducing electrical demand, not necessarily energy. This is applied to facilities with utility demand charges and is usually designed to reduce peak (max) demand thereby reducing utility demand charges.

Derivative Control Mode: Also called rate or anticipatory control, it is used to increase response time af-

ter a disturbance or step change. For a given rate of change of error, there is a unique value of controller output. This control mode reacts to the rate of change of error, not the magnitude of error. This mode cannot be used alone since there is no output when the error is zero.

Direct Acting: Control action that increases its output as the measured variable increases above setpoint.

Discrete: Synonymous with on-off or binary (input or output). Characterized by having two possible states. Typical example is an open-close contact input or relay output.

Dry Contact: A discrete signal, output or input, that is made with a contact closure such as a mechanical relay or switch, which has no voltage at the terminals and depends upon voltage from the initiating circuit to read its two-state resistance. In contrasting, when the device has a voltage at the terminals (as if to light a light bulb when the contacts close) it is termed the wetting voltage.

Duty Cycling: A demand-limiting strategy where multiple electric points of use, often motors that otherwise run continuously, are controlled to run some fraction of time, e.g. 40 minutes on and 20 minutes off. By coordinating the run and off times of multiple items, the aggregate utility electric demand can be reduced, generating demand savings.

Energy Monitoring and Control System (EMCS): Also called Energy Monitoring and Control Systems. See Building Automation System (BAS). See also Chapter 12.

Measure	Measure Savings	Other Process or Measure that Subtracts from Savings	Remarks
Reduce VAV Supply Air Duct Static Pressure according to Demand *Or* *Reduce Variable Flow Hydronic System Differential Pressure according to Demand*	Basis for savings is shown in the remarks column. Load profile is needed to determine pct flow needed in the building, and then compare incremental flows at the two static pressures. This calculation uses the difference in downstream duct/pipe pressure (SP1 and SP2), not the total pressure, so the "tsp" or "head" are the difference between the two. For summer operation, the extra fan / pump horsepower also creates new cooling load and ADDS to the measure savings.	This approach assumes a constant supply temperature. Anything that raises supply temperatures will increase mass flows (air flow or water flow) and reduce measure savings.	$HP2 = HP1 \times \left(^{SP2}/_{SP1}\right)$ (note 1) Example: a 50 bhp fan load at 1.3 in. w.c. static pressure is reduced to 1.0 in. w.c. by adjusting the control set point. New motor load = 50 bhp X $\left(^{1.0}/_{1.3}\right)$ = 38.5 bhp and savings would be 50-38.57 = 11.5 bhp Note 1: Using the fan laws directly, savings would be 50 Hp x $\left(^{1.0}/_{1.3}\right)^{y1.5}$ = 33.7 bhp. However, for standard VAV systems, the supply duct is held at a constant pressure. Assuming 2/3 of the duct system feels the benefit of the pressure reduction (1/3 does not) the exponent becomes 1.0 and can be dropped. Circulating fan and pump motor loads will change throughout the day and by season, so a load profile should be used to quantify savings. Make two identical load profiles in a spreadsheet, using the same volumetric flow rates and different static pressures (in. w.c.) for the existing and proposed systems. **Fan bhp =** cfm * tsp/$_{6356}$ * eff **Pump bhp =** gpm * head * sp. Grav/$_{3960}$ * eff
Demand Controlled Ventilation	Savings are from avoided ventilation due to reduced occupancy. Calculate the summer and winter ventilation cost per person, using the OA CFM per person and seasonal weather data profiles. With an occupancy profile, compare costs for constant vs. variable ventilation.	Building exhaust requirements and pressurization requirements will limit how far down the ventilation rate can go, regardless of the number of people.	Not all local building departments allow this. Savings are climate dependent. In high humidity areas, ventilation cost is very high in summer. In moderate and dry climates, it is much lower in summer and may only be a cost during winter.

Figure 22-24. Quantifying Control System Savings (*Concluded*)

Energy Management System (EMS): EMS is the term given to a computerized automatic control system when the primary focus "...is on saving energy by specific automatic control programs" [7: pp 38.10,11]. See also Building Automation System.

Facility Management System (FMS): FMS is the term given to a computerized automatic control system when the primary focus "...goes beyond HVAC controls and/or beyond a single building, such as including fire, security, or manufacturing systems" [7: pp 38.11]. See also Building Automation System.

Feedback: The measurement of the process output, which is returned to the process controller that is also influenced by the controlled device. Feedback is the differentiating factor for a closed-loop control system, in contrast to an open-loop control system.

Floating Control Mode: A form of discrete (on-off) control with a null position whereby the controller holds the last output when setpoint is achieved, rather than returning to zero output.

Gain: A tuning constant that multiplies an input or output parameter, usually to increase sensitivity and improve control.

GUI: Graphical User Interface. The software that overlays the machine coding, adding the user-friendliness look and feel to a digital control system work station.

Hunting: Chronic, repeating oscillation (overshoot and undershoot) of a modulating control system, usually indicating a poorly tuned control loop.

Hysteresis: A physical phenomenon best explained as inertia, or a body's tendency to stay at rest. In control systems this can affect input and output instruments, as well as controlled devices (valves and dampers). The hysteresis effect will require a different value of control to cause a change or movement, depending upon whether approaching the value from higher or lower values. Generally, hysteresis is an issue for heavier controlled devices or with minute I/O changes.

Indicating Transmitter: A transmitter with an auxiliary display or meter to provide local indication of the measurement and/or output signal.

Input/Output Points (I/O): The general term for the identifiable instruments (points) used to relay the control system information into (input points) and out from (output points) the controller, linking the controller to the actual process.

Integral Control Mode: Also called reset control, it is used to return the process to zero error by producing an output any time the error is other than zero. It is significant that the output increases over time and is useful to return the process to a zero error state.

Interlock: A control strategy that requires one process (discrete or analog) to depend upon the on-off state of a separate but related process.

Examples:

A boiler firing interlocked with combustion air damper requires the damper to be proven open prior to firing. This interlock would be for safety and is normally "hard-wired," so as not to depend upon any software intervention.

A cooling coil control routine interlocked with supply fan operation requires the supply fan to be proven running prior to engaging in control of the cooling coil. This interlock would be for normal control since the feedback measurement is downstream of the cooling coil and would not be sensed without air flow, it is usually a software interlock.

Ladder Diagram: Also called an elementary wiring diagram, this is an electrical circuit representation where the high and low voltage terminals (120VAC, 24VAC, 24VDC, etc) are shown as vertical lines on opposite sides of the page. Each circuit sharing this power source, with its control contacts and load, becomes a "rung" on the ladder.

M&V: Measurement and verification

Master-Sub Master Control: See "Cascade Control"

Measured Variable: The measurement representative of the process that is the basis for controller action, relative to setpoint.

Measurement and Verification (M&V): The after-the-fact activities that are used to verify whether and to what extent energy savings have occurred, to gage the actual energy savings and cost savings benefit of a project. Often, these are compared to estimated savings. In some cases, M&V is part of a contract stipulation for guaranteed savings and may form the basis of payment from one party to another. Depth and rigor of these activities varies depending upon need for accuracy and available funds.

Minimum Position: An adjustment parameter of controlled devices, referring to the lowest level of control action allowed, regardless of further reduction in controller signal.

Examples:

Minimum fan speed

Minimum damper position

Modulating: See also "Analog." Varying position in response to need, contrasting to two-position.

Most-open Valve: A control algorithm that senses damper or valve positions at the various points of use to determine demand. This strategy serves the most demanding area but strives to provide just enough process media to satisfy it, thereby reducing overall system energy use. Similar control routine applies to "most-open damper."

Open Loop Control: The controlled system has no impact on the measured variable. No amount of change in the controlled variable or in the output of the controller will cause a change in the measured variable.

Optimization: Actively monitoring and controlling each of the pertinent parameters of a process to maximize productivity and/or to minimize energy use. Often requires control of multiple dynamic variables. Optimization can apply to single equipment items or large systems.

Oscillation: Repeated over and over-shooting about the desired setpoint. For two-position control, this is normal. For modulating control, this is abnormal. See "Hunting."

Overshoot: A measure of control system response to changing loads. After a corrective action is taken, and before settling out at the new level, the controlled variable is usually driven beyond the setpoint as it tries to recover. The extent to which it temporarily goes beyond the mark is the overshoot.

Point—Controls context: Usually used with digital control systems, each unique identifiable input or output item. The number of system points represents the size of the system, e.g. the number of connected instruments that can be individually controlled. Most commonly used in reference to hardware items but also applies to software points.

Proportional Band: The range of error (+/-) that will cause the proportional controller output to vary between 0-100%. As the proportional gain increases, the proportional band decreases.

Proportional Control Mode: Produces a linear proportional output that is a direct response to the measured error, configured to counteract the error and provide basic regulation. A residual offset error is characteristic of this control mode, e.g. a true zero-error state can never be achieved using proportional-only control.

Proportional Offset: The amount of residual error remaining in the process using proportional-only control.

Pulse Width Modulation: An output signal conditioning strategy that converts a modulating 0-100% output to a percentage of on time output, to impose the proportional control onto a discrete device. Often used in the control of electric resistance heating or incremental control electric motor actuators.

Real Time (general context): A computer system that responds to inputs without delay. (Computer science context: a computer system that updates information at the same rate it receives information.)

Relay: A discrete output interface device used to control a large current (via heavy contacts) with a small current (the relay coil current) and/or to isolate controller low voltage power (DC) from the controlled equipment.

Relay Logic: The term given to discrete logical control (if-then, and, or) accomplished by arranging relay contact in series and/or parallel electrical circuits.

Repeatability: A performance parameter that measures the ability of a process or instrument to produce the same value of output during repeated trials. Instrument repeatability is synonymous with instrument precision.

Reset or Reset Schedule: A control algorithm whereby one setpoint is varied, usually linearly, between two values, as a function of another analog value. Example: Resetting hot water temperature setpoint (HW) from outside air temperature (OA):

Reset Schedule

OA	HW
70	120
30	180

Resolution: The minimum measurable value of an input variable or the minimum incremental change of the output variable. In digital control systems, this is often a measurement criterion of the A/D or D/A transformation, e.g. the number of steps provided and how closely it emulates true analog control.

Reverse Acting: Control action that decreases its output as the measured variable increases above setpoint.

RTD: Resistance Temperature Detector: A transducer that leverages the physical properties of certain materials that predictably change resistance as a function of temperature. RTD response is characterized as being nearly linear and stable over time.

SCADA: Supervisory Control and Data Acquisition. These are industrial control systems with very large point counts, lots going on, often widespread and remote, and often critical in nature. Thousands of point systems are common. Example: Utility energy and water metering and control.

SCR: Silicon Controlled Rectifier. A device used to proportionally control an electric load, often a resistance heater or motor. The rectifier is an electronic switch so it is technically an on-off device. However, SCRs are normally controlled at very high speeds in time-proportioning fashion to create a modulating effect.

Sensor: The general name for a device whose purpose is to sense some media and translate the measurement into a convenient and predictable form, often electrical or pneumatic.

Setpoint: The desired steady state of a controlled process. The goal of the control activity.

Settling Time: "…the time required for the process-control loop to bring the dynamic variable back to within the allowable range…" [3: pp 10].

Short Cycling: A dysfunction of some two-position control applications where the cut-in and cut-out settings are too narrow or the process rate is too fast, with the result being rapid cycling of equipment, often to the detriment of the equipment.

Soft Start: Any of several methods of reducing start-up demand upon a motor, usually by reduced voltage or mechanically unloading the driven load.

Step Change: Also known as a disturbance or bump, this describes a sudden and significant event process output change. In the context of control loop tuning, a step change is a valuable testing function that will demonstrate the ability of the tuned components to react properly to sudden process changes without undue loss of control, excess drift from setpoint, hunting, or other control anomalies.

System Capacitance: A general term that describes the

relative strength or capacity of the controlled equipment to effect a change to the process measured variable; a good indicator of system controllability and control mode choice. In lay terms, a low system capacitance acts like the equipment is over-sized with the tendency to short-cycle as a result. In technical terms this is measured by the pct process change resulting from a step change in output. Systems with high capacitance can usually be controlled with any control mode. Systems with low capacitance are often troublesome. Systems with low capacitance may experience short-cycling and instability using on-off control unless excessively wide range is used. Modulated systems with low system capacitance may require careful tuning and complex proportional control (PI, PID) to maintain stable control.

Thermistor: A semi-conductor device used to measure temperature. Lightweight and inexpensive, thermistors are popular for low cost or non-critical applications such as residential or light commercial temperature control, where minor errors and long term drift are acceptable. Thermistor outputs are not linear with respect to changes in temperature and usually require some form of signal conditioning to avoid errors.

Thermocouple: A basic transducer for temperature measurement that uses the physical principal of galvanic action between dissimilar materials. The galvanic response is a function of temperature, and so, with proper signal conditioning, it can be used to measure temperature.

Throttling Range: "…the amount of change in the controlled variable that causes the controlled device to move from one extreme to the other, from full-open to full-closed" [2: pp 1:21].

Transducer: A device that performs the initial input or output conversion of a dynamic variable into proportional electrical or pneumatic information, often a very low level signal requiring further signal conditioning.

Examples include: thermocouple, thermistor, RTD, variable capacitance dP cell, strain gage vibra-

tion transducer, flow element (orifice plate, etc.).

Transmitter: The general name for an input device that provides signal conditioning from a transducer or other basic measurement signal, transforming it into some proportional information in a useful form, often with the ability to send the information over long distance without loss of accuracy (transmit).

Tuning: Adjustment of gains and various parameters of a dynamic control system to achieve acceptable and stable control.

Two-Position Control Mode: Also called On-Off control, this is the most elementary control mode. Output is either 100% if sufficiently below setpoint, or 0% if sufficiently above setpoint. Over and undershoot is normal for this control mode, the least costly of all control methods. Regarding two position control: "Generally, the two-position control mode is best adapted to large scale systems with relatively slow process rates" [3: pp 290].

Wax Motor: A self-powered actuator whose motive force is thermal expansion of a wax pellet/bellows assembly. The expansion is translated into a linear movement to vary the position of another device. Common uses include small thermally compensated devices such as shower mixing valves and thermally compensated air diffusers.

References
1. *Optimization of Unit Operations*, Liptak, B., Chilton Book Company, 1987.
2. *Fundamentals of HVAC Control Systems*, Taylor, S., ASHRAE, Atlanta Georgia, 2004.
3. *Process Control Instrumentation Technology*, second ed, Johnson, C., John Wiley & Sons, 1982.
4. "Hierarchy of HVAC Design Needs," Schwaller, D., *ASHRAE Journal* August 2003.
5. *Productivity Benefits Due to Improved Indoor Air Quality*, NEMI, National Energy Management Institute, August 1995.
6. "Quantifying The Energy Benefits Of HVAC Maintenance Training and Preventive Maintenance," AEE, *Energy Engineering*; Vol. 96; Issue 2; 1999.
7. "Computer Applications," 1999 *ASHRAE Applications Handbook*, ASHRAE, Atlanta Georgia, 1999.
8. *Commercial Energy Auditing Reference Handbook*, first ed., Doty, S., Fairmont Press, 2008.

CHAPTER 23

SUSTAINABILITY AND HIGH PERFORMANCE GREEN BUILDINGS

LEED FOR NEW CONSTRUCTION AND EXISTING BUILDINGS

NICK STECKY
NJS Associates, LLC

23.1 BEGINNINGS

The publication of Rachel Carson's *Silent Spring* in 1962 alerted the general public to the dangers of pesticides, in particular the dangers to humans. This helped precipitate the rise of an environmental movement, and associated politics and laws in the United States during the sixties and seventies. New laws were passed to protect the environment. These included:

- The National Environmental Policy Act of 1969—this created the Federal Environmental Protection Agency.
- The Clean Air Act was passed in 1970. This greatly expanded the protection of two previous laws, the Air Pollution Control Act of 1955 and the first Clean Air Act of 1963.
- The Water Pollution Control Act of 1972.
- The Endangered Species Act of 1973.
- The formation of the Federal Department of Energy in the late seventies.

But a problem with environmentalism was beginning to brew. There came a tension, an apparent conflict between the need to preserve the environment and the need to grow and expand the economy and jobs. Environmentalists began to be seen as opponents of growth and industry. There appeared to be a contradiction between business and protection of the environment. Environmentalism began to be seen as just another "special interest" group which simply added cost to running a business, with very little added value.

In addition to legislation, other events were occurring during the seventies, eighties, and nineties that encouraged the development of sustainability:

- Earth Day, launched April 22, 1970.
- Nuclear power suffered major setbacks with the incidents at Three Mile Island and Chernobyl.
- Major oceanic oil spills, including the EXXON Valdez.
- The OPEC oil crises of the mid-seventies and early eighties.
- The mid-seventies natural gas shortages that caused many plants to close during the winter to preserve gas for home heating.
- Discovery that ozone depleting compounds, such as CFC refrigerants, were destroying the ozone layer of the atmosphere.

During the eighties, in reaction to the forces of high energy costs, inadequate energy supplies, environmentalism, and pollution control, a new approach to designing, building, and operating buildings began to develop. It was recognized that buildings consume significant percentages of our resources, open space, and energy. However, some of the new approaches were not without problems. For example, architects experimented with half or even fully buried homes, but many times these homes had problems, such as humidity that led to mold. As a result of the desire to save energy on ventilation air, ASHRAE modified Standard 62 and reduced ventilation rates to five cubic feet per minute (CFM) for offices. This contributed to what was later called sick building syndrome, or SBS. SBS was a catchall phrase for any building that provided an uncomfortable, irritating, and possibly unhealthy indoor environment for the occupants. Employees in these buildings generally had higher rates of absenteeism, lower productivity, and higher incidences of lawsuits against building owners and employers.

23.1.1 The Sustainability Movement

While buildings and occupants were suffering through difficulties as described above, the nineties gave rise to the sustainability movement. It was in 1999 that the book *Natural Capitalism* by P. Hawken, A. Lovins, and

L.H. Lovins was released. The authors present the connection between economics and environmentalism, and they argue that these are mutually supportive, not mutually exclusive. *Natural Capitalism* points out that, as in its title, nature itself is capital. For example, what is the value of a clean lake? It is drinking water that needs less expensive water treatment before being potable. It is the recreational value and income for some, through swimmers and boaters. If the lake were destroyed through pollution, water treatment costs would rise and lake use revenues go down. Seems simple enough, but it has mostly been ignored until now.

Historically, the environmental movement consisted primarily of regulation, legislation, and mandates. However, this regulatory approach has often been seen by the business community as an obstacle to growth. But a new form of green revolution has been emerging. This one may succeed where traditional legislative environmentalism has had limited success. This new form is a larger view, taking economic, community, and technological considerations into account, as well as environmental ones. This new sustainability recognizes the need to consider the cost-benefit analysis in evaluating and/or promoting various programs. And the themes presented in *Natural Capitalism* help us identify and quantify the complete costs and benefits associated with sustainability.

23.1.2 Sustainability Defined

Some of the many definitions of sustainability that exist include:

Design Ecology Project

Sustainability is a state or process that can be maintained indefinitely. The principles of sustainability integrate three closely intertwined elements—the environment, the economy, and the social system—into a system that can be maintained in a healthy state indefinitely.

Brundtland Commission of the UN

Development is sustainable "if it meets the needs of the present without compromising the ability of future generations to meet their own needs."

ASHRAE defines sustainability as

"providing for the needs of the present without detracting from the ability to fulfill the needs of the future."

Note the commonality of theme. The concept is rooted in maintaining our current standards of living without jeopardizing future generation's standards of living.

23.2 SUSTAINABILITY GIVES RISE TO THE GREEN BUILDING MOVEMENT

The late eighties and early nineties were a crucial developmental period for the green design movement. Leaders of green design included William McDonough, Paul Hawken, John Picard, Bill Browning, and David Gottfried, who later was one of the co-founders of the USGBC. The movement acknowledged that buildings represent a very significant usage of resources, land, and energy, and that improvements to the ways in which we design, construct, operate, and decommission buildings could make significant contributions to improvement of the environment and overall sustainability.

Today, owners, occupants, and communities are beginning to hold buildings to higher standards. Industry leaders are responding by creating physical assets that save energy and resources and are more satisfying and productive while being economical as well as environmentally accountable. Building owners who want the greatest return on investment can take a path that is green both economically and environmentally. How are they doing it? They are doing it through integrated solutions for the design, construction, maintenance, and operations, as well as the ultimate disposal of a building.

An integrated, or whole building approach may mean that up-front costs may be no more than conventional construction, and life cycle costs over the life of the asset will be lower. By designing, building, and operating in an integrated way, owners can expect high performance buildings that offer:

- Increased efficiencies of systems and use of resources and energy.
- Quality indoor environments that are healthy, secure, pleasing, and productive for occupants and operators.
- Optimal economic and environmental performance.
- Wise use of building sites, assets, and materials.
- Landscaping, material use, and recycling efforts inspired by the natural environment.
- Lessened human impact upon the natural environment.

23.2.1 Formation of the Unites States Green Building Council, USGBC

In 1993, David Gottfried, a developer, Mike Italiano, an environmental attorney, and Rick Fedrizzi of Carrier Corporation got together to form the United States Green Building Council, USGBC. They had become concerned about the fragmentation of the building industry as it relates to sustainable design. At the time, there was no clear consensus among industry professionals as to what constituted a "green design." The USGBC was initially formed to create an educational organization that would bring building professionals together to promote sustainable design. In 1997, the USGBC was awarded a $200,000 grant from the Federal Department of Energy, and the USGBC was off and running. The USGBC has made a large impact on the design and building industry. There are more than 5,500 members consisting of individuals, large and leading corporations, governmental entities, universities, educational entities, consultants, product manufacturers, trade associations, and more.

23.2.2 Making It All Come Together: The USGBC

The U.S. Green Building Council (USGBC.org), a balanced consensus coalition representing every sector of the building industry, spent five years developing, testing, and refining the LEED Green Building Rating System. LEED stands for "leadership in energy & environmental design," and when adopted from the start of a project it helps facilitate integration throughout the design, construction, and operation of buildings.

MISSION STATEMENT

The U.S. Green Building Council is the nation's foremost coalition of leaders from across the building industry working to promote buildings that are environmentally responsible, profitable, and healthy places to live and work.

The mission of this unprecedented coalition is to accelerate the adoption of green building practices, technologies, policies, and standards. The USGBC is a committee-based organization endeavoring to move the green building industry forward with market-based solutions. Another vital function of the council is linking industry and government. The council has formed effective relationships and priority programs with key federal agencies, including the U.S. DOE, EPA, and GSA.

The council's membership is open and balanced. It is comprised of leading and visionary representation from all segments of the building industry, including product manufacturers, environmental groups, building owners, building professionals, utilities, city governments, research institutions, professional societies, and universities. This type of representation provides a unique, integrated platform for carrying out important programs and activities.

LEED™ Overview

The LEED Green Building Rating System™ is a pri-

ority program of the US Green Building Council. It is a voluntary, consensus-based, market-driven building rating system based on existing, proven technology. It evaluates environmental performance from an integrated or "whole building," perspective over a building's life cycle, providing a definitive standard for what constitutes a "green building."

LEED™ is based on accepted energy and environmental principles and strikes a balance between known effective practices and emerging concepts. Unlike other rating systems currently in existence, the development of LEED Green Building Rating System™ was initiated by the US Green Council Membership, representing all segments of the building industry. It has also been open to public scrutiny.

LEED™ is an assessment system that incorporates third party verification and is designed for rating new and existing commercial, institutional, and high-rise residential buildings. It is a feature-oriented system where credits, also called points, are earned for satisfying each criterion. Different levels of green building certification are awarded, based on the total credits earned. The system is designed to be comprehensive in scope, yet simple in operation.

23.2.3 General Introduction & Discussion

There are several key points in this section:

- LEED is becoming nationally accepted at local, state, and federal levels.
- The engineering community, such as ASHRAE, AEE, IESNA, has the knowledge and skills that can add significant value to a LEED design team.
- LEED has value to the engineering community, both as owners/operators of facilities and as design team members.
- It is a WIN-WIN proposition for us all.
- Green buildings are a process, not a collection of technologies.
- Encourage engineering participation/membership in USGBC and its local chapters.
- Encourage professional accreditation and attendance at LEED workshops.
- Encourage building owners/operators to register and certify projects.

There are many different terms for sustainable buildings, but basically they all convey the same message:

- Sustainable design
- High performance buildings
- High efficiency buildings
- Integrated building design
- Green buildings

Sustainable buildings may be our goal, but the most common term used for these high-performance buildings is green buildings. For some, the term green buildings may sound too environmentally focused, but the way it is used here it represents the all-inclusive idea of sustainable buildings.

Characteristics of Sustainable Green Buildings
- Optimal environmental and economic performance.
- Increased efficiencies, saving energy and resources.
- Satisfying, productive, quality indoor spaces.
- Whole building design, construction, and operation over the entire life cycle.
- A fully integrated approach—teams, processes, and systems.

23.2.4 LEED Green Buildings

Green buildings are designed and constructed in accordance with practices that significantly reduce or eliminate the negative impact of buildings on the environment and its occupants. This includes design, construction, operations, AND, ultimately, demolition. Five fundamental categories constitute the USGBC green building designation. They are:

- Sustainable site planning
- Safeguarding water and water efficiency
- Energy efficiency and renewable energy
- Conservation of materials and resources
- Indoor environmental quality

All relate back to the previous definitions and discussions of sustainability, and all are contained in the LEED standard (leadership in energy and environmental design). As mentioned earlier, this is the trademark rating system developed by the United States Green Building Council, USGBC.

Besides the LEED rating system to define and describe sustainable buildings, there are others, such as:

- The British Research Establishment Environmental Assessment Method (BREEAM) was launched in 1990, and its use is increasing.
- Canada's Building Environmental Performance Assessment Criteria (BEPAC) began in 1994 but was never fully implemented due to its complexity. Canada has now licensed use of LEED from the USGBC.
- The Hong Kong Building Environmental Assessment Method (HK-BEAM) is currently in pilot form.

- The USGBC LEED family of programs.
- State and regional guides include high performance building guidelines in NY and PA for creating high performance buildings, as well as California's programs for school construction.
- Green Globes is a web-based self-assessment program that guides the integration of green principles into a building's design.

LEED strives to encompass a wide band of sustainability that includes:
- Society and Community—recognizes that buildings exist to serve the needs of the community but that their impacts must also be minimized.
- Environment—again, striving to minimize negative impacts on the environment.
- Economics—recognizes that adoption of sustainability initiatives by business will require economic benefits that can be delivered by green buildings.
- Energy—recognizes that energy plays a key role in building operating costs as well as in a sustainable energy future.

23.2.5 Benefits of LEED Buildings
The Environment

At the onset, LEED was created to standardize the concept of building green to offer the building industry a universal program that provided concrete guidelines for the design and construction of sustainable buildings for a livable future. As such, it is firmly rooted in the conservation of our world's resources. Each credit point awarded through the rating system reduces our demand and footprint upon the natural environment.

Economics

Conventional wisdom says that the construction of an environmentally friendly, energy efficient building brings with it a substantial price tag and extended timetables. This need not be the case. Breakthroughs in building materials, operating systems, and integrated technologies have made building green not only a timely, cost effective alternative but a preferred method of construction among the nation's leading professionals.

From the USGBC

"Smart business people recognize that high performance green buildings produce more than just a cleaner, healthier environment. They also positively impact the bottom line. Benefits include better use of building materials, significant operational savings, and increased workplace productivity."

In many instances, green alternatives to conventional building methods are less expensive to purchase and install. An even larger number provide tremendous operational savings. The USGBC and LEED offer the building industry a fiscally sound platform on which to build their case for whole-building design and construction. Green buildings can show a positive return on investment for owners and builders.

Economic benefits can also include:

- Improved occupant performance—employee productivity rises, students' grades improve. California schools have analyses that show children in high performance facilities have improved test scores.
- Absenteeism is reduced.
- Retail stores have observed measurable sales improvements in stores with daylighting.

The savings associated with productivity gains can be the single largest category of savings in an office building. For example, salaries per square foot can be in the range of ten times (or more) than the cost of energy per square foot in a typical office building. So while the energy engineering community agonizes over extracting each cent in energy costs, there are greater savings potentials by leveraging the increased employee productivity through high performance buildings. As described earlier, these buildings provide a superior indoor environment with regard to lighting, noise control, temperature and humidity control, ventilation, and fresh air.

Building green also enhances asset value. According to organizations such as the International Facilities Management Association (IFMA) and the Building Owners and Managers Association (BOMA), the asset value of a property rises at a rate of ten times the value of the operational savings. For example, if green building efficiency reduces operating costs by $1/sq ft per year, the asset value of that property rises by ten times that amount, or $10. For a 300,000 sq ft office complex, annual savings could be $300,000, and the asset value increase would be approximately $3,000,000. It pays to be efficient!

Societal Implications

In addition to the environmental and economic benefits of building green, there are also significant societal benefits. These include increased productivity, a healthy work environment, comfort, and satisfaction, to name a few. The good news about green buildings can be leveraged with local and trade media through press releases, ceremonies, or events. By calling attention to the building and its certification status, owners speak volumes about themselves. The USGBC writes:

"Like a strong prospectus, building green sends the right message about a company or organization: it's well run, responsible, and committed to the future."

This, too, has a direct effect on the bottom line.

23.2.6 Benefits to the Architectural and Engineering Community

All too often, market forces drive a building design team to focus on minimum first cost regardless of what the overall life cycle costs of such a design may be. The end result is that all elements of society—owner, occupants, and the community—end up with a building that is less than what it could have or should have been. Too many resources were in construction, its energy and operating costs are high, and it does not provide the optimum indoor environment for employee productivity.

However, installation of the LEED process of forming an integrated design team (on conceptual design day one) promotes the formation of a creative solution to the particular building needs being planned for. Within this creative roundtable of equals, the team is able to maximize use of the collective wisdom of the members and develop a design concept that can be both green and economic. For example, increasing the use of natural light to displace some artificial light can result in a reduced load on air conditioning systems. These AC systems can then be downsized, resulting in reduced equipment first costs and reduction in electricity use, both through the artificial lighting reduction and reduced cooling loads. Small savings multiply and reverberate throughout the design, ultimately having significant impacts on overall first costs and operating costs.

When building "green," the sum is larger than the individual parts only when all components are integrated into a single unified system. Integration draws upon every aspect of the building to realize efficiencies, cost savings, and continuous returns on investment.

The most significant benefit to the architectural/ engineering (A/E) design community is that LEED promotes and rewards creative solutions. Sustainable and green designs are not yet commodity skills that all firms can lay claim to. For those A/E firms seeking to provide more value to their clients, LEED is a way to achieve this. LEED can become a standard for design excellence and provides an A/E with a brand differentiator versus the competition. LEED can become an outstanding competitive edge in the marketplace for firms seeking a leadership position of excellence.

LEED Acceptance

There's a groundswell of acceptance taking place. When LEED was first introduced in 2000, there was reluctance to accept it. It represented a new way of looking at the design, construction, operations, and disposal of facilities. It called for raising the bar, which many have been slow to accept. There were many questions about higher first costs, paybacks, overall benefits, ability of the design community to deliver, doubts about the technologies involved, etc. The perception was that green buildings were too futuristic, involved unobtainable goals, and required too many tradeoffs to be able to deliver a practical, efficient facility where people could live, work, and play.

Some elements of these doubts remain, especially as to the financial benefits question. However, as we gain experience with green buildings, we are developing the experience and the data needed to resolve these doubts.

The First Cost "Premium" of LEED

Conventional wisdom says that green buildings cost more. However, as the industry becomes more experienced with the actual delivery of green buildings, green is becoming more cost-neutral. In an article titled "The Costs and Financial Benefits of High Performance Buildings," Greg Kats of Capital E analyzed 40 California LEED buildings for the "cost premium" of LEED. The study consisted of 32 office buildings and eight schools:

- The eight LEED-certified buildings (the basic level of LEED certification) cost an average of 0.7% more.
- The twenty-one Silver-rated buildings cost an average 1.9% more.
- The nine Gold-rated buildings cost an average of 2.2% more.
- The two Platinum-rated buildings cost an average of 6.8% more.

23.2.7 LEED Described

Why was LEED created?
- Provides a way to define and quantify what constitutes a sustainable green design.
- Defines "green" by providing a standard for measurement.
- Addresses "greenwashing" issues, such as false or exaggerated claims.
- Facilitates positive results for the environment, occupant health, and financial return.
- Is useful as a design guideline.
- Recognizes leaders.
- Stimulates green competition.
- Establishes market value with a recognizable national "brand."
- Raises consumer awareness.
- Transforms the marketplace.
- Promotes a whole-building, integrated design process.

What is LEED
- Consists of performance-based and prescriptive-based criteria.
- Focuses on the whole building system instead of components.
- Is life cycle based, not first cost based.
- Promotes architectural and engineering innovation, i.e., innovation LEED credits.
- Provides a third-party verification process to ensure quality and compliance.

The growing family of LEED building rating systems includes:

LEED NC for new construction
LEED EB for existing buildings
LEED CI for commercial interiors
LEED CS for commercial core & shell
LEED H for homes, currently in pilot.

Other LEED programs are being developed.

23.3 INTRODUCING THE LEED NC RATING SYSTEM: A TECHNICAL REVIEW

The LEED format for rating a green building consists of two categories:
- Prerequisites—These are mandatory requirements, and all must be satisfied before a building can be certified.
- Credits—Each credit is optional, with each contributing to the overall total of credits. This will determine the level at which a building will be rated—Certified, Silver, Gold, or Platinum.

23.3.1 Sustainable Sites—14 Possible Points
Prerequisite:
- Erosion & Sedimentation Control—Control erosion to reduce negative impacts on water and air quality by complying with the EPA storm water management requirements for construction activities.

Credits:
- Site Selection—Avoid development of inappropriate sites and reduce the environmental impact from the location of a building on a site. One point.
- Urban Redevelopment—Channel development to urban areas with existing infrastructures, protecting green fields and preserving habitat and natural resources. One point.
- Brownfield Redevelopment—Rehabilitate damaged

sites where development is complicated by real or perceived environmental contamination, thereby reducing pressure on undeveloped land. One point.

- Alternative Transportation—Four points available, one each for: public transportation access, bicycle storage, alternative fuel vehicles, and parking capacity.
- Reduced Site Disturbance—Conserve existing natural areas and restore damaged areas to provide habitat and promote biodiversity. Two credits available: one for protecting or restoring open space and one for reducing the development footprint.
- Storm Water Management—Limit disruption of natural water flows by minimizing storm water runoff, increasing on-site infiltration and reducing contaminants. Two credits available: one for no increase in the rate or quantity of runoff and one for treatment systems designed to remove total dissolved solids and phosphorous, complying with EPA guidelines.
- Heat Island Effect—Reduce heat islands, which are thermal gradient differences between developed and underdeveloped areas. There is one point each for roof and non-roof applications, up to two points.
- Light Pollution Reduction—one point. The intent is to eliminate light trespass from the building site, improve night sky access, and reduce development impact on nocturnal environments.

23.3.2 Water Efficiency—5 possible points

No prerequisites in this category.
Credits:

- Water Efficient Landscaping—Limit or eliminate the use of potable water for landscape irrigation. Up to two points available. One for use of high efficiency irrigation technology and one for using captured rain or recycled water for irrigation.
- Innovative Wastewater Technologies—one point. Reduce the generation of wastewater and potable water demand, while increasing the local aquifer recharge.
- Water Use Reduction—20% reduction is one point, 30% reduction is two points. Maximize water efficiency within the building to reduce the burden on municipal water supply and wastewater systems.

23.3.3 Energy & Atmosphere—17 possible points

Three prerequisites:

- Fundamental Building Systems Commissioning—Verify and ensure that fundamental building elements and systems such as HVAC are designed, installed, and calibrated to operate as intended.
- Minimum Energy Performance—Establish the minimum level of energy for the base building systems to comply with ASHRAE/IES 90.1-1999.
- CFC Reduction in HVAC&R Equipment—This requires zero use of CFC based refrigerants (such as R11 and R12) in new buildings.

Credits:

- Optimize Energy Efficiency—ten possible points. Achieve increasing levels of energy performance above the prerequisite standard (ASHRAE 90.1-2004) to reduce environmental impacts associated with excessive energy use. 20% better is two points, 30% is four points, and on up to 60% better being worth ten points.
- Renewable Energy—three points possible, one point each for 5%, 10%, 20% of total energy. Encourage and recognize increasing levels of self-supply through renewable technologies to reduce environmental impacts associated with fossil fuel energy use.
- Additional Commissioning—one point. Verify and ensure that the *entire* building, including the building envelope, is designed, constructed, and calibrated to operate as intended. This as opposed to the prerequisite which called for only *fundamental* systems.
- Elimination of HCFC s and HALONS—one point. Reduce ozone depletion and support early compliance with the Montreal Protocol. This applies to refrigerants such as R22 and R123.
- Measurement & Verification—one point. Provide for the ongoing accountability and optimization of building energy and water consumption performance over time.
- Green Power—one point. Encourage the development and use of grid-source energy technologies on a net zero pollution basis by the purchase of green power that meets the Center for Resource Solutions Green-E products.

23.3.4 Materials & Resources—13 possible points

Prerequisite:

- Storage & Collection of Recyclables—Facilitate the reduction of waste generated by building occupants that is hauled to and disposed of in landfills.

Credits:

- Building Reuse—two points possible at 75% and 100% reuse of building shell and non-shell. A third point is available by maintaining 100% of an existing building's structure and 50% of the non-shell, such as walls, floor coverings, and ceilings. The purpose is to extend the life cycle of existing building stock.

- Construction Waste Management—two points available. This is to divert construction, demolition, and land clearing debris from landfill disposal. Redirect recyclable material back into the manufacturing process: 50% recycled or salvaged materials by weight gets one point, 75% earns one more point.
- Resource Reuse—two possible points at 5% or 10% of using salvaged or refurbished materials. The intent is to extend the life cycle of targeted building materials by reducing environmental impacts related to materials manufacturing and transport.
- Recycled Content—two possible points. One point specifying a minimum of 25% building materials that contain post consumer recycled materials. An additional point is available if an additional 25% is recycled content.
- Local/Regional Materials—two possible points for materials either manufactured or harvested locally. The purpose is to increase the demand for building products that are manufactured locally (less than 500 miles), thereby reducing the environmental impacts resulting from long-distance transportation.
- Rapidly Renewable Materials—one point. Reduce the use and depletion of finite raw and long cycle renewable materials by replacing them with renewables.
- Forest Stewardship Council (FSC) certified wood. One point.

Note: Regarding "certified products," this does not mean "LEED certified," but certified by other entities such as the Forest Stewardship Council (FSC). Beware of manufactures claiming "LEED certified" products. The USGBC and LEED do not certify products. What they do is adopt industry standards as applicable, such as FSC certified wood.

23.3.5 Indoor Environmental Quality (IEQ)— 15 possible points

Two prerequisites:
- Minimum IAQ Performance—Comply with the ASHRAE 62-2004 indoor air quality standard to prevent the development of air quality problems.
- Environmental Tobacco Smoke (ETS) Control—Prevent exposure of building occupants and systems to ETS.

Credits:
- Carbon Dioxide Monitoring—one point. Provide an HVAC system which can monitor and ventilate a building based upon CO_2 levels.
- Ventilation Effectiveness—one point. Provide for the effective delivery of mixed and outdoor air to support health, safety, and comfort of occupants. Uses ASHRAE 129 methodology.
- Construction IAQ Management Plan—two possible points. Develop an IAQ management plan for during construction and before occupancy, one point. An additional point is available by conducting a two-week building flushout prior to occupancy, using 100% outside air.
- Low Emitting Materials—four possible points. One point for low VOC adhesives. One point for low VOC paints. One point for carpet exceeding the Carpet & Rug Institute Green Label IAQ Program. One point for composite wood and agrifiber products containing no additional urea formaldehyde resins.
- Indoor Chemical & Pollutant Source Control—one point. This is to avoid exposing occupants to potentially hazardous chemicals that adversely affect IAQ.
- Controllability of Systems—two possible points, one each for perimeter and non-perimeter. This is to provide a high level of individual occupant control of thermal, ventilation, and lighting systems.
- Thermal Comfort—comply with ASHRAE Standard 55-2004—one point. This relates to providing a thermally comfortable environment that supports healthy and productive performance of occupants.
- Permanent Monitoring System—one point. These are permanent systems to monitor temperature and humidity that also allow occupants to have partial control over these parameters.
- Daylighting & Views—two points available. This is to provide a connection between indoor spaces and outdoor environments through the introduction of sunlight and views into the occupied areas of the building. For one point, achieve a daylight factor of 2% in 75% of all space occupied for critical visual tasks, but not including the likes of laundry rooms, copying rooms etc. For an additional point, achieve the 2% rating in 90% of spaces.

23.3.6 Innovation & Design Process—5 possible points
- Use of LEED-accredited professional—one point.
- Innovation in design—four possible points.

23.3.7 Discussion
Note: Energy and engineering skills are applicable to as many as 58% of total available points on water, energy, and IAQ. Energy related credits are the largest category of available credits.

Of particular interest is the May '04 *Energy User News* article by Peter D'Antonio, entitled "The LEEDing

Way." The article analyzes the activity in energy & atmosphere, E&A, and indoor environmental quality, IEQ, for the first 53 LEED-certified buildings:

- Regarding E&A, average points earned is only 5.3 out of the possible 17!
- This is the lowest % achieved in any of the five categories! So although E & A is the largest plum, few appear to be taking advantage of it.
- Renewable energy points are earned in fewer than 10% of the certified buildings.
- Regarding IEQ points, ventilation effectiveness and controllability points are achieved in less than one third of buildings.

This EUN article provides support for the proposition that engineers, AEE, ASHRAE, IESNA, and others are not maximizing the potential contributions to LEED buildings. Hence, there is a significant opportunity to take on a larger role in the design, construction, and operation of green buildings.

Other factors for the design team to consider are the forces that drive the LEED points on a project. Many times, for the design team of an LEED project, it boils down to, "How many points can we get?" This becomes especially the case for the mechanical, electrical, and plumbing (MEP) team members. They control or influence approximately 75% of the total LEED credits on a job. Commonly called "pointchasing," it is an effort by the design team to achieve the maximum available points at the minimum cost and effort. And although it is a rather ugly approach to green building design, it has become a matter of fact that teams will focus on points. It is even possible that the acquisition of points is one of the elements in the design contract, with possibly a bonus linked to points achieved.

But this can and should be "managed" by the owner. It gets back to the integrated design process and the setting of goals during the design charette. Does the owner want very high energy efficiency, or does he want to make an environmental statement with a green roof?

23.4 LEED FOR EXISTING BUILDINGS RATING SYSTEM (LEED-EB) ADOPTED IN 2004

LEED for existing buildings

- Pilot: 2002—2004.
- Addresses:
  —Operation and upgrades of existing buildings.
  —Initial certification and ongoing re-certification.
- Achievements:
  —More than 95 registered buildings.

  —4 certified buildings.
- Range of Users:
  —Federal, state, and local governments; schools, colleges and universities, commercial buildings.

LEED-EB approval was completed and released for use, Oct. 2004.

Why LEED-EB Is So Important

Drawing on similarities to LEED NC, LEED EB has a larger potential impact and resulting benefits to society simply because there are many times more existing buildings than new construction. LEED EB focuses on where the greatest impact potential is.

23.4.1 LEED for Existing Buildings Rating System LEED-EB Rating System Goals

- Help building owners upgrade and operate their buildings in a sustainable way over the long term. Avoid the "saw tooth" approach (upgrade, decline, upgrade, decline).
- Support high productivity by building occupants.
- Operations:
  —Help building owners upgrade and operate their buildings in a sustainable way over the long term.
  —Reduce building operating costs.
  —Solve building operating problems.
  —Improve indoor environment.
  —Support higher productivity of building occupants.

- Communications:
  —Help building managers, operators, and service providers communicate the importance of effective, ongoing building operation and maintenance to decision makers in their organization.
  —Help building managers and operators make sustainability part of the culture of their organization.
  —Help CEOs and CFOs make sustainability part of the culture of their organization.
  —Help communicate the organization's sustainability commitments and achievements to its customers and the community.

Prerequisites and Credits

Same categories as for other LEED Rating Systems:

- Sustainable Sites
- Water Efficiency
- Energy and Atmosphere
- Materials and Resources
- Indoor Environmental Quality

- Innovation and Accredited Professional

LEED-EB Rating System
Four Levels of Certification:

- LEED-EB Certified   32-39 points
- Silver Level   40-47 points
- Gold Level   48-63 points
- Platinum Level   64-85 points

23.4.2 LEED for EB Technical Review

Similar to LEED NC, all prerequisites must be satisfied, and the credits are optional depending upon the final points and certification level desired.

23.4.2.1 Sustainable Sites—14 possible points
Two prerequisites:
- Erosion and Sedimentation Control—Control erosion to reduce negative impacts on water and air quality.
- Age of Building—two years old or more.

Credits:
- Plan for Green Site & Bldg Exterior Management—up to two points. Encourage grounds/site/building exterior management practices that have the lowest environmental impact possible and preserve ecological integrity, enhance diversity, and protect wildlife while supporting building performance.
- Hi Development Density Building & Area—one point. Channel development to urban areas with existing infrastructure, protect greenfields, and preserve habitat and natural resources.
- Environmentally Preferable Alt Transportation—up to four points available. One point each for: public transportation access, bicycle storage and changing rooms, alternative fueled vehicles/car pooling, and telecommuting.
- Reduced Site Disturbance—up to two points. One point for protecting or restoring 50% of the site area. An additional point to protect or restore open space at 75% of the site area.
- Storm water Management—up to two points. One point for measures that mitigate at least 25% of the annual storm water falling on the site. An additional point for mitigation of at least 50% of storm water.
- Reduce Heat Islands Effect (roof and non-roof)—up to two points. One point for reduction of heat islands. An additional point is available for an ENERGYSTAR-compliant roof.
- Light Pollution Reduction—one point. Eliminate light trespass from the building and site, improve the night sky, and reduce developmental impact on nocturnal environments.

23.4.2.2 Water Use and Water Efficiency— 5 possible points
Two Prerequisites:
- Minimum Water Efficiency—maximize fixture water efficiency within buildings to reduce the burden on potable water supply and wastewater systems.
- Discharge Water Compliance—protect natural habitat, waterways, and water supply from pollutants carried by building discharge water.

Credits:
- Water Efficient Landscaping—up to two points. Requires the use of water efficient irrigation technologies or captured rain and recycled water to reduce potable water consumption for irrigation. The first point is based on a 50% reduction, and an additional point is available for 95% reduction in potable water use.
- Innovative Wastewater Technology—one point. Reduce the generation of wastewater and potable water demand, while increasing the local aquifer recharge.
- Water Use Reduction—up to two points. Maximize fixture potable water efficiency to reduce burdens on potable and wastewater municipal systems. The first point is for a 10% reduction, and an additional point is available for a 20% reduction.

23.4.2.3 Energy and Atmosphere—23 possible points
Three Prerequisites:
- Existing Building Commissioning—Verify that fundamental buildings systems are performing as intended.
- Minimum Energy Performance—Satisfy the minimum level of energy efficiency, using the ENERGYSTAR portfolio manager. Needs a rating of 60 or more.
- Ozone Protection—Reduce ozone depletion potentials by not using CFC refrigerants such as R11 and R12.

Credits:
- Optimize Energy Performance—up to ten points available. Achieve increasing levels of energy efficiency above the ENERGYSTAR prerequisite score of 60. Thus, a score 63 earns one point and 79 earns five points, up to a maximum of ten points for a 99 rating.
- On-site & Off-site Renewable Energy—up to four

points. The first point is for 5% on-site renewable OR 25% off-site renewables, up to a maximum of four points for 30% on-site renewable energy OR 100% off-site renewable energy.

- Building Operations & Maintenance—up to three points. One point each for maintenance staff education, building systems maintenance, and building systems monitoring.
- Additional Ozone Protection—one point. Reduce ozone depletion potential in compliance with the Montreal Protocol. Thus, HCFC refrigerants such as R22 and R123 are not used.
- Performance Measurement—up to four points. Have in place a continuous metering system for a number of facilities functions: lighting systems, electric and gas metering, cooling load, chilled water system efficiency, irrigation water metering, boiler efficiencies, HVAC systems such as economizers, variable speed pumps and fans, air distribution, and emissions monitoring. Note these can all be incorporated into the building automation system (BAS).
- Documenting Sustainable Building Cost Impacts—one point. Document overall building operating costs for the previous five years and track changes in the overall operating costs.

23.4.2.4 Materials and Resources -16 possible points

Two prerequisites:

- Source Reduction and Waste Management—establish minimum source reduction and recycling program elements.
- Toxic Material Source Reduction—reduced mercury in lamps.

Credits:

- Construction, Demolition and Renovation Waste Management—up to two points. First point for diverting 50% or more of construction, demolition and land clearing waste from landfills. An additional point if 75% or more is diverted.
- Optimize Use of Alternative Materials—up to five points available. Maintain a sustainable purchasing program covering at least office paper, office equipment, furniture, furnishings, and building materials. One point is awarded for each 10% of total purchases that achieve criteria such as 70% salvaged materials, 10% post consumer recycled, 50% rapidly renewables, FSC-certified wood, and materials manufactured within 500 miles of the site.
- Optimize Use of IAQ Compliant Products—up to two points. These relate to the purchase of products using low emitting materials, such as carpets, seal-

ants, paints and coatings, composite materials, and agrifiber products with no added urea formaldehyde.

- Sustainable Cleaning Products and Materials—up to three points. Points accumulate based upon quantities of products that meet the Green Seal GS-37 or comply with the California Code of Regulations for VOCs. Disposable janitorial paper products and trash bags meeting the requirements of the EPA comprehensive procurement guidelines are also considered.
- Occupant Recycling—up to three points. Set up divert/recycle programs for occupants. 30% is one point, 40% another; the third point is given if 50% of total waste stream is diverted or recycled.
- Additional Toxic Material Source Reduction—one point. Establish a program to reduce the potential amounts of mercury brought into the building through lamps.

23.4.2.5 Indoor Environmental Quality— 22 points available

Four prerequisites:

- Outside Air Introduction and Exhaust Systems—Satisfy ASHRAE 62-2004 for IAQ.
- Environmental Tobacco Smoke (ETS) Control—Prevent or minimize occupant exposure to ETS.
- Asbestos Removal or Encapsulation—Establish an asbestos remediation and control management plan.
- PCB Removal—Establish a PCB management plan, including a facility survey for PCBs.

Credits:

- Outside Air Delivery Monitoring—one point. Provide permanent monitoring systems on ventilation system performance, measuring outdoor air and CO_2.
- Increased Ventilation—one point. Increase ventilation rates to exceed ASHRAE 62-2004 by 30%.
- Construction IAQ Management Plan—one point. Prevent any IAQ problems from arising due to construction/renovation work. Isolate occupied areas from dust, noise, and other irritants.
- Documenting Productivity Impacts—up to two points. Document the history of absenteeism, productivity, and health care costs, and submit to the USGBC.
- Indoor Chemical & Pollution Source Control—up to two points. Reduce the exposure of occupants to dusts and particulates by using filters of effectiveness of MERV 13 or greater. An additional point is

earned by reducing occupants' exposures to contaminants that may arise from operations such as copying, faxing, etc.

- Controllability of Systems—up to two points. One point is available for occupant control of lighting systems, another for HVAC and temperature control.
- Thermal Comfort—ASHRAE Standard 55-2004—up to two points. The first point is for compliance with the standard and an additional point is available for a permanent monitoring system to ensure compliance.
- Daylighting and Views—up to four points. Provide a connection between indoor spaces and the outdoor environment through the introduction of sunlight and views. Points are available for 50% and 75% of spaces that have a 2% daylight factor. Two more points are available for 45% of spaces (1 point) and 90% of spaces (1 point) that have direct line of sight vision to the outdoors.
- Contemporary IAQ Practice—one point. Enhance IAQ performance by optimizing practices and developing procedures to prevent the development of IAQ problems.
- Green Cleaning—up to six points. Points are available for cleaning entryway systems, isolation of janitorial closets, low environmental impact cleaning policy, low environmental impact pest management policy, and low environmental impact cleaning equipment policy.

23.4.2.6 Innovation and Accredited Professional— 5 possible points

- LEED EB Innovation in Operation, Upgrades and Maintenance—up to four points.
- LEED-accredited Professional—one point.

23.5 SUMMARY DISCUSSION OF TWO NEW LEED PROGRAMS:

LEED-CI for Commercial Interiors and LEED-CS for Core and Shell

LEED-CI addresses the specifics of tenant spaces, primarily in office, retail, and industrial buildings. It was formally adopted in the fall of 2004. A companion rating is LEED for core & shell, which is currently under development and in its pilot phase. Adoption is expected in the fall of 2005. Together, LEED-CI, and LEED-CS will establish green building criteria for commercial office real estate, for use by both developers and tenants.

LEED-CI serves building owners and occupants,

as well as the interior designers and architects who design building interiors and the teams of professionals who install them. It addresses performance areas including water efficiency, energy efficiency, HVAC systems & equipment, resource utilization, furnishings, and indoor environmental quality.

23.5.1 LEED for Commercial Interiors (CI)

- Pilot: 2002—2004.
- Addresses the design and construction of interiors in existing buildings and tenant fit-outs in new core and shell buildings.
- Achievements: More than 45 projects in pilot.
- LEED CI adopted in fall 2004.

23.5.1.1 LEED-CI Point Distribution

The same five basic categories as the other LEED rating systems are used.

Possible Points

Sustainable Sites	7
Water Efficiency	2
Energy & Atmosphere	12
Materials & Resources	14
Indoor Environmental Quality	17
Innovation & Design Process	4
LEED Accredited Professional	1
Total Points Available	57

4 Levels of Certification

Certified	21-26
Silver	27-31
Gold 2-41	
Platinum	42-57

Because of its nature, i.e., the interior parts of a building, the energy engineer has less opportunity to aid in earning points in this program. This is the only program to date wherein the E & A credits are not the largest category. However, engineers in general do have the potential to aid points, especially in the indoor environmental quality category.

23.5.2 LEED for Core & Shell, CS

Based upon the LEED NC rating system for new construction and major renovation, LEED CS was developed in recognition of the unique nature of core and shell developments. In particular, there is the lack of developer control over key aspects, such as interior finishes, lighting, and HVAC distribution. Thus, the scope of CS is limited to those elements of the project under the direct control of the developer.

With its standards for CS and CI, the USGBC addresses the entire building—core, shell, and interiors. The responsibilities for particular sections, however, are assigned to those parties having direct control over them. LEED CS was still in its pilot phase as of early 2006.

23.5.2.1 LEED CS Credit Categories

Below is a summary of where the points will be for LEED CS. Note its similarities to LEED NC, and that energy & atmosphere is the largest points category.

Possible Points

Sustainable Sites	15
Water Efficiency	5
Energy & Atmosphere	16
Materials & Resources	11
Indoor Environmental Quality	13
Innovation and Design Process	4
LEED Accredited Professional	1
Total Points Available	65

4 Levels of Certification

Certified	24-29 points
Silver	30-35 points
Gold	36-47 points
Platinum	48-64 points

You can download all four of the LEED rating systems. The rating systems download are free. However, other tools and workbooks such as reference guides do have a fee associated with them, with discounts given to members.

- Visit U.S. Green Building Council Web Site at www.usgbc.org/leed.
- Choose rating system.
- Click on rating system you would like to download.

23.6 THE LEED PROCESS

- Design Team Integration
- Project Registration
- Project Certification
- Documentation

The LEED Design Process:

When does a Green Design Begin?

"It Begins in the Beginning"

Critical to success is the integration of the design TEAM on Day 1 of design. LEED is a marketplace transformer. It is a paradigm shift away from top down, minimum first cost emphasis. The hierarchical old-fashioned way was design, bid, build.

But the LEED design process is one of integrated, holistic building design, construction, operations and maintenance. All involved participate as equals on a construction roundtable:

- Owner operations personnel
- Owner
- Architect
- Engineer
- Construction manager
- Contractors & subcontractors
- Equipment suppliers & manufacturers
- Commissioning authority—watchdog role

During the very early stages of a green building's development, a design charette should be held. This refers to meetings that are held over the course of a day or two, wherein the entire team, the construction roundtable group, gets together to develop the roadmap to successful green building.

- The ENTIRE team joins in—all stakeholders (including), the owner, designers, the commissioning authority, and operations personnel—to collectively:
- Gain buy-in and consensus.
- Explore environmental issues.
- Propose alternatives.
- Identify modeling and resource allocation.
- Use the LEED checklists as a guide for the level of green desired.
- Use an outside facilitator who specializes in integrated design.
- Present examples of resources and ways to trace costs and benefits of modeling.
- Establish a task-responsible team to track and manage compliance with the process.
- Determine an LEED leader who will be the "watchdog" over points. (The commissioning authority can be a good choice for this.)

23.6.1 The Energy Engineer's Goal: Get Invited!

It is during the design charette, which occurs at the earliest moments of a project, that the energy engineer can provide the maximum overall benefit to the project. It is during this time that key choices are made about the lighting, HVAC, and building envelope. The energy engineer can help guide the team to the most appropriate energy efficient design strategies, based upon the team's energy goals and the available energy sources.

On an LEED project, a major concern of the design team is simply, "How do I get the points?" Commonly called "points chasing," it is an effort by the design team to achieve the maximum available points at the minimum cost and effort. However, this can be ameliorated or driven by the owner. During the design charette, it is the owner's responsibility to clearly identify the goals and objectives of the project. If it is simply, "get me the most points," the team will point chase. If the goals are, for example, to have the highest energy efficiency possible, or the most daylighting possible, then other choices, alternatives, and evaluations may be examined.

23.6.2 Example and Discussion on Obtaining LEED Points on a Project

The following is an example of how in an integrated green building design, a "simple" decision such as to have a computerized building automation system (BAS), can have substantial overall impact on acquiring points.

Knowing that LEED buildings will generally require sophisticated controls, the BAS can have a major role in obtaining and facilitating points. The following is a summary of the impact a BAS may have on LEED points. The design team can use this example to guide them in the same process for other techniques and/or technologies. Again, the emphasis is on integrated design. We are designing an *integrated* building, not a collection of parts and systems.

23.6.2.1 Some Examples of BAS Influence on LEED Credits:

- Sustainable Sites—Light pollution reduction through use of controls.
- Water Efficiency—Use of metering to document water consumption; although not a credit itself, it can facilitate credits in this category.

Energy & Atmosphere—many credits and much influence here:

- Energy Prerequisite and Optimized Energy Performance—ten credits are in play here. The BAS is an integral part of the energy-consuming system including lighting, HVAC, load management, etc., and it helps earn credits through performance improvements that will be quantified in the building energy simulations required by ASHRAE 90.1.

- Commissioning—BAS aids the commissioning authority in performing their duties, serving as a time saver.

- Measurement and Verification (one prescriptive credit)—Provide for accountability & optimization of energy and water consumption over time.

- Optimize Energy Performance—In LEED CI, up to four prescriptive credits are available for lighting and power controls. Other credits in energy performance, can provide an additional four.

- Energy Submetering—In LEED CI, measure for energy accountability. Up to two prescriptive credits.

- Building Operations & Maintenance (three prescriptive credits)—These relate to staff education, building systems maintenance, and building systems monitoring.

- Performance Measurement—Enhanced metering and emission reduction reporting gain up to four prescriptive credits.

Indoor Environmental Quality

- CO_2 Monitoring—one prescriptive credit.
- Increase Ventilation Effectiveness—BAS can aid in earning this credit.
- Controllability of Systems—up to two prescriptive credits.
- Thermal Comfort—up to two prescriptive credits—Comply with ASHRAE Standard 55 and permanently monitor temperature and humidity.
- Outdoor Air Delivery Monitoring—one prescriptive credit.

The conclusion is that a BAS can directly add to points accumulation but indirectly has a great deal of influence on other points. The intent of these examples is to demonstrate how design decisions flow through the entire integrated design, having direct effects on some credits and indirect effects on others.

23.6.3 Marketing LEED and Sustainability to the Community, Owners, and Designers

LEED offers a great deal of value to various members of the community. The owners benefit by having high performance buildings that are cost efficient and provide for better employee productivity. The design community benefits by now having a way to craft a stronger value message for superior architecture and design. LEED provides a way for designers to qualify and quantify their competitive advantage over other non-green designers. The community benefits by having a program such as LEED that promotes urban and brownfield development, reduces demands on infrastructure such as roads and waste disposal, improves the environment, and provides for a healthy living style.

Elaborating as to why a design team should be promoting LEED design, let us examine the overall life cycle cost of a typical commercial office building.

Ownership Cost Breakdown—40 Year Life Cycle Costs

Construction or First Cost is 11%
Financing is 14%
Alterations are 25%
Operations are 50%

It is interesting that the cost that most design teams grapple with, to keep low first costs, is actually the least significant cost element in the overall life cycle costs of a building. Thus, those decisions to keep first costs low by specifying cheaper designs and equipment can have a serious negative impact on the overall life cycle performance of a facility.

Additionally, considering that first cost is only 11% of the total life cycle cost (and that A/E fees are only 6% to 8% of that 11%, or .88% of the total costs), could it be beneficial to the owner to pay more for superior architecture and engineering? The answer is yes, because the designs and selections made by the design team have a great deal of leverage on the total life cycle ownership costs.

23.6.4 Credits that Engineers can help in acquiring

Many times it is believed that the architectural profession has the most potential to aid in acquiring LEED credits. However, it is the engineering profession that, in fact, influences the most points. For example:

- Energy & atmosphere and indoor environments are responsible for 40% of available points.
- Minimum energy performance is a prerequisite!
- Energy modeling is required.
- Energy measurement and verification is a credit.
- Commissioning is required.

Other benefits of bringing engineers to the design team

- Provides required creativity vs. CAD commodity design. Creativity is desired and rewarded.
- Promotes investment in A/E design $ to value engineers before, not after, beginning the process.
- Promotes the collective wisdom of the integrated design team.
- Enhances the interplay between professions that occurs during the design charette.

- Provides a catalyst for the design team to do the high performance job that it is capable of.
- Provides a value message of premium engineering and design to the owner. (This may be the path to higher fees and/or more work?)

23.6.5 Impediments to Green Acceptance

Typically, the first and possibly most serious impediment to the wide-scale adoption of LEED is the perception that it costs more. The facts are that it may add cost, from 1% to 5%, depending upon the level of green the ownership team has identified in that design charette. But it does not necessarily cost more if the design team is clever about making design decisions and using all available resources that may be at hand.

Hint: If trying to promote LEED, look to identify market conditions in the project's locale that supports LEED. Many states have various programs to incentivize energy efficiency, as well as other marketplace conditions which can affect the viability of a green project. The following uses New Jersey as an example of "market conditions" that can drive LEED adoption:

The Case for LEED in New Jersey:

First cost is less of an issue because of high efficiency equipment incentives. There is a program through the NJ Board of Public Utilities called NJ Smart Start Buildings, which provides rebates for high efficiency equipment, such as lighting, HVAC, boilers, and chillers, as well as for commissioning and design team meetings. Essentially much of the cost differential between cheap inefficient equipment and high efficiency equipment is offset by the rebates.

Renewable energy sources are promoted by statewide programs such as the Clean Energy Program. Similar to Smart Start, renewables such as wind, solar PV, and biomass projects are rebated up to 50% of the initial installed cost.

ASHRAE Std 90.1-2004 is the State Energy Code, as well as the prerequisite for energy and atmosphere LEED credits. So there is no additional cost for NJ buildings to comply with this. In other states that may not have this code requirement, compliance with ASHRAE 90.1 would add cost.

Many brownfields are available for development, with incentives from the NJ Economic Development Administration, NJEDA. This can be a simple prescriptive credit.

Mass transportation is generally adjacent to the brownfields, which aids in acquiring more points. NJ, as the most densely populated state in the nation, has many former industrial sites that are in inner city areas close to mass transit and part of urban renewal. Thus, a brownfield site can facilitate a number of other credits.

Voluntary LEED adoption decreases need for additional environmental regulation. NJ is one of the most regulated and legislated states, but if we have more adoption of LEED, many of the goals of environmental legislation can be achieved voluntarily.

High energy costs in NJ promote equipment and operational cost efficiencies. Paybacks on high efficiency equipment are quicker than in other states, which helps drive the recognized value of energy efficiency versus low first cost equipment.

Other states and regions may have similar incentives and programs. New York and California are two that come to mind. In addition, the Energy Policy Act of 2005 promises to encourage energy efficiency through various programs.

23.7 ASHRAE GUIDES DEVELOPED TO SUPPORT LEED

Introduction to ASHRAE's GreenGuide and the Advanced Energy Design Guide

Besides market conditions which may favor LEED adoption, organizations, such as ASHRAE, are developing tools to help design teams accomplish their green design goals.

During the year 2000, the USGBC released its first green building rating system. Called "Leadership in Energy and Environmental Design for New Construction," LEED 2.0 was later revised and reissued as LEED 2.1 NC. Although the LEED rating system incorporated many ASHRAE standards into the rating system, the architectural community, not the engineering community, was the vanguard of LEED. ASHRAE recognized the potential value of LEED to the community and the building industry, as well as the value that ASHRAE could bring to the LEED program. Under then president-elect Bill Coad, Tech Committee 1.10, Energy Resources, was tasked with developing a handbook or guide for sustainable engineering design, specifically targeted for ASHRAE members. In addition, during this time, after dialog between both organizations, the USGBC and ASHRAE entered into a partnering agreement. The result of these efforts was the development of the ASHRAE *GreenGuide*, released in December 2003, to assist the USGBC in its efforts at promoting sustainable design. Additionally, in 2005, ASHRAE released the *Advanced Energy Design Guide for Small Office Buildings* (less than 20,000 square feet). This guide provides a prescriptive description and requirements for various building components and systems that would be energy efficient.

23.7.1 ASHRAE GreenGuide

The ASHRAE *GreenGuide* aids designers of sustainable, high performance green facilities. It offers various "green tips" to aid the integrated design team in developing a green building. In 2002, the American Society of Refrigerating and Air-Conditioning Engineers, ASHRAE, and the United States Green Building Council (USGBC) entered into a partnering agreement to team together to promote green buildings. This *GreenGuide* was developed by ASHRAE to assist the USGBC in efforts at promoting sustainable design.

The guide was developed to provide guidance on how to apply green design techniques. Its purpose is to help the designer of a "green design" with the question of "What do I do next?" It is organized to be relevant to the audience, useful, and practical, and to encourage innovative ideas from the design team. A key component of the guide is the "green tips," which will be covered in some detail later in this section.

However, the guide is not a consensus document, and one does not have to agree with all elements of the guide for it to be helpful. It was not developed to motivate the use of green design.

23.7.1.1 Green Design, Sustainability and Good Design

"Green" has become one of those words that can have too many possible meanings. One of the USGBC's initial goals was to provide a definition of green through the development and release of the LEED rating system. It was here that we had a measurable, quantifiable way of determining how green a building was. It also addresses the "greenwashing" issue, wherein all types of green technologies and techniques could be employed, some valid and others questionable, all in the effort to be able to label a building green. The conclusion is that green buildings are LEED buildings. This message is almost universally accepted in the USA, as well as internationally. But do be aware that there are other strong green rating systems that have been developed in Canada and Europe. For example, Canada has the Green Building Challenge (GBC), and Britain has the Building Research Establishment Environmental Assessment Method (BREEAM). According to the guide, the consensus on green buildings is that they achieve a high level of performance over the full life cycle in the following areas:

• Minimal consumption of nonrenewable, depletable resources, such as water, energy, land, and materials.

• Minimal atmospheric emissions that have negative environmental effects.

- Minimal discharge of harmful liquid and solid materials, including demolition debris at the end of a building's life.
- Minimal negative impact on site ecosystems.
- Maximum quality of indoor environment, including lighting, air quality, temperature, and humidity.

23.7.1.2 "Good" Design

ASHRAE asks whether good design is intrinsically green design. The *GreenGuide* authors make the distinction between green and good. Good design includes:

- Meeting the purpose and needs of the building's owners and occupants.
- Meets the requirements of health and safety.
- Achieves a good indoor environment consisting of thermal comfort, indoor air quality, acoustical comfort, and compatibility with the surrounding buildings.
- Creates the intended emotional impact on building's occupants.

But this is not green design in the sense that it does not address energy conservation, environmental impact, low impact emissions, and waste disposal. So integrating good and green design, such as with the LEED rating system, helps us achieve the optimum building for the owners' needs as well as the needs of society. Hence the authors of *GreenGuide* "strongly advocate that buildings should strive to achieve both."

The *GreenGuide* emphasizes the design process. This process is the first crucial element in producing a green building. There needs to be an integrated design team created in the beginning. This team should include: the owner, project manager, representative of the end user, architect, mechanical engineer, plumbing and fire protection engineer, electrical engineer, lighting designer, structural engineer, landscaping specialist, civil engineer, energy analyst, environmental consultant, commissioning authority, construction manager, cost estimator, building operator, and code enforcement representative.

Each individual professional works together in a team environment to establish the building's goals and the manner in which these goals will be achieved. Each professional must be able to recognize the impact of one another on others' designs and process.

For example, during conceptual design, the architect will determine the size and number of floors of the building. The building envelope will determine the size of HVAC equipment, as well as the types of systems being evaluated. The energy analyst will advise the team on the energy cost implications of its selections. Everyone is interdependent upon the others. What should come at the end of this iterative process is a single, integrated design that functions a unit and not as a collection of parts. Integration and interdependency of the design team professionals are the keys to a successful green design.

23.7.2 Conceptual Engineering Design

The principle intent of the *GreenGuide* is to assist the design-engineering professionals in integrating their skills and bringing value into the green design. The guide discusses a number of design responsibilities and suggests a number of "green tips" to the design team. You may find that these suggestions are not new to you and that many of these concepts have been in use, or at least in consideration, for years. However, with the advent of the desire to build green, which requires high performance systems, these techniques have a much better chance of being incorporated into the green building than in the past when first cost was likely the primary concern of an owner and the design team. The tips are arranged so that they are listed after the design section responsibility to which they are most closely linked.

23.7.2.1 Load Determination

Loads are determined by summing up internal and external gains and losses. The HVAC engineer can assist the architect in determining necessary characteristics of the building envelope. Working together with the energy analyst and others, they will aid the architect in selecting building orientation, insulation, fenestration, roofing, lighting, day lighting, systems sizes, efficiencies, etc. A key element to strive for is the initial reduction of loads resulting from an efficient building envelope, such as lighting loads, power loads, and A/C tonnages.

Green Tip #1: Night Precooling

This involves the circulation of cool air in the nighttime hours during the cooling season with the intent of cooling the structure. There are two variations on this theme. First is the use of fans only to bring in cool ventilation air; this is a somewhat passive technique. The other is to actively run the HVAC plants to precool the facility, potentially below daytime occupied temperatures, to take advantage of the building's thermal mass. Parameters to consider when evaluating this strategy are local diurnal temperature variations, humidity levels, and thermal coupling of the circulated air to the building's thermal mass.

23.7.2.2 Space Thermal/Comfort Delivery Systems

Occupant comfort and health are important, and the quality of the indoor environment promotes satisfied, productive workers within the building. Green buildings provide a more productive workplace environment.

Thermal comfort is primarily concerned with satisfying ASHRAE Standard 55 for temperature and humidity requirements.

Indoor air quality is primarily concerned with ASHRAE Standard 62 for fresh air and ventilation. However, both of these can negatively impact energy consumption; thus, the following green tips are offered:

Green Tip #2 Air-to-air Heat Recovery— Heat Exchange Enthalpy Wheels

This is a rotary cylinder filled with an air-permeable medium with a large internal surface area. Intake and exhaust air streams pass through opposite ends of the wheel in a reverse flow configuration. Latent and sensible heat are then transferred from exhaust to inlet air, thereby recovering some of the conditioning energy that was invested in the exhaust air.

Green Tip #3 Air-to-Air Heat Recovery— Heat Pipe Systems

These are passive devices, usually configured as tubes with fins for maximum surface area. They contain a thermal fluid that transports sensible heat only between exhausts and inlet air streams.

Green Tip #4 Air-to-Air Heat Recovery— Run Around Systems

This consists of energy recovery coils in the exhaust and inlet air streams and a circulating loop containing a heat transfer fluid. These systems do not transfer latent energy. An option added to this system and the heat pipe system is the use of an indirect evaporative water process that can reduce cooling loads in addition to the heat recovery process.

Green Tip #5 Displacement Ventilation

This technique supplies conditioned air at a temperature slightly lower than the desired room temperature, at low velocities horizontally at the floor level. Returns are located in the ceiling. This supply air rises by convection, picks up the room load, and exits through the ceiling returns.

Green Tip #6 Dedicated Outdoor Air Systems (DOAS)

This uses a dedicated, separate air handler to condition the outdoor air before delivering it directly to the occupants. The air delivered should be conditioned sufficiently to not adversely affect the thermal comfort of the occupants. The only absolute with this system is that the ventilation air must be delivered directly to the space from a separate system. Control strategy, energy recovery, and leaving air conditions are all variables that can be fixed by the designer.

Green Tip #7 Ventilation Demand Control Using CO_2

CO_2 concentrations are measured in a space, and ventilation rates are automatically adjusted by the BAS to maintain CO_2 concentrations within predetermined limits. This system is best used in buildings and spaces with variable occupancies, such as public spaces, theaters, meeting rooms etc.

Green Tip #8 Hybrid Ventilation

This allows the controlled introduction of outside air ventilation into a building by both mechanical and passive means. It is sometimes called "mixed mode ventilation." It has built-in strategies to allow the mechanical and passive portions to work with one another so as not to cause additional ventilation loads, as would occur using mechanical ventilation alone. This is a non- conventional technique that has the promise of reducing operating expenses as well as providing a healthier stimulating environment.

23.7.2.3 Energy Distribution Systems

These provide the heating, cooling, lighting, and electric power throughout the building. The most common media to distribute energy are steam, hydronics (water), air, and electricity. Steam supply, because of its pressure characteristics, does not need to be pumped, although generally the steam condensate is pumped back to the boilers. Water and air are the principle media that require mechanical intervention for distribution and hence can be major consumers of electric power for pumps and fans.

Green Tip #9 Variable Flow/Variable Speed Pumping Systems

Pumps and fans can be significant users of electrical power. In a conventional application, the pumps and fans operate at a fixed rate based on maximum design conditions, regardless of actual loads. Adding variability to the pumping and fans, to allow modulation of flows based upon actual systems needs as opposed to design conditions, can provide significant electrical savings.

23.7.2.4 Energy Conservation Systems

This section focuses on the equipment that generates electricity, steam, hot water, and chilled water. These include distributed electrical generation, boilers, furnaces, electrically driven water chillers, and thermally driven absorption chillers.

Green Tip #10 Low-NO_x Burners

These are natural gas burners that improve energy efficiency and lower emissions of oxides of nitrogen, NO_x. They can be purchased as an option for new equipment or retrofitted to existing equipment.

Green Tip #11 Combustion Air Preheating

This tip refers to preheating combustion air by using waste heat from the exhaust stack, increasing energy efficiency of equipment such as boilers and furnaces. The principle is to introduce preheated hot air into the combustor instead of cold air, thereby reducing energy consumption.

Green Tip #12 Combustion Space/Water Heaters

These consist of a storage water heater, a heat delivery system such as fan coils or baseboards, and associated pumps and controls. The single unit does dual duty, both as a water heater for domestic hot water heater and as a source of hot water for the hydronic heating system.

Green Tip #13 Ground Source Heat Pumps, GSHP

These extract heat stored in the ground when in the heating mode and reject heat removed from the building into the ground in the cooling season. They consist of a loop of piping, or a well in the ground, and indoor units consisting of evaporators and condensers connected into the ground water loop. GSHPs can reduce the energy required for space heating, cooling, and service water in commercial/institutional buildings by as much as 50%.

Green Tip #14 Water Loop Heat Pump Systems

These consist of multiple water source heat pumps within a building and tied into a neutral temperature loop that serves as a heat source and a heat sink. The loop temperature in turn is maintained at this neutral point by supplementing with heat from a boiler, or cooling from a cooling tower.

Green Tip #15 Thermal Energy Storage for Cooling

This is an active storage system that uses the building's cooling equipment to remove heat, usually during the night and off-peak periods, to take advantage of lower-cost electricity during those periods and make ice slurry or chilled water. This enables a number of control and operational strategies. For example, smaller chillers can be purchased, and the building peak loads are satisfied with ice made during the off- peak periods.

Green Tip #16 Double Effect Absorption Chillers

Chilled water for facility cooling can be driven by electricity or thermal energy such as steam. In absorption chillers, thermal energy such as steam is used to drive a process using water and a salt solution such as lithium bromide in a vacuum-sealed shell to produce chilled water. There are no ozone depleting refrigerants used in this process. Absorbers come in single and double effect types, the double effect having a COP of about 1.2 versus the single effect COP of about .8.

Green Tip #17 Gas Engine Driven Chillers

Chilled water systems that use energy sources besides electricity can help offset high electricity costs. Generally, these are engines run on natural gas and are coupled to a chiller compressor section. Essentially, it is an engine replacing the electric motor of an electrically driven chiller. Gasoline and diesel fuel can also be used, depending upon engine types selected.

Green Tip #18 Gas-Fired Chiller/Heater

Gas-fired absorption chillers are a special type of absorption chiller wherein the thermal energy source is a direct burner typically firing natural gas, although other fuels could be used. This is as opposed to the conventional absorber operated with steam. The chiller/heater can do double duty to both make cold and hot water simultaneously.

Green Tip #19 Desiccant Cooling and Dehumidification

Rotary desiccant dehumidifiers use solid desiccants such as silica gel to attract water from the air. Humid air is dehumidified in one part of the desiccant bed while a different part of the bed is dried for reuse by a second air stream.

Green Tip #20 Evaporative Cooling

This technique can be used to reduce the amount of energy consumed by mechanical cooling equipment. There are two types of evaporation coolers, direct and indirect. Direct introduces water directly into the air stream, and the water evaporates, reducing the dry bulb temperature of the air while raising the relative humidity. Indirect systems spray water onto a coil and, through evaporation from the fins of the coils, reduce the dry bulb temperature also. However, no water is added to the air stream with this method.

23.7.2.5 Energy and Water Sources

The designer may not have much option in the selections of energy sources for the building being designed. Typically, energy is provided by the area utilities in the form of gas and electric. However, the designer can consider options to supplement the conventional energy sources with renewables such as wind, solar photovoltaics, PV, and hydro, as the site permits. PV is generally the most applicable renewable energy source for buildings. The others are site specific.

Green Tip #21 Solar Energy System—Photovoltaics

Sunlight shines on solid state crystals of silicon and

generates low voltage direct current electricity. Applications simply require full access to the sun and sufficient space, typically on the roof, to generate useful amounts of electricity. The low voltage DC can be inverted and voltage boosted to be supplied directly into the building's electrical distribution system. Thus, there is no need to locate a specific low voltage DC load for the power produced.

23.7.2.6 Lighting Systems

The *GreenGuide* section here is designed to familiarize the HVAC&R engineer with lighting systems and their potential impact on the equipment sizes and designs. However, it states that it is best to engage a lighting designer who will design according to IESNA standards (Illuminating Engineering Society-North America). However, the guide does make one suggestion for lighting.

Green Tip #22 Light Conveyor

A light conveyor is a large pipe or duct with reflective sides that transmits artificial or natural light along its length. There are occasions wherein designers have used light tracking to enhance the performance of the light conveyors.

23.7.2.7 Plumbing and Fire Protection Systems

Although not usually within the purview of the HVAC designer's expertise, both must be able to work together in developing a fully integrated design. There are several tips that include aspects of both skill sets.

Green Tip #23 Water Conserving Plumbing Fixtures

Water conserving strategies can save owners both consumption and demand charges. Options to be considered for water conservation are:
* Infrared faucet sensors
* Delayed action shut off valves
* Low flow toilets
* Faucets with flow restrictors
* Metering faucets
* Water efficient dishwashers
* Waterless urinals
* Closed cooling towers to eliminate drift, and filters for cleaning tower water

Green Tip #24 Graywater Systems

Graywater is defined as wastewater coming from operations such as showers, bathtubs, washing machines and sinks. This is separate from blackwater, which is wastewater from toilets and sinks that contain organic or toxic materials. Where allowed by code, graywater can be filtered, treated, stored, and later used for nonpotable

uses such as irrigation of landscaping and flushing of toilets. Distribution would be through a piping system clearly separated from all others.

Green Tip #25 Point-of-Use Domestic Hot Water Heaters

These provide small quantities of hot water at the point of use, without a tie into a central hot water source. Generally electrically heated, savings are obtained through the avoidance of large amounts of hot water storage and thermal losses of hot water distribution piping.

Green Tip #26 Direct Contact Water Heaters

This consists of a heat exchanger in which flue gases are in direct contact with the water to be heated. Cold water enters the top of the heat exchanger; natural gas is burned and flows up through the heat exchanger wherein the water is cascading down, acquiring the heat of the burned gas. Although there is direct contact between exhaust gases and the water, there is very little contamination of the water. These are suitable for dairy and food processing uses, as well as many other processes.

Green Tip #27 Rainwater Harvesting

Although this is not a new concept and has been around for thousands of years, it is a simple and effective technology to apply. Rainwater is channeled into cisterns for nonpotable uses as needed for irrigation, toilet flushing, etc.

23.7.2.8 Controls

These can be thought of as the "nervous system" of the building's mechanical and electrical infrastructure. Controls can be used for temperature and humidity control, ventilation control, energy management and analysis, etc. (See also Chapter 22.)

Green Tip #28 Mixed Air Temperature Reset

This refers to the mix of outside and return air that exists on an operating system supply air handling unit prior to any "new" energy being applied to it. The concept is to reset the mixed air temperature (MAT) to a temperature that just satisfies the lowest cold air demand. Reset controls involve raising the setpoint of the MAT controls based on input that indicates the demand of the zone needing the coldest air, limited by the minimum amount of outside air for IAQ purposes.

Green Tip #29 Cold Deck Temperature Reset, CDT, with Humidity Override

CDT is similar to MAT, but it applies to air leaving a cooling coil. The concept here is to allow the discharge temperature from the cooling coil to go to higher temperatures when demand for cooling is low. However, this strategy can create high humidity conditions indoors on occasion. Thus, humidity sensors are located in the spaces to override the rising cold deck temperatures and drive temperatures back down to extract more moisture, based upon demands for temperature and humidity control.

Other elements of green, good design covered by the *GreenGuide,* but with no specific green tips suggested, are:

- Expressing and testing concepts
- Completing design and documentation for construction
- Post Design—construction to demolition
- Builder/contractor selection
- Construction
- Commissioning
- Operation/maintenance/performance evaluation

23.7.3 ASHRAE Advanced Energy Design Guide

This ASHRAE manual for small office buildings is intended for buildings up to 20,000 square feet in size. It is a prescriptive description of how designers and contractors can achieve up to 30% reduction in energy consumption in comparison to ASHRAE Standard 90.1-1999. This would be worth four credits in LEED NC. The manual, although primarily of use for new construction (NC) design, can also be used as a planning tool for LEED EB, because the measures given are a form of best energy practices for new or existing buildings.

The guide is divided into three basic chapters:

(I) Integrated Process to Achieve Energy Savings
1. Pre-design phase—Prioritize goals
2. Design phase
3. Construction
4. Acceptance
5. Occupancy
6. Operation

(II) The guide provides recommendations by climate. It divides the USA into 8 zones, from #1, the warmest zone (the southern tip of Florida), to #8 for Alaska.

(III) For each climate zone there is a table providing prescriptive recommendations for the building components listed below:
1. Roofs
2. Walls
3. Floors
4. Slabs
5. Doors
6. Vertical glazing
7. Skylighting
8. Interior lighting
9. HVAC category, including air-conditioning, heating, air side economizers, ventilation, and air handling ducts.
10. Service water heating

References:
ASHRAE *GreenGuide,* 2003—Editor David L. Grumman
ASHRAE Advanced Energy Design Guide
ASHRAE Website *www.ASHRAE.ORG*
Energy User News, May 2004 by Peter D'Antonio, entitled "The LEED-ing Way."
The United States Green Building Council, www.USGBC.ORG

CHAPTER 24

ELECTRIC DEREGULATION

GEORGE R. OWENS, P.E. C.E.M.
Energy and Engineering Solutions, Inc.

24.0 INTRODUCTION

Utility deregulation, customer choice, unbundled rates, re-regulation, universal service charge, off tariff gas, stranded costs, competitive transition charge (CTC), caps and floors, load profiles and on and on are the new energy buzzwords. They are all the jargon being used as customers, utilities, and the new energy service suppliers become proficient in doing the business of utility deregulation.

Add to that the California energy shortages and rolling blackouts, the Northeast and Midwest outages of 2003, scandal, rising energy prices, and loss of price protection in deregulated states, and you can see why utility deregulation is increasingly on the mind of utility customers throughout the United States and abroad.

With individual state actions on deregulating natural gas in the late 80's and then the passage of the Energy Policy Act (EPACT) of 1992, the process of deregulating the gas and electric industry was begun. Because of this historic change toward a competitive arena, the utilities, their customers, and the new energy service providers have begun to reexamine their relationships.

How will utility customers, each with varying degrees of sophistication, choose their suppliers of these services? Who will supply them? What will it cost? How will it impact comfort, production, tenants, and occupants? How will the successful new players bring forward the right product to the marketplace to stay profitable? And how will more and better energy purchases improve the bottom line?

This chapter reviews the historic relationships between utilities, their customers, and the new energy service providers, as well as the tremendous possibilities for doing business in new and different ways.

The following figure portrays how power is generated and how it is ultimately delivered to the end customer.

1. Generator—Undergoing deregulation
2. Generator Substation—See 1

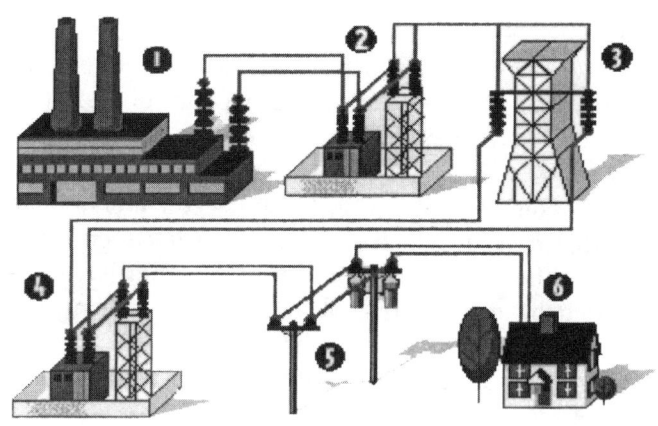

The Power Flow Diagram

3. Transmission System—Continues to be regulated by the Federal Energy Regulatory Commission (FERC) for interstate systems and by the individual states for in-state ones
4. Distribution Substation—Continues to be regulated by individual states
5. Distribution Lines—See 4
6. End Use Customer—As a result of deregulation, will be able to purchase power from a number of generators. Will still be served by the local "wires" distribution utility, which is regulated by the state.

24.1 AN HISTORICAL PERSPECTIVE OF THE ELECTRIC POWER INDUSTRY

At the turn of the century, vertically integrated electric utilities produced approximately two-fifths of the nation's electricity. At the time, many businesses (nonutilities) generated their own electricity. When utilities began to install larger and more efficient generators and more transmission lines, the associated increase in convenience and economical service prompted many industrial consumers to shift to the utilities for their electricity needs. With the invention of the electric motor came the inevitable use of more and more home appliances. Consumption of electricity skyrocketed, along with the utility's share of the nation's generation.

The early structure of the electric utility industry was predicated on the concept that a central source of power supplied by efficient, low-cost utility generation, transmission, and distribution was a natural monopoly. In addition to its intrinsic design to protect consumers, regulation generally provided reliability and a fair rate of return to the utility. The result was traditional rate base regulation.

For decades, utilities were able to meet increasing demand at decreasing prices. Economies of scale were achieved through capacity additions, technological advances, and declining costs, even during periods when the economy was suffering. Of course, the monopolistic environment in which they operated left them virtually unhindered by the worries that would have been created by competitors. This overall trend continued until the late 1960s, when the electric utility industry saw decreasing unit costs and rapid growth give way to increasing unit costs and slower growth.

The passage of EPACT-1992 began the process of drastically changing the way that utilities, their customers, and the energy services sector deal (or do not deal) with each other. Regulated monopolies are out and customer choice is in. The future will require knowledge, flexibility, and maybe even size to parlay this changing environment into profit and cost-saving opportunities.

One of the provisions of EPACT-1992 mandates open access on the transmission system to "wholesale" customers. It also provides for open access to "exempt wholesale generators" to provide power in direct competition with the regulated utilities. This provision fostered bilateral contracts (those directly between a generator and a customer) in the wholesale power market.

The regulated utilities then continue to transport the power over the transmission grid and ultimately, through the distribution grid, directly to the customer. This process of transmitting power across utilities is called "electric wheeling." Each utility that participates in the wheeling process adds a transmission charge to the electric power, based upon the tariffs.

What EPACT-1992 did not do was to allow for "retail" open access. Unless you are a wholesale customer, power can only be purchased from the regulated utility. However, EPACT-1992 made provisions for the states to investigate retail wheeling. ("Wheeling" and "open access" are other terms used to describe deregulation.) Many states have held or are currently holding hearings. Several states either have or will soon have pilot programs for retail wheeling. The model being used is that the electric generation component (typically 60-70% of the total bill) will be deregulated and subject to full competition. The transmission and distribution systems will remain regulated and subject to FERC and state Public Service Commission (PSC) control.

ELECTRIC INDUSTRY DEREGULATION TIME LINE

1992 - Passage of EPACT and the start of the debate.

1995 & 1996 - The first pilot projects and the start of special deals. Examples are: The automakers in Detroit; New Hampshire programs for direct purchase, including industrial, commercial and residential; and large user pilots in Illinois and Massachusetts.

1997 - Continuation of more pilots in many states and almost every state had deregulation on the legislative and regulatory commission agenda.

1998 - Full deregulation in a few states for large users (i.e., California and Massachusetts). Many states converged upon 1/1/98 as the start of their deregulation efforts, with more pilots and the first 5% roll-in of users, such as Pennsylvania and New York.

2000 - Deregulation of electricity became common for most industrial and commercial users and began to penetrate the residential market in several states. These included Maryland, New Jersey, New York, and Pennsylvania, among others. See Figure 24-1.

2001/2-California experienced rolling blackouts and high prices due to reduced power availability from other states and manipulation of the power trading system by some of the energy suppliers working in that market. Enron, who at the time was one of the largest suppliers of electricity in the nation, collapsed due to an accounting scandal. This action lowered public confidence in electric deregulation. California effectively shut down portions of the electric deregulation program.

2002/3-Customers had always had a "backstop" of regulated pricing. Since the transition periods were nearing their end, customers were faced with the option of buying electricity on the open market without a regulated default price.

2003 - During the summer, parts of the northeast and upper Midwest experienced a massive blackout that shut down businesses and residential customers. The adequacy of the transmission system was blamed.

2005 - EPACT-2005 became law, expanding transmission links that should reduce certain charges.

2007 - The Energy Independence and Security Act of 2007 was signed into law. This act addresses renewables, miles per gallon standards for cars, and efficiency improvements. It does not appear to affect electric deregulation.

24.2 THE TRANSMISSION SYSTEM AND THE FEDERAL ENERGY REGULATORY COMMISSION'S (FERC) ROLE IN PROMOTING COMPETITION IN WHOLESALE POWER

Even before the passage of EPACT in 1992, FERC played a critical role in the competitive transformation of wholesale power generation in the electric power industry. Specific initiatives include notices of proposed rulemaking steps toward the expansion of competitive wholesale electricity markets. FERC's Order 888, which was issued in 1996, required public utilities that own, operate, or control transmission lines to file tariffs that were non-discriminatory at rates no higher than what the utility charges itself. These actions essentially opened up the national transmission grid to non-discretionary access on the wholesale level (to public utilities, municipalities and rural cooperatives). This order did not give access to the transmission grid for retail customers.

Another initiative undertaken by the states is the disaggregation of the various electric charges shown on the typical electric bill. Previously, electric utility bills had the fixed, distribution, transmission, generation, tax, and other charges aggregated into one or two line items. In general, current billing practice is to show each cost item on a separate line item.

In an effort to ensure that the transmission grid is open to competition on a non-discriminatory basis, independent system operators (ISOs) are being formed in many regions of the country. An ISO is an independent operator of the transmission grid and is primarily responsible for reliability, maintenance (even if the day-to-day maintenance is performed by others), and security. In addition, ISO's generally provide the following functions: congestion management, administering transmission and ancillary pricing, making transmission information publicly available, etc.

24.3 STRANDED COSTS

Stranded costs are generally described as legitimate, prudent, and verifiable costs incurred by a public utility or a transmitting utility to provide a service to a customer that subsequently is no longer used. Since the asset or capacity is generally paid for through rates, ceasing to use the service leaves the asset, and its cost, stranded. In the case of de-regulation, stranded costs are created when the utility service or asset is provided, in whole or in part, to a deregulated customer of another public utility or transmitting utility. Stranded costs emerge because new generating capacity can currently be built and operated at costs that are lower than many utilities' embedded costs. Wholesale and retail customers have, therefore, an incentive to turn to lower cost producers. Such actions make it difficult for utilities to recover all their prudently incurred costs in generating facilities.

Stranded costs can occur during the transition to a fully competitive wholesale power market as some wholesale customers leave a utility's system to buy power from other sources. This may idle the utility's existing generating plants, imperil its fuel contracts, and inhibit its capability to undertake planned system expansion, leading to the creation of "stranded costs." During the transition to a fully competitive wholesale power market, some utilities may incur stranded costs as customers switch to other suppliers. If power previously sold to a departing customer cannot be sold to an alternative buyer, or if other means of mitigating the stranded costs cannot be found, the options for recovering stranded costs are limited.

The issue of stranded costs has become contentious in state proceedings on electric deregulation. Utilities have argued vehemently that they are justified in recovering their stranded costs. Customer advocacy groups, on the other hand, have argued that the stranded costs proposed by the utilities are excessive. This is being worked out in the state utility commissions. Often, in exchange for recovering stranded costs, utilities are joining in settlement agreements that offer guaranteed rate reductions and opening up their territories to deregulation.

24.4 STATUS OF STATE ELECTRIC INDUSTRY RESTRUCTURING ACTIVITY

Electric deregulation on the retail level is determined by state activity. Many states have or are in the process of enacting legislation and/or conducting proceedings. See Figure 24-1. Although this chart was last updated in 2003, little change has occurred since then. Because of rising fuel prices and the removal of price caps, electric prices have gone up faster than inflation. Some states have considered a partial reregulation of electricity in response to consumer complaints over rising prices.

Retail access is either currently available to all or some customers or will soon be available. Those states are Arizona, Connecticut, Delaware, District of Columbia, Illinois, Maine, Maryland, Massachusetts, Michigan, New Hampshire, New Jersey, New York, Ohio, Oregon, Pennsylvania, Rhode Island, Texas, and Virginia. In Oregon, no customers are currently participating in the State's retail access program, but the law allows nonresidential customers access. Yellow colored states are not actively pursuing restructuring. Those states are Alabama, Alaska, Colorado, Florida, Georgia, Hawaii, Idaho, Indiana, Iowa, Kansas, Kentucky, Louisiana, Minnesota, Mississippi, Missouri, Nebraska, North Carolina, North Dakota, South Carolina, South Dakota, Tennessee, Utah, Vermont, Washington, West Virginia, Wisconsin, and Wyoming. In West Virginia, the Legislature and Governor have not approved the Public Service Commission's restructuring plan, authorized by HB 4277. The Legislature has not passed a resolution resolving the tax issues of the PSC's plan, and no activity has occurred since early in 2001. A green colored state signifies a delay in the restructuring process or the implementation of retail access. Those states are Arkansas, Montana, Nevada, New Mexico, and Oklahoma. California is the only blue colored state because direct retail access has been suspended.

*As of January 30, 2003, Department of Energy, Energy Information Administration

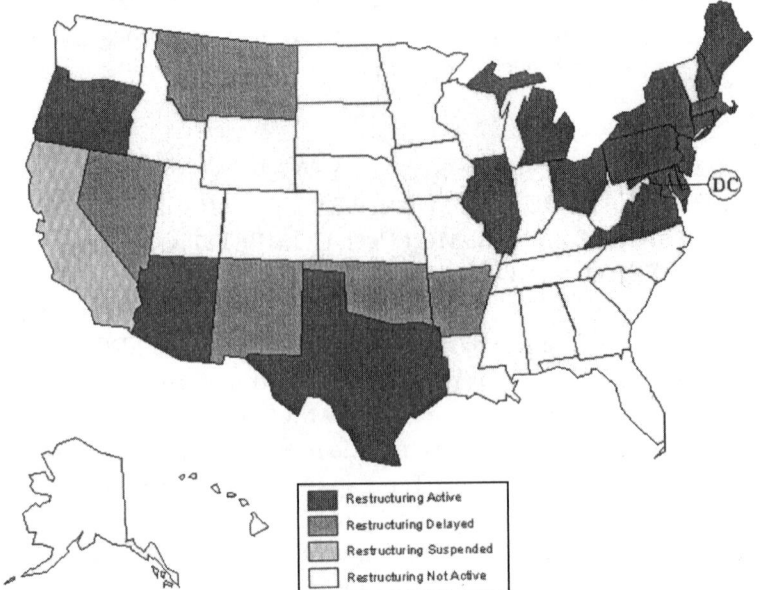

Figure 24-1. Status of State Electric Industry Restructuring Activity*

24.5 TRADING ENERGY - MARKETERS AND BROKERS

With the opening of retail electricity markets in several states, new suppliers of electricity have developed beyond the traditional vertically integrated electric utility. Energy marketers and brokers are the new companies that are being formed to fill this need. An energy marketer is one that buys electricity or gas commodity and transmission services from traditional utilities or other suppliers, then resells these products. An energy broker, like a real estate broker, arranges for sales but does not take title to the product. There are independent energy marketers and brokers, as well as unregulated subsidiaries of the regulated utility.

According to The Edison Electric Institute, the energy and energy services market was $360 billion in 1996 and was expected to grow to $425 billion in 2000. To help put these numbers in perspective, this market is over six times the telecommunications marketplace. As more states open for competition, the energy marketers and brokers are anticipating strong growth. Energy suppliers have been in a merger and consolidation mode for the past few years. This will probably continue at the same pace as the energy industry redefines itself even further. Guidance on how to choose the right supplier for your business or clients will be offered later on in this chapter.

The trading of electricity on the commodities market is a rather new phenomenon. It has been recognized that the marketers, brokers, utilities and end users need to have vehicles that are available for the managing of risk in the sometimes-volatile electricity market. The New York Mercantile Exchange (NYMEX) has instituted the trading of electricity along with its more traditional commodities. A standard model for an electricity futures contract has been established and is traded for delivery at several points around the country. As these contracts become more actively traded, their usefulness will increase as a means to mitigate risk. An example of a risk management play would be when a power supplier locks in a future price via a futures or options contract to protect its position at that point in time. Then if the prices rise dramatically, the supplier's price will be protected.

24.6 THE IMPACT OF DEREGULATION

Historically, electricity prices have varied by a factor of two to one or greater, depending upon where in the county the power is purchased. See Figure 24-2. These major differences even occur in utility jurisdictions that are joined. The cost of power has varied because of several factors, some of which are under the utilities control and some that are not, such as:

• Decisions on projected load growth

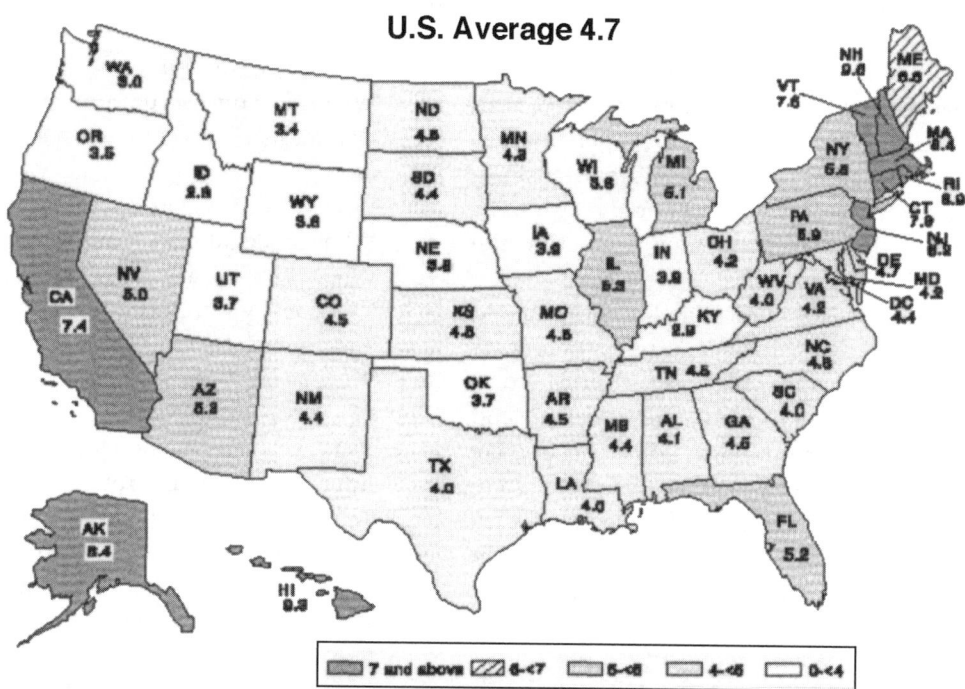

Source: Energy Information Administration, Form EIA-861, "Annual Electric Utility Report," (1995).

Figure 24-2a. Electricity Cost by State, 1995

Average Revenue from Electric Sales to Industrial Consumers by State, 1995 (Cents per Kilowatt-hour)

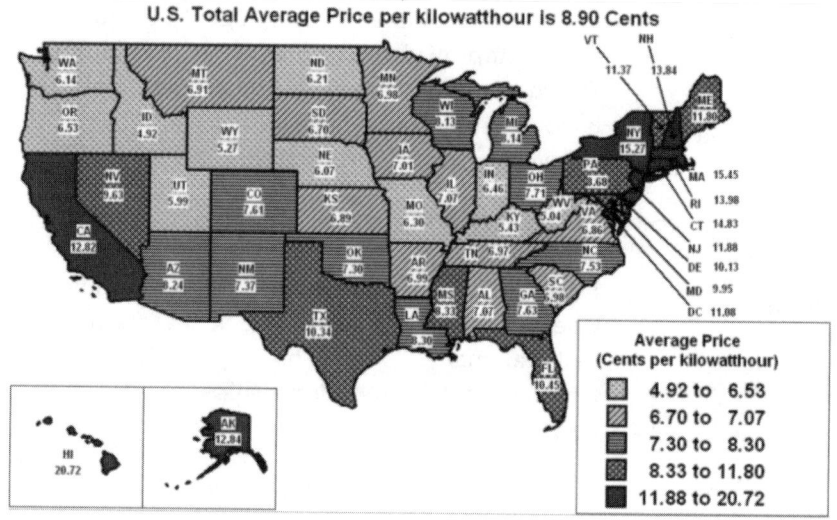

Figure 24-2b. Electricity Cost by State 2006
http://www.eia.doe.gov/cneaf/electricity/epa/
fig7p4.html

The original chart in the first edition of this book was from 1995. The most current chart, 2006, shows that the average price has nearly doubled since then. It is anticipated that the results of a similar chart in 2007 and 2008 would show significant increases over 2006. This is not what was supposed to happen with the introduction of electric deregulation.

- The type of generation
- Fuel selections
- Cost of labor and taxes
- The regulatory climate

All of these factors contribute to the range of pricing. Customers have been clamoring for the right to choose the supplier and gain access to cheaper power for quite some time. This has driven regulators to impose utility deregulation, often with opposition from the incumbent

utilities.

Many believe that electric deregulation will even out this difference *and* bring down the total average price through competition. There are others that do not share that opinion. Most utilities are already taking actions to reduce costs. Consolidations, layoffs, and mergers are occurring with increased frequency. As part of the transition to deregulation, many utilities are requesting and receiving rate freezes and reductions in exchange for stranded costs.

One factor remained a constant until the early 2000's. Customers have always had a "backstop" of regulated pricing until recently. Now that the transition periods are nearing their end, customers are faced with the option of buying electricity on the open market without a regulated default price. The risks to customers have increased dramatically. And, energy consultants and ESCOs are having a difficult time predicting the direction of electricity costs.

All of this provides for interesting background and statistics, but what does it mean to energy managers interested in providing and procuring utilities, commissioning, O&M (operations and maintenance), and the other energy services required to build and operate buildings effectively? Just as almost every business enterprise has experienced changes in the way that they operate in the 90's and 2000 and beyond, the electric utilities, their customers, and the energy service sector must also transform. Only well-prepared companies will be in a position to take advantage of the opportunities that will present themselves after deregulation. Building owners and managers need to be in a position to actively participate in the early opening states. The following questions will have to be answered by each and every company if they are to be prepared:

- Will they participate in the deregulated electric market?
- Is it better to do a national account style supply arrangement or divide the properties by region and/or by building type?
- How will electric deregulation affect their relationships with tenants in commercial, governmental, and institutional properties?
- Would there be a benefit for multi-site facilities to partake in purchasing power on their own?
- Should the analysis and operation of electric deregulation efforts be performed in-house or by consultants or a combination?
- What criteria should be used to select the energy suppliers when the future is uncertain?

24.7 THE TEN-STEP PROGRAM TO SUCCESSFUL UTILITY DEREGULATION

In order for the building sector to get ready for the new order and answer the questions raised above, a ten-step program has been developed to ease the transition and take advantage of the new opportunities. This ten-step program is ideally suited to building owners and managers, as well as to energy engineers in the process of developing their utility deregulation program.

Step #1 - Know Thyself
- When do you use the power?
- Distinguish between summer vs. winter, night vs. day.
- What load can you control/change?
- What $$$ goal does your business have?
- What is your 24 hr. load profile?
- What are your in-house engineering, monitoring and financial strengths?

Step #2 - Keep Informed
- Read, read, read—network, network, network.
- Interact with your professional organizations.
- Talk to vendors, consultants, and contractors.
- Subscribe to trade publications.
- Attend seminars and conferences.
- Utilize internet resources—news groups, WWW, and E-mail.
- Investigate buyer's groups.

Step #3 - Talk to Your Utilities (all energy types)
- Recognize customer relations are improving.
- Discuss alternate contract terms or other energy services.
- Find out if they are "for" or "agin" deregulation.
- Obtain improved service items (i.e., reliability).
- Tell them your position and what you want. Now is not the time to be bashful!
- Renegotiate existing contracts.

Step #4 - Talk to Your Future Utility(ies)
- See Step #3.
- Find out who is actively pursuing your market.
- Check the neighborhood, check the region, and look nationally.
- Develop your future relationships.
- Partner with energy service companies (ESCOs), power marketers, financial reps, vendors, and others for your energy services needs.

Step #5 - Explore Energy Services Now
(Why wait for deregulation?)
- Implement "standard" energy projects such as lighting, HVAC, etc.
- Investigate district cooling/heating.
- Explore selling your central plant.
- Calculate square foot pricing.
- Buy comfort, Btus or GPMs, not kWhs.
- Outsource your operations and maintenance.
- Consider other work on the customer side of the meter.

Step #6 - Understand the Risks

- Realize that times will be more complicated in the future.
- Consider the length of a contract term in uncertain times.
- Identify whether you want immediate reductions now, larger reductions later, or prices tied to some other index.
- Determine the value of a flat price for utilities.
- Be wary of losing control of your destiny by turning over some of the operational controls of your energy systems.
- Realize the possibility that some companies will not be around in a few years.
- Determine how much risk you are willing to take in order to achieve higher rewards.

Step #7 - Solicit Proposals

- Meet with the bidders prior to issuing the request for proposal (RFP).
- Prepare the RFP for the services you need.
- Identify qualified players.
- Make commissioning a requirement to achieve the results.

Step #8 - Evaluate Options

- Enlist the aid of internal resources and outside consultants.
- Narrow the playing field and interview the finalists prior to awarding.
- Prepare a financial analysis of the results over the life of the project—return on investment (ROI) and net present value (NPV).
- Remember that the least first cost may or may not be the best value.
- Pick someone that has financial and technical strengths for the long term.
- Evaluate financial options such as leasing or sharing.

Step #9 - Negotiate Contracts

Remember the following guidelines when negotiating a contract:

- The longer the contract, the more important are the escalation clauses due to compounding.
- Since you may be losing some control, the contract document is your only protection.
- The supplying of energy is not regulated like the supplying of kWhs are now.
- The clauses that identify the party taking responsibility for an action, or "Who Struck John" clauses, are often the most difficult to negotiate.
- Include monitoring and evaluation of results.

- Understand how the contract can be terminated and the penalties for early termination.

Step #10 - Sit Back and Reap the Rewards

- Monitor, measure, and compare.
- Don't forget operations and maintenance for the long term.
- Keep looking, there are more opportunities out there.
- Get off your duff and go to step #1 for the next round of reductions.

24.8 PRICING OPTIONS FOR ELECTRIC SUPPLY

One thing that electric deregulation has delivered is the many pricing options available to electric suppliers and their customers. Depending upon the customers risk tolerance and their special needs for electricity pricing, an electric supply product is probably available to meet them. It has become so complicated that many customers choose to use utility consultants to help them sort out the many options and choose the best supplier. Such options include:

- Fixed price per kWh for the duration of the contract.

- Index pricing. This pricing shall be the index price plus a supplier adder.

- A discount off what the customer would have paid to the regulated utility.

- Other pricing options that would be advantageous to the customer. The methodology should be described in detail. Examples include "block and index," "caps," and "collars."

- Terms. Provide a recommendation and pricing for multiple terms: 6 month, 12 months, 24 months, 36 months, etc.

- Customer has the option to lock into a fixed price per kwh contract to reflect then current pricing at any time during the contract term during a variable price contract.

- Demand response programs, with the limitations of the customer taken into consideration, such as availability of loads to be shed and maintaining a comfortable environment.

- Percentage Purchasing—i.e. purchasing a percentage of the load at different times and lengths of contracts to take advantage of pricing opportunities and provide averaging of prices over time (similar to a CD ladder).

24.9 AGGREGATION

Aggregation is the grouping of utility customers to jointly purchase commodities and/or other energy services. There are many aggregations already formed or being formed in the states where utility deregulation is occurring. There are two basic forms of aggregation:

1. Similar Customers with Similar Needs
   Similar customers may be better served via aggregation, even if they have the same load profiles.
   — Pricing and risk can be tailored to similar customers needs.
   — Similar billing needs can be met.
   — Cross subsidization would be eliminated.
   — Trust in the aggregator, i.e. BOMA (office building managers membership).

2. Complementary Customers that May Enhance the Total
   Different load profiles can benefit the aggregated group by combining different load profiles.
   — Match a manufacturing facility having a flat or inverted load profile to an office building that has a peaky load profile, etc.
   — Combining of load profiles is more attractive to a supplier than either would be individually.

   Why Aggregate?
   Some potential advantages to aggregating are:
- Reduction of internal administration expense
- Shared consulting expenses
- More supplier attention resulting from a larger bid
- Lower rates may be the result of a larger bid.
- Lower average rates resulting from combining dissimilar user profiles

   Why Not Aggregate?
   Some potential disadvantages from aggregating are:
- If you are big enough, you are your own aggregation.
- Good load factor customers may subsidize poor load factor customers.
- The average price of an aggregation may be lower

than your unique price.
- An aggregation cannot meet "unique" customer requirements.

Factors that affect the decision on joining an aggregation
Determine if an aggregation is right for your situation by considering the following factors. An understanding of how these factors apply to your operation will result in an informed decision.
- Size of load
- Load profile
- Risk tolerance
- Internal abilities (or via consulting)
- Contract length flexibility
- Contract terms and conditions flexibility
- Regulatory restrictions

24.10 IN-HOUSE VS. OUTSOURCING ENERGY SERVICES

The end user sector has always used a combination of in-house and outsourced energy services. Many large managers and owners have a talented and capable staff to analyze energy costs, develop capital programs, and operate and maintain the in-place energy systems. Others (particularly the smaller players who cannot justify an in-house staff) have outsourced these functions to a team of consultants, contractors, and utilities. These relationships have evolved recently due to downsizing and returning to the core businesse,s. In the new era of deregulation, the complexion of how energy services are delivered will evolve further.

Customers and energy services companies are already getting into the utility business of generating and delivering power. Utilities are also getting into the act by going beyond the meter and supplying chilled/hot water, conditioned air, and comfort. In doing so, many utilities are setting up unregulated subsidiaries to provide commissioning, O&M, and many other energy services to customers located within their territory, and nationwide as well.

A variety of terms are often used: performance contracting, energy system outsourcing, utility plant outsourcing, guaranteed savings, shared savings, sell/leaseback of the central plant, chauffage (used in Europe), energy services performance contract (ESPC), etc. Definitions are as follows:
- Performance Contracting
  This is the process of providing a specific improvement such as a lighting retrofit or a chiller

change-out, usually using the contractor's capital and then paying for the project via the savings over a specific period of time. Often the contractor guarantees a level of savings. The contractor supplies capital, engineering, equipment, installation, commissioning, and often the maintenance and repair.

• Energy System Outsourcing
This the process of divesting of the responsibilities and often the assets of the energy systems to a third party. The third party then supplies the commodity, whether it be chilled water, steam, hot water, electricity, etc., at a per unit cost. The third party supplier is then responsible for the improvement capital and operations and maintenance of the energy system for the duration of the contract.

See Chapter 25 for a more detailed discussion of performance contracting.

24.11 SUMMARY

This chapter presented information on the changing world of the utility industry in the new millennium. Starting in the 80's with gas deregulation and the passage of the Energy Policy Act of 1992 for electricity, the method of providing and purchasing energy was changed forever. Utilities began a slow change from vertically integrated monopolies to providers of regulated wires and transmission services. Some utilities continued to supply generation services, through their unregulated enterprises and by independent power producers in the deregulated markets, while others sold their generation assets and became "wires" companies. Customers became confused in the early stages of deregulation, but by the end of the 1990s some became more knowledgeable and successful in buying deregulated natural gas and electricity.

In the early 2000s, difficulties developed in the deregulated utility arena. California rescinded deregulation (except for existing contracts) after shortages, rolling blackouts, and price increases sent the utilities into a tailspin. The great blackout of 2003 raised concerns about the reliability of the transmission system. And the loss of regulated rates provides more challenges to customers and their consultants. However, many customers continue to participate in the deregulated markets to obtain reduced (or stable) prices, reduce their risk of big price swings, and incorporate energy reduction programs with energy procurement programs.

Another result of deregulation has been a re-examination by customers of outsourcing their energy needs. Some customers have "sold" their energy systems to energy suppliers and are now purchasing Btus instead of kWhs. The energy industry responded with energy service business units to meet this new demand for outsourcing. Performance contracting and energy system outsourcing can be advantageous when the organization does not have internal expertise to execute these projects and when other sources of capital are needed. However, performance contracting and energy system outsourcing is not without peril if the risks are not understood and mitigated. Before undertaking a performance contract or energy system outsourcing project, the owner or manager first needs to define the financial, technical, legal, and operational issues of importance. Next, the proper resources, whether internal or outsourced, need to be marshaled to define the project, prepare the request for proposal, evaluate the suppliers and bids, negotiate a contract, and monitor the results, often over a long period. If these factors are properly considered and executed, the performance contract or energy system outsourcing often produces results that could not be obtained via other project methods.

Bibliography

Power Shopping and *Power Shopping II*, A publication of the Building Owners and Managers Association (BOMA) International, 1201 New York Avenue, N.W., N.W., Suite 300, Washington, DC 20005.

The Changing Structure of the Electric Power Industry: Historical Overview, United States Department of Energy, Energy Information Administration, Washington, DC.

The Ten Step Program to Successful Utility Deregulation for Building Owners and Managers, George R Owens PE CEM, President Energy and Engineering Solutions, Inc. (EESI), 9449 Penfield Ct., Columbia, MD 21045.

Performance Contracting and Energy System Outsourcing, George R Owens PE CEM, President Energy and Engineering Solutions, Inc. (EESI), 9449 Penfield Ct., Columbia, MD 21045.

Generating Power and Getting It to The Consumer, Edison Electric Institute, 701 Pennsylvania Ave NW, Washington, DC, 20004.

The Changing Structure of the Electric Power Industry: An Update, US Department of Energy, Energy Information Administration, DOE/EIA-0562(96)

PJM Electricity Futures, New York Mercantile Exchange (NYMEX) web page, www.nymex.com

Wikipedia—California electricity Crises http://en.wikipedia.org/wiki/California_electricity_crisis

SOME USEFUL INTERNET RESOURCES

10 Step paper - www.eesienergy.com
State activities - www.eia.doe.gov/fuelelectricHTML
State regulatory commissions www.naruc.org
Utilities - www.utilityconnection.com

CHAPTER 25

FINANCING AND PERFORMANCE CONTRACTING

ERIC A. WOODROOF, PH.D., CEM, CRM
ProfitableGreenSolutions.com

25.1 INTRODUCTION

Financing can be a key success factor for projects. This chapter's purpose is to help facility managers understand and apply the financial arrangements available to them. Hopefully, this approach will increase the implementation rate of good energy management projects that would have otherwise been cancelled or postponed due to lack of funds.

Most facility managers agree that energy management projects (EMPs) are good investments. Generally, EMPs reduce operational costs, have a low risk/reward ratio, usually improve productivity, and have even been shown to improve a firm's stock price.[1] Despite these benefits, many cost-effective EMPs are not implemented due to financial constraints. A study of manufacturing facilities revealed that first-cost and capital constraints represented over 35% of the reasons cost-effective EMPs were not implemented.[2] Often, the facility manager does not have enough cash to allocate funding or cannot get budget approval to cover initial costs. Financial arrangements can mitigate a facility's funding constraints,[3] allowing additional energy savings to be reaped without delay.

Alternative finance arrangements can overcome the "initial cost" obstacle, allowing firms to implement more EMPs. However, many facility managers are either unaware or have difficulty understanding the variety of financial arrangements available to them. Most facility managers use simple payback analyses to evaluate projects, which do not reveal the added value of after-tax benefits.[4] Sometimes facility managers do not implement an EMP because financial terminology and contractual details intimidate them.[5]

To meet the growing demand, there has been a dramatic increase in the number of finance companies specializing in EMPs. These financiers are introducing new payment arrangements to implement EMPs. Often, the financier's innovation will satisfy the unique customer needs of a large facility. This is a great service; however, most financiers are not attracted to small facilities with EMPs requiring less than $100,000. Thus, many facility managers remain unaware or confused about the common financial arrangements that could help them implement EMPs.

Numerous papers and government programs have been developed to show facility managers how to use quantitative (economic) analysis to evaluate financial arrangements.[4,5,6] (Refer to Chapter 4 of this book.) *Quantitative analysis includes computing the simple payback, net present value (NPV), internal rate of return (IRR), or life-cycle cost of a project with or without financing.* Although these books and programs show how to evaluate the economic aspects of projects, they do not incorporate qualitative factors like strategic company objectives (which can impact the financial arrangement selection). Without incorporating a facility manager's qualitative objectives, it is hard to select an arrangement that meets all of the facility's needs. Qualitative objectives can be at least as important as quantitative objectives.[9]

This chapter hopes to provide some valuable information, which can be used to overcome barriers that keep good projects from getting implemented. The chapter is divided into several sections to accomplish three objectives. Sections 2 and 3 *introduce the basic financial arrangements* via a simple example. In sections 4 and 5, financial terminology is defined and each arrangement is explained in greater detail while applied to a case study. The remaining sections show *how to match financial arrangements to different projects and facilities.*

25.2 FINANCIAL ARRANGEMENTS: A SIMPLE EXAMPLE

Consider a small company, "PizzaCo," that makes frozen pizzas and distributes them regionally. PizzaCo uses an old delivery truck that breaks down frequently and is inefficient. Assume the old truck has no salvage value and is fully depreciated. PizzaCo's management would like to obtain a new and more efficient truck to reduce expenses and improve reliability. However, they do not have the cash on hand to purchase the truck. Thus, they consider their financing options.

25.2.1 Purchase the Truck with a Loan or Bond

Just like most car purchases, PizzaCo borrows money from a lender (a bank) and agrees to a monthly re-payment plan. Figure 25-1 shows PizzaCo's annual cash flows for a loan. The solid arrows represent the financing cash flows between PizzaCo and the bank. Each year, PizzaCo makes payments (on the principal, plus interest based on the unpaid balance) until the balance owed is zero. The payments are the negative cash flows. Thus, at time zero when PizzaCo borrows the money, it receives a large sum of money from the bank, which is a positive cash flow (be used to purchase the truck).

The *dashed* arrows represent the truck purchase as well as savings cash flows. Thus, at time zero, PizzaCo purchases the truck (a negative cash flow) with the money from the bank. Due to the new truck's greater efficiency, PizzaCo's annual expenses are reduced (which is a savings). The annual savings are the positive cash flows. The remaining cash flow diagrams in this chapter utilize the same format.

PizzaCo could also purchase the truck by selling a bond. This arrangement is similar to a loan, except investors (not a bank) give PizzaCo a large sum of money (called the bond's "par value"). Periodically, PizzaCo would pay the investors only the interest accumulated. As Figure 25-2 shows, when the bond reaches maturity, PizzaCo returns the par value to the investors. The equipment purchase and savings cash flows are the same as with the loan.

25.2.2 Sell Stock to Purchase the Truck

In this arrangement, PizzaCo sells its stock to raise money to purchase the truck. In return, PizzaCo is expected to pay dividends back to shareholders. Selling stock has a similar cash flow pattern as a bond, with a few subtle differences. Instead of interest payments to bondholders, PizzaCo would pay dividends to share-

holders until some future date when PizzaCo could buy the stock back. However, these dividend payments are not mandatory, and if PizzaCo is experiencing financial strain, it does not need to distribute dividends. On the other hand, if PizzaCo's profits increase, this wealth will be shared with the new stockholders, because they now own a part of the company.

25.2.3 Rent/Lease the Truck

Just like renting a car, PizzaCo could rent a truck for an annual fee. This would be equivalent to a "true lease," or "operating lease." The rental company (lessor) owns and maintains the truck for PizzaCo (the lessee). PizzaCo pays the rental fees (lease payments), which are considered tax-deductible business expenses.

Figure 25-3 shows that the lease payments (solid arrows) start as soon as the equipment is leased (year zero) to account for lease payments paid in advance. Lease payments "in arrears" (starting at the end of the first year) could also be arranged. However, the leasing company may require a security deposit as collateral. Notice that the savings cash flows are essentially the same as the previous arrangements, except there is no equipment purchase, which is a large negative cash flow at year zero.

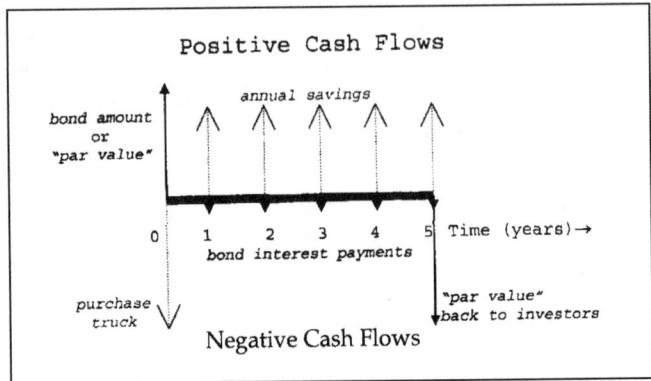

Figure 25-2 PizzaCo's Cash Flows for a Bond.

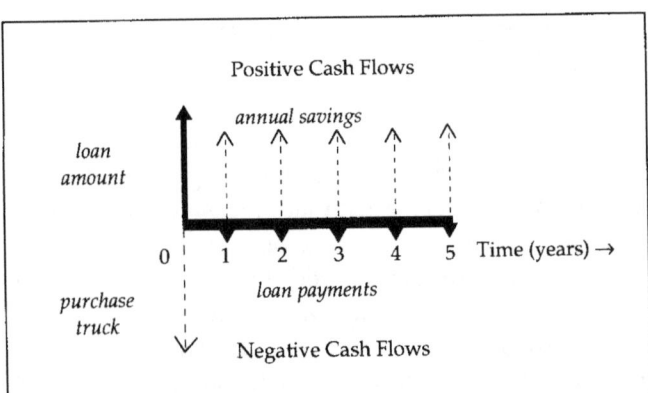

Figure 25-1. PizzaCo's Cash Flows for a Loan.

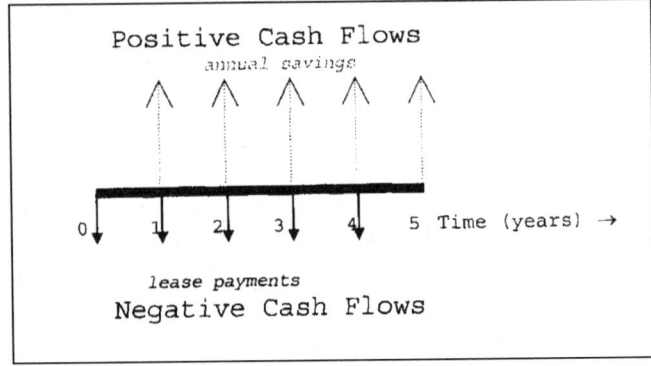

Figure 25-3. PizzaCo's Cash Flows for a True Lease.

In a true lease, the contract period should be shorter than the equipment's useful life. The lease is cancelable because the truck can be leased easily to someone else. At the end of the lease, PizzaCo can either return the truck or renew the lease. In a separate transaction, PizzaCo could also negotiate to buy the truck at the fair market value.

If PizzaCo wanted to secure the option to buy the truck (for a bargain price) at the end of the lease, then it would use a capital lease. A capital lease can be structured like an installment loan, however ownership is not transferred until the end of the lease. The lessor retains ownership as security in case the lessee (PizzaCo) defaults on payments. Because the entire cost of the truck is eventually paid, the lease payments are larger than the payments in a true lease (assuming similar lease periods). Figure 25-4 shows the cash flows for a capital lease with advance payments and a bargain purchase option at the end of year five.

There are some additional scenarios for lease arrangements. A "vendor-financed" agreement is when the lessor (or lender) is the equipment manufacturer. Alternatively, a third party could serve as a financing source. With "third party financing," a finance company would purchase a new truck and lease it to PizzaCo. In either case, there are two primary ways to repay the lessor.

1. With a "fixed payment plan," where payments are due whether or not the new truck actually saves money.

2. With a "flexible payment plan," where the savings from the new truck are shared with the third party until the truck's purchase cost is recouped with interest. This is basically a "shared savings" arrangement.

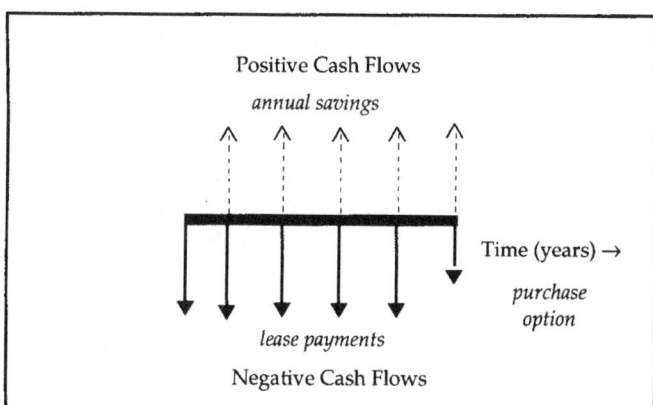

Figure 25-4. PizzaCo's Cash Flows for a Capital Lease.

25.2.4 Subcontract Pizza Delivery to a Third Party

Since PizzaCo's primary business is not delivery, it could subcontract that responsibility to another company. Let's say that a delivery service company would provide a truck and deliver the pizzas at a reduced cost. Each month, PizzaCo would pay the delivery service company a fee. However, this fee is guaranteed to be less than what PizzaCo would have spent on delivery. Thus, PizzaCo would obtain savings without investing any money or risk in a new truck. This arrangement is analogous to a performance contract.

This arrangement is very similar to a third-party lease. However, with a performance contract, the contractor assumes most of the risk, and the contractor also is responsible for ensuring that the delivery fee is less than what PizzaCo would have spent. For the PizzaCo example, the arrangement would be designed under the conditions below.

*   The delivery company owns and maintains the truck. It is also responsible for all operations related to delivering the pizzas.

*   The monthly fee is related to the number of pizzas delivered. This is the performance aspect of the contract; if PizzaCo doesn't sell many pizzas, the fee is reduced. *A minimum amount of pizzas may be required by the delivery company (performance contractor) to cover costs.* Thus, the delivery company assumes these risks:
    1. PizzaCo will remain solvent.
    2. PizzaCo will sell enough pizzas to cover costs.
    3. The new truck will operate as expected and will actually reduce expenses per pizza.
    4. The external financial risk, such as inflation and interest rate changes, are acceptable.

*   The delivery company is an expert in delivery; it has specially skilled personnel and uses efficient equipment. Thus, the delivery company can deliver the pizzas at a lower cost (even after adding a profit) than PizzaCo.

Figure 25-5 shows the net cash flows according to PizzaCo. Since the delivery company simply reduces PizzaCo's operational expenses, there is only a net savings. There are no negative financing cash flows. Unlike the other arrangements, the delivery company's fee is a less expensive substitute for PizzaCo's in-house delivery expenses. With the other arrangements, PizzaCo had to pay a specific financing cost (loan, bond or lease payments, or dividends) associated with the truck, whether or not the truck actually saved money. In addition, PizzaCo would have to spend time maintaining the truck, which would detract from its core focus—making pizzas. With a performance contract, the delivery company is paid from the operational savings it generates. Because the savings are greater than the fee, there is a net savings. Often, the contractor guarantees the savings.

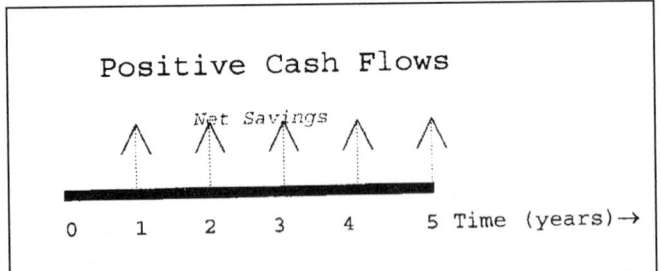

Figure 25-5. PizzaCo's Cash Flows for a Performance Contract.

Supplementary Note: Combinations of the basic finance arrangements are possible. For example, a lease arrangement can be structured within a performance contract. Also, performance contracts are often designed so that the facility owner (PizzaCo) would own the asset at the end of the contract.

25.3 FINANCIAL ARRANGEMENTS: DETAILS AND TERMINOLOGY

To explain the basic financial arrangements in more detail, each one is applied to an energy management-related case study. To understand the economics behind each arrangement, some finance terminology is presented below.

25.3.1 Finance Terminology

Equipment can be purchased with cash on-hand (officially labeled "retained earnings"), a loan, a bond, a capital lease, or by selling stock. Alternatively, equipment can be utilized with a true lease or with a performance contract.

Note that with performance contracting, the building owner is not paying for the equipment itself but the benefits provided by the equipment. *In the Simple Example, the benefit was the pizza delivery. PizzaCo was not concerned with what type of truck was used.*

The decision to purchase or utilize equipment is partly dependent on the company's strategic focus. If a company wants to delegate some or all of the responsibility of managing a project, it should use a true lease or a performance contact.[10] However, if the company

wants to be intricately involved with the EMP, purchasing and self-managing the equipment could yield the greatest profits. When the building owner purchases equipment, he/she usually maintains the equipment and lists it as an asset on the balance sheet so it can be depreciated.

Financing for purchases has two categories:

1. *Debt Financing*, which is borrowing money from someone else or another firm (using loans, bonds and capital leases).

2. *Equity Financing*, which is using money from your company or your stockholders (using retained earnings or issuing common stock).

In all cases, the borrower will pay an interest charge to borrow money. The interest rate is called the "cost of capital." The cost of capital is essentially dependent on three factors: (1) the borrower's credit rating, (2) project risk, and (3) external risk. External risk can include energy price volatility and industry-specific economic performance, as well as global economic conditions and trends. The cost of capital (or "cost of borrowing") influences the return on investment. If the cost of capital increases, then the return on investment decreases.

The minimum attractive rate of return (MARR) is a company's "hurdle rate" for projects. *Because many organizations have numerous projects competing for funding, the MARR can be much higher than interest earned from a bank or other risk-free investment.* Only projects with a return on investment greater than the MARR should be accepted. The MARR is also used as the discount rate to determine the net present value (NPV).

25.3.2 Explanation of Figures and Tables

Throughout this chapter's case study, figures are presented to illustrate the transactions of each arrangement. Tables are also presented to show how to perform the economic analyses of the different arrangements. The NPV is calculated for each arrangement.

It is important to note that the NPV of a particular arrangement can change significantly if the cost of capital, MARR, equipment residual value, or project life is adjusted. Thus, the examples within this chapter are provided only to illustrate how to perform the analyses. The cash flows and interest rates are estimates, which can vary from project to project. To keep the calculations simple, end-of-year cash flows are used throughout this chapter.

Within the tables, the following abbreviations and equations are used:

EOY = End of Year
Savings = pre-Tax Cash Flow
Depr. = Depreciation
Taxable Income = Savings - Depreciation - Interest
Payment
Tax = (Taxable Income)*(Tax Rate)
ATCF = After Tax Cash Flow =
Savings – Total Payments – Taxes

Table 25-1 shows the basic equations that are used to calculate the values under each column heading within the economic analysis tables.

$2.5 million. The expected equipment life is 15 years; however, the process will only be needed for 5 years, after which the chilled water system will be sold at an estimated market value of $1,200,000 (book value at year five = $669,375). The chilled water system should save PizzaCo about $1 million/year in energy savings. PizzaCo's tax rate is 34%. The equipment's annual maintenance and insurance cost is $50,000. PizzaCo's MARR is 18%. Since, at the end of year 5, PizzaCo expects to sell the asset for an amount greater than its book value, the additional revenues are called a "capital gain" (which equals the market value – book value) and are taxed.

Table 25-1. Table of Sample Equations used in Economic Analyses.

A	B	C	D	E	F	G	H	I	J
			Payments			Principal	Taxable		
EOY	Savings	Depreciation	Principal	Interest	Total	Outstanding	Income	Tax	ATCF
n									
n+1		= (MACRS %)*			=(D) +(E)	=(G at year n)	=(B)–(C)–(E)	=(H)*(tax rate)	=(B)–(F)–(I)
n+2		(Purchase Price)				–(D at year n+1)			

Regarding depreciation, the modified accelerated cost recovery system (MACRS) is used in the economic analyses. This system indicates the percent depreciation claimable year-by-year after the equipment is purchased. Table 25-2 shows the MACRS percentages for seven-year property. *For example, after the first year, an owner could depreciate 14.29% of an equipment's value. The equipment's "book value" equals the remaining unrecovered depreciation. Thus, after the first year, the book value would be 100%-14.29%, which equals 85.71% of the original value. If the owner sells the property before it has been fully depreciated, he/she can claim the book value as a tax-deduction.**

25.4 APPLYING FINANCIAL ARRANGEMENTS: A CASE STUDY

Suppose PizzaCo (*the "host" facility*) needs a new chilled water system for a specific process in its manufacturing plant. The installed cost of the new system is

If PizzaCo sells the asset for less than its book value, PizzaCo incurs a "capital loss."

PizzaCo does not have $2.5 million to pay for the new system; thus, it considers its finance options. PizzaCo is a small company with an average credit rating, which means that it will pay a higher cost of capital than a larger company with an excellent credit rating. As usual, if investors believe that an investment is risky, they will demand a higher interest rate.

Table 25-2. MACRS Depreciation Percentages.

EOY	MACRS Depreciation Percentages for 7-Year Property
0	0
1	14.29%
2	24.49%
3	17.49%
4	12.49%
5	8.93%
6	8.92%
7	8.93%
8	4.46%

To be precise, the IRS uses a "half-year convention" for equipment that is sold before it has been completely depreciated. In the tax year that the equipment is sold (say year "x"), the owner claims only Ω of the MACRS depreciation percent for that year. (This is because the owner has only used the equipment for a fraction of the final year.) Then on a separate line entry (in the year "x"), the remaining unclaimed depreciation is claimed as "book value." The x* year is presented as a separate line item to show the book value treatment; however, x* entries occur in the same tax year as "x."

25.4.1 Purchase Equipment with Retained Earnings (Cash)

If PizzaCo did have enough retained earnings (cash on-hand) available, it could purchase the equipment without external financing. Although external finance expenses would be zero, the benefit of tax-deductions (from interest expenses) is also zero. Also, any cash used to purchase the equipment would carry an "opportunity cost," because that cash could have been used to earn a return somewhere else. This opportunity cost rate is usually set equal to the MARR. In other words, the company lost the opportunity to invest the cash and gain at least the MARR from another investment.

Of all the arrangements described in this chapter, purchasing equipment with retained earnings is probably the simplest to understand. For this reason, it will serve as a brief example and introduction to the economic analysis tables that are used throughout this chapter.

25.4.1.1 Application to the Case Study

Figure 25-6 illustrates the resource flows between the parties. In this arrangement, PizzaCo purchases the chilled water system directly from the equipment manufacturer.

Once the equipment is installed, PizzaCo recovers the full $1 million/year in savings for the entire five years, but it must spend $50,000/year on maintenance and insurance. At the end of the five-year project, PizzaCo expects to sell the equipment for its market value of $1,200,000. Assume MARR is 18% and that the equipment is classified as 7-year property for MACRS depreciation. Table 25-3 shows the economic analysis for purchasing the equipment with retained earnings.

Reading Table 25-3 from left to right, and top to bottom, at EOY 0, the single payment is entered into the table. Each year thereafter, the savings as well as the depreciation (which equals the equipment purchase price multiplied by the appropriate MACRS % for each year) are entered into the table. Year by year, the taxable income = savings – depreciation. The taxable income is

then taxed at 34% to obtain the tax for each year. The after-tax cash flow = savings – tax for each year.

At EOY 5, the equipment is sold before the entire value was depreciated. EOY 5* shows how the equipment sale and book value are claimed. In summary, the NPV of all the ATCFs would be $320,675.

25.4.2 Loans

Loans have been the traditional financial arrangement for many types of equipment purchases. A bank's willingness to loan depends on the borrower's financial health, experience in energy management, and number of years in business. Obtaining a bank loan can be difficult if the loan officer is unfamiliar with EMPs. Loan officers and financiers may not understand energy-related terminology (demand charges, kVAR, etc.). In addition, facility managers may not be comfortable with the financier's language. Thus, to save time, a bank that can understand EMPs should be chosen.

Most banks will require a down payment and collateral to secure a loan. However, securing assets can be difficult with EMPs because the equipment often becomes part of the real estate of the plant. *For example, it would be very difficult for a bank to repossess lighting fixtures from a retrofit.* In these scenarios, lenders may be willing to secure other assets as collateral.

25.4.2.1 Application to the Case Study

Figure 25-7 illustrates the resource flows between the parties. In this arrangement, PizzaCo purchases the chilled water system with a loan from a bank. PizzaCo makes equal payments (principal + interest) to the bank for five years to retire the debt. Due to PizzaCo's small size, credibility, and inexperience in managing chilled water systems, PizzaCo is likely to pay a relatively high cost of capital. For example, let's assume 15%.

PizzaCo recovers the full $1 million/year in savings for the entire five years, but it must spend $50,000/year on maintenance and insurance. At the end of the five-year project, PizzaCo expects to sell the equipment

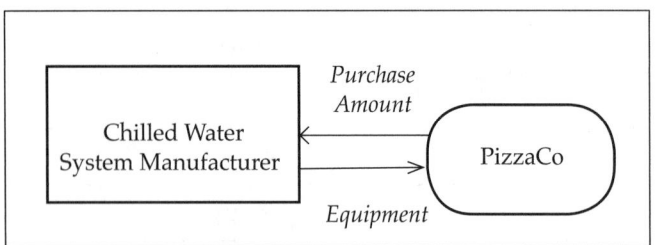

Figure 25-6. Resource Flows for Using Retained Earnings

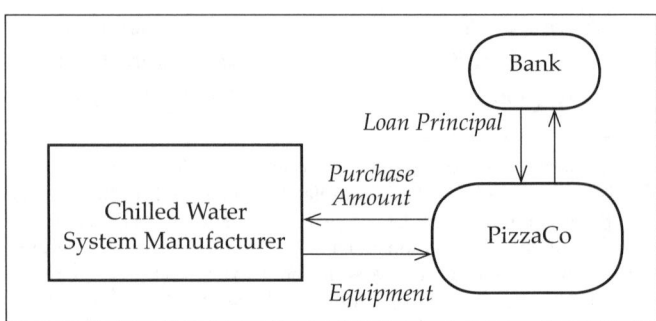

Figure 25-7. Resource Flow Diagram for a Loan.

Table 25-3. Economic Analysis for Using Retained Earnings.

EOY	Savings	Depr.	Payments Principal	Payments Interest	Total	Principal Outstanding	Taxable Income	Tax	ATCF
0					2,500,000				-2,500,000
1	950,000	357,250					592,750	201,535	748,465
2	950,000	612,250					337,750	114,835	835,165
3	950,000	437,250					512,750	174,335	775,665
4	950,000	312,250					637,750	216,835	733,165
5	950,000	111,625					838,375	285,048	664,953
5*	1,200,000	669,375					530,625	180,413	1,019,588
		2,500,000							

Net Present Value at 18%: $320,675

Notes: Loan Amount: 0
Loan Finance Rate: 0% MARR 18%
Tax Rate 34%

MACRS Depreciation for 7-Year Property, with half-year convention at EOY 5
Accounting Book Value at end of year 5: 669,375
Estimated Market Value at end of year 5: 1,200,000
EOY 5* illustrates the Equipment Sale and *Book* Value
Taxable Income: =(Market Value - Book Value)
=(1,200,000 - 669,375) = $530,625

Table 25-4. Economic Analysis for a Loan with No Down Payment.

EOY	Savings	Depr.	Payments Principal	Payments Interest	Total	Principal Outstanding	Taxable Income	Tax	ATCF
0				2,500,000					
1	950,000	357,250	370,789	375,000	745,789	2,129,211	217,750	74,035	130,176
2	950,000	612,250	426,407	319,382	745,789	1,702,804	18,368	6,245	197,966
3	950,000	437,250	490,368	255,421	745,789	1,212,435	257,329	187,492	116,719
4	950,000	312,250	563,924	181,865	745,789	648,511	455,885	55,001	49,210
5	950,000	111,625	648,511	97,277	745,789	0	741,098	251,973	-47,761
5*	1,200,000	669,375			530,625	180,413	1,019,588		
		2,500,000							

Net Present Value at 18%: $757,121

Notes: Loan Amount: 2,500,000 (used to purchase equipment at year 0)
Loan Finance Rate: 15% MARR 18%
Tax Rate 34%

MACRS Depreciation for 7-Year Property, with half-year convention at EOY 5
Accounting Book Value at end of year 5: 669,375
Estimated Market Value at end of year 5: 1,200,000
EOY 5* illustrates the Equipment Sale and *Book* Value
Taxable Income: =(Market Value - Book Value)
=(1,200,000 - 669,375) = $530,625

Table 25-5. Economic Analysis for a Loan with a 20% Down-Payment,

EOY	Savings	Depr.	Payments Principal	Interest	Total	Principal Outstanding	Taxable Income	Tax	ATCF
0					500,000	2,000,000			−500,000
1	950,000	357,250	302,567	280,000	582,567	1,697,433	312,750	106,335	261,098
2	950,000	612,250	344,926	237,641	582,567	1,352,507	100,109	34,037	333,396
3	950,000	437,250	393,216	189,351	582,567	959,291	323,399	109,956	257,477
4	950,000	312,250	448,266	134,301	582,567	511,0241	503,449	171,173	196,260
5	950,000	111,625	511,024	71,543	582,567	0	766,832	260,723	106,710
5*	1,200,000	669,375					530,625	180,413	1,019,588
	2,500,000								

Net Present Value at 18%: $710,962

Notes: Loan Amount: 2,000,000 (used to purchase equipment at year 0)
Loan Finance Rate: 14%   MARR   18%
500,000   Tax Rate   34%

MACRS Depreciation for 7-Year Property, with half-year convention at EOY 5
Accounting Book Value at end of year 5:   669,375
Estimated Market Value at end of year 5:   1,200,000
EOY 5* illustrates the Equipment Sale and *Book* Value
Taxable Income: =(Market Value - Book Value)
=(1,200,000 - 669,375) = $530,625

for its market value of $1,200,000. Tables 25-4 and 25-5 show the economic analysis for loans with a zero down payment and a 20% down payment, respectively. Assume that the bank reduces the interest rate to 14% for the loan with the 20% down payment. Since the asset is listed on PizzaCo's balance sheet, PizzaCo can use depreciation benefits to reduce the after-tax cost. In addition, all loan interest expenses are tax-deductible.

25.4.3 Bonds

Bonds are very similar to loans; a sum of money is borrowed and repaid with interest over a period of time. The primary difference is that with a bond, the issuer (PizzaCo) periodically pays the investors only the interest earned. This periodic payment is called the "coupon interest payment." *For example, a $1,000 bond with a 10% coupon will pay $100 per year. When the bond matures, the issuer returns the face value ($1,000) to the investors.*

Bonds are issued by corporations and government entities. Government bonds generate tax-free income for investors; thus, these bonds can be issued at lower rates than corporate bonds. This benefit provides government facilities an economic advantage to use bonds to finance projects.

25.4.3.1 Application to the Case Study

Although PizzaCo (a private company) would not be able to obtain the low rates of a government bond, it could issue bonds with coupon interest rates competitive with the loan interest rate of 15%.

In this arrangement, PizzaCo receives the investors' cash (bond par value) and purchases the equipment. PizzaCo uses part of the energy savings to pay the coupon interest payments to the investors. When the bond matures, PizzaCo must then return the par value to the investors. See Figure 25-8.

As with a loan, PizzaCo owns, maintains, and depreciates the equipment throughout the project's life. All coupon interest payments are tax-deductible. At the end of the five-year project, PizzaCo expects to sell the equipment for its market value of $1,200,000. Table 25-6 shows the economic analysis of this finance arrangement.

25.4.4 Selling Stock

Although less popular, selling company stock is an equity financing option that can raise capital for projects. For the host, selling stock offers a flexible repayment schedule, because dividend payments to shareholders aren't absolutely mandatory. Selling stock is also often used to help a company attain its desired capital structure. However, selling new shares of stock dilutes the power of existing shares and may send an inaccurate "signal" to investors about the company's financial strength. If the company is selling stock, investors may

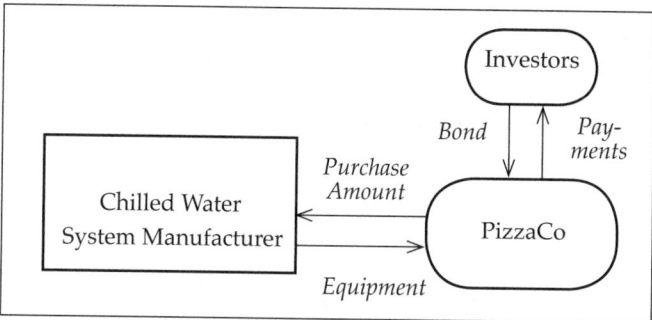

Figure 25-8. Resource Flow Diagram for a Bond.

think that it is desperate for cash and in a poor financial condition. Under this belief, the company's stock price could decrease. However, recent research indicates that when a firm announces an EMP, investors react favorably.[11] On average, stock prices were shown to increase by 21.33%, independent of overall market growth.

By definition, the cost of capital (rate) for selling stock is:

$$\text{cost of capital}_{\text{selling stock}} = D/P$$

where D = annual dividend payment

P = company stock price

However, in most cases, the after-tax cost of capital for selling stock is higher than the after-tax cost of debt financing (using loans, bonds and capital leases). This is because interest expenses (on debt) are tax deductible,

but dividend payments to shareholders are not.

In addition to tax considerations, there are other reasons why the cost of debt financing is less than the financing cost of selling stock. Lenders and bond buyers (creditors) will accept a lower rate of return, because they are in a less risky position due to the reasons below.

- Creditors have a contract to receive money at a certain time and future value. (Stockholders have no such guarantee with dividends.)

- Creditors have first claim on earnings. (Interest is paid before shareholder dividends are allocated.)

- Creditors usually have secured assets as collateral and have first claim on assets in the event of bankruptcy.

Despite the high cost of capital, selling stock does have some advantages. This arrangement does not bind the host to a rigid payment plan (like debt financing agreements), because dividend payments are not mandatory. The host has control over when it will pay dividends. Thus, when selling stock, the host receives greater payment flexibility, but at a higher cost of capital.

25.4.4.1 Application to the Case Study

As Figure 25-9 shows, the financial arrangement is very similar to a bond; at year zero the firm receives $2.5

Table 25-6. Economic Analysis for a Bond.

EOY	Savings	Depr.	Principal	Payments Interest	Total	Principal Outstanding	Taxable Income	Tax	ATCF
0						2,500,000			
1	950,000	357,250		375,000	375,000	2,500,000	217,750	74,035	500,965
2	950,000	612,250		375,000	375,000	2,500,000	-37,250	-12,665	587,665
3	950,000	437,250		375,000	375,000	2,500,000	137,750	46,835	528,165
4	950,0 0	312,250		375,000	375,000	2,500,000	262,750	89,335	485,665
5	950,000	111,625	2,500,000	375,000	2,875,000	0	463,375	157,548	-2,082,548
5*	1,200,000	669,375					530,625	180,413	1,019,588
		2,500,000							

Net Present Value at 18%: 953,927

Notes: Loan Amount: 2,500,000 (used to purchase equipment at year 0)

Loan Finance Rate: 0%   MARR   18%

Tax Rate   34%

MACRS Depreciation for 7-Year Property, with half-year convention at EOY 5

Accounting Book Value at end of year 5:   669,375

Estimated Market Value at end of year 5:   1,200,000

EOY 5* illustrates the Equipment Sale and *Book* Value

Taxable Income: =(Market Value - Book Value)

=(1,200,000 - 669,375) = $530,625

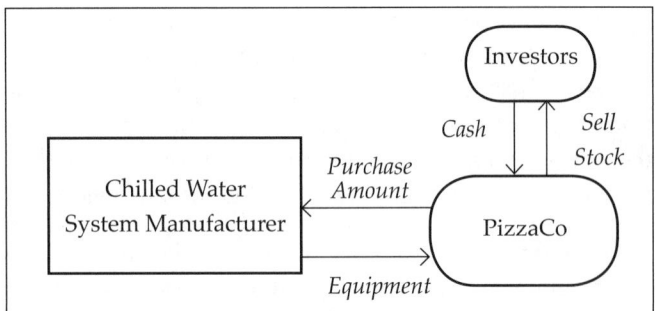

Figure 25-9. Resource Flow Diagram for Selling Stock.

million, except the funds come from the sale of stock. Instead of coupon interest payments, the firm distributes dividends. At the end of year five, PizzaCo repurchases the stock. Alternatively, PizzaCo could capitalize the dividend payments, which means setting aside enough money so that the dividends could be paid with the interest generated.

Table 25-7 shows the economic analysis for issuing stock at a 16% cost of equity capital, that repurchasing the stock at the end of year five. (For consistency of comparison to the other arrangements, the stock price does not change during the contract.) Like a loan or bond, PizzaCo owns and maintains the asset. Thus, the annual savings are only $950,000. PizzaCo pays annual dividends worth $400,000. At the end of year 5, PizzaCo

expects to sell the asset for $1,200,000.

Note that Table 25-7 is slightly different from the other tables in this chapter:

Taxable Income = Savings – Depreciation, and
ATCF = Savings – Stock Repurchases – Dividends – Tax

25.4.5 Leases

Firms generally own assets; however, it is the use of these assets that is important, not the ownership. Leasing is another way of obtaining the use of assets. There are numerous types of leasing arrangements, ranging from basic rental agreements to extended payment plans for purchases. Leasing is used for nearly one-third of all equipment utilization.[12] Leases can be structured and approved very quickly, even within 48 hours. Table 25-8 lists some additional reasons why leasing can be an attractive arrangement for the lessee.

Basically, there are two types of leases: the "true lease" (a.k.a. "operating" or "guideline lease") and the "capital lease." One of the primary differences between a true lease and a capital lease is the tax treatment. In a true lease, the lessor owns the equipment and receives the depreciation benefits. However, the lessee can claim the entire lease payment as a tax-deductible business expense. In a capital lease, the lessee (PizzaCo) owns and depreciates the equipment. However, only the in-

Table 25-7. Economic Analysis of Selling Stock.

EOY	Savings	Depr.	Stock Transactions		Dividend	Taxable	Tax	ATCF
			Sale of Stock	Repurchase	Payments	Income		
0			$2,500,000 from Stock Sale is used to purchase equipment, thur ATCF = 0					
1	950,000	357,250			400,000	592,750	201,535	348,465
2	950,000	612,250			400,000	337,750	114,835	435,165
3	950,000	437,250			400,000	512,750	174,335	375,665
4	950,000	312,250			400,000	637,750	216,835	333,165
5	950,000	111,625		2,500,000	400,000	838,375	285,048	2,235,048
5*	1,200,000	669,375				530,625	180,413	1,019,588
		2,500,000						

Net Present Value at 18%:   477,033

Notes: Value of Stock Sold (which is repurchased after year 5) 2,500,000 (used to purchase equipment at year 0)
      Cost of Capital = Annual Dividend Rate: 16%   MARR = 18%   Tax Rate = 34%
   MACRS Depreciation for 7-Year Property, with half-year convention at EOY 5
   Accounting Book Value at end of year 5: 669,375
   Estimated Market Value at end of year 5: 1,200,000
   EOY 5* illustrates the Equipment Sale and Book Value
       Taxable Income: = (Market Value - Book Value)
               = (1,200,000 - 669,375) = $530,625

Table 25-8. Good Reasons to Lease.

Financial Reasons

- With some leases, the entire lease payment is tax-deductible.
- Some leases allow "off-balance sheet" financing, preserving credit lines.

Risk Sharing

- Leasing is good for short-term asset use, and it reduces the risk of getting stuck with obsolete equipment.
- Leasing offers less risk and responsibility.

terest portion of the lease payment is tax-deductible. In general, a true lease is effective for a short-term project where the company does not plan to use the equipment when the project ends. A capital lease is effective for long-term equipment.

25.4.5.1 The True Lease

Figure 25-10 illustrates the legal differences between a true lease and a capital lease.[13] A true lease (or operating lease) is strictly a rental agreement. The word "strict" is appropriate, because the Internal Revenue Service will only recognize a true lease if it satisfies all of the following criteria:

1. The lease period must be less than 80% of the equipment's life.

2. The equipment's estimated residual value must be $\geq 20\%$ of its value at the beginning of the lease.

3. There is no "bargain purchase option."

4. There is no planned transfer of ownership.

5. The equipment must not be custom-made and only useful in a particular facility.

25.4.5.2 Application to the Case Study

It is unlikely that PizzaCo could find a lessor that would be willing to lease a sophisticated chilled water system and, after five years, move the system to another facility. Thus, obtaining a true lease would be unlikely. However, Figure 25-11 shows the basic relationship between the lessor and lessee in a true lease. A third-party leasing company could also be involved by purchasing the equipment and leasing to PizzaCo. Such a resource flow diagram is shown for the capital lease.

Table 25-9 shows the economic analysis for a true lease. Notice that the lessor pays the maintenance and

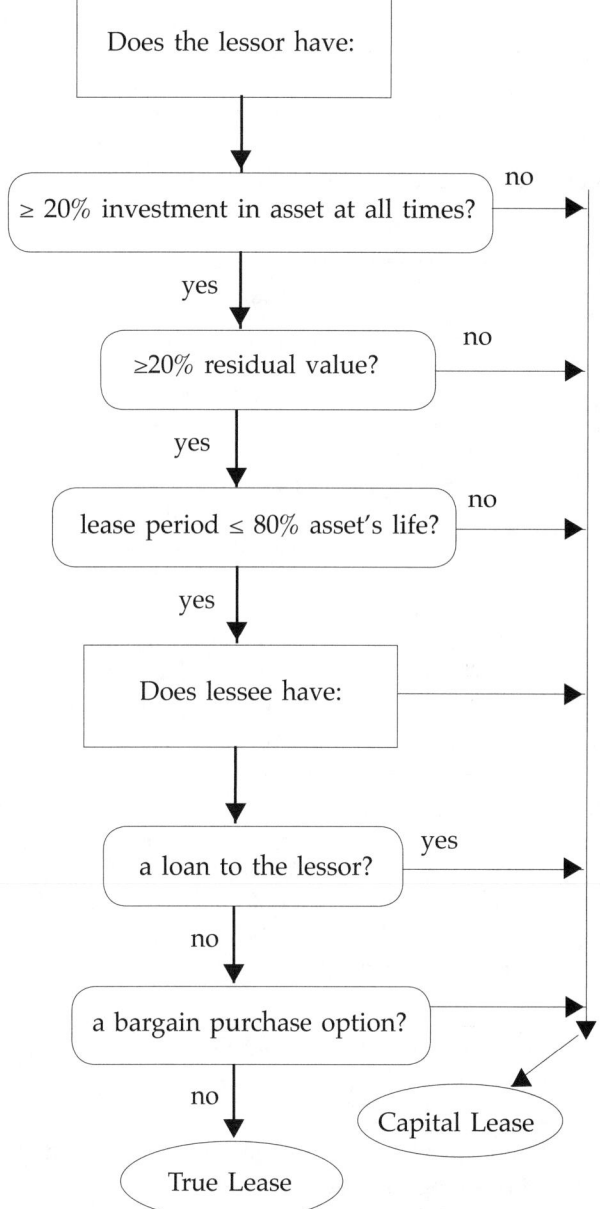

Figure 25-10. Classification for a True Lease.

insurance costs, so PizzaCo saves the full $1 million per year. PizzaCo can deduct the entire lease payment of $400,000 as a business expense. However PizzaCo does not obtain ownership, so it can't depreciate the asset.

25.4.5.3 The Capital Lease

The capital lease has a much broader definition than a true lease. A capital lease fulfills any one of the following criteria:

1. The lease term $\geq 80\%$ of the equipment's life.
2. The present value of the lease payments $\geq 80\%$ of the initial value of the equipment.
3. The lease transfers ownership.

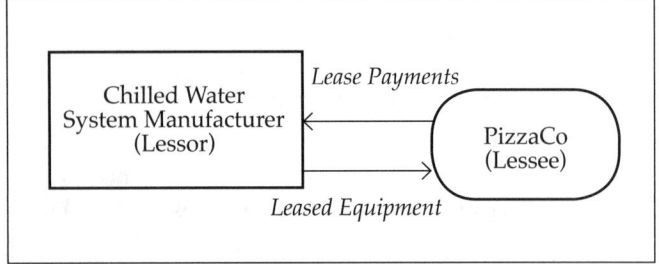

Figure 25-11. Resource Flow Diagram for a True Lease.

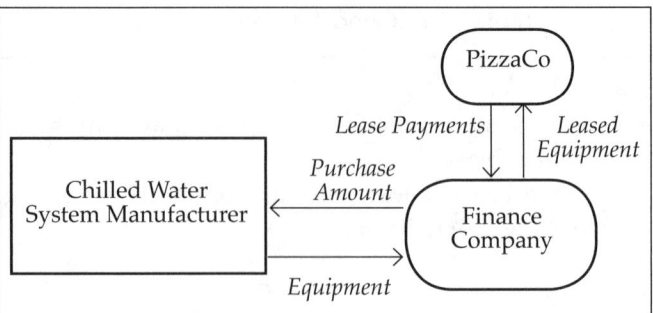

Figure 25-12. Resource Flow Diagram for a Capital Lease.

Table 25-9 Economic Analysis for a True Lease

EOY	Savings	Depr.	Lease Payments	Total	Principal Outstanding	Taxable Income	Tax	ATCF
0			400,000	400,000		-400,000		-400,000
1	1,000,000	0	400,000	400,000		600,000	204,000	396,000
2	1,000,000	0	400,000	400,000		600,000	204,000	396,000
3	1,000,000	0	400,000	400,000		600,000	204,000	396,000
4	1,000,000	0	400,000	400,000		600,000	204,000	396,000
5	1,000,000	0				1,000,000	340,000	660,000
					Net Present Value at 18%:			$953,757

Notes:	Annual Lease Payment:	400,000
	MARR = 18%	
	Tax Rate 34%	

4. The lease contains a "bargain purchase option." negotiated at the inception of the lease.

Most capital leases are basically extended payment plans, except ownership is usually not transferred until the end of the contract. This arrangement is common for large EMPs, because the equipment (such as a chilled water system) is usually difficult to reuse at another facility. With this arrangement, the lessee eventually pays for the entire asset (plus interest). In most capital leases, the lessee pays the maintenance and insurance costs.

The capital lease has some interesting tax implications, because the lessee must list the asset on its balance sheet from the beginning of the contract. Thus, like a loan, the lessee gets to depreciate the asset, and only the interest portion of the lease payment is tax deductible.

25.4.5.4 Application to the Case Study
Figure 25-12 shows the basic third-party financing relationship between the equipment manufacturer, les-sor, and lessee in a capital lease. The finance company (lessor) is shown as a third party, although it also could be a division of the equipment manufacturer. Because the finance company (with excellent credit) is involved, a lower cost of capital (12%) is possible due to reduced risk of payment default.

Like an installment loan, PizzaCo's lease payments cover the entire equipment cost. However, the lease payments are made in advance. Because PizzaCo is considered the owner, it pays the $50,000 annual maintenance expenses, which reduces the annual savings to $950,000. PizzaCo receives the benefits of depreciation and tax-deductible interest payments. To be consistent with the analyses of the other arrangements, PizzaCo would sell the equipment at the end of the lease for its market value. Table 25-10 shows the economic analysis for a capital lease.

25.4.5.5 The Synthetic Lease
A synthetic lease is a "hybrid" lease that combines aspects of a true lease and a capital lease. Through careful structuring and planning, the synthetic lease appears

Table 25-10. Economic Analysis for a Capital Lease.

EOY	Savings	Depr.	Payments Principal	Interest	Total	Principal Outstanding	Taxable Income	Tax	ATCF
0			619,218	0	619,218	1,880,782		-619,218	
1	950,000	357,250	393,524	225,694	619,218	1,487,258	367,056	124,799	205,983
2	950,000	612,250	440,747	178,471	619,218	1,046,511	159,279	54,155	276,627
3	950,000	437,250	493,637	125,581	619,218	552,874	387,169	131,637	199,145
4	950,000	312,250	552,874	66,345	619,218	0	571,405	194,278	136,503
5	950,000	111,625					838,375	285,048	664,953
5*	1,200,000	669,375					530,625	180,413	1,019,588

2,500,000

Net Present Value at 18%: $681,953

Notes: Total lease amount: 2,500,000

However, since the payments are in advance, the first payment is analogous to a downpayment

Thus the actual amount borrowed is only = 2,500,000 – 619,218 = 1,880,782

Lease finance rate: 12%   MARR 18%   Tax Rate 34%

MACRS depreciation for 7-year property, with half-year convention at EOY 5

Accounting book value at end of year 5: 669,375

Estimated market value at end of year 5: 1,200,000

EOY 5* illustrates the equipment sale and book value

Taxable income: = (Market Value – Book Value)

= (1,200,000 – 669,375) = $530,625

as an operating lease for accounting purposes (enables the host to have off-balance sheet financing), yet also appears as a capital lease for tax purposes (to obtain depreciation for tax benefits). Consult your local financing expert to learn more about synthetic leases; they must be carefully structured to maintain compliance with the associated tax laws.

With most types of leases, loans, and bonds the monthly payments are fixed, regardless of the equipment's utilization or performance. However, shared savings agreements can be incorporated into certain types of leases.

25.4.6 Performance Contracting

Performance contracting can be an arrangement that allows the building owner to make necessary improvements while investing very little money upfront. The contractor usually assumes responsibility for purchasing and installing the equipment, as well as maintenance throughout the contract. But the unique aspect of performance contracting is that the contractor is responsible for the performance of the installed equipment. In some cases, only after the installed equipment actually reduces expenses does the contractor get paid. Energy service companies (ESCOs) typically serve as contractors within this line of business.

Unlike most loans, leases, and other fixed payment arrangements, the ESCO is responsible for the performance of the equipment. In other words, if the finished product doesn't save energy or operational costs, the host doesn't pay. This aspect removes the incentive to "cut corners" on construction or other phases of the project. as with bid/spec contracting. In fact, often there is an incentive to exceed savings estimates. For this reason, performance contracting usually entails a more "facility-wide" scope of work (to find extra energy savings) than loans or leases on particular pieces of equipment.

With a facility-wide scope, many improvements can occur at the same time. For example, lighting and air conditioning systems can be upgraded at the same time. In addition, the indoor air quality can be improved. With a comprehensive facility management approach, a "domino-effect" on cost reduction is possible. For example, if facility improvements create a safer and higher quality environment for workers, productivity could increase. As a result of decreased employee absenteeism, the workman's compensation cost could also be reduced. These are additional benefits to the facility.

Depending on the host's capability to manage the risks (equipment performance, financing, etc.), the host will delegate some of these responsibilities to the ESCO. In general, the amount of risk assigned to the ESCO is directly related to the percent savings that must be shared with the ESCO.

For facilities that are not in a good position to manage the risks of an energy project, performance contracting may be the only economically feasible implementation method. *For example, the US federal government used performance contracting to upgrade facilities when budgets were being dramatically cut. In essence, they "sold" some of their future energy savings to an ESCO in return for receiving new equipment and efficiency benefits.*

In general, performance contracting may be the best option for facilities that:

- are severely constrained by their cash flows;

- have a high cost of capital;

- don't have sufficient resources (such as a lack of in-house energy management expertise or an inadequate maintenance capacity*);

- are seeking to reduce in-house responsibilities and focus more on their core business objectives; or

- are attempting a complex project with uncertain reliability or a host not fully capable of managing the project. *For example, a lighting retrofit has a high probability of producing the expected cash flows, whereas a completely new process does not have the same "time-tested" reliability. If the in-house energy management team cannot manage this risk, performance contracting may be an attractive alternative.*

Performance contracting does have some drawbacks. In addition to sharing the savings with an ESCO, the tax benefits of depreciation and other economic benefits must be negotiated. Whenever large contracts are involved, there is reason for concern. One study found that 11% of customers considering EMPs felt that dealing with an ESCO was too confusing or complicated.[14] Another reference claims, "with complex contracts, there may be more options and more room for error."[15] Therefore, it is critical to choose an ESCO with a good reputation and experience within the types of facilities that are involved.

There are a few common types of contracts. The ESCO will usually offer the following options:

- guaranteed fixed dollar savings or "avoided costs"

- guaranteed fixed energy (MMBtu) savings
- a percent of energy savings
- a combination of the above

Obviously, some facility managers would prefer the options with "guaranteed savings." However, this extra security (and risk to the ESCO) usually costs more. The primary difference between the two guaranteed options is that guaranteed fixed dollar savings contracts ensure dollar savings, even if energy prices fall. *For example, if energy prices drop and the equipment does not save as much money as predicted, the ESCO must pay (out of its own pocket) the contracted savings to the host.*

Percent energy savings contracts are agreements that basically share energy savings between the host and the ESCO. The more energy saved, the higher the revenues to both parties. However, the host has less predictable savings and must also periodically negotiate with the ESCO to determine "who saved what" when sharing savings. There are numerous hybrid contracts available that combine the positive aspects of the above options.

25.4.6.1 Application to the Case Study

PizzaCo would enter into a hybrid contract, *percent energy savings/guaranteed arrangement.* The ESCO would purchase, install and operate a highly efficient chilled water system. The ESCO would guarantee that PizzaCo would save the $1,000,000 per year, but PizzaCo would pay the ESCO 80% of the savings. In this way, PizzaCo would not need to invest any money and would simply collect the net savings of $200,000 each year. To avoid periodic negotiations associated with shared savings agreements, the contract could be worded such that the ESCO will provide guaranteed energy savings worth $200,000 each year. Additional savings would be kept by PizzaCo.

With this arrangement, there is no depreciation, interest payment or tax-benefit for PizzaCo. However, PizzaCo receives a positive cash flow with no investment and little risk. At the end of the contract, the ESCO removes the equipment. (At the end of most performance contracts, the host usually acquires or purchases the equipment for fair market value. However, for this case study, the equipment was removed to make a consistent comparison with the other financial arrangements.)

Figure 25-13 illustrates the transactions between the parties. Table 25-11 presents the economic analysis for performance contracting.

Note that Table 25-11 is slightly different from the other tables in this chapter: Taxable Income = Savings – Depreciation – ESCO Payments.

---

*Maintenance capacity represents the ability that the maintenance personnel will be able to maintain the new system. It has been shown that systems fail and are replaced when maintenance concerns are not incorporated into the planning process. See Woodroof, E. (1997) "Lighting Retrofits: Don't Forget About Maintenance," Energy Engineering, 94(1) pp. 59-68.

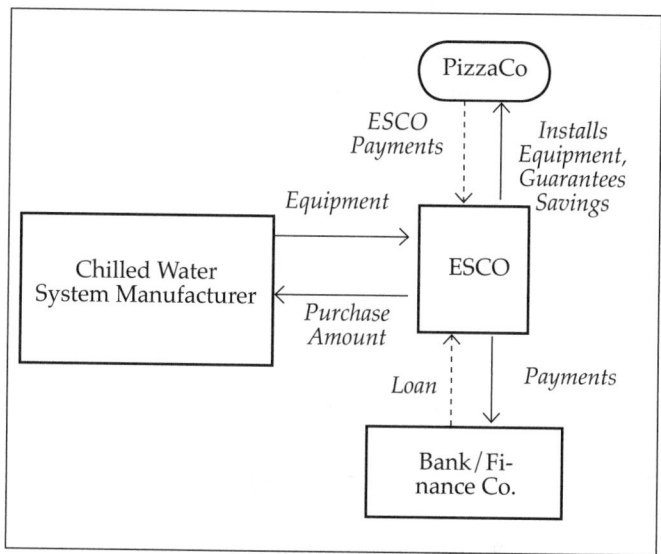

Figure 25-13. Transactions for a Performance Contract.

25.4.7 Summary Of Tax Benefits

Table 25-12 summarizes the tax benefits of each financial arrangement presented in this chapter.

25.4.8 Additional Options

Combinations of the basic financial arrangements can be created to enhance the value of a project. A sample of the possible combinations are described below.

- Third party financiers often cooperate with performance contracting firms to implement EMPs.

- Utility rebates and government programs may provide additional benefits for particular projects.

- Tax-exempt leases are available to government facilities.

- Insurance can be purchased to protect against risks relating to equipment performance, energy savings, etc.

- Some financial arrangements can be structured as non-recourse to the host. Thus, the ESCO or lessor would assume the risks of payment default. However, as mentioned before, profit sharing increases with risk sharing.

Attempting to identify the absolute best financial arrangement is a rewarding goal, unless it takes too long. As every minute passes, potential dollar savings are lost forever. When considering special grant funds, rebate programs or other unique opportunities, it is important to consider the lost savings due to delay.

25.5 "PROS" & "CONS" OF EACH FINANCIAL ARRANGEMENT

This section presents a brief summary of the "pros" and "cons" of each financial arrangement from the host's perspective.

Loan

Pros:

- Host keeps all savings.
- Depreciation and interest payments are tax-deductible.
- Host owns the equipment.
- The arrangement is good for long-term use of equipment.

Cons:

- Host takes all the risk and must install and manage project.

Table 25-11. Economic Analysis of a Performance Contract.

EOY	Savings	Depr.	ESCO Payments	Total	Principal Outstanding	Taxable Income	Tax	ATCF
0								
1	1,000,000	0	800,000	800,000		200,000	68,000	132,000
2	1,000,000	0	800,000	800,000		200,000	68,000	132,000
3	1,000,000	0	800,000	800,000		200,000	68,000	132,000
4	1,000,000	0	800,000	800,000		200,000	68,000	132,000
5	1,000,000	0	800,000	800,000		200,000	68,000	132,000
					Net Present Value at 18%:			$412,787

Notes: ESCO purchases/operates equipment. Host pays ESCO 80% of the savings = $800,000.
The contract could also be designed so that PizzaCo can buy the equipment at the end of year 5.

Table 25-12. Host's Tax Benefits for each Arrangement.

ARRANGEMENT	Depreciation Benefits	Interest Payments are Tax-Deductible	Total Payments are Tax-Deductible
Retained Earnings	X		
Loan	X	X	
Bond	X	X	
Sell Stock	X		
Capital Lease	X	X	
True Lease			X
Performance Contract			X

Bond

Has the same pros/cons as loan, and:

Pro:

- Good for government facilities, because they can offer a tax-free rate (that is lower, but considered favorable by investors)

Sell Stock

Has the same Pros/Cons as loan, and

Pro:

- Selling stock could help the host achieve its target capital structure.

Con:

- Dividend payments (unlike interest payments) are not tax-deductible.
- Dilutes company control.

Use Retained Earnings

Has the same pros/cons as loan and:

Pro:

- Host pays no external interest charges. However, retained earnings do carry an opportunity cost, because such funds could be invested somewhere at the MARR.

Con:

- Host loses tax-deductible benefits of interest charges.

Capital Lease

Has the same pros/cons as loan, and:

Pro:

- Greater flexibility in financing (possible lower cost of capital with third-party participation).

True Lease

Pros:

- Allows use of equipment, without ownership risks.

- Reduced risk of poor performance, service, equipment obsolescence, etc.,
- Good for short-term use of equipment.
- An entire lease payment is tax-deductible.

Cons:

- No ownership at end of lease contract.
- No depreciation tax benefits.

Performance Contract

Pros:

- Allows use of equipment, with reduced installment/operational risks.
- Reduced risk of poor performance, service, equipment obsolescence, etc.
- Allows host to focus on its core business objectives.

Cons:

- Potentially binding contracts, legal expenses, and increased administrative costs.
- Host must share project savings.

25.5.1 Rules of Thumb

When investigating financing options, consider the following generalities:

Loans, bonds and other host-managed arrangements should be used when a customer has the resources (experience, financial support, and time) to handle the risks. Performance contracting (ESCO assumes most of the risk) is usually best when a customer doesn't have the resources to properly manage the project. Remember that with any arrangement where the host delegates risk to another firm, the host must also share the savings.

Leases are the "middle ground" between owning and delegating risks. Leases are very popular due to their tax benefits.

True leases tend to be preferred when:

- The equipment is needed on a short-term basis.
- The equipment has unusual service problems that cannot be handled by the host.
- Technological advances cause equipment to become obsolete quickly.
- Depreciation benefits are not useful to the lessee.

Capital Leases are preferred when:

- The installation and removal of equipment is costly.
- The equipment is needed for a long time.
- The equipment user desires to secure a "bargain purchase option."

25.6 CHARACTERISTICS THAT INFLUENCE WHICH FINANCIAL ARRANGEMENT IS BEST

There are at least three types of characteristics that can influence which financial arrangement should be used for a particular EMP. These include facility characteristics, project characteristics, and financial arrangement characteristics. In this section, quantitative characteristics are bulleted with the symbol "$." The qualitative characteristics are bulleted with the symbol ☺. Note that qualitative characteristics are generally "strategic" and are not associated with an exact dollar value.

A few of the facility characteristics include:

☺ The long-term plans of facility. For example, is the facility trying to focus on core business objectives and outsourcing other tasks, such as EMPs?

$ The facility's current financial condition. Credit ratings and ability to obtain loans can determine whether certain financial arrangements are feasible.

☺ The experience and technical capabilities of in-house personnel. Will additional resources (personnel, consultants, technologies, etc.) be needed to successfully implement the project?

$ The facility's ability to obtain rebates from the government, utilities, or other organizations. For example, there are Dept. of Energy subsidies available for DOE facilities.

$ The facility's ability to obtain tax benefits. For example, government facilities can offer tax-exempt interest rates on bonds.

A few of the Project Characteristics include:

$ The project's economic benefits: net present value, internal rate of return and simple payback.

☺ The project's complexity and overall risk. For example, a complex project that has never been done before has a different level of risk than a standard lighting retrofit.

☺ The project's alignment with the facility's long-term objectives. Will this project's equipment be needed for long-term goals?

☺ The project's cash flow schedule and the variance between cash flows. For example, there may be significant differences in the acceptability of a project based on when revenues are received.

A few of the financial arrangement characteristics include:

$ The economic benefit of a project using a particular financial arrangement. The net present value and internal rate of return can be influenced by the financial arrangement selected.

☺ The impact on the corporate capital structure. For example, will additional debt be required to finance the project? Will additional liabilities appear on the firm's balance sheet and impact the image of the company to investors?

☺ The flexibility of the financial arrangement. For example, can the facility manager alter the contract and payment terms in the event of revenue shortfall or changes in operational hours?

25.7 INCORPORATING STRATEGIC ISSUES WHEN SELECTING FINANCIAL ARRANGEMENTS

Because strategic issues can be important when selecting financial arrangements, the facility manager should include them in the selection process. The following questions can help assess a facility manager's needs.

- Does the facility manager want to manage projects or outsource?
- Are net positive cash flows required?
- Will the equipment be needed for long-term needs?

- Is the facility government or private?
- If private, does the facility manager want the project's assets on or off the balance sheet?
- Will operations be changing?

From the research experience, a Strategic Issues Financing Decision Tree was developed to guide facility managers to the financial arrangement that is most likely optimal. Figure 25-14 illustrates the decision tree, which is by no means a rule but embodies some general observations from the industry.

Working the tree from the top to bottom, the facility manager should assess the project and facility characteristics to decide whether it is strategic to manage the project or outsource it. If outsourced, the performance contract would be the logical choice.* If the facility manager wants to manage the project, the next step (moving down the tree) is to evaluate whether the project's equipment will be needed for long or short-term purposes. If short-term, the true lease is logical. If it is a long-term project, in a government facility, the bond is likely to be the best option. If the facility is in the private sector,

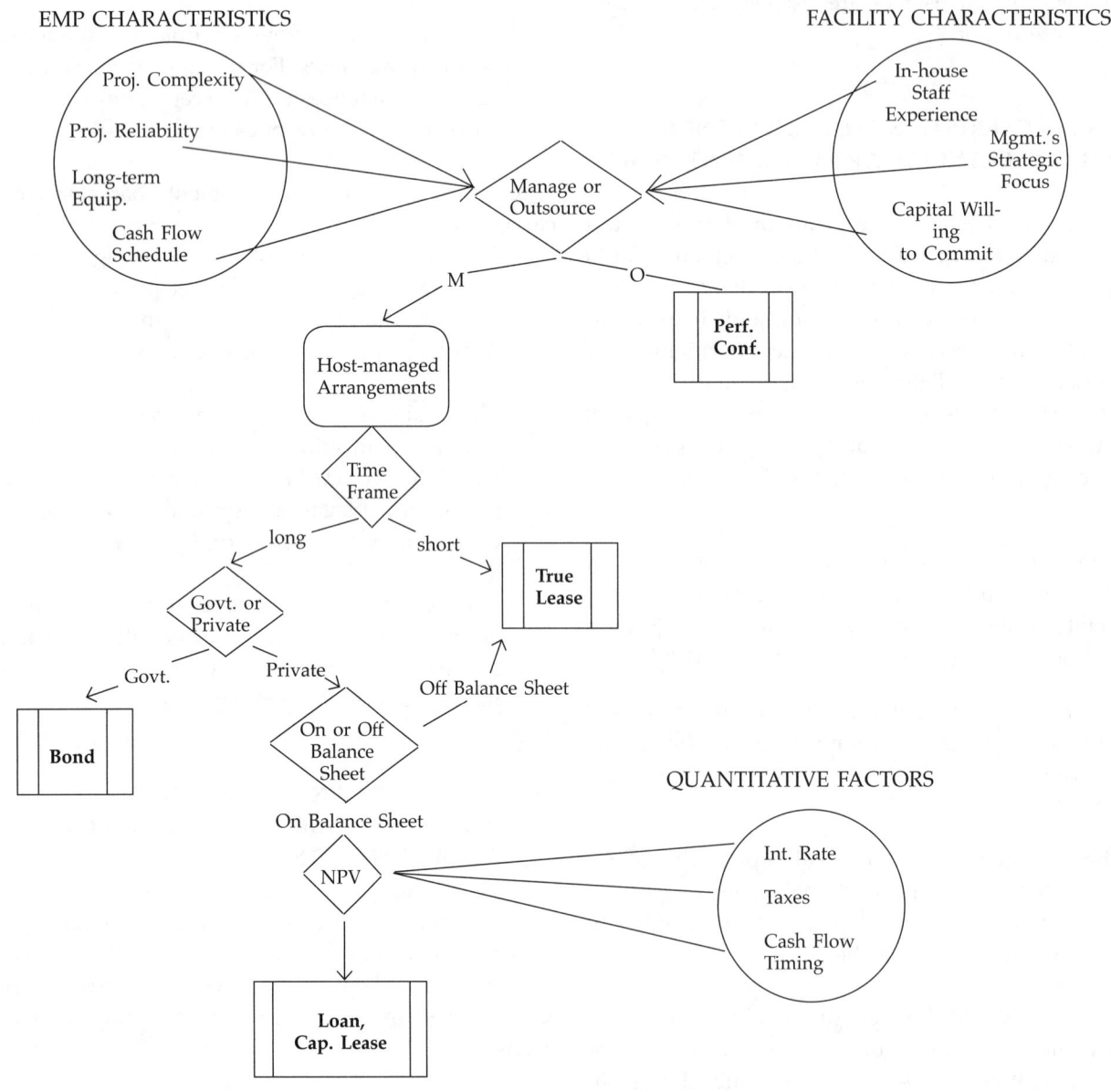

Figure 25-14. Strategic Issues Financing Decision Tree.

---

*It should be noted that a performance contract could be structured using leases and bonds.

the facility manager should decide whether the project should be on or off the balance sheet. An off-balance sheet preference would lead back to the true lease. If the facility manager wants the project's assets on the balance sheet, the net present value (or other economic benefit indicator) can help determine which host-managed arrangement (loan, capital lease, or cash) would be most lucrative.

25.8 PERSPECTIVES OF OWNERS AND PERFORMANCE CONTRACTORS... BOTH SIDES OF THE STORY

Strategic considerations from the owner's perspective for performance contracting success.

1. Loan payback period should not exceed the life of the equipment or systems prompting the loan. This would be analogous to a 20-year loan on an automobile purchase. You do not want to end up still paying for it when it is gone. Tables of expected equipment life can be provided for your use.

2. If the guaranteed savings term is much shorter than the contract total term length (e.g. 3 years guaranteed vs. 15 years total), there is a risk to the owner; if savings degrade, the cash flow will look different and unanticipated costs may arise. If the contract is designed to be "revenue-neutral." then any year when the savings do not come to pass will constitute a new bill, for which there needs to a be a method of payment. To guard against this, encourage the ESCO to provide conservative savings estimates so that the actual savings are likely to be better than the contract stipulations, or build-in a degradation factor on the savings that increases over time. Some of the degradation could come from lack of maintenance, so identify maintenance expectations and commit to these.

3. If design defects are identified during the guarantee period, this can cost the owner money over the life of the loan after the guarantee time is up. Unless the guaranteed savings extends for the full loan term, this is a risk to the owner. A contract stipulation can guard against this by providing for long term annual payment equivalent to the missed savings, with the option to do additional measures that create equal savings in lieu of a long term penalty payment.

4. Measurement and verification method options are many. To guard the bottom line, consider using actual service utility costs and IPMV Option C, adjusted for utility rate changes, adjusted for weather variations. This will eliminate controversy and excuses. It either saves money or it doesn't. This will help assure the "revenue neutral" project intent.

5. Any evaporative cooling considerations should include the water and waste costs. Depending on rates, these can erode half or more of the energy savings.

6. Savings from utility tariff changes should not be accepted for project "savings."

7. Contractually, the single biggest item to focus on is the burden of proof for savings. In advance of the work, there must be a clear and unambiguous definition of how the savings will be quantified and agreed upon, as well as the recourse for both sides if the savings are more or less than estimated. After the agreement is tentatively met, it is strongly suggested that a third party consultant review and comment on the arrangement. Without this firmly and comfortably in place, the PC delivery system may become adversarial.

8. Of all the construction delivery methods, performance contracting and design-build methods both rely on some non-traditional attributes. Design-build requires a significant level of trust between owner and vendor. Performance contracting, especially during the audit stage, requires a good comfort level between owner and vendor to discuss pros and cons of the plethora of options that are being explored. The vendor must be willing to teach and the owner must be willing to learn during this phase. The owner needs to fully comprehend the scope of services and payments and be comfortable asking questions at any time. For all these reasons, the partner chosen should include considerations of soft skills like communication and personality. For interviews, a stipulation should be made that the people who show up are the actual people that would do the work and that the contract may be void if this promise is not kept.

9. Bear in mind that the focus of the performance contractor with a guaranteed savings provision is to achieve the savings. This will be the prime objective in all activities, and other considerations, while not ignored, may be taken less seriously than the owner is used to. These linclude adherence to facility guidelines and preferences, reliability and redundancy, maintenance provisions, aesthetics, detailed design and documentation, and attention to related systems.)

10. For guaranteed savings contracts, the performance contractor essentially holds the checkbook and has a greater amount of control over the project. Cost containment is the other half of the savings equation, because it is the ratio of the two (cost and savings) that make the cash flow work out. Things like owner review, design comments, and general owner participation may be conceded along with the control of the project. In order to guarantee the savings and maintain the cash flow expectations, the ESCO will require the controls to achieve the stated savings objectives and to stay in budget.

11. Like all contracts, what is not said is as important as what is said. Find someone who has experience with this type of contract and get review comments from them.

Strategic considerations from the performance contractor's perspective for project success.

1. The customer must be around longer than the life of the loan period. Permanent institutions such as government buildings, schools, etc. are very good candidates. More volatile business segments, such as leased office buildings, are poor candidates.

2. The longer the contract terms, the higher the risk to the ESCO. Out of 10 projects that extend more than 10 years, the chances of at least one of them going "bad" are quite good, and the other 9 must absorb the loss for the ESCO to remain solvent. Margins are always proportional to risk, and thus the ESCO long term guaranteed savings margins will be high.

3. For the case of a customer wishing to miraculously find financing for old and expensive equipment, there must be significant energy savings in many areas to fund it if it is to be revenue neutral. Savings from energy improvements alone cannot usually justify normal equipment replacement costs, i.e. the car needs a steering wheel anyway. Energy savings has a good chance at showing financial merit for upgrades but not for whole projects. Thus, the lighting savings may be subsidizing the roof replacement, and a package proposal of work is the usual solution.

4. Unforeseen conditions always exist, and establishing a contingency is very helpful in preserving the original cash flow expectations and "riding through" the little things that come up.

5. Regardless of the energy estimating model, there are factors and dynamics that cannot be included and surprises will come up. Estimating conservatively is a good way to cover some of these uncertainties and project risks. Undersell and over-deliver. If one area saves more than advertised, it can cover for something else that does not perform as anticipated. It is the bottom line savings that matter most.

6. How equipment is operated and maintained has an effect on energy use. It is essential to establish the "rules" for O/M, such as indoor temperatures, hours of operation, and frequency of servicing—and then provide the necessary monitoring to assure the promises are kept. Excess run time, cranking thermostats up or down, changing reset schedules, dirty heat exchangers, etc. are items that, if the owner chooses to do, will affect savings at their expense. For this to be enforceable and fair, it must be easy to demonstrate that system performance (e.g. comfort) can be achieved using the assumption parameters and that the choices made by owner staff were not necessary. If the owner must make the changes just to make the system work, then it really isn't at their expense since the expectation, in addition to cash flow and financing, is that the new systems work.

7. M/V and baselines are the crux of the savings contract and need to be clearly defined. (Refer to chapter 27 measurement and verification of Energy Savings.) Baseline definitions must include provisions for adjusting up or down with external influences, like colder or warmer than usual seasonal weather, building additions, occupancy, production rates, etc.

8. Savings based on cost have the risk of volatile utility pricing. Savings based on energy use instead of cost are less volatile, but also less likely since the project premise is cost control. Agree on assumptions of utility cost and increases, including escalation, during the life of the loan period.

9. Projects will tend to be large and, except for work within large cities, will usually be remote. Remote project management introduces new risks from unfamiliar contractors, local codes, etc.

10. The project delivery is different from conventional design-bid-build and may be new to the customer. Newness will bring a lack of confidence and some apprehension, and project management should anticipate this. Thus, an educational component is part of the work. Building customer confidence with high skilled technical staff, combined with excellent interpersonal skills and regular customer contact, are effective. The proj-

ect is contractual but relies heavily on the built relationship to go smoothly and assure satisfaction, as well as a referral at the end.

11. For customers with large groups of stakeholders, in-house technical staff, etc. a challenge will be to identify project goals and limits and to condition expectations. A single point of customer contact is important. Excessive input from customer groups, what-if scenarios, etc. will increase project cost, but this requires a delicate balance. Turning this off with contract clauses and extra services should be a last resort since it may alienate the customer, but too much is also too much. Project management experience is needed here, and a conservative savings estimate can help to roll through some of the small things that the customer will appreciate and remember.

12. The guaranteed savings contract provision has the potential to create an adversarial relationship with the customer, which should be anticipated and avoided. Establishing clear cut rules, with owner buy-in on how it is determined who owes who what, are essential. Project management, interpersonal skills, and a conservative estimate with a little wiggle room are other tools, but they do not replace clear contract language that is, from the onset, fair to both parties.

APPENDIX

Overcoming the Three Main Barriers to Energy Efficiency or "Green" Projects

INTRODUCTION

Although the popularity of energy management and "green" projects is improving, there are MANY good projects that are postponed or cancelled due to common barriers. However, there are some common and cutting-edge strategies that leverage marketing, educational resources, and financing approaches to make your projects irresistible. The goal of this section is to help organizations get more "good" energy management/ green projects approved.

Why is this important? The polar ice is melting, and as far as the planet is concerned, engineers are wasting their time if the projects they so carefully develop are not implemented and deliver no value. This article refers to "good" projects as those with a 3 year payback or less. Why don't good projects get implemented? There are a variety of reasons and a few common barriers:

1. *Marketing* (under-marketing a project's value)

2. *Education & collaboration* (not expanding the value of a project)

3. *Money* (not having a positive cash flow solution)

If a project can't satisfy these criteria, it probably won't be implemented anyway... so focus on the ones that will!

PROBLEM #1: MARKETING

People often ask me why marketing is first on the list. Answer: Because NOTHING HAPPENS WITHOUT A SALE. For example, your first job (or your first date) began with you "selling yourself" on a resume or during an interview. In fact, the development of every product/service begins with someone selling a solution to some type of problem. Now I am not saying that selling/convincing is "bad" or un-ethical. Convincing someone when it improves their lives is good and can be done with passion. When something (like an energy management project) is great, we should sell the benefits **with all the passion in the world**. *You would do the same when talking to your kids about "getting a good education," or "learning good manners."* Passion can also emerge from fear, such as from the chaos and the violence that occurs during an electrical blackout. *Most of the time, humans are more passionate and action-oriented when they are at risk of losing something, versus gaining something.*

Thus, we must communicate in a way so the audience (the buyers or project approvers) can understand the problem/pain that they are in now. After they agree that they are "in pain," then they will want to hear about potential solutions.

Attention

It starts with getting the buyer's attention on the problem, the pain it is causing, and a sense of urgency to solve the problem. Only then will a solution seem to be logical. In addition, after they understand the problem/pain, they will be able to become passionate about the solution.

If you fail to get the attention of the approver, you are actually doing them a disservice; they won't know they are in trouble and wasting money. It is like allowing someone to bleed to death when they don't even know they are cut. So, don't be shy. You have a duty to perform.

Warning! Some approvers personalities' won't like to hear about problems/pain. Some approvers may "put their head in the sand" (like an ostrich) when problems are discussed. Don't blame them; it is their personality (which has strengths in other areas). Discover ways to communicate in a way that will prompt them to respond. FYI, it can take 7 impressions (explanations/presentations) before some people will agree on the problem and take action on a solution. Don't give up, and don't be surprised or depressed when they don't take action after the first impression.

Below are a few examples of effective headlines that can help get the attention of an approver. Feel free to use these in executive summaries:

- "How will the shareholders feel about us throwing money away every month?"

- "A way to make money while reducing emissions…"

- "What will we do with the yearly savings?"

- "We are paying for energy-efficiency projects, *whether or not* we do them!"

- "Guaranteed, high-yield investments…"

- "If you enjoy throwing money away every month, don't read this…"

- "4.6 billion years of reliability… solar energy"

- This project could improve our stock price by over 20%![1]

- "Good planets are hard to find."

There are many other great proven examples that are available.[2] However, you can experiment by looking around for "marketing copy" in magazine advertisements, commercials, etc. *There is a reason they call it "copy"; some of the principles are thousands of years old, and they still work!* Just change the words to relate to your problem/solution. Try a few versions and test, test, test to see which ones are most effective. Go for it!

Benefits[3]

After you have their attention, be sure that you include compelling benefits that "take away the pain" the audience is feeling. As engineers, we are good at mentioning the typical benefits:

- Saves energy, money, waste, and emissions.
- Offsets the cost of a planned capital project.
- Improves cost-competitiveness, productivity, etc.
- Is a relatively low-risk, high-profit investment that directly impacts the bottom line[4].

In today's green-minded economy, we could also demonstrate that "green" projects are a very effective marketing tool (which could get the client's marketing department behind your project), because these benefits have also been proven[5]:

- Improves the client's "green" image.
- Differentiates the client from the competition.[6]
- Introduces them to new markets, suppliers and clients.[7,8]
- *Helps them grow sales/revenue.*

However, we should also mention the *passionate, global, and moral* reasons behind a good green project:

- Slows global warming; reduces acid rain.
- Reduces mercury pollution, which allows us to eat healthy.
- Improves our national energy independence.
- Reduces security/disaster risks, etc.

Dollar values for these benefits can often be calculated and should be included in your proposal. To calculate the "green benefit equivalencies," such as "number of trees planted" (from reduced power plant emissions), see the "Money" section of this article.

The list above can be expanded, refined, and optimized for any project. To build a list like the one above, one technique is "WSGAT" (What is So Good About That?") Ask that question for every project feature, and you will develop a long list of passionate benefits. By the way, this approach has been used in TV sales and has helped sell billions of dollars of material[9]. *If they can sell that much junk on TV, we should have no problem selling green projects that are factually saving the planet!* Add the emotional benefits of "going green," and you will have a project that touches the hearts of leaders in your organization.

Call to Action

The call to action becomes easy and logical when all of the benefits have been quantified, and they are aligned with the client's strategic objectives. Tell the approvers what you want them to do and why. Be sure to include the "cost of delay" in your executive summary. Remind them that they are "in pain" and the project/solution will solve it. Visual aids can be helpful.

For example, during one presentation, buckets of dollars were shoveled out a window to demonstrate the losses that were occurring every minute. The executives were literally in pain watching those dollars fly away. They couldn't stand it, and they took action. It is OK to get creative and have some fun in your presentation!

BUT WAIT... THERE IS MORE!

"Configuring" your presentation can make the difference between immediate approval and further delay. There are many ways to configure or "package" your product/project so that it is IRRESISTABLE. One way to do this is to find a way to make a project's performance guaranteed or "risk-free." Another way is to separate (or add) one part of the project and introduce it as a "free bonus." Everyone likes a "FREE" bonus—it helps them understand that they are getting a good deal. *For example, on a recent green, facility-related project, carbon offsets for a company's fleet were included as a free bonus. The bonus delighted the client and differentiated the project (it was extra value), yet the additional costs were less than $1,000.*

Engineers can be two, three or ten times more productive by developing sales and marketing skills. However, there is another reason for developing these sales/marketing skills—your career! The skills you learn will be valuable to your organization, as well as to other organizations. These skills are transferable to other industries too. So keep this in mind when you are investing in yourself... There will almost always be a fantastic payoff.

Finally, there are two prerequisites that a buyer must see in you before any sale is made: "trust" and "value." As far as trust goes, it must be earned and once it is earned, it must be cherished. To accelerate the buyer's trust in you, be an advocate for the client and put their needs ahead of your own. Assume the role of their "most trusted advisor," and then deliver. Value comes from applying knowledge, tools, resources, partners, etc., in the best way for the client, which is why education and collaboration is such an important component of success. This is discussed further in the next section. Be sure to read the sub-sections on reciprocal business agreements, and joint ventures and incentives/rebates—great ideas!

PROBLEM #2: EDUCATION & COLLABORATION

Knowing how to deliver the value is an area that requires continuous updating. Today, with the proliferation of energy/green technologies, it is impossible for one person to know all the ways to add value to a project. Green specialties are expanding every day, for example: energy efficiency, water efficiency, green janitorial, LEED[10], recycling, transportation, etc.

Learn all you can, then collaborate with other professionals who are also actively learning, and the value available to your clients increases exponentially. It is important to be open to new ideas and fresh perspectives in this process. *"Mind-sharing" or brain-storming techniques can facilitate the process and maximize the number of useful ideas.*[11]

Fortunately, education is a low-cost investment. Collaboration and even joint-ventures/partnering can be done inexpensively as well, and the returns can be huge!

Free Sources of green/energy efficiency education include:

https://www.aeecenter.org/seminars
http://www.eere.energy.gov
http://www.ashrae.org/education
www.usgbc.org
www.ase.org
www.energystar.org
http://greeninginterior.doi.gov

In addition, there are many innovative ways to bring more value to a project. Some include:
* Reciprocal business agreements
* Joint ventures
* Free tax and utility incentives/rebates

Reciprocal Business Agreements

For example, after presenting a $1,000,000 service contract for a global car rental company, the deal was sweetened with an agreement on our part to choose that car company while traveling, which generated over $1,000,000 in extra car rentals for them. To the client, they were getting an extra $1,000,000 in revenue by working with us versus the competitors. *With what suppliers, partners, colleagues, professionals, etc. could you develop reciprocal business agreements? How could you help two clients (or a supplier) benefit from each other? How could you help them become more green?*

Another example: We helped client #1 supply green solutions to client #2. Both clients were extremely happy to generate more sales and save money. When it was time to approve our next round of projects, there was little resistance, because we had helped them earn/save far more than the costs of the proposed projects. This illustrates the value of being the "trusted advisor."

Joint Ventures

For example, a green travel agency gives 50% of its commissions back to its clients in exchange for their travel business.[12] The client can use this extra, free money to fund green initiatives, scholarships, or other social programs. The travel agency guarantees the lowest prices and easily doubles its business, because it delivers more value to its clients via joint ventures.

Free Tax and Utility Incentives/Rebates

For example, in California, 50% of a solar project was funded by federal and state rebates. Utility incentives lowered the installation costs even further. There are numerous free tax and utility incentives available, and some are discussed in the next section.

In addition to the options above, many utilities and third parties are offering "green power purchase agreements," which are essentially "wind and solar performance contracts." For example, if you want to put solar panels on your roof, a third party (often a utility or solar contractor) finances the project installation and then sells you the renewable energy produced from your roof (at a known price) for 15-25 years. So you get "green" power at no upfront cost and a known future energy cost (lowers your risk to energy price volatility). The financier wins because the project will pay back their investment within 10 years, and the rest is profit.

There are an unlimited number of creative "win-win" contracts available. However, before finalizing or even developing your solution, be sure that you understand the client's strategic and financial goals, then align the value to support the client's larger objectives.

PROBLEM #3: MONEY

If you do a good job tapping into the passion behind the project and are satisfying the emotional, financial, and other approval criteria, you should have enough benefits to get the project approved, especially if the project is above the client's MARR.[13] However, if your organization is capitally-constrained, you can finance a project and have positive cash flow. *CFOs like positive cash flow projects!* On the contrary, cash flow constraints (not having the upfront capital to install a project) represent over 35% of the reasons why projects are not implemented[14].

Financing does not have to be complicated. In fact, financing energy efficiency/green projects can be very similar to your mortgage or car payment, fixed payments for a length of time. However, with a good project, you can finance the project such that the annual savings are greater than the finance payments, which means the project becomes "cash flow positive" and does not impact the capital budget! This can allow the approver to move forward without sacrificing any other budget line item.

Table 1 shows the cash flow for a non-financed project[15]. Assume the project costs $100,000 and saves $28,000 per year for 15 years. This project could get approved *if the client has $100,000 in cash to fund it*. The project has a net present value of $102,700 and an internal rate of return of 27%.

Now, let's look at financing the project with a simple loan. Let's say the client finances the $100,000 for 15 years at 10% per year. That means instead of investing $100,000 upfront (the bank provides these funds), the client pays $13,147 each year to the bank for 15 years.

EOY	Savings	Cost	Cash Flow
0	-	(100,000)	$ (100,000)
1	28,000		$ 28,000
2	28,000		$ 28,000
3	28,000		$ 28,000
4	28,000		$ 28,000
5	28,000		$ 28,000
6	28,000		$ 28,000
7	28,000		$ 28,000
8	28,000		$ 28,000
9	28,000		$ 28,000
10	28,000		$ 28,000
11	28,000		$ 28,000
12	28,000		$ 28,000
13	28,000		$ 28,000
14	28,000		$ 28,000
15	28,000		$ 28,000
NPV i=10%			$102,700
IRR			27%

Table 1. Project Cash Flow (paid with Cash)

EOY	Savings	Finance Cost	Cash Flow
0	-	-	$ -
1	28,000	13,147	$ 14,853
2	28,000	13,147	$ 14,853
3	28,000	13,147	$ 14,853
4	28,000	13,147	$ 14,853
5	28,000	13,147	$ 14,853
6	28,000	13,147	$ 14,853
7	28,000	13,147	$ 14,853
8	28,000	13,147	$ 14,853
9	28,000	13,147	$ 14,853
10	28,000	13,147	$ 14,853
11	28,000	13,147	$ 14,853
12	28,000	13,147	$ 14,853
13	28,000	13,147	$ 14,853
14	28,000	13,147	$ 14,853
15	28,000	13,147	$ 14,853
NPV i=10%			$112,970
IRR			n/a

Table 2. Financed Project Cash Flow

At the end of 15 years, the bank loan is paid off (just like a mortgage or car payment—just a different time period). *To keep this simple, ignore interest tax deductions and depreciation, both of which would likely improve the financials even further.*

In this case, the project generates $14,853 each year for the client. Because there is no upfront investment required, the IRR value becomes infinity.

Adding in the green benefits could further illustrate the project's benefits. Table 3 shows what some of these benefits could include. *Note that it can be easier for the audience to visualize equivalencies ("car miles not driven," or "trees planted") instead of lbs of CO_2.*

However, there are even more benefits… when you consider the following impacts the project could have on:

- Shareholders in the annual report
- Community morale & green image
- Productivity improvements
- Legal risk reduction
- LEED points, white certificates, RECs[17]
- FREE public press[18]

FREE Money

In addition, there are utility rebates, tax refunds, credits, and other sources of free money that will improve a project's financial return. Here are some useful websites that allow you to see utility and tax benefits in your state:

www.dsireusa.org
www.energytaxincentives.org
http://www.efficientbuildings.org
http://www.lightingtaxdeduction.org

But don't just rely on websites. Use professionals; they should know what techniques, technologies, and rebates are best for your geographic area.

SUMMARY

This section has described the 3 common barriers (marketing, education and money) as well as a start on how to overcome them. To get a project approved:

1. Articulate the problem/pain.
2. Collaborate to add value in the solution.
3. Quantify all the benefits.
4. Minimize financial risk.
5. Develop/configure an executive summary that "sings" to the hearts of the approver.

Hopefully, these techniques will help you get your next project approved. Why is this important? because the ice is melting! We are counting on you.

Footnotes

1. Wingender, J. and Woodroof, E., (1997) "When Firms Publicize Energy Management Projects: Their Stock Prices Go Up"—How much—21.33% on Average! *Strategic Planning for Energy and the Environment*, Summer Issue 1997.
2. The "Vault Files," www.ProfitableGreenSolutions.com
3. Download the FREE emissions calculator from www.ProfitableGreenSolutions.com
4. For Example: an energy-efficient project that saves $100,000 in operating costs is equivalent to generating $1,000,000 in new sales (assuming the company has a 10% profit margin). It can be more difficult to add $1,000,000 in sales, and would require more infrastructure, etc.
5. Several examples include: Patagonia, Google, GE, Home Depot, etc. Other examples can be downloaded from the "Resource Vault" at www.ProfitableGreenSolutions.com
6. For example: a construction firm switched to hybrid vehicles and also offset the carbon emissions. The firm's name is prominently displayed on each vehicle. They get tons of new business because they are seen and known as the "greenest construction firm" in the city. Plus, they charge a premium for their services!
7. For Example: a law firm renovated their office in a "green" manner and attracted a new client (who chose the firm due to its "green" emphasis). The new client was worth an extra $100,000 in revenue in the first month.
8. Additional Examples: "Green" networking groups such as "greendrinks.org" and can supplement the traditional business networking clubs like Rotary Club, Kiwanis, Chamber of Commerce, etc. Also, when joining groups such as the Climate Action Registry, companies are exposed to other members, who could be superior suppliers, clients and partners.
9. Marketing to Millions Manual, Bob Circosta Communications, LLC.
10. LEED = Leadership in Energy & Environmental Design
11. Results from the Profitable Green Strategies Course, www.ProfitableGreenSolutions.com
12. www.GreenTravelPartners.com
13. MARR= Minimum Attractive Rate of Return. For more info on this topic see: Woodroof, E., Thumann, A.(2005) *Handbook for Financing Energy Projects*, Fairmont Press, Atlanta.
14. U.S. Department of Energy, (1996) "Analysis of Energy-Efficiency Investment Decisions by Small and Medium-Sized Manufacturers," U.S. DOE, Office of Policy and Office of Energy Efficiency and Renewable Energy, pp. 37-38.
15. Advanced Project Financing Course, www.ProfitableGreenSolutions.com
16. Download the FREE emissions calculator from www.ProfitableGreenSolutions.com
17. REC = Renewable Energy Credit
18. Press release samples from the "Vault" at www.ProfitableGreenSolutions.com

kWh Saved per Year	260,000	
# of Years	15	
kWh Saved during Project	3,900,000	
Equivalent Environmental Benefits — Barrels of Oil Not Consumed	7,917	
Car Miles Not Driven	6,924,450	
Acid Rain Emission Reduction	29,250	lbs of Sox
Smog Emission Reductions	14,040	lbs of Nox
GreenHouse Gas Reduction	6,045,000	lbs of CO2
Mature Trees Planted	13,260	

Table 3. Green Benefits[16]

25.9 CHAPTER SUMMARY

It is clear that knowing the strategic needs of the facility manager is critical to selecting the best arrangement. There are practically an infinite number of financial alternatives to consider. This chapter has provided some information on the basic financial arrangements. Combining these arrangements to construct the best contract for your facility is only limited by your creativity.

25.10 GLOSSARY

Capitalize—To convert a schedule of cash flows into a principal amount, called capitalized value, by dividing by a rate of interest. In other words, to set aside an amount large enough to generate (via interest) the desired cash flows forever.

Capital or Financial Lease—Lease that under Statement 13 of the Financial Accounting Standards Board must be reflected on a company's balance sheet as an asset and corresponding liability. Generally, this applies to leases where the lessee acquires essentially all of the economic benefits and risks of the leased property.

Depreciation—The amortization of fixed assets, such as plant and equipment, so as to allocate the cost over their depreciable life. Depreciation reduces taxable income, but it is not an actual cash flow.

Energy Service Company (ESCO)—Company that provides energy services (and possibly financial services) to an energy consumer.

Host—The building owner or facility that uses the equipment.

Lender—Individual or firm that extends money to a borrower with the expectation of being repaid, usually with interest. Lenders create debt in the form of loans or bonds. If the borrower is liquidated, the lender is paid off before stockholders receive distributions.

Lessee—The renter. The party that buys the right to use equipment by making lease payments to the lessor.

Lessor—The owner of the leased equipment.

Line of Credit—An informal agreement between a bank and a borrower indicating the maximum credit the bank will extend. A line of credit is popular because it allows numerous borrowing transactions to be approved without the re-application paperwork.

Liquidity—Ability of a company to convert assets into cash or cash equivalents without significant loss. For example, investments in money market funds are much more liquid than investments in real estate.

Leveraged Lease—Lease that involves a lender in addition to the lessor and lessee. The lender, usually a bank or insurance company, puts up a percentage of the cash required to purchase the asset, usually more than half. The balance is put up by the lessor, who is both the equity participant and the borrower. With the cash the lessor acquires the asset, giving the lender (1) a mortgage on the asset and (2) an assignment of the lease and lease payments. The lessee then makes periodic payments to the lessor, who in turn pays the lender. As owner of the asset, the lessor is entitled to tax deductions for depreciation on the asset and interest on the loan.

MARR (Minimum Attractive Rate of Return)—MARR is the "hurdle rate" for projects within a company. MARR is used to determine the NPV; the annual after-tax cash flow is discounted at MARR (which represents the rate the company could have received with a different project).

Net Present Value (NPV)—As the saying goes, "a dollar received next year is not worth as much as a dollar today." The NPV converts the worth of that future dollar into what it is worth today. NPV converts future cash flows by using a given discount rate. For example, at 10%, $1,000 dollars received one year from now is worth only $909.09 dollars today. In other words, if you invested $909.09 dollars today at 10%, in one year it would be worth $1,000.

NPV is useful because you can convert future savings cash flows back to "time zero" (present), and then compare to the cost of a project. If the NPV is positive, the investment is acceptable. In capital budgeting, the discount rate used is called the hurdle rate and is usually equal to the incremental cost of capital.

"Off-balance Sheet" Financing—Typically refers to a true lease, in which the assets are not listed on the balance sheet. Because the liability is not on the balance sheet, the host appears to be financially stronger. However, most large leases must be listed in the footnotes of financial statements, which reveal the "hidden assets."

Par Value or Face Value—Equals the value of the bond at maturity. For example, a bond with a $1,000 dollar par value will pay $1,000 to the issuer at the maturity date.

Preferred Stock—A hybrid type of stock that pays dividends at a specified rate (like a bond) and has preference over common stock in the payment of dividends and liquidation of assets. However, if the firm is financially strained, it can avoid paying the preferred dividend as it would the common stock dividends. Preferred stock doesn't ordinarily carry voting rights.

Project Financing—A type of arrangement, typically meaning that a single purpose entity (SPE) is constructed. The SPE serves as a special bank account. All funds are sent to the SPE, from which all construction costs are paid. Then all savings cash flows are also distributed from the SPE. The SPE is essentially a mini-company, with the sole purpose of funding a project.

Secured Loan—Loan that pledges assets as collateral. Thus, in the event that the borrower defaults on payments, the lender has the legal right to seize the collateral and sell it to pay off the loan.

True Lease, Operating Lease, or Tax-oriented Lease—Type of lease, normally involving equipment, whereby the contract is written for considerably less time than the equipment's life, and the lessor handles all maintenance and servicing; also called service lease. Operating leases are the opposite of capital leases, where the lessee acquires essentially all the economic benefits and risks of ownership. Common examples of equipment financed with operating leases are office copiers, computers, automobiles, and trucks. Most operating leases are cancelable.

WACC (Weighted Average Cost of Capital)—The firm's average cost of capital, as a function of the proportion of different sources of capital: Equity, Debt, preferred stock, etc. *For example, a firm's target capital structure is:*

Capital Source	Weight (w_i)
Debt	30%
Common Equity	60%
Preferred Stock	10%

and the firm's costs of capital are:

before tax cost of debt = k_d = 10%
cost of common equity = k_s = 15%
cost of preferred stock = k_{ps} = 12%

Then the weighted average cost of capital will be:

$$WACC = w_d k_d (1-T) + w_s k_s + w_{ps} k_{ps}$$

where w_i = weight of Capital Source$_i$
T = tax rate = 34%
After-tax cost of debt = $k_d(1-T)$

Thus,

$$WACC = (.3)(.1)(1-.34) + (.6)(.15) + (.1)(.12)$$

$$WACC = 12.18\%$$

References

1. Wingender, J. and Woodroof, E., (1997) "When Firms Publicize Energy Management Projects Their Stock Prices Go Up: How High?—As Much as 21.33% within 150 days of an Announcement," *Strategic Planning for Energy and the Environment*, Vol. 17(1), pp. 38-51.
2. U.S. Department of Energy, (1996) "Analysis of Energy-Efficiency Investment Decisions by Small and Medium-Sized Manufacturers," U.S. DOE, Office of Policy and Office of Energy Efficiency and Renewable Energy, pp. 37-38.
3. Woodroof, E. and Turner, W. (1998), "Financial Arrangements for Energy Management Projects," *Energy Engineering* 95(3) pp. 23-71.
4. Sullivan, A. and Smith, K. (1993) "Investment Justification for U.S. Factory Automation Projects," *Journal of the Midwest Finance Association*, Vol. 22, p. 24.
5. Fretty, J. (1996), "Financing Energy-Efficient Upgraded Equipment," Proceedings of the 1996 International Energy and Environmental Congress, Chapter 10, Association of Energy Engineers.
6. Pennsylvania Energy Office, (1987) The Pennsylvania Life Cycle Costing Manual.
7. United States Environmental Protection Agency (1994). ProjectKalc, Green Lights Program, Washington DC
8. Tellus Institute, (1996), P2/Finance version 3.0 for Microsoft Excel Version 5, Boston MA.
9. Woodroof, E. And Turner, W. (1999) "Best Ways to Finance Your Energy Management Projects," *Strategic Planning for Energy and the Environment*, Summer 1999, Vol. 19(1) pp. 65-79.
10. Cooke, G.W., and Bomeli, E.C., (1967), *Business Financial Management*, Houghton Mifflin Co., New York.
11. Wingender, J. and Woodroof, E., (1997) "When Firms Publicize Energy Management Projects: Their Stock Prices Go Up," *Strategic Planning for Energy and the Environment*, 17 (1) pp. 38-51.
12. Sharpe, S. and Nguyen, H. (1995) "Capital Market Imperfections and the Incentive to Lease," *Journal of Financial Economics*, 39(2), p. 271-294.
13. Schallheim, J. (1994), *Lease or Buy?*, Harvard Business School Press, Boston, p. 45.
14. Hines, V. (1996),"EUN Survey: 32% of Users Have Signed ESCO Contracts," *Energy User News* 21(11), p.26.
15. Coates, D.F. and DelPonti, J.D. (1996), "Performance Contracting: a Financial Perspective" *Energy Business and Technology Sourcebook*, Proceedings of the 1996 World Energy Engineering Congress, Atlanta. p. 539-543.

CHAPTER 26

COMMISSIONING

DAVID E. CLARIDGE
Leland Jordan Professor
Mechanical Engineering Department
Texas A&M University

MINGSHENG LIU
Professor
Architectural Engineering
University of Nebraska, Lincoln

W.D. TURNER
Professor
Mechanical Engineering Department
Texas A&M University

26.1 INTRODUCTION TO COMMISSIONING FOR ENERGY MANAGEMENT

Commissioning an existing building has been shown to be a key energy management activity over the last decade, often resulting in energy savings of 10%, 20%, or sometimes 30% without significant capital investment. Commissioning is more often applied to new buildings today than to existing buildings, but the energy manager will have more opportunities to apply the process to an existing building as part of the overall energy management program. Hence, this chapter emphasizes commissioning applied to existing buildings, but it also provides some commissioning guidance for the energy manager who is involved in a construction project.

Commissioning an existing building provides several benefits in addition to being an extremely effective energy management strategy. It typically provides an energy payback of one to three years. In addition, building comfort is improved, systems operate better, and maintenance cost is reduced. Commissioning measures typically require no capital investment, though the process often identifies maintenance that is required before the commissioning can be completed. Potential capital upgrades or retrofits are often identified during the commissioning activities, and knowledge gained during the process permits more accurate quantification of benefits than is possible with a typical audit. Involve-

ment of facilities personnel in the process can also lead to improved staff technical skills.

The energy manager's effort is generally directed toward improving the efficiency of existing buildings. However, whenever the organization initiates design and construction of a new building that will become part of the energy manager's portfolio of buildings, it is extremely important that the energy manager become an active part of the design and construction team to ensure that the building incorporates all appropriate energy efficiency technologies. It is just as important that the perspective of operational personnel be included in the design process so it will be possible to effectively and efficiently operate the building. One of the best ways to accomplish these objectives is to commission the building as it is designed and built.

This chapter is intended to provide the energy manager with the information needed to make the decision whether to conduct an in-house commissioning program or to select and work with an outside commissioning provider. There is no single definition of commissioning for an existing building, or for new buildings, so several widely used commissioning definitions are given. The commissioning process used by the authors in existing buildings is described in some detail, and common commissioning measures and commissioning resources are described so the energy manager can choose how to implement a commissioning program. Measurement and verification is very important to a successful commissioning program. Some commissioning specific M&V issues are discussed, particularly the role of M&V in identifying the need for follow-up commissioning activities. Commissioning a new building is described from the perspective of the energy manager. Three case studies illustrate different applications of the commissioning process as part of the overall energy management program.

26.2 COMMISSIONING DEFINITIONS

To commission a navy ship refers to the order or process that makes it completely ready for active duty. Over the last two decades, the term has come to refer to the process that makes a building or some of its systems completely ready for use. In the case of existing buildings, it generally refers to a restoration or improvement in the operation or function of the building systems.

26.2.1 New Building Commissioning

ASHRAE defines building commissioning as *"the process of ensuring systems are designed, installed, functionally tested, and operated in conformance with the design intent. Commissioning begins with planning and includes design, construction, start-up, acceptance, and training and can be applied throughout the life of the building. Furthermore, the commissioning process encompasses and coordinates the traditionally separate functions of systems documentation, equipment start-up, control system calibration, testing and balancing, and performance testing."*[1]

This guideline was restricted to new buildings, but it later became evident that, while initial start-up problems were not an issue in older buildings, most of the other problems that commissioning resolved were even more prevalent in older systems.

26.2.2 Recommissioning

Recommissioning refers to commissioning a building that has already been commissioned at least once. After a building has been commissioned during the construction process, recommissioning ensures that the building continues to operate effectively and efficiently. Buildings, even if perfectly commissioned, will normally drift away from optimum performance over time, due to system degradation, usage changes, or failure to correctly diagnose the root cause of comfort complaints. Therefore, recommissioning normally reapplies the original commissioning procedures in order to keep the building operating according to design intent, or it may modify them for current operating needs.

Optimally, recommissioning becomes part of a facility's continuing O&M program. There is not yet a consensus on recommissioning frequency, but some consider that it should occur every 3 to 5 years. If there are frequent build-outs or changes in building use, re-commissioning should be applied more often. [2]

26.2.3 Retro Commissioning (RcX)

Retro commissioning is the first-time commissioning of an existing building. Many of the steps in the retro commissioning process are similar to those for commissioning. Retro commissioning, however, occurs after construction as an independent process, and its focus is usually on energy-using equipment such as mechanical equipment and related controls. Retro commissioning may or may not bring the building back to its original design intent since the usage may have changed or the original design documentation may no longer exist. [3]

26.2.4 Ongoing Commissioning[4]

Ongoing commissioning is a commissioning process conducted continually for the purpose of maintaining, improving, and optimizing the performance of building systems after new building commissioning, or "RCx."

26.3. THE COMMISSIONING PROCESS IN EXISTING BUILDINGS

There are multiple terms that describe the commissioning process for existing buildings, as noted in the previous section. Likewise, there are many adaptations of the process itself. The same practitioner will implement the process differently in different buildings, based on the budget and the owner requirements. The commissioning process described in this chapter is the *retro* **commissioning (RCx) process.** The model described assumes that a commissioning provider is involved since that is normally the case. Some (or all) of the steps may be implemented by the facility staff if they have the expertise and adequate staffing levels to take on the work.

26.3.1 Commissioning Team

The RCx team consists of a project manager, one or more RCx professionals and RCx technicians, and one or more designated members of the facility operating team. The primary responsibilities of the team members are shown in Table 1. The project manager can be an owner representative or an RCx provider representative. It is essential that the RCx professionals have the qualifications and experience to perform the work specified in the table. The designated facility team members generally include at least one lead HVAC technician and an EMCS operator or engineer. It is also essential that the designated members of the facility operating team actively participate in the process and be convinced of the value of the measures proposed and implemented, or operation will rapidly revert to old practices.

26.3.2 RCx Process

The RCx process, or "RCx" consists of two phases. The first phase is the project development phase that identifies the buildings and facilities to be included in the project and develops the project scope. At the end of this phase, the RCx scope is clearly defined and an RCx contract is signed as described in Section 26.3.2.1. The second phase implements RCx and verifies project performance through the six steps outlined in Figure 26-1 and described in Section 26.3.2.2.

Table 26-1. Commissioning team members and their primary responsibilities.

Team Member(s)	Primary Responsibilities
Project Manager	1. Coordinate the activities of building personnel and the commissioning team 2. Schedule project activities
RCx professional(s)	1. Develop metering and field measurement plans 2. Develop improved operational and control schedules 3. Work with building staff to develop mutually acceptable implementation plans 4. Make necessary programming changes to the building automation system 5. Supervise technicians implementing mechanical systems changes 6. Project potential performance changes and energy savings 7. Conduct an engineering analysis of the system changes 8. Write the project report
Designated Facility Staff	1. Participate in the initial facility survey 2. Provide information about problems with facility operation 3. Suggest commissioning measures for evaluation 4. Approve all RCx measures before implementation 5. Actively participate in the implementation process
RCx Technicians	1. Conduct field measurements 2. Implement mechanical, electrical, and control system program modifications and changes, under the direction of the project engineer

26.3.2.1 Phase 1: Project Development
Step 1: Identify buildings or facilities

Objective: Screen potential RCx candidates with minimal effort to identify buildings or facilities that will receive an RCx assessment. The RCx candidate can be a building, an entire facility, or a piece of equipment. If the building is part of a complex or campus, it is desirable to select the entire facility as the RCx candidate since one mechanical problem may be rooted in another part of the building or facility.

Approach: The RCx candidates can be selected based on one or more of the following criteria:
• The candidate provides poor thermal comfort.
• The candidate consumes excessive energy.
• The design features of the facility HVAC systems are not fully used.

If one or more of the above criteria fits the description of the facility, it is likely to be a good candidate for RCx. RCx can be effectively implemented in buildings that have received energy efficiency retrofits, in newer buildings, and in existing buildings that have not received energy efficiency upgrades. In other words, virtually any building is a potential RCx candidate.

The RCx candidates can be selected by the building owner or the RCx provider. However, the building owner is usually in the best position to select the most promising candidates because of his or her knowledge of the facility operation and costs. The RCx provider

should, then, perform a preliminary analysis to check the feasibility of using the RCx process on candidate facilities before performing an RCx assessment.

The following information is needed for the preliminary assessment:
• Monthly utility bills (both electricity and gas) for at least 12 months. (Actual bills are preferable to a table of historic energy and demand data because meter reading dates are needed.)
• General building information: size, function, major equipment, and occupancy schedules.
• O&M records, if available.
• Description of any problems in the building, such as thermal comfort, indoor air quality, moisture, or mildew.

An experienced engineer should review this information and determine the potential of the RCx process to improve comfort and reduce energy cost. The RCx projects often improve building comfort and reduce building energy consumption at the same time. However, some of the RCx measures may increase building energy consumption in order to satisfy room comfort and indoor air requirements. For example, providing building minimum outside air will certainly increase the cooling energy consumption during summer and heating consumption during winter, compared to operating the building with no outside air. If the potential justifies an RCx assessment, a list of preliminary commissioning measures for evaluation in an RCx assessment should be developed. If the owner is interested in proceeding at this point, an RCx assessment may be performed.

Step 2: Perform RCx assessment and develop project scope

Objectives: The objectives of this step are to:
- Define owner's requirements.
- Check the availability of in-house technical support such as RCx technicians.
- Identify major RCx measures.
- Estimate potential savings from RCx measures and cost to implement.

Approach: The owner's representative, the RCx project manager, and the RCx project engineer will meet. The expectations and interest of the building owner in comfort improvements, utility cost reductions, and maintenance cost reductions will be discussed and documented. The availability and technical skills of in-house technicians will be discussed. After this discussion, a walkthrough must be conducted to identify the feasibility of the owner expectations for comfort performance and improved energy performance. During the walkthrough, the RCx professional and project manager will identify major RCx measures applicable to the building. An in-house technician should participate in this walk-through to provide a local operational perspective and input. The project engineer estimates the potential savings and the commissioning cost and, together with the project manager, prepares the RCx assessment report that documents these findings as well as the owner expectations.

Special Considerations
- A complete set of mechanical and control system design documentation is needed.
- The RCx professional and technician will make preliminary measurements of key equipment operating parameters during the walk-through.

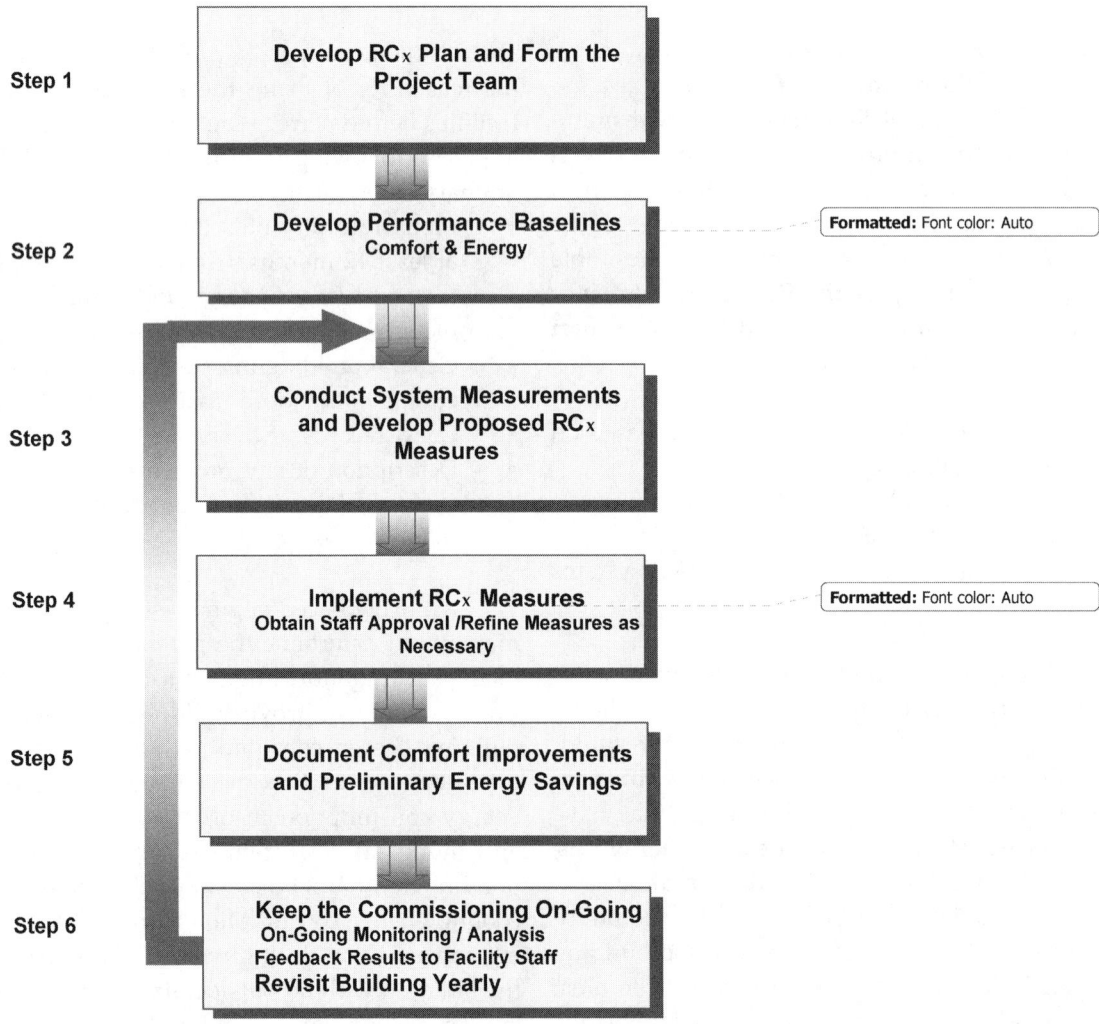

Figure 26-1. Outline of Phase II of the Retro Commissioning (RCx) Process

- Any available measured whole building level or sub-metered energy consumption data from stand-alone meters or the building automation system should be utilized while preparing the report.

An RCx assessment report must be completed that lists and describes preliminary RCx measures, the estimated energy savings from implementation, and the cost of carrying out the RCx process on the building(s) evaluated in the RCx assessment.

There may be more than one iteration or variation at each step described here, but once a contract is signed, the process moves to phase 2 as detailed below.

Phase 2: RCx Implementation and Verification
Step 1: Develop RCx plan and form the project team
Objectives:
- Develop a detailed work plan.
- Identify the entire project team.
- Clarify the duties of each team member.

Approach: The RCx project manager and project engineer develop a detailed work plan for the project that includes major tasks, their sequence, time requirements, and technical requirements. The work plan is then presented to the building owner or representative(s) at a meeting attended by any additional RCx professionals and/or technicians on the project team. During the meeting, the owner contact personnel and in-house technicians who will work on the project should be identified. If in-house technicians are going to conduct measurements and system adjustments, additional time should be included in the schedule unless they are going to be dedicated full time to the RCx project. Typically, in-house technicians must continue their existing duties and cannot devote full time to the RCx effort, which results in project delays. In-house staff may also require additional training. The work plan may need to be modified, depending on the availability and skill levels of in-house staff utilized.

Special Issues
- Availability of funding to replace/repair parts found broken.
- Time commitment of in-house staff.
- Training needs of in-house staff.

Deliverable: RCx Report Part I: RCx plan that includes project scope and schedule, project team, and task duties of each team member.

Step 2: Develop performance baselines
Objectives:
- Document existing comfort conditions.

- Document existing system conditions.
- Document existing energy performance.

Approach: Document all known comfort problems in individual rooms resulting from issues related to heating, cooling, noise, humidity, odors (especially from mold or mildew), or lack of outside air. Also, identify and document any HVAC system problems, including:
- Valve and damper hunting.
- Disabled systems or components.
- Operational problems.
- Frequently replaced parts.

An interview and walk-through may be required although most of this information is collected during the RCx assessment and Step 1. Room comfort problems should be quantified using hand held meters or portable data loggers. System and/or component problems should be documented, based on interviews with occupants and technical staff in combination with field observations and measurements.

Baseline energy models of building performance are necessary to document the energy savings after commissioning. The baseline energy models can be developed using one or more of the following types of data:
- Short-term measured data obtained from data loggers or the EMCS system,
- Long-term, hourly, or 15-minute whole-building energy data such as whole-building electricity and cooling and heating consumption.
- Utility bills for electricity, gas, and/or chilled or hot water.

The whole-building energy baseline models normally include whole building electricity, cooling energy, and heating energy models. These models are generally expressed as functions of outside air temperature since both heating and cooling energy are normally weather dependent.

Any component baseline models should be represented using the most relevant physical parameter(s) as the independent variable(s). For example, the fan motor power should be correlated with the fan airflow, and pump motor energy consumption should be correlated with water flow.

Short-term measured data are often the most cost effective and accurate if the potential savings of RCx measures are independent of the weather. For example, a single true power measurement can be used to develop the baseline fan energy consumption if the pulley is to be changed in a constant air volume system. Short-term data are useful for determining the baseline for specific pieces of equipment but are not reliable for baselining overall building energy use. They may be used with calibrated simulation to obtain plausible baselines when no longer-term data are available.

Long-term measurements are normally required since potential savings of RCx measures are weather-dependent. These measurements provide the most convincing evidence of the impact of RCx projects. Long-term data also help in continuing to diagnose system faults during ongoing RCx. Although more costly than short-term measured data, long-term data often produces additional savings, making them the preferred data type. For example, unusual energy consumption patterns can be easily identified using long-term, short-interval measured data. "Fixing" these unusual patterns can result in significant energy savings. Generally speaking, long-term interval data for electricity, gas, and thermal usage are preferred.

Utility bills may be used to develop the energy use baselines if the RCx process will result in energy savings that are a significant fraction (>15%) of baseline use, and if the building functions and use patterns will remain the same throughout the project.

The RCx professionals should provide the metering options that meet the project requirements to the building owner or representative. A metering method should be selected from the options presented by the RCx professional and a detailed metering implementation plan developed. It may be necessary to hire a metering subcontractor if an energy information system is installed prior to implementation of the RCx measures. More detailed information on savings determination is contained in the measurement and verification chapter of this handbook (Chapter 27).

Special Considerations
- Use the maintenance log to help identify major system problems.
- Select a metering plan that suits the RCx goals and facility needs.
- Always measure or obtain weather data as part of the metering plan.
- Keep meters calibrated. When the EMCS system is used for metering, both sensors and transmitters should be calibrated using field measurements.

Deliverables: RCx Report Part II: Report on current building performance that includes current energy performance, current comfort and system problems, and metering plans if new meters are to be installed. Alternatively, if utility bills are used to develop the base-

line energy models, the report should include baseline energy models.

Step 3: Conduct system measurements and develop proposed RCx measures
Objectives:
- Identify current operational schedules & problems.
- Develop solutions to existing problems.
- Develop improved operation and control schedules and setpoints.
- Identify potential cost effective energy retrofit measures.

Approach: The RCx professional should develop a detailed measurement cut-sheet for each major system. The cut-sheet should list all the parameters to be measured and all mechanical and electrical parts to be checked. The RCx professional should also provide measurement training to the technician if a local technician is used to perform system measurements. The RCx technicians should follow the procedures on the cut-sheets to obtain the measurements, using appropriate equipment.

The RCx professional conducts an engineering analysis to develop solutions for the existing problems and develops improved operation and control schedules and setpoints for terminal boxes, air handling units (AHUs), exhaust systems, water and steam distribution systems, heat exchangers, chillers, boilers, and other components or systems as appropriate. Cost effective energy retrofit measures can also be identified and documented during this step if desired by the building owner.

Special Considerations
Trend main operational parameters using the EMCS and compare with the measurements from hand meters.
- Print out EMCS control sequences and schedules.
- Verify system operation in the building and compare to EMCS schedules.

Deliverable: RCx Report Part III: Existing system conditions. This report includes:
- Existing control sequences and setpoints for all major equipment, such as AHU supply air temperature, AHU supply static pressures, terminal box minimum airflow and maximum airflow values, water loop differential pressure setpoints, and equipment on/off schedules.
- List of disabled control sequences and schedules.
- List of malfunctioning equipment and control devices.

- Engineering solutions to the existing problems and a list of repairs required.
- Proposed improved control and operation sequences and schedules.

Step 4: Implement RCx measures
Objectives:
- Obtain approval for each RCx measure from the building owner's representative prior to implementation.
- Implement solutions to existing operational and comfort problems.
- Implement and refine improved operation and control schedules.

Approach: The RCx project manager and project engineer should present the engineering solutions to existing problems and the improved operational and control schedules to the building owner's representative in one or more meetings. The in-house operating staff should be invited to meeting(s). All critical questions should be answered. It is important at this point to get "buy-in" and approval from both the building owner's representative and the operating staff. The meeting(s) will decide the following issues:
- Approval, modification or disapproval of each RCx measure.
- Implementation sequence of RCx measures.
- Implementation schedules.

RCx implementation should start by solving existing problems. The existing comfort and difficult control problems are the first priority of the occupants, operating staff, and facility owner. Solving these problems improves occupant comfort and increases productivity. The economic benefits from comfort improvements are sometimes higher than the energy cost savings, though less easily quantified. The successful resolution of existing problems can also earn trust in the RCx professional from facility operating staff, facility management. and the occupants.

Implementation of the improved operation and control schedules should start at the end of the comfort delivery system, such as at the terminal boxes, and end with the central plant. This procedure provides benefits to the building occupants as quickly as possible. It also reduces the overall load on the system. If the process is reversed, the chiller plant is commissioned first. The chiller sequences are developed based on the current load. After the rest of the commissioning is complete, the chiller load may decrease by 30%, resulting in a need to revise the chiller operating schedules.

The RCx professionals should develop a detailed implementation plan that lists each major activity. The RCx technician should follow this plan in implementing the measures. The RCx professionals should closely supervise the implementation and refine the operational and control schedules as necessary. The RCx professionals should also be responsible for the key software changes as necessary.

Following implementation, the new operation and control sequences must be documented in a way that helps the building staff understand why they were implemented. Maintenance procedures for these measures should be provided. If any measures have not been implemented due to temporary impediments such as an out of stock part, recommendations for their future implementation should be included.

In many RCx projects, this portion of the project is contracted to an external service provider rather than being implemented by the RCx provider in cooperation with the building staff. If this approach is taken, careful coordination with the RCx provider is needed to ensure that measures are properly and completely implemented. The external provider often does not fully understand the measures being implemented and hence is often unable to modify them slightly to make them completely successful, as can be done when the commissioning professional is part of the implementation process.

Special Considerations:
Ensure that the owner's technical representative understands each major measure.
- Encourage involvement of in-house technicians in the implementation and/or have them implement as many measures as possible.
- Document improvements in a timely manner.

Deliverable: RCx Report Part IV: RCx implementation. This report includes detailed documentation of implemented operation and control sequences, maintenance procedures for these measures, and recommendations for measures to be implemented in the future.

**Step 5: Document comfort improvements
and preliminary energy savings**
Objectives:
- Document improved comfort conditions.
- Document improved system conditions.
- Document preliminary energy savings.

Approach: The comfort measurements taken in Step 2 (Phase 2) should be repeated at the same locations under comparable conditions to determine impact on room conditions. The measured parameters, such as temperature and humidity, should be compared with the measurements from Step 2.

The M&V procedures adopted in Step 2 should be used to determine the early post-RCx energy performance. Energy performance should be compared under the same occupancy conditions and weather normalized.

Special Considerations:
• Savings analyses should follow accepted measurement and verification protocols such as the IPMVP.
• Comfort conditions should conform to appropriate guidelines/design documents such as ASHRAE standards.

Deliverable: RCx Report, Part 5: Preliminary measurement and verification report. This report includes results of detailed measurements of room conditions and energy consumption after RCx activities, and any retrofit recommendations that may be provided. The room conditions should be compared with those from the pre-RCx period. The projected annual energy savings should be determined according to the M&V approach adopted in Step 2.

Step 6: Ongoing commissioning
Objectives:
• Maintain improved comfort and energy performance.
• Provide measured annual energy savings.

Approach: The RCx professionals should review the system operation periodically to identify any operating problems and develop improved operation and control schedules as described below.

The RCx professionals should provide follow-up phone consultation to the operating staff as needed, supplemented by site visits. This will allow the operating staff to make wise decisions and maintain the savings and comfort in years to come. If long-term measured data are available, the RCx professionals should review the energy data quarterly to evaluate the need for a site visit. If the building energy consumption has increased, the RCx professionals determine possible reasons and verify with facility operating staff. Once the problem(s) is identified, an RCx professional should visit the site, develop measures to restore the building performance, and supervise the facility staff in implementing the measures. If the RCx professional can remotely log onto

the EMCS system, the RCx professional can check the existing system operation quarterly using the EMCS system. When a large number of operation and control measures are disabled, a site visit is necessary. If the RCx professional cannot evaluate the facility using long-term measured energy data and EMCS system information, the RCx professional should visit the facility semi-annually.

One year after RCx implementation is complete, the RCx professional should write a project follow-up report that documents the first-year savings, recommendations or changes resulting from any consultation or site visits provided, and any recommendations to further improve building operations.

The importance of ongoing commissioning is discussed in more detail, and a case study is provided in Section 26.5.6.

Special Considerations: Operating personnel often have a high turnover rate, and it is important to train new staff members in the RCx process and make sure they are aware of the reasons the RCx measures were implemented. Ongoing follow-up is essential if the savings are to be maintained at high levels over time.

Deliverable: Special RCx report that documents measured first-year energy savings, results from first-year follow-up, recommendations for ongoing staff training, and a schedule of follow-up RCx activities.

26.3.3 Uses of Commissioning in the Energy Management Process

Commissioning can be used as a part of the energy management program in several different ways. It can be used:
• As a stand-alone measure.
• As a follow-up to the retrofit process.
• As an energy conservation measure (ECM) in a retrofit program.
• To ensure that a new building meets or exceeds its energy performance goals.

26.3.3.1 A stand-alone measure

Commissioning is probably most often implemented in existing buildings, because it is the most cost-effective step the owner can take to increase the energy efficiency of the building, generally offering a pay-back under three years, and often 1-2 years[5]. The RCx process also provides a high level of understanding of the building and its operation, enabling retrofit recommendations developed as part of the RCx process to be made with a high level of certainty. The load reductions resulting from implementation of the RCx process may also enable other changes, for example the use of a smaller high efficiency chiller.

26.3.3.2 A follow-up to the retrofit process

RCx has often been used to provide additional savings after a successful retrofit.[6] It has also been used numerous times to make an under-performing retrofit meet or exceed the original expectations. The process was initially developed for these purposes as part of the Texas LoanSTAR program.

26.3.3.3 As an Energy Conservation Measure (ECM) in a retrofit program

The rapid payback that generally results from RCx may be used to lower the payback of a package of measures to enable inclusion of a desired equipment replacement that has a longer payback in a retrofit package[8]. This is illustrated by a case study in Section 26.3.4. In this approach the RCx professionals conduct the RCx assessment in parallel with the retrofit audit conducted by the design engineering firm. Because the two approaches are different and look at different opportunities, it is very important to closely coordinate these two audits. For example, the RCx professional may determine a need for a variable frequency drive on a chilled water pump. This is a retrofit opportunity for the audit engineer and should be written up as a retrofit ECM. Similarly, the audit engineer may encounter an RCx opportunity during the building walk-through audit, that should be reported to the RCx professional. Similarly, RCx may be included as a highly cost effective measure within an energy service company contract or "guaranteed savings" contract.

26.3.3.4 To ensure that a new building meets or exceeds its energy performance goals

Commissioning is generally used for a new building to ensure that the systems work and provide comfort for the occupants with minimum start-up problems. It also has been found to reduce expensive change orders and other construction problems. It may also be used to significantly improve the efficiency of a new building by optimizing operation to meet its actual loads and uses instead of working to design assumptions[9,10].

The commissioning process has been described using an outside provider. It is certainly possible to perform commissioning using internal personnel when the needed skills are available on staff and these engineers and technicians can be assigned to the commissioning process. This is directly analogous to the retrofit process. Most energy audits and retrofit designs are performed by external consultants, but they can and are provided by internal personnel on occasion.

26.3.4 Case Study With RCx As An ECM[11]

Prairie View A&M University is a 1.7 million square foot campus, with most buildings served by a central thermal plant. Electricity is purchased from a local electric co-op.

University staff identified the need for major plant equipment replacements on campus. They wished to use the Texas LoanSTAR program to finance the project. The LoanSTAR program finances energy efficiency upgrades for public buildings, requiring that the aggregate energy payback of all energy conservation measures (ECMs) financed be ten years or less. The program requires that participating state agencies meter all buildings/plants receiving the ECMs and implement a comprehensive M&V program. The cost of the detailed investment grade audit and the mandatory M&V can be rolled into the loan, but the simple payback must still meet the ten-year criterion. This typically means that the aggregate payback of the ECMs must be 8 to 8-1/2 years, without the audit and M&V costs included. Replacement of items such as chillers, cooling towers, and building automation systems typically have paybacks of considerably more than 10 years. Hence, they can only be included in a loan if packaged with low payback measures that bring the aggregate payback below 10 years.

The university administration wanted to maximize the loan amount to get as much equipment replacement as possible. They also wanted to ensure that the retrofits work properly after being installed. To maximize their loan dollars, they chose to include RCx as an ECM. They also chose to include the audit and M&V costs in the loan to minimize up front costs.

The LoanSTAR Program provides a brief walk-through audit of the candidate buildings and plants. This audit is performed to determine whether there is sufficient retrofit potential to justify a more thorough investment grade audit.

When RCx is to be included as an ECM, the RCx assessment is conducted in parallel with the retrofit audit conducted by the engineering design firm. The two approaches look at different opportunities, but there can be some overlap, so it is very important to closely coordinate both assessments. For example, the RCx professional may determine a need for a variable frequency drive on a chilled water pump. This is a retrofit opportunity for the audit engineer and should be written up as a retrofit ECM. Similarly, the audit engineer may encounter an RCx opportunity during the building audit that should be reported to the RCx professional. It is particularly important that the savings estimated by the audit team are not "double counted." The area of greatest overlap in this case was the building automation system. Considerable care was taken not to mix improved EMCS operation with operational improvements determined by the RCx professional, so both measures received proper credit.

The same design engineering firm conducted both the initial walk-through audit and the detailed, investment grade audit. ESL RCx professionals likewise conducted an initial RCx walk-through assessment as well as the detailed RCx assessment.

The RCx measures identified included:

- Hot and cold deck temperature resets.
- Extensive EMCS programming to avoid simultaneous heating and cooling.
- Air and water balancing.
- Duct static pressure resets.
- Sensor calibration/repair.
- Improved start/stop/warm-up/shutdown schedules.

The RCx professionals took the measurements required and collected adequate data on building operation during the RCx assessment to perform a calibrated simulation on the major buildings. Available metered data and building EMCS data were also used. The RCx energy savings were then written as an ECM and discussed with the design engineer. Any potential overlaps were removed. The combined ECMs were then listed and the total savings determined.

Table 26-2 summarizes the ECMs identified from the two assessments.

The RCx savings were calculated to be $204,563, as determined by conducting calibrated simulation of 16 campus buildings and by engineering calculations

of savings from improved loop pumping. No RCx savings were claimed for central plant optimization. Those savings were all applied to ECM #7, although it seems likely that additional RCx savings will accrue from this measure. The simple payback from RCx is slightly under three years, making it by far the most cost effective of the ECMs to be implemented. The RCx savings represent nearly 30% of the total project savings.

Perhaps more importantly, RCx accounted for 2/3 of the "surplus" savings dollars available to buy down the payback of the chillers and EMCS upgrade. Without RCx as an ECM, the University would have had to choose which ECMs to delete—one chiller and the EMCS upgrades, or some combination of chillers and limited building EMCS upgrades—to meet the ten-year payback criteria. With RCx, however, the university was able to include all these hardware items and still meet the ten-year payback.

26.4. COMMISSIONING MEASURES

RCx measures can be placed in two basic categories. The first category includes a number of long-time energy management measures that eliminate operation when it isn't needed, or put simply, "Shut it off if it isn't needed." A number of these measures are a bit more complex than simply turning it off, but all are widely recognized and practiced. However, opportunities to implement some of these measures are often found, even in well-run facilities. These are discussed in some detail since most facility personnel can implement these measures. The second category of measures can broadly be categorized as implementing control practices that are optimized to the facility. These measures require a relatively high level of knowledge and skill to analyze the operation of a building, develop the optimal control

ECM #	ECM	Annual Savings				Cost to Implement	Simple Payback
		Electric kWh/yr	Electric Demand kW/yr	Gas MCF/yr	Cost Savings		
#1	Lighting	1,565,342	5,221	(820)	$94,669	$ 561,301	6.0
#2	Replace Chiller #3	596,891	1,250	-0-	$33,707	$ 668,549	19.8
#3	Repair Steam System	-0-	-0-	13,251	$58,616	$ 422,693	7.2
#4	Install Motion Sensors	81,616	-0-	(44.6)	$ 3,567	$ 26,087	7.3
#5	Add 2 Bldgs. to CW Loop	557,676	7,050	-0-	$ 60,903	$ 508,565	8.4
#6	Add Chiller #4	599,891	1,250	-0-	$ 33,707	$ 668,549	19.8
#7	Primary/Secondary Pumping	1,070,207	-0-	-0-	$49,230	$ 441,880	9.0
#8	Replace DX Systems	38,237	233	-0-	$ 2,923	$ 37,929	13.0
#9	Replace DDC/EMCS	2,969,962	670	2,736	$151,488	$2,071,932	13.7
#10	RcX Commissioning	2,129,855	-0-	25,318	$204,563	$ 605,000	3.0
	Assessment Reports					$ 102,775	
	Metering					$ 157,700	
	M&V					$ 197,500	
		9,606,677	15,674	40,440	$693,373	$6,470,460	9.3

Table 26-2. Summary of Energy Cost Measures (ECMs) [12]

sequences, and then implement them. These measures are presented in less detail, but references are provided for the reader who wishes to learn more. Some measures include the implementation of retrofits or new hardware in ways that are relatively new and innovative to provide rapid payback comparable to the other RCx measures. These measures are sometimes considered as a separate category but are not discussed to that level of detail in this chapter.

26.4.1 Eliminating Unnecessary Operation

Commissioning begins with simple measures that are included in any good energy management program. Simple rules like shutting off any system that isn't needed are the beginning point of a good commissioning program as well as a good energy management program.

26.4.1.1 Remove Foot Heaters and Turn Off Desk Fans

The presence of foot heaters and desk fans indicates an unsuitable working environment and wastes energy as well. To turn off foot heaters and desk fans, the following actions should be taken:

- Adjust the individual zone temperature setpoint according to the occupant's desires.
- Balance zone airflow if foot heaters are used in a portion of the zone.
- Adjust AHU supply air temperature and static pressure if the entire building is too cold or too hot.
- Repair existing mechanical and control problems, such as replacing diffusers of the wrong type and relocating return air grilles, to maintain a comfortable zone temperature.

Different people require different temperatures to feel comfortable. Some organizations, however, mandate the zone temperature setpoint for both summer and winter. This often leads to comfort complaints and negatively impacts productivity. The operating staff must place comfort as a priority and adjust the room temperature setpoint as necessary. Workers should be asked to dress appropriately during the summer and winter to maintain their individual comfort if setpoints are centrally mandated for a facility. Most complaints can be eliminated when the room temperature is within the range of ASHRAE's recommended comfort zone.

26.4.1.2 Turn off Heating Systems During Summer

Heating is not needed for most buildings during the summer. When the heating system is on, hot water or steam often leaks through control valves, causes ther-

mal comfort problems, and consumes excessive cooling and heating energy. To improve building comfort and decrease heating and cooling energy consumption, the following actions should be taken:

- Turn off boilers or heat exchangers if the entire building does not need heating.
- Manually valve off heating and preheating coils if the heating system has to be on for other systems.
- Reset differential pressure of the hot water loop to a lower value to prevent excessive pressure on control valves during the summer.
- Troubleshoot individual zones or systems that have too many cold complaints.
- Do not turn heating off too early in the spring (to avoid having to turn the system back on repeatedly).

This measure may be applied in constant air volume systems in dry climates. When the reheat system is shut off, room comfort may be maintained by increasing supply air temperature. This measure is not suitable for other climates where the cooling coil leaving air temperature has to be controlled below 57°F to control room humidity levels.

This simple measure results in significant energy savings, as well as improved comfort, in most buildings. Figure 26-2 compares the measured heating energy consumption before and after manually shutting off AHU heating valves in a building in Austin, Texas.

This building has a floor area of 147,000 square feet with two dual duct VAV systems. Before closing the heating coil manual valves, the average daily steam consumption varied from a low of 0.2 MMBtu/hr to a high of about 0.28 MMBtu/hr. After the manual valves were closed, steam leakage was eliminated through the heating coil. The steam consumption immediately dropped to slightly above 0.10 MMBtu/hr. Since the manual valves in this building can stay closed for more than seven months, the annual steam savings are 756 MMBtu/yr. The same amount of chilled water will also be saved if the building remains at the same temperature, so the cooling energy savings will be 756 MMBtu/yr as well. The annual energy cost savings is $15,120 at an energy price of $10/MMBtu. This savings is not particularly large, but the only action required was shutting two manual valves.

26.4.1.3 Turn Off Systems During Unoccupied Hours

If a building is not occupied at nights or on weekends, the HVAC system may often be turned off completely during these periods. With a properly designed

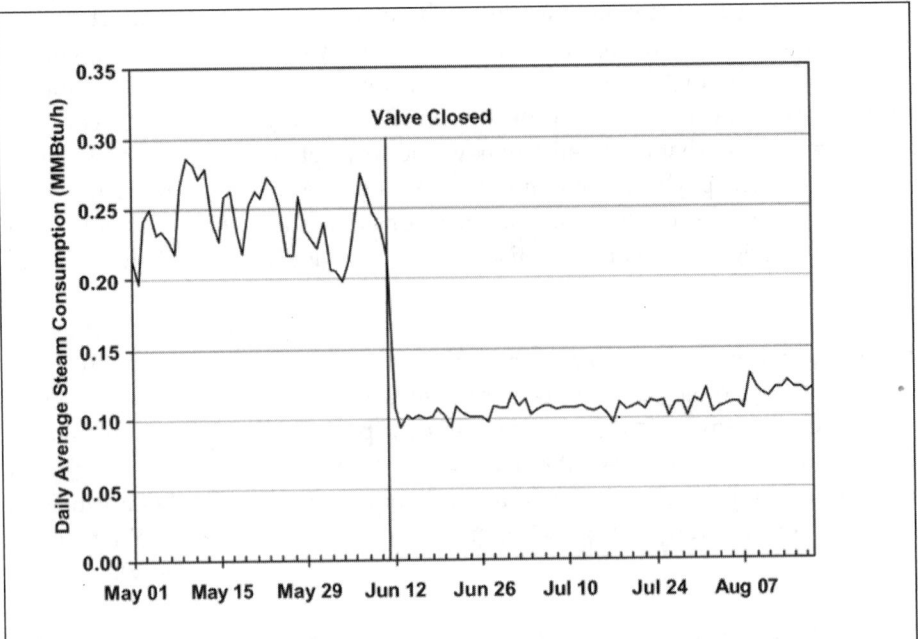

Figure 26-2. Comparison of measured daily average steam consumption before and after manually shutting off heating coil valves in the business administration building at the University of Texas at Austin [13]

warm-up/cool-down, building comfort can be maintained with significant energy savings. In a commercial or institutional building, office equipment and lighting make up a large portion (often 50% or more) of the electrical requirements. However, a significant portion of a building (15% or more) is normally unoccupied during office hours due to travel, meetings, vacations, and sick leave. Turning off systems during unoccupied hours results in significant energy savings without degrading occupant comfort. This measure can be achieved by the following actions:

- Turn off lights, computers, printers, fax machines, desk fans, and other office equipment when leaving the office.
- Turn off lights and set back room thermostats after cleaning.
- Turn off AHUs at nights and on weekends. A schedule needs to be developed for each zone or air handling unit. *Turning off the system too early in the evening or turning the system on too late in the morning may cause comfort problems. Conversely, turning off a system too late in the evening and turning the system on too early in the morning may lose considerable savings.*
- Turn off the boiler hot water pump at night during the summer when AHUs are turned off.
- Turn off chillers and chilled water pumps when free cooling is available or when AHUs are turned off.

Figure 26-3 presents the measured building electricity consumption, excluding chiller consumption, before and after implementation of AHU and office equipment turn-off on nights and weekends in the Stephen F. Austin Building (SFA) in Austin, Texas.

The Stephen F. Austin Building has 470,000 square feet of floor area with 22 dual duct AHUs. During the first phase of implementation, 16 AHUs were turned off from midnight to 4 a.m., weekdays and weekends. During the second phase, 22 AHUs were turned off from 11:00 p.m. to 5 a.m. during weekdays and weekends. In addition, during the second phase, all occupants were asked to turn off office equipment when leaving their office.

The measured results show that the nighttime whole-building electricity use decreased from 1,250 kW to 900 kW during the first phase. During the second phase, the nighttime minimum electricity decreased to 800 kW.

It was observed that the daily peak electricity consumption after night shutdowns began is significantly lower than the base peak. For example, the lowest peak during the second phase is 1,833 kW, which is 8% lower than the base peak. The lower electricity peak indicates that some office equipment remained off during the daytime or that the employees were more conscientious in turning off lights and equipment when they left the office. The annual energy cost savings, including electricity, heating, and cooling, were determined to be $100,000/yr using measured hourly data.

26.4.1.4 Slow Down Systems During Unoccupied/Lightly-Occupied Hours

Most large buildings are never completely unoccupied. It is not uncommon to find a few people working,

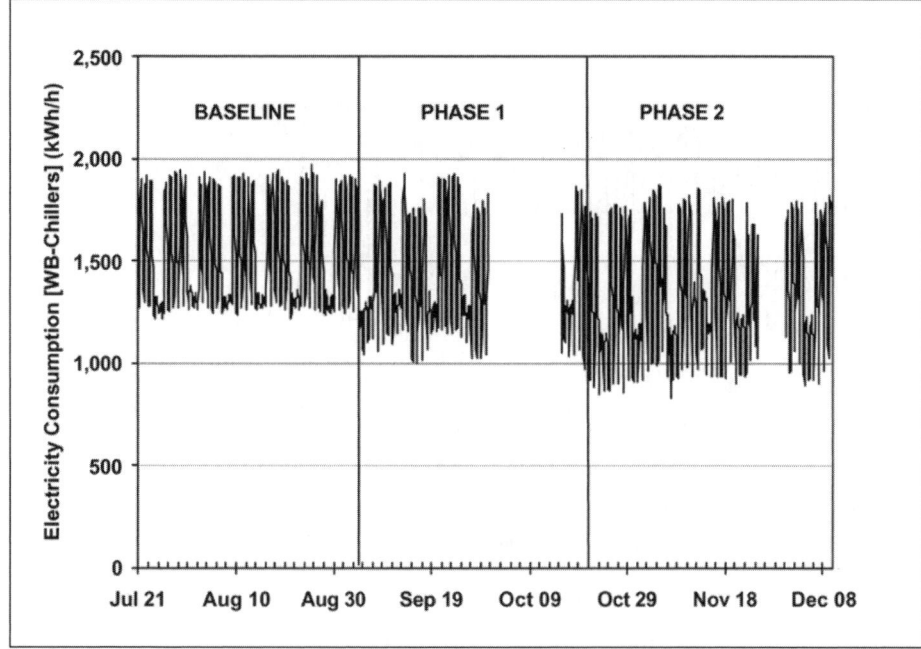

Figure 26-3. Hourly whole building electricity consumption at the SFA Building before and after night shutdown of AHUs was initiated [14]

regardless of the time of day. The zones that may be used during the weekends or at night are also unpredictable. System shutdown often results in complaints. Substantial savings can be achieved while maintaining comfort conditions in a building by an appropriate combination of the following actions:

*   Reset outside air intake to a lower level (0.05 cfm/ft²) during these hours in hot summer and cold winter weather. Outside air can be reduced since there will be very few people in the building. Check outside and exhaust air balance to maintain positive building pressure.
*   Reset the minimum airflow to a lower value, possibly zero, for VAV terminal boxes.
*   Program constant volume terminal boxes as VAV boxes and reset the minimum flow from the maximum to a lower value, possibly zero during unoccupied hours.
*   Reset AHU static pressure and water loop differential pressure to lower values.
*   Set supply air fan at a lower speed.

These measures maintain building comfort while minimizing energy consumption. The savings are often comparable with the shutdown option. Figure 26-4 presents the measured hourly fan energy consumption in the Education Building at the University of Texas at Austin.

The education building has 251,000 ft² of floor area with eight 50-hp AHUs that are operated on VFDs. Prior to the introduction of this measure, the motor control center (MCC) energy consumption was almost constant.

The RCx measure implemented was to set the fan speed at 30% at night and on weekends. The nighttime slow roll decreased the fan power from approximately 50 kW to approximately 25 kW while maintaining building comfort.

26.4.1.5 Limit Fan Speed During Warm-up and Cool-down Periods

If nighttime shutdown is implemented, warm-up is necessary during the winter and cool-down is required during the summer. During warm-up and cool-down periods, fan systems are often run at maximum speed since all terminal boxes require either maximum heating or maximum cooling. A simple fan speed limit can reduce fan power significantly. This principle may also be used in other systems, such as pumps. The following actions should be taken to achieve the fan energy savings:

*   Determine the optimal start up time using 80% (adjustable) fan capacity if automatic optimal start-up is used.
*   Set the fan speed limit at 80% (adjustable) manually and extend the warm-up or cool-down period by 25%. If the speed limit is set at another fractional value (x), determine the warm up period using the following equation:

$$T_n = \frac{T_{exist}}{x}$$

*   Keep outside air damper closed during warm-up and cool-down periods.

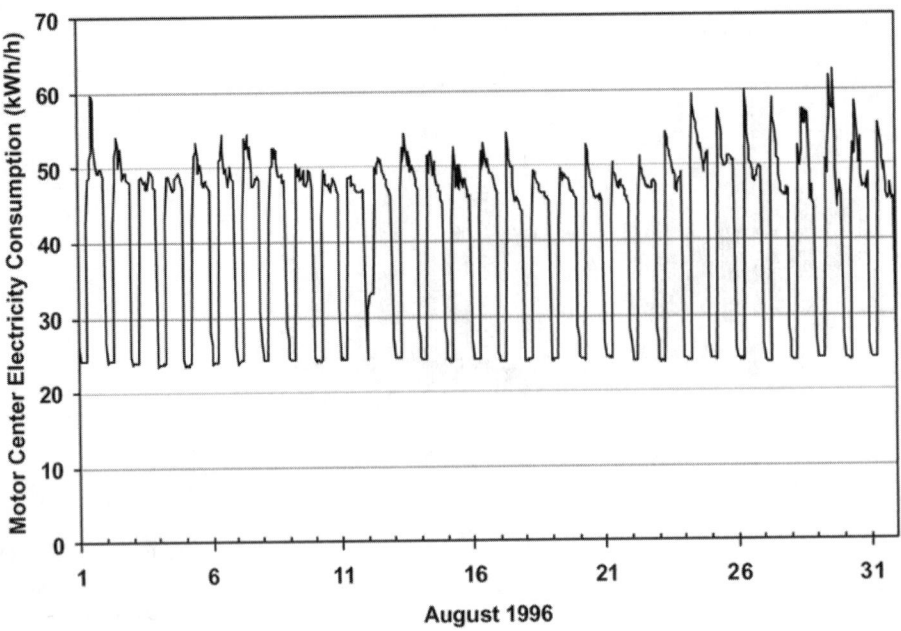

Figure 26-4. Measured Post-RCx hourly supply fan electricity consumption in the education building[15]

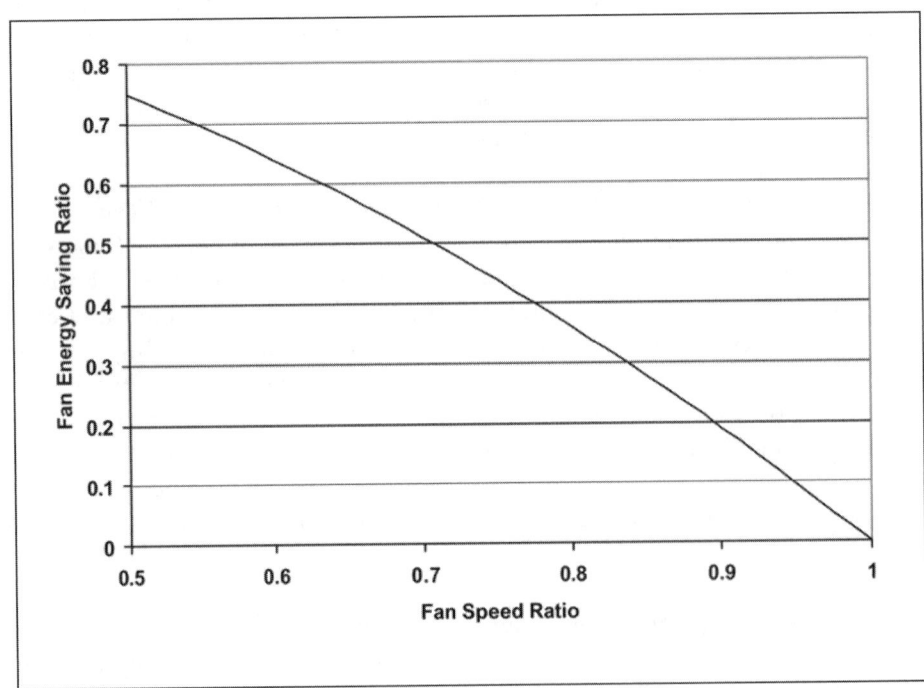

Figure 26-5. Potential fan energy savings using fan speed limiting[16]

The fan energy savings increase significantly as the fan speed limit decreases. Figure 26-5 presents the theoretical fan power savings. When the fan speed limit is 50% of design fan speed, the potential fan energy savings are 75% of the fan energy, even if the fan runs twice as long. The theoretical model did not consider the variable speed drive loss. The actual energy savings will normally be somewhat lower than the model projected value.

Note that if the outside air damper cannot be closed tightly, extra thermal energy may be required to cool or warm up outside air that leaks through the damper. This factor should be considered when this measure is used.

26.4.2 Operational Efficiency Measures for AHU Systems

Air handler systems normally condition and distribute air inside buildings. A typical AHU system consists of some combination of heating and cooling coils, supply and return air fans, filters, humidifiers, dampers, ductwork, terminal boxes, and associated safety and control devices, and it may include an economizer. As the building load changes, AHUs change one or more of the following parameters to maintain building comfort: outside air intake, total airflow, static pressure, and supply air temperature and humidity. Both operating schedules and initial system set up, such as total airflow and outside airflow, significantly impact building energy consumption and comfort. The following ten major RCx measures should be used to optimize AHU operation and control schedules:

- Adjust total airflow for constant air volume systems.
- Set minimum outside air intake correctly.
- Improve static pressure set-point and schedule.
- Optimize supply air temperatures.
- Improve economizer operation and control.
- Improve coupled control AHU operation.
- Valve off hot air flow for dual duct AHUs during summer.
- Install VFD on constant air volume systems.
- Install airflow control for VAV systems.
- Improve terminal box operation.

26.4.2.1 Adjust Total Air Flow and Fan Head for Constant Air Volume Systems

Air flow rates are significantly higher than required in most buildings, primarily due to system over-sizing. In some large systems, an oversized fan causes over-pressurization in terminal boxes. This excessive pressurization is the primary cause of room noise. The excessive airflow often causes excessive fan energy consumption, excessive heating and cooling energy consumption, humidity control problems, and excessive noise in terminal boxes[17].

26.4.2.2 Set Minimum Outside Air Intake Correctly

Outside air intake rates are often significantly higher than design values in existing buildings due to lack of accurate measurements, incorrect design calculations and balancing, and operation and maintenance problems. Excessive outside air intake is caused by the mixed air chamber pressure being lower than the design value, significant outside air leakage through the maximum outside air damper on systems with an economizer, the minimum outside air intake being set to use minimum total airflow for a VAV system, or lower than expected/designed occupancy.

26.4.2.3 Improve Static Pressure Setpoint and Schedule

The supply air static pressure is often used to control fan speed and ensure adequate airflow to each zone. If the static pressure setpoint is lower than required, some zones may experience comfort problems due to lack of airflow. If the static pressure setpoint is too high, fan power will be excessive. In most existing terminal boxes, proportional controllers are used to maintain the airflow setpoint. When the static pressure is too high, the actual airflow is higher than its setpoint. The additional airflow depends on the setting of the control band. Field measurements[18] have found that the excessive airflow can be as high as 20%. Excessive airflow can also occur when terminal box controllers are malfunctioning. For pressure dependent terminal boxes, high static pressure causes significant excessive airflow. Consequently, high static pressure often causes unnecessary heating and cooling energy consumption. A higher than necessary static pressure setpoint is also the primary reason for noise problems in buildings.

26.4.2.4 Optimize Supply Air Temperatures

Supply air temperatures (cooling coil discharge air temperature for single duct systems; cold deck and hot deck temperatures for dual duct systems) are the most important operation and control parameters for AHUs. If the cold air supply temperature is too low, the AHU may remove excessive moisture during the summer using mechanical cooling. The terminal boxes must then warm the over-cooled air before sending it to each individual diffuser for a single duct AHU. More hot air is required in dual duct air handlers. The lower air temperature consumes more thermal energy in either system. If the cold air supply temperature is too high, the building may lose comfort control. The fan must supply more air to the building during the cooling season, so fan power will be higher than necessary. The goal of optimal supply air temperature schedules is to minimize combined fan power and thermal energy consumption or cost. Although developing optimal reset schedules requires a comprehensive engineering analysis, improved (near optimal) schedules can be developed based on several simple rules. Guidelines for developing improved supply air temperature reset schedules are available for four major types of AHU systems[19].

26.4.2.5 Improve Economizer Operation and Control

An economizer is designed to eliminate mechanical cooling when the outside air temperature is lower than the supply air temperature setpoint and to decrease mechanical cooling when the outside air temperature is between the cold deck temperature and a high temperature limit, which is typically less than 70°F. Economizer control is often implemented so it controls mixed air temperature at the cold deck temperature, or simply 55°F. This control algorithm is far from optimum. It may, in fact, actually increase the building energy consumption. The economizer operation can be improved using the following steps:

1. Integrate economizer control with optimal cold deck temperature reset. It is tempting to ignore cold deck reset when the economizer is operating since the cooling is free. However, cold deck reset normally saves significant heating.

2. For a draw-through AHU, set the mixed air temperature 1°F lower than the cold deck temperature setpoint. For a blow-through unit, set the mixed air temperature at least 2°F lower than the supply air temperature setpoint. This will eliminate chilled water valve hunting and unnecessary mechanical cooling.

3. For a dual duct AHU, the economizer should be disabled if the hot air flow is higher than the cold air flow since the heating energy penalty is then typically higher than cooling energy savings.

4. Set the economizer operating range as wide as possible. For dry climates, the economizer should be activated when the outside air temperature is between 30°F and 75°F, between 30°F and 65°F for normal climates, and between 30°F and 60°F for humid climates. When proper return and outside air mixing can be achieved, the economizer can be activated even when the outside air temperature is below 30°F.

5. Measure the true mixed air temperature. Most mixing chambers do not achieve complete mixing of the return air and outside air before reaching the cooling coil. It is particularly important that mixed air temperature be measured accurately when an economizer is being used. An averaging temperature sensor should be used for the mixed air temperature measurement.

26.4.2.6 Improve Coupled Control AHU Operation

Coupled control is often used in single-zone, single-duct, constant volume systems. Conceptually, this system provides cooling or heating as needed to maintain the setpoint temperature in the zone and uses simultaneous heating and cooling only when the humidistat indicates that additional cooling (followed by reheat) is needed to provide humidity control. However, the humidistat is often disabled for a number of reasons. To control room relative humidity level, the control signals or spring ranges are overlapped. Simultaneous heating

and cooling often occurs almost continuously.

26.4.2.7 Valve Off Hot Air Flow for Dual Duct AHUs During Summer

During the summer, most commercial buildings do not need heating. Theoretically, hot air should be zero for dual duct VAV systems. However, hot air leakage through terminal boxes is often significant due to excessive static pressure on the hot air damper. For constant air volume systems, hot air flow is often up to 30% of the total airflow. During summer months, hot air temperatures as high as 140°F have been observed due to hot water leakage through valves. The excessively high hot air temperature often causes heat complaints in some locations. Eliminating this hot air flow can improve building thermal comfort, reduce fan power, cooling consumption, and heating consumption.

26.4.2.8 Install VFD on Constant Air Volume Systems

The building heating load and cooling load varies significantly with both weather and internal occupancy conditions. In constant air volume systems, a significant amount of energy is consumed unnecessarily due to humidity control requirements. Most of this energy waste can be avoided by simply installing a VFD on the fan without a major retrofit effort. Guidelines for VFD installation are available for dual duct, multi-zone, and single duct systems[20].

26.4.2.9. Airflow Control for VAV Systems

Airflow control of VAV systems has been an important design and research subject since the VAV system was introduced. An airflow control method should: (1) ensure sufficient airflow to each space or zone; (2) control outside air intake properly; and (3) maintain a positive building pressure. These goals can be achieved using the variable speed drive volume tracking (VSDVT) method[21,22].

26.4.2.10. Improve Terminal Box Operation

The terminal box is the end device of the AHU system. It directly controls room temperature and airflow. Improving the set up and operation are critical for room comfort and energy efficiency. The following measures are suggested:

*   Set minimum air damper position properly for pressure dependent terminal boxes.
*   Use VAV control algorithm for constant air volume terminal boxes.
*   Use airflow reset.
*   Integrate lighting and terminal box control.
*   Integrate airflow and temperature reset

- Improve Terminal Box Control Performance

26.4.3 Case Study— AHU RCx[23]
26.4.3.1 Facility Description and Energy Use

The case study building is a 2-story building with a basement and an HVAC penthouse. (See Figure 26-6.) The total conditioned space is 99,579 ft^2 of which the mechanical rooms in the basement and penthouse account for approximately 20%. Heating and cooling for the building are provided by two 4 MMBtu/hr hot water boilers and three 225 ton electric screw chillers. Five single duct, variable air volume (VAV) systems, with reheat at the terminal boxes, serve the building. Air handler units 1 and 4 serve the exterior zones, which are primarily office space; air handler unit 5 provides conditioned air for a conference room; and the remaining air handlers, units 2 and 3, were originally designed to use 100% outside air and serve laboratory areas, which are interior zones. Sometime near the end of the design phase or just prior to the construction phase, a decision was made to install returns with dampers for AHUs 2 and 3. This was done because it was anticipated that

significant areas would not be used as laboratories. Each of the AHUs contains both a chilled water coil and a cooling coil that is connected to the cooling tower sump through a heat exchanger. The cooling towers are used to provide building cooling for about six months per year. The energy management control system (EMCS) is a Siemens Apogee system. The direct digital control (DDC) hierarchical level is capable of monitoring and controlling down to the VAV terminal box.

Energy use in the building is measured using two electrical meters and a gas meter. The electrical meters provide hourly readings, while the gas meter is normally read on the first day of each month. When the RCx project began, daily readings were made for several months. One of the electric meters primarily monitors a number of computer servers, so the nominal load of 100 kW on this meter shows very little variation throughout the day or the year. The other electric meter monitors all remaining electricity use in the building, including the chillers and distribution systems.

Gas use from November 2003-November 2004 is shown in Figure 26-7. Total use for the 12-month period beginning with December, 2003, which was used as the baseline period for this project, was 3464 MCF. An average gas cost of $7.00/MCF was used. Electricity use was 3,070,189 kWh during the same December-November period, for an average use level of approximately 350 kW as shown in Figure 26-8. Figure 26-8 shows both total electricity consumption in the top series and process electricity consumption in the bottom series. The electricity price paid for September-November, 2004 averaged $0.0407/kWh, so this value was used as the basis for this project. Using these prices, annual baseline energy costs for the building totaled $149,205, including $24,248 for gas and $124,957 for electricity as shown in Table 26-3.

Examination of the pattern of gas consumption shows that summer use is typically half of the winter consumption, indicating that there is significant reheat in the building. Similar examination of the electricity consumption pattern shows that base process consumption increased by 15-20 kW during May and that chillers were used beginning in early May and continuing through late October, with scattered use during November. Chiller operation appears to have been continuous from about mid-June through mid-August. Analysis of the consumption data, observations of AHUs and

Figure 26-6. The case study building

Use	Annual Consumption	Cost ($/year)
Electricity	3,070,189 kWh	$124,957
Gas	3464 MCF	$24,248
Baseline Energy Cost		$149,205
Estimated HVAC Use/Cost		
Heating	3464 MCF	$24,248
Chiller cooling	224,400 kWh	$9,135
Fans and pumps	475,000 kWh	$19,337
Total HVAC Cost		$52,720

Table 26-3. Electricity and gas consumption for the building from December 2003-November 2004, including estimated HVAC use.

pumps during the site visit (including selected pressure measurements and flow measurements); and use of EMCS schedule information lead to the estimates of HVAC consumption provided in Table 26-3.

26.4.3.2 HVAC Systems and Operation

On each single duct VAV unit, the supply fan speed is controlled by the duct static pressure setpoint. The control algorithms indicate that constant value

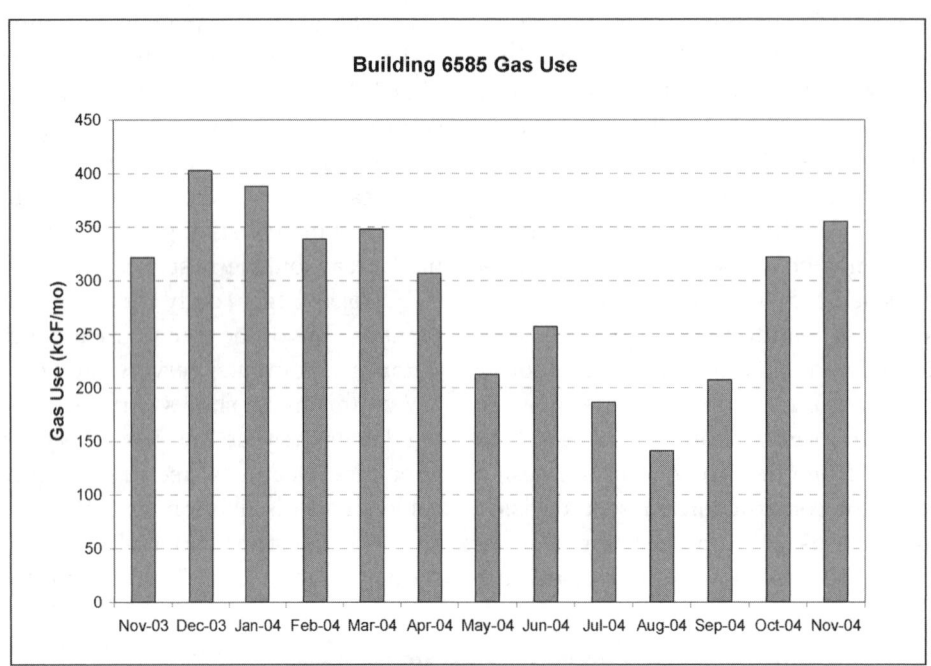

Figure 26-7. Gas use for the building from December, 2003 through November, 2004 in thousands of cubic feet per month.

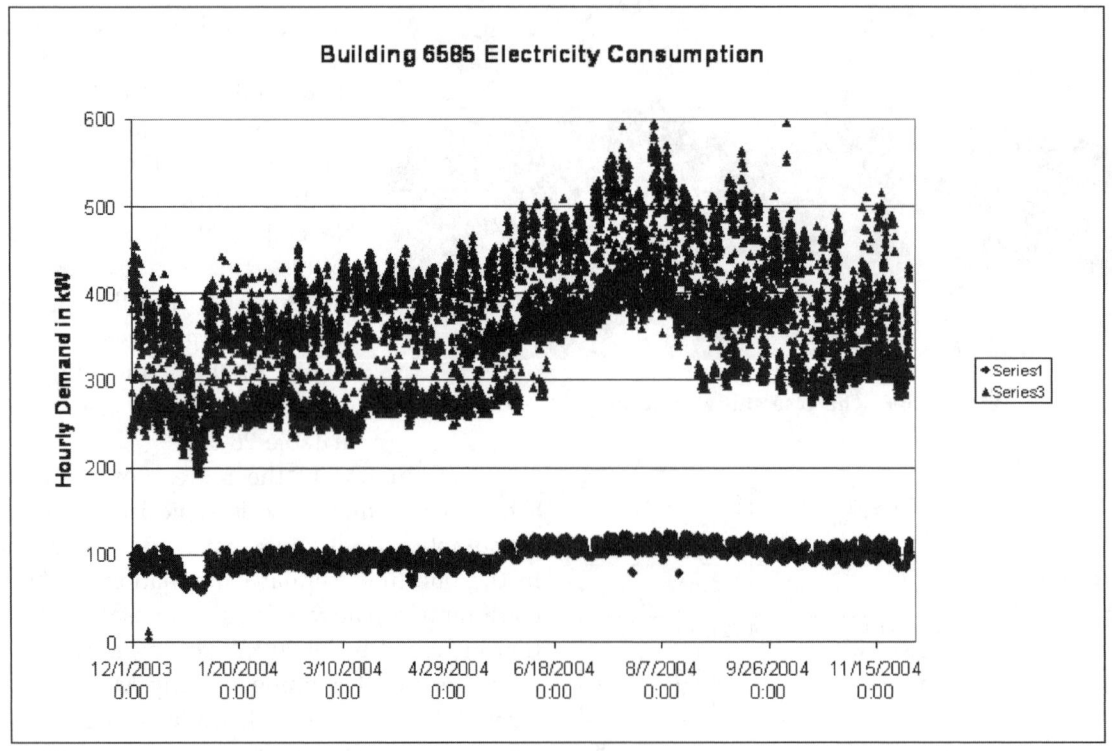

Figure 26-8. Total hourly electricity use for the building from December, 2003 through November, 2004 in kW (upper data). Bottom data are process consumption.

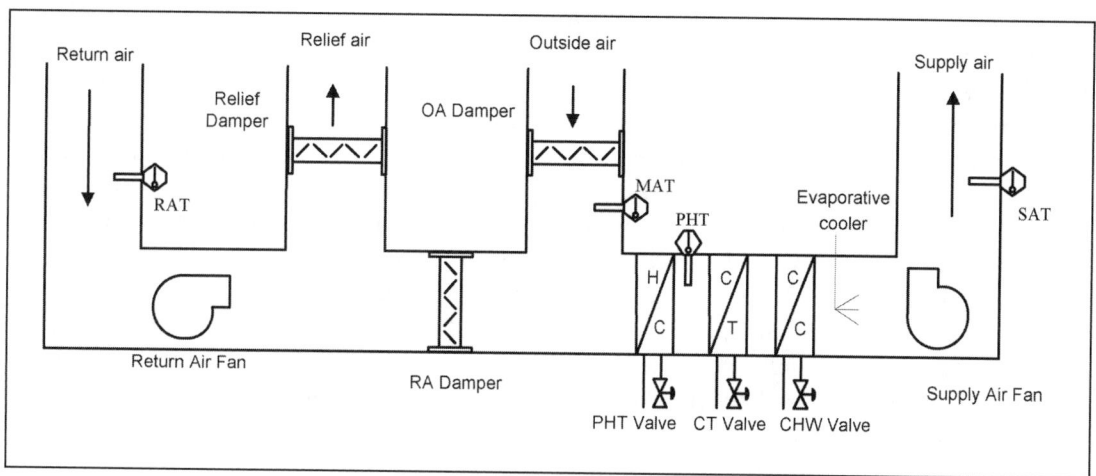

Figure 26-9. Air handler schematic for units 1, 4, and 5

setpoints are used. The static pressure setpoints range from 1.0-1.8 in. H_2O.

A dedicated chilled water and hot water system is provided for the building cooling and heating loads. Three chillers (with R-22, each at 225 tons) are operating in a parallel configuration supplying chilled water to a primary/secondary distribution system. The logged runtime hours of the chillers have averaged 2621 hours/year since the building was built. This corresponds to 24-hour operation of a single chiller for 2 months with an average of 33 hours per week of chiller operation for an additional four months. An additional source of cooling for the building is provided by the cooling tower system. Three cooling towers are located on the south side of the building. Each tower has a nominal cooling capacity of 235 tons.

The building hot water system is located in the penthouse. Heating is provided by two Cleaver Brooks natural gas water tube boilers (4 MMBTUH max. output/boiler). The hot water system is configured in a primary/secondary distribution system. The secondary hot water distribution system is driven by two variable speed hot water pumps (85 GPM each). In addition to the secondary distribution system, two variable speed pumps (41 GPM each) deliver hot water to the laboratory section.

RCx Assessment

A retro-commissioning (RCx) assessment of the building was conducted in January 2005. The assessment began with a meeting between two RCx professionals, the energy manager, two other site engineers, and the building controls technician (energy management team or EMT). At this meeting the HVAC system characteristics and control as currently implemented

in the building were discussed, and current gas and electricity consumption data were reviewed. They revealed that the building basically works well, but site staff indicated that they expected opportunities to improve operational characteristics of the building since the basic building operation was determined and set up as part of a turn-key project in the mid-1990s and has not been optimized since then. This was followed by examination and printing of EMCS screens providing current operating status for all major air handlers, chillers, boilers, and water side distribution systems in the building.

A walkthrough of the building was conducted in the afternoon and on the morning of the next day by the RCx personnel and EMT personnel. This walkthrough was primarily devoted to a detailed examination of the systems in the basement and penthouse mechanical rooms, supplemented by visits to several offices on the first floor. Measurements of key temperatures, flows and pressures were made during the walkthrough. Information obtained during the walkthrough, supplemented by building drawings, energy consumption data, and additional information supplied by the EMT was subsequently analyzed to identify a preliminary list of RCx measures recommended for implementation in the building.

26.4.3.3 Observations and Findings of the Walkthrough

Air handler units 3, 4, and 5 were started at approximately 4:00 a.m. and stopped at 7:00 p.m., Monday through Friday. Air handler unit 2 was started at approximately 5:00 a.m. and stopped at 6:00 p.m. Air Handler unit 1 was operating on a continuous basis. During the periods when an air handler unit is scheduled off, occupants can use the override buttons located on the thermostats to activate the air handler unit for a two hour period. Static pressure setpoints on all five AHUs were constant values.

In evaluating the chilled water and hot water systems, special attention was given to the positioning of all manual valves. The manual valves located on the discharge side of the secondary chilled water pumps were 50% closed. The manual valves located on the chilled water return lines for each air handler were also 50% closed. Rebalancing was recommended to reduce the pumping power needed to supply the loop.

Based on information gathered through discussion with facility personnel, sensor calibration was not performed except when a problem was noted. VAV box calibration plays an important role in the reduction of fan power. Because of time constraints, verification of maximum and minimum air flows for individual terminal boxes was not possible. Because functions within the building change, minimum design flow settings may exceed the necessary airflow requirements. The combined minimum supply flow of 46,800 cfm currently set on the terminal boxes lead to requirements for reheat during a significant portion of the year and contributed to the relatively high reheat observed in the building. It was recommended that minimum flow requirements for each box be evaluated and minimum flow settings be reduced where appropriate.

The preliminary list of RCx measures recommended for adoption were:

1. Spot check calibration of existing sensors.
2. Spot check VAV boxes, determine required minimum flows, and reduce where appropriate.
3. Develop and implement optimum start-stop strategy for each AHU.
4. Develop and implement a static pressure reset schedule based on outside air temperature for each AHU.
5. Optimize chilled and hot water secondary loop performance. Installation of additional temperature sensors may be required to monitor the ΔT for each loop.

Measure minimum outside air flow settings and reduce when they exceed the amount needed to meet Standard 62, or increase if necessary.

Examine and optimize combined economizer/tower cooling control strategy and operation.

Optimize the preheat control strategy. Reheat the supply air at the terminal box only. Use preheat for coil freeze protection.

Evaluate the supply temperature reset strategy and optimize to minimize fan power and heating and cooling energy.

It was conservatively estimated that implementation of the RCx measures would provide annual HVAC operational savings of $23,086 and would cost $50,000.

26.4.3.4 Implementation of Retro Commissioning (RCx) Measures

To begin implementation of the RCx measures, each system (AHUs, hot water, and cooling system) was set up for trending on the EMCS. All analog input and output points, in addition to on/off and status points were trended. Time series plots were developed for each system and analyzed. System performance problems, as well as physical component problems, were identified.

Optimization of Air Handler Operation Schedule

Normal operating hours for the building were from 0600 hours to 1800 hours Monday-Friday. Normally, the air handler equipment was scheduled off for Saturday and Sunday.

It was learned that AHU 1 had been temporarily set to operate continuously in December, 2004. However, when the normal operating schedule was put in place, it still operated continuously due to a problem in the control program. Hence it had apparently been operating continuously for an unknown period, perhaps since the building was built. This problem was located and corrected, and operation then returned to the normal schedule.

The trend data showed that the building reached occupied conditions within 15 minutes of startup and that the optimum start/stop programs were not functioning properly. It was determined that the optimum start algorithms were complex and apparently contained one or more bugs. The RCx team recommended that the optimum start/stop programs be temporarily removed until the programming could be corrected. The start times for each AHU were pushed back to 0600 hours. The stop time of 1800 hours was not altered. Building personnel that choose to work outside the normal operating time schedule can use the occupancy override button located at the thermostat to run the corresponding AHU for a two hour period. After the two hour period the AHU will shut off.

Preheat Control Strategy Optimization

Preheat control for AH01-AH04 was modified from its original discharge air temperature control strategy. It is based on mixed air temperature and outside air temperature.

Static Pressure Optimization

Static pressure sensors for each AHU were located

and spot checked for accuracy by the RCx engineers. The static pressure transducer for AH02 was found to be faulty. The EMCS system showed that AH02 was supplying 1.5 in. H_2O to the system, but field measurements indicated that 0.6 in. H_2O was actually being supplied. The building HVAC maintenance technician replaced the faulty transducer. Once sensor verification was complete, static pressure measurements were taken at each box located at the end-of-line for each duct system. These measurements revealed an excessive amount of static pressure for all the AHU systems.

The terminal boxes used in these AHU systems require a minimum static pressure of 0.17-0.22 in. H_2O in order to operate properly. Field measurements taken by RCx engineers indicated that a 0.5 in. H_2O static pressure setpoint for each AHU system would satisfy the maximum airflow requirements for the most remote box in each system. In each system the adjusted static pressure setpoint could have been reduced further. However, limitations of the variable frequency drives (VFDs) prohibit this from happening.

Assuming that building cooling and heating loads are linear functions of outside air temperature, static pressure reset schedules based on outside air temperature were implemented for each AHU system serving exterior zones. This reset schedule includes three stages. When the outside air is below 40°F the static pressure setpoint will maintain a constant minimum value (0.5 in H_2O); when the outside air temperature is above 90°F the static pressure setpoint will not exceed its maximum value (0.8 in. H_2O). Between 40°F and 90°F the setpoint varies linearly between the maximum and minimum settings.

Optimization of Hot Water Loop

All the manual valves on the preheat coils were opened (office loop). The RCx engineers attached gages at the end-of-line location, AH01 preheat coil, and then used the control system to place all the preheat valve positions on each AHU at 100%. Based on information provided from the AHU schedule, the pressure drop across the preheat coils was approximately 1.2 psi. This was the target differential pressure needed for the coil. Being conservative and adjusting for possible weather abnormalities that may develop, the differential setpoint for the office loop was set at 5 psi.

The lab heating loop serves only the reheat coils for the interior sections. Typically, internal spaces don't require heat and outside weather conditions don't influence the system. The differential pressure for the lab loop was reduced to 5 psi. However, the heating demand was minimal for this loop, and setpoint could not be achieved with the secondary pump running at minimum speed. The lowest obtainable differential pressure for this loop was 10 psi. Without the secondary pumps running, the differential pressure of 8.0 psi was still above the 5 psi setpoint. The boiler constant speed pump was capable of meeting the needs to the system.

The existing control strategy for each secondary loop system was modified. Secondary pumps were turned off and allowed to cycle on as required to maintain the 5 psi setpoint. Deadbands were used to eliminate unnecessary cycling of the secondary pumps. The manual bypass between the primary and secondary loop could not be completely closed, because each secondary loop does not contain a bypass or three-way valve to prevent pump dead heading. It was closed approximately 75%, forcing supply hot water into the secondary loops instead of re-circulating the water back to the return side of the primary loop.

Terminal Box Optimization

Measurements of airflow were taken and compared to flow values reported by the DDC control system for several boxes. Typically, these boxes were the ones used to determine the required static pressure for the system. Some boxes required that the flow coefficient be recalculated to correct the reading from the control system. Concluding that no major problems exist with the box flow stations, investigation into possible minimum airflow reduction was pursued.

The original minimum flow settings for most terminal boxes were found to be approximately 30% of their maximum settings. The design maximum cooling flow for this building was greater than 1.5 CFM per square foot, resulting in minimum flow settings of approximately 0.5 CFM per square foot. This is a fairly typical minimum flow setting, but it was causing terminal boxes to use significant reheat in this building. To reduce the amount of reheat, minimum air flow settings were lowered. The airflow reductions were based on lighting density, plug loads, and observed space loads. The majority of minimum flow settings were reduced by 50% and in some cases were reduced even further. Design sizing of the terminal boxes prevents further reduction of airflow in many cases.

At the beginning of commissioning, each box's temperature setpoint was controlled separately by the occupants of the space. Space setpoints varied throughout the building to satisfy the comfort needs of the occupants. This means that adjacent zones could be in different modes (heating or cooling). A standard of 74°F for cooling and 70°F for heating has been established for the building. Occupants will have the capability to make minor adjustments to the thermostat as needed.

26.4.3.5 Energy Impact of RCx Implementation

RCx measures were implemented, beginning on February 14 and continuing through March. Figure 26-10 shows that electric consumption immediately dropped by about 50 kW. The figure shows (from top to bottom) a time series plot of the total electricity consumption in the building, the non-process consumption, and the process consumption.

It is clear that consumption has dropped by about 50 kW during both weekdays and on weekends.

Figure 26-11 shows the baseline monthly gas consumption data with a 3-parameter model of the monthly average baseline use as a function of temperature, a common behavior for gas consumption (IPMVP 2001). Daily gas consumption data are also shown for the two-week period while measures were being implemented and for five subsequent periods of about four weeks each. It is clear that consumption dropped immediately, with consumption during the last two periods shown to be less than 50% of the baseline consumption. The large circles show the average consumption for each of the three periods plotted.

The amount of post-RCx data shown are still quite limited, but it is clear that savings will exceed the projected savings of $23,086 per year. Table 26-4 shows the actual savings for the February/March-July periods shown, with annual projections based on use of

3-parameter models of the gas and electricity use during the baseline period and models of the March-July data for the post-RCx period. Weather data for August 2004-February 2005 was used to estimate savings for the remainder of the year. It appears the savings exceeded 50% of the baseline HVAC consumption in this building.

26.4.3.5 Lessons Learned

- Oversized VAV AHUs tend to have minimum flow values that cause excess reheat.
- Static pressure setpoints should be determined by measurement in the hydraulically remote terminal boxes. Design values tend to waste fan power.
- Trend AHU operation periodically to be sure schedules haven't been changed.
- Track consumption to be sure efficiency is maintained.

26.4.4   RCx Measures for Water/Steam Distribution Systems

Distribution systems include central chilled water, hot water, and steam systems that deliver thermal energy from central plants to buildings. In turn, the system distributes the chilled water, hot water, and steam to AHU coils and terminal boxes. Distribution systems mainly consist of pumps, pipes, control valves, and variable speed pumping devices. This section focuses on the

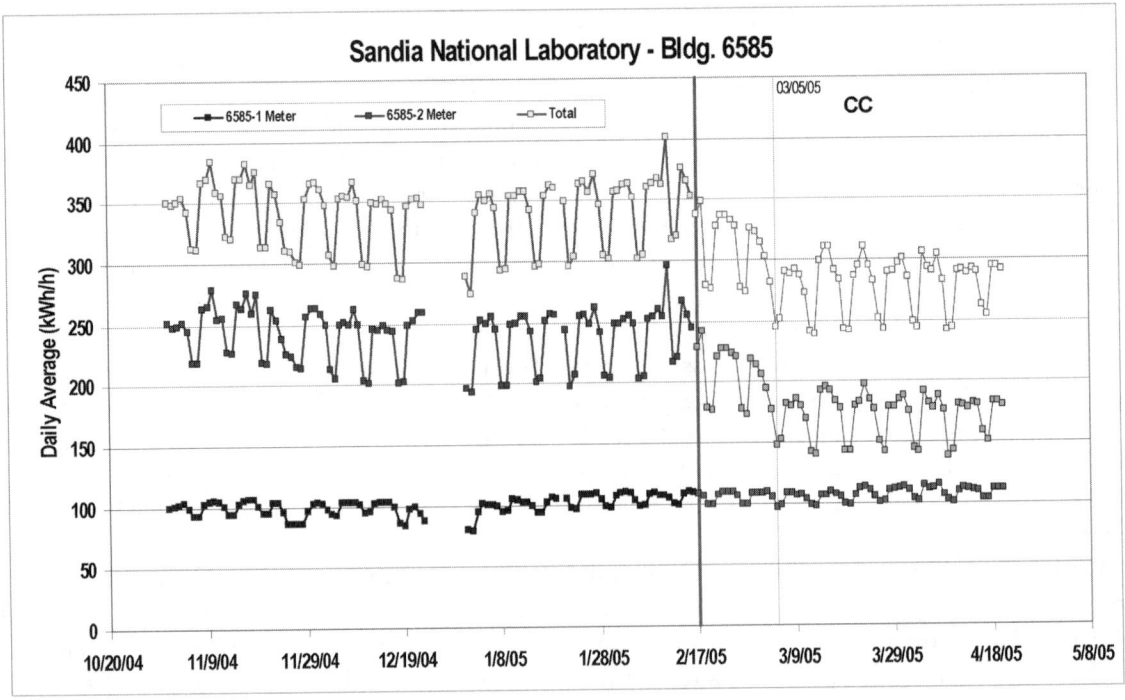

Figure 26-10. Time Series of the Electricity Use for the Building for the Period Immediately Before and After the RCx Measures were Implemented

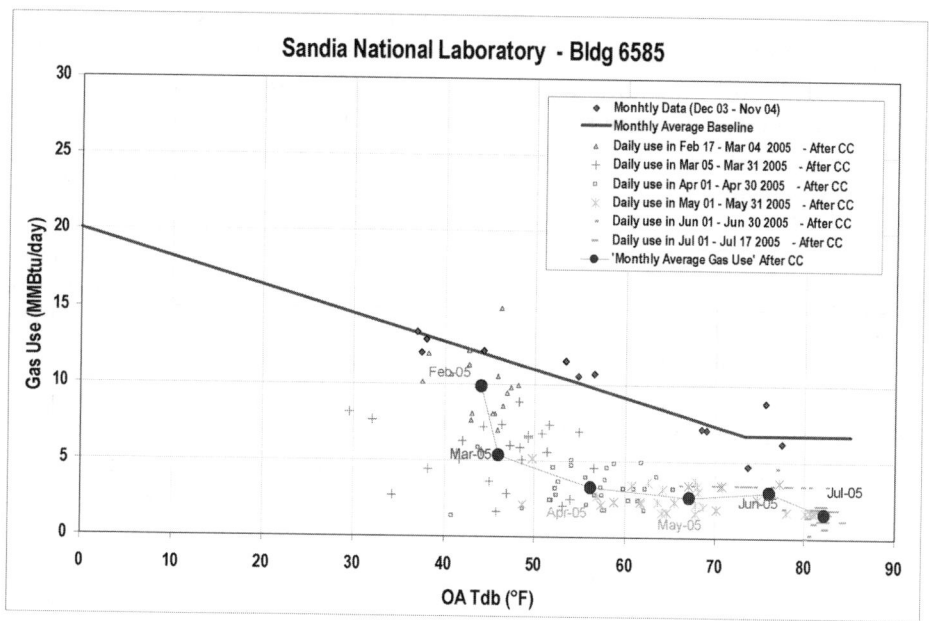

Figure 26-11. Baseline and daily gas use after RCx measures implemented at the building

Table 26-4. Savings from implementation of RCx measures in the building

Energy Type	Savings Period	Savings	Annual Projected
Electricity (kWh)	Mar 1 – July 17, 2005	$5,200 (127,756)	$16,018
Gas use (MCF)	Feb 17 – July 17, 2005	$5,295 (756)	$14,251 (2,036)
		TOTAL	$30,269

RCx measures for optimal pressure control, water flow control, and general optimization.

26.4.4.1 Improve Building Chilled Water Pump Operation

Most building chilled water pumping systems are equipped with variable speed devices (VSDs). If a VSD is not installed, retrofit of a VSD is generally recommended. The discussion here is limited to systems where a VSD is installed. The goal of pumping optimization is to avoid excessive differential pressures across the control valves, while providing enough water to each building, coil, or other end use. An optimal pump differential pressure schedule should be developed that provides adequate pressure across the hydraulically most remote coil in the system under all operating conditions but does not provide excess head.

26.4.4.2 Improve Secondary Loop Operation

For buildings supplied by a secondary loop from a central plant, building loop optimization should be performed before the secondary loop optimization.

Source Distributed Systems

If there are no building pumps, the secondary pumps must provide the pressure head required to overcome both the secondary loop and the building loop pressure losses. In this case, the secondary loop is called a source distributed system. The secondary loop pumps should be controlled to provide enough pressure head for the most remote coil. If VFDs are installed, the differential pressure can be controlled by modulating pump speed. Otherwise, the differential pressure can be modulated by changing the number of pumps in operation.

Source Distributed Systems with Building Pumps

In most campus settings, each building has a pump. The optimal differential pressure setpoint should then be determined by optimizing the secondary loop pressure setpoint so the combined secondary pump and building pumping power is minimized. This can be done by developing a pressure reset schedule that requires maximum building pump power at the most hydraulically remote building on the secondary loop. This may occur with a negative differential pressure across the most remote building.

26.4.4.3 Improving Central Plant Water Loop Operation

The central plant loop optimization should be performed after secondary loop optimization.

Single Loop Systems

For most heating distribution systems and some chilled water systems, a single loop is used instead of primary and secondary systems. Under partial load conditions, fewer pumps can be used for both chillers and heat exchangers. This can result in less pump power consumption.

Primary and Secondary Loop Systems

Primary and secondary systems are the most common chilled water distribution systems used with central chiller plants. This design is based on the assumption that the chilled water flow through the chiller must be maintained at the design level. This is seldom needed. Due to this incorrect assumption, a significant amount of pumping power is wasted in numerous central plants. Design engineers may or may not include an isolation valve on the by-pass line of the primary loop. Procedures are available to optimize pump operation for both cases[23,24,25,26].

26.4.5 RCx Measures for Central Chiller Plants

The central chiller plant includes chillers, cooling towers, a primary water distribution system, and the condenser water distribution system. Although a secondary pumping system may be physically located inside the central plant, commissioning issues dealing with secondary loops are discussed in the previous section. The central chiller plant produces chilled water using electricity, steam, hot water, or gas. The detailed commissioning measures vary with the type of chiller, and this section gives general commissioning measures that apply to a typical central cooling plant that can produce significant energy savings.

Use the Most Efficient Chillers

Most central chiller plants have several chillers with different performance factors or efficiencies. The differences in performance may be due to the design, performance degradation, age, or operational problems. One chiller may have a higher efficiency at a high load ratio, while another may have a higher efficiency at a lower load ratio. Running chillers with the highest performance can result in significant energy savings and will also reduce the number of complaints, because you will be providing the greatest output for the least input.

Reset the Supply Water Temperature

Increasing the chilled water supply temperature can decrease chiller electricity consumption significantly. The general rule-of-thumb is that a one degree Fahrenheit increase corresponds to a decrease in compressor electricity consumption of 1.7%. The chilled water supply temperature can be reset based on either cooling load or ambient conditions.

Reset Condenser Return Water Temperature

Decreasing cooling tower return water temperature has the same effect as increasing the chilled water supply temperature. The cooling tower return temperature should be reset based on weather conditions. The following provides general guidelines:

- The cooling tower return water temperature setpoint should be at least 5°F (adjustable based on tower design) higher than the ambient wet bulb temperature. This prevents excessive cooling tower fan power consumption.
- The cooling tower water return temperature should not be lower than 65°F for chillers made before 1999, and it should not be lower than 55°F for newer chillers. It is also recommended that you consult the chiller manufacturer's manual for more information.

The cooling tower return water temperature reset can often be implemented using the BAS. If it cannot be implemented using the BAS, operators can reset the setpoint daily, using the daily maximum wet bulb or dry bulb temperature.

Decreasing the cooling tower return temperature may increase fan power consumption. However, fan power may not necessarily increase with lower cooling tower return water temperature. The following tips can help.

- Use all towers. For example, use all three towers when one of the three chillers is used. This may eliminate fan power consumption entirely. The pump power may actually stay the same. Be sure to keep the other two tower pumps off.
- Never turn on the cooling tower fan before the bypass valve is totally closed. If the bypass valve is not totally closed, the additional cooling provided by the fan is not needed and will not be used. Save the fan power!
- Balance the water distribution to the towers and within the towers. Towers are often seen where water is flowing down only one side of the tower, or one tower may have twice the flow of another. This significantly increases the water return temperature from the towers.

Increase Chilled Water Return Temperature

Increasing chilled water return temperature has

the same effect as increasing chilled water supply temperature. It can also significantly decrease the secondary pump power since the higher the return water temperature (for a given supply temperature), the lower the chilled water flow. Maximizing chilled water return temperature is much more important than optimizing supply water temperature since it often provides much more savings potential. It is hard to increase supply temperature 5°F above the design setpoint. It is often easy to increase the return water temperature as much as 7°F by conducting water balance and shutting off bypass and three way valves.

Use Variable Flow under Partial Load Conditions

Typical central plants use primary and secondary loops. A constant speed primary pump is often dedicated to a particular chiller. When the chiller is turned on, the pump is on. Chilled water flow through each chiller is maintained at the design flow rate by this operating schedule. When the building-loop flow is less than the chiller loop flow, part of the chiller flow bypasses the building and returns to the chiller. This practice causes excessive primary pump power consumption and excessively low entering water temperature to the chiller, which increases the compressor power consumption.

It is generally perceived that chilled water flows must remain constant for chiller operational safety. Actually, most new chillers allow chilled water flow as low as 30% of the design value. The chilled water flow can be decreased to be as low as 50% for most existing chillers if the proper procedures are followed[27].

Varying chilled water flow through a chiller can result in significant pump power savings. Although the primary pumps are kept on all the time, the secondary pump power consumption is decreased significantly when compared to the conventional primary and secondary system operation. Varying chilled water flow through the chillers will also increase the chiller efficiency when compared to constant water flow with chilled water bypass. More information can be found in a paper by Liu[28].

Optimize Chiller Staging

For most chillers, the kW/ton decreases (COP increases) as the load ratio increases from 40% to 80%. When the load ratio is too low, the capacity modulation device in the chiller lowers the chiller efficiency. When the chiller has a moderate load, the capacity modulation device has reasonable efficiency. The condenser and evaporator are oversized for the load under this condition, so the chiller efficiency is higher. When the chiller is at maximum load, the evaporator and condenser

have a smaller load ratio, reducing the chiller efficiency below its maximum value. Running chillers in the high efficiency range can result in significant electrical energy savings and can improve the reliability of plant operation. Optimal chiller staging should be developed using the following procedures.

- Determine and understand the optimal load range for each chiller. This information should be available from the chiller manufacturer. For example, chiller kW/ton typically has a minimum value when the chiller load is somewhere between 50% and 70% of the design value.
- Turn on the most efficient chiller first. Optimize the pump and fan operation accordingly.
- Turn on more chillers to maintain the load ratio (chiller load over the design load) within the optimal efficiency range for each chiller. This assumes that the building bypass is closed. If the building bypass cannot be closed, the minimum chiller load ratio should be maintained at 50% or higher to limit primary pumping power increases

Maintain Good Operating Practices

The operating procedures recommended by the manufacturer should be followed. It is important to calibrate the temperature, pressure, current sensors, and the flow switches periodically. The temperature sensors are especially important for maintaining efficient operation. Control parameters must be set properly, particularly the time delay relay.

26.4.6 RCx Measures for Central Heating Plants

Central heating plants produce hot water, steam, or both, typically using either natural gas, coal, or oil as fuel. Steam, hot water, or both are distributed to buildings for HVAC systems and other end uses, such as cooking, cleaning, sterilization, and experiments. Boiler plant operation involves complex chemical, mechanical, and control processes. Energy performance and operational reliability can be improved through numerous measures. However, the RCx measures discussed in this section are limited to those that can be implemented by an operating technician, operating engineers, and RCx professionals.

26.4.6.1 Optimize Supply Water Temperature and Steam Pressure

Steam pressure and hot water temperature are the most important safety parameters for a central heating plant. Reducing the boiler steam pressure and hot water temperature has the following advantages:

- Improves plant safety.
- Increases boiler efficiency and decreases fuel consumption.
- Increases condensate return from buildings and improves building automation system performance.
- Reduces hot water and steam leakage through malfunctioning valves.

26.4.6.2 Optimize Feedwater Pump Operation

The feedwater pump is sized based on boiler design pressure. Since most boilers operate below the design pressure, the feedwater pump head is often significantly higher than required. This excessive pump head is often dropped across pressure reducing valves and manual valves. Installing a VSD on the feedwater pump in such cases can decrease pump power consumption and improve control performance. Trimming the impeller or changing feedwater pumps may also be feasible, and the cost may be lower. However, the VSD provides more flexibility, and it can be adjusted to any level. Consequently, it maximizes the savings and can be adjusted to future changes as well.

26.4.6.3 Optimize Airside Operation

The key issues are excessive airflow and flu gas temperature control. Some excess airflow is required to improve the combustion efficiency and avoid having insufficient combustion air during fluctuations in airflow. However, excessive airflow will consume more thermal energy since it has to be heated from the outside air temperature to the flue gas temperature. The boiler efficiency goes down as excessive airflow increases. The flue gas temperature should be controlled properly. If the flue gas temperature is too low, acid condensation can occur in the flue. If the flue gas temperature is too high, it carries out too much thermal energy. The airside optimization starts with a combustion analysis that determines the combustion efficiency based on the flu gas composition, flu gas temperature, and fuel composition. The typical combustion efficiency should be higher than 80%. If the combustion efficiency is lower than this value, available procedures[29,30] should be used to determine the reasons.

26.4.6.4 Optimize Boiler Staging

Most central plants have more than one boiler. Using optimal staging can improve plant energy efficiency and reduce maintenance cost. The optimal staging should be developed using the following guidelines.
- Measure boiler efficiency.
- Run the higher efficiency boiler as the primary system, and run the lower efficiency boiler as the

back up system.
- Avoid running any boiler at a load ratio less than 40% or higher than 90%.
- If two boilers are running at average load ratios less than 60%, no stand-by boiler is necessary. If three boilers are running at loads of less than 80%, no stand by boiler is necessary.

Boiler staging involves boiler shut off, start up, and standby. Realizing the large thermal inertial and the temperature changes between shut off, standby, and normal operation, precautions must be taken to prevent corrosion damage and expansion damage. Generally speaking, short-term (monthly) turn on/off should be avoided for steam boilers. Hot water boilers are sometimes operated to provide water temperatures as low as 80°F. This improves distribution efficiency but may lead to acid condensate in the flue. The hot water temperature must be kept high enough to prevent this condensation.

26.4.6.5 Improve Multiple Heat Exchanger Operation

Heat exchangers are often used in central plants or buildings to convert steam to hot water, or high temperature hot water to lower temperature hot water. If more than one heat exchanger is installed, use as many heat exchangers as possible, provided the average load ratio is 30% or higher. This approach provides the following benefits.
- Lower pumping power. For example, if two heat exchangers are used instead of one under 100% load, the pressure loss through the heat exchanger system will be decreased by 75%. The pumping power will also be decreased by 75%.
- Lower leaving temperature on the heat source. The condensate should be super-cooled when the heat exchangers are operated at low load ratio. The exit hot water temperature will be lower than the design value under the partial load condition. This will result in less water or steam flow and more energy extracted from each pound of water or steam. For example, the condensate water may be sub-cooled from 215°F to 150°F under low heat exchanger loads. Compared with leaving the heat exchanger at 215°F, each pound of steam delivers 65 Btu more thermal energy to the heat exchanger.

Using more heat exchangers will result in more heat loss. If the load ratio is higher than 30%, the benefits mentioned above normally outweigh the heat loss. More information can be found in a paper by Liu et al.[31]

26.4.6.6 Maintain Good Operating Practices

Central plant operation involves both energy efficiency and safety issues. Proper safety and maintenance guidelines should be followed. The following maintenance issues should be carefully addressed.

- Blowdown: Check blowdown setup if a boiler is operating at partial load most of the time. The purpose of blowdown is to remove the mineral deposits in the drum. This deposit is proportional to the cumulative make-up water flow, which is then proportional to the steam or hot water production. The blowdown can often be set back significantly. If the load ratio is 40% or higher, the blowdown can be reset proportional to the load ratio. If the load ratio is less than 40%, keep the blowdown rate at 40% of the design blowdown rate.

- Steam traps: Check steam traps frequently. Steam traps still have a tendency to fail, and leakage costs can be significant. A steam trap maintenance program is recommended. Consult the manufacturer and other manuals for proper procedures and methods.

- Condensate return: Inspect the condensate return frequently. Make sure you are returning as much condensate as possible. This is very expensive water. It has high energy content and is treated water. When you lose condensate, you have to pay for the make-up water, chemicals, fuel, and, in some cases, sewage costs.

26.5 ENSURING OPTIMUM BUILDING PERFORMANCE

The RCx activities described in the previous sections will optimize building system operation and reduce energy consumption. To ensure excellent long-term performance, the following activities should be conducted.

- Document RCx activities.
- Measure energy and maintenance cost savings.
- Train operating and maintenance staff.
- Measure energy data and continuously measure energy performance.
- Obtain ongoing assistance from RCx professionals.

This section discusses guidelines to perform these tasks.

26.5.1 Document the RCx Project

The documentation should be brief and accurate. *The operating sequences should be documented accurately and carefully.* This documentation should not repeat the existing building documentation. It should describe the procedures implemented, including control algorithms, and briefly give the reasons behind these procedures. The emphasis is on accurate and usable documentation. The documentation should be easily used by operating staff. For example, operating staff should be able to create operating manuals and procedures from the document.

The RCx project report should include accurate documentation of current energy performance, building data, AHUs and terminal boxes, water loops and pumps, control system, and performance improvements.

26.5.2 Measure Energy Savings

Most building owners expect the RCx project to pay for itself through energy savings. Measurement of energy savings is one of the most important issues for RCx projects. The measurements should follow the procedures described in Chapter 27, Measurement and Verification, of this handbook. Chapter 27 describes procedures from the *International Performance Measurement and Verification Protocol*[32] (IPMVP). This section will provide a very brief description of these procedures, emphasizing issues that are important in M&V for RCx projects.

The process for determining savings as adopted in the IPMVP defines energy savings, E_{save}, as:

$$E_{Save} = E_{base} - E_{post}$$

where E_{base} is the "baseline" energy consumption before the RCx measures were implemented, and E_{post} is the measured consumption following implementation of the RCx measures.

Figure 26-12 shows the daily electricity consumption of the air handlers in a large building in which the HVAC systems were converted from constant volume systems to VAV systems using variable frequency drives. Consumption is shown for slightly over a year before the VFDs were installed (Pre), for about three months of construction (Con), and for about two years after installation (Post). In this case, the base daily electricity consumption is 8,300 kWh/day. The post-retrofit electricity consumption is 4,000 kWh/day, corresponding to electricity savings of 4,300 kWh/day. During the construction period, the savings are slightly lower.

However, in most cases, consumption shows more variation from day to day and month to month than that shown by the fan power for these constant speed fans. Hence, determination of the baseline must consider a number of factors, including weather changes, changes

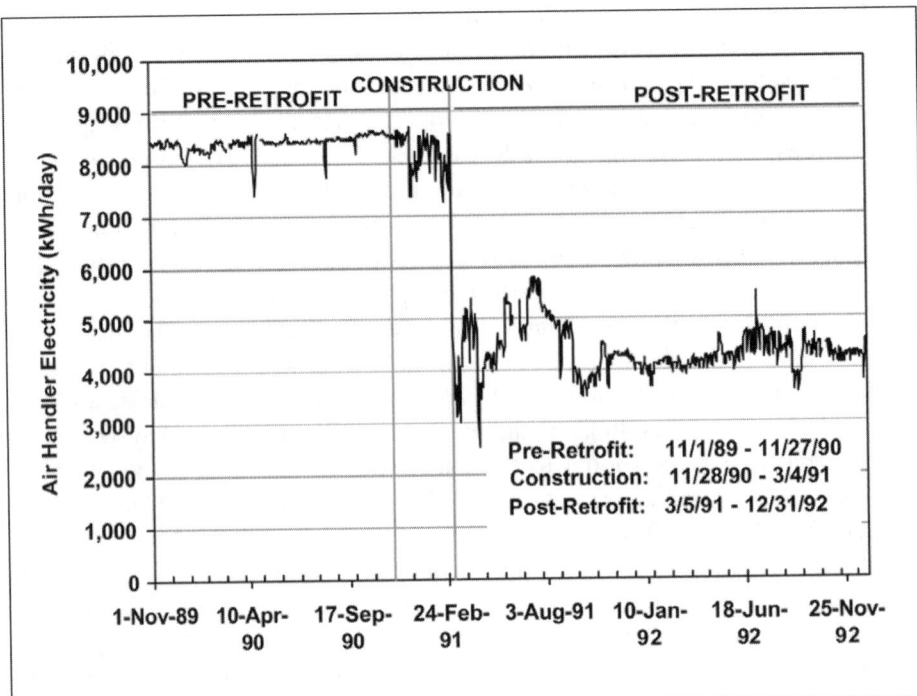

Figure 26-12. Daily electricity consumption for approximately one year before a retrofit and two years after the retrofit[33]

in occupancy schedule, changes in number of occupants, remodeling of the spaces, equipment changes, etc.

In the IPMVP, the baseline energy use, E_{base}, is determined from a model of the building operation before the retrofit (or commissioning) that uses post-installation operating conditions (e.g. weather, occupancy, etc.). The post-installation energy use is generally simply the measured energy use, but it may be determined from a model if measured data are not available.

The IPMVP includes four different M&V techniques or options. These options, may be summarized as Option A—some measurements, but mostly stipulated savings, Option B—measurement at the system or device level, Option C—measurement at the whole-building or facility level, and Option D—determination from calibrated simulation. Each option has its advantages for some special applications. Refer to Chapter 27, Measurement and Verification, for further information.

26.5.2.1 Data Used to Determine Savings

Note that monthly bills may be used to estimate the energy savings. This method is one version of Option C described above. It is typically the least expensive method of verification. It will work fine if the following conditions are met.

1. Significant savings are expected at the utility meter level.
2. Savings are too small to cost-justify more data.
3. There will be no changes in:
    a. Equipment

b. Schedules
c. Occupancy
d. Space utilization

The case shown in Figure 26-13 is an example where monthly bills clearly show the savings. The savings were large and consistent following the retrofit until June. At this point, a major deviation occurred. The presence of other metering at this site showed that the utility bill was incorrect. Further investigation showed that the utility meter had been changed, and this had not been considered in the bill sent. The consumption included in this bill was greater than the site would have used if it had used the peak demand recorded on the utility meter for every hour of the billing period!

However, daily or hourly data will show the results of commissioning measures much more quickly and are an extremely valuable diagnostic tool when problems arise as described in Section 26.5.4. Hence, it is recommended that such data be used for savings determination and follow-up tracking whenever possible.

26.5.3 Trained Operating and Maintenance Staff

Efficient building operation begins with a qualified and committed staff. Since the RCx process generally makes changes in the way a building is operated to improve comfort and efficiency, it is essential that the operators be a part of the commissioning team. They need to work with the RCx professionals, propose RCx measures, and implement or help implement them. In addition to actively participating in the RCx process, formal technical

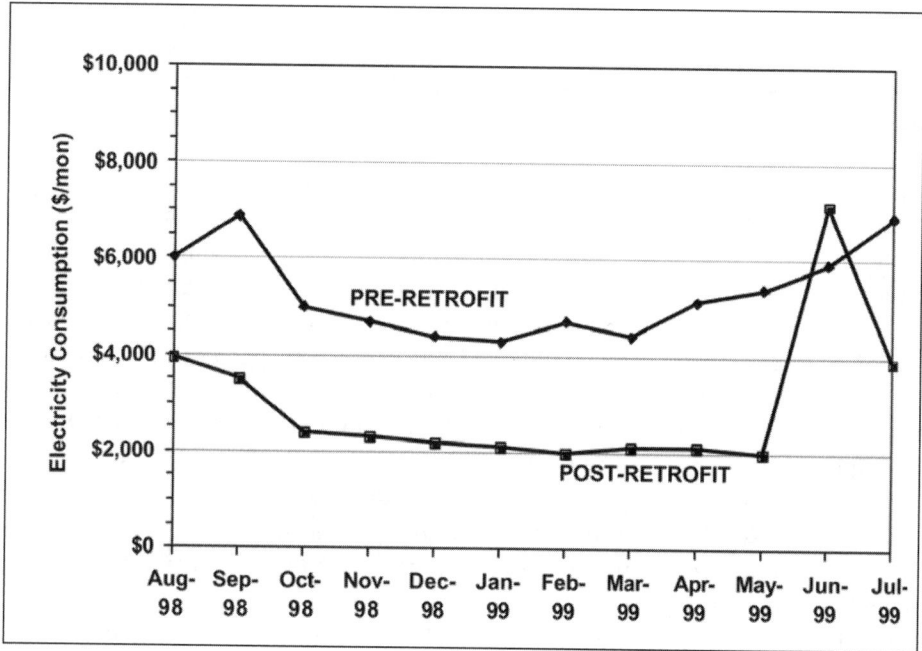

Figure 26-13. Comparison of monthly utility bills before (top line) and after (bottom line) a retrofit[34]

training should be provided to ensure that the operating staff understands the procedures implemented so they can perform troubleshooting properly.

26.5.4 Continuously Measure Energy Performance

The measurement of energy consumption data is very important to maintain building performance and maintain RCx savings. The metered data can be used to:

1. *Identify and solve problems.* Metered consumption data are needed to be sure that the building is still operating properly. If there is a component failure or an operating change that makes such a small change in comfort or operating efficiency that it is not visible in metered consumption data, it generally isn't worth worrying about. If it does show up as even a marginal increase in consumption, troubleshooting should be initiated.

2. *Trend/measure energy consumption data.* This continuing activity is the first line of defense against declining performance. The same procedures used to establish a pre-RCx baseline can be used to establish a baseline for post-RCx performance, and this post-RCx baseline can be used as a standard against which future performance is compared. Consumption that exceeds this baseline for a few days, or even a month may not be significant, but if it persists much more than a month, troubleshooting should be used to find out what has led to the increase. If it is the result of a malfunctioning valve, you can fix it. If it is the result of 100 new computers added to the building, you will adjust

the baseline accordingly.

3. *Trend and check major operating parameters.* Parameters such as cold-deck temperatures, zone supply temperatures, etc. should be trended periodically for comparison with historic levels. This can be extremely valuable when troubleshooting and when investigating consumption above the post-RCx baseline.

4. *Find the real problems when the system needs to be repaired or fixed.* It is essential that the same fundamental approach used to find and fix problems while the RCx process was initiated be used whenever new hot calls or cold calls are received.

26.5.5 Utilize Expert Support as Needed

It is inevitable that a problem will come up which, even after careful troubleshooting, points toward a problem with one or more of the RCx measures that have been implemented. Ask the RCx providers for help in solving such problems before undoing an implemented RCx measure. Sometimes it will be necessary to modify a measure that has been implemented. The RCx professionals will often be able to help with finding the most efficient solution, and they will sometimes be able to help you find another explanation so the problem can be remedied without changing the measure.

Ask for help from the RCx providers when you run into a new problem or situation. Problems occasionally crop up that defy logical explanation. These are the problems that generally get resolved by trying one of three things that seem like possible solutions and playing with system settings until the problem goes away.

This is one of the most important situations in which expert help is needed. These are precisely the kind of problems—and the trial and error solutions—that often lead to major operating cost increases.

26.5.6 How Well Do Commissioning Savings Persist?

The Energy Systems Laboratory at Texas A&M has conducted a study of 10 buildings on the Texas A&M campus that had RCx measures implemented in 1996-97[35,36]. Table 26-5 shows the baseline cost of combined heating, cooling, and electricity use of each building and the commissioning savings for 1998 and 2000. The baseline consumption and savings for each year were normalized to remove any differences due to weather.

Looking at the totals for the group of 10 buildings, heating and cooling consumption increased by $207,258 (12.1%) from 1998 to 2000, but savings from the earlier commissioning work were still $985,626. However, it may also be observed that almost three-fourths of this consumption increase occurred in two buildings, the Kleberg Building and G. Rollie White Coliseum. The increased consumption of the Kleberg Building was due to a combination of component failures and control problems, as described in the case study in Section 26.5.6.1. The increased consumption in G. Rollie White Coliseum was due to different specific failures and changes, but it was qualitatively similar to Kleberg since it resulted from a combination of component failures and control changes. The five buildings that showed consumption changes of more than 5% from 1998 to 2000 were all found to have different control settings that appear to account for the changed consumption (including the decrease in the Wehner Business Building).

These data suggest that commissioning savings

generally persist, but tracking can subsequently uncover problems that did not cause comfort problems but have increased consumption by $10,000-$100,000 per year in large buildings.

26.5.6.1 *Commissioning persistence case study—Kleberg Building[38]*

The Kleberg Building is a teaching/research facility on the Texas A&M campus consisting of classrooms, offices, and laboratories, with a total floor area of approximately 165,030 ft[2]. Ninety percent of the building is heated and cooled by two (2) single duct variable air volume (VAV) air handling units (AHU), each having a pre-heat coil, a cooling coil, one supply air fan (100 hp), and a return air fan (25 hp). Two smaller constant volume units handle the teaching/lecture rooms in the building. The campus plant provides chilled water and hot water to the building. The two (2) parallel chilled water pumps (2×20 hp) have variable frequency drive control. There are 120 fan-powered VAV boxes with terminal reheat in 12 laboratory zones and 100 fan-powered VAV boxes with terminal reheat in the offices. There are six exhaust fans (10-20 hp, total 90 hp) for fume hoods and laboratory general exhaust. The air handling units, chilled water pumps and 12 laboratory zones are controlled by a direct digital control (DDC) system. DDC controllers modulate dampers to control exhaust airflow from fume hoods and laboratory general exhaust.

An RCx investigation was initiated in the summer of 1996 due to the extremely high level of simultaneous heating and cooling observed in the building. Figures 26-14 and 26-15 show daily heating and cooling consumption (expressed in average kBtu/hr) as functions of daily average temperature. The Pre-RCx heating con-

Table 26-5. Commissioning savings in 1998 and 2000 for 10 buildings on the Texas A&M campus[37]

Building	Baseline Use ($/yr)	1998 Savings ($/yr)	2000 Savings($/yr)
Kleberg Building	$ 484,899	$ 313,958	$ 247,415
G.R. White Coliseum	$ 229,881	$ 154,973	$ 71,809
Blocker Building	$ 283,407	$ 76,003	$ 56,738
Eller O&M Building	$ 315,404	$ 120,339	$ 89,934
Harrington Tower	$ 145,420	$ 64,498	$ 48,816
Koldus Building	$ 192,019	$ 57,076	$ 61,540
Richardson Petroleum Building	$ 273,687	$ 120,745	$120,666
Veterinary Medical Center Addition	$ 324,624	$ 87,059	$ 92,942
Wehner Business Building	$ 224,481	$ 47,834	$ 68,145
Zachry Engineering Center	$ 436,265	$ 150,400	$127,620
Totals	$ 2,910,087	$ 1,192,884	$ 985,626

sumption data given in Figure 26-14 shows very little temperature dependence, as indicted by the regression line derived from the data. Data values were typically between 5 and 6 MMBtu/hr, with occasional lower values. The cooling data (Figure 26-15) shows more tem- perature dependence, and the regression line indicates that average consumption on a design day would exceed 10 MMBtu/hr. This corresponds to only 198 sq.ft./ton based on average load.

It was soon found that the preheat was operating

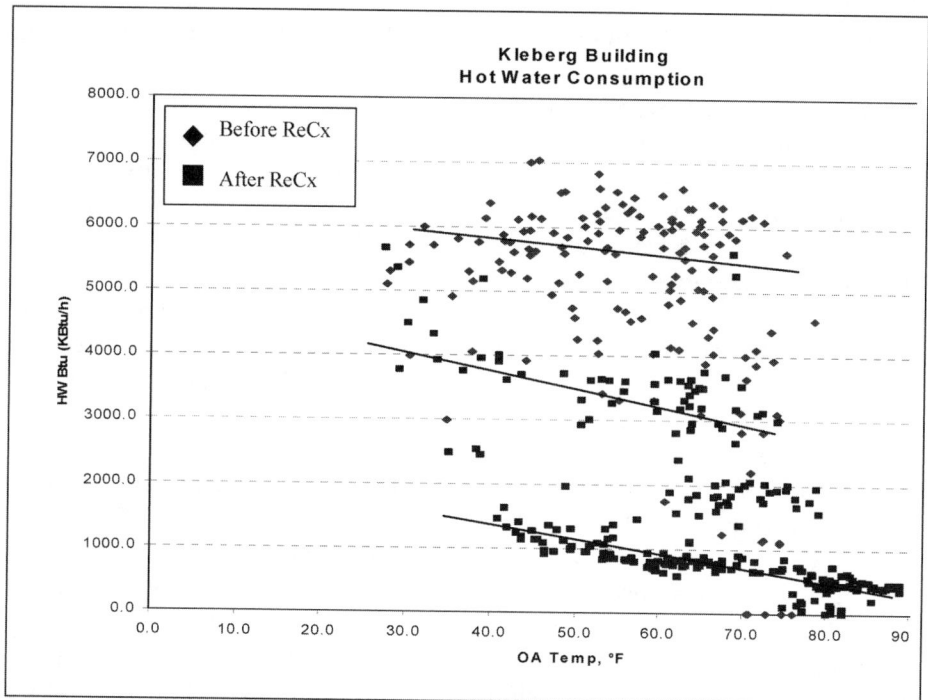

Figure 26-14. Pre-RCx and post-RCx heating water consumption at the Kleberg Building vs. daily average outdoor temperature[39]

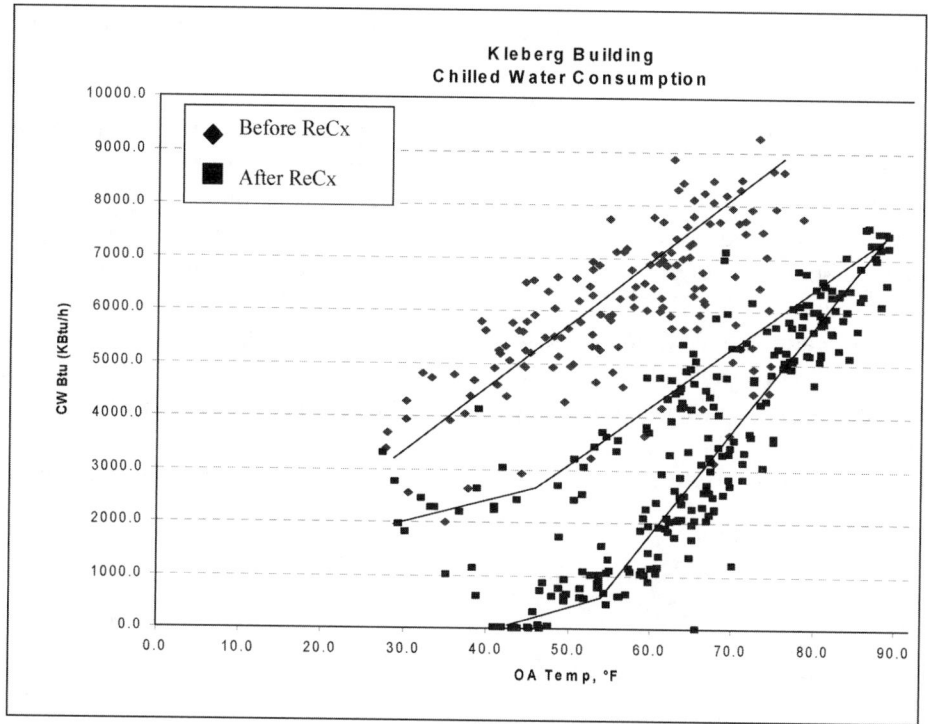

Figure 26-15. Pre-RCx and post-RCx chilled water consumption at the Kleberg Building vs. daily average outdoor temperature[40]

continuously, heating the mixed air entering the cooling coil to approximately 105°F, instituted in response to a humidity problem in the building. The preheat was turned off, and heating and cooling consumption both dropped by about 2 MMBtu/hour as shown by the middle clouds of data in Figures 26-14 and 26-15. Subsequently, the building was thoroughly examined, and a comprehensive list of commissioning measures was developed and implemented. The principal measures implemented that led to reduced heating and cooling consumption were:

- Preheat to 105°F was changed to preheat to 40°F.
- Cold deck was schedule changed from 55°F fixed to vary from 62°F to 57°F as ambient temperature varies from 40°F to 60°F.
- Economizer was set to maintain mixed air at 57°F whenever outside air is below 60°F.
- Static pressure control was reduced from 1.5 inH$_2$O to 1.0 in H$_2$O and implemented night-time set back to 0.5 inH$_2$O.
- Replaced or repaired a number of broken VFD boxes.
- Chilled water pump VFDs were turned on.

Additional measures implemented included changes in CHW pump control (changed so one pump modulates to full speed before the second pump comes on instead of operating both pumps in parallel at all times), building static pressure was reduced from 0.05 in H$_2$O to 0.02 in H$_2$O, and control changes were made to eliminate hunting in several valves. It was also observed that there was a vibration at a particular frequency in the pump VFDs that influenced the operators to place these VFDs in the manual mode, so it was recommended that the mountings be modified to solve this problem.

These changes further reduced chilled water and heating hot water use as shown in Figures 26-14 and 26-15 for a total annualized reduction of 63% in chilled water use and 84% in hot water use. Additional follow-up conducted from June 1998 through April 1999 focused on air balance in the 12 laboratory zones, general exhaust system rescheduling, VAV terminal box calibration, adjusting the actuators and dampers, and calibrating fume

hoods and return bypass devices to remote DDC control[41]. These changes reduced electricity consumption by about 7% or 30,000 kWh/mo.

In 2001 it was observed that chilled water savings for 2000 had declined to 38% and hot water savings to 62%, as shown in Table 26-6. Chilled water data for 2001 and the first three months of 2002 are shown in Figure 26-16. The two lines shown are the regression fits to the chilled water data before RCx implementation and after implementation of RCx measures in 1996 as shown in Figure 26-15. It is evident that consumption during 2001 is generally appreciably higher than immediately following implementation of RCx measures. The RCx group performed field tests and analyses that soon focused on two SDVAV AHU systems, two chilled water pumps, and the energy management control system (EMCS) control algorithms, as described in Chen et al. [42]. Several problems were observed as noted below.

26.5.6.2 Problems Identified

- The majority of the VFDs were running at a constant speed, near 100%.
- VFD control on two chilled water pumps was again bypassed to run at full speed.
- Two chilled water control valves were leaking badly. Combined with a failed electronic to pneumatic switch and the high water pressure noted above, this resulted in discharge air temperatures of 50°F and lower and activated preheat continuously.
- A failed pressure sensor and two failed CO$_2$ sensors put all outside air dampers to the full open position.
- The damper actuators were leaking and unable to maintain pressure in some of the VAV boxes. This caused cold air to flow through the boxes even when they were in the heating mode, resulting in simultaneous heating and cooling. Furthermore, some of the reheat valves were malfunctioning. This caused the reheat to remain on continuously in some cases.
- Additional problems identified from the field survey included: (1) high air resistance from the filters and coils, (2) errors in a temperature sensor and

Type	Pre-RCx Baseline (MMBtu/yr)	Post-RCx Use/Savings		2000 Use/Savings	
		Use (MMBtu/yr)	Savings (%)	Use (MMBtu/yr)	Savings (%)
CHW	72935	26537	63.6%	45431	37.7%
HW	43296	6841	84.2%	16351	62.2%

Table 26-6. Chilled water and heating water usage and saving in the Kleberg Building for three different years normalized to 1995 weather[43]

static pressure sensor, and (3) high static pressure setpoints in AHU1 and AHU2.

This combination of equipment failure, compounded by control changes that returned several pumps and fans to constant speed operation, had the consequence of increasing chilled water use by 18,894 MMBtu and hot water use by 9,510 MMBtu. This amounted to an increase of 71% in chilled water use and more than doubled hot water use from two years earlier.

These problems were corrected, and building performance returned to previously low levels, as illustrated by the data for April-June 2002 in Figure 26-16. These data are all below the lower of the two regression lines and is comparable to the level achieved after additional RCx measures were implemented in 1998-99.

26.6. COMMISSIONING NEW BUILDINGS FOR ENERGY MANAGEMENT

The energy manager's effort is generally directed toward improving the efficiency of existing buildings. However, whenever the organization initiates design and construction of a new building that will become part of the energy manager's portfolio of buildings, it is extremely important that the energy manager become an active part of the design and construction team to ensure that the building incorporates all appropriate energy efficiency technologies. It is just as important that the perspective of operational personnel be included in the design process so it will be possible to effectively and

efficiently operate the building. One of the best ways to accomplish these objectives is to commission the building as it is designed and built.

The primary motivation for commissioning HVAC systems is generally to achieve HVAC systems that work properly to provide comfort to building occupants at low cost. In principle, all building systems should be designed, installed, documented, and tested—and building staff trained in their use. In practice, competitive pressures, fee structures, and financial pressures to occupy new buildings as quickly as possible generally result in buildings that are handed over to the owners with minimal contact between designers and operators, minimal functional testing of systems, documentation that largely consists of manufacturers' system component manuals, and little or no training for operators. This in turn leads to problems such as mold growth in walls of new buildings, rooms that never cool properly, air quality problems, etc. Such experiences were doubtless the motivation of the facility manager for a large university medical center who stated a few years ago that he didn't want to get any new buildings. He only wanted three-year old buildings in which the problems had been fixed.

Although commissioning provides higher quality buildings and results in fewer initial and subsequent operational problems, most owners will include commissioning in the design and construction process only if they believe they will benefit financially from commissioning. It is much more difficult to document the energy cost savings from commissioning a new building than an existing building. There is no historical use pat-

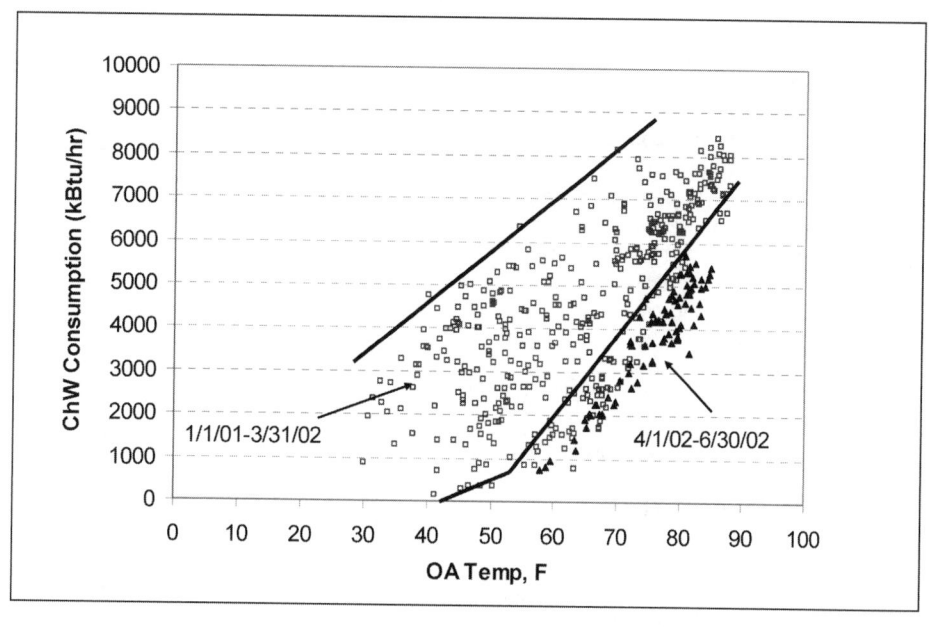

Figure 26-16. CHW data for the Kleberg Building for January 2001- June 2002[44]

tern to use as a baseline. However, it has been estimated that new building commissioning will save 8% in energy cost alone compared with the average building which is not commissioned[45]. This offers a payback for the cost of commissioning in just over four years from the energy savings alone and also provides improved comfort and air quality.

Commissioning is often considered to be a punch-list process that ensures that the systems in a building function before the building is turned over to the owner. However, the process outlined in Table 26-7 shows the process beginning in the pre-design phase. It is most effective if allowed to influence both design and construction. It is essential that the energy manager be involved in the commissioning process on the owner's team no later than the design phase of the construction process. This permits input into the design process that can have major impact on the efficiency of the building as built and can lead to a building that has far fewer operational problems.

26.7 SUMMARY

Commissioning of existing buildings is emerging as one of the most cost effective ways for an energy manager to lower operating costs, and it typically does so with no capital investment, or with a very minimal amount. It has been successfully implemented in several hundred buildings and provides typical paybacks of one to three years.

It is much more than the typical O&M program. It does not ensure that the systems function as originally designed but focuses on improving overall system control and operations for the building as it is currently utilized and on meeting existing facility needs. During the RCx process, a comprehensive engineering evaluation is conducted for both building functionality and system functions. The optimal operational parameters and schedules are developed based on actual building conditions. An integrated approach is used to implement these optimal schedules to ensure practical local and global system optimization and to ensure persistence of the improved operational schedules.

The approach presented in this chapter begins by conducting a thorough examination of all problem areas or operating problems in the building, diagnoses these problems, and develops solutions that solve these problems while almost always reducing operating costs at the same time. Equipment upgrades or retrofits may be implemented as well but have not been a factor in the case studies presented, except where the commission-

Table 26-7. The commissioning process for new buildings[46]

1. Conception or pre-design phase
(a) Develop commissioning objectives
(b) Hire commissioning provider
(c) Develop design phase commissioning requirements
(d) Choose the design team
2. Design phase
(a) Commissioning review of design intent
(b) Write commissioning specifications for bid documents
(c) Award job to contractor
(d) Develop commissioning plan
3. Construction/Installation phase
(a) Gather and review documentation
(b) Hold commissioning scoping meeting and finalize plan
(c) Develop pre-test checklists
(d) Start up equipment or perform pre-test checklists to ensure readiness for functional testing during acceptance
4. Acceptance phase
(a) Execute functional test and diagnostics
(b) Fix deficiencies
(c) Retest and monitor as needed
(d) Verify operator training
(e) Review O&M manuals
(f) Building accepted by owner
5. Post-acceptance phase
(a) Prepare and submit final report
(b) Perform deferred tests (if needed)
(c) Develop recommissioning plan/schedule

ing was used to finance equipment upgrades. This is in sharp contrast to the more usual approach to improving the efficiency of HVAC systems and cutting operating costs that primarily emphasizes system upgrades or retrofits to improve efficiency.

Commissioning of new buildings is also an important option for the energy manager, offering an opportunity to help ensure that new buildings have the energy efficiency and operational features that are most needed.

26.8 FOR ADDITIONAL INFORMATION

Two major sources of information on commissioning existing buildings are:

A Practical Guide for Commissioning Existing Buildings (Haasl, T. and Sharp, T., Portland Energy Conservation, Inc. and Oak Ridge National Laboratory for U.S. DOE, ORNL/TM-1999/34, 69 pp. + App., 1999) and

*Continuous Commissioning*SM *Guidebook: Maximizing Building Energy Efficiency and Comfort* (Liu, M., Claridge, D.E. and Turner, W.D., Federal Energy Management Program, U.S. Dept. of Energy, 144 pp., 2002)

There are a much wider range of materials available that treat commissioning of new buildings. Two documents that provide a good starting point are:

ASHRAE Guideline 1-1996: The HVAC Commissioning Process (American Society of Heating, Refrigerating and Air-Conditioning Engineers, Atlanta, GA, 1996) and

Building Commissioning Guide—Version 2.2 (U.S. GSA and U.S. DOE, 1998 by McNeil Technologies, Inc. and Enviro-Management & Research, Inc. available at http://www.eere.energy.gov/femp/techassist/bldguide.pdf)

The case studies in this chapter have been largely abridged and adapted from the following three papers:

Evans, C., Cordero, J., Atencio. M., Claridge, D., Martinez, J., Oberle, C., Baltazar, J.C. and Zhu, Y., "Continuous Commissioning® of an Office/Laboratory Building," *Proc. of 6th International Conference for Enhanced Building Operation*, Pittsburgh, PA, October 11-13, 2005, CD.

Turner, W.D., Claridge, D.E., Deng, S. and Wei, G., "The Use of Continuous CommissioningSM As An Energy Conservation Measure (ECM) for En-

ergy Efficiency Retrofits," Proc. of 11th National Conference on Building Commissioning, Palm Springs, CA, CD, May 20-22, 2003.

Claridge, D.E., Turner, W.D., Liu, M., Deng, S., Wei, G., Culp, C., Chen, H. and Cho, S.Y., "Is Commissioning Once Enough?," *Solutions for Energy Security & Facility Management Challenges: Proc. of the 25th WEEC*, Atlanta, GA, pp. 29-36, Oct. 9-11, 2002.

References

1. ASHRAE, *ASHRAE Guideline 1-1996: The HVAC Commissioning Process*, American Society of Heating, Refrigerating and Air-Conditioning Engineers, Atlanta, GA, 1996.
2. U.S. Department of Energy, Building Commissioning: The Key to Quality Assurance, Washington, DC. 1999.
3. ibid.
4. Visier, J.C., ed., Commissioning Tools for Improved Energy Performance: Results of IEA ECBCS ANNEX 40, International Energy Agency ECBCS, 2005.
5. Liu, M., Claridge, D.E. and Turner, W.D., *Continuous Commissioning SM Guidebook: Maximizing Building Energy Efficiency and Comfort*, Federal Energy Management Program, U.S. Dept. of Energy, 144 pp., 2002.
6. Claridge, D.E., Haberl, J., Liu, M., Houcek, J., and Athar, A., "Can You Achieve 150% of Predicted Retrofit Savings: Is It Time for Recommissioning?" *ACEEE 1994 Summer Study on Energy Efficiency In Buildings Proceedings: Commissioning, Operation and Maintenance*, Vol. 5, American Council for an Energy Efficient Economy, Washington, D.C., pp. 73-87, 1994.
7. Evans, C., Cordero, J., Atencio. M., Claridge, D., Martinez, J., Oberle, C., Baltazar, J.C. and Zhu, Y., "Continuous Commissioning® of an Office/Laboratory Building," *Proc. of 6th International Conference for Enhanced Building Operation*, Pittsburgh, PA, October 11-13, 2005, CD.
8. Turner, W.D., Claridge, D.E., Deng, S. and Wei, G., "The Use of Continuous CommissioningSM As An Energy Conservation Measure (ECM) for Energy Efficiency Retrofits," Proc. of 11th National Conference on Building Commissioning, Palm Springs, CA, CD, May 20-22, 2003.
9. Zhu, Y., Liu, M., Claridge, D.E., Feary, D. and Smith, T., "A Continuous Commissioning Case Study of a State-of-the-Art Building," *Proceedings of the 5th National Commissioning Conference*, Huntington Beach, CA, pp. 13.1-13.10, April, 1997.
10. Liu, M., Zhu, Y., Powell, T., and Claridge, D.E., "System Optimization Saves $195,000/yr. in a New Medical Facility," *Proceedings of the 6th National Conference on Building Commissioning*, Lake Buena Vista, FL, pp. 14.2.1-14.2.11, May 18-20, 1998.
11. Turner, et al., op. cit.
12. Source: Adapted from Turner et al. op. cit.
13. Source: Liu, Claridge, and Turner, op. cit.
14. Source: ibid.
15. Source: ibid.
16. Source: ibid.
17. Liu, M., Zhu, Y., Park, B.Y., Claridge, D.E., Feary, D.K. and Gain, J., "Airflow Reduction to Improve Building Comfort and Reduce Building Energy Consumption—A Case Study," *ASHRAE Transactions-Research*, Vol. 105, Part I, pp. 384—390, 1999.
18. Liu, M., Zhu, Y., Claridge, D. and White, E., "Impacts of Static Pressure Set Level on the HVAC Energy Consumption and Indoor Conditions," *ASHRAE Transactions-Research*. Volume 103, Part 2, pp. 221-228, 1997.
19. Liu, Claridge, and Turner, op. cit.
20. ibid.

21. ibid.

22. Liu, M., "Variable Speed Drive Volumetric Tracking (VSDVT) for Airflow Control in Variable Air Volume (VAV) Systems," Proceedings of Thirteenth Symposium on Improving Building Systems in Hot and Humid Climates, San Antonio, TX, pp. 195-198, May 15-16, 2002.

23. This case study is adapted from Evans, C., Cordero, J., Atencio. M., Claridge, D., Martinez, J., Oberle, C., Baltazar, J.C. and Zhu, Y., "Continuous Commissioning® of an Office/Laboratory Building," Proc. of 6th International Conference for Enhanced Building Operation, Pittsburgh, PA, October 11-13, 2005, CD.

24. Liu, M., "Variable Water Flow Pumping for Central Chilled Water Systems," ASME Journal of Solar Energy Engineering, Vol. 124, pp. 300-304, 2002.

25. Deng, S., Turner, W.D., Batten, T., and Liu, M., "Continuous Commmissioning^SM of a Central Chilled Water and Heating Hot Water System," Proc. Twelfth Symposium on Improving Building Systems in Hot and Humid Climates, San Antonio, TX, pp. 199-206, May 15-16, 2000.

26. Deng, S. Turner, W.D., Claridge, D.E., Liu, M., Bruner, H., Chen, H. and Wei, G. "Retrocommissioning of Central Chilled/Hot Water Systems," ASHRAE Transactions-Research, Vol. 108, Part 2, pp. 75-81, 2002.

27. Liu, Claridge, and Turner, op. cit.

28. Liu, M., op. cit.

29. Liu, Claridge, and Turner, op. cit.

30. Wei, G., Liu, M., and Claridge, D.E., "In-situ Calibration of Boiler Instrumentation Using Analytic Redundancy," International Journal of Energy Research, Vol. 25, pp. 375-387, 2001.

31. Liu, M. et al., 1998, op. cit.

32. IPMVP Committee, International Performance Measurement & Verification Protocol: Concepts and Options for Determining Energy and Water Savings, Vol. 1, U.S. Dept. of Energy, DOE/GO-102001-1187, January, 2001, 86 pp..

33. Source: Liu, Claridge, and Turner, op. cit.

34. Source: Liu, Claridge, and Turner, op. cit.

35. Turner, W.D., Claridge, D.E., Deng, S., Cho, S., Liu, M., Hagge, T., Darnell, C., Jr., and Bruner, H., Jr., "Persistence of Savings Obtained from Continuous Commissioning^SM," Proc. of 9th National Conference on Building Commissioning, Cherry Hill, NJ, pp. 20-1.1-20-1.13, May 9-11, 2001.

36. Claridge, D.E., Turner, W.D., Liu, M., Deng, S., Wei, G., Culp, C., Chen, H. and Cho, S.Y., "Is Commissioning Once Enough?," Solutions for Energy Security & Facility Management Challenges: Proc. of the 25th WEEC, Atlanta, GA, pp. 29-36, Oct. 9-11, 2002.

37. Source: ibid.

38. ibid.

39. Source: ibid.

40. Source: ibid.

41. Lewis, T., H. Chen, and M. Abbas, "CC summary for the Kleberg Building," Internal ESL Report, July, 1999.

42. Chen, H., Deng, S., Bruner, H., Claridge, D. and Turner, W.D., "Continuous Commissioning^SM Results Verification And Follow-Up For An Institutional Building—A Case Study," Proc. 13th Symposium on Improving Building Systems in Hot and Humid Climates, Houston, TX, pp. 87-95, May 20-23, 2002.

43. Source: Claridge, D.E. et al. 2002, op. cit.

44. Source: ibid.

45. PECI, "National Strategy for Building Commissioning," Portland Energy Conservation, Inc., Portland, OR, 1999.

46. Adapted from Haasl, T. and Sharp, T., A Practical Guide for Commissioning Existing Buildings, Portland Energy Conservation, Inc. and Oak Ridge National Laboratory for U.S. DOE, ORNL/TM-1999/34, 69 pp. + App., 1999.

CHAPTER 27

MEASUREMENT AND VERIFICATION OF ENERGY SAVINGS

JEFF S. HABERL, PH.D., P.E.
CHARLES C. CULP, PH.D., P.E.
Energy Systems Laboratory
Texas A&M University

27.1 INTRODUCTION— M&V METHOD SELECTION

M&V has a dual role. First, M&V quantifies the savings being obtained. This applies to the initial savings and the long-term savings. Since the persistence of savings has been shown to decrease with time,[1] long-term M&V provides data to make these savings sustainable. Second, M&V must be cost effective so that the cost of measurement and the analysis does not consume the savings[2,3]. The 1997 International Performance Measurement and Verification Protocol (IPMVP) set the target costs for M&V to be in the range of 1% to 10% of the construction cost for the life of the ECM, depending upon the option selected. Most approaches fall in the recommended range of 3% to 10% of the construction cost. The IPMVP 2001 removed this guidance on the recommended costs for M&V. Currently, a goal of about 5% of the savings per year has evolved as a preferred criterion for costing M&V, since the cost justification directly results from the calculation.

A general procedure for selecting an approach can be summarized by the following five steps.

- a. *In general one wants to try to* perform monthly utility bill before/after analysis.
- b. *If this does not work, then* perform daily or hourly before/after analysis.
- c. *If this does not work, then* perform component isolation analysis.
- d. *If this does not work, then* perform calibrated simulation analysis.
- e. *Then* report savings and finish analysis.

27.2 HISTORY OF M&V

27.2.1 History of Building Energy Measurement
The history of the measurement of building energy use can be traced back to the 19th century for electricity and earlier for fuels such as coal and wood, which were used to heat buildings[4,5,6,7]. By the 1890s, although electricity was common in many new commercial buildings, its use was primarily for incandescent lighting and, to a lesser extent, for the electric motors associated with ventilating buildings since most of the work in office buildings was carried out during daylight hours. The metering of electricity closely paralleled the spread of electricity into cities, as its inventors needed to recover the cost of its production through the collection of payments from electric utility customers[8,9]. Commercial meters for the measurement of flowing liquids in pipes can be traced back to the same period, beginning with the invention of the first commercial flowmeter by Clemens Herschel in 1887, which used principles based on the Pitot tube and venturi flowmeter, invented in 1732 and 1797 by their respective namesakes[10]. Commercial meters for the measurement of natural gas can likewise be traced to the sale and distribution of natural gas, which paralleled the development of the electric meters.

27.2.2 History of Measurement and Verification (M&V) in the U.S.
The history of the measurement and verification of building energy use parallels the development and use of computerized energy calculations in the 1960s, with a much accelerated awareness in 1973 when the embargo on Mideast oil made energy a front page issue[11,12]. During the 1950s and 1960s most engineering calculations were performed using slide rules, engineering tables, and desktop calculators that could only add, subtract, multiply and divide. Since the public was lead to believe energy was cheap and abundant[13], the measurement and verification of the energy use in a building was limited for the most part to simple, unadjusted comparisons of monthly utility bills.

In the 1960s several efforts were initiated to formulate and codify equations that could predict dynamic heating and cooling loads, including efforts at the National Bureau of Standards to predict temperatures in fallout shelters[14] and the 1967 HCC program developed by a group of mechanical engineering consultants[15] that used the total equivalent temperature difference/time averaging (TETD/TA) method. The popularity of this program prompted the American Society of Heating, Refrigeration and Air-Conditioning Engineers (ASHRAE) to embark on a series of efforts that eventually delivered today's modern, general purpose simulation programs[16] (i.e., DOE-2, BLAST, EnergyPlus, etc.), which utilize thermal response factors,[17,18] as well as algorithms, for simulation of the quasi-steady-state performance of primary and secondary equipment[19]. One of these efforts was to validate the hourly calculations with field measurements at the Legal Aid Building on the Ohio State University campus[20], which was probably the first application of a calibrated, building energy simulation program.

Developing standardized methods for the M&V of building energy use began with efforts to normalize residential heating energy use in single-family and multi-family buildings,[21] which include the Princeton Scorekeeping Method[22] (PRISM), a forerunner to ASHRAE's variable-based degree day (VBDD) calculation method. In commercial buildings, numerous methods reported over the years[23,24,25] varied from weather normalization, using monthly utility billing data[26,27,28] daily and hourly methods,[29] and even dynamic inverse models using resistance-capacitance (RC) networks.[30] Procedures and methodologies to baseline energy use in commercial buildings began being published in the 1980s[31,32,33] and the early 1990s[34,35,36]. Modeling toolkits and software have been developed that are useful in developing performance metrics for buildings, as well as for HVAC system components. These efforts include:

- The Princeton Scorekeeping Software (PRISM),[37] useful for developing variable-based degree day models of monthly or daily data.

- ASHRAE's HVAC01 software for modeling primary HVAC systems such as boilers and chillers[38].

- ASHRAE's HVACO$_2$ software for modeling secondary systems, including air-handlers, blowers, cooling coils, and terminal boxes[39].

Also included are ASHRAE research projects, which include: 827-RP for *in-situ* measurement of chillers, pumps, and blowers,[40] 1004-RP for *in-situ* measurement of thermal storage systems[41], 1050-RP toolkit for calculating linear, change-point linear and multiple-linear inverse building energy analysis models,[42,43,44] and 1093-RP toolkit for calculating diversity factors for energy and cooling loads[45,46].

In 1989, a report by Oak Ridge National Laboratory[47] classified the diverse commercial building analysis methods into five categories, including: annual total energy and energy intensity comparisons, linear regression and component models, multiple linear regression, building simulation, and dynamic (inverse) thermal performance models. In 1997 a reorganized and expanded version of this classification appeared in the *ASHRAE*

Handbook of Fundamentals, and it is shown Table 27-1 and Table 27-2. In Table 27-1 different methods of analyzing building energy are presented, which have been classified according to model type, including: forward, inverse, and hybrid models[48].

In the first method, forward modeling, a thermodynamic model is created of a building using fundamental engineering principles to predict the hypothetical energy use of a building for 8,760 hours of the year, given the location and weather conditions. This requires a complete description of the building, system, or component of interest, as well as the physical description of the building geometry, geographic location, system type, wall insulation value, etc. Forward models are normally used to design and size HVAC systems, and they have begun to be used to model existing building, using a technique referred to as calibrated simulation.

In the second method, inverse modeling, an empirical analysis is conducted on the behavior of the building as it relates to one or more driving forces or parameters. This approach is referred to as a system identification, parameter identification, or inverse modeling. To develop an inverse model, one must assume a physical configuration of the building or system and then identify the parameter of interest using statistical analysis[49]. Two primary types of inverse models have been reported in the literature, including steady state inverse models and dynamic inverse models. A third category, hybrid models, consists of models that have characteristics of both forward and inverse models[50].

The simplest steady-state inverse model regresses monthly utility consumption data against average billing period temperatures. More robust methods include multiple linear regression, change-point linear regression, and variable-based degree day regressions, as indicated in Table 27-1. The advantage of steady-state inverse models is that their use can be automated and applied to large datasets where monthly utility billing data and average daily temperatures for the billing periods are available. Steady-state inverse models can also be applied to daily data, which allows one to compensate for differences in weekday and weekend use[51]. Unfortunately, steady state inverse models are insensitive to dynamic effects (i.e., thermal mass) and other variables (for example humidity and solar gain), and they are difficult to apply to certain building types, for example buildings that have strong on/off schedule dependent loads or buildings that display multiple change-points.

Dynamic inverse models include: equivalent thermal network analysis,[52] ARMA models,[53,54] Fourier series models,[55,56] machine learning[57], and artificial neural networks.[58,59] Unlike steady-state, inverse models, dynamic models are capable of capturing dynamic effects such as thermal mass, which traditionally has required the solution of a set of differential equations. The disadvantages of dynamic inverse models are that they are increasingly complex and need more detailed measurements to "tune" the model.

Hybrid models are models that contain forward and inverse properties. For example, when a traditional fixed-schematic simulation program such as DOE-2 or BLAST (or even a component-based simplified model) is used to simulate the energy use of an existing building then one has a forward analysis method that is being used in an inverse application, i.e., the forward simulation model is being calibrated or fit to the actual energy consumption data from a building in much the same way that one fits a linear regression of energy use to temperature.

Table 27-2 presents information that is useful for selecting an inverse model where usage of the model (diagnostics—D, energy savings calculations—ES, design—De, and control—C), degree of difficulty in understanding and applying the model, time scale for the data used by the model (hourly—H, daily—D, monthly—M, and sub-hourly—S), calculation time, input variables used by the models (temperature—T, humidity—H, solar—S, wind—W, time—t, thermal mass—tm), and accuracy are used to determine the choice of a particular model.

27.2.3 History of M&V protocols in the United States

The history of measurement and verification protocols in the United States can be traced to independent M&V efforts in different regions of the country as shown in Table 27-3, with states such as New Jersey, California, and Texas developing protocols that contained varying procedures for measuring the energy and demand savings from retrofits to existing buildings. These efforts culminated in the development of the USDOE's 1996 North American Measurement and Verification Protocol (NEMVP)[62], which was accompanied by the USDOE's 1996 FEMP guidelines[63]; both relied on analysis methods developed in the Texas LoanSTAR program[64]. In 1997 the NEMVP was updated and republished as the International Performance Measurement and Verification Protocols (IPMVP)[65]. The IPMVP was then expanded in 2001 into two volumes: Volume I covering Energy and Water Savings[66], and Volume II covering Indoor Environmental Quality[67]. In 2003 Volume III of the IPMVP was published, which covers protocols for new construction[68]. Finally, in 2002 the American Society of Heating Refrigeration Air-conditioning Engineers (ASHRAE) released Guideline 14-2002: Measurement of Energy and

Table 27-1. ASHRAE's 1997 Classification of Methods for the Thermal Analysis of Buildings.[60]

METHOD:	FORWARD	INVERSE	HYBRID	COMMENTS:
Steady State Methods				
Simple linear regression		X		One dependent parameter, one independent parameter. May have slope and y-intercept.
Multiple linear regression		X	X	One dependent parameter, multiple independent parameters.
Modified degree day method	X			Based on fixed reference temperature of 65F.
Variable base degree day method	X			Variable reference temperatures.
Traditional ASHRAE bin method and inverse bin method	X	X	X	Hours in temperature bin times load for that bin.
Change point models: 3-parameter (PRISM CO,HO), 4-parameter, 5-parameter (PRISM HC).		X	X	Uses daily or monthly utility billing data and average period temperatures.
ASHRAE TC 4.7 modified bin method	X		X	Modified bin method with cooling load factors.
Dynamic Methods				
Thermal network	X	X	X	Uses equivalent thermal parameters (inverse mode).
Response factors	X			Tabulated or as used in simulation programs.
Fourier Analysis	X	X	X	Frequency domain analysis convertible to time domain.
ARMA Model		X		Autoregressive Moving Average model.
ARMA Model		X		Multiple-input autoregressive moving average model.
BEVA, PSTAR	X	X	X	Combination of ARMA and Fourier series, includes loads in time domain.
Modal analysis	X	X	X	Bldg. described by diagonalized differential equation using nodes.
Differential equation		X		Analytical linear differential equation.
Computer simulation (DOE-2, BLAST)	X		X	Hourly simulation programs with system models.
Computer emulation (HVACSIM+, TRNSYS)	X		X	Sub-hourly simulation programs.
Artificial Neural Networks		X	X	Connectionist models.

Table 27-2. ASHRAE's 1997 Decision Diagram for Selection of Model.[61]

METHOD:	USAGE: (1)	DIFFICULTY:	TIME SCALE: (2)	CALC. TIME:	VARIABLES:	ACCURACY:
Simple linear regression	ES	Simple	D,M	Very Fast	T	Low
Multiple linear regression	D,ES	Moderate	D,M	Fast	T,H,S,W,t	Medium
Inverse ASHRAE bin method	ES	Moderate	H	Fast	T	Medium
Change point models.	D,ES	Moderate	H,D,M	Fast	T	Medium
ASHRAE TC 4.7 modified bin method	ES, De	Moderate	H	Medium	T,S,tm	Medium
Thermal network	D,ES,C	Complex	S,H	Fast	T,S,tm	High
Fourier Series Analysis	D,ES,C	Complex	S,H	Medium	T,H,S,W,t,tm	High
ARMA Model	D,ES,C	Complex	H	Medium	T,H,S,W,t,tm	High
Modal analysis	D,ES,C	Complex	H	Medium	T,H,S,W,t,tm	High
Differential equation	D,ES,C	Very Complex	S,H	Fast	T,H,S,W,t,tm	High
Computer Simulation (Component-based)	D,ES,C, De	Very Complex	S,H	Slow	T,H,S,W,t,tm	Medium
Computer simulation (Fixed schematic)	D,ES,De	Very Complex	H	Slow	T,H,S,W,t,tm	Medium
Computer emulation	D,C	Very Complex	S,H	Very Slow	T,H,S,W,t,tm	High
Artificial Neural Networks	D,ES,C	Complex	S,H	Medium	T,H,S,W,t,tm	High

NOTE: (1) Usage shown includes: diagnostics (D), energy savings calculations (ES), design (De), and control (C).
(2) Time scales shown are hourly (H), daily (D), monthly (M), and sub-hourly (S).
(3) Variables include: temperature (T), humidity (H), solar (S), wind (W), time (t), thermal mass (TM).

2003—IPMVP-2003 Volume III (new construction)
2002—ASHRAE Guideline 14-2002
2001—IPMVP-2001 Volume I & II (revised and expanded IPMVP)
1998—Texas State Performance Contracting Guidelines
1997—IPMVP (revised NEMVP)
1996—FEMP Guidelines
1996—NEMVP
1995—ASHRAE Handbook—Ch. 37 "Building Energy Monitoring"
1994—PG&E Power Saving Partner "Blue Book"
1993—NAESCO M&V Protocols
1993—New England AEE M&V Protocols
1992—California CPUC M&V Protocols
1989—Texas LoanSTAR Program
1988—New Jersey M&V Protocols
1985—First Utility Sponsored Large Scale Programs to Include M&V
1985—ORNLs "Field Data Acq. For Bld & Eqp Energy Use Monitoring"
1983—International Energy Agency "Guiding Principles for Measurement"
1980s—USDOE funds the End-Use Load and Consumer Assessment Program (ELCAP)
1980s—First Utility Sponsored Large Scale Programs to Include M&V
1970s—First Validation of Simulations
1960s—First Building Energy Simulations on Mainframe Computers

Table 27-3. History of M&V Protocols.

Demand Savings[69], which is intended to serve as the technical document for the IPMVP.

27.3 PERFORMANCE CONTRACTS

To reduce costs and improve the HVAC and lighting systems in its buildings, the U.S. federal government has turned to the private energy efficiency sector to develop methods to finance and deliver energy ef-

ficiency to the government. One of these arrangements, the performance contract, often includes a guarantee of performance, which benefits from accurate, reliable measurement and verification. In such a contract all costs of the project (i.e., administration, measurement and verification, overhead and profit) are paid for by the energy saved from the energy or water conservation projects. In principle, this is a very attractive option for the government since it avoids paying the initial costs of the retrofits, which would have to come from shrinking taxpayer revenues. Instead, the costs are paid over a series of years, because the government agrees to pay the Energy Service Company (ESCO) an annual fee that equals the annual normalized costs savings of the retrofit (plus other charges). This allows the government to finance the retrofits by paying a pre-determined annual utility bill over a series of years, which equals the utility bill during the base year had the retrofit not occurred. In reality, because the building has received an energy conservation retrofit, the actual utility bill is reduced, which allows funding the annual fee to the ESCO without realizing any increase in the total annual utility costs (i.e., utility costs plus the ESCO fee). Once the performance contract is paid off, the total annual utility bills for the government are reduced, and the government receives the full savings amount of the retrofit.

27.3.1   Definitions, Roles and Participants

There are many different types of performance contracts, that vary according to risk and financing, including guaranteed savings, and shared savings[70]. In a guaranteed savings performance contract, a fixed payment is established that repays the ESCO's debt financing of the energy conservation retrofit and any fees associated with the project. In return, the ESCO guarantees that the energy savings will cover the fixed payment to the ESCO. Hence, in a guaranteed savings contract the ESCO is responsible for the majority of the project risks. In a shared savings performance contract, payments to the ESCO are based on an agreed-upon portion of the estimated savings generated by the retrofit.

In such contracts the M&V methods selected determine the level of risk, as well as the responsibilities of the ESCO and building owner. In both types of contracts the measurement and verification of the energy savings plays a crucial role in determining payment amounts.

27.4 OVERVIEW OF MEASUREMENT AND VERIFICATION METHODS

Nationally recognized protocols for measurement and verification have evolved since the publication of the 1996 NEMVP as shown in Table 27-4. This evolution reflects the consensus process that the Department of Energy has chosen as a basis for the protocols. This process was chosen to produce methods that all parties agree can be used by the industry to determine savings from performance contracts, varying in accuracy and cost from partial stipulation to complete measurement. In 1996 three M&V methods were included in the NEMVP: Option A: measured capacity with stipulated consumption; Option B: end-use retrofits, which utilized measured capacity and measured consumption; and Option C: whole-facility or main meter measurements, which utilize before/after regression models.

In 1997, Options A, B and C were modified and relabeled, and Option D, calibrated simulation, was added. Also included in the 1997 IPMVP was a chapter on measuring the performance of new construction, which primarily utilized calibrated simulation and a discussion of the measurement of savings due to water conservation efforts. In 2001 the IPMVP was published in two volumes: Volume I, which covers Options A, B, and C, which were redefined and relabeled from the 1997 IPMVP, and Volume II, which covers indoor

Table 27-4. Evolution of M&V Protocols in the United States.

1996 NEMVP	1997 IPMVP	2001/2003 IPMVP	2002 ASHRAE GUIDELINE 14
OPTION A: Measured Capacity Stipulated Consumption	OPTION A: End-use Retrofits: Measured Capacity, Stipulated Consumption	VOLUME I: OPTION A: Partially Measured Retrofit Isolation	
OPTION B: End-use Retrofits: Measured Capacity, Measured Consumption	OPTION B: End-use Retrofits: Measured Capacity, Measured Consumption	VOLUME I: OPTION B: Retrofit Isolation	RETROFIT ISOLATION APPROACH
OPTION C: Whole-facility or Main Meter Measurement	OPTION C: Whole-facility or Main Meter Measurement	VOLUME I: OPTION C: Whole-building	WHOLE-BUILDING APPROACH
	OPTION D: Calibrated Simulation	VOLUME I: OPTION D: Calibrated Simulation	WHOLE-BUILDING CALIBRATED SIMULATION APPROACH
		VOLUME II: IEQ M&V 5 Approaches	
	Measurement and Verification of New Buildings	VOLUME III: New Construction	
	EXAMPLE: Water Projects		

environmental quality (IEQ) and includes five M&V approaches for IEQ, including: no IEQ M&V, M&V based on modeling, short-term measurements, long-term measurements, and a method based on occupant perceptions of IEQ. In 2003 the IPMVP released Volume III, which contains four M&V methods: Option A: partially measured energy conservation measure (ECM) isolation, Option B: isolation, Option C: whole-building comparisons, and Option D: whole-building calibrated simulation.

In 2002 ASHRAE released Guideline 14-2002: Measurement of Energy and Demand Savings, which is intended to serve as the technical document for the IPMVP. As the name implies, Guideline 14 contains approaches for measuring energy and demand savings from energy conservation retrofits to buildings. This includes three methods: a retrofit isolation approach, which parallels Option B of the IPMVP; a whole-building approach, which parallels Option C of the IPMVP; and a whole-building calibrated simulation approach, which parallels Option D of the 1997 and 2001 IPMVP. ASHRAE's Guideline 14 does not explicitly contain an approach parallels Option A in the IPMVP, although several of the retrofit isolation approaches use partial measurement procedures, as will be discussed in a following section.

27.4.1 Role of M&V

Each energy conservation measure (ECM) presents particular requirements. These can be grouped in functional sections as shown in Table 27-5. Unfortunately, in most projects, numerous variables exist so the assessments can be easily disputed. In general, the low risk (L)—reasonable payback ECMs exhibit steady performance characteristics that tend not to degrade or become easily noticed when savings degradation occurs. These include lighting, constant speed motors, two-speed motors and IR radiant heating. The high risk (H)—reasonable payback ECMs include EMCSs, variable speed drives and control retrofits. The savings from these ECMs can be overridden by building operators and not be noticed until years later. Most other ECMs fall in the category of "it depends." The attention that the operations and maintenance directs at these dramatically impacts the sustainability of the operation and the savings. With an EMCS, operators can set up trend reports to measure and track occupancy schedule overrides, the various reset schedule overrides, and variable speed drive controls—and even monitor critical parameters that track mechanical systems performance. Table 27-5 illustrates a "most likely" range of ratings for the various categories.[71]

Table 27-5. Overview of Risks and Costs for ECMs.

ECM Strategies	ECM			M&V		
	Implement Cost	Payback	Savings	Implement Cost	Short Term Savings Risk	Long Term Savings Risk
Boiler Replace/Rebuild	M	M	M	H	M	H
Building Envelope Upgrades	H	L	M	H	L	M
Central Heating Plant Decentralization	VH	M	M		H	M
Chiller Plant Decentralization	VH	L	TBD			
Chiller Replace/Rebuild	H	L	M			
Constant Speed Motors	L	L	L		L	L
Cooling Tower Replace/Rebuild	H	L	M			
DDC Controls	M-H	M	H			
EMCS	H	L-M	H			
Ground Source HP	M	M	L-M			
HP Replace	L	M	L			
IR Radiant Heating	M-H	M	H			
Lighting	M	H	TBD			
Continuous Commissioning	H	VH	H			
Propane Air Plant	H	L-M				
Steam Traps	L	TBD				
Thermal Storage	H	L				
Tower Free Cooling	M-H	M				
Varaiable Speed Motors	L	H				
Very High	>$1M	<3 years	>$100K	>$100K	>15%/yr	>50%
High	>$100K	<5 years	>$10K	>$10K	>10%/yr	>35%
Medium	>$10K	<7 years	>$1K	>$1K	>5%/yr	>25%
Low	>$1K	> 8 years	<$1K	<$1K	<4%/yr	<10%

Often, building envelope or mechanical systems need to be replaced. Mechanical systems have finite lifetimes, ranging from two to five years for most light bulbs to 10-20+ years for chillers and boilers. Building envelop replacements like insulation, siding, roof, windows and doors can have lifetimes from 10 to 50 years. In these instances, life cycle costing should be done to compare the total cost of upgrading to more efficient technology. Also, the cost of M&V should be considered when determining how to sustain the savings and performance of the replacement. In many cases, the upgraded efficiency will have a payback of less than 10 years when compared to the current efficiency of the existing equipment. Current technology high efficiency upgrades normally use controls to acquire the high efficiency. These controls often connect to standard interfaces so that they communicate with today's state of the art energy management and control systems (EMCSs).

27.4.2 M&V Methods: Existing Buildings

In general, a common theme between the NEMVP, IPMVP and ASHRAE's Guideline 14-2002, is that M&V methods for measuring energy and demand savings in existing building are best represented by the following three approaches: retrofit isolation approach, a whole-building approach, and a whole-building calibrated simulation approach. Similarly, the measurement of the performance of new construction, renewables, and water use utilize one or more of these same methods.

27.4.2.1 Retrofit Isolation Approach

The retrofit isolation approach is best used when end use capacity, demand, or power can be measured during the baseline period and after the retrofit for short-term period(s), or continuously over the life of the project. This approach can use continuous measurement of energy use, both before and after the retrofit. Likewise, periodic, short-term measurements can be used to during the baseline and after the retrofit to determine the retrofit savings. Often such short-term measurements are accompanied by periodic inspections of the equipment to assure that the equipment is operating as specified. In most cases energy use is calculated by developing representative models of the isolated component or energy end use (i.e., the kW or Btu/hr) and use (i.e., the kWh or Btu).

27.4.2.1.1 Classifications of Retrofits

According to ASHRAE's Guideline 14-2002 retrofit isolation approach, components or end-uses can be classified according to the following definitions.[72]

1. *Constant Load, Constant Use.* Constant load, constant use systems consist of systems where the energy used by the system is constant (i.e., varies by less than 5%), and the use of the system is constant (i.e., varies by less than 5%) through either the baseline or post-retrofit period.

2. *Constant Load, Variable Use.* Constant load, variable use systems consist of systems where the energy used by the system is constant (i.e., varies by less than 5%), but the use of the system is variable (i.e., varies by more than 5%) through either the baseline or post-retrofit period.

3. *Variable Load, Constant Use.* Variable load, constant use systems consist of systems where the energy used by the system is variable (i.e., varies by more than 5%), but the use of the system is constant (i.e., varies by less than 5%) through either the baseline or post-retrofit period.

4. *Variable Load, Variable Use.* Variable load, variable use systems consist of systems where the energy used by the system is variable (i.e., varies by more than 5%), and the use of the system is variable (i.e., varies by more than 5%) through either the baseline or post-retrofit period.

Use of these classifications then allows for a simplified decision table (Table 27-6) to be used in determining which type of retrofit-isolation procedure to use. For example, in the first row (i.e., a CL/TS-pre-retrofit to CL/TS-post-retrofit) if a constant load with a known or timed schedule is replaced with a new device that has a reduced constant load and a known or constant schedule, then the pre-retrofit and post-retrofit metering can be performed with one-time load measurement(s). Contrast this with the last row (i.e., a VL/VS-pre-retrofit to VL/VS-post-retrofit) if a variable load with a timed or variable schedule is replaced with a new device that has a reduced variable load and a variable schedule; then the pre-retrofit and post-retrofit metering should use continuous or short-term measurement that are sufficient in length to allow for the characterization of the performance of the component to be accomplished with a model (e.g., regression, or engineering model).

27.4.2.1.2 Detailed Retrofit Isolation Measurement and Verification Procedures

Appendix E of ASHRAE's Guideline 14-2002 contains detailed retrofit isolation procedures for the measurement and verification of savings, including:

Table 27-6. Metering Requirements to Calculate Energy and Demand Savings From the ASHRAE Guideline 14-2002[73].

Pre-Retrofit	Retrofit changes	Required metering	
		Pre-retrofit	Post-retrofit
CL/TS	Load but still CL	One time load measurement	One time load measurement
CL/TS	Load to VL	One time load measurement	Sufficient load measurements to characterize load
CL/TS	Schedule but still TS	One time load measurement (either pre- or post-retrofit)	
CL/TS	Schedule to VS	One time load measurement (either pre- or post-retrofit)	Sufficient measurement of runtime
CL/TS	Load but still CL and schedule but still TS	One time load measurement	One time load measurement
CL/TS	Load to VL and schedule but still TS	One time load measurement	Sufficient load measurements to characterize load
CL/TS	Load but still CL and schedule to VS	One time load measurement	One time load measurement and sufficient measurement of runtime
CL/TS	Load to VL and schedule to VS	One time load measurement	Sufficient load measurements to characterize load
CL/VS	Load but still CL	One time load measurement and sufficient measurement of runtime	One time load measurement and sufficient measurement of runtime
CL/VS	Load to VL	One time load measurement and sufficient measurement of runtime	Sufficient load measurements to characterize load
CL/VS	Schedule to TS	One time load measurement (either pre- or post-retrofit) and sufficient measurement of runtime	
CL/VS	Schedule but still VS	One time load measurement (either pre- or post-retrofit) and sufficient measurement of runtime	Sufficient measurement of runtime
CL/VS	Load but still CL and schedule to TS	One time load measurement and sufficient measurement of runtime	One time load measurement
CL/VS	Load to VL and schedule but still TS	One time load measurement and sufficient measurement of runtime	Sufficient load measurements to characterize load
CL/VS	Load but still CL and schedule to VS	One time load measurement and sufficient measurement of runtime	One time load measurement and sufficient measurement of runtime
CL/VS	Load to VL and schedule but still VS	One time load measurement and sufficient measurement of runtime	Sufficient load measurements to characterize load
VL/TS or VS	Load to CL	Sufficient load measurements to characterize load	One time load measurement and sufficient measurement of runtime
VL/TS or VS	Load but still VL	Sufficient load measurements to characterize load	Sufficient load measurements to characterize load
VL/TS or VS	Schedule still or to TS	Sufficient load measurements to characterize load	Sufficient load measurements to characterize load
VL/TS or VS	Schedule to or still VS	Sufficient load measurements to characterize load	Sufficient load measurements to characterize load
VL/TS or VS	Load to CL and schedule still or to TS	Sufficient load measurements to characterize load	One time load measurement
VL/TS or VS	Load but still VL and schedule still or to TS	Sufficient load measurements to characterize load	Sufficient load measurements to characterize load
VL/TS or VS	Load to CL and schedule to or still VS	Sufficient load measurements to characterize load	One time load measurement and sufficient measurement of runtime
VL/TS or VS	Load but still VL and schedule to or still VS	Sufficient load measurements to characterize load	Sufficient load measurements to characterize load

CL = constant load
TS = timed (known) schedule
VL = variable load
VS = variable (unknown) schedule

pumps, fans, chillers, boilers and furnaces, lighting, and large and unitary HVAC systems. In general, the procedures were drawn from the previous literature, including ASHRAE's Research Project 827-RP[74] (i.e., pumps, fans, chillers), various published procedures for boilers and furnaces,[75,76,77,78,79,80,81] lighting procedures, and calibrated HVAC calibration simulations.[82,83] A review of these procedures, which vary from simple one-time measurements to complex, calibrated air-side psychrometric models, is described it the following sections.

A. PUMPS

Most large HVAC systems utilize electric pumps for moving heating/cooling water from the building's primary systems (i.e., boiler or chiller) to the building's secondary systems (i.e., air-handling units, radiators, etc.) where it can condition the building's interior. Such pumping systems use different types of pumps, varying control strategies, and piping layouts. Therefore, the characterization of pumping electric power depends on the system design and control method used. Pumping systems can be characterized by the three categories shown in Table 27-7.[84] Table 27-8 shows the six pump testing methods, including the required measurements, applications, and procedures steps.

ASHRAE's Research Project 827-RP[85] developed six in-situ methods for measuring the performance of pumps of varying types and controls. To select a method the user needs to determine the pump system type and control and the desired level of uncertainty, cost, and degree of intrusion. The user also needs to record the pump and motor data (i.e., manufacturer, model and serial number), fluid characteristics, and operating conditions. The first two methods (i.e., single-point and single-point with a manufacturer's curve) involve testing at a single operating point. The third and fourth procedures involve testing at multiple operating points under imposed system loading. The fifth method also involves multiple operating points, in this case obtained through short-term monitoring of the system without imposed loading. The sixth procedure operates the pump with the fluid flow path completely blocked. While the sixth procedure is not useful for generating a power versus load relationship, it can be used to confirm manufacturer's data or to identify pump impeller diameter. A summary of the methods is provided below. Additional details can be found by consulting ASHRAE's Guideline 14-2002.

A-1. Constant Speed and Constant Volume Pumps

Constant volume pumping systems use three way valves and bypass loops at the end-use or at the pump. As the load varies in the system, pump pressure and flow are held relatively constant, and the pump input power remains nearly constant. Because pump motor speed is constant, constant volume pumping systems have a single operating point. Therefore, measuring the power use at the operating point (i.e., a single point measurement) and the total operating hours are enough to determine annual energy use.

A-2. Constant Speed and Variable Volume Pumps

Variable pumping systems with constant speed pumps use two-way control valves to modulate flow to the end-use as required. In constant speed variable volume pumping systems, the flow varies along the pump curve as the system pressure drop changes in response to the load. In some cases, a bypass valve may be modulated if system differential pressure becomes too large. Such systems have a single possible operating point for any given flow, as determined by the pump curve at that

flow rate. In such systems the second and third testing methods can be used to characterize the pumps energy use at varying conditions. In the second procedure, measurements of in-situ power use is performed at one flow rate and manufacturer's data on the pump, motor, and drive system are used to create a part load power use curve. In the third testing method in-situ measurements are made of the electricity use of the pump with varying loads imposed on the system using existing control, discharge, or balancing valves. The fourth and fifth methods can also be used to characterize the pump electricity use. Using one of these methods, the part load power use curve and a representative flow load frequency distribution are used to determine annual energy use.

A-3. Variable Speed and Variable Volume Pumps

Like the constant speed variable volume system, flow to the zone loads is typically modulated using two-way control valves. However, in variable speed variable volume pumping systems, a static pressure controller is used to adjust pump speed to match the flow load requirements. In such systems the operating point cannot be determined solely from the pump curve and flow load, because a given flow can be provided at various pressures or speeds. Furthermore, the system design and control strategy place constraints on either the pressure or flow. Such systems have a range of system curves which call for the same flow rate, depending on the pumping load. 827-RP provides two options (i.e., multiple point with imposed loads and short-term monitoring) for accurately determining the in-situ part load power use. In both cases, the characteristics of the in-situ test include the pump and piping system (piping, valves, and controllers); therefore, the control strategy is included within the data set. In the fourth method (i.e., multiple point with imposed load at the zone), the pump power use is measured at a range of imposed loads. These imposed loads are done at the zone level to account for the in-situ control strategy and system design. In the fifth method (i.e., multiple point through short-term monitoring), the pump system is monitored as the building experiences a range of thermal loads, with no artificial imposition of loads. If the monitored loads reflect the full range of loads, then an accurate part load power curve can be developed that represents the full range of annual load characteristics. For methods #4 and #5, the measured part load power use curves and flow load frequency distribution are used to determine annual energy use.

Table 27-7. Applicability of Test Methods to Common Pumping Systems
From the ASHRAE Guideline 14-2002[86].

Test Method:	Pumping System:		
	Constant Speed, Constant Volume	Constant Speed, Variable Volume	Variable Speed, Variable Volume
1. Single Point	✓		
2. Single Point with Manufacturer's Pump Curve		✓	
3. Multiple Point with Imposed Loads at Pump		✓	
4. Multiple Point with Imposed Loads at Zone		✓	✓
5. Multiple Point through Short Term Monitoring		✓	✓
6. No-Flow Test for Pump Characteristics	✓	✓	✓

A-4. Calculation of Annual Energy Use

Once the pump performance has been measured, the annual energy use can be calculated using the following procedures, depending upon whether the system is a constant volume or variable volume pumping system. Savings are then calculated by comparing the annual energy use of the baseline with the annual energy use of the post-retrofit period.

Constant Volume Constant Speed Pumping Systems. In a constant volume constant speed pumping system, the volume of the water moving through the pump is almost constant, and therefore the power load of the pump is virtually constant. The annual energy calculation is therefore a constant times the frequency of the operating hours of the pump:

$$E_{annual} = T * P$$

where:

T = annual operating hours
P = equipment power input

Variable Volume Pumping Systems. For variable volume pumping systems the volume of water moving through the pump varies over time; hence, the power demand of the pump and motor varies. The annual energy use then becomes a frequency distribution of the load times the power associated with each of the bins of operating hours. In-situ testing is used to determining the power associated with the part load power use.

$$E_{annual} = \sum_i (T_i * P_i)$$

where:

i = bin index, as defined by the load frequency distribution
T_i = number of hours in bin i
P_i = equipment power use at load bin i

B.   FANS

Most large HVAC systems utilize fans or air-handling units to deliver heating and cooling to the building's interior. Such air-handling systems use different types of fans, varying control strategies, and duct layouts. Therefore, the characterization of fan electric power depends on the system design and control method used. Fan systems can be characterized by the three categories shown in Table 27-9.[88] Table 27-10 shows the five fan testing methods, including the required measurements, applications, and procedures steps.

In a similar fashion as pumping systems, ASHRAE's Research Project 827-RP developed five in-situ methods for measuring the performance of fans of varying types and controls. To select a method the user needs to determine the system type and control and the desired level of uncertainty, cost, and degree of intrusion. The user also needs to record the fan and motor data (i.e., manufacturer, model and serial number), as well as the operating conditions (i.e., temperature, pressure and humidity of the air stream). The first two methods (i.e., single-point and single-point with a manufacturer's curve) involve testing at a single operating point. The third and fourth procedures involve testing at multiple operating points under imposed system loading. The fifth method also involves multiple operating points, in this case obtained through short-term monitoring of the system without imposed loading. Additional details about fan testing procedures can be found by consulting ASHRAE's Guideline 14-2002.

B-1. Calculation of Annual Energy Use

Once the fan performance has been measured, the annual energy use can be calculated using the following procedures, depending upon whether the system is a constant volume or variable volume system. Savings are then calculated by comparing the annual energy use of the baseline with the annual energy use of the post-

Table 27-8. Pump Testing Methods from ASHRAE Guideline 14-2002[87].

METHOD	PUMPS
Method #1: Single Point Test	Measure: i) volumetric flow rate, ii) coincident RMS power, iii) differential pressure, iv) and rotational speed while the pump is at typical operating conditions. Applications: Constant volume constant speed pumping systems. Used to confirm design operating conditions and pump and system curves. Steps: • Operate pump at typical existing operating conditions for the system. • Measure pump suction and discharge pressure, or differential pressure. • Measure pump capacity. • Measure motor RMS power input. • Measure speed. • Calculate pump and energy characteristics.
Method #2: Single Point Test with Manufacturer's Curve	Measure: i) volumetric flow rate, ii) coincident RMS power, iii) differential pressure, iv) rotational speed while the pump is at typical operating conditions. Applications: Variable volume constant speed pumping systems. Used with manufacturer's data on the pump, motor, and drive system to determine power at other operating points. Steps: • Obtain manufacturer's pump performance curves. • Operate pump at typical existing operating conditions. • Measure pump suction and discharge pressure, or differential pressure. • Measure pump capacity. • Measure motor RMS power input. • Measure speed. • Calculate pump and energy characteristics.
Method #3: Multiple Point Test with Imposed Loads at Pump or Fan	Measure: i) volumetric flow rate and ii) coincident RMS power while the pump is at operated at a range of flow load conditions as prescribed in the test procedures. Applications: Variable volume constant speed pumping systems. The loads are imposed downstream of the pump with existing control valves. Pump operation follows the pump curve. Pump differential pressure and rotational speed may also be measured for more complete pump system evaluation. Steps: • Operate pump with system configuration set for maximum flow. • Measure pump capacity. • Measure motor RMS power input. • Change system configuration to reduce flow and repeat measurement steps 2 and 3. • Calculate pump and energy characteristics.
Method #4: Multiple Point Test with Imposed Loads at Zone	Measure: i) volumetric flow rate and ii) coincident RMS power for a range of building or zone thermal loads as prescribed in the test procedures. Applications: Variable volume systems. The loads are imposed on the building or zones such that the system will experience a broad range of flow rates. The existing pump variable speed control strategy is allowed to operate. Pump differential pressure and rotational speed may also be measured for more complete pump system evaluation. Steps: • Operate pump with system configured for maximum flow rate. • Measure pump capacity. • Measure motor RMS power input. • Change system configuration and repeat measurement steps 2 and 3. • Calculate pump and energy characteristics.
Method #5: Multiple Point Test through Short Term Monitoring	Measure: i) volumetric flow rate and ii) coincident RMS power for a range of building or zone thermal loads as prescribed in the test procedures. Applications: Variable volume variable speed systems. A monitoring period must be selected such that the system will experience a broad range of loads and pump flow rates. Pump differential pressure and rotational speed may also be measured for more complete pump system evaluation. Steps: • Choose appropriate time period for test. • Monitor pump operation and record data values for pump capacity and motor RMS power input. • Calculate pump and energy characteristics.
Method #6: No-Flow Test for Pump Characteristics	Measure: i) differential pressure at zero flow conditions (shut-off head) and compare to manufacturer's pump curves to determine impeller size. Applications: All types of centrifugal pumps (not recommended for use on positive displacement pumps) Steps: • Run pump at design operating conditions and close discharge valve completely. • Measure pump suction and discharge pressure, or differential pressure. • Measure speed. • Calculate shut-off head. • Compare shut-off head with manufacturer's pump performance curve to determine and/or verify impeller diameter.

Table 27-9. Applicability of Test Methods to Common Fan Systems From the ASHRAE Guideline 14-2002[89].

Test Method:	Fan System:		
	Constant Speed, Constant Volume	Constant Speed, Variable Volume	Variable Speed, Variable Volume
1. Single Point Test	✓		
2. Single Point with Manufacturer's Fan Data		✓	
3. Multiple Point with Imposed Loads at Fan		✓	
4. Multiple Point with Imposed Loads at Zone		✓	✓
5. Multiple Point through Short Term Monitoring		✓	✓

retrofit period.

Constant Volume Fan Systems. In a constant volume system the volume of the air moving across the fan is almost constant, and therefore the power load of the fan is virtually constant. The annual energy calculation is therefore a constant times the frequency of the operating hours of the fan.

$$E_{annual} = T * P$$

where:

T = annual operating hours
P = equipment power input

Variable Volume Systems. For variable volume systems the volume of the air being moved by the fan varies over time; hence, the power demand of the fan and motor varies. The annual energy use then becomes a frequency distribution of the load times the power associated with each of the bins of operating hours. In-situ testing is used to determine the power associated with the part load power use.

$$E_{annual} = \sum_i (T_i * P_i)$$

where:

i = bin index, as defined by the load frequency distribution
T_i = number of hours in bin i
P_i = equipment power use at load bin i

C.   CHILLERS

In a similar fashions as pumps and fans, in-situ chiller performance measurements have been also been developed as part of ASHRAE research project 827-RP. These models provide useful performance testing methods to evaluate annual energy use and peak demand characteristics for installed water-cooled chillers and selected air-cooled chillers. These procedures require short-term testing of the part load performance of an installed chiller system over a range of building thermal loads and coincident ambient conditions. The test methods determine chiller power use at varying thermal loads, using thermodynamic models or statistical models with inputs from direct measurements, or manufacturer's data. With these models annual energy use can be determined using the resultant part load power use curve with a load frequency distribution. Such models are capable of calculating the chiller power use as a function of the building thermal load, evaporator and condenser flow rates, entering and leaving chilled water temperatures, entering condenser water temperatures, and internal chiller controls. ASHRAE's Guideline 14-2002 describes two models for calculating the power input of a chiller, including simple and temperature dependent thermodynamic models.[91,92,93] A third method, which uses a tri-quadratic regression model such as those found in the DOE-2 simulation program,[94,95,96,97,98] also provides acceptable performance models, provided that measurements are made over the full operating range.

C-1. Simple Thermodynamic Model

Both the simple thermodynamic model and the temperature-dependent thermodynamic model express chiller efficiency as 1/COP, because it has a linear relationship with 1/(evaporator load). The simpler version of the chiller model developed predicts a linear relationship between 1/COP and $1/Q_{evap}$, which is independent of the evaporator supply temperature or condenser temperature returning to the chiller. The full form of simple thermodynamic model is shown in the equation below.

$$\frac{1}{COP} = -1 + \left(\frac{T_{cwRT}}{T_{chwST}} \right) + \left(\frac{1}{Q_{evap}} \right) \left(\frac{q_{evap} T_{cwRT}}{T_{chwST}} - q_{cond} \right) + f_{HX}$$

where:

T_{cwRT} = Entering (return) condenser water temperature (Kelvin)

T_{chwST} = Leaving (supply) evaporator water temperature (Kelvin)

Table 27-10. Fan Testing Methods from ASHRAE Guideline 14-2002[90].

METHOD	FANS
Method #1: Single Point Test	Measure: i) volumetric flow rate, ii) coincident RMS power use, iii) fan differential pressure, and iiii) fan rotational speed while the fan is at typical operating conditions. Applications: Constant volume fan systems. Data are used to confirm design operating conditions and fan and system curves. Steps: • Operate fan at typical existing operating conditions for the system. • Measure fan inlet and discharge pressure or (preferably) differential pressure. • Measure fan flow capacity. • Measure motor RMS power input. • Measure fan speed. • Calculate fan and energy characteristics.
Method #2: Single Point Test with Manufacturer's Curve	Measure: i) volumetric flow rate, ii) coincident RMS power use, iii) fan differential pressure, and iv) fan rotational speed while the fan is at typical operating conditions. Applications: Variable volume systems without fan control. Data are used with manufacturer's data on the fan, motor, and drive system and engineering principles to determine power at other operating points. Steps: • Obtain manufacturer's fan performance curves. • Operate fan at typical existing operating conditions. • Measure fan inlet and discharge pressure, or differential pressure. • Measure fan flow capacity. • Measure motor RMS power input. • Measure fan speed. • Calculate fan and energy characteristics.
Method #3: Multiple Point Test with Imposed Loads at Pump or Fan	Measure: i) volumetric flow rate and ii) coincident RMS power while the fan is at operated at a range of flow rate conditions as prescribed in the test procedures. Applications: Variable volume systems without fan control. The loads are imposed downstream of the fan with existing dampers. Fan operation follows the fan curve. Fan differential pressure and rotational speed may also be measured for more complete fan system evaluation. Steps: • Operate fan with system configuration set for maximum flow. • Measure fan flow capacity. • Measure motor RMS power input. • Change system configuration to reduce flow and repeat measurement steps 2 and 3. • Calculate fan and energy characteristics.
Method #4: Multiple Point Test with Imposed Loads at Zone	Measure: i) volumetric flow rate and ii) coincident RMS power use while the fan is operated at a range of flow rate conditions as prescribed in the test procedures. Applications: Variable volume systems. Thermal loads are imposed at the building or zone level such that the system will experience a broad range of flow rates. The existing fan variable speed control strategy is allowed to operate. Fan differential pressure and rotational speed may also be measured for more complete fan system evaluation. Steps: • Operate fan with system configured for maximum flow rate. • Measure fan capacity. • Measure motor RMS power input. • Change system configuration and repeat measurement steps 2 and 3. • Calculate fan and energy characteristics.
Method #5: Multiple Point Test through Short Term Monitoring	Measure: i) volumetric flow rate and ii) coincident RMS power while the fan operates at a range of flow rates. Applications: Variable volume systems. The range of flow rates will depend on the building or zones experiencing a wide range of thermal loads. A time period must be selected such that the system will experience a broad range of loads and fan flow rates. The existing fan variable speed control strategy is allowed to operate. Fan differential pressure and rotational speed may also be measured for more complete fan system evaluation. Steps: • Choose appropriate time period for test. • Monitor fan operation and record data values for fan capacity and motor RMS power input. • Calculate fan and energy characteristics.

Q_{evap} = Evaporator load

q_{evap} = rate of internal losses in evaporator

q_{cond} = rate of internal losses in condenser

f_{HX} = dimensionless term[99]

This equation reduces to a simple form that allows for the determination of two coefficients using linear regression, which is shown in the following equation.

$$\frac{1}{COP} = c_1 \left(\frac{1}{Q_{evap}} \right) + c_0$$

In this simplified form, the coefficient c_1 characterizes the internal chiller losses, while the coefficient c_0 combines the other terms of the simple model. The COP figure of merit can be converted into conventional efficiency measures of COP or kW per ton using the following relationships:

Coefficient of Performance (COP):

$$COP = \frac{kW\ refrigeration\ effect}{kW\ input}$$

Energy Efficiency Ratio (EER):

$$EER = \frac{Btu/hr\ refrigeration\ effect}{Watt\ input} = 3.412\ COP$$

Power per Ton (kW/ton):

$$kW/ton = \frac{kW\ input}{tons\ refrigeration\ effect} = 12/EER$$

The simple thermodynamic model can be determined with relatively few measurements of the chiller load (evaporator flow rate, entering and leaving chilled water temperatures) and coincident RMS power use. Unfortunately, variations in the chilled water supply (i.e., the temperature of the chilled water leaving the evaporator) and the condenser water return temperature are not considered. Hence, this model is best used with chiller systems that maintain constant temperature control of evaporator and condenser temperatures. In systems with varying temperatures, a temperature-dependent thermodynamic model or a tri-quadratic model yields a more accurate performance prediction.

C-2. Temperature-dependent
Thermodynamic Model

The temperature-dependent thermodynamic model includes the losses in the heat exchangers of the evaporator and condenser, which are expressed as a function of the chilled water supply and condenser water return temperatures. The resulting expression uses three coefficients (A_0, A_1, A_2), which are found with linear regression, as shown in the equation that follows.

$$\frac{1}{COP} = -1 + \left(T_{cwRT} \Big/ T_{chwST} \right) + \frac{-A_0 + A_1 \left(T_{cwRT} \right) - A_2 \left(T_{cwRT} \Big/ T_{chwST} \right)}{Q_{evap}}$$

Use of this temperature-dependent thermodynamic model requires the measurement of the chiller load (i.e., evaporator flow rate, entering and leaving chilled water temperatures), coincident RMS power use, and condenser water return temperature. Since this model is sensitive to varying temperatures it is applicable to a wider range of chiller systems.

To use the temperature dependent model, measured chiller thermal load, coincident RMS power use, chilled water supply temperature, and condenser water return temperatures are used to calculate the three coefficients (A_0, A_1, and A2).

To determine A2 the following equation is plotted against T_{cwRT}/T_{chwST} (Kelvin), with value of A_2. Being determined from the regression lines, which should resemble a series of straight parallel lines, one for each condenser temperature setting.

$$\alpha = \left(\frac{1}{COP} + 1 - \left(T_{cwRT} \Big/ T_{chwST} \right) \right) Q_{evap}$$

The coefficients A_0 and A_1 are determined by plotting β from the next equation, using the already determined value of A_2 versus the condenser water return temperature T_{cwRT} (Kelvin). This should result in a group of data points forming a single straight line. The slope of the regression line determines the value of coefficient A_1, while the intercept determines the value of coefficient A_0.

After A_0, A_1, and A_2 have been determined using αa and βb from the equations above, the 1/COP can be calculated and used to determine the chiller perfor-

$$\beta = \left(\frac{1}{COP} + 1 - \left(\frac{T_{cwRT}}{T_{chwST}} \right) \right) Q_{evap} + A_2 \left(\frac{T_{cwRT}}{T_{chwST}} \right)$$

mance over a wide range of measured input parameters of chiller load, chilled water supply temperature, and condenser water return temperature.

C-3. Quadratic Chiller Models

Chiller performance models can also be calculated with quadratic models, which can include models that express the chiller power use as a function of the chiller load (quadratic), as a function of the chiller load and chilled water supply temperature (bi-quadratic), or as a function of the chiller load, evaporator supply temperature and condenser return temperature (tri-quadratic). Such models use the quadratic functional form used in the DOE-2 energy simulation program to model part-load equipment and plant performance characteristics. Two examples of quadratic models are shown below, one for a monitoring project where chiller electricity use, chilled water production, chilled water supply temperature, and condenser water temperature returning to the chiller were available, which uses a tri-quadratic model as follows:

$$
\begin{aligned}
kW/ton = \ & a + b \ x \ Tons + c \ x \ T_{cond} + d \ x \ T_{evap} + \\
& e \ x \ Tons^2 + f \ x \ T_{cond}^2 \\
& + g \ x \ T_{evap}^2 + h \ x \ Tons \ x \ T_{cond} + I \ x \ T_{evap} \ x \ Tons \\
& + j \ x \ T_{cond} \ x \ T_{evap} + k \ x \ Tons \ x \ T_{cond} \ x \ T_{evap}.
\end{aligned}
$$

In a second example, chiller electricity use is modeled with a bi-quadratic model that includes only the chilled water production and chilled water supply temperature, which reduces to the following form. Either model can easily be calculated from field data in a spreadsheet using multiple linearized regression.

$$kW/ton = a + b \ x \ Tons + c \ x \ T_{evap} + d \ x \ Tons^2 + e \ x \ T_{evap}^2 + f \ x \ T_{evap} \ x \ Tons$$

C-4. Example: Quadratic Chiller Models

An example of a quadratic chiller performance analysis model is provided from hourly measurements that were taken to determine the baseline model of a cooling plant at an Army base in Texas. Figure 27-1 shows the time series data that were recorded during June and August of 2002. The upper trace is the chiller thermal load (tons), and the lower trace is the ambient temperature during this period. Figure 27-2 shows a time series plot of the recorded temperatures of the condenser water returning to the chiller (upper trace) and the chilled water supply temperatures (lower trace). In Figure 27-3 and Figure 27-4 the performance of the chiller is shown as the chiller efficiency (i.e., kW/ton) versus the chiller load (tons). In Figure 27-3 linear ($R^2 = 34.3\%$) and quadratic ($R^2 = 53.4\%$) models of the chiller are superimposed over the measured data from the chiller to illustrate how a quadratic model fits the chiller data. In Figure 27-4 a tri-quadratic model ($R^2 = 83.7\%$) is shown superimposed over the measured data.

A quick inspection of the R^2 goodness-of-fit indicators for the linear, quadratic, and tri-quadratic models begins to shed some light on how well the models are fitting the data. However, one must also inspect how well the model is predicting the chiller performance at the intended operation points. For example, although a linear model has an inferior R^2 when compared to a quadratic model, for this particular chiller it gives similar performance values for cooling loads ranging from 200 to 450 tons. Choosing the quadratic model improves the prediction of the chiller performance for values below 200 tons. However, it significantly under predicts the kW/ton at 350 tons and over-predicts the kW/ton at values over 500 tons. Hence, both models should be used with caution.

The tri-quadratic model has an improved R^2 of 83.7% and does not seem to contain any ranges where the model's bias is significant from the measured data (excluding the few stray points which are caused by transient data). Therefore, in the case of this chiller, the additional effort to gather and analyze the chiller load against the chilled water supply temperature and condenser water return temperature is well justified.

C-5. Calculation of Annual Energy Use

Once the chiller performance has been determined, the annual energy use can be calculated using the simple or temperature dependent models to determine the power demand of the chiller at each bin of the cooling load distribution. For chillers with varying temperatures, a load frequency distribution, which contains the two water temperatures, provides the operating hours of the chiller at each bin level. The energy use E_i and power level P_i are given by the equations below. The total annual energy use is then the sum of the product of the number of hours in each bin times the chiller power associated with that bin. Savings are then calculated by comparing the annual energy use of the baseline with the annual energy use of the post-retrofit period.

$$E_i = T_i * P_i$$

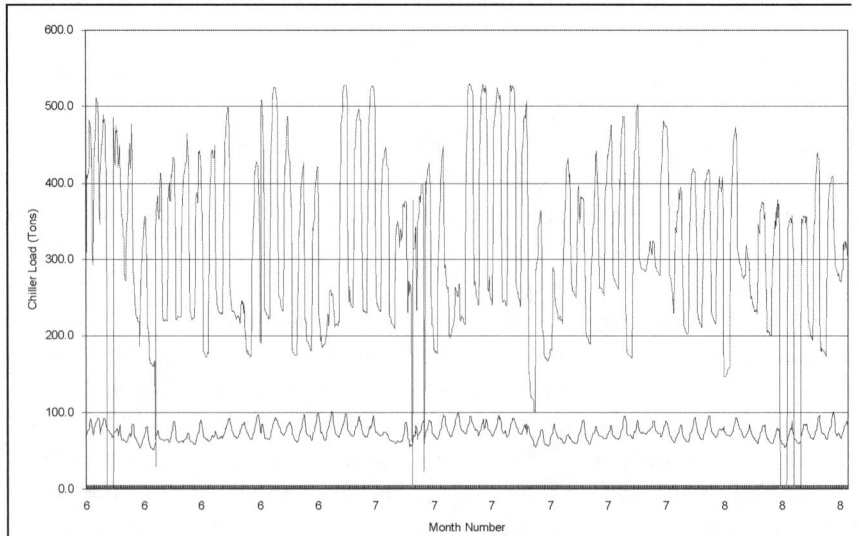

Figure 27-1. Example chiller analysis. Time series plot of chiller load (upper trace, tons) and ambient temperature (lower trace, degrees F).

Figure 27-2. Example chiller analysis. Time series plot of condenser water return temperature (upper trace, degrees F) and chilled water supply temperature (lower trace, degrees F).

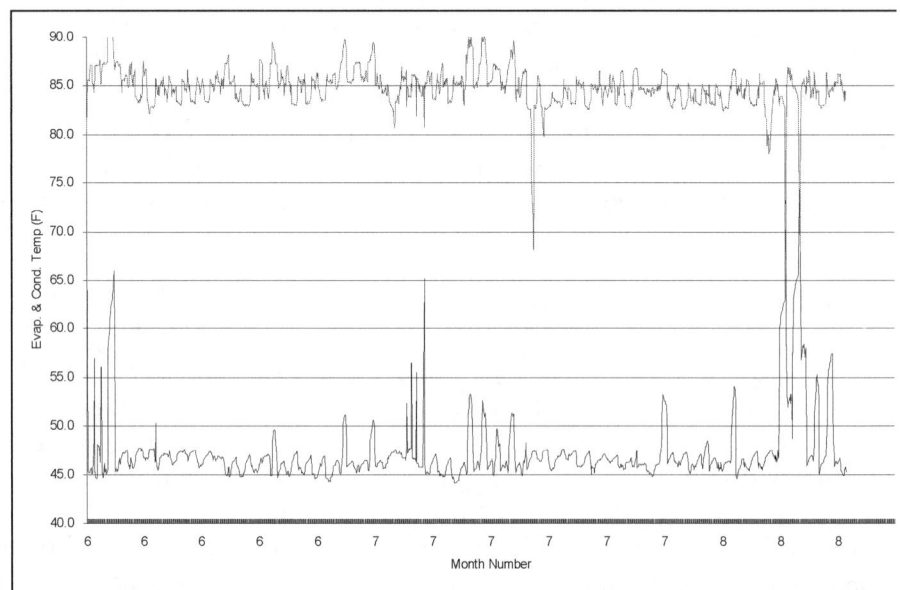

$$P_i = (1 / Eff_i) * (Q_{evap,i})$$

$$E_{annual} = \sum_i (T_i * P_i)$$

where:

i = bin index, as defined by load frequency distribution

T_i = number of hours in bin i

P_i = equipment power use at load bin i

Eff_i = chiller 1/COP in bin i

$Q_{evap,i}$ = chiller load in bin i

D. BOILERS AND FURNACES

In-situ boiler and furnace performance measurements, for non-reheat boilers and furnaces, are listed in Appendix E of ASHRAE Guideline 14-2002. These procedures, which were obtained from the previously noted published literature on performance measurements of boilers and furnaces,[100,101,102,103,104,105,106] are grouped into four methods (i.e., single-point, single-point with manufacturer's data, multiple point with imposed loads, and multiple point tests using short-term monitoring) that use three measurement techniques (i.e., direct method, direct heat loss method, and indirect combustion method), for a total of twelve methods.

The choice of method depends on boiler type (i.e., constant fire boiler or variable-fire boilers) and availability of measurements (i.e., fuel meters, steam meters, etc.). For constant fire boilers, the boiler load is virtually constant. Therefore, a single measurement or series of measurements of a full load will characterize the boiler

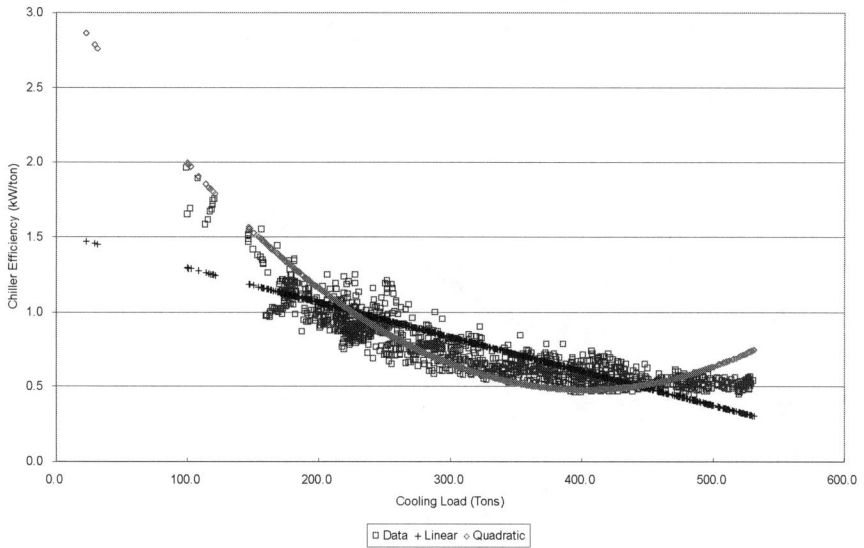

Figure 27-3. Example chiller analysis. Chiller performance plot of chiller efficiency (kW/ton) versus the chiller cooling load. Comparisons of linear (R2 34.3%) and quadratic (R2 = 53.4.3%) chiller models are shown.

Figure 27-4. Example chiller analysis. Chiller performance plot of chiller efficiency (kW/ton) versus the chiller cooling load. In this figure a tri-quadratic chiller model (R2 = 83.7%) is shown.

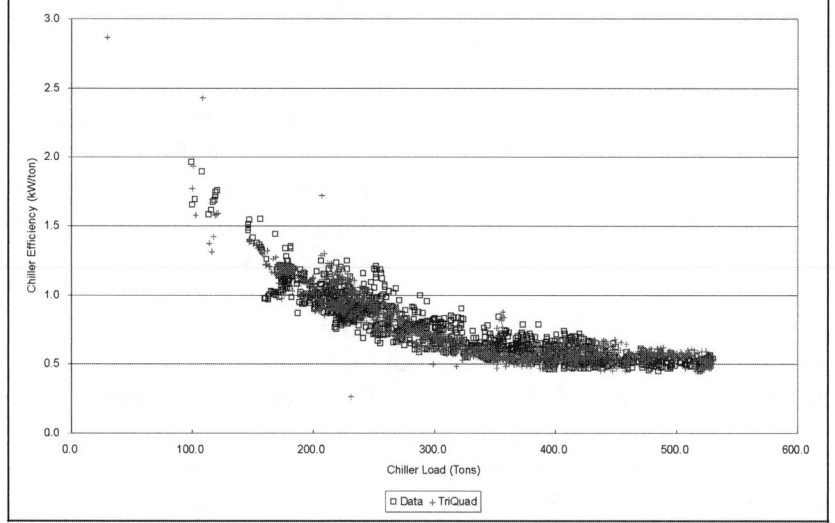

or furnace efficiency at a given set of ambient conditions. For variable fire boilers the fuel use and output of the boiler varies. Therefore, the efficiency of the boiler will vary depending upon the load of the boiler as described by the manufacturer's efficiency curve. Figure 27-5 shows an example of the measured performance of a variable-fire, low pressure steam boiler installed at an army base in Texas.[107]

D-1. Boiler Efficiency Measurements

There are three principal methods for determining boiler efficiency, the direct method (i.e., input-output method); the direct heat loss method, also known as the indirect method; and the indirect combustion efficiency method. The first two are recognized by the American Society of Mechanical Engineers (ASME) and are mathematically equivalent. They give identical results

if all the heat balance factors are considered and the boiler measurements performed without error. ASME has formed committees from members of the industry and developed the performance test codes[109] that detail procedures of determining boiler efficiency by the first two methods mentioned above. The accuracy of boiler performance calculations is dependent on the quantities measured and the method used to determine the efficiency. In the direct efficiency method, these quantities are directly related to the overall efficiency. For example, if the measured boiler efficiency is 80%, then an error of 1% in one of the quantities measured will result in a 0.8% error in the efficiency. Conversely, in the direct heat loss method the measured parameters are related to the boiler losses. Therefore, for the same boiler which had an 80% efficiency, a measurement error of 1% in any quantity affects the overall efficiency by only 0.2% (i.e.,

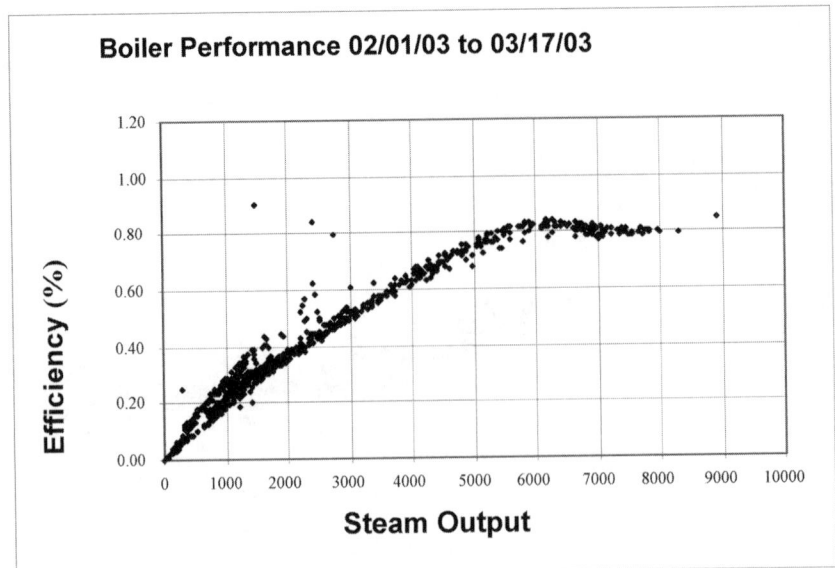

Figure 27-5. Example Boiler Performance Curve from Short-term Monitoring[108].

1% of the measured losses of 20%). As a result, the direct heat loss method is inherently more accurate than the direct method for boilers. However, the direct heat loss method requires more measurement and calculation. In general, boiler efficiencies range from 75% to 95% for utility boilers; for industrial and commercial boilers, the average efficiency ranges from 76% to 83% on gas, 78% to 89% on oil, and 85% to 88% for coal.[110,111]

- **Direct Method**

    The direct method (i.e., the input-output method) is the simplest method to determine boiler efficiency. In this method, the heat supplied to the boiler and the heat absorbed by the water in the boiler in a given time period are directly measured. Using the direct method, the efficiency of a non-reheat boiler is given by[112]:

$$\eta_b = \frac{Q_a}{Qi} \times 100$$

where

Q_a = heat absorbed (Btu/hr) = $\Sigma m_o h_o - \Sigma m_i h_i$

$m_o h_o$ = mass flow-enthalpy products of working fluid streams leaving boiler envelope, including main steam, blowdown, soot blowing steam, etc.

$m_i h_i$ = mass flow-enthalpy products of working fluid streams entering boiler envelope, including feedwater, desuperheating sprays, etc.

Q_i = heat inputs (Btu/hr) = $V_{fuel} \times HHV + Q_c$

V_{fuel} = volumetric flow of fuel into boiler (SCF/hr)

HHV = fuel higher heating value (Btu/SCF)

Q_c = heat credits (Btu/hr). Heat credits are defined as the heat added to the envelope

of the steam generating unit other than the chemical heat in the fuel "as fired." These credits include quantities such as sensible heat in the fuel, the entering air, and the atomizing steam. Other credits include heat from power conversion in the pulverizer or crusher, circulating pump, primary air fan, and recirculating gas fan.

- **Direct Heat Loss Method**

    In the direct heat loss method the boiler efficiency equals 100% minus the boiler losses. The direct heat loss method tends to be more accurate than the direct method, because the direct heat loss method focuses on determining the heat lost from the boiler, rather than on the heat absorbed by the working fluid. The direct heat loss method determines efficiency using the following[113]:

$$\eta_b = \frac{Q_a}{Q_i} \times 100 = \frac{Q_i - Q_{loss}}{Q_i} \times 100$$

$$= 100 - L_{df} - L_{fh} - L_{am} - L_{rad} - L_{conv} - L_{bd} - L_{inc} - L_{u\text{-}nacct}$$

where

Q_{loss} = heat losses (Btu/hr)

L_{df} = dry flue gas heat loss (%)

L_{fh} = fuel hydrogen heat loss (%)

L_{am} = combustion air moisture heat loss (%)

L_{rad} = radiation heat loss (%)

L_{conv} = convection heat loss (%)

L_{inc} = uncombusted fuel loss (%)

L_{bd} = blowdown heat loss (%)

L_{unacct} = unaccounted for heat losses (%)

Using this method the flue gas loss (sensible and latent heat), radiation and convection loss, fuel losses due to incomplete combustion, and blowdown loss are accounted for. In most boilers the flue gas loss is the largest loss, which can be determined by a flue gas analysis. Flue gas losses vary with flue gas exit temperature, fuel composition, and type of firing.[114] Radiation and convection loss can be obtained from the standard curves.[115] Unaccounted for heat losses can also be obtained from published industry sources,[116] which cite losses of 1.5% for solid fuels and 1% for gaseous or liquid fuel boilers. Losses from boiler blowdown should also be measured. Typical values can be found in various sources.[117,118]

- **Indirect Combustion Method**

The indirect combustion method can also be used to measure boiler efficiency. The combustion efficiency is the measure of the fraction of fuel-air energy available during the combustion process, calculated from the following[119,120]:

$$\eta_c = \frac{|h_p| - |h_f + h_a|}{HHV} \times 100$$

where

η_c = combustion efficiency (%)
h_p = enthalpy of products (Btu/lb)
h_f = enthalpy of fuel (Btu/lb)
h_a = enthalpy of combustion air (Btu/lb)
HHV = higher heating value of fuel (Btu/lb)

Indirect combustion efficiency can be related to direct efficiency or direct heat loss efficiency measurements using the following[121,122]:

$$\eta_b = \eta_c - L_{rad} - L_{conv} - L_{unacct}$$

On the right side of the equation the loss terms are usually small for well insulated boilers. These terms must be accounted for when boilers are poorly insulated or operated poorly (i.e., excessive blowdown control, etc.).

Table 27-11 provides a summary of the performance measurement methods (i.e., single-point, single-point with manufacturer's data, multiple point with imposed loads, and multiple point tests using short-term monitoring), which use three efficiency measurement techniques (i.e., direct method, direct heat loss method,

and indirect combustion method) that are listed in Appendix E of ASHRAE Guideline 14-2002. For each method the pertinent measurements are listed, along with the steps that should be taken to calculate the efficiency of the boiler or furnace being measured.

D-2. Calculation of Annual Energy Use

Once the boiler performance has been measured, the annual energy use can be calculated using the following procedures, depending upon whether the system is a constant fire boiler or variable fire boiler. Savings are then calculated by comparing the annual energy use of the baseline with the annual energy use of the post-retrofit period.

- **Constant Fire boilers**

In constant fire boilers the method assumes the load and fuel use are constant when the boiler is operating. Therefore, the annual fuel input is simply the full-load operating hours of the boiler times the fuel input. The total annual energy use is given by:

$$E_{annual} = T * P$$

where: T = annual operating hours under full load
P = equipment power use

- **Variable Fire Boilers**

For variable fire boilers the output of the boiler and fuel input vary according to load. Hence, a frequency distribution of the load is needed that provides the operating hours of the boiler at each bin level. *In-situ* testing is then used to determine the efficiency of the boiler or furnace for each bin. The total annual energy use for variable fire boilers is given by:

$$E_{annual} = \Sigma(T_i^* P_i)$$

where:

i = bin index, as defined by load variable frequency distribution
T_i = number of hours in bin i
P_i = equipment fuel input (& efficiency) at load bin (i)

E. LIGHTING

One of the most common retrofits to commercial buildings is to replace inefficient T-12 fluorescents and magnetic ballasts with T-8 fluorescents and electronic ballasts. This type of retrofit saves electricity associated with the use of the more efficient lighting and, depending on system type, can reduce cooling energy use

Table 27-11. Boiler and Furnace Performance Testing Methods from ASHRAE Guideline 14-2002[123].

METHOD	DESCRIPTION OF THE METHOD
Method #1a: Single Point Test (direct method)	Measure: i) mass flow and enthalpy of fluid streams leaving the boiler (main steam, blowdown, etc.) ii) mass flow and enthalpy of fluid streams entering the boiler (feedwater, desuperheating sprays, etc.), iii) heat inputs. Applications: Non-reheat boilers and furnaces. Steps: • Operate boiler at typical existing operating conditions for the system. • Measure mass flow and enthalpy of fluid streams leaving the boiler (main steam, blowdown, etc.). • Measure mass flow and enthalpy of fluid streams entering the boiler (feedwater desuperheating sprays, etc.). • Measure heat inputs. • Calculate efficiency using the direct efficiency method. • Calculate boiler and efficiency characteristics.
Method #1b: Single Point Test (direct heat loss method)	Measure: i) all boiler losses (dry flue gas loss, fuel hydrogen heat loss, combustion air moisture heat loss, radiation heat loss, convection heat loss, uncombusted fuel loss, blowdown loss & unaccounted for losses), ii) heat inputs. Applications: Non-reheat boilers and furnaces. Steps: • Operate boiler at typical existing operating conditions for the system. • Measure all boiler losses (dry flue gas loss, fuel hydrogen heat loss, combustion air moisture heat loss, radiation heat loss, convection heat loss, uncombusted fuel loss, blowdown loss & unaccounted for losses). • Measure heat inputs. • Calculate efficiency using direct heat loss method. • Calculate boiler and efficiency characteristics.
Method #1c: Single Point Test (indirect combustion method)	Measure: i) enthalpy of all combustion products, ii) enthalpy of fuel, iii) enthalpy of combustion air, iv) heat inputs. Applications: Non-reheat boilers and furnaces. Steps: • Operate boiler at typical existing operating conditions for system. • Measure enthalpy of all combustion products, the enthalpy of the fuel, the enthalpy of combustion air. • Measure heat inputs. • Calculate efficiency using the indirect combustion method. • Calculate boiler and efficiency characteristics.
Method #2a: Single Point Test with Manufacturer's Data (direct method)	Measure: i) mass flow and enthalpy of fluid streams leaving the boiler (main steam, blowdown, etc.) ii) mass flow and enthalpy of fluid streams entering the boiler (feedwater, desuperheating sprays, etc.), iii) heat inputs. Applications: Non-reheat boilers and furnaces. Data are used with manufacturer's published boiler efficiency curves and engineering principles to determine efficiency at other operating points. Boiler efficiency at other operating points is assumed to follow the manufacturer's curve. Steps: • Operate boiler at typical existing operating conditions for the system. • Obtain manufacturer's boiler efficiency curve. • Measure mass flow and enthalpy of fluid streams leaving the boiler (main steam, blowdown, etc.). • Measure mass flow and enthalpy of fluid streams entering the boiler (feedwater desuperheating sprays, etc.). • Measure heat inputs. • Calculate efficiency using the direct efficiency method for the single point and compare to manufacturer's curve. • Calculate boiler and efficiency characteristics.
Method #2b: Single Point Test with Manufacturer's Data (direct heat loss method)	Measure: i) all boiler losses (dry flue gas loss, fuel hydrogen heat loss, combustion air moisture heat loss, radiation heat loss, convection heat loss, uncombusted fuel loss, blowdown loss & unaccounted for losses), ii) heat inputs. Applications: Non-reheat boilers and furnaces. Data are used with manufacturer's published boiler efficiency curves and engineering principles to determine efficiency at other operating points. Boiler efficiency at other operating points is assumed to follow the manufacturer's curve. If single point does not confirm manufacturer's curve within 5% another boiler efficiency method will need to be used. Steps: • Operate boiler at typical existing operating conditions for the system. • Obtain manufacturer's boiler efficiency curve. • Measure all boiler losses (dry flue gas loss, fuel hydrogen heat loss, combustion air moisture heat loss, radiation heat loss, convection heat loss, uncombusted fuel loss, blowdown loss & unaccounted for losses). • Measure heat inputs. • Calculate efficiency using direct heat loss method for a single point and compare to manufacturer's curve. • Calculate boiler and efficiency characteristics.
Method #2c: Single Point Test with Manufacturer's Data (indirect combustion method)	Measure: i) enthalpy of all combustion products, ii) enthalpy of fuel, iii) enthalpy of combustion air, iv) heat inputs.

Table 27-11 (Cont'd). Boiler and Furnace Performance Testing Methods from ASHRAE Guideline 14-2002[123].

	Applications: Non-reheat boilers and furnaces. Data are used with manufacturer's published boiler efficiency curves and engineering principles to determine efficiency at other operating points. Boiler efficiency at other operating points is assumed to follow the manufacturer's curve. If single point does not confirm manufacturer's curve within 5% another boiler efficiency method will need to be used. Steps: • Operate boiler at typical existing operating conditions for system. • Obtain manufacturer's boiler efficiency curve. • Measure enthalpy of all combustion products, the enthalpy of the fuel, the enthalpy of combustion air. • Measure heat inputs. • Calculate efficiency using the indirect combustion method, compare to manf. curve. • Calculate boiler and efficiency characteristics.
Method #3a: Multiple Point Test with Imposed Loads (direct method)	Measure over a range of operating conditions: i) mass flow and enthalpy of fluid streams leaving the boiler (main steam, blowdown, etc.) ii) mass flow and enthalpy of fluid streams entering the boiler (feedwater, desuperheating sprays, etc.), iii) heat inputs. Different loads are imposed on the boiler and measurements repeated. Boiler operation is assumed to follow manufacturer's efficiency curve. Applications: Non-reheat boilers and furnaces. Steps: • Obtain manufacturer's efficiency curves. • Operate boiler at a given load. • Measure mass flow and enthalpy of fluid streams leaving the boiler (main steam, blowdown, etc.). • Measure mass flow and enthalpy of fluid streams entering the boiler (feedwater desuperheating sprays, etc.). • Measure heat inputs. • Calculate efficiency using the direct efficiency method. • Change load on boiler and repeat steps 2 through 6. • Calculate boiler and efficiency characteristics.
Method #3b: Multiple Point Test with Imposed Loads (direct heat loss method)	Measure over a range of operating conditions: i) all boiler losses (dry flue gas loss, fuel hydrogen heat loss, combustion air moisture heat loss, radiation heat loss, convection heat loss, uncombusted fuel loss, blowdown loss & unaccounted for losses), ii) heat inputs. Different loads are imposed on the boiler and measurements repeated. Boiler operation is assumed to follow manufacturer's efficiency curve. Applications: Non-reheat boilers and furnaces. Steps: • Obtain manufacturer's boiler efficiency curve. • Operate boiler at a given load. • Measure all boiler losses (dry flue gas loss, fuel hydrogen heat loss, combustion air moisture heat loss, radiation heat loss, convection heat loss, uncombusted fuel loss, blowdown loss & unaccounted for losses). • Measure heat inputs. • Calculate efficiency using direct heat loss method for a single point and compare to manufacturer's curve. • Change load on boiler and repeat steps 2 through 5. • Calculate boiler and efficiency characteristics.
Method #3c: Multiple Point Test with Imposed Loads (indirect combustion method)	Measure over a range of operating conditions: i) enthalpy of all combustion products, ii) enthalpy of fuel, iii) enthalpy of combustion air, iv) heat inputs. Different loads are imposed on the boiler and measurements are repeated. Boiler operation is assumed to follow the manufacturer's efficiency curve. Applications: Non-reheat boilers and furnaces. Steps: • Obtain manufacturer's boiler efficiency curve. • Operate boiler at a given load. • Measure enthalpy of all combustion products, the enthalpy of the fuel, the enthalpy of combustion air. • Measure heat inputs. • Calculate efficiency using the indirect combustion method and compare to manufacturer's curve. • Change load on boiler and repeat steps 2 through 5. • Calculate boiler and efficiency characteristics.
Method #4a: Multiple Point Test through Short Term Monitoring (direct method)	Monitor over a range of operating conditions: i) mass flow and enthalpy of fluid streams leaving the boiler (main steam, blowdown, etc.) ii) mass flow and enthalpy of fluid streams entering the boiler (feedwater, desuperheating sprays, etc.), iii) heat inputs. The range of boiler loads should cover the normally expected loads that the boiler will experience (low and high). Applications: Non-reheat boilers and furnaces. Steps: • Choose appropriate time period for test. • Monitor boiler operation and record data values for mass flow and enthalpy of fluid streams leaving the boiler (main steam, blowdown, etc.), mass flow and enthalpy of fluid streams entering the boiler (feedwater desuperheating sprays, etc.), and heat inputs. • Calculate efficiency using the direct efficiency method. • Calculate boiler and efficiency characteristics.

Table 27-11 (Cont'd). Boiler and Furnace Performance Testing Methods from ASHRAE Guideline 14-2002[123].

Method #4b: Multiple Point Test through Short Term Monitoring (direct heat loss method)	Monitor over a range of operating conditions: i) all boiler losses (dry flue gas loss, fuel hydrogen heat loss, combustion air moisture heat loss, radiation heat loss, convection heat loss, uncombusted fuel loss, blowdown loss & unaccounted for losses), ii) heat inputs. The range of boiler loads should cover the normally expected loads that the boiler will experience (low and high). Applications: Non-reheat boilers and furnaces. Steps: • Choose appropriate period for the test. • Monitor all boiler losses (dry flue gas loss, fuel hydrogen heat loss, combustion air moisture heat loss, radiation heat loss, convection heat loss, uncombusted fuel loss, blowdown loss & unaccounted for losses), and monitor heat inputs. • Calculate efficiency using direct heat loss method for a single point and compare to manufacturer's curve. • Calculate boiler and efficiency characteristics.
Method #4c: Multiple Point Test through Short Term Monitoring (indirect combustion efficiency method)	Monitor over a range of operating conditions: i) enthalpy of all combustion products, ii) enthalpy of fuel, iii) enthalpy of combustion air, iv) heat inputs. The range of boiler loads should cover the normally expected loads that the boiler will experience (low and high). Applications: Non-reheat boilers and furnaces. Steps: • Choose appropriate time period for test. • Monitor: enthalpy of all combustion products, the enthalpy of the fuel, the enthalpy of combustion air, and monitor heat inputs. • Calculate efficiency using the indirect combustion method and compare to manufacturer's curve. • Calculate boiler and efficiency characteristics.

because of reduced internal loads from the removal of the inefficient lighting. In certain climates, depending on system type, this can also mean an increase in heating loads that are required to offset the heat from the inefficient lighting. Previously published studies show the cooling interaction can increase savings by 10 to 20%. The increased heating requirements can reduce savings by 5 to 20%.[124] Therefore, where the costs can be justified, accurate measurement of total energy savings can involve before/after measurements of the lighting loads, cooling loads, and heating loads.

E-1. Lighting Methods

ASHRAE Guideline 14-2002 provides six measurement methods to account for the electricity and thermal savings, varying from methods that utilize sampled before-after measurements to methods that use submetered before-after lighting measurement with measurements of increases or decreases to the heating and cooling systems from the removal of the internal lighting load. In general, the calculation of savings from lighting retrofits involves ascertaining the wattage or power reduction associated with the new fixtures, which is then multiplied times the hours per day (i.e., lighting usage profiles) that the lights are used. The lighting usage profiles can be calculated based on appropriate estimates of use, measured at the electrical distribution panel, or

sampled with lighting loggers. Figure 27-6 shows an example of weekday-weekend profiles calculated with ASHRAE's Diversity Factor Toolkit.[125]

Some lighting retrofits involve the installation of daylighting sensors to dim fixtures near the perimeter of the building or below skylights when lighting levels can be maintained with daylighting, thus reducing the electricity used for supplemental lighting. Measuring the savings from such daylighting retrofits usually involves before-after measurements of electrical power and lighting usage profiles.

Any lighting retrofit should include an assessment of the existing lighting levels measured during daytime and nighttime conditions. All lighting retrofits should achieve and maintain lighting levels recommended by the Illuminating Engineering Society of North America (IESNA)[126]. Any pre-retrofit lighting levels not maintaining IESNA lighting levels should be brought to the attention of the building owner or administrator. In the following section, the six methods described in the ASHRAE Guideline 14-2002 are summarized. Table 27-12 contains the lighting performance measurement methods from ASHRAE's Guideline 14-2002.

Method #1: Baseline and post-retrofit measured lighting power levels and stipulated diversity profiles.

In Method #1 before-after lighting power levels for

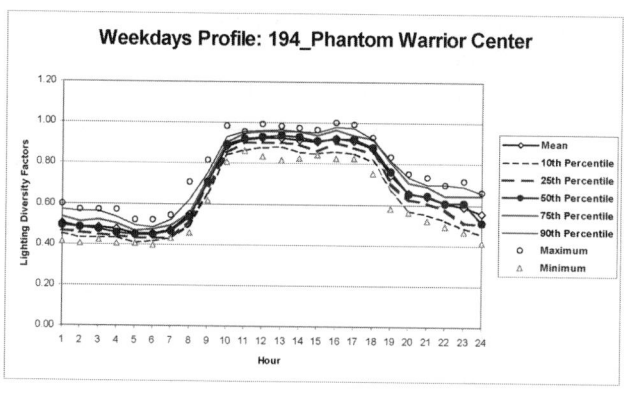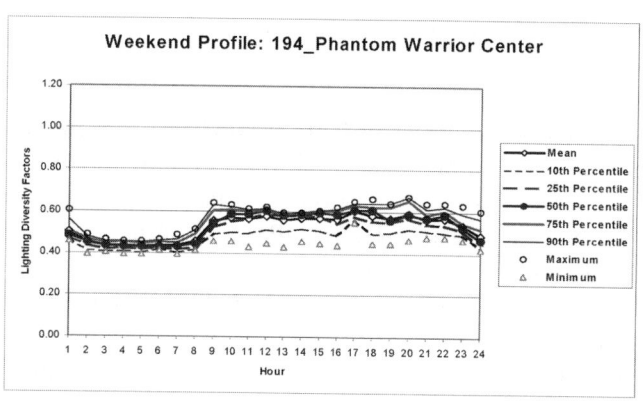

Figure 27-6. Example Weekday-Weekend Lighting Profiles.

a representative sample of lighting fixtures are measured using a Wattmeter, yielding an average Watt/fixture measurement for the pre-retrofit fixtures and post-retrofit fixtures. Lighting usage profiles are estimated or stipulated using the best available information, which represents the lighting usage profiles for the fixtures. This method works best for exterior lighting fixtures or lighting fixtures controlled by a timer or photocell. Lighting fixtures located in hallways, or any interior lighting fixtures that are operated 24 hours per day, 7 days per week or controlled by a timer is also suitable for this method. Savings benefits or penalties from thermal interactions are not included in this method.

Method #2: Baseline and post-retrofit measured lighting power levels and sampled baseline and post-retrofit diversity profiles.

In Method #2 before-after lighting power levels for a representative sample of lighting fixtures are measured using a Wattmeter, yielding an average Watt/fixture measurement for the pre-retrofit fixtures and post-retrofit fixtures. Lighting usage profiles are measured with portable lighting loggers or portable current meters attached to lighting circuits to determine the lighting usage profiles for the fixtures. This method is appropriate for any interior or exterior lighting circuit that has predictable usage profiles. Savings benefits or penalties from thermal interactions are not included in this method.

Method #3: Baseline measured lighting power levels with baseline sampled diversity profiles and post-retrofit power levels with post-retrofit continuous diversity profile measurements.

In Method #3 pre-retrofit lighting power levels for a representative sample of lighting fixtures are measured using a Wattmeter, yielding an average Watt/fixture measurement for the pre-retrofit fixtures. Pre-retrofit lighting usage profiles are measured with portable

lighting loggers or portable current meters attached to lighting circuits to determine the lighting usage profiles for the fixtures. Post-retrofit lighting usage is measured continuously using either sub-metered lighting electricity measurements or post-retrofit lighting power levels for a representative sample of lighting fixtures times a continuously measured diversity profile (i.e., using lighting loggers or current measurements on lighting circuits). This method is appropriate for any interior or exterior lighting circuit that has predictable usage profiles. Savings benefits or penalties from thermal interactions are not included in this method.

Method #4: Baseline measured lighting power levels with baseline sampled diversity profiles and post-retrofit continuous sub-metered lighting.

In Method #4 pre-retrofit lighting power levels for a representative sample of lighting fixtures are measured using a Wattmeter, yielding an average Watt/fixture measurement for the pre-retrofit fixtures. Pre-retrofit lighting usage profiles are measured with portable lighting loggers or portable current meters attached to lighting circuits to determine the lighting usage profiles for the fixtures. Post-retrofit lighting usage is measured continuously using sub-metered lighting electricity measurements. This method is appropriate for any interior or exterior lighting circuit that has predictable usage profiles. Savings benefits or penalties from thermal interactions are not included in this method.

Method #5: Includes methods #1, #2, or #3 with measured thermal effect (heating & cooling).

In Method #5, pre-retrofit and post-retrofit lighting electricity use is measured with Methods #1, #2, #3, or #4, and the thermal effect is measured using the component isolation method for the cooling or heating system. This method is appropriate for any interior lighting circuit that has predictable usage profiles. Savings benefits or penalties from thermal interactions are included in this method.

Method #6: Baseline and post-retrofit sub-metered lighting measurements and thermal measurements.

In Method #6, pre-retrofit and post-retrofit lighting electricity use is measured continuously using sub-metering, and the thermal effect is measured using whole-building cooling and heating sub-metered measurements. This method is appropriate for any interior lighting circuit. Savings benefits or penalties from thermal interactions are included in this method.

E-2. Calculation of Annual Energy Use

The calculation of annual energy use varies according to lighting calculation method as shown in Table 27-13. The savings are determined by comparing the annual lighting energy use during the baseline period to the annual lighting energy use during the post-retrofit period. In Methods #5 and #6 the thermal energy effect can either be calculated using the component efficiency methods, or it can be measured using whole-building, before-after cooling and heating measurements. Electric demand savings can be calculated using Methods #5 and #6 using diversity factor profiles from the pre-retrofit period and continuous measurement in the post-retrofit period. Peak electric demand reductions attributable to reduced chiller loads can be calculated using the component efficiency tests for the chillers. Savings are then calculated by comparing the annual energy use of the baseline with the annual energy use of the post-retrofit period.

F. HVAC SYSTEMS

As mentioned previously, during the 1950s and 1960s most engineering calculations were performed using slide rules, engineering tables, and desktop calculators that could only add, subtract, multiply, and divide. In the 1960s efforts were initiated to formulate and codify equations that could predict dynamic heating and cooling loads, including efforts to simulate HVAC systems. In 1965 ASHRAE recognized that there was a need to develop public-domain procedures for calculating the energy use of HVAC equipment and formed the Presidential Committee on Energy Consumption, which became the Task Group on Energy Requirements (TGER) for heating and cooling in 1969.[129] TGER commissioned two reports that detailed the public domain procedures for calculating the dynamic heat transfer through the building envelopes,[130] and procedures for simulating the performance and energy use of HVAC systems.[131] These procedures became the basis for today's public-domain

building energy simulation programs such as BLAST, DOE-2, and EnergyPlus.[132,133]

In addition, ASHRAE has produced several additional efforts to assist with the analysis of building energy use, including a modified bin method,[134] the HVAC01[135] and HVAC-02[136] toolkits, and HVAC simulation accuracy tests,[137] which contain detailed algorithms and computer source code for simulating secondary and primary HVAC equipment. Studies have also demonstrated that properly calibrated simplified HVAC system models can be used for measuring the performance of commercial HVAC systems.[138,139,140,141]

F-1. HVAC System Types

To facilitate the description of measurement methods that are applicable to a wide range of HVAC systems, it is necessary to categorize HVAC systems into groups, ranging from single zone, steady state systems to the more complex systems such as multi-zone systems with simultaneous heating and cooling. To accomplish this, two layers of classification are proposed; in the first layer, systems are classified into two categories (Table 27-14): systems that provide heating or cooling under separate thermostatic control and systems that provide heating and cooling under a combined control. In the second classification, systems are grouped according to those that provide constant heating rates, systems that provide varying heating rates, those that provide constant cooling rates, and those that provide varying cooling rates.

- HVAC systems that provide heating or cooling at a constant rate include: single zone, 2-pipe fan coil units, ventilating and heating units, window air conditioners, and evaporative cooling. Systems that provide heating or cooling at a constant rate can be measured using: single-point tests, multi-point tests, short-term monitoring techniques, or in-situ measurement combined with calibrated, simplified simulation.

- HVAC systems that provide heating or cooling at a varying rate include: 2-pipe induction units, single zone with variable speed fan and/or compressors, variable speed ventilating and heating units, variable speed, and selected window air conditioners. Systems that provide heating or cooling at a varying rate can be measured using: single-point tests, multi-point tests, short-term monitoring techniques, or short-term monitoring combined with calibrated, simplified simulation.

- HVAC systems that provide simultaneous heating and cooling include: multi-zone, dual duct constant volume dual duct variable volume, single duct constant volume w/reheat, single duct variable volume

Table 27-12. Lighting Performance Measurement Methods from ASHRAE Guideline 14-2002[127].

METHOD	DESCRIPTION OF THE METHOD
Method #1: Baseline and post-retrofit measured lighting power levels and stipulated diversity profiles.	Description: i) Obtain before-after lighting power levels using RMS watt/fixture measurements for the pre-retrofit fixtures and the post-retrofit fixtures, ii) stipulate the lighting usage profiles using the best available information that represents lighting usage profiles for the facility. Application: Exterior lighting on a timer or photocell. Interior hallway lighting or any interior lighting used continuously or on a timer. Steps: • Obtain measured RMS watt/fixture data for pre-retrofit and post-retrofit fixtures. • Count the fixtures associated with each functional area in the building (e,g,, areas that have different usage profiles). • Define the lighting usage profiles for each functional area using the appropriate information that represents lighting usage profiles (e.g., continuously on, on during evening hours, etc.). • Calculate lighting energy usage characteristics.
Method #2: Before/after measured lighting power levels with sampled before/after diversity profiles.	Description: i) Measure lighting power levels using RMS watt meter for a sample of the pre-retrofit fixtures and the post-retrofit fixtures, ii) measure the lighting usage profiles using light loggers or portable metering attached to the lighting circuits. Application: Any exterior lighting or interior lighting with predictable usage profiles. Steps: • Measure watt/fixture using RMS watt meter for pre-retrofit and post-retrofit fixtures. • Count the fixtures associated with each functional area in the building (i.e., areas that have different usage profiles). • Sample lighting usage profiles for each functional area using lighting loggers and/or portable submetered RMS watt meters on lighting circuits. • Calculate lighting energy usage characteristics.
Method #3: Baseline measured lighting power levels with baseline sampled diversity profiles and post-retrofit power levels with continuous diversity profile measurements.	Description: i) Obtain lighting power levels using RMS watt/fixture measurements for the pre-retrofit fixtures and the post-retrofit fixtures, ii) sample the baseline lighting usage profiles using light loggers or RMS watt measurements on submetered lighting circuits, iii) continuously measure the post-retrofit lighting usage profiles using light loggers or RMS watt measurements on submetered lighting circuits. Application: Any exterior lighting or interior lighting. Steps: • Obtain lighting power levels using RMS watt/fixture measurements for the pre-retrofit fixtures and the post-retrofit fixtures. • Sample the baseline lighting usage profiles using light loggers or RMS watt measurements on submetered lighting circuits. • Continuously measure the post-retrofit lighting usage profiles using light loggers or RMS watt measurements on submetered lighting circuits. • Calculate lighting energy usage characteristics.
Method #4: Baseline measured lighting power levels with baseline sampled diversity profiles and post-retrofit continuous sub-metered lighting.	Description: i) Obtain lighting power levels using RMS watt/fixture measurements for the pre-retrofit fixtures, ii) sample the baseline lighting usage profiles using light loggers or RMS watt measurements on submetered lighting circuits, iii) continuously measure the post-retrofit lighting power usage using RMS watt measurements on submetered lighting circuits. Application: Any exterior lighting or interior lighting. Steps: • Obtain lighting power levels using RMS watt/fixture measurements for the pre-retrofit fixtures. • Sample the baseline lighting usage profiles using light loggers or RMS watt measurements on submetered lighting circuits. • Continuously measure the post-retrofit lighting usage using RMS watt measurements on submetered lighting circuits. • Calculate lighting energy usage characteristics.
Method #5: Method #1, #2, or #3 with stipulated thermal effects.	Description: i) Obtain lighting power profiles and usage using Method(s) #1, #2, #3, or #4 ii) Calculate the heating or cooling system efficiency using HVAC component isolation methods described in this document, iii) Calculate decrease in cooling load and increase in heating load. Application: Any interior lighting. Steps: • Obtain lighting power profiles and usage using Method(s) #1, #2, or #3, • Calculate the heating or cooling system efficiency using HVAC component isolation methods described in this document, • Calculate lighting energy usage characteristics. • Calculate decrease in cooling load and increase in heating load.
Method #6: Before/after sub-metered lighting and thermal measurements.	Description: i) Obtain lighting energy usage by measuring RMS lighting use continuously at the sub-metered level for pre-retrofit and post-retrofit conditions, ii) Obtain thermal energy use data by measuring sub-metered cooling or heating energy use for pre-retrofit and post-retrofit conditions, iii) develop representative lighting usage profiles from the sub-metered lighting data. Application: Any interior lighting projects. Any exterior lighting projects (no thermal interaction). Steps: • Obtain measured sub-metered lighting data for pre-retrofit and post-retrofit periods. • Develop representative lighting usage profiles from the sub-metered lighting data. • Calculate lighting energy usage characteristics. • Calculate decrease in cooling load and increase in heating load.

Table 27-13. Lighting Calculations Methods from ASHRAE Guideline 14-2002[128].

TYPE OF MEASUREMENT	PRE-RETROFIT ELECTRICITY USAGE CALCULATIONS	POST-RETROFIT ELECTRICITY USAGE CALCULATIONS	THERMAL ENERGY USAGE CALCULATIONS
Method #1: Baseline and post-retrofit measured lighting power levels and stipulated diversity profiles.	For each lighting circuit: Annual energy use = (Power levels) x (24-hr *stipulated* profiles) x (number of days assigned to each profile)	For each lighting circuit: Annual energy use = (Power levels) x (24-hr *stipulated* profiles) x (number of days assigned to each profile)	None.
Method #2: Before/after measured lighting power levels with sampled before/after diversity profiles.	For each lighting circuit: Annual energy use = (Power levels) x (24-hr *sampled* profiles) x (number of days assigned to each profile)	For each lighting circuit: Annual energy use = (Power levels) x (24-hr *sampled* profiles) x (number of days assigned to each profile)	None.
Method #3: Baseline measured lighting power levels with baseline sampled diversity profiles and post-retrofit power levels with continuous diversity profile measurements.	For each lighting circuit: Annual energy use = (Power levels) x (24-hr *sampled* profiles) x (number of days assigned to each profile)	For each lighting circuit: Annual energy use = (Power levels) x (continuous diversity profile measurements)	None.
Method #4: Baseline measured lighting power levels with baseline sampled diversity profiles and post-retrofit continuous sub-metered lighting.	For each lighting circuit: Annual energy use = (Power levels) x (24-hr *sampled* profiles) x (number of days assigned to each profile)	For each lighting circuit: Annual use = sub-metered lighting energy use.	None.
Method #5: Method #1, #2, #3, or #4 with calculated thermal effect.	Annual use = method #1, #2, #3, or #4 as appropriate.	Annual use = method #1, #2, #3, or #4 as appropriate.	Pre and post thermal load from the lighting is calculated using the component efficiency measurement methods for HVAC systems.
Method #6: Before/after sub-metered lighting and thermal measurements.	For each lighting circuit: Annual use = sub-metered lighting energy use.		Pre and post thermal load is calculated using before-after whole-building cooling and heating sub-metered measurements.

Table 27-14. Relationship of HVAC Test Methods to Type of System.

Test Method	HVAC System		
	Constant Heating or Cooling	Variable Heating or Cooling	Simultaneous Heating and Cooling
1. Single Point with Manufacturer's Performance Data	✓		
2. Multiple Point with Manufacturer's Performance Data.	✓	✓	
3. Multiple Point through Short Term Monitoring with Manufacturer's Data.	✓	✓	
4. Short Term Monitoring and Calibrated, Simplified Simulation.	✓	✓	✓

w/reheat, dual path systems (i.e., with main and preconditioning coils), 4-pipe fan coil units, and 4-pipe induction units. Such systems can be measured using *in-situ* measurement combined with calibrated, simplified simulation.

F-2. HVAC System Testing Methods

In this section, four methods are described for the in-situ performance testing of HVAC systems as shown in Table 27-15, including: a single point method that uses manufacturer's performance data, a multiple point method that includes manufacturer's performance data, a multiple point that uses short-term data and manufacturer's performance data, and a multiple point that uses short-term data and manufacturer's performance data. Each of these methods are explained in the sections that follow.

Method #1: Single point with manufacturer's performance data

In this method the efficiency of the HVAC system is measured with a single-point (or a series) of field measurements at steady operating conditions. On-site measurements include: the energy input to the system (e.g., electricity, natural gas, hot water or steam), the thermal output of system, and the temperature of the surrounding environment. The efficiency is calculated as the measured output/input. This method can be used in the following constant systems: single zone systems, 2-pipe fan coil units, ventilating and heating units, single speed window air conditioners, and evaporative coolers.

Method #2: Multiple point with manufacturer's performance data

In this method the efficiency of the HVAC system is measured with multiple points on the manufacturer's performance curve. On-site measurements include: the energy input to the system (e.g., electricity, natural gas, hot water or steam), the thermal output of system, the system temperatures, and the temperature of surrounding environment. The efficiency is calculated as the measured output/input, which varies according to the manufacturer's performance curve. This method can be used in the following systems: single zone (constant or varying), 2-pipe fan coil units, ventilating and heating units (constant or varying), window air conditioners

(constant or varying), evaporative cooling (constant or varying), 2-pipe induction units (varying), single zone with variable speed fan and/or compressors, variable speed ventilating and heating units, and variable speed window air conditioners.

Method #3: Multiple point using short-term data and manufacturer's performance data, continuous

In this method the efficiency of the HVAC system is measured continuously over a short-term period, with data covering the manufacturer's performance curve. On-site measurements include: the energy input to the system (e.g., electricity, natural gas, hot water or steam), the thermal output of system, the system temperatures, and the temperature of the surrounding environment. The efficiency is calculated as the measured output/input, which varies according to the manufacturer's performance curve. This method can be used in the following systems: single zone (constant or varying), 2-pipe fan coil units, ventilating and heating units (constant or varying), window air conditioners (constant or varying), evaporative cooling (constant or varying) 2-pipe induction units (varying), single zone with variable speed fan and/or compressors, variable

Table 27-15. HVAC System Testing Methods[142,143].

METHOD	DESCRIPTION OF THE METHOD
Method #1: Single point with manufacturer's performance data	Measure: i) energy input to system (e.g., electricity, natural gas, hot water or steam), ii) thermal output of system, iii) temperature of surrounding environment (adjust for differences in efficiency with manufacturer's data). Applications: • single zone (constant mode) • 2-pipe fan coil units (constant mode) • ventilating and heating units, • single speed window air conditioners • evaporative coolers Steps: • Measure energy input to system (i.e, electricity, natural gas, hot water or steam). • Measure thermal output of system. • Measure temperature of space where system is operating. • Calculate efficiency as output/input. If efficiency does not agree to within 5% of manufacturer's performance data then use method #2, #3 or #4. • Adjust efficiency, using manufacturer's data if surrounding environmental conditions vary significantly over the year (i.e., ambient temperatures that the HVAC unit is exposed to). • Calculate savings by comparing differences in before-after component efficiency calculations applied to continuously measured post-retrofit thermal output energy requirements or sampled thermal load profiles.
Method #2: Multiple point with manufacturer's performance data	Measure at multiple points: i) energy input to system (e.g., electricity, natural gas, hot water or steam), ii) thermal output of system, iii) temperature of surrounding environment (adjust for differences in efficiency with manufacturer's data). Applications: • single zone (constant or varying) • 2-pipe fan coil units • ventilating and heating units (constant or varying) • window air conditioners (constant or varying) • evaporative cooling (constant or varying) • 2-pipe induction units (varying) • single zone with variable speed fan and/or compressors • variable speed ventilating and heating units • variable speed window air conditioners

(Continued)

Table 27-15 (Cont'd). HVAC System Testing Methods[142,143].

	Steps: • Measure energy input to system at multiple points (i.e., electricity, natural gas, hot water or steam). • Measure corresponding thermal output of system at multiple points. • Measure temperature of space where system is operating during all tests. • Calculate efficiency as output/input. If efficiency does not agree to within 5% of manufacturer's performance data then use method #3 or #4. • Adjust efficiency, using manufacturer's data if surrounding environmental conditions vary significantly over the year (i.e., ambient temperatures that the HVAC unit is exposed to). • Calculate savings by comparing differences in before-after component efficiency calculations applied to continuously measured post-retrofit thermal output energy requirements or sampled thermal load profiles.
Method #3: Multiple point using short-term data and manufacturer's performance data, continuous	Continuously measure over a short-term period: i) energy input to system (e.g., electricity, natural gas, hot water or steam), ii) thermal output of system, iii) temperature of surrounding environment (adjust for differences in efficiency with manufacturer's data). Applications: • single zone (constant or varying) • 2-pipe fan coil units • ventilating and heating units (constant or varying) • window air conditioners (constant or varying) • evaporative cooling (constant or varying) • 2-pipe induction units (varying) • single zone with variable speed fan and/or compressors • variable speed ventilating and heating units • variable speed window air conditioners Steps: • Continuously measure over a short-term period energy input to system (e.g., electricity, natural gas, hot water or steam). • Measure corresponding thermal output of system at multiple points. • Measure temperature of space where system is operating during all tests. • Calculate efficiency as output/input. If efficiency does not agree to within 5% of manufacturer's performance data then use method #4. • Adjust efficiency, using manufacturer's data if surrounding environmental conditions vary significantly over the year (i.e., ambient temperatures that the HVAC unit is exposed to). • Calculate savings by comparing differences in before-after component efficiency calculations applied to continuously measured post-retrofit thermal output energy requirements or sampled thermal load profiles.
Method #4: Multiple point using short-term data and manufacturer's performance data, range	Measure over a representative range: i) thermal and electric energy input to system (e.g., electricity, natural gas, hot water or steam), ii) thermal output of system, iii) temperature of surrounding environment (may be used to adjust for losses to space), iv) develop an air-side simulation model that is representative of the system (Knebel 1983), v) calibrate the air-side model to the measured data for both pre-retrofit and post-retrofit conditions. Applications: • single zone (constant or varying) • 2-pipe fan coil units • ventilating and heating units (constant or varying) • window air conditioners (constant or varying) • evaporative cooling (constant or varying) • 2-pipe induction units (varying) • single zone with variable speed fan and/or compressors • variable speed ventilating and heating units • variable speed window air conditioners • multi-zone • dual duct constant volume • dual duct variable volume • single duct constant volume w/reheat • single duct variable volume w/reheat • dual path systems (i.e., with main and preconditioning coils) • 4-pipe fan coil units • 4-pipe induction units Steps: • Measure thermal input to system over a representative range (e.g., electricity, natural gas, hot water or steam). • Measure thermal output of system. • Measure temperature of space where system is operating. • Measure important system operation characteristics (e.g., cooling coil setpoint, heating coil setpoint, mixture of outside air to returning air, etc.) • For systems using chilled water, calculate efficiency as output/input over a range of conditions representative of operating conditions. • For systems using direct expansion of refrigerant, calculate efficiency as output/input over a range of varying cooling supply temperatures and heat rejection supply temperatures (i.e., to capture the efficiency of the A/C unit over varying outside conditions). • Develop an airside simulation model that is representative of the system (Knebel 1983) and calibrate the air-side model to the measured data for both pre-retrofit and post-retrofit conditions. • Calculate savings by applying the calibrated simulation models for the baseline and post-retrofit system to continuously measured post-retrofit cooling requirements or sampled cooling load profiles.

speed ventilating and heating units, and variable speed window air conditioners.

Method #4: Multiple point using short-term data and manufacturer's performance data, range

In this method the efficiency of the HVAC system is measured continuously over a short-term period with data covering the manufacturer's performance curve. On-site measurements include: the energy input to the system (e.g., electricity, natural gas, hot water or steam), the thermal output of the system, the system temperatures, and the temperature of the surrounding environment. The efficiency is calculated using a calibrated air-side simulation of the system, which can include manufacturer's performance curves for various components. Similar measurements are repeated after the retrofit. This method can be used in the following systems: single zone (constant or varying), 2-pipe fan coil units, ventilating and heating units (constant or varying), window air conditioners (constant or varying), evaporative cooling (constant or varying), 2-pipe induction units (varying), single zone with variable speed fan and/or compressors, variable speed ventilating and heating units, variable speed window air conditioners, multi-zone, dual duct constant volume, dual duct variable volume, single duct constant volume w/reheat, single duct variable volume w/reheat, dual path systems (i.e., with main and preconditioning coils), and 4-pipe fan coil units, 4-pipe induction units.

F-3. Calculation of Annual Energy Use

The calculation of annual energy use varies according to HVAC calculation method as shown in Table 27-16. The savings are determined by comparing the annual HVAC energy use and demand during the baseline period to the annual HVAC energy use and demand during the post-retrofit period.

27.4.2.2 Whole-building or Main-meter Approach Overview

The whole-building approach, also called the main-meter approach, includes procedures that measure the performance of retrofits for those projects where whole-building pre-retrofit and post-retrofit data are available to determine the savings, and where the savings are expected to be significant enough that the difference between pre-retrofit and post-retrofit usage can be measured using a whole-building approach. Whole-building methods can use monthly utility billing data (i.e., demand or usage) or continuous measurements of the whole-building energy use after the retrofit on a more detailed measurement level (weekly, daily or

hourly). Sub-metering measurements can also be used to develop the whole-building models, providing that the measurements are available for the pre-retrofit and post-retrofit period and that meter(s) measures that portion of the building where the retrofit was applied. Each sub-metered measurement then requires a separate model. Whole-building measurements can also be used on stored energy sources, such as oil or coal inventories. In such cases, the energy used during a period needs to be calculated (i.e., any deliveries during the period minus measured reductions in stored fuel).

In most cases, the energy use and/or electric demand are dependent on one or more independent variables. The most common independent variable is outdoor temperature, which affects the building's heating and cooling energy use. Other independent variables can also affect a building's energy use and peak electric demand, including: the building's occupancy (i.e., often expressed as weekday or weekend models), parking or exterior lighting loads, special events (i.e., Friday night football games), etc.

27.4.2.2.1 Whole-building Energy Use Models

Whole-building models usually involve the use of a regression model that relates the energy use and peak demand to one or more independent variables. The most widely accepted technique uses linear regression or change-point linear regression to correlate energy use or peak demand as the dependent variable with weather data and/or other independent variables. In most cases the whole-building model has the form:

$$E = C + B_1V_1 + B_2V_2 + B_3V_3 + \dots$$

Where

E = the energy use or demand estimated by the equation

C = a constant term in energy units/day or demand units/billing period

B_n = the regression coefficient of an independent variable V_n

V_n = the independent driving variable

In general, when creating a whole-building model, a number of different regression models are tried for a particular building, the results are compared, and the best model is selected using R^2 and CV(RMSE). Table 27-17 and Figure 27-7 contain models listed in ASHRAE's Guideline 14-2002, which include steady-state: constant or mean models, models adjusted for the days in the billing period, two parameter models, three parameter models or variable-based degree-day models, four

Table 27-16. HVAC Performance Measurement Methods from ASHRAE Guideline 14-2002,[144].

TYPE OF MEASUREMENT	PRE-RETROFIT ENERGY USAGE AND ELECTRIC DEMAND CALCULATIONS	POST-RETROFIT ENERGY USAGE AND ELECTRIC DEMAND CALCULATIONS
Method #1: Single point with manufacturer's performance data	For each HVAC system: Annual energy use = (Measured Energy Use) x (Full Load Runtime Hours) Monthly demand use = (Measured Peak Electric Demand Use for System) x (System on/off Status for Each Month)	For each HVAC system: Annual energy use = (Measured Energy Use) x (Full Load Runtime Hours) Monthly demand use = (Measured Peak Electric Demand Use for System) x (System on/off Status for Each Month)
Method #2: Multiple point with manufacturer's performance data	For each HVAC system: Annual energy use = (Measured Energy Use for Each Operating Point) x (Full Load Runtime Hours for Each Operating Point) Monthly demand use = (Measured Peak Electric Demand Use for System for Each Operating Point) x (Maximum Operating Point for Each Month)	For each HVAC system: Annual energy use = (Measured Energy Use for Each Operating Point) x (Full Load Runtime Hours for Each Operating Point) Monthly demand use = (Measured Peak Electric Demand Use for System for Each Operating Point) x (Maximum Operating Point for Each Month)
Method #3: Multiple point using short-term data and manufacturer's performance data	For each HVAC system: Annual energy use = (Measured Energy Use for Each Operating Point) x (Full Load Runtime Hours for Each Operating Point) Monthly demand use = (Measured Peak Electric Demand Use for System for Each Operating Point) x (Maximum Operating Point for Each Month)	For each HVAC system: Annual energy use = (Measured Energy Use for Each Operating Point) x (Full Load Runtime Hours for Each Operating Point) Monthly demand use = (Measured Peak Electric Demand Use for System for Each Operating Point) x (Maximum Operating Point for Each Month)
Method #4: Multiple point using short-term data and manufacturer's performance data	For each HVAC system: Annual energy use = (Simulated Energy Use Using Calibrated Air-side Model) x (Binned Weather Data) Monthly demand use = (Simulated Peak Electric Demand Use for System) x (Maximum Bin Temperature for Month)	For each HVAC system: Annual energy use = (Simulated Energy Use Using Calibrated Air-side Model) x (Binned Weather Data) Monthly demand use = (Simulated Peak Electric Demand Use for System) x (Maximum Bin Temperature for Month)

parameter models, five parameter models, and multivariate models. All of these models can be calculated with ASHRAE Inverse Model Toolkit (IMT), which was developed from Research Project 1050-RP.[145]

The steady-state, linear, change-point linear, variable-based degree-day, and multivariate inverse models contained in ASHRAE's IMT have advantages over other types of models. First, since the models are simple, and their use with a given dataset requires no human intervention, the application of the models can be automated and applied to large numbers of buildings, such as those contained in utility databases. Such a procedure can assist a utility, or an owner of a large number of buildings, to identify which buildings have abnormally high energy use. Second, several studies have shown that linear and change-point linear model coefficients have physical significance to operation of heating and cooling equipment that is controlled by a thermostat.[146,147,148,149] Finally, numerous studies have reported the successful use of these models on a variety

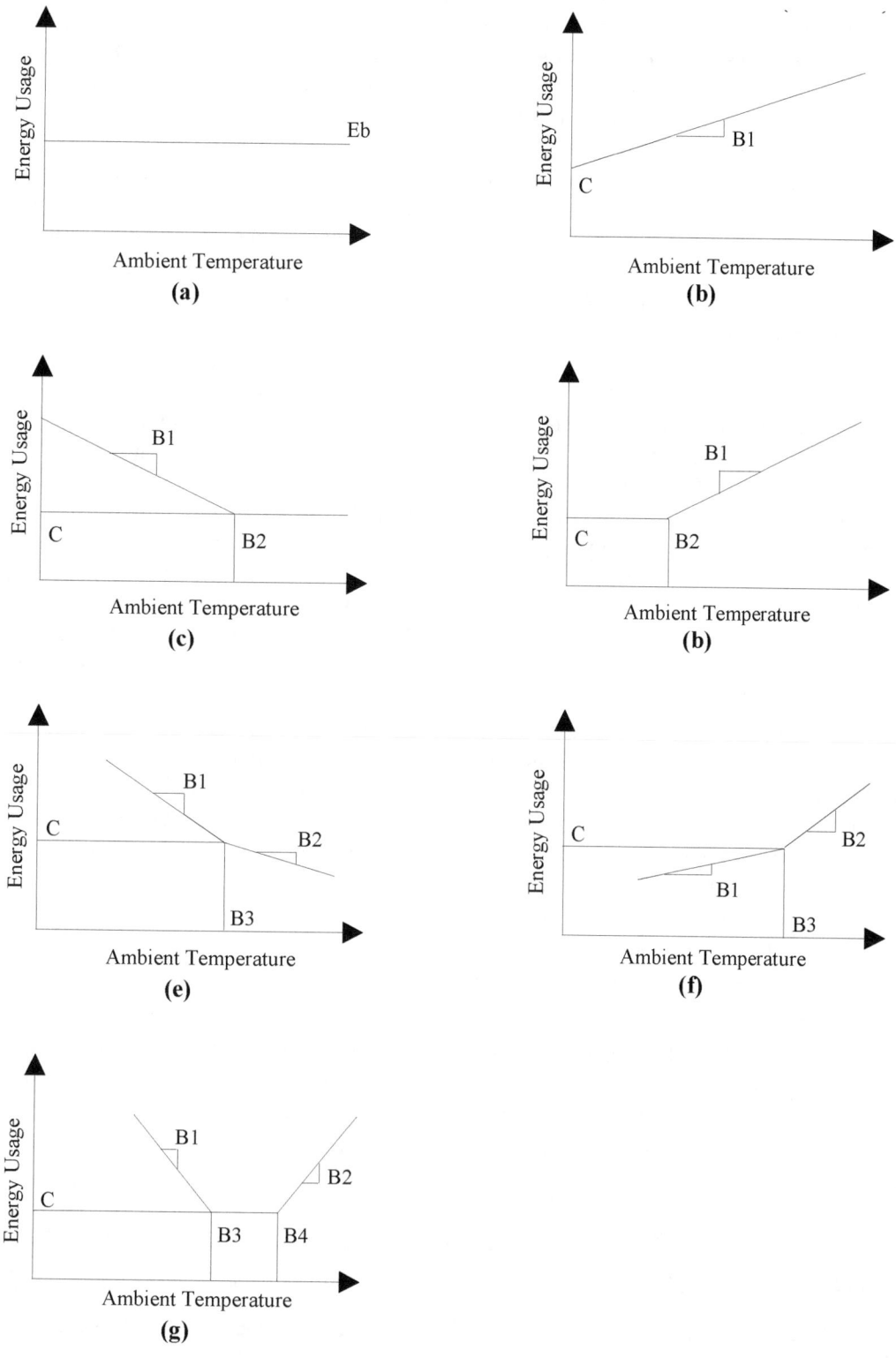

Figure 27-7. Sample Models for the Whole-building Approach. Included in this figure is: (a) mean or 1 parameter model, (b) 2 parameter model, (c) 3 parameter heating model (similar to a variable based degree-day model (VBDD) for heating), (d) 3 parameter cooling model (VBDD for cooling), (e) 4 parameter heating model, (f) 4 parameter cooling model, and (g) 5 parameter model.[157]

of different buildings.[150,151,152,153,154,155]

Steady-state models have disadvantages, including: an insensitivity to dynamic effects (e.g., thermal mass), insensitivity to variables other than temperature (e.g., humidity and solar), and inappropriateness for certain building types, for example buildings that have strong on/off schedule dependent loads or buildings that display multiple change-points. If whole-building models are required in such applications, alternative models will need to be developed.

A. ONE-PARAMETER OR CONSTANT MODEL

One-parameter, or constant models are models where the energy use is constant over a given period. Such models are appropriate for modeling buildings that consume electricity in a way that is independent of the outside weather conditions. For example, such models are appropriate for modeling electricity use in buildings that are on district heating and cooling systems since the electricity use can be well represented by a constant weekday-weekend model. Constant models are often used to model sub-metered data on lighting use that is controlled by a predictable schedule.

B. DAY-ADJUSTED MODEL

Day-adjusted models are similar to one-parameter constant models, with the exception that the final coefficient of the model is expressed as an energy use per day, which is then multiplied by the number of days in the billing period to adjust for variations in the utility billing cycle. Such day-adjusted models are often used with one, two, three, four, and five parameter linear or change-point linear monthly utility models, where the energy use per period is divided by the days in the bill-

ing period before the linear regression or change-point linear regression is performed.

C. TWO-PARAMETER MODEL

Two-parameter models are appropriate for modeling building heating or cooling energy use in extreme climates where a building is exposed to heating or cooling year-around, and the building has an HVAC system with constant controls that operates continuously. Examples include outside air pre-heating systems in arctic conditions or outside air pre-cooling systems in near-tropical climates. Dual-duct, single-fan, and constant-volume systems without economizers can also be modeled with two-parameter regression models. Constant use, domestic water heating loads can also be modeled with two-parameter models, which are based on the water supply temperature.

D. THREE-PARAMETER MODEL

Three parameter models, which include change-point linear models or variable-based degree day models, can be used on a wide range of building types, including residential heating and cooling loads, small commercial buildings, and models that describe the gas used by boiler thermal plants that serve one or more buildings. In Table 27-17, three parameter models have several formats, depending upon whether or not the model is a variable based degree-day model or three-parameter, change-point linear model for heating or cooling. The variable-based degree day model is defined as:

$$E = C + B_1 (DD_{BT})$$

Table 27-17. Sample Models for the Whole-Building Approach from ASHRAE Guideline 14-2002.[156]

Name	Independent Variable(s)	Form	Examples
No Adjustment /Constant Model	None	$E = E_b$	Non weather sensitive demand
Day Adjusted Model	None	$E = E_b \times \dfrac{day_b}{day_c}$	Non weather sensitive use (fuel in summer, electricity in summer)
Two Parameter Model	Temperature	$E = C + B_1(T)$	
Three Parameter Models	Degree days/Temperature	$E = C + B_1(DD_{BT})$ $E = C + B_1(B_2 - T)^+$ $E = C + B_1(T - B_2)^+$	Seasonal weather sensitive use (fuel in winter, electricity in summer for cooling) Seasonal weather sensitive demand
Four Parameter, Change Point Model	Temperature	$E = C + B_1(B_3 - T)^+ - B_2(T - B_3)^+$ $E = C - B_1(B_3 - T)^+ + B_2(T - B_3)^+$	
Five Parameter Models	Degree days/Temperature	$E = C - B_1(DD_{TH}) + B_2(DD_{TC})$ $E = C + B_1(B_3 - T)^+ + B_2(T - B_4)^+$	Heating and cooling supplied by same meter.
Multi-Variate Models	Degree days/Temperature, other independent variables	Combination form	Energy use dependent non-temperature based variables (occupancy, production, etc.).

Where

C = the constant energy use below (or above) the change point

B_1 = the coefficient or slope that describes the linear dependency on degree-days

DD_{BT} = the heating or cooling degree-days (or degree hours), which are based on the balance-point temperature.

The three-parameter change-point linear model for heating is described by[158]:

$$E = C + B_1 (B_2 - T)^+$$

Where

C = the constant energy use above the change point

B_1 = the coefficient or slope that describes the linear dependency on temperature

B_2 = the heating change point temperature

T = the ambient temperature for the period corresponding to the energy use

$+$ = positive values only inside the parenthesis

The three-parameter change-point linear model for cooling is described by:

$$E = C + B_1 (T - B_2)^+$$

Where

C = the constant energy use below the change point

B_1 = the coefficient or slope that describes the linear dependency on temperature

B_2 = the cooling change point temperature

T = the ambient temperature for the period corresponding to the energy use

$+$ = positive values only for the parenthetical expression

E. FOUR-PARAMETER MODEL

The four-parameter change-point linear heating model is typically applicable to heating usage in buildings with HVAC systems that have variable-air volume or whose output varies with the ambient temperature. Four-parameter models have also been shown to be useful for modeling the whole-building electricity use of grocery stores that have large refrigeration loads and significant cooling loads during the cooling season. Two types of four-parameter models are listed in Table 27-17, including a heating model and a cooling model. The four parameter change-point linear heating model is given by:

$$E = C + B_1 (B_3 - T)^+ - B_2 (T - B_3)^+$$

Where

C = the energy use at the change point

B_1 = the coefficient or slope that describes the linear dependency on temperature below the change point

B_2 = the coefficient or slope that describes the linear dependency on temperature above the change point

B_3 = the change-point temperature

T = the temperature for the period of interest

$+$ = positive values only for the parenthetical expression

The four parameter change-point linear cooling model is given by:

$$E = C - B_1 (B_3 - T)^+ + B_2 (T - B_3)^+$$

Where

C = the energy use at the change point

B_1 = the coefficient or slope that describes the linear dependency on temperature below the change point

B_2 = the coefficient or slope that describes the linear dependency on temperature above the change point

B_3 = the change-point temperature

T = the temperature for the period of interest

$+$ = positive values only for the parenthetical expression

F. FIVE-PARAMETER MODEL

Five parameter change-point linear models are useful for modeling the whole-building energy use in buildings that contain air conditioning and electric heating. Such models are also useful for modeling the weather-dependent performance of the electricity consumption of variable air volume air-handling units. The basic form for the weather dependency of either case is shown in Figure 27-7f, where there is an increase in electricity use below the change point associated with heating, an increase in the energy use above the change point associated with cooling, and constant energy use between the heating and cooling change points. Five parameter change-point linear models can be described, using variable-based degree day models or a five-parameter model. The equation for describing the energy use with variable-based degree days is:

$$E = C - B_1 (DD_{TH}) + B_2 (DD_{TC})$$

where

C = the constant energy use between the heating and cooling change points

B_1 = the coefficient or slope that describes the linear dependency on heating degree-days

B_2 = the coefficient or slope that describes the linear dependency on cooling degree-days

DD_{TH} = the heating degree-days (or degree hours), which are based on the balance-point temperature

DD_{TC} = the cooling degree-days (or degree hours), which are based on the balance-point temperature

The five parameter change-point linear model that is based on temperature is:

$$E = C + B_1 (B_3 - T)^+ + B_2 (T - B_4)^+$$

where

C = the energy use between the heating and cooling change points

B_1 = the coefficient or slope that describes the linear dependency on temperature below the heating change point

B_2 = the coefficient or slope that describes the linear dependency on temperature above the cooling change point

B_3 = the heating change-point temperature

B_4 = the cooling change-point temperature

T = the temperature for the period of interest

$+$ = positive values only for the parenthetical expression.

G. WHOLE-BUILDING PEAK DEMAND MODELS

Whole-building peak electric demand models differ from whole-building energy use models in several respects. First, the models are not adjusted for the days in the billing period since the model is meant to represent the peak electric demand. Second, the models are usually analyzed against the maximum ambient temperature during the billing period. Models for whole-building peak electric demand can be classified according to weather-dependent and weather-independent models.

G-1. Weather-dependent, Whole-building Peak Demand Models

Weather-dependent, whole-building peak demand models can be used to model the peak electricity use of a facility. Such models can be calculated with linear and change-point linear models regressed against maximum temperatures for the billing period, or calculated with an inverse bin model.[159,160]

G-2. Weather-independent, Whole-building Peak Demand Models

Weather-independent, whole-building peak demand models are used to measure the peak electric use in buildings or sub-metered data that do not show significant weather dependencies. ASHRAE has developed a diversity factor toolkit for calculating weather-independent whole-building peak demand models as part of Research Project 1093-RP. This toolkit calculates the 24-hour diversity factors using a quartile analysis. An example of the application of this approach is given in the following section.

27.4.2.2.2 Example: Whole-building Energy Use Models

Figure 27-8 presents an example of the typical data requirements for a whole-building analysis, including one year of daily average ambient temperatures and twelve months of utility billing data. In this example of a residence, the daily average ambient temperatures were obtained from the National Weather Service (i.e., the average of the published min/max data), and the utility bill readings represent the actual readings from the customer's utility bill. To analyze these data several calculations need to be performed. First, the monthly electricity use (kWh/month) needs to be divided by the days in the billing period to obtain the average daily electricity use for that month (kWh/day). Second, the average daily temperatures need to be calculated from the published NWS min/max data. From these average daily temperatures the average billing period temperature needs to be calculated for each monthly utility bill.

The data set containing average billing period temperatures and average daily electricity use is then analyzed with ASHRAE's Inverse Model Toolkit (IMT)[161] to determine a weather normalized consumption, as shown in Figure 27-9 and Figure 27-10. In Figure 27-9 the twelve monthly utility bills (kWh/period) are shown plotted against the average billing period temperature, along with a three-parameter change-point model calculated with the IMT. In Figure 27-10 the twelve monthly utility bills, which were adjusted for days in the billing period (i.e., kWh/day), are shown plotted against the average billing period temperature, along with a three-parameter change-point model calculated with the IMT. In the analysis for this house, the use of an average daily model improved the accuracy of the unadjusted model (i.e., Figure 27-9) from an R^2 of 0.78 and CV(RMSE) of 24.0% to an R^2 of 0.83 and a CV(RMSE) of 19.5% for

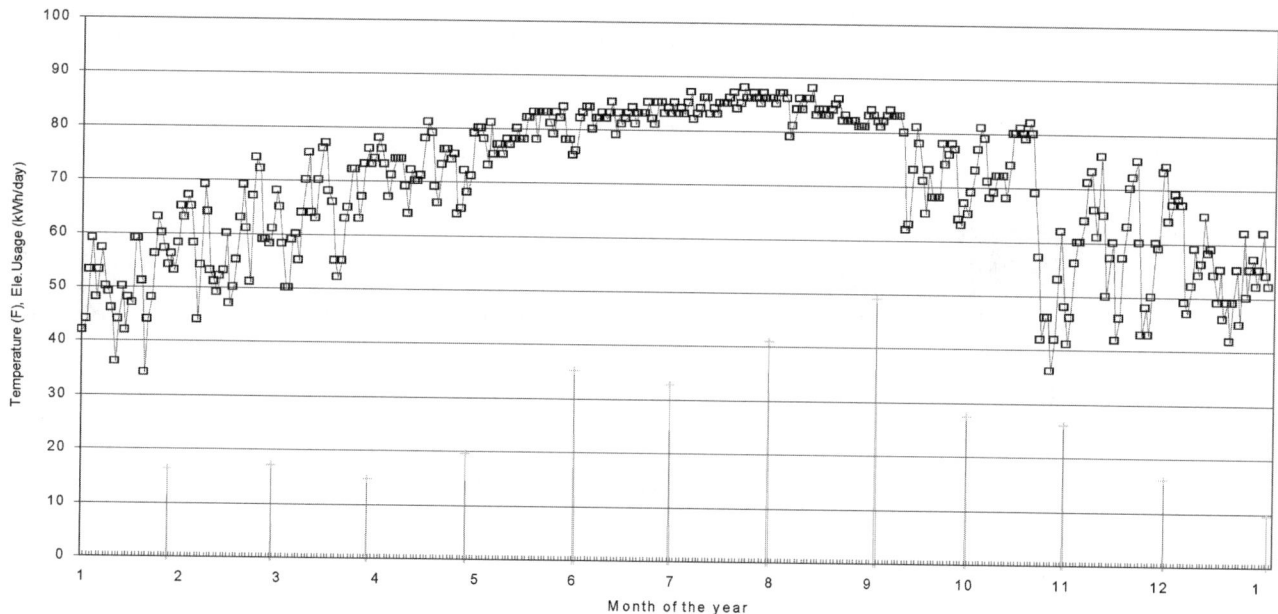

Figure 27-8. Example Data for Monthly Whole-building Analysis (upper trace, daily average temperature, F, lower points, monthly electricity use, kWh/day).

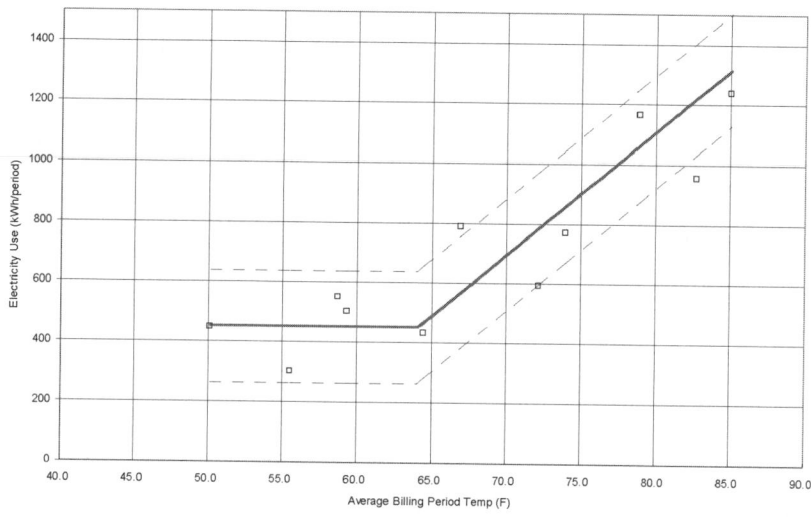

Figure 27-9. Example Unadjusted Monthly Whole-building Analysis (3P Model) for kWh/period ($R^2 = 0.78$, CV(RMSE) = 24.0%).

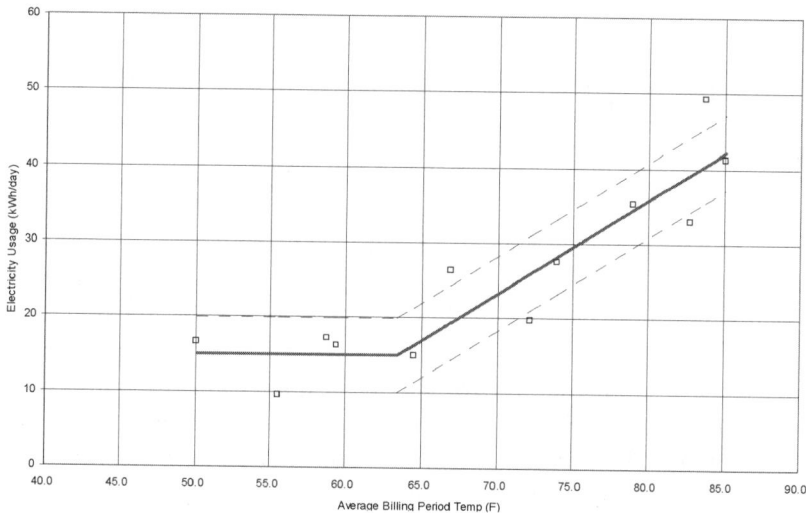

Figure 27-10. Example Adjusted Whole-building Analysis (3P Model) for kWh/day ($R^2 = 0.83$, CV(RMSE) = 19.5%).

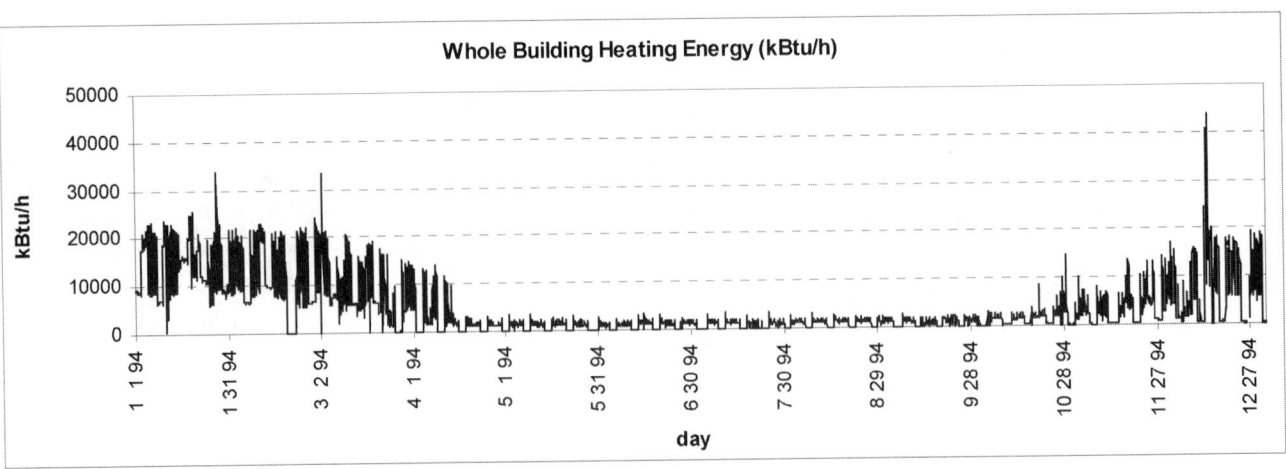

Figure 27-11. Example Heating Data for Daily Whole-building Analysis.

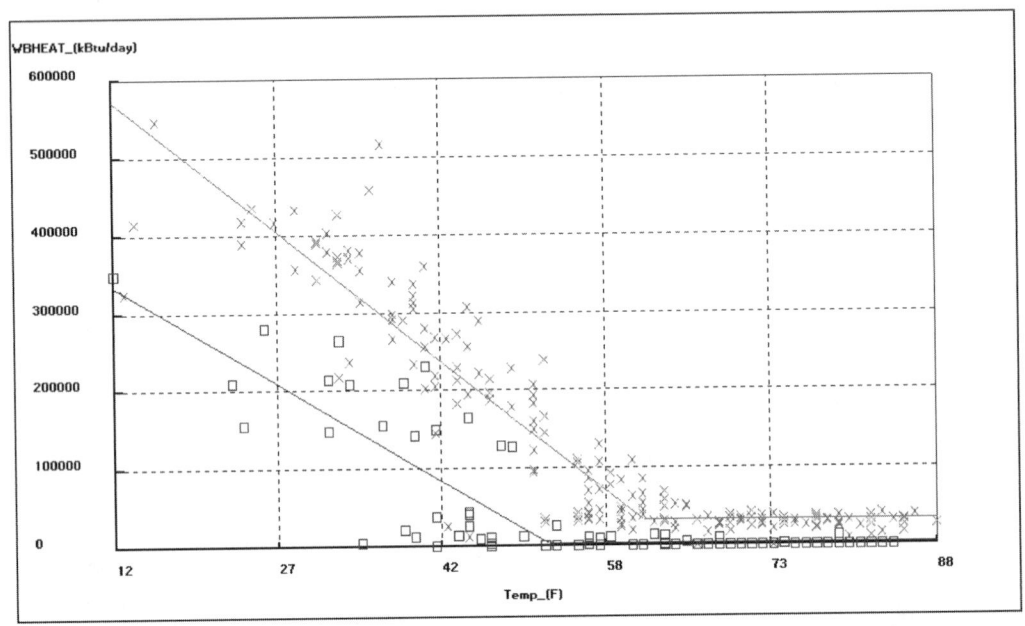

Figure 27-12. Example Daily Weekday-weekend Whole-building Analysis (3P Model) for Steam Use (kBtu/day, R^2 = 0.87, RMSE = 50,085.95, CV(RMSE) = 37.1%). Weekday use (×), weekend use (□).

the adjusted model (i.e., Figure 27-10), which indicates a significant improvement in the model.

In another example, the hourly steam use (Figure 27-11) and hourly electricity use (Figure 27-13) for the U.S.D.O.E. Forrestal Building is modeled with a daily weekday-weekend three parameter, change-point model for the steam use (Figure 27-12), and an hourly weekday-weekend demand model for the electricity use (Figure 27-14). To develop the weather-normalized model for the steam use, the hourly steam data and hourly weather data were first converted into average daily data, then a three parameter, weekday-weekend model was calculated using the EModel software[162], which contains similar al-

gorithms as ASHRAE's IMT. The resultant model, which is shown in Figure 27-12, along with the daily steam, is well described with *an R^2 of 0.87 an RMSE of 50,085.95 kBtu/day and a CV (RMSE) of 37.1%.*

In Figure 27-14 hourly weather-independent 24-hour weekday-weekend profiles have been created for the whole-building electricity use using ASHRAE's 1093-RP Diversity Factor Toolkit.[163] These profiles can be used to calculate the baseline whole-building electricity use (i.e., using the mean hourly use) by multiplying times the expected weekdays and weekends in the year. The profiles can also be used to calculate the peak electricity use (i.e., using the 90th percentile).

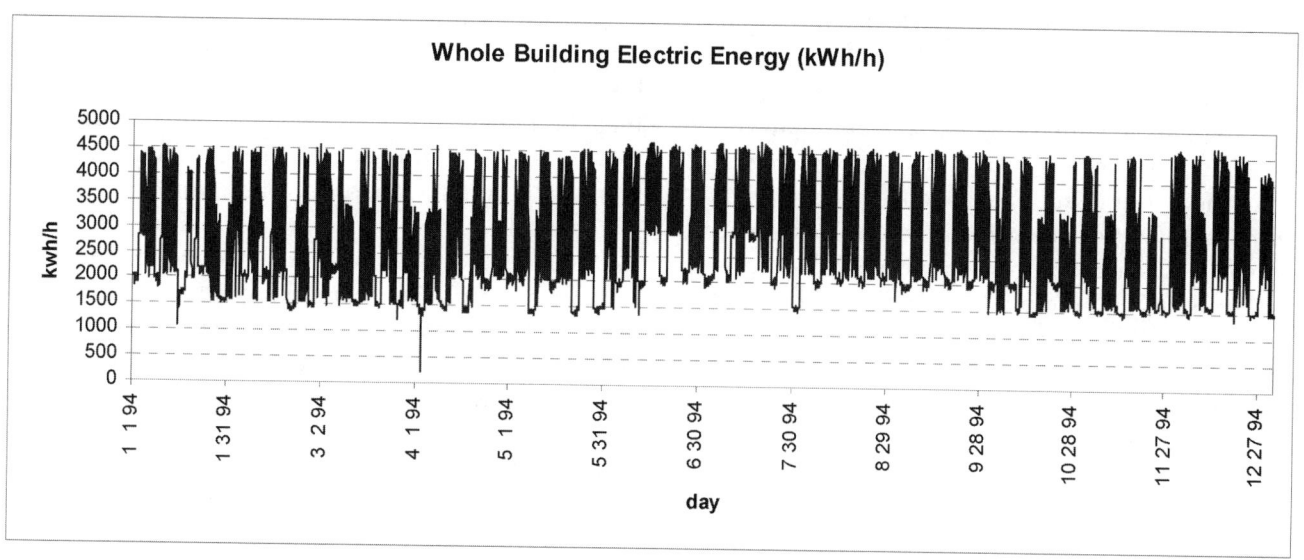

Figure 27-13. Example Electricity Data for Hourly Whole-building Demand Analysis.

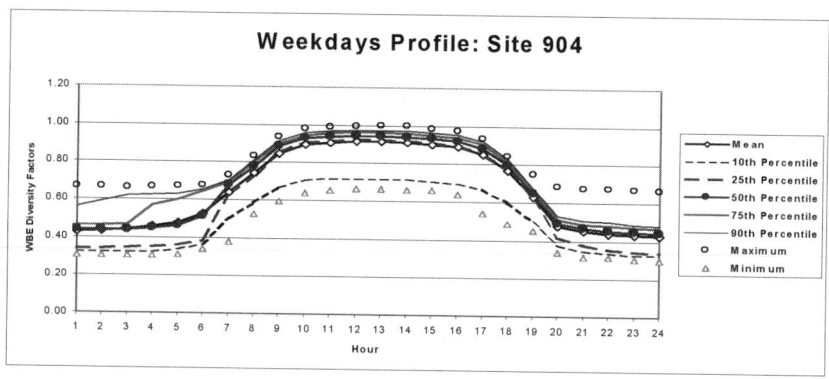

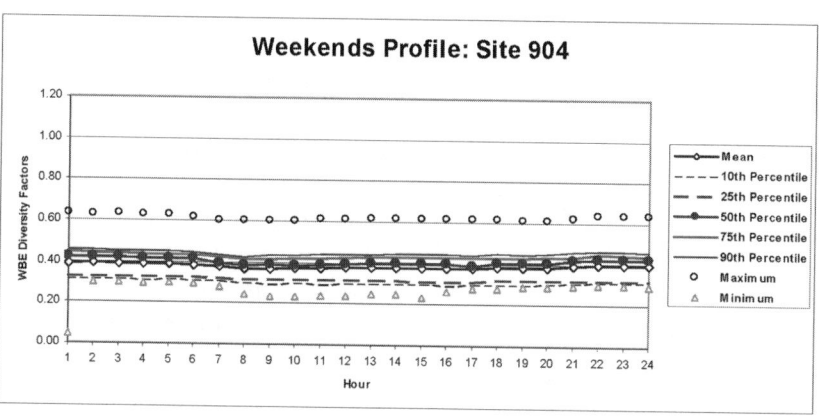

Figure 27-14. Example Weekday-weekend Hourly Whole-building Demand Analysis (1093-RP Model) for Electricity Use.

27.4.2.2.3 Calculation of Annual Energy Use

Once the appropriate whole-building model has been chosen and applied to the baseline data, the annual energy use for the baseline period and the post-retrofit period are then calculated. Savings are then calculated by comparing the annual energy use of the baseline with the annual energy use of the post-retrofit period.

27.4.2.3 Whole-building Calibrated Simulation Approach

Whole-building calibrated simulation normally requires the hourly simulation of an entire building, including the thermal envelope, interior and occupant loads, secondary HVAC systems (i.e., air handling units), and the primary HVAC systems (i.e., chillers, boilers). This is usually accomplished with a general purpose simulation program such as BLAST, DOE-2, EnergyPlus or E-Quest, or similar proprietary programs. Such programs require an hourly weather input file for the location in which the building is being simulated. Calibrating the simulation refers to the process whereby selected outputs from the simulation are compared and eventually matched with measurements taken from an actual building. A number of papers in the literature have addressed techniques for accomplishing these calibrations and include results from case study buildings where calibrated simulations have been developed for various purposes.[164,165,166,167,168,169,170,171,172,173,174, 175 176,177,178,179,180,181,182]

27.4.2.3.1 Applications of Calibrated Whole-building Simulation

Calibrated whole-building simulation can be a useful approach for measuring the savings from energy conservation retrofits to buildings. However, it is generally more expensive than other methods, and therefore it is best reserved for applications where other, less costly approaches cannot be used. For example, calibrated simulation is useful in projects where either pre-retrofit or post-retrofit whole-building metered electrical data are not available (i.e., new buildings or buildings without meters such as many college campuses with central facilities). Calibrated simulation is desired in projects where there are significant interactions between retrofits, for example lighting retrofits combined with changes to HVAC systems, or chiller retrofits. In such cases the whole-building simulation program can account for the interactions, and in certain cases actually isolate interactions to allow for end-use energy allocations. It is useful in projects where there are significant changes in the facility's energy use during or after a retrofit has been installed, where it may be necessary to account for additions to a building that add or subtract thermal loads from the HVAC system. In other cases, demand may change over time, where the changes are not related to the energy conservation measures. Therefore, adjustments to account for these changes will be also be needed. Finally, in many newer buildings, as-built design simulations are being delivered as a part of the building's final documents. In cases where such simulations are properly documented, they can be calibrated to the baseline conditions and then used to calculate and measure retrofit savings.

Unfortunately, calibrated, whole-building simulation is not useful in all buildings. For example, if a building cannot be readily simulated with available simulation programs, significant costs may be incurred

in modifying a program or developing a new program to simulate only one building (e.g., atriums, underground buildings, and buildings with complex HVAC systems that are not included in a simulation program's system library). Additional information about calibrated, whole-building simulation can be found in ASHRAE's Guideline 14-2002.

Figure 27-15 provides an example of the use of calibrated simulation to measure retrofit savings in a project where pre-retrofit measurements were not available. In this figure both the before-after whole-building approach and the calibrated simulation approach are illustrated. On the left side of the figure the traditional whole-building, before-after approach is shown for a building that had a dual-duct, constant volume system (DDCV) replaced with a variable air volume (VAV) system. In such a case where baseline data are available, the energy use for the building is regressed against the coincident weather conditions to obtain the representative baseline regression coefficients. After the retrofit is installed, the energy savings are calculated by comparing the projected pre-retrofit energy use against the measured post-retrofit energy use, where the projected pre-retrofit energy use is calculated with the regression model (or empirical model), which was determined with the facility's baseline DDCV data.

In cases where the baseline data are not available (i.e., the right side of the figure), a simulation of the building can be developed and calibrated to the post-retrofit conditions (i.e., the VAV system). Then, using the calibrated simulation program, the pre-retrofit energy use (i.e., DDCV system) can be calculated for conditions in the post-retrofit period and the savings calculated by comparing the simulated pre-retrofit energy use against the measured post-retrofit energy use. In such a case the calibrated post-retrofit simulation can also be used to fill-in any missing post-retrofit energy use, which is a common occurrence in projects that measure hourly energy and environmental conditions. The accuracy of the post-retrofit model depends on numerous factors.

27.4.2.3.2 Methodology for Calibrated Whole-building Simulation

Calibrated simulation requires a systematic approach that includes the development of the whole-building simulation model, collection of data from the building being retrofitted, and the coincident weather data. The calibration process then involves the comparison of selected simulation outputs against measured data from the systems being simulated and the adjustment of the simulation model to improve the comparison of the simulated output against the corresponding

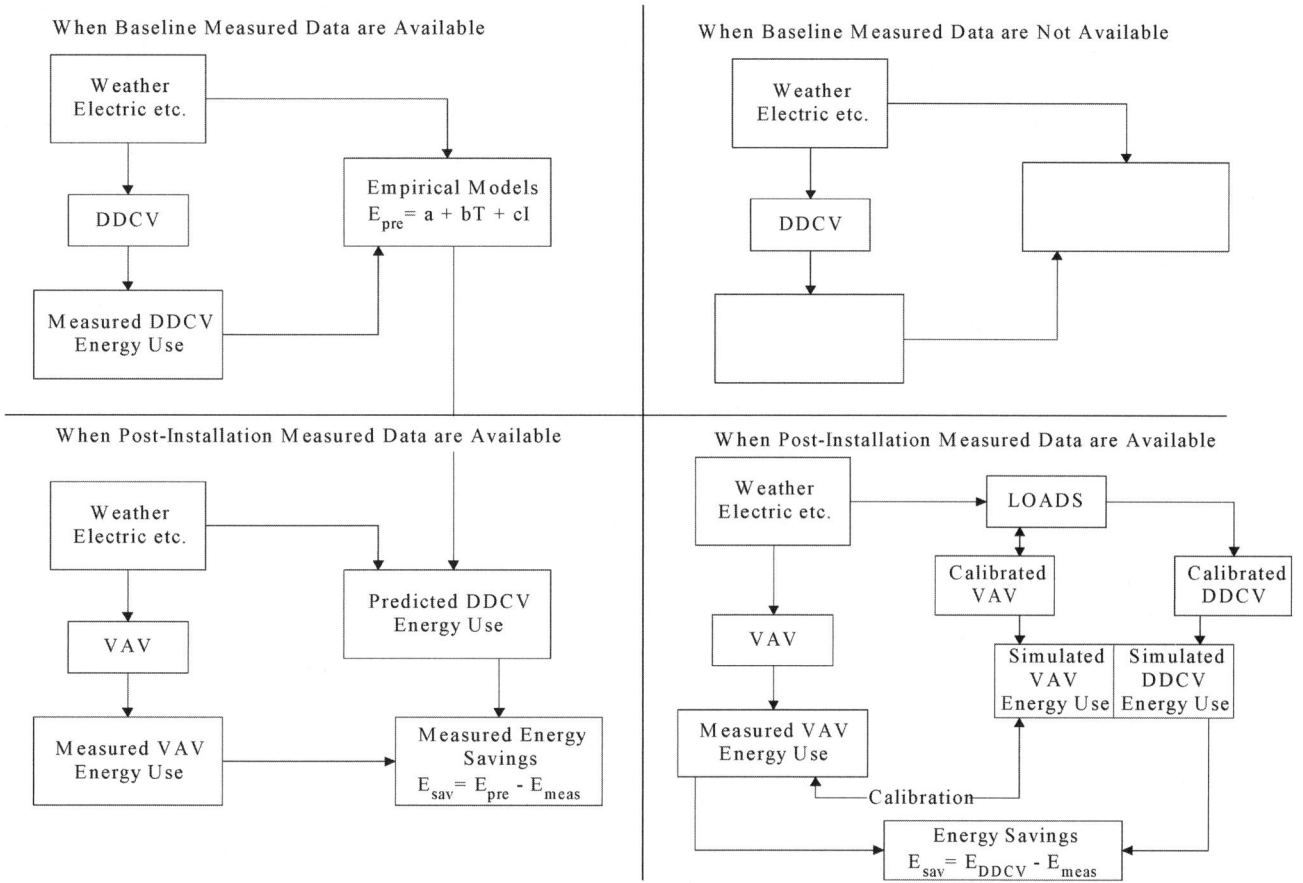

Figure 27-15. Flow Diagram for Calibrated Simulation Analysis of Air-Side HVAC System[183].

measurements. The choice of simulation program is a critical step in the process, which must balance the model appropriateness, algorithmic complexity, user expertise, and degree of accuracy against the resources available to perform the modeling.

Data collection from the building include the collection of data from the baseline and post-retrofit periods, which can cover several years of time. Building data to be gathered include such information as the building location, building geometry, materials characteristics, equipment nameplate data, operations schedules, temperature settings and, at a minimum, whole-building utility billing data. If the budget allows, hourly whole-building energy use and environmental data can be gathered to improve the calibration process, which can be done over short-term or long-term period.

Figure 27-16 provides an illustration of a calibration process that used hourly graphical and statistical comparisons of the simulated versus measured energy use and environmental conditions. In this example, the site-specific information was gathered and used to develop a simulation input file, including the use of measured weather data, which was then used by the DOE-2

program to simulate the case study building. Hourly data from the simulation program was then extracted and used in a series of special-purpose graphical plots to help guide the calibration process (i.e., time series, bin and 3-D plots). After changes were made to the input file, DOE-2 was then run again, and the output compared against the measured data for a specific period. This process was then repeated until the desired level of calibration was reached, at which point the simulation was proclaimed to be "calibrated." The calibrated model was then used to evaluate how the new building was performing compared to the design intent.

A number of different calibration tools have been reported by various investigators, ranging from simple X-Y scatter plots to more elaborate statistical plots and indices. Figure 27-17, Figure 27-18, and Figure 27-19 provide examples of several of these calibration tools. In Figure 27-17, an example of an architectural rendering tool is shown that assists the simulator with viewing the exact placement of surfaces in the building, as well as shading from nearby buildings and north-south orientation. In Figure 27-18, temperature binned calibration plots are shown comparing the weather dependency of an hourly

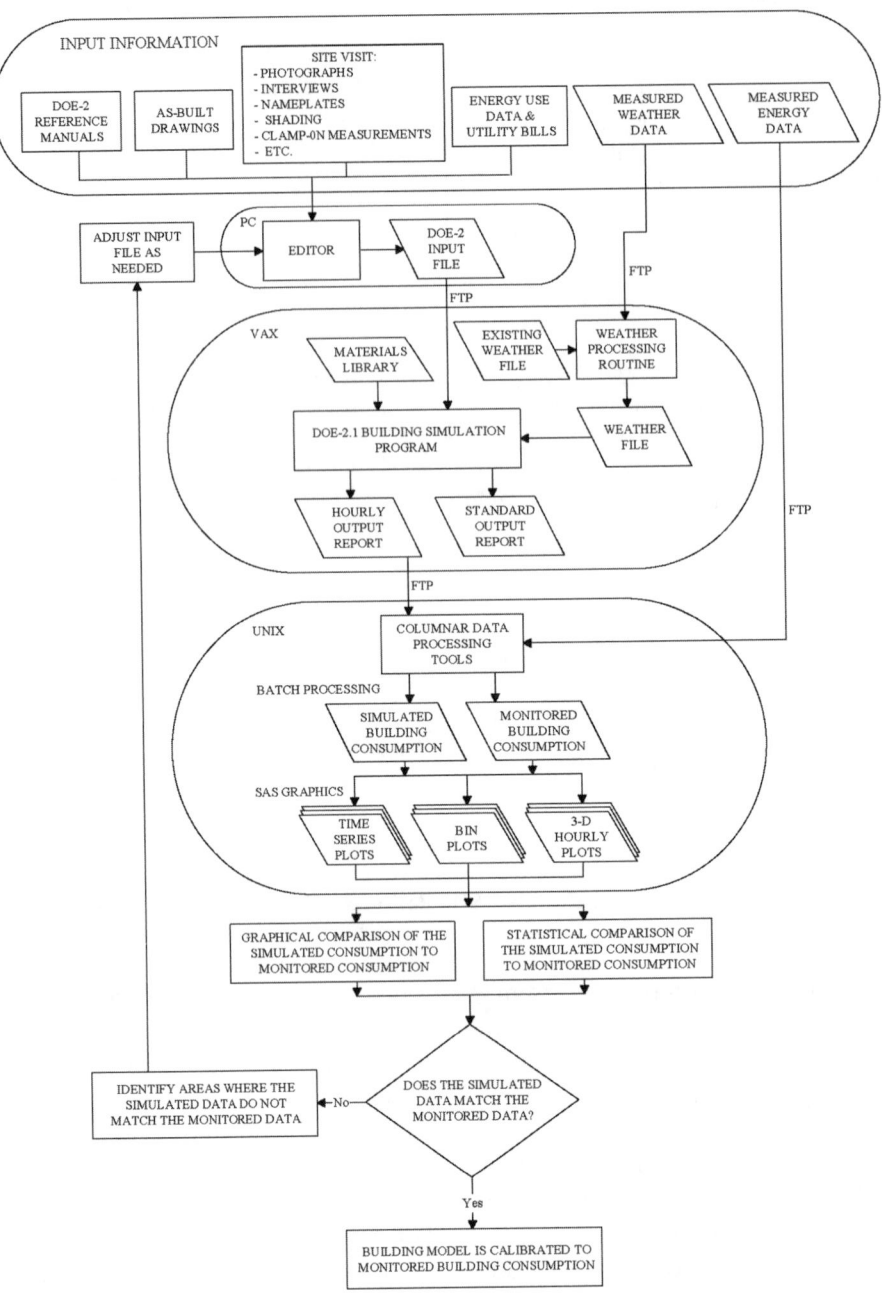

Figure 27-16. Calibration Flowchart. This figure shows the sequence of processing routines that were used to develop graphical calibration procedures[185].

simulation against measured data. In this figure the upper plots show the data as scatter plots against temperature. The lower plots are statistical, temperature-binned box-whisker-mean plots, which include the superpositioning of measured mean line onto the simulated mean line to facilitate a detailed evaluation. In Figure 27-19, comparative three-dimensional plots are shown that show measured data (top plot), simulated data (second plot from the top), simulated minus measured data (second plot from the bottom), and measured minus simulated

data (bottom plot). In these plots the day-of-the-year is the scale across the page (y axis), the hour-of-the-day is the scale projecting into the page (x axis), and the hourly electricity use is the vertical scale of the surface above the x-y plane. These plots are useful for determining how well the hourly schedules of the simulation match the schedules of the real building, and they can be used to identify other certain schedule-related features. For example, in the front of plot (b) the saw-toothed feature is indicating on/off cycling of the HVAC system, which

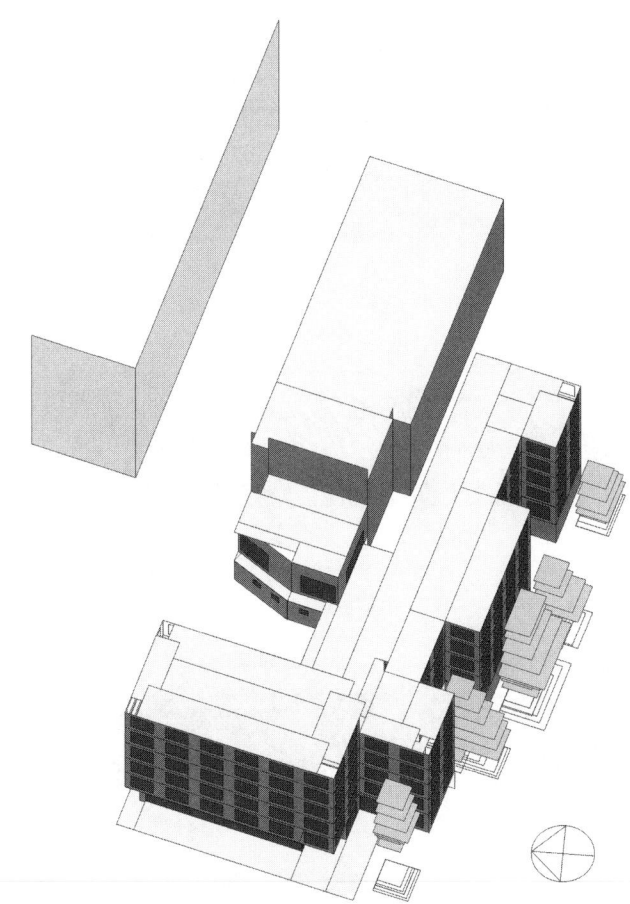

Figure 27-17. Example Architectural Rendering of the Robert E. Johnson Building, Austin, Texas[186,187].

is not occurring in the actual building.

Table 27-18 contains a summary of the procedures used for developing a calibrated, whole-building simulation program, as defined in ASHRAE's Guideline 14-2002. In general, to develop a calibrated simulation, detailed information is required for a building, including information about the building's thermal envelope (i.e., the walls, windows, roof, etc.), and information about the building's operation, including temperature settings, HVAC systems, and heating-cooling equipment that existed both during the baseline and post-retrofit period. This information is input into two simulation files, one for the baseline and one for the post-retrofit conditions. Savings are then calculated by comparing the two simulations of the same building, one that represents the baseline building, and one that represents the building's operations during the post-retrofit period.

27.4.3 Cost/Benefit Analysis

The target for work for the USAF has been 5% of the savings.[190] The cost of the M&V can exceed 5% if the risk of losing savings exceeds predefined limits. The variable speed drive ECM illustrates these opportunities and risks. VSD equipment exhibits high reliability. Equipment type of failures normally happen when connection breaks occur with the control input, the remote sensor. Operator-induced failures occur when the operator sets the unit to 100% speed and does not re-enable the control. Setting the unit to 100% can occur

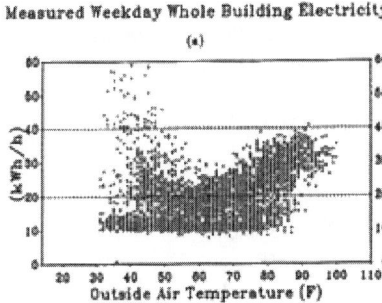

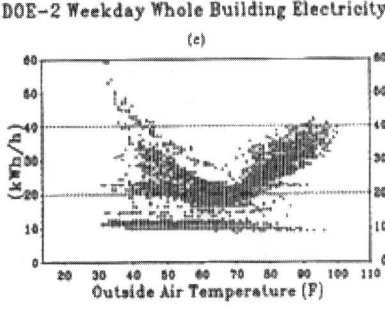

Figure 27-18. Temperature Bin Calibration Plots. This figure shows the measured and simulated hourly weekday data as scatter plots against temperature in the upper plots and as statistical binned box- whisker-mean plots in the lower plots.[188]

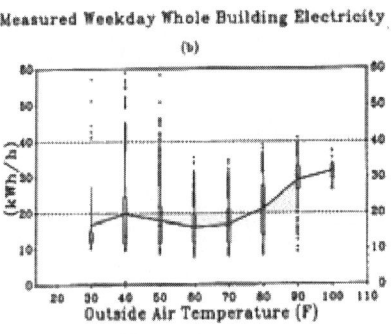

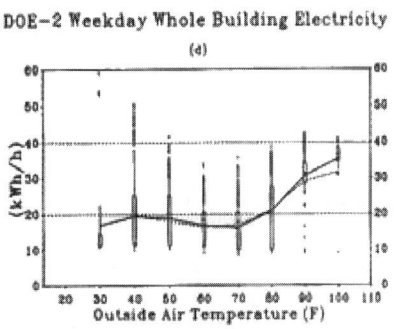

for legitimate reasons. These reasons include running a test, overriding a control program that does not provide adequate speed under specific and typically unusual, circumstances, or requiring 100% operation for a limited time. The savings disappear if the VSD remains at 100% operating speed.

For example, consider a VSD ECM with ten motors, with each motor on an different air handling unit. Each motor has fifty horsepower (HP). The base case measured these motors running 8760 hours per year at full speed. Assume that the loads on the motors matched the nameplate 50 HP at peak loads. Although the actual load on an AHU fan varies with the state of the terminal boxes, assume that the load average equates to 80% of the full load since the duct pressure will rise as the terminal boxes reduce flow at the higher speed. Table 27-19 contains the remaining assumptions. To correctly determine the average power load, either the average power must be integrated over the period of consumption or the bin method must be used. For the purposes of this example, the 14.4% value will be used.

The equation below shows the relationship between the fan speed and the power consumed. The exponent has been observed to vary between 2.8 (at high flow) and 2.7 (at reduced flow) for most duct systems. This includes the loss term from pressure increases at a given fan speed. Changing the exponent from 2.8 to 2.7 reduces the savings by less than 5%.

Demand savings will not be considered in this example. Demand savings will likely be very low if the demand has a 13 month ratchet and the summer load requires some full speed operation during peak times. Assuming a $12.00/kW per month demand charge, demand savings could be high for off-season months if the demand billing resets monthly. Without a ratchet clause, rough estimates have yearly demand savings ranging up to $17,000 if the fan speed stays under 70% for 6 months per year. Yearly demand savings jump to over $20,000 if the fan speed stays under 60% for 6 months per year.

Figure 27-20 illustrates the savings expected from the VSD ECM by hours of use per year. The 5% and 10% of Savings lines define the amount available for M&V expenditures at these levels. In this example, the ECM savings exceeds $253,000 per year. Five percent of

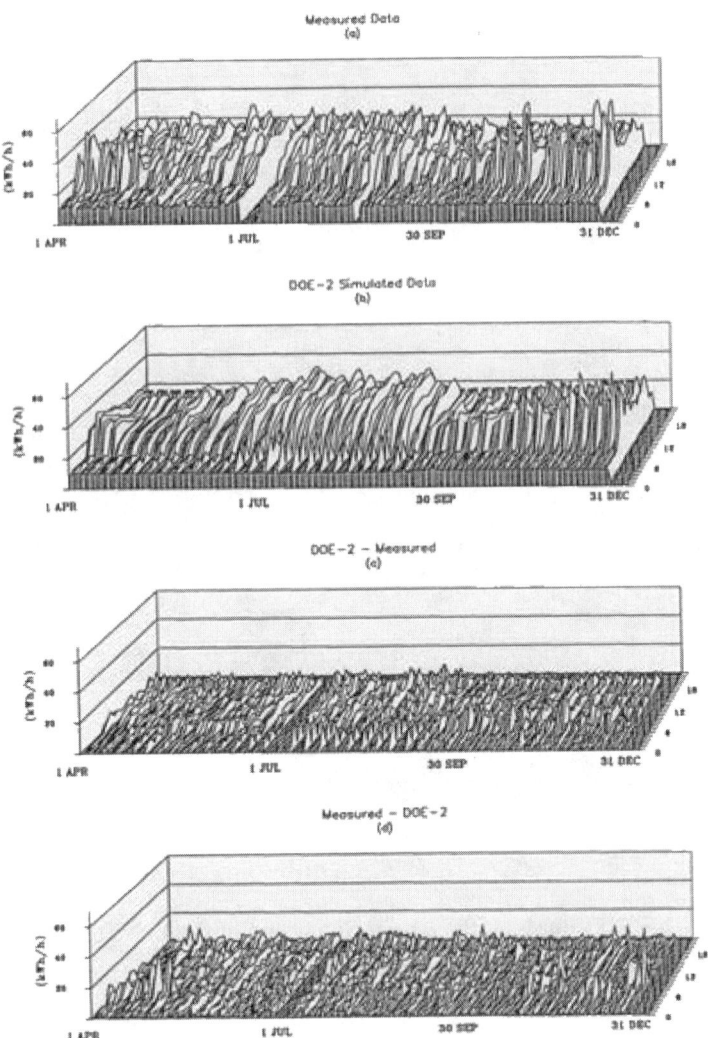

Figure 27-19. Comparative Three-dimensional Plots. (a) Measured Data. (b) Simulated Data. (c) Simulated-Measured Data. (d) Measured-Simulated Data[189].

savings over a 20 year project life makes $253K available for M&V, and ten percent of the savings makes $506K available over the 20-year period. If the motors run less frequently than continuous, savings decrease as shown in Figure 27-20. Setting up the M&V program to monitor the VSDs on an hourly basis and report savings on a monthly report requires monitoring the VSD inverter with an EMCS to poll the data and create reports.

To provide the impact of the potential losses from losing the savings, assume the savings degrades at a loss of 10% of the total yearly savings per year. Studies have shown that control ECMs like the VSD example can expect to see 20% to 30% degradation in savings in 2 to 3 years. Figure 27-21 illustrates what happens to the savings in 20 years with 10% of the savings spent on

Table 27-18. Calibrated Whole-building Simulation Procedures from ASHRAE Guideline 14-2002[184].

STEP	PROCEDURES	DATA OR INFORMATION REQUIRED
Step 1: Calibrated Simulation Plan	• Develop Baseline Scenario • Develop Post-retrofit Scenario • Select simulation package • Select calibration tools	• Information about existing conditions, building location, surroundings, etc. • Information about post-retrofit conditions.
Step 2: Data Collection From Existing Building	• Visit site and collect data about baseline and post-retrofit conditions, characteristics, plans, • Collect energy use data, and information about building operations, and systems • Determine internal heat loads • Prepare data for input into simulation	• Utility billing data for baseline and for post-retrofit periods • Interval energy use and environmental data from building • Equipment schedules, thermostat settings, HVAC system types, performance measurements
Step 3: Data Input Into Simulation Program	• Develop whole-building simulation or building, including envelope, loads, systems, plant • Use architectural rendering to confirm inputs	• All building characteristics, system types, schedules, setpoints, etc., needed for developing simulation.
Step 4: Baseline Simulation Run and Comparison	• Run simulation. • Compare with measured data • Repeat process until simulation matches actual building	• All building characteristics, system types, schedules, setpoints, etc., needed for developing simulation. • On-sites measurements of energy and environmental data
Step 5: Post-retrofit Simulation Run and Comparison.	• Run simulation. • Compare with measured data • Repeat process until simulation matches actual building	• All building characteristics, system types, schedules, setpoints, etc., needed for developing simulation. • On-sites measurements of energy and environmental data
Step 6: Calculate Savings	• Compare baseline and post-retrofit simulations	
Step 7: Report findings	• Provide baseline & post-retrofit building descriptions. • Document building measurements • Include simulation plan • Present results • Provide input/output files from simulation	

Table 27-19. VSD Example Assumptions.

	Pre-ECM	Post-ECM	Comments
Hours / Yr	8760	8760	
$ / kWh	$0.06	$0.06	
$ / kW / month	$12.00	$12.00	
Average Speed	100%	50%	
PowerCurveExponent	2.8	2.8	Contains duct loss impact
Average Power Load	80%	14.4%	See Equation XXX
Total Motor HP	500	500	
Total kW	373	384	At full speed, ~3% loss in inverter

M&V. Note that the losses exceed the M&V cost during the first year, resulting in a net loss of almost $3,000,000 over the 20 year period. Figure 27-22 shows the savings per year with a 10% loss of savings. M&V costs remain at 10% of savings. At the end of the 20 year period, the savings drop to almost $30,000 per year out of a potential savings of over $250,000 per year.

This example shows the cumulative impact of losing savings on a year by year basis. The actual savings amounts will vary, depending upon the specific factors in an ECM, and can be scaled to reflect a specific application. Increasing the M&V cost to reduce the loss of savings often makes sense and must be carefully thought through.

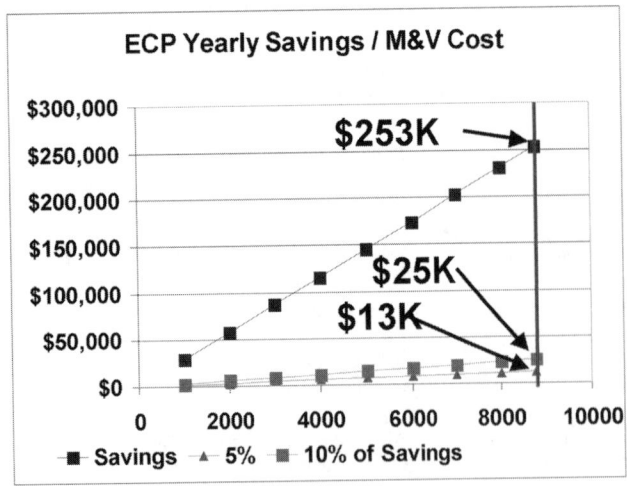

Figure 27-20. ECP Yearly Savings/M&V Cost.

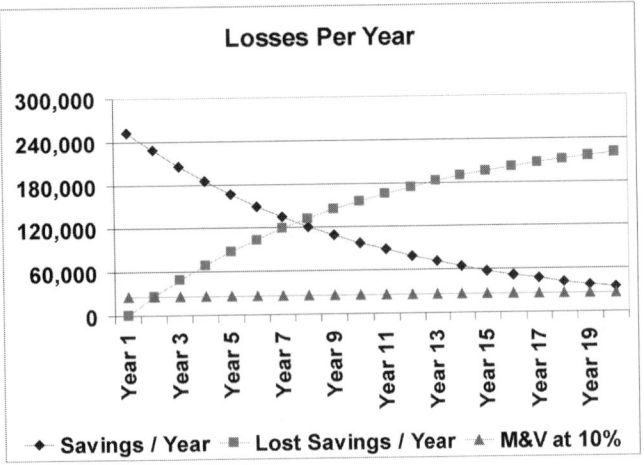

Figure 27-21. Yearly Impact of Ongoing Losses.

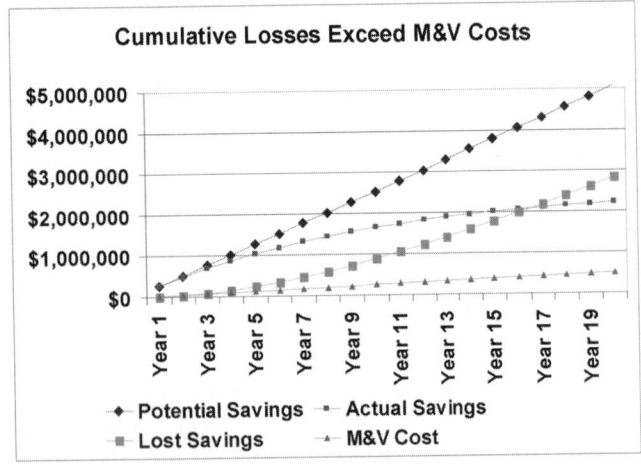

Figure 27-22. Cumulative Impact of Savings Loss.

27.5 COST REDUCTION STRATEGIES

M&V strategies can be cost reduced by lowering the requirements for M&V or by statistical sampling. Reducing requirements involves performing tradeoffs with the risks and benefits of having reliable numbers to determine the savings and the costs for these measurements.

27.5.1 Constant Load ECMs

Lighting ECMs can save 30% of the pre-ECM energy and have a payback in the range of 3 to 6 years. Assuming that the lighting ECM was designed and implemented per the specifications and the savings were verified to be occurring, just verifying that the storeroom has the correct ballasts and lamps may constitute acceptable M&V on a yearly basis. This costs far less than performing a yearly set of measurements, analyzing them, and then creating reports. In this case, other safeguards should be implemented to assure that the bulb and ballast replacement occurs and meets the requirements specified.

High efficiency motor replacements provide another example of constant load ECMs. The key short-term risks with motor replacements involve installing the right motor with all mechanical linkages and electrical components installed correctly. Once verified, the long-term risks for maintaining savings occur when the motor fails. The replacement motor must be the correct motor or savings can be lost. A sampled inspection reduces this risk. Make sure to inspect all motors at least once every five (5) years.

27.5.2 Major Mechanical Systems

Boilers, chillers, air handler units, and cooling towers comprise the category of major mechanical equipment in buildings. The need to be considered separately as each carries its own set of short-term and long-term risks. In general, measurements provide necessary risk reduction. The question becomes, "What measurements reduce the risk of savings loss by an acceptable amount?"

First a risk assessment needs to be performed. The short-term risks for boilers involve installing the wrong size or installing the boiler improperly (not to specifications). Long-term savings sustainability risks tend to focus on the water side and the fire side. Water deposits (K^+, Ca^{++}, Mg^+) will form on the inside of the tubes and add a thermal barrier to the heat flow. The fire side can add a layer of soot if the O_2 level drops too low. Either of these reduces the efficiency of the boiler over the long haul. Generally, this can take several years to impact the efficiency if regular tune-ups and water treatment occur.

Boilers come in a wide variety of shapes and sizes. Boiler size can be used as a defining criterion for measurements. Assume that natural gas or other boiler fuels cost about $5.00 per MMBtu. (Although fuel price constantly changes, it provides a reference point for this analysis.) Thus, a boiler with 1MMBtu per hour output, an efficiency of 80%, and operating at 50% load 3500 hours per year, consumes about $11,000 per year. If this boiler replaced a less efficient boiler, say at 65%, then the net savings amounts to about $2,500 per year, assuming the same load from the building. At 5% of the savings, $125 per year can be used for M&V. This does not allow much M&V. At 10% of the annual savings, $250 per year can be used. At this level of cost, a combustion efficiency measurement could be performed, either yearly or bi-yearly, depending on the local costs. In 2003, the ASME's Power Test Code 4.1 (PTC-4.1) [191] was replaced with PTC4. Either of these codes allow two methods to measure boiler efficiency. The first method uses the energy in equals energy out—using the first law of thermodynamics. This requires measuring the Btu input via the gas flow and the Btu output via the steam (or water) flow and temperatures. The second method measures the energy loss due to the content and temperature of the exhausted gases, radiated energy from the shell and piping, and other loss terms (like blowdown). The energy loss method can be performed in less than a couple of hours. The technician performing these measurements must be skilled, or significant errors will result in the calculated efficiency. The equation below shows the calculations required.

Efficiency = 100% – Losses + Credits

The losses term includes the temperature of the exhaust gas and a measure of the unburned hydrocarbons by measuring CO_2 or O_2 levels, the loss due to excess CO and a radiated term. Credits seldom occur but could arise from solar heating the makeup water or similar contributions. The Greek letter "η" usually denotes efficiency.

As with boilers, a risk assessment needs to be performed. The short-term risks for chillers involve sizing or improper installation. Long-term savings sustainability risks focus on the condenser water system, as circulation occurs in an open system. Water deposits (K^+, Ca^{++}, Mg^+, organics) will form on the inside of the condenser tubes and add a barrier to the thermal flow. These reduce the efficiency of the chiller over the long haul. Generally, this can take several years to impact the efficiency if proper water treatment occurs. Depending on the environmental conditions, the quality of the makeup water, and the water treatment, condenser tube fouling should be checked every year, or at least every other year.

Chillers consume electricity in the case of most centrifugal, screw, scroll, and reciprocating compressors. Absorbers and engine-driven compressors use a petroleum based fuel. As with boilers, chiller size and application sets the basic energy consumption levels. Assume, for the purpose of this example, that electricity provides the chiller energy. Older chillers with water towers often operate at the 0.8 to 1.3 kW per ton level of efficiency. New chillers with water towers can operate in the 0.55 to 0.7 range of efficiency. Note that the efficiency of any chiller depends upon the specific operating conditions. Also assume: 500 tons centrifugal chiller with the specifications shown in Table 27-20. Under these conditions the chiller produces 400 tons of chilled water and requires an expenditure of $ 38,000 per year, considering both energy use and demand charges. Some utilities only charge demand charges on the transmission and

Table 27-20. Example of Savings with a 500 Ton Chiller.

	No Rachet-4 Mo/Yr		13 Month Rachet		
	Pre-ECM	Post-ECM	Pre-ECM	Post-ECM	Comments
Hours / Yr	2000	2000	2000	2000	
Tons Cooling	400	400	400	400	
$ / kWh	$0.06	$0.06	$0.06	$0.06	
$ / kW / month	$12.00	$12.00	$12.00	$12.00	
Average Efficiency	0.90	0.57	0.90	0.57	kW / Ton
Yearly kWh	720000	456000	720000	456000	
Yearly kWh Cost	$43,200	$27,360	$43,200	$27,360	Energy use cost
Yearly kW Cost	$17,280	$10,944	$51,840	$32,832	Demand cost
Yearly Cost	$60,480	$38,304	$95,040	$60,192	Total
Savings		$22,176		$34,848	

delivery (T&D) parts of the rate structure. In that case, the cost at $0.06/kWh would be closer to $28,000. Using the 5% (10%) guideline for M&V costs as a percentage of savings leaves almost $1,100 ($2,200) per year to spend on M&V. This creates an allowable expenditure over a 20 year project of $22,000 ($44,000) for M&V. If the utility has a ratchet clause in the rate structure, the amount for M&V increases to $1,700 ($3,400) per year. At $1,100 per year, tradeoffs will need to be made to stay within that "budget." The risks need to be weighed and decisions made as to what level of M&V costs will be allowed.

To determine the actual efficiency of a chiller requires accurate measurements of the chilled water flow, the difference between the chilled water supply and return temperatures, and the electrical power provided to the chiller. Costs can be reduced using an EMCS if only the temperature, flow, and power sensors need to be installed.

Cooling tower replacement requires knowledge of the risks and costs involved. As with boilers and chillers, the primary risks involve the water treatment. Controls can be used to improve the efficiency of a chiller-tower combination by as much as 15% to 20%. As has been previously stated, control ECMs often get overridden and the savings disappear.

27.5.3   Control Systems

Control ECMs encompass a wide spectrum of capabilities and costs. Upgrading a pneumatic control system and installing EP (electronic to pneumatic) sensors involves the simple end. The complex side could span installing a complete EMCS will sophisticated controls, with various reset, pressurization and control strategies. Generally, EMCSs function as on/off controls and do not get widely used in sophisticated applications.

Savings due to EMCS controls bear high sustainability risks. When an operator overrides a strategy and forgets to re-enable it, the savings disappear. A common EMCS ECM requires the installation of equipment and programs used to set back temperatures or turn off equipment. Short-term risks involve setting up the controls so that performance enhances, or at least does not degrade, the comfort of the occupants. When discomfort occurs, either occupants set up "portable electric reheat units" or operators override the control program. For example, when the night set-back control does not get the space to comfort by occupancy, operators typically override instead of adjusting the parameters in the program. These actions tend to occur during peak loading times and then not get re-enabled during milder times. Long-term risks cover the same area as short-term risks. A new operator or a failure in remote

equipment that does not get fixed will likely cause the loss of savings. Estimating the savings cost for various projects can be done when the specifics are known.

Risk abatement can be as simple as requiring a trend report weekly, or at least monthly. M&V costs can generally be easily held under 5% when using an EMCS and creating trend reports.

27.6   M&V SAMPLING STRATEGIES

M&V can be made significantly lower-cost by sampling. Sampling also reduces the timeliness of obtaining specific data on specific equipment. The benefits of sampling arise when the population of items increases. Table 27-21 (M&V Guidelines: Measurement and Verification for Federal Energy Projects, Version 2.2, Appendix D) illustrates how confidence and precision impact the number of samples required in a given population for items.

For example, lighting ECMs typically involve thousands of fixtures. To obtain a savings estimate with a confidence of 80% and a precision of 20%, 11 fixtures would need to be sampled. If the requirements increased to a confidence of 90% and a precision of 10%, 68 fixtures would need to be sampled.

The boiler ECM also represents opportunity for

Table 27-21. Sampling Requirements.

Precision	20%	20%	10%
Confidence	80%	90%	90%
Population Size, N	Sample Size, n		
4	3	4	4
12	6	8	11
20	8	10	16
30	9	11	21
40	9	12	26
50	10	13	29
60	10	14	32
70	10	15	35
80	10	15	37
90	10	15	39
100	10	15	41
200	11	16	51
300	11	17	56
400	11	17	59
500	11	17	60
Infinite	11	17	68

M&V cost reduction using sampling. Assume that the ECM included replacing 50 boilers. If a confidence of 80% and a precision of 20% satisfy the requirements, 10 boilers would need to be sampled. The cost is then reduced to 20% of the cost of measuring all boilers, a significant savings. A random sampling to select the sample set can easily be implemented.

SUMMARY

This chapter covers various measurement and verification (M&V) methods that can be directly applicable in practice in order to reduce both energy consumption and cost in buildings. M&V methods for measuring energy and demand savings in existing building are best represented by the following three approaches: retrofit isolation approach, whole-building/main meter approach, and whole-building calibrated simulation approach (Table 27-22). A number of examples and diagrams have been presented to help understand the detailed procedures of each M&V method. In addition to the M&V methods, the history of M&V and its protocols in the United States has been introduced, and the issues related to the cost/benefit analysis of M&V and several cost reduction strategies have been discussed together.

Table 27-22. Summary of the M&V Methods.

M&V Methods	Description
Retrofit Isolation Approach	This approach decides savings by comparing the actual energy use data of specific equipment. Generally, statistically representative models of each isolated component are developed for the calculation of savings.
Whole-building or Main Meter Approach	This approach decides savings by comparing the actual energy use data of an entire building. A number of regression models that relates the energy use and peak demand to one or more independent variables such as weather data are developed for the calculation.
Whole-building Calibrated Simulation Approach	This approach decides savings by comparing the calibrated simulation results of an entire building. Calibrating the simulation refers to the process whereby selected outputs from the simulation are compared and eventually matched with measurements taken from an actual building.

FUTURE STUDY

This chapter is designed to assist readers to apply various M&V methods in practice. Readers can choose the appropriate M&V method according to their intentions of conducting M&V. There are still some topics related to M&V that were not covered in this chapter but would be helpful to readers, including:

- Measurement of site and source energy.
- Calculation of air pollution savings from energy efficiency: SO_x, NO_x, and CO_2 (carbon credit).
- Procedures for the normalization of baseline data.
- Additional case studies, including whole-building before and after retrofit, retrofit isolation, and calibrated simulation approach.

References

1. Claridge D.E., Turner, W.D., Liu, M., Deng, S., Wei, G., Culp, C., Chen, H., and Cho, S. 2002. "Is Commissioning Once Enough?," Solutions for Energy Security and Facility Management Challenges: *Proceedings of the 25th WEEC*, Atlanta, GA, October 19-11, 2002, pp.29-36.
2. Haberl, J., Lynn, B., Underwood, D., Reasoner, J., Rury, K. 2003a. "Development an M&V Plan and Baseline for the Ft. Hood ESPC Project," *ASHRAE Seminar Presentation*, (June).
3. C. Culp, K.Q. Hart, B. Turner, S. Berry-Lewis, 2003. "Energy Consumption Baseline: Fairchild AFB's Major Boiler Retrofit," *ASHRAE Seminar Presentation* (January).
4. Arnold, D., 1999. "The Evolution of Modern Office Buildings and Air Conditioning," *ASHRAE Journal*, American Society of Heating Refrigeration Air-conditioning Engineers, Atlanta, GA, p. 40-54, (June).
5. Donaldson, B., Nagengast. 1994. *Heat and Cold: Mastering the Great Indoors*. American Society of Heating Refrigeration Air-conditioning Engineers, Atlanta, GA.
6. Cheney, M., Uth, R. 1999. *Tesla: Master of Lightning*. Barnes and Noble Books, New York, N.Y.
7. Will, H. 1999. *The First Century of Air Conditioning*. American Society of Heating Refrigeration Air-conditioning Engineers, Atlanta, GA.
8. Israel, P. 1998. *Edison: A Life of Invention*. John Wiley and Sons, New York, N.Y.
9. EEI 1981. *Handbook for Electricity Metering*, 8th Edition with Appendix, Edition Electric Institute, Washington D.C.
10. Miller, R. 1989. *Flow Measurement Engineering Handbook*. McGraw Hill, New York, N.Y.
11. American Institute of Physics, 1975. *Efficient Use of Energy: The APS Studies on the Technical Aspects of the More Efficient Use of Energy*, American Physical Society, New York, N.Y., (A report on the 1973 summer study at Princeton University).
12. *National Geographic*, February 1981. Special Report on Energy: Facing up to the Problem, Getting Down to Solutions, National Geographic Society, Washington, D.C.
13. *Scientific American* 1971. Energy and Power. W.H. Freeman and Company, San Francisco, CA. (A reprint of eleven articles that appeared in the September 1971 Scientific American).
14. Kusuda, T. 1999. "Early History and Future Prospects of Building System Simulation," *Proceedings of the Sixth International Building Performance Simulation Association* (IBPSA BS' 99), Kyoto, Japan, (September).
15. APEC 1967. *HCC-heating/cooling load calculations program*. Dayton, Ohio, Automated Procedures for Engineering Consultants.
16. Ayres, M., Stamper, E. 1995. "Historical Development of Building Energy Calculations," *ASHRAE Transactions*, Vol. 101, Pt. 1. American Society of Heating Refrigeration Air-conditioning Engineers, Atlanta, GA.
17. Stephenson, D., and Mitalas, G. 1967. "Cooling Load Calculations by Thermal Response Factor Method," *ASHRAE Transactions*, Vol. 73, pt. 1.
18. Mitalas, G. and Stephenson, D. 1967. "Room Thermal Response Factors," *ASHRAE Transactions*, Vol. 73, pt. 2.
19. Stoecker, W. 1971. *Proposed Procedures for Simulating the Per-*

formance of Components and Systems for Energy Calculations, 2nd Edition, American Society of Heating Refrigeration Air-conditioning Engineers, Atlanta, GA.

20. Sepsy, C. 1969. "Energy Requirements for Heating, and Cooling Buildings (ASHRAE RP 66-OS), Ohio State University.

21. Socolow, R. 1978. *Saving Energy in the Home: Princeton's Experiments at Twin Rivers*, Ballinger Publishing Company, Cambridge, Massachusetts, (This book contains a collection of papers that were also published in Energy and Buildings, Vol.1, No. 3., (April)).

22. Fels, M. 1986. Special Issue Devoted to Measuring Energy Savings: The Scorekeeping Approach, *Energy and Buildings*, Vol. 9, Nos. 1 &2, Elsevier Press, Lausanne, Switzerland, (February/May).

23. DOE 1985. *Proceedings of the DOE/ORNL Data Acquisition Workshop*, Oak Ridge National Laboratory, Oak Ridge, TN, (October).

24. Lyberg, M. 1987. *Source Book for Energy Auditors*: Vols. 1&2, International Energy Agency, Stockholm, Sweden, (Report on IEA Task XI).

25. IEA 1990. *Field Monitoring For a Purpose*. International Energy Agency Workshop, Chalmers University, Gothenburg, Sweden, (April).

26. Omnicomp 1984. Faser Software, Omnicomp, Inc., State College, PA, (monthly accounting software with VBDD capability).

27. Eto, J. 1988. "On Using Degree-days to Account for the Effects of Weather on Annual Energy Use in Office Buildings," *Energy and Buildings*, Vol. 12, No. 2, pp. 113-127.

28. SRC Systems 1996. Metrix: Utility Accounting System, Berkeley, CA, (monthly accounting software with combined VBDD/multiple regression capabilities).

29. Haberl, J. and Vajda. E. 1988. "Use of Metered Data Analysis to Improve Building Operation and Maintenance: Early Results From Two Federal Complexes," *Proceedings of the ACEEE 1988 Summer Study on Energy Efficient Buildings*, Pacific Grove, CA, pp. 3.98-3.111, (August).

30. Sonderegger, R. 1977. Dynamic Models of House Heating Based on Equivalent Thermal Parameters, Ph.D. Thesis, Center for Energy and Environmental Studies, Report No. 57, Princeton University.

31. DOE 1985. op.cit.

32. Lyberg, M. 1987. op.cit.

33. IEA 1990. op.cit.

34. ASHRAE 1991. *Handbook of HVAC Applications*, Chapter 37: Building Energy Monitoring, American Society of Heating Refrigeration Air-conditioning Engineers, Atlanta, GA.

35. Haberl, J., and Lopez, R. 1992. "LoanSTAR Monitoring Workbook: Workbook and Software for Monitoring Energy in Buildings," submitted to the Texas Governor's Energy Office, *Energy Systems Laboratory Report ESL-TR-92-06-03*, Texas A&M University, (August).

36. Claridge, D., Haberl, J., O'Neal, D., Heffington, W., Turner, D., Tombari, C., Roberts, M., Jaeger, S. 1991. "Improving Energy Conservation Retrofits with Measured Savings." *ASHRAE Journal*, Volume 33, Number 10, pp. 14-22, (October).

37. Feis, M., Kissock, K., Marean, M., and Reynolds, C. 1995. *PRISM, Advanced Version 1.0 User's Guide*, Center for Energy and Environmental Studies, Princeton University, Princeton, N.J., (January).

38. ASHRAE 1999, HVAC01 Toolkit: A Toolkit for Primary HVAC System Energy Calculation, ASHRAE Research Project—RP 665, Lebrun, J., Bourdouxhe, J-P, and Grodent, M., American Society of Heating Refrigeration Air-conditioning Engineers, Atlanta, GA.

39. ASHRAE 1993. HVAC02 Toolkit: Algorithms and Subroutines for Secondary HVAC System Energy Calculations, ASHRAE Research Project-827-RP, Authors: Brandemuehl, M., Gabel,

S.,Andresen, American Society of Heating Refrigeration Air-conditioning Engineers, Atlanta, GA.

40. Brandemuehl, M., Krarti, M., Phelan, J. 1996. "827-RP Final Report: Methodology Development to Measure In-Situ Chiller, Fan, and Pump Performance," *ASHRAE Research*, ASHRAE, Atlanta, GA, (March).

41. Haberl, J., Reddy, A., and Elleson, J. 2000a. "Determining Long-Term performance Of Cool Storage Systems From Short-Term Tests, Final Report," submitted to ASHRAE under Research Project 1004-RP, *Energy Systems Laboratory Report ESL-TR-00/08-01*, Texas A&M University, 163 pages, (August).

42. Kissock, K., Haberl, J., and Claridge, D. 2001. "Development of a Toolkit for Calculating Linear, Change-point Linear and Multiple-Linear Inverse Building Energy Analysis Models: Final Report," submitted to ASHRAE under Research Project 1050-RP, *University of Dayton and Energy Systems Laboratory*, (December).

43. Kissock, K., Haberl, J., Claridge, D. 2003. "Inverse Model Toolkit (1050-RP): Numerical Algorithms for Best-Fit Variable-Base Degree-Day and Change-Point Models," *ASHRAE Transactions-Research*, Vol. 109, Pt. 2, pp. 425-434.

44. Haberl, J., Claridge, D., Kissock, K. 2003b. "Inverse Model Toolkit (1050-RP): Application and Testing," *ASHRAE Transactions-Research*, Vol. 109, Pt. 2, pp. 435-448.

45. Abushakra, B., Haberl, J., Claridge, D., and Sreshthaputra, A. 2001 "Compilation Of Diversity Factors And Schedules For Energy And Cooling Load Calculations; ASHRAE Research Project 1093: Final Report," submitted to ASHRAE under Research Project 1093-RP, *Energy Systems Lab Report ESL-TR-00/06-01*, Texas A&M University, 150 pages, (June).

46. Claridge, D., Abushakra, B., Haberl, J. 2003. "Electricity Diversity Profiles for Energy Simulation of Office Buildings (1093-RP)," *ASHRAE Transactions-Research*, Vol. 110, Pt. 1, pp. 365-377 (February).

47. MacDonald, J. and Wasserman, D. 1989. Investigation of Metered Data Analysis Methods for Commercial and Related Buildings, *Oak Ridge National Laboratory Report No. ORNL/CON-279*, (May).

48. Rabl, A. 1988. "Parameter Estimation in Buildings: Methods for Dynamic Analysis of Measured Energy Use," *Journal of Solar Energy Engineering*, Vol. 110, pp. 52-66.

49. Rabl, A., Riahle, A. 1992. "Energy Signature Model for Commercial Buildings: Test With Measured Data and Interpretation," *Energy and Buildings*, Vol. 19, pp. 143-154.

50. Gordon, J.M. and Ng, K.C. 1994. "Thermodynamic Modeling of Reciprocating Chillers," *Journal of Applied Physics*, Volume 75, No. 6, March 15, 1994, pp. 2769-2774.

51. Claridge, D. E., Haberl, J. S., Sparks, R., Lopez, R., Kissock, K. 1992. "Monitored Commercial Building Energy Data: Reporting the Results." 1992 *ASHRAE Transactions-Research*, Vol. 98, Part 1, pp. 881-889.

52. Sonderegger, R. 1977, op.cit.

53. Subbarao, K., Burch, J., Hancock, C. E. 1990. "How to accurately measure the load coefficient of a residential building," *Journal of Solar Energy Engineering*, in preparation.

54. Reddy, A. 1989. "Application of Dynamic Building Inverse Models to Three Occupied Residences Monitored Non-intrusively," *Proceedings of the Thermal Performance of Exterior Envelopes of Buildings IV*, ASHRAE/DOE/BTECC/CIBSE.

55. Shurcliff, W.A. 1984. *Frequency Method of Analyzing a Building's Dynamic Thermal Performance*, W.A. Shurcliff, 19 Appleton St., Cambridge, MA.

56. Dhar, A. 1995, "Development of Fourier Series and Artificial Neural Networks Approaches to Model Hourly Energy Use in Commercial Buildings," Ph.D. Dissertation, Mechanical Engineering Department, Texas A&M University, May.

57. Miller, R., and Seem, J. 1991. "Comparison of Artificial Neural Networks with Traditional Methods of Predicting Return Time

from Night Setback," *ASHRAE Transactions*, Vol. 97, Pt.2, pp. 500-508.

58. J.F. Kreider and X.A. Wang, (1991). "Artificial Neural Networks Demonstration for Automated Generation of Energy Use Predictors for Commercial Buildings." *ASHRAE Transactions*, Vol. 97, part 1.

59. Kreider, J. and Haberl, J. 1994. "Predicting Hourly Building Energy Usage: The Great Energy Predictor Shootout: Overview and Discussion of Results," *ASHRAE Transactions-Research*, Volume 100, Part 2, pp. 1104-1118, (June).

60. ASHRAE 1997. *Handbook of Fundamentals*, Chapter 30: Energy Estimating and Modeling Methods, American Society of Heating Refrigeration Air-conditioning Engineers, Atlanta, GA., p. 30.27 (Copied with permission).

61. ASHRAE 1997. op.cit., p. 30.28 (Copied with permission).

62. USDOE 1996. *North American Energy Measurement and Verification Protocol (NEMVP)*, United States Department of Energy DOE/EE-0081, (March).

63. FEMP 1996. *Standard Procedures and Guidelines for Verification of Energy Savings Obtained Under Federal Savings Performance Contracting Programs*, USDOE Federal Energy Management Program (FEMP).

64. Haberl, J., Claridge, D., Turner, D., O'Neal, D., Heffington, W., Verdict, M. 2002. "LoanSTAR After 11 Years: A Report on the Successes and Lessons Learned From the LoanSTAR Program," *Proceedings of the 2nd International Conference for Enhanced Building Operation*, Richardson, Texas, pp. 131-138, (October).

65. USDOE 1997. *International Performance Measurement and Verification Protocol (IPMVP)*, United States Department of Energy DOE/EE-0157, (December).

66. USDOE 2001. *International Performance Measurement and Verification Protocol (IPMVP): Volume I: Concepts and Options for Determining Energy and Water Savings*, United States Department of Energy DOE/GO-102001-1187 (January).

67. USDOE 2001. *International Performance Measurement and Verification Protocol (IPMVP): Volume II: Concepts and Practices for Improved Indoor Environmental Quality*, United States Department of Energy DOE/GO-102001-1188 (January).

68. USDOE 2003. *International Performance Measurement and Verification Protocol (IPMVP): Volume III: Concepts and Options for Determining Energy Savings in New Construction*, United States Department of Energy (April).

69. ASHRAE 2002. *Guideline 14: Measurement of Energy and Demand Savings*, American Society of Heating Refrigeration Air-conditioning Engineers, Atlanta, GA (September).

70. Hansen, S. 1993. *Performance Contracting for Energy and Environmental Systems*, Fairmont Press, Lilburn, GA, pp. 99-100.

71. C. Culp, K.Q. Hart, B. Turner, S. Berry-Lewis, 2003. "Cost Effective Measurement and Verification at Fairchild AFB, International Conference on Enhance Building Operation," *Energy Systems Laboratory Report*, Texas A&M University, (October).

72. ASHRAE 2002. op.cit., pp. 27-30.

73. Ibid, p. 30 (Copied with permission).

74. Brandemuehl et al. 1996. op.cit.

75. Wei, G. 1997. "A Methodology for In-situ Calibration of Steam Boiler Instrumentation," MS Thesis, Mechanical Engineering Department, Texas A&M University, August.

76. Dukelow, S.G. 1991. *The Control of Boilers*. Research Triangle Park, NC: Instrument Society of America.

77. Dyer, F.D. and Maples, G. 1981. *Boiler Efficiency Improvement*. Boiler Efficiency Institute. Auburn: AL.

78. Garcia-Borras, T. 1983. *Manual for Improving Boiler and Furnace Performance*. Houston, TX: Gulf Publishing Company.

79. Aschner, F.S. 1977. *Planning Fundamentals of Thermal Power Plants*. Jerusalem, Israel: Israel Universities Press.

80. ASME 1974 *Performance Test for Steam Units—PTC 4.1a*, 1974.

81. Babcock and Wilcox. 1992. *Steam: Its generation and Use*, Babcock and Wilcox, Barberton, Ohio, ISBN 0-9634570-0-4.

82. Katipamula, S., and Claridge, D. 1992. "Monitored Air Handler Performance and Comparison with a Simplified System Model," *ASHRAE Transactions*, Vol. 98, Pt 2., pp. 341-351.

83. Liu, M., and Claridge, D. 1995 "Application of Calibrated HVAC Systems to Identify Component Malfunctions and to Optimize the Operation and Control Schedules," *ASME/JSME International Solar Energy Conference*, pp. 209-217.

84. ASHRAE 2002. op.cit., p. 144.

85. Brandemuehl, et al. 1996. op. cit.

86. ASHRAE 2002, op cit., p. 144, (Copied with permission).

87. ASHRAE 2002. op.cit., pp.144-147, (Copied with permission).

88. ASHRAE 2002. op.cit., p. 144.

89. Ibid, p. 148, (Copied with permission).

90. ASHRAE 2002. op.cit., pp.147-149, (Copied with permission).

91. Gordon, J.M. and Ng, K.C. 1994. op.cit.

92. Gordon, J.M. and Ng, K.C., 1995. "Predictive and diagnostic aspects of a universal thermodynamic model for chillers," *International Journal of Heat Mass Transfer*, 38(5), p.807.

93. Gordon, J.M., Ng, K.C., and Chua, H.T., 1995. "Centrifugal chillers: thermodynamic modeling and a diagnostic case study," *International Journal of Refrigeration*, 18(4), p.253.

94. LBL. 1980. *DOE-2 User Guide, Ver. 2.1*. Lawrence Berkeley Laboratory and Los Alamos National Laboratory, Rpt No. LBL-8689 Rev. 2; DOE-2 User Coordination Office, LBL, Berkeley, CA.

95. LBL. 1981. *DOE-2 Engineers Manual, Ver. 2.1A*, Lawrence Berkeley Laboratory and Los Alamos National Laboratory, Rpt No. LBL-11353; DOE-2 User Coordination Office, LBL, Berkeley, CA.

96. LBL. 1982. *DOE-2.1 Reference Manual Rev. 2.1A*. Lawrence Berkeley Laboratory and Los Alamos National Laboratory, Rpt No. LBL-8706 Rev. 2; DOE-2 User Coordination Office, LBL, Berkeley, CA.

97. LBL. 1989. DOE-2 *Supplement, Ver 2.1D*. Lawrence Berkeley Laboratory, Rpt No. LBL-8706 Rev. 5 Supplement. DOE-2 User Coordination Office, LBL, Berkeley, CA.

98. Haberl, J. S., Reddy, T. A., Figueroa, I., Medina, M. 1997. "Overview of LoanSTAR Chiller Monitoring and Analysis of In-Situ Chiller Diagnostics Using ASHRAE RP827 Test Method," *Proceedings of the PG&E Cool Sense National Integrated Chiller Retrofit Forum* (September).

99. According to Gordon et al. 1995, f_{HX} is a dimensionless term that is normally negligible.

100. Wei, G. 1997. op.cit.

101. Dukelow, S.G. 1991. op.cit.

102. Dyer, F.D. and Maples, G. 1981. op.cit.

103. Garcia-Borras, T. 1983. op.cit.

104. Aschner, F.S. 1977. op.cit.

105. ASME 1974. op.cit.

106. Babcock and Wilcox. 1992. op.cit.

107. Haberl, J., Lynn, B., Underwood, D., Reasoner, J., Rury, K. 2003a. op.cit.

108. Haberl, et al. 2003a. ibid.

109. ASME, 1974. *Power Test Codes (PTC) 4.1a, Steam Generating Units*. New York: ASME.

110. Stallard, G.S. and Jonas, T.S. 1996. *Power Plant Engineering: Combustion Processes*. New York: Chapman & Hall.

111. Payne, F.W. 1985. *Efficient Boiler Operations Sourcebook*. Atlanta, GA: The Fairmont Press.

112. Wei 1997. op.cit.

113. Wei 1997. ibid.

114. Aschner 1977. op.cit.

115. Babcock and Wilcox 1992. op.cit.

116. Dukelow 1991. op.cit.

117. Witte, L.C., Schmidt, P.S., and Brown, D.R. 1988. *Industrial Energy Management and Utilization*. New York: Hemisphere Publishing Corporation.

118. Aschner 1977. op.cit.

119. Thumann, A. 1988. *Guide to Improving Efficiency of Combustion*

Systems. Lilburn, GA: The Fairmont Press.

120. Wei 1997. op.cit.

121. Garcia-Borras, T. 1983. *Manual for Improving Boiler and Furnace Performance.* Houston, TX: Gulf Publishing Company.

122. Wei 1997. op.cit.

123. ASHRAE 2002, op.cit., pp. 154-156, (Copied with permission).

124. Bou Saada, T., Haberl, J., Vajda, J., and Harris, L. 1996. "Total Utility Savings From the 37,000 Fixture Lighting Retrofit to the USDOE Forrestal Building," *Proceedings of the 1996 ACEEE Summery Study,* (August).

125. Abushakra, B., Sreshthaputra, A., Haberl, J., and Claridge, D. 2001. "Compilation of Diversity Factors and Schedules for Energy and Cooling Load Calculations-Final Report," submitted to ASHRAE under Research Project 1093-RP, *Energy Systems Lab Report ESL-TR-01/04-01,* Texas A&M University, (April).

126. IESNA 2003. *Lighting Handbook,* 9th Edition, Illuminating Engineering Society of North America, New York, N.Y.

127. ASHRAE 2002, op. cit., pp. 156-159, (Copied with permission).

128. ASHRAE 2002, op. cit., p. 160, (Copied with permission).

129. Ayres, M., Stamper, E. 1995, op.cit.

130. ASHRAE 1969. *Procedures for Determining Heating and Cooling Loads for Computerized Energy Calculations: Algorithms for Building Heat Transfer Sub-routines.* M. Lokmanhekim, Editor, American Society of Heating Refrigeration Air-conditioning Engineers, Atlanta, GA.

131. ASHRAE 1971. *Procedures for Simulating the Performance of Components and Systems for Energy Calculations.* Stoecker, W.F. Stoecker, editor, 2nd edition, American Society of Heating Refrigeration Air-conditioning Engineers, Atlanta, GA

132. BLAST. 1993. *BLAST Users Manual.* BLAST Support Office, University of Illinois Urbana-Champaign.

133. LBL 1980, 1981, 1982, 1989, op.cit.

134. Knebel, D.E., 1983. *Simplified Energy Analysis Using the Modified Bin Method,* American Society of Heating, Refrigerating and Air-Conditioning Engineers, Inc., Atlanta, Georgia.

135. ASHRAE 1999. op.cit.

136. ASHRAE 1993. op.cit.

137. Yuill, G., K., Haberl, J.S. 2002. Development of Accuracy Tests For Mechanical System Simulation. Final Report for ASHRAE Research Project 865-RP, The University of Nebraska at Lincoln, (July).

138. Katipamula, S. and Claridge, D.E., 1993. " Use of Simplified Systems Models to Measure Retrofit Savings," *ASME Journal of Solar Energy Engineering,* Vol.115, pp.57-68, May.

139. Liu, M. and Claridge, D.E., 1995. "Application of Calibrated HVAC System Models to Identify Component Malfunctions and to Optimize the Operation and Control Schedules," *Solar Engineering* 1995, W.B. Stine, T. Tanaka and D.E. Claridge (Eds.), ASME/JSME/JSES International Solar Energy Conference, Maui, Hawaii, March.

140. Liu, M. and Claridge, D. E., 1998. "Use of Calibrated HVAC System Models to Optimize System Operation," *Journal of Solar Energy Engineering,* May 1998, Vol.120.

141. Liu, M., Wei, G., Claridge, D., E., 1998, "Calibrating AHU Models Using Whole Building Cooling and Heating Energy Consumption Data," *Proceedings of 1998 ACEEE Summer Study on Energy Efficiency in Buildings.* Vol. 3.

142. Haberl, J., Claridge, D., Turner. D. 2000b. "Workshop on Energy Measurement, Verification and Analysis Technology," Energy Conservation Task Force, Federal Reserve Bank, Dallas, Texas (April).

143. This table contains material adapted from proposed HVAC System Testing Methods for ASHRAE Guideline 14-2002, which were not included in the published ASHRAE Guideline 14-2002.

144. Haberl et al. 2000b, op. cit.

145. Kissock et al. 2001. op.cit.

146. Fels 1986. op.cit.

147. Rabl 1988. op.cit.

148. Rabl and Raihle 1992. op.cit.

149. Claridge et al. 1992. op.cit.

150. Reddy, T. A., Haberl, J. S., Saman, N.F., Turner, W. D., Claridge, D.E., Chalifoux, A. T. 1997. "Baselining Methodology for Facility-Level Monthly Energy Use—Part 1: Theoretical Aspects," *ASHRAE Transactions-Research,* Volume 103, Part 2, pp. 336-347, (June).

151. Reddy, T.A., Haberl, J.S., Saman, N.F., Turner, W.D., Claridge, D.E., Chalifoux, A. T. 1997 "Baselining Methodology for Facility-Level Monthly Energy Use—Part 2: Application to Eight Army Installations," *ASHRAE Transactions-Research,* Volume 103, Part 2, pp. 348-359, (June).

152. Haberl, J., Thamilseran, S., Reddy, A., Claridge, D., O'Neal, D., Turner, D. 1998. "Baseline Calculations for Measuring and Verification of Energy and Demand Savings in a Revolving Loan Program in Texas," *ASHRAE Transactions-Research,* Volume 104, Part 2, pp. 841-858, (June).

153. Turner, D., Claridge, D., O'Neal, D., Haberl, J., Heffington, W., Taylor, D., Sifuentes, T. 2000. "Program Overview: The Texas LoanSTAR Program: 1989-1999 A 10-year Experience," *Proceedings of the 2000 ACEEE Summery Study on Energy Efficiency in Buildings,* Volume 4, pp. 4.365-4.376, (August).

154. Haberl, J., Sreshthatputra, A., Claridge, D., Turner, D. 2001. "Measured Energy Indices for 27 Office Buildings," *Proceedings of the 1st International Conference for Enhanced Building Operation,* Austin, Texas, pp. 185-200, (July).

155. Beasley, R., Haberl, J. 2002. "Development of a Methodology for Baselining The Energy Use of Large Multi-building Central Plants," *ASHRAE Transactions-Research,* Volume 108, Part 1, pp. 251-259, (January).

156. ASHRAE 2002. op.cit. p. 25, (Copied with permission).

157. Haberl et al. 2000b, op. cit.

158. Temperatures below zero are calculated as positive increases away from the change point temperature.

159. Thamilseran, S., Haberl, J. 1995. "A Bin Method for Calculating Energy Conservation Retrofits Savings in Commercial Buildings," *Proceedings of the 1995 ASME/JSME/JSES International Solar Energy Conference,* Lahaina, Maui, Hawaii, pp. 111-124 (March).

160. Thamilseran, S. 1999. "An Inverse Bin Methodology to Measure the Savings from Energy Conservation Retrofits in Commercial Buildings," Ph.D. Thesis, Mechanical Engineering Department, Texas A&M University, (May).

161. Kissock et al. 2001. op.cit.

162. Kissock, J.K, Xun,W., Sparks, R., Claridge, D., Mahoney, J. and Haberl, J., 1994. "EModel Version 1.4de," Texas A&M University, Energy Systems Laboratory, Department of Mechanical Engineering, Texas A&M University, College Station, TX, December.

163. Abushakra et al. 2001. op.cit.

164. Haberl, J., Bou-Saada, T. 1998. "Procedures for Calibrating Hourly Simulation Models to Measured Building Energy and Environmental Data," *ASME Journal of Solar Energy Engineering,* Volume 120, pp. 193-204, (August).

165. Clarke, J.A, Strachan, P.A. and Pernot, C.. 1993. An Approach to the Calibration of Building Energy Simulation Models. *ASHRAE Transactions.* 99(2): 917-927.

166. Diamond, S.C. and Hunn, B.D.. 1981. Comparison of DOE-2 Computer Program Simulations to Metered Data for Seven Commercial Buildings. *ASHRAE Transactions.* 87(1): 1222-1231.

167. Haberl, J., Bronson, D., Hinchey, S. and O'Neal, D. 1993. "Graphical Tools to Help Calibrate the DOE-2 Simulation Program to Non-weather Dependent Measured Loads," 1993 *ASHRAE Journal,* Vol. 35, No.1, pp. 27-32, (January).

168. Haberl, J., Bronson, D. and O'Neal, D. 1995. "An Evaluation of the Impact of Using Measured Weather Data Versus TMY

Weather Data in a DOE-2 Simulation of an Existing Building in Central Texas." *ASHRAE Transactions-Research,*Vol. 101, Pt. 2, pp 558-576 (June).

169. Hinchey, S.B. 1991. Influence of Thermal Zone Assumptions on DOE-2 Energy Use Estimations of a Commercial Building. M.S. Thesis, Energy Systems Report No. ESL-TH-91/09-06, Texas A&M University, College Station, TX.

170. Hsieh, E.S. 1988. Calibrated Computer Models of Commercial Buildings and Their Role in Building Design and Operation. M.S. Thesis, PU/CEES Report No. 230, Princeton University, Princeton, NJ.

171. Hunn, B.D., Banks, J.A. and Reddy, S.N. 1992. Energy Analysis of the Texas Capitol Restoration. *The DOE-2 User News.* 13(4): 2-10.

172. Kaplan, M.B., Jones, B. and Jansen, J. 1990a. DOE-2.1C Model Calibration with Monitored End-use Data. *Proceedings from the ACEEE 1990 Summer Study on Energy Efficiency in Buildings,* Vol. 10, pp. 10.115-10.125.

173. Kaplan, M.B., Caner, P. and Vincent, G.W. 1992. Guidelines for Energy Simulation of Commercial Buildings. *Proceedings from the ACEEE 1992 Summer Study on Energy Efficiency in Buildings,* Vol. 1, pp. 1.137-1.147.

174. Katipamula, S. and Claridge, D.E., 1993. " Use of Simplified Systems Models to Measure Retrofit Savings," ASME *Journal of Solar Energy Engineering,* Vol.115, pp.57-68, May.

175. Liu, M. and Claridge, D.E., 1995. "Application of Calibrated HVAC System Models to Identify Component Malfunctions and to Optimize the Operation and Control Schedules," *Solar Engineering 1995,* W.B. Stine, T. Tanaka and D.E. Claridge (Eds.), ASME/JSME/JSES International Solar Energy Conference, Maui, Hawaii, March.

176. Liu, M. and Claridge, D. E., 1998. "Use of Calibrated HVAC System Models to Optimize System Operation," *Journal of Solar Energy Engineering,* May 1998, Vol.120.

177. Liu, M., Wei, G., Claridge, D., E., 1998, "Calibrating AHU Models Using Whole Building Cooling and Heating Energy Consumption Data," *Proceedings of 1998 ACEEE Summer Study on Energy Efficiency in Buildings.* Vol. 3.

178. Manke, J., Hittle, D. and Hancock 1996. "Calibrating Building Energy Analysis Models Using Short-term Test Data," *Proceedings of the 1996 International ASME Solar Energy Conference,* p.369, San Antonio, TX.

179. McLain, H.A., Leigh, S.B., and MacDonald, J.M.. 1993. Analysis of Savings Due to Multiple Energy Retrofits in a Large Office Building. Oak Ridge National Laboratory, *ORNL Report No. ORNL/CON-363,* Oak Ridge, TN.

180. Sreshthaputra, A., Haberl, J., Andrews, M. 2004. "Improving Building Design and Operation of a Thai Buddhist Temple," *Energy and Buildings,* Vol. 36, pp. 481-494.

181. Song, S., Haberl, J. 2008. "A Procedure for the Performance Evaluation of a New Commercial Building, Part 1: Calibrated As-Built Simulation," *ASHRAE Transactions-Research,* Vol. 114, Pt. 2 (May).

182. Song, S., Haberl, J. 2008. "A Procedure for the Performance Evaluation of a New Commercial Building, Part 2: Overall Methodology and Comparison of Methods," *ASHRAE Transactions-Research,* Vol. 114, Pt. 2 (May).

183. Haberl et al. 2000b, op. cit.

184. ASHRAE 2002. op.cit. p. 35-43, (Copied with permission).

185. Bou-Saada, T. 1994. An Improved Procedure for Developing A Calibrated Hourly Simulation Model of an Electrically Heated and Cooled Commecial Building, Master's Thesis, Mechanical Engineering Department, Texas A&M University, (December), p. 54.

186. Sylvester, K., Song, S., Haberl, J., and Turner, D. 2002. Case Study: Energy Savings Assessment for the Robert E. Johnson State Office Building in Austin, Texas," *IBPSA Newsletter,* Vol. 12, Number 2, pp. 22-28, (Summer).

187. Huang & Associates. 1993. *DrawBDL user's guide.* 6720 Potrero Ave., El Cerrito, California, 94530.

188. Bou-Saada 1994. op. cit. p. 150.

189. Bou-Saada 1994. op. cit. p. 144.

190. Culp et al. 2003, ibid.

191. ASME 1974 op cit.

CHAPTER 28

GROUND-SOURCE HEAT PUMPS APPLIED TO COMMERCIAL BUILDINGS

STEVEN A. PARKER, P.E., C.E.M.
DONALD L. HADLEY
Energy Science and Technology Directorate
Pacific Northwest National Laboratory[1]
Richland, WA

28.1 ABSTRACT

Ground-source heat pumps can provide an energy-efficient, cost-effective way to heat and cool commercial facilities. While ground-source heat pumps are well established in the residential sector, their application in larger, commercial-style facilities is lagging, in part because of limited experience with the technology by those in decision-making positions. Through the use of a ground-coupling system, a conventional water-source heat pump design is transformed to a unique means of utilizing thermodynamic properties of earth and groundwater for efficient operation throughout the year in most climates. In essence, the ground (or groundwater) serves as a heat source during winter operation and a heat sink for summer cooling. Many varieties in design are available, so the technology can be adapted to almost any site. Ground-source heat pump systems can be used widely in commercial building applications and, with proper installation, offer great potential for the commercial sector, where increased efficiency and reduced heating and cooling costs are important. Ground-source heat pump systems require less refrigerant than conventional air-source heat pumps or air-conditioning systems, with the exception of direct expansion type ground-source heat pump systems.

Installation costs are relatively high but are offset by low maintenance and operating expenses and efficient energy use. The greatest barrier to effective use is improper design and installation; well-trained, experienced, and responsible designers and installers are of critical importance.

This chapter provides information and procedures that an energy manager can use to evaluate most ground-source heat pump applications. Ground-source heat pump operation, system types, design variations, energy savings, and other benefits are explained. Guidelines are provided for appropriate application and installation. Two case studies are presented to give the reader a sense of the actual costs and energy savings. A list of manufacturers and references for further reading are included for prospective users who have specific or highly technical questions not fully addressed in this chapter. Sample case spreadsheets are also provided.

28.2 BACKGROUND

This chapter is based on a Federal Technology Alert sponsored by the U.S. Department of Energy (DOE), Federal Energy Management Program (FEMP). The original Federal Technology Alert was published in 1995 and updated in 2001. The material was updated in 2005 and 2008 to develop this chapter.

28.3 INTRODUCTION TO GROUND-SOURCE HEAT PUMPS

Ground-source heat pumps are known by a variety of names: geoexchange heat pumps, ground-coupled heat pumps, geothermal heat pumps, earth-coupled heat pumps, ground-source systems, groundwater source heat pumps, well water heat pumps, solar energy heat pumps, and a few other variations. Some names are used to describe more accurately the specific application; however, most are the result of marketing efforts and the need to associate (or disassociate) the heat pump systems from other systems. This chapter refers to them as ground-source heat pumps except when it is necessary to distinguish a specific design or application of the technology. A typical ground-source heat pump system design applied to a commercial facility is illustrated in Figure 28-1.

It is important to remember that the primary equip-

[1]Pacific Northwest National Laboratory is operated for the U.S. Department of Energy by Battelle Memorial Institute under contract DE-AC05-76RL01830.

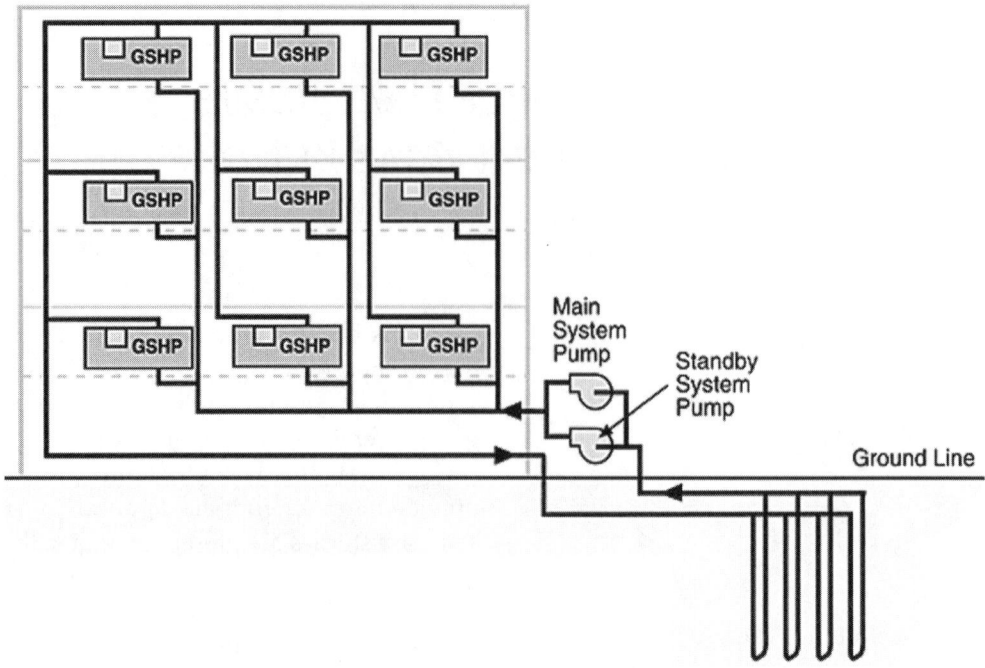

Figure 28-1. Typical ground-source heat pump system applied to a commercial facility

ment used for ground-source heat pumps are water-source heat pumps. What makes a ground-source heat pump different (unique, efficient, and usually more expensive to install) is the ground-coupling system. In addition, most manufacturers have developed extended-range water-source heat pumps for use as ground-source heat pumps.[2]

A conventionally designed water-source heat pump system would incorporate a boiler as a heat source during the winter heating operation and a cooling tower to reject heat (heat sink) during the summer cooling operation. This system type is also sometimes called a boiler/tower water-loop heat pump system. The water loop circulates to all the water-source heat pumps connected to the system. The boiler (for winter operation) and the cooling tower (for summer operation) provide a fairly constant water-loop temperature, which allows the water-source heat pumps to operate at high efficiency.

A conventional air-source heat pump uses the outdoor ambient air as a heat source during the winter heating operation and as a heat sink during the summer cooling operation. Air-source heat pumps are subject to higher temperature fluctuations of the heat source and heat sink. They become much less effective—and less

efficient—at extreme ambient air temperatures. This is particularly true at low temperatures. In addition, heat transfer using air as a transfer medium is not as effective as water systems because of air's lower thermal mass.

A ground-source heat pump uses the ground (or in some cases groundwater) as the heat source during the winter heating operation and as the heat sink during the summer cooling operation. Ground-source heat pumps may be subject to higher temperature fluctuations than conventional water-source heat pumps but not as high as air-source heat pumps. Consequently, most manufacturers have developed extended-range systems. The extended-range systems operate more efficiently while subject to the extended-temperature range of the water loop. Like water-source heat pumps, ground-source heat pumps use a water loop between the heat pumps and the heat source/heat sink (the earth). The primary exception is the direct expansion ground-source heat pump, which is described in more detail later in this chapter.

Ground-source heat pumps take advantage of the thermodynamic properties of the earth and groundwater. Temperatures below the ground surface do not fluctuate significantly through the day or the year as do ambient air temperatures. Ground temperatures a few feet below the surface stay relatively constant throughout the year. For this reason, ground-source heat pumps remain extremely efficient throughout the year in virtually any climate.

---

[2]The extended-range designation is important. Conventional water-source heat pumps are designed to operate with a water-loop as a heat sink that maintains a narrow temperature range. Ground-source heat pumps, however, are typically required to operate with a water-loop heat sink under a wider range of temperatures.

28.4 ABOUT THE TECHNOLOGY

In 1999, an estimated 400,000 ground-source heat pumps were operating in residential and commercial applications, up from 100,000 in 1990. In 1985, it was estimated that only around 14,000 ground-source heat pump systems were installed in the United States. Annual sales of approximately 45,000 units were reported in 1997. With a projected annual growth rate of 10%, 120,000 new units would be installed in 2010, for a total of 1.5 million units in 2010 (Lund and Boyd 2000). In Europe, the estimated total number of installed ground-source heat pumps at the end of 1998 was 100,000 to 120,000 (Rybach and Sanner 2000). Nearly 10,000 ground-source heat pumps have been installed in U.S. federal buildings, over 400 schools, and thousands of low-income houses and apartments (ORNL/SERDP, no date).

Although ground-source heat pumps are used throughout the United States, the majority of new ground-source heat pump installations in the United States are in the southern and mid-western states (from North Dakota to Florida). Oklahoma, Texas, and the East Coast have been particularly active with new ground-source heat pump installations. Environmental concerns, particularly from the potential for groundwater contamination with a leaking ground loop, and a general lack of understanding of the technology by HVAC companies and installers have limited installations in the West (Lund and Boyd 2000). Usually the technology does well in an area where it has been actively promoted by a local utility or the manufacturer.

Ground-source heat pumps are not a new idea. Patents on the technology date back to 1912 in Switzerland (Calm 1987). One of the oldest ground-source heat pump systems, in the United Illuminating headquarters building in New Haven, Connecticut, has been operating since the 1930s (Pratsch 1990). Although ground-source heat pump systems are probably better established today in rural and suburban residential areas because of the land area available for the ground loop, the market has expanded to urban and commercial applications.

The vast majority of ground-source heat pump installations utilize unitary equipment consisting of multiple water-source heat pumps connected to a common ground-coupled loop. Most individual units range from 1 to 10 tons (3.5 to 35.2 kW), but some equipment is available in sizes up to 50 tons (176 kW). Large-tonnage commercial systems are achieved by using several unitary water-source heat pumps, each responsible for an individual control zone.

One of the largest commercial ground-source heat pump systems is at Stockton College in Pomona, New Jersey, where 63 ground-source heat pumps totaling 1,655 tons (5,825 kW) are connected to a ground-coupled loop consisting of 400 wells, each 425 feet (129 m) deep (Gahran 1993).

Public schools are another good application for the ground-source heat pump technology with over 400 installations nationwide. In 1995, the Lincoln, Nebraska, public school district built four new 70,000 square foot elementary schools. Space conditioning loads are met by 54 ground-source heat pumps ranging in size from 1.4 to 15 tons, with a total cooling capacity of 204 tons. Gas-fired boilers provided hot water for pre-heating of the outside air and for terminal re-heating. Compared with other similar new schools, these four ground-source heat pump conditioned facilities used approximately 26% less source energy per square foot of floor area (Shonder et al. 1999).

Multiple unitary systems are not the only arrangements suitable for large commercial applications. It is also possible to design large centralized heat-pump systems consisting of reciprocating and centrifugal compressors (up to 19.5 million Btu/h) and to use these systems to support central-air-handling units, variable air-volume systems, or distributive two-pipe fan coil units.

28.4.1 How the Technology Works

Heat normally flows from a warmer medium to a colder one. This basic physical law can only be reversed with the addition of energy. A heat pump is a device that does so by essentially "pumping" heat up the temperature scale, then transferring it from a cold material to a warmer one by adding energy, usually in the form of electricity. A heat pump functions by using a refrigerant cycle similar to the household refrigerator. In the heating mode, a heat pump removes the heat from a low temperature source, such as the ground or air, and supplies that heat to a higher temperature sink, such as the heated interior of a building. In the cooling mode, the process is reversed and the heat is extracted from the cooler inside air and rejected to the warmer outdoor air or other heat sink. For space conditioning of buildings, heat pumps that remove heat from outdoor air in the heating mode and reject it to outdoor air in the cooling mode are common. These are normally called air-source or air-to-air heat pumps. Air-source heat pumps have the disadvantage that the greatest requirement for building heating or cooling is necessarily coincident with the times when the outdoor air is least effective as a heat source or sink. Below about 37°F (2.8°C), supplemental heating is required to meet the heating load. For this reason, air-source heat pumps are essentially unfeasible in cold climates with outdoor temperatures below 37°F (2.8°C) for extended periods of time.

The efficiency of any heat pump is inversely proportional to the temperature difference between the conditioned space and the heat source (heating mode) or heat sink (cooling mode), as can be easily shown by a simple thermodynamic analysis (Reynolds and Perkins 1977). For this reason, air-source heat pumps are less efficient and have a lower heating capacity in the heating mode at low outdoor air temperatures. Conversely, air-source heat pumps are also less efficient and have a lower cooling capacity in the cooling mode at high outdoor air temperatures. Ground-source heat pumps, however, are not impacted directly by outdoor air temperatures. Ground-source heat pumps use the ground, groundwater, or surface water, which are all more thermally stable and not subject to large annual swings of temperature as a heat source or sink.

28.4.2 Other benefits

The primary benefit of ground-source heat pumps is the increase in operating efficiency, which translates to a reduction in heating and cooling costs, but there are additional advantages. One notable benefit is that ground-source heat pumps, although electrically driven, are classified as renewable-energy technology. The justification for this classification is that the ground acts as an effective collector of solar energy. The renewable-energy classification can affect federal goals and potential federal funding opportunities.

An environmental benefit is that ground-source heat pumps typically use 25% less refrigerant than split-system air-source heat pumps or air-conditioning systems. Ground-source heat pumps generally do not require tampering with the refrigerant during installation. Systems are generally sealed at the factory, reducing the potential for leaking refrigerant in the field during assembly.

Ground-source heat pumps also require less space than conventional heating and cooling systems. While the requirements for the indoor unit are about the same as conventional systems, the exterior system (the ground coil) is underground, and there are no space requirements for cooling towers or air-cooled condensers. In addition, the ground-coupling system does not necessarily limit future use of the land area over the ground loop, with the exception of siting a building. Interior space requirements are also reduced. There are no floor space requirements for boilers or furnaces, just the unitary systems and circulation pumps. Furthermore, many distributed ground-source heat pump systems are designed to fit in ceiling plenums, reducing the floor space requirement of central mechanical rooms.

Compared with air-source heat pumps that use outdoor air coils, ground-source heat pumps do not require defrost cycles or crankcase heaters, and there is virtually no concern for coil freezing. Cooling tower systems require electric resistance or steam heaters to prevent freezing in the tower basin—also not necessary with ground-source heat pumps.

It is generally accepted that maintenance requirements are also reduced, although research continues to be directed toward verifying this claim. It is clear, however, that ground-source heat pumps eliminate the exterior fin-coil condensers of air-cooled refrigeration systems and eliminate the need for cooling towers and their associated maintenance and chemical requirements. This is a primary benefit cited by facilities in highly corrosive areas, such as near the ocean where salt spray can significantly reduce outdoor equipment life.

Ground-source heat pump technology offers further benefits: less need for supplemental resistance heaters, no exterior coil freezing (requiring defrost cycles) such as that associated with air-source heat pumps, improved comfort during the heating season (compared with air-source heat pumps, the supply air temperature does not drop when recovering from the defrost cycle), significantly reduced fire hazard over that associated with fossil fuel-fired systems, reduced space requirements and hazards by eliminating fossil-fuel storage, and reduced local emissions from those associated with other fossil fuel-fired heating systems.

Another benefit is quieter operation, because ground-source heat pumps have no outside air fans. Finally, ground-source heat pumps are reliable and long-lived, because the heat pumps are generally installed in climate-controlled environments and therefore are not subject to the stresses of extreme temperatures. Because of the materials and joining techniques, the ground-coupling systems are also typically reliable and long-lived. For these reasons, ground-source heat pumps are expected to have a longer life and require less maintenance than alternative (more conventional) technologies.

28.4.3 Ground-Coupled System Types

The ground-coupling systems used in ground-source heat pumps fall under three main categories: closed-loop, open-loop and direct expansion. These are illustrated in Figure 28-2 and discussed in the following sections. The type of ground coupling employed will affect heat pump system performance (therefore the heat pump energy consumption), auxiliary pumping energy requirements, and installation costs. Choice of the most appropriate type of ground coupling for a site is usually a function of specific geography, available land area, and life-cycle cost economics.

Closed-loop Systems

Closed-loop systems consist of an underground network of sealed, high-strength plastic pipe[3] acting as a heat exchanger. The loop is filled with a heat transfer fluid, typically water or a water-antifreeze[3] solution, although other heat transfer fluids may be used.[4] When cooling requirements cause the closed-loop liquid temperature to rise, heat is transferred to the cooler earth. Conversely, when heating requirements cause the closed-loop fluid temperature to drop, heat is absorbed from the warmer earth. Closed-loop systems use pumps to circulate the heat transfer fluid between the heat pump and the ground loop. Because the loops are closed and sealed, the heat pump heat exchanger is not subject to mineral buildup and there is no direct interaction (mixing) with groundwater.

There are several varieties of closed-loop configurations, including horizontal, spiral, vertical, and submerged.

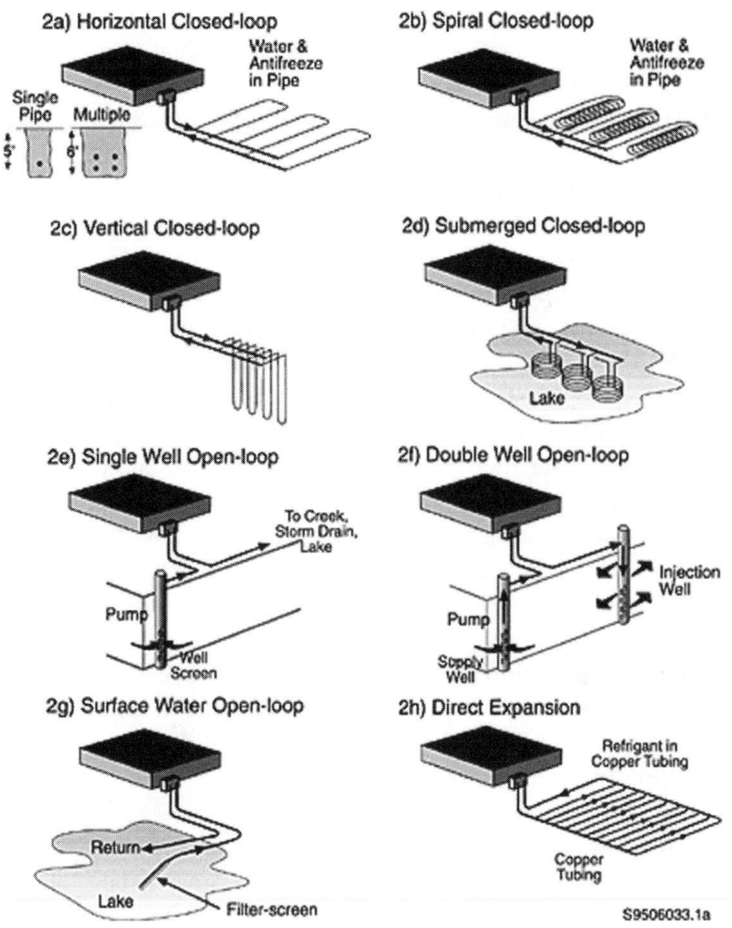

Horizontal Loops

Horizontal loops, illustrated in Figure 28-2a, are often considered when adequate land surface is available. The pipes are placed in trenches, typically at a depth of 4 to 10 feet (1.2 to 3.0 m). Depending on the specific design, from one to six pipes may be installed in each trench. Although requiring more linear feet of pipe, multiple-pipe configurations conserve land space, require less trenching, and therefore frequently cost less to install than single-pipe configurations. Trench lengths can range from 100 to 400 feet per system cooling ton (8.7 to 34.6 m/kW), depending on soil characteristics and moisture content and the number of pipes in the trench. Trenches are usually spaced from 6 to 12 feet (1.8 to 3.7 m) apart.

These systems are common in residential applications but are not frequently applied to large-tonnage commercial applications because of the significant land

Figure 28-2. Ground-coupling system types

area required for adequate heat transfer. The horizontal-loop systems can be buried beneath lawns, landscaping, and parking lots. Horizontal systems tend to be more popular where there is ample land area with a high water table.

- **Advantages**: Trenching costs typically lower than well-drilling costs; flexible installation options.

- **Disadvantages**: Large ground area required; ground temperature subject to seasonal variance at shallow depths; thermal properties of soil fluctuate with season, rainfall, and burial depth; soil dryness must be properly accounted for in designing the required pipe length, especially in sandy soils and on hilltops that may dry out during the summer; pipe system could be damaged during backfill process; longer pipe lengths are required than for vertical wells; antifreeze solution viscosity increases pumping energy, and decreases the heat transfer rate, thus reducing overall efficiency; lower system efficiencies.

---

[3]Acceptable piping includes high quality polyethylene or polybutylene. PVC is not acceptable in either heat transfer characteristics or strength.

[4]Common heat transfer fluids include water or water mixed with an antifreeze, such as: sodium chloride, calcium chloride, potassium carbonate, potassium acetate, ethylene glycol, propylene gycol, methyl alcohol, or ethyl alcohol.

[6]Note that various heat transfer fluids have different densities and thermodynamic properties. Therefore, the heat transfer fluid selected will affect the required pumping power and the amount of heat transfer pipe. Furthermore, some local regulations may limit the selection and use of certain antifreeze solutions.

Spiral Loops

A variation on the multiple pipe horizontal-loop configuration is the spiral loop, commonly referred to as the "slinky." The spiral loop, illustrated in Figure 28-2b, consists of pipe unrolled in circular loops in trenches; the horizontal configuration is shown.

Another variation of the spiral-loop system involves placing the loops upright in narrow vertical trenches. The spiral-loop configuration generally requires more piping, typically 500 to 1,000 feet per system cooling ton (43.3 to 86.6 m/kW) but less total trenching than the multiple horizontal-loop systems described above. For the horizontal spiral-loop layout, trenches are generally 3 to 6 feet (0.9 to 1.8 m) wide; multiple trenches are typically spaced about 12 feet (3.7 m) apart. For the vertical spiral-loop layout, trenches are generally 6 inches (15.2 cm) wide; the pipe loops stand vertically in the narrow trenches. In cases where trenching is a large component of the overall installation costs, spiral-loop systems are a means of reducing the installation cost. As noted with horizontal systems, slinky systems are also generally associated with lower-tonnage systems where land area requirements are not a limiting factor.

- **Advantages**: Requires less ground area and less trenching than other horizontal loop designs; installation costs sometimes less than other horizontal loop designs.

- **Disadvantages**: Requires more total pipe length than other ground-coupled designs; relatively large ground area required; ground temperature subject to seasonal variance; larger pumping energy requirements than other horizontal loops defined above; backfilling the trench can be difficult with certain soil types, and the pipe system could be damaged during backfill process.

Vertical Loops

Vertical loops, illustrated in Figure 28-2c, are generally considered when land surface is limited. Wells are bored to depths that typically range from 75 to 300 feet (22.9 to 91.4 m) deep. The closed-loop pipes are inserted into the vertical well. Typical piping requirements range from 200 to 600 feet per system cooling ton (17.4 to 52.2 m/kW), depending on soil and temperature conditions. Multiple wells are typically required with well spacing not less than 15 feet (4.6 m) in the northern climates and not less than 20 feet (6.1 m) in southern climates to achieve the total heat transfer requirements. A 300- to 500-ton capacity system can be installed on one acre of land, depending on soil condi-

tions and ground temperature.

There are three basic types of vertical-system heat exchangers: U-tube, divided-tube, and concentric-tube (pipe-in-pipe) system configurations.

- **Advantages**: Requires less total pipe length than most closed-loop designs; requires the least pumping energy of closed-loop systems; requires least amount of surface ground area; ground temperature typically not subject to seasonal variation.

- **Disadvantage**: Requires drilling equipment; drilling costs frequently higher than horizontal trenching costs; some potential for long-term heat buildup underground with inadequately spaced bore holes.

Submerged Loops

If a moderately sized pond or lake is available, the closed-loop piping system can be submerged, as illustrated in Figure 28-2d. Some companies have installed ponds on facility grounds to act as ground-coupled systems. (Ponds also serve to improve facility aesthetics.) Submerged-loop applications require some special considerations, and it is best to discuss these directly with an engineer experienced in the design applications. This type of system requires adequate surface area and depth to function satisfactorily in response to heating or cooling requirements under local weather conditions. In general, the submerged piping system is installed in loops attached to concrete anchors. Typical installations require around 300 feet of heat transfer piping per system cooling ton (26.0 m/kW) and around 3,000 square feet of pond surface area per ton (79.2 m²/kW), with a recommended minimum one-half acre total surface area. The concrete anchors act to secure the piping, restricting movement, but also hold the piping 9 to 18 inches (22.9 to 45.7 cm) above the pond floor, allowing for good convective flow of water around the heat transfer surface area. It is also recommended that the heat-transfer loops be at least 6 to 8 feet (1.8 to 2.4 m) below the pond surface, preferably deeper. This maintains adequate thermal mass even in times of extended drought or other low-water conditions. Rivers are typically not used, because they are subject to drought and flooding, both of which may damage the system.

- **Advantages**: Can require the least total pipe length of closed-loop designs; can be less expensive than other closed-loop designs if body of water available.

- **Disadvantage**: Requires a large body of water and may restrict lake use (i.e., boat anchors).

Open-Loop Systems

Open-loop systems use local groundwater or surface water (i.e., lakes) as a direct heat transfer medium instead of the heat transfer fluid described for the closed-loop systems. These systems are sometimes referred to specifically as "ground-water-source heat pumps" to distinguish them from other ground-source heat pumps. Open-loop systems consist primarily of extraction wells, extraction and reinjection wells, or surface water systems. These three types are illustrated in Figures 28-2e, 28-2f, and 28-2g, respectively.

A variation on the extraction well system is the standing column well. This system reinjects the majority of the return water back into the source well, minimizing the need for a reinjection well and the amount of surface discharge water.

There are several special factors to consider in open-loop systems. One major factor is water quality. In open-loop systems, the primary heat exchanger between the refrigerant and the groundwater is subject to fouling, corrosion, and blockage. A second major factor is the adequacy of available water. The required flow rate through the primary heat exchanger between the refrigerant and the groundwater is typically between 1.5 and 3.0 gallons per minute per system cooling ton (0.027 and 0.054 L/s-kW). This can add up to a significant amount of water and can be affected by local water resource regulations. A third major factor is what to do with the discharge stream. The groundwater must either be re-injected into the ground by separate wells or discharged to a surface system such as a river or lake. Local codes and regulations may affect the feasibility of open-loop systems.

Depending on the well configuration, open-loop systems can have the highest pumping load requirements of any of the ground-coupled configurations. In ideal conditions, however, an open-loop application can be the most economical type of ground-coupling system.

- **Advantages**: Simple design; lower drilling requirements than closed-loop designs; subject to better thermodynamic performance than closed-loop systems because well(s) are used to deliver groundwater at ground temperature rather than as a heat exchanger delivering heat transfer fluid at temperatures other than ground temperature; typically lowest cost; can be combined with potable water supply well; low operating cost if water already pumped for other purposes, such as irrigation.

- **Disadvantages**: Subject to various local, state, and federal clean water and surface water codes and regulations; large water flow requirements; water availability may be limited or not always available; heat pump heat exchanger subject to suspended matter, corrosive agents, scaling, and bacterial contents; typically subject to highest pumping power requirements; pumping energy may be excessive if the pump is oversized or poorly controlled; may require well permits or be restricted for extraction; water disposal can limit or preclude some installations; high cost if reinjection well required.

Direct-Expansion Systems

Each of the ground-coupling systems described above uses an intermediate heat transfer fluid to transfer heat between the earth and the refrigerant. Use of an intermediate heat transfer fluid necessitates a higher compression ratio in the heat pump to achieve sufficient temperature differences in the heat transfer chain (refrigerant to fluid to earth). Each also requires a pump to circulate water between the heat pump and the ground-couple. Direct expansion systems, illustrated in Figure 28-2h, remove the need for an intermediate heat transfer fluid, the fluid-refrigerant heat exchanger, and the circulation pump. Copper coils are installed underground for a direct exchange of heat between refrigerant and earth. The result is improved heat transfer characteristics and thermodynamic performance.

The coils can be buried either in deep vertical trenches or wide horizontal excavations. Vertical trenches typically require from 100 to 150 square feet of land surface area per system cooling ton (2.6 to 4.0 m^2/kW) and are typically 9 to 12 feet (2.7 to 3.7 m) deep. Horizontal installations typically require from 450 to 550 square feet of land area per system cooling ton (11.9 to 14.5 m^2/kW) and are typically 5 to 10 feet (1.5 to 3.0 m) deep. Vertical trenching is not recommended in sandy, clay, or dry soils because of the poor heat transfer.

Because the ground coil is metal, it is subject to corrosion. (The pH level of the soil should be between 5.5 and 10, although this is normally not a problem.) If the ground is subject to stray electric currents and/or galvanic action, a cathodic protection system may be required. Because the ground is subject to larger temperature extremes from the direct expansion system, there are additional design considerations. In winter heating operation, the lower ground coil temperature may cause the ground moisture to freeze. Expansion of the ice buildup may cause the ground to buckle. Also, because of the freezing potential, the ground coil should not be located near water lines. In the summer cooling operation, the higher coil temperatures may drive moisture from the soil. Low moisture content will change soil heat transfer characteristics.

At the time this chapter was initially drafted (1995), only one U.S. manufacturer offered direct expansion ground-source heat pump systems. However, new companies have released similar direct expansion systems. In November 2005, the Geothermal Heat Pump Consortium web site identified four manufacturers of direct exchange systems. Systems were available from 16,000 to 83,000 Btu/h (heating/cooling capacity) (4.7 to 24.3 kW). Larger commercial applications would require multiple units with individual ground coils.

- **Advantages**: Higher system efficiency; no circulation pump required.

- **Disadvantages**: Large trenching requirements for effective heat transfer area; ground around the coil subject to freezing (may cause surface ground to buckle and can freeze nearby water pipes); copper coil should not be buried near large trees where root system may damage the coil; compressor oil return can be complicated, particularly for vertical heat exchanger coils or when used for both heating and cooling; leaks can be catastrophic; higher skilled installation required; installation costs typically higher; this system type requires more refrigerant than most other systems; smaller infrastructure in the industry.

28.4.4 Variables Affecting Design and Performance

Among the variables that have a major impact on the sizing and effectiveness of a ground-coupling system, the importance of underground soil temperatures and soil type deserve special mention.

Underground Soil Temperature

The soil temperature is of major importance in the design and operation of a ground-source heat pump. In an open-loop system, the temperature of groundwater entering the heat pump has a direct impact on the efficiency of the system. In a closed-loop system and in the direct expansion system, the underground temperature will affect the size of the required ground-coupling system and the resulting operational effectiveness of the underground heat exchanger. Therefore, it is important to determine the underground soil temperature before selecting a system design.

Annual air temperatures, moisture content, soil type, and ground cover all have an impact on underground soil temperature. In addition, underground temperature varies annually as a function of the ambient surface air temperature swing, soil type, depth, and time lag. Figure 28-3 contains a map of the United States indicating mean annual underground soil temperatures and amplitudes of annual surface ground temperature swings. Figure 28-4, though for a specific location, illustrates how the annual soil temperature varies with depth, soil type, and season. For vertical ground-loop systems, the mean annual earth temperature (Figure 28-3a) is an important factor in the ground-loop design. With horizontal ground-loop systems, the ground surface annual temperature variation (Figure 28-3b and Figure 28-4b) becomes an important design consideration.

Soil and Rock Classification

The most important factor in the design and successful operation of a closed-loop ground-source heat pump system is the rate of heat transfer between the closed-loop ground-coupling system and the surrounding soil and rock. The thermal conductivity of the soil and rock is the critical value that determines the length of pipe required. The pipe length, in turn, affects the installation cost as well as the operational effectiveness, which in turn affects the operating cost. Because of local variations in soil type and moisture conditions, economic designs may vary by location. Soil classifications include coarse-grained sands and gravels, fine-grained silts and clays, and loam (equal mixtures of sand, silt, and clay). Rock classifications are broken down into nine different petrologic groups. Thermal conductivity values vary significantly within each of the nine groups. Each of these classifications plays a role in determining the thermal conductivity and thereby affects the design of the ground-coupling system. For more information on the thermal properties of soils and rocks and how to identify the different types of soils and rocks, see Soil and Rock Classification for the Design of Ground-Coupled Heat Pump Systems (STS Consultants 1989).

Series versus Parallel Flow

Closed-loop ground-coupled heat exchangers may be designed in series, parallel, or a combination of both. In series systems, the heat transfer fluid can take only one path through the loop, whereas in parallel systems the fluid can take two or more paths through the circuit. The selection will affect performance, pumping requirements, and cost. Small-scale ground-coupling systems can use either series or parallel-flow design, but most large ground-coupling systems use parallel-flow systems. The advantages and disadvantages of series and parallel systems are summarized below. In large systems, pressure drop and pumping costs need to be carefully considered, or they will be very high. Variable-speed drives can be used to reduce pumping energy and costs during part-load conditions. Total life-cycle cost

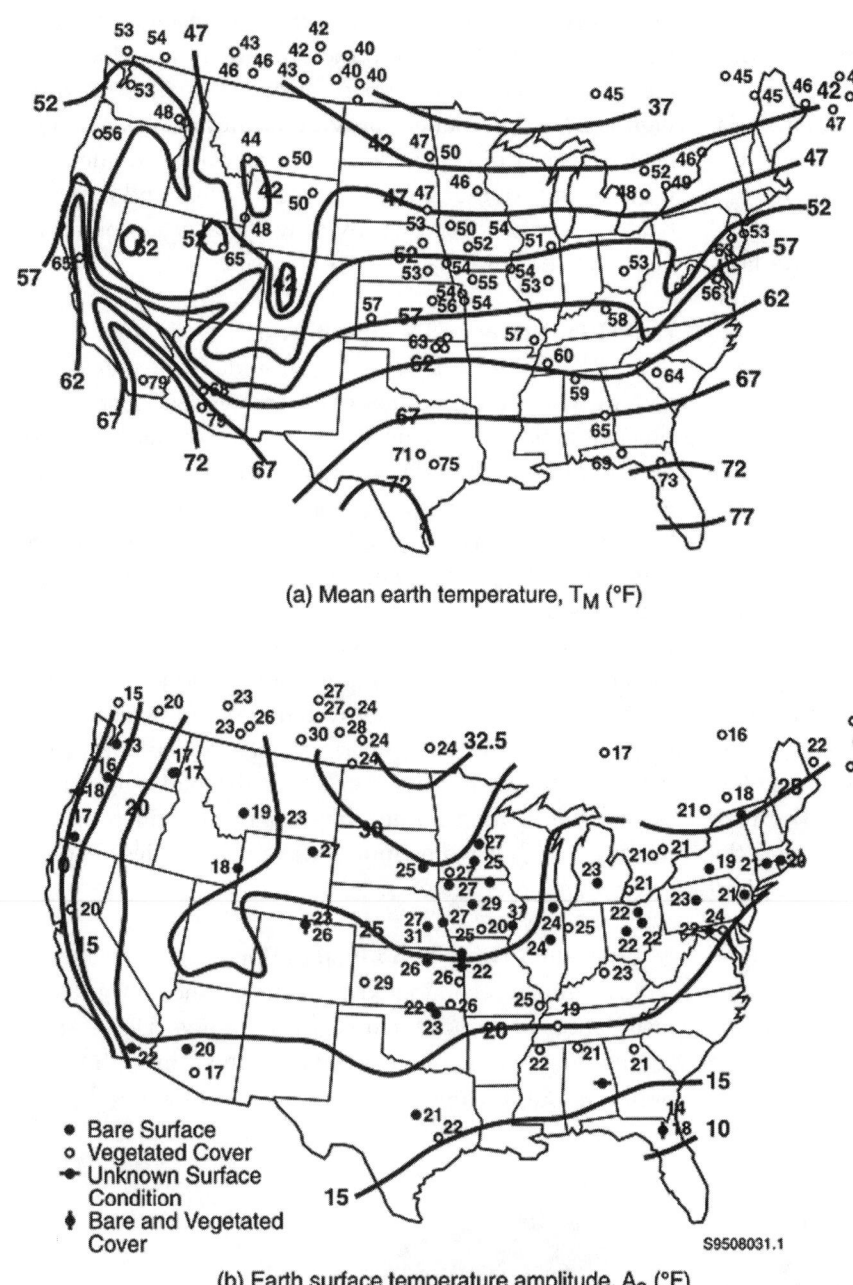

(a) Mean earth temperature, T_M (°F)

- Bare Surface
- Vegetated Cover
- Unknown Surface Condition
- Bare and Vegetated Cover

S9508031.1

(b) Earth surface temperature amplitude, A_s (°F)

Figure 28-3. Mean annual soil temperatures. Source: OSU (1988)

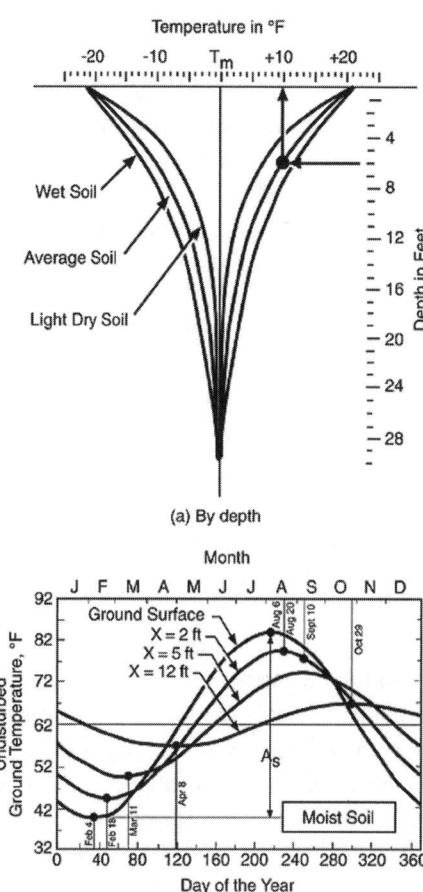

(a) By depth

(b) Annual

S9508031.2

Figure 28-4. Soil temperature variation Source: OSU (1988)

volumes; higher pipe cost per unit of performance; increased installed labor cost; limited capacity (length) caused by fluid pressure drop characteristics; larger pressure drop resulting in larger pumping load; requires larger purge system to remove air from the piping network.

- **Parallel-System Advantages**: Smaller pipe diameter has lower unit cost; lower volume requires less antifreeze; smaller pressure drop resulting in smaller pump-ing load; lower installation labor cost.

- **Parallel-System Disadvantages**: Special attention required to ensure air removal and flow balancing between each parallel path to result in equal length loops.

28.4.5 Variations

The ground-coupling system is what makes the ground-source heat pump unique among heating and

and design limitations should be used to design a specific system.

- **Series-System Advantages**: Single path flow and pipe size; easier air removal from the system; slightly higher thermal performance per linear foot of pipe because larger pipe size required in the series system.

- **Series-System Disadvantages**: Larger fluid volume of larger pipe in series requires greater antifreeze air-conditioning systems and, as described above, there are several types of ground-coupling systems. In addition, variations to ground-source heat pump design and installation can save additional energy or reduce installation costs. Three notable variations are described below.

Cooling-Tower-Supplemented System

The ground-coupling system is typically the largest component of the total installation cost of a ground-source heat pump. In southern climates or in thermally heavy commercial applications where the cooling load is the driving design factor, supplementing the system with a cooling tower or other supplemental heat rejection system can reduce the required size of a closed-loop ground-coupling system. The supplemental heat rejection system is installed in the loop by means of a heat exchanger (typically a plate and frame heat exchanger) between the facility load and the ground couple. A cooling tower system is illustrated in Figure 28-5. The cooling tower acts to precool the loop's heat transfer fluid upstream of the ground couple, which lowers the cooling-load requirement on the ground-coupling system. By significantly reducing the required size of the ground-coupling system, using a cooling tower can lower the overall installation cost. This type of system is operating successfully in several commercial facilities, including some mission-critical facilities at Fort Polk in Louisiana.

Solar-Assisted System

In northern climates where the heating load is the driving design factor, supplementing the system with solar heat can reduce the required size of a closed-loop ground-coupling system. Solar panels, designed to heat water, can be installed into the ground-coupled loop (by means of a heat exchanger or directly), as illustrated in Figure 28-6. The panels provide additional heat to the heat transfer fluid. This type of variation can reduce the required size of the ground-coupled system and increase heat pump efficiency by providing a higher temperature heat transfer fluid.

Hot Water Recovery/Desuperheating

The use of heat pumps to provide hot water is becoming common. Because of their high efficiency, this practice makes economic sense. Most manufacturers offer an option to include desuperheating heat exchangers to provide hot water from a heat pump. These dual-wall heat exchangers are installed in the refrigerant loop to recover high temperature heat from the superheated refrigerant gas. Hot-water recovery systems can supplement, or sometimes replace, conventional facility water-heating systems. With the heat pump in cooling mode, hot-water recovery systems increase system operating efficiency while acting as a waste-heat-recovery device—and provide essentially free hot water. When the load is increased during the heating mode, the heat pump still provides heating and hot water more efficiently and less expensively than other systems.

28.4.6 System Design and Installation

More is becoming known about the design and installation of ground-source heat pumps. Design-day cooling and heating loads are determined through traditional design practices such as those documented by the

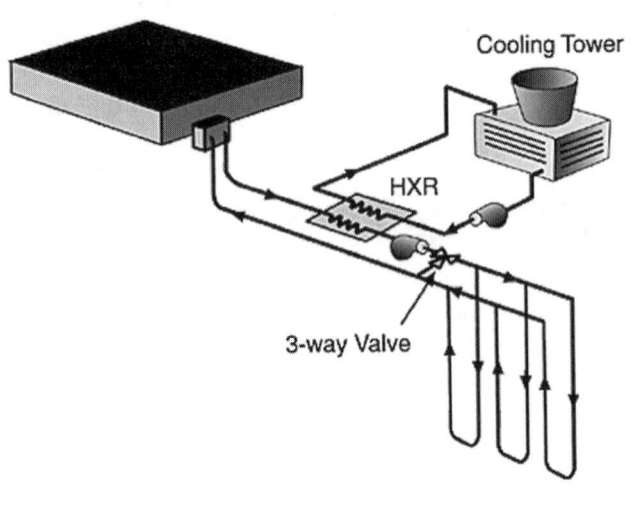

S9506033.2

Figure 28-5. Cooling-tower-supplemented system for cooling-dominated loads

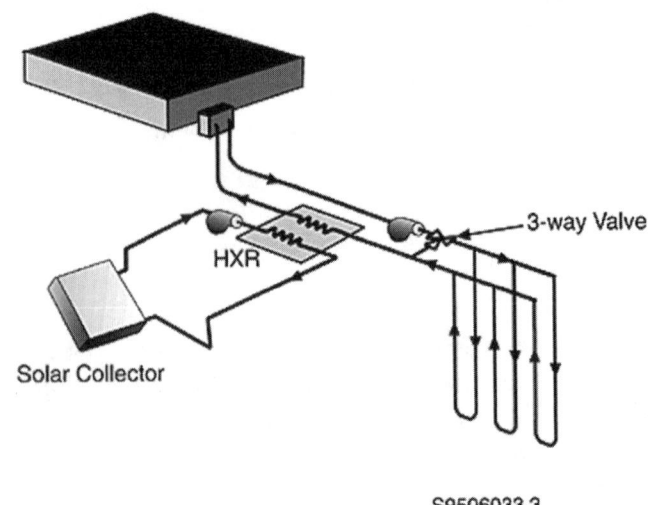

S9506033.3

Figure 28-6. Solar-assisted system for heating-dominated loads

American Society of Heating, Refrigerating, and Air-Conditioning Engineers (ASHRAE). Systems are also zoned using commonly accepted design practices.

The key issue that makes ground-source heat pumps unique is the design of the ground-coupling system. Most operational problems with ground-source heat pumps stem from the performance of the ground-coupling system. Today, software tools are available to support the design of the ground-coupling systems that meet the needs of designers and installers. These tools are available from several sources, including the International Ground-Source Heat Pump Association (IGSHPA). In addition, several manufacturers have designed their own proprietary tools more closely tuned to their particular system requirements.

Ground loops can be placed just about anywhere—under landscaping, parking lots, or ponds. Selection of a particular ground-coupling system (vertical, horizontal, spiral, etc.) should be based on life-cycle cost of the entire system, in addition to practical constraints. Horizontal closed-loop ground-coupling systems can be installed using a chain-type trenching machine, horizontal boring machine, backhoe, bulldozer, or other earth-moving heavy equipment. Vertical applications (for both open and closed systems) require a drilling rig and qualified operators. Most applications of ground-source heat pumps to large facilities use vertical closed-loop ground-coupling systems primarily because of land constraints. Submerged-loop applications require some special considerations and, as noted earlier, it is best to discuss these directly with an experienced design engineer.

It is important to assign overall responsibility for the entire ground-source heat pump system to a single individual or contractor. Installation of the system, however, will involve several trades and contractors, many of whom may not have worked together in previous efforts. In addition to refrigeration/air-conditioning and sheet metal contractors, installation involves plumbers and (in the case of vertical systems) well drillers. Designating a singular responsible party and coordinating activities will significantly reduce the potential for problems with installation, startup, and proper operation.

In heating-dominated climates, a mixture of antifreeze and water must be used in the ground-coupling loops if loop temperatures are expected to fall below about 41°F (5°C). A study by Heinonen (1997) establishes the important considerations for antifreeze solutions for ground-source heat pump systems and provides guidance on selection.

One note of caution to the designer: some regulations, installation manuals, and/or local practices call for partial or full grouting of the borehole. The thermal conductivity of materials normally used for grouting is very low compared with the thermal conductivity of most native soil formations. Thus, grouting tends to act as insulation and hinders heat transfer to the ground. Some experimental work by Spilker (1998) has confirmed the negative impact of grout on borehole heat transfer. Under heat rejection loading, average water temperature was nearly 11°F (6°C) higher for a 6.5-in. (16.5-cm) diameter borehole backfilled with standard bentonite grout than for a 4.75-in. (12.1-cm) diameter borehole backfilled with thermally enhanced bentonite grout. Using fine sand as backfill in a 6.5-in. (16.5-cm) diameter borehole lowered the average water temperature over 14°F (8°C) compared with the same diameter bore backfilled with standard bentonite grout. For a typical system (Spilker 1998) with a 6.5-in. (16.5-cm) diameter borehole, the use of standard bentonite grout would increase the required bore length by 49% over fine sand backfill in the same borehole. By using thermally enhanced grout in a smaller 4.75-in. (12.1-cm) borehole, the bore length is increased by only 10% over fine sand backfill in the larger 6.5-in. (16.5-cm) diameter borehole. Thus, the results of this study (Spilker 1998) suggest three steps that may be taken to reduce the impact of grout on vertical borehole system performance:

- Reduce the amount of grout used to the bare minimum. Sand or cuttings may be used where allowed, but take care to ensure that the entire interstitial space between the piping and the borehole diameter is filled.

- Use thermally enhanced grout wherever possible. For information on thermally enhanced grout, consult ASHRAE (1997) and Spilker (1998).

- Reduce the borehole diameter as much as possible to mitigate the effects of the grout or backfill used. The regulatory requirements for vertical boreholes used for ground-coupling heat exchangers vary widely by state. Current state and federal regulations, as well as related building codes, are summarized at the Geothermal Heat Pump Consortium web site (www.geoexchange.org/publications/regs.htm).

28.4.7 Summary of Ground-loop Design Software

Because of the diversity in loads in multi-zone buildings, the design of the ground-coupling heat exchanger (the ground loop) must be based on peak block load rather than the installed capacity. This is of paramount importance, because ground coupling is usually a major portion of the total ground-source heat pump system cost, and over-sizing will render a project economically unattractive.

In the residential sector, many systems have been designed using rules-of-thumb and local experience, but for commercial-scale systems such practices are ill advised. For all but the most northern climates, commercial-scale buildings will have significantly more heat rejection than extraction. This imbalance in heat rejection/extraction can cause heat buildup in the ground to the point where heat pump performance is adversely affected and hence system efficiency and possibly occupant comfort suffer. (This is an important consideration in producintg accurate life-cycle cost estimates of energy use.) Proper design for commercial-scale systems almost always benefits from the use of design software. Software for commercial-scale ground-source heat pump system design should consider the interaction of adjacent loops and predict the potential for long-term heat buildup in the soil. The heat rejection/extraction imbalance and the long term soil temperature effect (and efficiency impact) can be slowed with bore hole spacing or increased loop field size. To achieve long term soil temperature equilibrium there would need to be equal seasonal deposit and extraction of heat. Minor differences in the annual heat accounting are inconsequential, but where the imbalance is significant and its long term effect are unacceptable, intervention strategies can be employed. For example, where heat rejection dominates and ground temperature is expected to increase unacceptably, the design could include a fluid cooler in the sealed loop to reduce the heat rejection to the soil in cooling season. Some sources of PC-based design software packages that address this need are:

*   **GchpCalc**, Version 3.1, Energy Information Services, Tel: (205) 799-4591. This program includes built-in tables for heat pump equipment from most manufacturers. Input is in the form of heat loss/gain during a design day and the approximate equivalent full-load heating hours and equivalent full-load cooling hours. Primary output from the program is the ground loop length required. This program will also calculate the optimal size for a supplemental fluid cooler for hybrid systems, as discussed later.

*   **GLHEPRO**, International Ground Source Heat Pump Association (IGSHPA), Tel: (800) 626-4747. Input required is monthly heating/cooling loads on heat pumps and monthly peak loads either entered directly by user or read from BLAST or Trane System Analyzer and Trane Trace output files. Output includes long-term soil temperature effect from rejection/extraction imbalance. The current

configuration of the program has some constraints on selection of borehole spacing, depth, and overall layout that will be removed from a future version now being prepared.

*   **GS2000**, Version 2.0, Caneta Research Inc., Tel: (905) 542-2890, email: caneta@compuserve.com. Heating/cooling loads are input as monthly totals on heat pumps or, alternatively, monthly loads on the ground loop may be input. Equipment performance is input at ARI/ISO rating conditions. For operating conditions other than the rating conditions, the equipment performance is adjusted based on generic heat pump performance relationships.

Each of these programs requires input about the soil thermal properties, borehole resistance, type of piping and borehole arrangement, fluid to be used, and other design parameters. Many of the required inputs will be available from tables of default values. The designer should be careful to ensure that the values chosen are representative of the actual conditions to be encountered to ensure efficient and cost-effective designs. Test borings and *in situ* thermal conductivity analysis to determine the type of soil formations and aquifer locations will substantially improve design accuracy and may help reduce costs. Even with the information from test borings, some uncertainty will remain with respect to the soil thermal properties. These programs make it possible to vary design parameters easily within the range of anticipated values and determine the sensitivity of the design to a particular parameter (OTL 1999). In some instances, particularly very large projects, it may be advisable to obtain specific information on ground-loop performance by thermal testing of a sample borehole (Shonder and Beck 2000).

28.5 APPLICATION

This section addresses technical aspects of applying ground-source heat pumps. The range of applications and climates in which the technology has been installed are discussed. The advantages, limitations, and benefits are enumerated. Design and integration considerations for ground-source heat pumps are highlighted, including energy savings estimates, equipment warranties, relevant codes and standards, equipment and installation costs, and utility incentives.

28.5.1 Application Screening

A ground-source heat pump system is one of the most efficient technologies available for heating and cooling. It can be applied in virtually any climate or building category. Although local site conditions may dictate the type of ground-coupling system employed, the high first cost and its impact on the overall life-cycle cost are typically the constraining factors.

The operating efficiency of ground-source heat pumps is very dependent on the entering water temperature, which, in turn, depends on ground temperature, system load, and size of ground loop. As with any HVAC system, the system load is a function of the facility, internal activities, and the local weather. Furthermore, with ground-source heat pumps, the load on the ground-coupling system may impact the underground temperature. Therefore, energy consumption will be closely tied to the relationship between the annual load distribution and the annual ground loop-temperature distribution (e.g., their joint frequency distribution).

There are several techniques to estimate the annual energy consumption of ground-source heat pump systems. The most accurate methodologies use computer simulation, and several software systems now support the analysis of ground-source heat pumps. These methods, while more accurate than hand techniques, are also difficult and expensive to employ and are therefore more appropriate when additional detail is required rather than as an initial screening tool.

The bin method is another analytical tool for screening technology applications. In general, a bin method is a simple computational procedure that is readily adaptable to a spreadsheet-type analysis and can be used to estimate the energy consumption of a given application and climate. Bin methods rely on load and ambient wet and dry bulb temperature distributions. This methodology is used in the case study presented later in this chapter.

28.5.2 Where to Apply Ground-source Heat Pumps

Ground-source heat pumps are generally applied to air-conditioning and heating systems but may also be used in any refrigeration application. The decision whether to use a ground-source heat pump system is driven primarily by economics. Almost any HVAC system can be designed using a ground-source heat pump. The primary technical limitation is a suitable location for the ground-coupling system. The following list identifies some of the best applications of ground-source heat pumps.

- Ground-source heat pumps are probably least cost-prohibitive in new construction; the technology is relatively easy to incorporate.

- Ground-source heat pumps can also be cost effective to replace an existing system at the end of its useful life, or as a retrofit, particularly if existing ductwork can be reused with minimal modification.

- In climates with either cold winters or hot summers, ground-source heat pumps can operate much more efficiently than air-source heat pumps or other air-conditioning systems. Ground-source heat pumps are also considerably more efficient than other electric heating systems and, depending on the heating fuel cost, may be less expensive to operate than other heating systems.

- In climates with high daily temperature swings, ground-source heat pumps show superior efficiency. In addition, in climates characterized by large daily temperature swings, the ground-coupling system also offers some thermal storage capability, which may benefit the operational coefficient of performance.

- In areas where natural gas is not available or where the cost of natural gas or other fuel is high compared with electricity, ground-source heat pumps are economical. They operate with a heating coefficient of performance in the range of 3.0 to 4.5, compared with conventional heating efficiencies in the range of 80% to 97%. Therefore, when the cost of electricity (per Btu) is less than 3.5 times that of conventional heating fuels (per Btu), ground-source heat pumps have lower energy costs.

- Areas of high natural gas (or fuel oil) costs will favor ground-source heat pumps over conventional gas (or fuel oil) heating systems. High electricity costs will favor ground-source heat pumps over air-source heat pumps.

- In facilities where multiple temperature control zones or individual load control is beneficial, ground-source heat pumps provide tremendous capability for individual zone temperature control, because they are primarily designed using multiple unitary systems.

- In areas where drilling costs are low, vertical-loop systems may be especially attractive. The initial cost of the ground-source heat pump system is one of the prime barriers to the economics. In locations with a significant ground-source heat pump industry infrastructure (such as Oklahoma, Louisiana, Florida, Texas, and Indiana), installation costs may be lower and the contractors more experienced. This, however, is changing as the market for ground-source heat pumps grows.

28.5.3 What to Avoid

The following precautions should be followed when the application of ground-source heat pump technology is considered:

- Avoid threaded plastic pipe connections in the ground loop. Specify thermal fusion welding. Unlike conventional water-source heat pump systems where the water loop temperature ranges from 60° to 90°F (15.6° to 32.2°C), ground-source systems are subject to wider temperature ranges (20° to 110°F [-6.7° to 43.3°C]), and the resulting expansion and contraction may result in leaks at the threaded connection. It is also generally recommended to specify piping and joining methods approved by International Ground-Source Heat Pump Association (IGSHPA).

- Check local water and well regulations. Regulations affecting open-loop systems are common, and local regulations can vary significantly. Some local regulations may require reinjection wells rather than surface drainage. Some states require permits to use even private ponds as a heat source/sink.

- Have the ground-source heat pump system installed as a complete and balanced assemblage of components, each of which must be properly designed, sized, and installed (Giddings 1988). Also, have the system installed under the responsibility of a single party. If the entire system is installed by three different professionals, none of whom understands or appreciates the other two parts of the system, then the system may not perform satisfactorily.

- One of the most frequent problems cited is improper sizing of the heat pump or the ground-coupling system. Approved calculation procedures should be used in the sizing process—as is the case with any heating or air-conditioning system regardless of technology. ASHRAE has established one of the most widely known and accepted standards for the determination of design heating and cooling loads. Sizing the ground-coupling system is just as critical. Because of the uncertainty of soil conditions, a site analysis to determine the thermal conductivity and other heat transfer properties of the local soil may be required. This should be the responsibility

of the designing contractor, because it can significantly affect the final design.

- Avoid inexperience; check on the previous experience of potential designers and installers. (See above.) It is also generally recommended to specify IGSHPA certified designers and installers.

28.5.4 Design and Equipment Integration

The purpose of this chapter is to familiarize the energy manager and facility engineer with the benefits and liabilities of ground-source heat pumps in their application to commercial buildings. It is beyond the scope of this chapter to fully explain the design requirements of a ground-source heat pump system. It is, however, important that the reader know the basic steps in the design process.

The design of a ground-source heat pump system will generally follow the following sequence:

1. Determine local design conditions, including climatic and soil thermal characteristics.

2. Determine local water, well, and grouting requirements.

3. Determine building heating and cooling loads at design conditions.

4. Select the alternative HVAC system components, including the indoor air-distribution system type; size the alternatives as required; and select equipment that will meet the demands calculated in Step 2 (using the preliminary estimate of the entering water temperatures to determine the heat pump's heating and cooling capacities and efficiencies).

5. Determine the monthly and annual building heating and cooling energy requirements.

6. Make preliminary selection of a ground-coupling system type.

7. Determine a preliminary design of the ground-coupling system. This often includes soil testing.

8. Determine the thermal resistance of the ground-coupling system.

9. Determine the required length of the ground-coupling system; recalculate the entering and exiting water temperatures on the basis of system loads and the ground-coupling system design.

10. Redesign the ground-coupling system as required to balance the requirements of the system load (heating and cooling) with the effectiveness of the ground-coupling system. Note that designing and sizing the ground-coupling system for one season (such as cooling) will impact its effectiveness and ability to meet system load requirements during the other season (such as heating).

11. Depending upon whether cooling or heating demand is greatest, the heat pump selected may be oversized for the other season. Over sizing is usually problematic in HVAC design. Capacity modulation, such as a two-speed compressor, may be advisable.

12. Perform life-cycle cost analysis on the system design (or system design alternatives).

Although the design procedure for the ground-coupling system is an iterative and sometimes difficult process, several sources are available to simplify the task. First, an experienced designer should be assigned responsibility for the heat pump and ground-coupling system designs. Several manufacturers of ground-source heat pump equipment have their own software tools to support the design of large, commercial-type systems. However, for those who typically design systems in-house, there are support tools available. Software programs are available to support the design of ground-source heat pump HVAC systems and the ground-coupling system. Several software tools are available through the IGSHPA, including an Earth-Coupled Analysis Program and a Ground-Loop Heat Exchanger Design Program. In addition, several technical design manuals also are available through IGSHPA, ASHRAE, and equipment manufacturers. (Refer to earlier section for an introduction to ground-loop design software.)

There are several different approaches for incorporating ground-source heat pumps into the HVAC design. However, most applications in large facilities involve multiple smaller heat pump units (<20 ton) applied in a modular zone control system and connected to a common water loop and associated ground-coupling system. Although some agencies are experimenting with larger equipment sizes, most manufacturers are supporting the development of efficient smaller systems (1/2 ton to 15 tons [1.8 to 52.8 kW]).

Equipment efficiencies will vary. Energy consumption of the GSHP system will be proportional to the COP (heating) and EER (cooling). Care must be taken when specifying or comparing equipment efficiencies, because the entering water temperature impacts the efficiency directly. For example, a high efficiency heat pump system (combination of compressor and heat exchangers) may be specified:

- Efficiency in heating mode: COP 4.0 with 32°F entering water temperature
- Efficiency in cooling mode: EER-17 with 77°F entering water temperature

Note that the loop temperature is warmer in the cooling mode and cooler in the heating mode. This is to be expected because the function of the heat exchanger between the refrigerant and the ground loop change from a condensing coil during the cooling mode to an evaporating coil during the heating mode. Thus, heat is released to or absorbed from the soil, depending on the season. In real operating conditions, however, some parts of the building may be cooling (such as the building core) while other parts are heating (such as the outer perimeter). Accuracy of seasonal energy use, and thus system payback calculation, are dependent upon the accuracy of predicted loop temperature and associated heat pump power requirement.

28.5.5 Equipment Warranties

The prospective user should ask potential suppliers, contractors, and installers about equipment warranties. The heat pump equipment is typically guaranteed free from manufacturer defects from 1 to 5 years. Some manufacturers offer extended warranties up to 10 years. Residential applications have been found to have longer warranties than commercial applications.

Warranties should also be requested for the ground-coupling system, which is less common. Some installers and pipe manufacturers have offered limited ground-coupling system warranties as long as 50 years. Quality control in the installation of the ground-coupling system has been a concern in the industry. The IGSHPA now offers training and certification programs for installers and designers. In addition, the IGSHPA also administers a registration program for those organizations in the industry. These services have gone far to improve quality control and customer satisfaction.

The track records reported in the literature for the ground-coupling systems are good. In cases where pipe joints were thermally welded (fused) and followed the standards recommended by IGSHPA, systems have proven reliable and resistant to system leaks.

28.5.6 Energy Codes and Standards

Applications of ground-source heat pumps are subject to building and facility energy codes and standards.

In addition, the equipment used is subject to commercial equipment energy codes and standards. Most energy regulations that impact commercial buildings derive from ASHRAE standards, specifically ASHRAE Standard 90.1. Minimum equipment efficiency standards, as identified in ASHRAE Standard 90.1-2004 for commercial equipment relative to ground-source heat pumps, are shown in Table 28-1.

Although codes and standards identify minimum efficiencies, such as those identified above, they do not fully communicate the energy efficiency that is achieved by today's heat pumps. A review of manufacturers' literature on commercially available equipment indicates that cooling efficiencies[5] (EERs) of 13.4 to 20 Btu/W-h and heating efficiencies (COPs) of 3.1 to 4.3 are readily available.[6] When comparing equipment efficiencies, it is important to make an appropriate comparison. The efficiency of any heating or cooling equipment varies with application, load, and related heat-source and heat-sink temperatures.

As partially illustrated in Table 28-1, ISO Standard 13256-1 provides separate equipment rating conditions for water-source heat pumps, ground-water-source heat pumps, and closed-loop ground-source heat pumps. The difference among these rating conditions is in the application, not necessarily the equipment. Therefore, a given heat pump rated under closed-loop ground-source heat pump conditions would have a different EER (cooling) and COP (heating) than if the same heat pump were rated under water-source heat pump rating conditions, and different still if the equipment were rated under the ground-water-source heat pump rating conditions.

[5]Rating based on ANSI/ARI/ASHRAE ISO Standard 13256-1-2005 for closed-loop ground-source heat pumps.
[6]For more information on the various terms used to define efficiency in HVAC systems (see Appendix).

Also, standard ratings are for specific temperatures and operating conditions. Standard ratings do not necessarily reflect the efficiency of systems under true seasonal operating conditions. In determining the efficiency of heating or cooling equipment for estimating energy consumption or potential energy savings, the efficiency should be adjusted for the appropriate operating conditions.

28.5.7 Utility Incentives and Support

Many utilities are promoters of ground-source heat pumps, and many offer incentive programs and support. The Geothermal Heat Pump Consortium web site (www. geoexchange.org/incentives/incentives.htm) identifies federal and state incentives for ground-source heat pump systems for both residential and commercial applications. Incentives identified include rebates (by ton, by unit, and/or by kWh energy saved), low interest loans, sales tax exemptions, tax credits, and/or tax deductions. Rebates for ground-source heat pumps identified ranged from $150 to $600 per ton. For commercial facilities, some incentives ran into several thousand dollars. In addition to the technology-specific rebate programs, some utilities also offer custom rebate programs or programs that are not technology-specific but are based on the energy savings regardless of the technology employed. Readers are encouraged to contact their local utility to find out more about what programs, services, and rebates may be offered by the utility to promote energy management and new energy efficient technologies.

28.6 TECHNOLOGY PERFORMANCE

In 1999, an estimated 400,000 ground-source heat pumps were operating in residential and commercial applications, up from 100,000 in 1990. With a projected

Table 28-1. Minimum Commercial Equipment Efficiency Ratings Standards (ASHRAE 90.1-2004)

Technology Application	Category (capacity)	Rating Condition* (entering water temperature)	Minimum Performance (ASHRAE 90.1-2004)	
			Cooling	Heating
Water-source heat pumps	< 17 kBtuh	86°F (30°C)	11.2 EER	- - -
		68°F (20°C)	- - -	4.2 COP
	≥ 17 kBtuh and < 65 kBtuh	86°F (30°C)	12.0 EER	- - -
		68°F (20°C)	- - -	4.2 COP
	≥ 65 kBtuh and < 135 kBtuh	86°F (30°C)	12.0 EER	- - -
		68°F (20°C)	- - -	4.2 COP
Groundwater-source heat pumps	< 135 kBtuh	59°F (15°C)	16.2 EER	- - -
		50°F (10°C)	- - -	3.6 COP
Ground-source heat pumps	< 135 kBtuh	77°F (25°C)	13.4 EER	- - -
		32°F (0°C)	- - -	3.1 COP

*Reference Standard: ANSI/ARI/ASHRAE ISO Standard 13256-1: Water-source Heat Pumps – Testing and Rating for Performance – Part 1: Water-to-Air and Brine-to-Air Heat Pumps

annual growth rate of 10%, 120,000 new units would be installed in 2010, for a total of 1.5 million units in 2010 (Lund and Boyd 2000).

28.6.2 Energy Savings

The most important reason to consider the application of ground-source heat pumps to commercial building is the potential energy savings and its impact on overall life-cycle cost of the heating and cooling system. Ground-source heat pumps save energy and money, because the equipment operates more efficiently than conventional systems, the maintenance costs are lower (see next section for maintenance benefits), and the equipment has a longer life expectancy than conventional unitary equipment. In addition, a ground-source heat pump does not require a defrost cycle, or (in most situations) backup electric resistance heat, as do air-source heat pumps.

By comparison, the average cooling efficiency at commercial-type facilities is estimated to be an EER of 8.0 (2.33 COP) for existing facilities and an EER of 10.0 (2.93 COP) for new facilities. Ground-source heat pump systems have the potential to reduce consumption of cooling energy by 30% to 50% and to reduce heating energy by 20% to 40%, compared with typical air-source heat pumps.

A review of manufacture's literature on commercially available systems indicates that cooling efficiencies (EERs) of 13.4 to 20 Btu/W-h and heating efficiencies (COPs) of 3.1 to 4.3 are readily available.[7] A study prepared by DOE estimates energy-saving ranges of 17% to 42% comparing ground-source heat pumps to air-source heat pumps, depending on region (Calm 1987). Figure 28-7 illustrates the range of efficiencies typical of various heating and cooling equipment.

28.6.3 Maintenance

The ground-source heat pump technology is mature and reliable. Systems have standard warranties ranging from 1 to 5 years. The heat pump units are self-contained, and maintenance requirements are relatively straightforward; no new maintenance skills are necessary. Because heat pump equipment is not exposed to outdoor elements, the units actually require less maintenance than typical air-source heat pumps. One site reported a problem with cottonwood trees clogging outside condenser units of the previous air-conditioning systems. This maintenance problem was eliminated with the application of ground-source heat pumps.

---

[7]Rating based on ANSI/ARI/ASHRAE ISO Standard 13256-1-2005 for closed-loop ground-source heat pumps.

In closed-loop systems, the ground loop is virtually maintenance free. The circulating pump(s) requires routine maintenance, as with any pump and motor system, and the water loop (a closed system) should be routinely monitored for temperature, pressure, flow, and antifreeze concentration. Unless there is a leak, no action is generally required.

In open-loop systems, the well requires maintenance similar to any water well. The system should be routinely monitored for temperature, pressure, and flow. Because groundwater is being supplied to the heat pump, the heat exchangers should be routinely inspected for potential fouling and scale buildup.

Maintenance costs for ground-source heat pumps are about the same or less than for conventional equipment. For example, maintenance costs for four schools in Lincoln, Nebraska, equipped with ground-source heat pump systems, have been well documented (Martin et al. 1999, 2000). The results of these studies are summarized in Table 28-2.

Table 28-2 Maintenance Costs for Public Schools in Lincoln, Nebraska

HVAC System Type	Preventive Maintenance (cents/yr-ft^2)	Repair (cents/hr-ft^2)	Total (cents/hr-ft^2)
GSHP System	7.1	2.1	9.2
Conventional system	5.9 to 12.6	2.9 to 6.1	8.8 to 18.7

Source: Martin et al. 1999, 2000

Another analysis, this one by the Geothermal Heat Pump Consortium, found that the average total preventive and corrective maintenance costs for 25 ground-source-heat-pump-equipped buildings were approximately 11 cents/ft^2, compared to 30 to 40 cents/ft^2 for conventional systems (Cane 1998).

28.6.4 Installation Costs

Application-specific parameters, such as equipment capacity, type, refrigerant, air-distribution system, control system, plumbing configuration, and ground-coupling system type, significantly affect the total cost of the overall system. While equipment costs are competitive, installation costs vary significantly.

To illustrate the potential range of installation costs, the following examples have been reported.

- Stockton State College, in Pomona, New Jersey, retrofit a system that totaled 1,655 tons (5,826 kW) at a total cost of $5,246,000 (Gahran 1993). The facility re-

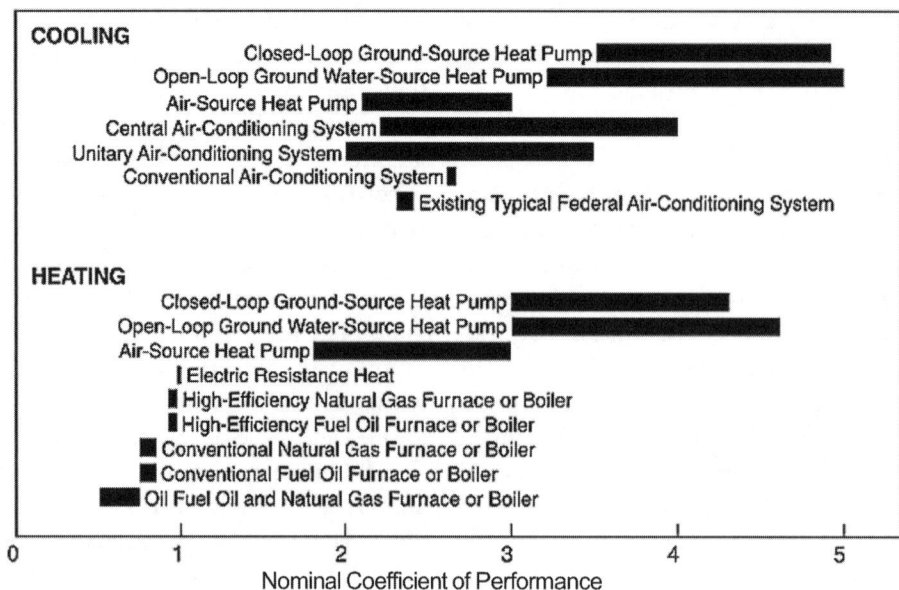

Figure 28-7. Typical heating and cooling equipment efficiencies

ceived grants and rebates reducing the capital outlay to $135,000. The system included 63 rooftop units, 500 variable-air-volume (VAV) boxes, and a 3,500- point energy management control system (EMCS). The unit cost was $3,170/ton in 1993 dollars.

• WaterFurnace, a ground-source heat pump manufacturer, designed the technology using a submerged pond closed-loop system into its new office building located in Fort Wayne, Indiana. The system totaled 134 tons (471.7 kW) at a cost of $239,800 (WaterFurnace WF639). The unit cost was $1,790/ton in 1991 dollars.

• Salem Community College, in Carney's Point, New Jersey, retrofit a system that totaled 160 tons at a total cost of $284,000 (Gahran 1994). The system included 32 heat pumps. The unit cost was $1,775/ton in 1993 dollars.

• Paint Lick Elementary School, in Garrard County, Kentucky, designed the technology into new school building. The system totaled 123 tons (433 kW) at a total cost of $380,010 (WaterFurnace WF666). The unit cost was $3,090/ton in 1992 dollars.

• Maywood Elementary School, in Hammond, Indiana, designed the technology into a new school building. The unit totaled 250 tons (880 kW) at a total cost of $1,277,190 (WaterFurnace WF925). The system consisted of 74 heat pumps and a ground-coupling system consisting of 244 vertical wells. The unit cost was $5,110/ton in 1994 dollars.

• The Lincoln, Nebraska, school district installed ground-source heat pump systems in four new elementary schools. In each school, the system consisted of 54 heat pumps ranging in size from 1.4 tons to 15 tons, with a total cooling capacity of 180 tons (630 kW). Four gas-fired boilers with a capacity of 330,000 Btuh each provide hot water for preheat and terminal reheat. The total heat pump cost per school was approximately $657,000 ($3,650/ton) in 1995 dollars (Shonder et al. 1999).

• Kavanaugh (1995) concluded that the average cost of ground-source heat pumps (including unit, loop, duct, and installation) ranged from $2360/ton for a 5-ton horizontal loop to $3000/ton for a 3-ton vertical loop system. Compared to a 3-ton conventional system, the added cost was $1250 to $1550 per ton.

Installation costs are expected to drop as the ground-source heat pump industry infrastructure grows and designers and installers become more experienced. Reducing installation costs is one of the prime goals of the International Ground-Source Heat Pump Association and the Geothermal Heat Pump Consortium.

28.6.5 Other Impacts

There are no significant negative environmental impacts associated with ground-source heat pumps. There is, however, the potential for systems to be affected by some local codes and regulations. The most likely source of conflict, if any, lies with the installation of the ground-coupling system. Working with an experienced installer is the best advice. However, local electric utilities and other local sites with existing ground-source heat pump installations are other sources of information about local permit and regulation issues.

With the application of any electrotechnology, there is a potential environmental benefit. Installing a ground-source heat pump system in lieu of a fossil-fuel heating system will reduce local emissions. Furthermore, installing a more efficient electrotechnology such as a ground-source heat pump system for cooling will reduce source emissions at the utility power plant. Typical emission reductions per MWh of energy conserved are 0.3 pounds (0.14 kg) of particulates, 3.3 pounds (1.5 kg) of sulfur oxides, 5.3 pounds (2.4 kg) of nitrogen oxides, and 1,720 pounds (780 kg) of carbon dioxide. These numbers vary with time and region, depending on the power generation fuel mix (EPA 1994; Nemeth 1993).

28.7 HYPOTHETICAL CASE STUDY

The example in this section is designed to assist the energy manager or facility engineer in estimating the energy consumption and costs associated with the construction and operation of ground-source heat pump systems and comparing them with those for conventional HVAC technologies. The goal is to estimate energy consumption and savings, not to design systems. There are several methods for estimating energy consumption of HVAC technologies, from simplistic degree-day calculations to sophisticated hour-by-hour energy modeling and simulation systems supported by computer programs. The examples used in this chapter are based on an outdoor temperature bin method. This method is described in more detail in *Closed-Loop/ Closed-Source Heat Pump Systems: Installation Guide* (OSU 1988).

28.7.1 Example: Oklahoma City Facility

The facility is a hypothetical administrative building of typical single-story construction that operates continuously (no night setback temperature control). Economizer-based control has also been removed from this analysis for simplicity. The facility is located in Oklahoma City, Oklahoma. Typical local weather conditions are 3,588°F-day/yr (base 65°F) heating degree-days (1,993 heating Celsius degree-days) and 2,068 (base 65°F) cooling degree-days (1,149 cooling Celsius degree-days). The 97.5% heating design temperature is 13°F (–10.6°C) and the 2.5% cooling design temperature is 96°F (35.6°C). The mean annual earth temperature is 62°F (16.7°C). The electric rate tariff for this example consists of a summer demand charge of $11.94/kW-month (5-months per year),

a winter demand charge of $5.72/kW-month (7 months per year), and a fixed energy charge of $0.0556/kWh. The gas rate schedule consists of an energy charge of $5.75/mcf ($0.575/therm). The corresponding heating load for this facility during the design day is estimated to be 2,392 kBtu/h (701.7 kW). The design cooling load is 1,832 kBtu/h (537.4 kW). The same balance temperature of 60°F (15.6°C) is estimated for the facility.

Technology Description

Three technologies are compared in this simplified example. The number of equipment units varies to match commercially available equipment capacities.

The first alternative uses rooftop air-conditioning equipment with integral natural gas furnace heaters. To meet the cooling load requirements during the cooling design day, 10 rooftop units are required. This number also meets the heating load requirements during the design winter conditions. The estimated life of this equipment, according to ASHRAE, is 15 years (*ASHRAE HVAC Applications Handbook*, 2005, pp 36.3).

The second alternative uses rooftop air-source heat pumps with electric resistance supplemental heaters. To meet the cooling requirements during the cooling design day, 10 rooftop units are required. To meet the heating requirements during the heating design day, each rooftop unit is equipped with a 50-kW supplemental heater. The estimated equipment life of this alternative, according to ASHRAE, is 15 years (*ASHRAE HVAC Applications Handbook*, 2005, pp 36.3).

The third alternative uses a ground-source heat pump system using extended range water-source heat pumps and a vertical bore closed-loop ground-coupling system. To meet the cooling requirements during the cooling design day, 35 units are required. To meet the heating requirement during the heating design day, each unit is equipped with a 10-kW supplemental heater. The water-loop system employs a variable-speed drive for added energy savings. The estimated equipment life of this alternative, according to ASHRAE, is 19 years (*ASHRAE HVAC Applications Handbook*, 2005, pp 36.3).

Savings Potential

Energy consumption for the three alternatives is estimated by an outdoor temperature bin method. The calculations are performed in a spreadsheet using 5°F bin data, because this is a form in which data are readily available (TM 5-785).

The main assumption underlying this type of analysis is that the facility is "thermally light." This implies that the heating and cooling loads on the building are proportional to the outside air temperature. A "thermally heavy" facility would be a building in which the cooling load on the building is not very proportional to the outside air temperature because of significant internal heat loads or a thermally massive structure, which tends to "hold" the heat or cold. In the case of a thermally heavy facility, the building heating and cooling loads would have to be calculated using a methodology other than the one utilized in the following analysis.

The analysis for the ground-source heat pump option shown in Table 28-3 begins with an estimate of the building load for each bin. Column 1 indicates the midpoint of each temperature bin, e.g., the bin from 95° to 99°F has a midpoint temperature of 97°F. Column 2 gives the number of hours that occur in each bin over a "typical" year for that location. Column 3 estimates the corresponding building load. This is calculated by linear interpolation between no heating or cooling load at the facility balance temperature and the design heating or cooling load at the design heating or cooling temperature. The actual equations used in this analysis are listed in Table 28-4 The entering water temperature (EWT) is estimated in column 4 of Table 28-3. This is the temperature of the water entering the heat pump from the ground-coupling system. For initial estimating, the entering water temperature is a linear interpolation between two points for the heating cycle, and again for the cooling cycle.

During zero building load conditions, which occur at the facility balance temperature, the entering water temperature is assumed to be equal to the ground temperature. During the peak load, the entering EWT is set to the maximum temperature desired during the peak cooling season load; similarly, it is set to the minimum temperature desired during the peak heating season load. Publications by the International Ground-Source Heat Pump Association recommend for initial calculations that these temperatures be set to 100°F maximum (37.8°C) and 37°F (2.8°C) minimum, although in this example, the minimum temperature was set to 35°F (1.7°C). Today, most commercial systems are commonly designed at 90°F (32.2°C) maximum EWT. The estimated entering water temperatures are refined later in the design process as the ground-coupling system design is finalized.

Column 5 is the net capacity of the ground-source heat pump equipment. This is taken from equipment specifications based on the entering water temperature. Columns 6, 7, and 8 are used to determine the loss in efficiency caused by part-load operating characteristics of equipment, which result in increased run time on the units. The part-load factor is dependent on the equipment "degradation factor." The degradation factor is typically assumed to be 0.25 unless the equipment manufacturer has tested the unit and determined a lower value.

Column 10 is the input power of the ground-source heat pump equipment. Like the net capacity, this is taken from equipment specifications based on the entering water temperature. Column 9 is the efficiency of the ground source heat pump alone and is determined by dividing the net capacity by the input power and correcting for the appropriate units of measure. The cooling efficiency is typically expressed as the energy-efficiency ratio (EER) and has units of Btu/watt-h. Heating efficiency is typically expressed as the coefficient of performance (COP) and is a dimensionless measure. Column 11 is the average input power required by the supplemental heaters to meet the estimated building heating load. It is estimated by taking the difference between the building heating load and the capacity of the heat pumps. When the ground-source heat pumps can meet the building load, there is no load on the supplemental heaters. The maximum load does not exceed the installed capacity of the supplemental heaters. Not included in this simplified example is an air-to-air heat recovery system that would reduce the need for supplemental heating. These recovery systems are common in cold region applications and can be specified to be part of the basic rooftop unit.

The average pump flow rate in column 12 is estimated based on the number of heat pumps, the actual run time (column 8), and the rated flow per unit. Most ground-source heat pumps are rated at around 3 gpm per ton of cooling (0.054 L/s-kW), although this varies by manufacturer.

The ground-source heat pump energy consumption, column 13, is the result of the input power multiplied by the percent run time multiplied by the number of hours in each bin (column 10 x column 8 x column 2). The supplemental heater energy consumption is similarly estimated as the result of the supplemental heater power multiplied by the run time multiplied by the number of hours in each bin (column 11 x column 8 x column 2). The energy consumption of the ground-coupling loop is dependent on the configuration of the ground-coupling and control systems. For this example, it is assumed the net pressure rise of the circulation pump at full design flow is 48 psi and, as a result of the control system, the variable speed drive reduces the energy requirement as a square function of the flow rate (assumes automatic water shut-off valves at each heat pump that close when the unit is off). The total energy consumption (column 16) is the sum of the component energy consumptions (column 13 + column 14 + column 15). The monthly-billed demand is estimated by examining the corresponding temperature bin during the time the monthly peak demand is typically set for the facility. Not included in the sample equipment design is an air-to-air heat recovery system that would reduce the

Table 28-3. Oklahoma City Facility Ground-source Heat Pump Energy Consumption Bin Method Analysis Example

Avg. Bin Temp. (°F)	Bin Hours (h/yr)	Bldg. Load (kBtuh)	EWT (°F)	Net GSHP Capacity (kBtuh)	Theor. Run Time (%)	Partial Load Factor	Actual Run Time (%)	GSHP Eff'y (EER or COP)	GSHP Input Power (kW)	Suppl. Heater Power (kW)	Pump Flow Rate (gpm)	GSHP Electric Energy (kWh)	Suppl. Heater Energy (kWh)	Circ. Pump Energy (kWh)	Circ. Pump Energy (kWh)	Total Electric Energy (kWh)
(1)	(2)	(3)	(4)	(5)	(6)	(7)	(8)	(9)	(10)	(11)	(12)	(13)	(15)	(16cv)	(16vsd)	(17)
Cooling Mode:																
112	0	2,646	117	1,699.8	1.00	1.00	1.00	9.55	177.93	n/a	494	0	0	0	0	0
107	1	2,392	112	1,751.3	1.00	1.00	1.00	10.20	171.63	n/a	494	172	0	14	14	186
102	18	2,137	106	1,802.8	1.00	1.00	1.00	10.90	165.32	n/a	494	2,976	0	256	256	3,231
97	94	1,883	101	1,854.3	1.00	1.00	1.00	11.66	159.02	n/a	494	14,948	0	1,335	1,335	16,282
92	240	1,628	96	1,905.8	0.85	0.96	0.89	12.48	152.71	n/a	438	32,500	0	3,407	2,679	35,180
87	393	1,374	91	1,957.2	0.70	0.93	0.76	13.37	146.41	n/a	374	43,644	0	5,580	3,210	46,854
82	603	1,120	85	2,008.7	0.56	0.86	0.63	14.34	140.10	n/a	309	52,945	0	8,561	3,362	56,308
77	829	829	80	2,060.2	0.42	0.85	0.49	15.40	133.80	n/a	242	54,477	0	11,770	2,839	57,316
72	948	611	75	2,111.7	0.29	0.82	0.35	16.56	127.49	n/a	174	42,505	0	13,459	1,665	44,170
67	819	356	69	2,163.1	0.16	0.79	0.21	17.85	121.19	n/a	103	20,659	0	11,628	504	21,163
62	729	102	64	2,214.6	0.05	0.76	0.06	19.28	114.88	n/a	30	5,054	0	10,350	38	5,092
Heating Mode:																
57	654	153	61	1,931.3	0.08	0.77	0.10	3.88	146.07	0.00	51	9,811	0	9,285	98	9,909
52	627	407	58	1,872.9	0.22	0.80	0.27	3.82	143.53	0.00	133	24,323	0	8,902	650	24,973
47	608	622	56	1,814.5	0.36	0.84	0.43	3.77	140.99	0.00	214	37,160	0	8,632	1,622	38,782
42	592	916	54	1,756.1	0.52	0.88	0.59	3.72	138.45	0.00	292	48,566	0	8,405	2,951	51,516
37	537	1,171	51	1,697.7	0.69	0.92	0.75	3.66	135.91	0.00	369	54,559	0	7,624	4,260	58,820
32	446	1,425	49	1,639.3	0.87	0.97	0.90	3.60	133.37	0.00	443	53,457	0	6,332	5,114	58,571
27	282	1,679	47	1,580.9	1.00	1.00	1.00	3.54	130.84	28.91	494	36,896	8,152	4,004	4,004	49,051
22	174	1,934	44	1,522.5	1.00	1.00	1.00	3.48	128.30	120.60	494	22,323	20,985	2,470	2,470	45,779
17	85	2,188	42	1,464.1	1.00	1.00	1.00	3.41	125.76	212.30	494	10,689	18,046	1,207	1,207	29,942
12	49	2,443	40	1,405.6	1.00	1.00	1.00	3.34	123.22	304.00	494	6,038	14,896	696	696	21,629
7	16	2,697	37	1,347.2	1.00	1.00	1.00	3.27	120.68	350.00	494	1,931	5,600	227	227	7,758
2	9	2,952	35	1,288.8	1.00	1.00	1.00	3.20	118.14	350.00	494	1,063	3,150	128	128	4,341
-3	1	3,206	33	1,230.4	1.00	1.00	1.00	3.12	115.60	350.00	494	116	350	14	14	480
-8	0	3,461	30		1.00	1.00	1.00	3.04	113.06	350.00	494	0	0	0	0	0
Cooling												269,880	0	66,359	15,902	285,782
Heating												306,932	71,179	57,926	23,441	401,551
Total												576,812	71,179	124,284	39,342	687,333

need for supplemental electric resistance heating. Heat pump manufacturers make this equipment available as part of the rooftop unit for installations in colder northern climates. Similar spreadsheets were developed for each of the other technology alternatives using a standard bin method. Table 28-5 shows the analysis for a conventional air-conditioning with natural-gas furnace option. The analysis for an air-source heat pump option is shown in Table 28-6.

Life-cycle Cost

The costs for energy, maintenance, and installation for each of the three alternatives are shown in Table 28-7. The GSHP system has the higher installation cost, not surprising given the added cost of the ground-coupled bore field. However, the greater efficiency results in the lowest energy consumption and lowest annual energy cost. The life expectancy of the equipment is also greatest. When life-cycle costs are considered in the assessment, the GSHP alternative proves to have the lowest cost. Because the equipment lives for the alternatives in this example are different, the present values for the alternatives can not be directly compared. As noted in Chapter 4, the project life in the comparison must be equivalent. To accomplish this, Table 28-7 includes the annualized cost for

each of the alternatives, which can be directly compared. In this example, the GSHP alternative has the lowest annualized cost and, therefore, would be the most life-cycle cost effective solution.

Life-cycle cost can be calculated manually (see Chapter 4), or it can be determined through software. One such software tool is the Building Life-Cycle Cost (BLCC) software developed by the National Institute for Standards and Technology (NIST). BLCC is available at no charge from the DOE Federal Energy Management Program (FEMP) website at www.eere.energy.gov/femp (use the search feature for BLCC). BLCC 5.3-08 was used to calculate the life-cycle and annualized cost shown in Table 28-7.

If only first costs are considered, GSHP systems may frequently appear more expensive than other alternatives. Utility incentives may reduce the cost premium but, as illustrated, energy costs over the life cycle are typically more dominant relative to the initial cost.

The economic merit of GSHP systems vary by location. First, the higher the energy cost, the more benefit the GSHP system will show, given the increase in efficiency. (This is true for any energy reduction measures.) Also, a greater heating and cooling season duration will result in greater system utilization, which will result in

Table 28-4. Summary of Equations for Bin Method Energy Consumption Analysis

Name	Column	Season	Description/Equation
Average temp (°F)	(1)	Both	Average bin temperature = Midpoint of temperature bin from weather data
Bin hours (hr/yr)	(2)	Both	Bin hours = Number of hours in temperature bin from weather data
Building load	(3)	Cooling Heating	Building load = (avg. bin temp. - fac. balance temp.) * [(fac. design load) / (fac. design temp. - fac. balance temp.)] = (fac. balance temp. - avg. bin temp.) * [(fac. design load) / (fac. balance temp. - fac. design temp.)]
Entering water temperature	(4)	Cooling Heating	Entering water temperature = (ground temp.) + [(max. entering water temp. – ground temp.) / (max. bin temp. - fac. balance temp.)] * (avg. bin temp. - fac. balance temp.) = (min. entering water temp.) + [(ground temp. - min. entering water temp.) / (fac. balance temp. - min. bin temp.)] * (avg. bin temp. - min. bin temp.)
Net equipment capacity	(5)	Both	Net equipment capacity = base on equipment specifications corrected for temperature
Theoretical run time	(6)	Both	Theoretical run time = (column 3) / (column 5); maximum = 1.00
Partial load factor	(7)	Both	Partial load factor = 1.00 - { degradation factor * [1.00 - (column 6)] }; maximum = 1.00
Actual run time	(8)	Both	Actual run time = (column 6) / (column 7)
Efficiency	(9)	Cooling Heating	Efficiency = (column 5) / (column 10) = (column 5) / [(column 10) * 3.412]
HVAC equipment input power	(10)	Both	HVAC equipment input power = based on equipment specifications corrected for temperature, may also include indoor and outdoor fan power
Supplemental heater power	(11)	Heating	Supplemental heater power = [(column 3) - (column 5)] /3.412; minimum = 0, maximum = capacity
Average water-loop pump flow rate	(12)	Both	Average water-loop pump flow rate = [no. of units * flow rate per unit * (column 8)]
HVAC equipment energy	(13)	Both	HVAC equipment energy = (column 2) * (column 8) * (column 10)
Fan energy	(14)	Cooling Heating	Fan energy = (column 2) * (column 8) * (total indoor air fan load + total outdoor air fan load); ignore if included in column 10 = (column 2) * (column 8) * (total indoor air fan load); ignore if included in column 10
Supplemental heater energy	(15)	Heating	Supplemental heater energy = (column 2) * (column 11)
Constant-volume water-loop pump energy	(16cv)	Both	Constant-volume water-loop pump energy = [no. of units * flow rate per unit * pump pressure rise in p.s.i. * 0.746 * (column 2)] / [1715 * pump eff. * motor eff.]
Variable-speed drive water-loop pump energy	(16vsd)	Both	Variable-speed drive water-loop pump energy = (column 16cv) * [(column 8)^2]
Total electric energy	(17)	Both	Total electric energy = (column 13) + (column 14) + (column 15) + (column 16)
Total electric gas furnace energy	(18)	Heating	Total natural gas furnace energy = (column 2) * (column 5) * (column 8) / [AFUE * 100]

a greater energy saving benefit. A high efficiency HVAC system installed in a mild climate with fewer heating or cooling hours will provide less annual energy savings compared to the same system installed in location with a longer, more dominant, heating and cooling season. Finally, in the case of GSHP technology, the cost of installation can be a significant local variable. Where installation conditions are favorable and installers are prevalent and experienced, the installation costs will be lower than in locations where conditions are difficult and the technology is unfamiliar to local installers. The availability of a well trained and experienced infrastructure can have the greatest impact on the installed cost of a GSHP system.

28.8 THE TECHNOLOGY IN PERSPECTIVE

The future of ground-source heat pump technology looks good, because there are many potential commercial applications. Although installation costs are typically higher for ground-source heat pumps than for other technologies, the decision criteria should be based on life-cycle costs rather than first costs; thus, a ground source heat pump system can be the most cost-effective alternative. According to the EPA study, Space Conditioning: The Next Frontier, ground-source heat umps are consistently the most energy-efficient, least polluting of all space conditioning technologies throughout the country

Table 28-5. Oklahoma City Facility Ground-Source Heat Pump Energy Consumption Bin Method Analysis Example

Avg. Bin Temp (°F)	Bin Hours (h/yr)	Bldg. Load (kBtuh)	Net Equip. Capacity (kBtuh)	Theor. Run Time (%)	Partial Load Factor	Actual Run Time (%)	A/C Eff'y (EER)	A/C Input Power (kW)	A/C Electric Energy (kWh)	Fan Energy (kWh)	Total Electric Energy (kWh)	Nat. Gas Furnace Energy (therms)
(1)	(2)	(3)	(5)	(6)	(7)	(8)	(9)	(10)	(13)	(14)	(17)	(18)
Cooling Mode:												
112	0	2,646	1,621.0	1.00	1.00	1.00	8.63	187.90	0	0	0	0
107	1	2,392	1,706.0	1.00	1.00	1.00	9.25	184.40	184	52	236	0
102	18	2,137	1,776.0	1.00	1.00	1.00	9.95	178.50	3,213	931	4,144	0
97	94	1,883	1,836.0	1.00	1.00	1.00	10.74	171.00	16,074	4,860	20,934	0
92	240	1,628	1,887.0	0.86	0.97	0.89	11.29	167.10	35,837	11,088	46,924	0
87	393	1,374	1,932.0	0.71	0.93	0.77	11.67	165.60	49,886	15,574	65,461	0
82	603	1,120	1,977.0	0.57	0.89	0.64	12.05	164.10	62,850	19,801	82,652	0
77	829	829	2,022.0	0.43	0.86	0.50	12.44	162.10	67,298	21,398	88,696	0
72	948	611	2,067.0	0.30	0.82	0.36	12.83	161.10	54,766	17,576	72,342	0
67	819	356	2,112.0	0.17	0.79	0.21	13.23	159.60	27,831	9,015	36,846	0
62	729	102	2,157.0	0.05	0.76	0.06	13.64	158.10	7,139	2,334	9,473	0
Heating Mode:												
57	654	153	2,700.0	0.06	0.76	0.07	n/a	n/a	0	1,210	1,210	1,633
52	627	407	2,700.0	0.15	0.79	0.19	n/a	n/a	0	3,001	3,001	4,051
47	608	622	2,700.0	0.25	0.81	0.30	n/a	n/a	0	4,591	4,591	6,198
42	592	916	2,700.0	0.34	0.83	0.41	n/a	n/a	0	6,015	6,015	8,120
37	537	1,171	2,700.0	0.43	0.86	0.51	n/a	n/a	0	6,780	6,780	9,154
32	446	1,425	2,700.0	0.53	0.88	0.60	n/a	n/a	0	6,673	6,673	9,008
27	282	1,679	2,700.0	0.62	0.91	0.69	n/a	n/a	0	4,843	4,843	6,538
22	174	1,934	2,700.0	0.72	0.93	0.77	n/a	n/a	0	3,354	3,354	4,527
17	85	2,188	2,700.0	0.81	0.95	0.85	n/a	n/a	0	1,808	1,808	2,441
12	49	2,443	2,700.0	0.90	0.98	0.93	n/a	n/a	0	1,135	1,135	1,533
7	16	2,697	2,700.0	1.00	1.00	1.00	n/a	n/a	0	400	400	540
2	9	2,952	2,700.0	1.00	1.00	1.00	n/a	n/a	0	225	225	304
-3	1	3,206	2,700.0	1.00	1.00	1.00	n/a	n/a	0	25	25	34
-8	0	3,461	2,700.0	1.00	1.00	1.00	n/a	n/a	0	0	0	0
Cooling									325,079	102,629	427,708	0
Heating									0	40,060	40,060	54,081
Total									325,079	142,688	467,768	54,081

(EPA 1993). The limited industry volume has been one of the factors holding back the development of a broader contractor base. The technology has not enjoyed broad national promotion. According to the results of one survey on the barriers to ground-source heat pumps, HVAC, plumbing, and architectural/engineering contractors are conservative, and are therefore reluctant to commit to what they regard as innovative and (to some) unproven technology and equipment (Technical Marketing Associates 1988). Installation of ground-source heat pump systems requires skills beyond those of most HVAC or plumbing contractors. Until recently, only a small number of contractors have performed enough installations to develop an extensive base of experience and expertise in ground-source heat pumps.

28.8.1 The Technology's Development

The ground-source heat pump technology has been shown through laboratory testing, field testing, and theo-retical analysis to be technically valid and economically attractive in many applications. Energy savings have been verified in a large number of field tests over the past 30 years. In several installations, reductions in maintenance costs have also been verified. The technology has gained rapid acceptance from users. Remaining barriers to rapid implementation include: (1) acceptance from new users and engineers unfamiliar with the technology, (2) out-of-date cost estimating guides for construction and maintenance, (3) general lack of engineers able to design ground-source heat pump systems, and (4) lack of contractors to install and service ground-source heat pump systems. This chapter is intended to address some of these concerns by reporting on the collective experience of ground-source heat pump users and evaluators and by providing application guidance. For actual design guidance, readers should refer to any of the many publications available, some of which are listed at the end of this chapter.

Table 28-6. Oklahoma City Facility Air-source Heat Pump System Bin Method Analysis Example

Avg. Bin Temp (°F)	Bin Hours (h/yr)	Bldg. Load (kBtuh)	Net ASHP Capacity (kBtuh)	Theor. Run Time (%)	Partial Load Factor	Actual Run Time (%)	ASHP Eff'y (EER or COP)	ASHP Input Power (kW)	Suppl. Heater Power (kW)	ASHP Electric Energy (kWh)	Suppl. Heater Energy (kWh)	Total Electric Energy (kWh)
(1)	(2)	(3)	(5)	(6)	(7)	(8)	(9)	(10)	(11)	(13)	(15)	(17)
Cooling Mode:												
112	0	2,646	1,647.8	1.00	1.00	1.00	9.48	173.90	n/a	0	0	0
107	1	2,392	1,702.5	1.00	1.00	1.00	10.09	168.73	n/a	169	0	169
102	18	2,137	1,757.1	1.00	1.00	1.00	10.74	163.57	n/a	2,944	0	2,944
97	94	1,883	1,811.8	1.00	1.00	1.00	11.44	158.40	n/a	14,890	0	14,890
92	240	1,628	1,866.5	0.87	0.97	0.90	12.18	153.23	n/a	33,143	0	33,143
87	393	1,374	1,921.1	0.72	0.93	0.77	12.97	148.07	n/a	44,808	0	44,808
82	603	1,120	1,975.8	0.57	0.89	0.64	13.83	142.90	n/a	54,759	0	54,759
77	829	829	2,030.5	0.43	0.86	0.50	14.74	137.73	n/a	56,798	0	56,798
72	948	611	2,085.1	0.29	0.82	0.36	15.73	132.57	n/a	44,709	0	44,709
67	819	356	2,139.8	0.17	0.79	0.21	16.80	127.40	n/a	21,942	0	21,942
62	729	102	2,194.5	0.05	0.76	0.06	17.95	122.23	n/a	5,426	0	5,426
Heating Mode:												
57	654	153	2,019.0	0.08	0.77	0.10	3.11	190.20	0.00	12,234	0	12,234
52	627	407	1,884.0	0.22	0.80	0.27	3.03	182.20	0.00	30,706	0	30,706
47	608	622	1,752.0	0.38	0.84	0.45	2.93	175.10	0.00	47,611	0	47,611
42	592	916	1,622.0	0.56	0.89	0.63	2.82	168.60	0.00	63,254	0	63,254
37	537	1,171	1,447.0	0.81	0.95	0.85	2.61	162.70	0.00	74,223	0	74,223
32	446	1,425	1,242.0	1.00	1.00	1.00	2.32	157.20	53.64	70,111	23,924	94,035
27	282	1,679	1,118.0	1.00	1.00	1.00	2.14	152.90	164.56	43,118	46,407	89,525
22	174	1,934	1,048.0	1.00	1.00	1.00	2.06	149.40	259.66	25,996	45,181	71,176
17	85	2,188	975.0	1.00	1.00	1.00	1.96	145.40	355.63	12,376	30,229	42,605
12	49	2,443	900.0	1.00	1.00	1.00	1.86	141.60	452.20	6,938	22,158	29,096
7	16	2,697	810.0	1.00	1.00	1.00	1.76	134.90	500.00	2,158	8,000	10,158
2	9	2,952	710.0	1.00	1.00	1.00	1.65	126.40	500.00	1,138	4,500	5,638
-3	1	3,206	619.0	1.00	1.00	1.00	1.52	119.40	500.00	119	500	619
-8	0	3,461	534.0	1.00	1.00	1.00	1.38	113.40	500.00	0	0	0
Cooling										279,589	0	279,589
Heating										389,983	180,898	570,880
Total										669,571	180,898	850,469

In addition to the U.S. Department of Energy and the Environmental Protection Agency, two associations are working to address the barriers to increased utilization of ground-source heat pumps: the International Ground Source Heat Pump Association (www.igshpa. okstate.edu) and the Geothermal Heat Pump Consortium (www.geoexchange.org). The International Ground Source Heat Pump Association (IGSHPA) is a non-profit, member-driven organization established in 1987 to advance ground-source heat pump technology on local, state, national, and international levels. IGSHPA conducts ground-source heat pump system installation training and geothermal research. The mission of the International Ground Source Heat Pump Association and its membership is to promote the use of ground-source heat pump technology worldwide through education and communication. The government-industry-utility consortium, Geothermal Heat Pump Consortium, Inc., was formed in 1994 to create a self-sustaining ground-source heat pump market. The Geothermal Heat Pump Consortium is a nonprofit organization working to raise awareness and increase the use of ground-source heat pump technology throughout the United States. As a cooperative venture, the Consortium counts among its partners electric utilities, equipment manufacturers, architects, designers, engineers, contractors, builders, drillers, energy service companies, and other private sector companies that operate in the ground-source heat pump market, as well as national, state and local organizations and public agencies.

Table 28-7. Oklahoma City Example: Summary of Results

Oklahoma City Facility	First Alternative Rooftop AC and Gas Heat	Second Alternative ASHP	Third Alternative GSHP
Number of equipment	10	10	35
Nominal capacity (tons) Each	13.5	13.5	4.8
Total	135	135	168
Supplemental heaters (kW) Each	n/a	50	10
Total	n/a	500	350
Equipment capacity (kBtuh) (at design conditions) Summer	1,848.0	1,822.7	1,864.6
Winter	2,700.0	2,621.0	2,611.5
Energy consumption (/yr) Electricity (kWh)	467,708	850,469	687,333
Demand* (kW-mo) - Winter	455	2,748	1,931
- Summer	842	770	770
Natural gas (therm)	54,081	0	0
Total energy (10^6 Btu)	7,004	2,902	2,345
Energy cost ($/yr) Electricity	26,005	47,286	38,216
Demand	12,656	24,912	20,239
Natural gas	31,097	0	0
Total energy	69,758	72,198	58,455
O&M costs ($/yr)	6,750	4,725	5,880
Installed cost ($)	349,300	236,000	451,500
Equipment life (yr)	15	15	19
Total life-cycle cost[†] ($)	1,236,558	1,132,706	1,349,276
Annualized cost[†] ($/yr)	103,608	94,907	94,216

* Assumptions are required to estimate the monthly billing demand. In this case, the authors matched a temperature bin to when the peak demand window was likely to occur. A sensitivity analysis could be performed by changing the assumptions.
[†] Life-cycle cost and annualized costs were determined using BLCC 5.3-08.

28.8 FOR FURTHER INFORMATION

Trade Associations

American Society of Heating, Refrigerating, Refrigerating and Air Conditioning Engineers (ASHRAE)
1791 Tullie Circle, NE
Atlanta, GA 30329
Tel: (404) 636-8400
Fax: (404) 321-5478
www.ashrae.org

Canadian Earth Energy Association
130 Slater Street, Suite 1050
Ottawa, ON Canada K1P6E2
Tel: (613) 230-2332
Fax: (613) 237-1480 Fax
www.earthenergy.ca

Electric Power Research Institute (EPRI)
P.O. Box 50490
3412 Hillview Ave.
Palo Alto, CA 94303
www.epri.com

Geothermal Heat Pump Consortium, Inc. (Geoexchange)
1050 Connecticut Ave, NW, Suite 1000
Washington, D.C.
Tel: (202) 558-7175
Fax: (202) 558-6759
www.geoexchange.org

Geothermal Resources Council
P.O. Box 1350
Davis, CA 95617
www.geothermal.org

International Ground-Source Heat Pump Association (IGSHPA)
379 Cordell South
Oklahoma State University
Stillwater, OK 74078
Tel: (800) 626-4747
Fax: (405) 744-5283
www.igshpa.okstate.edu

National Rural Electric Cooperative Assn. (NRECA)
1800 Massachusetts Ave., NW
Washington, D.C. 20036
www.nreca.org

National Ground Water Association
601 Dempsey Road
Westerville, OH 43801
Tel: (800) 551-7379
www.ngwa.org

Michigan Geothermal Energy Association
2859 West Jolly Road
Okemos, MI 48864
Tel: (800) 417-5555
www.earthcomfort.com

Newsletters

"Heat Pump News Exchange," Electric Power Research Institute (EPRI), published quarterly and distributed by EPRI, Palo Alto, California.

"The Source," International Ground Source Heat Pump Association, published quarterly and distributed by Oklahoma State University Ground Source Heat Pump Publications, Stillwater, Oklahoma.

"Geo-Heat Center Quarterly Bulletin," Geo-Heat Center, published quarterly and distributed by Geo-Heat Center, Oregon Institute of Technology, Klamath Falls, OR.

"Down to Earth," published by the Michigan Geothermal Energy Association, Okemos, MI.

"Outside the Loop," published quarterly by the University of Alabama, Tuscaloosa, Alabama (www.oit.edu/~geoheat/otl/index.htm).

User and Third-Party Field and Lab Test Reports[14]

Collie, M.J. (editor) 1979. "Comparison of Water-Source and Air-Source Heat Pumps in Northern Environment," a chapter in Heat Pump Technology for Saving Energy. Noyes Data Corporation, Park Ridge, New Jersey.

*Hughes, P.J. and J.A. Shonder. 1998. The Evaluation of a 4000-Home Geothermal Heat Pump Retrofit at Fort Polk, Louisiana: Final Report. ORNL/CON-460, Oak Ridge National Laboratory, Oak Ridge, Tennessee.

Phetteplace, G., H. Ueda, and D. Carbee. 1992. "Performance of Ground-Coupled Heat Pumps in Military Family Housing Units." Solar Engineering. G0656A-1992, American Society of Mechanical Engineers, New York, NY.

Phetteplace, G. 1995. Ground-Coupled Heat Pumps for Family Housing Units. FEAP-UG-CRREL-95/01, U.S. Army Center for Public Works, Alexandria, Virginia.

Svec, O.J. 1987. "Potential of Ground Heat Source Systems." International Journal of Energy Research, Vol. 11, pp. 573-581.

Design and Installation Guides

*ASHRAE. 1995. "Chapter 29: Geothermal Energy." ASHRAE Handbook: Heating, Ventilating, and Air-Conditioning Applications. American Society of Heating, Refrigerating, and Air- Conditioning Engineers, Inc. Atlanta, Georgia.

Bose, J.E., J.D. Parker, and F.C. McQuiston. 1985. Design/Data Manual for Closed-Loop Ground-Coupled Heat Pump Systems. American Society of Heating, Refrigerating, and Air-Conditioning Engineers, Inc., Atlanta, Georgia.

Caneta Research, Inc. 1995. Commercial/Institutional Ground-Source Heat Pump Engineering Manual. American Society of Heating, Refrigerating, and Air-Conditioning Engineers, Inc., Atlanta, Georgia.

*International Ground-Source Heat Pump Association (not dated). Ground Source Systems: Design and Installation Standards. Oklahoma State University, Stillwater, Oklahoma.

*Oklahoma State University. 1988. Closed-Loop/Ground-Source Heat Pump Systems: Installation Guide. International Ground-Source Heat Pump Association, Stillwater, Oklahoma.

---

[14]*Denotes literature cited in the technical body of this chapter.

*STS Consultants. 1989. Soil and Rock Classification for the Design of Ground-Coupled Heat Pump Systems: Field Manual. CU-6600, Electric Power Research Institute, Palo Alto, California.

*WaterFurnace. October 1991. Polyethylene Pond/Lake Loop, Mat Design. WF390, WaterFurnace International, Fort Wayne, Indiana.

Manufacturer's Application Notes

Command-Aire. 1988. Earth Energy Heat Pumps: Commercial Water Source Heat Pump Systems; Application, Installation, Operation Manual. Bulletin C-A10-0191.

Guarino, D.S. March 1993. "WaterFurnace: From the Ground Up." Reprint from Contracting Business.

Utility, Information Service, or
Government Agency Tech-Transfer Literature

Bas, E. February 13, 1995. "Geothermal Market Receives Hefty Boost from Consortium." The Air Conditioning, Heating, and Refrigeration News, p.1.

*Calm, J.M. September 1987. Proceedings of the Workshop on Ground-Source Heat Pumps. From the workshop held in Albany, New York, October 27 through November 1, 1986. HPC-WR-2. International Energy Agency Heat Pump Center.

Goldfish, L.H., and R.A. Simonelli. 1988. Ground-Source and Hydronic Heat Pump Market Study. EM-6062, Electric Power Research Institute, Palo Alto, California.

Ground Source Systems: Educational and Marketing Material Catalog. International Ground Source Heat Pump Association, Stillwater, Oklahoma.

GS-Systems: An Answer to U.S. Energy and Environmental Concerns: A comprehensive report on ground-source heat pumps and their benefits (not dated). International Ground-Source Heat Pump Association, Oklahoma State University, Stillwater, Oklahoma.

*Pratsch, L.W. 1990. "Geothermal Heat Pumps: A Major Opportunity for the Utility Industry." Presented at the Geothermal Energy Conference, Columbus, Ohio, November 14, 1990.

Scofield, M., and P. Joyner. September 1991. "Heat Pumps for Northern Climates." EPRI Journal 16(6), pp. 28-33.

Case Studies

Duffy, G. March 22, 1993. "120-ton geothermal system replaces 30-year-old relics." The Air Conditioning, Heating, and Refrigeration News.

*Gahran, A. September 1993. "Grants, Util. Rebate Pay 97% of Ground-Source Heat Pump Project Cost." Energy User News 18(9) pp. 10, 51, 72.

*Gahran, A. January 1994. "New Ground-Source Heat Pumps Cut Energy, Maintenance Costs $68K." Energy User News 19(1) p. 16.

Greenleaf, J. July-September 1982. "College invests in an energy-efficient future." Brown and Caldwell Quarterly. KCPL. 1994. A Model of Efficiency from the Ground Up. Commercial case study 2M/3-135, Kansas City Power and Light, Kansas City, Missouri.

KCPL. 1994. Innovative Heating and Cooling System Allows Increased Efficiencies to Surface. Commercial case study 4-132, Kansas City Power and Light, Kansas City, Missouri.

KCPL. 1994. Station's Ground Source Heat Pump System Finds Efficiency in Unusual Places. Commercial case study 4-118, Kansas City Power and Light, Kansas City, Missouri.

PSO. "Facility Type: Family Fitness Center, Technology Application: Ground- Source Heat Pump." Power Profile. Number 1, Public Service Company of Oklahoma, Tulsa, Oklahoma.

Randazzo, M. November 1994. "Retrofit Cut Elec. Use by 32MMkWh/Yr." Energy User News 19(11) pp. 1,14.

WaterFurnace. Dallas' "Hope of the Sun" goes underground for affordable, efficient, environmental housing. Case Study 5, WF915, WaterFurnace International, Fort Wayne, Indiana.

*WaterFurnace. Elementary school teaches lesson in efficiency. Case Study 2, WF666, WaterFurnace International, Fort Wayne, Indiana.

WaterFurnace. Geothermal comfort goes low-rise in Kisilano, British Columbia. Case Study 8, WF923, WaterFurnace International, Fort Wayne, Indiana.

WaterFurnace. John Deere dealership comforts its clientele from the floor up. Case Study 11, WF922, WaterFurnace International, Fort Wayne, Indiana.

WaterFurnace. Last century's structure retrofitted for the next. Case Study 7, WF924, WaterFurnace International, Fort Wayne, Indiana.

WaterFurnace. Patients treated to geothermal comfort at Dupont Medical Center. Case Study 12, WF942, WaterFurnace International, Fort Wayne, Indiana.

WaterFurnace. Salem Community college enhances environment and education. Case Study 4, WF907, WaterFurnace International, Fort Wayne, Indiana.

*WaterFurnace. Three Indiana schools compare HVAC notes. Case Study 9, WF925, WaterFurnace International, Fort Wayne, Indiana.

*WaterFurnace. WaterFurnace builds state-of-the-art facility. Case Study 1, WF639, WaterFurnace International, Fort Wayne, Indiana.

WaterFurnace. WaterFurnace goes down under, down under. Case Study 10, WF921, WaterFurnace International, Fort Wayne, Indiana.

WaterFurnace. Whitehawk Ranch looks earthward for heating and cooling. Case Study 6, WF916, WaterFurnace International, Fort Wayne, Indiana.

WaterFurnace. York County contributes to a healthier environment. Case Study 3, WaterFurnace International, Fort Wayne, Indiana.

Codes and Standards

*Energy-Efficient Design of New Buildings Except Low-Rise Residential Buildings. 2004. ASHRAE/IESNA Standard 90.1-2004, American Society of Heating, Refrigerating, and Air- Conditioning Engineers, Inc., Atlanta, Georgia.

*Water-source Heat Pumps – Testing and Rating for Performance – Part 1: Water-to-air and Brine-to-air Heat Pumps. 2005. ANSI/ARI/ASHRAE ISO Standard 13256-1-2005, American Society of Heating, Refrigerating, and Air- Conditioning Engineers, Inc., Atlanta.

*Ground Source Closed-Loop Heat Pumps. 1998. ANSI/ARI Standard 330-98, Air-Conditioning and Refrigeration Institute, Inc., Arlington, Virginia.

*Ground Water-Source Heat Pumps. 1998. ARI Standard 325-98. Air-Conditioning and Refrigeration Institute, Inc., Arlington, Virginia.

*Water-Source Heat Pumps. 1998. ANSI/ARI Standard 320-98. Air-Conditioning and Refrigeration Institute, Inc., Arlington, Virginia.

Other Sources

American Society of Heating, Refrigerating, and Air-Conditioning Engineers (ASHRAE). 1997. Ground Source Heat Pumps — Design of Geothermal Systems for Commercial and Institutional Buildings. ASHRAE, Atlanta, GA.

Cane, D., A. Morrison, and C. Ireland. 1998. "Maintenance and Service Costs of Commercial Building Ground-Source Heat Pump Systems." ASHRAE Transactions, Vol 104(2): 699-706.

Duffy, G. July/August 1989. "Commercial Earth-Coupled Heat Pumps Are Ready To Roll." Engineered Systems, pp. 40-47.

*Electric Power Research Institute (EPRI). May 1993. 1992 Survey of Utility Demand-Side Management Programs. EPRI-TR 102193, Palo Alto, California.

*EPA. 1993. Space Conditioning: The Next Frontier. Office of Air and Radiation, 430-R-93-004 (4/93), EPA, Washington, D.C.

*EPA. 1994. Voluntary Reporting Formats for Greenhouse Gas Emissions and Carbon Sequestration. (Public Review Draft, May 1994). EPA, Washington, D.C.

*Geothermal Heat Pumps." February 1990. Custom Builder. 5(2) pp. 32-35.

*Giddings, T. February 1988. "Ground Water Heat Pumps: Why Doesn't Everybody Want One?" Well Water Journal 42(2) pp. 32-33.

Gilmore, V.E. June 1988. "Neo-geo heat pump." Popular Science 232(6) pp. 88-89,112.

*Heinonen, E.W., R.E. Tapscott, M.W. Wildin, and A.N. Beall. February 1997. "Assessment of Antifreeze Solutions for Ground-Source Heat Pump Systems." Report 908RP, American Society of Heating, Refrigerating, and Air-Conditioning Engineers (ASHRAE), Atlanta.

Hurlburt, S. February 1988. "Ground Water Heat Pumps - The Second Wave." Well Water Journal 42(2) pp. 34-40.

Kavanaugh, S., C Gilbreath, J. Kilpatrick. 1995. Cost Containment for Ground-Source Heat Pumps, Final Report. University of Alabama, Tuscaloosa.

Kavanaugh, S. September 1992. "Ground-coupled heat pumps for commercial buildings." ASHRAE Journal 34(9), 30-37.

*Lund, J.W. and T. L. Boyd. 2000. "Geothermal Direct-Use in the United States in 2000." Geo-Heat Center Quarterly Bulletin, Vol. 21, No. 1 (March), Klamath Falls, Oregon, pp. 1-5.

Martin, M.A., D.J. Durfee, and P.J. Hughes. 1999. "Comparing Maintenance Costs of Geothermal Heat Pump Systems with Other HVAC Systems in Lincoln Public Schools: Repair, Service, and Corrective Action." ASHRAE Transactions, Vol. 105(2): 1208-1216.

Martin, M.A., M.G. Madgett, and P.J. Hughes. 2000. "Comparing Maintenance Costs of Geothermal Heat Pump Systems with Other HVAC Systems: Preventive Maintenance Actions and Total Maintenance Costs." ASHRAE Transactions, Vol. 106(1): 1-15.

*Nemeth, R.J., D.E. Fornier, and L.A. Edgar. 1993. Renewables and Energy- Efficiency Planning. U.S. Army Construction Engineering Research Laboratory, Champaign, Illinois.

Oak Ridge National Laboratory (not dated). "Big Savings from the World's Largest Installation of Geothermal Heat Pumps at Fort Polk, Louisiana." Oak Ridge National Laboratory, Oak Ridge, Tennessee.

Outside the Loop. 1999. "Impact of Conductivity Error on Design Results." Outside the Loop - Volume 2, Number 3, University of Alabama, Tuscaloosa, Alabama.

Phetteplace, G., H. Ueda, and D. Carbee. April 1992. "Performance of Ground-Coupled Heat Pumps in Military Family Housing Units." Proceedings of the 1992 ASME International Solar Energy Conference. Maui, Hawaii, 4-8 April, 1992, pp. 377-383.

Reynolds, William C., and Herry C. Perkins. 1977. Engineering Thermodynamics, pp. 287-289. McGraw-Hill Book Co., New York.

Rybach, L., and B. Sanner. March 2000. "Ground-Source Heat Pump Systems the European Experience." Geo-Heat Center Quarterly Bulletin 21(1):16-26, Klamath Falls, OR.

*Shonder J.A., D. Durfee, M.A. Martin, P.J. Hughes, and T.R. Sharp. 1999. "Benchmark for Performance: Geothermal Applications in Lincoln Public Schools." ASHRAE Transactions 105(2): 1199-1207.

Shonder, J. A., M. A. Martin, T. R. Sharp, D. J. Durfee, and P. J. Hughes. 1999. "Benchmark for Performance: Geothermal Applications in Lincoln Public Schools." ASHRAE SE-99-20-03, Atlanta.

Shonder, J. A., V. Baxter, J. Thorton, and P. Hughes. June 1999. "A New Comparison of Vertical Ground Heat Exchanger Design Methods for Residential Applications." ASHRAE Transactions.

Shonder, J. A., V. Baxter, P. Hughes, and J. Thorton (draft - no date). "A Comparison of Vertical Ground Heat Exchanger Design Software for Commercial Applications." ASHRAE Transactions.

Shonder, J.A. and J.V. Beck. 2000. "Field Test of a New Method for Determining Soil Formation Thermal Conductivity and Borehole Resistance from In Situ Measurements." ASHRAE Transactions.

*Spilker, E. H. January 1998. "Ground-Coupled Heat Pump Loop Design Using Thermal Conductivity Testing and the Effect of Different Backfill Materials on Vertical Bore Length." Proceedings of the American Society of Heating, Refrigerating, and Air-Conditioning Engineers (ASHRAE). January 1998 meeting, San Francisco, paper SF98-1-3.

Swanson, G.L. August 1986. "Heating and Cooling with GWHPs." Well Water Journal 40(8) pp. 44-46.

*Technical Marketing Associates, Inc. 1988. Ground-Source and Hydronic Heat Pump Market Study. EM-6062, Electric Power Research Institute, Palo Alto, California.

*TM 5-785. 1978. Engineering Weather Data. Department of the Army Technical Manual. TM 5-785, U.S. Government Printing Office, Washington, D.C.

Appendix

PERFORMANCE AND EFFICIENCY TERMINOLOGY

There are many terms used in the heating, air-conditioning, and refrigeration industry that convey performance and efficiency. Many of these terms are synonymous; others are not. When comparing various systems, it is important to understand what the assorted terms are, how they are determined, and their relationship. The following provides a brief description of many (but not all) of the terms used to convey efficiency. Use of the terms is also summarized in Table 28-A.1.

Annual Fuel Utilization Efficiency (AFUE): For fuel-fired systems such as boilers and furnaces, AFUE is defined as the ratio of annual output energy to annual input energy. This term is generally applied to systems <=300,000 Btu/h input. AFUE is a weighted average efficiency under standard rated conditions at various part-load conditions and also includes any non-heating-season pilot input losses. AFUE is a unitless term generally expressed as a percentage.

Coefficient of Performance (COP): The COP is the basic parameter used to report the efficiency of refrigerant-based systems. It is a unitless term. This term is universal in its use but not in its meaning. COP can be used to define both cooling efficiency or heating efficiency, such as for a heat pump. For cooling, COP is defined as the ratio of the rate of heat removal to the rate of energy input to the compressor, in consistent units. For heating, COP is defined as the ratio of rate of heat delivered to the rate of energy input to the compressor, in consistent units. COP can be used to define the efficiency at a single (standard or nonstandard) rated condition or a weighted average (seasonal) condition. Depending on its use, the term may or may not include the energy consumption of auxiliary systems such as indoor or outdoor fans, chilled water pumps, or cooling tower systems. For purposes of comparison, the higher the COP the more efficient the system. For mathematical purposes, COP can be treated as an efficiency. (COP of 2.00 = 200% efficient.) For unitary heat pumps, ratings at two standard (outdoor) temperatures (47°F and 17°F [8.3°C and -8.3°C]) may be reported based on old ARI standards.

Combustion Efficiency (n_c or E_c): For fuel-fired systems, this efficiency term is defined as the ratio of the fuel energy input minus the flue gas losses (dry flue gas, incomplete combustion and moisture formed by combustion of hydrogen) to the fuel energy input. In the U.S., fuel-fired efficiencies are reported based on the higher heating value of the fuel. Other countries report fuel-fired efficiencies based on the lower heating value of the fuel. The combustion efficiency is calculated by determining the fuel gas losses as a percent of fuel burned. [$E_c = 1 -$ (flue gas losses).]

Energy Efficiency Ratio (EER): The EER is a term generally used to define the cooling efficiency of unitary air-conditioning and heat pump systems. The term implies that the efficiency is determined at a single rated condition specified by the appropriate equipment standard and is defined as the ratio of net cooling capacity (heat removed in Btu/h) to the total input rate of electric energy required (watt). The units of EER are Btu/w-h. It is important to note that this efficiency term typically includes the energy requirement of auxiliary systems such as the indoor and outdoor fans. For purposes of comparison, the higher the EER, the more efficient the system. To convert to a COP, divide the EER by 3.412. [COP = EER/3.412.]

Heating Seasonal Performance Factor (HSPF): The term HSPF is similar to the term SEER, except it is used to signify the seasonal heating efficiency of heat pumps. The HSPF is a weighted average efficiency over a range of outside air conditions following a specific standard test method. The term is generally applied to heat pump systems less than 60,000 Btu/h (rated cooling capacity). The units of HSPF are Btu/w-h. It is important to note that this efficiency term typically includes the energy requirement of auxiliary systems such as the indoor and outdoor fans. For purposes of comparison, the higher the HSPF, the more efficient the system.

Integrated Part-Load Value (IPLV): The term IPLV is used to signify the cooling efficiency related to a typical (hypothetical) season rather than a single rated condition. The IPLV is calculated by determining the weighted average efficiency at part-load capacities specified by an accepted standard. It is also important to note that IPLVs are typically calculated using the same condensing temperature for each part-load condition and that IPLVs do not include cycling or load/unload losses. The units of IPLV are not consistent in the literature; therefore, it is important to confirm which units are implied when the term IPLV is used. ASHRAE Standard 90.1 (using ARI reference standards) uses the term IPLV to report seasonal cooling efficiencies for both seasonal COPs (unitless) and seasonal EERs (Btu/w-h), depending on the equipment capacity category; most chiller manufacturers report seasonal efficiencies for large chillers as IPLV using units of kW/ton. Depending on how a cooling system loads and unloads (or cycles), the IPLV can be between 5 and 50% higher than the EER at the standard rated condition.

kW/ton: The term kW/ton is generally used for large commercial and industrial air-conditioning, heat pump, and refrigeration systems. The term is defined as the ratio of the rate of energy consumption (kW) to the rate of heat removal (ton) at a rated condition. As the term suggests, the units are kW/ton. Because refrigeration systems of this capacity are typically custom designed, the reported kW/ton generally implies only the compressor and does not include the auxiliaries. However, for specific references, auxiliaries can be added to report the overall system efficiency using this term. It is important to note that this term is inverse to the other performance and efficiency terminology. Therefore, for purposes of comparison, the lower the kW/ton, the more efficient the system. To convert to a COP, divide 12,000 by the product of 3,412 multiplied by the kW/ton. [$COP = 12,000/\{(3,412) * (kW/ton)\}$.]

Seasonal Energy Efficiency Ratio (SEER): The term SEER is used to define the average annual cooling efficiency of an air conditioning or heat pump system. The term SEER is similar to the term EER but is related to a typical (hypothetical) season rather than for a single rated condition. The SEER is a weighted average of EERs over a range of rated outside air conditions following a specific standard test method. The term is generally applied to systems less than 60,000 Btu/h. The units of SEER are Btu/W-h. It is important to note that this efficiency term typically includes the energy requirements of auxiliary systems such as the indoor and outdoor fans. For purposes of comparison, the higher the SEER, the more efficient the system. Although SEERs and EERs cannot be directly compared, the SEERs usually range from 0.5 to 1.0 higher than corresponding EERs.

Thermal Efficiency (n_t or E_t): This efficiency term is generally defined as the ratio of the heat absorbed by the water (or the water and steam) to the heat value of the energy consumed. The combustion efficiency of a fuel-fired system will be higher than its thermal efficiency. (See ASME Power Test Code 4.1 for more details on determining the thermal efficiency of boilers and other fuel-fired systems.) In the U.S., fuel-fired efficiencies are typically reported based on the higher heating value of the fuel. Other countries typically report fuel-fired efficiencies based on the fuel's lower heating value. The difference between a fuel's higher heating value and its lower heating value is the latent energy contained in the water vapor (in the exhaust gas) that results when hydrogen (from the fuel) is burned. The efficiency of a system based on a fuel's lower heating value can be 10 to 15% higher than its efficiency based on a fuel's higher heating value.

Table 28-A.1.
Summary of Performance and Efficiency Terminology

Operating Mode	Design Rated Conditions	Seasonal Average Conditions
Cooling	COP	COP
	EER	SEER
	kW/ton	IPLV
Heating	COP	AFUE
	E_c	COP
	E_t	HSPF

Appendix I
Thermal Sciences Review

L.C. WITTE
Professor of Mechanical Engineering
University of Houston
Houston, Texas

I.1. INTRODUCTION

Many technical aspects of energy management involve the relationships that result from the thermal sciences: classical thermodynamics, heat transfer, and fluid mechanics. For the convenience of the user of this handbook, brief reviews of some applicable topics in the thermal sciences are presented. Derivation of equations are omitted; for readers needing those details, references to readily available literature are given.

I.2. THERMODYNAMICS

Classical thermodynamics represents our understanding of the relationships of energy transport to properties and characteristics of various types of systems. This science allows us to describe the global behavior of energy-sensitive devices. The relationships that can be developed will find application in both fluid-flow and heat-transfer systems.

A thermodynamic system is a region in space that we select for analysis. The boundary of the system must be defined. Usually, the system boundary will coincide with the physical shell of a piece of hardware. A closed system is one where no mass may cross the boundary, whereas an open system, sometimes called a control volume, will generally have mass flowing through it.

We generally divide energy into two categories: stored and transient types of energy. The stored forms are potential, kinetic, internal, chemical, and nuclear. These terms are fairly self-descriptive in that they relate to ways in which the energy is stored. Chemical and nuclear energy represent the energy tied up in the structure of the molecular and atomic compounds themselves. These two types of stored energy presently form the prime energy sources for most industrial and utility applications and thus are of great importance to us.

Potential, kinetic, and internal energy forms generally are nonchemical and nonnuclear in nature. They relate to the position, velocity, and state of material in a thermodynamic system. More detailed representations of these energy forms will be shown later.

I.2.1 Properties and States

Thermodynamic systems are a practical necessity for the calculation of energy transformations. But to do this, certain characteristics of the system must be defined in a quantifiable way. These characteristics are usually called properties of the system. The properties form the basis of the state of the system. A state is the overall nature of the system defined by a unique relationship between properties.

Properties are described in terms of four fundamental quantities: length, mass, time, and temperature. Mass, length, and time are related to a force through Newton's second law.

In addition to the fundamental quantities of a system there are other properties of thermodynamic importance. They are pressure, volume, internal energy, enthalpy, and entropy—P, V, U, H, and S, respectively.

Equation of State. Returning to the concept of a state, we use an equation of state to relate the pertinent properties of a system. Generally, we use $P = P(m, V, T)$ as the functional equation of state. The most familiar form is the ideal gas equation of state written as

$$PV = mRT \tag{I.1}$$

or

$$PV = n\bar{R}T \tag{I.2}$$

Equation I.1 is based on the mass in a system, whereas equation I.2 is molal-based. R is called the gas constant and is unique to a particular gas. $\bar{R}$, on the other hand, is called the universal gas constant and retains the same value regardless of the gas (i.e., $\bar{R} = 1545$ ft $lb_f/lb_m \cdot mol \cdot °R = 1.9865$ Btu/$lb_m \cdot mol \cdot °R$). It can be easily shown that R is simply $\bar{R}/M$, where M is the molecular weight of the gas.

The ideal gas equation is useful for superheated but not for saturated vapors. A vexing question that often arises is, "What is the range of application of the ideal gas equation?" Figure I-1 shows the limits of appli-

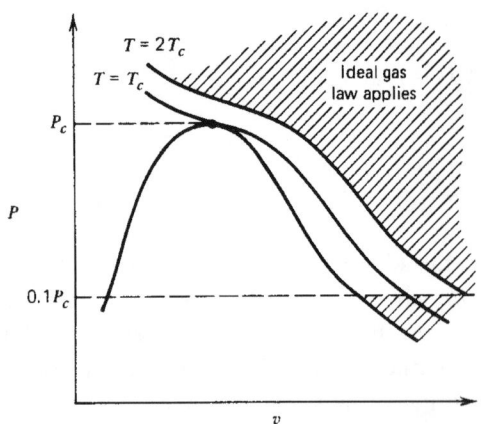

Figure I-1 Applicability of ideal gas equation of state.

cability. The shaded areas demonstrate where the ideal gas equation applies to within 10% accuracy.

If high pressures or vapors near saturation are involved, other means of representing the equation of state are available. The compressibility factor Z, for example $Z = PV/RT$, is a means of accounting for nonideal gas conditions. Several other techniques are available; see Ref. 1, for example.

Changes of state for materials that do not behave in an ideal way can be calculated by use of generalized charts for property changes. For example, changes in enthalpy and entropy can be presented in terms of reduced pressure and temperature, P_r and T_r. The reduced properties are the ratio of the actual to the critical properties. These charts (see Appendix II) can be used to calculate the property changes for any change in state for those substances whose thermophysical properties are well documented.

Ideal Gas Mixtures. Where several gases are mixed, a way to conveniently represent the properties of the mixture is needed. The simplest way is to treat the system as an ideal gas mixture. In combustion systems where fuel vapor and air are mixed, and in the atmosphere where oxygen nitrogen and water vapor form the essential elements of air, the concept of the ideal gas mixture is very useful.

There are two ways to represent gas mixtures. One is to base properties on mass, called the gravimetric approach. The second is based on the number of moles in a system, called a molal analysis. This leads to the definition of a mass fraction, $x_i = m_i/m_t$, where m_i is the mass of the ith component in the mixture, m_t is the total mass of the system, and the mole fraction, $y_i = n_i/n_t$, where n represents the number of moles.

Commonly, the equation of state for an ideal mixture involves the use of Dalton's law of additive pressures. This uses the concept that the volume of a system is occupied by all the components. Using a two-component system as an example, we write

$$P_1 V_1 = n_1 \bar{R} T_1$$

and $$P_2 V_2 = n_2 \bar{R} T_2$$

Also $$P_1 V_1 = n_1 \bar{R} T_2$$

Since V, $\bar{R}$, and T are the same for all three equations above, we see that

$$P_t = P_1 + P_2$$

and $$(P_1 P_2)V = (n_1 n_1)\bar{R}T$$

for a mixture. Each gas in the mixture of ideal gases then behaves in all respects as though it exists alone at the volume and temperature of the total system.

I.2.2 Thermodynamic Processes

A transformation of a system from one state to another is called a process. A cycle is a set of processes by which a system is returned to its initial state. Thermodynamically, it is required that a process be quantifiable by relations between properties if an analysis is to be possible.

A process is said to be reversible if a system can be returned to its initial state along a reversed process line with no change in the surroundings of the system. In actual practice a reversible process is not possible. All processes contain effects that render them irreversible. For example, friction, nonelastic deformation, turbulence, mixing, and heat transfer are all effects that cause a process to be irreversible. The reversible process, although impossible, is valuable, because it serves as a reference value. That is, we know that the ideal process is a theoretical limit toward which we can strive by minimizing the irreversible effects listed above.

Many processes can be described by a phrase which indicates that one of its properties or characteristics remains constant during the process. Table I-1 shows the more common of these processes, together with expressions for work, heat transfer, and entropy change for ideal gases.

I.2.3 Thermodynamic Laws

Thermodynamic laws are relationships between mass and energy quantities for both open and closed systems. In classical form they are based on the conservation of mass for a system with no relativistic effects. Table I-2 shows the conservation-of-mass relations for

Table I-1 Ideal Gas Processes[a]

Process	Describing Equations	$_1W_2$	$_1Q_2$	$S_2 - S_1$
Isometric or constant volume	$V = c,^{(1)}\ v = c, \dfrac{v}{t} = c$	0	$U_2 - U_1 = m(u_2 - u_1)$ $= mc_v(T_2 - T_1)^{(2)}$	$m\left(s_2^0 - s_1^0 - R\ln\dfrac{p_2}{p_1}\right)^{(3)}$ $= mc_v \ln\dfrac{T_2}{T_1}$
Isobaric, isopiestic, or constant pressure	$P = c,\ V/T = c, \dfrac{v}{t} = c$	$p(V_2 - V_1)$ $= mR(T_2 - T_1)$	$H_2 - H_1 = m(h_2 - h_1)$ $= mc_p(T_2 - T_1)$	$m(s_2^0 - s_1^0)$ $= mc_p \ln\dfrac{T_2}{T_1}$
Isothermal or constant temperature	$T = c,\ pV = pv = c$	$p_1V_1 \ln r^{(4)} = (p_2 - V_2)\ln r$ $= mRT \ln r$	$_1W_2$	$mR \ln r$
Isentropic or reversible adiabatic	$s = c$ $pV^{k(5)} = c,\ TV^{k-1}$ $= c,\ p^{k-1}T^{k-1} = c$	$U_1 - U_2 = m(u_1 - u_2)$ $\dfrac{P_1V_1 - P_2V_2}{k-1} = \dfrac{mR(T_1 - T_2)}{k-1}$	0	0
Polytropic	$pV^{k(6)} = c,\ TV^{n-1}$ $= c,\ p^{n-1}T^{n-1} = c$	$\dfrac{P_1V_1 - P_2V_2}{n-1} = \dfrac{mR(T_1 T_2)}{n-1}$	Use first law	$m\left(s_2^0 - s_1^0 - R\ln\dfrac{p_2}{p_1}\right)$ $= \dfrac{k-n}{k-1} \ln r$

[a](1) c stands for an unspecified constant; (2) the second line of each entry applies when c_p and c_v are independent of temperature; (3) $s_2 - s_1 = s^0(T_2) - s^0(T_1) - R\ln(p_2/p_1)$; (4) r = volume ratio or compression ratio $= V_2/V_1$; (5) $k = c_p/c_v > 1$; (6) n = polytropic exponent.

various types of systems. In energy conservation, the first and second laws of thermodynamics form the basis of most technical analysis.

For open systems two approaches to analysis can be taken, depending upon the nature of the process. For steady systems, the steady-state steady-flow assumption

Table I-2 Law of Conservation of Mass

Closed system, any process: m = constant

Open system, SSSF: $\displaystyle\sum_{in} \dot{m} = \sum_{out} \dot{m}$

Open system, USUF: $(m_2 - m_2)_{c.v.} = \left[\displaystyle\sum_{in} \dot{m} = \sum_{out} \dot{m}\right] t$

Open system, general case:

$\dfrac{d}{dt} m_{c.v.} = \displaystyle\sum_{in} \dot{m} = \sum_{in} \dot{m}\ \text{or}\ \dfrac{d}{dt} \int_V \rho\, dV = \int_A \rho\, dV_{rn}\, dA$

(SSSF) is adequate. This approach assumes that the state of the material is constant at any point in the system. For transient processes, the uniform flow uniform state (UFUS) assumption fits most situations. This involves the assumption that at points where mass crosses the system boundary, its state is constant with time. Also, the state of the mass in the system may vary with time but is uniform at any time.

Tables I.3 and I.4 give listings of the conservation-of-energy (first-law) and the second-law relations for various systems.

The first law simply gives a balance of energy during a process. The second law, however, extends the utility of thermodynamics to the prediction of both the possibility of a proposed process or the direction of a system change following a perturbation of the system. Although the first law is perhaps more directly valuable in energy conservation, the implications of the second law can be equally illuminating.

Table I-3 First Law of Thermodynamics

Closed system, cyclic process:

$$\oint dQ = \oint dW$$

Closed system, state 1 to state 2:

$$_1Q_2 = E_2 - E_1 + _1W_2$$

E = internal energy + kinetic energy + potential energy = $m(u + V^2/2 + gz)$

Open system:

$$\dot{Q}_{c.v.} = \frac{d}{dt}\int_v \rho\,(u + V^2/2 + gz)dV + \int_v \rho\,(h + V^2/2 + gz)\,V_{rn}dA + \dot{W}_{c.v.,}$$

where enthalpy per unit mass $h = u + pv$;

alternative form:

$$\dot{Q}_{c.v.} + \sum_{in} in\,(h + V^2/2 + gz) = \dot{W}_{c.v.} + \sum_{out} in\,(h + V^2/2 + gz) + \dot{E}_{c.v.,}$$

where

$$\dot{E}_{c.v.} = \frac{d}{dt}\int_v r\,(u + V^2/2 + gz)dV$$

Open system, steady state steady flow (SSSF):

$$\dot{Q}_{c.v.} = \sum_{in} in\,(h + V^2/2 + gz) = \dot{W}_{c.v.} + \sum_{out} in\,(h + V^2/2 + gz)$$

Open system, uniform state uniform flow (USUF):

$$_1\dot{Q}_{c.v.} = \sum_{in} m\,(h + V^2/2 + gz) = _1\dot{W}_{2c.v.} + \sum_{out} m\,(h + V^2/2 + gz) + [m\,(u + V^2/2 + gz)]_{1c.v.}^{2c.v.}$$

There are two statements of the second law. The two, although appearing to be different, actually can be shown to be equivalent. Therefore, we state only one of them, the Kelvin-Planck version:

It is impossible for any device to operate in a cycle and produce work while exchanging heat only with bodies at a single fixed temperature.

The other statement is called the Clausius statement.

The implications of the second law are many. For example, it allows us to (1) determine the maximum possible efficiency of a heat engine, (2) determine the maximum coefficient of performance for a refrigerator, (3) determine the feasibility of a proposed process, (4) predict the direction of a chemical or other type of

process, and (5) correlate physical properties. So we see that the second law is quite valuable.

I.2.4 Efficiency

Efficiency is a concept used to describe the effectiveness of energy conversion devices that operate in cycles as well as in individual system components that operate in processes. Thermodynamic efficiency, η, and coefficient of performance, COP, are used for devices that operate in cycles. The following definitions apply:

$$\eta = \frac{W_{net}}{Q_H} \quad COP = \frac{Q_L}{W}$$

where Q_L and Q_H represent heat transferred from cold and hot regions, respectively; W_{net} is useful work pro-

Table I-4 Second Law of Thermodynamics

Closed system, cyclic process:

$$\oint \frac{dQ}{T} \leq 0$$

Closed system, state 1 to state 2:

$$\oint_1^2 \frac{dQ}{T} \leq S_2 - S_1 = m\left(S_2 - S_1\right)$$

Open system:

$$\frac{d}{dt}\int_V \rho s \, dV + \sum_{out} \dot{m}s \geq \sum_{in} \dot{m}s + \int_A \frac{\dot{Q}}{T} \, dA$$

or

$$\frac{d}{dt}\int_V \rho s \, dV + \int_V \rho s \, V_m dA \geq \int_A \frac{Q}{T} \, dA$$

Open system, SSSF:

$$\sum_{out} \dot{m}s + \sum_{in} \dot{m}s \int_A \frac{Q}{T} \, dA$$

Open system, USUF:

$$\left(m_2 s_2 - m_1 s_1\right)_{c.v.} + \sum_{out} \dot{m}s \geq \sum_{in} \dot{m}s \int_A \frac{\dot{Q}_{c.v.}}{T} \, dA$$

duced in a heat engine; and W is the work required to drive the refrigerator. A heat engine produces useful work, while a refrigerator uses work to transfer heat from a cold to a hot region. There is an ideal cycle, called the Carnot cycle, which yields the maximum efficiency for heat engines and refrigerators. It is composed of four ideal reversible processes; the efficiency of this cycle is

$$\eta_c = 1 - \frac{T_L}{T_H}$$

and the COP is

$$[COP]_c = \frac{TL/TH}{1 - TL/TH}$$

These represent the best possible performance of cyclic energy conversion devices operating between temperature extremes T_H and T_L. The thermodynamic efficiency

should not be confused with efficiencies applied to devices that operate along a process line. This efficiency is defined as

$$\eta_{device} = \frac{\text{actual energy transfer}}{\text{ideal energy transfer}}$$

for a work-producing device and

$$\eta_{device} = \frac{\text{ideal energy transfer}}{\text{actual energy transfer}}$$

for a work-consuming device. Note these definitions are such that $\eta < 1$. These efficiencies are convenience factors in that the actual performance can be calculated from an ideal process line and the efficiency, which generally must be experimentally determined. Table I-5 shows the most commonly encountered versions of efficiencies.

I.2.5 Power and Refrigeration Cycles

Many cycles have been devised to convert heat into work, and vice versa. Several of these take advantage of the phase change of the working fluid: for example the Rankine, the vapor compression, and the absorption cycles. Others involve approximations of thermodynamic processes to mechanical processes and are called air-standard cycles.

Table I-5 Thermodynamic Efficiency

Heat engines and refrigerators:

Engine efficiency $\eta \equiv W/Q_H \leq \eta_{Carnot} = (T_H - T_L)/T_H < 1$

Heat pump c.o.p. $\beta' \equiv Q_H/W \leq \beta'_{Carnot} = T_H/(T_H - T_L) > 1$

Refrigerator c.o.p. $\beta \equiv Q_L/W \leq \beta_{Carnot} = T_L/(T_H - T_L)$,

$\quad 0 < \beta < \infty, (Q_H/Q_L)_{Carnot} = T_H - T_L$

Process efficiencies

$\quad \eta_{ad, turbine} = w_{actual, adiabatic}/w_{isentropic}$
$\quad \eta_{ad, compressor} = w_{isentropic}/w_{actual, adiabatic}$
$\quad \eta_{ad, nozzle} = \text{K.E.}_{actual, adiabatic}/\text{K.E.}_{isentropic}$

$$\eta_{nozzle} = \frac{V_a^2/2gc}{V_s^2/2gc}$$

$\eta_{cooled\ nozzle} = w_{isentropic\ rev.}/w_{actual}$

The Rankine cycle is probably the most frequently encountered cycle in thermodynamics. It is used in almost all large-scale electric generation plants, regardless of the energy source (gas, coal, oil, or nuclear). Many modern steam-electric power plants operate at supercritical pressures and temperatures during the boiler heat addition process. This leads to the necessity of reheating between high- and lower-pressure turbines to prevent excess moisture in the latter stages of turbine expansion (prevents blade erosion). Feedwater heating is also extensively used to increase the efficiency of the basic Rankine cycle. (See Ref. 1 for details.)

The vapor compression cycle is almost a reversed Rankine cycle. The major difference is that a simple expansion valve is used to reduce the pressure between the condensor and the evaporator rather than being a work-producing device. The reliability of operation of the expansion valve is a valuable trade-off compared to the small amount of work that could be reclaimed. The vapor compression cycle can be used for refrigeration or heating (heat pump).

In the energy conservation area, applications of the heat pump are taking on added emphasis. The device is useful for heating from an electrical source (compressor) in situations where direct combustion is not available. Additionally, the device can be used to upgrade the temperature level of waste heat recovered at a lower temperature.

Air-standard cycles, useful both for power generation and heating/cooling applications, are the thermodynamic approximations to the processes occurring in the actual devices. In the actual cases, a thermodynamic cycle is not completed, necessitating the approximations. Air-standard cycles are analyzed by using the following approximations.

1. Air is the working fluid and behaves as an ideal gas.

2. Combustion and exhaust processes are replaced by heat exchangers.

Other devices must be analyzed component by component using property data for the working fluids. (See Appendix II.) Figure I-2 gives a listing of various power systems with their corresponding thermodynamic cycle and other pertinent information.

I.2.6 Combustion Processes

The combustion process continues to be the most prevalent means of energy conversion. Natural and manufactured gases, coal, liquid fuel/air mixtures, and even wood and peat are examples of energy sources requiring combustion.

There are two overriding principles of importance in analyzing combustion processes. They are the combustion equation and the first law for the combustion chamber. The combustion equation is simply a mass balance between reactants and products of the chemical reaction combustion process. The first law is the energy balance for the same process, using the results of the combustion equation as input.

In practice, we can restrict our discussion to hydrocarbon fuels, meaning that the combustion equation (chemical balance) is written as

Table I-6 Characteristics of Some of the Hydrocarbon Families

Family	Formula	Structure	Saturated
Paraffin	C_nH_{2n+2}	Chain	Yes
Olefin	C_nH_{2n}	Chain	No
Diolefin	C_nH_{2n-2}	Chain	No
Naphthene	C_nH_{2n}	Ring	Yes
Aromatic Benzene	C_nH_{2n-6}	Ring	No
Aromatic Naphthalene	C_nH_{2n-12}	Ring	No

Molecular structure of some hydrocarbon fuels:

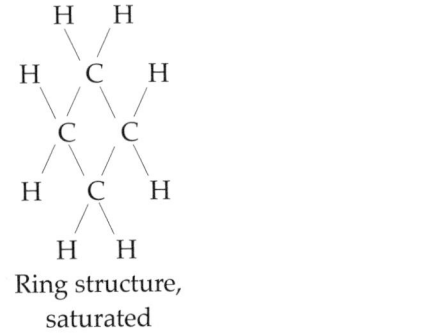

Chain structure,
saturated

Chain structure,
unsaturated

Ring structure,
saturated

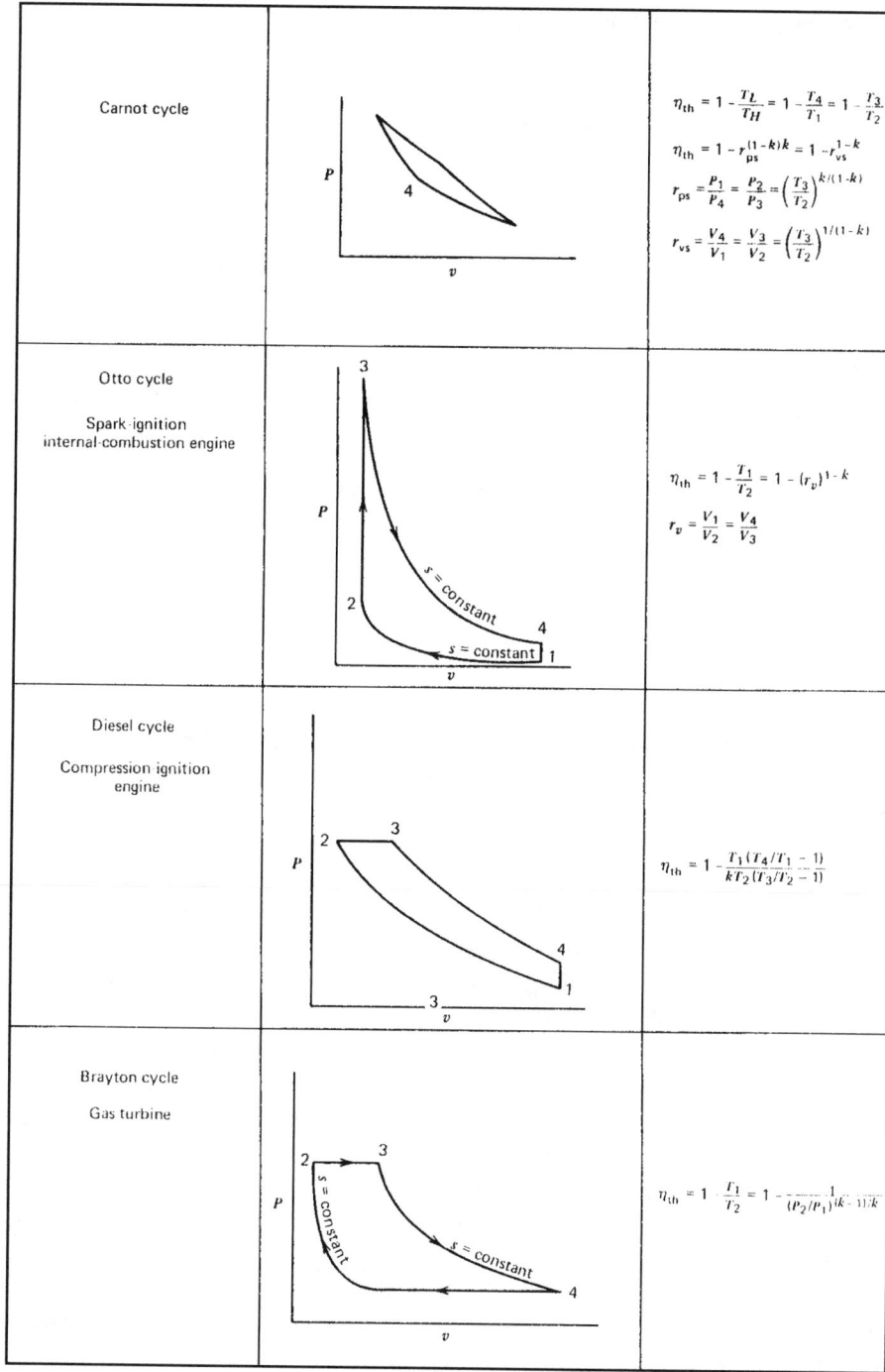

Figure I-2 Air standard cycles.

The equations shown in the figure are:

Carnot cycle:
$$\eta_{th} = 1 - \frac{T_L}{T_H} = 1 - \frac{T_4}{T_1} = 1 - \frac{T_3}{T_2}$$
$$\eta_{th} = 1 - r_{ps}^{(1-k)k} = 1 - r_{vs}^{1-k}$$
$$r_{ps} = \frac{P_1}{P_4} = \frac{P_2}{P_3} = \left(\frac{T_3}{T_2}\right)^{k/(1-k)}$$
$$r_{vs} = \frac{V_4}{V_1} = \frac{V_3}{V_2} = \left(\frac{T_3}{T_2}\right)^{1/(1-k)}$$

Otto cycle Spark-ignition internal-combustion engine:
$$\eta_{th} = 1 - \frac{T_1}{T_2} = 1 - (r_v)^{1-k}$$
$$r_v = \frac{V_1}{V_2} = \frac{V_4}{V_3}$$

Diesel cycle Compression-ignition engine:
$$\eta_{th} = 1 - \frac{T_1(T_4/T_1 - 1)}{kT_2(T_3/T_2 - 1)}$$

Brayton cycle Gas turbine:
$$\eta_{th} = 1 - \frac{T_1}{T_2} = 1 - \frac{1}{(P_2/P_1)^{(k-1)/k}}$$

ticular application. Table I-6 gives the characteristics of some of the hydrocarbons. Table I-7 shows the volumetric analyses of several gaseous fuels.

Once a combustion process is decided upon (i.e., the fuel to be used and the heat transfer/combustion chamber are selected), the relative amount of fuel and air become of prime importance. This is because the air/fuel ratio (AF) controls the temperature of the combustion zone and the energy available to be transferred to a working fluid or converted to work. Stoichiometric air is that quantity of air required such that no oxygen would appear in the products. Excess air occurs when more than enough air is provided to the combustion process. Ideal combustion implies perfect mixing and complete reactions. In this case theoretical air (TA) would yield no free oxygen in the products. Excess air, then, is actual air less theoretical air.

Most industrial combustion processes conform closely to a steady-state, steady-flow case. The first law for an open control volume surrounding the combustion zone can then be written. If we assume that Q and W are zero and that ΔK.E. and ΔP.E. are negligible, then the following equation results:

$$\sum_{products} (H_e - H_{ref}) = \sum_{reactants} (H_i - H_{ref}) + H_{comb} \qquad (I.3)$$

Subscripts i and e refer to inlet and exit conditions, respectively. H_{ref} is the enthalpy of each component at some reference temperature. $\Delta H comb$ represents the heat of combustion for the fuel and, in general, carries a negative value, meaning that heat would have to be transferred out of the system to maintain inlet and exit temperatures at the same level.

The adiabatic flame temperature occurs when the

$$C_xH_y + \alpha(O_2 + 3.76N_2) \rightarrow b\ CO_2 + c\ CO_2$$
$$+ e\ H_2O + d\ O_2 + 3.76a\ N_2$$

This equation neglects the minor components of air; that is, air is assumed to be 1 mol of O_2 mixed with 3.76 mol of N_2. The balance is based on 1 mol of fuel C_xH_y. The unknowns are determined for each par-

Table I-7 Volumetric Analyses of Some Typical Gaseous Fuels

Constituent	Various Natural Gases				Producer Gas from Bituminous Coal	Carbureted Water Gas	Coke Oven Gas
	A	B	C	D			
Methane	93.9	60.1	67.4	54.3	3.0	10.2	32.1
Ethane	3.6	14.8	16.8	16.			
Propane	1.2	13.4	15.8	16.2			
Butanes plus[a]	1.3	4.2		7.4			
Ethene						6.1	3.5
Benzene						2.8	0.5
Hydrogen					14.0	40.5	46.5
Nitrogen		7.5		5.8	50.9	2.9	8.1
Oxygen					0.6	0.5	0.8
Carbon monoxide					27.0	34.0	6.3
Carbon dioxide					4.5	3.0	2.2

[a]This includes butane and all heavier hydrocarbons.

combustion zone is perfectly insulated. The solution of equation I.3 would give the adiabatic flame temperature for any particular case. The maximum adiabatic flame temperature would occur when complete combustion occurs with a minimum of excess O_2 appearing in the products.

Appendix II gives tabulated values for the important thermophysical properties of substances important in combustion.

Gas Analysis. During combustion in heaters and boilers, the information required for control of the burner settings is the amount of excess air in the fuel gas. This percentage can be a direct reflection of the efficiency of combustion.

The most accurate technique for determining the volumetric makeup of combustion by-products is the Orsat analyzer. The Orsat analysis depends upon the fact that for hydrocarbon combustion the products may contain CO_2, O_2, CO, N_2, and water vapor. If enough excess air is used to obtain complete combustion, no CO will be present. Further, if the water vapor is removed, only CO_2, O_2, and N_2 remain.

The Orsat analyzer operates on the following principles. A sample of fuel gas is first passed over a desiccant to remove the moisture. (The amount of water vapor can be found later from the combustion equation.) Then the sample is exposed in turn to materials that absorb first the CO_2, then the O_2, and finally the CO (if present). After each absorption the volumetric change is carefully measured in a graduated pipette system. The

remaining gas is assumed to be N_2. Of course, it could contain some trace of gases and pollutants.

I.2.7 Psychrometry

Psychrometry is the science of air/water vapor mixtures. Knowledge of the behavior of such systems is important, both in meteorology and industrial processes, especially heating and air conditioning. The concepts can be applied to other ideal gas/water vapor mixtures.

Air and water vapor mixed together at a total pressure of 1 atm is called atmospheric air. Usually, the amount of water vapor in atmospheric air is so minute that the vapor and air can be treated as an ideal gas. The air existing in the mixture is often called dry air, indicating that it is separate from the water vapor coexisting with it.

Two terms frequently encountered in psychrometry are relative humidity and humidity ratio. Relative humidity, ϕ, is defined as the ratio of the water vapor pressure to the saturated vapor pressure at the temperature of the mixture. Figure I-3 shows the relation between points on the T–s diagram that yield the relative humidity. Relative humidity cannot be greater than unity, or 100%, as is normally stated.

The humidity ratio, ω, on the other hand, is defined as the ratio of the mass of water vapor to the mass of dry air in atmospheric air, $\omega = m_v/m_a$. This can be shown to be $\omega = v_a/v_v$, and a relationship between ω and ϕ exists, $\omega = (va/v_g)\phi$, where v_g refers to the specific volume of saturated water vapor at the temperature of the mixture.

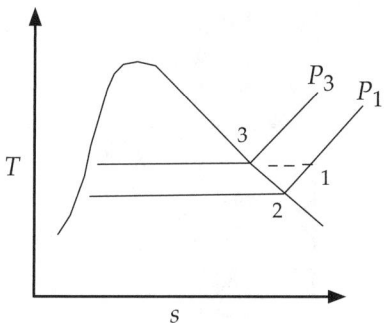

Figure I-3 Behavior of water in air: $\phi = P_1/P_3$; T_2 = dew point.

A convenient way of describing the condition of atmospheric air is to define four temperatures: dry-bulb, wet-bulb, dew-point, and adiabatic saturation temperatures. The dry-bulb temperature is simply that temperature which would be measured by any of several types of ordinary thermometers placed in atmospheric air.

The dew-point temperature (point 2 on Figure I-3) is the saturation temperature of the water vapor at its existing partial pressure. In physical terms it is the mixture temperature where water vapor would begin to condense if cooled at constant pressure. If the relative humidity is 100%, the dew-point and dry-bulb temperatures are identical.

In atmospheric air with relative humidity less than 100%, the water vapor exists at a pressure lower than saturation pressure. Therefore, if the air is placed in contact with liquid water, some of the water would be evaporated into the mixture, and the vapor pressure would be increased. If this evaporation were done in an insulated container, the air temperature would decrease since part of the energy to vaporize the water must come from the sensible energy in the air. If the air is brought to the saturated condition, it is at the adiabatic saturation temperature.

A psychrometric chart is a plot of the properties of atmospheric air at a fixed total pressure, usually 14.7 psia. The chart can be used to quickly determine the properties of atmospheric air in terms of two independent properties, for example dry-bulb temperature and relative humidity. Also, certain types of processes can be described on the chart. Appendix II contains a psychrometric chart for 14.7-psia atmospheric air. Psychrometric charts can also be constructed for pressures other than 14.7 psia.

I.3 HEAT TRANSFER

Heat transfer is the branch of engineering science that deals with the prediction of energy transport caused by temperature differences. Generally, the field is broken down into three basic categories: conduction, convection, and radiation heat transfer.

Conduction is characterized by energy transfer by internal microscopic motion, such as lattice vibration and electron movement. Conduction will occur in any region where mass is contained and across which a temperature difference exists.

Convection is characterized by motion of a fluid region. In general, the effect of the convective motion is to augment the conductive effect caused by the existing temperature difference.

Radiation is an electromagnetic wave transport phenomenon and requires no medium for transport. In fact, radiative transport is generally more effective in a vacuum since there is attenuation in a medium.

I.3.1 Conduction Heat Transfer

The basic tenet of conduction is called Fourier's law,

$$\dot{Q} = -kA\frac{dT}{dx}$$

The heat flux is dependent upon the area across which energy flows and the temperature gradient at that plane. The coefficient of proportionality is a material property, called thermal conductivity k. This relationship always applies, both for steady and transient cases. If the gradient can be found at any point and time, the heat flux density, $\dot{Q}/A$, can be calculated.

Conduction Equation. The control volume approach from thermodynamics can be applied to give an energy balance, which we call the conduction equation. For brevity we omit the details of this development; see Refs. 2 and 3 for these derivations. The result is

$$G + K\nabla^2 T = -\rho C \frac{\partial T}{\partial \tau} \tag{I.4}$$

This equation gives the temperature distribution in space and time; G is a heat-generation term, caused by chemical, electrical, or nuclear effects in the control volume. Equation I.4 can be written

$$\nabla^2 T + \frac{G}{K} = \frac{\rho C}{k}\frac{\partial T}{\partial \tau}$$

The ratio $k/\rho C$ is also a material property called thermal diffusivity u. Appendix II gives thermophysical properties of many common engineering materials.

For steady, one-dimensional conduction with no heat generation,

$$\frac{D^2T}{dx^2} = 0$$

This will give $T = ax + b$, a simple linear relationship between temperature and distance. Then the application of Fourier's law gives

$$\dot{Q} = kA\frac{T}{x}$$

a simple expression for heat transfer across the Δx distance. If we apply this concept to insulation, for example, we get the concept of the R value. R is just the resistance to conduction heat transfer per inch of insulation thickness (i.e., $R = 1/k$).

Multilayered, One-Dimensional Systems. In practical applications, there are many systems that can be treated as one-dimensional, but they are composed of layers of materials with different conductivities. For example, building walls and pipes with outer insulation fit this category. This leads to the concept of overall heat-transfer coefficient, U. This concept is based on the definition of a convective heat-transfer coefficient,

$$\dot{Q} = hA\ T$$

This is a simplified way of handling convection at a boundary between solid and fluid regions. The heat-transfer coefficient h represents the influence of flow conditions, geometry, and thermophysical properties on the heat transfer at a solid-fluid boundary. Further discussion of the concept of the h factor will be presented later.

Figure I-4 represents a typical one-dimensional, multilayered application. We define an overall heat-transfer coefficient U as

$$Q = UA\ (T_i - T_o)$$

We find that the expression for U must be

$$U = \frac{1}{\frac{1}{h_1} + \frac{x_1}{k_1} + \frac{x_2}{k_2} + \frac{x_3}{k_3} + \frac{1}{h_0}}$$

This expression results from the application of the conduction equation across the wall components and the convection equation at the wall boundaries. Then, by noting that in steady state each expression for heat must be equal, we can write the expression for U, which contains both convection and conduction effects. The U factor is extremely useful to engineers and architects in a wide variety of applications.

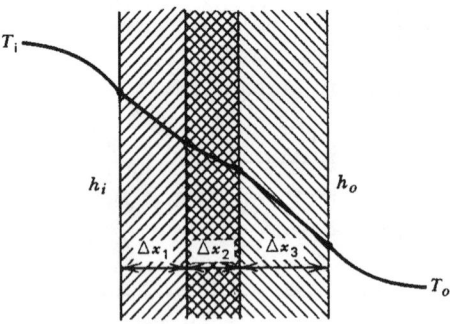

Figure I-4 Multilayered wall with convection at the inner and outer surfaces.

The U factor for a multilayered tube with convection at the inside and outside surfaces can be developed in the same manner as for the plane wall. The result is

$$U = \frac{1}{\frac{1}{h_0} + \sum_j \frac{r_0\ln\left(r_j + 1/r_j\right)}{k_j} + \frac{1r_0}{h_ir_i}}$$

where r_i and r_o are inside and outside radii.

Caution: The value of U depends upon which radius you choose (i.e., the inner or outer surface).

If the inner surface were chosen, we would get

$$U = \frac{1}{\frac{1r_i}{h_0r_0} + \sum_j \frac{r_i\ln\left(r_j + 1/r_j\right)}{k_j} + \frac{1}{h_i}}$$

However, there is no difference in heat-transfer rate; that is,

$$Q_0 = U_iA_iT_{overall} = U_0A_0T_{overall}$$

so it is apparent that

$$U_iA_i = U_0A_0$$

for cylindrical systems.

Finned Surfaces. Many heat-exchange surfaces experience inadequate heat transfer because of low heat-transfer coefficients between the surface and the adjacent fluid. A remedy for this is to add material to the surface. The added material in some cases resembles a fish "fin," thereby giving rise to the expression "a finned surface." The performance of fins and arrays of fins is an important item in the analysis of many heat-exchange devices. Figure I-5 shows some possible shapes for fins.

The analysis of fins is based on a simple energy balance between one-dimensional conduction down the length of the fin and the heat convected from the exposed surface to the surrounding fluid. The basic equation that applies to most fins is

$$\frac{d_2\theta}{dx^2} + \frac{1}{A}\frac{dA}{dx}\frac{d\theta}{dx} - \frac{h}{k}\frac{1}{A}\frac{dS}{dx}\theta = 0 \qquad (I.5)$$

when θ is $(T - T_\infty)$, the temperature difference between fin and fluid at any point; A is the cross-sectional area of the fin; S is the exposed area; and x is the distance along the fin. Chapman[2] gives an excellent discussion of the development of this equation.

The application of equation I.5 to the myriad of possible fin shapes could consume a volume in itself. Several shapes are relatively easy to analyze; for example, fins of uniform cross section and annular fins can be treated so that the temperature distribution in the fin and the heat rate from the fin can be written. Of more utility, especially for fin arrays, are the concepts of fin efficiency and fin surface effectiveness (see Holman[3]).

Fin efficiency η_f is defined as the ratio of actual heat loss from the fin to the ideal heat loss that would occur if the fin were isothermal at the base temperature. Using this concept, we could write

$$\dot{Q}_{fin} = A_{fin}\frac{h}{}(T_b - T_\infty)\eta_f$$

η_f is the factor that is required for each case. Figure I-6 shows the fin efficiency for several cases.

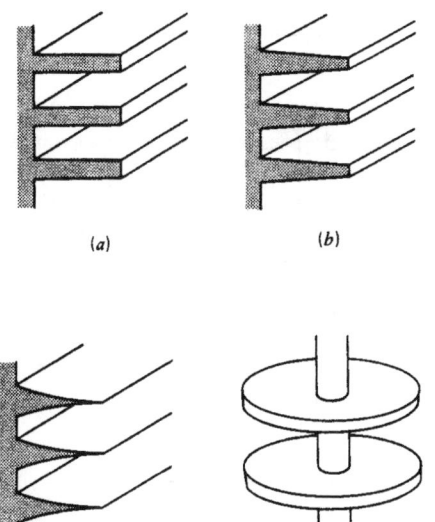

Figure I-5 Fins of various shapes. (a) Rectangular, (b) Trapezoidal, (c) Arbitrary profile, (d) Circumferential.

Surface effectiveness K is defined as the actual heat transfer from a finned surface to that which would occur if the surface were isothermal at the base temperature. Taking advantage of fin efficiency, we can write

$$K = \frac{(A - A_f)h\,\theta_0 + \eta_f A_f \theta_0}{A_h h\theta_0} \qquad (I.6)$$

Equation I.6 reduces to

$$K = 1 - \frac{A_f}{A}(1 - \eta_f)$$

which is a function only of geometry and single fin efficiency. To get the heat rate from a fin array, we write

$$Q_{array} = Kh\,(T_b - T_\infty)\,A$$

where A is the total area exposed.

Transient Conduction. Heating and cooling problems involve the solution of the time-dependent conduction equation. Most problems of industrial significance occur when a body at a known initial temperature is suddenly exposed to a fluid at a different temperature. The temperature behavior for such unsteady problems can be characterized by two dimensionless quantities, the Biot number, Bi = hL/k, and the Fourier modulus, Fo = $\alpha\tau/L^2$. The Biot number is a measure of the effectiveness of conduction within the body. The Fourier modulus is simply a dimensionless time.

If Bi is a small, say Bi $\leq$ 0.1, the body undergoing the temperature change can be assumed to be at a uniform temperature at any time. For this case,

$$\frac{T - T_f}{T_i - T_f} = \exp\left[-\left(\frac{hA}{\rho CV}\right)\tau\right]$$

where T_f and T_i are the fluid temperature and initial body temperature, respectively. The term $(\rho CV/hA)$ takes on the characteristics of a time constant.

If Bi $\geq$ 0.1, the conduction equation must be solved in terms of position and time. Heisler[4] solved the equation for infinite slabs, infinite cylinders, and spheres. For convenience he plotted the results so that the temperature at any point within the body and the amount of heat transferred can be quickly found in terms of Bi and Fo. Figures I-7 to I-10 show the Heisler charts for slabs and cylinders. These can be used if h and the properties of the material are constant.

I.3.2 Convection Heat Transfer

Convective heat transfer is considerably more complicated than conduction because motion of the medium is involved. In contrast to conduction, where many geometrical configurations can be solved analytically, there are only limited cases where theory alone will give convective heat-transfer relationships. Consequently, convection is largely what we call a semi-empirical science. That is, actual equations for heat transfer are based strongly on the results of experimentation.

Convection Modes. Convection can be split into several subcategories. For example, forced convection refers to the case where the velocity of the fluid is completely independent of the temperature of the fluid. On the other hand, natural (or free) convection occurs when the temperature field actually causes the fluid motion through buoyancy effects.

We can further separate convection by geometry into external and internal flows. Internal refers to channel, duct, and pipe flow, and external refers to unbounded fluid flow cases. There are other specialized forms of convection, for example the change-of-phase phenomena: boiling, condensation, melting, freezing, etc. Change-of-phase heat transfer is difficult to predict analytically. Tongs[5] gives many of the correlations for boiling and two-phase flow.

Dimensional Heat-Transfer Parameters. Because experimentation has been required to develop appropriate correlations for convective heat transfer, the use of generalized dimensionless quantities in these correlations is preferred. In this way, the applicability of experimental data covers a wider range of conditions and fluids. Some of these parameters, which we generally call "numbers," are given below:

$$\text{Nusselt number: Nu} = \frac{hL}{k}$$

where k is the fluid conductivity and L is measured along the appropriate boundary between liquid and solid; the Nu is a nondimensional heat-transfer coefficient.

$$\text{Reynolds number: Re} = \frac{Lu}{\upsilon}$$

defined in Section I.4; it controls the character of the flow.

$$\text{Prandtl number: Pr} = \frac{C\mu}{k}$$

ratio of momentum transport to heat-transport characteristics for a fluid; it is important in all convective cases, and is a material property.

$$\text{Grashof number: Gr} = \frac{g\ \beta(T - T_\infty)L^3}{\upsilon^2}$$

serves in natural convection the same role as Re in forced convection; that is, it controls the character of the flow.

$$\text{Stanton number: St} = \frac{h}{\rho\ uC_p}$$

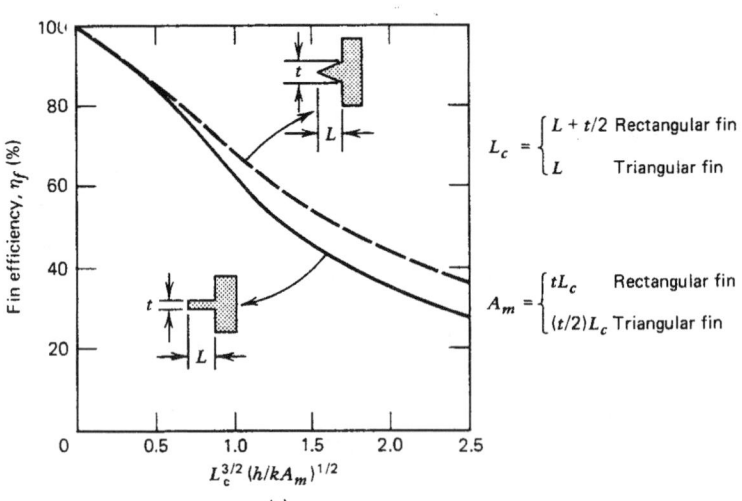

(a)

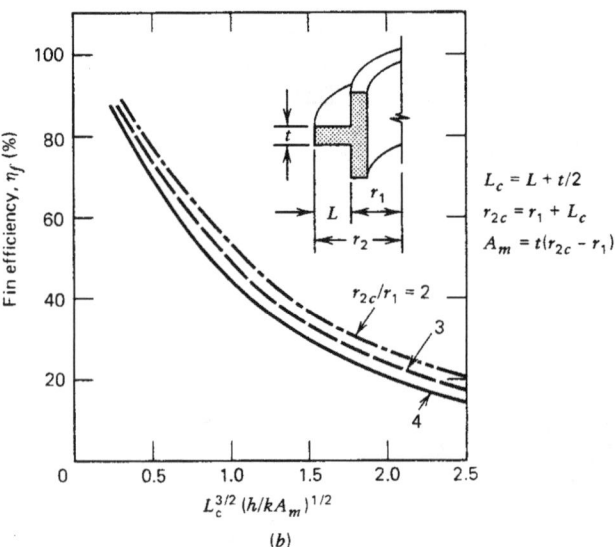

(b)

Figure I-6 (a) Efficiencies of rectangular and triangular fins, (b) Efficiencies of circumferential fins of rectangular profile.

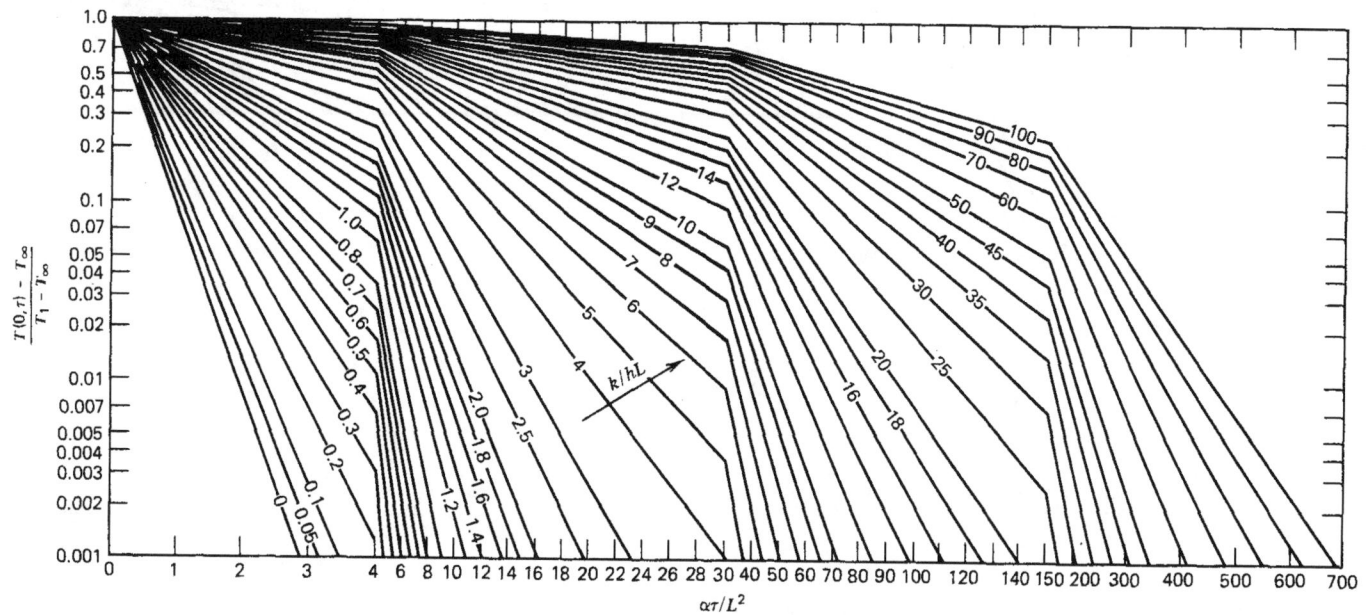

Figure I-7 Midplane temperature for an infinite plate of thickness 2L. (From Ref. 4.)

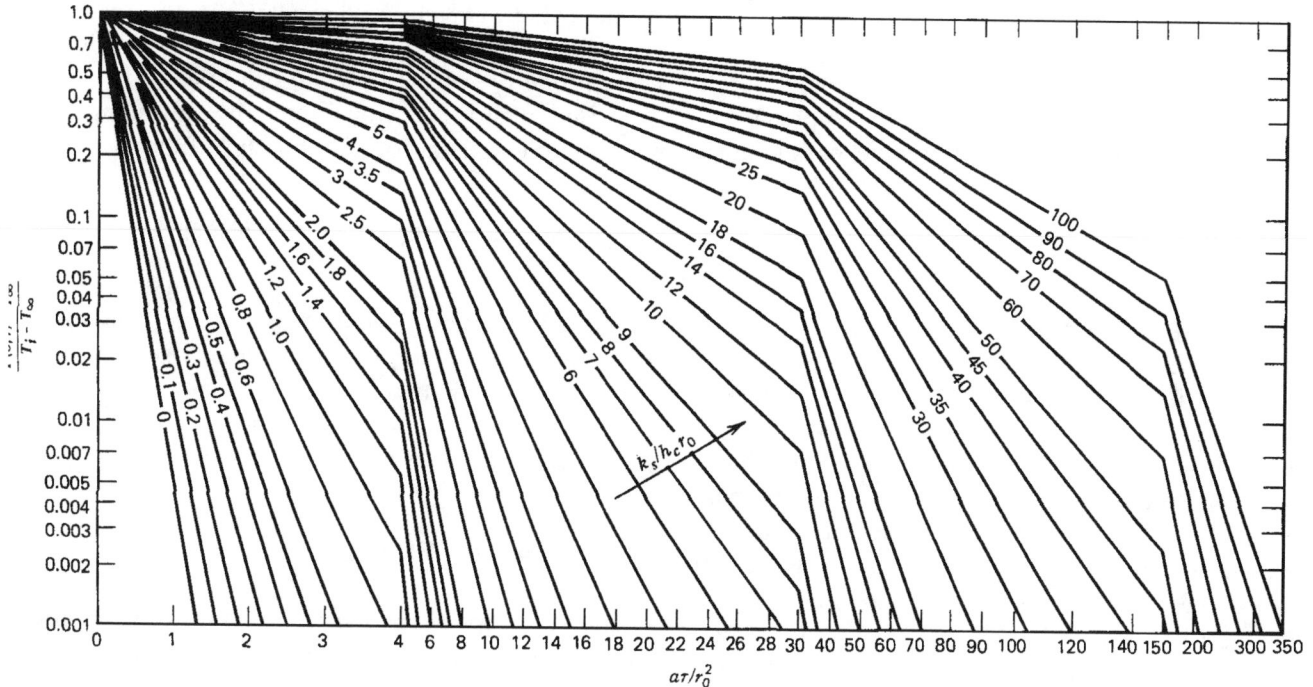

Figure I-8 Axis temperature for an infinite cylinder of radius r_0. (From Ref. 4.)

also a nondimensional heat-transfer coefficient; it is very useful in pipe flow heat transfer.

In general, we attempt to correlate data by using relationships between dimensionless numbers: for example, in many convection cases we could write Nu = Nu(Re, Pr) as a functional relationship. Then it is possible, either from analysis, experimentation, or both, to write an equation that can be used for design calculations. These are generally called working formulas.

Forced Convection Past Plane Surfaces. The average heat-transfer coefficient for a plate of length L may be calculated from

$$Nu_L = 0.664 \, (Re_L)^{1/2}(Pr)^{1/3}$$

if the flow is laminar (i.e., if $Re_L \leq 4{,}000$). For this case the fluid properties should be evaluated at the mean film temperature T_m, which is simply the arithmetic

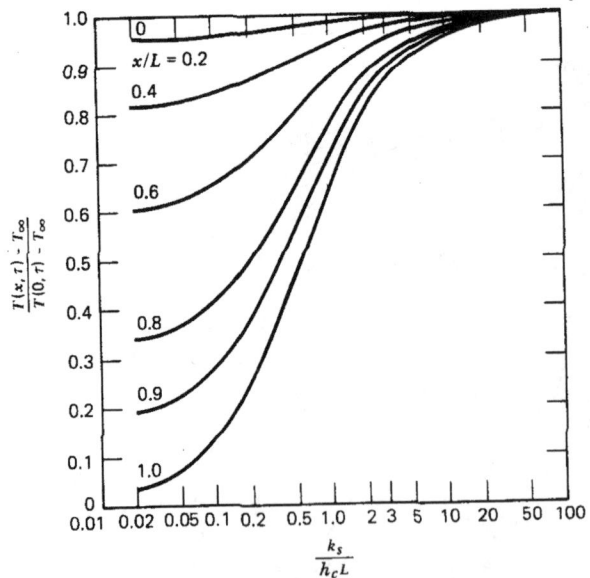

Figure I-9 Temperature as a function of center temperature in an infinite plate of thickness 2L. (From Ref. 4.)

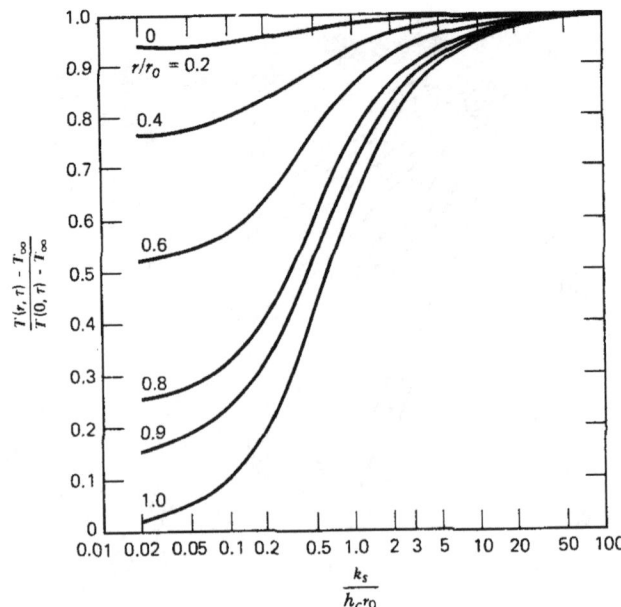

Figure I-10 Temperature as a function of axis temperature in an infinite cylinder of radius r_0. (From Ref. 4.)

average of the fluid and the surface temperature.

For turbulent flow, there are several acceptable correlations. Perhaps the most useful includes both laminar leading edge effects and turbulent effects. It is

$$\text{Nu} = 0.0036 \, (\text{Pr})^{1/3} \, [(\text{Re}_L)^{0.8} - 18.700]$$

where the transition Re is 4,000.

Forced Convection Inside Cylindrical Pipes or Tubes. This particular type of convective heat transfer is of special engineering significance. Fluid flows through pipes, tubes, and ducts are very prevalent, both in laminar and turbulent flow situations. For example, most heat exchangers involve the cooling or heating of fluids in tubes. Single pipes and/or tubes are also used to transport hot or cold liquids in industrial processes. Most of the formulas listed here are for the $0.5 \leq \text{Pr} \leq 100$ range.

Laminar Flow. For the case where $\text{Re}_D < 2300$, Nusselt showed that $\text{Nu}_D = 3.66$ for long tubes at a constant tube-wall temperature. For forced convection cases (laminar and turbulent) the fluid properties are evaluated at the bulk temperature T_b. This temperature, also called the mixing-cup temperature, is defined by

$$T_b = \frac{\int_0^R uTr \, dr}{\int_0^R ur \, dr}$$

if the properties of the flow are constant.

Sieder and Tate developed the following more convenient empirical formula for short tubes.

$$\text{Nu}_D = 1.86 \, (\text{Re}_D)^{1/3} (\text{Pr})^{1/3} \left(\frac{D}{L}\right)^{1/3} \left(\frac{\mu}{\mu_s}\right)^{0.14}$$

The fluid properties are to be evaluated at T_b except for the quantity μ_s, which is the dynamic viscosity evaluated at the temperature of the wall.

Turbulent Flow. McAdams suggests the empirical relation

$$\text{Nu}_D = 0.023 \, (\text{Pr}_D)^{0.8} (\text{Pr})^n \qquad (\text{I.7})$$

where $n = 0.4$ for heating and $n = 0.3$ for cooling. Equation I.7 applies as long as the difference between the pipe surface temperature and the bulk fluid temperature is not greater than 10°F for liquids or 100°F for gases.

For temperature differences greater then the limits specified for equation I.7, or for fluids more viscous than water, the following expression from Sieder and Tate will give better results.

$$\text{NU}_D = 0.027 \, (\text{Pr}_D)^{0.8} (\text{Pr})^{1/3} \left(\frac{\mu}{\mu_s}\right)^{0.14}$$

Note that the McAdams equation requires only a knowledge of the bulk temperature, whereas the Sieder-Tate expression also requires the wall temperature. Many people prefer equation I.7 for that reason.

Nusselt found that short tubes could be represented by the expression

$$\text{Nu}_D = 0.036\,(\text{Pe}_D)^{0.8}(\text{Pr})^{1/3}\left(\frac{\mu}{\mu_s}\right)^{0.14}\left(\frac{D}{L}\right)^{1/18}$$

For noncircular ducts, the concept of equivalent diameter can be employed so that all the correlations for circular systems can be used.

Forced Convection in Flow Normal to Single Tubes and Banks. This circumstance is encountered frequently, for example air flow over a tube or pipe carrying hot or cold fluid. Correlations of this phenomenon are called semi-empirical and take the form $\text{Nu}_D = C(\text{Re}_D)^m$. Hilpert, for example, recommends the values given in Table I-8. These values have been in use for many years and are considered accurate.

Flows across arrays of tubes (tube banks) may be even more prevalent than single tubes. Care must be exercised in selecting the appropriate expression for the tube bank. For example, a staggered array and an in-line array could have considerably different heat-transfer characteristics. Kays and London[6] have documented many of these cases for heat-exchanger applications. For a general estimate of order-of-magnitude heat-transfer coefficients, Colburn's equation

$$\text{Nu}_D = 0.33\,(\text{Re}_D)^{0.6}\,(\text{Pr})^{1/3}$$

is acceptable.

Free Convection Around Plates and Cylinders. In free convection phenomena, the basic relationships take on the functional form $\text{Nu} = f(\text{Gr}, \text{Pr})$. The Grashof number replaces the Reynolds number as the driving function for flow.

In all free convection correlations it is customary to evaluate the fluid properties at the mean film temperature T_m, except for the coefficient of volume expansion β, which is normally evaluated at the temperature of the undisturbed fluid far removed from the surface—namely, T_f. Unless otherwise noted, this convention should be used in the application of all relations quoted here.

Table I-9 gives the recommended constants and exponents for correlations of natural convection for vertical plates and horizontal cylinders of the form $\text{Nu} = C \cdot \text{Ra}^m$. The product $\text{Gr} \cdot \text{Pr}$ is called the Rayleigh number (Ra) and is clearly a dimensionless quantity associated with any specific free convective situation.

I.3.3 Radiation Heat Transfer

Radiation heat transfer is the most mathematically complicated type of heat transfer. This is caused primarily by the electromagnetic wave nature of thermal radiation. However, in certain applications, primarily high-temperature, radiation is the dominant mode of heat transfer. So it is imperative that a basic understanding of radiative heat transport be available. Heat transfer in boiler and fired-heater enclosures is highly dependent upon the radiative characteristics of the surface and the hot combustion gases. It is known that for a body radiating to its surroundings, the heat rate is

$$\dot{Q} = \varepsilon\sigma A\left(T^4 - T_s^4\right)$$

where ε is the emissivity of the surface, σ is the Stefan-Boltzmann constant, and $\sigma = 0.1713 \times 10^{-8}$ Btu/hr ft^2 $\cdot$ R^4. Temperature must be in absolute units, R or K. If $\varepsilon = 1$ for a surface, it is called a "blackbody," a perfect emitter of thermal energy. Radiative properties of various surfaces are given in Appendix II. In many cases, the heat exchange between bodies when all the radiation emitted by one does not strike the other is of interest. In this case we employ a shape factor F_{ij} to modify the basic transport equation. For two blackbodies we would write

$$\dot{Q}_{12} = F_{12}\sigma A\left(T_1^4 - T_2^4\right)$$

Table I-8 Values of C and m for Hilpert's Equation

Range of $N_{\text{Re}D}$	C	m
1-4	0.891	0.330
4-40	0.821	0.385
40-4000	0.615	0.466
4000-40,000	0.175	0.618
40,000-250,000	0.0239	0.805

Table I-9 Constants and Exponents for Natural Convection Correlations

Ra	Vertical Plate[a]		Horizontal Cylinders[b]	
	c	m	c	m
$10^4 < \text{Ra} < 10^9$	0.59	1/4	0.525	1/4
$10^9 < \text{Ra} < 10^{12}$	0.129	1/3	0.129	1/3

[a]Nu and Ra based on vertical height L.
[b]Nu and Ra based on diameter D.

for the heat transport from body 1 to body 2. Figures I-11 to I-14 show the shape factors for some commonly encountered cases. Note that the shape factor is a function of geometry only.

Gaseous radiation that occurs in luminous combustion zones is difficult to treat theoretically. It is too complex to be treated here; the interested reader is referred to Siegel and Howell[7] for a detailed discussion.

I.4   FLUID MECHANICS

In industrial processes we deal with materials that can be made to flow in a conduit of some sort. The laws that govern the flow of materials form the science that is called fluid mechanics. The behavior of the flowing fluid controls pressure drop (pumping power), mixing efficiency, and in some cases the efficiency of heat transfer. So it is an integral portion of an energy conservation program.

I.4.1 Fluid Dynamics

When a fluid is caused to flow, certain governing laws must be used. For example, mass flows in and out of control volumes must always be balanced. In other words, conservation of mass must be satisfied.

In its most basic form the continuity equation (conservation of mass) is

$$\iint_{\text{c.s.}} \rho(\bar{v}\cdot\bar{n})\, dA + \frac{\partial}{\partial_t}\iiint_{\text{c.v.}} \rho\, dV = 0$$

In words, this is simply a balance between mass entering and leaving a control volume and the rate of mass storage. The $\rho(\bar{v}\cdot\bar{n})$ terms are integrated over the control surface, whereas the $\rho\, dV$ term is dependent upon an integration over the control volume.

For a steady flow in a constant-area duct, the continuity equation simplifies to

$$\dot{m} = \rho_f A_c \bar{u} = \text{constant}$$

That is, the mass flow rate $\dot{m}$ is constant and is equal to the product of the fluid density ρ_f, the duct cross section A_c, and the average fluid velocity $\bar{u}$.

If the fluid is compressible and the flow is steady, one gets

$$\frac{\dot{m}}{\rho_f} = \text{constant} = (\bar{u}A_c)(\bar{u}A_c)_2$$

where 1 and 2 refer to different points in a variable area duct.

I.4.2 First Law—Fluid Dynamics

The first law of thermodynamics can be directly applied to fluid dynamical systems, such as duct flows. If there is no heat transfer or chemical reaction and if the internal energy of the fluid stream remains unchanged, the first law is

$$\frac{V_i^2 - V_e^2}{2g_c} + \frac{z_i - z_e}{g_c}g + \frac{p_i - p_e}{\rho} + \left(w_p - w_f\right) = 0 \qquad (I.8)$$

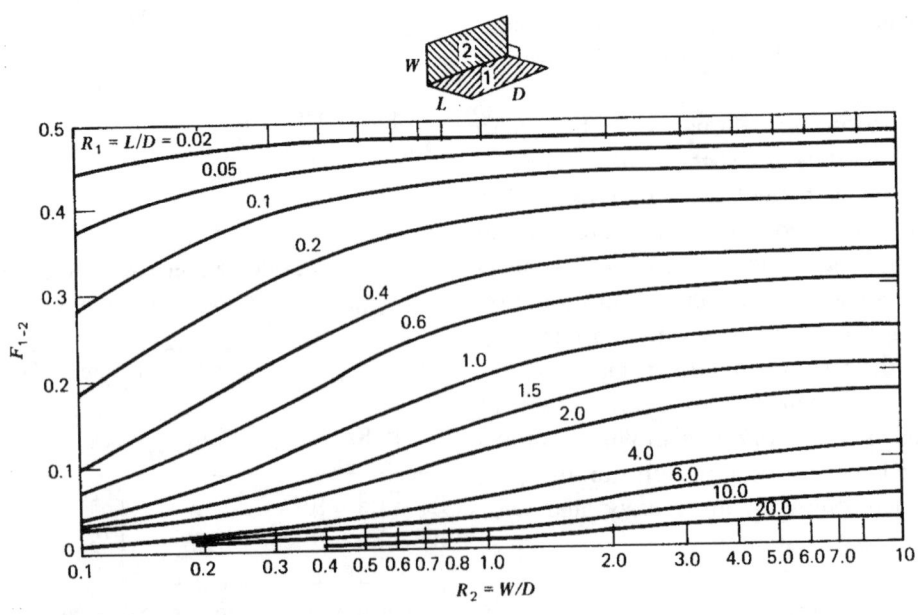

Figure I-11 Radiation shape factor for perpendicular rectangles with a common edge.

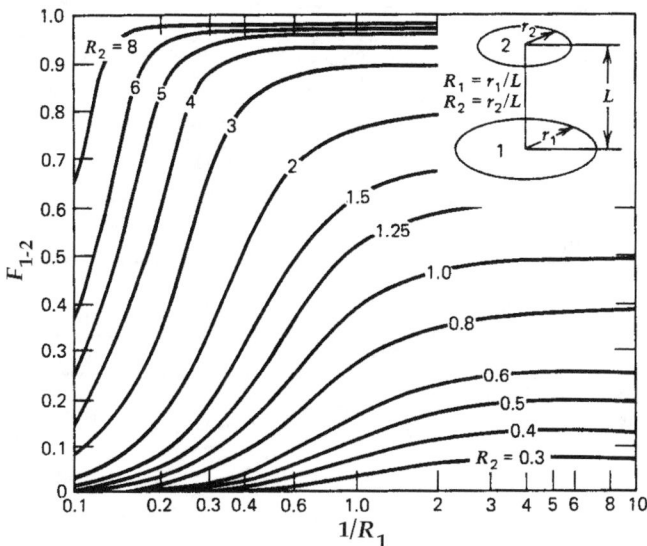

Figure I-12 Radiation shape factor for parallel, concentric disks.

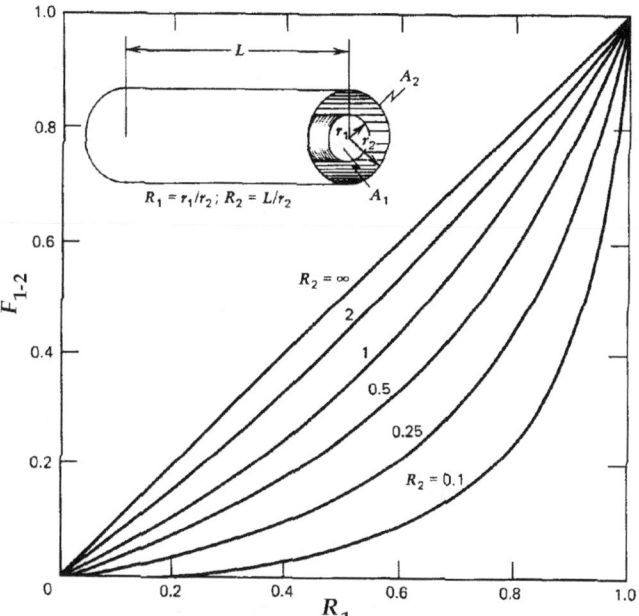

Figure I-13 Radiation shape factor for concentric cylinders of finite length.

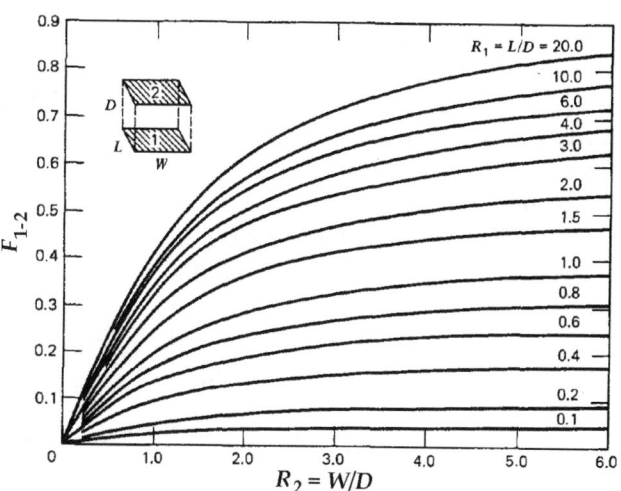

Figure I-14 Radiation shape factor for parallel, directly opposed rectangles.

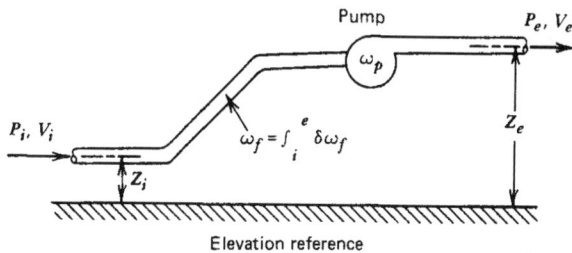

Figure I-15 The first law applied to adiabatic flow system.

In the English system, horsepower is

$$hp = \dot{m}\left(\frac{lb_m}{sec}\right) w_p = \left(\frac{ft \cdot lb_f}{lb_m}\right) \times \left(\frac{1\ hp - sec}{500\ ft - lb}\right) = \left(\frac{\dot{m}w_p}{550}\right)$$

Referring back to equation I.8, the most difficult term to determine is usually the frictional work term w_f. This is a term that depends upon the fluid viscosity, the flow conditions, and the duct geometry. For simplicity, w_f is generally represented as

$$w_f = \frac{p_f}{\rho}$$

when Δp_f is the frictional pressure drop in the duct. Further, we say that

$$\frac{p_f}{\rho} = \frac{2\ f\bar{u}^2 L}{g_c\ D}$$

in a duct of length L and diameter D. The friction factor f is a convenient way to represent the differing influence of laminar and turbulent flows on the friction pressure drop.

where the subscripts i and e refer to inlet and exit conditions, and w_p and w_f are pump work and work required to overcome friction in the duct. Figure I-15 shows schematically a system illustrating this equation.

Any term in equation I.8 can be converted to a rate expression by simply multiplying by the mass flow rate. Take, for example, the pump horsepower,

$$W\left(\frac{energy}{time}\right) = \dot{m}w_p\left(\frac{mass}{time}\right)\left(\frac{energy}{mass}\right)$$

The character of the flow is determined through the Reynolds number, $Re = \rho u D / \mu$, where μ is the viscosity of the fluid. This nondimensional grouping represents the ratio of dynamic to viscous forces acting on the fluid.

Experiments have shown that if $Re \leq 2300$, the flow is laminar. For larger Re the flow is turbulent. Figure I-16 shows how the friction factor depends upon the Re of the flow. Note that for laminar flow the f vs. Re curve is single-valued and is simply equal to $16/Re$. In the turbulent regime, the wall roughness e can affect the friction factor because of its effect on the velocity profile near the duct surface.

If a duct is not circular, the equivalent diameter D_e can be used so that all the relationships developed for circular systems can still be used. D_e is defined as

$$D_e = \frac{4A_c}{P}$$

P is the "wetted" perimeter, that part of the flow cross section that touches the duct surfaces. For a circular system $D_e = 4(\pi D^2/4\pi D) = D$, as it should. For an annular duct, we get

$$D_e = \frac{(\pi D_o^2/4 - \pi D_i^2/4)4}{\pi D_o + \pi D_i} = \frac{\pi(D_o + D_i)(D_o + D_i)}{\pi D_o + \pi D_i}$$

$$= D_o + D_i$$

Pressure Drop in Ducts. In practical applications, the essential need is to predict pressure drops in piping and duct networks. The friction factor approach is adequate for straight runs of constant area ducts. But valves, nozzles, elbows, and many other types of fittings are necessarily included in a network. This can be accounted for by defining an equivalent length L_e for the fitting. Table I-10 shows L_e/D values for many different fittings.

Pressure Drop across Tube Banks. Another commonly encountered application of fluid dynamics is the pressure drop caused by transverse flow across arrays of heat-transfer tubes. One technique to calculate this effect is to find the velocity head loss through the tube bank:

$$N_v = fNF_d$$

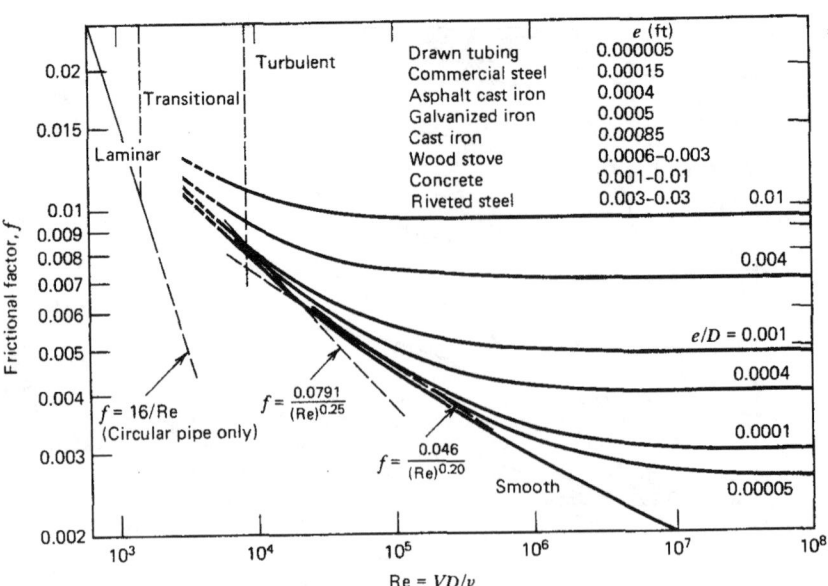

Figure I-16 Friction factors for straight pipes.

where f is the friction factor for the tubes (a function of the Re), N the number of tube rows crossed by the flow, and F_d is the "depth factor." Figures I-17 and I-18 show the f factor and F_d relationship that can be used in pressure-drop calculations. If the fluid is air, the pressure drop can be calculated by the equation

$$p = N\left(\frac{30}{B}\right)\frac{T}{1.73 \times 10^5}\left(\frac{G}{10^3}\right)^2$$

where B is the atmospheric pressure (in. Hg), T is temperature (°R), and G is the mass velocity (lbm/ft² hr).

Bernoulli's Equation. There are some cases where the equation

$$\frac{p}{\rho} + \frac{u^2}{2} + gz = \text{constant}$$

which is called Bernoulli's equation, is useful. Strictly speaking, this equation applies for inviscid, incompressible, steady flow along a streamline. However, even in pipe flow where the flow is viscous, the equation can be applied because of the confined nature of the flow. That is, the flow is forced to behave in a streamlined manner. Note that the first law equation (I.8) yields Bernoulli's equation if the friction drop exactly equals the pump work.

I.4.3 Fluid-Handling Equipment

For industrial processes, another prime application of fluid dynamics lies in fluid-handling equipment.

Table I-10 L_e/D for Screwed Fittings, Turbulent Flow Only[a]

Fitting	L_e/D
45° elbow	15
90° elbow, standard radius	31
90° elbow, medium radius	26
90° elbow, long sweep	20
90° square elbow	65
180° close return bend	75
Swing check valve, open	77
Tee (as el, entering run)	65
Tee (as el, entering branch)	90
Couplings, unions	Negligible
Gate valve, open	7
Gate valve, 1/4 closed	40
Gate valve, 1/2 closed	190
Gate valve, 3/4 closed	840
Globe valve, open	340
Angle valve, open	170

[a]Calculated from Crane Co. Tech. Paper 409, May 1942.

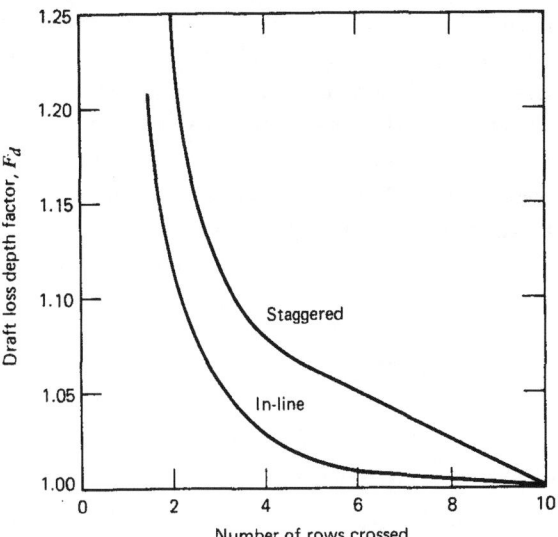

Figure I-17 Depth factor for number of tube rows crossed in convection banks.

Pumps, compressors, fans, and blowers are extensively used to move gases and liquids through the process network and over heat-exchanger surfaces. The general constraint in equipment selection is a matching of fluid handler capacity to pressure drop in the circuit connected to the fluid handler.

Pumps are used to transport liquids, whereas compressors, fans, and blowers apply to gases. There are features of performance common to all of them. For purposes of illustration, a centrifugal pump will be used to discuss performance characteristics.

Centrifugal Machines. Centrifugal machines operate on the principle of centrifugal acceleration of a fluid element in a rotating impeller/housing system to achieve a pressure gain and circulation.

The characteristics that are important are flow rate (capacity), head, efficiency, and durability. Q_f (capacity), h_p (head), and η_p (efficiency) are related quantities, dependent basically on the fluid behavior in the pump and the flow circuit. Durability is related to the wear, corrosion, and other factors that bear on a pump's reliability and lifetime.

Figure I-19 shows the relation between flow rate and related characteristics for a centrifugal pump at constant speed. Graphs of this type are called performance curves; fhp and bhp are fluid and brake horsepower, respectively. The primary design constraint is a matching

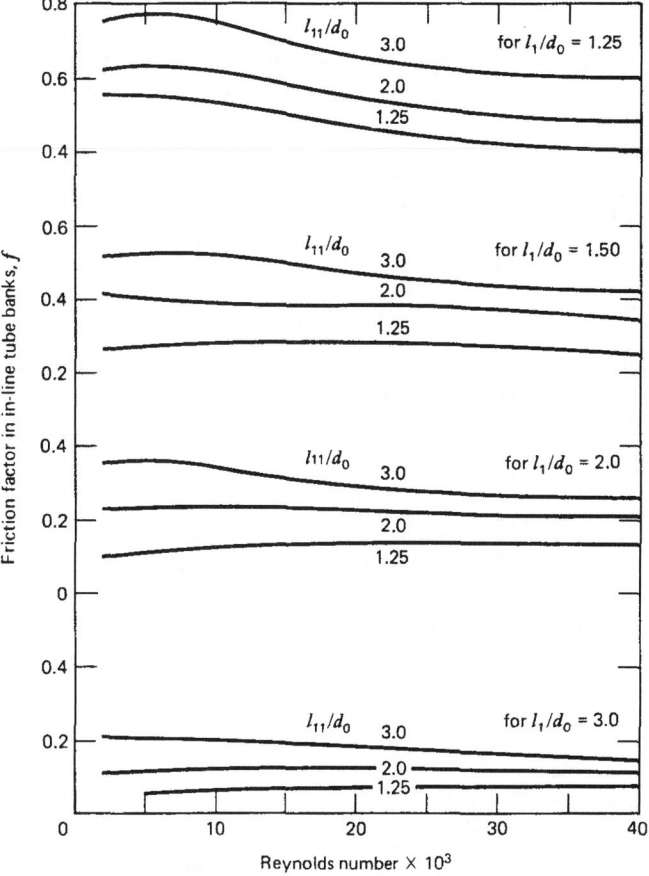

Figure I-18 Friction factor f as affected by Reynolds number for various in-line tube patterns, crossflow gas or air, d_o, tube diameter; $l_\perp$, gap distance perpendicular to the flow; $l_\parallel$, gap distance parallel to the flow.

of flow rate to head. Note that as the flow-rate requirement is increased, the allowable head must be reduced if other pump parameters are unchanged.

Analysis and experience has shown that there are scaling laws for centrifugal pump performance that give the trends for a change in certain performance parameters. Basically, they are:

Efficiency:

$$\eta_p = f_1\left(\frac{Q_f}{D^3 n}\right)$$

Dimensionless head:

$$\frac{h_p g}{D^2 n^2} = f^2\left(\frac{Q_f}{D^3 n}\right)$$

Dimensionless brake horsepower:

$$\frac{\text{bhp} \cdot g}{\gamma D^2 n^3} = f^3\left(\frac{Q_f}{D^3 n}\right)$$

where D is the impeller diameter, n is the rotary impeller speed, g is gravity, and γ is the specific weight of fluid.

The basic relationships yield specific proportionalities such as $Q_f \propto n$ (rpm), $h_p \propto n^2$, $fhp \propto n^3$, $Q_f \propto \frac{1}{D^2}$, $h_p \propto \frac{1}{D^4}$, and $fhp \propto \frac{1}{D^4}$.

For pumps, density variations are generally negligible since liquids are incompressible. But for gas-handling equipment, density changes are very important. The scaling laws will give the following rules for changing density:

$$h_p \propto \rho$$
$$fhp \propto \rho \qquad (Q_f,\ n\ \text{constant})$$

$$\left\{\begin{array}{c} n \\ fhp \\ Q_f \end{array}\right\} \propto \rho^{-1/2} \qquad (h_p\ \text{constant}$$

$$\left\{\begin{array}{c} n \\ Q_f \\ h_p \end{array}\right\} \propto \frac{1}{\rho} \qquad (\dot{m}\ \text{constant})$$

$$fhp \propto \frac{1}{\rho^2}$$

For centrifugal pumps, the following equations hold:

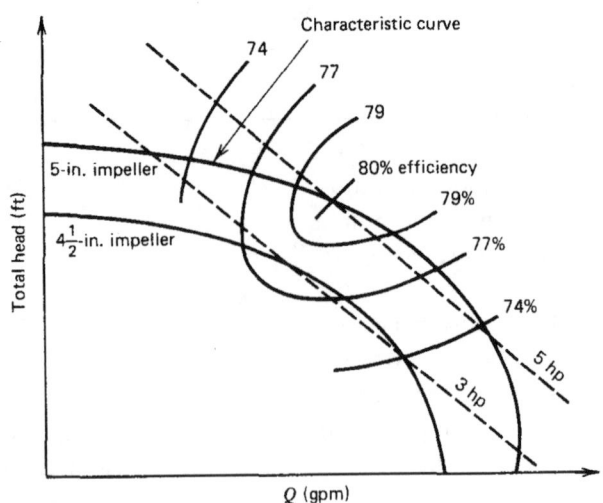

Figure I-19 Performance curve for a centrifugal pump.

$$fhp = \frac{Q_f \rho g h_p}{550 g_c}$$

$$\eta_p p = \frac{Q_f \rho g h_p 550 g_c}{\text{bhp}} = \frac{fhp}{\text{bhp}}$$

system efficiency $\eta_s = \eta_p \times \eta_m$ (motor efficiency)

It is important to select the motor and pump so that at nominal operating conditions, the pump and motor operate at near their maximum efficiency.

For systems where two or more pumps are present, the following rules are helpful. To analyze pumps in parallel, add capacities at the same head. For pumps in series, simply add heads at the same capacity.

There is one notable difference between blowers and pump performance. This is shown in Figure I-20. Note that the bhp continues to increase as permissible head goes to zero, in contrast to the pump curve when bhp approaches zero. This is because the kinetic energy imparted to the fluid at high flow rates is quite significant for blowers.

Manufactures of fluid-handling equipment provide excellent performance data for all types of equipment. Anyone considering replacement or a new installation should take full advantage of these data.

Fluid-handling equipment that operates on a principle other than centrifugal does not follow the centrifugal scaling laws. Evans[8] gives a thorough treatment of most types of equipment that would be encountered in industrial application.

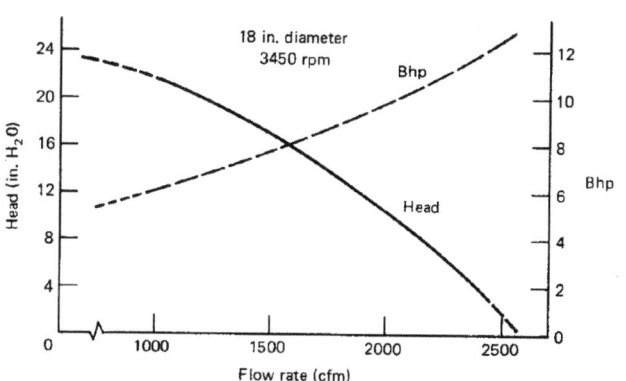

Figure I-20 Variation of head and bhp with flow rate for a typical blower at constant speed.

References

1. G.J. Van Wylen and R.E. Sonntag, *Fundamentals of Classical Thermodynamics*, 2nd ed., Wiley, New York, 1973.
2. A.S. Chapman, *Heat Transfer*, 3rd ed., Macmillan, New York, 1974.
3. J.P. Holman, *Heat Transfer*, 4th ed., McGraw-Hill, New York, 1976.
4. M.P. Heisler, *Trans. ASME*, Vol. 69 (1947), p. 227.
5. L.S. Tong, *Boiling Heat Transfer and Two-Phase Flow*, Wiley, New York, 1965.
6. W.M. Kays and A.L. London, *Compact Heat Exchangers*, 2nd ed., McGraw-Hill, New York, 1963.
7. R. Siegel and J.R. Howell, *Thermal Radiation Heat Transfer*, McGraw-Hill, New York, 1972.
8. FRANK L. Evans, JR., *Equipment Design Handbook for Refineries and Chemical Plants*, Vols. 1 and 2, Gulf Publishing, Houston, Tex., 1974.

SYMBOLS

Thermodynamics

AF	air/fuel ratio
C_p	constant-pressure specific heat
C_v	constant-volume specific heat
Cp_0	zero-pressure constant-pressure specific heat
C_{v0}	zero-pressure constant-volume specufic heat
e, E	specific energy and total energy
g	acceleration due to gravity
g, G	specific Gibbs function and total Gibbs function
g_e	a constant that relates force, mass, length, and time
h, H	specific enthalpy and total enthalpy
k	specific heat ratio: C_p/C_v

K.E.	kinetic energy
lb_f	pound force
lb_m	pound mass
lb mol	pound mole
m	mass
$\dot{m}$	mass rate of flow
M	molecular weight
n	number of moles
n	polytropic exponent
P	pressure
P_i	partial pressure of component i in a mixture
P.E.	potential energy
P_r	relative pressure as used in gas tables
q, Q	heat transfer per unit mass and total heat transfer
$\dot{Q}$	rate of heat transfer
QH, QL	heat transfer from high- and low-temperature bodies
R	gas constant
$\bar{R}$	universal gas constant
s, S	specific entropy and total entropy
t	time
T	temperature
u, U	specific internal energy and total internal energy
v, V	specific volume and total volume
V	velocity
V_r	relative velocity
w, W	work per unit mass and total work
W	rate of work, or power
w_{rev}	reversible work between two states assuming heat transfer with surroundings
x	mass fraction
Z	elevation
Z	compressibility factor

Greek Letters

β	coefficient of performance for a refrigerator
β'	coefficient of performance for a heat pump
η	efficiency
ρ	density
φ	relative humidity
ω	humidity ratio or specific humidity

Subscripts

c	property at the critical point
c.v.	control volume
e	state of a substance leaving a control volume
f	formation
f	property of saturated liquid
fg	difference in property for saturated vapor

	and saturated liquid
g	property of saturated vapor
r	reduced property
s	isentropic process

Superscripts

-	bar over symbol denotes property on a molal basis (over V, H, S, U, A, G, the bar denotes partial molal property)
$\circ$	property at standard-state condition
*	ideal gas
L	liquid phase
S	solid phase
V	vapor phase

Heat Transfer—Fluid Flow

A	surface area
A_m	profile area for a fin
Bi	Biot number, (hL/k)
c_p	specific heat at constant pressure
c	specific heat
D	diameter
D_e	hydraulic diameter
$F_{i\text{-}j}$	shape factor of area i with respect to area j
f	friction factor
Gr	Grashof number, $g\,\beta\Delta T L_c^3/v^2$
g	acceleration due to gravity
g_c	gravitational constant
h	convective heat-transfer coefficient
k	thermal conductivity
m	mass
$\dot{m}$	mass rate of flow
N	number of rows
Nu	Nusselt number, hL/k
Pr	Prandtl number, $\mu C_p/k$
p	pressure
Q	volumetric flow rate
Q	rate of heat flow
Ra	Rayleigh number, $g\,\beta\Delta T L_c^3/v \propto$

Re	Reynolds number, $\rho u_{av}\, L_c/\mu$
r	radius
St	Stanton number, $h/Cp\,\rho u_\infty$
T	temperature
U	overall heat-transfer coefficient
u	velocity
ux	free-stream velocity
V	volume
V	velocity
W	rate of work done

Greek Symbols

α	thermal diffusivity
β	coefficient of thermal expansion
Δ	difference, change
ε	surface emissivity
η_f	fin effectiveness
μ	viscosity
v	kinematic viscosity
ρ	density
σ	Stefan-Boltzmann constant
τ	time

Subscripts

b	bulk conditions
cr	critical condition
c	convection
cond	conduction
conv	convection
e	entrance, effective
f	fin, fluid
i	inlet conditions
o	exterior condition
0	centerline conditions in a tube at $r = 0$
o	outlet condition
p	pipe, pump
s	surface condition
∞	free-stream condition

APPENDIX II
CONVERSION FACTORS

Compiled by
L.C. WITTE

Professor of Mechanical Engineering
University of Houston
Houston, Texas

Table II.1 Conversion Factors

To Obtain:	Multiply:	By:
Acres	Sq miles	640.0
Atmospheres	Cm of Hg @ 0 deg C	0.013158
Atmospheres	Ft of H_2O @ 39.2 F.	0.029499
Atmospheres	Grams/sq cm	0.00096784
Atmospheres	In. Hg @ 32 F	0.033421
Atmospheres	In. H_2O @ 39.2 F	0.0024583
Atmospheres	Pounds/sq ft	0.00047254
Atmospheres	Pounds/sq in.	0.068046
Btu	Ft-lb	0.0012854
Btu	Hp-hr	2545.1
Btu	Kg-cal.	3.9685
Btu	kW-hr	3413
Btu	Watt-hr	3.4130
Btu/(cu ft) (hr)	kW/liter	96,650.6
Btu/hr	Mech. hp	2545.1
Btu/hr	kW	3413
Btu/hr	Tons of refrigeration	12,000
Btu/hr	Watts	3.4127
Btu/kW hr	Kg cal/kW hr	3.9685
Btu/(hr) (ft) (deg F)	Cal/(sec) (cm) (deg C)	241.90
Btu/(hr) (ft) (deg F)	Joules/(sec) (cm) (deg C)	57.803
Btu/(hr) (ft) (deg F)	Watts/(cm) (deg C)	57.803
Btu/(hr) (sq ft)	Cal/(sec) (sq cm)	13,273.0
Btu/min	Ft-lb/min	0.0012854
Btu/min	Mech. hp	42.418
Btu/min	kW	56.896
Btu/lb	Cal/gram	1.8
Btu/lb	Kg cal/kg	1.8
Btu/(lb) (deg F)	Cal/(gram) (deg C)	1.0
Btu/(lb) (deg F)	Joules/(gram) (deg C)	0.23889
Btu/sec	Mech. hp	0.70696
Btu/sec	Mech. hp (metric)	0.6971
Btu/sec	Kg-cal/hr	0.0011024
Btu/sec	kW	0.94827
Btu/sq ft	Kg-cal/sq meter	0.36867

Table II.1 Continued

To Obtain:	Multiply:	By:
Calories	Ft-lb	0.32389
Calories	Joules	0.23889
Calories	Watt-hr	860.01
Cal/(cu cm) (sec)	kW/liter	0.23888
Cal/gram	Btu/lb	0.55556
Cal/(gram) (deg C)	Btu/(lb) (deg F)	1.0
Cal/(sec) (cm) (deg C)	Btu/(hr) (ft) (deg F)	0.0041336
Cal/(sec) (sq cm)	Btu/(hr) (sq ft)	0.000075341
Cal/(sec) (sq cm) (deg C)	Btu/(hr) (sq ft) (deg F)	0.0001355
Centimeters	Inches	2.540
Centimeters	Microns	0.0001
Centimeters	Mils	0.002540
Cm of Hg @ 0 deg C	Atmospheres	76.0
Cm of Hg @ 0 deg C	Ft of H_2O @ 39.2 F	2.242
Cm of Hg @ 0 deg C	Grams/sq cm	0.07356
Cm of Hg @ 0 deg C	In. of H_2O @ 4 C	0.1868
Cm of Hg @ 0 deg C	Lb/sq in.	5.1715
Cm of Hg @ 0 deg C	Lb/sq ft	0.035913
Cm/deg C	In./deg F	4.5720
Cm/sec	Ft/min	0.508
Cm/sec	Ft/sec	30.48
Cm/(sec) (sec)	Gravity	980.665
Cm of H_2O @39.2 F	Atmospheres	1033.24
Cm of H_2O @39.2 F	Lb/sq in.	70.31
Centipoises	Centistokes	Density
Centistokes	Centipoises	1/density
Cu cm	Cu ft	28,317
Cu cm	Cu in.	16.387
Cu cm	Gal. (USA, liq.)	3785.43
Cu cm	Liters	1000 03
Cu cm	Ounces (USA, liq.)	29.573730
Cu cm	Quarts (USA, liq.)	946.358
Cu cm/sec	Cu ft/min	472.0
Cu ft	Cords (wood)	128.0
Cu ft	Cu meters	35.314
Cu ft	Cu yards	27.0
Cu ft	Gal. (USA, liq.)	0.13368
Cu ft	Liters	0.03532
Cu ft/min	Cu meters/sec	2118.9
Cu ft/min	Gal. (USA, liq./sec)	8.0192
Cu ft/lb	Cu meters/kg	16.02
Cu ft/lb	Liters/kg	0.01602
Cu ft/sec	Cu meters/min	0.5886
Cu ft/sec	Gal. (USA, liq.)/min	0.0022280
Cu ft/sec	Liters/min	0.0005886
Cu in.	Cu centimeters	0.061023
Cu in.	Gal. (USA, liq.)	231.0
Cu in.	Liters	61.03
Cu in.	Ounces (USA. liq.)	1.805

Table II.1 Continued

To Obtain:	Multiply:	By:
Cu meters	Cu ft	0.028317
Cu meters	Cu yards	0.7646
Cu meters	Gal. (USA. liq.)	0.0037854
Cu meters	Liters	0.001000028
Cu meters/hr	Gal./min	0.22712
Cu meters/kg	Cu ft/lb	0.062428
Cu meters/min	Cu ft/min	0.02832
Cu meters/min	Gal./sec	0.22712
Cu meters/sec	Gal./min	0.000063088
Cu yards	Cu meters	1.3079
Dynes	Grams	980.66
Dynes	Pounds (avoir.)	444820.0
Dyne-centimeters	Ft-lb	13,558,000
Dynes/sq cm	Lb/sq in.	68947
Ergs	Joules	10,000,000
Feet	Meters	3.281
Ft of H_2O @ 39.2 F	Atmospheres	33.899
Ft of H_2O @ 39.2 F	Cm of Hg @ 0 deg C	0.44604
Ft of H_2O @ 39.2 F	In. of Hg @ 32 deg F	1.1330
Ft of H_2O @ 39.2 F	Lb/sq ft	0.016018
Ft of H_2O @ 39.2 F	Lb/sq in.	2.3066
Ft/min	Cm/sec	1.9685
Ft/min	Miles (USA. statute)/hr	88.0
Ft/sec	Knots	1.6889
Ft/sec	Meters/sec	3.2808
Ft/sec	Miles (USA, statute)/hr	1.4667
Ft/(sec) (sec)	Gravity (sea level)	32.174
Ft/(sec) (sec)	Meters/(sec) (sec)	3.2808
Ft-lb	Btu	778.0
Ft-lb	Joules	0.73756
Ft-lb	Kg-calories	3087.4
Ft-lb	kW-hr	2,655,200
Ft-lb	Mech. hp-hr	1,980,000
Ft-lb/min	Btu/min	778.0
Ft-lb/min	Kg cal/min	3087.4
Ft-lb/min	kW	44,254.0
Ft-lb/min	Mech. hp	33,000
Ft-lb/sec	Btu/min	12.96
Ft-lb/sec	kW	737.56
Ft-lb/sec	Mech. hp	550.0
Gal. (Imperial, liq.)	Gal. (USA. Liq.)	0.83268
Gal. (USA, liq.)	Barrels (petroleum, USA)	42
Gal. (USA. liq.)	Cu ft	7.4805
Gal. (USA. liq.)	Cu meters	264.173
Gal. (USA, liq.)	Cu yards	202.2
Gal. (USA. liq.)	Gal. (Imperial, liq.)	1.2010
Gal. (USA. liq.)	Liters	0.2642
Gal. (USA. liq.)/min	Cu ft/sec	448.83
Gal. (USA, liq.)/min	Cu meters/hr	4.4029

Table II.1 Continued

To Obtain:	Multiply:	By:
Gal. (USA. liq.)/sec	Cu ft/min	0.12468
Gal. (USA. liq.)/sec	Liters/min	0.0044028
Grains	Grams	15.432
Grains	Ounces (avoir.)	437.5
Grains	Pounds (avoir.)	7000
Grains/gal. (USA. liq.)	Parts/million	0.0584
Grams	Grains	0.0648
Grams	Ounces (avoir.)	28.350
Grams	Pounds (avoir.)	453.5924
Grams/cm	Pounds/in.	178.579
Grams/(cm) (sec)	Centipoises	0.01
Grams/cu cm	Lb/cu ft	0 .016018
Grams/cu cm	Lb/cu in.	27.680
Grams/cu cm	Lb/gal.	0.119826
Gravity (at sea level)	Ft/(sec) (sec)	0.03108
Inches	Centimeters	0.3937
Inches	Microns	0.00003937
Inches of Hg @ 32 F	Atmospheres	29.921
Inches of Hg @ 32 F	Ft of H_2O @ 39.2 F	0.88265
Inches of Hg @ 32 F	Lb/sq in.	2.0360
Inches of Hg @ 32 F	In. of H_2O @ 4 C	0.07355
Inches of H_2O@ 4 C	In. of Hg @ 32 F	13.60
Inches of H_2O @ 39.2 F	Lb/sq in.	27.673
Inches/deg F	Cm/deg C	0.21872
Joules	Btu	1054.8
Joules	Calories	4.186
Joules	Ft-lb	1.35582
Joules	Kg-meters	9.807
Joules	kW-hr	3,600,000
Joules	Mech. hp-hr	2,684,500
Kg	Pounds (avoir.)	0.45359
Kg-cal	Btu	0.2520
Kg-cal	Ft-lb	0.00032389
Kg-cal	Joules	0.0002389
Kg-cal	kW-hr	860.01
Kg-cal	Mech. hp-hr	641.3
Kg-cal/kg	Btu/lb	0.5556
Kg-cal/kW hr	Btu/kW hr	0.2520
Kg-cal/min	Ft-lb/min	0.0003239
Kg-cal/min	kW	14,33
Kg-cal/min	Mech. hp	10.70
Kg-cal/sq meter	Btu/sq ft	2.712
Kg/cu meter	Lb/cu ft	16.018
Kg/(hr) (meter)	Centipoises	3.60
Kg/liter	Lb/gal. (USA, liq.)	0.11983
Kg/meter	Lbm	1.488
Kg/sq cm	Atmospheres	1.0332
Kg sq cm	Lb/sq in .	0.0703
Kg/sq meter	Lb/sq ft	4.8824

Table II.1 Continued

To Obtain:	Multiply:	By:
Kg/sq meter	Lb/sq in.	703.07
Km	Miles (USA, statute)	1.6093
kW	Btu/min	0.01758
kW	Ft-lb/min	0.00002259
kW	Ft-lb/sec	0.00135582
kW	Kg-cal/hr	0.0011628
kW	Kg-cal/min	0.069767
kW	Mech. hp	0.7457
kW-hr	Btu	0.000293
kW-hr	Ft-lb	0.0000003766
kW-hr	Kg-cal	0.0011628
kW-hr	Mech. hp-hr	0.7457
Knots	Ft/sec	0.5921
Knots	Miles/hr	0.8684
Liters	Cu ft	28 . 316
Liters	Cu in.	0.01639
Liters	Cu centimeters	999.973
Liters	Gal. (Imperial. liq.)	4.546
Liters	Gal. (USA, liq.)	3.78533
Liters/kg	Cu ft/lb	62.42621
Liters/min	Cu ft/sec	1699.3
Liters/min	Gal. (USA. liq.)/min	3.785
Liters/sec	Cu ft/min	0.47193
Liters/sec	Gal./min	0.063088
Mech. hp	Btu/hr	0.0003929
Mech. hp	Btu/min	0.023575
Mech. hp	Ft-lb/sec	0.0018182
Mech. hp	Kg-cal/min	0.093557
Mech. hp	kW	1.3410
Mech. hp-hr	Btu	0.00039292
Mech. hp-hr	Ft-lb	0.00000050505
Mech. hp-hr	Kg-calories	0.0015593
Mech. hp-hr	kW-hr	1.3410
Meters	Feet	0.3048
Meters	Inches	0.0254
Meters	Miles (Int., nautical)	1852.0
Meters	Miles (USA, statute)	1609.344
Meters/min	Ft/min	0.3048
Meters/min	Miles (USA. statute)/hr	26.82
Meters/sec	Ft/sec	0.3048
Meters/sec	Km/hr	0.2778
Meters/sec	Knots	0.5148
Meters/sec	Miles (USA, statute)/hr	0.44704
Meters/(sec) (sec)	Ft/(sec) (sec)	0.3048
Microns	Inches	25,400
Microns	Mils	25.4
Miles (Int., nautical)	Km	0.54
Miles (Int., nautical)	Miles (USA, statute)	0.8690
Miles (Int., nautical)/hr	Knots	1.0

Table II.1 Continued

To Obtain:	Multiply:	By:
Miles (USA, statute)	Km	0.6214
Miles (USA, statute)	Meters	0.0006214
Miles (USA, statute)	Miles (Int., nautical)	1.151
Miles (USA, statute)/hr	Knots	1.151
Miles (USA, statute)/hr	Ft/min	0.011364
Miles (USA, statute)/hr	Ft/sec	0.68182
Miles (USA, statute)/hr	Meters/min	0.03728
Miles (USA, statute)/hr	Meters/sec	2.2369
Milliliters/gram	Cu ft/lb	62.42621
Millimeters	Microns	0.001
Mils	Centimeters	393.7
Mils	Inches	1000
Mils	Microns	0.03937
Minutes	Radians	3437.75
Ounces (avoir.)	Grains (avoir.)	0.0022857
Ounces (avoir.)	Grams	0.035274
Ounces (USA, liq.)	Gal. (USA, liq.)	128.0
Parts/million	Gr/gal. (USA, liq.)	17.118
Percent grade	Ft/100 ft	1.0
Pounds (avoir.)	Grains	0.0001429
Pounds (avoir.)	Grams	0.0022046
Pounds (avoir.)	Kg	2.2046
Pounds (avoir.)	Tons, long	2240
Pounds (avoir.)	Tons, metric	2204.6
Pounds (avoir.)	Tons, short	2000
Pounds/cu ft	Grams/cu cm	62.428
Pounds/cu ft	Kg/cu meter	0.062428
Pounds/cu ft	Pounds/gal.	7.48
Pounds/cu in .	Grams/cu cm	0.036127
Pounds/ft	Kg/meter	0.67197
Pounds/hr	Kg/min	132.28
Pounds/(hr) (ft)	Centipoises	2.42
Pounds/inch	Grams/cm	0.0056
Pounds/(sec) (ft)	Centipoises	0.000672
Pounds/sq inch	Atmospheres	14.696
Pounds/sq inch	Cm of Hg @ 0 deg C	0.19337
Pounds/sq inch	Ft of H_2O @ 39.2 F	0.43352
Pounds/sq inch	In. Hg @ l 32 F	0.491
Pounds/sq inch	In. H_2O @ 39.2 F	0.0361
Pounds/sq inch	Kg/sq cm	14 . 223
Pounds/sq inch	Kg/sq meter	0.0014223
Pounds/gal. (USA, liq.)	Kg/liter	8.3452
Pounds/gal. (USA, liq.)	Pounds/cu ft	0.1337
Pounds/gal. (USA, liq.)	Pounds/cu inch	231
Quarts (USA, liq.)	Cu cm	0.0010567
Quarts (USA, liq.)	Cu in.	0.01732
Quarts (USA, liq.)	Liters	1.057
Sq centimeters	Sq ft	929.0
Sq centimeters	Sq inches	6.4516

Table II.1 Continued

To Obtain:	Multiply:	By:
Sq ft	Acres	43,560
Sq ft	Sq meters	10.764
Sq inches	Sq centimeters	0.155
Sq meters	Acres	4046.9
Sq meters	Sq ft	0.0929
Sq mlles (USA. statute)	Acres	0.001562
Sq mils	Sq cm	155.000
Sq mils	Sq inches	1,000.000
Tons (metric)	Tons (short)	0.9072
Tons (short)	Tons (metric)	1.1023
Watts	Btu/sec	1054.8
Yards	Meters	1.0936

I-P/SI CONVERSION FACTORS

When making conversions, remember that a converted value is no more precise than the original value. Round off the final value to the same number of significant figures as those in the original value.

CAUTION: The conversion values are rounded to three or four significant figures, which is sufficiently accurate for most applications. See ANSI SI 10 for additional conversions with more significant figures.

Multiply	By	To Obtain
acre	0.4047	ha
atmosphere, standard	*101.325	kPa
bar	*100	kPa
barrel (42 US gal, petroleum)	159	L
Btu, (International Table)	1.055	kJ
Btu/ft^2	11.36	kJ/m^2
Btu·ft/h·ft^2·°F	1.731	W/(m·K)
Btu·in/h·ft^2·°F		
(thermal conductivity, k)	0.1442	W/(m·K)
Btu/h	0.2931	W
Btu/h·ft	0.9615	W/m
Btu/h·ft^2	3.155	W/m^2
Btu/h·ft^2·°F		
(heat transfer coefficient, U)	5.678	W/(m^2·K)
Btu/lb	*2.326	kJ/kg
Btu/lb·°F (specific heat, c_p)	4.184	kJ/(kg·K)
bushel	0.03524	m^3
calorie, (thermochemical)	*4.184	J
calorie, nutrition (kilocalorie)	*4.184	kJ
candle, candlepower	*1.0	cd
centipoise, dynamic vicosity, μ	*1.00	mPa·s
centistokes, kinematic viscosity, ν	*1.00	mm^2/s
clo	0.155	m^2·K/W
dyne/cm^2	*0.100	Pa
EDR hot water (150 Btu/h)	44.0	W
EDR steam (240 Btu/h)	70.3	W
fuel cost comparison at 100% eff.		
cents per gallon (no. 2 fuel oil)	0.0677	$/GJ
cents per gallon (no. 6 fuel oil)	0.0632	$/GJ
cents per gallon (propane)	0.113	$/GJ
cent per kWh	2.78	$/GJ
cents per therm	0.0948	$/GJ
ft	*0.3048	m
ft	*304.8	mm
ft/min, fpm	*0.00508	m/s
ft/s, fps	*0.3048	m/s
ft of water	2.99	kPa
ft of water per 100 ft of pipe	0.0981	kPa/m
ft^2	0.09290	m^2
ft^2·h·°F/Btu (thermal resistance, R)	0.176	m^2·K/W
ft^2/s, kinematic viscosity, ν	92 900	mm^2/s
ft^3	28.32	L
ft^3	0.02832	m^3
ft^3/h, cfh	7.866	mL/s
ft^3/min, cfm	0.4719	L/s
ft^3/s, cfs	28.32	L/s
footcandle	10.76	lx
ft·lb$_f$ (torque or moment)	1.36	N·m
ft·lb$_f$ (work)	1.36	J
ft·lb$_f$/lb (specific energy)	2.99	J/kg
ft·lb$_f$/min (power)	0.0226	W
gallon, US (*231 in^3)	3.785	L
gph	1.05	mL/s
gpm	0.0631	L/s
gpm/ft^2	0.6791	L/(s·m^2)
gpm/ton refrigeration	0.0179	mL/J
grain (1/7000 lb)	0.0648	g
gr/gal	17.1	g/m^3
horsepower (boiler)(33,470 Btu/h)	9.81	kW
horsepower (550 ft·lb$_f$/s)	0.746	kW
inch	*25.4	mm
inch of mercury (60°F)	3.377	kPa
inch of water (60°F)	248.8	Pa
To Obtain	**By**	**Divide**

Multiply	By	To Obtain
in/100 ft (thermal expansion)	0.833	mm/m
in·lb$_f$ (torque or moment)	113	mN·m
in^2	645	mm^2
in^3 (volume)	16.4	mL
in^3/min (SCIM)	0.273	mL/s
in^3 (section modulus)	16 400	mm^3
in^4 (section moment)	416 200	mm^4
km/h	0.278	m/s
kWh	*3.60	MJ
kW/1000 cfm	2.12	kJ/m^3
kilopond (kg force)	9.81	N
kip (1000 lb$_f$)	4.45	kN
kip/in^2 (ksi)	6.895	MPa
litre	*0.001	m^3
MBtuh (1000 Btu/h)	0.2931	kW
met	58.15	W/m^2
micron (μm) of mercury (60°F)	133	mPa
mil (0.001 in.)	*25.4	mm
mile	1.61	km
mile, nautical	1.85	km
mph	1.61	km/h
mph	0.447	m/s
millibar	*0.100	kPa
mm of mercury (60°F)	0.133	kPa
mm of water (60°F)	9.80	Pa
ounce (mass, avoirdupois)	28.35	g
ounce (force of thrust)	0.278	N
ounce (liquid, US)	29.6	mL
ounce (avoirdupois) per gallon	7.49	kg/m^3
perm (permeance)	57.45	ng/(s·m^2·Pa)
perm inch (permeability)	1.46	ng/(s·m·Pa)
pint (liquid, US)	473	mL
pound		
lb (mass)	0.4536	kg
lb (mass)	453.6	g
lb$_f$ (force or thrust)	4.45	N
lb/ft (uniform load)	1.49	kg/m
lb$_m$/(ft·h) (dynamic viscosity, μ)	0.413	mPa·s
lb$_m$/(ft·s) (dynamic viscosity, μ)	1490	mPa·s
lb$_f$·s/ft^2 (dynamic viscosity, μ)	47 880	mPa·s
lb/min	0.00756	kg/s
lb/h	0.126	g/s
lb/h (steam at 212°F)(970 Btu/h)	0.284	kW
lb$_f$/ft^2	47.9	Pa
lb/ft^2	4.88	kg/m^2
lb/ft^3 (density, ρ)	16.0	kg/m^3
lb/gallon	120	kg/m^3
ppm (by mass)	*1.00	mg/kg
psi	6.895	kPa
quad (10^{15} Btu)	1.06	EJ
quart (liquid, US)	0.946	L
revolutions per minute (rpm)	*1/60	Hz
square (100 ft^2)	9.29	m^2
tablespoon (approx.)	15	mL
teaspoon (approx.)	5	mL
therm (100,000 Btu)	105.5	MJ
ton, short (2000 lb)	0.907	Mg; t (tonne)
ton, refrigeration (12,000 Btu/h)	3.517	kW
torr (1 mm Hg at 0°C)	133	Pa
watt per square foot	10.8	W/m^2
yd	*0.9144	m
yd^2	0.836	m^2
yd^3	0.7646	m^3
To Obtain	**By**	**Divide**

Note: In this list the kelvin (K) expresses temperature intervals. The degree Celsius symbol (°C) is often used for this purpose as well.

*Conversion factor is exact.

Source: Writing for ASHRAE Journal, 2008, pp 9, ©American Society of Heating, Refrigerating and Air-Conditioning Engineers

PROPERTY TABLES

Table III-1. Saturated Steam: Temperature

Temp. t (°F)	Abs. Press. p (psi)	Specific Volume			Enthalpy			Entropy			Temp. t (°F)
		Sat. Liquid v_f	Evap v_{fg}	Sat. Vapor v_g	Sat. Liquid h_f	Evap h_{fg}	Sat. Vapor h_g	Sat. Liquid s_f	Evap s_{fg}	Sat. Vapor s_g	
32.0"	0.08859	0.016022	3304.7	3304.7	−0.0179	1075.5	1075.5	0.0000	2.1873	2.1873	32.0"
34.0	0.09600	0.016021	3061.9	3061.9	1.996	1074.4	1076.4	0.0041	2.1762	2.1802	34.0
36.0	0.10395	0.016020	2839.0	2839.0	4.008	1073.2	1077.2	0.0081	2.1651	2.1732	36.0
38.0	0.11249	0.016019	2634.1	2634.2	6.018	1072.1	1078.1	0.0122	2.1541	2.1663	38.0
40.0	0.12163	0.016019	2445.8	2445.8	8.027	1071.0	1079.0	0.0162	2.1432	2.1594	40.0
42.0	0.13143	0.016019	2272.4	2272.4	10.035	1069.8	1079.9	0.0202	2.1325	2.1527	42.0
44.0	0.14192	0.016019	2112.8	2112.8	12.041	1068.7	1080.7	0.0242	2.1217	2.1459	44.0
46.0	0.15314	0.016020	1965.7	1965.7	14.047	1067.6	1081.6	0.0282	2.1111	2.1393	46.0
48.0	0.16514	0.016021	1830.0	1830.0	16.051	1066.4	1082.5	0.0321	2.1006	2.1327	48.0
50.0	0.17796	0.016023	1704.8	1704.8	18.054	1065.3	1083.4	0.0361	2.0901	2.1262	50.0
52.0	0.19165	0.016024	1589.2	1589.2	20.057	1064.2	1084.2	0.0400	2.0798	2.1197	52.0
54.0	0.20625	0.016026	1482.4	1482.4	22.058	1063.1	1085.1	0.0439	2.0695	2.1134	54.0
56.0	0.22183	0.016028	1383.6	1383.6	24.059	1061.9	1086.0	0.0478	2.0593	2.1070	56.0
58.0	0.23843	0.016031	1292.2	1292.2	26.060	1060.8	1086.9	0.0516	2.0491	2.1008	58.0
60.0	0.25611	0.016033	1207.6	1207.6	28.060	1059.7	1087.7	0.0555	2.0391	2.0946	60.0
62.0	0.27494	0.016036	1129.2	1129.2	30.059	1058.5	1088.6	0.0593	2.0291	2.0885	62.0
64.0	0.29497	0.016039	1056.5	1056.5	32.058	1057.4	1089.5	0.0632	2.0192	2.0824	64.0
66.0	0.31626	0.016043	989.0	989.1	34.056	1056.3	1090.4	0.0670	2.0094	2.0764	66.0
68.0	0.33889	0.016046	926.5	926.5	36.054	1055.2	1091.2	0.0708	1.9996	2.0704	68.0
70.0	0.36292	0.016050	868.3	868.4	38.052	1054.0	1092.1	0.0745	1.9900	2.0645	70.0
72.0	0.38844	0.016054	814.3	814.3	40.049	1052.9	1093.0	0.0783	1.9804	2.0587	72.0
74.0	0.41550	0.016058	764.1	764.1	42.046	1051.8	1093.8	0.0821	1.9708	2.0529	74.0
76.0	0.44420	0.016063	717.4	717.4	44.043	1050.7	1094.7	0.0858	1.9614	2.0472	76.0
78.0	0.47461	0.016067	673.8	673.9	46.040	1049.5	1095.6	0.0895	1.9520	2.0415	78.0
80.0	0.50683	0.016072	633.3	633.3	48.037	1048.4	1096.4	0.0932	1.9426	2.0359	80.0
82.0	0.54093	0.016077	595.5	595.5	50.033	1047.3	1097.3	0.0969	1.9334	2.0303	82.0
84.0	0.57702	0.016082	560.3	560.3	52.029	1046.1	1098.2	0.1006	1.9242	2.0248	84.0
86.0	0.61518	0.016087	227.5	527.5	54.026	1045.0	1099.0	0.1043	1.9151	2.0193	86.0
88.0	0.65551	0.016093	496.8	496.8	56.022	1043.9	1099.9	0.1079	1.9060	2.0139	88.0
90.0	0.69813	0.016099	468.1	468.1	58.018	1042.7	1100.8	0.1115	1.8970	2.0086	90.0
92.0	0.74313	0.016105	441.3	441.3	60.014	1041.6	1101.6	0.1152	1.8881	2.0033	92.0
94.0	0.79062	0.016111	416.3	416.3	62.010	1040.5	1102.5	0.1188	1.8792	1.9980	94.0
96.0	0.84072	0.016117	392.8	392.9	64.006	1039.3	1103.3	0.1224	1.8704	1.9928	96.0
98.0	0.89356	0.016123	370.9	370.9	66.003	1038.2	1104.2	0.1260	1.8617	1.9876	98.0
100.0	0.94924	0.016130	350.4	350.4	67.999	1037.1	1105.1	0.1295	1.8530	1.9825	100.0
102.0	1.00789	0.016137	331.1	331.1	69.995	1035.9	1105.9	0.1331	1.8444	1.9775	102.0
104.0	1.06965	0.016144	313.1	313.1	71.992	1034.8	1106.8	0.1366	1.8358	1.9725	104.0
106.0	1.1347	0.016151	296.16	296.18	73.99	1033.6	1107.6	0.1402	1.8273	1.9675	106.0
108.0	1.2030	0.016158	280.28	280.30	75.98	1032.5	1108.5	0.1437	1.8188	1.9626	108.0
110.0	1.2750	0.016165	265.37	265.39	77.98	1031.4	1109.3	0.1472	1.8105	1.9577	110.0
112.0	1.3505	0.016173	251.37	251.38	79.98	1030.2	1110.2	0.1507	1.8021	1.9528	112.0
114.0	1.4299	0.016180	238.21	238.22	81.97	1029.1	1111.0	0.1542	1.7938	1.9480	114.0
116.0	1.5133	0.016188	225.84	225.85	83.97	1027.9	1111.9	0.1577	1.7856	1.9433	116.0
118.0	1.6009	0.016196	214.20	214.21	85.97	1026.8	1112.7	0.1611	1.7774	1.9386	118.0
120.0	1.6927	0.016204	203.25	203.26	87.97	1025.6	1113.6	0.1646	1.7693	1.9339	120.0
122.0	1.7891	0.016213	192.94	192.95	89.96	1024.5	1114.4	0.1680	1.7613	1.9293	122.0
124.0	1.8901	0.016221	183.23	183.24	91.96	1023.3	1115.3	0.1715	1.7533	1.9247	124.0
126.0	1.9959	0.016229	174.08	174.09	93.96	1022.2	1116.1	0.1749	1.7453	1.9202	126.0
128.0	2.1068	0.016238	165.45	165.47	95.96	1021.0	1117.0	0.1783	1.7374	1.9157	128.0
130.0	2.2230	0.016247	157.32	157.33	97.96	1019.8	1117.8	0.1817	1.7295	1.9112	130.0
132.0	2.3445	0.016256	149.64	149.66	99.95	1018.7	1118.6	0.1851	1.7217	1.9068	132.0
134.0	2.4717	0.016265	142.40	142.41	101.95	1017.5	1119.5	0.1884	1.7140	1.9024	134.0
136.0	2.6047	0.016274	135.55	135.57	103.95	1016.4	1120.3	0.1918	1.7063	1.8980	136.0
138.0	2.7438	0.016284	129.09	129.11	105.95	1015.2	1121.1	0.1951	1.6986	1.8937	138.0
140.0	2.8892	0.016293	122.98	123.00	107.95	1014.0	1122.0	0.1985	1.6910	1.8895	140.0
142.0	3.0411	0.016303	117.21	117.22	109.95	1012.9	1122.8	0.2018	1.6534	1.8852	142.0
144.0	3.1997	0.016312	111.74	111.76	111.95	1011.7	1123.6	0.2051	1.6759	1.8810	144.0
146.0	3.3653	0.016322	106.58	106.59	113.95	1010.5	1124.5	0.2084	1.6684	1.8769	146.0
148.0	3.5381	0.016332	101.68	101.70	115.95	1009.3	1125.3	0.2117	1.6610	1.8727	148.0

Table III-1. Saturated Steam: Temperature

Temp. t (°F)	Abs. Press. p (psi)	Specific Volume			Enthalpy			Entropy			Temp. t (°F)
		Sat. Liquid v_f	Evap v_{fg}	Sat. Vapor v_g	Sat. Liquid h_f	Evap h_{fg}	Sat. Vapor h_g	Sat. Liquid s_f	Evap s_{fg}	Sat. Vapor s_g	
150.0	3.7184	0.016343	97.05	97.07	117.95	1008.2	1126.1	0.2150	1.6536	1.8686	150.0
152.0	3.9065	0.016353	92.66	92.68	119.95	1007.0	1126.9	0.2183	1.6463	1.8646	152.0
154.0	4.1025	0.016363	88.50	88.52	121.95	1005.8	1127.7	0.2216	1.6390	1.8606	154.0
156.0	4.3068	0.016374	84.56	84.57	123.95	1004.6	1128.6	0.2248	1.6318	1.8566	156.0
158.0	4.5197	0.016384	80.82	80.83	125.96	1003.4	1129.4	0.2281	1.6245	1.8526	158.0
160.0	4.7414	0.016395	77.27	77.29	127.96	1002.2	1130.2	0.2313	1.6174	1.8487	160.0
162.0	4.9722	0.016406	73.90	73.92	129.96	1001.0	1131.0	0.2345	1.6103	1.8448	162.0
164.0	5.2124	0.016417	70.70	70.72	131.96	999.8	1131.8	0.2377	1.6032	1.8409	164.0
166.0	5.4623	0.016428	67.67	67.68	133.97	998.6	1132.6	0.2409	1.5961	1.8371	166.0
168.0	5.7223	0.016440	64.78	64.80	135.97	997.4	1133.4	0.2441	1.5892	1.8333	168.0
170.0	5.9926	0.016451	62.04	62.06	137.97	996.2	1134.2	0.2473	1.5822	1.8295	170.0
172.0	6.2736	0.016463	59.43	59.45	139.98	995.0	1135.0	0.2505	1.5753	1.8258	172.0
174.0	6.5656	0.016474	56.95	56.97	141.98	993.8	1135.8	0.2537	1.5684	1.8221	174.0
176.0	6.8690	0.016486	54.59	54.61	143.99	992.6	1136.6	0.2568	1.5616	1.8184	176.0
178.0	7.1840	0.016498	52.35	52.36	145.99	991.4	1137.4	0.2600	1.5548	1.8147	178.0
180.0	7.5110	0.016510	50.21	50.22	148.00	990.2	1138.2	0.2631	1.5480	1.8111	180.0
182.0	7.850	0.016522	48.172	18.189	150.01	989.0	1139.0	0.2662	1.5413	1.8075	182.0
184.0	8.203	0.016534	46.232	46.249	152.01	987.8	1139.8	0.2694	1.5346	1.8040	184.0
186.0	8.568	0.016547	44.383	44.400	154.02	986.5	1140.5	0.2725	1.5279	1.8004	186.0
188.0	8.947	0.016559	42.621	42.638	156.03	985.3	1141.3	0.2756	1.5213	1.7969	188.0
190.0	9.340	0.016572	40.941	40.957	158.04	984.1	1142.1	0.2787	1.5148	1.7934	190.0
192.0	9.747	0.016585	39.337	39.354	160.05	982.8	1142.9	0.2818	1.5082	1.7900	192.0
194.0	10.168	0.016598	37.808	37.824	162.05	981.6	1143.7	0.2848	1.5017	1.7865	194.0
196.0	10.605	0.016611	36.348	36.364	164.06	980.4	1144.4	0.2879	1.4952	1.7831	196.0
198.0	11.058	0.016624	34.954	34.970	166.08	979.1	1145.2	0.2910	1.4888	1.7798	198.0
200.0	11.526	0.016637	33.622	33.639	168.09	977.9	1146.0	0.2940	1.4824	1.7764	200.0
204.0	12.512	0.016664	31.135	31.151	172.11	975.4	1147.5	0.3001	1.4697	1.7698	204.0
208.0	13.568	0.016691	28.862	28.878	176.14	972.8	1149.0	0.3061	1.4571	1.7632	208.0
212.0	14.696	0.016719	26.782	26.799	180.17	970.3	1150.5	0.3121	1.4447	1.7568	212.0
216.0	15.901	0.016747	24.878	24.894	184.20	967.8	1152.0	0.3181	1.4323	1.7505	216.0
220.0	17.186	0.016775	23.131	23.148	188.23	965.2	1153.4	0.3241	1.4201	1.7442	220.0
224.0	18.556	0.016805	21.529	21.545	192.27	962.6	1154.9	0.3300	1.4081	1.7380	224.0
228.0	20.015	0.016834	20.056	20.073	196.31	960.0	1156.3	0.3359	1.3961	1.7320	228.0
232.0	21.567	0.016864	18.701	18.718	200.35	957.4	1157.8	0.3417	1.3842	1.7260	232.0
236.0	23.216	0.016895	17.454	17.471	204.40	954.8	1159.2	0.3476	1.3725	1.7201	236.0
240.0	24.968	0.016926	16.304	16.321	208.45	952.1	1160.6	0.3533	1.3609	1.7142	240.0
244.0	26.826	0.016958	15.243	15.260	212.50	949.5	1162.0	0.3591	1.3494	1.7085	244.0
248.0	28.796	0.016990	14.264	14.281	216.56	946.8	1163.4	0.3649	1.3379	1.7028	248.0
252.0	30.883	0.017022	13.358	13.375	220.62	944.1	1164.7	0.3706	1.3266	1.6972	252.0
256.0	33.091	0.017055	12.520	12.538	224.69	941.4	1166.1	0.3763	1.3154	1.6917	256.0
260.0	35.427	0.017089	11.745	11.762	228.76	938.6	1167.4	0.3819	1.3043	1.6862	260.0
264.0	37.894	0.017123	11.025	11.042	232.83	935.9	1168.7	0.3876	1.2933	1.6808	264.0
268.0	40.500	0.017157	10.358	10.375	236.91	933.1	1170.0	0.3932	1.2823	1.6755	268.0
272.0	43.249	0.017193	9.738	9.755	240.99	930.3	1171.3	0.3987	1.2715	1.6702	272.0
276.0	46.147	0.017228	9.162	9.180	245.08	927.5	1172.5	0.4043	1.2607	1.6650	276.0
280.0	49.200	0.017264	8.627	8.644	249.17	924.6	1173.8	0.4098	1.2501	1.6599	280.0
284.0	52.414	0.01730	8.1280	8.1453	253.3	921.7	1175.0	0.4154	1.2395	1.6548	284.0
288.0	55.795	0.01734	7.6634	7.6807	257.4	918.8	1176.2	0.4208	1.2290	1.6498	288.0
292.0	59.350	0.01738	7.2301	7.2475	261.5	915.9	1177.4	0.4263	1.2186	1.6449	292.0
296.0	63.084	0.01741	6.8259	6.8433	265.6	913.0	1178.6	0.4317	1.2082	1.6400	296.0
300.0	67.005	0.01745	6.4483	6.4658	269.7	910.0	1179.7	0.4372	1.1979	1.6351	300.0
304.0	71.119	0.01749	6.0955	6.1130	273.8	907.0	1180.9	0.4426	1.1877	1.6303	304.0
308.0	75.433	0.01753	5.7655	5.7830	278.0	904.0	1182.0	0.4479	1.1776	1.6256	308.0
312.0	79.953	0.01757	5.4566	5.4742	282.1	901.0	1183.1	0.4533	1.1676	1.6209	312.0
316.0	84.688	0.01761	5.1673	5.1849	286.3	897.9	1184.1	0.4586	1.1576	1.6162	316.0
320.0	89.643	0.01766	4.8961	4.9138	290.4	894.8	1185.2	0.4640	1.1477	1.6116	320.0
324.0	94.826	0.01770	4.6418	4.6595	294.6	891.6	1186.2	0.4692	1.1378	1.6071	324.0
328.0	100.245	0.01774	4.4030	4.4208	298.7	888.5	1187.2	0.4745	1.1280	1.6025	328.0
332.0	105.907	0.01779	4.1788	4.1966	302.9	885.3	1188.2	0.4798	1.1183	1.5981	332.0
336.0	111.820	0.01783	3.9681	3.9859	307.1	882.1	1189.1	0.4850	1.1086	1.5936	336.0

Table III-1. Saturated Steam: Temperature

Temp. t (°F)	Abs. Press. p (psi)	Specific Volume			Enthalpy			Entropy			Temp. t (°F)
		Sat. Liquid v_f	Evap v_{fg}	Sat. Vapor v_g	Sat. Liquid h_f	Evap h_{fg}	Sat. Vapor h_g	Sat. Liquid s_f	Evap s_{fg}	Sat. Vapor s_g	
340.0	117.992	0.01787	3.7699	3.7878	311.3	878.8	1190.1	0.4902	1.0990	1.5892	340.0
344.0	124.430	0.01792	3.5834	3.6013	315.5	875.5	1191.0	0.4954	1.0894	1.5849	344.0
348.0	131.142	0.01797	3.4078	3.4258	319.7	872.2	1191.1	0.5006	1.0799	1.5806	348.0
352.0	138.138	0.01801	3.2423	3.2603	323.9	868.9	1192.7	0.5058	1.0705	1.5763	352.0
356.0	145.424	0.01806	3.0863	3.1044	328.1	865.5	1193.6	0.5110	1.0611	1.5721	356.0
360.0	153.010	0.01811	2.9392	2.9573	332.3	862.1	1194.4	0.5161	1.0517	1.5678	360.0
364.0	160.903	0.01816	2.8002	2.8184	336.5	858.6	1195.2	0.5212	1.0424	1.5637	364.0
368.0	169.113	0.01821	2.6691	2.6873	340.8	855.1	1195.9	0.5263	1.0332	1.5595	368.0
372.0	177.648	0.01826	2.5451	2.5633	345.0	851.6	1196.7	0.5314	1.0240	1.5554	372.0
376.0	186.517	0.01831	2.4279	2.4462	349.3	848.1	1197.4	0.5365	1.0148	1.5513	376.0
380.0	195.729	0.01836	2.3170	2.3353	353.6	844.5	1198.0	0.5416	1.0057	1.5473	380.0
384.0	205.294	0.01842	2.2120	2.2304	357.9	840.8	1198.7	0.5466	0.9966	1.5432	384.0
388.0	215.220	0.01847	2.1126	2.1311	362.2	837.2	1199.3	0.5516	0.9876	1.5392	388.0
392.0	225.516	0.01853	2.0184	2.0369	366.5	833.4	1199.9	0.5567	0.9786	1.5352	392.0
396.0	236.193	0.01858	1.9291	1.9477	370.8	829.7	1200.4	0.5617	0.9696	1.5313	396.0
400.0	247.259	0.01864	1.8444	1.8630	375.1	825.9	1201.0	0.5667	0.9607	1.5274	400.0
404.0	258.725	0.01870	1.7640	1.7827	379.4	822.0	1201.5	0.5717	0.9518	1.5234	404.0
408.0	270.600	0.01875	1.6877	1.7064	383.8	818.2	1201.9	0.5766	0.9429	1.5195	408.0
412.0	282.894	0.01881	1.6152	1.6340	388.1	814.2	1202.4	0.5816	0.9341	1.5157	412.0
416.0	295.617	0.01887	1.5463	1.5651	392.5	810.2	1202.8	0.5866	0.9253	1.5118	416.0
420.0	308.780	0.01894	1.4808	1.4997	396.9	806.2	1203.1	0.5915	0.9165	1.5080	420.0
424.0	322.391	0.01900	1.4184	1.4374	401.3	802.2	1203.5	0.5964	0.9077	1.5042	424.0
428.0	336.463	0.01906	1.3591	1.3782	405.7	798.0	1203.7	0.6014	0.8990	1.5004	428.0
432.0	351.00	0.01913	1.30266	1.32179	410.1	793.9	1204.0	0.6063	0.8903	1.4966	432.0
436.0	366.03	0.01919	1.24887	1.26806	414.6	789.7	1204.2	0.6112	0.8816	1.4928	436.0
440.0	381.54	0.01926	1.19761	1.21687	419.0	785.4	1204.4	0.6161	0.8729	1.4890	440.0
444.0	397.56	0.01933	1.14874	1.16806	423.5	781.1	1204.6	0.6210	0.8643	1.4853	444.0
448.0	414.09	0.01940	1.10212	1.12152	428.0	776.7	1204.7	0.6259	0.8557	1.4815	448.0
452.0	431.14	0.01947	1.05764	1.07711	432.5	772.3	1204.8	0.6308	0.8471	1.4778	452.0
456.0	448.73	0.01954	1.01518	1.03472	437.0	767.8	1204.8	0.6356	0.8385	1.4741	456.0
460.0	466.87	0.01961	0.97463	0.99424	441.5	763.2	1204.8	0.6405	0.8299	1.4704	460.0
464.0	485.56	0.01969	0.93588	0.95557	446.1	758.6	1204.7	0.6454	0.8213	1.4667	464.0
468.0	504.83	0.01976	0.89885	0.91862	450.7	754.0	1204.6	0.6502	0.8127	1.4629	468.0
472.0	524.67	0.01984	0.86345	0.88329	455.2	749.3	1204.5	0.6551	0.8042	1.4592	472.0
476.0	545.11	0.01992	0.82958	0.84950	459.9	744.5	1204.3	0.6599	0.7956	1.4555	476.0
480.0	566.15	0.02000	0.79716	0.81717	464.5	739.6	1204.1	0.6648	0.7871	1.4518	480.0
484.0	587.81	0.02009	0.76613	0.78622	469.1	734.7	1203.8	0.6696	0.7785	1.4481	484.0
488.0	610.10	0.02017	0.73641	0.75658	473.8	729.7	1203.5	0.6745	0.7700	1.4444	488.0
492.0	633.03	0.02026	0.70794	0.72820	478.5	724.6	1203.1	0.6793	0.7614	1.4407	492.0
496.0	656.61	0.02034	0.68065	0.70100	483.2	719.5	1202.7	0.6842	0.7528	1.4370	496.0
500.0	680.86	0.02043	0.65448	0.67492	487.9	714.3	1202.2	0.6890	0.7443	1.4333	500.0
504.0	705.78	0.02053	0.62938	0.64991	492.7	709.0	1201.7	0.6939	0.7357	1.4296	504.0
508.0	731.40	0.02062	0.60530	0.62592	497.5	703.7	1201.1	0.6987	0.7271	1.4258	508.0
512.0	757.72	0.02072	0.58218	0.60289	502.3	698.2	1200.5	0.7036	0.7185	1.4221	512.0
516.0	784.76	0.02081	0.55997	0.58079	507.1	692.7	1199.8	0.7085	0.7099	1.4183	516.0
520.0	812.53	0.02091	0.53864	0.55956	512.0	687.0	1199.0	0.7133	0.7013	1.4146	520.0
524.0	841.04	0.02102	0.51814	0.53916	516.9	681.3	1198.2	0.7182	0.6926	1.4108	524.0
528.0	870.31	0.02112	0.49843	0.51955	521.8	675.5	1197.3	0.7231	0.6839	1.4070	528.0
532.0	900.34	0.02123	0.47947	0.50070	526.8	669.6	1196.4	0.7280	0.6752	1.4032	532.0
536.0	931.17	0.02134	0.46123	0.48257	531.7	663.6	1195.4	0.7329	0.6665	1.3993	536.0
540.0	962.79	0.02146	0.44367	0.46513	536.8	657.5	1194.3	0.7378	0.6577	1.3954	540.0
544.0	995.22	0.02157	0.42677	0.44834	541.8	651.3	1193.1	0.7427	0.6489	1.3915	544.0
548.0	1028.49	0.02169	0.41048	0.43217	546.9	645.0	1191.9	0.7476	0.6400	1.3876	548.0
552.0	1062.59	0.02182	0.39479	0.41660	552.0	638.5	1190.6	0.7525	0.6311	1.3837	552.0
556.0	1097.55	0.02194	0.37966	0.40160	557.2	632.0	1189.2	0.7575	0.6222	1.3797	556.0
560.0	1133.38	0.02207	0.36507	0.38714	562.4	625.3	1187.7	0.7625	0.6132	1.3757	560.0
564.0	1170.10	0.02221	0.35099	0.37320	567.6	618.5	1186.1	0.7674	0.6041	1.3716	564.0
568.0	1207.72	0.02235	0.33741	0.35975	572.9	611.5	1184.5	0.7725	0.5950	1.3675	568.0
572.0	1246.26	0.02249	0.32429	0.34678	578.3	604.5	1182.7	0.7775	0.5859	1.3634	572.0
576.0	1285.74	0.02264	0.31162	0.33426	583.7	597.2	1180.9	0.7825	0.5766	1.3592	576.0

Table III-1. Saturated Steam: Temperature

Temp. t (°F)	Abs. Press. p (psi)	Specific Volume			Enthalpy			Entropy			Temp. t (°F)
		Sat. Liquid v_f	Evap v_{fg}	Sat. Vapor v_g	Sat. Liquid h_f	Evap h_{fg}	Sat. Vapor h_g	Sat. Liquid s_f	Evap s_{fg}	Sat. Vapor s_g	
580.0	1326.17	0.02279	0.29937	0.32216	589.1	589.9	1179.0	0.7876	0.5673	1.3550	**580.0**
584.0	1367.7	0.02295	0.28753	0.31048	594.6	582.4	1176.9	0.7927	0.5580	1.3507	**584.0**
588.0	1410.0	0.02311	0.27608	0.29919	600.1	574.7	1174.8	0.7978	0.5485	1.3464	**588.0**
592.0	1453.3	0.02328	0.26499	0.28827	605.7	566.8	1172.6	0.8030	0.5390	1.3420	**592.0**
596.0	1497.8	0.02345	0.25425	0.27770	611.4	558.8	1170.2	0.8082	0.5293	1.3375	**596.0**
600.0	1543.2	0.02364	0.24384	0.26747	617.1	550.6	1167.7	0.8134	0.5196	1.3330	**600.0**
604.0	1589.7	0.02382	0.23374	0.25757	622.9	542.2	1165.1	0.8187	0.5097	1.3284	**604.0**
608.0	1637.3	0.02402	0.22394	0.24796	628.8	533.6	1162.4	0.8240	0.4997	1.3238	**608.0**
612.0	1686.1	0.02422	0.21442	0.23865	634.8	524.7	1159.5	0.8294	0.4896	1.3190	**612.0**
616.0	1735.9	0.02444	0.20516	0.22960	640.8	515.6	1156.4	0.8348	0.4794	1.3141	**616.0**
620.0	1786.9	0.02466	0.19615	0.22081	646.9	506.3	1153.2	0.8403	0.4689	1.3092	**620.0**
624.0	1839.0	0.02489	0.18737	0.21226	653.1	496.6	1149.8	0.8458	0.4583	1.3041	**624.0**
628.0	1892.4	0.02514	0.17880	0.20394	659.5	486.7	1146.1	0.8514	0.4474	1.2988	**628.0**
632.0	1947.0	0.02539	0.17044	0.19583	665.9	476.4	1142.2	0.8571	0.4364	1.2934	**632.0**
636.0	2002.8	0.02566	0.16226	0.18792	672.4	465.7	1138.1	0.8628	0.4251	1.2879	**636.0**
640.0	2059.9	0.02595	0.15427	0.18021	679.1	454.6	1133.7	0.8686	0.4134	1.2821	**640.0**
644.0	2118.3	0.02625	0.14644	0.17269	685.9	443.1	1129.0	0.8746	0.4015	1.2761	**644.0**
648.0	2178.1	0.02657	0.13876	0.16534	692.9	431.1	1124.0	0.8806	0.3893	1.2699	**648.0**
652.0	2239.2	0.02691	0.13124	0.15816	700.0	418.7	1118.7	0.8868	0.3767	1.2634	**652.0**
656.0	2301.7	0.02728	0.12387	0.15115	707.4	405.7	1113.1	0.8931	0.3637	1.2567	**656.0**
660.0	2365.7	0.02768	0.11663	0.14431	714.9	392.1	1107.0	0.8995	0.3502	1.2498	**660.0**
664.0	2431.1	0.02811	0.10947	0.13757	722.9	377.7	1100.6	0.9064	0.3361	1.2425	**664.0**
668.0	2498.1	0.02858	0.10229	0.13087	731.5	362.1	1093.5	0.9137	0.3210	1.2347	**668.0**
672.0	2566.6	0.02911	0.09514	0.12424	740.2	345.7	1085.9	0.9212	0.3054	1.2266	**672.0**
676.0	2636.8	0.02970	0.08799	0.11769	749.2	328.5	1077.6	0.9287	0.2892	1.2179	**676.0**
680.0	2708.6	0.03037	0.08080	0.11117	758.5	310.1	1068.5	0.9365	0.2720	1.2086	**680.0**
684.0	2782.1	0.03114	0.07349	0.10463	768.2	290.2	1058.4	0.9447	0.2537	1.1984	**684.0**
688.0	2857.4	0.03204	0.06595	0.09799	778.8	268.2	1047.0	0.9535	0.2337	1.1872	**688.0**
692.0	2934.5	0.03313	0.05797	0.09110	790.5	243.1	1033.6	0.9634	0.2110	1.1744	**692.0**
696.0	3013.4	0.03455	0.04916	0.08371	804.4	212.8	1017.2	0.9749	0.1841	1.1591	**696.0**
700.0	3094.3	0.03662	0.03857	0.07519	822.4	172.7	995.2	0.9901	0.1490	1.1390	**700.0**
702.0	3135.5	0.03824	0.03173	0.06997	835.0	144.7	979.7	1.0006	0.1246	1.1252	**702.0**
704.0	3177.2	0.04108	0.02192	0.06300	854.2	102.0	956.2	1.0169	0.0876	1.1046	**704.0**
705.0	3198.3	0.04427	0.01304	0.05730	873.0	61.4	934.4	1.0329	0.0527	1.0856	**705.0**
705.47[b]	3208.2	0.05078	0.00000	0.05078	906.0	0.0	906.0	1.0612	0.0000	1.0612	**705.47**[b]

Table III-2. Saturated Steam: Pressure

Abs. Press. p (psi)	Temp. t (°F)	Specific Volume			Enthalpy			Entropy			Abs. Press. P (psi)
		Sat. Liquid V_f	Evap v_{fx}	Sat. Vapor v_x	Sat. Liquid h_f	Evap h_{fx}	Sat. Vapor h_x	Sat. Liquid s_f	Evap s_{fx}	Sat. Vapor s_f	

Pressure Table

0.08865	32.018	0.016022	3302.4	3302.4	0.0003	1075.5	1075.5	0.0000	2.1872	2.1872	0.08865
0.25	59.323	0.016032	1235.5	1235.5	27.382	1060.1	1087.4	0.0542	2.0425	2.0967	0.25
0.50	79.586	0.016071	641.5	641.5	47.623	1048.6	1096.3	0.0925	1.9446	2.0370	0.50
1.0	101.74	0.016136	333.59	333.60	69.73	1036.1	1105.8	0.1326	1.8455	1.9781	1.0
5.0	162.24	0.016407	73.515	73.532	130.20	1000.9	1131.1	0.2349	1.6094	1.8443	5.0
10.0	193.21	0.016592	38.404	38.420	161.26	982.1	1143.3	0.2836	1.5043	1.7879	10.0
14.696	212.00	0.016719	26.782	26.799	180.17	970.3	1150.5	0.3121	1.4447	1.7568	14.696
15.0	213.03	0.016726	26.274	26.290	181.21	969.7	1150.9	0.3137	1.4415	1.7552	15.0
20.0	227.96	0.016834	20.070	20.087	196.27	960.1	1156.3	0.3358	1.3962	1.7320	20.0
30.0	250.34	0.017009	13.7266	13.7436	218.9	945.2	1164.1	0.3682	1.3313	1.6995	30.0
40.0	267.25	0.017151	10.4794	10.4965	236.1	933.6	1169.8	0.3921	1.2844	1.6765	40.0
50.0	281.02	0.017274	8.4967	8.5140	250.2	923.9	1174.1	0.4112	1.2474	1.6586	50.0
60.0	292.71	0.017383	7.1562	7.1736	262.2	915.4	1177.6	0.4273	1.2167	1.6440	60.0
70.0	302.93	0.017482	6.1875	6.2050	272.7	907.8	1180.6	0.4411	1.1905	1.6316	70.0
80.0	312.04	0.017573	5.4536	5.4711	282.1	900.9	1183.1	0.4534	1.1675	1.6208	80.0
90.0	320.28	0.017659	4.8779	4.8953	290.7	894.6	1185.3	0.4643	1.1470	1.6113	90.0
100.0	327.82	0.017740	4.4133	4.4310	298.5	888.6	1187.2	0.4743	1.1284	1.6027	100.0
110.0	334.79	0.01782	4.0306	4.0484	305.8	883.1	1188.9	0.4834	1.1115	1.5950	110.0
120.0	341.27	0.01789	3.7097	3.7275	312.6	877.8	1190.4	0.4919	1.0960	1.5879	120.0
130.0	347.33	0.01796	3.4364	3.4544	319.0	872.8	1191.7	0.4998	1.0815	1.5813	130.0
140.0	353.04	0.01803	3.2010	3.2190	325.0	868.0	1193.0	0.5071	1.0681	1.5752	140.0
150.0	358.43	0.01809	2.9958	3.0139	330.6	863.4	1194.1	0.5141	1.0554	1.5695	150.0
160.0	363.55	0.01815	2.8155	2.8336	336.1	859.0	1195.1	0.5206	1.0435	1.5641	160.0
170.0	368.42	0.01821	2.6556	2.6738	341.2	854.8	1196.0	0.5269	1.0322	1.5591	170.0
180.0	373.08	0.01827	2.5129	2.5312	346.2	850.7	1196.9	0.5328	1.0215	1.5543	180.0
190.0	377.53	0.01833	2.3847	2.4030	350.9	846.7	1197.6	0.5384	1.0113	1.5498	190.0
200.0	381.80	0.01839	2.2689	2.2873	355.5	842.8	1198.3	0.5438	1.0016	1.5454	200.0
210.0	385.91	0.01844	2.16373	2.18217	359.9	839.1	1199.0	0.5490	0.9923	1.5413	210.0
220.0	389.88	0.01850	2.06779	2.08629	364.2	835.4	1199.6	0.5540	0.9834	1.5374	220.0
230.0	393.70	0.01855	1.97991	1.99846	368.3	831.8	1200.1	0.5588	0.9748	1.5336	230.0
240.0	397.39	0.01860	1.89909	1.91769	372.3	828.4	1200.6	0.5634	0.9665	1.5299	240.0
250.0	400.97	0.01865	1.82452	1.84317	376.1	825.0	1201.1	0.5679	0.9585	1.5264	250.0
260.0	404.44	0.01870	1.75548	1.77418	379.9	821.6	1201.5	0.5722	0.9508	1.5230	260.0
270.0	407.80	0.01875	1.69137	1.71013	383.6	818.3	1201.9	0.5764	0.9433	1.5197	270.0
280.0	411.07	0.01880	1.63169	1.65049	387.1	815.1	1202.3	0.5805	0.9361	1.5166	280.0
290.0	414.25	0.01885	1.57597	1.59482	390.6	812.0	1202.6	0.5844	0.9291	1.5135	290.0
300.0	417.35	0.01889	1.52384	1.54274	394.0	808.9	1202.9	0.5882	0.9223	1.5105	300.0
350.0	431.73	0.01912	1.30642	1.32554	409.8	794.2	1204.0	0.6059	0.8909	1.4968	350.0
400.0	444.60	0.01934	1.14162	1.16095	424.2	780.4	1204.6	0.6217	0.8630	1.4847	400.0
450.0	456.28	0.01954	1.01224	1.03179	437.3	767.5	1204.8	0.6360	0.8378	1.4738	450.0
500.0	467.01	0.01975	0.90787	0.92762	449.5	755.1	1204.7	0.6490	0.8148	1.4639	500.0
550.0	476.94	0.01994	0.82183	0.84177	460.9	743.3	1204.3	0.6611	0.7936	1.4547	550.0
600.0	486.20	0.02013	0.74962	0.76975	471.7	732.0	1203.7	0.6723	0.7738	1.4461	600.0
650.0	494.89	0.02032	0.68811	0.70843	481.9	720.9	1202.8	0.6828	0.7552	1.4381	650.0
700.0	503.08	0.02050	0.63505	0.65556	491.6	710.2	1201.8	0.6928	0.7377	1.4304	700.0
750.0	510.84	0.02069	0.58880	0.60949	500.9	699.8	1200.7	0.7022	0.7210	1.4232	750.0
800.0	518.21	0.02087	0.54809	0.56896	509.8	689.6	1199.4	0.7111	0.7051	1.4163	800.0
850.0	525.24	0.02105	0.51197	0.53302	518.4	679.5	1198.0	0.7197	0.6899	1.4096	850.0
900.0	531.95	0.02123	0.47968	0.50091	526.7	669.7	1196.4	0.7279	0.6753	1.4032	900.0
950.0	538.39	0.02141	0.45064	0.47205	534.7	660.0	1194.7	0.7358	0.6612	1.3970	950.0
1000.0	544.58	0.02159	0.42436	0.44596	542.6	650.4	1192.9	0.7434	0.6476	1.3910	1000.0
1050.0	550.53	0.02177	0.40047	0.42224	550.1	640.9	1191.0	0.7507	0.6344	1.3851	1050.0
1100.0	556.28	0.02195	0.37863	0.40058	557.5	631.5	1189.1	0.7578	0.6216	1.3794	1100.0
1150.0	561.82	0.02214	0.35859	0.38073	564.8	622.2	1187.0	0.7647	0.6091	1.3738	1150.0
1200.0	567.19	0.02232	0.34013	0.36245	571.9	613.0	1184.8	0.7714	0.5969	1.3683	1200.0

Table III-2. Saturated Steam: Pressure (Cont'd)

Abs. Press. p (psi)	Temp. t (°F)	Specific Volume			Enthalpy			Entropy			Abs. Press. P (psi)
		Sat. Liquid V_f	Evap v_{fx}	Sat. Vapor v_x	Sat. Liquid h_f	Evap h_{fx}	Sat. Vapor h_x	Sat. Liquid s_f	Evap s_{fx}	Sat. Vapor s_f	
1250.0	572.38	0.02250	0.32306	0.34556	578.8	603.8	1182.6	0.7780	0.5850	1.3630	1250.0
1300.0	577.42	0.02269	0.30722	0.32991	585.6	594.6	1180.2	0.7843	0.5733	1.3577	1300.0
1350.0	582.32	0.02288	0.29250	0.31537	592.3	585.4	1177.8	0.7906	0.5620	1.3525	1350.0
1400.0	587.07	0.02307	0.27871	0.30178	598.8	576.5	1175.3	0.7966	0.5507	1.3474	1400.0
1450.0	591.70	0.02327	0.26584	0.28911	605.3	567.4	1172.8	0.8026	0.5397	1.3423	1450.0
1500.0	596.20	0.02346	0.25372	0.27719	611.7	558.4	1170.1	0.8085	0.5288	1.3373	1500.0
1550.0	600.59	0.02366	0.24235	0.26601	618.0	549.4	1167.4	0.8142	0.5182	1.3324	1550.0
1600.0	604.87	0.02387	0.23159	0.25545	624.2	540.3	1164.5	0.8199	0.5076	1.3274	1600.0
1650.0	609.05	0.02407	0.22143	0.24551	630.4	531.3	1161.6	0.8254	0.4971	1.3225	1650.0
1700.0	613.13	0.02428	0.21178	0.23607	636.5	522.2	1158.6	0.8309	0.4867	1.3176	1700.0
1750.0	617.12	0.02450	0.20263	0.22713	642.5	513.1	1155.6	0.8363	0.4765	1.3128	1750.0
1800.0	621.02	0.02472	0.19390	0.21861	648.5	503.8	1152.3	0.8417	0.4662	1.3079	1800.0
1850.0	624.83	0.02495	0.18558	0.21052	654.5	494.6	1149.0	0.8470	0.4561	1.3030	1850.0
1900.0	628.56	0.02517	0.17761	0.20278	660.4	485.2	1145.6	0.8522	0.4459	1.2981	1900.0
1950.0	632.22	0.02541	0.16999	0.19540	666.3	475.8	1142.0	0.8574	0.4358	1.2931	1950.0
2000.0	635.80	0.02565	0.16266	0.18831	672.1	466.2	1138.3	0.8625	0.4256	1.2881	2000.0
2100.0	642.76	0.02615	0.14885	0.17501	683.8	446.7	1130.5	0.8727	0.4053	1.2780	2100.0
2200.0	649.45	0.02669	0.13603	0.16272	695.5	426.7	1122.2	0.8828	0.3848	1.2676	2200.0
2300.0	655.89	0.02727	0.12406	0.15133	707.2	406.0	1113.2	0.8929	0.3640	1.2569	2300.0
2400.0	662.11	0.02790	0.11287	0.14076	719.0	384.8	1103.7	0.9031	0.3430	1.2460	2400.0
2500.0	668.11	0.02859	0.10209	0.13068	731.7	361.6	1093.3	0.9139	0.3206	1.2345	2500.0
2600.0	673.91	0.02938	0.09172	0.12110	744.5	337.6	1082.0	0.9247	0.2977	1.2225	2600.0
2700.0	679.53	0.03029	0.08165	0.11194	757.3	312.3	1069.7	0.9356	0.2741	1.2097	2700.0
2800.0	684.96	0.03134	0.07171	0.10305	770.7	285.1	1055.8	0.9468	0.2491	1.1958	2800.0
2900.0	690.22	0.03262	0.06158	0.09420	785.1	254.7	1039.8	0.9588	0.2215	1.1803	2900.0
3000.0	695.33	0.03428	0.05073	0.08500	801.8	218.4	1020.3	0.9728	0.1891	1.1619	3000.0
3100.0	700.28	0.03681	0.03771	0.07452	824.0	169.3	993.3	0.9914	0.1460	1.1373	3100.0
3200.0	705.08	0.04472	0.01191	0.05663	875.5	56.1	931.6	1.0351	0.0482	1.0832	3200.0
3208.2^c	705.47	0.05078	0.00000	0.05078	906.0	0.0	906.0	1.0612	0.0000	1.0612	3208.2^c

^{a} The states shown are metastable.
^{b} Critical temperature.
^{c} Critical pressure.
Source: Copyright 1967 ASME (Abridged); reprinted by permission.

Table III-2. Superheated Steam[a]

Abs. Press. (psi) (Sat. Temp.)		Sat. Water	Sat. Steam	Temperature (°F)													
				200	250	300	350	400	450	500	600	700	800	900	1000	1100	1200
1	Sh			98.26	148.26	198.26	248.26	298.26	348.26	398.26	498.26	598.26	698.26	798.26	898.26	998.26	1098.26
(101.74)	v	0.01614	333.6	392.5	422.4	452.3	482.1	511.9	541.7	571.5	631.1	690.7	750.3	809.8	869.4	929.0	988.6
	h	69.73	1105.8	1150.2	1172.9	1195.7	1218.7	1241.8	1265.1	1288.6	1336.1	1384.5	1433.7	1483.8	1534.9	1586.8	1639.7
	s	0.1326	1.9781	2.0509	2.0841	2.1152	2.1445	2.1722	2.1985	2.2237	2.2708	2.3144	2.3551	2.3934	2.4296	2.4640	2.4969
5	Sh			37.76	87.76	137.76	187.76	237.76	287.76	337.76	437.76	537.76	637.76	737.76	837.76	937.76	1037.76
(162.24)	v	0.01641	73.53	78.14	84.21	90.24	96.25	102.24	108.23	114.21	126.15	138.08	150.01	161.94	173.86	185.78	197.70
	h	130.20	1131.1	1148.6	1171.7	1194.8	1218.0	1241.3	1264.7	1288.2	1335.9	1384.3	1433.6	1483.7	1534.7	1586.7	1639.6
	s	0.2349	1.8443	1.8716	1.9054	1.9369	1.9664	1.9943	2.0208	2.0460	2.0932	2.1369	2.1776	2.2159	2.2521	2.2866	2.3194
10	Sh			6.79	56.79	106.79	156.79	206.79	256.79	306.79	406.79	506.79	606.79	706.79	806.79	906.79	1006.79
(193.21)	v	0.01659	38.42	38.84	41.93	44.98	48.02	51.03	54.04	57.04	63.03	69.00	74.98	80.94	86.91	92.87	98.84
	h	161.26	1143.3	1146.6	1170.2	1193.7	1217.1	1240.6	1264.1	1287.8	1335.5	1384.0	1433.4	1483.5	1534.6	1586.6	1639.5
	s	0.2836	1.7879	1.7928	1.8273	1.8593	1.8892	1.9173	1.9439	1.9692	2.0166	2.0603	2.1011	2.1394	2.1757	2.2101	2.2430
14.696	Sh				38.00	88.00	138.00	188.00	238.00	288.00	388.00	488.00	588.00	688.00	788.00	888.00	988.00
(212.00)	v	.0167	26.799		28.42	30.52	32.60	34.67	36.72	38.77	42.86	46.93	51.00	55.06	59.13	63.19	67.25
	h	180.17	1150.5		1168.8	1192.6	1216.3	1239.9	1263.6	1287.4	1335.2	1383.8	1433.2	1483.4	1534.5	1586.5	1639.4
	s	.3121	1.7568		1.7833	1.8158	1.8459	1.8743	1.9010	1.9265	1.9739	2.0177	2.0585	2.0969	2.1332	2.1676	2.2005
15	Sh				36.97	86.97	136.97	186.97	236.97	286.97	386.97	486.97	586.97	686.97	786.97	886.97	986.97
(213.03)	v	0.01673	26.290		27.837	29.899	31.939	33.963	35.977	37.985	41.986	45.978	49.964	53.946	57.926	61.905	65.882
	h	181.21	1150.9		1168.7	1192.5	1216.2	1239.9	1263.6	1287.3	1335.2	1383.8	1433.2	1483.4	1534.5	1586.5	1639.4
	s	0.3137	1.7552		1.7809	1.8134	1.8437	1.8720	1.8988	1.9242	1.9717	2.0155	2.0563	2.0946	2.1309	2.1653	2.1982
20	Sh				22.04	72.04	122.04	172.04	222.04	272.04	372.04	472.04	572.04	672.04	772.04	872.04	972.04
(227.96)	v	0.01683	20.087		20.788	22.356	23.900	25.428	26.946	28.457	31.466	34.465	37.458	40.447	43.435	46.420	49.405
	h	196.27	1156.3		1167.1	1191.4	1215.4	1239.2	1263.0	1286.9	1334.9	1383.5	1432.9	1483.2	1534.3	1586.3	1639.3
	s	0.3358	1.7320		1.7475	1.7805	1.8111	1.8397	1.8666	1.8921	1.9397	1.9836	2.0244	2.0628	2.0991	2.1336	2.1982
25	Sh				9.93	59.93	109.93	159.93	209.93	259.93	359.93	459.93	559.93	659.93	759.93	859.93	959.93
(240.07)	v	0.01693	16.301		16.558	17.829	19.076	20.307	21.527	22.740	25.153	27.557	29.954	32.348	34.740	37.130	39.518
	h	208.52	1160.6		1165.6	1190.2	1214.5	1238.5	1262.5	1286.4	1334.6	1383.3	1432.7	1483.0	1534.2	1586.2	1639.2
	s	0.3535	1.7141		1.7212	1.7547	1.7856	1.8145	1.8415	1.8672	1.9149	1.9588	1.9997	2.0381	2.0744	2.1089	2.1418
30	Sh					49.66	99.66	149.66	199.66	249.66	349.66	449.66	549.66	649.66	749.66	849.66	949.66
(250.34)	v	0.01701	13.744			14.810	15.859	16.892	17.914	18.929	20.945	22.951	24.952	26.949	28.943	30.936	32.927
	h	218.93	1164.1			1189.0	1213.6	1237.8	1261.9	1286.0	1334.2	1383.0	1432.5	1482.8	1534.0	1586.1	1639.0
	s	0.3682	1.6995			1.7334	1.7647	1.7937	1.8210	1.8467	1.8946	1.9386	1.9795	2.0179	2.0543	2.0888	2.1217
35	Sh					40.71	90.71	140.71	190.71	240.71	340.71	440.71	540.71	640.71	740.71	840.71	940.71
(259.29)	v	0.01708	11.896			12.654	13.562	14.453	15.334	16.207	17.939	19.662	21.379	23.092	24.803	26.512	28.220
	h	228.03	1167.1			1187.8	1212.7	1237.1	1261.3	1285.5	1333.9	1382.8	1432.3	1482.7	1533.9	1586.0	1638.9
	s	0.3809	1.6872			1.7152	1.7468	1.7761	1.8035	1.8294	1.8774	1.9214	1.9624	2.0009	2.0372	2.0717	2.1046
40	Sh					32.75	82.75	132.75	182.75	232.75	332.75	432.75	532.75	632.75	732.75	832.75	932.75
(267.25)	v	0.01715	10.497			11.036	11.838	12.624	13.398	14.165	15.685	17.195	18.699	20.199	21.697	23.194	24.689
	h	236.14	1169.8			1186.6	1211.7	1236.4	1260.8	1285.0	1333.6	1382.5	1432.1	1482.5	1533.7	1585.8	1638.8
	s	0.3921	1.6765			1.6992	1.7312	1.7608	1.7883	1.8143	1.8624	1.9065	1.9476	1.9860	2.0224	2.0569	2.0899
45	Sh					25.56	75.56	125.56	175.56	225.56	325.56	425.56	525.56	625.56	725.56	825.56	925.56
(274.44)	v	0.01721	9.399			9.777	10.497	11.201	11.892	12.577	13.932	15.276	16.614	17.950	19.282	20.613	21.943
	h	243.49	1172.1			1185.4	1210.4	1235.7	1260.2	1284.6	1333.3	1382.3	1431.9	1482.3	1533.6	1585.7	1638.7
	s	0.4021	1.6671			1.6849	1.7173	1.7471	1.7748	1.8010	1.8492	1.8934	1.9345	1.9730	2.0093	2.0439	2.0768
50	Sh					18.98	68.98	118.98	168.98	218.98	318.98	418.98	518.98	618.98	718.98	818.98	918.98
(281.02)	v	0.01727	8.514			8.769	9.424	10.062	10.688	11.306	12.529	13.741	14.947	16.150	17.350	18.549	19.746
	h	250.21	1174.1			1184.1	1209.9	1234.9	1259.6	1284.1	1332.9	1382.0	1431.7	1482.2	1533.4	1585.6	1638.6
	s	0.4112	1.6586			1.6720	1.7048	1.7349	1.7628	1.7890	1.8374	1.8816	1.9227	1.9613	1.9977	2.0322	2.0652
55	Sh					12.93	62.93	112.93	162.93	212.93	312.93	412.93	512.93	612.93	712.93	812.93	912.93
(287.07)	v	0.01733	7.787			7.945	8.546	9.130	9.702	10.267	11.381	12.485	13.583	14.677	15.769	16.859	17.948
	h	256.43	1175.9			1182.9	1208.9	1234.2	1259.1	1283.6	1332.6	1381.8	1431.5	1482.0	1533.3	1585.5	1638.5
	s	0.4196	1.6509			1.6601	1.6933	1.7237	1.7518	1.7781	1.8266	1.8710	1.9121	1.9507	1.987	2.022	2.055
60	Sh					7.29	57.29	107.29	157.29	207.29	307.29	407.29	507.29	607.29	707.29	807.29	907.29
(292.71)	v	0.01738	7.174			7.257	7.815	8.354	8.881	9.400	10.425	11.438	12.446	13.450	14.452	15.452	16.450
	h	262.21	1177.6			1181.6	1208.0	1233.5	1258.5	1283.2	1332.3	1381.5	1431.3	1481.8	1533.2	1585.3	1638.4
	s	0.4273	1.6440			1.6492	1.6934	1.7134	1.7417	1.7681	1.8168	1.8612	1.9024	1.9410	1.9774	2.0120	2.0450
65	Sh					2.02	52.02	102.02	152.02	202.02	302.02	402.02	502.02	602.02	702.02	802.02	902.02
(297.98)	v	0.01743	6.653			6.675	7.195	7.697	8.186	8.667	9.615	10.552	11.484	12.412	13.337	14.261	15.183
	h	267.63	1179.1			1180.3	1207.0	1232.7	1257.9	1282.7	1331.9	1381.3	1431.1	1481.6	1533.0	1585.2	1638.3
	s	0.4344	1.6375			1.6390	1.6731	1.7040	1.7324	1.7590	1.8077	1.8522	1.8935	1.9321	1.9685	2.0031	2.0361
70	Sh						47.07	97.07	147.07	197.07	297.07	397.07	497.07	597.07	697.07	797.07	897.07
(302.93)	v	0.01748	6.205				6.664	7.133	7.590	8.039	8.922	9.793	10.659	11.522	12.382	13.240	14.097
	h	272.74	1180.6				1206.0	1232.0	1257.3	1282.2	1331.6	1381.0	1430.9	1481.5	1532.9	1585.1	1638.2
	s	0.4411	1.6316				1.6640	1.6951	1.7237	1.7504	1.7993	1.8439	1.8852	1.9238	1.9603	1.9949	2.0279
75	Sh						42.39	92.39	142.39	192.39	292.39	392.39	492.39	592.39	692.39	792.39	892.39
(307.61)	v	0.01753	5.814				6.204	6.645	7.074	7.494	8.320	9.135	9.945	10.750	11.553	12.355	13.155
	h	277.56	1181.9				1205.0	1231.2	1256.7	1281.7	1331.3	1380.7	1430.7	1481.3	1532.7	1585.0	1638.1
	s	0.4474	1.6260				1.6554	1.6868	1.7156	1.7424	1.7915	1.8361	1.8774	1.9161	1.9526	1.9872	2.0202

Table III-2.   Continued

Abs. Press. (psi) (Sat. Temp.)		Sat. Water	Sat. Steam	350	400	450	500	550	600	700	800	900	1000	1100	1200	1300	1400
80 (312.04)	Sh			37.96	87.96	137.96	187.96	237.96	287.96	387.96	487.96	587.96	687.96	787.96	887.96	987.96	1087.96
	v	0.01757	5.471	5.801	6.218	6.622	7.018	7.408	7.794	8.560	9.319	10.075	10.829	11.581	12.331	13.081	13.829
	h	282.15	1183.1	1204.0	1230.5	1256.1	1281.3	1306.2	1330.9	1380.5	1430.5	1481.1	1532.6	1584.9	1638.0	1692.0	1746.8
	s	0.4534	1.6208	1.6473	1.6790	1.7080	1.7349	1.7602	1.7842	1.8289	1.8702	1.9089	1.9454	1.9800	2.0131	2.0446	2.0750
85 (316.26)	Sh			33.74	83.74	133.74	183.74	233.74	283.74	383.74	483.74	583.74	683.74	783.74	883.74	983.74	1083.74
	v	0.01762	5.167	5.445	5.840	6.223	6.597	6.966	7.330	8.052	8.768	9.480	10.190	10.898	11.604	12.310	13.014
	h	286.52	1184.2	1203.0	1229.7	1255.5	1280.8	1305.8	1330.6	1380.2	1430.3	1481.0	1532.4	1584.7	1637.9	1691.9	1746.8
	s	0.4590	1.6159	1.6396	1.6716	1.7008	1.7279	1.7532	1.7772	1.8220	1.8634	1.9021	1.9386	1.9733	2.0063	2.0379	2.0682
90 (320.28)	Sh			29.72	79.72	129.72	179.72	229.72	279.72	379.72	479.72	579.72	679.72	779.72	879.72	979.72	1079.72
	v	0.01766	4.895	5.128	5.505	5.869	6.223	6.572	6.917	7.600	8.277	8.950	9.621	10.290	10.958	11.625	12.290
	h	290.69	1185.3	1202.0	1228.9	1254.9	1280.3	1305.4	1330.2	1380.0	1430.1	1480.8	1532.3	1584.6	1637.8	1691.8	1746.7
	s	0.4643	1.6113	1.6323	1.6646	1.6940	1.7212	1.7467	1.7707	1.8156	1.8570	1.8957	1.9323	1.9669	2.0000	2.0316	2.0619
95 (324.13)	Sh			25.87	75.87	125.87	175.87	225.87	275.87	375.87	475.87	575.87	675.87	775.87	875.87	975.87	1075.87
	v	0.01770	4.651	4.845	5.205	5.551	5.889	6.221	6.548	7.196	7.838	8.477	9.113	9.747	10.380	11.012	11.643
	h	294.70	1186.2	1200.9	1228.1	1254.3	1279.8	1305.0	1329.9	1379.7	1429.9	1480.6	1532.1	1584.5	1637.7	1691.7	1746.6
	s	0.4694	1.6069	1.6253	1.6580	1.6876	1.7149	1.7404	1.7645	1.8094	1.8509	1.8897	1.9262	1.9609	1.9940	2.0256	2.0559
100 (327.82)	Sh			22.18	72.18	122.18	172.18	222.18	272.18	372.18	472.18	572.18	672.18	772.18	872.18	972.18	1072.18
	v	0.01774	4.431	4.590	4.935	5.266	5.588	5.904	6.216	6.833	7.443	8.050	8.655	9.258	9.860	10.460	11.060
	h	298.54	1187.2	1199.9	1227.4	1253.7	1279.3	1304.6	1329.6	1379.5	1429.7	1480.4	1532.0	1584.4	1637.6	1691.6	1746.5
	s	0.4743	1.6027	1.6187	1.6516	1.6814	1.7088	1.7344	1.7586	1.8036	1.8451	1.8839	1.9205	1.9552	1.9883	2.0199	2.0502
105 (331.37)	Sh			18.63	68.63	118.63	168.63	218.63	268.63	368.63	468.63	568.63	668.63	768.63	868.63	968.63	1068.63
	v	0.01778	4.231	4.359	4.690	5.007	5.315	5.617	5.915	6.504	7.086	7.665	8.241	8.816	9.389	9.961	10.532
	h	302.24	1188.0	1198.8	1226.6	1253.1	1278.8	1304.2	1329.2	1379.2	1429.4	1480.3	1531.8	1584.2	1637.5	1691.5	1746.4
	s	0.4790	1.5988	1.6122	1.6455	1.6755	1.7031	1.7288	1.7530	1.7981	1.8396	1.8785	1.9151	1.9498	1.9828	2.0145	2.0448
110 (334.79)	Sh			15.21	65.21	115.21	165.21	215.21	265.21	365.21	465.21	565.21	665.21	765.21	865.21	965.21	1065.21
	v	0.01782	4.048	4.149	4.468	4.772	5.068	5.357	5.642	6.205	6.761	7.314	7.865	8.413	8.961	9.507	10.053
	h	305.80	1188.9	1197.7	1225.8	1252.5	1278.3	1303.8	1328.9	1379.0	1429.2	1480.1	1531.7	1584.1	1637.4	1691.4	1746.4
	s	0.4834	1.5950	1.6061	1.6396	1.6698	1.6975	1.7233	1.7476	1.7928	1.8344	1.8732	1.9099	1.9446	1.9777	2.0093	2.0397
115 (338.08)	Sh			11.92	61.92	111.92	161.92	211.92	261.92	361.92	461.92	561.92	661.92	761.92	861.92	961.92	1061.92
	v	0.01785	3.881	3.957	4.265	4.558	4.841	5.119	5.392	5.932	6.465	6.994	7.521	8.046	8.570	9.093	9.615
	h	309.25	1189.6	1196.7	1225.0	1251.8	1277.9	1303.3	1328.6	1378.7	1429.0	1479.9	1531.6	1584.0	1637.2	1691.4	1746.3
	s	0.4877	1.5913	1.6001	1.6340	1.6644	1.6922	1.7181	1.7425	1.7877	1.8294	1.8682	1.9049	1.9396	1.9727	2.0044	2.0347
120 (341.27)	Sh			8.73	58.73	108.73	158.73	208.73	258.73	358.73	458.73	558.73	658.73	758.73	858.73	958.73	1058.73
	v	0.01789	3.7275	3.7815	4.0786	4.3610	4.6341	4.9009	5.1637	5.6813	6.1928	6.7006	7.2060	7.7096	8.2119	8.7130	9.2134
	h	312.58	1190.4	1195.6	1224.1	1251.2	1277.4	1302.9	1328.2	1378.4	1428.8	1479.8	1531.4	1583.9	1637.1	1691.3	1746.2
	s	0.4919	1.5879	1.5943	1.6286	1.6592	1.6872	1.7132	1.7376	1.7829	1.8246	1.8635	1.9001	1.9349	1.9680	1.9996	2.0300
130 (347.33)	Sh			2.67	52.67	102.67	152.67	202.67	252.67	352.67	452.67	552.67	652.67	752.67	852.67	952.67	1052.67
	v	0.01796	3.4544	3.4699	3.7489	4.0129	4.2672	4.5151	4.7589	5.2384	5.7118	6.1814	6.6486	7.1140	7.5781	8.0411	8.5033
	h	318.95	1191.7	1193.4	1222.5	1249.9	1276.4	1302.1	1327.5	1377.9	1428.4	1479.4	1531.1	1583.6	1636.9	1691.1	1746.1
	s	0.4998	1.5813	1.5833	1.6182	1.6493	1.6775	1.7037	1.7283	1.7737	1.8155	1.8545	1.8911	1.9259	1.9591	1.9907	2.0211
140 (353.04)	Sh				46.96	96.96	146.96	196.96	246.96	346.96	446.96	546.96	646.96	746.96	846.96	946.96	1046.96
	v	0.01803	3.2190		3.4661	3.7143	3.9526	4.1844	4.4119	4.8588	5.2995	5.7364	6.1709	6.6036	7.0349	7.4652	7.8946
	h	324.96	1193.0		1220.8	1248.7	1275.3	1301.3	1326.8	1377.4	1428.0	1479.1	1530.8	1583.4	1636.7	1690.9	1745.9
	s	0.5071	1.5752		1.6085	1.6400	1.6686	1.6949	1.7196	1.7652	1.8071	1.8461	1.8828	1.9176	1.9508	1.9825	2.0129
150 (358.43)	Sh				41.57	91.57	141.57	191.57	241.57	341.57	441.57	541.57	641.57	741.57	841.57	941.57	1041.57
	v	0.01809	3.0139		3.2208	3.4555	3.6799	3.8978	4.1112	4.5298	4.9421	5.3507	5.7568	6.1612	6.5642	6.9661	7.3671
	h	330.65	1194.1		1219.1	1247.4	1274.3	1300.5	1326.1	1376.9	1427.6	1478.7	1530.5	1583.1	1636.5	1690.7	1745.7
	s	0.5141	1.5695		1.5993	1.6313	1.6602	1.6867	1.7115	1.7573	1.7992	1.8383	1.8751	1.9099	1.9431	1.9748	2.0052
160 (363.55)	Sh				36.45	86.45	136.45	186.45	236.45	336.45	436.45	536.45	636.45	736.45	836.45	936.45	1036.45
	v	0.01815	2.8336		3.0060	3.2288	3.4413	3.6469	3.8480	4.2420	4.6295	5.0132	5.3945	5.7741	6.1522	6.5293	6.9055
	h	336.07	1195.1		1217.4	1246.0	1273.3	1299.6	1325.4	1376.4	1427.2	1478.4	1530.3	1582.9	1636.3	1690.5	1745.6
	s	0.5206	1.5641		1.5906	1.6231	1.6522	1.6790	1.7039	1.7499	1.7919	1.8310	1.8678	1.9027	1.9359	1.9676	1.9980
170 (368.42)	Sh				31.58	81.58	131.58	181.58	231.58	331.58	431.58	531.58	631.58	731.58	831.58	931.58	1031.58
	v	0.01821	2.6738		2.8162	3.0288	3.2306	3.4255	3.6158	3.9879	4.3536	4.7155	5.0749	5.4325	5.7888	6.1440	6.4983
	h	341.24	1196.0		1215.6	1244.7	1272.2	1298.8	1324.7	1375.8	1426.8	1478.0	1530.0	1582.6	1636.1	1690.4	1745.4
	s	0.5269	1.5591		1.5823	1.6152	1.6447	1.6717	1.6968	1.7428	1.7850	1.8241	1.8610	1.8959	1.9291	1.9608	1.9913
180 (373.08)	Sh				26.92	76.92	126.92	176.92	226.92	326.92	426.92	526.92	626.92	726.92	826.92	926.92	1026.92
	v	0.01827	2.5312		2.6474	2.8508	3.0433	3.2286	3.4093	3.7621	4.1084	4.4508	4.7907	5.1289	5.4657	5.8014	6.1363
	h	346.19	1196.9		1213.8	1243.4	1271.2	1297.9	1324.0	1375.3	1426.3	1477.7	1529.7	1582.4	1635.9	1690.2	1745.3
	s	0.5328	1.5543		1.5743	1.6078	1.6376	1.6647	1.6900	1.7362	1.7784	1.8176	1.8545	1.8894	1.9227	1.9545	1.9849
190 (377.53)	Sh				22.47	72.47	122.47	172.47	222.47	322.47	422.47	522.47	622.47	722.47	822.47	922.47	1022.47
	v	0.01833	2.4030		2.4961	2.6915	2.8756	3.0525	3.2246	3.5601	3.8889	4.2140	4.5365	4.8572	5.1766	5.4949	5.8124
	h	350.94	1197.6		1212.0	1242.0	1270.1	1297.1	1323.3	1374.8	1425.9	1477.4	1529.4	1582.1	1635.7	1690.0	1745.1
	s	0.5384	1.5498		1.5667	1.6006	1.6307	1.6581	1.6835	1.7299	1.7722	1.8115	1.8484	1.8834	1.9166	1.9484	1.9789
200 (381.80)	Sh				18.20	68.20	118.20	168.20	218.20	318.20	418.20	518.20	618.20	748.20	818.20	918.20	1018.20
	v	0.01839	2.2873		2.3598	2.5480	2.7247	2.8939	3.0583	3.3783	3.6915	4.0008	4.3077	4.6128	4.9165	5.2191	5.5209
	h	355.51	1198.3		1210.1	1240.6	1269.0	1296.2	1322.6	1374.3	1425.5	1477.0	1529.1	1581.9	1635.4	1689.8	1745.0
	s	0.5438	1.5454		1.5593	1.5938	1.6242	1.6518	1.6773	1.7239	1.7663	1.8057	1.8426	1.8776	1.9109	1.9427	1.9732

Table III-2. Continued

Abs. Press. (psi) (Sat. Temp.)		Sat. Water	Sat. Steam	Temperature (°F)													
				400	450	500	550	600	700	800	900	1000	1100	1200	1300	1400	1500
210 (385.91)	Sh			14.09	64.09	114.09	164.09	214.09	314.09	414.09	514.09	614.09	714.09	814.09	914.09	1014.09	1114.09
	v	0.01844	2.1822	2.2364	2.4181	2.5880	2.7504	2.9078	3.2137	3.5128	3.8080	4.1007	4.3915	4.6811	4.9695	5.2571	5.5440
	h	359.91	1199.0	1208.02	1239.2	1268.0	1295.3	1321.9	1373.7	1425.1	1476.7	1528.8	1581.6	1635.2	1689.6	1744.8	1800.8
	s	0.5490	1.5413	1.5522	1.5872	1.6180	1.6458	1.6715	1.7182	1.7607	1.8001	1.8371	1.8721	1.9054	1.9372	1.9677	1.9970
220 (389.88)	Sh			10.12	60.12	110.12	160.12	210.12	310.12	410.12	510.12	610.12	710.12	810.12	910.12	1010.12	1110.12
	v	0.01850	2.0863	2.1240	2.2999	2.4638	2.6199	2.7710	3.0642	3.3504	3.6327	3.9125	4.1905	4.4671	4.7426	5.0173	5.2913
	h	364.17	1199.6	1206.3	1237.8	1266.9	1294.5	1321.2	1373.2	1424.7	1476.3	1528.5	1581.4	1635.0	1689.4	1744.7	1800.6
	s	0.5540	1.5374	1.5453	1.5808	1.6120	1.6400	1.6658	1.7128	1.7553	1.7948	1.8318	1.8668	1.9002	1.9320	1.9625	1.9919
230 (393.70)	Sh			6.30	56.30	106.30	156.30	206.30	306.30	406.30	506.30	606.30	706.30	806.30	906.30	1006.30	1106.30
	v	0.01855	1.9985	2.0212	2.1919	2.3503	2.5008	2.6461	2.9276	3.2020	3.4726	3.7406	4.0068	4.2717	4.5355	4.7984	5.0606
	h	368.28	1200.1	1204.4	1236.3	1265.7	1293.6	1320.4	1372.7	1424.2	1476.0	1528.2	1581.1	1634.8	1689.3	1744.5	1800.5
	s	0.5588	1.5336	1.5385	1.5747	1.6062	1.6344	1.6604	1.7075	1.7502	1.7897	1.8268	1.8618	1.8952	1.9270	1.9576	1.9869
240 (397.39)	Sh			2.61	52.61	102.61	152.61	202.61	302.61	402.61	502.61	602.61	702.61	802.61	902.61	1002.61	1102.61
	v	0.01860	1.9177	1.9268	2.0928	2.2462	2.3915	2.5316	2.8024	3.0661	3.3259	3.5831	3.8385	4.0926	4.3456	4.5977	4.8492
	h	372.27	1200.6	1202.4	1234.9	1264.6	1292.7	1319.7	1372.1	1423.8	1475.6	1527.9	1580.9	1634.6	1689.1	1744.3	1800.4
	s	0.5634	1.5299	1.5320	1.5687	1.6006	1.6291	1.6552	1.7025	1.7452	1.7848	1.8219	1.8570	1.8904	1.9223	1.9528	1.9822
250 (400.97)	Sh				49.03	99.03	149.03	199.03	299.03	399.03	499.03	599.03	699.03	799.03	899.03	999.03	1099.03
	v	0.01865	1.8432		2.0016	2.1504	2.2909	2.4262	2.6872	2.9410	3.1909	3.4382	3.6837	3.9278	4.1709	4.4131	4.6546
	h	376.14	1201.1		1233.4	1263.5	1291.8	1319.0	1371.6	1423.4	1475.3	1527.6	1580.6	1634.4	1688.9	1744.2	1800.2
	s	0.5679	1.5264		1.5629	1.5951	1.6239	1.6502	1.6976	1.7405	1.7801	1.8173	1.8524	1.8858	1.9177	1.9482	1.9776
260 (404.44)	Sh				45.56	95.56	145.56	195.56	295.56	395.56	495.56	595.56	695.56	795.56	895.56	995.56	1095.56
	v	0.01870	1.7742		1.9173	2.0619	2.1981	2.3289	2.5808	2.8256	3.0663	3.3044	3.5408	3.7758	4.0097	4.2427	4.4750
	h	379.90	1201.5		1231.9	1262.4	1290.5	1318.2	1371.1	1423.0	1474.9	1527.3	1580.4	1634.2	1688.7	1744.0	1800.1
	s	0.5722	1.5230		1.5573	1.5899	1.6189	1.6453	1.6930	1.7359	1.7756	1.8128	1.8480	1.8814	1.9133	1.9439	1.9732
270 (407.80)	Sh				42.20	92.20	142.20	192.20	292.20	392.20	492.20	592.20	692.20	792.20	892.20	992.20	1092.20
	v	0.01875	1.7101		1.8391	1.9799	2.1121	2.2388	2.4824	2.7186	2.9509	3.1806	3.4084	3.6349	3.8603	4.0849	4.3087
	h	383.56	1201.9		1230.4	1261.2	1290.0	1317.5	1370.5	1422.6	1474.6	1527.1	1580.1	1634.0	1688.5	1743.9	1800.0
	s	0.5764	1.5197		1.5518	1.5848	1.6140	1.6406	1.6885	1.7315	1.7713	1.8085	1.8437	1.8771	1.9090	1.9396	1.9690
280 (411.07)	Sh				38.93	88.93	138.93	188.93	288.93	388.93	488.93	588.93	688.93	788.93	888.93	988.93	1088.93
	v	0.01880	1.6505		1.7665	1.9037	2.0322	2.1551	2.3909	2.6194	2.8437	3.0655	3.2855	3.5042	3.7217	3.9384	4.1543
	h	387.12	1202.3		1228.8	1260.0	1289.1	1316.8	1370.0	1422.1	1474.2	1526.8	1579.9	1633.8	1688.4	1743.7	1799.8
	s	0.5805	1.5166		1.5464	1.5798	1.6093	1.6361	1.6841	1.7273	1.7671	1.8043	1.8395	1.8730	1.9050	1.9356	1.9649
290 (414.25)	Sh				35.75	85.75	135.75	185.75	285.75	385.75	485.75	585.75	685.75	785.75	885.75	985.75	1085.75
	v	0.01885	1.5948		1.6988	1.8327	1.9578	2.0772	2.3058	2.5269	2.7440	2.9585	3.1711	3.3824	3.5926	3.8019	4.0106
	h	390.60	1202.6		1227.3	1258.9	1288.1	1316.0	1369.5	1421.7	1473.9	1526.5	1579.6	1633.5	1688.2	1743.6	1799.7
	s	0.5844	1.5135		1.5412	1.5750	1.6048	1.6317	1.6799	1.7232	1.7630	1.8003	1.8356	1.8690	1.9010	1.9316	1.9610
300 (417.35)	Sh				32.65	82.65	132.65	182.65	282.65	382.65	482.65	582.65	682.65	782.65	882.65	982.65	1082.65
	v	0.01889	1.5427		1.6356	1.7665	1.8883	2.0044	2.2263	2.4407	2.6509	2.8585	3.0643	3.2688	3.4721	3.6746	3.8764
	h	393.99	1202.9		1225.7	1257.7	1287.2	1315.2	1368.9	1421.3	1473.6	1526.2	1579.4	1633.3	1688.0	1743.4	1799.6
	s	0.5882	1.5105		1.5361	1.5703	1.6003	1.6274	1.6758	1.7192	1.7591	1.7964	1.8317	1.8652	1.8972	1.9278	1.9572
310 (420.36)	Sh				29.64	79.64	129.64	179.64	279.64	379.64	479.64	579.64	679.64	779.64	879.64	979.64	1079.64
	v	0.01894	1.4939		1.5763	1.7044	1.8233	1.9363	2.1520	2.3600	2.5638	2.7650	2.9644	3.1625	3.3594	3.5555	3.7509
	h	397.30	1203.2		1224.1	1256.5	1286.3	1314.5	1368.4	1420.9	1473.2	1525.9	1579.2	1633.1	1687.8	1743.3	1799.4
	s	0.5920	1.5076		1.5311	1.5657	1.5960	1.6233	1.6719	1.7153	1.7553	1.7927	1.8280	1.8615	1.8935	1.9241	1.9536
320 (423.31)	Sh				26.69	76.69	126.69	176.69	276.69	376.69	476.69	576.69	676.69	776.69	876.69	976.69	1076.69
	v	0.01899	1.4480		1.5207	1.6462	1.7623	1.8725	2.0823	2.2843	2.4821	2.6774	2.8708	3.0628	3.2538	3.4438	3.6332
	h	400.53	1203.4		1222.5	1255.2	1285.3	1313.7	1367.8	1420.5	1472.9	1525.6	1578.9	1632.9	1687.6	1743.1	1799.3
	s	0.5956	1.5048		1.5261	1.5612	1.5918	1.6192	1.6680	1.7116	1.7516	1.7890	1.8243	1.8579	1.8899	1.9206	1.9500
330 (426.18)	Sh				23.82	73.82	123.82	173.82	273.82	373.82	473.82	573.82	673.82	773.82	873.82	973.82	1073.82
	v	0.01903	1.4048		1.4684	1.5915	1.7050	1.8125	2.0168	2.2132	2.4054	2.5950	2.7828	2.9692	3.1545	3.3389	3.5227
	h	403.70	1203.6		1220.9	1254.0	1284.4	1313.0	1367.3	1420.0	1472.5	1525.3	1578.7	1632.7	1687.5	1742.9	1799.2
	s	0.5991	1.5021		1.5213	1.5568	1.5876	1.6153	1.6643	1.7079	1.7480	1.7855	1.8208	1.8544	1.8864	1.9171	1.9466
340 (428.99)	Sh				21.01	71.01	121.01	171.01	271.01	371.01	471.01	571.01	671.01	771.01	871.01	971.01	1071.01
	v	0.01908	1.3640		1.4191	1.5399	1.6511	1.7561	1.9552	2.1463	2.3333	2.5175	2.7000	2.8811	3.0611	3.2402	3.4186
	h	406.80	1203.8		1219.2	1252.8	1283.4	1312.2	1366.7	1419.6	1472.2	1525.0	1578.4	1632.5	1687.3	1742.8	1799.0
	s	0.6026	1.4994		1.5165	1.5525	1.5836	1.6114	1.6606	1.7044	1.7445	1.7820	1.8174	1.8510	1.8831	1.9138	1.9432
350 (431.73)	Sh				18.27	68.27	118.27	168.27	268.27	368.27	468.27	568.27	668.27	768.27	868.27	968.27	1068.27
	v	0.01912	1.3255		1.3725	1.4913	1.6002	1.7028	1.8970	2.0832	2.2652	2.4445	2.6219	2.7980	2.9730	3.1471	3.3205
	h	409.83	1204.0		1217.5	1251.5	1282.4	1311.4	1366.2	1419.2	1471.8	1524.7	1578.2	1632.3	1687.1	1742.6	1798.9
	s	0.6059	1.4968		1.5119	1.5483	1.5797	1.6077	1.6571	1.7009	1.7411	1.7787	1.8141	1.8477	1.8798	1.9105	1.9400
360 (434.41)	Sh				15.59	65.59	115.59	165.59	265.59	365.59	465.59	565.59	665.59	765.59	865.59	965.59	1065.59
	v	0.01917	1.2891		1.3285	1.4454	1.5521	1.6525	1.8421	2.0237	2.2009	2.3755	2.5482	2.7196	2.8898	3.0592	3.2279
	h	412.81	1204.1		1215.8	1250.3	1281.5	1310.6	1365.6	1418.7	1471.5	1524.4	1577.9	1632.1	1686.9	1742.5	1798.8
	s	0.6092	1.4943		1.5073	1.5441	1.5758	1.6040	1.6536	1.6976	1.7379	1.7754	1.8109	1.8445	1.8766	1.9073	1.9368
380 (439.61)	Sh				10.39	60.39	110.39	160.39	260.39	360.39	460.39	560.39	660.39	760.39	860.39	960.39	1060.39
	v	0.01925	1.2218		1.2472	1.3606	1.4635	1.5598	1.7410	1.9139	2.0825	2.2484	2.4124	2.5750	2.7366	2.8973	3.0572
	h	418.59	1204.4		1212.4	1247.7	1279.5	1309.0	1364.5	1417.9	1470.8	1523.8	1577.4	1631.6	1686.5	1742.2	1798.5
	s	0.6156	1.4894		1.4982	1.5360	1.5683	1.5969	1.6470	1.6911	1.7315	1.7692	1.8047	1.8384	1.8705	1.9012	1.9307

Table III-2.   Continued

Abs. Press. (psi) (Sat. Temp.)		Sat. Water	Sat. Steam	450	500	550	600	650	700	800	900	1000	1100	1200	1300	1400	1500
400 (444.60)	Sh			5.40	55.40	105.40	155.40	205.40	255.40	355.40	455.40	555.40	655.40	755.40	855.40	955.40	1055.40
	v	0.01934	1.1610	1.1738	1.2841	1.3836	1.4763	1.5646	1.6499	1.8151	1.9759	2.1339	2.2901	2.4450	2.5987	2.7515	2.9037
	h	424.17	1204.6	1208.8	1245.1	1277.5	1307.4	1335.9	1363.4	1417.0	1470.1	1523.3	1576.9	1631.2	1686.2	1741.9	1798.2
	s	0.6217	1.4847	1.4894	1.5282	1.5611	1.5901	1.6163	1.6406	1.6850	1.7255	1.7632	1.7988	1.8325	1.8647	1.8955	1.9250
420 (449.40)	Sh			.60	50.60	100.60	150.60	200.60	250.60	350.60	450.60	550.60	650.60	750.60	850.60	950.60	1050.60
	v	0.01942	1.1057	1.1071	1.2148	1.3113	1.4007	1.4856	1.5676	1.7258	1.8795	2.0304	2.1795	2.3273	2.4739	2.6196	2.7647
	h	429.56	1204.7	1205.2	1242.4	1275.4	1305.8	1334.5	1362.3	1416.2	1469.4	1522.7	1576.4	1630.8	1685.8	1741.6	1798.0
	s	0.6276	1.4802	1.4808	1.5206	1.5542	1.5835	1.6100	1.6345	1.6791	1.7197	1.7575	1.7932	1.8269	1.8591	1.8899	1.9195
440 (454.03)	Sh				45.97	95.97	145.97	195.97	245.97	345.97	445.97	545.97	645.97	745.97	845.97	945.97	1045.97
	v	0.01950	1.0554		1.1517	1.2454	1.3319	1.4138	1.4926	1.6445	1.7918	1.9363	2.0790	2.2203	2.3605	2.4998	2.6384
	h	434.77	1204.8		1239.7	1273.4	1304.2	1333.2	1361.1	1415.3	1468.7	1522.1	1575.9	1630.4	1685.5	1741.2	1797.7
	s	0.6332	1.4759		1.5132	1.5474	1.5772	1.6040	1.6286	1.6734	1.7142	1.7521	1.7878	1.8216	1.8538	1.8847	1.9143
460 (458.50)	Sh				41.50	91.50	141.50	191.50	241.50	341.50	441.50	541.50	641.50	741.50	841.50	941.50	1041.50
	v	0.01959	1.0092		1.0939	1.1852	1.2691	1.3482	1.4242	1.5703	1.7117	1.8504	1.9872	2.1226	2.2569	2.3903	2.5230
	h	439.83	1204.8		1236.9	1271.3	1302.5	1331.8	1360.0	1414.4	1468.0	1521.5	1575.4	1629.9	1685.1	1740.9	1797.4
	s	0.6387	1.4718		1.5060	1.5409	1.5711	1.5982	1.6230	1.6680	1.7089	1.7469	1.7826	1.8165	1.8488	1.8797	1.9093
480 (462.82)	Sh				37.18	87.18	137.18	187.18	237.18	337.18	437.18	537.18	637.18	737.18	837.18	937.18	1037.18
	v	0.01967	0.9668		1.0409	1.1300	1.2115	1.2881	1.3615	1.5023	1.6384	1.7716	1.9030	2.0330	2.1619	2.2900	2.4173
	h	444.75	1204.8		1234.1	1269.1	1300.8	1330.5	1358.8	1413.6	1467.3	1520.9	1574.9	1629.5	1684.7	1740.6	1797.2
	s	0.6439	1.4677		1.4990	1.5346	1.5652	1.5925	1.6176	1.6628	1.7038	1.7419	1.7777	1.8116	1.8439	1.8748	1.9045
500 (467.01)	Sh				32.99	82.99	132.99	182.99	232.99	332.99	432.99	532.99	632.99	732.99	832.99	932.99	1032.99
	v	0.01975	0.9276		0.9919	1.0791	1.1584	1.2327	1.3037	1.4397	1.5708	1.6992	1.8256	1.9507	2.0746	2.1977	2.3200
	h	449.52	1204.7		1231.2	1267.0	1299.1	1329.1	1357.7	1412.7	1466.6	1520.3	1574.4	1629.1	1684.4	1740.3	1796.9
	s	0.6490	1.4639		1.4921	1.5284	1.5595	1.5871	1.6123	1.6578	1.6990	1.7371	1.7730	1.8069	1.8393	1.8702	1.8998
520 (471.07)	Sh				28.93	78.93	128.93	178.93	228.93	328.93	428.93	528.93	628.93	728.93	828.93	928.93	1028.93
	v	0.01982	0.8914		0.9466	1.0321	1.1094	1.1816	1.2504	1.3819	1.5085	1.6323	1.7542	1.8746	1.9940	2.1125	2.2302
	h	454.18	1204.5		1228.3	1264.8	1297.4	1327.7	1356.5	1411.8	1465.9	1519.7	1573.9	1628.7	1684.0	1740.0	1796.7
	s	0.6540	1.4601		1.4853	1.5223	1.5539	1.5818	1.6072	1.6530	1.6943	1.7325	1.7684	1.8024	1.8348	1.8657	1.8954
540 (475.01)	Sh				24.99	74.99	124.99	174.99	224.99	324.99	424.99	524.99	624.99	724.99	824.99	924.99	1024.99
	v	0.01990	0.8577		0.9045	0.9884	1.0640	1.1342	1.2010	1.3284	1.4508	1.5704	1.6880	1.8042	1.9193	2.0336	2.1471
	h	458.71	1204.4		1225.3	1262.5	1295.7	1326.3	1355.3	1410.9	1465.1	1519.1	1573.4	1628.2	1683.6	1739.7	1796.4
	s	0.6587	1.4565		1.4786	1.5164	1.5485	1.5767	1.6023	1.6483	1.6897	1.7280	1.7640	1.7981	1.8305	1.8615	1.8911
560 (478.84)	Sh				21.16	71.16	121.16	171.16	221.16	321.16	421.16	521.16	621.16	721.16	821.16	921.16	1021.16
	v	0.01998	0.8264		0.8653	0.9479	1.0217	1.0902	1.1552	1.2787	1.3972	1.5129	1.6266	1.7388	1.8500	1.9603	2.0699
	h	463.14	1204.2		1222.2	1260.3	1293.9	1324.9	1354.2	1410.0	1464.4	1518.6	1572.9	1627.8	1683.3	1739.4	1796.1
	s	0.6634	1.4529		1.4720	1.5106	1.5431	1.5717	1.5975	1.6438	1.6853	1.7237	1.7598	1.7939	1.8263	1.8573	1.8870
580 (482.57)	Sh				17.43	67.43	117.43	167.43	217.43	317.43	417.43	517.43	617.43	717.43	817.43	917.43	1017.43
	v	0.02006	0.7971		0.8287	0.9100	0.9824	1.0492	1.1125	1.2324	1.3473	1.4593	1.5693	1.6780	1.7855	1.8921	1.9980
	h	467.47	1203.9		1219.1	1258.0	1292.1	1323.4	1353.0	1409.2	1463.7	1518.0	1572.4	1627.4	1682.9	1739.1	1795.9
	s	0.6679	1.4495		1.4654	1.5049	1.5380	1.5668	1.5929	1.6394	1.6811	1.7196	1.7556	1.7898	1.8223	1.8533	1.8831
600 (486.20)	Sh				13.80	63.80	113.80	163.80	213.80	313.80	413.80	513.80	613.80	713.80	813.80	913.80	1013.80
	v	0.02013	0.7697		0.7944	0.8746	0.9456	1.0109	1.0726	1.1892	1.3008	1.4093	1.5160	1.6211	1.7252	1.8284	1.9309
	h	471.70	1203.7		1215.9	1255.6	1290.3	1322.0	1351.8	1408.3	1463.0	1517.4	1571.9	1627.0	1682.6	1738.8	1795.6
	s	0.6723	1.4461		1.4590	1.4993	1.5329	1.5621	1.5884	1.6351	1.6769	1.7155	1.7517	1.7859	1.8184	1.8494	1.8792
650 (494.89)	Sh				5.11	55.11	105.11	155.11	205.11	305.11	405.11	505.11	605.11	705.11	805.11	905.11	1005.11
	v	0.02032	0.7084		0.7173	0.7954	0.8634	0.9254	0.9835	1.0929	1.1969	1.2979	1.3969	1.4944	1.5909	1.6864	1.7813
	h	481.89	1202.8		1207.6	1249.6	1285.7	1318.3	1348.7	1406.0	1461.2	1515.9	1570.7	1625.9	1681.6	1738.0	1794.9
	s	1.6828	1.4381		1.4430	1.4858	1.5207	1.5507	1.5775	1.6249	1.6671	1.7059	1.7422	1.7765	1.8092	1.8403	1.8701
700 (503.08)	Sh					46.92	96.92	146.92	196.92	296.92	396.92	496.92	596.92	696.92	796.92	896.92	996.92
	v	0.02050	0.6556			0.7271	0.7928	0.8520	0.9072	1.0102	1.1078	1.2023	1.2948	1.3858	1.4757	1.5647	1.6530
	h	491.60	1201.8			1243.4	1281.0	1314.6	1345.6	1403.7	1459.4	1514.4	1569.4	1624.8	1680.7	1737.2	1794.3
	s	1.6928	1.4304			1.4726	1.5090	1.5399	1.5673	1.6154	1.6580	1.6970	1.7335	1.7679	1.8006	1.8318	1.8617
750 (510.84)	Sh					39.16	89.16	139.16	189.16	289.16	389.16	489.16	589.16	689.16	789.16	889.16	989.16
	v	0.02069	0.6095			0.6676	0.7313	0.7882	0.8409	0.9386	1.0306	1.1195	1.2063	1.2916	1.3759	1.4592	1.5419
	h	500.89	1200.7			1236.9	1276.1	1310.7	1342.5	1401.5	1457.6	1512.9	1568.2	1623.8	1679.8	1736.4	1793.6
	s	0.7022	1.4232			1.4598	1.4977	1.5296	1.5577	1.6065	1.6494	1.6886	1.7252	1.7598	1.7926	1.8239	1.8538
800 (518.21)	Sh					31.79	81.79	131.79	181.79	281.79	381.79	481.79	581.79	681.79	781.79	881.79	981.79
	v	0.02087	0.5690			0.6151	0.6774	0.7323	0.7828	0.8759	0.9631	1.0470	1.1289	1.2093	1.2885	1.3669	1.4446
	h	509.81	1199.4			1230.1	1271.1	1306.8	1339.3	1399.1	1455.8	1511.4	1566.9	1622.7	1678.9	1735.7	1792.9
	s	0.7111	1.4163			1.4472	1.4869	1.5198	1.5484	1.5980	1.6413	1.6807	1.7175	1.7522	1.7851	1.8164	1.8464
850 (525.24)	Sh					24.76	74.76	124.76	174.76	274.76	374.76	474.76	574.76	674.76	774.76	874.76	974.76
	v	0.02105	0.5330			0.5683	0.6296	0.6829	0.7315	0.8205	0.9034	0.9830	1.0606	1.1366	1.2115	1.2855	1.3588
	h	518.40	1198.0			1223.0	1265.9	1302.8	1336.0	1396.8	1454.0	1510.0	1565.7	1621.6	1678.0	1734.9	1792.3
	s	0.7197	1.4096			1.4347	1.4763	1.5102	1.5396	1.5899	1.6336	1.6733	1.7102	1.7450	1.7780	1.8094	1.8395
900 (531.95)	Sh					18.05	68.05	118.05	168.05	268.05	368.05	468.05	568.05	668.05	768.05	868.05	968.05
	v	0.02123	0.5009			0.5263	0.5869	0.6388	0.6858	0.7713	0.8504	0.9262	0.9998	1.0720	1.1430	1.2131	1.2825
	h	526.70	1196.4			1215.5	1260.6	1298.6	1332.7	1394.4	1452.2	1508.5	1564.4	1620.6	1677.1	1734.1	1791.6
	s	0.7279	1.4032			1.4223	1.4659	1.5010	1.5311	1.5822	1.6263	1.6662	1.7033	1.7382	1.7713	1.8028	1.8329

Table III-2. Continued

Abs. Press. (psi) (Sat. Temp.)		Sat. Water	Sat. Steam	550	600	650	700	750	800	850	900	1000	1100	1200	1300	1400	1500
									Temperature (°F)								
950	Sh			11.61	61.61	111.61	161.61	211.61	261.61	311.61	361.61	461.61	561.61	661.61	761.61	861.61	961.61
(538.39)	v	0.02141	0.4721	0.4883	0.5485	0.5993	0.6449	0.6871	0.7272	0.7656	0.8030	0.8753	0.9455	1.0142	1.0817	1.1484	1.2143
	h	534.74	1194.7	1207.6	1255.1	1294.4	1329.3	1361.5	1392.0	1421.5	1450.3	1507.0	1563.2	1619.5	1676.2	1733.3	1791.0
	s	0.7358	1.3970	1.4098	1.4557	1.4921	1.5228	1.5500	1.5748	1.5977	1.6193	1.6595	1.6967	1.7317	1.7649	1.7965	1.8267
1000	Sh			5.42	55.42	105.42	155.42	205.42	255.42	305.42	355.42	455.42	555.42	655.42	755.42	855.42	955.42
(544.58)	v	0.02159	0.4460	0.4535	0.5137	0.5636	0.6080	0.6489	0.6875	0.7245	0.7603	0.8295	0.8966	0.9622	1.0266	1.0901	1.1529
	h	542.55	1192.9	1199.3	1249.3	1290.1	1325.9	1358.7	1389.6	1419.4	1448.5	1505.4	1561.9	1618.4	1675.3	1732.5	1790.3
	s	0.7434	1.3910	1.3973	1.4457	1.4833	1.5149	1.5426	1.5677	1.5908	1.6126	1.6530	1.6905	1.7256	1.7589	1.7905	1.8207
1050	Sh				49.47	99.47	149.47	199.47	249.47	299.47	349.47	449.47	549.47	649.47	749.47	849.47	949.47
(550.53)	v	0.02177	0.4222		0.4821	0.5312	0.5745	0.6142	0.6515	0.6872	0.7216	0.7881	0.8524	0.9151	0.9767	1.0373	1.0973
	h	550.15	1191.0		1243.4	1285.7	1322.4	1355.8	1387.2	1417.3	1446.6	1503.9	1560.7	1617.4	1674.4	1731.8	1789.6
	s	0.7507	1.3851		1.4358	1.4748	1.5072	1.5354	1.5608	1.5842	1.6062	1.6469	1.6845	1.7197	1.7531	1.7848	1.8151
1100	Sh				43.72	93.72	143.72	193.72	243.72	293.72	343.72	443.72	543.72	643.72	743.72	843.72	943.72
(556.28)	v	0.02195	0.4006		0.4531	0.5017	0.5440	0.5826	0.6188	0.6533	0.6865	0.7505	0.8121	0.8723	0.9313	0.9894	1.0468
	h	557.55	1189.1		1237.3	1281.2	1318.8	1352.9	1384.7	1415.2	1444.7	1502.4	1559.4	1616.3	1673.5	1731.0	1789.0
	s	0.7578	1.3794		1.4259	1.4664	1.4996	1.5284	1.5542	1.5779	1.6000	1.6410	1.6787	1.7141	1.7475	1.7793	1.8097
1150	Sh				39.18	89.18	139.18	189.18	239.18	289.18	339.18	439.18	539.18	639.18	739.18	839.18	939.18
(561.82)	v	0.02214	0.3807		0.4263	0.4746	0.5162	0.5538	0.5889	0.6223	0.6544	0.7161	0.7754	0.8332	0.8899	0.9456	1.0007
	h	564.78	1187.0		1230.9	1276.6	1315.2	1349.9	1382.2	1413.0	1442.8	1500.9	1558.1	1615.2	1672.6	1730.2	1788.3
	s	0.7647	1.3738		1.4160	1.4582	1.4923	1.5216	1.5478	1.5717	1.5941	1.6353	1.6732	1.7087	1.7422	1.7741	1.8045
1200	Sh				32.81	82.81	132.81	182.81	232.81	282.81	332.81	432.81	532.81	632.81	732.81	832.81	932.81
(567.19)	v	0.02232	0.3624		0.4016	0.4497	0.4905	0.5273	0.5615	0.5939	0.6250	0.6845	0.7418	0.7974	0.8519	0.9055	0.9584
	h	571.85	1184.8		1224.2	1271.8	1311.5	1346.9	1379.7	1410.8	1440.9	1499.4	1556.9	1614.2	1671.6	1729.4	1787.6
	s	0.7714	1.3683		1.4061	1.4501	1.4851	1.5150	1.5415	1.5658	1.5883	1.6298	1.6679	1.7035	1.7371	1.7691	1.7996
1300	Sh				22.58	72.58	122.58	172.58	222.58	272.58	322.58	422.58	522.58	622.58	722.58	822.58	922.58
(577.42)	v	0.02269	0.3299		0.3570	0.4052	0.4451	0.4804	0.5129	0.5436	0.5729	0.6287	0.6822	0.7341	0.7847	0.8345	0.8836
	h	585.58	1180.2		1209.9	1261.9	1303.9	1340.8	1374.6	1406.4	1437.1	1496.3	1554.3	1612.0	1669.8	1727.9	1786.3
	s	0.7843	1.3577		1.3860	1.4340	1.4711	1.5022	1.5296	1.5544	1.5773	1.6194	1.6578	1.6937	1.7275	1.7596	1.7902
1400	Sh				12.93	62.93	112.93	162.93	212.93	262.93	312.93	412.93	512.93	612.93	712.93	812.93	912.93
(587.07)	v	0.02307	0.3018		0.3176	0.3667	0.4059	0.4400	0.4712	0.5004	0.5282	0.5809	0.6311	0.6798	0.7272	0.7737	0.8195
	h	598.83	1175.3		1194.1	1251.4	1296.1	1334.5	1369.3	1402.0	1433.2	1493.2	1551.8	1609.9	1668.0	1726.3	1785.0
	s	0.7966	1.3474		1.3652	1.4181	1.4575	1.4900	1.5182	1.5436	1.5670	1.6096	1.6484	1.6845	1.7185	1.7508	1.7815
1500	Sh				3.80	53.80	103.80	153.80	203.80	253.80	303.80	403.80	503.80	603.80	703.80	803.80	903.80
(596.20)	v	0.02346	0.2772		0.2820	0.3328	0.3717	0.4049	0.4350	0.4629	0.4894	0.5394	0.5869	0.6327	0.6773	0.7210	0.7639
	h	611.68	1170.1		1176.3	1240.2	1287.9	1328.0	1364.0	1397.4	1429.2	1490.1	1549.2	1607.7	1666.2	1724.8	1783.7
	s	0.8085	1.3373		1.3431	1.4022	1.4443	1.4782	1.5073	1.5333	1.5572	1.6004	1.6395	1.6759	1.7101	1.7425	1.7734
1600	Sh					45.13	95.13	145.13	195.13	245.13	295.13	395.13	495.13	595.13	695.13	795.13	895.13
(604.87)	v	0.02387	0.2555			0.3026	0.3415	0.3741	0.4032	0.4301	0.4555	0.5031	0.5482	0.5915	0.6336	0.6748	0.7153
	h	624.20	1164.5			1228.3	1279.4	1321.4	1358.5	1392.8	1425.2	1486.9	1546.6	1605.6	1664.3	1723.2	1782.3
	s	0.8199	1.3274			1.3861	1.4312	1.4667	1.4968	1.5235	1.5478	1.5916	1.6312	1.6678	1.7022	1.7347	1.7657
1700	Sh					36.87	86.87	136.87	186.87	236.87	286.87	386.87	486.87	586.87	686.87	786.87	886.87
(613.13)	v	0.02428	0.2361			0.2754	0.3147	0.3468	0.3751	0.4011	0.4255	0.4711	0.5140	0.5552	0.5951	0.6341	0.6724
	h	636.45	1158.6			1215.3	1270.5	1314.5	1352.9	1388.1	1421.2	1483.8	1544.0	1603.4	1662.5	1721.7	1781.0
	s	0.8309	1.3176			1.3697	1.4183	1.4555	1.4867	1.5140	1.5388	1.5833	1.6232	1.6601	1.6947	1.7274	1.7585
1800	Sh					28.98	78.98	128.98	178.98	228.98	278.98	378.98	478.98	578.98	678.98	778.98	878.98
(621.02)	v	0.02472	0.2186			0.2505	0.2906	0.3223	0.3500	0.3752	0.3988	0.4426	0.4836	0.5229	0.5609	0.5980	0.6343
	h	648.49	1152.3			1201.2	1261.1	1307.4	1347.2	1383.3	1417.1	1480.6	1541.4	1601.2	1660.7	1720.1	1779.7
	s	0.8417	1.3079			1.3526	1.4054	1.4446	1.4768	1.5049	1.5302	1.5753	1.6156	1.6528	1.6876	1.7204	1.7516
1900	Sh					21.44	71.44	121.44	171.44	221.44	271.44	371.44	471.44	571.44	671.44	771.44	871.44
(628.56)	v	0.02517	0.2028			0.2274	0.2687	0.3004	0.3275	0.3521	0.3749	0.4171	0.4565	0.4940	0.5303	0.5656	0.6002
	h	660.36	1145.6			1185.7	1251.3	1300.2	1341.4	1378.4	1412.9	1477.4	1538.8	1599.1	1658.8	1718.6	1778.4
	s	0.8522	1.2981			1.3346	1.3925	1.4338	1.4672	1.4960	1.5219	1.5677	1.6084	1.6458	1.6808	1.7138	1.7451
2000	Sh					14.20	64.20	114.20	164.20	214.20	264.20	364.20	464.20	564.20	664.20	764.20	864.20
(635.80)	v	0.02565	0.1883			0.2056	0.2488	0.2805	0.3072	0.3312	0.3534	0.3942	0.4320	0.4680	0.5027	0.5365	0.5695
	h	672.11	1138.3			1168.3	1240.9	1292.6	1335.4	1373.5	1408.7	1474.1	1536.2	1596.9	1657.0	1717.0	1777.1
	s	0.8625	1.2881			1.3154	1.3794	1.4231	1.4578	1.4874	1.5138	1.5603	1.6014	1.6391	1.6743	1.7075	1.7389
2100	Sh					7.24	57.24	107.24	157.24	207.24	257.24	357.24	457.24	557.24	657.24	757.24	857.24
(642.76)	v	0.02615	0.1750			0.1847	0.2304	0.2624	0.2888	0.3123	0.3339	0.3734	0.4099	0.4445	0.4778	0.5101	0.5418
	h	683.79	1130.5			1148.5	1229.8	1284.9	1329.3	1368.4	1404.4	1470.9	1533.6	1594.7	1655.2	1715.4	1775.7
	s	0.8727	1.2780			1.2942	1.3661	1.4125	1.4486	1.4790	1.5060	1.5532	1.5948	1.6327	1.6681	1.7014	1.7330
2200	Sh					.55	50.55	100.55	150.55	200.55	250.55	350.55	450.55	550.55	650.55	750.55	850.55
(649.45)	v	0.02669	0.1627			0.1636	0.2134	0.2458	0.2720	0.2950	0.3161	0.3545	0.3897	0.4231	0.4551	0.4862	0.5165
	h	695.46	1122.2			1123.9	1218.0	1276.8	1323.1	1363.3	1400.0	1467.6	1530.9	1592.5	1653.3	1713.9	1774.4
	s	0.8828	1.2676			1.2691	1.3523	1.4020	1.4395	1.4708	1.4984	1.5463	1.5883	1.6266	1.6622	1.6956	1.7273
2300	Sh						44.11	94.11	144.11	194.11	244.11	344.11	444.11	544.11	644.11	744.11	844.11
(655.89)	v	0.02727	0.1513				0.1975	0.2305	0.2566	0.2793	0.2999	0.3372	0.3714	0.4035	0.4344	0.4643	0.4935
	h	707.18	1113.2				1205.3	1268.4	1316.7	1358.1	1395.7	1464.2	1528.3	1590.3	1651.5	1712.3	1773.1
	s	0.8929	1.2569				1.3381	1.3914	1.4305	1.4628	1.4910	1.5397	1.5821	1.6207	1.6565	1.6901	1.7219

Table III-2. Continued

Abs. Press. (psi) (Sat. Temp.)		Sat. Water	Sat. Steam	700	750	800	850	900	950	1000	1050	1100	1150	1200	1300	1400	1500
2400 (662.11)	Sh			37.89	87.89	137.89	187.89	237.89	287.89	337.89	387.89	437.89	487.89	537.89	637.89	737.89	837.89
	v	0.02790	0.1408	0.1824	0.2164	0.2424	0.2648	0.2850	0.3037	0.3214	0.3382	0.3545	0.3703	0.3856	0.4155	0.4443	0.4724
	h	718.95	1103.7	1191.6	1259.7	1310.1	1352.8	1391.2	1426.9	1460.9	1493.7	1525.6	1557.0	1588.1	1649.6	1710.8	1771.8
	s	0.9031	1.2460	1.3232	1.3808	1.4217	1.4549	1.4837	1.5095	1.5332	1.5553	1.5761	1.5959	1.6149	1.6509	1.6847	1.7167
2500 (668.11)	Sh			31.89	81.89	131.89	181.89	231.89	281.89	331.89	381.89	431.89	481.89	531.89	631.89	731.89	831.89
	v	0.02859	0.1307	0.1681	0.2032	0.2293	0.2514	0.2712	0.2896	0.3068	0.3232	0.3390	0.3543	0.3692	0.3980	0.4259	0.4529
	h	731.71	1093.3	1176.7	1250.6	1303.4	1347.4	1386.7	1423.1	1457.5	1490.7	1522.9	1554.6	1585.9	1647.8	1709.2	1770.4
	s	0.9139	1.2345	1.3076	1.3701	1.4129	1.4472	1.4766	1.5029	1.5269	1.5492	1.5703	1.5903	1.6094	1.6456	1.6796	1.7116
2600 (673.91)	Sh			26.09	76.09	126.09	176.09	226.09	276.09	326.09	376.09	426.09	476.09	526.09	626.09	726.09	826.09
	v	0.02938	0.1211	0.1544	0.1909	0.2171	0.2390	0.2585	0.2765	0.2933	0.3093	0.3247	0.3395	0.3540	0.3819	0.4088	0.4350
	h	744.47	1082.0	1160.2	1241.1	1296.5	1341.9	1382.1	1419.2	1454.1	1487.7	1520.2	1552.2	1583.7	1646.0	1707.7	1769.1
	s	0.9247	1.2225	1.2908	1.3592	1.4042	1.4395	1.4696	1.4964	1.5208	1.5434	1.5646	1.5848	1.6040	1.6405	1.6746	1.7068
2700 (679.53)	Sh	·		20.47	70.47	120.47	170.47	220.47	270.47	320.47	370.47	420.47	470.47	520.47	620.47	720.47	820.47
	v	0.03029	0.1119	0.1411	0.1794	0.2058	0.2275	0.2468	0.2644	0.2809	0.2965	0.3114	0.3259	0.3399	0.3670	0.3931	0.4184
	h	757.34	1069.7	1142.0	1231.1	1289.5	1336.3	1377.5	1415.2	1450.7	1484.6	1517.5	1549.8	1581.5	1644.1	1706.1	1767.8
	s	0.9356	1.2097	1.2727	1.3481	1.3954	1.4319	1.4628	1.4900	1.5148	1.5376	1.5591	1.5794	1.5988	1.6355	1.6697	1.7021
2800 (684.96)	Sh			15.04	65.04	115.04	165.04	215.04	265.04	315.04	365.04	415.04	465.04	515.04	615.04	715.04	815.04
	v	0.03134	0.1030	0.1278	0.1685	0.1952	0.2168	0.2358	0.2531	0.2693	0.2845·	0.2991	0.3132	0.3268	0.3532	0.3785	0.4030
	h	770.69	1055.8	1121.2	1220.6	1282.2	1330.7	1372.8	1411.2	1447.2	1481.6	1514.8	1547.3	1579.3	1642.2	1704.5	1766.5
	s	0.9468	1.1958	1.2527	1.3368	1.3867	1.4245	1.4561	1.4838	1.5089	1.5321	1.5537	1.5742	1.5938	1.6306	1.6651	1.6975
2900 (690.22)	Sh			9.78	59.78	109.78	159.78	209.78	259.78	309.78	359.78	409.78	459.78	509.78	609.78	709.78	809.78
	v	0.03262	0.0942	0.1138	0.1581	0.1853	0.2068	0.2256	0.2427	0.2585	0.2734	0.2877	0.3014	0.3147	0.3403	0.3649	0.3887
	h	785.13	1039.8	1095.3	1209.6	1274.7	1324.9	1368.0	1407.2	1443.7	1478.5	1512.1	1544.9	1577.0	1640.4	1703.0	1765.2
	s	0.9588	1.1803	1.2283	1.3251	1.3780	1.4171	1.4494	1.4777	1.5032	1.5266	1.5485	1.5692	1.5889	1.6259	1.6605	1.6931
3000 (695.33)	Sh			4.67	54.67	104.67	154.67	204.67	254.67	304.67	354.67	404.67	454.67	504.67	604.67	704.67	804.67
	v	0.03428	0.0850	0.0982	0.1483	0.1759	0.1975	0.2161	0.2329	0.2484	0.2630	0.2770	0.2904	0.3033	0.3282	0.3522	0.3753
	h	801.84	1020.3	1060.5	1197.9	1267.0	1319.0	1363.2	1403.1	1440.2	1475.4	1509.4	1542.4	1574.8	1638.5	1701.4	1763.8
	s	0.9728	1.1619	1.1966	1.3131	1.3692	1.4097	1.4429	1.4717	1.4976	1.5213	1.5434	1.5642	1.5841	1.6214	1.6561	1.6888
3100 (700.28)	Sh				49.72	99.72	149.72	199.72	249.72	299.72	349.72	399.72	449.72	499.72	599.72	699.72	799.72
	v	0.03681	0.0745		0.1389	0.1671	0.1887	0.2071	0.2237	0.2390	0.2533	0.2670	0.2800	0.2927	0.3170	0.3403	0.3628
	h	823.97	993.3		1185.4	1259.1	1313.0	1358.4	1399.0	1436.7	1472.3	1506.6	1539.9	1572.6	1636.7	1699.8	1762.5
	s	0.9914	1.1373		1.3007	1.3604	1.4024	1.4364	1.4658	1.4920	1.5161	1.5384	1.5594	1.5794	1.6169	1.6518	1.6847
3200 (705.08)	Sh				44.92	94.92	144.92	194.92	244.92	294.92	344.92	394.92	444.92	494.92	594.92	694.92	794.92
	v	0.04472	0.0566		0.1300	0.1588	0.1804	0.1987	0.2151	0.2301	0.2442	0.2576	0.2704	0.2827	0.3065	0.3291	0.3510
	h	875.54	931.6		1172.3	1250.9	1306.9	1353.4	1394.9	1433.1	1469.2	1503.8	1537.4	1570.3	1634.8	1698.3	1761.2
	s	1.0351	1.0832		1.2877	1.3515	1.3951	1.4300	1.4600	1.4866	1.5110	1.5335	1.5547	1.5749	1.6126	1.6477	1.6806
3300	Sh																
	v				0.1213	0.1510	0.1727	0.1908	0.2070	0.2218	0.2357	0.2488	0.2613	0.2734	0.2966	0.3187	0.3400
	h				1158.2	1242.5	1300.7	1348.4	1390.7	1429.5	1466.1	1501.0	1534.9	1568.1	1632.9	1696.7	1759.9
	s				1.2742	1.3425	1.3879	1.4237	1.4542	1.4813	1.5059	1.5287	1.5501	1.5704	1.6084	1.6436	1.6767
3400	Sh																
	v				0.1129	0.1435	0.1653	0.1834	0.1994	0.2140	0.2276	0.2405	0.2528	0.2646	0.2872	0.3088	0.3296
	h				1143.2	1233.7	1294.3	1343.4	1386.4	1425.9	1462.9	1498.3	1532.4	1565.8	1631.1	1695.1	1758.2
	s				1.2600	1.3334	1.3807	1.4174	1.4486	1.4761	1.5010	1.5240	1.5456	1.5660	1.6042	1.6396	1.6728
3500	Sh																
	v				0.1048	0.1364	0.1583	0.1764	0.1922	0.2066	0.2200	0.2326	0.2447	0.2563	0.2784	0.2995	0.3198
	h				1127.1	1224.6	1287.8	1338.2	1382.2	1422.2	1459.7	1495.5	1529.9	1563.6	1629.2	1693.6	1757.2
	s				1.2450	1.3242	1.3734	1.4112	1.4430	1.4709	1.4962	1.5194	1.5412	1.5618	1.6002	1.6358	1.6691
3600	Sh																
	v				0.0966	0.1296	0.1517	0.1697	0.1854	0.1996	0.2128	0.2252	0.2371	0.2485	0.2702	0.2908	0.3106
	h.				1108.6	1215.3	1281.2	1333.0	1377.9	1418.6	1456.5	1492.6	1527.4	1561.3	1627.3	1692.0	1755.9
	s				1.2281	1.3148	1.3662	1.4050	1.4374	1.4658	1.4914	1.5149	1.5369	1.5576	1.5962	1.6320	1.6654
3800	Sh																
	v				0.0799	0.1169	0.1395	0.1574	0.1729	0.1868	0.1996	0.2116	0.2231	0.2340	0.2549	0.2746	0.2936
	h				1064.2	1195.5	1267.6	1322.4	1369.1	1411.2	1450.1	1487.0	1522.4	1556.8	1623.6	1688.9	1753.2
	s				1.1888	1.2955	1.3517	1.3928	1.4265	1.4558	1.4821	1.5061	1.5284	1.5495	1.5886	1.6247	1.6584
4000	Sh																
	v				0.0631	0.1052	0.1284	0.1463	0.1616	0.1752	0.1877	0.1994	0.2105	0.2210	0.2411	0.2601	0.2783
	h				1007.4	1174.3	1253.4	1311.6	1360.2	1403.6	1443.6	1481.3	1517.3	1552.2	1619.8	1685.7	1750.6
	s				1.1396	1.2754	1.3371	1.3807	1.4158	1.4461	1.4730	1.4976	1.5203	1.5417	1.5812	1.6177	1.6516
4200	Sh																
	v				0.0498	0.0945	0.1183	0.1362	0.1513	0.1647	0.1769	0.1883	0.1991	0.2093	0.2287	0.2470	0.2645
	h				950.1	1151.6	1238.6	1300.4	1351.2	1396.0	1437.1	1475.5	1512.2	1547.6	1616.1	1682.6	1748.0
	s				1.0905	1.2544	1.3223	1.3686	1.4053	1.4366	1.4642	1.4893	1.5124	1.5341	1.5742	1.6109	1.6452
4400	Sh																
	v				0.0421	0.0846	0.1090	0.1270	0.1420	0.1552	0.1671	0.1782	0.1887	0.1986	0.2174	0.2351	0.2519
	h				909.5	1127.3	1223.3	1289.0	1342.0	1388.3	1430.4	1469.7	1507.1	1543.0	1612.3	1679.4	1745.3
	s				1.0556	1.2325	1.3073	1.3566	1.3949	1.4272	1.4556	1.4812	1.5048	1.5268	1.5673	1.6044	1.6389

Table III-2. Continued

Abs. Press. (psi) (Sat. Temp.)	Sat. Water	Sat. Steam	Temperature (°F)													
			750	800	850	900	950	1000	1050	1100	1150	1200	1250	1300	1400	1500
4600 Sh																
v			0.0380	0.0751	0.1005	0.1186	0.1335	0.1465	0.1582	0.1691	0.1792	0.1889	0.1982	0.2071	0.2242	0.2404
h			883.8	1100.0	1207.3	1277.2	1332.6	1380.5	1423.7	1463.9	1501.9	1538.4	1573.8	1608.5	1676.3	1742.7
s			1.0331	1.2084	1.2922	1.3446	1.3847	1.4181	1.4472	1.4734	1.4974	1.5197	1.5407	1.5607	1.5982	1.6330
4800 Sh																
v			0.0355	0.0665	0.0927	0.1109	0.1257	0.1385	0.1500	0.1606	0.1706	0.1800	0.1890	0.1977	0.2142	0.2299
h			866.9	1071.2	1190.7	1265.2	1323.1	1372.6	1417.0	1458.0	1496.7	1533.8	1569.7	1604.7	1673.1	1740.0
s			1.0180	1.1835	1.2768	1.3327	1.3745	1.4090	1.4390	1.4657	1.4901	1.5128	1.5341	1.5543	1.5921	1.6272
5000 Sh																
v			0.0338	0.591	0.0855	0.1038	0.1185	0.1312	0.1425	0.1529	0.1626	0.1718	0.1806	0.1890	0.2050	0.2203
h			854.9	1042.9	1173.6	1252.9	1313.5	1364.6	1410.2	1452.1	1491.5	1529.1	1565.5	1600.9	1670.0	1737.4
s			1.0070	1.1593	1.2612	1.3207	1.3645	1.4001	1.4309	1.4582	1.4831	1.5061	1.5277	1.5481	1.5863	1.6216
5200 Sh																
v			0.0326	0.0531	0.0789	0.0973	0.1119	0.1244	0.1356	0.1458	0.1553	0.1642	0.1728	0.1810	0.1966	0.2114
h			845.8	1016.9	1156.0	1240.4	1303.7	1356.6	1403.4	1446.2	1486.3	1524.5	1561.3	1597.2	1666.8	1734.7
s			0.9985	1.1370	1.2455	1.3088	1.3545	1.3914	1.4229	1.4509	1.4762	1.4995	1.5214	1.5420	1.5806	1.6161
5400 Sh																
v			0.0317	0.0483	0.0728	0.0912	0.1058	0.1182	0.1292	0.1392	0.1485	0.1572	0.1656	0.1736	0.1888	0.2031
h			838.5	994.3	1138.1	1227.7	1293.7	1348.4	1396.5	1440.3	1481.1	1519.8	1557.1	1593.4	1663.7	1732.1
s			0.9915	1.1175	1.2296	1.2969	1.3446	1.3827	1.4151	1.4437	1.4694	1.4931	1.5153	1.5362	1.5750	1.6109
5600 Sh																
v			0.0309	0.0447	0.0672	0.0856	0.1001	0.1124	0.1232	0.1331	0.1422	0.1508	0.1589	0.1667	0.1815	0.1954
h			832.4	975.0	1119.9	1214.8	1283.7	1340.2	1389.6	1434.3	1475.9	1515.2	1552.9	1589.6	1660.5	1729.5
s			0.9855	1.1008	1.2137	1.2850	1.3348	1.3742	1.4075	1.4366	1.4628	1.4869	1.5093	1.5304	1.5697	1.6058
5800 Sh																
v			0.0303	0.0419	0.0622	0.0805	0.0949	0.1070	0.1177	0.1274	0.1363	0.1447	0.1527	0.1603	0.1747	0.1883
h			827.3	958.8	1101.8	1201.8	1273.6	1332.0	1382.6	1428.3	1470.6	1510.5	1548.7	1585.8	1657.4	1726.8
s			0.9803	1.0867	1.1981	1.2732	1.3250	1.3658	1.3999	1.4297	1.4564	1.4808	1.5035	1.5248	1.5644	1.6008
6000 Sh																
v			0.0298	0.0397	0.0579	0.0757	0.0900	0.1020	0.1126	0.1221	0.1309	0.1391	0.1469	0.1544	0.1684	0.1817
h			822.9	945.1	1084.6	1188.8	1263.6	1323.6	1375.7	1422.3	1465.4	1505.9	1544.6	1582.0	1654.2	1724.2
s			0.9758	1.0746	1.1833	1.2615	1.3154	1.3574	1.3925	1.4229	1.4500	1.4748	1.4978	1.5194	1.5593	1.5960
6500 Sh																
v			0.0287	0.0358	0.0495	0.0655	0.0793	0.0909	0.1012	0.1104	0.1188	0.1266	0.1340	0.1411	0.1544	0.1669
h			813.9	919.5	1046.7	1156.3	1237.8	1302.7	1358.1	1407.3	1452.2	1494.2	1534.1	1572.5	1646.4	1717.6
s			0.9661	1.0515	1.1506	1.2328	1.2917	1.3370	1.3743	1.4064	1.4347	1.4604	1.4841	1.5062	1.5471	1.5844
7000 Sh																
v			0.0279	0.0334	0.0438	0.0573	0.0704	0.0816	0.0915	0.1004	0.1085	0.1160	0.1231	0.1298	0.1424	0.1542
h			806.9	901.8	1016.5	1124.9	1212.6	1281.7	1340.5	1392.2	1439.1	1482.6	1523.7	1563.1	1638.6	1711.1
s			0.9582	1.0350	1.1243	1.2055	1.2689	1.3171	1.3567	1.3904	1.4200	1.4466	1.4710	1.4938	1.5355	1.5735
7500 Sh																
v			0.0272	0.0318	0.0399	0.0512	0.0631	0.0737	0.0833	0.0918	0.0996	0.1068	0.1136	0.1200	0.1321	0.1433
h			801.3	889.0	992.9	1097.7	1188.3	1261.0	1322.9	1377.2	1426.0	1471.0	1513.3	1553.7	1630.8	1704.6
s			0.9514	1.0224	1.1033	1.1818	1.2473	1.2980	1.3397	1.3751	1.4059	1.4335	1.4586	1.4819	1.5245	1.5632
8000 Sh																
v			0.0267	0.0306	0.0371	0.0465	0.0571	0.0671	0.0762	0.0845	0.0920	0.0989	0.1054	0.1115	0.1230	0.1338
h			796.6	879.1	974.4	1074.3	1165.4	1241.0	1305.5	1362.2	1413.0	1459.6	1503.1	1544.5	1623.1	1698.1
s			0.9455	1.0122	1.0864	1.1613	1.2271	1.2798	1.3233	1.3603	1.3924	1.4208	1.4467	1.4705	1.5140	1.5533
8500 Sh																
v			0.0262	0.0296	0.0350	0.0429	0.0522	0.0615	0.0701	0.0780	0.0853	0.0919	0.0982	0.1041	0.1151	0.1254
h			792.7	871.2	959.8	1054.5	1144.0	1221.9	1288.5	1347.5	1400.2	1448.2	1492.9	1535.3	1615.4	1691.7
s			0.9402	1.0037	1.0727	1.1437	1.2084	1.2627	1.3076	1.3460	1.3793	1.4087	1.4352	1.4597	1.5040	1.5439
9000 Sh																
v			0.0258	0.0288	0.0335	0.0402	0.0483	0.0568	0.0649	0.0724	0.0794	0.0858	0.0918	0.0975	0.1081	0.1179
h			789.3	864.7	948.0	1037.6	1125.4	1204.1	1272.1	1333.0	1387.5	1437.1	1482.9	1526.3	1607.9	1685.3
s			0.9354	0.9964	1.0613	1.1285	1.1918	1.2468	1.2926	1.3323	1.3667	1.3970	1.4243	1.4492	1.4944	1.5349
9500 Sh																
v			0.0254	0.0282	0.0322	0.0380	0.0451	0.0528	0.0603	0.0675	0.0742	0.0804	0.0862	0.0917	0.1019	0.1113
h			786.4	859.2	938.3	1023.4	1108.9	1187.7	1256.6	1318.9	1375.1	1426.1	1473.1	1517.3	1600.4	1679.0
s			0.9310	0.9900	1.0516	1.1153	1.1771	1.2320	1.2785	1.3191	1.3546	1.3858	1.4137	1.4392	1.4851	1.5263
10000 Sh																
v			0.0251	0.0276	0.0312	0.0362	0.0425	0.0495	0.0565	0.0633	0.0697	0.0757	0.0812	0.0865	0.0963	0.1054
h			783.8	854.5	930.2	1011.3	1094.2	1172.6	1242.0	1305.3	1362.9	1415.3	1463.4	1508.6	1593.1	1672.8
s			0.9270	0.9842	1.0432	1.1039	1.1638	1.2185	1.2652	1.3065	1.3429	1.3749	1.4035	1.4295	1.4763	1.5180
10500 Sh																
v			0.0248	0.0271	0.0303	0.0347	0.0404	0.0467	0.0532	0.0595	0.0656	0.0714	0.0768	0.0818	0.0913	0.1001
h			781.5	850.5	923.4	1001.0	1081.3	1158.9	1228.4	1292.4	1351.1	1404.7	1453.9	1500.0	1585.8	1666.7
s			0.9232	0.9790	1.0358	1.0939	1.1519	1.2060	1.2529	1.2946	1.3371	1.3644	1.3937	1.4202	1.4677	1.5100

Table III-2. Continued

Abs. Press. (psi) (Sat. Temp.)		Sat. Water	Sat. Steam	Temperature (°F)													
				750	800	850	900	950	1000	1050	1100	1150	1200	1250	1300	1400	1500
11000	v			0.0245	0.0267	0.0296	0.0335	0.0386	0.0443	0.0503	0.0562	0.0620	0.0676	0.0727	0.0776	0.0868	0.0952
	h			779.5	846.9	917.5	992.1	1069.9	1146.3	1215.9	1280.2	1339.7	1394.4	1444.6	1491.5	1578.7	1660.6
	s			0.9196	0.9742	1.0292	1.0851	1.1412	1.1945	1.2414	1.2833	1.3209	1.3544	1.3842	1.4112	1.4595	1.5023
11500	v			0.0243	0.0263	0.0290	0.0325	0.0370	0.0423	0.0478	0.0534	0.0588	0.0641	0.0691	0.0739	0.0827	0.0909
	h			777.7	843.8	912.4	984.5	1059.8	1134.9	1204.3	1268.7	1328.8	1384.4	1435.5	1483.2	1571.8	1654.7
	s			0.9163	0.9698	1.0232	1.0772	1.1316	1.1840	1.2308	1.2727	1.3107	1.3446	1.3750	1.4025	1.4515	1.4949
12000	v			0.0241	0.0260	0.0284	0.0317	0.0357	0.0405	0.0456	0.0508	0.0560	0.0610	0.0659	0.0704	0.0790	0.0869
	h			776.1	841.0	907.9	977.8	1050.9	1124.5	1193.7	1258.0	1318.5	1374.7	1426.6	1475.1	1564.9	1648.8
	s			0.9131	0.9657	1.0177	1.0701	1.1229	1.1742	1.2209	1.2627	1.3010	1.3353	1.3662	1.3941	1.4438	1.4877
12500	v			0.0238	0.0256	0.0279	0.0309	0.0346	0.0390	0.0437	0.0486	0.0535	0.0583	0.0629	0.0673	0.0756	0.0832
	h			774.7	838.6	903.9	971.9	1043.1	1115.2	1184.1	1247.9	1308.8	1365.4	1418.0	1467.2	1558.2	1643.1
	s			0.9101	0.9618	1.0127	1.0637	1.1151	1.1653	1.2117	1.2534	1.2918	1.3264	1.3576	1.3860	1.4363	1.4808
13000	v			0.0236	0.0253	0.0275	0.0302	0.0336	0.0376	0.0420	0.0466	0.0512	0.0558	0.0602	0.0645	0.0725	0.0799
	h			773.5	836.3	900.4	966.8	1036.2	1106.7	1174.8	1238.5	1299.6	1356.5	1409.6	1459.4	1551.6	1637.4
	s			0.9073	0.9582	1.0080	1.0578	1.1079	1.1571	1.2030	1.2445	1.2831	1.3179	1.3494	1.3781	1.4291	1.4741
13500	v			0.0235	0.0251	0.0271	0.0297	0.0328	0.0364	0.0405	0.0448	0.0492	0.0535	0.0577	0.0619	0.0696	0.0768
	h			772.3	834.4	897.2	962.2	1030.0	1099.1	1166.5	1229.7	1291.0	1348.1	1401.5	1451.8	1545.2	1631.9
	s			0.9045	0.9548	1.0037	1.0524	1.1014	1.1495	1.1948	1.2361	1.2749	1.3098	1.3415	1.3705	1.4221	1.4675
14000	v			0.0233	0.0248	0.0267	0.0291	0.0320	0.0354	0.0392	0.0432	0.0474	0.0515	0.0555	0.0595	0.0670	0.0740
	h			771.3	832.6	894.3	958.0	1024.5	1092.3	1158.5	1221.4	1283.0	1340.2	1393.8	1444.4	1538.8	1626.5
	s			0.9019	0.9515	0.9996	1.0473	1.0953	1.1426	1.1872	1.2282	1.2671	1.3021	1.3339	1.3631	1.4153	1.4612
14500	v			0.0231	0.0246	0.0264	0.0287	0.0314	0.0345	0.0380	0.0418	0.0458	0.0496	0.0534	0.0573	0.0646	0.0714
	h			770.4	831.0	891.7	954.3	1019.6	1086.2	1151.4	1213.8	1275.4	1332.9	1386.4	1437.3	1532.6	1621.1
	s			0.8994	0.9484	0.9957	1.0426	1.0897	1.1362	1.1801	1.2208	1.2597	1.2949	1.3266	1.3560	1.4087	1.4551
15000	v			0.0230	0.0244	0.0261	0.0282	0.0308	0.0337	0.0369	0.0405	0.0443	0.0479	0.0516	0.0552	0.0624	0.0690
	h			769.6	829.5	889.3	950.9	1015.1	1080.6	1144.9	1206.8	1268.1	1326.0	1379.4	1430.3	1526.4	1615.9
	s			0.8970	0.9455	0.9920	1.0382	1.0846	1.1302	1.1735	1.2139	1.2525	1.2880	1.3197	1.3491	1.4022	1.4491
15500	v			0.0228	0.0242	0.0258	0.0278	0.0302	0.0329	0.0360	0.0393	0.0429	0.0464	0.0499	0.0534	0.0603	0.0668
	h			768.9	828.2	887.2	947.8	1011.1	1075.7	1139.0	1200.3	1261.1	1319.6	1372.8	1423.6	1520.4	1610.8
	s			0.8946	0.9427	0.9886	1.0340	1.0797	1.1247	1.1674	1.2073	1.2457	1.2815	1.3131	1.3424	1.3959	1.4433

^{a} Sh-superheat, °F; v specific volume, ft^3/lb; h enthalpy, Btu/lb; s entropy, Btu/°F · lb.
Source: Copyright 1967 ASME (Abridged); reprinted by permission.

Table III-3   Mollier Diagram for Steam

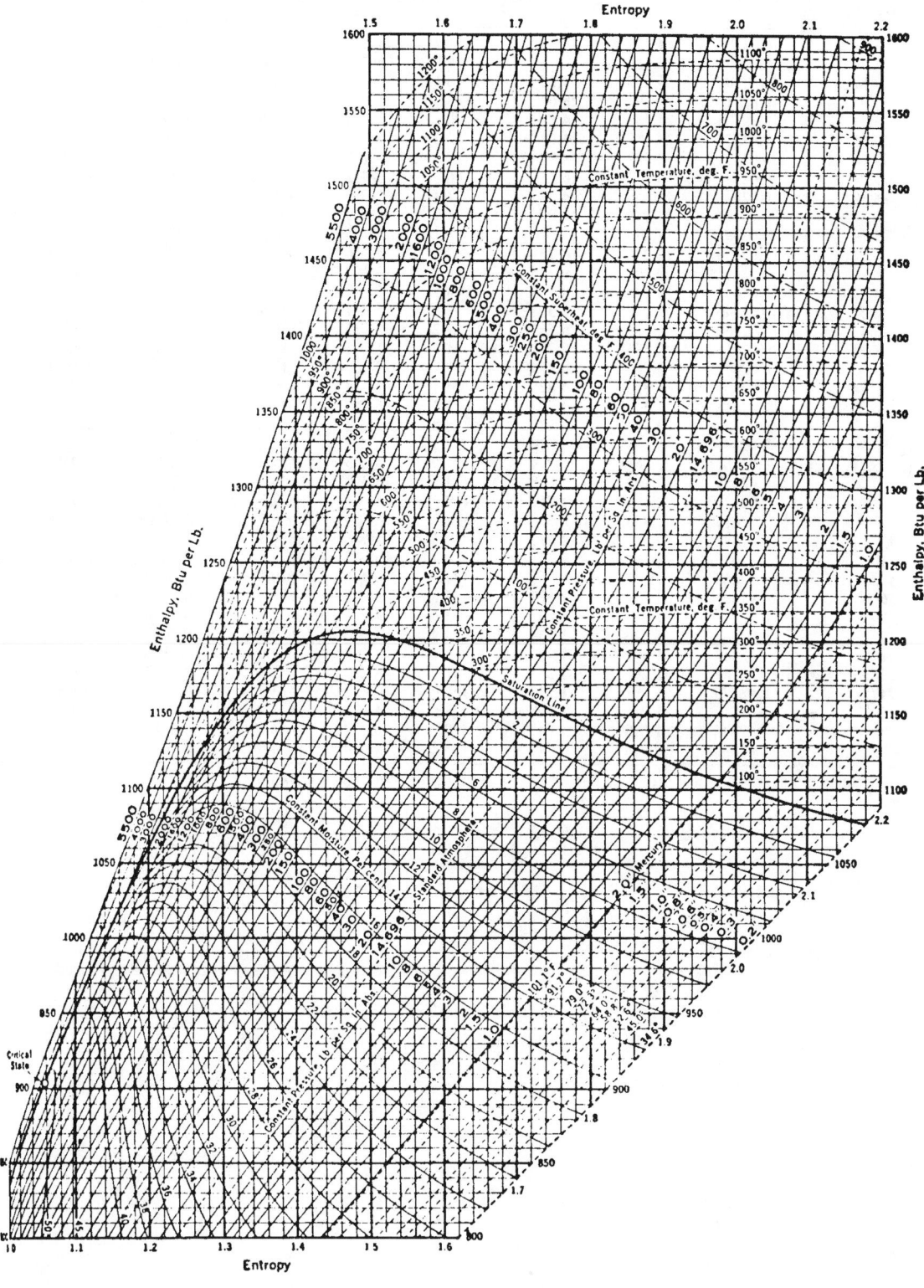

Source: Modified and greatly reduced from J.H. Keenan and F.G. Keyes, *Thermodynamic Properties of Steam*, John Wiley & Sons Inc., New York, 1936; reproduced by permission of the publishers.

Table II.2-6 Thermodynamic Properties of Saturated Ammonia

Temp. (°F)	Abs. Press. P (lb$_f$/in.²)	Specific Volume (ft³/lb$_m$)			Enthalpy (Btu/lb$_m$)			Entropy (Btu/lb$_m$ · °R)		
		Sat. Liquid v_f	Evap. v_{fg}	Sat. Vapor v_g	Sat. Liquid h_f	Evap. h_{fg}	Sat. Vapor h_g	Sat. Liquid s_f	Evap. s_{fg}	Sat. Vapor s_g
−60	5.55	0.0228	44.707	44.73	−21.2	610.8	589.6	−0.0517	1.5286	1.4769
−55	6.54	0.0229	38.357	38.38	−15.9	607.5	591.6	−0.0386	1.5017	1.4631
−50	7.67	0.0230	33.057	33.08	−10.6	604.3	593.7	−0.0256	1.4753	1.4497
−45	8.95	0.0231	28.597	28.62	−5.3	600.9	595.6	−0.0127	1.4495	1.4368
−40	10.41	0.02322	24.837	24.86	0	597.6	597.6	0.000	1.4242	1.4242
−35	12.05	0.02333	21.657	21.68	5.3	594.2	599.5	0.0126	1.3994	1.4120
−30	13.90	0.0235	18.947	18.97	10.7	590.7	601.4	0.0250	1.3751	1.4001
−25	15.98	0.0236	16.636	16.66	16.0	587.2	603.2	0.0374	1.3512	1.3886
−20	18.30	0.0237	14.656	14.68	21.4	583.6	605.0	0.0497	1.3277	1.3774
−15	20.88	0.02381	12.946	12.97	26.7	580.0	606.7	0.0618	1.3044	1.3664
−10	23.74	0.02393	11.476	11.50	32.1	576.4	608.5	0.0738	1.2820	1.3558
−5	26.92	0.02406	10.206	10.23	37.5	572.6	610.1	0.0857	1.2597	1.3454
0	30.42	0.02419	9.092	9.116	42.9	568.9	611.8	0.0975	1.2377	1.3352
5	34.27	0.02432	8.1257	8.150	48.3	565.0	613.3	0.1092	1.2161	1.3253
10	38.51	0.02446	7.2795	7.304	53.8	561.1	614.9	0.1208	1.1949	1.3157
15	43.14	0.02460	6.5374	6.562	59.2	557.1	616.3	0.1323	1.1739	1.3062
20	48.21	0.02474	5.8853	5.910	64.7	553.1	617.8	0.1437	1.1532	1.2969
25	53.73	0.02488	5.3091	5.334	70.2	548.9	619.1	0.1551	1.1328	1.2879
30	59.74	0.02503	4.8000	4.825	75.7	544.8	620.5	0.1663	1.1127	1.2790
35	66.26	0.02518	4.3478	4.373	81.2	540.5	621.7	0.1775	1.0929	1.2704
40	73.32	0.02533	3.9457	3.971	86.8	536.2	623.0	0.1885	1.0733	1.2618
45	80.96	0.02548	3.5885	3.614	92.3	531.8	624.1	0.1996	1.0539	1.2535
50	89.19	0.02564	3.2684	3.294	97.9	527.3	625.2	0.2105	1.0348	1.2453
55	98.06	0.02581	2.9822	3.008	103.5	522.8	626.3	0.2214	1.0159	1.2373
60	107.6	0.02597	2.7250	2.751	109.2	518.1	627.3	0.2322	0.9972	1.2294
65	117.8	0.02614	2.4939	2.520	114.8	513.4	628.2	0.2430	0.9786	1.2216
70	128.8	0.02632	2.2857	2.312	120.5	508.6	629.1	0.2537	0.9603	1.2140
75	140.5	0.02650	2.0985	2.125	126.2	503.7	629.9	0.2643	0.9422	1.2065
80	153.0	0.02668	1.9283	1.955	132.0	498.7	630.7	0.2749	0.9242	1.1991
85	166.4	0.02687	1.7741	1.801	137.8	493.6	631.4	0.2854	0.9064	1.1918
90	180.6	0.02707	1.6339	1.661	143.5	488.5	632.0	0.2958	0.8888	1.1846
95	195.8	0.02727	1.5067	1.534	149.4	483.2	632.6	0.3062	0.8713	1.1775
100	211.9	0.02747	1.3915	1.419	155.2	477.8	633.0	0.3166	0.8539	1.1705
105	228.9	0.02769	1.2853	1.313	161.1	472.3	633.4	0.3269	0.8366	1.1635
110	247.0	0.02790	1.1891	1.217	167.0	466.7	633.7	0.3372	0.8194	1.1566
115	266.2	0.02813	1.0999	1.128	173.0	460.9	633.9	0.3474	0.8023	1.1497
120	286.4	0.02836	1.0186	1.047	179.0	455.0	634.0	0.3576	0.7851	1.1427
125	307.8	0.02860	0.9444	0.973	185.1	448.9	634.0	0.3679	0.7679	1.1358

Source: National Bureau of Standards Circular No. 142, *Tables of Thermodynamic Properties of Ammonia*.

Table II.2-7 Thermodynamic Properties of Superheated Ammonia

Abs. Press. (Sat. Temp.) (lb$_f$/in.^{2})		Temperature (°F)											
		0	20	40	60	80	100	120	140	160	180	200	220
10 (−41.34)	v	28.58	29.90	31.20	32.49	33.78	35.07	36.35	37.62	38.90	40.17	41.45	
	h	618.9	629.1	639.3	649.5	659.7	670.0	680.3	690.6	701.1	711.6	722.2	
	s	1.477	1.499	1.520	1.540	1.559	1.578	1.596	1.614	1.631	1.647	1.664	
15 (−27.29)	v	18.92	19.82	20.70	21.58	22.44	23.31	24.17	25.03	25.88	26.74	27.59	
	h	617.2	627.8	638.2	648.5	658.9	669.2	679.6	690.0	700.5	711.1	721.7	
	s	1.427	1.450	1.471	1.491	1.511	1.529	1.548	1.566	1.583	1.599	1.616	
20 (−16.64)	v	14.09	14.78	15.45	16.12	16.78	17.43	18.08	18.73	19.37	20.02	20.66	21.3
	h	615.5	626.4	637.0	647.5	658.0	668.5	678.9	689.4	700.0	710.6	721.2	732.0
	s	1.391	1.414	1.436	1.456	1.476	1.495	1.513	1.531	1.549	1.565	1.582	1.598
25 (−7.96)	v	11.19	11.75	12.30	12.84	13.37	13.90	14.43	14.95	15.47	15.99	16.50	17.02
	h	613.8	625.0	635.8	646.5	657.1	667.7	678.2	688.8	699.4	710.1	720.8	731.6
	s	1.362	1.386	1.408	1.429	1.449	1.468	1.486	1.504	1.522	1.539	1.555	1.571
30 (−.57)	v	9.25	9.731	10.20	10.65	11.10	11.55	11.99	12.43	12.87	13.30	13.73	14.16
	h	611.9	623.5	634.6	645.5	656.2	666.9	677.5	688.2	698.8	709.6	720.3	731.1
	s	1.337	1.362	1.385	1.406	1.426	1.446	1.464	1.482	1.500	1.517	1.533	1.550
35 (5.89)	v		8.287	8.695	9.093	9.484	9.869	10.25	10.63	11.00	11.38	11.75	12.12
	h		622.0	633.4	644.4	655.3	666.1	676.8	687.6	698.3	709.1	719.9	730.7
	s		1.341	1.365	1.386	1.407	1.427	1.445	1.464	1.481	1.498	1.515	1.531
40 (11.66)	v		7.203	7.568	7.922	8.268	8.609	8.945	9.278	9.609	9.938	10.27	10.59
	h		620.4	632.1	643.4	654.4	665.3	676.1	686.9	697.7	708.5	719.4	730.3
	s		1.323	1.347	1.369	1.390	1.410	1.429	1.447	1.465	1.482	1.499	1.515
45 (16.87)	v		6.363	6.694	7.014	7.326	7.632	7.934	8.232	8.528	8.822	9.115	9.406
	h		618.8	630.8	642.3	653.5	664.6	675.5	686.3	697.2	708.0	718.9	729.9
	s		1.307	1.331	1.354	1.375	1.395	1.414	1.433	1.450	1.468	1.485	1.501
50 (21.67)	v			5.988	6.280	6.564	6.843	7.117	7.387	7.655	7.921	8.185	8.448
	h			629.5	641.2	652.6	663.7	674.7	685.7	696.6	707.5	718.5	729.4
	s			1.317	1.340	1.361	1.382	1.401	1.420	1.437	1.455	1.472	1.488
60 (30.21)	v			4.933	5.184	5.428	5.665	5.897	6.126	6.352	6.576	6.798	7.019
	h			626.8	639.0	650.7	662.1	673.3	684.4	695.5	706.5	717.5	728.6
	s			1.291	1.315	1.337	1.358	1.378	1.397	1.415	1.432	1.449	1.466
70 (37.7)	v	4.401	4.615	4.822	5.025	5.224	5.420	5.615	5.807	6.187	6.563		
	h	636.6	648.7	660.4	671.8	683.1	694.3	705.5	716.6	738.9	761.4		
	s	1.294	1.317	1.338	1.358	1.377	1.395	1.413	1.430	1.463	1.494		
80 (44.4)	v	3.812	4.005	4.190	4.371	4.548	4.722	4.893	5.063	5.398	5.73		
	h	634.3	646.7	658.7	670.4	681.8	693.2	704.4	715.6	738.1	760.7		
	s	1.275	1.298	1.320	1.340	1.360	1.378	1.396	1.414	1.447	1.478		
90 (50.47)	v	3.353	3.529	3.698	3.862	4.021	4.178	4.332	4.484	4.785	5.081		
	h	631.8	644.7	657.0	668.9	680.5	692.0	703.4	714.7	737.3	760.0		
	s	1.257	1.281	1.304	1.325	1.344	1.363	1.381	1.400	1.432	1.464		
100 (56.05)	v	2.985	3.149	3.304	3.454	3.600	3.743	3.883	4.021	4.294	4.562		
	h	629.3	642.6	655.2	667.3	679.2	690.8	702.3	713.7	736.5	759.4		
	s	1.241	1.266	1.289	1.310	1.331	1.349	1.368	1.385	1.419	1.451		
140 (74.79)	v		2.166	2.288	2.404	2.515	2.622	2.727	2.830	3.030	3.227	3.420	
	h		633.8	647.8	661.1	673.7	686.0	698.0	709.9	733.3	756.7	780.0	
	s		1.214	1.240	1.263	1.284	1.305	1.324	1.342	1.376	1.409	1.440	
180 (89.78)	v			1.720	1.818	1.910	1.999	2.084	2.167	2.328	2.484	2.637	
	h			639.9	654.4	668.0	681.0	693.6	705.9	730.1	753.9	777.7	
	s			1.199	1.225	1.248	1.269	1.289	1.308	1.344	1.377	1.408	
220 (102.42)	v				1.443	1.525	1.601	1.675	1.745	1.881	2.012	2.140	2.265
	h				647.3	662.0	675.8	689.1	701.9	726.8	751.1	775.3	799.5
	s				1.192	1.217	1.239	1.260	1.280	1.317	1.351	1.383	1.413
240 (108.09)	v				1.302	1.380	1.452	1.521	1.587	1.714	1.835	1.954	2.069
	h				643.5	658.8	673.1	686.7	699.8	725.1	749.8	774.1	798.4
	s				1.176	1.203	1.226	1.248	1.268	1.305	1.339	1.371	1.402
260 (113.42)	v				1.182	1.257	1.326	1.391	1.453	1.572	1.686	1.796	1.904
	h				639.5	655.6	670.4	684.4	697.7	723.4	748.4	772.9	797.4
	s				1.162	1.189	1.213	1.235	1.256	1.294	1.329	1.361	1.391
280 (118.45)	v				1.078	1.151	1.217	1.279	1.339	1.451	1.558	1.661	1.762
	h				635.4	652.2	667.6	681.9	695.6	721.8	747.0	771.7	796.3
	s				1.147	1.176	1.201	1.224	1.245	1.283	1.318	1.351	1.382

Source: National Bureau of Standards Circular No. 142, *Tables of Thermodynamic Properties of Ammonia.*

Table II.2-8 Thermodynamic Properties of Saturated Nitrogen

Temp. (°R)	Abs. Press. P (lb$_f$/in.²)	Specific Volume (ft³/lb$_m$)			Enthalpy (Btu lb$_m$)			Entropy (Btu/lb$_m$ · °R)		
		Sat. Liquid v_f	Evap. v_{fg}	Sat. Vapor v_g	Sat. Liquid h_f	Evap. h_{fg}	Sat. Vapor h_g	Sat. Liquid s_f	Evap. s_{fg}	Sat. Vapor s_g
113.670	1.813	0.01845	23.793	23.812	0.000	92.891	92.891	0.00000	0.81720	0.81720
120.000	3.337	0.01875	13.570	13.589	3.113	91.224	94.337	0.02661	0.76020	0.78681
130.000	7.654	0.01929	6.3208	6.3401	8.062	88.432	96.494	0.06610	0.68025	0.74634
139.255	14.696	0.01984	3.4592	3.4791	12.639	85.668	98.306	0.09992	0.61518	0.71510
140.000	15.425	0.01989	3.3072	3.3271	13.006	85.436	98.443	0.10253	0.61026	0.71279
150.000	28.120	0.02056	1.8865	1.9071	17.945	82.179	100.124	0.13628	0.54786	0.68414
160.000	47.383	0.02132	1.1469	1.1682	22.928	78.458	101.476	0.16795	0.49093	0.65888
170.000	74.991	0.02219	0.7299	0.7521	28.045	74.383	102.427	0.19829	0.43754	0.63584
180.000	112.808	0.02323	0.4789	0.5021	33.411	69.478	102.889	0.22805	0.38599	0.61404
190.000	162.761	0.02449	0.3190	0.3435	39.153	63.582	102.735	0.25789	0.33464	0.59254
200.000	226.853	0.02613	0.2119	0.2380	45.283	56.474	101.757	0.28780	0.28237	0.57017
210.000	307.276	0.02845	0.1354	0.1639	52.061	47.474	99.536	0.31894	0.22607	0.54501
220.000	406.739	0.03249	0.0750	0.1075	60.336	34.536	94.872	0.35494	0.15698	0.51192
226.000	477.104	0.03806	0.0374	0.0755	68.123	20.423	88.546	0.38789	0.09037	0.47826

Source: Abstracted from National Bureau of Standards Technical Note 129A, *The Thermodynamic Properties of Nitrogen from 114 to 540 R between 1.0 and 3000 psia, Supplement A (British Units)*, by Thomas R. Strobridge.

Table II.2-9 Thermodynamic Properties of Superheated Nitrogen

Temp. (°R)	v (ft³/lb$_m$)	h (Btu/lb$_m$)	s (Btu/lb$_m$ · °R)	v (ft³/lb$_m$)	h (Btu/lb$_m$)	s (Btu/lb$_m$ · °R)	v (ft³/lb$_m$)	h (Btu/lb$_m$)	s (Btu/lb$_m$ · °R)
	14.7 lb$_f$/in.²			20 lb$_f$/in.²			50 lb$_f$/in.²		
150	3.7782	101.086	0.7343	2.7395	100.715	0.7109			
200	5.1366	113.849	0.8078	3.7538	113.625	0.7852	1.4534	112.315	0.7159
250	6.4680	126.443	0.8640	4.7397	126.293	0.8418	1.8663	125.432	0.7744
300	7.7876	138.958	0.9096	5.7138	138.850	0.8875	2.2662	138.239	0.8212
350	9.1015	151.432	0.9481	6.6820	151.351	0.9261	2.6599	150.896	0.8602
400	10.412	163.882	0.9814	7.6469	163.821	0.9594	3.0502	163.471	0.8938
450	11.721	176.319	1.0107	8.6098	176.271	0.9887	3.4385	175.997	0.9233
500	13.028	188.748	1.0368	9.5714	188.710	1.0149	3.8255	188.492	0.9496
540	14.073	198.690	1.0560	10.340	198.657	1.0341	4.1344	198.474	0.9688
	100 lb$_f$/in.²			200 lb$_f$/in.²			500 lb$_f$/in.²		
200	0.6834	109.931	0.6585	0.2884	103.911	0.5875	0.1321	108.378	0.5608
250	0.9078	123.948	0.7212	0.4272	120.763	0.6631	0.1966	128.168	0.6335
300	1.1169	137.205	0.7696	0.5420	135.076	0.7153	0.2473	143.838	0.6819
350	1.3192	150.133	0.8094	0.6490	148.589	0.7570	0.2932	158.205	0.7202
400	1.5181	162.888	0.8435	0.7522	161.718	0.7921	0.3368	171.933	0.7526
450	1.7149	175.540	0.8733	0.8532	174.630	0.8225	0.3790	185.292	0.7807
500	1.9103	188.129	0.8998	0.9529	187.408	0.8494	0.4120	195.807	0.8010
540	2.0660	198.170	0.9192	1.0319	197.567	0.8690			
	1000 lb$_f$/in.²			2000 lb$_f$/in.²			3000 lb$_f$/in.²		
250	0.0384	78.126	0.4145	0.0286	70.290	0.3596	0.0261	69.719	0.3371
300	0.0828	115.224	0.5514	0.0398	97.820	0.4599	0.0321	93.216	0.4228
350	0.1150	135.789	0.6150	0.0552	122.614	0.5366	0.0403	116.066	0.4933
400	0.1417	152.487	0.6597	0.0699	142.869	0.5908	0.0493	136.883	0.5490
450	0.1659	167.637	0.6954	0.0833	160.406	0.6321	0.0582	155.522	0.5930
500	0.1887	181.969	0.7256	0.0958	176.411	0.6659	0.0667	172.551	0.6289
540	0.2063	193.069	0.7470	0.1053	188.526	0.6892	0.0732	185.361	0.6535

Source: Abstracted from National Bureau of Standards Technical Note 129A, *The Thermodynamic Properties of Nitrogen from 114 to 540 R between 1.0 and 3000 psia, Supplement A (British Units)*, by Thomas R. Strobridge.

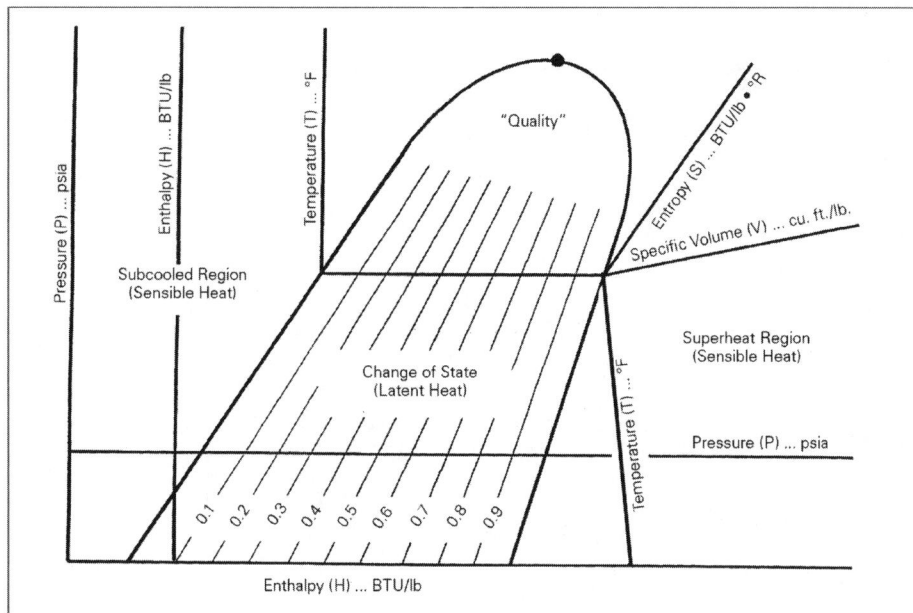

Table III-8. Use of Refrigerant Pressure/Enthalpy Diagrams

Table III-9. Pressure/Enthalpy Diagram. R-22.

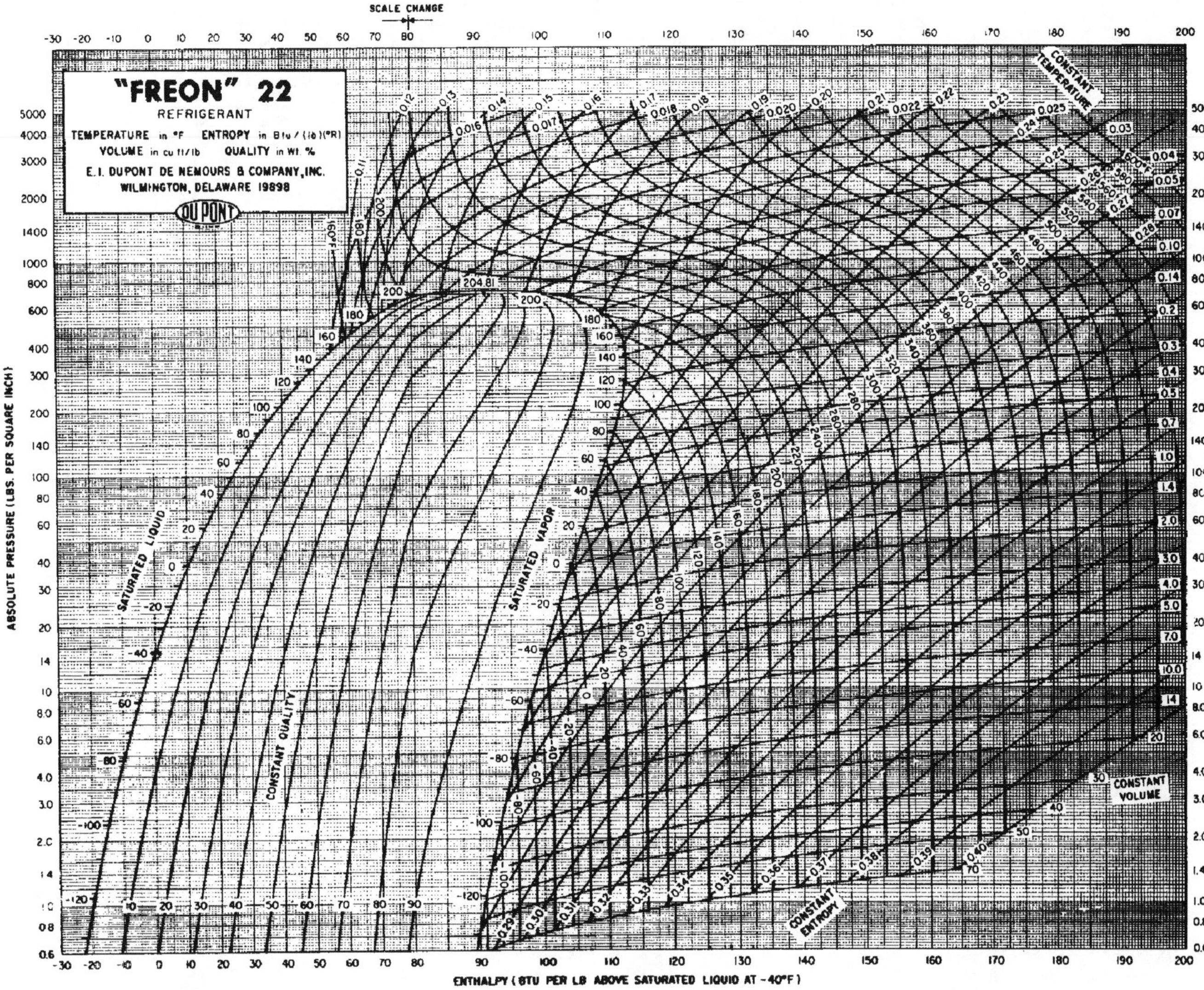

Table III-10. Pressure/Enthalpy Diagram. R-134a.

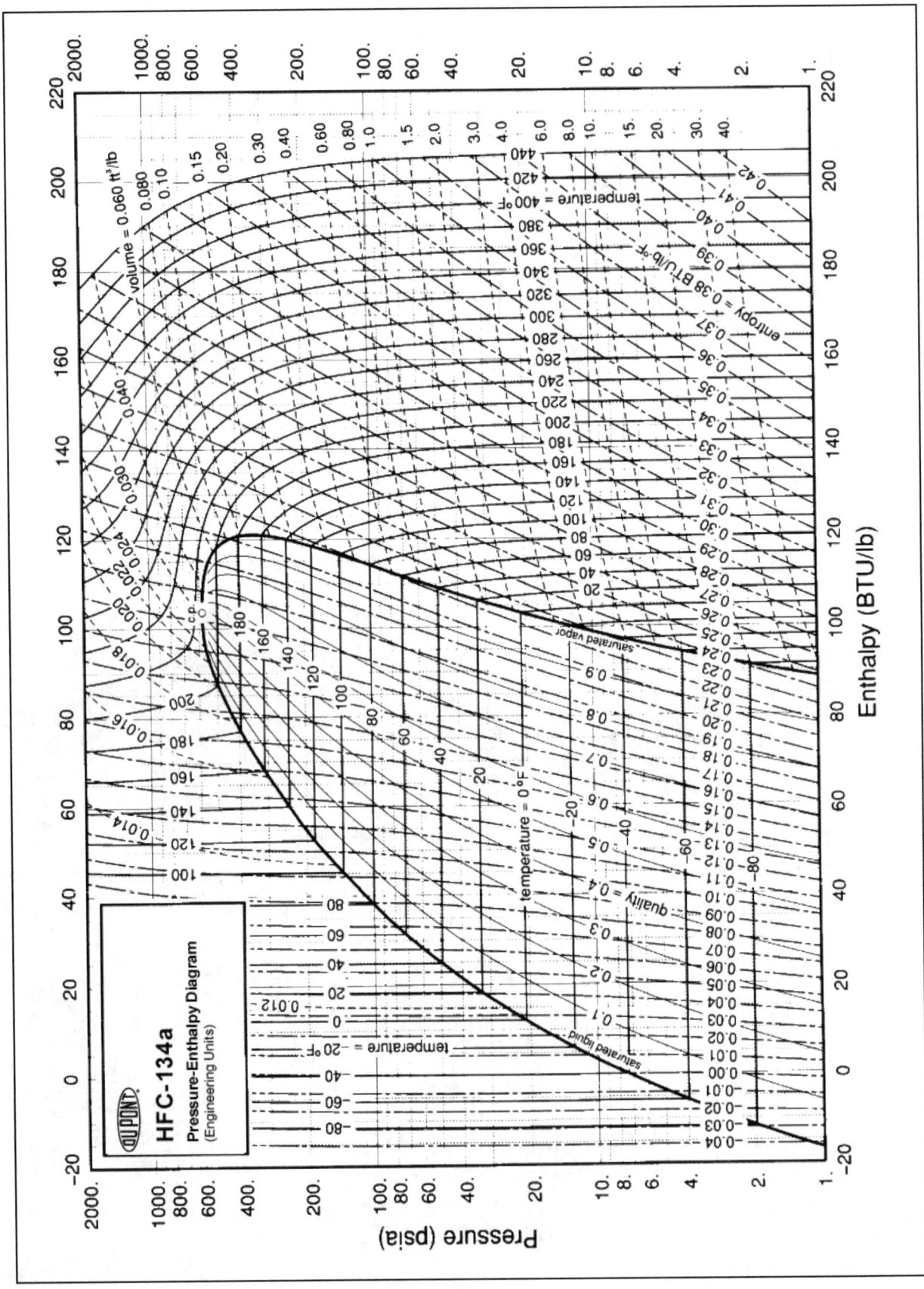

Table III-11. Pressure/Enthalpy Diagram. R-123.

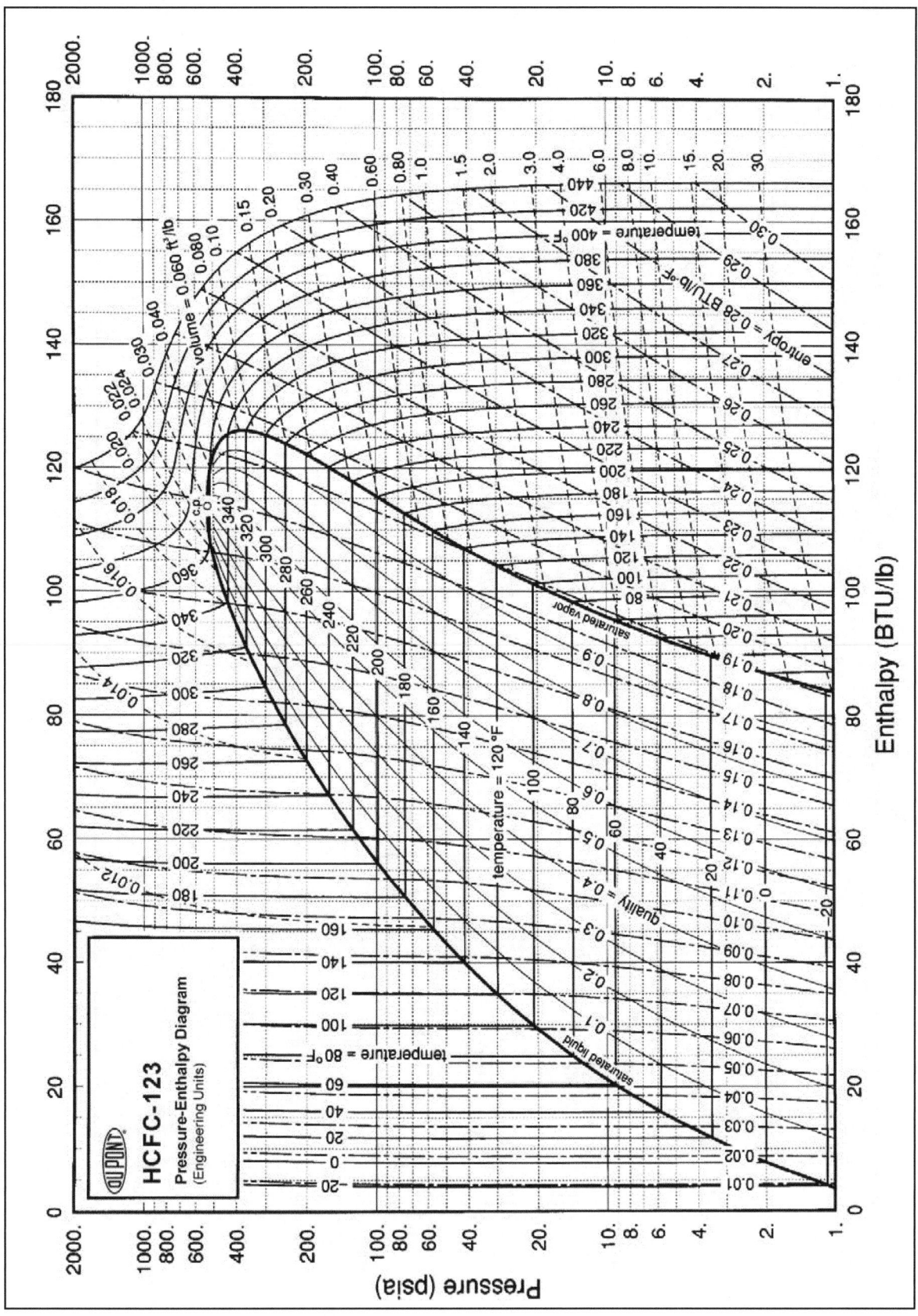

Table III-12. Pressure/Enthalpy Diagram. R-404a.

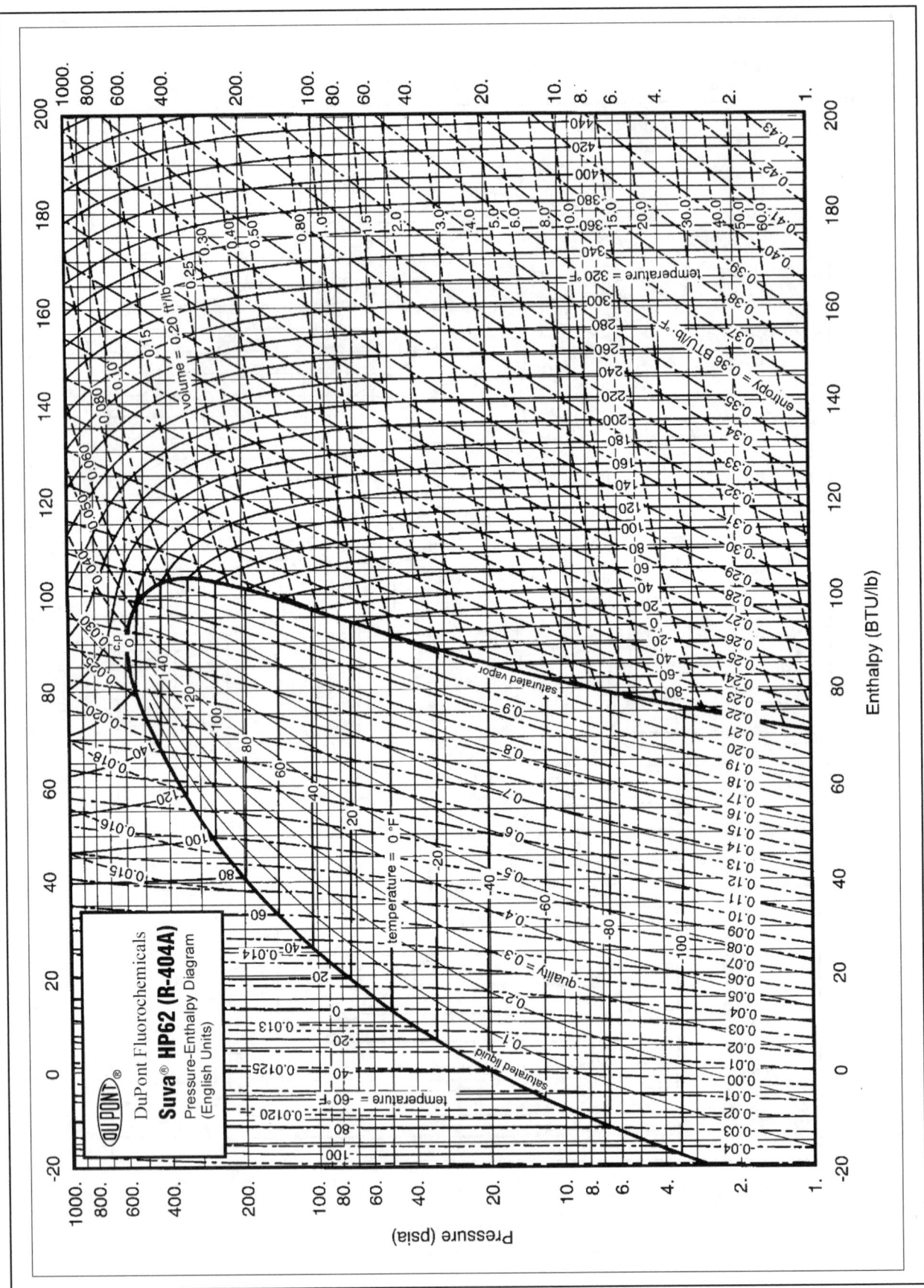

Table III-13. Pressure/Enthalpy Diagram. R-410a.

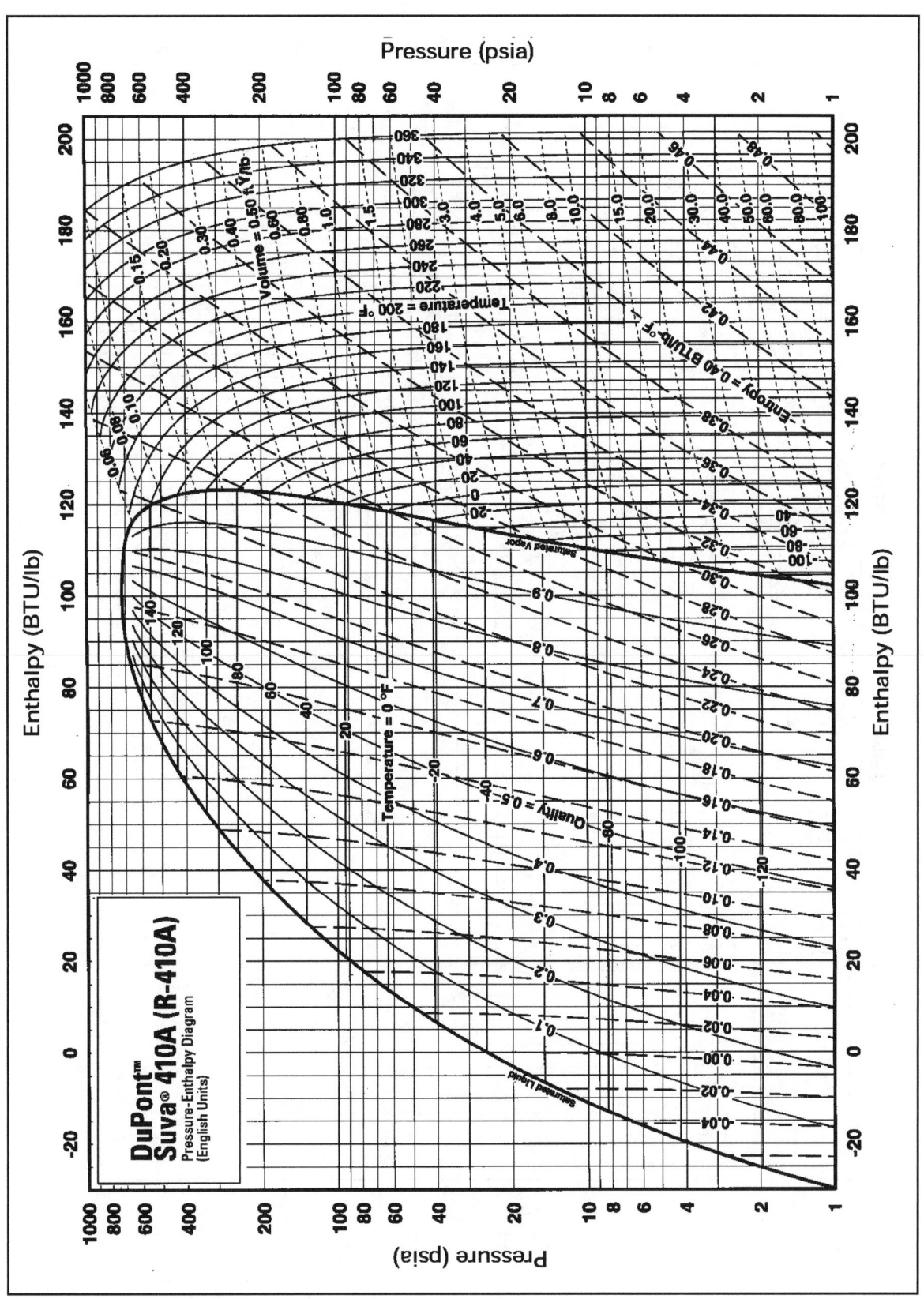

Table III-14. Pressure/Enthalpy Diagram.   R-717 (ammonia).

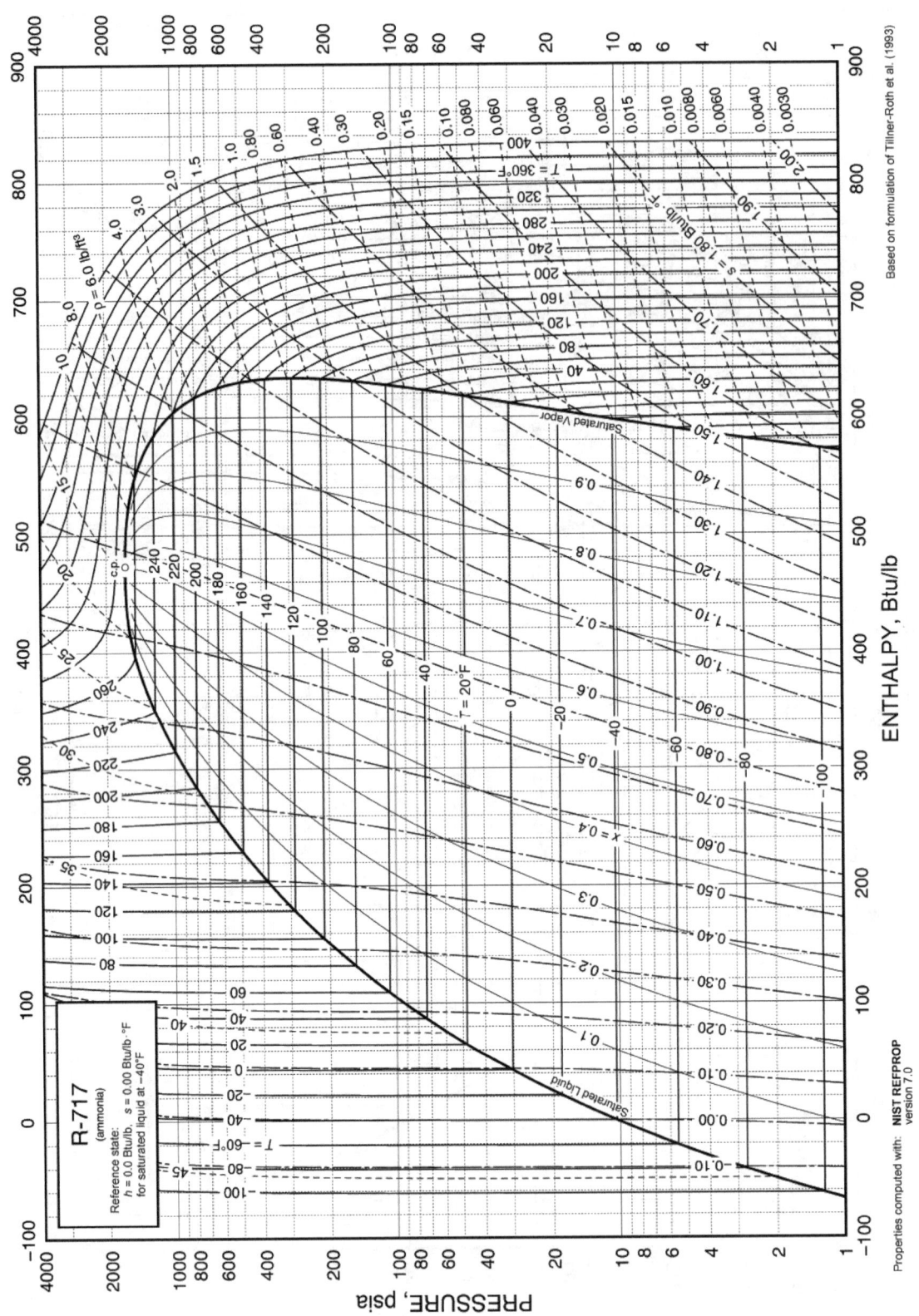

Source: ASHRAE Fundamentals Handbook, 2005, © American Society of Heating, Refrigerating and Air-Conditioning Engineers

Table III-15. Critical Constants

Substance	Formula	Molecular Weight	Temperature		Pressure		Volume (ft³/lb mol)
			°K	°R	atm	lb_f/in.²	
Ammonia	NH_3	17.03	405.5	729.8	111.3	1636	1.16
Argon	Ar	39.944	151	272	48.0	705	1.20
Bromine	Br_2	159.832	584	1052	102	1500	2.17
Carbon dioxide	CO_2	44.01	304.2	547.5	72.9	1071	1.51
Carbon monoxide	CO	28.01	133	240	34.5	507	1.49
Chlorine	Cl_2	70.914	417	751	76.1	1120	1.99
Deuterium (normal)	D_2	4.00	38.4	69.1	16.4	241	—
Helium	He	4.003	5.3	9.5	2.26	33.2	0.926
Helium	He	3.00	3.34	6.01	1.15	16.9	—
Hydrogen (normal)	H_2	2.016	33.3	59.9	12.8	188.1	1.04
Krypton	Kr	83.7	209.4	376.9	54.3	798	1.48
Neon	Ne	20.183	44.5	80.1	26.9	395	0.668
Nitrogen	N_2	28.016	126.2	227.1	33.5	492	1.44
Nitrous oxide	N_2O	44.02	309.7	557.1	71.7	1054	1.54
Oxygen	O_2	32.00	154.8	278.6	50.4	736	1.25
Sulfur dioxide	SO_2	64.06	430.7	775.2	77.8	1143	1.95
Water	H_2O	18.016	647.4	1165.3	218.3	3204	0.90
Xenon	Xe	131.3	289.75	521.55	58.0	852	1.90
Benzene	C_6H_6	78.11	562	1012	48.6	714	4.17
n-Butane	C_4H_{10}	58.120	425.2	765.2	37.5	551	4.08
Carbon tetrachloride	CCl_4	153.81	556.4	1001.5	45.0	661	4.42
Chloroform	$CHCl_3$	119.39	536.6	965.8	54.0	794	3.85
Dichlorodifluoromethane	CCl_2F_2	120.92	384.7	692.4	39.6	582	3.49
Dichlorofluoromethane	$CHCl_2F$	102.93	451.7	813.0	51.0	749	3.16
Ethane	C_2H_6	30.068	305.5	549.8	48.2	708	2.37
Ethyl alcohol	C_2H_5OH	46.07	516.0	929.0	63.0	926	2.68
Ethylene	C_2H_4	28.052	282.4	508.3	50.5	742	1.99
n-Hexane	C_6H_{14}	86.172	507.9	914.2	29.9	439	5.89
Methane	CH_4	16.012	191.1	343.9	45.8	673	1.59
Methyl alcohol	CH_3OH	32.04	513.2	923.7	78.5	1154	1.89
Methyl chloride	CH_3Cl	50.49	416.3	749.3	65.9	968	2.29
Propane	C_3H_8	44.094	370.0	665.9	42.0	617	3.20
Propene	C_3H_6	42.078	365.0	656.9	45.6	670	2.90
Propyne	C_3H_4	40.062	401	722	52.8	776	—
Trichlorofluoromethane	CCl_3F	137.38	471.2	848.1	43.2	635	3.97

Source: K. A. Kobe and R. E. Lynn, Jr., *Chemical Reviews*, Vol. 52 (1953), pp. 117–236.

Table III-16. Thermal Properties of Metals

Metal	Properties at 68°F				k(Btu/hr·ft·°F)									
	ρ (lbm/ft³)	C_P (Btu/lbm·°F)	k (Btu/hr·ft·°F)	α (ft²/hr)	−148°F −100°C	32°F 0°C	212°F 100°C	392°F 200°C	572°F 300°C	752°F 400°C	1112°F 600°C	1472°F 800°C	1832°F 1000°C	2192°F 1200°C
Aluminum														
Pure	169	0.214	132	3.665	134	132	132	132	132					
Al-Cu (Duralumin): 94–96 Al, 3–5 Cu, trace Mg	174	0.211	95	2.580	73	92	105	112						
Al-Mg (Hydronalium): 91–95 Al, 5–9 Mg	163	0.216	65	1.860	54	63	73	82						
Al-Si (Silumin): 87 Al, 13 Si	166	0.208	95	2.773	86	94	101	107						
Al-Si (Silumin, copper bearing): 86.5 Al, 12.5 Si, 1 Cu	166	0.207	79	2.311	69	79	83	88	93					
Al-Si (Alusil): 78–80 Al, 20–22 Si	164	0.204	93	2.762	83	91	97	101	103					
Al-Mg-Si: 97 Al, 1 Mg, 1 Si, 1 Mn	169	0.213	102	2.859	—	101	109	118						
Lead	710	0.031	20	0.924	21.3	20.3	19.3	18.2	17.2					
Iron														
Pure	493	0.108	42	0.785	50	42	39	36	32	28	23	21	20	21
Wrought iron (C < 0.50%)	490	0.11	34	0.634	—	34	33	30	28	26	21	19	19	19
Cast iron (C ≈ 4%)	454	0.10	30	0.666										
Steel (C_{max} ≈ 1.5%)														
Carbon steel (C ≈ 0.5%)	489	0.111	31	0.570	—	32	30	28	26	24	20	17	17	18
1.0%	487	0.113	25	0.452	—	25	25	24	23	21	19	17	16	17
1.5%	484	0.116	21	0.376	—	21	21	21	20	19	18	16	16	17
Nickel steel (Ni ≈ 0%)	493	0.108	42	0.785										
10%	496	0.11	15	0.279										
20%	499	0.11	11	0.204										
30%	504	0.11	7	0.118										
40%	510	0.11	6	0.108										
50%	516	0.11	8	0.140										
60%	523	0.11	11	0.182										
70%	531	0.11	15	0.258										
80%	538	0.11	20	0.344										
90%	547	0.11	27	0.452										
100%	556	0.106	52	0.892										
Invar (Ni ≈ 36%)	508	0.11	6.2	0.108										
Chrome steel (Cr = 0%)	493	0.108	42	0.785	50	42	39	36	32	28	23	21	20	21
1%	491	0.11	35	0.645	—	36	32	30	27	24	21	19	19	
2%	491	0.11	30	0.559	—	31	28	26	24	22	19	18	18	
5%	489	0.11	23	0.430	—	23	22	21	21	19	17	17	17	17
10%	486	0.11	18	0.344	—	18	18	18	17	17	16	16	17	
20%	480	0.11	13	0.258	—	13	13	13	13	14	14	15	17	
30%	476	0.11	11	0.204										
Cr-Ni (chrome-nickel)														
15 Cr, 10 Ni	491	0.11	11	0.204										
18 Cr, 8 Ni (V2A)	488	0.11	9.4	0.172	—	9.4	10	10	11	11	13	15	18	
20 Cr, 15 Ni	489	0.11	8.7	0.161										
25 Cr, 20 Ni	491	0.11	7.4	0.140										
Ni-Cr (nickel-chrome)														
80 Ni, 15 Cr	532	0.11	10	0.172										
60 Ni, 15 Cr	516	0.11	7.4	0.129										
40 Ni, 15 Cr	504	0.11	6.7	0.118										
20 Ni, 15 Cr	491	0.11	8.1	0.151	—	8.1	8.7	8.7	9.4	10	11	13		
Cr-Ni-Al: 6 Cr, 1.5 Al, 0.5 Si (Sicromal 8)	482	0.117	13	0.237										
24 Cr, 2.5 Al, 0.5 Si (Sicromal 12)	479	0.118	11	0.194										
Manganese steel (Ma = 0%)	493	0.118	42	0.784										
1%	491	0.11	29	0.538										
2%	491	0.11	22	0.376	—	22	21	21	21	20	19			
5%	490	0.11	13	0.247										
10%	487	0.11	10	0.194										
Tungsten steel (W = 0%)	493	0.108	42	0.785										
1%	494	0.107	38	0.720										
2%	497	0.106	36	0.677	—	36	34	31	28	26	21			
5%	504	0.104	31	0.591										
10%	519	0.100	28	0.527										
20%	551	0.093	25	0.484										
Silicon steel (Si = 0%)	493	0.108	42	0.785										
1%	485	0.11	24	0.451										
2%	479	0.11	18	0.344										
5%	463	0.11	11	0.215										

Table III-16. Continued

Metal	Properties at 68°F ρ (lb_m/ft³)	C_p (Btu/lb_m•°F)	k (Btu/hr•ft•°F)	α (ft²/hr)	k(Btu/hr-ft-.F) −148°F −100°C	32°F 0°C	212°F 100°C	392°F 200°C	572°F 300°C	752°F 400°C	1112°F 600°C	1472°F 800°C	1832°F 1000°C	2192°F 1200°C
Copper															
Pure	559	0.0915	223	4.353	235	223	219	216	—	210	204			
Aluminum bronze: 95 Cu, 5 Al	541	0.098	48	0.903										
Bronze: 75 Cu. 25 Sn	541	0.082	15	0.333										
Red brass: 85 Cu. 9 Sn. 6 Zn	544	0.092	35	0.699	—	34	41							
Brass: 70 Cu. 30 Zn	532	0.092	64	1.322	51	—	74	83	85	85				
German silver 62 Cu, 15 Ni. 22 Zn	538	0.094	14.4	0.290	11.1	—	18	23	26	28				
Constantan: 60 Cu, 40 Ni	557	0.098	13.1	0.237	12	—	12.8	15						
Magnesium															
Pure	109	0.242	99	3.762	103	99	97	94	91					
Mg-Al (electrolytic) 6-8% Al, 1-2% Zn	113	0.24	38	1.397	—	30	36	43	48					
Mg-Mn: 2% Mn	111	0.24	66	2.473	54	64	72	75						
Molybdenum	638	0.060	79	2.074	80	79	79							
Nickel															
Pure (99.9%)	556	0.1065	52	0.882	60	54	48	42	37	34				
Impure (99.2%)	556	0.106	40	0.677	—	40	37	34	32	30	32	36	39	40
Ni-Cr: 90 Ni. 10 Cr	541	0.106	10	0.172	—	9.9	10.9	12.1	13.2	14.2				
80 Ni, 20 Cr	519	0.106	7.3	0.129	—	7.1	8.0	9.0	9.9	10.9	13.0			
Silver															
Purest	657	0.0559	242	6.601	242	241	240	238						
Pure (99.9%)	657	0.0559	235	6.418	242	237	240	216	209	208				
Tungsten	1208	0.0321	94	2.430	—	96	87	82	77	73.	65	44		
Zinc. Pure	446	0.0918	64.8	1.591	66	65	63	61	58	54				
Tin. pure	456	0.0541	37	1.505	43	38.1	34	33						

Source: From E.R.G. Eckert and R.M. Drake, *Heat and Mass Transfer-,* copyright 1959 McGraw-Hill; used with the permission of McGraw-Hill Book Company.

Table III-17. Property Values of Fluids in Saturated State

t (°F)	ρ (lb/ft³)	C_p (Btu/ lb·°F)	ν (ft²/sec)	k (Btu/ hr·ft·°F)	α (ft²/hr)	Pr	β (1/R)
			Water (H₂O)				
32	62.57	1.0074	1.925×10^{-5}	0.319	5.07×10^{-3}	13.6	
68	62.46	0.9988	1.083	0.345	5.54	7.02	0.10×10^{-3}
104	62.09	0.9980	0.708	0.363	5.86	4.34	
140	61.52	0.9994	0.514	0.376	6.02	3.02	
176	60.81	1.0023	0.392	0.386	6.34	2.22	
212	59.97	1.0070	0.316	0.393	6.51	1.74	
248	59.01	1.015	0.266	0.396	6.62	1.446	
284	57.95	1.023	0.230	0.395	6.68	1.241	
320	56.79	1.037	0.204	0.393	6.70	1.099	
356	55.50	1.055	0.186	0.390	6.68	1.004	
392	54.11	1.076	0.172	0.384	6.61	0.937	
428	52.59	1.101	0.161	0.377	6.51	0.891	
464	50.92	1.136	0.154	0.367	6.35	0.871	
500	49.06	1.182	0.148	0.353	6.11	0.874	
537	46.98	1.244	0.145	0.335	5.74	0.910	
572	44.59	1.368	0.145	0.312	5.13	1.019	
			Ammonia (NH₃)				
−58	43.93	1.066	0.468×10^{-5}	0.316	6.75×10^{-3}	2.60	
−40	43.18	1.067	0.437	0.316	6.88	2.28	
−22	42.41	1.069	0.417	0.317	6.98	2.15	
−4	41.62	1.077	0.410	0.316	7.05	2.09	
14	40.80	1.090	0.407	0.314	7.07	2.07	
32	39.96	1.107	0.402	0.312	7.05	2.05	
50	39.09	1.126	0.396	0.307	6.98	2.04	
68	38.19	1.146	0.386	0.301	6.88	2.02	1.36×10^{-3}
86	37.23	1.168	0.376	0.293	6.75	2.01	
104	36.27	1.194	0.366	0.285	6.59	2.00	
122	35.23	1.222	0.355	0.275	6.41	1.99	
			Carbon Dioxide (CO₂)				
−58	72.19	0.44	0.128×10^{-5}	0.0494	1.558×10^{-3}	2.96	
−40	69.78	0.45	0.127	0.0584	1.864	2.46	
−22	67.22	0.47	0.126	0.0645	2.043	2.22	
−4	64.45	0.49	0.124	0.0665	2.110	2.12	
14	61.39	0.52	0.122	0.0635	1.989	2.20	
			Eutectic Calcium Chloride Solution (29.9% CaCl₂)				
−58	82.39	0.623	39.13×10^{-5}	0.232	4.52×10^{-3}	312	
−40	82.09	0.6295	26.88	0.240	4.65	208	
−22	81.79	0.6356	18.49	0.248	4.78	139	
−4	81.50	0.642	11.88	0.257	4.91	87.1	
14	81.20	0.648	7.49	0.265	5.04	53.6	
32	80.91	0.654	4.73	0.273	5.16	33.0	
50	80.62	0.660	3.61	0.280	5.28	24.6	
68	80.32	0.666	2.93	0.288	5.40	19.6	
86	80.03	0.672	2.44	0.295	5.50	16.0	
104	79.73	0.678	2.07	0.302	5.60	13.3	
122	79.44	0.685	1.78	0.309	5.69	11.3	
			Glycerin [C₃H₅(OH)₃]				
32	79.66	0.540	0.0895	0.163	3.81×10^{-3}	84.7×10^3	
50	79.29	0.554	0.0323	0.164	3.74	31.0	
68	78.91	0.570	0.0127	0.165	3.67	12.5	0.28×10^{-3}
86	78.54	0.584	0.0054	0.165	3.60	5.38	
104	78.16	0.600	0.0024	0.165	3.54	2.45	
122	77.72	0.617	0.0016	0.166	3.46	1.63	

Table III-17. Continued

t (°F)	ρ (lb/ft³)	C_p (Btu/ lb·°F)	ν (ft²/sec)	k (Btu/ hr·ft·°F)	α (ft²/hr)	Pr	β (1/R)
			Ethylene Glycol [$C_2H_4(OH_2)$]				
32	70.59	0.548	61.92×10^{-5}	0.140	3.62×10^{-3}	615	
68	69.71	0.569	20.64	0.144	3.64	204	0.36×10^{-3}
104	68.76	0.591	9.35	0.148	3.64	93	
140	67.90	0.612	5.11	0.150	3.61	51	
176	67.27	0.633	3.21	0.151	3.57	32.4	
212	66.08	0.655	2.18	0.152	3.52	22.4	
			Engine Oil (Unused)				
32	56.13	0.429	0.0461	0.085	3.53×10^{-3}	47100	
68	55.45	0.449	0.0097	0.084	3.38	10400	0.39×10^{-3}
104	54.69	0.469	0.0026	0.083	3.23	2870	
140	53.94	0.489	0.903×10^{-3}	0.081	3.10	1050	
176	53.19	0.509	0.404	0.080	2.98	490	
212	52.44	0.530	0.219	0.079	2.86	276	
248	51.75	0.551	0.133	0.078	2.75	175	
284	51.00	0.572	0.086	0.077	2.66	116	
320	50.31	0.593	0.060	0.076	2.57	84	
			Mercury (Hg)				
32	850.78	0.0335	0.133×10^{-5}	4.74	166.6×10^{-3}	0.0288	
68	847.71	0.0333	0.123	5.02	178.5	0.0249	1.01×10^{-4}
122	843.14	0.0331	0.112	5.43	194.6	0.0207	
212	835.57	0.0328	0.0999	6.07	221.5	0.0162	
302	828.06	0.0326	0.0918	6.64	246.2	0.0134	
392	820.61	0.0375	0.0863	7.13	267.7	0.0116	
482	813.16	0.0324	0.0823	7.55	287.0	0.0103	
600	802	0.032	0.0724	8.10	316	0.0083	
			Carbon Dioxide (CO₂)				
32	57.87	0.59	0.117	0.0604	1.774	2.38	
50	53.69	0.75	0.109	0.0561	1.398	2.80	
68	48.23	1.2	0.098	0.0504	0.860	4.10	7.78×10^{-3}
86	37.32	8.7	0.086	0.0406	0.108	28.7	
			Sulfur Dioxide (SO₂)				
−58	97.44	0.3247	0.521×10^{-5}	0.140	4.42×10^{-3}	4.24	
−40	95.94	0.3250	0.456	0.136	4.38	3.74	
−22	94.43	0.3252	0.399	0.133	4.33	3.31	
−4	92.93	0.3254	0.349	0.130	4.29	2.93	
14	91.37	0.3255	0.310	0.126	4.25	2.62	
32	89.80	0.3257	0.277	0.122	4.19	2.38	
50	88.18	0.3259	0.250	0.118	4.13	2.18	
68	86.55	0.3261	0.226	0.115	4.07	2.00	1.08×10^{-3}
86	84.86	0.3263	0.204	0.111	4.01	1.83	
104	82.98	0.3266	0.186	0.107	3.95	1.70	
122	81.10	0.3268	0.174	0.102	3.87	1.61	
			Methyl Chloride (CH₃Cl)				
−58	65.71	0.3525	0.344×10^{-5}	0.124	5.38×10^{-3}	2.31	
−40	64.51	0.3541	0.342	0.121	5.30	2.32	
−22	63.46	0.3561	0.338	0.117	5.18	2.35	
−4	62.39	0.3593	0.333	0.113	5.04	2.38	
14	61.27	0.3629	0.329	0.108	4.87	2.43	
32	60.08	0.3673	0.325	0.103	4.70	2.49	
50	58.83	0.3726	0.320	0.099	4.52	2.55	
68	57.64	0.3788	0.315	0.094	4.31	2.63	
86	56.38	0.3860	0.310	0.089	4.10	2.72	
104	55.13	0.3942	0.303	0.083	3.86	2.83	
122	53.76	0.4034	0.295	0.077	3.57	2.97	

Source: From E. R. G. Eckert and R. M. Drake, *Heat and Mass Transfer;* copyright 1959 McGraw-Hill; Used with the permission of McGraw-Hill Book Company.

Table III-18. Property Values of Gases at Atmospheric Pressure

T (°F)	ρ (lb/ft³)	C_p (Btu/ lb·°F)	μ (lb/sec·ft)	ν (ft²/sec)	k (Btu/ hr·ft·°F)	α (ft²/hr)	Pr
Air							
−280	0.2248	0.2452	0.4653×10^{-5}	2.070×10^{-5}	0.005342	0.09691	0.770
−190	0.1478	0.2412	0.6910	4.675	0.007936	0.2226	0.753
−100	0.1104	0.2403	0.8930	8.062	0.01045	0.3939	0.739
−10	0.0882	0.2401	1.074	10.22	0.01287	0.5100	0.722
80	0.0735	0.2402	1.241	16.88	0.01516	0.8587	0.708
170	0.0623	0.2410	1.394	22.38	0.01735	1.156	0.697
260	0.0551	0.2422	1.536	27.88	0.01944	1.457	0.689
350	0.0489	0.2438	1.669	31.06	0.02142	1.636	0.683
440	0.0440	0.2459	1.795	40.80	0.02333	2.156	0.680
530	0.0401	0.2482	1.914	47.73	0.02519	2.531	0.680
620	0.0367	0.2520	2.028	55.26	0.02692	2.911	0.680
710	0.0339	0.2540	2.135	62.98	0.02862	3.324	0.682
800	0.0314	0.2568	2.239	71.31	0.03022	3.748	0.684
890	0.0294	0.2593	2.339	79.56	0.03183	4.175	0.686
980	0.0275	0.2622	2.436	88.58	0.03339	4.631	0.689
1070	0.0259	0.2650	2.530	97.68	0.03483	5.075	0.692
1160	0.0245	0.2678	2.620	106.9	0.03628	5.530	0.696
1250	0.0232	0.2704	2.703	116.5	0.03770	6.010	0.699
1340	0.0220	0.2727	2.790	126.8	0.03901	6.502	0.702
1520	0.0200	0.2772	2.955	147.8	0.04178	7.536	0.706
1700	0.0184	0.2815	3.109	169.0	0.04410	8.514	0.714
1880	0.0169	0.2860	3.258	192.8	0.04641	9.602	0.722
2060	0.0157	0.2900	3.398	216.4	0.04880	10.72	0.726
2240	0.0147	0.2939	3.533	240.3	0.05098	11.80	0.734
2420	0.0138	0.2982	3.668	265.8	0.05348	12.88	0.741
2600	0.0130	0.3028	3.792	291.7	0.05550	14.00	0.749
2780	0.0123	0.3075	3.915	318.3	0.05750	15.09	0.759
2960	0.0116	0.3128	4.029	347.1	0.0591	16.40	0.767
3140	0.0110	0.3196	4.168	378.8	0.0612	17.41	0.783
3320	0.0105	0.3278	4.301	409.9	0.0632	18.36	0.803
3500	0.0100	0.3390	4.398	439.8	0.0646	19.05	0.831
3680	0.0096	0.3541	4.513	470.1	0.0663	19.61	0.863
3860	0.0091	0.3759	4.611	506.9	0.0681	19.92	0.916
4160	0.0087	0.4031	4.750	546.0	0.0709	20.21	0.972
Helium							
−456		1.242	5.66×10^{-7}		0.0061	0.1792	0.74
−400	0.0915	1.242	33.7	3.68×10^{-5}	0.0204	2.044	0.70
−200	0.211	1.242	84.3	39.95	0.0536	3.599	0.694
−100	0.0152	1.242	105.2	69.30	0.0680	5.299	0.70
0	0.0119	1.242	122.1	102.8	0.0784	9.490	0.71
200	0.00829	1.242	154.9	186.9	0.0977	14.40	0.72
400	0.00637	1.242	184.8	289.9	0.114	20.21	0.72
600	0.00517	1.242	209.2	404.5	0.130	25.81	0.72
800	0.00439	1.242	233.5	531.9	0.145	34.00	0.72
1000	0.00376	1.242	256.5	682.5	0.159	41.98	0.72
1200	0.00330	1.242	277.9	841.0	0.172		0.72
Hydrogen							
−406	0.05289	2.589	1.079×10^{-6}	2.040×10^{-5}	0.0132	0.0966	0.759
−370	0.03181	2.508	1.691	5.253	0.0209	0.262	0.721
−280	0.01534	2.682	2.830	18.45	0.0384	0.933	0.712
−190	0.01022	3.010	3.760	36.79	0.0567	1.84	0.718
−100	0.00766	3.234	4.578	59.77	0.0741	2.99	0.719
−10	0.00613	3.358	5.321	86.80	0.0902	4.38	0.713
80	0.00511	3.419	6.023	117.9	0.105	6.02	0.706
170	0.00438	3.448	6.689	152.7	0.119	7.87	0.697
260	0.00383	3.461	7.300	190.6	0.132	9.95	0.690
350	0.00341	3.463	7.915	232.1	0.145	12.26	0.682
440	0.00307	3.465	8.491	276.6	0.157	14.79	0.675
530	0.00279	3.471	9.055	324.6	0.169	17.50	0.668
620	0.00255	3.472	9.599	376.4	0.182	20.56	0.664
800	0.00218	3.481	10.68	489.9	0.203	26.75	0.659
980	0.00191	3.505	11.69	612	0.222	33.18	0.664
1160	0.00170	3.540	12.62	743	0.238	39.59	0.676
1340	0.00153	3.575	13.55	885	0.254	46.49	0.686
1520	0.00139	3.622	14.42	1039	0.268	53.19	0.703
1700	0.00128	3.670	15.29	1192	0.282	60.00	0.715
1880	0.00118	3.720	16.18	1370	0.296	67.40	0.733
1940	0.00115	3.735	16.42	1429	0.300	69.80	0.736

Table III-18. Continued

T (°F)	ρ (lb/ft³)	C_p (Btu/ lb·°F)	μ (lb/sec·ft)	ν (ft²/sec)	k (Btu/ hr·ft·°F)	α (ft²/hr)	Pr
			Oxygen				
−280	0.2492	0.2264	5.220×10^{-6}	2.095×10^{-5}	0.00522	0.09252	0.815
−190	0.1635	0.2192	7.721	4.722	0.00790	0.2204	0.773
−100	0.1221	0.2181	9.979	8.173	0.01054	0.3958	0.745
−10	0.0975	0.2187	12.01	12.32	0.01305	0.6120	0.725
80	0.0812	0.2198	13.86	17.07	0.01546	0.8662	0.709
170	0.0695	0.2219	15.56	22.39	0.01774	1.150	0.702
260	0.0609	0.2250	17.16	28.18	0.02000	1.460	0.695
350	0.0542	0.2285	18.66	34.43	0.02212	1.786	0.694
440	0.0487	0.2322	20.10	41.27	0.02411	2.132	0.697
530	0.0443	0.2360	21.48	48.49	0.02610	2.496	0.700
620	0.0406	0.2399	22.79	56.13	0.02792	2.867	0.704
			Nitrogen				
−280	0.2173	0.2561	4.611×10^{-6}	2.122×10^{-5}	0.005460	0.09811	0.786
−100	0.1068	0.2491	8.700	8.146	0.01054	0.3962	0.747
80	0.0713	0.2486	11.99	16.82	0.01514	0.8542	0.713
260	0.0533	0.2498	14.77	27.71	0.01927	1.447	0.691
440	0.0426	0.2521	17.27	40.54	0.02302	2.143	0.684
620	0.0355	0.2569	19.56	55.10	0.02646	2.901	0.686
800	0.0308	0.2620	21.59	70.10	0.02960	3.668	0.691
980	0.:267	0.2681	23.41	87.68	0.03241	4.528	0.700
1160	0.0237	0.2738	25.19	98.02	0.03507	5.404	0.711
1340	0.0213	0.2789	26.88·	126.2	0.03741	6.297	0.724
1520	0.0194	0.2832	28.41	146.4	0.03958	7.204	0.736
1700	0.0178	0.2875	29.90	168.0	0.04151	8.111	0.748
			Carbon Dioxide				
−64	0.1544	0.187	7.462×10^{-6}	4.833×10^{-5}	0.006243	0.2294	0.818
−10	0.1352	0.192	8.460	6.257	0.007444	0.2868	0.793
80	0.1122	0.208	10.051	8.957	0.009575	0.4103	0.770
170	0.0959	0.215	11.561	12.05	0.01183	0.5738	0.755
260	0.0838	0.225	12.98	15.49	0.01422	0.7542	0.738
350	0.0744	0.234	14.34	19.27	0.01674	0.9615	0.721
440	0.0670	0.242	15.63	23.33	0.01937	1.195	0.702
530	0.0608	0.250	16.85	27.71	0.02208	1.453	0.685
620	0.0558	0.257	18.03	32.31	0.02491	1.737	0.668
			Carbon Monoxide				
−64	0.09699	0.2491	9.295×10^{-6}	9.583×10^{-5}	0.01101	0.4557	0.758
−10	0.0525	0.2490	10.35	12.14	0.01239	0.5837	0.750
80	0.07109	0.2489	11.990	16.87	0.01459	0.8246	0.737
170	0.06082	0.2492	13.50	22.20	0.01666	1.099	0.728
260	0.05329	0.2504	14.91	27.98	0.01864	1.397	0.722
350	0.04735	0.2520	16.25	34.32	0.0252	1.720	0.718
440	0.04259	0.2540	17.51	41.11	0.02232	2.063	0.718
530	0.03872	0.2569	18.74	48.40	0.02405	2.418	0.721
620	0.03549	0.2598	19.89	56.04	0.02569	2.786	0.724
			Ammonia (NH₃)				
−58	0.0239	0.525	4.875×10^{-6}	2.04×10^{-4}	0.0099	0.796	0.93
32	0.0495	0.520	6.285	1.27	0.0127	0.507	0.90
122	0.0405	0.520	7.415	1.83	0.0156	0.744	0.88
212	0.0349	0.534	8.659	2.48	0.0189	1.015	0.87
302	0.0308	0.553	9.859	3.20	0.0226	1.330	0.87
392	0.0275	0.572	11.08	4.03	0.0270	1.713	0.84
			Steam (H₂O Vapor)				
224	0.0366	0.492	8.54×10^{-6}	2.33×10^{-4}	0.0142	0.789	1.060
260	0.0346	0.481	9.03	2.61	0.0151	0.906	1.040
350	0.0306	0.473	10.25	3.35	0.0173	1.19	1.010
440	0.0275	0.474	11.45	4.16	0.0196	1.50	0.996
530	0.0250	0.477	12.66	5.06	0.0219	1.84	0.991
620	0.0228	0.484	13.89	6.09	0.0244	2.22	0.986
710	0.0211	0.491	15.10	7.15	0.0268	2.58	0.995
800	0.0196	0.498	16.30	8.31	0.0292	2.99	1.000
890	0.0183	0.506	17.50	9.56	0.0317	3.42	1.005
980	0.0171	0.514	18.72	10.98	0.0342	3.88	1.010
1070	0.0161	0.522	19.95	12.40	0.0368	4.38	1.019

Source: From E. R. G. Eckert and R. M. Drake, *Heat and Mass Transfer;* copyright 1959 by McGraw-Hill; used with permission of McGraw-Hill Book Company.

Table III-19.　Zero-pressure Properties of Gases

Gas	Chemical Formula	Molecular Weight	R (ft lb_f/$lb_m \cdot °R$)	C_{p0} (Btu/$lb_m \cdot °R$)[a]	C_{v0} (Btu/$lb_m \cdot °R$)[a]	k[a]
Air	—	28.97	53.34	0.240	0.171	1.400
Argon	Ar	39.94	38.66	0.1253	0.0756	1.667
Carbon dioxide	CO_2	44.01	35.10	0.203	0.158	1.285
Carbon monoxide	CO	28.01	55.16	0.249	0.178	1.399
Helium	He	4.003	386.0	1.25	0.753	1.667
Hydrogen	H_2	2.016	766.4	3.43	2.44	1.404
Methane	CH_4	16.04	96.35	0.532	0.403	1.32
Nitrogen	N_2	28.016	55.15	0.248	0.177	1.400
Oxygen	O_2	32.000	48.28	0.219	0.157	1.395
Steam	H_2O	18.016	85.76	0.445	0.335	1.329

[a] C_{p0}, C_{v0}, and k are at 80°F.

Table III-20.　Constant-pressure specific Heats of Various Substances at Zero Pressure

Gas or Vapor	Equation: $\tilde{C}_{p0}$ in Btu/lb mol · °R, T in °R	Range (°R)	Max. Error (%)
O_2	$\tilde{C}_{p0} = 11.515 - \dfrac{172}{\sqrt{T}} + \dfrac{1530}{T}$	540–5000	1.1
	$= 11.515 - \dfrac{172}{\sqrt{T}} + \dfrac{1530}{T}$ $+ \dfrac{0.05}{1000}(T - 4000)$	5000–9000	0.3
N_2	$\tilde{C}_{p0} = 9.47 - \dfrac{3.47 \times 10^3}{T} + \dfrac{1.16 \times 10^6}{T^2}$	540–9000	1.7
CO	$\tilde{C}_{p0} = 9.46 - \dfrac{3.29 \times 10^3}{T} + \dfrac{1.07 \times 10^6}{T^2}$	540–9000	1.1
H_2	$\tilde{C}_{p0} = 5.76 + \dfrac{0.578}{1000}T + \dfrac{20}{\sqrt{T}}$	540–4000	0.8
	$= 5.76 + \dfrac{0.578}{1000}T + \dfrac{20}{\sqrt{T}}$ $- \dfrac{0.33}{1000}(T - 4000)$	4000–9000	1.4
H_2O	$\tilde{C}_{p0} = 19.86 \dfrac{597}{\sqrt{T}} + \dfrac{7500}{T}$	540–5400	1.8
CO_2	$\tilde{C}_{p0} = 16.2 - \dfrac{6.53 \times 10^3}{T} + \dfrac{1.41 \times 10^6}{T^2}$	540–6300	0.8
CH_4	$\tilde{C}_{p0} = 4.52 + 0.00737T$	540–1500	1.2
C_2H_4	$\tilde{C}_{p0} = 4.23 + 0.01177T$	350–1100	1.5
C_2H_6	$\tilde{C}_{p0} = 4.01 + 0.01636T$	400–1100	1.5
C_3H_8	$\tilde{C}_{p0} = 2.258 + 0.0320T - 5.43 \times 10^{-6}T^2$	415–2700	1.8
C_4H_{10}	$\tilde{C}_{p0} = 4.36 + 0.0403T - 6.83 \times 10^{-6}T^2$	540–2700	1.7
C_8H_{18}	$\tilde{C}_{p0} = 7.92 + 0.0601T$	400–1100	est. 4
$C_{12}H_{26}$	$\tilde{C}_{p0} = 8.68 + 0.0889T$	400–1100	est. 4

Source: From R. L. Sweigert and M. W. Beardsley, Bulletin No. 2, Georgia School of Technology, 1938, except C_3H_8 and C_4H_{10}, which are from H. M. Spencer, Journal of the American Chemical Society, Vol. 67 (1945), p. 1859.

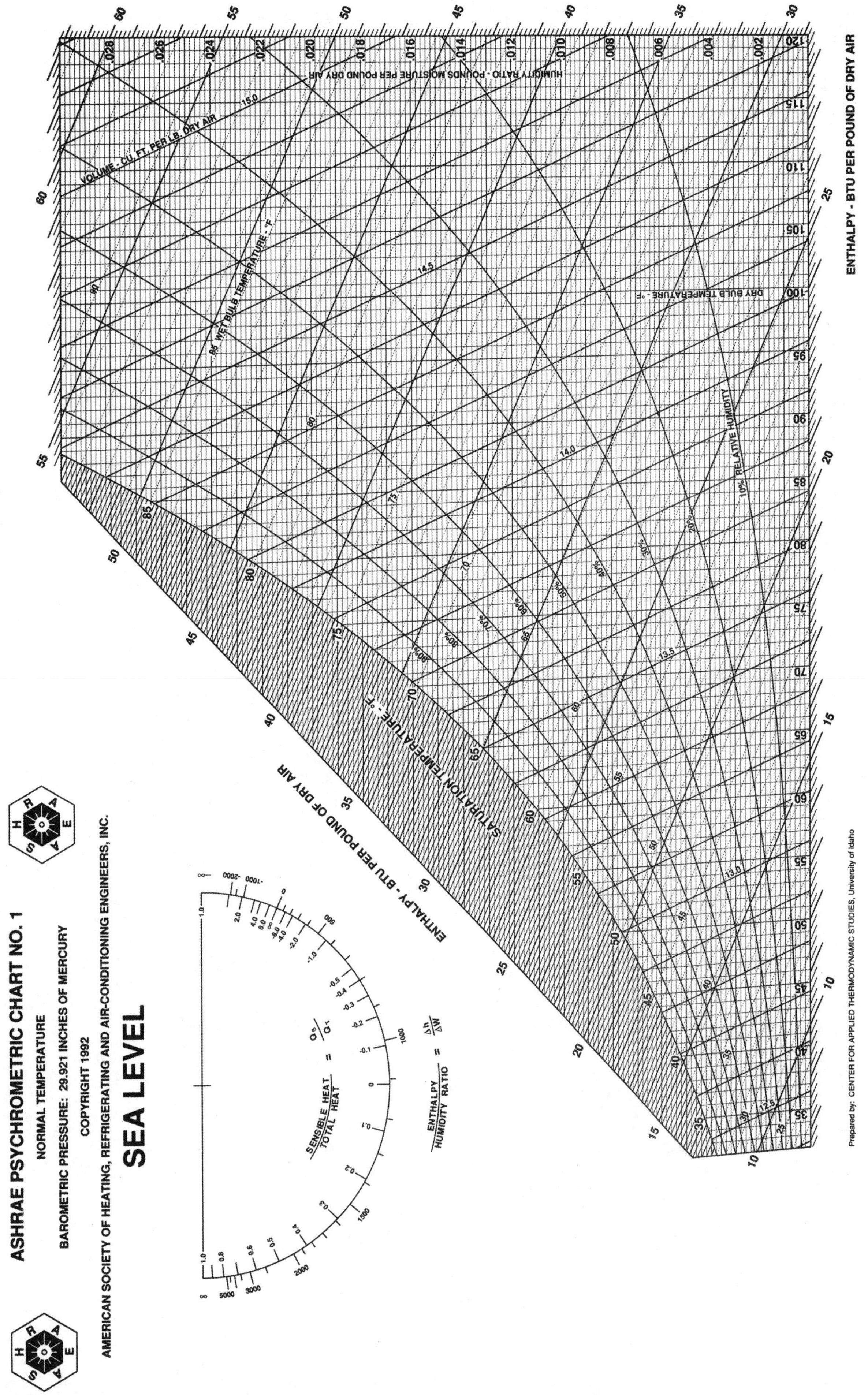
ASHRAE PSYCHROMETRIC CHART NO. 1
NORMAL TEMPERATURE
BAROMETRIC PRESSURE: 29.921 INCHES OF MERCURY
COPYRIGHT 1992
AMERICAN SOCIETY OF HEATING, REFRIGERATING AND AIR-CONDITIONING ENGINEERS, INC.
SEA LEVEL
SENSIBLE HEAT / TOTAL HEAT = Qs / Qt
ENTHALPY / HUMIDITY RATIO = Δh / ΔW
ENTHALPY - BTU PER POUND OF DRY AIR
SATURATION TEMPERATURE - °F
WET BULB TEMPERATURE - °F
VOLUME - CU. FT. PER LB. DRY AIR
HUMIDITY RATIO - POUNDS MOISTURE PER POUND DRY AIR
DRY BULB TEMPERATURE - °F
RELATIVE HUMIDITY
Prepared by: CENTER FOR APPLIED THERMODYNAMIC STUDIES, University of Idaho

Table III-21. Psychrometric Chart.

Table III-22. Total Emissivity Data

Surface	°C	°F	ε
Metals			
Aluminum			
Polished, 98% pure	200–600	400–1100	0.04–0.06
Commercial sheet	100	200	0.09
Rough plate	40	100	0.07
Heavily oxidized	100–550	200–1000	0.20–0.33
Antimony			
Polished	40–250	100–500	0.28–0.31
Bismuth			
Bright	100	200	0.34
Brass			
Highly polished	250	500	0.03
Polished	40	100	0.07
Dull plate	40–250	100–500	0.22
Oxidized	40–250	100–500	0.46–0.56
Chromium			
Polished sheet	40–550	100–1000	0.08–0.27
Cobalt			
Unoxidized	250–550	500–1000	0.13–0.23
Copper			
Highly polished electrolytic	100	200	0.02
Polished	40	100	0.04
Slightly polished	40	100	0.12
Polished, lightly tarnished	40	100	0.05
Dull	40	100	0.15
Black oxidized	40	100	0.76
Gold			
Pure, highly polished	100–600	200–1100	0.02–0.035
Inconel			
X, stably oxidized	230–900	450–1600	0.55–0.78
B, stably oxidized	230–1000	450–1750	0.32–0.55
X and B, polished	150–300	300–600	0.20
Iron and Steel			
Mild steel, polished	150–500	300–900	0.14–0.32
Steel, polished	40–250	100–500	0.07–0.10
Sheet steel, ground	1000	1700	0.55
Sheet steel, rolled	40	100	0.66
Sheet steel, strong rough oxide	40	100	0.80
Steel, oxidized at 1100°F	250	500	0.79
Cast iron, with skin	40	100	0.70–0.80
Cast iron, newly turned	40	100	0.44
Cast iron, polished	200	400	0.21
Cast iron, oxidized	40–250	100–500	0.57–0.66
Iron, red rusted	40	100	0.61
Iron, heavily rusted	40	100	0.85
Wrought iron, smooth	40	100	0.35
Wrought iron, dull oxidized	20–360	70–680	0.94
Stainless, polished	40	100	0.07–0.17
Stainless, after repeated heating and cooling	230–930	450–1650	0.50–0.70
Lead			
Polished	40–250	100–500	0.05–0.08
Gray, oxidized	40	100	0.28
Oxidized at 390°F	200	400	0.63
Oxidized at 1100°F	40	100	0.63
Magnesium			
Polished	40–250	100–500	0.07–0.13
Manganin			
Bright rolled	100	200	0.05
Mercury			
Pure, clean	40–100	100–200	0.10–0.12

Table III-22. Continued

Surface	°C	°F	ε
Molybdenum			
Polished	40–250	100–500	0.06–0.08
Polished	550–1100	1000–2000	0.11–0.18
Filament	550–2800	1000–5000	0.08–0.29
Monel			
After repeated heating and cooling	230–930	450–1650	0.45–0.70
Oxidized at 1100°F	200–600	400–1100	0.41–0.46
Polished	40	100	0.17
Nickel			
Polished	40–250	100–500	0.05–0.07
Oxidized	40–250	100–500	0.35–0.49
Wire	250–1100	500–2000	0.10–0.19
Platinum			
Pure, polished plate	200–600	400–1100	0.05–0.10
Oxidized at 1100°F	250–550	500–1000	0.07–0.11
Electrolytic	250–550	500–1000	0.06–0.10
Strip	550–1100	1000–2000	0.12–0.14
Filament	40–1100	100–2000	0.04–0.19
Wire	200–1370	400–2500	0.07–0.18
Silver			
Polished or deposited	40–550	100–1000	0.01–0.03
Oxidized	40–550	100–1000	0.02–0.04
German silver," polished	250–550	500–1000	0.07–0.09
Tin			
Bright tinned iron	40	100	0.04–0.06
Bright	40	100	0.06
Polished sheet	100	200	0.05
Tungsten			
Filament	550–1100	1000–2000	0.11–0.16
Filament	2800	5000	0.39
Filament, aged	40–3300	100–6000	0.03–0.35
Polished	40–550	100–1000	0.04–0.08
Zinc			
Pure polished	40–250	100–500	0.02–0.03
Oxidized at 750°F	400	750	0.11
Galvanized, gray	40	100	0.28
Galvanized, fairly bright	40	100	0.23
Dull	40–250	100–500	0.21
Nonmetals			
Asbestos			
Board	40	100	0.96
Cement	40	100	0.96
Paper	40	100	0.93–0.95
Slate	40	100	0.97
Brick			
Red, rough	40	100	0.93
Silica	1000	1800	0.80–0.85
Fireclay	1000	1800	0.75
Ordinary refractory	1100	2000	0.59
Magnesite refractory	1000	1800	0.38
White refractory	1100	2000	0.29
Gray, glazed	1100	2000	0.75
Carbon			
Filament	1050–1420	1900–2600	0.53
Lampsoot	40	100	0.95
Clay			
Fired	100	200	0.91
Concrete			
Rough	40	100	0.94

Table III-22. Continued

Surface	°C	°F	ϵ
Corundum			
Emery rough	100	200	0.86
Glass			
Smooth	40	100	0.94
Quartz glass (2 mm)	250–550	500–1000	0.96–0.66
Pyrex	250–550	500–1000	0.94–0.75
Gypsum	40	100	0.80–0.90
Ice			
Smooth	0	32	0.97
Rough crystals	0	32	0.99
Hoarfrost	–18	0	0.99
Limestone	40–250	100–500	0.95–0.83
Marble			
Light gray, polished	40	100	0.93
White	40	100	0.95
Mica	40	100	0.75
Paints			
Aluminum, various ages and compositions	100	200	0.27–0.62
Black gloss	40	100	0.90
Black lacquer	40	100	0.80–0.93
White paint	40	100	0.89–0.97
White lacquer	40	100	0.80–0.95
Various oil paints	40	100	0.92–0.96
Red lead	100	200	0.93
Paper			
White	40	100	0.95
Writing paper	40	100	0.98
Any color	40	100	0.92–0.94
Roofing	40	100	0.91
Plaster			
Lime, rough	40–250	100–500	0.92
Porcelain			
Glazed	40	100	0.93
Quartz	40–550	100–1000	0.89–0.58
Rubber			
Hard	40	100	0.94
Soft, gray rough	40	100	0.86
Sandstone	40–250	100–500	0.83–0.90
Snow	(–12) – (–6)	10–20	0.82
Water			
0.1 mm or more thick	40	100	0.96
Wood			
Oak, planed	40	100	0.90
Walnut, sanded	40	100	0.83
Spruce, sanded	40	100	0.82
Beech	40	100	0.94
Planed	40	100	0.78
Various	40	100	0.80–0.90
Sawdust	40	100	0.75

Source: From E. M. Sparrow and R. D. Cess, *Radiation Heat Transfer*, rev. ed.; copyright ©
1970 by Wadsworth Publishing Company, Inc.; reprinted by permission of the publisher, Brooks/
Cole Publishing Company, Monterey, Calif.
[a] German silver is actually an alloy of copper, nickel, and zinc.

REVIEW OF ELECTRICAL SCIENCE

RUSSELL L. HEISERMAN, Ed.D.
School of Technology
Oklahoma State University
Stillwater, Oklahoma

IV. 1 INTRODUCTION

This brief review of electrical science is intended for those readers who may use electrical principles only on occasion and is intended to be supportive of the material found in those chapters of the handbook based on electrical science. The review consists of selected topics in basic ac circuit theory presented at a nominal analytical level. Much of the material deals with power in ac circuits and principles of power-factor improvement.

IV.2 REVIEW OF VECTOR ALGEBRA

Vector algebra is the mathematics most appropriate for ac circuit problems. Most often electric quantities, voltage, and current are not in phase in ac circuits, so phase relationships as well as magnitude have to be considered. This brief review will cover the basic idea of a vector quantity and then refresh the process of adding, subtracting, multiplying, and dividing vectors.

IV.2.1 Review

A vector is a quantity having both direction and magnitude. Familiar vector quantities are velocity and force. Other familiar quantities, such as speed, volume, area, and mass, have magnitude only.

A vector quantity is expressed as having both magnitude and direction, such as

$$Ae^{\pm j\theta}$$

where A is the magnitude and $e^{\pm j\theta}$ expresses the direction in the complex plane (Figure IV-1).

The important feature of this vector notation is to note that the angle of displacement is in fact an exponent. This feature is significant since it will allow the use of the law of exponents when multiplying, dividing, or raising to a power.

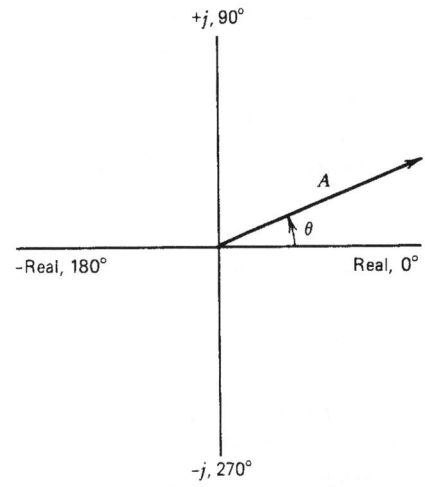

Figure IV-1. The generalized vector $Ae^{j\theta}$ shown in the complex plane. If j is positive, it is referenced to the positive real axis with a counterclockwise displacement. If j is negative, it is referenced to the positive real axis with a clockwise displacement.

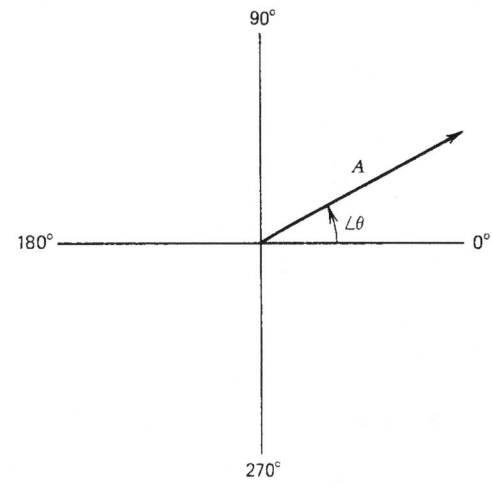

Figure IV-2. The vector A $\angle\theta$ shown in the polar coordinate system. A vector expressed as A $\angle\theta$ is said to be in polar form.

Common practice has created a shorthand for expressing vectors. This method is quicker to write and for many, more clearly expresses the idea of a vector:

$$A \angle \theta$$

This shorthand is read as a vector magnitude A operating or pointing in the direction θ. It is termed the polar representation of a vector, as shown in Figure IV-2.

Now the function $e^{j\theta}$ may be expressed or resolved into its horizontal and vertical components in the complex plane:

$$e^{j\theta} = \cos\theta + j\sin\theta$$

The vector has been resolved and expressed in rectangular form. Using the shorthand notation

$$A\angle\theta = A\cos\theta + jA\sin\theta$$

where $A\cos\theta$ is the vector projection on the real axis and $jA\sin\theta$ is the vector projection on the imaginary axis, as shown in Figure IV-3.

Both rectangular and polar expressions of a vector quantity are useful when performing mathematical operations.

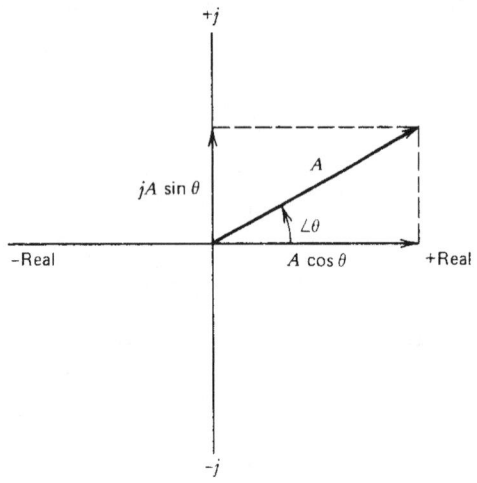

Figure IV-3. The vector A $\angle\theta$ shown together with its rectangular components.

IV.2.2 Addition and Subtraction of Vectors

When adding or subtracting vectors, it is most convenient to use the rectangular form. This is best demonstrated through an example. Suppose that we have two vectors, $20\angle30°$ and $25\angle-45°$, and these vectors are to be added. The quickest way to accomplish this is to resolve each vector into its rectangular components, add the real components, then add the imaginary components, and, if needed, express the results in polar form:

$$20\angle30° = 20\cos30° + j20\sin30°$$

$$= 17.3 + j10$$

$$25\angle-45° = 25\cos45° - j25\sin45°$$

$$= 17.7 - j17.7$$

A calculator is a handy tool for resolving vectors. Many calculators have automatic programs for converting vectors from one form to another.

Now adding, we obtain

$$\begin{array}{r} 17.3 + j10 \\ (+)\quad \underline{17.7 - j17.7} \\ 35.0 - j7.7 \end{array}$$

By inspection, this vector is seen to be slightly greater in magnitude than 35.0 and at a small angle below the positive real axis. Again using a calculator to express the vector in polar form: $35.8\angle-12.4°$, an answer in agreement with what was anticipated. Figure IV-4 shows roughly the same result using a graphical technique. Subtraction is accomplished in much the same way. Suppose that the vector $25\angle-45°$ is to be subtracted from the vector $20\angle30°$:

$$20\angle30° = 17.3 + j10$$

$$25\angle-45° = 17.7 - j17.7$$

To subtract

$$\begin{array}{r} 17.3 + j10 \\ \underline{17.7 - j17.7} \end{array}$$

first change sign of the subtrahend and then add:

$$\begin{array}{r} 17.3 + j10 \\ \underline{-17.7 + j17.7} \\ -0.4 + j27.7 \end{array}$$

The effect of changing the sign of the subtrahend is to push the vector back through the origin. as shown in Figure IV-5.

The resulting vector appears to be about 28 units long and barely in the second quadrant. The calculator gives $27.7\angle90.8°$.

IV.2.3 Multiplication and Division of Vectors

Vectors are expressed in polar form for multiplication and division. The magnitudes are multiplied or divided, and the angles follow the rules governing

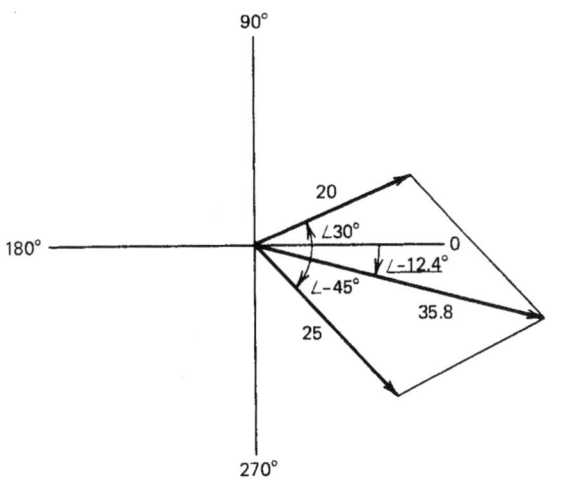

Figure IV-4. Use of the graphical parallelogram method for adding two vectors. The result or sum is the diagonal originating at the origin of the coordinate system.

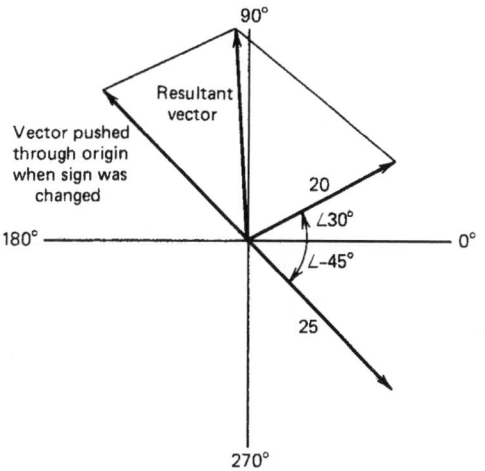

Figure IV-5. Graphical solution to subtraction of vectors.

exponents, added when multiplying, subtracted when dividing. Consider

$$20\angle 30° \times 25\angle{-45°} = 500\angle{-15°}$$

The magnitudes are multiplied and the angles are added. Consider

$$\frac{20\angle 30°}{25\angle{-45°}} = 0.8\angle 75°$$

The magnitudes are divided and the angle of the divisor is subtracted from the angle of dividend.

Raising to powers is a special case of multiplication. The magnitude is raised to the power and the angle is multiplied by the power. Consider

$$(20\angle 30)^3 = 8000\angle 90°$$

or consider

$$(20\angle 30°)^{1/2} = 4.47\angle 15°$$

IV.2.4 Summary

Vector manipulation is straightforward and easy to do. This presentation is intended to refresh those techniques most commonly used by those working at a practical level with ac electrical circuits. It has been the author's intent to exclude material on dot and cross products in favor of techniques that tend to allow the user more of a feeling for what is going on.

IV.3 RESISTANCE, INDUCTANCE, AND CAPACITANCE

The three types of electric circuit elements having distinct characteristics are resistance, inductance, and capacitance. This brief review will focus on the characteristics of these circuit elements in ac circuits to support later discussions on circuit impedance and power-factor-improvement principles.

IV.3. 1 Resistance

Resistance R in an ac circuit is the name given to circuit elements that consume real power in the form of heat, light. mechanical work, and so on. Resistance is a physical property of the wire used in a distribution system that results in power loss commonly called I^2R loss. Resistance can be thought of as a name given that portion of a circuit load that performs real work, that is, the portion of the power fed to a motor that results in measurable mechanical work being accomplished.

If resistance is the only circuit element in an ac circuit, the physical properties of that circuit are easily summarized, as shown in Figure IV-6. The important property is that the voltage and current are in phase.

Since the current and voltage are in phase and the ac source is a sine wave, the power used by the resistor is easily computed from root-mean-square (rms) (effective) voltage and current readings taken with a typical multimeter. The power is computed by taking the product of the measured voltage in volts or kilovolts and the measured current in amperes:

$$P(\text{watts}) = V(\text{volts}) \times I \text{ (amperes)}$$

where V is the voltage measured in volts and I is the current measured in amps. Both quantities are measured with an rms reading meter.

In many industrial settings the voltage may be

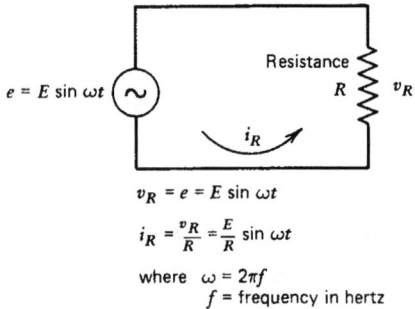

$$v_R = e = E \sin \omega t$$

$$i_R = \frac{v_R}{R} = \frac{E}{R} \sin \omega t$$

where $\omega = 2\pi f$
           f = frequency in hertz

Figure IV-6. Circuit showing an ac source with radian frequency ω. The current through the resistor is in phase with the voltage across the resistor.

measured in kilovolts and the current in amperes. The power is computed as the product of current and voltage and expressed as kilowatts:

$$P \text{ (kilowatts)} = V \text{ (kilovolts)} \times I \text{ (amps)}$$

If it is inconvenient to measure both voltage and current, one can compute power using only voltage or current if the resistance R is known:

$$P \text{ (watts)} = I^2 \text{(amps)} \times R \text{ (ohms)}$$

or

$$P \text{(watts)} = \frac{V^2 \text{(volts)}}{R \text{(ohms)}}$$

IV.3.2 Inductance

Inductance L in an ac circuit is usually formed as coils of wire, such as those found in motor windings, solenoids, or inductors. In a real circuit it is impossible to have only pure inductance, but for purposes of establishing background we will take the theoretical case of a pure inductance so that its circuit properties can be isolated and presented.

An inductor is a circuit element that uses no real power; it simply stores energy in the form of a magnetic field and will give up this stored energy, alternately storing energy and giving it up every half-cycle. The result of this storing and giving up energy when an inductor is driven by a sine-wave source is to put the measured magnetizing current (i_c.) 90° out of phase with the driving voltage. The magnetizing current lags behind the driving voltage by 90°. If pure inductance were the load of a sine-wave generator, we could summarize its characteristics as in Figure IV-7.

An inductor limits the current flowing through it by reacting with the voltage change across it. This property is called inductive reactance X_L. The inductive

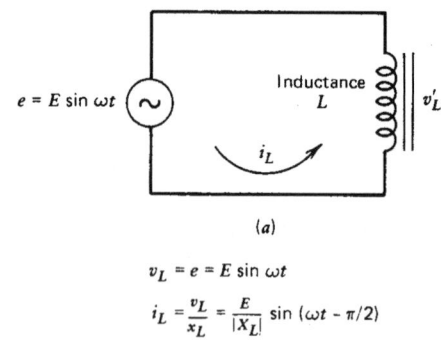

(a)

$$v_L = e = E \sin \omega t$$

$$i_L = \frac{v_L}{x_L} = \frac{E}{|X_L|} \sin (\omega t - \pi/2)$$

where $\pi/2$ is 90°; this expresses
the idea that the current lags
the voltage by 90°

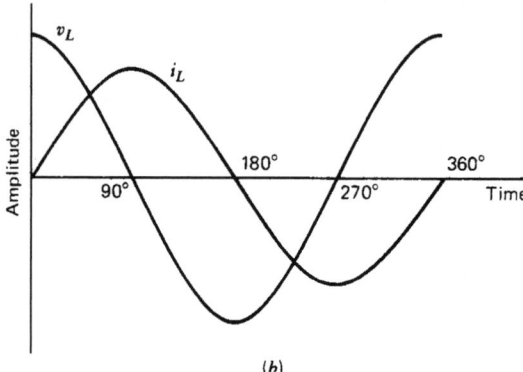

(b)

Figure IV-7. (a) Ac circuit with pure inductance. (b) Plot of the voltage across the inductor v_L, and the current i_L through it. The plot shows a 90° displacement between the current and the voltage.

reactance of a coil whose inductance is known in henrys (H) may be computed using the expression

$$X_L = 2\pi f L$$

where f is the frequency in hertz and L is the coil's inductance in henrys.

IV.3.3 Capacitance

Capacitance C, like inductance, only stores and gives up energy. However, the voltage and current phasing is exactly opposite that of an inductor in an ac circuit. The current in an ac circuit containing only capacitance leads the voltage by ∠90°. Figure IV-8 summarizes the characteristics of an ac circuit with a pure capacity load. A capacitor also reacts to changes. This property is called capacitive reactance X_c. The capacity reactance may be computed by using the expression

$$X_c = \frac{1}{2\pi f C}$$

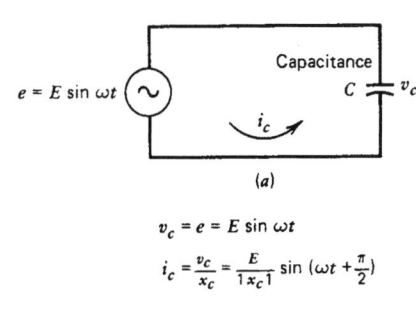

$$v_c = e = E \sin \omega t$$

$$i_c = \frac{v_c}{x_c} = \frac{E}{1 x_c 1} \sin \left(\omega t + \frac{\pi}{2}\right)$$

(a)

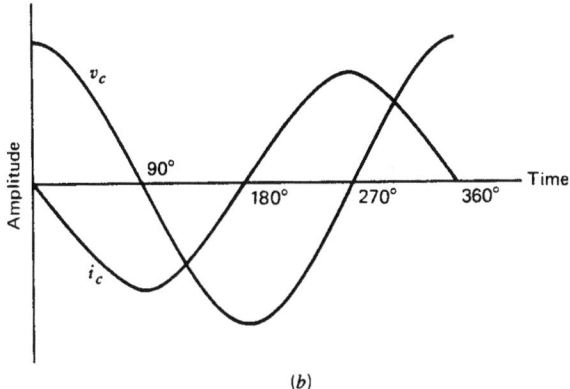

(b)

Figure IV-8. (*a*) Ac circuit with pure capacitance. (*b*) Plot of the voltage across the capacitor v_c, and the current i_c through it. The plot shows a 90° displacement between the current and the voltage.

where f is the frequency in hertz and C is the capacity in farads.

IV.3.4 Summary

Circuit elements are resistance that consumes real power and two reactive elements that only store and give up energy. These two reactive elements, capacitors and inductors, have opposite effects on the phase displacement between the current and voltage in ac circuits. These opposite effects are the key to adding capacitors in an otherwise inductive circuit for purposes of reducing the current-voltage phase displacement. Reducing the phase displacement improves the power factor of the circuit. (Power factor is defined and discussed later.)

IV.4 IMPEDANCE

In the preceding section it was mentioned that pure inductance does not occur in a real-world circuit. This is because the wire that is used to form the most carefully made coil still has resistance. This section considers circuits containing resistance and inductive reactance and circuits containing resistance and capacity reactance. At-

tention will be given to the notation used to describe such circuits since vector algebra must be used exclusively.

IV.4.1 Circuits with Resistance and Inductive Reactance

Figure IV-9 shows a circuit that has both resistive and inductive elements. Such a circuit might represent a real inductor, with the resistance representing the wire resistance, or such a circuit might be a simple model of a motor, with the inductance reflecting the inductive characteristics of the motor's windings and the resistance representing both the wire resistance and the real power consumed and converted to mechanical work performed by the motor.

In Figure IV-9 the current is common to both circuit elements. Recall that the voltage across the resistor is in phase with this current, while the voltage across the inductor leads the current. This idea is shown by plotting these quantities in the complex plane. Since i is the reference, it is plotted on the positive real axis as shown in Figure IV-10.

The voltage across the resistor is in phase with the current, so it is also on the positive real axis, whereas the voltage across the inductor is on the positive j axis

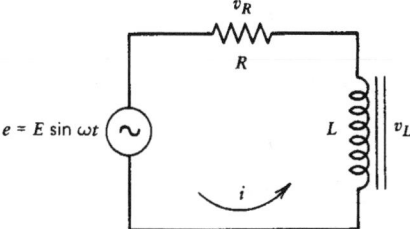

Figure IV-9. Circuit with both resistance and inductance. The circuit current i is common to both elements.

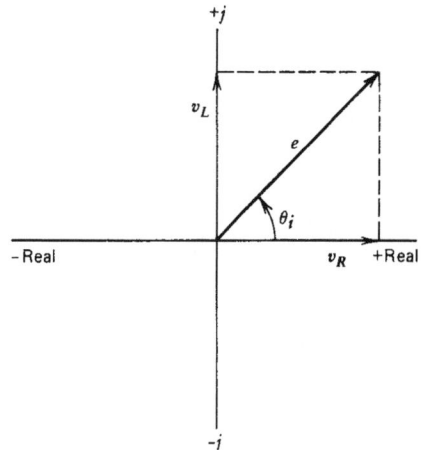

Figure IV-10. Circuit voltages and current plotted in the complex plane. Both i and V_R are on the positive real axis since they are in phase; V_L is on the positive j axis since it leads the current by 90°.

since it leads the current by 90°. However, the sum of the voltages must be the source voltage e. Figure IV-10 shows that the two voltages must be added as vectors:

$$e = v_r + jv_L$$

$$e = i_R + jiX_L$$

If we call the ratio of voltage to current the circuit impedance, then

$$Z = \frac{e}{i} = R + jX_L$$

Z, the circuit impedance, is a complex quantity and may be expressed in either polar or rectangular form:

$$Z = R + jX_L$$

or

$$Z = |Z| \underline{\theta}$$

In circuits with resistance and inductance the complex impedance will have a positive phase angle, and if R and X_L are plotted in the complex plane, X_L is plotted on the positive j axis, as shown in Figure IV-11.

IV.4.2 Circuits with Resistance and Capacity Reactance

Circuits containing resistance and capacitance are approached about the same way. Going through a similar analysis and looking at the relationship among R, X_c, and Z would show that X_c is plotted on the negative j axis, as shown in Figure IV-12.

IV.4.3 Summary

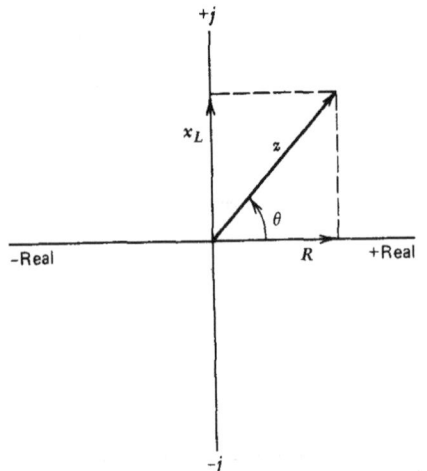

Figure IV-11. Plot in the complex plane showing the complex relationship of R, X_L, and Z.

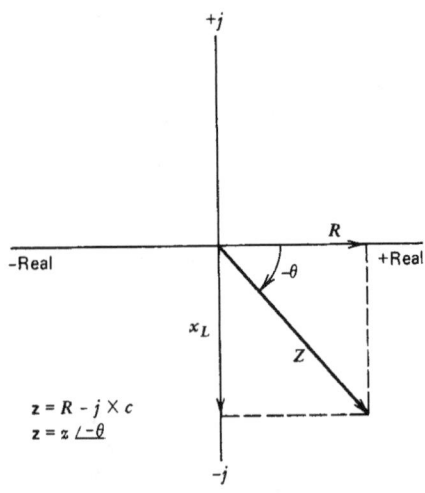

Figure IV-12. Summary of the relationship among R, X_c, and Z shown in the complex plane.

In circuits containing both resistive and reactive elements, the resistance is plotted on the positive real axis, while the reactances are plotted on the imaginary axis. The fact that inductive and capacitive reactance causes opposite phase displacements (has opposite effects in ac circuits) is further emphasized by plotting their reactance effects in opposite directions on the imaginary axis of the complex plane. The case is building for why capacitors might be used in an ac circuit with inductive loading to improve the circuit's power factor.

IV.5 POWER IN AC CIRCUITS

This section considers three aspects of power in ac circuits. First, the case of a circuit containing resistance and inductance is discussed, followed by the introduction of the power triangle for circuits containing resistance and inductance. Finally, power-factor improvement by the use of capacitors is presented.

IV.5.1 Power in a Circuit Containing Both Resistance and Inductance

Figure IV-13 reviews this situation through a circuit drawing and the voltages and currents shown in the complex plane. Meters are in place that read the effective or rms voltage V across the complex load and the effective or rms line current I.

Power is usually thought of as the product of voltage and the current in a circuit. The question is: The current I times *which* voltage will yield the correct or true power? This is an important question since Figure IV-13b shows three voltages in the complex plane.

Each of the three products may be taken, and each

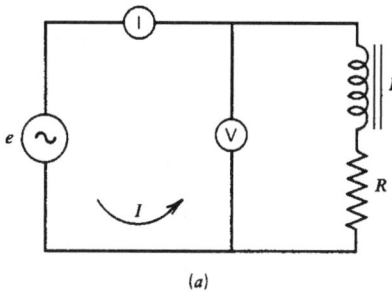

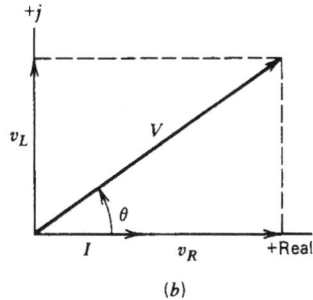

Figure IV-13. (*a*) Circuit having resistance and inductance; meters are in place to measure the line current *I* and the voltage V. (*b*) Relationship between the various voltages and the line current for this circuit.

has a name and a meaning. Taking the ammeter reading *I* times the voltmeter reading yields the *apparent power*. The apparent power is the load current-load voltage product without regard to the phase relationship of the current and voltage. This figure by itself is meaningless:

$$P(\text{apparent}) = IV$$

If the voltmeter could be connected across the resistor only, to measure v_R, then the line current-voltage product would yield the *true power* since the current and voltage are in phase:

$$P(\text{true}) = Iv_R$$

Usually, this connection cannot be made, so the true power of a load is measured with a special meter called a wattmeter that automatically performs the following calculation.

$$P(\text{true}) = IV \cos \theta$$

Note that in Figure IV-13b, the circuit voltage *V* and the resistance voltage V_R are related through the cosine of θ. The third product that could be taken is called *imaginary power* or VAR, the voltampere reactive product:

$$P(\text{imaginary}) = Iv_L$$

This is the power that is alternately stored and given up by the inductor to maintain its magnetic field. None of this reactive power is actually used.

If the voltages in the foregoing examples were measured in kilovolts, the three values computed would be the more familiar:

$$P(\text{apparent}) = \text{kVA}$$

$$P(\text{real}) = \text{kW}$$

$$P(\text{imaginary}) = \text{kVAR}$$

This discussion, together with Figure IV-13b, leads to the power triangle.

IV.5.2 The Power Triangle

The power triangle consists of three values, kVA, kW, and kVAR, arranged in a right triangle. The angle between the line current and voltage, Θ, becomes an important factor in this triangle. Figure IV-14 shows the power triangle.

To emphasize the relationship between these three quantities, an example may be helpful. Suppose that we have a circuit with inductive characteristics, and using a voltmeter, ammeter, and wattmeter, the following values are measured:

$$\text{watts} = 1.5 \text{ kW}$$

$$\text{line current} = 10 \text{ A}$$

$$\text{line voltage} = 240 \text{ V}$$

From this information we should be able to determine the kVA, Θ, and the kVAR.

The kVA can be computed directly from the voltmeter and ammeter readings:

$$\text{kVA} = (10 \text{ A})(0.24) \text{ kV} = 2.4 \text{ kVA}$$

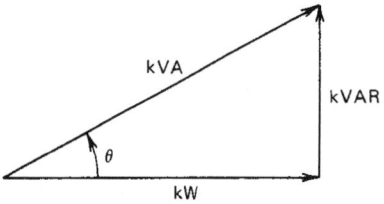

Figure IV-14. Power triangle for an inductive load. The angle θ is the angle of displacement between the line voltage and the line current.

Looking at the triangle in Figure IV-14 and recalling some basic trigonometry, we have

$$\cos \theta = \frac{kW}{kVA} = \frac{1.5}{2.4} = 0.625$$

and θ is the angle whose cosine equals 0.625. This can be looked up in a table or calculated using a hand calculator that computes trig functions:

$$\theta = \cos^{-1} 0.625 = 51.3°$$

Again referring to the power triangle and a little trig, we see that

$$kVAR = kVA \sin \theta$$

$$= 2.4 \ kVA \sin 51.3°$$

$$= 1.87 \ kVAR$$

Figure IV-15 puts all these measured and calculated data together in a power triangle.

Of particular interest is the ratio kW/kVA. This ratio is called the *power factor* (PF) of the circuit. So the power factor is the ratio of true power to apparent power in a circuit. This is also the cosine of the angle θ, the angle of displacement between the line voltage and the line current. To improve the power factor, the angle θ must be reduced. This could be accomplished by reducing the kVAR side of the triangle.

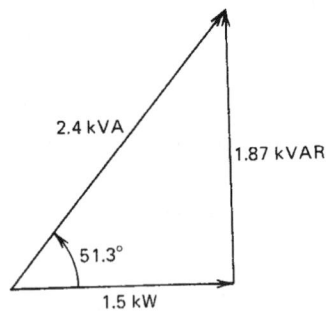

Figure IV-15. Organization of the measured and computed data of the example into a power triangle.

IV.5.3 Power-Factor Improvement

Recall that inductive reactance and capacity reactance are plotted in opposite directions on the imaginary axis, *j*. Thus, it should be no surprise to consider that kVAR produced by a capacitive load behaves in an opposite way to kVAR produced by inductive loads. This is the case and is the reason capacitors are commonly added to circuits having inductive loads to improve power factor (reduce the angle θ).

Suppose in the example being considered that enough capacity is added across the load to offset the effects of 90% of the inductive load. That is, we will try

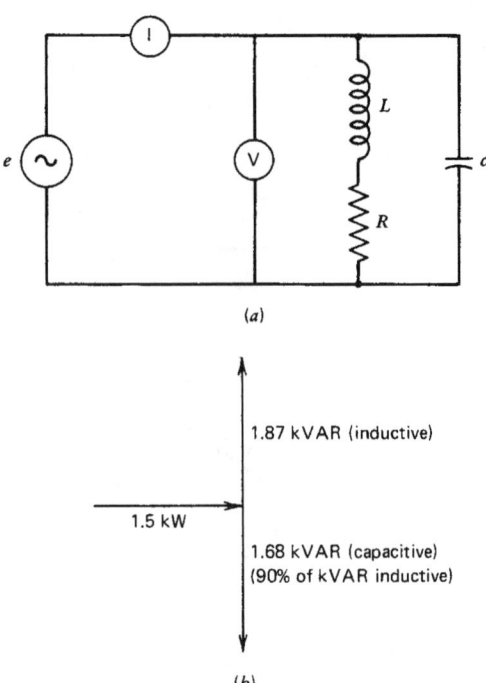

Figure IV-16. (*a*) Inductive circuit with capacity added to correct power factor. (*b*) Power vectors showing the relationship among kW, kVAR inductive, and kVAR capacitive.

to improve the power factor by better than 90%. Figure IV-16 shows the circuit arrangement with the kW and kVAR vectors drawn to show their relationship.

Following the example through, consider Figure IV-17, where 90% of the kVAR inductive load has been neutralized by adding the capacitor.

Working with the modified triangle in Figure IV-17, we can compute the new θ, call it θ_2:

$$\theta_2 = \tan^{-1} \frac{0.19}{1.5}$$
$$= 7.2$$

Again, a calculator comes in handy.

Since the new power factor is the cosine of θ_2, we compute

$$PF \ new = cosine \ 7.2 = 0.99$$

certainly an improvement.

Recall that the power factor can be expressed as a ratio of kW to kVA. From this idea we can compute a new kVA value:

$$PF = 0.99 = \frac{1.5 \ kW}{kVA \ new}$$

or

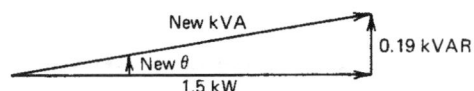

Figure IV-17. Resulting net power triangle when the capacitor is added. A new kVA can be calculated as well as a new θ.

$$\text{kVA new} = \frac{1.5 \text{ kW}}{0.99}$$

$$= 1.52 \text{ kVA}$$

The line voltage did not change, so the line current must be lower:

$$1.52 \text{ kVA new} = 0.24 \text{ kV} \times I \text{ new}$$

$$I \text{ new} = \frac{1.52 \text{ kVA}}{0.24} = 6.3 \text{ A}$$

Comparing the original circuit to the circuit after adding capacity, we have:

	Inductive Circuit	Improved Circuit
Line voltage	240 V	240 V
Line current	10 A	6.3 A
PF	62.5%	99%
kVA	2.4 kVA	1.52 kVA
kW	1.5 kW	1.5 kW
kVAR	1.87 kVAR	0.19 kVAR

The big improvement noted is the reduction of line current by 37%, with no decrease in real power, kW, used by the load. Also note the big change in kVA; less generating capacity is used to meet the same real power demand (generator input power is determined by KVA output).

IV.5.4 Summary

Through an example it has been demonstrated how the addition of a capacitor across an inductive load can improve power factor, reduce line current, and reduce the amount of generating capacity required to supply the load. The way this comes about is by having the capacitor supply the inductive magnetizing current locally. Since inductive and capacitive elements store and release power at different times in each cycle, this reactive current simply flows back and forth between the capacitor and inductor of the load. This idea is

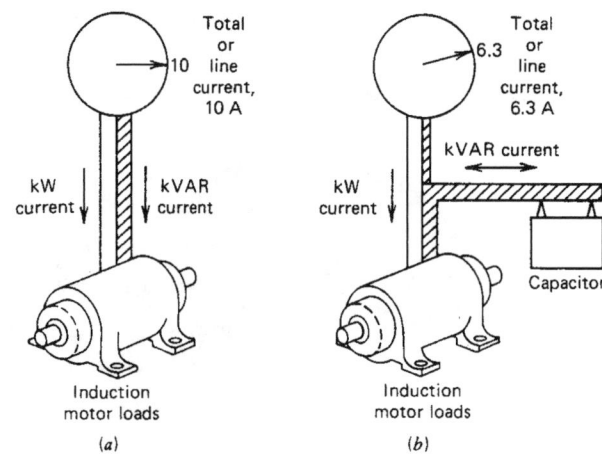

Figure IV-18. (a) Pictorial showing the inductive load of the example in this section. (b) The load with a capacitor added. With the exchange of the kVAR current between the capacitor and inductive load, very little kVAR current is supplied by the generator.

reinforced by Figure IV-18. Adding capacitors to inductive loads can free generating capacity, reduce line loss, improve power factor, and in general be cost effective in controlling energy bills.

IV.6 THREE-PHASE POWER

Three-phase power is the form of power most often distributed to industrial users. This form of transmission has three advantages over single-phase systems: (1) less copper is required to supply a given power at given voltage; (2) if the load of each phase of the three-phase source is identical, the instantaneous output of the alternator is constant; and (3) a three-phase system produces a magnetic field of constant density that rotates at the line frequency—which greatly reduces the complexity of motor construction.

The author realizes that both delta systems and wye systems exist but will concentrate on four-wire wye systems as being representative of internal distribution systems. This type of internal distribution system allows the customer both single-phase and three-phase service. Our focus will be on measuring power and determining power factor in four-wire, three-phase wye-connected systems.

IV.6.1 The Four-Wire Wye-Connected System

Figure IV-19 shows a generalized four-wire wye-connected system. The coils represent the secondary windings of the transformers at the site substation, while the generalized loads represent phase loads that

Table IV-1. How to Select Capacitor Ratings for Induction Motors/*Source: 1.*

Reference No. 1: For motor designs pre-dating TRI-CLAD 700® and CUSTOM 8000® Motors
(See GED-6063-02, Reference No. 2, for TRI-CLAD and CUSTOM 8000 motor designs)

Now it's easy to choose the right capacitors for your induction motors. Just refer to the following tables to find the kvar required by your particular motors. Locate motors by horsepower, rpm, and number of poles. All ratings are based on General Electric motor designs.

Tables I and V are also applicable to standard wound-rotor, open-type, three-phase, 60-cycle motors, provided the kvar values in the table are multiplied by a factor of 1.1, and the reduction in line current is increased by multiplying the values in the table by 1.05.

When selecting and installing capacitors, keep in mind the following: A capacitor located at the motor releases the maximum system capacity and is most effective in reducing system losses. Also, for a motor that runs continuously, or nearly so, it is usually most economical to locate the capacitor right at the motor terminals and switch it with the motor.

TABLE I—220-, 440-, AND 550-VOLT MOTORS, ENCLOSURE OPEN—INCLUDING DRIPPROOF AND SPLASHPROOF, GENERAL ELECTRIC TYPE K (NEMA DESIGN "B"), NORMAL STARTING TORQUE AND CURRENT

Induction Motor Horsepower Rating	Nominal Motor Speed in Rpm and Number of Poles											
	3600 2		1800 4		1200 6		900 8		720 10		600 12	
	Kvar	%AR	Kvar	%AR	Kvar	%AR	Kvar	%AR	Kvar	%AR	Kvar	%AR
2	1	16	1	20	1	22	1	24	..	..	..	..
3	1	10	1	16	1	21	2	24	..	..	..	..
5	1	9	2	16	2	21	2	21	..	..	..	..
7½	1	8	2	13	2	15	4	21	5	29	..	..
10	2	8	2	13	4	15	5	21	5	25	7.5	34
15	4	8	4	13	5	15	5	15	7.5	23	7.5	25
20	4	5	5	9	5	12	7.5	15	10	23	10	24
25	4	7	5	9	5	11	7.5	12	10	23	10	23
30	5	7	7.5	9	5	11	10	12	10	15	10	19
40	5	5	7.5	9	10	11	10	12	10	15	15	19
50	7.5	5	10	7	10	9	15	12	20	15	25	19
60	7.5	5	10	7	10	9	15	11	20	15	30	19
75	10	5	10	7	15	9	15	10	30	15	40	19
100	15	5	20	7	25	9	30	10	40	15	45	17
125	15	5	20	7	30	9	35	10	45	15	50	17
150	15	5	25	6	30	9	40	9	50	13	60	17
200	40	5	40	6	45	8	50	9	70	13	75	17
250	45	5	50	6	50	8	70	9	75	12	90	17
300	50	5	50	5	70	8	75	9	75	11	105	17
350	50	5	50	5	75	8	80	9	80	11	105	17
400	60	5	60	5	75	6	100	9	100	11	110	17
450	60	5	75	5	75	6	100	9	100	11	110	17
500	70	5	90	5	90	6	110	9	120	11	120	17

TABLE II—220-, 440-, 550-VOLT MOTOR, TOTALLY-ENCLOSED, FAN-COOLED, GENERAL ELECTRIC TYPE K (NEMA DESIGN "B"), NORMAL STARTING TORQUE, NORMAL STARTING CURRENT

Induction Motor Horsepower Rating	Nominal Motor Speed in Rpm and Number of Poles											
	3600 2		1800 4		1200 6		900 8		720 10		600 12	
	Kvar	%AR	Kvar	%AR	Kvar	%AR	Kvar	%AR	Kvar	%AR	Kvar	%AR
2	1	17	1	20	1	23	1	24	..	..	..	..
3	1	11	1	16	1	19	2	24	..	..	..	..
5	1	9	2	15	2	19	2	20	..	..	..	..
7½	1	6	2	13	4	19	4	20	..	..	..	..
10	2	6	2	11	4	16	5	15	7.5	20	7.5	27
15	4	6	4	11	4	13	5	15	7.5	20	7.5	24
20	4	6	5	11	5	13	7.5	15	10	20	10	24
25	5	6	5	8	5	9	7.5	15	10	17	10	18
30	5	6	7.5	8	7.5	9	10	15	10	15	10	18
40	7.5	6	10	8	10	9	10	15	10	15	15	17
50	7.5	6	10	8	10	9	15	12	15	12	20	17
60	10	6	10	8	10	9	15	12	20	12	25	17
75	15	6	15	8	15	9	20	11	25	12	35	17
100	15	6	20	8	25	9	25	11	40	12	45	17
125	20	6	25	7	30	9	30	11	45	12	45	15
150	25	6	30	7	40	9	40	11	45	12	50	15
200	35	6	40	7	60	9	60	11	55	11	60	13
250	40	5	50	6	60	9	80	11	60	11	100	13
300	50	5	45	6	80	8	80	10	80	10	125	13
350	60	5	70	6	80	8	80	9	..	..	..	..
400	60	5	80	6	80	6	160	..	..	..	..	..
450	70	5	100	6	..	..	..	..	..	..	..	..
500	70	5	..	..	..	..	..	..	..	..	..	..

TABLE III—220-, 440-, 550-VOLT MOTORS, ENCLOSURE OPEN—INCLUDING DRIPPROOF AND SPLASHPROOF, GENERAL ELECTRIC TYPE KG (NEMA DESIGN "C"), HIGH STARTING TORQUE AND NORMAL STARTING CURRENT

Induction Motor Horsepower Rating	Nominal Motor Speed in Rpm and Number of Poles							
	1800 4		1200 6		900 8		720 10	
	Kvar	%AR	Kvar	%AR	Kvar	%AR	Kvar	%AR
3	..	..	2	19	2	30	..	..
5	1	15	2	19	4	26	..	..
7½	2	15	2	15	4	22	..	..
10	2	12	4	12	5	17	..	..
15	4	12	4	12	5	17	..	..
20	4	12	5	12	7.5	17	..	..
25	7.5	9	7.5	11	10	17	..	..
30	7.5	9	7.5	11	10	15	..	..
40	10	9	10	11	10	13	..	..
50	10	9	10	9	15	13	..	..
60	10	8	15	9	20	13	20	18
75	15	8	15	9	20	12	30	18
100	20	8	25	9	30	12	40	18
125	20	7	30	8	30	12	50	18
150	35	7	35	8	45	12	..	..
200	40	7	50	8	70	12	..	..
250	50	7	50	8	80	12	..	..
300	60	7	80	8	80	10	..	..
350	60	7	80	8	80	9	..	..

NOTE: A capacitor located on the motor side of the overload relay reduces current through the relay, and therefore, a smaller relay may be necessary. The motor-overload relay should be selected on the basis of the motor full-load nameplate current reduced by the percent reduction in line current (% AR) due to capacitors.

TABLE IV—220-, 440-, 550-VOLT MOTORS, TOTALLY-ENCLOSED, FAN-COOLED, GENERAL ELECTRIC TYPE KG (NEMA DESIGN "C"), HIGH STARTING TORQUE, NORMAL STARTING CURRENT

Induction Motor Horsepower Rating	Nominal Motor Speed in Rpm and Number of Poles									
	1800 / 4		1200 / 6		900 / 8		720 / 10		600 / 12	
	Kvar	%AR	Kvar	%AR	Kvar	%AR	Kvar	%AR	Kvar	%AR
3	...	...	2	20	2	27	...	...	...	...
5	2	10	2	19	2	21	...	...	...	...
7½	2	10	2	17	4	21	...	...	...	...
10	2	9	3	17	5	21	...	...	...	...
15	4	9	3	14	5	16	...	...	...	...
20	5	9	3	11	7.5	16	...	...	...	...
25	7.5	9	5	10	10	15	10	15	10	18
30	7.5	9	7.5	10	10	15	10	15	10	18
40	10	9	10	9	10	12	15	15	15	18
50	10	9	10	9	15	12	15	15	25	18
60	15	9	15	9	20	12	30	15	30	18
75	15	9	15	8	20	11	30	15	40	16
100	20	7	25	8	25	11	40	15	45	16
125	25	7	35	8	35	11	50	15	50	15
150	35	7	35	7	50	11	60	14	50	13
200	40	7	45	7	50	10	70	14	70	13

TABLE VII—2300- AND 4000-VOLT MOTORS, ENCLOSURE OPEN—INCLUDING DRIPPROOF AND SPLASHPROOF, GENERAL ELECTRIC TYPE KG (NEMA DESIGN "C"), HIGH STARTING TORQUE AND NORMAL STARTING CURRENT

Induction Motor Horsepower Rating	Nominal Motor Speed in Rpm and Number of Poles							
	1800 / 4		1200 / 6		900 / 8		720 / 10	
	Kvar	%AR	Kvar	%AR	Kvar	%AR	Kvar	%AR
100	25	7	25	10	25	10	25	10
125	25	7	25	9	25	9	25	10
150	25	8	25	8	25	9	...	...
200	50	9	50	9	50	10	...	...
250	50	8	50	8	50	9	...	...
300	50	7	75	9	75	10	...	...
350	50	6	75	8	75	9	...	...

TABLE V—2300- AND 4000-VOLT MOTORS, ENCLOSURE OPEN—INCLUDING DRIPPROOF AND SPLASHPROOF, GENERAL ELECTRIC TYPE K (NEMA DESIGN "B"), NORMAL STARTING TORQUE AND CURRENT

Induction Motor Horsepower Rating	Nominal Motor Speed in Rpm and Number of Poles											
	3600 / 2		1800 / 4		1200 / 6		900 / 8		720 / 10		600 / 12	
	Kvar	%AR	Kvar	%AR	Kvar	%AR	Kvar	%AR	Kvar	%AR	Kvar	%AR
100	...	...	25	10	25	9	25	11	25	11	25	11
125	25	7	25	8	25	9	25	9	25	10	50	15
150	25	6	25	7	25	7	25	9	50	11	50	14
200	25	6	25	6	50	8	50	9	50	11	75	14
250	50	7	50	6	50	8	50	8	75	11	100	14
300	50	6	50	6	75	8	75	9	75	9	100	13
350	50	5	50	5	75	8	75	9	75	9	100	12
400	50	5	50	5	75	6	100	9	100	9	100	12
450	75	5	75	5	75	6	100	8	100	8	100	8
500	75	5	75	5	100	6	125	8	125	8	100	8
600	75	5	100	5	125	6	125	7	125	8	150	8
700	100	5	100	5	125	5	125	7	150	8	150	8
800	100	5	125	5	125	5	150	7	150	8	150	8

TABLE VIII—2300- AND 4000-VOLT MOTORS, TOTALLY-ENCLOSED, FAN-COOLED, GENERAL ELECTRIC TYPE KG (NEMA DESIGN "C"), HIGH-STARTING TORQUE, NORMAL STARTING CURRENT

Induction Motor Horsepower Rating	Nominal Motor Speed in Rpm and Number of Poles							
	1200 / 6		900 / 8		720 / 10		600 / 12	
	Kvar	%AR	Kvar	%AR	Kvar	%AR	Kvar	%AR
75	...	...	...	...	25	16	25	12
100	25	9	25	12	50	12	50	12
125	25	9	25	8	50	18	50	17
150	25	8	50	14	50	15	50	15
200	25	7	50	10	75	15	74	14

TABLE VI—2300- AND 4000-VOLT MOTORS, TOTALLY-ENCLOSED, FAN-COOLED, GENERAL ELECTRIC TYPE K (NEMA DESIGN "B"), NORMAL STARTING TORQUE, NORMAL STARTING CURRENT

Induction Motor Horsepower Rating	Nominal Motor Speed in Rpm and Number of Poles											
	3600 / 2		1800 / 4		1200 / 6		900 / 8		720 / 10		600 / 12	
	Kvar	%AR	Kvar	%AR	Kvar	%AR	Kvar	%AR	Kvar	%AR	Kvar	%AR
100	...	...	25	8	25	8	25	9	25	10	25	12
125	25	5	25	7	25	8	25	9	25	9	50	14
150	25	5	25	6	25	8	50	10	50	10	50	14
200	25	5	25	6	50	9	50	11	50	10	75	14
250	25	4	50	7	50	8	75	10	75	12	75	13
300	50	5	50	7	50	7	75	10	100	12	100	14
350	50	5	50	6	75	8	75	9	100	11	125	15
400	75	5	75	7	125	9	100	9	100	10	125	13
450	50	4	100	8	100	8	100	8	100	10	125	13
500	75	5	125	8	125	8	125	8	125	10	150	13

TABLE IX—440-VOLT OPEN, DRIPPROOF, OIL-FIELD MOTORS, 1200-RPM, GENERAL ELECTRIC TYPE KG, KR AND KOF

Induction Motor Horsepower Rating	High Starting Torque—Low Starting Current					
	Type KG Motor		Type KR Motor		Type KOF Motor	
	Kvar	% AR	Kvar	% AR	Kvar	% AR
5	3	25½	3	14½	4	24
7½	5	25½	4	20½	6	26
10	5	13½	4	12½	6	19
15	6	17½	6	14	6	13
20	8	17	10	20½	10	16
25	10	15	10	14½	10	15
30	10	16	10	14½	10	13
40	10	13½	10	9	10	10
50	10	10½	10	7½	15	10
60	10	9	15	9½	15	9
75	15	10½	15	8½	...	...
100	30	13	...	...	...	...

CAPACITORS BENEFIT YOUR DISTRIBUTION SYSTEM BY

- ● Reducing power costs
- ● Releasing system capacity
- ● Improving voltage levels
- ● Reducing system losses

INDUSTRIAL AND POWER CAPACITOR PRODUCTS DEPT. ● GENERAL ELECTRIC CO. ● HUDSON FALLS, N.Y.

GENERAL ⓖ ELECTRIC

GED-6063.01A
12-72 (10M) 6200

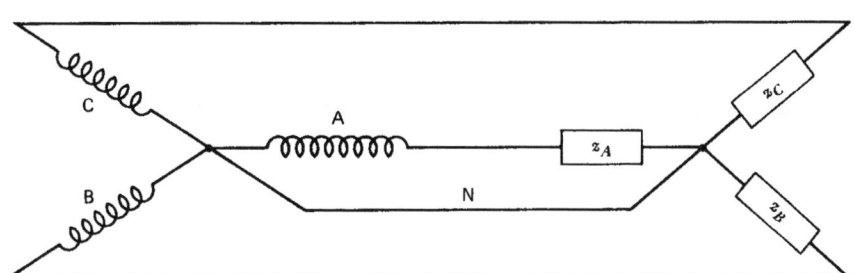

Figure IV-19. Generalized four-wire wye-connected system. The coils A, B, and C represent the three transformer secondaries at the site substation, while Z_A, Z_B, and Z_C are the generalized loads seen by each phase.

Table IV-2. How to Select Capacitor Ratings for Induction Motors/*Source: 2.*

how to select capacitor ratings for induction motors

Reference No. 2: For TRI-CLAD 700® and CUSTOM 8000® motors only.
(See GED-6063-01, Reference No. 1, for motor designs pre-dating TRI-CLAD 700 and CUSTOM 8000 Line)

Now it's easy to choose the right capacitors for your induction motors. Just refer to the following tables to find the kvar required by your particular motors. Locate motors by horsepower, rpm, and number of poles. All ratings are based on General Electric motor designs.

Tables I and V are also applicable to standard, wound-rotor, open-type, three-phase, 60-cycle motors, provided the kvar values in the table are multiplied by a factor of 1.1, and the reduction in line current is increased by multiplying the values in the table by 1.05.

When selecting and installing capacitors, keep in mind the following: A capacitor located at the motor releases the maximum system capacity and is most effective in reducing system losses. Also, for a motor that runs continuously, or nearly so, it is usually most economical to locate the capacitor right at the motor terminals and switch it with the motor.

TABLE I—230-, 460-, AND 575-VOLT MOTORS, ENCLOSURE OPEN—INCLUDING DRIPPROOF GENERAL ELECTRIC TYPE K (NEMA DESIGN "B"), NORMAL STARTING TORQUE AND CURRENT

Induction Motor Horsepower Rating	Nominal Motor Speed in Rpm and Number of Poles											
	3600 2		1800 4		1200 6		900 8		720 10		600 12	
	Kvar	%AR	Kvar	%AR	Kvar	%AR	Kvar	%AR	Kvar	%AR	Kvar	%AR
2	1	14	1	24	1	28	2	42	...	...	3	50
3	1	14	2	28	3	26	4	42	4	40	4	49
5	2	14	2	21	3	26	4	31	5	40	5	49
7½	2	14	4	21	4	21	5	26	7.5	40	10	49
10	4	14	4	17	5	21	5	26	7.5	36	10	41
15	5	12	5	17	5	19	10	26	7.5	31	10	34
20	5	11	7.5	17	7.5	19	10	23	10	29	15	34
25	7.5	10	7.5	17	10	19	10	23	10	24	20	34
30	7.5	10	7.5	17	15	19	15	23	15	24	25	32
40	10	10	15	17	20	19	20	23	20	24	30	32
50	10	10	20	17	25	19	20	23	24	24	35	32
60	10	10	20	17	30	19	30	23	30	22	45	32
75	15	10	25	14	30	16	30	17	35	21	40	19
100	15	10	30	14	30	12	40	16	45	15	50	17
125	30	10	35	12	30	12	50	16	45	15	50	17
150	30	10	35	12	35	12	50	14	60	13	60	17
200	35	10	50	11	55	12	70	14	70	13	90	17
250	35	10	55	9	70	12	85	14	90	13	100	17
300	35	10	65	9	75	12	95	14	100	13	110	17
350	40	10	80	9	85	12	125	14	120	13	150	17
400	100	10	80	8	100	12	140	14	150	13	150	17
450	100	10	90	8	140	12	150	13	150	13	175	17
500	100	8	115	8	150	12	150	12	175	13	175	17

TABLE II—230-, 460-, 575-VOLT MOTOR, TOTALLY ENCLOSED, FAN-COOLED GENERAL ELECTRIC TYPE K (NEMA DESIGN "B"), NORMAL STARTING TORQUE, NORMAL STARTING CURRENT

Induction Motor Horsepower Rating	Nominal Motor Speed in Rpm and Number of Poles											
	3600		1800		1200		900		720		600	
	Kvar	%AR	Kvar	%AR	Kvar	%AR	Kvar	%AR	Kvar	%AR	Kvar	%AR
2	1	15	1	24	1	28	2	42	...	...	3	50
3	1	15	2	28	2	26	4	32	3	41	4	49
5	2	15	2	20	3	22	4	29	4	35	5	49
7½	3	15	4	20	4	22	5	29	7.5	35	10	49
10	3	11	4	17	5	22	5	29	7.5	35	10	41
15	4	10	5	17	7.5	22	10	29	7.5	24	10	34
20	5	10	5	17	10	21	10	23	10	23	15	34
25	5	10	7.5	17	10	21	10	22	10	22	20	31
30	5	10	7.5	17	10	21	15	22	15	22	20	3½
40	10	10	10	12	15	21	20	22	20	22	30	3½
50	10	10	15	12	25	21	23	21	25	22	35	31
60	12	9	15	12	25	19	25	19	30	22	40	30
75	15	9	20	12	25	15	30	19	35	21	40	30
100	20	9	30	12	25	13	40	19	40	12	50	30
125	20	9	35	12	30	13	45	18	50	12	50	30
150	25	9	40	11	25	13	55	18	50	12	70	30
200	30	9	40	8	60	13	60	18	70	12	75	30
250	60	9	50	8	60	13	115	18	100	12	125	30
300	65	7	50	8	60	13	140	18	150	12	150	30
350	70	7	55	8	80	13	160	18	150	12	150	30
400	70	7	60	8	130	13	160	17	175	12	175	30
450	90	7	95	8	145	13	160	17	175	12	200	30
500	100	7	110	7	170	13	210	17	...	...	...	...

TABLE III—230-, 460-, 575-VOLT MOTORS, ENCLOSURE OPEN—INCLUDING DRIPPROOF, GENERAL ELECTRIC TYPE KG (NEMA DESIGN "C") HIGH STARTING TORQUE AND NORMAL STARTING CURRENT

Induction Motor Horsepower Rating	Nominal Motor Speed in Rpm and Number of Poles									
	1800 4		1200 6		900 8		720 10		600 12	
	Kvar	%AR	Kvar	%AR	Kvar	%AR	Kvar	%AR	Kvar	%AR
3	...	...	2	28	4	42	...	...	...	...
5	2	21	3	26	4	32	...	...	...	...
7.5	4	21	4	22	4	29	...	...	...	...
10	4	17	5	22	5	29	...	...	20	40
15	5	17	7.5	22	10	29	...	...	20	40
20	5	17	7.5	21	10	25	...	...	30	40
25	7.5	17	7.5	21	10	23	20	28	35	39
30	7.5	17	10	21	15	23	...	...	45	39
40	10	17	15	21	20	23	30	28	...	...
50	20	17	15	21	25	23	35	28	...	...
60	20	15	30	21	30	23	45	28	...	...
75	23	14	30	17	40	23	40	15	45	17
100	30	13	30	14	50	23	40	15	50	17
125	35	12	40	14	50	14	45	15	60	17
150	35	10	45	13	50	14	50	13	60	17
200	50	10	55	11	70	14	70	13	90	17
250	55	9	70	11	85	14	90	13	100	17
300	65	9	75	10	95	14	100	13	110	17
350	80	9	85	9	125	14	120	13	150	17

NOTE: A capacitor located on the motor side of the overload relay reduces current through the relay, and therefore, a smaller relay may be necessary. The motor-overload relay should be selected on the basis of the motor full-load nameplate current reduced by the percent reduction in line current (% AR) due to capacitors.

TABLE IV—230-, 460-, 575-VOLT MOTORS, TOTALLY-ENCLOSED, FAN-COOLED, GENERAL ELECTRIC TYPE KG (NEMA DESIGN "C"), HIGH STARTING TORQUE, NORMAL STARTING CURRENT

Induction Motor Horse-power Rating	1800 / 4		1200 / 6		900 / 8		720 / 10		600 / 12	
	Kvar	%AR	Kvar	%AR	Kvar	%AR	Kvar	%AR	Kvar	%AR
3	...	...	2	28	4	42	...	...	...	...
5	2	20	3	26	4	32	...	...	...	...
7.5	4	19	4	22	4	30	...	...	...	...
10	4	19	5	22	5	30	...	...	...	...
15	5	19	7.5	22	10	30	...	...	...	...
20	7.5	19	10	21	10	29	...	...	20	40
25	7.5	19	10	19	10	29	...	...	...	...
30	7.5	19	10	19	20	29	25	31	30	40
40	15	19	15	19	20	24	...	...	35	38
50	15	19	20	19	20	22	30	28	40	37
60	25	19	25	18	25	22	35	28	40	37
75	25	13	25	15	35	22	45	28	50	37
100	30	13	30	12	45	20	40	28	50	37
125	35	12	30	12	45	20	50	28	50	37
150	40	11	40	12	50	20	50	28	70	37
200	40	8	60	12	60	20	70	28	75	38

TABLE V—2400- AND 4160-VOLT MOTORS, ENCLOSURE OPEN—INCLUDING DRIPPROOF AND SPLASHPROOF, GENERAL ELECTRIC TYPE K (NEMA DESIGN "B") NORMAL STARTING TORQUE AND CURRENT

Induction Motor Horse-power Rating	3600 / 2		1800 / 4		1200 / 6		900 / 8		720 / 10		600 / 12	
	Kvar	%AR	Kvar	%AR	Kvar	%AR	Kvar	%AR	Kvar	%AR	Kvar	%AR
100	...	...	25	11	25	12	50	24	25	14	25	20
125	...	...	25	9	25	12	25	13	25	14	50	20
150	25	9	25	9	25	13	50	13	50	14	50	20
200	25	9	50	9	50	12	75	13	75	14	100	20
250	25	9	50	8	50	12	75	13	75	14	100	20
300	50	9	50	8	75	12	100	12	100	14	125	20
350	75	9	50	8	75	12	100	12	100	14	125	19
400	75	9	50	8	100	12	100	12	125	14	150	19
450	100	9	75	8	100	12	125	11	125	14	150	19
500	100	8	100	8	125	12	125	11	150	14	150	19
600	125	8	125	8	175	12	150	11	150	14	200	17
700	150	8	150	8	200	11	150	10	200	14	200	15
800	175	8	150	7	175	10	175	10	225	13	250	15

TABLE VI—2400- AND 4160-VOLT MOTORS, TOTALLY-ENCLOSED, FAN-COOLED, GENERAL ELECTRIC TYPE K (NEMA DESIGN "B"), NORMAL STARTING TORQUE, NORMAL STARTING CURRENT

Induction Motor Horse-power Rating	3600 / 2		1800 / 4		1200 / 6		900 / 8		720 / 10		600 / 12	
	Kvar	%AR	Kvar	%AR	Kvar	%AR	Kvar	%AR	Kvar	%AR	Kvar	%AR
100	...	...	25	17	...	...	50	22	25	12	50	15
125	...	...	50	17	25	15	50	17	25	12	50	15
150	25	6	25	12	50	15	50	17	50	12	75	15
200	25	6	50	12	75	15	50	17	50	12	100	15
250	25	6	50	11	75	15	75	17	75	12	100	15
300	50	6	50	11	75	13	125	17	100	12	125	15
350	50	6	50	11	75	13	125	17	125	12	150	15
400	75	6	125	11	125	13	150	17	150	12	200	15
450	75	6	125	10	150	13	175	17	200	12	225	15
500	75	6	125	8	175	13	225	17	225	12	225	13

TABLE VII—2400- AND 4160-VOLT MOTORS, ENCLOSURE OPEN—INCLUDING DRIPPROOF AND SPLASHPROOF, GENERAL ELECTRIC TYPE KG (NEMA DESIGN "C"), HIGH STARTING TORQUE AND NORMAL STARTING CURRENT

Induction Motor Horse-power Rating	1800 / 4		1200 / 6		900 / 8		720 / 10	
	Kvar	%AR	Kvar	%AR	Kvar	%AR	Kvar	%AR
100	25	10	25	11	25	13	25	14
125	25	8	25	9	50	13	25	14
150	25	7	50	12	50	13	50	14
200	25	8	50	12	50	13	75	14
250	25	8	50	12	75	13	75	14
300	50	8	75	12	100	13	100	14
350	50	8	75	12	100	12	100	14

TABLE VIII—2400- AND 4160-VOLT MOTORS, TOTALLY-ENCLOSED, FAN-COOLED, GENERAL ELECTRIC TYPE KG (NEMA DESIGN "C"), HIGH STARTING TORQUE, NORMAL STARTING CURRENT

Induction Motor Horse-power Rating	1200 / 6		900 / 8		720 / 10		600 / 12	
	Kvar	%AR	Kvar	%AR	Kvar	%AR	Kvar	%AR
75	...	...	...	...	...	...	...	...
100	...	...	...	...	25	12	50	15
125	25	10	50	17	25	12	50	15
150	...	...	50	17	50	12	75	15
200	75	15	50	17	50	12	100	15

CAPACITORS BENEFIT YOUR DISTRIBUTION SYSTEM BY

- Reducing power costs
- Releasing system capacity
- Improving voltage levels
- Reducing system losses

INDUSTRIAL AND POWER CAPACITOR PRODUCTS DEPT. • GENERAL ELECTRIC CO. • HUDSON FALLS, N.Y.

GENERAL ⊕ ELECTRIC

GED-6063.02A
12-72 (10M) 6200

are the sum loads on each phase. These loads may be composites of single-phase services and three-phase motors being fed by the distribution system. N is the neutral or return.

To determine the power and power factor of any phase A, B, or C, consider that phase as if it were a single-phase system. Measure the real power, kW, delivered by the phase by use of a wattmeter and measure and compute the volt-ampere product, apparent power, kVA, using a voltmeter and ammeter.

The power factor of the phase can then be determined and corrected as needed. Each phase can be treated independently in turn. The only caution to note is to make the measurements during nominal load periods, as this will allow power-factor correction for the most common loading.

If heavy motors are subject to intermittent duty, additional power and power-factor information can be gathered while they are operating. Capacitors used to correct power factor for these intermittent loads should be connected to relays so that they are across the motors and on phase only when the motor is on; otherwise, overcorrection can occur.

In the special case of a four-wire wye-connected system with balanced loading, two wattmeters may be used to monitor the power consumed on the service and also allow computation of the power factor from the two wattmeter readings.

IV.6.1.1 Balanced Four-Wire Wye-Connected System

Figure IV-20 shows a balanced system containing two wattmeters. The sum of these two wattmeter readings are the total real power being used by the service:

$$P_T = P_1 + P_2$$

Further, the angle of displacement between each line current and voltage can be computed from P_1 and P_2:

$$\theta = \tan^{-1} \sqrt{3} \frac{P_2 - P_1}{P_2 + P_1}$$

and the power factor PF = cos θ.

This quick method for monitoring power and power factor is useful in determining both fixed capacitors to be tied across each phase for the nominal load and the capacitors that are switched in only when intermittent loads come on-line.

The two-wattmeter method is useful for determining real power consumed in either wye- or delta-connected systems with or without balanced loads:

$$P_T = P_1 + P_2$$

However, the use of these readings for determining phase power factor as well is restricted to the case of balanced loads.

IV.6.2 Summary

This brief coverage of power and power-factor determination in three-phase systems covers only the very basic ideas in this important area. It is the aim of this brief coverage to recall or refresh ideas once learned but seldom used.

Tables IV-1 and IV-2 were supplied by General Electric, who gave permission for the reproduction of materials in this handbook.

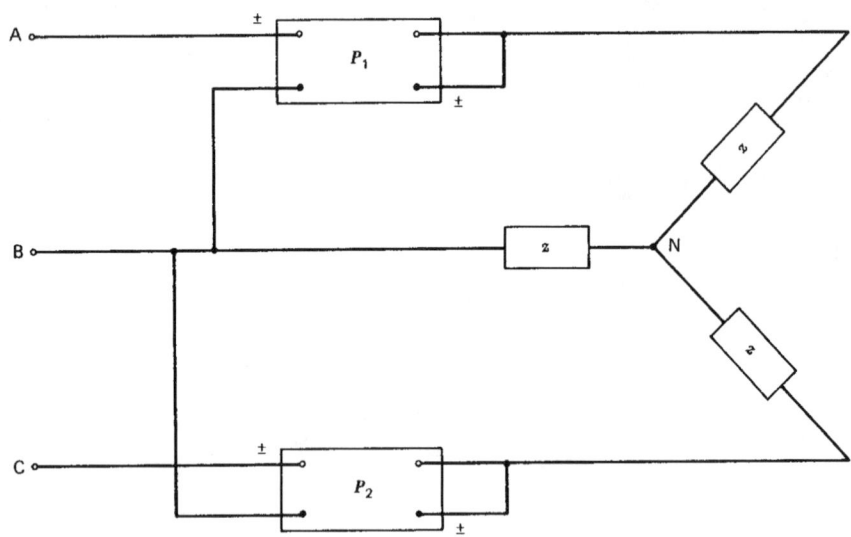

Figure IV-20. Four-wire wye-connected system with wattmeter connections detailed. Solid circle voltage connections to wattmeter; open circle, current connections to wattmeter.

INDEX